MICHIGAN MAMMALS

MICHIGAN MAMMALS

by

Rollin H. Baker

Michigan State University Press 1983

Library of Congress Catalog Card Number: 82-61821
ISBN: 087013-299-7
Printed in the United States of America
Michigan State University Press
East Lansing, Michigan 48823-5202

The paper used in this publication meets the minimum requirements of
American National Standard for Information Sciences—
Permanence of Paper for Printed Library Materials
ANSI Z39.48—1984

To Mary

A thousand hours out of the treasury
of time that yet remains for me to live
would I yield up if that small, fugitive,
alert-eyed chipmunk to his granary
would lead me as a friend, and there disclose,
with confident winkings and a secret smile,
the mysteries of the woodland that he knows,
and be my intimate for a little while.
I tempt him toward me; half-way he will come,
but then whisks bushy tail, and is away.
My gentlest language to his ears is dumb.
A million years have passed since that fair day
when in these woods we two, alike at home,
could smile and go each unperturbed his way.

Arthur Davison Ficke

Contents

Figures

COLOR PLATES
following page 16

Maps

Tables

Foreword

When I look carefully at the countryside, I see splashes of color—browns, yellows, greens, blues, reds. These are obvious, but do I also perceive the environment beyond? Do I distinguish cultivated plants growing in their symmetrical rows? Do I see the arrangement of fields in relation to soil types, to exposures on north or south hillsides, to valleys? Do I detect previously-cultivated fields and recognize that the types of plants growing there are influenced by the length of time since the field was last plowed and sowed? Can I envision such a piece of land in a successional stage of natural plant growth from bare ground through annual weeds and grasses, perennial weeds and grasses, shrubs, low trees and ultimately forest. Further, can I predict what kinds of birds, mammals, and other animals might have lived there during each of these successional stages?

Can I also interpret the details in the topography—hills, valleys, and drainage patterns? In road cuts, say, or stream-banks can I recognize the subsurface layers of soils, gravels, and rocks and what they might suggest about the geologic history of the area? If I can do more than comprehend, and perceive something of the drama of my surroundings, then I have attained a degree of "natural literacy."

This book concerns mammals, a dominant but often inconspicuous assemblage in the Michigan outdoors. It is hoped that readers will come away, perhaps as I have, with a greater appreciation and knowledge of the diversity and eccentricities of these creatures and the prominent roles which they play in the Michigan environment.

Acknowledgements

Most specimens referred to in text are preserved in Michigan institutions. I am therefore indebted to colleagues for allowing me to record data from collections at Adrian College, Andrews University, Central Michigan University, Michigan State University, Michigan Tech University, Northern Michigan University, Northwestern Michigan College, Olivet College, The University of Michigan, Wayne State University, and Western Michigan University.

The conscientious effort by wildlife biologists of the Michigan Department of Natural Resources to acquire distributional and ecological data on Michigan mammals has added significantly to this report. I acknowledge especially the cooperation and generous support of David A. Arnold, Ralph E. Bailey, Glenn Y. Belyea, Carl L. Bennett, Jr., Charles T. "Ted" Black, Ralph I. Blouch, Marvin E. Cooley, Tom Cooley, Lawrence Dayton, Jerry P. Duvendeck, L. Dale Fay, Elsworth M. Harger, Louis J. Hawn, Robert S. Huff, Victor S. Janson, David H. Jenkins, Ford Kellum, John M. Lerg, Edward S. Mikula, Herb Miller, Richard J. Moran, John Ozoga Jr., Harry Ruhl, Lawrence A. Ryel, Raymond D. Schofield, Charles Shick, John N. Studt, Frederick W. "Fritz" Stuewer, Donald Switzenberg, Sylvia M. Taylor, Louis J. Verme, Joseph E. Vogt, Oscar "Ozz" Warbach and William G. Youatt. Help from several present and former colleagues at Michigan State University is also recognized: Alwynelle Ahl, Richard J. Aulerich, Charles E. Cleland, William Cooper, Leslie C. Drew, Glen Dudderar, J. Keever Greer, Leslie Gysel, Donald O. Hayne, Max Hensley, Richard Hill, Victor Hogg, J. Alan Holman, Joe Johnson, Kenneth Keahey, John King, William Lovis, Richard Manville, Bert Ostenson, George Petrides, James Sikarskie, Donald O. Straney, Chester Trout, Roswell D. VanDuesen, and George Wallace. William H. Burt must be singled out as a major authority for data in these accounts. Other important contributors include Karl D. Bailey, Richard Brewer, Larry Caldwell, John F. Douglass, Lowell Getz, E. Raymond Hall, Margaret Herman, Emmet Hooper, G. William Irwin, Russell Jameson, Kenneth Kram, James P. Ludwig, Lawrence L. Master, Jack Mell, Philip Myers, R. A. Powell, William Prychodko, James M. Ryan, William Robinson, William Scharf, Norman Sloan, Robert Stones, George Taft, Jens Touborg, and Harold Walter.

The author thanks the many persons who read and edited sections of this report with special acknowledgement made to Byron L. Baker, Jean Busfield, Lawrence L. Master, Lynne Taft, and particularly David H. Jenkins. Finally, I must recognize the invaluable assistance of the many advanced students at Michigan State University whose devotion to mammalian study and whose superior, inquisitive minds provided much of the catalytic action which made this all possible: Thom Alcoze, Larry Bowdre, James Bowers, Christopher Carmichael, Don Christian, Steven Collett, Peter Dalby, Gary Dawson, James Dietz, James Drake, John Enders, Ronald J. Field, John Fitch, Richard Fitzner, Robert A. Goertner, Nancy M. Gosling, J. Keever Greer, Jerry Hall, Melvin Hathaway, Gary Heidt, John Helm, Danny Herman, Larry Holcomb, Richard Hoppe, David I. Johnson, Gordon Kirkland, James Koschmann, Allen Kurta, Albert Manville, Terrance Martin, John Matson, Douglas McWhirter, Alan Muchlinski, Donald M. Osterburg, John Ozoga Jr., Michael Petersen, Carleton Phillips, Mary Lou Rabe, Steven Rogers, Jacquelyn Shier, Henry

Short, Ann and Karl Shump, Max Terman, William Teska, Robert Tuck, and N. Geoffroy Weilert. Suggestions about Michigan plantlife were generously provided by botanists John Beaman, John Cantlon, Peter Murphy, and Steve Stephenson of Michigan State University. The author is indebted to John A. Hannah and Herman King for their encouragement and to Walter B. Barrows (1855–1923), whose classic "Michigan Bird Life" published by Michigan Agricultural College in 1912, provided inspiration for the present work. I must also acknowledge the patient assistance rendered by Museum colleagues Judy DeJaegher, Laurena Jenkins, and Bernice Rochon without whose help this project might not have been initiated.

Drawings of mammals are the work of artists Dirk Gringhuis, Don Dickerson, Bonnie Marris, Patricia Stinson, and Oscar Warbach. Most of the skulls and lower jaws were drawn by James Zablotny; some by Bonnie Marris. Maps showing distributions of mammals were prepared by Jane Kaminski. Photographs were generously provided by Bruce R. Baker, Robert Brown, William H. Burt, Robert P. Carr, Robert Harrington, Dikran Kashkashian, William Mitcham, Roger A. Powell, and Lawrence A. Ryel. I am also flattered that Gijsbert van Frankenhuyzen, Bonnie Marris, David Mohrhardt, Charles Schwartz, and James Zablotny kindly allowed me to reproduce special examples of their wildlife artistry. These outstanding and talented persons have my sincere thanks. I have made every effort to contact the executors of the estate of Arthur Davison Ficke for permission to reproduce the poem which appears here as an epigraph.

Finally, I am most indebted to the Michigan State University Foundation for generous support toward publication of this work, and to Mary W. Baker for her patience and encouragement.

Introduction

In 1946, Professor William Henry Burt published his classic summarization of the mammals of Michigan (Burt, 1946). He compiled records of specimens preserved in museums, summarized results of his own extensive investigations, gleaned biological information from the literature of the day, and also noted the contributions of nineteenth century Michigan observers. He referred to the writings of John T. Blois (1839), Adolphe B. Covert (1881), Morris Gibbs (1895), Manly Miles (1861), Abram Sager (1839), and James J. Strang (1855). When it became known that a revision of this book was not forthcoming, I asked for and received Dr. Burt's hearty endorsement to bring out this updated report. I soon found the project to be a major task, simply because in the post-1946 years abundant new information on Michigan mammals has come to light. Since World War II, ecologists have been busy utilizing remote sensing devices and other kinds of space age technology to probe the life styles of inconspicuous mammals. Conservationists, physiologists, geneticists, behaviorists, and even biochemists have published original contributions to our overall knowledge of how mammals live. Even so, in terms of diversity, only two new mammals, the smoky shrew (*Sorex fumeus*) and the eastern pipistrelle (*Pipistrellus subflavus*), have been added to the 64 species of mammals Burt reported as native to Michigan in 1946. Using his work as a base, a wealth of data was also extracted from subsequent compilations on mammals of North America (Hall, 1981), Canada (Banfield, 1974), eastern Canada (Peterson, 1966), the Upper Great Lakes (Burt, 1972) Wisconsin (Jackson, 1961), Illinois (Hoffmeister and Mohr, 1957), Indiana (Mumford, 1969; Mumford and Whitaker, 1982), Ohio (Gottschang, 1981) and Minnesota (Hazard, 1982).

My aim has been to place on record our present knowledge of the distribution and life habits of 66 native mammals known from Michigan within the historical period (since 1700). Conspicuous by its absence herein is a formal account of the sixty-seventh native species, the human (*Homo sapiens*). However, the profound influence of this ingenious, creative, adaptable, but often destructive inhabitant on other mammals is the subject of considerable in-text discussion.

Generally, descriptive accounts pertain specifically to WHERE, HOW, and WHY each mammal lives within the political boundaries of Michigan. Two non-residents, the small-footed bat (*Myotis leibii*) and Franklin's ground squirrel (*Spermophilus franklinii*), are also included because each has been recorded in localities very close to the Michigan border. Hopefully, their inclusion will alert readers to watch for them and perhaps obtain the first state records—always a thrill for animal observers. Accounts of the introduced European or cape hare (*Lepus capensis*) and of the Old World commensal rodents, the Norway rat (*Rattus norvegicus*) and the house mouse (*Mus musculus*), complete the species documentation.

The Mammalian Fauna of Michigan

Mammals are surely among the most prominent creatures on earth. On land, they range in size from tiny shrews, bats, and mice to massive elephants and rhinoceroses. In the aquatic environment, whales are conspicuous by their bulk, with the endangered blue or sulphur-bottomed whale (*Balaenoptera musculus*) being the largest of all known animals, living or extinct. Despite their adaptive radiation into all types of global environments, which according to the paleontologists has characterized their spectacular emergence from reptilian stock, mammals actually rank low in numbers of living species. Ornithologists can boast of more than twice as many bird species (8,600) and there are over 750,000 insect species. However, in at least one major feature, biomass (*i.e.*, pounds per acre or kilograms per hectare) the mammalian community takes a commanding position among terrestrial animal life. Nevertheless, mammals are shy and attempt to remain unobtrusive. The average Michigan observer is probably most familiar with tree squirrels and bats as examples of smaller mammals, and the white-tailed deer among the larger ones. Many secretive and nocturnal species are viewed only briefly as they scurry across highways at night or are left on doorsteps by foraging house cats.

POSTGLACIAL HISTORY

The 66 species of mammals known to be native to Michigan within historic times belong to 17 families in seven orders (Table 1). They are all newcomers because Michigan's land surface during the last geologic epoch, the Pleistocene, was covered on several occasions by massive, life-obliterating glaciers (Kelley and Farrand, 1967). According to studies of the stratified layers of rocks and soils, warm weather conditions alternated with cold periods during the so-called Ice Age which began almost two million years ago (Dorr and Eschman, 1971). About 30,000 years ago the last of these southward thrusts of ice—called the late Wisconsin glacial stage—engulfed the area which is now the State of Michigan.

This final ice sheet was slow to melt. It would recede slowly and then surge southward again to envelop the terrain while the churning action and

Table 1. Native Mammals of Michigan.

ORDERS	FAMILIES	NUMBER OF SPECIES
MARSUPIALIA	DIDELPHIDAE	1
INSECTIVORA	SORCIDAE	7
	TALPIDAE	2
CHIROPTERA	VESPERTILIONIDAE	9
LAGOMORPHA	LEPORIDAE	2
RODENTIA	SCIURIDAE	9
	CASTORIDAE	1
	CRICETIDAE	8
	ZAPODIDAE	2
	ERETHIZONTIDAE	1
CARNIVORA	CANIDAE	4
	URSIDAE	1
	PROCYONIDAE	1
	MUSTELIDAE	10
	FELIDAE	3
ARTIODACTYLA	CERVIDAE	4
	BOVIDAE	1
		66*

*7 Extirpated, of which 3 are re-established.

sheer weight of the motile ice gouged out vestiges of the substrate deposited in earlier times. The ice ultimately departed completely from what is now the southern part of the Lower Peninsula about 15,000 years ago, not receding until 12,000 years ago in northern areas, with parts of the Superior Basin remaining icebound until as recently as 10,000 years ago. Geological treatises such as Dorr and Eschman (1971) give details of this melt-back, the glacial effects on the landscape, and the formation of the modern lake basins and stream flow patterns.

The environment at the receding edge of the glacier must have been bleak and probably barren. The top layers of soil were likely saturated with moisture resulting from the lack of seepage into the still frozen subsoils and the absence of developed drainage patterns to allow melt water to flow away. Nevertheless, plant and animal life were quick to invade suitable sites near the edge of this retreating ice. The first were species highly tolerant of the cold and moist environment—no doubt reminiscent of conditions today on the tundra bordering the Arctic Ocean. Plants characteristic of this habitat grew about 13,770 years ago in what is now Lapeer County (Farrand and Eschman, 1974). Remains of the "tundra" form of the caribou (*Rangifer tarandus*) found in Macomb County date from 11,200 years ago (Cleland, 1965). The Jefferson mammoth (*Mammuthus jeffersoni*) and muskox (*Symbos cavifrons*), both species now extinct, also lived in this arctic-like environment, closely following the slow retreat of the ice (see Table 2; Fig. 1).

Table 2. Postglacial Mammals—Now Extinct. In Michigan

SPECIES REPORTED	LATEST RECORDED DATE IN MICHIGAN
Giant Beaver, *Castoroides ohioensis*	11,400 ± 160 years B.P.
American Mastodon, *Mammut americanus*	6,000 ± 150 years B.P.
Jefferson Mammoth, *Mammuthus jeffersoni*	11,400 ± 400 years B.P.
Peccary, *Platygonus compressus*	4,290 ± 150 years B.P.
Giant Moose, *Cervalces scotti*	no dated record
Woodland Muskox, *Symbos cavifrons*	11,150 ± 400 years B.P.

Figure 1. The artist's conception of how the Michigan landscape in Lapeer County may have looked about 13,770 years ago. The tundra-like environment, with the glacial edge in the background harbored the Jefferson mammoth, the muskox, and the "tundra" form of the caribou. Sketch by Dirk Gringhuis.

The climate gradually warmed and became less humid, causing the ice sheet to disappear beyond the northern horizon. The lichens, mosses, grasses, and forbs, which had first occupied the bare soils, rocky rubble, and other glacial debris, found the milder, dryer conditions unacceptable. Eventually they gave way to the northward advance of shrub growth, followed by boreal conifer forests (Amundson and Wright, 1979), and finally to mixed conifers and hardwoods and associated animal life (Handley, 1971). This wave-like succession may have taken as long as 4,000 years. At any rate, some 7,000 years ago, plant and animal communities slowly began to take on a modern appearance (Farrand and Eschman, 1974). With the early forest environments (see Fig. 2) came the now-extinct giant beaver (*Castoroides ohioensis*), American mastodon (*Mammut americanus*), Scott's moose (*Cervalces scotti*), and the flat-headed peccary (*Platygonus compressus*). These species (see Table 2) associated at the time with such survivors as white-tailed deer (*Odocoileus virginianus*), wapiti (*Cervus elaphus*), moose (*Alces alces*), and the "forest" form of the caribou (*Rangifer tarandus*); (see Fig. 3).

Early in these postglacial times, marine mammals from the Atlantic Ocean entered the basins which were later to become the Great Lakes (Handley, 1953). Remains of these creatures uncovered in old beach deposits have been identified as those of walrus (*Odobenus*), sperm whale (*Physeter*), finback whale (*Balaenoptera*), and bowhead whale (*Balaena*). Causes for the demise of these large terrestrial and aquatic creatures remain obscure (Dorr and Eschman, 1971; Hibbard, 1951; Holman, 1975; Wilson, 1967).

Figure 2. The boreal forests of the early postglacial period were inhabited by such now-extinct creatures as the giant beaver, Scott's moose, the American mastodon, as well as the modern "forest" form of the caribou.
Sketch by Dirk Gringhuis.

THE HUMAN INTRUSION

The first postglacial peoples, the Paleo-Indian hunters (see Fig. 4), followed the northward retreat of the ice sheet into what is now Michigan between 10,000 and 11,000 years ago (Cleland, 1975). A long series of cultural adaptations to changing environments of Michigan resulted in the Archaic, or hunting and gathering, way of life which evolved from the earlier Paleo-Indian Culture. This was followed by the Woodland Culture, based on the cultivation of corn, beans, and squash. There is some evidence that these prehistoric people lived in association with the mastodon and other now departed creatures, leading some authorities to suspect (although with little actual evidence) that these early hunters influenced the extinction of some of these animals. This first wave of human intervention appeared to make no immediate major changes in the mammalian fauna; major habitat alteration was possibly the setting of fires. Small clearings in the forest of southern Michigan for village sites and the agricultural crops probably constituted a minor Indian impact on early Michigan habitats. Evidence obtained from studying the remains of food stuffs found at prehistoric campsites leads archaeologists to believe that the Indians maintained a non-depleting co-existence with the mammalian biota.

These prehistoric peoples were the sole human inhabitants of the Michigan area for perhaps 10,000 years. The second wave of human encroachment—the French, followed by the British, and then by people from other Old World areas—began about 350 years ago. These new intruders spent most of their first 150 years in Michigan as

Figure 3. The mixed coniferous and deciduous forests in the northern part of what is now Michigan's Lower Peninsula 500–1,000 years ago was home to the four modern cervids, the "forest" form of the caribou, the moose, the wapiti, and the white-tailed deer, with the latter in much smaller populations than today. Sketch by Dirk Gringhuis.

traders exchanging European weaponry, tools, religion, and other goods and services for exportable pelts of beaver and other animals. Commercial demand for these natural resources was the beginning of a new and drastic change in human regard for the Michigan environment. Furs, edible fish and wildlife, lumber, mining products, and other natural resources, until then primarily home-use commodities, suddenly became marketable. The new settlers accelerated their effect on the environment with the use of the axe, trap, musket, shovel, torch, and plow. As one resource became depleted, another seemed ripe for the taking.

Modern views on natural resource management in the early twentieth century allowed for the successful return, by natural propagation and controlled harvest, of the beaver (*Castor canadensis*), white-tailed deer (*Odocoileus virginianus*), and other depleted resident mammals. Some, like the extirpated marten (*Martes americana*), fisher (*Martes pennanti*), and wapiti (*Cervus elaphus*), were reintroduced. Others, however, like the wolverine (*Gulo gulo*), mountain lion (*Felis concolor*), caribou (*Rangifer tarandus*), and bison (*Bison bison*), were lost as free-living native species (Table 3).

The small and reclusive shrews and rodents received little attention from early scientific observers. Modern mammalogists can only surmise how these species increased or decreased in efforts to adjust to forest clearing, reforestation and fire prevention, propagation of domestic livestock, cultivation of new agricultural crops, introduction of noxious weeds, free use of such environmental additives as fertilizers and pesticides, drainage projects and lowering of water tables, and the ultimate formation, especially in southern Michigan, of a mosaic pattern of woodlots and cleared lands in various stages of plant succession. Small native mammals not only had to contend with the introduction of the aggressive, adaptable Old World house mouse (*Mus musculus*) but also with such midwestern prairie mammals as the thirteen-lined ground squirrel (*Spermophilus tridecemlineatus*), the prairie subspecies of the deer mouse (*Peromyscus maniculatus*), and the badger (*Taxidea taxus*) which spread naturally northeastward into the lands cleared of forests. Southern species like the fox squirrel (*Sciurus niger*) populated second-growth forest lands in place of the gray squirrel (*Sciurus carolinensis*) which had formerly thrived in the mature hardwoods (Handley, 1971).

Figure 4. The artist's conception of the lifestyle of the Paleo-Indian hunters, the first postglacial people to inhabit what is now Michigan. Sketch by Dirk Gringhuis.

Michigan mammals found in abundance in various environments today are generally ones which

Table 3. Michigan Mammals—Extinct, 1800 to 1937.

SPECIES	APPROXIMATE DATE OF EXTINCTION
Marten, *Martes americana**	*circa* 1911
Fisher, *Martes pennanti**	*circa* 1936
Wolverine, *Gulo gulo*	*circa* 1880
Mountain Lion, *Felis concolor*	*circa* 1937
Wapiti or Elk, *Cervus elaphus**	*circa* 1877
Caribou, *Rangifer tarandus***	*circa* 1910
Bison, *Bison bison*	*circa* 1800

*Reintroduced by man
**On Isle Royale as late as 1925

Table 4. Mammals Considered as Endangered, Threatened, Rare, or Peripheral in Michigan.

ENDANGERED SPECIES*
Indiana Bat, *Myotis sodalis*
Gray Wolf (eastern subspecies), *Canis lupus lycaon*
THREATENED SPECIES**
Marten (introduced population), *Martes americana*
Lynx, *Felis lynx*
RARE SPECIES***
Hoary Bat, *Lasiurus cinereus*
Woodland Vole, *Microtus pinetorum*
Wapiti or Elk (introduced population), *Cervus elaphus*
PERIPHERAL SPECIES***
Smoky Shrew, *Sorex fumeus*
Least Shrew, *Cryptotis parva*
Eastern Pipistrelle, *Pipistrellus subflavus*
Evening Bat, *Nycticeius humeralis*
Prairie Vole, *Microtus ochrogaster*
Moose, *Alces alces*

*Determined under the Endangered Species Act of 1973 (P.L. 93-205).

**Determined under Michigan's Comprehensive Endangered Species Law, Act No. 203, Public Acts of 1974.

***Lack legal status under the Michigan Endangered Species Act.

have been successful in adapting to human land-use practices. Those endangered and threatened species (see Table 4) have often not thrived because of these environmental changes.

GEOGRAPHICAL AFFINITIES OF MICHIGAN MAMMALS

The 66 Michigan mammals are arranged in Table 5 in terms of their specific distributions within both continental and local political areas. Fully three-quarters of the species (48) have distributions which are entirely North American, including Central America. Of these, 29 species (43.9%) occur only in the United States and Canada while 21 species (31.8%) also occur in Mexico and in some cases other Central American countries. Some seven species (10.6%)—among them two important game species, the eastern cottontail and the white-tailed deer—include at least northern South America within their distributions. Finally, nine species (13.6%) have circumboreal distributions and occur both in North America and in Eurasia.

Table 5. Geographic Affinities of Michigan Mammals.

GEOGRAPHICAL AREAS	ASSOCIATED MAMMALS	PERCENT
NORTH AMERICA and EURASIA	Gray Wolf, *Canis lupus* Red Fox, *Vulpes vulpes* Ermine, *Mustela erminea* Least Weasel, *Mustela nivalis* Wolverine, *Gulo gulo* Lynx, *Felis lynx* Wapiti, *Cervus elaphus* Moose, *Alces alces* Caribou, *Rangifer tarandus*	13.6
NORTH AMERICA and SOUTH AMERICA	Red Bat, *Lasiurus borealis* Hoary Bat, *Lasiurus cinereus* Eastern Cottontail, *Sylvilagus floridanus* Gray Fox, *Urocyon cinereoargenteus* Long-Tailed Weasel, *Mustela frenata* Mountain Lion, *Felis concolor* White-Tailed Deer, *Odocoileus virginianus*	10.6

Table 5. Cont'd.

GEOGRAPHICAL AREAS	ASSOCIATED MAMMALS	PERCENT
NORTH AMERICA—Canada to Mexico	Virginia Opossum, *Didelphis virginiana* Least Shrew, *Cryptotis parva* Eastern Mole, *Scalopus aquaticus* Little Brown Bat, *Myotis lucifugus* Eastern Pipistrelle, *Pipistrellus subflavus* Big Brown Bat, *Eptesicus fuscus* Evening Bat, *Nycticeius humeralis* Fox Squirrel, *Sciurus niger* Southern Flying Squirrel, *Glaucomys volans* Beaver, *Castor canadensis* White-Footed Mouse, *Peromyscus leucopus* Deer Mouse, *Peromyscus maniculatus* Meadow Vole, *Microtus pennsylvanicus* Porcupine, *Erethizon dorsatum* Coyote, *Canis latrans* Black Bear, *Ursus americanus* Raccoon, *Procyon lotor* Badger, *Taxidea taxus* Striped Skunk, *Mephitis mephitis* Bobcat, *Felis rufus* Bison, *Bison bison*	31.8
NORTH AMERICA—Canada and United States	Arctic Shrew, *Sorex arcticus* Masked Shrew, *Sorex cinereus* Smoky Shrew, *Sorex fumeus* Pygmy Shrew, *Sorex hoyi* Water Shrew, *Sorex palustris* Short-Tailed Shrew, *Blarina brevicauda* Star-Nosed Mole, *Condylura cristata* Keen's Bat, *Myotis keenii* Indiana Bat, *Myotis sodalis* Silver-Haired Bat, *Lasionycteris noctivagans* Snowshoe Hare, *Lepus americanus* Eastern Chipmunk, *Tamias striatus* Least Chipmunk, *Eutamias minimus* Woodchuck, *Marmota monax* Thirteen-Lined Ground Squirrel, *Spermophilus tridecemlineatus* Gray Squirrel, *Sciurus carolinensis* Red Squirrel, *Tamiasciurus hudsonicus* Northern Flying Squirrel, *Glaucomys sabrinus* Southern Red-Backed Vole, *Clethrionomys gapperi* Prairie Vole, *Microtus ochrogaster* Woodland Vole, *Microtus pinetorum* Muskrat, *Ondatra zibethicus* Southern Bog Lemming, *Synaptomys cooperi* Meadow jumping Mouse, *Zapus hudsonius* Woodland Jumping Mouse, *Napaeozapus insignis* Marten, *Martes americana* Fisher, *Martes pennanti* Mink, *Mustela vison* River otter, *Lutra canadensis*	43.9

ENVIRONMENTAL AFFINITIES OF MICHIGAN MAMMALS

When the Michigan environment began to stabilize about 5,500 years ago (Cleland, 1966; Dorr and Eschman, 1971), two major environments for mammals became evident (Barrows, 1912; Burt, 1946; Dice, 1943; Odum, 1953; Somers, 1977; Veatch, 1953, 1959). In the Upper Peninsula and the northern part of the Lower Peninsula, the cool climate and Spodosol (Podzol) soils nourished northern conifer and hardwood forests belonging to the Northern Coniferous Forest Biome, which stretches as a broad belt across boreal North America. In the southern part of the Lower Peninsula, the mild climate and gray-brown Alfisol (Podzolic) soils encouraged the growth of southern hardwoods (beech, maples, oaks, and hickories) belonging to the Temperate Deciduous Forest Biome, which occurs throughout southeastern United States. The distributional picture was complicated, however, by a broad transition zone between these two biomes, allowing for interdigitation of characteristic plants and animals. This transition habitat occurs both in the central sector of the Lower Peninsula as well as in the extreme south-central part of the Upper Peninsula along Green Bay in the Menominee County area (Map 1). Some inroads of the Grassland Biome (from the Midwest) occur in glades in southwestern Michigan (Butler, 1978; Transeau, 1935); the clearing of forest lands following settlement encouraged mammals characteristic of this biome to spread northeastward.

Twenty-one of Michigan's 66 native mammals (31.8%) have close association with conifer-hardwood environments of the Northern Coniferous Forest Biome. Fifteen (22.7%) are characteristic of the hardwood environments of the Southern Deciduous Forest Biome. Seven (10.6%) are primarily open land mammals with affinities with the Grassland Biome. The other 23 species (34.8%) are sufficiently well adapted to more than one biome to be considered non specific in terms of environmental association in the Upper Great Lakes Region (see Table 6.)

Of special interest is the deer mouse, *Peromyscus maniculatus*. A woodland-adapted subspecies, *P. m.*

Table 6. Ecologic Affinities of Michigan Mammals with Biomes.

MAJOR BIOMES	ASSOCIATED MAMMALS*	PERCENT
NORTHERN CONIFEROUS FORESTS	Arctic Shrew, *Sorex arcticus* Masked Shrew, *Sorex cinereus* Smoky Shrew, *Sorex fumeus* Pygmy Shrew, *Sorex hoyi* Water Shrew, *Sorex palustris* Star-Nosed Mole, *Condylura cristata* Snowshoe Hare, *Lepus americanus* Least Chipmunk, *Eutamias minimus* Red Squirrel, *Tamiasciurus hudsonicus* Northern Flying Squirrel *Glaucomys sabrinus* Deer Mouse, *Peromyscus maniculatus gracilis* (woodland subspecies)* Southern Red-Backed Vole, *Clethrionomys gapperi* Meadow Vole, *Microtus pennsylvanicus* Woodland Jumping Mouse, *Napaeozapus insignis* Porcupine, *Erethizon dorsatum* Marten, *Martes americana* Fisher, *Martes pennanti* Ermine, *Mustela erminea* Wolverine, *Gulo gulo* Lynx, *Felis lynx* Moose, *Alces alces* Caribou, *Rangifer tarandus*	31.8

Table 6. Cont'd.

MAJOR BIOMES	ASSOCIATED MAMMALS*	PERCENT
SOUTHERN DECIDUOUS FORESTS	Virginia Opossum, *Didelphis virginiana* Short-Tailed Shrew, *Blarina brevicauda* Least Shrew, *Cryptotis parva* Eastern Mole, *Scalopus aquaticus* Indiana Bat, *Myotis sodalis* Eastern Pipistrelle, *Pipistrellus subflavus* Evening Bat, *Nycticeius humeralis* Eastern Cottontail, *Sylvilagus floridanus* Eastern Chipmunk, *Tamias striatus* Gray Squirrel, *Sciurus carolinensis* Fox Squirrel, *Sciurus niger* Southern Flying Squirrel, *Glaucomys volans* White-Footed Mouse, *Peromyscus leucopus* Woodland Vole, *Microtus pinetorum* Gray Fox, *Urocyon cinereoargenteus*	22.7
GREAT PLAINS GRASSLANDS	Thirteen-Lined Ground Squirrel, *Spermophilius tridecemlineatus* Deer Mouse, *Peromyscus maniculatus bairdii* (prairie subspecies)* Prairie Vole, *Microtus ochrogaster* Least Weasel, *Mustela nivalis* Badger, *Taxidea taxus* Wapiti, *Cervus elaphus* Bison, *Bison bison*	10.6
INTERBIOME IN DISTRIBUTION	Keen's Bat, *Myotis keenii* Little Brown Bat, *Myotis lucifugus* Silver-Haired Bat, *Lasionycteris noctivagans* Big Brown Bat, *Eptesicus fuscus* Red Bat, *Lasiurus borealis* Hoary Bat, *Lasiurus cinereus* Woodchuck, *Marmota monax* Beaver, *Castor canadensis* Muskrat, *Ondatra zibethicus* Southern Bog Lemming, *Synaptomys cooperi* Meadow Jumping Mouse, *Zapus hudsonius* Coyote, *Canis latrans* Gray Wolf, *Canis lupus* Red Fox, *Vulpes vulpes* Black Bear, *Ursus americanus* Raccoon, *Procyon lotor* Long-Tailed Weasel, *Mustela frenata* Mink, *Mustela vison* Striped Skunk, *Mephitis mephitis* River Otter, *Lutra canadensis* Mountain Lion, *Felis concolor* Bobcat, *Felis rufus* White-Tailed Deer, *Odocoileus virginianus*	34.8

*The deer mouse, *Peromyscus maniculatus*, is counted both in the Northern Coniferous Forest Biome and the Southern Deciduous Forest Biome (see text).

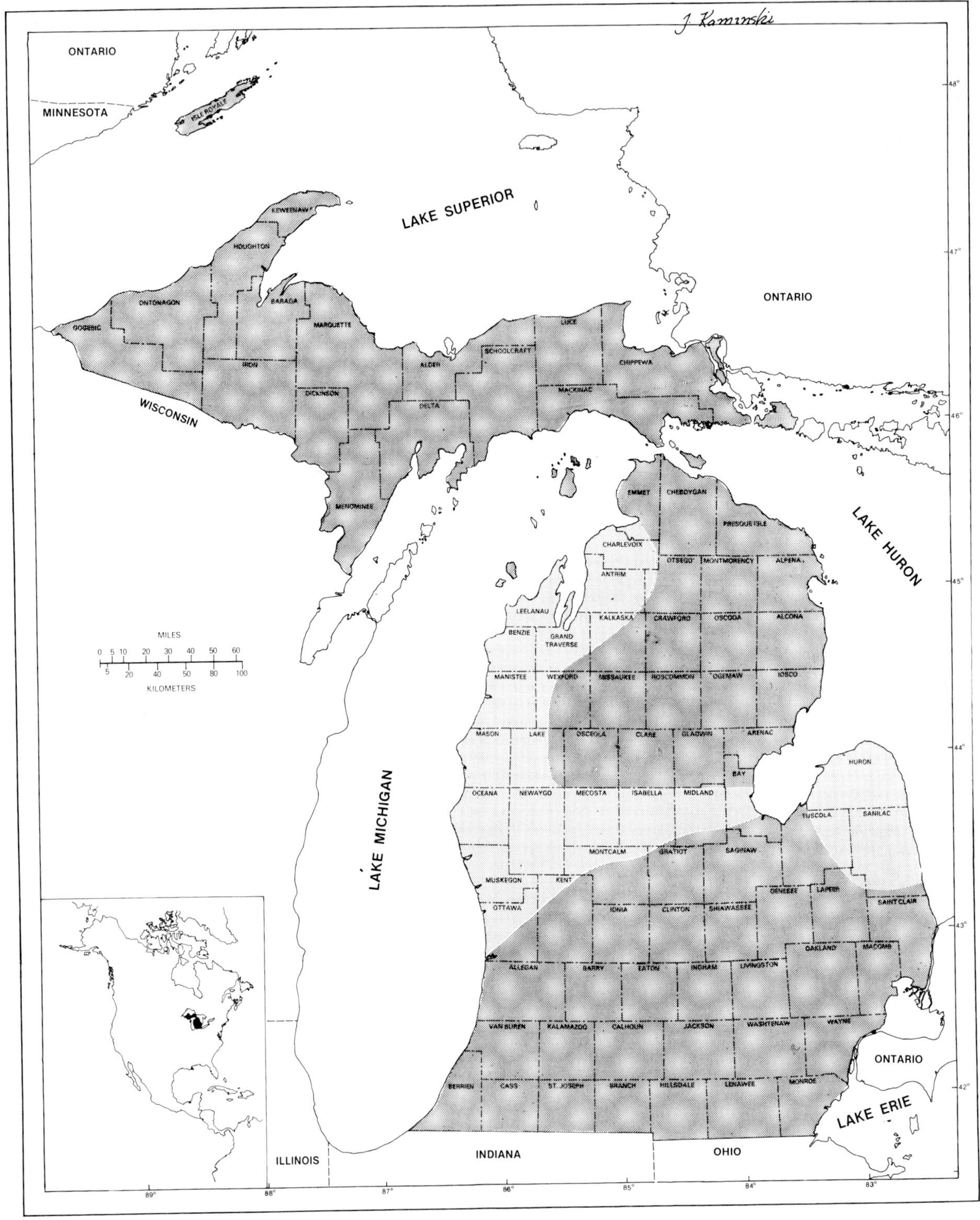

Map 1. Geographic distribution of biomes. Areas in the Upper Peninsula and the northern part of the Lower Peninsula marked with dark shading are within the Northern Coniferous Forest Biome. The area in the southern part of the Lower Peninsula marked with dark shading is within the Temperate Deciduous Forest Biome. A zone of transition in the middle part of the Lower Peninsula is marked with pale shading.

gracilis, has apparently been a longtime resident of Michigan's northern forests. An openland-adapted subspecies, *P. m. bairdii,* is suspected of being a recent arrival to southern Michigan from the western prairies, thriving mostly in cultivated and recently fallowed lands (Handley, 1971). A third subspecies, *P. m. maniculatus,* on Isle Royale shows major affinity with the deer mouse population on the adjacent Ontario mainland to the north.

On close inspection of local environments, there is a great diversity of situations hospitable to mammals. These habitats include bogs, swamps, streamside (riparian) growth, cleared or forested uplands, cleared or forested poorly drained lowlands, and cultivated or fallow farmland (Dice, 1932; Stearns and Kobriger, 1975). The ecological distributions of six common small mammals in relation to the various plant successional stages of southern Michigan farmland are shown in Table 7.

EFFECTS OF THE UPPER GREAT LAKES WATERSHED ON MAMMALS

Each of Michigan's two peninsulas is surrounded on three sides by bodies of water and interconnecting streamways of the Upper Great Lakes system. The Upper Peninsula projects eastward as a boreal forested extension of northeastern Wisconsin to which the elongated land mass bears its closest faunal affinities. To the north, Lake Superior and the fast-moving waters of the St. Marys River separate the Upper Peninsula from Ontario. To the south and east, lakes Michigan and Huron (connected by the deep Straits of Mackinac) separate it from the Lower Peninsula.

The Lower Peninsula, likewise, projects northward as a southern deciduous forested extension of the habitat characteristic of southeastern United States. However, its northern half is covered with boreal vegetation, presumably isolated there since early postglacial times, and inhabited by a number of mammals characteristic of the Northern Coniferous Biome. To the west, the Lower Peninsula is bordered by Lake Michigan, to the north by the Straits of Mackinac, to the northeast by Lake Huron, and to the southeast by the fast-moving St. Clair and Detroit rivers with Lake St. Clair in between. It is well to emphasize that the boreal mammals in the northern half of the Lower Peninsula are completely surrounded either by water or by alien southern deciduous habitat.

These water barriers (Map 2), especially the narrower interlake connectors, have influenced the distribution and subspeciation of non-volant mammals living in the Upper and Lower Pen-

Table 7. Habitat preferences of six common small mammals in plant successional stages in upland areas in southern Michigan. After abandonment, cultivated cropland passes through several vegetative phases as it reverts to forest. This process is rapid at first with bare ground covered by annual weeds in the first growing season. Perennial grasses and forbs appear in two years, remaining for as long as ten years depending upon the rate of shrub and tree seedling growth, and after 20 to 40 years a woodland canopy forms.

SMALL MAMMALS	CULTIVATED → FIELDS	ANNUAL WEEDS GRASSES AND → FORBS	PERENNIAL GRASSES → AND FORBS	SHRUB AND TREE → SEEDLINGS	MATURE HARDWOOD FORESTS
Deer Mouse (*Peromyscus maniculatus bairdii*)	——	——	——		
Short-Tailed Shrew (*Blarina brevicauda*)	——	——	——	——	——
Meadow Jumping Mouse (*Zapus hudsonius*)		——	——	——	——
Masked Shrew (*Sorex cinereus*)		——	——	——	——
Meadow Vole (*Microtus pennsylvanicus*)		——	——	——	
White-Footed Mouse (*Peromyscus leucopus*)			——	——	——

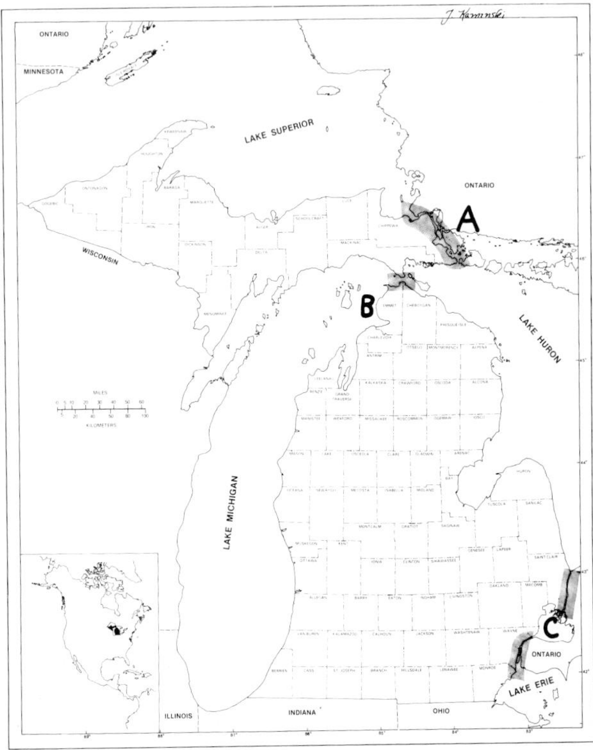

Map 2. Aquatic barriers influencing distribution and subspeciation of non-volant mammals. A. Superior-Huron Waterway Connection; B. Michigan-Huron Waterway Connection; C. Huron-Erie Waterway Connection.

insulas and adjacent parts of Ontario. To move across these water barriers (in search of suitable habitats or to interbreed with relatives on the other side), the resident mammals would have to either be strong swimmers or excel at ice-crossing. Rafting on floating debris could also conceivably be a means of crossing (McCabe and Cowan, 1945), and there is always a possibility that mammals are transported (purposely or not) as a result of human activities. However, mammalian distribution in this area is not primarily due to any of these water-crossing techniques. Since glacial times and the subsequent formation of the Upper Great Lakes basins (Dorr and Eschman, 1971), the only means by which non-volant mammals could enter the Upper Peninsula has been from the west through what is now Wisconsin. Likewise, to enter the Lower Peninsula, terrestrial mammals had to move from the south through what is now Ohio and Indiana. Even so, of the 57 non-flying mammals native to Michigan within historic times (nine species of bats excluded), all but nine have been recorded in both peninsulas (see Table 8).

In the Upper Peninsula, 43 mammals (86% of its mammalian fauna) are widespread in both western and eastern counties. Seven others occur only in the extreme southern part, Menominee County mostly, and have affinities with southern forest or open habitats (see Table 9). The gray fox (*Urocyon cinereoargenteus*) is an example of a southern species which has extended its range to the eastern part of the Upper Peninsula as well. In the Lower Peninsula, 40 species (74% of its mammalian fauna) occur in both the northern boreal and the southern temperate environments. The eight species restricted to the northern part of the Lower Peninsula (see Table 10) are boreal in affinity and show close relationship to populations of the same species in the Upper Peninsula across the Straits of Mackinac. With the exception of powerful-swimming moose and caribou, these species may well have been residents of this isolated pocket since the

Table 8. Non-volant mammals with distribution restricted to only one of Michigan's two peninsulas.

FOUND ONLY IN THE UPPER PENINSULA	FOUND ONLY IN THE LOWER PENINSULA
Arctic Shrew, *Sorex arcticus*	Least Shrew, *Cryptotis parva*
Smoky Shrew, *Sorex fumeus**	Eastern Mole, *Scalopus aquaticus*
Least Chipmunk, *Eutamias striatus*	Prairie Vole, *Microtus ochrogaster*
	Woodland Vole, *Microtus pinetorum*
	Wapiti, *Cervus elaphus*
	Bison, *Bison bison*

*Only on Sugar Island, Chippewa County.

Table 9. Non-volant mammals occurring in the Upper Peninsula, only in Menominee County and vicinity.

SPECIES
Virginia Opossum, *Didelphis virginiana*
Eastern Cottontail, *Sylvilagus floridanus*
Thirteen-Lined Ground Squirrel, *Spermophilus tridecemlineatus*
Fox Squirrel, *Sciurus niger**
Southern Flying Squirrel, *Glaucomys volans*
White-Footed Mouse, *Peromyscus leucopus*
Least Weasel, *Mustela nivalis*

*Populations introduced through human agency in Chippewa and Mackinac counties not considered here.

Table 10. Non-volant mammalian distributional patterns in the Lower Peninsula.

SPECIES CONFINED TO THE NORTHERN PART OF THE LOWER PENINSULA	SPECIES CONFINED TO THE SOUTHERN PART OF THE LOWER PENINSULA
Water Shrew, *Sorex palustris*	Virginia Opossum, *Didelphis virginiana***
Snowshoe Hare, *Lepus americanus**	Least Shrew, *Cryptotis parva*
Northern Flying Squirrel, *Glaucomys sabrinus*	Southern Flying Squirrel, *Glaucomys volans***
Southern Red-Backed Vole, *Clethrionomys gapperi*	Prairie Vole, *Microtus ochrogaster*
Woodland Jumping Mouse, *Napaeozapus insignis*	Least Weasel, *Mustela nivalis*
Ermine, *Mustela erminea**	Bison, *Bison bison*
Moose, *Alces alces*	
Caribou, *Rangifer tarandus**	

*A few records from southern counties are known
**A few records from northern counties are known

time several thousand years ago when boreal habitats shifted northward following glacial melt-back (Handley, 1971). The arctic shrew (*Sorex arcticus*) and the least chipmunk (*Eutamias minimus*) are the only two boreal species occurring in the Upper Peninsula which are absent from suitable habitats south of the Straits of Mackinac. These species are derived from the west instead of directly from the south, and are possibly rather recent arrivals to Michigan (Handley, 1971).

The six species occurring in the temperate habitats of the southern part of the Lower Peninsula (see Table 10) have either southern or southwestern environmental affinities (Handley, 1971). On the other hand, several other such species, notably the eastern mole (*Scalopus aquaticus*), eastern cottontail (*Sylvilagus floridanus*), fox squirrel (*Sciurus niger*), thirteen-lined ground squirrel (*Spermophilus tridecemlineatus*), and the white-footed mouse (*Peromyscus leucopus*), have—apparently in response to environmental changes resulting from the mosaic pattern of human settlement—spread northward through the entire Lower Peninsula, although not necessarily using areas where boreal environments have been little altered.

Current knowledge of the distribution and subspeciation of non-volant mammals influenced by the drainage pattern on the Upper Great Lakes Region is summarized as follows in terms of the barrier effect on the three narrow waterway connections between lakes Superior and Huron, Michigan and Huron, and Huron and Erie (Map 2).

The Superior-Huron Waterway Connection—The St. Marys River and the island-filled downstream expansion form a narrow drain for Lake Superior into Lake Huron and separate the eastern end of the Upper Peninsula from adjacent Ontario (see Map 2). Of the 47 species of non-volant mammals living on one or both sides of this interlake connection, seven have apparently been barred from crossing while ten others occur on both sides but are classified as being subspecifically distinctive (see Table 11). This indicates that 38% of this mammalian community is influenced by this aquatic obstruction. Recent studies (W. L. Robinson, pers. comm.) disclose that the highly mobile coyote, gray wolf, red fox, white-tailed deer, and moose regularly cross on the winter ice from island to island in this waterway connection. But Upper Peninsula mammals generally have a greater affinity for those species in Wisconsin to the west than those in Ontario to the east (Banfield, 1974; Handley, 1971; Jackson, 1961).

The Michigan-Huron Waterway Connection—The Straits of Mackinac, through which water from Lake Michigan flows into Lake Huron, separates by almost 6 mi (8 km) the Upper and Lower Peninsulas (see Map 2). Of the 51 species of non-volant mammals living on one or both sides of this water barrier, at least nine are found on only one side, while nine others occur on both sides but are classified as belonging to different subspecies (Table 12). These data show that 35% of the mammals living in the area of the Michigan-Huron Waterway Connection have found this water gap

Table 11. Influence of Superior-Huron waterway connection on the distribution and subspeciation of non-volant mammals of Michigan's Upper and Lower Peninsulas.

MAMMALS FOUND ON ONE SIDE OF THE CONNECTION BUT NOT ON THE OTHER	MAMMALS FOUND ON BOTH SIDES OF THE CONNECTION BUT DISTINGUISHABLE AS DIFFERENT SUBSPECIES
On the Ontario side only	Short-Tailed Shrew, *Blarina brevicauda*
Hairy-Tailed Mole, *Parascalops breweri*	Snowshoe Hare, *Lepus americanus*
Heather Vole, *Phenacomys intermedius*	Eastern Chipmunk, *Tamias striatus*
Rock Vole, *Microtus chrotorrhinus*	Least Chipmunk, *Eutamias minimus*
	Woodchuck, *Marmota monax*
On the Upper Peninsula side only	Meadow Jumping Mouse, *Zapus hudsonius*
Arctic Shrew, *Sorex arcticus*	Woodland Jumping Mouse, *Napaeozapus insignis*
Gray Squirrel, *Sciurus carolinensis*	Red Fox, *Vulpes vulpes*
Gray Fox, *Urocyon cinereoargenteus*	Striped Skunk, *Mephitis mephitis*
Badger, *Taxidea taxus*	Bobcat, *Felis rufus*

an effective barrier to north-south dispersal. As mentioned previously, the eight boreal species (see Table 10) confined to the northern part of the Lower Peninsula include six (moose and caribou excluded) which were undoubtedly trapped in this isolated area when boreal habitats shifted northward in early postglacial times. A distinctive subspecies of the eastern chipmunk (*Tamias striatus peninsulae*) also occurs in this restricted area and is considered part of this group of early arrivals. The extent of the intergradation of this subspecies with eastern chipmunks of the subspecies *T. s. rufescens* in the southern part of the Lower Peninsula has not yet been fully determined.

The Huron-Erie Waterway Connection—The St. Clair and Detroit rivers form narrow streamways connecting lakes Huron, St. Clair, and Erie and separating the southeastern part of the Lower Peninsula from adjacent southeastern Ontario (see Map 2). Of the 51 species of non-volant mammals living on one or both sides of this waterway, at least ten have

Table 12. Influence of the Michigan-Huron waterway connection on the distribution and subspeciation of non-volant mammals of Michigan's Upper and Lower Peninsulas.

MAMMALS FOUND ON ONE SIDE OF THE CONNECTION BUT NOT ON THE OTHER*	MAMMALS FOUND ON BOTH SIDES OF THE CONNECTION BUT DISTINGUISHABLE AS DIFFERENT SUBSPECIES*
On the Upper Peninsula side only	
Arctic Shrew, *Sorex arcticus*	Pygmy Shrew, *Sorex hoyi*
Least Chipmunk, *Eutamias minimus*	Eastern Chipmunk, *Tamias striatus*
	Woodchuck, *Marmota monax*
	Gray Squirrel, *Sciurus carolinensis*
On the Lower Peninsula side only	Meadow Jumping Mouse, *Zapus hudsonius*
Eastern Mole, *Scalopus aquaticus*	Mink, *Mustela vison*
Eastern Cottontail, *Sylvilagus floridanus*	Striped Skunk, *Mephitis mephitis*
Thirteen-Lined Ground Squirrel, *Spermophilus tridecemlineatus*	Mountain Lion, *Felis concolor*
Fox Squirrel, *Sciurus niger***	Bobcat, *Felis rufus*
White-Footed Mouse, *Peromyscus leucopus*	
Woodland Vole, *Microtus pinetorum*	
Wapiti, *Cervus elaphus*	

*Species of mammals found only in the Menominee County area of southwestern Upper Peninsula (Table 9) and the smoky shrew (*Sorex fumeus*), known only from Sugar Island, Chippewa County, not included.
**Introduction in Chippewa and Mackinac counties not considered.

failed to cross while two occur on both sides but are classified as belonging to distinctive subspecies (see Table 13). Thus, the distribution of about 24% of the local mammals are influenced by this barrier to east-west dispersal. This aquatic obstruction to mammalian movements may gain meaning to Detroit-Windsor area citizens when they note fox squirrels in Detroit's city parks but not in Windsor's.

The extent to which these water barriers isolate the non-volant mammals of the Upper Great Lakes Region cannot be fully understood until detailed biochemical studies are made. Such studies could be designed to compare the genetic characteristics of populations of the same species living on each side of these waterways. White-tailed deer investigations by Manlove (1979) indicate a genetic gap between these populations. His examination of protein phenotypes, using starch-gel electrophoresis of muscle tissue extracts, found population differences attributable to at least partial isolation by the barrier effect of the Straits of Mackinac.

In summary, the fascinating mammals of Michigan are (1) newcomers in the geological sense, being derived from hardy post-glacial ancestors; (2) diversified by being derived from two major environments, the northern coniferous and the temperate hardwood forest, broadly overlapping within Michigan, with most boreal species in the Upper Peninsula being derived from the west (Wisconsin) and many of those in the northern part of the Lower Peninsula being isolated there since early postglacial times; (3) profoundly influenced in terms of distributional patterns and subspeciation by the isolating effects of the drainage pattern of the Upper Great Lakes; and (4) co-inhabitants of Michigan with the human species from the postglacial beginnings, compatibly at first but seriously disturbed by the environmental changes of the second wave of human intrusion, mostly in the past 150 years.

LITERATURE CITED

Amundson, D. C., and H. E. Wright, Jr.
1979 Forest changes in Minnesota at the end of the Pleistocene. Ecol. Monogr., 49(1):1–16.

Banfield, A. W. F.
1974 The mammals of Canada. Univ. Toronto Press, Toronto. xxiv+438 pp.

Barrows, W. B.
1912 Michigan bird life. Michigan Agric. College, xiv+822 pp.

Blois, J. T.
1839 Gazetteer of the State of Michigan. Sydney L. Rood and Co., Detroit. xi+418 pp.

Burt, W. H.
1946 The Mammals of Michigan. Univ. Michigan Press, Ann Arbor. xv+288 pp.

Table 13. Influence of the Huron-Erie waterway connection on the distribution and subspeciation of non-volant mammals of Michigan's Upper and Lower Peninsulas.

MAMMALS FOUND ON ONE SIDE OF THE CONNECTION BUT NOT ON THE OTHER*	MAMMALS FOUND ON BOTH SIDES OF THE CONNECTION BUT DISTINGUISHABLE AS DIFFERENT SUBSPECIES*
On the Lower Peninsula side only	
Least Shrew, *Cryptotis parva*	Eastern Chipmunk, *Tamias striatus*
Thirteen-Lined Ground Squirrel, *Spermophilus tridecemlineatus*	Mink, *Mustela vison*
Fox Squirrel, *Sciurus niger*	
Least Weasel, *Mustela nivalis*	
*On the Ontario side only**	
Smoky Shrew, *Sorex fumeus*	
Pygmy Shrew, *Sorex hoyi*	
Hairy-Tailed Mole, *Parascalops breweri*	
Northern Flying Squirrel, *Glaucomys sabrinus*	
Woodland Jumping Mouse, *Napaeozapus insignis*	
Ermine, *Mustela erminea***	

*Advice of Randolph L. Peterson is acknowledged.

**Although there are scattered records of this species in southern Michigan, its distribution in southern counties is inconsistent and marginal.

1972 Mammals of the Great Lakes Region. Ann Arbor Paperbacks, Univ. Michigan Press, Ann Arbor. xv+246 pp.

Butler, A. F.
1978 Prairies lost, prairies found. Michigan Nat. Res. Mag., 47(3):32–39.

Cleland, C. E.
1965 Barren ground caribou (*Rangifer arcticus*) from an early man site in southeastern Michigan. American Antiquity, 30(3):350–351.
1966 The prehistoric animal ecology and ethnozoology of the Upper Great Lakes Region. Univ. Michigan, Mus. Anthro., Anthro. Papers No. 29, x+294 pp.
1975 A brief history of Michigan Indians. Michigan Dept. State, Hist. Div., 38 pp.

Covert, A. B.
1881 Natural history. Chapter V, pp. 173–194 *in* History of Washtenaw County, Michigan. Chas C. Chapman & Co., Chicago. 1,452 pp.

Dice, L. R.
1932 A preliminary classification of the major terrestrial ecologic communities of Michigan, exclusive of Isle Royale. Michigan Acad. Sci., Arts and Ltrs., Papers, 16:217–239.
1943 The biotic provinces of North America. Univ. Michigan Press, Ann Arbor. viii+78 pp.

Dorr, J. A., Jr., and D. F. Eschman
1971 Geology of Michigan. Univ. Michigan Press, Ann Arbor. viii+576 pp.

Farrand, W. R., and D. F. Eschman
1974 Glaciation of the Southern Peninsula of Michigan: A review. Michigan Acad., 7(1):31–56.

Gibbs, M.
1895 The rodents of Michigan. The Museum, 1:145–152.

Gottschang, J. L.
1981 A guide to the mammals of Ohio. Ohio State Univ. Press, Columbus. 181 pp.

Hall, E. R.
1981 The mammals of North America. Second Edition. John Wiley and Sons, New York. Vol. 1, xv+600+90 pp.; Vol. 2, vi+601–1181+90 pp.

Handley, C. O., Jr.
1953 Marine mammals in Michigan Pleistocene beaches. Jour. Mammalogy, 34(2):252–253.
1971 Appalachian mammalian geography—Recent Epoch. Virginia Poly. Inst. and State Univ., Research Div. Monogr. 4:263–303.

Hazard, E. B.
1982 The mammals of Minnesota. Univ. Minnesota Press, Minneapolis. xii+280 pp.

Hibbard, C. W.
1951 Animal life of Michigan during the Ice Age. Michigan Alumnus Quart. Rev., 57(18):200–208.

Hoffmeister, D. F., and C. O. Mohr
1957 Fieldbook of Illinois mammals. Illinois Nat. Hist. Surv. Div., Manual 4, xi+233 pp.

Holman, J. A.
1975 Michigan's fossil vertebrates. Michigan State Univ., Publ. Mus., Educ. Bull. No. 2, 54 pp.

Jackson, H. H. T.
1961 Mammals of Wisconsin. Univ. of Wisconsin Press, Madison. xii+504 pp.

Kelley, R. W., and W. R. Farrand
1967 The glacial lakes around Michigan. Michigan Dept. Conserv., Geol. Surv., Bull. 4, 23 pp.

Manlove, M. N.
1979 Genetic similarity among contiguous and isolated populations of white-tailed deer in Michigan. Michigan State Univ., unpubl. M.S. thesis, 49 pp.

McCabe, T. T., and I. McT. Cowan
1945 *Peromyscus maniculatus macrorhinus* and the problem of insularity. Trans. Royal Canadian Inst., No. 54, Vol. 25, pt. 2, pp. 117–215.

Miles, M.
1861 A catalogue of the mammals, birds, reptiles and mollusks of Michigan. Geol. Surv. Michigan, First Biennial Rept., pp. 219–241.

Mumford, R. E.
1969 Distribution of the mammals of Indiana. Indiana Acad. Sci., Monogr. No. 1, vii+114 pp.

Mumford, R. E., and J. O. Whitaker, Jr.
1982 Mammals of Indiana. Indiana Univ. Press, Bloomington. 537 pp.

Odum, E. P.
1953 Fundamentals of ecology. W. B. Saunders Co., Philadelphia. xii+384 pp.

Peterson, R. L.
1966 The mammals of eastern Canada. Oxford Univ. Press, Toronto. xxxii+465 pp.

Rausch, R. L.
1963 A review of the distribution of Holarctic Recent mammals. Tenth Pacific Sci. Congr., Bishop Museum Press, Honolulu, pp. 29–43.

Sager, A.,
1839 Report of zoologist of geological survey. Pp. 410–421 *in* Second Annual Report of State Geologist, Michigan House Document No. 23.

Somers, L. M., ed.
1977 Atlas of Michigan. Michigan State Univ. Press, East Lansing. xi+242 pp.

Stearns, F., and N. Kobriger
1975 Environmental status of the Lake Michigan region. Vegetation of the Lake Michigan drainage basin. Argonne Natl. Lab., ANL/ES-40, Vol. 10, 113 pp.

Strang, J. J.
1855 Some remarks on the natural history of Beaver Islands, Michigan. Smithsonian Inst., Ninth Ann. Rept., 1854, pp. 282–288.

Transeau, E. N.
1935 The prairie peninsula. Ecology, 16:423–437.

Veatch, J. O.
1953 Soils and land of Michigan. Michigan State Col. Press, East Lansing, xi+241 pp.
1959 Presettlement forest in Michigan. Michigan State Univ., Dept. Resource Develop.

Wilson, R. L.
1967 The Pleistocene vertebrates of Michigan. Michigan Acad. Sci., Arts and Ltrs., Papers, 52:197–234.

Plate 1. Michigan mammals photographed by Robert Carr. Virginia opossum, *Didelphis virginiana* (above), and short-tailed shrew, *Blarina brevicauda* (below).

Plate 2. The snowshoe hare, *Lepus americanus,* in winter (above) and summer (below) pelages.

Paintings by David Mohrhardt.

Plate 3. Michigan mammals photographed by Robert Carr. White-footed mouse, *Peromyscus leucopus* (above), and badger, *Taxidea taxus* (below).

Plate 4a. The deer mouse, *Peromyscus maniculatus* (left) and the woodland jumping mouse, *Napaeozapus insignis* (right).

Plate 4b. Widespread Michigan voles: southern red-backed vole, *Clethrionomys gapperi* (above), meadow vole, *Microtus pennsylvanicus* (middle), southern bog lemming, *Synaptomys cooperi* (below).

Paintings by James Zablotny.

Plate 5. Profiles of the red fox, *Vulpes vulpes* (above), and the gray fox, *Urocyon cineroargenteus* (below).
Paintings by Gijsbert van Frankenhuyzen.

Plate 6. Portraits of the lynx, *Felis lynx* (above), and the bobcat, *Felis rufus* (below).

Paintings by David Mohrhardt.

Accounts of Mammals

The mammals discussed in the following accounts are arranged in orders, families, and genera following Simpson (1945) and in species following Hall (1981). Common names are from Jones *et al.* (1979). In the heading for each account, the scientific name (listed there and elsewhere in italics) is followed by the name of the person who originally described the species. Some of the names of the authorities are in parentheses because in the original description the mammal was named as belonging to a genus different from that currently recognized. For purposes of the accounts, each mammal is designated by generic and specific names only, with subspecific names given in text as appropriate. For several species, geographic variation within the state has been sufficient that Upper Peninsula populations may be classed as belonging to different subspecies from Lower Peninsula populations. The eastern chipmunk (*Tamias striatus*) has distinctive subspecies named from the southern part of the Lower Peninsula, from the northern part of the Lower Peninsula, and from the Upper Peninsula.

For each species of mammal, measurements and weights of adults are given in both millimeters and grams or kilograms and in inches or feet and ounces or pounds. Listed measurements and weights are taken from Michigan specimens when possible. Technical words used in describing mammals and their habits are explained in the text or defined in the Glossary. This book is not intended to be an all-inclusive account of Michigan mammals. For further details, works such as Cockrum (1962), Gunderson (1976), and Vaughan (1978) and specific articles reporting original research findings in such technical periodicals as the Journal of Mammalogy may be consulted.

LITERATURE CITED

Cockrum, E. L.
1962 Introduction to mammalogy. Ronald Press Co., New York. viii+455 pp.

Gunderson, H. L.
1976 Mammalogy. McGraw-Hill Book Co., New York. viii+483 pp.

Vaughan, T. A.
1978 Mammalogy. Second Edition. W. B. Saunders Co., Philadelphia. x+522 pp.

Jones, J. K., Jr., D. C. Carter, and H. H. Genoways
1979 Revised checklist of North American mammals north of Mexico, 1979. Texas Tech Univ., Mus., Occas. Papers No. 62, 17 pp.

Hall, E. R.
1981 The mammals of North America. Second Edition. John Wiley and Sons, New York, Vol. 1, xv+600+90 pp.; Vol. 2, vi+601–1181+90 pp.

Simpson, G. G.
1945 The principles of classification and a classification of mammals. American Mus. Nat. Hist., Bull., Vol. 85, xvi+350 pp.

KEY TO THE ORDER OF NATIVE MAMMALS IN MICHIGAN
Using External and Cranial Characters of Adult Animals

1a. Elongated fingers support flight membrane (see Fig. 15); skull less than ¾ in (20 mm) long, wide U-shaped notch at anterior end ..... Order CHIROPTERA, page 81.

1b. Fingers not elongated or supporting flight membrane; skull more than ¾ in (20 mm) long, no U-shaped notch at anterior end..... 2a

2a. Inner toe (hallux) thumb-like, apposable to other hind toes; tail long, nearly naked, terminally whitish, prehensile; teeth numerous, totalling 50, 26 in upper and 24 in lower jaws .... Order MARSUPIALIA, page 19.

2b. Inner hind toe (hallux) not thumb-like or apposable; tail haired or naked, neither terminally white nor prehensile; teeth numbering 44 or less, never more than 22 in either upper or lower jaws .... 3a.

3a. Each foot terminating in 2 large, hard hoofs; skull massive more than 11⅞ in (300 mm) long; upper jaws bearing no incisor teeth ...... Order ARTIODACTYLA, page 563.

3b. Each foot possessing either 4 or 5 digits, each with terminal claws; skull small to large but less than 11⅞ in (300 mm) long; upper jaws bearing front teeth (incisors) .... 4a.

4a. Total length (head, body, tail) variable, from 7¼ in (185 mm) to more than 61 in (1,600 mm); canines markedly longer than adjacent teeth .... Order CARNIVORA, page 383.

4b. Total length (head, body, tail) small to medium, usually much less than 61 in (1,600 mm); canines, when present, not markedly longer than adjacent teeth.................... 5a.

5a. Total length (head, body, tail) small, not more than 8⅛ in (205 mm); eyes minute; external ears reduced or absent; toothrow continuous, gaps between individual teeth no more than length of a single grinding tooth; canines present.... Order INSECTIVORA, page 29.

5b. Total length (head, body, tail) small to medium, often more than 8⅛ in (205 mm); eyes large, often beady; external ears large, conspicuous; tooth-row discontinuous, distinct gap (diastema) between incisors and cheek teeth; canines absent .... 6a.

6a. Total length (head, body, tail) medium, more than 15 in (380 mm); tail a cotton-like tuft, much shorter than length of external ear; small peg-like tooth directly behind each large, upper, chisel-shaped incisor .... Order LAGMORPHA, page 137.

6b. Total length (head, body, tail) small to medium, less to more than 15 in (380 mm); tail elongate, not a cotton-like tuft, usually longer than length of external ear; no small peg-like tooth directly behind each large, upper, chisel-like incisor .... Order RODENTIA, page 161.

Pouched Mammals

ORDER MARSUPIALIA

This ancient group of pouched mammals was once widespread in many continental areas. Their early start—paleontologists report their remains from the fossil record as far back in geologic time as the Cretaceous Period of the Mesozoic Era—gave them opportunity to disperse to most major land surfaces. Their later decline in many of these areas has been attributed to competition with the higher and perhaps more aggressive placental mammals. Ultimately the placental mammals replaced the pouched mammals except in Australia and adjacent islands (cut off from the later-developing placentals of Asia by a barrier of sea water) and in South America (also cut off for a long time by a water gap at what is now the Isthmus of Panama). Nevertheless, when the land-bridge between South and North America was reestablished, some marsupials proved able to compete effectively with the "advanced" southward-invading herbivores and carnivores. In fact, a most successful one is represented by the modern Virginia opossum, whose ancestors moved northward amid the established "higher" mammals into what is now the eastern United States and even parts of southern Canada. Although marsupials have shown spectacular ability to diversify into a variety of species types, especially in Australia, they retain certain reptilian features, including small brains.

Today, most marsupials are found in Australia; the koala, kangaroo, wombat, and bandicoot are well-known. Less well-known species survive in Latin America. On occasion, however, Michigan residents are confronted with a startled tiny mouse opossum which has stowed away in a stalk of bananas in tropical America only to arrive in a produce house or grocery store in Michigan.

The ordinal name, Marsupialia, is derived from a Latin word meaning "pouch" and refers to the marsupium or pouch formed by a fold of skin on the abdomen of most females. Following conception and a brief gestation period, the premature, almost larval-like young venture out of the birth canal and enter the pouch (perhaps with maternal assistance) to nurse for an extended period before being weaned.

In contrast, the more advanced (placental) mammals—including the other 65 in Michigan—carry their young interuterine much longer with nutrients supplied by a highly developed placenta, allowing the young to be much further developed at birth. In some, the young may be fully haired and in other ways precocial, allowing for remarkable mobility within a few hours of birth.

FAMILY DIDELPHIDAE

American Opossums

This family is composed of several Neotropical (Middle and South American) marsupials, with only one sufficiently adapted to withstand the non-tropical climates of the United States and southern Canada. The lineage of this family dates back to late Mesozoic times, with some fossil forms closely resembling the existing species. Perhaps this conservative development has, in some way, aided this group in surviving for these millions of y ars.

VIRGINIA OPOSSUM

Didelphis virginiana Kerr

NAMES. The generic name *Didelphis,* proposed by Linneaus in 1758, is derived from the Greek and means "two uteri" (*di,* double; *delphys,* uterus or womb). Although there is a double uterus present, the original describer may well have referred to the presence of the uterus and the pouch (marsupium). The specific name, *virginiana,* proposed by Kerr in 1792, refers to the state of Virginia from which the animal first became known to science. The common name originated from the Algonquian language and was spelled out as *apasum* by early colonists. In Michigan, the animal is universally known as possum or opossum (sometimes spelled oppossum). The term possum generally refers to Australian marsupials of certain genera such as *Trichosurus.*

In many published reports, the opossum in Michigan is classified as *Didelphis marsupialis,* with *D. virginiana* considered a subspecies. However, Gardner (1973) showed that *D. virginiana* is distinct from *D. marsupialis* and concluded that animals belonging to this latter species occur widespread in South and Central America but no further north than Mexico. The species *Didelphis virginiana,* on the other hand, is found mostly in the United States. The subspecies, *Didelphis virginiana virginiana* Kerr, is found in Michigan.

RECOGNITION. Body equal to that of a large house cat (see Plate 1 and Fig. 5); length of head and body averaging approximately 17 in (432 mm); length of tail averaging 12½ in (318 mm); head elongate with pointed snout; eyes small; ears rounded, thin, leathery (sometimes reduced by frostbite); whiskers (vibrissae) long; body somewhat heavyset; female with fur-lined pouch on abdomen, conspicuous when filled with young; tail long and tapering, distally naked, scaled and prehensile (sometimes shortened by frostbite); legs short and stocky; feet flat with five toes, all bearing claws except first hind toe (hallux) which is thumblike, extending sideways, and opposing the other four hind toes (see Fig. 6 for this distinctive track).

The pelage is composed of long and rather coarse guard hairs all white or black-tipped and short, thick, soft, and wooly underfur grayish in color. The front and hind quarters are often darker than the overall grayish-white body. The head may be yellowish-white; eyes black and beady encircled by dusky hairs; nose pink; ears blackish sometimes with white margins; tail basally black and terminally white for more than one-half its length; legs and feet dark; and toes whitish. The opossum may attain the size of a half-grown woodchuck, but the combination of grayish-white color; sharply-pointed snout; thin, leathery ears; long

Figure 5. The Virginia opossum (*Didelphis virginiana*). Sketches from life by Bonnie Marris.

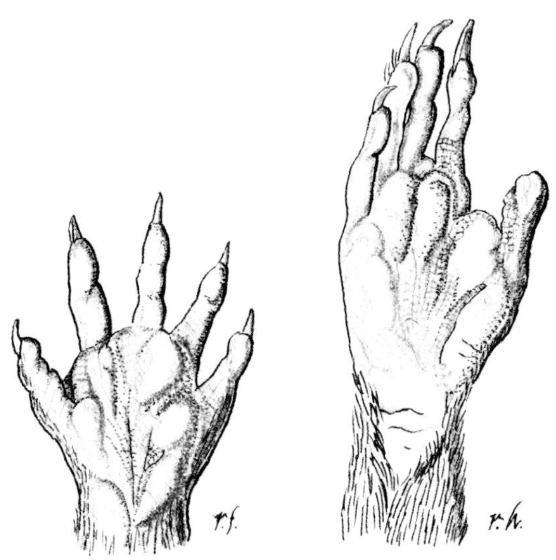

Figure 6. Right front (l.) and right hind (r.) foot of the Virginia opossum. Note space separating the big toe (hallux) from the other four digits on the hind foot.

and ungroomed-looking pelage; and tapering, scaly tail distinguishes this unique mammal from other Michigan species.

MEASUREMENTS AND WEIGHTS. Adult Virginia opossums have the following dimensions: length from tip of nose to end of tail vertebrae, 26 to 33 in (660 to 840 mm); length of tail vertebrae, 10 to 15 in (255 to 380 mm); length of hind foot, 2 to 3 in (50 to 80 mm); height of ear from notch, 1⅝ to 2¼ in (40 to 60 mm); weight, 5 to 12 lbs (2.2 to 5.4 kg). Females tend to be slightly larger than males.

DISTINCTIVE CRANIAL AND DENTAL CHARACTERISTICS. The skull of the adult opossum (see Fig. 7) is characterized by a prominent rostrum and a distinctly narrow and small braincase; nasal bones basally expanded forward of the orbits; zygomatic arches heavy; smallness of the cranium emphasized by the high sagittal crest at midline; sutures between individual cranial bones comparatively open even in mature animals. The angular process on each side of the lower jaw is characteristically inwardly turned. The skull of a mature animal averages 4¾ in (120 mm) long and 2½ in (65 mm) wide (across zygomatic arches).

The adult opossum has 50 teeth, more than any other Michigan mammal. Most of these teeth are small except for the long and somewhat curved canines. The conspicuously diminutive and narrow braincase and the abundance of teeth are the most obvious means of distinguishing skulls of this species from those of any other Michigan mammals. The dental formula is:

$$\text{I (incisors)}\ \frac{5\text{-}5}{4\text{-}4},\ \text{C (canines)}\ \frac{1\text{-}1}{1\text{-}1},\ \text{P (premolars)}\ \frac{3\text{-}3}{3\text{-}3},\ \text{M (molars)}\ \frac{4\text{-}4}{4\text{-}4} = 50$$

DISTRIBUTION. The Virginia opossum occurs from the Central American tropics northward into

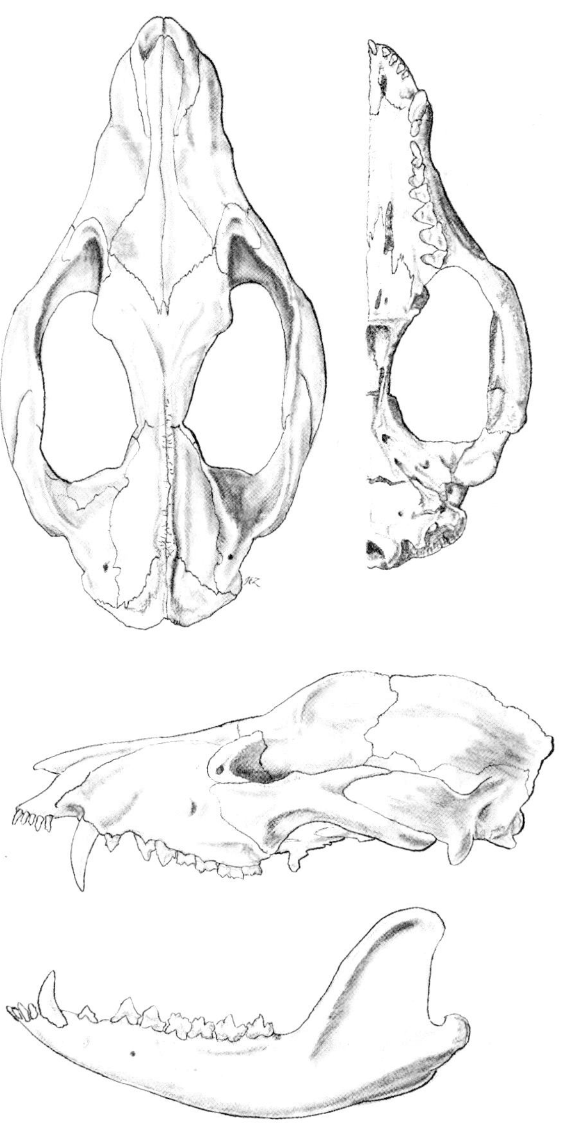

Figure 7. Cranium and lower jaw of the Virginia opossum (*Didelphis virginiana*).

eastern United States and parts of extreme southern Canada (Gardner, 1973). It also occurs along the western coast of Mexico and has been successfully introduced into California and other Pacific states. The species is derived from ancestral stock originating in South America (Neotropica) from which it must have dispersed slowly northward after the formation of the land connection between the two continents in the late Tertiary period of the geologic time scale. In spreading into Michigan and adjacent parts of southern Ontario, the opossum had to adapt to cold north temperate winters, as well as compete with numerous already-adapted mammals for food and living space and as predators. Presumably the opossum's success in surviving the Michigan climate is due in part to its ability to live in close proximity to rural and urban human habitations (Map 3).

There are few records of the opossum from archaeological sites in what is now northcentral United States (Guilday, 1958). It is suspected, however, that in prehistoric times "oak openings"

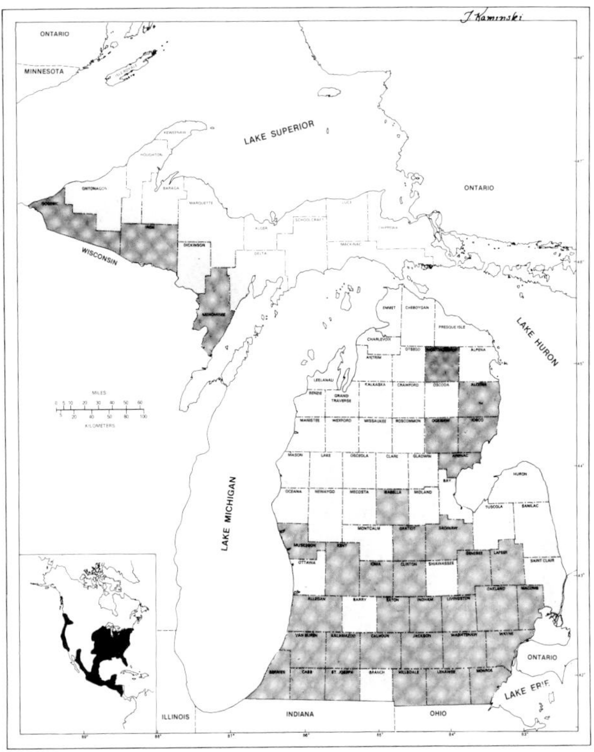

Map 3. Geographic distribution of the Virginia opossum (*Didelphis virginiana*). Actual records exist for counties with dark shading.

in extreme southwestern Michigan and forest and marsh habitat in southeastern areas may have been within the range of the animal. The earliest written reference to this marsupial in Michigan is the account by Antoine de la Mothe Cadillac at the beginning of the eighteenth century. The founder of Detroit described the animals as follows (Burton, 1904:135): "There are wood rats which are as large as rabbits; most of them are grey, but there are some seen which are as white as snow. The female has a pouch under her belly which opens and shuts as she requires, so that sometimes when her little ones are playing, if the mother finds herself pressed, she quickly shuts them up in her pouch and carries them all away with her at once and gains her retreat."

In Berrien County, an archaeological site from the vicinity of Fort St. Joseph and dating from approximately 1750 contained opossum bones (Baker, 1899). The land clearing by settlers in southern Michigan in the 1820s and 1830s is thought to have produced a "disturbed" habitat conducive to the continued northward spread of the opossum. By the mid-1850s, newspaper articles about this mammal began to appear, records being mostly from the first three or four tiers of southern counties. The opossum was reported in Lodi Township of Washtenaw County in 1845 (Wood, 1922), and in the same year at New Richmond in Allegan County (Wood and Dice, 1924). In 1850, there were recorded occurrences of the species in Monroe and Oakland counties and in 1857 and 1859 in Genesee County (Miles, 1861; Wood and Dice, 1924). On November 10, 1851, Perry (1899: 140; Burroughs, 1958) wrote in his diary that near the Little St. Joseph River, 21 mi from Hillsdale in Hillsdale County, "Opossums are very numerous in the woods, and I kill numbers of them for Canada, who, in account of their skin and oil, considers them quite a prize."

By the turn of the century, the opossum was widely known in most of the southern half of the Lower peninsula (Wood and Dice, 1924)—in Wayne County (1898), Ingham County (1899), Eaton County (1903), Gratiot County (1905), Cass County (1908), and Ottawa County (1911). However, it was not until after 1920 that the animal became less sporadic and more evenly distributed as a conspicuous member of the southern Michigan mammalian fauna (Conger, 1920; Dice, 1927; Allen, 1940; DeVos, 1964). Populations in southern

Ontario have been similarly uneven until recently (Banfield, 1974; Peterson, 1966; Peterson and Downing, 1956). This marsupial's distribution in the Lower Peninsula has expanded (see Map 3), with positive records as far north as Montmorency County, where a road kill was reported on highway M-32 in 1949 (David H. Jenkins, pers. comm.).

Within the past 40 years, the opossum has also appeared in three Upper Peninsula counties bordering Wisconsin, from which it undoubtedly entered Michigan (Long and Copes, 1968). The first record of its presence was obtained by Otto Dewaard at Hermansville in Menominee County in about 1942. By 1962, Ozoga and Goertner (1963) reported the animal in both Iron and Gogebic counties. Observers conclude that environmental changes produced by lumbering and agriculture provide favorable conditions for oppossum survival despite cold winters.

HABITAT PREFERENCES. The ability of the opossum to sustain itself at the northern edge of its natural distribution in Michigan is surely the result of its ability to use a variety of habitats. Wooded areas with surface water nearby are natural abodes, hardwoods being preferred to stands of pines and other evergreens. This mammal's characteristic tracks in soft ground at the edges of swamps and marshes indicate its liking for these areas for foraging. However, opossums perhaps occur in maximum densities in the disturbed habitats produced by small farming operations and human habitations both in rural and urban settings. The animals are even seen occasionally in the inner cities.

DENSITY AND MOVEMENTS. Mammalogists have estimated populations of the opossum by: counting tracks along streambanks or roadways; the number of pelts taken by trappers and hunters; trapping, marking, releasing and re-trapping individuals in live traps set in grids; and monitoring movements of individuals fitted with radio collars (Fitch and Shirer, 1970; Gillette, 1980; Lay, 1942). Estimations of densities of opossums in the Upper Great Lakes Region show marked fluctuations, probably the result of differences in habitat quality (carrying capacity of the particular area for the animals). In farmland habitat in Illinois, for example, Holmes and Sanderson (1965) calculated that midsummer populations were as dense as about one opossum per two acres (one per ha). In southern Michigan, in contrast, it was Linduska's (1950) estimate that autumn populations of opossums in Clinton County were about one animal per 50 to 80 acres (one per 20 to 32 ha). In Washtenaw County, Craighead and Craighead (1956) considered that the opossum was no denser than one per 270 acres (one per 108 ha).

Field studies have shown that the Virginia opossum normally concentrates its movements in search of food and mates, or to avoid enemies, around a preferred den site or food source, be it fruiting vegetation, field crops, or even garbage dumps (Fitch and Sandidge, 1953; Verts, 1963; Wiseman and Hendrickson, 1950). Tracking studies in recent years show that these home ranges are rather discrete, with male individuals occupying large areas which may overlap those of two or more females, which have smaller and apparently more stable ranges (Fitch and Shirer, 1970; Gillette, 1980). In Wisconsin, Gillette's (1980) study of radio-tagged opossums during the warm seasons of the year showed that males had home ranges averaging 270 acres (108 ha) and females, 127 acres (51 ha). In addition to resident populations, there are also young individuals, as well as some unsettled adults, in search of homesites of their own. Because they may be excluded from territories defended by residents, such individuals may wander considerable distances. In Missouri, Reynolds (1945) recorded movements as great as 7 mi (11 km). However, in Illinois, Holmes and Sanderson (1965) found that marked individuals traveled an average of 748 ft (190 m) between captures in live traps on successive nights. Presumably these animals were merely moving about in their home areas rather than traveling across country looking for new homesites.

BEHAVIOR. The Virginia opossum's short legs and flat feet contribute to its awkward, deliberate, and plodding gait, which adds to the impression that the animal is a melancholy loner. In truth, the opossum is prone to a rather solitary existence, associating with the opposite sex in times of courtship and in the case of females with their young. The animals may gather at food concentrations on occasion but apparently usually keep to themselves in home ranges from which they discourage others of the same sex. Even so, on occasion, two opossums of the same sex may occupy the same winter den (Shirer and Fitch, 1970). As yet, our knowledge of the year-around social habits and spatial relationships of opossums is rudimentary.

Although surviving well in the cold winter weather of Michigan, the opossum has kept its rather sparse pelage (no doubt reflecting its Neotropical ancestry) with its poor insulative properties (Scholander *et al.*, 1950). This animal, however, like other marsupials does not maintain a precise body temperature. According to McManus (1969), the opossum holds a body temperature of about 96°F (35.5°C) when surrounding temperatures range from 96°F (35.5°C) down to 37.5°F (3°C). When the ambient temperature drops lower, the body temperature declines but usually not to below 91°F (33°C). Otherwise, the animal does not seem to have made conspicuous adjustments in its usual life-style except to refrain from activity during severe winter weather (Brocke, 1970) when its tropical-adapted tail and naked, parchment-like ears are subject to frostbite (Hamilton, 1958).

The Virginia opossum is essentially a nocturnal animal, although it may occasionally be active on sunny days in winter (Fitch and Shirer, 1970; McManus, 1971; Reynolds, 1952). The animal primarily uses the senses of hearing and smell, with sight less well developed (Langley, 1978). Although generally silent, the opossum does utter low growls and faint hissing calls. When disturbed, the animal may turn to run or climb a tree but usually stands its ground with teeth bared. If captured, the opossum is more than apt to feign death, with body relaxed, eyes glassy, and mouth gaping. This condition, known as "playing possum" in American folklore, is apparently caused by a nervous reaction, although there are no significant changes in rate of heart beat or other electrocardiogram patterns (Francq, 1970). Supposedly this performance, although involuntary and not appearing to be the result of shock, has survival value. An attacking domestic dog, for example, might shake the limp creature a few times and depart (Fig. 5).

The opossum does not excavate its own ground den but uses one dug by a groundhog or striped skunk (Grizzell, 1955). A well-insulated underground burrow is perhaps the winter refuge choice. In addition, opossums use a variety of surface and arboreal sites which can be in thickets, brush piles, and tree cavities. The animal appears to take special effort in providing abundant insulating materials for nest chambers. These materials are usually leaves and as many as six bushels have been found in a single den. Leaves are gathered and carried in a twist of the opossum's prehensile tail (Fig. 5). M. E. Nestell of Flint (*in litt.*) observed (by flashlight at night in Grand Blanc Township in Genesee County in February, 1969) an opossum collect dry leaves in its mouth, then push them backward between the forelimbs while dragging the body forward a bit to roll the leaves into a compressed bundle. The tail was then wrapped firmly about the mass allowing the animal to carry it slowly to its homesite.

ASSOCIATES. The opossum will join other mammalian scavengers at such food sources as garbage dumps. On occasion, opossums have been found occupying the same underground dens with eastern cottontails, groundhogs, raccoons, and striped skunks (Shirer and Fitch, 1970). As a rule, however, they live apart from the other mammals which inhabit their homesites.

REPRODUCTIVE ACTIVITIES. For a seemingly uncomplicated animal, the opossum appears beset with difficulties throughout its life span (Allen, 1940; Hartman, 1923), due in part to its unusual reproductive habits. Courtship and mating occur as normal mammalian activities, usually twice each year from late January to March and from late May to July. In Ohio, a litter was reported as early as February 23 (Grote and Dalby, 1973). In Michigan, the opossum generally has only one litter annually, as early as March. Two litters are common in more southern latitudes. A second litter occasionally is produced in east-central Illinois (Holmes and Sanderson, 1965).

After conception, however, the gestation period of this marsupial is remarkably short, no more than 13 days. A litter may include as many as 23 offspring. The premature individual is no longer than a navy bean and weighs only about one ten-thousandth as much as the female parent. The entire litter may weigh no more than a penny with 16 young filling an ordinary tablespoon. These newly-born opossums are equivalent in development to a five-week old human embryo.

The larval-like young must now journey about 3 in (76 mm) from the birth canal to the pouch. Prior to birth, the expectant mother has carefully licked the inner surface of the pouch and the fur leading to it from the opening of the birth canal. In a sitting position, the female grooms each young by licking as it appears. The offspring then squirms in a zig-zag action, propelled in part by movement of the clawed forelimbs, along the moistened pathway to the pouch, where there are about 13 teats arranged in a horseshoe shape with one centered. The young grasp these teats and hold on tightly and, as there are usually more young than teats, late arivals are lost. The babies grow rapidly in the female's protective pouch, with about eight individuals usually surviving to become weaned in ten to twelve weeks (Holmes and Sanderson, 1965). The young remain with the female parent for a while after weaning, often riding on her back with their prehensile tails entwined in her fur (Fig. 5).

The young mature sexually nine months after birth. In nature, the life span is rarely more than 24 months (Gillette, 1980); in captivity, opossums have been known to live for as long as six years.

FOOD HABITS. The opossum's ability to eat a variety of animal and plant foods has probably aided in its successful occupation of most of temperate North America. It is truly an omnivore. Fruits, seeds, grains, grasses, leaves, insects, worms, mice, carrion, and sticks were found in examining the contents of stomachs of opossums (Dearborn, 1932; Allen, 1940; Taube, 1947; Dexter, 1951). A fondness for carrion makes garbage dumps attractive. The opossum does not hesitate to eat salamanders, toads, frogs, snakes, nestling birds, and eggs, including those from the henhouse. Some complaints from bird lovers and poultrymen may be justified (Allen, 1940), as well as nature enthusiasts' displeasure because of the opossum's taste for mushrooms and wild fruits.

ENEMIES. Carnivorous animals equal in size or larger than the opossum constitute threats to young and adult individuals. Michigan observers point to the domestic dog, both the gray and red foxes, coyote, bobcat, and lynx as major predators. Larger hawks and owls also eat opossums, especially the young. Many of these carnivores may have greater opportunity to devour highway-killed animals (Davis, 1940; Haugen, 1944). Such carcasses also provide food for crows, ravens, and vultures (Harlow *et al.*, 1975).

PARASITES AND DISEASES. An array of both internal and external parasites infest Michigan opossums. Some of these host/parasite relationships can be traced to the subtropical and tropical areas from which the opossum originated. Fleas, ticks, mites, and lice (Mallophaga) have been collected from the opossum's exterior while tapeworms, flatworms, thorny-headed worms, and roundworms are known to infest various organ systems (summarized by Blumenthal and Kirkland, 1976).

MOLT AND COLOR ABERRATIONS. The opossum appears to have no regular molt. Both guard hair and underfur seem to be replaced on a rather continuous basis throughout the year. Animals with all white pelage (albinistic) are reported occasionally. Individuals with dark and reddish fur are also known.

ECONOMIC IMPORTANCE. Numerous youthful trappers have probably had their first experiences with fur animals by capturing and pelting opossums. These youngsters learn quickly that care in preparation of the pelts mean higher prices-especially important in winter when farm cash crops are often short. Fur prices have varied depending on current fashion, yet opossum fur has never brought the spectacular prices of mink, beaver, or otter. It has always been at the bottom of the fur price list—usually under $1.00 per pelt.

According to the Biennial Reports of the Michigan Department of Natural Resources (previously the Department of Conservation), the opossum first appeared as a commercial fur-producing mammal in the winter season of 1928–29 with an estimated take of 994 pelts from the southern part of the Lower Peninsula. This small catch undoubtedly reflects both low fur prices and the small opossum population at the time. The catch increased to 3,460 in 1929–30 and ultimately to a record high of 58,979 in the 1953–54 season. Low market demand and lack of interest apparently caused a sharp drop in the catch after the 1955–56 season. In the season of 1970–71, for example, Michigan trappers sold 8,690 opossum pelts at an average price of 50¢. In 1973–74, opossum pelts brought up to $1.50; in 1976–77, a high of $4.75.

The opossum's sporting qualities attract some hunting dog enthusiasts. Trail dogs bring the wily opossum to bay in brush or tree. The flesh is tender and palatable, especially that of the young. It is probably advisable to parboil a dressed carcass and remove excessive fat prior to frying or baking the meat. The Possum Growers and Breeders Association of America publishes a booklet of recipes for tasty dishes using opossum meat.

The opossum occasionally invades the henhouse (Schofield, 1957); it may have to be removed from a garbage can or when seeking denning quarters in a garage or barn. Otherwise, the opossum remains as a rather innocuous member of the outdoor community.

COUNTY RECORDS FROM MICHIGAN. Records of specimens in collections of museums and from the literature show that the Virginia opossum occurs in the extreme western part of the Upper Peninsula and throughout the southern half of Michigan's Lower Peninsula (see Map 3). Records are from the following counties in the Upper Peninsula: Gogebic, Iron, and Menominee; in the Lower Peninsula: Allegan, Berrien, Calhoun, Cass, Clinton, Eaton, Genesee, Gratiot, Hillsdale, Ingham, Ionia, Isabella, Jackson, Kalamazoo, Kent, Lapeer, Livingston, Macomb, Monroe, Montmorency, Muskegon, Oakland, Saginaw, Van Buren, Washtenaw, and Wayne.

LITERATURE CITED

Allen, D.
1940 Nobody loves the 'possum. Michigan Conservation, March, pp. 1–4.

Baker, G. A.
1899 The St. Joseph-Kankakee Portage. Its location and use by Marquette, LaSalle and the French voyageurs. Northern Indiana Historical Society, Publ. No. 1, 48 pp.

Banfield, A. W. F.
1974 The mammals of Canada. Univ. Toronto Press, Toronto. xxiv+438 pp.

Blumenthal, E. M., and G. L. Kirkland, Jr.
1976 The biology of the opossum, *Didelphis virginiana,* in southcentral Pennsylvania. Proc. Pennsylvania Acad. Sci., 50:81–85.

Brocke, R. H.
1970 The winter ecology and bioenergetics of the opossum, *Didelphis marsupialis,* as distributional factors in Michigan. Michigan State Univ., unpubl. Ph.D. dissertation, 215 pp.

Burroughs, R. D.
1958 Perry's deer hunting in Michigan, 1838–1855. Michigan History, 42(1):35–58.

Burton, C. M., ed.
1904 Cadillac Papers. Description of Detroit, advantages found there. Michigan Pioneer and Hist. Coll., 33: 131–151.

Conger, A. C.
1920 A key to Michigan vertebrates, except birds. Michigan Agri. College, Zoology Department and Teacher-Training Department, 76 pp.

Craighead, J. J., and F. C. Craighead, Jr.
1956 Hawks, owls and wildlife. The Stackpole Co., Harrisburg, Pennsylvania. xix+443 pp.

Davis, W. B.
1940 Mortality of wildlife on a Texas highway. Jour. Wildlife Mgmt., 4:90–91.

Dearborn, N.
1932 Foods of some predatory fur-bearing animals in Michigan. Univ. of Michigan, School of Forestry and Conserv. Bulletin No. 1, 52 pp.

De Vos, A.
1964 Range changes of mammals in the Great Lakes region. American Midl. Nat., 7(1):210–231.

Dexter, R. W.
1951 Earthworms in the winter diet of the opossum and raccoon. Jour. Mammalogy, 32(3):464.

Dice, L. R.
1927 A manual of the recent wild mammals of Michigan. Univ. Michigan, Michigan Handbook Series No. 2, 63 pp.

Fitch, H. S., and L. L. Sandidge
1953 Ecology of the opossum on a natural area in northeastern Kansas. Univ. Kansas Publ., Mus. Nat. Hist., 7:305–338.

Fitch, H. S., and H. W. Shirer
1970 A radiotelemetric study of spatial relationships in the opossum. American Midl. Nat., 84:170–186.

Francq, E. N.
1970 Electrocardiograms of the opossum, *Didelphis marsupialis,* during feigned death. Jour. Mammalogy, 51(2):395.

Gardner, A. L.
1973 The systematics of the genus *Didelphis* (Marsupialia: Didelphidae) in North and Middle America. Texas Tech. Univ. Mus., Sp. Publ., 4:1–81.

Gillette, L. N.
1980 Movement patterns of radio-tagged opossums in Wisconsin. American Midl. Nat., 104(1):1–12.

Grizzell, R. A., Jr.
1955 A study of the southern woodchuck, *Marmota monax monax.* American Midl. Nat., 53(2):257–293.

Grote, J. C., and P. L. Dalby
1973 An early litter for the opossum (Didelphis marsupialis) in Ohio. Ohio Jour. Sci., 73(4):240.

Guilday, J. E.
1958 The prehistoric distribution of the opossum. Jour. Mammalogy, 39:39–43.

Hamilton, W. J., Jr.
1958 Life history and economic relations of the opossum *(Didelphis marsupialis virginiana)* in New York state. Cornell Univ., Agri. Exp. Sta., Memoir 354, 48 pp.

Harlow, R. F., R. G. Hooper, D. R. Chamberlain, and H. E. Crawford
1975 Some winter and nesting season foods of the common raven in Virginia. Auk, 92:298–306.

Hartman, C. G.
1923 Breeding habits, development and birth of the opossum. Smithsonian Rept. for 1921, pp. 347–363.
1962 Possums. Univ. Texas Press, Austin. xvi+174 pp.

Haugen, A. O.
1944 Highway mortality of wildlife in southern Michigan. Jour. Mammalogy, 25(2):177–184.

Holmes, A. C. W., and G. S. Sanderson
1965 Populations and movements of opossums in east-central Illinois. Jour. Wildlife Mgmt., 29(2):287–295.

Langley, W. M.
1978 Senses used in finding moving prey by skunks and opossums. Trans. Kansas Acad. Sci., 81(1):91.

Lay, D. W.
1942 Ecology of the opossum in eastern Texas. Jour. Mammalogy, 23:147–159.

Linduska, J. P.
1950 Ecology and land-use relationships of small mammals on a Michigan farm. Michigan Dept. Conservation, Game Div. ix+144 pp.

Long, C. A., and F. A. Copes
1968 Note on the rate of dispersion of the opossum in Wisconsin. American Midl. Nat., 80(1):283–284.

McManus, J. J.
1969 Temperature regulation in the opossum, *Didelphis marsupialis virginiana*. Jour. Mammalogy, 40(3):550–558.
1971 Activity of captive *Didelphis marsupialis*. Journ. Mammalogy, 52(4):846–848.

Miles, M.
1861 A catalogue of the mammals, birds, reptiles and mollusks of Michigan. Geol. Survey Michigan, First Biennial Rept., pp. 219–241.

Ozoga, J. and R. Goertner
1963 Noteworthy locality records for some Michigan mammals. Jack-Pine Warbler, 41(2):89–90.

Perry, O. H.
1899 Hunting expeditions of Oliver Hazard Perry of Cleveland verbatim from his diaries. Privately printed, Cleveland, viii+246 pp.

Peterson, R. L.
1966 The mammals of eastern Canada. Oxford Univ. Press, Toronto. xxxii+465 pp.

Peterson, R. L., and S. D. Downing
1956 Distributional records of the opossum in Ontario. Jour. Mammalogy, 37(3):431–435.

Reynolds, H. C.
1945 Some aspects of the life history and ecology of the oposum in central Missouri. Jour. Mammalogy, 26: 361–379.
1952 Studies on reproduction in the opossum *(Didelphis virginiana virginiana)*. Univ. California, Publ. Zool., 5:233–284.

Schofield, R. D.
1957 Livestock and poultry losses caused by wild animals in Michigan. Michigan Dept. Conserv., Game Div., Rept. No. 2118 (mimeo.), 6 pp.

Scholander, P. F., V. Walters, R. Hock, and L. Irving
1950 Body insulation of some arctic and tropical mammals and birds. Biol. Bull., 99:225–236.

Shirer, H. W., and H. S. Fitch
1970 Comparison from radiotracking of movements and denning habits of the raccoon, striped skunk, and opossum in northeastern Kansas. Jour. Mammalogy, 51(3):491–503.

Taube, C. M.
1947 Food habits of Michigan opossums. Jour. Wildlife Mgmt., 11:97–103.

Verts, B. J.
1963 Movements and populations of opossums in a cultivated area. Jour. Wildlife Mgt., 27(1):127–129.

Wiseman, G. L., and G. O. Hendrickson
1950 Notes on the life history and ecology of the opossum in southeast Iowa. Jour. Mammalogy, 31:331–337.

Wood, N. A.
1922 The mammals of Washtenaw County, Michigan. Univ. Michigan, Mus. Zool., Occas. Papers No. 123:1–23.

Wood, N. A., and L. R. Dice
1924 Records of the distribution of Michigan mammals. Michigan Acad. Sci., Arts and Ltrs., Papers, 3:426–469.

Insect-Eating Mammals

ORDER INSECTIVORA

Paleontologists have concluded that ancient representatives of the order Insectivora living in the Mesozoic (Cretaceous) time—perhaps one hundred million years ago—were the ancestoral stocks from which all modern placental mammals developed. The placental mammals are those grouped in the infraclass Eutheria ("true mammals") and have a highly developed, embryo-feeding mechanism called the placenta. Non-placentals living today include only the egg-laying Monotremata of Australia, Tasmania, and New Guinea and the pouch-bearing Marsupialia of the Australian area and Latin America with one example, the Virginia opossum, in Michigan. Today's Insectivora (only about 400 species) are remnants of this long ancestoral line. In fact, several of the living families contain only a few highly specialized species living in such places as the West Indies and the Malagasy Republic. Many of these forms may owe their very survival to the protection afforded by the isolation of their homes. So diverse are these families of Insectivora that it might be supposed they are classified together for convenience, instead of being derived from a remote yet common ancestral stock. The families containing Michigan moles and shrews show closest relationship (in suborder Lipotyhla) to those of the Old World hedgehogs, tenrecs of Africa and Malagasy, otter shrews of west-central Africa, solenodons of the West Indies, and golden moles of southern Africa. Other living Insectivora, grouped in the suborder Menotyphla, include the elephant shrews of Africa and the tree shrews of southeastern Asia and some adjacent islands. In distribution, this odd assortment occurs naturally on most continental areas except Australia. In South America, they inhabit only the extreme northern part. Aside from the insular solenodons of the West Indies, only two families of Insectivora, the moles (Talpidae) and the shrews (Soricidae), occur in the Western Hemisphere. Ten representatives of this order are found in Michigan.

KEY TO MICHIGAN FAMILIES OF INSECTIVORA
Using Characteristics of Adult Animals in the Flesh

1a. Size large, total length (head, body, tail) more than 5⅞ in (150 mm); weight more than 1.0 oz (30 gm); external ear (conch) absent; front feet broad, more than twice as wide (across palms) as hind feet... Family TALPIDAE (moles), page 67.

2b. Size small, total length (head, body, tail) usually less than 5⅞ in (150 mm); weight less than 1.0 oz (30 gm); external ear (conch) present although only a rim around ear opening in some species; front feet narrow, approximately same width (across palms) as hind feet... Family SORICIDAE (shrews), page 30.

USING CHARACTERISTICS OF THE SKULL OF ADULT ANIMALS

1a. Size large, length more than 1⅛ in (30 mm); both zygomatic arches and auditory bullae complete; front teeth not tipped with reddish or brownish coloring . . . Family TALPIDAE (moles), page 67.

1b. Size small, length less than 1⅛ in (30 mm); both zygomatic arches and auditory bullae incomplete; front teeth tipped with reddish or brownish coloring . . . . . Family SORICIDAE (shrews), page 30.

FAMILY SORICIDAE
Shrews

Shrews belong to the most widespread family of Insectivora; they occur in Eurasia, Africa, North America, and northern South America. Most are small and inconspicuous, leaving little evidence of their presence. To the casual observer, they resemble small mice, except that shrews have long and pointed noses and highly differentiated teeth. Shrews are voracious eaters, feeding on insects and their larvae, worms, other shrews, and mice. Being so small, the amount of body surface through which heat (into which much of their food must be converted) is lost is great in comparison with their size. As a result, over a period of 24 hours, shrews may consume food equivalent to their own body weight—so great are their bodily demands for energy and heat. Fortunately, as the size of the mammal increases, the body surface is relatively smaller. Thus an African elephant eats much less than a shrew per day in terms of body size.

Shrews are the world's smallest mammals, with the pygmy shrew *(Sorex hoyi)* found in Michigan being the most diminutive. Shrews are delicately constructed. Their skulls are fragile, primitive, and lack complete zygomatic arches and auditory bullae which are characteristic of most other mammals. The milk teeth are grown and shed before birth and never used. Glands located on the flanks of several species exude a musky odor. Both front and hind feet have five toes. Shrews are generally shy and secretive, with nervous, quick movements. They are active both in daylight and after dark, probably because of their need to ingest food at frequent intervals. Seven species are known to occur in Michigan.

KEY TO SPECIES OF THE FAMILY SORICIDAE (SHREWS) IN MICHIGAN
Using Characteristics of Adult Animals in the Flesh

1a. Tail long, more than 1¼ in (30 mm) and exceeding one-half the length of the head and body, external ear (conch) visible, not hidden in fur on side of head . . . . 2a.

1b. Tail short, less than 1½ in (30 mm) and always less than one-half the length of the head and body; external ear (conch) inconspicuous, poorly developed, and hidden in fur . . . 6a.

2a. Length of tail less than 1 5/16 in (33 mm); when viewed from the side, each upper jaw shows three unicuspid (single-pointed) teeth . . . . PYGMY SHREW (*Sorex hoyi*), page 45.

2b. Length of tail more than 1 5/16 in (33 mm); when viewed from the side, each upper jaw shows four or more unicuspid (single-pointed) teeth . . . . 3a.

3a. Size large, total length (head, body, tail) more than 5¾ in (145 mm), tail more than 1¾ in (38 mm); hind feet with fringe of short, stiff hairs along both inner and outer margins . . . . . . WATER SHREW (*Sorex palustris*), page 50.

3b. Size small, total length (head, body, tail) less than 5¾ in (145 mm), tail less than 1¾ in (38 mm); hind foot without marginal fringe . . . . 4a.

4a. Size larger, total length (head, body, tail), usually more than 4⅛ in (104 mm); color of pelage strongly tricolor or gray brown . . 5a.

4b. Size smaller, total length (head, body, tail) usually less than 4⅛ in (104 mm); color of pelage dark brown . . . . . MASKED SHREW (*Sorex cinereus*), page 35.

5a. Color strongly tricolored, dark brown on midback, light brown on sides, grayish on belly (most pronounced in winter); tail tapering and not swollen in spring and summer . . ARCTIC or SADDLE-BACK SHREW (*Sorex arcticus*), page 32.

5b. Color not strongly tricolored but gray-brown (darker in summer, paler in winter); tail conspicuously swollen in spring and summer; in Michigan only known from Sugar Island, Chippewa County . . SMOKY SHREW (*Sorex fumeus*), page 41.

6a. Size large, total length (head, body, tail) more than 3½ in (88 mm), tail more than ¾ in (19 mm) long; color of pelage strongly blackish; visible in side view of open mouth are four small teeth (sometimes five) on each side of the upper jaw between the first large upper incisor and the first large upper cheek tooth ..... SHORT-TAILED SHREW (*Blarina brevicauda*), pp. 53.

6b. Size small, total length (head, body, tail) less than 3½ in (88 mm), tail less than ¾ in (19 mm) long; color of pelage gray-brown; visible in side view of open mouth are three small teeth on each side of the upper jaw between the first large upper incisor and the first large upper cheek tooth ...... LEAST SHREW (*Cryptotis parva*), pp. 62.

KEY TO SPECIES OF THE FAMILY SORICIDAE (SHREWS) IN MICHIGAN Using Characteristics of the Skull of Adult Animals

1. Skull small, fragile, smooth and narrow, no more than 7/16 in (11 mm) wide, with long pointed rostrum; five unicuspid teeth (one or two may be minute) on each side of upper jaws (see Fig. 8) .... 2a.

1b. Skull large, less fragile, angular and either broader than 7/16 in (11 mm) with five unicuspid teeth on each side of the upper jaws or narower than 7/16 in (11 mm) with only four unicuspid teeth on each side of the upper jaws (see Fig. 9) .... 6a.

2a. Four unicuspid teeth plainly visible on each side of the upper jaws; one other, the fifth, minute and inconspicuous when viewed from the size (Fig. 8) .... 3a.

2b. Three unicuspid teeth plainly visible on each side of the upper jaws; two others, the third and fifth, minute and inconspicuous when viewed from the side (Fig. 8) ..... PYGMY SHREW (*Sorex hoyi*), page 45.

3a. Skull shorter, less than 11/16 in (17.5 mm) long; braincase narrow, less than 5/16 in (9 mm) wide; first four unicuspid teeth on each side of the upper jaws near equal in size, although the first and second are slightly

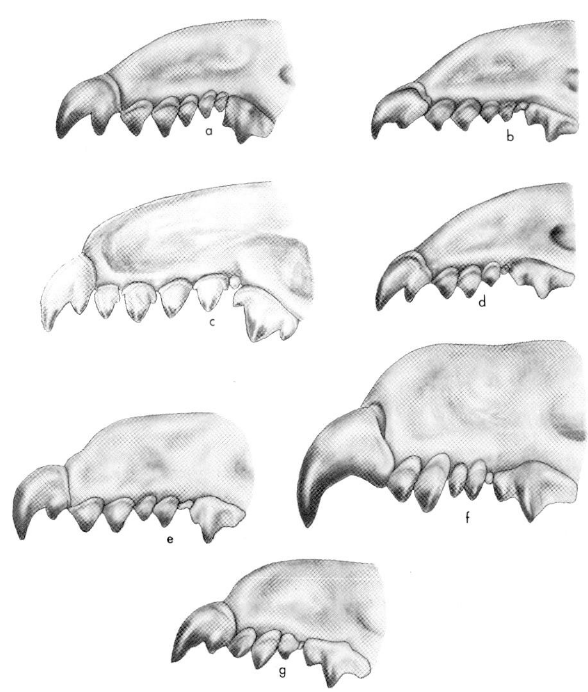

Figure 8. Side views of teeth in left upper jaws of Michigan soricids: (a) arctic shrew (*Sorex arcticus*); (b) masked shrew (*Sorex cinereus*); (c) smoky shrew (*Sorex fumeus*); (d) pygmy shrew (*Sorex hoyi*); (e) water shrew (*Sorex palustris*); (f) short-tailed shrew (*Blarina brevicauda*); (g) least shrew (*Cryptotis parva*).

larger than the third and fourth (Fig. 8) ...... MASKED SHREW (*Sorex cinereus*), page 35.

3b. Skull longer, more than 11/16 in (17.5 mm) long; braincase wide, more than ⅜ in (9 mm) wide; first four unicuspid teeth on each side on the upper jaws distinctly unequal in size, especially the third and fourth .... 4a.

4a. Skull shorter, less than 13/16 in (21 mm) long; braincase narrow, less than ⅜ in (10 mm) wide .... 5a.

4b. Skull longer, more than 13/16 in (21 mm) long; braincase wide, more than ⅜ in (10 mm) wide (Fig. 9).......WATER SHREW (*Sorex palustris*), page 50.

5a. Skull flatter overall; braincase less deep, narrower and less angular; rostrum more slender; third unicuspid tooth on each side of the upper jaw only slightly larger than the fourth; in Michigan only known from Sugar Island,

Chippewa County (Fig. 8) . . SMOKY SHREW (*Sorex fumeus*), page 41.

5b. Skull more inflated overall; braincase flatter, wider and more angular; rostrum broader; third unicuspid on each side of the upper jaw definitely larger than the fourth; (Fig. 8) . . . . ARCTIC or SADDLE-BACK SHREW (*Sorex arcticus*), page 32.

6a. Braincase broad, more than ½ in (11 mm) in width; five unicuspid teeth on each side of the upper jaws, first and second equally large, third and fourth small, fifth minute (Fig. 8) . . . . SHORT-TAILED SHREW (*Blarina brevicauda*), page 53.

6b. Braincase narrow, less than ½ in (11 mm) in width; four unicuspid teeth on each side of the upper jaws, the first three large, the fourth minute (Fig. 8) . . LEAST SHREW (*Cryptotis parva*), page 62.

ARCTIC SHREW
Sorex arcticus Kerr

NAMES. The generic name *Sorex,* proposed by Linnaeus in 1758 for a European shrew, is derived from the Latin and means shrew. The specific name *arcticus,* proposed by Kerr in 1792, is also from the Latin and is a descriptive word referring to the northern distribution of the species. The animal was first described from a specimen obtained at Fort Severn on the southwest side of Hudson Bay. Saddleback shrew is a name also applied to this insectivore. Arctic shrews in the Upper Peninsula are classified as belonging to the subspecies *Sorex arcticus laricorum* Jackson.

RECOGNITION. Body medium for a shrew, cylindrical, robust, and covered with soft, short hair; head elongate with pointed nose; eyes small and partly concealed in facial fur; ears inconspicuous, pinnae slit-like and hidden by fur; tail long and slender; legs short and delicate; feet with five toes, front and back; slit-like flank glands larger in males, approximately ¼ in (7 mm) long in a male from Iron County. Three pairs of mammary glands are characteristic, one pair abdominal and two pairs inguinal (between hind limbs). Adults show a striking tricolor pattern; upper back broadly striped longitudinally with glossy grayish-black from nose to rump; sides and flank contrasting cinnamon brown; underparts paler and more grayish than sides; tail bicolored, dark above and pale below.

MEASUREMENTS AND WEIGHTS. Adults measure as follows: length from tip of nose to end of tail vertebrae, 4⅛ to 4⅞ in (104 to 125 mm); length of tail vertebrae, 1½ to 1⅞ in (38 to 46 mm); length of hind foot, ½ in (13 to 14 mm); weight, approximately ¼ oz (7 to 13 gm).

DISTINCTIVE CRANIAL AND DENTAL CHARACTERISTICS. The skull of the adult arctic shrew is long, slender, flattened, and delicate (see Fig. 9). On the inner surface and near the articulating end of each lower jaw is the distinctive post-mandibular canal. The skull of an adult measures approximately ¾ in (19.0 to 20.6 mm) in greatest length and ⅜ in (9.1 to 10.1 mm) in greatest width.

The arctic shrew has a continuous row of small, sharp, and brown-tipped teeth with the first ones large and forward-extending to facilitate grasping insects and other motile prey. In lateral view, the small unicuspid teeth in the upper jaw are distinctive from those of other Michigan shrews; the first two are larger and swollen; the fourth is smaller than the third (see Fig. 8). The skull is distinctly larger and more robust than those of the masked shrew and the pygmy shrew, and smaller in all dimensions than that of the water shrew. The dental formula for the arctic shrew is:

$$\text{I (incisors)}\frac{3\text{-}3}{1\text{-}1},\ \text{C (canines)}\frac{1\text{-}1}{1\text{-}1},\ \text{P (premolars)}\frac{3\text{-}3}{1\text{-}1},\ \text{M (molars)}\frac{3\text{-}3}{3\text{-}3} = 32$$

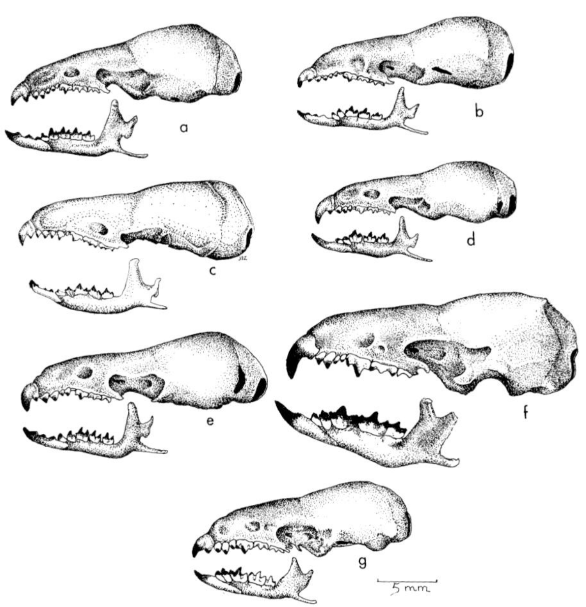

Figure 9. Skulls and lower jaws of Michigan soricids: (a) arctic shrew (*Sorex arcticus*); (b) masked shrew (*Sorex cinereus*); (c) smoky shrew (*Sorex fumeus*); (d) pygmy shrew (*Sorex hoyi*); (e) water shrew (*Sorex palustris*); (f) short-tailed shrew (*Blarina brevicauda*); (g) least shrew (*Cryptotis parva*).

DISTRIBUTION. The arctic shrew ranges widely in boreal and arctic regions of North America. The status of related shrews in Eurasia is yet to be fully resolved (Rausch, 1963). In Michigan (see Map 4), this insectivore is found throughout the Upper Peninsula. It reached this area by way of Wisconsin (Handley, 1971) and is near the most southern limit of its range. The arctic shrew has apparently failed to invade acceptable habitats in the northern part of the Upper Peninsula and adjacent parts of Ontario directly east of the Upper Peninsula, presumably being unable to cross the water barriers between these land areas (see Tables 11 and 12).

HABITAT PREFERENCES. As is often the case with northern mammals whose southernmost distribution is in the Michigan area, the arctic shrew seems to thrive best in the Upper Peninsula in spruce and tamarack swamps and in vegetation on lake and stream borders. Wilkinson (1980) found this soricid in the western part of the Upper Peninsula living chiefly in moist grassy areas bordering boreal woodland and swamps. Ozoga and Goertner (1963) reported the capture of a male in Iron County ". . . along the margin of an abandoned field, 12 ft from the edge of a cedar-spruce swamp, and about 15 ft from an intermittent spring-fed stream." Others have been taken in sedge openings in mixed conifer swamp in Alger County (Ozoga and Verme, 1968). In Minnesota, Whitaker and Pascal (1971) found the species in old fields; Timm (1975) took specimens in dense grass along road ditches, in mixed grasses, strawberries and ferns at forest openings, and in alder thickets. In Wisconsin, Clough (1960) caught arctic shrews in seasonally dry marsh containing grasses, sedge hammocks, forbs, clumps of cattail, and willow and red-osier shrubs. This evidence suggests that the arctic shrew may be found in many habitats in northern Michigan.

DENSITY AND MOVEMENTS. In favorable habitat, arctic shrews, according to Banfield (1974), may occur at densities as high as five individuals per acre (12.5 per ha). In Wisconsin, Clough (1960, 1963) estimated populations of these small insectivores as 3.5 shrews per acre (8.8 per ha). In

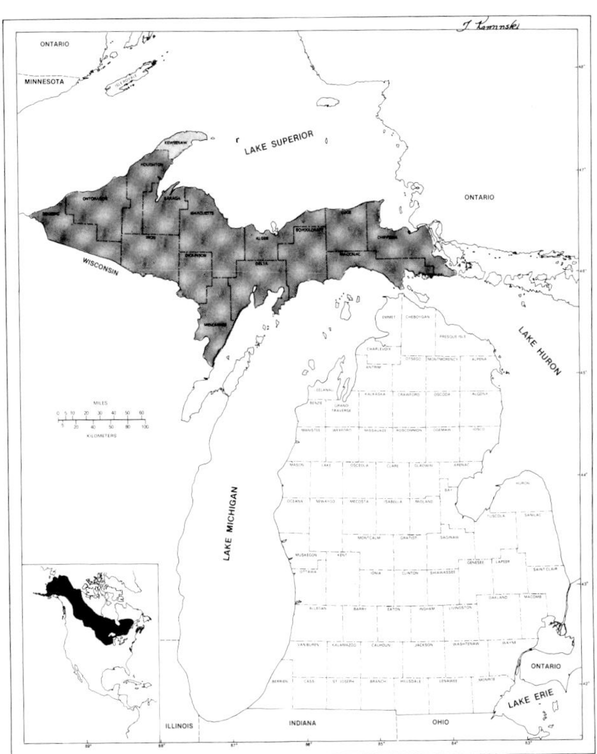

Map 4. Geographic distribution of the arctic shrew (*Sorex arcticus*). Actual records exist for counties with dark shading.

studies in the McCormick Experimental Forest in Marquette and Baraga counties, Haveman (1973) found densities of arctic shrews to be four per acre (10 per ha). Studies of individual movements indicate that arctic shrews maintain rather discrete areas of activity, perhaps no larger than 1/10 of an acre (.04 ha).

BEHAVIOR. This species has abundant energy and is active both day and night. Clough (1963) suspected that the arctic shrew forages more in daylight than after dark. There is some evidence that the arctic shrew may exclude others from its territory. Trapping indicates that these animals make use of small mammal runways in grass and leaf litter, and possibly also maintain their own trails. They construct small, globular surface nests of grasses and other vegetation.

ASSOCIATES. The arctic shrew lives in association with most other small mammals occupying the Upper Peninsula. In the McCormick Experimental Forest in Marquette and Baraga counties, Haveman (1973) found the arctic shrew in mature hardwood forests in close proximity to the masked shrew, pygmy shrew, short-tailed shrew, deer mouse, southern red-backed vole, meadow vole, southern bog lemming, and meadow jumping mouse. In bog environment, he found this same assemblage except for the deer mouse and southern red-backed vole. (The animals were caught in pitfall traps.) In three study areas in bog sites in Gogebic County, Wilkinson (1980) live-trapped 14 species of small mammals of which the arctic shrew was fourth most abundant. Of the total catch of 163 mammals (May 24 through August 10), meadow voles comprised 37%, masked shrews 23%, meadow jumping mice 12% and arctic shrews 10%. Clough (1960, 1963) found similar relationships in the Wisconsin habitats he studied, adding only the house mouse as an associate. Jackson (1961) thought that masked shrews outnumbered arctic shrews about 25 to one in Wisconsin environments.

REPRODUCTIVE ACTIVITIES. Courtship and mating activities may begin as early as February when the reproductive organs of the males enlarge, indicating sexual readiness. Even so, Clough (1963) found no pregnant females in his study area in Wisconsin until mid-May; elsewhere, they have been reported in late April (Banfield, 1974). Because the breeding season extends as late as September, there is opportunity for two or more litters to be produced annually by a single reproductive female (Jackson, 1961). Although litters are known to average about seven in number, with a high of ten, details about the birth and growth of the young are obscure. Clough (1963) noted the first young-of-the-year in the adult population in early July. First-year females rarely breed during their first summer (Buckner, 1966). Individual arctic shrews have been known to survive for at least 18 months in nature (Banfield, 1974).

FOOD HABITS. Little is known about the food eaten by the arctic shrew. Certainly larval, pupal, and adult insects and other invertebrates found in surface litter are consumed. Jackson (1961) suggested that this species, owing to its attraction to stream and bog banks, might tend to prey on aquatic insects.

ENEMIES. Like other shrews, the arctic shrew may discourage some predators because of the musky exudate from its flank glands. Nevertheless, both terrestrial and volant meat-eaters probably find this shrew fair game. Nelson (1934) reports the remains of an arctic shrew in a coughed-up pellet of a great horned owl.

PARASITES AND DISEASES. Fleas, mites, and ticks have been recorded as ectoparasites of the arctic shrew by Timm (1975) and Whitaker and Pascal (1971).

MOLT AND COLOR ABERRATIONS. The striking tricolor pattern characteristic of the arctic shrew is best observed in the dark, glossy winter coat beginning in October. The summer coat, which appears in June, is paler, shorter, and less insulative. Although no albinos are reported in the literature, Timm (1975) does describe an abnormally pale individual from Minnesota.

ECONOMIC IMPORTANCE. The arctic shrew is not known to transmit diseases to humans or domestic animals. It remains one of the many rather innocuous small mammals of northern areas.

COUNTY RECORDS FROM MICHIGAN. Records of specimens preserved in collections of museums and from the literature (Map 4) show that the arctic shrew is widespread in the Upper Peninsula and is known from the following counties: Alger, Baraga, Chippewa, Delta, Dickinson, Gogebic Houghton, Iron, Luce, Mackinac, Marquette, Menominee, Ontonagon, and Schoolcraft.

LITERATURE CITED

Banfield, A. W. F.
1974 The mammals of Canada. Univ. Toronto Press, Toronto. xxiv+438 pp.

Buckner, C. H.
1966 Populations and ecological relationships of shrews in tamarack bogs of southeastern Manitoba. Jour. Mammalogy, 47(2):181–194.

Clough, G. C.
1960 Arctic shrew in southern Wisconsin. Jour. Mammalogy, 41(2):263.
1963 Biology of the arctic shrew, *Sorex arcticus*. American Mid. Nat., 69(1):69–81.

Handley, C. O., Jr.
1971 Appalachian mammalian geography—Recent Epoch. Virginia Poly. Inst. and State Univ., Research Div. Monogr., 4:263–303.

Haveman, J. R.
1973 A study of population densities, habitats, and foods of four sympatric species of shrews. Northern Michigan Univ., unpubl. M.S. thesis, vii+70 pp.

Jackson, H. H. T.
1961 Mammals of Wisconsin. Univ. Wisconsin Press, Madison. xii+504 pp.

Nelson, A. L.
1934 Notes on Wisconsin mammals. Jour. Mammalogy, 15(3):252–253.

Ozoga, J., and R. Goertner
1963 Noteworthy locality records for some Michigan mammals. Jack-Pine Warbler, 41(2):89–90.

Ozoga, J. J., and L. J. Verme
1968 Small mammals of conifer swamp deeryards in northern Michigan. Michigan Acad. Sci., Arts and Ltrs., Papers, 53:37–49.

Rausch, R. L.
1963 A review of the distributions of Holarctic Recent mammals. Tenth Pacific Sci. Congr., Bishop Museum Press, Honolulu. Pp. 29–43.

Timm, R. M.
1975 Distribution, natural history, and parasites of mammals of Cook County, Minnesota. Univ. Minnesota, Bell Mus. Nat. Hist., Occas. Papers 14, 56 pp.

Whitaker, J. O., Jr., and D. D. Pascal, Jr.
1971 External parasites of arctic shrews *(Sorex arcticus)* taken in Minnesota. Jour. Mammalogy, 52(1):202.

Wilkinson, A. M.
1980 Status and distribution of threatened and rare mammals in certain wetland ecosystems in upper Michigan. Michigan Tech. Univ., unpubl. M.S. thesis, viii+85 pp.

MASKED SHREW
Sorex cinereus Kerr

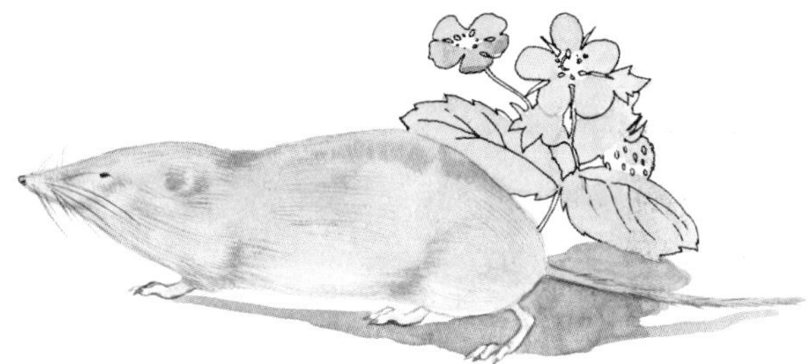

NAMES. The generic name *Sorex,* proposed by Linnaeus in 1758, is derived from the Latin and means shrew. The specific name *cinereus,* proposed by Kerr in 1792, is also of Latin origin and is a descriptive word referring to the ash-gray color of the pelage of this diminutive mammal. In Michigan, this animal is called shrew or sometimes cinereus shrew. Masked shrews in the Upper Peninsula and the northern part of the Lower Peninsula belong to the subspecies, *Sorex cinereus cinereus* Kerr. Those in the southern part of the Lower Peninsula are classified as belonging to the subspecies *Sorex cinereus lesueurii* Duvernoy.

RECOGNITION. Body small, cylindrical, and clothed in soft, short hair; head elongate with pointed nose and conspicuous whiskers; eyes minute and partly covered with facial hair; ears slit-like, inconspicuous, and almost hidden in fur on the sides of the head; neck short; tail long and slender with sparse hair covering; legs short and slender; feet delicate, each with five clawed toes; flank glands elongate; three pairs of mammary glands, one pair abdominal, two pairs inguinal (between hind limbs). Upperparts brownish, darker and more glossy in winter; underparts paler gray or buff; tail somewhat bicolored, darker above, paler below.

MEASUREMENTS AND WEIGHTS. Adults measure as follows: length from tip of nose and end of tail vertebrae, 3¼ to 4¾ in (80 to 115 mm); length of tail vertebrae, 1¼ to 1⅞ in (33 to 47 mm); length of hind foot, ⅜ to ½ in (10 to 13 mm); weight, less than 1/5 oz (3.5 to 5.5 g).

DISTINCTIVE CRANIAL AND DENTAL CHARACTERISTICS. The skull of the adult masked shrew is long with a slender rostrum and a broad and rather flattened cranium (Fig. 9). It measures approximately ⅝ in (16.0 to 17.5 mm) long and ¼ in (7.4 to 8.4 mm) wide. There is a continuous row of brown-tipped teeth, with the first incisor large and extending forward. The skull of the masked shrew is distinguished from those of other Michigan shrews by the nearly equal size of the first four upper unicuspid teeth, with the anterior two being slightly larger than the third and fourth (see Fig. 8). Except for the differences in tooth size, the masked shrew may be difficult to distinguish from the pygmy shrew. The dental formula for the masked shrew is:

$$\text{I (incisors)}\frac{3\text{-}3}{1\text{-}1}, \text{C (canines)}\frac{1\text{-}1}{1\text{-}1}, \text{P (premolars)}\frac{3\text{-}3}{1\text{-}1}, \text{M (molars)}\frac{3\text{-}3}{3\text{-}3} = 32$$

DISTRIBUTION. The masked shrew is the most widespread insectivore in north temperate, boreal, and arctic environments of North America. In both the Upper and Lower Peninsulas, this small, adaptable creature may be expected in any terrestrial situation (see Map 5). The masked shrew's presence on Beaver and Manitou islands in Lake Michigan and on Bois Blanc and Drummond islands in Lake Huron is best explained as the result of passive over-water transportation by human activities (Ozoga and Phillips, 1964). On Beaver Island, for example, the establishment of the masked shrew is fairly recent. Hatt *et al.* (1948) found no trace of the species there in 1938; while Michigan Department of Natural Resources wildlife biologists reported it in 1956.

HABITAT PREFERENCES. The masked shrew successfully inhabits all terrestrial environments in Michigan, with the possible exception of newly plowed fields (Getz, 1961). This diminutive mammal appears as much at home amid heavy leaf litter in mature woodlands as in grassy cover of fallow fields, in well-drained uplands or in moist swamp and waterway borders. There is also evidence that shrew populations increase due to major successional disturbances, even after the burning of swamp conifer cover (Verme and Ozoga, 1981). Many masked shrews are caught by house cats in densely settled urban areas (Toner, 1956).

DENSITY AND MOVEMENTS. Populations of masked shrews show marked multi-annual fluctuations, although their regularity is less well understood than those of such rodents as the meadow vole. In southern Michigan in 1955–1956, an extremely high population of masked shrews was observed. This was especially obvious to a mammalogy class at Michigan State University. In the course of their fieldwork, the students captured more of these shrews than any other species of small mammals. Because of this unexplained high density, the author collaborated with Anton de Vos in questioning other mammalogists about abund-

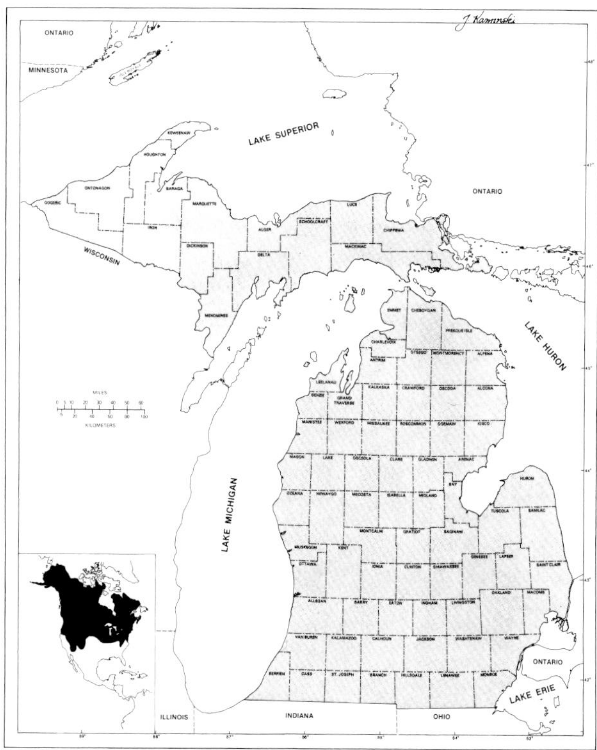

Map 5. Geographic distribution of the masked shrew (*Sorex cinereus*).

ance of shrews in their areas. We learned that unusually high numbers of shrews were reported to be widespread in many areas, including northern Ontario (de Vos, 1957), during that same period.

Most Michigan population estimates have been made in the Upper Peninsula. In Marquette County, Manville (1949) found approximately four masked shrews per acre (10 per ha) in northern white cedar swamp and three per acre (7.5 per ha) in black spruce habitat. In Alger County, Verme (1958) found populations of similar size in white cedar mixed with balsam fir, black spruce, tamarack, and swamp hardwoods. At the McCormick Experimental Forest in Baraga and Marquette counties, Haveman (1973) estimated that masked shrews occurred at the rate of 11 per acre (27.5 per ha) in spruce swamp; nine per acre (22.5 per ha) in mature hardwoods; six per acre (15 per ha) in bog habitat; and four per acre (10 per ha) in spruce barrens. In a subsequent study in the same area, Anderson (1977) found a considerably higher population of 93 masked shrews per acre (240 per ha) in mature hardwoods. These density figures are for warm season populations. In winter, due to the absence of reproductive effort, numbers will progressively decline until production resumes in the following spring.

The spatial relationships of masked shrews have been difficult to assess because these tiny mammals often appear reluctant to enter live-traps and are somewhat frail to withstand the ordeal of being trapped, examined, marked for release, and eventually re-trapped. Pit-traps, employing gallon-size containers buried so that the openings are level with the ground surface, make successful catching devices. If these are spaced in grids and masked shrews are successfully recaptured at more than one trap station in the grid, then both movements and density figures may be obtained. Individual masked shrews occupy rather defined territories of approximately 1/10 of an acre (0.04 ha) in size. Although there is some evidence that shrews maintain living spaces from which they may exclude other individuals (of the same sex perhaps), Hieshetter (1972) and Verme (1958) reported unexplained instances of aggregations. Both authors noted groups of shrews darting in and out of leaf litter, presumably attracted to a food source or perhaps involved in a courtship ritual as these were warm season observations.

BEHAVIOR. The masked shrew has the typical nervous energy characteristic of all small insectivores. Even when not moving about, the animal will constantly twitch its motile snout causing its whiskers to quiver. This shrew appears to be most active in early morning and just after dusk, although numerous observations show it may forage at all hours and during all seasons. Movements seem to be stimulated by the onset of rainfall and heavy cloud cover at night (Doucet and Bider, 1974). In Michigan, the masked shrew's restlessness is well known to hunters who are often entertained by the animals' hide-and-seek tactics as they move in leaf litter or between snow drifts in front of deer stands.

The masked shrew's reputation for savagery has been best demonstrated when the animal is in captivity, even though this excitable animal survives poorly in such situations. Not only will the masked shrew successfuly attack a larger adversary, such as an adult mouse, it will kill and eat other shrews, even litter mates. Merriam (1884) described this aggression: He placed three of these mammals under a single tumbler; within a few minutes one was dispatched and eaten; shortly thereafter the second was killed and eaten. The only evidence of the two victims was the slightly distended stomach of the cannibalistic victor. Such actions strengthen the belief that the masked shrew is indeed a solitary animal in nature except during breeding activities and, in the case of the female, when rearing (Jackson, 1961).

The masked shrew uses the senses of smell and hearing much more effectively than sight in probing its surroundings. Hearing acuity is demonstrated by its ability to maintain an awareness of the environment through echolocation (Gould *et al.*, 1964)—a means by which the shrew utters sounds which echo back to sensitive ears and provide it with some knowledge of the whereabouts of nearby objects. The masked shrew can make several calls audible to human ears, including nervous buzzing sounds and high-pitched squeaks (Blossom, 1932).

In moving about in its area of activity, the masked shrew often employs surface trails established by meadow voles or smaller trails of its own construction. Underground, it follows tunnels excavated by moles. In winter, this shrew probably has its greatest freedom of movement by easily tunneling in snowcover. The focal point of activity is presumably the nest which is generally constructed of grasses and leaves in a secluded place. Usual sites are in stumps, under logs, or in shallow burrows. On one occasion, a masked shrew constructed its nest in a small live-trap left in a locked-open position in a grassy field on the campus of Michigan State University (Muchlinski, 1977). The nest is usually a rounded or flattened ball of loose plant material about 3 in (75 mm) in diameter with an inner cavity no more than ½ in (13 mm) across. The inner chamber may be lined with finely shredded vegetable fibers with the opening at the side.

ASSOCIATES. Because the masked shrew can be expected in most terrestrial habitats in Michigan, it lives in association with all of the other resident small mammals. This is one reason to believe that the masked shrew, in sheer numbers, may be the most abundant mammal in Michigan (Lawrence Master, pers. comm.) In the McCormick Experimental Forest in Baraga and Marquette counties, Haveman (1973) caught masked shrews in the same environments with the arctic shrew, pygmy shrew, short-tailed shrew, deer mouse, southern red-backed vole, meadow vole, southern bog lemming, and meadow jumping mouse. In Alger County, Ozoga and Verme (1968) found masked shrews totaled 42% of the entire catch of 621 small mammals caught in swamp deeryards. Although evidence is scant, other small mammals probably keep their distance from the voracious masked shrew. Jackson (1961) estimated that masked shrews outnumbered arctic shrews in Wisconsin environments by 25 to 1 and water shrews and pygmy shrews by 200 to 1. Rabe (1981) obtained comparable ratios for the masked shrew, pygmy shrew, and water shrew in a leatherleaf-bog birch peatland community in Roscommon County using conventional snap traps. However, when using pitfall traps, catches in Michigan show that masked shrews outnumber pygmy shrews by only a 4 to 1 to a 10 to 1 ratio (Master, 1978; Ryan, 1982).

REPRODUCTIVE ACTIVITIES. In Michigan, the masked shrew breeds throughout the warm months, from March to October (Anderson, 1977; Moore, 1949), although moist conditions seem essential for success (Anderson, 1981). A lactating female was found as late as October 15 in Clinton County (Short, 1961). After a gestation period of about 18 days, litters containing from two to 10 individuals are born, with five to seven being usual (Timm, 1975). Kevin Murphy caught a pregnant female carrying eight near-term fetuses in Ingham County on May 15. The newly born young is helpless, hairless, pink in color, blind (eyelids fused), toothless, toeless (no digit separation), no more than ½ in (13 mm) long, and weighs 0.1 to 0.3 grams (Blossom, 1932; Kilham, 1951). Under the mother shrew's vigilant care and in the protection of the grass-lined nest, the young shrew exhibits rather slow growth. At 12 days of age, the young weighs as much as 2.8 grams; between 9 and 14 days, the incisors erupt; between 14 and 17 days, the ear slits open; at 18 days, the eyes open and the weight is as much as 3.4 grams; and at 19 days, the young is weaned and ready to begin life on its own (Forsyth, 1976).

Although the reproductive process is short, individual females usually produce only one litter during the breeding season. Females born early in the year may breed after reaching sexual maturity at about four months. Normally, however, young-of-the-year do not breed until their second year which may be their only opportunity because the animals rarely live beyond their second summer of life, perhaps no more than 16 months. Ozoga and Verme (1968) calculated that the September population of masked shrews in Alger County contained 40% young-of-the-year.

FOOD HABITS. Although the masked shrew is the smallest mammal in most Michigan habitats, it probably has no real difficulty in getting its share of the small animal foods. According to Getz (1961), it may not be in competition with the larger short-tailed shrew, the other common Michigan insectivore living as an associate, because the masked shrew selects smaller invertebrates as its major food supply. Even so, because of the large body surface of the masked shrew in relation to its small bulk, heat loss through the skin's surface is so great that feeding to replenish this energy loss must be a major daily activity. Most evidence of food preferences has come from examining the contents of shrew stomachs.

Whitaker and Mumford (1972) noted at least 26 categories of food in stomachs of 50 masked shrews captured in Indiana. Numerous small and obscure groups of insects are listed along with larval stages of moths and butterflies, crickets, adult and larval stages of beetles, and leaf hoppers (see Whitaker and Schmeltz, 1973). Also found in noticeable quantities were earthworms, spiders, slugs, centipedes, phalangids, and snails. Mouse flesh, seeds and other vegetable materials were at a minimum. Hamilton (1930) and Schmidt (1931) counted insects (mostly adult beetles, ants, and larval forms of flies, moths, and butterflies, and wasps) in stomachs of 62 masked shrews taken in summer. In pit traps, masked shrews attacked and partly ate such trapped associates as other shrews, meadow jumping mice, meadow voles and woodland voles. Two stomachs contained mammalian hair; two others contained remains of salamanders. In Wisconsin, Komarek (1932) watched a masked shrew dispatch and eat a grasshopper almost as large as the shrew.

Stomachs of masked shrews taken in various habitats in the McCormick Experimental Forest in Baraga and Marquette counties contained Diptera (pupae and adults), Coleoptera, Hymenoptera, Lepidoptera, spiders, annelid worms, unidentified mammalian hair, and other material (Haveman, 1973). Blossom (1932) kept a pair of Michigan masked shrews in captivity, feeding them various foods. Earthworms were killed at once and left whole or cut into pieces. The head and forelegs of a spotted salamander were eaten first, followed by the viscera. The base of the skull and the forelegs of a meadow vole were preferred to the posterior parts. Wings, legs, and heavy chitinous parts of insects were left uneaten. Small moths, sow-bugs, insect larval stages, grasshoppers, beetles, spiders, centipedes, and such items as hamburger and liver were relished. Blossom watched the shrews tear flesh by placing one or both front feet against the food while seizing it with the teeth and shaking their heads violently to pull away mouthfuls.

ENEMIES. Despite its smallness and quickness, the masked shrew seems a prime prey species for many carnivores. The strong-smelling exudate from the glands on each flank probably discourages some predators. House cats often catch masked shrews and other insectivores and leave them dead but otherwise unmolested on doorsteps (Toner, 1956). Four-footed hunters, including weasels, fishers, skunks, foxes (Murie, 1936), bobcats, even wolves, are known to catch and eat shrews. Nichols and Nichols (1935) watched a long-tailed weasel carrying a masked shrew, unmarked except for a small hole in the top of the head. Leopard frogs (Marshall, 1951), bullfrogs, snakes, and even other shrews (*e.g.,* the short-tailed shrew, Eadie, 1949) are also known to eat masked shrews. On September 4, 1977, Gary Dawson found a masked shrew in the stomach of a brown trout (*Salmo trutta*) caught in the Indian River in Schoolcraft County. However, perhaps equally important predators are hawks and owls (Gould *et al.,* 1964). Scientists know most about the food habits of owls, since they regurgitate, in the form of pellets, the undigestible hair, teeth and bones of their mammalian prey. Long-eared owls appear to eat few masked shrews (Armstrong, 1958; Voight and Glenn-Levin, 1978). In Michigan, Wallace (1948, 1950) found 6,815 prey animals in pellets found at barn owl roosts. Only 28 masked shrews (0.4% of the total) were recovered, with about 85% of the prey being meadow voles. Wallace suggested that when voles are scarce, shrews are more prevalent in this owl's diet.

PARASITES AND DISEASES. For a small mammal, the masked shrew appears overly infested with parasites. Sharf and Stewart (1980) found the fleas, *Corrodopsylla curvata curvata* and *Ctenophthalmus pseudagyrtes,* on masked shrews in Leelanau County. Other ectoparasites, reported include the fleas *Doratopsylla blarinae* and *Nearctopsylla genalis;* two ticks, *Isodes angustus* and *Ixodes muris;* and the mites *Labidophorus soricis, Androlaelaps fahrenholzi, Protomyobia claparedei, Myobia simplex,* and *Amorphacarus hengererorum* (Jackson, 1961; Timm, 1975).

MOLT AND COLOR ABERRATIONS. Despite its short natural life-span, the masked shrew molts and grows new pelage periodically. The winter coat is longer and glossier than the summer coat. A male in molt was recorded in July (Weaver, 1940). Occasionally, albino (or partly albino) animals appear. Husband (1963) reported a first-year white female captured in Ingham County in October. Undoubtedly these color mutations are of extremely negative survival value since conspicuous white pelage would contrast markedly with the drab browns, blacks, and greens of the shrew's natural habitat.

ECONOMIC IMPORTANCE. Despite its abundance, the masked shrew poses no problem in farming areas. Instead, its diet of insects and their larval forms may have some beneficial effect. This is not to encourage the notion that these small, always hungry, insectivores are by themselves effective controls of noxious farm and forest pests. Recent work in Canada, however, does point to their effective predation on the pupae of sawflies which attack conifers (Banfield, 1974).

COUNTY RECORDS FROM MICHIGAN. Records of specimens preserved in collections in museums and from the literature (Map 5) show that the subspecies, *Sorex cinereus cinereus* Kerr, is known from the following counties in the Upper Peninsula: Alger, Chippewa, Delta, Dickinson, Gogebic, Iron, Keweenaw, Luce, Marquette, Menominee, Ontonagan, Schoolcraft; in the northern part of the Lower Peninsula: Charlevoix, Cheboygan, Emmet, Gladwin, Huron, Leelanau, Montmorency, Ogemaw, Otsego, Presque Isle, Roscommon. The subspecies *Sorex cinereus lesuerii* Duvernoy, is known from the following counties in the southern part of the Lower Peninsula: Allegan, Berrien, Clinton, Eaton, Ingham, Ionia, Jackson, Kalamazoo, Lapeer, Livingston, Macomb, Oakland, Washtenaw, Wayne. The species is also known from Beaver, Garden, and North and South Manitou islands in Lake Michigan and from Bois Blanc and Drummond islands in Lake Huron (Manville, 1950; Ozoga and Phillips, 1964; Phillips *et al.*, 1965; Scharf, 1973; Scharf and Jorae, 1980).

LITERATURE CITED

Anderson, T. J.
1977 Population biology of the masked shrew, *Sorex cinereus*, in hardwood forest areas of the McCormick Experimental Forest, Marquette County, Michigan. Northern Michigan University. unpubl. M.S. thesis, viii+75 pp.
1981 Reproductive status of an upland population of masked shrews. Ohio Jour. Sci., 81(4):161–164.

Armstrong, W. H.
1958 Nesting and food habits of the long-eared owl in Michigan. Michigan State Univ., Publ. Mus., Biol. Ser., 1(2):61–96.

Banfield, A. W. F.
1974 The mammals of Canada. Univ. Toronto Press, Toronto. xxiv+438 pp.

Blossom, P. M.
1932 A pair of long-tailed shrews *(Sorex cinereus cinereus)* in captivity. Jour. Mammology, 13:136–143.

de Vos, A.
1957 Peak populations of the masked shrew in northern Ontario. Jour. Mammalogy, 38:256–258.

Doucet, G. J., and J. R. Bider
1974 The effects of weather on the activity of the masked shrew. Jour. Mammalogy, 55(2):348–363.

Eadie, W. R.
1949 Predation on *Sorex* by *Blarina*. Jour. Mammalogy, 30:308–309.

Forsyth, D. J.
1976 A field study of growth and development of nestling masked shrews *(Sorex cinereus)*. Jour. Mammalogy, 57(4):708–721.

Getz, L. L.
1961 Factors influencing the local distribution of shrews. American Mid. Nat., 65:67–88.

Gould, E., N. C. Negus and A. Novick
1964 Evidence of echolocation in shrews. Jour. Exp. Zool., 156:19–38.

Hamilton, W. J., Jr.
1930 The food of the Soricidae. Jour. Mammalogy, 11:26–39.

Hatt, R. T., J. Van Tyne, L. C. Stuart, C. H. Pope, and A. B. Grobman
1948 Island life: A study of the land vertebrates of the islands of eastern Lake Michigan. Cranbrook Inst. Sci., Bull. No. 27, xi+179 pp.

Haveman, J. R.
1973 A study of population densities, habitats and foods of four sympatric species of shrews. Northern Michigan Univ. unpubl. M.S. thesis, vii+70 pp.

Hieshetter, D.
1972 A concentration of masked shrews in Ingham County, Michigan. Jack-Pine Warbler, 40:63.

Husband, R. W.
1963 An albino shrew, *Sorex cinereus*, from Michigan. Jack-Pine Warbler, 41:67.

Jackson, H. H. T.
1961 Mammals of Wisconsin. Univ. Wisconsin Press, Madison. xii+504 pp.

Kilham, L.
1951 Mother and young of *Sorex cinereus fontenalis* in captivity. Jour. Mammalogy 32(1):115.

Komarek, E. V.
1932 Notes on mammals of Menominee Indian Reservation, Wisconsin. Jour. Mammalogy, 13:203–209.

Manville, R. H.
1949 A study of small mammal populations in northern Michigan. Univ. Michigan, Mus. Zool., Misc. Publ., No. 73, 83 pp.
1950 The mammals of Drummond Island, Michigan. Jour. Mammalogy, 31(3):358–359.

Marshall, W. H.
1951 Predation on shrews by frogs. Jour. Mammalogy, 32(2):219.

Master, L. L.
1978 A survey of the current distribution, abundance, and habitat requirements of threatened and potentially

threatened species of small mammals in Michigan. Univ. Michigan, Mus. Zool., iv+52 pp. (mimeo.).

Merriam, C. H.
1884 The mammals of the Adirondack region, northeastern New York. Publ. by author, New York. 316 pp.

Moore, J. C.
1949 Notes on the shrew, *Sorex cinereus*, in the southern Appalachians. Ecology, 30(2):234–237.

Muchlinski, A.
1977 Masked shrews use live traps as nest sites. Jack-Pine Warbler, 55(1):47.

Murie, A.
1936 Following fox trails. Univ. Michigan, Mus. Zool., Misc. Publ. No. 32, 45 pp.

Nichols, D. G., and J. T. Nichols
1935 Notes on the New York weasel (*Mustela novaboracensis*). Jour. Mammalogy, 16(4):297–199.

Ozoga, J. J., and C. J. Phillips
1964 Mammals of Beaver Island, Michigan. Michigan State Univ., Publ. Mus., Biol. Ser., 2:305–348.

Ozoga, J. J., and L. J. Verme
1968 Small mammals of conifer swamp deeryards in northern Michigan. Michigan Acad. Sci., Arts and Ltrs., Papers, 53:37–49.

Phillips, C. J., J. J. Ozoga, and L. C. Drew
1965 The land vertebrates of Garden Island. Jack-Pine Warbler, 43(1):20–25.

Rabe, M. L.
1981 New locations for pygmy (*Sorex hoyi*) and water (*Sorex palustris*) shrews in Michigan. Jack-Pine Warbler, 59(1):16–17.

Ryan, J. M.
1982 Distribution and habitat of the pygmy shrew, *Sorex (microsorex) hoyi*, in the Michigan. American Midl. Nat., (in press).

Scharf, W. C.
1973 Birds and land vertebrates of South Manitou Island. Jack-Pine Warbler, 51:2–19.

Scharf, W. C., and M. L. Jorae
1980 Birds and land vertebrates of North Manitou Island. Jack-Pine Warbler, 58(1):4–15.

Scharf, W. C., and K. R. Stewart
1980 New Records of Siphonaptera from northern Michigan. Great Lakes Ento., 13(3):165–167.

Schmidt, F. J. W.
1931 Mammals of western Clark County, Wisconsin. Jour. Mammalogy, 12:99–117.

Short, H. L.
1961 Fall breeding activity of a young shrew. Jack-Pine Warbler, 42:95.

Timm, R. M.
1975 Distribution, natural history, and parasites of mammals of Cook County, Minnesota. Univ. Minnesota, Bell Mus. Nat. Hist., Occas. Papers 14, 56 pp.

Toner, G. C.
1956 House cat predation on small animals. Jour. Mammalogy, 37(1):119.

Verme, L. J.
1958 Localized variation in masked shrew abundance. Jour. Mammalogy, 39:149–150.

Verme, L. J., and J. J. Ozoga
1981 Changes in small mammal populations following clear-cutting in upper Michigan conifer swamps. Canadian Field-Nat., 95(3):253–256.

Voight, J. and D. C. Glenn-Levin
1978 Prey availability and prey taken by long-eared owls in Iowa. American Midl. Nat., 99(1):162–171.

Wallace, G. J.
1948 The barn owl in Michigan. Its distribution, natural history and food habits. Michigan Agri. Exp. Sta., Tech. Bull. 208, 61 pp.
1950 In support of the barn owl. Michigan Agri. Exp. Stat. Quart. Bull., 33:96–105.

Weaver, R. L.
1940 Notes on a collection of mammals from the southern coast of the Laborador Peninsula. Jour. Mammalogy, 21:417–422.

Whitaker, J. O., Jr., and R. E. Mumford
1972 Food and ectoparasites of Indiana shrews. Jour. Mammalogy, 53:329–335.

Whitaker, J. O., Jr., and L. L. Schmeltz
1973 Food and external parasites of *Sorex palustris* and food of *Sorex cinereus* from St. Louis County, Minnesota. Jour. Mammalogy, 54(1):283–285.

SMOKY SHREW
Sorex fumeus Miller

NAMES. The generic name *Sorex,* proposed by Linneaus in 1758, is derived from the Latin and means shrew. The specific name *fumeus*, proposed by Miller in 1895, is also from the Latin and means smoky, referring to the grayish brown appearance of this insectivore. The best known common name for this species is smoky shrew. The subspecies of the smoky shrew occurring on Michigan's Sugar Island is *Sorex fumeus fumeus* Miller.

RECOGNITION. Body small, cylindrical, covered with soft, short hair; head elongate with pointed nose and conspicuous vibrissae, small eyes and slit-like external ears partly concealed in fur; tail long, slender, except in breeding season when swollen, with pencil of long hairs (in young adults) extending from tip; legs short, delicate; feet weak, each with five clawed toes. Both sexes are darker on the dorsal surface, with winter pelage (September to May) dark gray above and paler gray below and summer pelage (May to September) brown or grayish brown on the back side and only slightly paler, sometimes silvery, tail bicolored, dark above and pale below, tending to be hairless terminally in older adults. Three pairs of mammary glands are characteristic. Both sexes possess flank glands, most conspicuous in males during the courtship and mating season. The smoky shrew is larger and darker than the masked and pygmy shrews but smaller and less blackish above than the water shrew. In size it most resembles the arctic shrew, but the smoky shrew never displays the tricolor pattern so characteristic of the artic shrew.

MEASUREMENTS AND WEIGHTS. Adult smoky shrews measure as follows: length from tip of nose to end of tail vertebrae, 4¼ to 5 in (110 to 126 mm); tail, 1½ to 2 in (40 to 52 mm); hind foot, ½ in (13 to 15 mm); weight, ¼ oz (6.5 to 11.0 gm).

DISTINCTIVE CRANIAL AND DENTAL CHARACTERISTICS. The mature skull of the smoky shrew is long, fairly slender, flattened and delicate, with a total length of approximately ¾ in (18.0 to 19.2 mm) and a greatest width of approximately ⅜ in (8.5 to 9.3 mm). The form of the cranium and the teeth of this shrew compare closely with those of other Michigan representatives of the genus *Sorex* (see Fig. 9). The skull is larger and broader than those of the masked and pygmy shrews. Not only are the unicuspid teeth more massive in the smoky shrew, but they also differ in size, with the first and second almost being equal and the third being slightly larger than the fourth (Fig. 8). The skull of the smoky shrew is smaller in all dimensions than that of the water shrew. Although the skulls of the smoky shrew and the arctic shrew are similar in size, that of the former is flatter, with the braincase narrower and less angular. The dental formula for the smoky shrew is:

$$\text{I (incisors)}\frac{3\text{-}3}{1\text{-}1},\ \text{C (canines)}\frac{1\text{-}1}{1\text{-}1},\ \text{P (premolars)}\frac{3\text{-}3}{1\text{-}1},\ \text{M (molars)}\frac{3\text{-}3}{3\text{-}3} = 32$$

DISTRIBUTION. The smoky shrew is a distinctly eastern species, occurring in southern Canada from south of the Gulf of St. Lawrence westward to the northern edge of Lake Superior, southward in the eastern interior of the United States to northwestern South Carolina and adjacent parts of Georgia to central Kentucky (Map 6). The smoky shrew has also been found in a single isolated locality in southeastern Wisconsin, one being taken in this seemingly extralimital location in 1853 (Jackson, 1961). This most recently reported Michigan mammal (Master, 1982) is known only

Map 6. Geographic distribution of the smoky shrew (*Sorex fumeus*). Note that the species is known only from Sugar Island, Chippewa County.

from Sugar Island, located in the Superior-Huron Waterway Connection between mainland Ontario (from where the species has been previously known) and the Upper Peninsula where the species has not been recorded. Finding the smoky shrew on Sugar Island has prompted mammalogists to watch for evidence of this species in suitable habitat on the Michigan mainland of Chippewa County adjacent to the island. Since the smoky shrew crossed the water barrier between Ontario and Sugar Island—perhaps under snow cover on winter ice, on floating debris or passively, through human effort—there is a strong likelihood that its crossing of the water barrier between Sugar Island and mainland Michigan is a possibility. There are also records of the smoky shrew occuring in Middlesex County, Ontario, within 50 mi (80 km) of the St. Clair River at the border of the Lower Peninsula at Port Huron (R. L. Peterson, *in litt.*, February 24, 1981). Due chiefly to the lack of suitable streamside environment, there seems little likelihood of a smoky shrew crossing the Huron-Erie Waterway Connection.

HABITAT PREFERENCES. Cool, moist, shady boreal forest areas seem preferred by the smoky shrew within its rather restricted geographic range in eastern North America. The forest floor of favored habitat is underlain by black friable soils and covered with loose leaf mold. Characteristic also are growths of ferns, moss-covered logs and other litter (Hamilton, 1940). Field observers have found smoky shrews most commonly in mature maple, birch and hemlock woodlands (Doutt *et al.*, 1966; Hamilton and Whitaker, 1979). Despite more exacting environmental requirements than its common associate, the masked shrew, the smoky shrew has been found in such varied habitats as deep spruce forest under moss-covered logs on the Gaspe Peninsula (Goodwin, 1929), on a sparsely wooded hillside and in open woods in New York (Schoomaker, 1929), in wet, dense cedar swamp in Ontario (Soper, 1923), in clearcut and subsequent successional stages in northern deciduous and boreal coniferous forest habitats in West Virginia (Kirkland, 1977), and in northern deciduous forest treated with aerial application of the herbicide, 2,4,5-T, in West Virginia (Kirkland, 1978). In Michigan, the three smoky shrews obtained on Sugar Island were caught in pitfall (can)-traps set in open forest habitat (Master, 1982). The substrate consisted of a thick layer of leaf mold with sand below. Understory was mostly open with scattered seedlings of maple and oak and broken and fragile ferns. Mature trees making up the parklike forest stand included sugar maple and, to a lesser extent, striped maple, hemlock, and birch.

DENSITY AND MOVEMENTS. The few long-term studies of the smoky shrew show that this insectivore fluctuates in numbers from year to year as well as from season to season. It was Bole's (1939) contention that periods of drought were major factors in population declines. Hamilton (1940) found local populations of smoky shrews in New York to be low in winter when perhaps only 20% to 25% of the autumn numbers survive. He noted a spring increase with maximum densities from July to October. In populations living in favorable areas in beech, maple, and hemlock forests, Hamilton estimated smoky shrews to number from 25 to 50 individuals per acre (10 to 20 per ha). In similar habitat in Ohio, Bole (1939) found 15 smoky shrews per acre (6 per ha). However, in habitats other than mature deciduous-coniferous forest, Bole noted that shrew populations were much more sparse, averaging less than one per acre (0.4 per ha). Little is known about the spatial needs of individual smoky shrews. Perhaps like other members of the genus *Sorex*, individuals scurry about as solitary foragers in areas no more than one acre (0.4 ha) in size.

BEHAVIOR. Most data on the behavior of smoky shrews come from the careful observations of Hamilton (1940) in New York. He found them active at all hours of the day throughout the year. Most activity takes place below the forest floor in burrows constructed by moles and other mammals. Despite poor digging ability, the shrew can make its own trails through loosely packed leaf mold. Such secluded underground retreats may be a major reason why the smoky shrew's behavioral ecology is so little understood.

Apparently smoky shrews, except during the mating season, avoid direct contact with each other. Under captive conditions, the shrews are noisily aggressive toward one another, uttering high-pitched grating sounds. An almost indiscernible twitter can be detected when smoky shrews are foraging (Hamilton, 1940). Not only is hearing highly developed in this species but the sense of touch is of major importance in maintaining close contact with the environment. The vibrissae, as in other shrews, are constantly in motion as the snout twitches from side to side and up and down. The snout is also a major instrument for probing. The occurrence of smoky shrews in cedar swamps, along shaded brooks, and in sphagnum bogs shows its affinity for saturated substrate and even surface water areas. There is no evidence to show the smoky shrew can climb.

Smoky shrews apparently centralize their activities at a nest site which may connect to various underground runs. Nests are globular and fashioned from dried leaves, grass, and sometimes fur. They are secluded under logs or other forest floor debris.

ASSOCIATES. In contrast to the widespread masked shrew, the more restricted smoky shrew usually associates only with those species which venture into the moist, shady, mature woodlands where it thrives best. Small mammals using the same underground burrow systems as smoky shrews are masked shrews, short-tailed shrews, eastern chipmunks, white-footed mice, deer mice, southern red-backed voles, southern bog lemmings, and woodland jumping mice (Bole, 1939; Hamilton, 1940; Kirkland and Griffin, 1974). On Sugar Island in Michigan, Master (1982) trapped the smoky shrew in the same habitat with masked shrews, deer mice, southern red-backed voles, meadow voles, and woodland jumping mice.

REPRODUCTIVE ACTIVITIES. Breeding activity in the smoky shrew begins in late March with litters born usually in late April or early May and again in late August (Hamilton, 1940). There is also evidence that a third litter may appear in mid autumn. The gestation period is approximately 20 days in duration. The number of newly born averages four but may be as numerous as seven (McLaren and Kirkland, 1979). The tiny, grub-like, pink, helpless young become fully furred, alert, and near adult size by 30 days, at which time they also have been weaned. The young are easily identified from similar-sized adults by hairs extending from the tip of the tail. In older individuals, these tend to wear away. Adults, 12 months or more old, also show pronounced tooth wear. Not many smoky shrews survive beyond their first year of litter production.

FOOD HABITS. The diet of the smoky shrew consists of more than 70% animal matter (Hamilton and Hamilton, 1954). As with other shrews, this active species must forage for and eat food during much of its daily activity. Hamilton (1940) thought that this species required foods equal to at least one-half of its weight per 24 hour period in order to survive. In the warm months, the great bulk of food selections by smoky shrews is provided by ground-inhabiting insects, their pupae and larvae. In the cold months, the diet is apt to be more varied with snails, earthworms, and other invertebrates taken, as available. Some vegetable matter appears in stomach contents in autumn. However, the smoky shrew seems to prey less on earthworms and mice than do other soricids, especially the short-tailed shrew. Foods identified in stomach contents (Hamilton, 1930) include such insect groups as beetles (Coleoptera), flies (Diptera), bees and wasps (Hymenoptera), and moths and butterflies (Lepidoptera). Seasonal foods eaten in lesser quantities include centipedes, earthworms, sowbugs, salamanders, spiders, plant material, and an occasional mouse or bird. The importance of carrion in the smoky shrew's diet is unknown.

ENEMIES. Smoky shrews are similar to other insectivores in their ability to discourage some would-be predators by their musky, unpleasant odors. Even so, they can be fair game for both terrestrial and flying meat-eaters. Although Hamilton (1940) reported smoky shrews as being eaten only by the long-eared owl and short-tailed shrew, presumably other predators ingest them. However, the subsurface environment in which the smoky shrew spends much of its time undoubtedly shields it from many predators.

PARASITES AND DISEASES. Hamilton (1940) found fleas, *Ctenophthalmus pseudagyrtes* and *Doratopsylla blarinae,* and mites of the genera *Myobia, Haemogamasus* and *Protomyobia* as external parasites on smoky shrews. One roundworm belonging to the genus *Porrocaecum* infests muscle tissue and the viscera.

MOLT AND COLOR ABERRATIONS. As mentioned previously, the smoky shrew has a gray winter and a brown summer pelage. The spring molt occurs from late April to June. The autumn molt is in September and October. No albino individuals have been recorded.

ECONOMIC IMPORTANCE. The smoky shrew may make up 20% of the small mammalian fauna population of moist northern forests. By all reports, it appears to be an important predator of insect life, including some noxious species. Its underground actions no doubt have a favorable influence on soil formation and nutrient supplies. The species appears to have no major direct effect on human activities.

COUNTY RECORDS FROM MICHIGAN. Records of specimens in collections of museums show that the smoky shrew in Michigan occurs only on Sugar Island, which politically is a part of Chippewa County (see Map 6).

LITERATURE CITED

Bole, B. P., Jr.
1939 The quadrate method of studying small mammal populations. Cleveland Mus. Nat. Hist., Sci. Ser., 5(4):15–77.

Doutt, J. K., C. A. Heppenstall, and J. E. Guilday
1966 Mammals of Pennsylvania. Pennsylvania Game Comm., Harrisburg. 273 pp.

Goodwin, G. G.
1929 Mammals of the Cascapedia Valley, Quebec. Jour. Mammalogy, 10(3):239–246.

Hamilton, W. J., Jr.
1930 The food of the Soricidae. Jour. Mammalogy, 11(1): 26–39.
1940 The biology of the smoky shrew (*Sorex fumeus fumeus* Miller). Zoologica, 25(4):473–492.

Hamilton, W. J., Jr., and W. J. Hamilton, III
1954 The food of some small mammals from the Gaspe Peninsula, P. Q. Canadian Field-Nat., 68(3):108–109.

Hamilton, W. J., Jr., and John O. Whitaker, Jr.
1979 Mammals of the eastern United States. Second Ed. Cornell Univ. Press, Ithaca and London. 346 pp.

Jackson, H. H. T.
1961 Mammals of Wisconsin. Univ. Wisconsin Press, Madison. xii+504 pp.

Kirkland, G. L., Jr.
1977 Responses of small mammals to the clearcutting of northern Appalachian forests. Jour. Mammalogy, 58(4):600–609.
1978 Population and community responses of small mammals to 2,4,5-T. United States Dept. Agric., Forest Serv., Research Note PNW-314, 7 pp.

Kirkland, G. L., Jr., and R. J. Griffin
1974 Microdistribution of small mammals at the coniferous-deciduous forest ecotone in northern New York. Jour. Mammalogy, 55(2):417–427.

Master, L. L.
1982 The smoky shrew: A new mammal in Michigan. Jack-Pine Warbler, 60(1):28–29.

McLaren, S. B., and G. L. Kirkland, Jr.
1979 Geographic variation in litter size of small mammals in the central Appalachian region. Proc. Pennsylvania Acad. Sci., 53:123–126.

Schoomaker, W. J.
1929 Notes on some mammals of Alleghany State Park. Jour. Mammalogy, 10(3):246–249.

Soper, J. D.
1923 The mammals of Wellington and Waterloo counties, Ontario. Jour. Mammalogy, 4(4):244–252.

PYGMY SHREW
Sorex hoyi Baird

NAMES. The generic name *Sorex,* proposed by Linnaeus in 1758, is from the Latin and means shrew. The specific name *hoyi,* proposed by Baird in 1855, refers to P. R. Hoy, who obtained the specimen used in the original description. The use of the generic name *Sorex* is in accordance with the findings published in the revisionary study of Diersing (1980), who relegates the previously used generic name *Microsorex* to subgeneric status. The pygmy shrews in both the Upper and Lower Peninsulas have been classified as belonging to the subspecies *Sorex hoyi hoyi* Baird.

RECOGNITION. Body small and covered with soft, short hair; head elongate; nose pointed; whiskers conspicuous; eyes minute and mostly concealed in facial hair; ears obscure slits also hidden by fur on the sides of the head; neck short and not differentiated; body cylindrical; tail medium long, narrow, and sparsely haired; legs short and thin; feet delicate; flank glands slit-like and developed; three pairs of mammary glands, one abdominal, two inguinal. Upperparts gray-brown in summer and gray in winter; underparts lighter and more gray; tail indistinctly bicolored. The pygmy shrew resembles the masked shrew in external features but is smaller and slimmer in body proportions, has a shorter tail, and is grayer in overall color.

MEASUREMENTS AND WEIGHTS. Eighteen young adult pygmy shrews from the Lower Peninsula (MSU 29083–29100) measure as follows: length from tip of nose to end of tail vertebrae, 3⅛ to 3⅜ in (80 to 86 mm); length of tail vertebrae, 1 1/16 to 1¼ in(27 to 32 mm); length of hind foot, ⅜ in (8 to 9 mm); weight, approximately 1/10 oz (3.0 to 4.0 g).

DISTINCTIVE CRANIAL AND DENTAL CHARACTERISTICS. The skull of the adult pygmy shrew closely resembles those of other Michigan shrews of the genus *Sorex* (see Fig. 9). Compared with the skull of the masked shrew, the pygmy shrew's skull is only slightly smaller but the braincase is flatter and narrower with the rostrum shorter and broader. It measures approximately ⅝ in (15.0 to 16.5 mm) long and ¼ in (6.1 to 7.0 mm) wide. In side view, the skull of this smallest of shrews is easily distinguished from other Michigan shrews by the presence of only three obvious unicuspid (single-pointed) teeth plus a diminutive (peg-like) fourth tooth in the upper jaws (see Fig. 8). The dental formula for the pygmy shrew is:

$$\text{I (incisors)} \frac{3\text{-}3}{1\text{-}1}, \text{C (canines)} \frac{1\text{-}1}{1\text{-}1}, \text{P (premolars)} \frac{3\text{-}3}{1\text{-}1}, \text{M (molars)} \frac{3\text{-}3}{3\text{-}3} = 32$$

DISTRIBUTION. The pygmy shrew occurs widespread across boreal North America as far north as the tundra, and southward to the northern United States, in mountainous areas and in the Great Lakes Region. In Michigan, it has been recorded in 12 of the 15 counties in the Upper Peninsula and in six counties in the northern part of the Lower

Map 7. Geographic distribution of the pygmy shrew (*Sorex hoyi*). Actual records exist for counties with dark shading.

Peninsula (see Map 7). There is no evidence that the pygmy shrew occurs in the southern part of the Lower Peninsula; a record from Ann Arbor in Washtenaw County is questionable. However, the species is known from extreme southern Ontario, across the Huron-Erie Waterway Connection from the southeastern part of the Lower Peninsula (see Table 13).

HABITAT PREFERENCES. The pygmy shrew is characteristic of the northern coniferous and mixed coniferous and hardwood forests (Long, 1972a, 1972b, 1974). It has been reported in dry woodlands, thickets, and grassy clearings (Hoffmeister and Mohr, 1957); in moist alder clumps at the edge of a brook (Osgood, 1938); in partly submerged sphagnum and tall grass (Preble, 1937); in a garage in winter in Illinois (Sanborn and Tibbitts, 1949); in sub-climax beech-maple forest with clearings of chokecherry, *Viburnum*, sedges, and ferns (Miller, 1964); in leaf litter under hardwood pole timber (Svendsen, 1976); in moss mats with low spruce (Heinrich, 1953); in heavy spruce and pine at lakeside (Manville, 1942); in dry deciduous woods (Preble, 1910); in open woods and rocky areas, under logs, and among roots and stumps (Cahn, 1937); in stands of hemlock, white pine, hardwoods, and popular-soft maple-jack pine (Schmidt, 1931); in runways of southern red-backed voles (Goodwin, 1929); in sphagnum, tamarack swamp, heavy lakeside grass, and on dry sand ridges (Bailey, 1929); in bluegrass (Scott, 1939); in piles of driftwood along a stream (Goodwin, 1924); in an alpine bog (Spencer and Pettus, 1966); and in an old cabin (Dice, 1921). One pygmy shrew was even captured near noontime as it crossed a dirt road in Montana (Koford, 1938).

In the Upper Peninsula, William L. Robinson (*in litt.*, April 23, 1975) caught 11 pygmy shrews in several of these above listed habitats in the McCormick Experimental Forest in Marquette County. In Gogebic and Ontonagon counties, Dice and Sherman (1922) found this shrew in wet hardwood forest and in black spruce bog. Manville (1948) caught pygmy shrews in leatherleaf and white cedar swamps. In the Lower Peninsula, William H. Burt captured a female in beech-maple woodlands at Barnhart Lakes in Presque Isle County. In Roscommon County, Rabe (1981) caught 18 first-year pygmy shrews in swampy peatland dominated by leatherleaf (*Chamaedaphne calyculata*) and bog birch (*Betula pumila*), with sedges, grasses (mostly *Muhlenbergia glomerta*), and meadowsweet (*Spiraea alba*) interspersed, and *Sphagnum* forming thick matting.

DENSITY AND MOVEMENTS. Pygmy shrews have a strange way of appearing suddenly in large numbers and then apparently sharply decreasing or at least becoming inconspicuous to field observers. Although such population fluctuations may certainly be real, poor success in catching pygmy shrews and other small insectivores in standard mouse traps in the past may account for the lack of data on these fluctuations. Recent work in Michigan by L. L. Master, J. M. Ryan and others demonstrate that pit traps are much more efficient for capturing shrews than snap traps. Pit traps consist of wide-mouthed cans at least 7 in (18 cm) deep and 4 in (10 cm) in diameter sunk in the ground to surface level in favorable places (Hudson and Solf, 1959). Holes are punched in the upper part of each can so that it is never more than half full of water.

Two population estimates have been calculated for Michigan pygmy shrews. Manville (1948; 1949) found 0.21 individuals per acre (0.52 per ha) in a Marquette County cedar swamp in June. In the McCormick Experimental Forest in Baraga and Marquette counties, Haveman (1973) found pygmy shrews more numerous, about two individuals per acre (5 per ha). It was Manville's (1948) suggestion that individual pygmy shrews may move about in areas as large as ½ acre (0.2 ha). In terms of individual movement, Anderson (1977) re-captured a marked pygmy shrew after an 11 month interval; it had moved about 500 ft (150 m). If the two points of capture are considered the diameter of the individual's activity area, it would total 4.4 acres (1.8 ha), which might suggest a life-time area of movement rather than a seasonal one.

BEHAVIOR. Observations of captive individuals show that the pygmy shrew maintains an active pace, with occasional short rests throughout the entire 24 hour day. Preble (1938) reported catching pygmy shrews in early morning and in late afternoon. It appears to be an efficient digger, especially in soft, moist soil and leaf mold (Saunders, 1929). Prince (1940) observed a captive shrew darting about with tail held straight out and upwardly curved, climbing vertical cage walls, hanging upside down on the wire mesh ceiling, making jumps as high as 4½ in (115 mm), and continually probing its environment with a motile snout. During the shrew's brief sleeping periods, Prince noted that it curled up with legs drawn in under the body with the head and tail curled alongside. The pygmy shrew perceives its surroundings by means of its sensitive nose and ears. It is somewhat noisy as it moves about; Saunders (1929) reported a combination of whispering and whistling sounds, infinitely high on the musical scale. He also called attention to the animal's ability to discharge a potent musky odor from its flank glands when excited. Goodwin (1924) noted that captive shrews would attack and kill each other by biting the neck and chest.

ASSOCIATES. Because of the pygmy shrew's wide selection of habitats in the boreal environments, it may be expected to associate with all northern Michigan mammals. In Baraga and Marquette counties, Haveman (1973) caught pygmy shrews in pit traps along with the arctic shrew, masked shrew, short-tailed shrew, deer mouse, southern red-backed vole, meadow vole, and southern bog lemming. Ryan (1982) caught pygmy shrews in association with the southern bog lemming at three localities in Cheboygan County and at one in Otsego and Montmorency counties. In a leather-leaf-bog birch peatland community in Roscommon County, Rabe (1981) caught pygmy shrews in August in association with the masked shrew, water shrew, short-tailed shrew, star-nosed mole, meadow vole, and meadow jumping mouse. Her catch in snap traps set in spaced-out lines included 18 pygmy shrews and 233 masked shrews, a ratio of one to 18. However, the use of snap traps appears to discriminate against pygmy shrews for reasons unclear. Using pitfall traps instead, one pygmy shrew to only four to 10 masked shrews were taken in Michigan habitats (Master, 1978; Ryan, 1982). Parmalee and Munyer (1966) stressed the uncommon occurrence of the pygmy shrew when they identified, from post-glacial debris in a fissure in Monroe County, Illinois, an assortment of bones representing the remains of 1,000 short-tailed shrews but only one pygmy shrew.

REPRODUCTIVE ACTIVITIES. Knowledge of the breeding habits of the pygmy shrew is fragmentary. From what is known about other Michigan shrews of the genus *Sorex*, reproductive activities probably occur from late spring to perhaps late summer. Breeding readiness was noted in a female on June 25 (Palmer, 1947). Pregnant females carrying from three to seven embryos have been recorded as early as June 8 (Jackson, 1961), July (Long, 1976; Scott, 1939), and as late as August 3 (Osgood, 1938). The early life of young shrews has not been examined. It is generally thought, however, that females produce only one litter per year (Long, 1976). Young-of-the-year do not become sexually active until the second spring-summer, as demonstrated by the absence of any reproductive development in the 18 first-year pygmy shrews captured between August 19 and August 23 in Roscommon County. The only record indicating longevity is for a male re-captured 11 months after first being taken in the Upper Peninsula (Anderson, 1977).

FOOD HABITS. Insects probably make up the bulk of the pygmy shrew's diet, which also includes earthworms, spiders, mollusks, and other invertebrates (Cahn, 1937; Osgood, 1938). Digestive tracts from pygmy shrews taken in Baraga and Marquette counties contained flies (Diptera), beetles (Coleoptera), unidentified insects, spiders, and some *Sphagnum* (Haveman, 1973). Schmidt (1931) caught pygmy shrews apparently attracted by meat bait (small mammals, including the masked shrew and squirrel viscera) in pit traps. A captive pygmy shrew relished the flesh of the masked shrew, white-footed mouse, southern red-backed vole, and the liver of a meadow vole.

ENEMIES. The pygmy shrew can evidently avoid enemies successfully enough to survive its first winter. This is vital, since the young apparently do not breed until their second year. In addition, a female pygmy shrew produces only one litter per year. Supposedly the musky secretions from the flank glands do a fair job in discouraging some would-be attackers. Even so, there are reports that successful predators include a brook trout, garter snake, hawk, and house cat (Cahn, 1937; Long, 1970, 1972b; Manville, 1948).

PARASITES AND DISEASES. One flea, *Stenoponia americana*, unidentified mites, one tick of the genus *Ixodes*, and intestinal hymenolepidid tapeworms are known to infest pygmy shrews (Buckner and Blasko, 1969; Long, 1974).

MOLT AND COLOR ABERRATIONS. Pygmy shrews in the Michigan area molt their dull gray winter coats in late April and May and their browner summer pelage in October and November (Jackson, 1928; Long, 1974). No albino individuals have been reported.

ECONOMIC IMPORTANCE. These tiny creatures have no real positive or negative relationships with human affairs. Their importance in their natural environment certainly needs further appraisal.

COUNTY RECORDS FROM MICHIGAN. Records of specimens in collections in museums and from the literature show that the pygmy shrew is known from the following counties: in the Upper Peninsula: Baraga, Chippewa, Delta, Dickinson, Gogebic, Houghton, Iron, Luce, Marquette, Menominee, Ontonagon, Schoolcraft; in the Lower

Peninsula: Cheboygan, Crawford, Montmorency, Otsego, Presque Isle, Roscommon (see Map 7).

LITERATURE CITED

Anderson, T. J.
1977 Population biology of the masked shrew, *Sorex cinereus*, in hardwood forest areas of the McCormick Experimental Forest, Marquette County, Michigan. Northern Michigan Univ., unpubl. M.S. thesis, viii+75 pp.

Bailey, B.
1929 Mammals of Sherburne County, Minnesota. Jour. Mammalogy, 10(2):152–164.

Banfield, A. W. F.
1974 The mammals of Canada. Univ. Toronto Press, Toronto. xxiv+438 pp.

Buckner, C. H., and G. G. Blasko
1969 Additional range and host records of the fleas (Siphonaptera) of Manitoba. Manitoba Ent., 3:65–69.

Cahn, R.
1937 The mammals of the Quetico Provincial Park, Ontario. Jour. Mammalogy, 18(1):19–30.

Dice, L. R.
1921 Notes on the mammals of interior Alaska. Jour. Mammalogy, 2(1):20–28.

Dice, L. R., and H. B. Sherman
1922 Notes on the mammals of Gogebic and Ontonagon counties, Michigan, 1920. Univ. Michigan, Mus. Zool., Occas. Papers No. 109, 40 pp.

Diersing, V. E.
1980 Systematics and evolution of the pygmy shrew (subgenus *Microsorex*) of North America. Jour. Mammalogy, 6(1):76–101.

Goodwin, G. G.
1924 Mammals of the Gaspe Peninsula, Quebec. Jour. Mammalogy, 5(4):246–257.
1929 Mammals of the Cascapedia Valley, Quebec. Jour. Mammalogy, 10(3):231–246.

Haveman, J. R.
1973 A study of population densities, habitats and foods of four sympatric species of shrews. Northern Michigan Univ., unpubl. M.S. thesis, vii+70 pp.

Heinrich, G. H.
1953 *Microsorex, Sorex palustris,* and *Microtus chrotorrhinus* from Mt. Katahdin, Maine. Jour. Mammalogy, 34:382.

Hoffmeister, D. F., and C. O. Mohr
1957 Fieldbook of Illinois mammals. Illinois Nat. Hist. Survey Div., Manual 4, xi+233 pp.

Hudson, G. E. and J. D. Solf
1959 Control of small mammals with sunken-can pitfalls. Jour. Mammalogy, 40(3):455–457.

Jackson, H. H. T.
1928 A taxonomic review of the American long-tailed shrews (genera Sorex and Microsorex). American Fauna No. 51, 238 pp.
1961 Mammals of Wisconsin. Univ. Wisconsin Press, Madison. xii+504 pp.

Koford, C. B.
1938 *Microsorex hoyi washingtoni* in Montana. Jour. Mammalogy, 19(3):372.

Long, C. A.
1970 Mammals of central Wisconsin. Univ. Wisconsin, Stevens Point, Mus. Nat. Hist., Rpt. Fauna and Flora Wisconsin, No. 3, 59 pp.
1972a Taxonomic revision of the mammalian genus *Microsorex* Coues. Trans. Kansas Acad. Sci., 74:181–196.
1972 Notes on habitat preference and reproduction in pygmy shrews, *Microsorex*. Canadian Field-Nat., 86:155–160.
1974 *Microsorex hoyi* and *Microsorex thompsoni*. American Soc. Mammalogists, Mammalian Species No. 33, 4 pp.
1976 Notes on reproduction in pygmy shrews and observed ratios of mammae to body weights. Univ. Wisconsin, Stevens Point, Mus. Nat. Hist., Rpt. Fauna and Flora Wisconsin, No. 11:5–6.

Manville, R. H.
1942 Notes on the mammals of Mount Desert Island, Maine. Jour. Mammalogy, 33:391–398.
1948 The vertebrate fauna of the Huron Mountains, Michigan. American Mid. Nat., 39:615–640.
1949 A study of small mammal populations in northern Michigan. Univ. Michigan, Mus. Zool., Misc. Publ. No. 73, 83 pp.

Master, L. L.
1978 A survey of the current distribution, abundance, and habitat requirements of threatened and potentially threatened species of small mammals in Michigan. Univ. Michigan, Mus. Zool., iv+52 pp. (mimeo.).

Miller, D. H.
1964 Northern records of the pine mouse in Vermont. Jour. Mammalogy, 45:627–628.

Osgood, F. L., Jr.
1938 The mammals of Vermont. Jour. Mammalogy, 19: 435–441.

Palmer, R. S.
1947 Notes on some Maine shrews. Jour. Mammalogy, 28:13–16.

Parmalee, P. W., and E. A. Munyer
1966 Range extension of the least weasel and pygmy shrew in Illinois. Trans. Illinois Acad. Soc., 59:81–82.

Preble, F. A.
1910 A new *Microsorex* from the vicinity of Washington, D.C. Proc. Biol. Soc., Washington, 23:101–102.

Preble, N. A.
1937 Pygmy in New Hampshire. Jour. Mammalogy, 18: 362–363.
1938 Additional records of the pygmy shrew in New Hampshire. Jour. Mammalogy, 19(3):371–372.

Prince, L. A.
1940 Notes on the habits of the pygmy shrew (*Microsorex hoyi*) in captivity. Canadian Field-Nat., 54:97–100.

Rabe, M. L.
1981 New locations for pygmy (*Sorex hoyi*) and water (*Sorex palustris*) shrews in Michigan. Jack-pine Warbler, 59(1):16–17.

Ryan, J. M.
1982 Distribution and habitat of the pygmy shrew, *Sorex (Microsorex) hoyi*, in Michigan. Jack-Pine Warbler (in press).

Sanborn, C. C., and D. Tibbitts
1949 Hoy's pygmy shrew in Illinois. Chicago Acad. Sci., Nat. Hist. Misc., 36, 2 pp.

Sanders, P. B.
1929 *Microsorex hoyi* in captivity. Jour. Mammalogy, 10(1): 78–79.

Schmidt, F. J. W.
1931 Mammals of western Clark County, Wisconsin. Jour. Mammalogy, 12(2):99–117.

Scott, T. G.
1939 Number of fetuses in the Hoy pygmy shrew. Jour. Mammalogy, 20(2):251.

Spencer, A. W., and D. Pettus
1966 Habitat preferences of five sympatric species of long-tailed shrews. Ecology, 47:677–683.

Svendsen, G. E.
1976 *Microsorex hoyii* in southeastern Ohio. Ohio Jour. Sci., 76(3):102.

van Zyll de Jong, C. G.
1976 Are there two species of pygmy shrews (*Microsorex*)? Canadian Field Nat., 90:485–487.

WATER SHREW
Sorex palustris Richardson

NAMES. The generic name *Sorex,* proposed by Linnaeus in 1758, is derived from the Latin and means shrew. The specific name *palustris,* proposed by Richardson in 1828, is also from the Latin and means marshy, referring to a preferred habitat for this shrew. In Michigan, the animal is commonly called water shrew. The water shrews in both the Upper and Lower Peninsulas are classified as belonging to the subspecies *Sorex palustris hydrobadistes* Jackson.

RECOGNITION. Body large for a shrew of this genus and covered with soft, short hair; head long with pointed nose and conspicuous whiskers; eyes small and almost concealed by facial fur; ears with pinnae slit-like; neck short and not differentiated; body robust and cylindrical; tail long, slender, and clothed with short hairs; legs short and slender; forefeet delicate; hindfeet broad and fringed with stiff hairs; third and fourth hind toes joined by partial webbing; flank glands obvious in sexually-active males; three pairs of mammary glands, one abdominal, two inguinal (between hind limbs). Upperparts blackish; underparts silvery to brownish; tail dark above and silvery below. In bright light, the soft, velvet upperparts frequently show green or purple iridescence.

MEASUREMENTS AND WEIGHTS. Adult water shrews measure as follows: total length from tip of nose to end of tail vertebrae, 5¾ to 6½ in (138 to 164 mm); length of tail vertebrae, 2½ to 2⅞ in (63 to 72 mm); length of hind foot, ¾ in (19 to 20 mm); weight, approximately ⅓ oz (12 to 17 g).

DISTINCTIVE CRANIAL AND DENTAL CHARACTERISTICS. The skull of the adult water shrew is long, with a slender rostrum and a broad and flattened cranium (see Fig. 9). It measures approximately ⅝ in (21.1 to 22.5 mm) long and ⅜ in (10.0 to 10.9 mm) wide. The first two unicuspid teeth in the upper jaws are smaller than the next two with the third being smaller than the fourth (see Fig. 8). The skull of the water shrew is larger and heavier than those of the arctic, masked, pygmy, and least shrews. Although the water shrew's skull is almost as long as that of the short-tailed shrew, the former is narrower (width less than 11 mm instead of more), less angular and ridged, and lacks the mid-dorsal (sagittal) prominent crest toward the posterior end of the cranium in the short-tailed shrew (Fig. 8). The dental formula is as follows:

$$\text{I (incisors)}\frac{3\text{-}3}{1\text{-}1},\ \text{C (canines)}\frac{1\text{-}1}{1\text{-}1},\ \text{P (premolars)}\frac{3\text{-}3}{1\text{-}1},\ \text{M (molars)}\frac{3\text{-}3}{3\text{-}3} = 32$$

DISTRIBUTION. The water shrew occurs throughout the broad band of northern coniferous forest across North America, from the edge of the arctic tundra south to the states at the Canadian-American border except for those in the mid-continental plains. Within the United States, the water shrew also thrives in suitable boreal habitats in the Sierra Nevada, the southern Rockies, and the Appalachians. In Michigan, it is found throughout the Upper Peninsula. In the Lower Peninsula, the water shrew seems less common, with trapping records indicating a spotty distribution in the northern four or five tiers of counties (see Map 8).

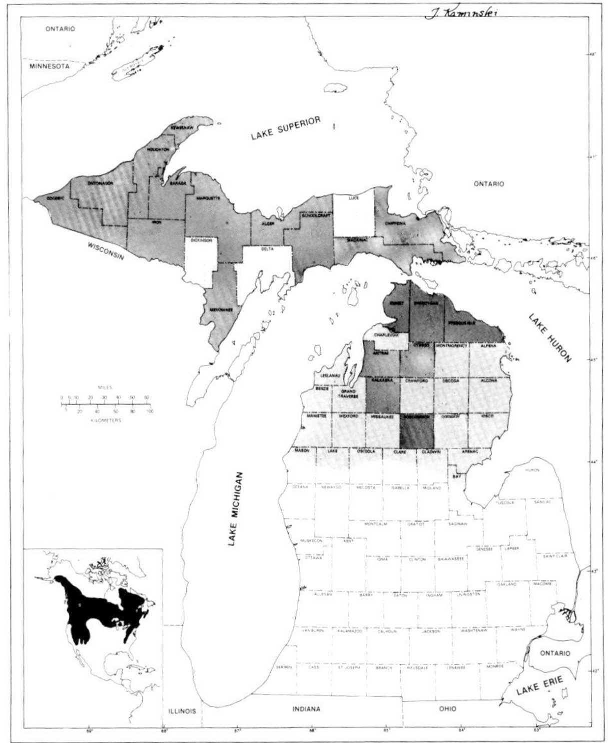

Map 8. Geographic distribution of the water shrew (*Sorex palustris*). Actual records exist for counties with dark shading.

HABITAT PREFERENCES. The water shrew, as its name implies, has a close affinity for water and is certainly the most aquatic of Michigan shrews. It has been found in moist situations along stone walls, along fence lines near cedar-spruce swamps, at bridge abutments, and on the wet forest floor, but especially at water's edge (Conaway, 1952; Kinsella, 1967; Ozoga and Gaertner, 1963; Rabe, 1981; Timm, 1975). Although it forages along swamps, bogs, and lakes, this shrew's favored habitat appears to be the edges of rock-filled and log-strewn cold-water streams bordered by woody vegetation (Borell and Ellis, 1934).

DENSITY AND MOVEMENTS. This close affinity for aquatic environments accounts for the water shrew's spotty distribution. Collectors have been most successful in catching individuals in traps placed adjacent to the water's edge or sometimes even floating in water. Even so, few population figures are available to show abundance of water shrews in their linear habitats along waterways. In Cheboygan County, for example, Master (1978) trapped seven water shrews along a 60-meter stretch of streamway in a 36-hour period. In Manitoba, Buckner and Ray (1968) re-trapped previously marked individuals to calculate home ranges of two individuals to be approximately 0.5 acres (0.2 ha) and 0.8 acres (0.3 ha), respectively.

BEHAVIOR. The water shrew's swimming ability is enhanced by broad hind feet fringed with stiff hairs and by the partial webbing between the third and fourth toes. Remarkable buoyancy is acquired, in part, by air bubbles trapped in body fur. Observations indicate that water shrews can literally run across pools of water 5 ft (1.5 m) wide without submerging (Jackson, 1961). Observations of animals in aquaria show the water shrew uses considerable energy in diving, especially with the vigorous paddling action of the feet (Howell, 1924). Its long nose is effective in probing bottom sediments. Water shrews may be active at any hour of the day and during any season, although major peaks of movements are just prior to daylight and after dusk.

Like other shrews, this species shows little inclination to be social except for breeding periods. Studies indicate that individuals move around in rather discrete areas, but there is no evidence that one animal will restrict others from entering its area. However, it can be supposed that individuals may very well stake out foraging areas along water courses for their own use. Little is known about either the sounds uttered by water shrews or how their senses are used to perceive their environments.

The nest may be as large as 4 in (102 mm) in diameter and constructed of vegetation including sticks and leaves. Bankside burrows are used as retreats and perhaps places to hoard food. Water shrews often move along their own surface runways under the protective covering of bank overhangs, fallen logs, and debris. The animals are known to seek refuge in beaver and muskrat houses (Bailey, 1929; Banfield, 1974).

ASSOCIATES. Water shrews appear to have the aggressive and voracious characteristics of other shrew species. Yet, Sorenson (1962) kept water shrews and deer mice (genus *Peromyscus*) in the same cage compatibly for more than ten months. In Iron County, Ozoga and Gaertner (1963) caught water shrews in close association with the masked shrew, short-tailed shrew, meadow vole, and meadow jumping mouse. All were taken in traps within a 15 ft (4.6 m) radius. In Emmet County, Hitchcock (1943) captured a water shrew and a deer mouse in traps set no more than 2 in (50 mm) apart. There may be considerably fewer water shrews than masked shrews, although most trapping to date has been with snap traps, which appear to be more successful in catching masked shrews than other soricids in the Michigan communities. For example, in Roscommon County Rabe (1981), using snap traps, caught only one water shrew along with 18 pygmy shrews and 233 masked shrews in a leatherleaf-bog birch peatland community.

REPRODUCTIVE ACTIVITIES. The breeding season begins as early as January and continues as late as August. Pregnant females have been found in late February and early March (Conaway, 1952). The gestation is approximately 21 days (Jackson, 1961). Adult females may produce two and possibly three litters in a single season. Females born early in the year occasionally produce a litter their first year. Litters of six individuals are usual with a maximum of eight known. Details of the growth of the nursing young are scant. By July, young-of-the-year are conspicuous in the population. Although Conaway (1952) suggested that the life span of the water shrew is no more than 18 months, this does give individuals the opportunity to develop their breeding potential in the second year of life.

FOOD HABITS. Although the water shrew requires almost its weight in food per day for sufficient energy to maintain correct body heat and organ function, foraging and hoarding occur just prior to dawn and just after sunset (Sorenson, 1962). The water shrew's diet consists chiefly of invertebrates which frequent the water's edge, including the aquatic nymphs of caddis flies, mayflies, and stoneflies. Hamilton (1930) found digestive tracts of water shrews to contain 78% insects including representatives of such orders as Hemiptera, Tricoptera, Ephemeroptera, Diptera, Coleoptera, and Hymenoptera. Slugs, earthworms, spiders, crickets, leafhoppers, fly larvae, *Endogone* and other vegetable material were found in stomachs of water shrews examined in Minnesota by Whitaker and Schmeltz (1973). Small fish have occasionally been identified as prey of water shrews, as well as a salamander, reported in Oregon (Nussbaum and Maser, 1969).

ENEMIES. In its semi-aquatic habitat, the water shrew is subject to predation from birds of prey, terrestrial meat-eaters, and predatory fish. Even so, records of predation on the water shrew are sparse. Jackson (1961) noted that weasels and garter snakes eat water shrews and suspected that hawks, owls, mink, and various large fish do also. The prevalence of water shrews' remains in trout stomachs, according to Doutt *et al.* (1966), is one of the better ways of recording the presence of the animal.

PARASITES AND DISEASES. There is little recorded evidence of parasites which infest the water shrew. A flea, *Doratopsylla blarinae,* was found on water shrews in New England (Jackson, 1961). Other fleas, ticks, and mites are reported by Timm (1975) and Whitaker and Schmeltz (1973).

MOLT AND COLOR ABERRATIONS. The winter pelt is shed between April and June. Summer growth begins at the animal's anterior and when complete, is dark but somewhat dull. In late July to September, the heavier winter pelt appears. It has a luxurious black coloring above with a silvery wash on the paler underparts.

COUNTY RECORDS FROM MICHIGAN. Records of specimens in collections in museums and from the literature (Map 8) show that the water shrew is known from the following counties; in the Upper Peninsula: Alger, Baraga, Chippewa, Gogebic, Houghton, Iron, Keweenaw, Mackinac, Marquette, Menominee, Ontonagon, Schoolcraft;

in the Lower Peninsula: Antrim, Cheboygan (Creaser, 1934), Emmet (Hitchcock, 1943), Kalkaska, Otsego, Presque Isle, and Roscommon (Rabe, 1981).

LITERATURE CITED

Bailey, B.
1929 Mammals of Sherburne County, Minnesota. Jour. Mammalogy, 10(2):153–164.

Banfield, A. W. F.
1974 The mammals of Canada. Univ. Toronto Press, Toronto. xxiv+438 pp.

Borell, A. E., and R. Ellis
1934 Mammals of the Ruby Mountains region of northeastern Nevada. Jour. Mammalogy, 15(1):12–44.

Buckner, C. H., and D. G. H. Ray
1968 Notes on the water shrew in bog habitats of southeastern Manitoba. Blue Jay, 26(2):95–96.

Conaway, C. H.
1952 Life history of the water shrew (*Sorex palustris navigator*). American Midl. Nat., 48(1):219–248.

Creaser, C. W.
1934 The moose (*Alces americanus*) and the water shrew (*Sorex palustris hydrobodistes*) rare mammals of the southern peninsula of Michigan. Michigan Acad. Sci., Arts and Ltrs., Papers, 20:597–598.

Doutt, J. K., C. A. Heppenstall, and J. E. Guilday
1966 Mammals of Pennsylvania. Pennsylvania Game Comm., Harrisburg. 273 pp.

Hamilton, W. J., Jr.
1930 The food of the Soricidae. Jour. Mammalogy, 11(1): 26–39.

Hitchcock, H. B.
1943 *Peromyscus maniculatus bairdii* and *Sorex palustris hydrobadistes* in the Lower Peninsula of Michigan. Jour. Mammalogy, 24(3):402–403.

Howell, A.
1924 The mammals of Mammoth, Mono County, California. Jour. Mammalogy, 5(1):24–36.

Jackson, H. H. T.
1961 Mammals of Wisconsin. Univ. Wisconsin Press, Madison. xii+504 pp.

Kinsella, J. M.
1967 Unusual habitat of the water shrew in western Montana. Jour. Mammalogy, 48(3):475–477.

Master, L. L.
1978 A survey of the current distribution, abundance and habitat requirements of threatened and potentially threatened species of small mammals in Michigan. Univ. Michigan, Mus. Zool., iv+52 pp. (mimeo.).

Nussbaum, R. A., and C. Maser
1969 Observations of *Sorex palustris* preying on *Dicamptodon ensatus*. Murrelet, 50(2):23–24.

Ozoga, J. and R. Gaertner
1963 Noteworthy locality records for some Michigan mammals. Jack-Pine Warbler, 41(2):89–90.

Rabe, M. L.
1981 New locations for pygmy (*Sorex hoyi*) and water (*Sorex palustris*) shrews in Michigan. Jack-Pine Warbler, 59(1):16–17.

Sorenson, M. W.
1962 Some aspects of water shrew behavior. American Midl. Nat., 68(2):445–462.

Timm, R. M.
1975 Distribution, natural history, and parasites of mammals of Cook County, Minnesota. Univ. Minnesota, Bell Mus. Nat. Hist., Occas. Papers 14, 56 pp.

Whitaker, J. O., Jr., and L. L. Schmeltz
1973 Food and external parasites of *Sorex palustris* and food of *Sorex cinereus* from St. Louis County, Minnesota. Jour. Mammalogy, 54(1):283–285.

SHORT-TAILED SHREW
Blarina brevicauda (Say)

NAMES. The generic name *Blarina,* proposed by British biologist J. E. Gray in 1838, has an uncertain origin, but it may have been the intent of the describer to name this shrew for Blair, the place in Nebraska near which the species was first described. The specific name *brevicauda,* proposed by Thomas Say in 1823, is of Latin derivation and refers to this shrew's characteristically short tail (*brevi* meaning short; *cauda* meaning tail). In some areas, this mammal is called mole shrew. Short-tailed shrews in Michigan's Upper and Lower Peninsulas belong to the subspecies *Blarina brevicauda kirtlandi* Bole and Moulthrop.

RECOGNITION. Body large for an American shrew, mouse-like, and covered with soft, short velvety fur; head elongate; nose somewhat blunt; eyes small and beady; ears slit-like, well concealed in the fur at the sides of the head (Plate 1 and Fig. 10); neck short and not differentiated; body robust and cylindrical; tail short and fairly thick; limbs

Figure 10. Side view of the head of a short-tailed shrew (*Blarina brevicauda*) showing details of the external ear and pinna with the fur parted. Sketch by Bonnie Marris.

short and sturdy; feet strongly constructed, especially the front ones; flank flands slit-like and well developed; midventral gland conspicuous by presence of a hairless area; three pairs of mammary glands, one on belly and two between hind limbs.

In color, the upperparts are slate black, slightly darker in the denser winter fur and paler in summer; underparts paler, slate gray both in winter and summer; well-haired tail not obviously bicolored. The large size and bobbed tail are distinctive when comparing the short-tailed shrew with other Michigan shrews. The dark slate coloring, soft velvet-like fur, inconspicuous ear pinnae, and long blunt nose serve to distinguish the short-tailed shrew from small rodent associates.

MEASUREMENTS AND WEIGHTS. Adult short-tailed shrews measure as follows: length from tip of nose to end of tail vertebrae, 4¼ to 5½ in (108 to 140 mm); length of tail vertebrae, ¾ to 1¼ in (18 to 30 mm); length of hind foot, 9/16 to 11/16 in (13 to 17 mm); weight, ½ to 1 oz (18 to 30 g).

DISTINCTIVE CRANIAL AND DENTAL CHARACTERISTICS. The skull of the adult short-tailed shrew is large, massive, and angular with prominent processes and ridges (see Fig. 9). In overview, the cranium is conspicuously wide and flat; with increased age the rostrum becomes heavier, with the large first incisors pointing progressively more ventrally. Behind these large front teeth, the unicuspid teeth on each side of the upper jaws are arranged in two pairs with the first pair (actually the second and third incisors) being much larger than the second pair (actually the canine and first premolar). The fifth unicuspid is much smaller and obscured when the skull is viewed from the side (see Fig. 8). These teeth are tipped with rich brown pigment. The skull measures approximately 1 in (23 to 25 mm) in greatest length and ½ in (11 to 13 mm) in greatest width. The dental formula is as follows:

$$\text{I (incisors)}\,\frac{3\text{-}3}{1\text{-}1},\ \text{C (canines)}\,\frac{1\text{-}1}{1\text{-}1},\ \text{P (premolars)}\,\frac{3\text{-}3}{1\text{-}1},\ \text{M (molars)}\,\frac{3\text{-}3}{3\text{-}3} = 32$$

DISTRIBUTION. The short-tailed shrew is a prominent member of the small mammalian community in eastern North America from southern Canada down to the Gulf coastal states and west to the Great Plains. In Michigan, this species occurs in all counties and most terrestrial communities (Map 9).

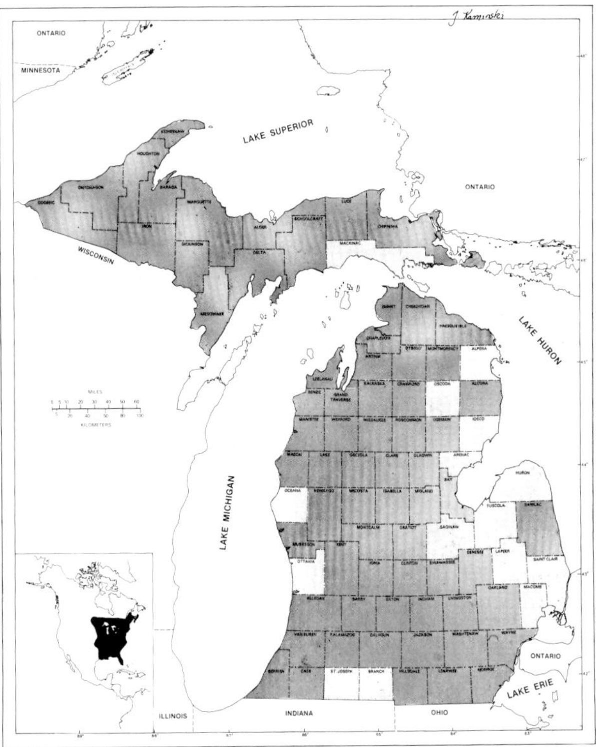

Map 9. Geographic distribution of the short-tailed shrew (*Blarina brevicauda*). Actual records exist for counties with dark shading.

HABITAT PREFERENCES. The short-tailed shrew is at home under varied living conditions in Michigan—from swamps and bogs in moist lowlands to dry uplands and even on sand dunes (Getz, 1961). Its ability to survive in vacant lots, city parks, and right-of-ways of streets and railroads makes this shrew a common urban and suburban resident. However, moist, litter-strewn forested areas are considered prime living places for the short-tailed shrew (Pruitt, 1959). Normally, it does not frequent areas completely devoid of vegetation, unless snow-covered (Schreiber and Graves, 1977). Shocked grain left in fields and surrounded by cleared stubble can serve as protective winter cover for the shrews (Linduska, 1942).

DENSITY AND MOVEMENTS. Many authors, notably Burt (1946) in Michigan and Jackson (1961) in Wisconsin, suggest that when all terrestrial habitats are considered, the short-tailed shrew is the most common of all mammals in the Upper Great Lakes Region. Some mammalogists feel the ubiquitous masked shrew may very well merit this distinction in Michigan (Lawrence L. Master, pers. comm.). The meadow vole will outnumber the short-tailed shrew, as well as the masked shrew, in many open grassy and marshy sites, but in brushy and forested habitats, the two shrews will be far more abundant. Although population estimates range as high as 50 short-tailed shrews per acre (125 per ha) in woodlands (Bole, 1939; Seton, 1909), with one New York study by Townsend (1935) showing a high of 104 per acre (260 per ha), most population studies in Michigan show lower densities. Estimates suggest that populations fluctuate considerably during the warm season, probably increasing when large numbers of young-of-the-year appear in the adult population. In the Upper Peninsula, Manville (1949) counted 0.3 to 3.0 short-tailed shrews per acre (.075 to 7.6 per ha) in late May and 8.5 to 11.6 per acre (21 to 29 per ha) in June in Marquette County. In another study in the McCormick Experimental Forest in Baraga and Marquette counties, Haveman (1973) estimated shrew populations at 19 per acre (47 per ha) in mature hardwood forest, seven per acre (17.5 per ha) in spruce swamp, and five per acre (12.5 per ha) in bog environment.

In the Lower Peninsula, Shull (1907) suggested that at least four short-tailed shrews (two pairs) occur per acre (10 per ha). Blair (1940) found shrew densities of 0.8 per acre (2 per ha) in August and 2.2 per acre (5.5 per ha) in September. There is abundant evidence to show that populations of short-tailed shrews fluctuate seasonally and multiannually, but reasons are obscure (Banfield, 1974; Manville, 1949).

Individual short-tailed shrews move around in rather restricted areas. In Marquette County, Manville (1949) found that in three days or less they travel as far as 200 ft (61 m) from first place of capture. It was Hamilton's (1931) opinion that normal movements are usually confined to areas no larger than an acre (0.4 ha). In the Upper Peninsula, Manville (1949) estimated, based on sparse data, that shrews in the Huron Mountain area of Marquette County may have used no more than 0.21 acres (less than 0.01 ha). Blair (1940), however, found cruising space of about 1.46 acres (0.59 ha) for females and 1.39 acres (0.56 ha) for males. In another study in the Lower Peninsula, Blair (1940) estimated that home ranges varied from 0.25 acres (0.01 ha) to 4.43 acres (1.8 ha), with those of males usually larger. Habitat quality which would influence density as well as the size of home range in the short-tailed shrew is, according to Getz (1961), profoundly affected by the moisture content of the environment, with prolonged dryness detrimental to the success of this species.

BEHAVIOR. The antics of the short-tailed shrew have been an attraction to many field observers (Lutz, 1964; Rood, 1958; etc.). By observing animals in captivity and in the field, it can be well-documented that this lively creature is active both day and night, alternating periods of intense movements and rest. Employing a device to photograph small mammals using a runway partly under a grass canopy in Ingham County, Osterberg (1962) found that the short-tailed shrew moved back and forth daily on this runway—most frequently (sometimes in bursts) from sunset to four to five hours afterwards. These data, obtained in April and May and again in October and November, showed that the shrew was generally more active in daylight hours in autumn than in spring.

As noted above, the short-tailed shrew uses surface runways of its own construction as well as larger ones made by the meadow vole, bog lemming, and other rodents. Less accurately known is its use of underground tunnels. In moist and spongy soils covered with leaf litter in a typical woodlot in southern Michigan, the short-tailed shrew population may reach maximum densities (Pruitt, 1959). The animal moves around just under the litter layer, in subsurface burrows of its own construction, or in those fashioned by the eastern mole. The network of underground galleries can be so thick that at times a hiker can not take a step without sinking slightly as a result of one or more of them caving in. Despite its apparent abundance, this insectivore is rarely observed by the average citizen unless the family house cat catches one and leaves it uneaten on the doorstep, although house cats will also eat this species (Nader and Martin, 1962).

Captive short-tailed shrews seem congenial to one another, sometimes huddling together during resting periods. Even so, it is thought that the animal is generally solitary in nature, except when attracted to the opposite sex for breeding, and that it moves about in a defined territory, and discourages intruders (Martin, 1981b). Behavioral actions (see Fig. 11) have been described by such workers as Olson (1969). In recent years, methods have been devised (Blus, 1971) for sustaining these nervous animals in captivity to allow for prolonged examination.

Figure 11. Agonistic behavior of the short-tailed shrew (*Blarina brevicauda*), after Olsen (1969).

In probing its immediate environment, the short-tailed shrew makes abundant use of its keen senses of touch and smell, the latter having to do also with aspects of attracting the opposite sex (Pearson, 1944). The snout and abundant whiskers seem most important in various probing actions, while sight seems poorly developed. An acute sense of hearing plus vocalizations combine to provide the animal with the ability to echolocate. Gould *et al.* (1964) showed that high frequency utterances bounce back from nearby objects to the sensitive ears of the short-tailed shrew and give it an awareness of the physical features of the surroundings. The utterances are pulsating calls having a frequency of 29 to 55 kilocycles and a duration of 1.5 to 20 milliseconds. These are emitted when the shrew searches about in its exploring activity. To the human ear, these shrew noises may be similar to the buzz of an electric arc (Pruitt, 1953). Jackson (1961) describes chattering with a high-pitched squeal followed by a rapid succession of short notes resembling *zee-zee-zee-zee-zee*. The variety of sounds made by the short-tailed shrew includes clicking sounds made by the courting male pursuing a female, twitters when the animal is feeding, and shrill squeals when individuals are fighting (Banfield, 1974).

The center of activity of the short-tailed shrew is the nest. Whether this structure is shared with other individuals, perhaps a member of the opposite sex, is undetermined. Surface trails and underground tunnels (see Fig. 12) lead into these nests (usually two entrances) which can be found in grass cover on the surface of the ground, under logs, in dead stumps, under debris, or as much as 1 ft (0.3 m) underground. The nests, according to Rapp and Rapp (1945), are round or oval about 8 in (200 mm) across and 4½ in (145 mm) deep with an inner living space 2 in (50 mm) by 2½ in (70 mm). Nesting materials, usually loosely arranged, consist of an array of plant and animal materials including grasses; sedges; nettle; goldenrod; leaves from such trees as ash, haw, maple,

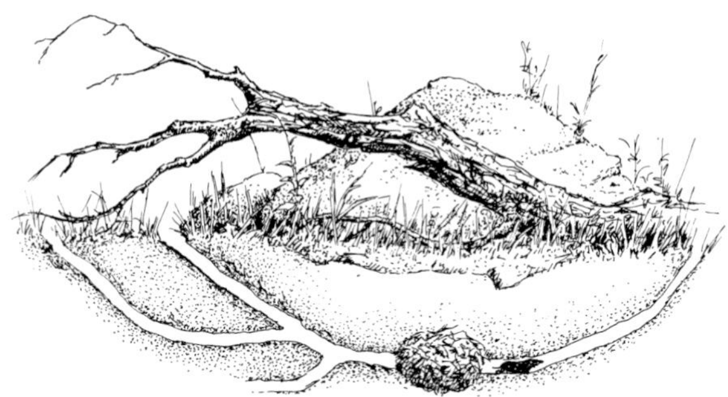

Figure 12. An underground nest and connecting tunnels of a short-tailed shrew (*Blarina brevicauda*), after Schwartz and Schwartz (1959).

sycamore, and oak; greeenbrier; and even meadow vole hair (Dusi, 1951; Hamilton, 1929; Nichols, 1936; Shull, 1907).

ASSOCIATES. The ubiquitous short-tailed shrew, through both its underground and surface activities, associates with most mammals in Michigan outdoor communities. In open habitats, perhaps this shrew's most common neighbor is the meadow vole. The shrew may have the advantage in this relationship as it makes use of runways, usually with grassy canopy, constructed by meadow voles and certainly preys on the meadow vole, especially in winter (Eadie, 1944). Based on his photographic study of trail-users, Osterberg (1962) showed that there are frequent encounters between these two small mammals. Of the 446 photographs obtained in Ingham County in spring (April and May) and in autumn (October and November), 62% were of short-tailed shrews and 36% of meadow vole with only two photographs of deer mice and least weasel and one each of eastern mole and eastern cottontail. In northern mammalian communities, Haveman (1973) caught the short-tailed shrew in pit traps in close association with the arctic shrew, masked shrew, pygmy shrew, deer mouse, southern red-backed vole, meadow vole, southern bog lemming, and meadow jumping mouse. In peatland grown to sedge-willow and leather-leaf bog birch in Roscommon County, Rabe (1981) caught short-tailed shrews with the masked shrew, pygmy shrew, water shrew, star-nosed mole, meadow vole, and meadow jumping mouse. It was Morris' (1979) contention that deer mice and short-tailed shrews are not close neighbors. Small rodents in general may avoid the short-tailed shrew, as shown by a study by Linduska (1942) of mammals living under corn shocks in winter fields of stubble. Meadow voles and prairie deer mice dominated in these food-filled refuges with short-tailed shrews using them only rarely. Interactions between the various species of Michigan shrews for living sites or, more importantly, for food, makes for interesting speculation, but little is known about whether such competition exists.

REPRODUCTIVE ACTIVITIES. Short-tailed shrews are reproductively active from March to September in Michigan. There is also some evidence of low intensity breeding through the winter as well (Christian, 1969; Dapson, 1968). Blus (1971) fostered year-around breeding in captive shrews when (a) light and temperature were kept at a level favorable for shrew activity, (b) food and water were adequate, (c) disturbance to females was minimal, and (d) males were separated from females prior to birth of the young.

Courtship generally reaches initial peaks in March with pregnancies in early April and first litters in late April. A single female may very well produce two and possibly three litters in a single season (Timm, 1975). However, according to Pearson (1944), only about 6% of the adults (including males) survive to produce a second litter. Young-of-the-year may come into breeding condition between one and two months after birth (Dapson, 1968).

After a gestation of 21 to 22 days, a litter averaging six with a maximum of ten is born (Hamilton, 1929). At birth the young are hairless, helpless, measure ⅞ in (22 mm) long, and weigh less than 1/10 oz (0.8 g). At about 17 days of age, the young are highly active, can move in and out of their grass-lined nest, and weigh about ⅓ oz (10 gm). At 25 days of age, the young are generally completely weaned (Blus, 1971). In comparison with shrews of the genus *Sorex,* the short-tailed shrew has marked fecundity but the turnover rate is high (Pearson, 1944). In nature, life expectancy is only a few months; in captivity, the animals are known to survive for 27 months (Jones, 1979).

FOOD HABITS. The short-tailed shrew must be a formidable opponent to many invertebrates and small vertebrates because of its quickness, stocky build and array of sharply-pointed teeth. Martinsen (1969) attributes the short-tailed shrew's dominance in the mammalian community to its proclivity for eating almost anything having nutritious content. This shrew has a heartbeat of 740 to 760 per minute (measured with the animal under sedation), a respiration rate of 164 breaths per minute, an average body temperature of 38.0°C (slightly more than 100°F), and a metabolic rate of 3.18 cm^3 0_2/gm/hr (Doremus, 1965; Neal and Lustick, 1973). These physiological data point to the need for the short-tailed shrew to eat between half and three times its body weight per day. Even so, this mammal is a methodical eater, taking as long as four minutes to consume one earthworm.

The short-tailed shrew is sufficiently vicious and aggressive to subdue prey larger than itself. Observations by Hamilton (1930) show that such attacks are prolonged; for example, it took about 10 minutes for a shrew to subdue a meadow vole and half an hour to dispatch a deer mouse. The shrew may begin the attack by catching the tail of the would-be victim, then maneuvering to grab the nape of the neck and ultimately making a fatal bite at the base of the brain.

In addition to its quickness and sharp teeth, the short-tailed shrew's salivary juice is toxic. Maynard (1889), who reported discomfort from shrew bites on the hand, first suspected this. Much later, Pearson (1942) found that saliva from the submaxillary glands produced a toxic effect when introduced into tooth wounds made by the short-tailed shrew in another small mammal. Insects have also been immobilized by this venom, Martin, (1981a). The small amount introduced was sufficient to stun or even paralyze the prey (Thomasi, 1978). When a victim such as a mouse was bitten, there would be an immediate local reaction in the tissue surrounding the wound; the blood pressure of the affected prey would drop, the heart beat would slow, and respiration would be inhibited. When as little as 0.4 mg of this toxic saliva enters venous blood, death results in a mouse weighing as much as ¾ oz (20 g).

This toxic saliva is also suspected to be a powerful digestive agent for the shrew's food (Lawrence, 1946). The enzyme glycosidase chitinase has the ability to aid in the digestion of hard, complex insect parts (Jeuniaux, 1961). Consequently, fecal deposits of the short-tailed shrews are not filled with undigested solids but consist of soft, often mushy, elongate and spindle-shaped masses, slightly twisted and as much as 1 in (25 mm) long. These fecal deposits are dark greenish-brown or black in color and often deposited on the floor of surface trails.

The short-tailed shrew feeds mostly on insects, other invertebrates, and small vertebrates. Some of the many foods eaten are crickets, moths and their larvae, flies (Diptera), ground-living bees and wasps, beetles and their larvae, spiders, millipedes, centipedes, sowbugs, snails, slugs, earthworms, and such vertebrates as salamanders, garter snakes, small birds, mice, other shrews, pine seeds, and even starchy vegetable material (DeByle, 1965; Eadie, 1949; Hamilton, 1930; Haveman, 1973; Ingram, 1942; Klugh, 1921). Shull (1907) found 40% of the prey of Michigan short-tailed shrews which he studied consisted of meadow vole flesh. He also noted that this insectivore relished pulmonate snails of the genus *Polygyra,* collecting and caching them in underground hoards in winter. Buckner (1964) noted that this shrew consumed the cocoons of pestiferous sawflies. Carter (1936) found a short-tailed shrew that climbed a small red oak to the height of 6 ft (1.8 m) to get to suet in a bird feeder. In captivity, Hamilton (1930) fed shrews assorted animal materials as well as sunflower seeds, raisins, apples, beechnuts, peanut butter, meat of walnuts, cattail root, soft corn, and mushrooms. Obviously the short-tailed shrew will rarely run out of things to eat in the Michigan countryside.

ENEMIES. Short-tailed shrews are bite-sized for many Michigan predators. However, the musky odor exuded by skin glands may very well discourage some of them. For example, Kirk (1921) noted that the tracks (in the snow) of an ermine showed that this mustelid approached no closer than 4 in (100 mm) to a trapped short-tailed shrew. However, in captivity, four female long-tailed weasels consumed 15 short-tailed shrews in 24 hours, and an ermine was observed carrying a short-tailed shrew in its mouth (Hamilton, 1933). On the other hand, there is also evidence in the literature that house cats and mink may pass up the chance to eat these shrews (Errington *et al.*, 1940). Even so, records based on the examination of the contents of the digestive tracts or of fecal droppings indicate that the following mammalian predators have eaten short-tailed shrews: domestic dog, red fox, gray fox, fisher, ermine, long-tailed weasel, striped skunk, bobcat, and house cat (Errington, 1935; Hamilton, 1933, 1936; Hamilton and Cook, 1955; Hamilton *et al.*, 1937; Hatfield, 1939; Jackson, 1961; Murie, 1936; Schofield, 1960; Van Hyning, 1931). Other predators finding short-tailed shrews edible include brown trout, green sunfish, large-mouthed bass, water snake, ring-billed gull, long-eared owl, great horned owl, barred owl, screech owl, snowy owl, great gray owl, barn owl, rough-legged hawk, and red-shouldered hawk (Allan, 1977; Armstrong, 1958; Blokpoel and Haymes, 1979; Chamberlain, 1980; Cope, 1949; Geis, 1952; Huish and Hoffmeister, 1947; Jackson, 1961; J. Krellwitz, pers. comm.; Linsdale,

1928; Master, 1979; Pettingill, 1976; Spiker, 1933; Wallace, 1948; Warthin and Van Tyne, 1922; Wilson, 1938). In a three-year study of the foods of the barn owl in southern Michigan, Wallace (1948) found remains of the short-tailed shrew made up 6.5% of the 6,815 prey animals found in 2,200 barn owl pellets examined. Of birds of prey, short-tailed shrews are probably more successfully captured by owls than by hawks. Gould *et al.* (1964) suggested that this success may be due in part to the owl's keen hearing which enables it to detect shrew calls.

PARASITES AND DISEASES. Fleas, mites, and beetles of the genus *Leptinus* are reported as ectoparasites on short-tailed shrews (Jameson, 1950; Moore, 1949; Timm, 1975; Whitaker and Mumford, 1972). Michigan shrews have been found to be infested with the chigger *Euschoengastia blarinae* and the fleas *Ctenophthalmus pseudogyrtes* and *Doratopsylla blarinae* (Scharf and Stewart, 1980; Wrenn, 1974).

MOLT AND COLOR ABERRATIONS. Short-tailed shrews molt twice annually. The spring molt (beginning as early as late winter) changes the darker and heavier winter pelage to the paler and less dense summer pelage. The new growth of fur appears first on the head and progresses backward. The autumn molt (as late as October and November) proceeds in reverse fashion from the rump forward (Hamilton, 1940). Albino individuals are found on occasion (Murray, 1939; Sime, 1940; Ulmer, 1940). One individual (MSU 24330) preserved from Ingham County has a conspicuous, narrow strip of white hair on the left side of the body from the top of the back to the belly. Blond and silvery lead-gray individuals have also been reported (Christian, 1947).

ECONOMIC IMPORTANCE. Short-tailed shrews are numerous in most Michigan terrestrial communities. Because they rarely enter human habitations or in any way damage agricultural crops, they have no adverse economic importance. Instead, the shrew's predation on meadow voles and noxious insects might endear them to farmers. These shrews may also be useful as monitors of chemical pollution. They tend, along highway right of ways, to show progressively higher levels of such environmental additives as lead, cadmium, and zinc in relation to increases in traffic volume (Blair *et al.*, 1977).

COUNTY RECORDS FROM MICHIGAN. Records of specimens preserved in collections in museums and from the literature show that the short-tailed shrew is known from almost all counties in the Upper and Lower Peninsulas (see Map 9). The species is also known on Little Summer Island in Lake Michigan and from Drummond Island in Lake Huron (Long, 1978; Manville, 1950).

LITERATURE CITED

Allan, T. A.
1977 Winter food of the snowy owl in northwestern lower Michigan. Jack-Pine Warbler, 55(1):42.

Armstrong, W. H.
1958 Nesting and food habits of the long-eared owl in Michigan. Michigan State Univ. Publ. Mus., Biol. Ser., 1(2):61–96.

Banfield, A. W. F.
1974 The mammals of Canada. Univ. Toronto Press, Toronto. xxiv+438 pp.

Blair, C. W., A. L. Hiller, and P. F. Scanlon
1977 Heavy metal concentrations in mammals associated with highways of different traffic densities. Virginia Jour. Sci., 18(2):61.

Blair, W. F.
1940 Notes on home ranges and population of the short-tailed shrew. Ecology, 21:284–288.
1941 Some data on the home ranges and general life history of the short-tailed shrew, red-backed vole, and woodland jumping mouse in northern Michigan. American Midl. Nat., 25(3):681–685.

Blokpoel, H., and G. T. Haymes
1979 Small mammals and birds as food items of ring-billed gulls on the lower Great Lakes. Wilson Bull., 91(4): 623–625.

Blus, L. J.
1971 Reproduction and survival of short-tailed shrews (*Blarina brevicauda*) in captivity. Lab. Anim. Sci., 21: 884–891.

Bole, B. P., Jr.
1939 The quadrat method of studying small mammal populations. Cleveland Mus. Nat. Hist., Sci. Publ., 5(4):15–77.

Buckner, C. H.
1964 Metabolism, food capacity, and feeding behavior in four species of shrews. Canadian Jour. Zool., 42:259–279.

Burt, W. H.
1946 The mammals of Michigan. Univ. Michigan Press, Ann Arbor. xxii+288 pp.

Carter, T. D.
1936 The short-tailed shrew as a tree climber. Jour. Mammalogy, 17:285.

Chamberlain, M. L.
1980 Winter hunting behavior of a snowy owl in Michigan. Wilson Bull., 92(1):116–120.

Christian, J. J.
1947 Note on light-colored *Blarina*. Jour. Mammalogy, 28(4):403.
1969 Maturation and breeding in *Blarina brevicauda* in winter. Jour. Mammalogy, 50(2):272–276.

Cope, J. B.
1949 Rough-legged hawk feeds on shrews. Jour. Mammalogy, 30(4):432.

Dapson, R. W.
1968 Reproduction and age structure in a population of short-tailed shrews, *Blarina brevicauda*. Jour. Mammalogy, 49:205–214.

DeByle, N. V.
1965 Short-tailed shrew attacks garter snake. Jour. Mammalogy, 46:329.

Doremus, H. M.
1965 Heart rate, temperature and respiration rate of the short-tailed shrew in captivity. Jour. Mammalogy, 46:424–425.

Dusi, J. L.
1951 The nest of a short-tailed shrew. Jour. Mammalogy, 32(1):115.

Eadie, W. R.
1944 The short-tailed shrew and field mouse predation. Jour. Mammalogy, 25(4):359–364.
1948 Shrew-mouse predation during low mouse abundance. Jour. Mammalogy, 29(1):35–37.
1949 Predation on *Sorex* by *Blarina*. Jour. Mammalogy, 30:308–309.

Errington, P. L.
1935 Food habits of mid-west foxes. Jour. Mammalogy, 16:192–200.

Errington, P. L., F. Hamerstrom, and F. N. Hamerstrom, Jr.
1940 The great horned owl and its prey in north-central United States. Iowa State Agric. Exp. Sta., Res. Bull. 277:759–850.

Geis, A. D.
1952 Winter food habits of a pair of long-eared owls. Jack-Pine Warbler, 30(3):93.

Getz, L. L.
1961 Factors influencing the local distribution of shrews. American Midl. Nat., 65:67–88.

Gould, E., N. C. Negus, and A. Novick
1964 Evidence for echolocation in shrews. Jour. Exp. Zool., 156:19–38.

Hamilton, W. J., Jr.
1929 Breeding habits of the short-tailed shrew, *Blarina brevicauda*. Jour. Mammalogy, 10:125–134.
1930 The food of the Soricidae. Jour. Mammalogy, 11:26–39.
1931 Habits of the short-tailed shrew, *Blarina brevicauda* (Say). Ohio Jour. Sci., 31:97–106.
1933 The weasels of New York. Their natural history and economic status. American Midl. Nat., 14:289–344.
1936 Seasonal food of skunks in New York. Jour. Mammalogy, 12:240–246.
1940 The molt in *Blarina brevicauda*. Jour. Mammalogy, 21:456–458.
1941 The food of small forest mammals in eastern United States. Jour. Mammalogy, 22:250–263.

Hamilton, W. J., Jr., and A. H. Cook
1955 The biology and management of the fisher in New York. New York Game and Fish Jour., 21(1):13–35.

Hamilton, W. J., Jr., N. W. Hosley and A. E. McGregor
1937 Late summer and early fall foods of the red fox in central Massachusetts. Jour. Mammalogy, 18: 366–367.

Hatfield, D. M.
1939 Winter food habits of foxes in Minnesota. Jour. Mammalogy, 10:202–206.

Haveman, J. R.
1973 A study of population densities, habitats and foods of four sympatric species of shrews. Northern Michigan Univ., unpubl. M.S. thesis, vii+70 pp.

Huish, M. T., and D. F. Hoffmeister
1947 The short-tailed shrew (*Blarina*) as a source of food for the green sunfish. Copeia, 3:198.

Ingram, W. M.
1942 Small associates of *Blarina brevicauda talpoides*. Jour. Mammalogy, 23:255–258.

Jackson, H. H. T.
1961 Mammals of Wisconsin. Univ. Wisconsin Press, Madison. xii+504 pp.

Jameson, E. W., Jr.
1950 The external parasites of the short-tailed shrew, *Blarina brevicauda* (Say). Jour. Mammalogy, 31(2): 138–145.

Jeuniaux, C.
1961 Chitinase: An addition to the list of hydrolases in the digestive tract of vertebrates. Nature, 192:135–136.

Jones, M. L.
1979 Longevity of mammals in captivity. Intern. Zoo News, 26(3):16–26.

Kirk, G. L.
1921 Shrews and weasels. Jour. Mammalogy, 2:111.

Klugh, A. B.
1921 Notes on the habits of *Blarina brevicauda*. Jour. Mammalogy, 2:35.

Lawrence, B.
1946 Brief comparison of the short-tailed shrew and reptile poisons. Jour. Mammalogy, 26:393–396.

Linduska, J. P.
1942 Winter rodent populations in field shocked corn. Jour. Wildlife Mgt., 6(4):353–363.

Linsdale, J.
1928 Mammals of a small area along the Missouri River. Jour. Mammalogy, 9:140–146.

Long, C. A.
1978 Mammals of the islands of Green Bay, Lake Michigan. Jack-Pine Warbler, 56(2):59–83.

Lutz, J. E.
1964 Natural history of the short-tailed shrew in southeastern Michigan. Univ. Michigan, unpubl. Ph.D. dissertation, 354 pp.

Manville, R. H.
1949 A study of small mammal populations in northern Michigan. Univ. Michigan, Mus. Zool., Misc. Publ., No. 73, 83 pp.
1950 The mammals of Drummond Island, Michigan. Jour. Mammalogy, 31(3):358–359.

Martin, I. G.
1981a Venom of the short-tailed shrew (*Blarina brevicauda*) as an insect immobilizing agent. Jour. Mammalogy, 62(1):189–192.
1982b Tolerance of conspecifics by short-tailed shrews (*Blarina brevicauda*) in simulated natural conditions. American Midl. Nat., 106(1):206–208.

Martinsen, D. L.
1969 Energetics and activity patterns of short-tailed shrews (*Blarina*) on restricted diets. Ecology, 50:505–510.

Master, L.
1979 Some observations on great gray owls and their prey in Michigan. Jack-Pine Warbler, 57(4):215–217.

Maynard, C. J.
1889 Singular effects produced by the bite of a short-tailed shrew, *Blarina brevicauda*. Contr. Science, 1(2):57–59.

Moore, J. C.
1949 Notes on the shrew, *Sorex cinereus*, in the southern Appalachians. Ecology, 30(2):234–237.

Morris, D. W.
1979 Microhabitat utilization and species distribution of sympatric small mammals in southwestern Ontario. American Midl. Nat., 101(2):373–384.

Murie, A.
1936 Following fox trails. Univ. Michigan, Mus. Zool., Misc. Publ. No. 32, 45 pp.

Murray, L. T.
1939 An albino shrew from Indiana. Jour. Mammalogy, 20:501.

Nader, I. A., and R. L. Martin
1962 The shrew as prey of the domestic cat. Jour. Mammalogy, 43(3):417.

Neal, C. M., and S. I. Lustick
1973 Energetics and evaporative water loss in the short-tailed shrew *Blarina brevicauda*. Physiol. Zool., 46:180–185.

Nichols, D. G.
1936 Note on the habits of *Blarina*. Jour. Mammalogy, 17:412.

Olsen, R. W.
1969 Agonistic behavior of the short-tailed shrew (*Blarina brevicauda*). Jour. Mammalogy, 50:494–500.

O'Reilly, R. A., Jr.
1949 Shrew preying on ribbon snake. Jour. Mammalogy, 30:309.

Osterberg, D. M.
1962 Activity of small mammals as recorded by a photographic device. Jour. Mammalogy, 43:219–229.

Pearson, O. P.
1942 On the cause and nature of a poisonous action produced by the bite of a shrew (*Blarina brevicauda*). Jour. Mammalogy, 23:159–166.
1944 Reproduction in the shrew (*Blarina brevicauda* Say). American Jour. Anatomy, 75:39–93.

Pettingill, O. S., Jr.
1976 The prey of six species of hawks in northern Lower Michigan. Jack-Pine Warbler, 54(2):70–74.

Pruitt, W. O., Jr.
1953 An analysis of some physical factors affecting the local distribution of the short-tailed shrew (*Blarina brevicauda*) in the northern part of the Lower Peninsula of Michigan. Univ. Michigan, Mus. Zool., Misc. Publ. No. 79, 39 pp.
1959 Microclimates and local distribution of small mammals on the George Reserve, Michigan. Univ. Michigan, Mus. Zool., Misc. Publ. No. 109, 27 pp.

Rabe, M. L.
1981 New locations for pygmy (*Sorex hoyi*) and water (*Sorex palustris*) shrews in Michigan. Jack-Pine Warbler, 59(1):16–17.

Rapp, J. L. C., and W. F. Rapp
1945 Resting nest of the short-tailed shrew. Jour. Mammalogy, 26(3):307.

Rood, J. P.
1958 Habits of the short-tailed shrew in captivity. Jour. Mammalogy, 39:499–507.

Scharf, W. C., and K. R. Stewart
1980 New records of Siphonaptera from northern Michigan. Great Lakes Ento., 13(3):165–167.

Schofield, R. D.
1960 A thousand miles of fox trails in Michigan's ruffed grouse range. Jour. Wildlife Mgt., 24(4):432–434..

Schreiber, R. K., and J. H. Graves
1977 Powerline corridors as possible barriers to the movements of small mammals. American Midl. Nat., 97(2):504–508.

Schwartz, C. W., and E. R. Schwartz
1959 The wild mammals of Missouri. Univ. Missouri Press and Missouri Conservation Comm., Columbia. 341 pp.

Seton, E. T.
1909 Life histories of northern animals. Charles Scribner's Sons, New York. 2:675–1267.

Shull, A. F.
1907 Habits of the short-tailed shrew, *Blarina brevicauda* (Say). American Nat., 41:495–522.

Sime, P. K.
1940 Albino short-tailed shrews from Connecticut. Jour. Mammalogy, 21(2):214–216.

Spiker, C. J.
1933 Analysis of two hundred long-eared owl pellets. Wilson Bull., 45(4):198.

Timm, R. M.
1975 Distribution, natural history and parasites of mammals of Cook County, Minnesota. Univ. Minnesota, Bell Mus. Nat. Hist., Occas. Papers 14, 56 pp.

Tomasi, T. E.
1978 Function of venom in the short-tailed shrew, *Blarina brevicauda*. Jour. Mammalogy, 59(4):852–854.

Townsend, M. T.
1935 Studies of some of the small mammals of central New York. Roosevelt Wildlife Ann., 4(1):1–120.

Ulmer, F. A.
1940 A Delaware record of albinism in *Blarina*. Jour. Mammalogy, 21:457.

Van Hyning, O. C.
1931 The house cat as a collector of mammals. Jour. Mammalogy, 12(2):164.

Wallace, G. J.
1948 The barn owl in Michigan: Its distribution, natural history and food habits. Michigan State Col., Agric. Exp. Sta., Tech. Bull. 208, 61 pp.

Warthin, A. S., Jr., and J. Van Tyne
1922 The food of long-eared owls. Auk, 39(3):417.

Whitaker, J. O., and R. E. Mumford
1972 Food and ectoparasites of Indiana shrews. Jour. Mammalogy, 53(2):329–335.

Wilson, K. A.
1938 Owl studies at Ann Arbor, Michigan. Auk, 55(2):187–197.

Wrenn, W. J.
1974 Notes on the ecology of chiggers (Acarina: Trombiculidae) from northern Michigan and the description of a new species of *Euschoengastia*. Jour. Kansas Ento. Soc., 47(2):227–238.

LEAST SHREW
Cryptotis parva (Say)

NAMES. The generic name *Cryptotis*, first proposed by Ponel in 1848, is of Latin derivation and refers to this shrew's inconspicuous external ear (*Crypt* means hidden; *otis* means ear). The specific name *parva*, proposed by Say in 1923, is also from the Latin and means small. Common names used for this species include oldfield shrew, little mole shrew, and little short-tailed shrew. The subspecies, *Cryptotis parva parva* (Say), occurs in southern Michigan.

RECOGNITION. Body size small for an American shrew, covered with soft, short hair; head elongate; nose pointed; eyes dark, small and beady; ears concealed in fur on sides of head; body heavily built, cylindrical; tail short, thin, finely haired; limbs short and delicate; feet heavy for a small shrew; flank, anal, and aural glands obvious; one pair of abdominal and two pairs of inguinal mammary glands present.

The least shrew is gray brown on the upperparts and silver gray on the underparts. The pelt is somewhat darker in winter than in summer. In size, the least shrew closely resembles Michigan shrews of the genus *Sorex* but has a more robust body and a decidedly shorter tail. Although its fur lacks the velvety appearance of that of the short-tailed shrew, in shape and shortness of tail the least shrew appears to be almost a small edition of the former.

MEASUREMENTS AND WEIGHTS. Adult least shrews measure as follows: length from tip of nose to end of tail vertebrae, 2½ to 3¼ in (64 to 82 mm); length of tail vertebrae, ½ to ¾ in (12 to 18 mm); length of hind foot, ⅜ in (9 to 11 mm); weight, 1/5 oz (4.0 to 6.5 g).

DISTINCTIVE CRANIAL AND DENTAL CHARACTERISTICS. The skull of the adult least shrew is small, compact, broad, high in the orbital area, with a relatively short and broad rostrum and widely spaced nasal openings (see Fig. 9). Behind the large and curved first upper incisors, there are three obvious unicuspid teeth plus a minute tooth wedged between the last large unicuspid and the first molariform tooth, as viewed from the side (see Fig. 8). These teeth are tipped with a rich brown coloring. The least shrew differs from other Michigan shrews in having only 30 teeth; the others have 32. Like the pygmy shrew, the least shrew has only three conspicuous unicuspid teeth plus a terminal diminutive one visible on each side of the upper jaw. In the case of the pygmy shrew, however, there is actually another small, peg-like tooth wedged obscurely between the second and third large unicuspids. The skull of the least shrew measures approximately ⅝ in (16.0 to 17.2 mm) long and ¼ in (7.5 to 8.2 mm) in greatest width. The dental formula is as follows:

$$\text{I (incisors)}\frac{3\text{-}3}{1\text{-}1}, \text{C (canines)}\frac{1\text{-}1}{1\text{-}1}, \text{P (premolars)}\frac{2\text{-}2}{1\text{-}1}, \text{M (molars)}\frac{3\text{-}3}{3\text{-}3} = 30$$

DISTRIBUTION. This southern shrew occurs in eastern North America from Nebraska to Mexico and Central America (in the west) and from southern New England and the Lake States south to peninsula Florida (in the east). The derivation of the Canadian record at Long Point, Ontario, is obscure (Banfield, 1974). Presumably, the species reached this area by rafting across Lake Erie or from the east by way of New York, although Peterson (1966) does consider this shrew's range to include the entire north shore area of Lake Erie. Even so, it is suggested herein that the Huron-Erie Waterway Connection has barred its entry into Ontario from Michigan (see Table 13).

In Michigan, the least shrew is at the northern edge of its distribution; perhaps the marginal nature of this habitat accounts for its patchy occurrence. To date, the least shrew is known from 12 of the 28 counties making up the southernmost four tiers of counties in the state (See Map 10). Most county records date from the first half of the present century, with the first record obtained in 1874 (in Ingham County, Wood and Dice, 1923).

In recent years, most efforts to capture least shrews have met with failure. Linduska (1950) caught two individuals in 1940–1941, at Rose Lake Wildlife Research Center in Clinton County. Shier (1981) repeated Linduska's trapping program in 1978–1980 and obtained no specimens. In a grid of live traps set in an open field in Washtenaw County, Getz (1962) caught 23 least shrews in five-day trapping periods in December and January of 1957–1958, even though he had been operating this trapping program for five days each month since the previous September. When the same area was intensively trapped again 20 years later, no least shrews were taken (Master, 1978). Annual classes (since the mid-1950s) in mammalogy at Michigan State University have captured no least shrews in the course of fieldwork in various habitats in and around Ingham County.

As of 1981, the last known specimen of the least shrew to be taken in Michigan was at the Edwin S. George Reserve in Livingston County in September 1960 (Master, 1978).

HABITAT PREFERENCES. The least shrew is reported from a variety of open and semi-open environments. It can be chiefly expected in grasslands and fallow fields which contain stands of annual and perennial grasses and forbs. It also thrives in marsh, shoreline areas, orchards, brushy edges of woods, shrubby fencerows, and even in moist woods (Hamilton, 1944; Whitaker, 1974). Efforts have been made to trap these small insectivores in all of the above Michigan habitats, especially in grassy areas, where many of the small number of Michigan specimens have been obtained (Blossom, 1931; Getz, 1962).

DENSITY AND MOVEMENTS. In sampling small mammal populations by using snap traps, live traps, or pit traps, small rodents usually greatly outnumber least shrews. The latter may actually be

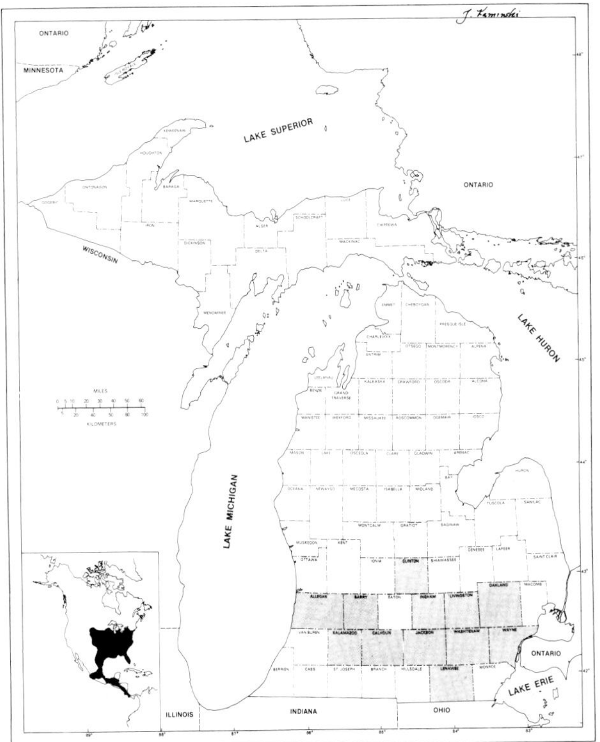

Map 10. Geographic distribution of the least shrew (*Cryptotis parva*). Actual records exist for counties with dark shading.

more abundant, but perhaps "trap shy," as is shown by the large numbers of least shrews found in the pellets of undigested items coughed up by owls at their roosts (Davis, 1938; Price, 1942). Kale (1972) estimated that least shrews may occur as abundantly as 12.6 per acre (31.7 per ha), as shown by his field studies in Florida. In Illinois, Hoffmeister and Mohr (1957) reported populations in fields as high as 10 or 15 per acre (25 to 37 per ha).

The few population studies involving least shrews indicate they live in rather discrete areas with perhaps a nest being the center of activity. A study in Kansas showed these small animals occupied living space as large as three acres (1.2 ha). One least shrew, captured, marked, and released, was recaptured six days later 885 ft (272 m) from the point of first capture (Choate and Fleharty, 1973).

BEHAVIOR. The habits of the least shrew most closely resemble those of the short-tailed shrew, although the former uses underground burrows to a lesser extent than the latter. The least shrew is most likely to be found in surface trails constructed by meadow voles. Major peaks of activity are at night and in the dim light of early morning and late afternoon, although Hamilton (1944) observed captives moving around at all hours without any special rhythmic sequence. As yet, there is no satisfactory explanation for the fact that most (80%) of the least shrews taken in Michigan were captured in the cold months of October through March (Lawrence Master, pers. comm.). Apparently this shrew is a poor climber and drowns easily (Jackson, 1961).

The sociability of individual shrews is shown by their compatibility as cage mates (Hamilton, 1944; Pfeiffer and Gass, 1963; Springer, 1937). In winter, these shrews tend to huddle together in communal nests, with as many as 31 individuals found in a single nest (Davis and Joeris, 1945; Hamilton, 1934; Jackson, 1961; McCarley, 1959). Their social habits seem to be in marked contrast to the intolerance shown toward one another by other Michigan shrews.

The least shrew, like related species, relies primarily on its senses of smell and touch for investigating its environment and finding food, with sight and hearing less developed (Hamilton, 1944). Smell is also important in courtship, as shown by studies of Kivett and Mock (1980) of the function of the aural glands. The motile snout and long whiskers play major roles in tactile sensitivity. Like other shrews, this species utters buzzes and squeaks. One call reminded Springer (1937) of the song of a flicker at a distance, rapidly repeated, high and thin. It is probable that these utterances serve for intraspecific communication.

It is suggested that there may be two kinds of nests used as refuge sites for the least shrew; a smaller one for the individual, especially a female with young, and a larger one for groups of individuals perhaps associating in the cold season of the year (Davis and Joeris, 1945). These refuges can be placed in slight depressions under a canopy of grasses but more usually are under logs, rocks, discarded boards or sheet metal, tin cans, and other such cover (Broadbrooks, 1952; Kilham, 1954; Snyder, 1929; Welter and Sollberger, 1939), but rarely underground. Nests are globular in shape and constructed of available plant materials, grasses and shrub or tree leaves, and sometimes even corn husks. Nests are roofed over and may have two side entrances. One such nest measured 2 in (50 mm) deep, 4 in (100 mm) wide and 6 in (150 mm) long (McCarley, 1959). Adjacent to this nest and under the same log was a pile of feces about ½ in (13 mm) high.

ASSOCIATES. In its range in southern Michigan, the least shrew has opportunity to associate with other small mammals characteristic of more open habitats. In Washtenaw County, for example, Blossom (1931) caught a least shrew at the same trap station with the short-tailed shrew, meadow vole, bog lemming, deer mouse, and meadow jumping mouse. Because the least shrew seems an unlikely predator on small rodents, its presence along runways of meadow voles may not elicit avoidance behavior in the mice.

REPRODUCTIVE ACTIVITIES. Studies in New York show that the least shrew may breed from early March to as late as November, with year-around breeding not unusual in more southern states (Brimley, 1923; Hamilton, 1944). The reproductive potential is high because a female is receptive to mating shortly after the birth of her litter, making it possible for her to have two or three litters per season (Hunt, 1951). In Ohio, Gottschang (1965) found a nursing female as late as November 4. Females can be sexually mature less than 40 days after birth, allowing them to produce litters in their first season (Mock, 1970).

After a gestation period of 21 to 23 days, a litter of three to six young are born (Conaway, 1958). A female weighed 6.3 g before and 4.0 g after the birth of six young (Hamilton, 1944). At birth, the young are hairless, wrinkled, helpless, pale in color except for dark spots marking places where whiskers will emerge, have closed eyes, measure about 20 mm in length, and weigh only about ⅓ g. At seven days of age, the young have grown a coat of short, dense fur. At 10 days of age, the young open their eyes. Growth is rapid and by about 21 to 22 days of age the young are weaned with a near adult weight of about ⅛ oz (3.4 g). At this time the total weight of the litter may exceed that of the mother by threefold (Hamilton, 1944). Mock (1982) reported that, under laboratory conditions, the least shrew can be remarkably productive, with a heterosexual pair weaning an average of 2.9 young per month. Individual least shrews may survive in nature for as long as 18 months (Banfield, 1974); in captivity animals have been maintained for 33 months (Pearson, 1945).

FOOD HABITS. Small size, a comparatively large surface area contributing to heat loss, and the necessity to maintain a constant high body temperature contribute the least shrew's need to eat nutritious foods as often as possible. In fact, this mammal may eat as much as its total body weight per 24 hours (Barrett, 1969). Perhaps 5% of the food of the least shrew is vegetable matter, with the remainder being edible invertebrates and small vertebrates. Digestive tracts examined show food items eaten to include caterpillars, grasshoppers, beetles and their larvae, chinch bugs, earthworms, snails, millipedes, sowbugs, and spiders (Hamilton, 1934, 1944; Jackson, 1961; Mohr, 1934; Whitaker and Mumford, 1972). In captivity, least shrews have consumed meal worms, crickets, ants and their pupae, caterpillers, grubs, fish, frogs, toads, lizards, snakes, mouse flesh, ground beef, horse meat, rabbit flesh, sparrow flesh, and cheese (Broadbrooks, 1952; Chamberlain, 1929; Conaway, 1958; Davis and Joeris, 1945; Hatt, 1938; Moseley, 1930; Peterson, 1936; Pfeiffer and Gass, 1963; Springer, 1937; Welter and Sollberger, 1939).

ENEMIES. The least shrew is sufficiently small to be fair game for most carnivorous mammals, reptiles, and birds. Known terrestrial predators on the least shrew include box turtle, puff adder, milk snake, domestic dog, red fox, gray fox, striped skunk, and house cat (Hamilton, 1934, 1936; McMurry, 1945; Moore, 1953; Murie, 1936; Peterson, 1936; Snyder, 1929; Welter and Sollberger, 1939). Predatory birds are probably as efficient as any group of shrew-eaters in obtaining these small mammals. There are more data on owls than hawks as enemies, probably because of the availability for study of pellets containing undigested food items coughed up by owls at their roosts. Even so, Cope (1949) counted the remains of 27 least shrews plus one short-tailed shrew in the digestive tract of one rough-legged hawk taken in January in Indiana.

There has been recovery of least shrew remains from the pellets of the great-horned owl, barred owl, screech owl, short-eared owl, long-eared owl, and barn owl (Armstrong, 1958; Davis, 1938; Jones, 1937; Linsdale, 1928; Master, 1978; Nelson, 1934; Voight and Glenn-Levin, 1978; Wallace, 1948; Wilson, 1938). The barn owl appears to be a highly efficient shrew-catcher. In a collection of barn owl pellets found in Texas, Davis (1938) counted cranial parts from 171 least shrews, making up 41% of the owl's total catch of mammals. In Wallace's (1948, 1950) study of the food habits of the barn owl in Ingham County, only four least shrews were counted in pellet remains which included 6,742 individual small mammals. This suggests that only 0.06% of the barn owl catch was least shrews.

PARASITES AND DISEASES. Whitaker and Mumford (1972) report that least shrews may be infested with eight kinds of mites and three kinds of fleas. In Wisconsin, Jackson (1961) found one flea on the least shrew.

MOLT AND COLOR ABERRATIONS. The least shrew changes its fur coat twice a year. In March, the dark winter coat is exchanged for a paler summer coat which lasts until late August or September, when the dark winter pelt reappears. Albino individuals also have been reported (Sealander, 1981).

ECONOMIC IMPORTANCE. The least shrew has little economic importance to human agricultural pursuits. They eat noxious insects, their larvae and pupae, which gains them favor with farmers even though their control efforts are not profound. On occasion, the least shrew may raid beehives and devour larval and pupal bees (Evermann and Butler, 1894).

COUNTY RECORDS FROM MICHIGAN. Records of specimens in collections of museums and from the literature, notably Master (1978), show that the least shrew is known from the following counties in the Lower Peninsula (see Map 10): Allegan, Barry, Calhoun, Clinton, Ingham, Jackson, Kalamazoo, Lenawee, Livingston, Oakland, Washtenaw, and Wayne.

LITERATURE CITED

Armstrong, W. L.
1958 Nesting and food habits of the long-eared owl in Michigan. Michigan State Univ., Publ. Mus., Biol. Ser., 1(2):61–96.

Banfield, A. W. F.
1974 The mammals of Canada. Univ. Toronto Press, Toronto. xxiv+438 pp.

Barrett, G. W.
1969 Bioenergetics of a captive least shrew, *Cryptotis parva.* Jour. Mammalogy, 50(3):629–630.

Blossom, P. M.
1931 Another record of *Cryptotis parva* for Michigan. Jour. Mammalogy, 12(4):429.

Brimley, C. S.
1923 Breeding dates of small mammals at Raleigh, North Carolina. Jour. Mammalogy, 4(4):263–264.

Broadbrooks, H. E.
1952 Nest and behavior of a short-tailed shrew, *Cryptotis parva.* Jour. Mammalogy, 33(2):241–243.

Chamberlain, E. B.
1929 Behavior of the least shrew. Jour. Mammalogy, 10(3): 250–251.

Choate, J. R., and E. D. Fleharty
1973 Habitat preference and spatial relations of shrews in a mixed grassland in Kansas. Southwestern Nat., 18(1): 110–112.

Conaway, C. H.
1958 Maintenance, reproduction and growth of the least shrew in captivity. Jour. Mammalogy, 39(4):507–512.

Cope, J. B.
1949 Rough-legged hawk feeds on shrews. Jour. Mammalogy, 30(4):432.

Davis, W. B.
1938 A heavy concentration of *Cryptotis.* Jour. Mammalogy, 19(4):499–500.

Davis, W. B., and L. Joeris
1945 Notes on the life-history of the little short-tailed shrew. Jour. Mammalogy, 26(2):136–138.

Evermann, B. W., and A. W. Butler
1894 Preliminary list of Indiana mammals. Proc. Indiana Acad. Sci. for 1893, pp. 124–139.

Getz, L. L.
1962 A local concentration of the least shrew. Jour. Mammalogy, 43(2):281–282.

Gottschang, J. L.
1965 Winter populations of small mammals in old fields of southwestern Ohio. Jour. Mammalogy, 46(1):44–52.

Hamilton, W. J., Jr.
1934 Habits of *Cryptotis parva* in New York. Jour. Mammalogy, 15(2):154–155.
1936 Seasonal food of skunks in New York. Jour. Mammalogy, 17(3):240–246.
1944 The biology of the little short-tailed shrew, *Cryptotis parva.* Jour. Mammalogy, 25(1):1–7.

Hatt, R. T.
1938 Feeding habits of the least shrew. Jour. Mammalogy, 19(2):247–248.

Hoffmeister, D. F., and C. O. Mohr
1957 Fieldbook of Illinois mammals. Illinois Nat. Hist. Surv. Div., Manual 4, xi+233 pp.

Hunt, T.
1951 Breeding of *Cryptotis parva* in Texas. Jour. Mammalogy, 32(1);115–116.

Jackson, H. H. T.
1961 Mammals of Wisconsin. Univ. Wisconsin Press, Madison. xii+504 pp.

Jones, J. C.
1937 *Cryptotis parva* in central New York, Jour. Mammalogy, 18(4):514.

Kale, H. W., II
1972 A high concentration of *Cryptotis parva* in a forest in Florida. Jour. Mammalogy, 53(1):216–218.

Kilham, L.
1954 Cow-pasture nests of *Cryptotis parva parva.* Jour. Mammalogy, 35(2):252.

Kivett, V. K., and O. B. Mock
1980 Reproductive behavior in the least shrew (*Cryptotis parva*) with special reference to the aural glandular region of the female. American Midl. Nat., 103(2): 339–345.

Linduska, J. P.
1950 Ecology and land-use relationships of small mammals on a Michigan farm. Michigan Dept. Conserv., Game Div., ix+144 pp.

Linsdale, J.
1928 Mammals of a small area along the Missouri River. Jour. Mammalogy, 9(2):140–146.

Master, L. L.
1978 A survey of the current distribution, abundance, and habitat requirements of threatened and potentially threatened species of small mammals in Michigan. Univ. Michigan, Mus. Zool., iv+52 pp. (mimeo.).

McCarley, W. H.
1959 An unusually large nest of *Cryptotis parva.* Jour. Mammalogy, 40(2):243.

McMurry, F. B.
1945 Three shrews, *Cryptotis parva,* eaten by a feral house cat. Jour. Mammalogy, 26(1):94.

Mock, O. B.
1970 Reproduction of the least shrew (*Cryptotis parva*) in captivity. Univ. of Missouri, unpubl. Ph.D. dissertation, 132 pp.
1982 The least shrew (*Cryptotis parva*) as a laboratory animal. Lab. Animal Sci., 32(2):177–179.

Mohr, C. O.
1934 Value of prey-individual analysis of stomach contents of predatory mammals. Jour. Mammalogy, 16(4):323–324.

Moore, J. C.
1953 Shrew on box turtle menu. Everglades Natural History, 1(3):129.

Moseley, E. L.
1930 Feeding a short-tailed shrew. Jour. Mammalogy, 11(2):224–225.

Murie, A.
1936 Following fox trails. Univ. Michigan, Mus. Zool., Misc. Publ. No. 32, 45 pp.

Nelson, A. L.
1934 Notes on Wisconsin mammals. Jour. Mammalogy, 15(3):252–253.

Pearson, O. P.
1945 Longevity of the short-tailed shrew. American Midl. Nat., 34(2):531–546.

Peterson, C. B.
1936 *Cryptotis parva* in western New York. Jour. Mammalogy, 17(3):284–285.

Peterson, R. L.
1966 The mammals of eastern Canada. Oxford Univ. Press, Toronto. xxxii+463 pp.

Pfeiffer, C. J., and C. H. Gass
1963 Notes on the longevity and habits of captive *Cryptotis parva.* Jour. Mammalogy, 44(3):427–428.

Price, H. F.
1942 Contents of owl pellets. American Midl. Nat., 28:524–525.

Sealander, J. A.
1981 Albino least shrews (*Cryptotis parva*) and a new locality record for the southeastern shrew (*Sorex longirostris*) from Arkansas. Southwestern Nat., 26(1):70.

Shier, J. L.
1981 Habitats of threatened small mammals on the Rose Lake Wildlife Research Area, Clinton County, Michigan—early forties and late seventies. Michigan State Univ., unpubl. M.S. thesis, vi+106 pp.

Snyder, L. L.
1929 *Cryptotis parva,* a new shrew for the Canadian list. Jour. Mammalogy, 10(1):79–80.

Springer, S.
1937 Observations on *Cryptotis floridana* in captivity. Jour. Mammalogy, 18(2):237–238.

Voight, J., and D. C. Glenn-Levin
1978 Prey availability and prey taken by long-eared owls in Iowa. American Midl. Nat., 99(1):162–171.

Wallace, G. J.
1948 The barn owl in Michigan: Its distribution, natural history and food habits. Michigan Agri. Exp. Sta., Tech. Bull. 208, 61 pp.
1950 In support of the barn owl. Michigan Agri. Exp. Sta., Quart. Bull., 33(2):96–105.

Welter, W. A., and S. E. Sollberger
1939 Notes on the mammals of Rowan and adjacent counties in eastern Kentucky. Jour. Mammalogy, 20(1):77–81.

Whitaker, J. O.
1974 *Cryptotis parva.* American Soc. Mammalogists, Mammalian Species, No. 43, 8 pp.

Whitaker, J. O., Jr. and R. E. Mumford
1972 Food and ectoparasites of Indiana shrews. Jour. Mammalogy, 53(2):329–335.

Wilson, K. A.
1938 Owl studies at Ann Arbor, Michigan. Auk, 55(2):187–197.

Wood, N. A., and L. R. Dice
1923 Records of the distribution of Michigan mammals. Michigan Acad. Sci. Arts and Ltrs., Papers, 3:425–469.

FAMILY TALPIDAE
Moles

The two Michigan moles are representatives of an insectivorous family occurring in parts of North America and Eurasia. These heavy-bodied, mouse-to-rat-sized mammals are fossorial and spend most of their time underground. At least three genera are also semiaquatic. Most are characterized by conspicuous snouts; dense, short fur; small or obscure eyes; usually no obvious ear pinnae; short legs with forefeet noticeably broad and adapted for digging or in some cases for swimming; and specialized bony structures (see Fig. 13).

Moles eat insects, worms, and other small animals but often accept vegetable matter as well. They construct a central underground chamber from which subsurface tunnels radiate. These tunnels are usually constructed to secure food in moist, loose soils in open lands, marshes, and forested areas. Moles are active day or night in their underground galleries. They are slow breeders; females usually have only one litter annually. Subsurface tunnels produced by moles cause ridges which are nuisances on lawns, in gardens, and on golf courses. When in fashion, the skins of moles, especially those of species living in Eurasia, have been used in the manufacture of moleskin garments, muffs, and dress and coat trimmings.

KEY TO THE SPECIES OF THE FAMILY TALPIDAE (MOLES) IN MICHIGAN
Using Characteristics of Adult Animals in the Flesh

1a. Tail short, less than 1¾ in (45 mm), thin; end of snout naked, lacking tentacles; nostrils open upwards; eyes obscured; forefoot conspicuously broad, approximately 1⅛ in (30 mm) across; color grayish with silver sheen . . . EASTERN MOLE (*Scalopus aquaticus*), page 68.

1b. Tail long, more than 1¾ in (45 mm), fleshy; end of snout bordered with 22 soft, slender, worm-like tentacles, motile in life; nostrils open toward the front, encircled by the tentacles; eyes visible; forefeet moderately broad, approximately ½ in (13 mm) across; color black . . . . . STAR-NOSED MOLE (*Condylura cristata*), page 74.

KEY TO THE SPECIES OF THE FAMILY TALPIDAE (MOLES) IN MICHIGAN
Using Characteristics of the Skull of Adult Animals

1a. Skull broad with short rostrum (snout); width of cranium more than 9/16 in (15 mm); bony palate terminates posterior to a line drawn between the hind ends of the last upper molars; teeth number 36, with 10 on each side of the upper jaws and 8 on each side of the lower jaws; first upper incisors curve downward and not visible from a dorsal view of the skull (see Fig. 14)) . . . . . . EASTERN MOLE (*Scalopus aquaticus*), page 68.

1b. Skull narrow with long, slender rostrum (snout); width of cranium less than 9/16 in (15 mm); bony palate terminates anterior to a line drawn between the hind ends of the last upper molars; teeth number 44 with 11 on each side of both the upper and lower jaws; second upper incisors minute and sometimes missing; first upper incisors project forward and highly visible from a dorsal view of the skull (See Fig. 14) . . . . STAR-NOSED MOLE (*Condylura cristata*), page 74.

EASTERN MOLE
Scalopus aquaticus (Linnaeus)

NAMES. The generic name *Scalopus,* proposed by E. Geoffroy Saint-Hilaire in 1803, is derived from the Greek work *skalops* and means to dig. The specific name *aquaticus,* proposed by Linnaeus in 1758, is from the Latin and means found in water, perhaps based on the misconception that the mole was aquatic and used its paddle-like forefeet in swimming. In Michigan this animal is called either mole or ground mole. The subspecies, *Scalopus aquaticus machrinus* (Rafinesque), occurs in the Lower Peninsula.

RECOGNITION. Body equal to that of a small, fat rat, covered with dense, short, velvety fur; head short; nose naked, elongated, and pointed; nostrils open upwards; eyes small, covered with a filmy, protective membrane resulting from the fusing of the eyelids and partly concealed in facial fur; ears obscured as small openings surrounded by fur; neck indistinct; body heavily-built and cylindrical; tail short, round, and so sparsely haired as to appear naked; forelimbs powerfully constructed (see Fig. 13); forefeet hand-like, with five broad, flat and massive claws, broader than long, sparsely haired above, hairless on palm side, palms facing outward; hind limbs less strongly built; hind feet small, narrow, scantily haired above, hairless on fleshy soles; one pair of pectoral, one pair of abdominal, and one pair of inguinal mammary glands present.

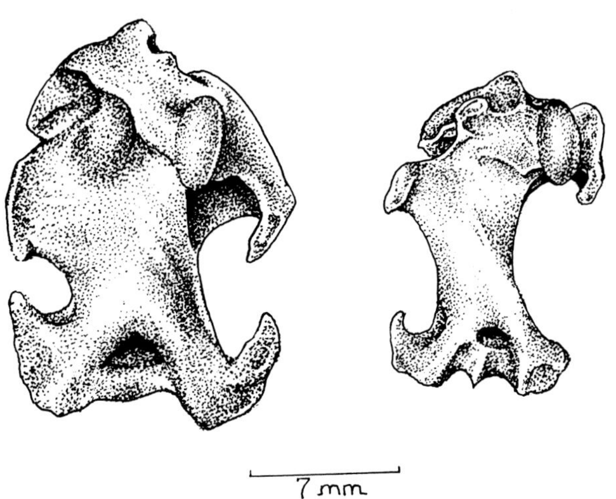

Figure 13. Upper right arm bones (humeri) of the eastern mole *(Scalopus aquaticus)*, left, and the star-nosed mole *(Condylura cristata)*, right.

The eastern mole is lead gray with a silky, silvery sheen on the upperparts, and light gray often with a tinge of reddish brown on the underparts. Staining on the throat and chest region results from the discharge of the skin glands. The hairless parts of the nose, tail, and feet are flesh-colored. The stout, cylindrical body covered with short, velvety fur, the pointed hairless nose, the short, flesh-colored tail, and the huge spade-like forefeet turned sideways serve to distinguish this mammal from all others in Michigan.

MEASUREMENTS AND WEIGHTS. Adult eastern moles measure as follows: length from tip of nose to end of tail vertebrae, 6 to 6⅞ in (150 to 200 mm); length of tail vertebrae, ⅞ to 1½ in (21 to 38 mm); length of hind foot, ¾ to 1⅛ in (22 to 26 mm); weight, 2¼ to 4¼ oz (65 to 120 g).

DISTINCTIVE CRANIAL AND DENTAL CHARACTERISTICS. The skull of the mature eastern mole is triangular in shape with a narrow rostrum, a wide, flattened braincase, and rather delicate zygomatic arches (see Fig. 14). The lower jaws are strongly constructed and curve upward both at the anterior and posterior ends. The jaw teeth possess pointed cusps useful in crushing the bodies of invertebrates used for food. The skull of the eastern mole measures about 1⅜ to 1 9/16 in (35 to 40 mm) long and ¾ in (19 to 21 mm) wide. The dental formula is:

$$\text{I (incisors) } \frac{3\text{-}3}{2\text{-}2}, \text{ C (canines) } \frac{1\text{-}1}{0\text{-}0}, \text{ P (premolars) } \frac{3\text{-}3}{3\text{-}3}, \text{ M (molars) } \frac{3\text{-}3}{3\text{-}3} = 36$$

DISTRIBUTION. The eastern mole is found throughout eastern United States as far west as the Great Plains and south to extreme northeastern Mexico. It also occurs in southwestern Ontario (Banfield, 1974; Peterson, 1966). This species ranges throughout Michigan's Lower Peninsula but is not known from suitable habitats in the Upper Peninsula (Map 11 and Table 12).

HABITAT PREFERENCES. The local distribution of the eastern mole depends on the presence of loose, moist loam and sandy soils (Bailey, 1929; Jackson, 1915, 1922). Most favorable living conditions are in alluvial soils along streams, especially where forest and shrub cover provide shade. Moist and friable soils under forest duff in woodlots in southern Michigan are also frequented by eastern moles. They tend to ex-

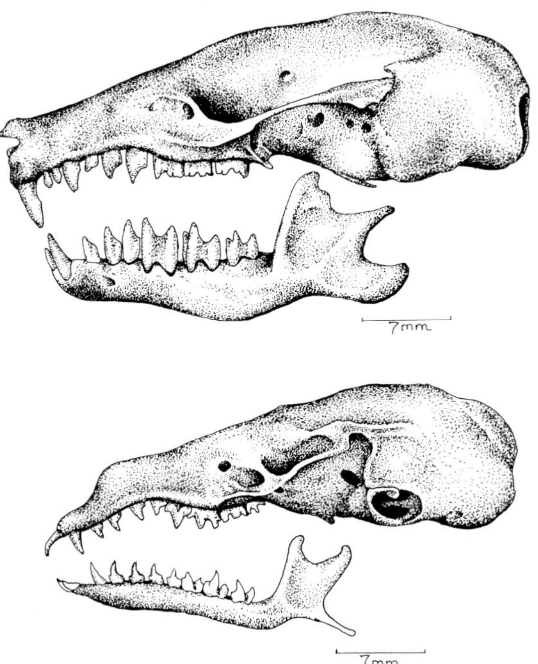

Figure 14. Skulls and lower jaws (viewed from the left sides) of the eastern mole *(Scalopus aquaticus)*, upper, and the star-nosed mole *(Condylura cristata)*, lower.

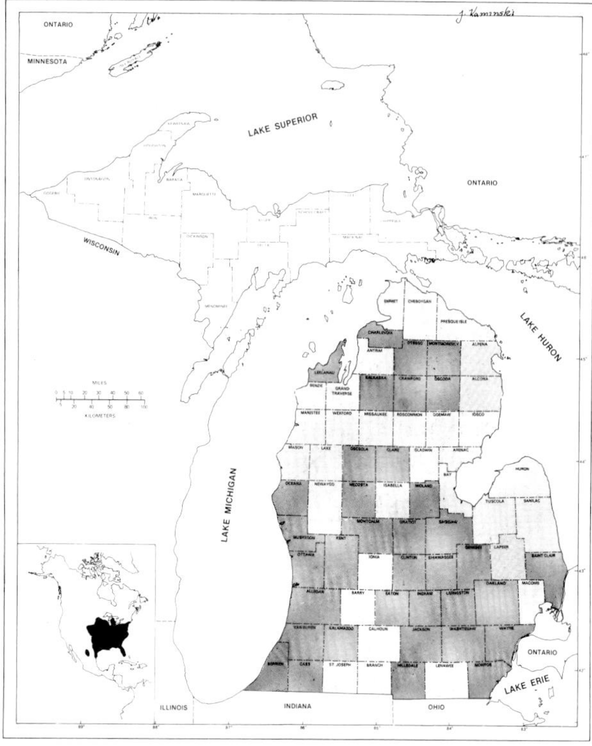

Map 11. Geographic distribution of the eastern mole (*Scalopus aquaticus*). Actual records exist for counties with dark shading.

cavate more deeply to attain moist conditions when surface dryness occurs. In addition to proper soil conditions, eastern moles require high populations of invertebrate foods. When fertilizers are used on pasture and crop lands, golf courses, and lawns and other green space in residential areas, cemeteries, and parks, these nutrients produce conditions for an increase in earthworms, grubs, and other invertebrate prey for the eastern mole. Producing environments for abundance of proper foods plus soil conditioning through tillage practices has undoubtedly improved habitats in southern Michigan for the eastern mole, presumably allowing the animal to spread generally throughout the entire Lower Peninsula.

DENSITY AND MOVEMENTS. Most observers who examine the crisscrossings of raised mole tunnels on lawns must wonder about the number of moles responsible. In truth, the work is very likely the product of a single individual. Populations as dense as one to two animals per acre (2 to 5 per ha) are probably normal in the Michigan area; records of densities of eight to 10 per acre (20 to 25 ha) may be unrealistically high (Jackson, 1961). Recent studies show that eastern moles occupy rather discrete areas, from 0.86 acres (0.35 ha) to two acres (0.75 ha) in size (Arlton, 1936; Harvey, 1976). Arlton also traced a raised tunnel along a fence line for a distance of about 3,280 ft (1,000 m).

BEHAVIOR. The small eye and ear openings protected by facial fur, the sensitive and mobile snout, the powerful forearms equipped with spadelike feet, and the cylindrical, fusiform body are fossorial adaptations of this most subterranean of American moles. In fact, the eastern mole spends almost its entire life in underground tunnels. Surface exposure time is probably short in duration, although long enough for individuals to be run down on a highway (Davis, 1940), photographed in a surface runway of a meadow vole (Osterberg, 1962), and consumed by hawks and owls (Craighead and Craighead, 1956; Wallace, 1950; Wilson, 1938). The animal moves around in its elaborate system of underground tunnels at all hours of the day but especially, according to Harvey (1976), from early morning to early afternoon (0800 to 1600) and at night (2300 to 0400). When resting, the eastern mole positions its body somewhat upright with nose tucked into the chest area. When aroused, the animal immediately forages for food; daily requirements call for food materials weighing as much as one-third of its own body weight.

The eastern mole excavates two types of underground passageways—near-surface tunnels and deep tunnels (Scheffer, 1913; Hisaw, 1923b). In Michigan, the mole generally constructs near-surface tunnels during the warm months, when soils are warm and insect larvae and earthworms are congregated among shallow rootstocks of grasses and forbs. These tunnels are dug by using a lateral and upward motion of one or the other of the flattened forefeet. The soil is not removed but merely pushed upward to loosen and crack the sod, causing conspicuous raised ridges on the surface. These tunnels may be used repeatedly or only once and then abandoned in the course of the mole's foraging activity.

The deep tunnels form a network of galleries approximately 2¼ in (55 mm) in diameter at depths of 10 to 24 in (25 to 60 cm). The mole constructs these for use when near-surface passageways become dry in midsummer or cold in winter. In both instances, the mole and its prey, earthworms and grubs, move to deeper soil layers. To construct these tunnels, the mole removes surplus soil by either storing it in abandoned tunnels or pushing it upwards and depositing it in mounds on the surface of the ground (Brown, 1972). These molehills are lumpy in consistency and crudely arranged in piles as wide as 12 in (300 mm) and as high as 6 in (150 mm). Sometimes it is difficult to distinguish between mounds of the eastern mole and the star-nosed mole. Usually those of the eastern mole are larger, flatter and more symmetrical in outline, and are rarely round as are those of the star-nosed mole in supersaturated or muck soils.

Several passageways lead to the central nest of the eastern mole. A large chamber, often 8 in (205 mm) in diameter, is filled with grasses and leaves. According to Scheffer (1914), plant material is pulled into the burrow by the roots. Nests may be located as close to the surface as 2 in (50 mm) or as deep as 12 in (300 mm), often under the protection of a stump or rock (Linsdale, 1928). Although the eastern mole remains active all winter, it does tend to spend periods of inactivity in nests at that season (Harvey, 1976).

The eastern mole is generally solitary and excludes other individuals from its tunnel system. Males use larger areas than females (Harvey, 1976). Tolerance is shown only during the mating season when the male enters the nearby tunnels of a female. Then as many as three breeding moles might be found in a single system of galleries (Arlton, 1936).

The eastern mole utters squeals in anguish and purrs when feeding (Jackson, 1961). Brown (1972) suggests that odors from exudates of skin glands serve as a means of communication, perhaps to attract the opposite sex during the breeding period or to warn and disperse intruders of the same sex. Certainly there is much more to be learned about the life habits of these subterranean insectivores.

ASSOCIATES. The eastern mole is continually excavating new tunnels and abandoning old ones in search of food. As a result, other ground-living and semi-fossorial small mammals also make use of these underground networks. Traps set directly across openings made in mole tunnels in woodlots on the campus of Michigan State University have captured masked shrews, short-tailed shrews, white-footed mice, meadow jumping mice, and house mice. Thirteen-lined ground squirrels and woodland voles also use mole burrows (Arlton, 1936; Jackson, 1961).

REPRODUCTIVE ACTIVITIES. Male eastern moles in the Michigan area become reproductively active before the end of January, although the mating season may not actually begin until females become receptive in March. In southern Wisconsin, Conaway (1959) found pregnant females as early as March 20 and as late as July 4. Most young are born in late April and May after a gestation period of about 28 days (Jackson, 1961). There is only one litter produced annually by a single female and no evidence that young become sexually mature during their first summer (Arlton, 1936; Conaway, 1959).

The usual number of young per litter is four; five is generally maximum. At birth, the young are hairless, have closed eye and ear openings, and measure about 2 in (50 mm) long. At seven to ten days of age, the young are clothed with fine hair, exhibit growing whiskers, show evidence of eyes and ears, develop well-formed forefeet, and possess bodily proportions much like the adult. At about 28 days of age, the young may be weaned and begin to forage on their own (Golley, 1962). At 35 days of age, the young are about one-half the size of the adult and leave the nest. Before autumn, the young-of-the-year are adult size and developing their own burrow systems. If indeed eastern moles are territorial, the dispersal to search for suitable, unoccupied burrowing areas may be a crucial period in the life of the maturing eastern mole, when considerable above-ground movement may expose the mole to predation, especially by hawks and owls.

Very likely individual eastern moles are long-lived; Jackson (1961) suggests they may live for three or four years. In captivity, one survived for 23 months (Jones, 1979).

FOOD HABITS. The sharp-cusped teeth of the eastern mole are effective for grasping and crushing the bodies of edible invertebrates. According to Jackson (1961), at least 80% of the mole's diet is made up of earthworms and the larval and adult stages of ground-living insects. The other 20% consists of vegetable material. Although moles will eat mouse flesh, there is no evidence that these insectivores actually can kill rodents. An examination of the contents of digestive tracts of moles by Hisaw (1923a) showed that foods consisted of 31% earthworms, 23% adult insects, and 29% larval insects. Besides the diet of earthworms, eastern moles are known to eat such specific animal items as mollusks, spiders, millipedes, centipedes, crickets, grasshoppers, and adult, larval, and pupal stages of yellow jackets, other wasps, wireworms, ground beetles, flies, and ants (Jackson, 1961; Scheffer, 1910; Schmidt, 1931). Brooks (1923) described an eastern mole extending a mound up to a low-hanging nest of the white-faced hornet (*Vespa maculata*) and then eating the young hornets.

Although vegetable material seems to be present on a regular basis in the diet, this use is minor. In the stomachs of 100 eastern moles taken in Kansas, Scheffer (1910) found remains of plant tissues including rootlets in 43 and seed pods or husks in eight. Captive eastern moles accepted walnuts, crackers, rolled oats, soaked corn, wheat, tomato, apple, potato, carrot and seeds of muskmelon and squash (Arlton, 1936; Hisaw, 1923a; Jackson, 1922; Rood, 1958).

ENEMIES. The usual underground activities of the eastern mole shield the species from many surface or flying predators. Of course, snakes and thin-bodied weasels, especially the least weasel, may easily invade its tunnel system (Jackson, 1961). As mentioned previously, aboveground activity, especially by dispersing young, may put the eastern mole in its most vulnerable position (Arlton, 1936). Some researchers however, felt that moles may sometimes be spared from predation becuase of their discouraging rank odor (Scheffer, 1923).

There is actually little evidence that terrestrial carnivores prey heavily on eastern moles. Red foxes, gray foxes, coyotes, domestic dogs, and house cats have been reported as taking eastern moles (Davis, 1951; Errington, 1935; Murie, 1936; Schofield, 1960; Van Hyning, 1931). In 1,175 fecal samples of red foxes in Iowa, the remains of only five eastern moles were found (Errington 1935).

Apparently winged predators are more apt to eat eastern moles than ground-hunting meat-eaters. Known to prey on eastern moles are red-tailed hawk, red-shouldered hawk, broad-winged hawk, screech owl, barred owl, and barn owl (Arlton, 1936; Davis, 1951; Scheffer, 1910; Wallace, 1950; Wilson, 1938). The eastern mole may not be a major item in the diet of the barn owl, according to Wallace's (1950) study in Ingham County. Of the 6,740 mammalian remains found in 2,200 pellets (coughed-up undigested remains of animals captured) Wallace found evidence of only two eastern moles.

Other records of mortality in the eastern mole are unusual. There are accounts of individuals being run over on highways (Davis, 1940; Kasul, 1976). An unusual mole death was noted by Charles Shick (pers. comm.) in the summer of 1957 in Hillsdale County when he saw a mole literally being stung to death by yellow jackets. It was his guess that the mole had brought on the attack by invading the wasps' underground nest to eat larval individuals.

PARASITES AND DISEASES. Although eastern moles and their parasites must have a close relationship in the confines of burrow systems, it was Jackson's (1961) opinion that moles are not heavily infested. External parasites reported from eastern moles include a louse (*Euhaematopinus abnormis*), fleas (*Ctenophthalmus pseudagrytes* and *Corypsylla ornata*), and mites (*Hirstionyssus blarinae, Androlaepus fahrenholzi*, and *Haemogamasus liponyssoides*), according to Arlton (1936), Jackson (1961), Scharf and Stewart (1980), and Yates *et al.* (1979). Intestinal worms identified from eastern moles include *Moniliformis moniliformis* and species of the genera *Filaria* and *Spiroptera* (Hanawalt, 1922; Scheffer, 1910).

MOLT AND COLOR ABERRATIONS. The eastern mole molts its winter pelt in spring and its summer pelt in late summer and early autumn (Jackson, 1961). There is little seasonal change in the dark coloring of the pelage. Secretions from skin glands present on the dorsal snout, chin, breast, abdomen, wrists and perineal areas tend to discolor parts of the fur (Eadie, 1954). Partial or complete albinism is found occasionally; five skins in the collections of the Michigan State University Museum show white streaks or blotches on the undersides, with one having a white spot on the nose.

ECONOMIC IMPORTANCE. The tunneling action of the eastern mole in pasture and forest lands in Michigan's Lower Peninsula is thought to be highly beneficial in soil mixing and aeration. This mammal's fondness for various noxious insect larvae is also a positive effect, although some fishing enthusiasts may be concerned to learn that the mole relishes earthworms. The mole's overall effects on the invertebrate populations, however, are not considered to be great. The eastern mole's mounds and raised tunnels are nuisances on lawns, golf courses, parks, green space in cemeteries, or formal gardens. Moles are generally attracted to such landscape areas, especially in suburbia, by an abundant food and water supply due to lawn fertilizers and frequent waterings.

Control practices might eliminate these unwanted creatures. However, if vacant lands adjoin, there is always opportunity for new invasions. The best way to discourage such encroachments probably is to install aluminium flashing to the depth of at least 1 ft (30 cm) entirely around the area to be protected.

Mole removal is often a difficult task. Unlike rats and mice, eastern moles are neither attracted to poisoned baits (Henning, 1952) nor discouraged by fumigants. The surest way to capture them is by trapping. Scheffer (1914, 1923) authored a series of governmental bulletins explaining methods of mole control. Others are available from state agencies (also see Jensen, 1982). There are two types of commercially-produced mole traps: the choker type and the harpoon type. To ensure success, the gardener should first roll the lawn where mole activity is detrimental, forcing the mole to push up its flattened passageways in the area currently occupied. This indicates where traps should be set.

Recently, animal control operators have suggested treating lawns and gardens with chemicals designed to reduce or eliminate the worm-insect foods of the eastern mole. It is sometimes recommended that this chemical treatment be given to a buffer zone of 20 to 40 ft (7 to 14 m) around the area. Although there are yet no data to substantiate the effectiveness of this control method, there is indication that once the food supply is decimated, the eastern mole will depart.

COUNTY RECORDS FROM MICHIGAN. Records of specimens preserved in collections in museums and from the literature show that the eastern mole is known from the following counties in the Lower Peninsula: Allegan, Berrien, Cass, Charlevoix, Clare, Clinton, Crawford, Eaton, Genesee, Gratiot, Hillsdale, Ingham, Jackson, Kalamazoo, Kalkaska, Kent, Leelanau, Livingston, Mecosta, Midland, Missaukee, Monroe, Montcalm, Montmorency, Muskegon, Oakland, Oceana, Osceola, Oscoda, Otsego, Ottawa, Saginaw, Shiawassee, St. Claire, Van Buren, Washtenaw, and Wayne.

LITERATURE CITED

Arlton, A. V.
1936 An ecological study of the mole. Jour. Mammalogy, 17(4):349–371.

Bailey, B.
1929 Mammals of Sherburne County, Minnesota. Jour. Mammalogy, 10(2):153–165.

Banfield, A. W. F.
1974 The mammals of Canada. Univ. Toronto Press, Toronto. xxiv+438 pp.

Brooks, Fred E.
1923 Moles destroy wasps' nest. Jour. Mammalogy, 4(3): 183.

Brown, L. N.
1972 Unique features of tunnel systems of the eastern mole in Florida. Jour. Mammalogy, 53(2):394–395.

Conaway, C. H.
1959 The reproductive cycle of the eastern mole. Jour. Mammalogy, 40(2):180–194.

Craighead, J. J., and F. C. Craighead, Jr.
1956 Hawks, owls and Wildlife. The Stackpole Co., Harrisburg, Pa. xix+443 pp.

Davis, W. B.
1940 Mortality of wildlife on a Texas highway. Jour. Wildlife Mgt., 4(1):90–91.
1951 Eastern moles eaten by cottonmouth and gray fox. Jour. Mammalogy, 32(1):114–115.

Eadie, W. R.
1954 Skin gland activity and pelage descriptions in moles. Jour. Mammalogy, 35(2):186–196.

Errington, P. L.
1935 Food habits of mid-west foxes. Jour. Mammalogy, 16(3):192–200.

Golley, F. B.
1962 Mammals of Georgia. Univ. Georgia Press. Athens. xii+218 pp.

Hanawalt, F. A.
1922 Habits of the common mole, Ohio Jour. Sci., 22:164–169.

Harvey, M. J.
1976 Home range, movements and diel activity of the eastern mole, *Scalopus aquaticus*. American Midl. Nat., 95(2):436–445.

Henning, W. L.
1952 Studies in control of the prairie mole *Scalopus aquaticus machrinus* (Rafinesque). Jour. Wildlife Mgt., 16(4): 419–424.

Hisaw, F. L.
1923a Feeding habits of moles. Jour. Mammalogy, 4(1):9–20.
1923b Observations on the burrowing habits of moles (*Scalopus aquaticus machrinoides*). Jour. Mammalogy, 4(2): 79–88.

Jackson, H. H. T.
1915 A review of the American moles. North Amer. Fauna No. 38, 100 pp.
1922 Some habits of the prairie mole, *Scalopus aquaticus machrinus*. Jour. Mammalogy, 3(2):115.
1961 Mammals of Wisconsin. Univ. Wisconsin Press. Madison. xii+504 pp.

Jensen, I. M.
1982 A new live trap for moles. Jour. Wildlife Mgt., 46(1): 249–252.

Jones, M. L.
1979 Longevity of mammals in captivity. Intern. Zoo News, 26(3):16–26.

Kasul, R. L.
1976 Mortality and movement of mammals and birds on a Michigan interstate highway. Michigan State Univ., unpubl. M.S. thesis, ix+39 pp.

Linsdale, J.
1928 Mammals of a small area along the Missouri River. Jour. Mammalogy, 9(2):140–146.

Murie, A.
1936 Following fox trails. Univ. Michigan Mus. Zool., Misc. Publ. No. 32, 45 pp.

Osterberg, D. M.
1962 Activity of small mammals as recorded by a photographic device. Jour. Mammalogy, 43(2):219–229.

Peterson, R. L.
1966 The mammals of eastern Canada. Oxford Univ. Press. Toronto. xxxii+465 pp.

Rood, J. F.
1958 Notes on a captive mole. Jour. Mammalogy, 39(4): 583–584.

Scharf, W. C., and K. R. Stewart
1980 New records of Siphonaptera from northern Michigan. Great Lakes Ento., 13(3):165–167.

Scheffer, T. H.
1910 The common mole, Kansas State Agri. Exp. Sta., Bul. 168, 36 pp.
1913 The common mole. Runway studies: Hours of activity. Trans. Kansas Acad. Sci., 25:160–163.
1914 The common mole of the eastern United States. U.S. Dept. Agri., Farmers' Bull. 583, 10 pp.
1923 American moles as agricultural pests and as fur producers. U.S. Dept. Agri., Farmers' Bull. No. 1247, 23 pp.

Schmidt, F. J. W.
1931 Mammals of western Clark County, Wisconsin. Jour. Mammalogy, 12(2):99–117.

Schofield, R. D.
1960 A thousand miles of fox trails in Michigan's ruffed grouse range. Jour. Wildlife Mgt., 24(4):432–434.

Van Hyning, O. C.
1931 The house cat as a collector of mammals. Jour. Mammalogy, 12(2):16.

Wallace, G. J.
1950 In support of the barn owl. Michigan Agri. Exp. Sta. Quart. Bull., 33(2):96–105.

Wilson, K. A.
1938 Owl studies at Ann Arbor, Michigan. Auk, 55(2):187–197.

Yates, T. L., D. B. Pence, and G. K. Launchbaugh
1979 Ectoparasites from seven species of North American moles (Insectivora: Talpidae). Jour. Med. Entomol., 16(2):166–168.

STAR-NOSED MOLE
Condylura cristata (Linnaeus)

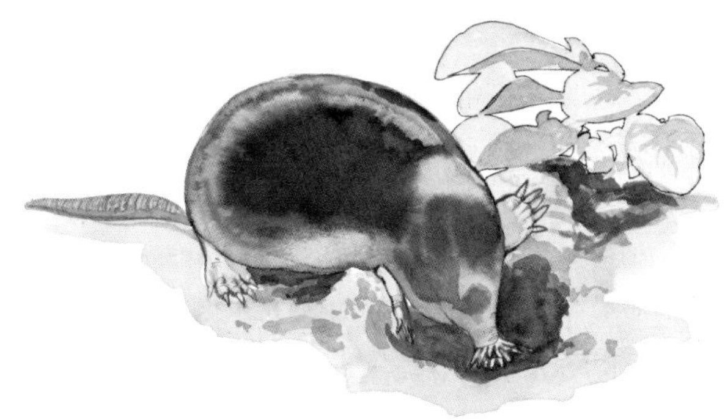

NAMES. The generic name *Condylura,* proposed by Illiger in 1811, is of Greek origin and means "knobbed tail." This inappropriate name was based on an incorrect figure of the star-nosed mole by De La Faille, who pictured the animal's tail as having a series of restrictions resembling beads. The specific name *cristata,* proposed by Linnaeus in 1758, is derived from the Latin and alludes to the "crested" appearance of the end of the animal's snout caused by the presence of star-shaped tentacles. The common name star-nosed mole is used throughout Michigan. The subspecies, *Condylura cristata cristata* (Linnaeus), occurs in both the Upper and Lower Peninsulas.

RECOGNITION. Body size equal to that of a small rat, covered with thin, long, coarse fur; head narrow, snout somewhat elongate terminating in a hairless disk bordered by 22 fleshy, finger-like feelers, symmetrically arranged with 11 on each side of a median ventral line; nostrils nearly circular and located on the anterior surface of the disk; eyes small (although larger than those of the eastern mole) and concealed in facial fur; ear openings lacking external pinnae and surrounded by facial fur; neck indistinct; body stout, cylindrical; tail long (almost half as long as head and body), distinctly annulated, scaly, sparsely covered with coarse hair, constricted at base, slender in summer, swollen with fatty deposits in winter; forelimbs strongly built (see Fig. 13); forefeet handlike with fleshy palms as broad as long (although smaller than those of the eastern mole) and edged with stiff bristles; foreclaws long and broad; hind limbs not strongly constructed; hind feet long and narrow, broadest at base; hind claws long and laterally compressed; mammary glands include two pairs pectoral, one pair abdominal, one pair inguinal.

The star-nosed mole is dark brown to black on the upperparts and paler brown on underparts; nasal disk and tentacles rose-colored; wrists tan; tail when fat showing pink where skin is tightly stretched. The peculiar snout with tentacles; the coarse, rather thin and black pelage; the small yet powerful forelimbs and feet; and the fat and basally constricted tail serve to distinguish the star-nosed mole from its only close Michigan relative, the eastern mole.

MEASUREMENTS AND WEIGHTS. Adult star-nosed moles measure as follows: length from tip of nose to end of tail vertebrae, 6½ to 8⅛ in (165 to 205 mm); length of tail vertebrae, 2⅜ to 3⅜ in (60 to 85 mm); length of hind foot, 1 to 1¼ in (25 to 30 mm); weight, 1.2 to 2.6 oz (35 to 75 g).

DISTINCTIVE CRANIAL AND DENTAL CHARACTERISTICS. The skull of the adult star-nosed mole is long, narrow, rather round, and delicately constructed (see Fig. 14). The braincase is high, not constricted in the region of orbits, and tapers anteriorly. The short, narrow zygomatic arches serve to distinguish this mole's skull from those of all shrews. The lower jaw is weakly constructed, curving upwards posteriorly and straight anteriorly. The upper molars are "W-shaped" and the lower ones "M-shaped." The star-nosed mole is the only Michigan mammal to possess 44 teeth, which include the full complement for modern placental mammals. The dental formula is:

$$\text{I (incisors)}\frac{3\text{-}3}{3\text{-}3}, \text{C (canines)}\frac{1\text{-}1}{1\text{-}1}, \text{P (premolars)}\frac{4\text{-}4}{4\text{-}4}, \text{M (molars)}\frac{3\text{-}3}{3\text{-}3} = 44$$

DISTRIBUTION. The geographic distribution of the star-nosed mole encompasses parts of southern Canada and northern United States in northeastern North America as far west as the Prairie Provinces (Petersen and Yates, 1980). In Michigan, this burrowing insectivore occurs in all counties in the Upper and Lower Peninsulas (see Map 12).

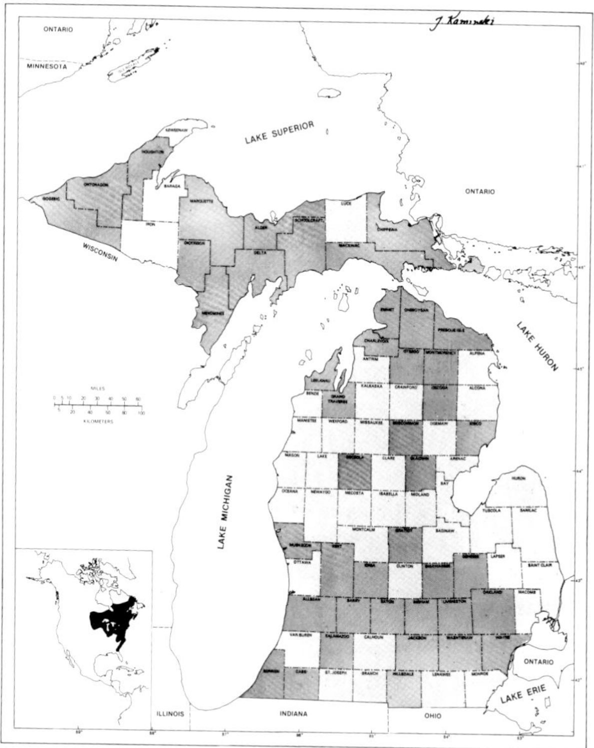

Map 12. Geographic distribution of the star-nosed mole (*Condylura cristata*). Actual records exist for counties with dark shading.

HABITAT PREFERENCES. This most aquatic of American moles requires moist soil conditions and is rarely found remote from poorly drained woodlands and openings, stream banks, and swamp, marsh, pond or lake borders. On rare occasions it has been found in rather dry situations; Paradiso (1969), for example, found one in a dry meadow ¼ mi (410 m) from the nearest water. By using snow tunnels in winter (Tenney, 1871), the star-nosed mole is also apt to move away from its normally damp haunts.

DENSITY AND MOVEMENTS. Surface evidence of the star-nosed mole's presence is much less conspicuous than that of the eastern mole because the former is less fossorial and digs fewer burrows which show as raised ridges. Also, its restriction to supersaturated soil areas makes it a less widespread member of the small mammal community. In narrow habitat along borders of streams and in low-lying areas, the star-nosed mole often lives in family groups, being much more congenial toward other individuals than is the eastern mole. Densities in ideal habitat have been estimated to be as high as 30 individuals per acre (75 per ha), although populations of five or even 10 per acre (12.5 to 25 per ha) are considered more usual (Banfield, 1974; Jackson, 1961). Although little evidence has been accumulated, it is suggested by Seton (1909) that individual star-nosed moles may move about in areas no larger than one acre (0.4 ha). Many questions concerning the ecology of this mole's life habits are yet to be answered.

BEHAVIOR. The star-nosed mole is apparently a less efficient digger than the eastern mole, possibly because of the difference in size of their forefeet. Even so, the loose, moist humus and water-saturated soils in which the former tunnels may be more easily excavated than the drier upland soils often preferred by the eastern mole. Sometimes it is difficult to distinguish the raised ridges and mounds produced by these two moles. On occasion, mounds of the star-nosed moles have also been mistaken for the mud chimneys constructed by crayfish. As a rule, the raised ridges produced by the star-nosed mole in near-surface tunnel excavations average 1¼ in (32 mm) high and 1¾ in (45 mm) wide, much lower and narrower than those produced by the eastern mole. The raised ridges of the star-nosed mole can also often be distinguished by their irregular directional patterns and intermittence because the burrow systems frequently extend deeper.

In digging near-surface tunnels, the star-nosed mole pushes the earth aside by action of its forefeet working forward at each side of the snout. Jackson (1961) indicated that raised-ridge tunnels are excavated at the rate of about 8 ft (2.4 m) per hour. When digging, the motile tentacles are curled up against the nostrils to exclude dirt (Hamilton, 1931).

Networks of shallow burrows connect with deeper galleries where ground water levels will permit and many descend to depths of more than 10 in (255 mm). To construct these, the mole must eliminate the unwanted earth by pushing it upwards and outward to form molehills. These mounds are generally more muddy and less lumpy than those of eastern moles. Those of the star-nosed mole may be as much as 25 in (635 mm) in diameter and 6 in (150 mm) high.

While the eastern mole uses near-surface tunnels in summer and deep galleries in winter, the star-nosed mole spends all seasons either in shallow burrows or on the surface. Above ground, it often follows runways constructed by the meadow vole and also tunnels freely in heavy snow cover (Merriam, 1884; Timm, 1975). Jackson (1961) noted that these moles can move in runways at speeds of 4 mi (6.8 km) per hour. Subsurface tunnels often open directly into such runways or lead to stream banks where fast-moving, ice-free water is accessible. Sites of this mole's activity in Michigan have been described in Alger (Wood, 1917), Berrien (Tenney, 1871), Montmorency (Green, 1925), and Marquette (Manville, 1948) counties.

Like the eastern mole, the star-nosed mole is prompt in repairing damaged runways. After pressing down places along 147 raised ridges produced by this mole, Hamilton (1931) found that the damage inflicted was repaired in all but 61 cases, with 57 tunnels re-opened between 9:00 a.m. and 6:00 p.m. and 29 between 6:00 p.m. and 8:00 a.m. On two occasions, flattened ridges were pushed up again within a few minutes of the time of damage.

The fusiform body, the powerful digging front feet, the protected nose, eye, and ear openings, and specialized sensory perceptors on the snout combine to provide the star-nosed mole with adaptations for fossorial life. Even so, in life actions the star-nosed mole reminds the observer of an agile short-tailed shrew rather than a ponderous eastern mole. The star-nosed mole scurries about both night and day, exhibiting nervous energy as it rapidly probes the environment with twitching snout. Close-up film sequences show the intricate motions of the individual feelers used effectively when the animal is foraging. Good hearing, a superor tactile sense, and probably a good sense of smell make up for the poor vision of this animal (Hamilton, 1931). Apparently its eyesight is only sufficiently acute to allow it to shy away from full daylight; Jackson (1961) found the animal most active on cloudy, overcast days.

Some observers have considered the star-nosed mole semi-aquatic. Burrows often open under the surface of water. In winter, individuals have been seen swimming under ice (Peterson, 1966). The star-nosed mole swims in a zig-zag fashion; the broad front feet act as oars, moving alternately as do the hind limbs. The tail, according to Jackson (1961), is not used in swimming, although Paradiso (1969) does suggest that it performs as a scull. The length of time a mole can remain submerged has not been determined.

Moles seem less noisy than shrews. Nestlings utter shrill cries and resting individuals have been known to make wheezing sounds (Schmidt, 1931; Wiegert, 1961). These animals probably have a repertoire of supersonic utterances yet to be categorized. No doubt, star-nosed moles communicate by such means as well as by the musky exudate from skin glands. Nestling moles have lower rates of oxygen consumption than adults (Pearson, 1947; Wiegert, 1961). It was Wiegert's suggestion that this was because the themoregulatory mechanism in the young had not yet been completely developed. Oxygen consumption rates of young star-nosed moles were found by Wiegert to be lower than those of meadow voles of similar age.

The nests of the star-nosed mole are centers of activity. They are placed above the high water line in an enlarged place in the burrow system at subsurface levels, from as shallow as 3 in (75 mm) to as deep as 12 in (300 mm). Nests may also be at the surface under the protection of an object such as a log. The nests themselves are flattened, spherical structures usually 5 or 6 in (125 to 150 mm) across and 4 to 5 in (100 to 125 mm) deep. These are made of grasses and other plant fibers (Hamilton, 1931); on occasion they are found in manure and compost heaps (Kennard, 1929; Simpson, 1923). A nest containing five young found on May 15 in Ingham County by Wiegert (1961) was 12 in (300 mm) below ground, measured about 5 in (130 mm) across, weighed 5 oz (142 g) air-dried, and was composed of leaves of cottonwood, willow, hawthorne, and oak. It is not known whether such nests are occupied, at least seasonally, by more than one adult star-nosed mole.

ASSOCIATES. In the course of the star-nosed mole's semi-aquatic existence, it frequents marsh and swamp habitat used by muskrats. In New York, Hamilton (1931) reported that these moles are caught in muskrat trails and feeding platforms by fur trappers in winter. Star-nosed moles use small mammal runways also frequented by deer mice, meadow voles, woodland voles, bog lemmings and short-tailed shrews (Eadie, 1937; Hamilton, 1931). In Delta County, the author caught the star-nosed mole, short-tailed shrew, and meadow vole in a single runway mostly covered by a canopy of grasses. In Roscommon County, Rabe (1981) caught star-nosed moles in a leatherleaf-bog birch peatland community in association with the masked shrew, pygmy shrew, water shrew, short-tailed shrew, meadow vole, and meadow jumping mouse.

REPRODUCTIVE ACTIVITIES. The male star-nosed mole shows evidence of sexual development beginning in February. Tail thickening, a phenomenon characteristic of this mole in winter, many also have some role in initiating the mating season. Hamilton (1931) suggested that adults may actually pair off in late autumn and remain together through the mating period which generally occurs from mid-March through April (Jackson, 1961). The female has one litter per year, which can be expected from late-April to mid-June, after a gestation period of about 45 days. Late litters—appearing sometime in July—result when first matings are unsuccessful and females subsequently again become receptive to males. The young do not mature sexually until the breeding season of the year following their birth.

Litters usually contain five individuals, with seven being maximum (Davis and Peak, 1970). At birth, the young are hairless, have wrinkled skin, are mostly pink in color, possess obvious tentacles somewhat compressed, have black spots where eyes will appear, possess a thickened tail, weigh about 1.5 g, and measure about 2 in (50 mm) long, of which the tail makes up about 1 in (25 mm). At seven days of age, the young have dark hair covering the back. Growth is rapid with the opening of the eyes and ears and the unfolding of the snout tentacles occurring during the first weeks after birth. At 15 days of age, the young weigh ¾ oz (21 g). At 30 days of age, the young weigh more than 1 oz (33 g) and venture out of the nest and on their own. In September the young are of adult size.

A long life is expected for individual star-nosed moles chiefly because of their slow reproductive rate of one litter per year. There is some evidence that animals may live three or four years (Jackson, 1961).

FOOD HABITS. Body agility and sharp-grasping front teeth enable the star-nosed mole to pursue and catch an assortment of prey species. Broad, convoluted jaw teeth are designed to crush the bodies of invertebrates. Like the eastern mole, the star-nosed mole eats larval, pupal, and adult insects and earthworms. In addition, the diet includes such aquatic foods as leeches; aquatic worms; immature stages (nymphs) of stone flies, caddis flies, and dragon and damsel flies; larval stages of midges, craneflies, horseflies, and water beetles; mollusks; crustacea; and minnows (Hamilton, 1931). Banfield (1974) noted that earthworms make up more than one-half of the diet, with aquatic species of annelids (including leeches) amounting to about two-thirds of the bulk. Schmidt (1931) fed captive star-nosed moles earthworms, white grubs, parts of a dead frog, and various organs from a meadow vole. Hamilton (1931) concluded that these moles eat at least one-quarter their weight per 24 hours.

ENEMIES. Terrestrial and flying predators probably have more opportunities to capture the star-nosed mole than the more subterranean eastern mole. Aquatic carnivores also catch them. Again, there is the typical insectivore's protection offered by unsavory odor. Terrestrial meat-eaters known to ingest star-nosed moles include domestic dog, red fox, fisher, striped skunk, and house cat (Hamilton, 1936; Murie, 1936; Schmidt, 1931; Schofield, 1960; Timm, 1975; Toner, 1956). Aquatic predators are largemouth bass, bullfrogs, and mink (Christian, 1977; Hamilton, 1936; Pine, 1975). Birds known to eat star-nosed moles are red-tailed hawk, broad-winged hawk, red-shouldered hawk, rough-legged hawk, screech owl, great-horned owl, barred owl, barn owl (De la Perriere, 1970; Hamilton, 1931; Jackson, 1961; Pettingill, 1976; Wallace, 1950). In his study of the food habits of the barn owl in Ingham County, Wallace (1950) counted only two star-nosed moles in the total remains of 6,742 mammals in 2,200 pellets examined.

PARASITES AND DISEASES. Ticks, mites and three species of fleas, *Ctenophthalmus pseudoagyrtes, Ctenophthalmus wenmanni* and *Ceratophyllus wickhami,* infest the surface of the star-nosed mole (Hamilton, 1931; Scharf and Stewart, 1980; Timm, 1975; Yates *et al.,* 1979). Tapeworms and roundworms are known to be internal parasites.

MOLT AND COLOR ABERRATIONS. The winter coat of the star-nosed mole is molted in June and early July. The summer pelage is changed in October. Except for the winter coat being slightly heavier, there seems to be little difference between the two seasonal pelts. There are records of partial and complete albinism in the star-nosed mole, including one completely white male from Michigan (Eadie, 1954; Jackson, 1961).

ECONOMIC IMPORTANCE. Because of the star-nosed mole's restriction to moist areas, it is seldom a problem in well-drained lawns, gardens, and golf greens. However, whenever such formal areas are maintained at the edge of swamps, marshes or streams, *Condylura* with its raised ridges and mounds can cause turf disfigurement. Trapping in a manner similar to the methods prescribed for the eastern mole seems the best way to discourage the inroads of this mole. The use of pesticides remains of questionable value (Henning, 1952; Polivka, 1963).

COUNTY RECORDS FROM MICHIGAN. Records of specimens preserved in collections in museums and from the literature show that the star-nosed mole is known from the following counties in the Upper Peninsula: Alger, Chippewa, Delta, Dickinson, Gogebic, Houghton, Mackinac, Marquette, Menominee, Ontonagon, and Schoolcraft; in the Lower Peninsula: Allegan, Barry, Berrien, Branch, Cass, Charlevoix, Cheboygan, Clinton, Eaton, Emmet, Genesee, Gladwin, Gratiot, Hillsdale, Ingham, Ionia, Iosco, Jackson, Kalamazoo, Kent, Leelanau, Livingston, Montmorency, Muskegon, Oakland, Osceola, Oscoda, Otsego, Presque Isle, Roscommon, Shiawassee, Washtenaw, and Wayne. The species also occurs on Big Summer Island in Lake Michigan (Long, 1978) and on Drummond Island in Lake Huron (Manville, 1950).

LITERATURE CITED

Banfield. A. W. F.
1974 The mammals of Canada. Univ. Toronto Press, Toronto. xxiv+438 pp.

Christian, D. P.
1977 An occurrence of fish predation on a star-nosed mole. Jack-Pine Warbler, 55(1):43.

Davis, D. E., and F. Peek
1970 Litter size of the star-nosed mole (*Condylura cristata*). Jour. Mammalogy, 51(1):156

De la Perriere, C. I.
1970 Food habits of red-tailed hawks, Summer Island, Michigan, 1969. Summer Sci. Jour., 2(2):72–73.

Eadie, W. R.
1937 The Cooper lemming mouse in southern New Hampshire. Jour. Mammalogy, 18(1);102–103.
1954 Skin gland activity and pelage descriptions in moles. Jour. Mammalogy, 35(2):186–196.

Green M. M.
1925 Notes on some mammals of Montmorency County, Michigan. Jour. Mammalogy, 6(3):173–178.

Hamilton, W. J., Jr.
1931 Habits of the star-nosed mole, *Condylura cristata*. Jour. Mammalogy, 12(4):345–355.
1936 Seasonal food of skunks in New York. Jour. Mammalogy, 17(3):240–246.

Henning, W. L.
1952 Studies in control of the prairie mole, *Scalopus aquaticus machrinus* (Rafinesque). Jour. Wildlife Mgt., 16(4):419–424.

Jackson, H. H. T.
1961 Mammals of Wisconsin. Univ. Wisconsin Press. Madison. xii+504 pp.

Kennard, F. H.
1929 A star-nosed mole's nest at Newton Centre, Massachusetts. Jour. Mammalogy, 10(1):77–78.

Long, C. A.
1978 Mammals of the islands of Green Bay, Lake Michigan. Jack-Pine Warbler, 56(2):59–83.

Manville, R. H.
1948 The vertebrate fauna of the Huron Mountains, Michigan. American Midl. Nat., 39(3):615–640.
1950 The mammals of Drummond Island, Michigan. Jour. Mammalogy, 31(3):358–359.

Merriam, C. H.
1884 The mammals of the Adirondack region, northeastern New York. Publ. by author. 316 pp.

Murie, A.
1936 Following fox trails. Univ. Michigan, Mus. Zool., Misc. Publ. No. 32, 45 pp.

Paradiso, J. L.
1969 Mammals of Maryland. North American Fauna No. 66, iv+193 pp.

Pearson, O. P.
1947 The rate of metabolism of some small mammals. Ecology, 28(2):127–145.

Petersen, K. E., and T. L. Yates
1980 *Condylura cristata*. American Soc. Mammalogists, Mammalian Species, No. 129, 4 pp.

Peterson, R. L.
1966 The mammals of eastern Canada. Oxford Univ. Press, Toronto. xxxii+465 pp.

Pettingill, O. S., Jr.
1976 The prey of six species of hawks in northern Lower Michigan. Jack-Pine Warbler, 54(2):70–74.

Pine, R. H.
1975 Star-nosed mole eaten by bullfrog. Mammalia, 39(4): 713–714.

Polivka, J. B.
1963 Effective mole control with chemicals. Ohio Farm & Home Res., 48(2):26.

Rabe, M. L.
1981 New locations for pygmy (*Sorex hoyi*) and water (*Sorex palustris*) shrews in Michigan. Jack-Pine Warbler, 59(1):16–17.

Scharf, W. C., and K. R. Stewart
1980 New records of Siphonaptera from northern Michigan. Great Lakes Ento., 13(3):165–167.

Schmidt, F. J. W.
1931 Mammals of western Clark County, Wisconsin. Jour. Mammalogy, 12(2):99–117.

Schofield, R. D.
1960 A thousand miles of fox trails in Michigan's ruffed grouse range. Jour. Wildlife Mgt., 24(4):432–434.

Seton, E. T.
1909 Life histories of northern animals. Chas. Schribner's Sons, New York. 2:675–1267.

Simpson, S. E.
1923 The nest and young of the star-nosed mole (*Condylura cristata*). Jour. Mammalogy, 4(3):167–171.

Tenney, S.
1871 On the appearance of the star-nosed mole on the snow at Niles, Michigan. American Nat., 5:314.

Timm, R. M.
1975 Distribution, natural history and parasites of mammals of Cook County, Minnesota. Univ. Minnesota, Bell Mus. Nat. Hist., Occas. Papers 14, 56 pp.

Toner, G. C.
1956 House cat predation on small animals. Jour. Mammalogy, 37(1):119.

Wallace, G. J.
1950 In support of the barn owl. Michigan Agri. Exp. Sta., Quart. Bull. 33(2):96–105.

Wiegert, R. G.
1961 Nest construction and oxygen consumption of *Condylura*. Jour. Mammalogy, 42(4):528–529.

Wood, N. A.
1917 Notes on the mammals of Alger County, Michigan. Univ. Michigan, Mus. Zool., Occ. Papers No. 36, 8 pp.

Yates, T. L., D. B. Pence, and G. K. Launchbaugh
1979 Ectoparasites from seven species of North American moles (Insectivora: Talpidae). Jour. Med. Entomol., 16(2):166–168.

Flying Mammals

ORDER CHIROPTERA

Bats fly—an ability unique among mammals—and this achievement is accomplished with no special added parts. Wings develop on a light wiry frame, with the humerus (bone of the upper arm) short and stout, the radius and ulna (bones of the lower arm) elongate, the carpals (bones of the wrist) small and compacted, the metacarpals (bones of the palm of the hand) and the digits (except the clawed thumb) greatly elongate. When the webbing grows and connects these digits with a thin parchment-like membrane and extends to the ankle area of the hind foot and ultimately to the tail, the flying mechanism is complete (see Figs. 15 and 16). Then, with the development of powerful breast muscles (Struhsaker, 1961), the bat can make the flapping, sculling motion which results in flight. This method of locomotion is not new; in fact, a remarkably complete fossil skeleton of a bat from the early Eocene of Wyoming (Jepsen, 1966) shows that chiropterans were capable of flight as long as 50 million years ago. How these creatures may have evolved from an insectivore-like terrestrial ancestor remains unexplained by the paleontologists.

Modern bats are usually divided into two suborders, Megachiroptera and Microchiroptera. The Megachiroptera occur only in the Old World tropics; are often large (wingspread as much as 1.5 m or almost 5 ft); usually have fox-like faces; have a smaller claw on the tip of the index finger in addition to the large claw on the thumb; and, among other distinctive features, have teeth with simple crown patterns. The single family, Pteropodidae, contains 149 species in Africa, and southern Eurasia to Australia. Smaller species may eat nectar and pollen while larger species are generally fruit eaters, with large congregations capable of damaging plantation fruit trees.

The Microchiroptera occur in all tropical, temperate and even subarctic regions. Members of this

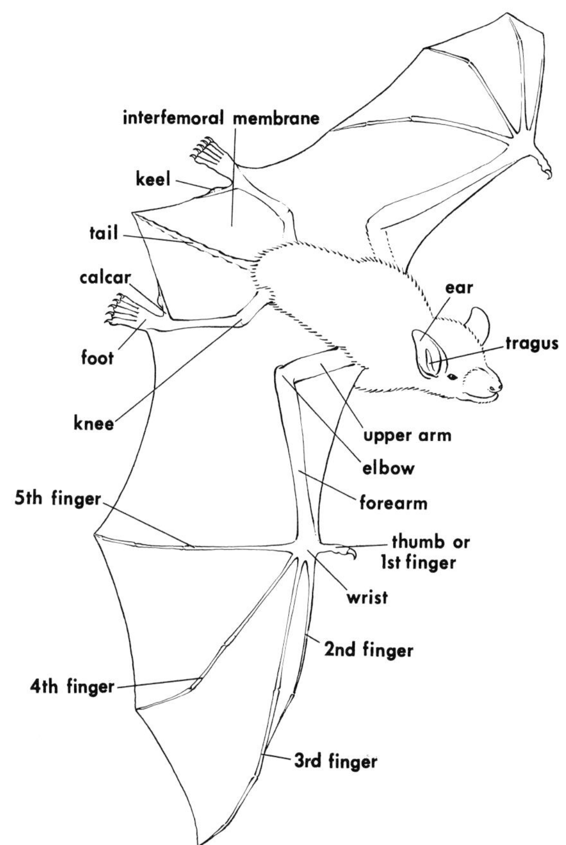

Figure 15. Diagram of the major features of the external anatomy of a vespertilionid bat. Sketch by Victor Hogg

Figure 16. Details of the wing of the big brown bat (*Eptesicus fuscus*). Note the four elongated and membrane-covered "fingers" and the short, hook-like thumb.
Photographed in Ingham County, Michigan, by Bob Brown

suborder are usually small; have a variety of facial forms, sometimes strange nose-leaves and irregularly-shaped ears; have only a single claw, on the thumb; and have highly specialized teeth, oftentimes with W-shaped crown patterns. There are 16 families and 704 species in this suborder of bats. Diversity is demonstrated by the wide variety of the facial configurations, the choice of foods which reflect the crown patterns of the teeth, the morphology of the digestive tract, and the mode of flight. Microchiropterans may catch and eat fish, bats, flying insects and other small animals or feed on blood; they have long tongues with which to probe flowers for nectar and pollen and strong jaws for eating fruits (Slaughter and Walton, 1970; Wimsatt, 1970). Each of these eating habits requires a special life style. Bats flying from a roosting area to fruit trees may not require the flight maneuvering necessary for a nectar feeder to hover, hummingbird-like, in front of a flower or for an insect-eater to pursue and catch flying insects. Although bats have functional eyes (see Fig. 17) and other senses, many rely on echolocation in flight to avoid striking obstacles and as a means of tracking insect prey (Griffin, 1958). The echoes resulting from the bat's ultrasonic pulses of sounds striking obstacles or food items provide the animal with the knowledge of its surroundings (see Vaughan, 1972; Simmons *et al.,* 1979). It has now been discovered that echolocation is used by such other mammals as shrews, whales, possibly some rodents, flying lemurs, and marsupials.

Most bats are nocturnal or crepuscular (active at twilight) and retire by daylight to refuges in caves, tree holes, man-made shelters, and various kinds of foliage. Some species are solitary; others are gregarious or colonial. Their ability to fly and forage in darkness allows bats to occupy environments which by day are filled by birds—thus sharing similar areas and resources without being direct competitors. Because of their small bodies and need for large amounts of energy for flight, bats have been able to ration stored nutrients (especially in the form of fat) through heterothermy (the ability to allow the body temperature to fluctuate). Thus, bats may enter hibernation in times of extended food shortages. They may also lower their energy output (metabolism) when resting in daylight retreats or even between early and late nightly foraging periods.

Bats which hibernate in winter, especially in north temperate regions (*e.g.,* Michigan), may reproduce by delayed fertilization. This has survival value in that the actual mating takes place in late summer or autumn, when both male and female bats are in excellent physical condition and insect foods abundant. The spermatozoa are stored in the uterus and remain viable until ovulation occurs at or near the end of the period of hibernation (dormancy) in spring. Fertilization occurs immediately after the female emerges from winter torpor allowing for the longest possible time for rearing of the young during late spring and summer, and before the next period of winter dormancy begins.

Bats are capable of flying great distances. In some cases colonial bats may move from a summer

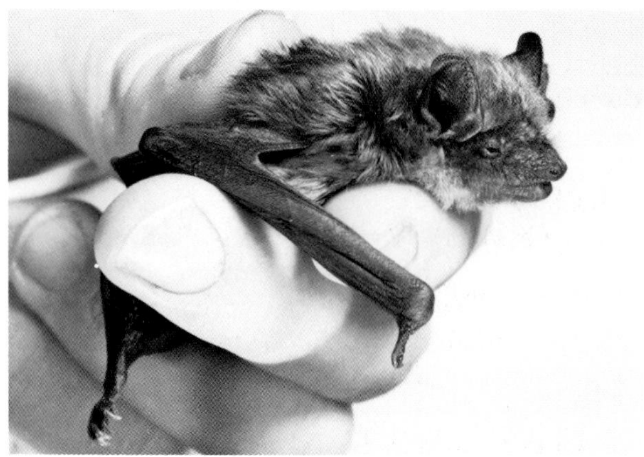

Figure 17. Details of head and forearm of the big brown bat, *Eptesicus fuscus.* Note small eyes, large naked ears and hook-like thumb at end of forearm.
Photographed in Ingham County, Michigan, by Bob Brown

roost to a winter roost a few miles away. Other bats, however, may fly from a summer habitat in a north temperate area southward to a more moderate winter habitation (Findley and Jones, 1964). Some knowledge of bat movements has been obtained from banding individuals and recovering them at later times and in different places (Greenhall and Paradiso, 1968; Humphrey and Cope, 1976). However, our knowledge of foliage-dwelling, solitary species which may be capable of flying more than 1,600 km (1,000 mi) on migrations is fragmentary.

FAMILY VESPERTILIONIDAE

The nine species of bats known to live in Michigan belong to this large (280 species) and widespread family. Although many members of this family are tropical, vespertilionids are the most common bats in temperate parts of the world. Some even occur northward to the treeline in Canada and Alaska (Barbour and Davis, 1969). Most members of this family are conservative in appearance, lacking the diverse facial morphology of some tropical groups. The nose has no complex nose leaf; the ears are generally shaped like those of mice; the wings are broad with the webbing between the hind legs and the tail (also called interfemoral membrane or uropatagium) large; the size is small, mostly from 0.15 to 16.0 oz (4 to 45 g); and the jaw teeth have a W-shaped crown pattern. These bats are insectivorous and mostly capture their prey by tracking it with echolocation and remarkable zig-zag flight capability. Vespertilionids in Michigan include colonial, and often highly sedentary species as well as solitary, foliage-dwelling, highly mobile species.

KEY TO SPECIES OF BATS OF THE FAMILY VESPERTILIONIDAE IN MICHIGAN
Using Characteristics of Adult Animals in the Flesh (See Table 14)

1a. Upper surface of tail (interfemoral) membrane (also called uropatagium) heavily furred outward from the body at least one-half or more the distance to the tip of the tail (see Fig. 18A); ears short and rounded; color of hair on back red, brown or black, often tipped with whitish . . . . 2a.

1b. Upper surface of tail membrane not heavily furred but nearly naked or only furred on the basal one-fourth (see Fig. 18B); ears not short and rounded; color of hair on back uniform brownish, never tipped with whitish . . . . 4a.

2a. Upper surface of tail membrane heavily furred outward from body for only slightly more than one-half the distance to the tip of the tail; color of hair on back black with silvery-white tips; first joints (metacarpals) of flight-digits of the third, fourth and fifth fingers almost the same length . . . SILVER-HAIRED BAT *(Lasionycteris noctivagans)*, page 108.

2b. Upper surface of tail membrane heavily furred outward from body and completely to tip of tail; color of hair on back not black; first joints of flight-digits of third, fourth, and fifth fingers not the same length but progressively (from third to fifth) much shorter . . . . . 3a.

3a. Body orange-red to yellowish-red with tips of hairs often frosted; throat without yellow band; ears without black rims; size small, head, body, and tail less than 4¾ in (120 mm) long; forearm less than 1¾ in (44 mm) long . . . . RED BAT *(Lasiurus borealis)*, page 122.

3b. Body yellowish to drab brown with tips of hairs often frosted; throat yellowish-tan; ears with conspicuous black rims; size large, head, body and tail more than 4¾ in (120 mm) long; forearm more than 1¾ in (44 mm) long . . . HOARY BAT *(Lasiurus cinereus)*, page 127.

4a. Length of bone of forearm more than 1½ in (44 mm) long . . . . BIG BROWN BAT *(Eptesicus fuscus)*, page 116.

4b. Length of bone of forearm less than 1½ in (40 mm) . . . . 5a.

5a. Tragus (elongated structure in front of ear openings, see Fig. 15) short, less than one-half the length of the external ear (pinna), slightly curved forward with blunt or broadly rounded tip; first visible tooth (premolar) behind the large canine (dog tooth), when viewing the head from the side, one-half as high as the canine . . . . EVENING BAT *(Nycticeius humeralis)*, page 132.

5b. Tragus in ear opening long, one-half or more the length of the external ear (see Fig. 15), with pointed or narrowly rounded tip; first visible tooth behind the large canine, when

Table 14. *Digest of Characteristics of Michigan Bats.*

Scientific Name	Common Name	Average Total Body Length- Nose to tip of Tail (in and mm)	Average Length of Forearm- Elbow to Wrist (in and mm)	Presence of Dense Fur on Upper Surface of Tail Membrane	Tragus—long, slender, straight and at least ½ the length of the ear	Calcar extends from heel with a definite keel	Ears long—when laid forward they project 1/16 inch beyond nostrils	Number of teeth	Color of the Fur on the Back
Pipistrellus subflavus	Eastern Pipistrelle	3¼ in, 82 mm	1½ in, 38 mm	none	yes	no	no	34	Yellow brown to golden brown, tricolored hair
Myotis leibii (recorded in Ontario and Ohio but not in Michigan)	Small-Footed Myotis	3 in, 77 mm	1¼ in, 32 mm	none	yes	yes	no	38	Yellow brown with shiny tips, blackish mask across face
Myotis sodalis	Indiana Myotis.	3¼ in, 82 mm	1½ in, 38 mm	none	yes rounded asymetrical	yes	no	38	Dull grayish brown tricolored hair
Myotis keenii	Keen's Myotis	3¼ in, 82 mm	1½ in, 38 mm	none	yes pointed	yes	yes	38	Light reddish brown
Myotis lucifugus	Little Brown Myotis	3½ in, 90 mm	1½ in, 38 mm	none	yes rounded asymetrical	no	no	38	Yellow brown to olive brown
Nycticeus humeralis	Evening Bat	3¾ in, 95 mm	1½ in, 38 mm	none	no	no	no	30	Medium brown to dark brown
Lasionycteris noctivagans	Silver-Haired Bat	4 in, 102 mm	1 5/6 in, 42 mm	basal ½ to ⅔ furred	no	no	no	36	Dark brown to black washed with white
Lasiurus borealis	Red Bat	4¼ in, 108 mm	1¾ in, 45 mm	densely furred to tail tip	no	yes	no	32	Red washed with white
Eptesicus fuscus	Big Brown Bat	4¼ in, 108 mm	1¾ in, 45 mm	none	no	yes	no	32	Brown
Lasiurus cinereus	Hoary Bat	5½ in, 140 mm	2 in, 51 mm	densely furred to tail tip	no	yes	no	32	Yellow brown to dark brown washed with white

viewing head from the side, only about one-third as high as the canine .... 6a.

6a. Ear long, pinna from notch more than ⅝ in (16 mm); when gently laid forward extends at least ⅛ in (3 mm) beyond tip of nose ..... KEEN'S BAT *(Myotis keenii)*, page 89.

6b. Ear short, pinna from notch less than ⅝ in (16 mm), when gently laid forward extends no more than 1/32 in (1 mm) beyond tip of nose .... 7a.

7a. Length of bone of forearm averages less than 1⅜ in (35 mm); fur on dorsal surface with long shiny tips, giving burnished appearance; blackish mask on face ... SMALL-FOOTED BAT *(Myotis leibii)* unreported in Michigan, page 94.

7b. Length of bone of forearm averaging more than 1⅜ in (35 mm); fur on dorsal surface without shiny tips; no blackish mask on face .... 8a.

8a. Upper surface of body yellowish brown to golden brown; hairs on back tricolored, dark at base and tip but paler in the middle; tragus in ear opening (Fig. 15) only about one-half of

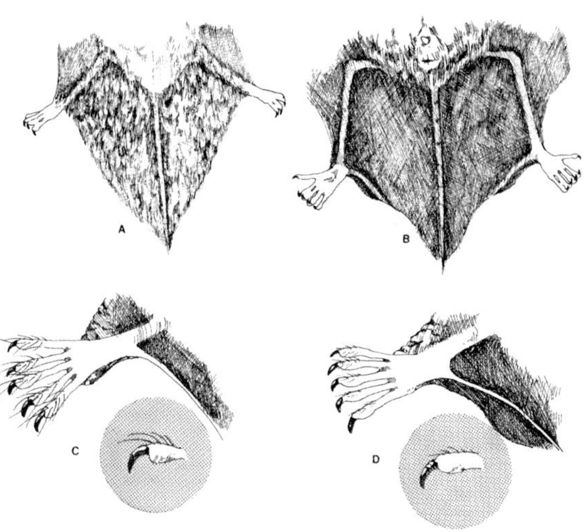

Figure 18. Features of the tail membrane and hind feet of vespertilionid bats.

length of the external ear (pinna) . . . . EASTERN PIPISTRELLE *(Pipistrellus subflavus),* page 112.

8b. Upper surface of body brownish; hairs on back not tricolored (except in *Myotis sodalis*); tragus (Fig. 15) more than one-half the length of the external ear . . . . 9a.

9a. Hairs on back tricolored, basally black, grayish in the middle and yellow brown on tips; fur dull not glossy; calcar (extension from heel, see Fig. 18D) with keel-shaped structure; no long hairs on toes (Fig. 18D) . . . INDIANA BAT *(Myotis sodalis),* page 104.

9b. Hairs on back not tricolored; fur glossy not dull; calcar (Fig. 18C) lacking keel-shaped structure; long hairs on toes (Fig. 18C) . . . LITTLE BROWN BAT *(Myotis lucifugus),* page 96.

KEY TO THE SPECIES OF BATS OF THE FAMILY VESPERTILIONIDAE IN MICHIGAN

Using Characteristics of the Skulls of Adult Animals (see Table 14 and Fig. 19)

1a. Teeth total 38 in number, nine on each side of upper and 10 on each side of lower jaw . . . 2a.

1b. Teeth total less than 38 in number, either 32, 34 or 36 (see Table 14) . . . . 5a.

2a. Skull smaller, usually less than ½ in (14 mm) long and less than 5/16 in (8.4 mm) wide; braincase distinctively flattened with gradually rising profile . . . . SMALL-FOOTED BAT *(Myotis leibii)* unreported in Michigan, page 94.

2b. Skull larger, usually more than ½ in (14 mm) long and more than 5/16 in (8.4 mm) wide; braincase arched and rounded, not flattened when viewed from side . . . . 3a.

3a. Skull comparatively narrow although bony space between orbits (interorbital constriction) 3/16 in (4 mm) or more . . . . LITTLE BROWN BAT *(Myotis lucifugus),* page 96.

3b. Skull comparatively wide although bony space between orbits less than 3/16 in (4 mm) . . 4a.

4a. Skull with median crest; length from front of canine (dog tooth) to back of last molar less than width across molars in upper jaw . . . INDIANA BAT *(Myotis sodalis),* page 104.

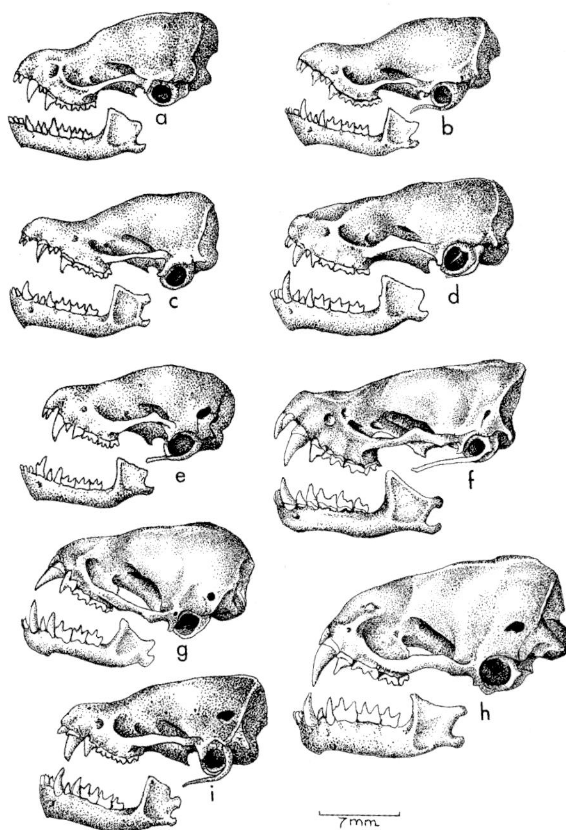

Figure 19. Profiles of the skulls and lower jaws of Michigan bats: (a) Keen's bat, (b) little brown bat, (c) Indiana bat, (d) silver-haired bat, (e) eastern pipistrelle, (f) big brown bat, (g) red bat, (h) hoary bat, and (i) evening bat.

4b. Skull without median crest; length from front of canine to back of last molar greater than width across molars in upper jaw .... KEEN'S BAT *(Myotis keenii)*, page 89.

5a. Teeth total 36 in number; eight on each side of upper and 10 on each side of lower jaw; skull comparatively long and low with tapered rostrum (nasal region) ... SILVER-HAIRED BAT *(Lasionycteris noctivagans)*, page 108.

5b. Teeth total 30, 32 or 34 in number ... 6a.

6a. Teeth total 34 in number, eight on each side of upper and nine on each side of lower jaw ... EASTERN PIPISTRELLE *(Pipistrellus subflavus)*, page 112.

6b. Teeth total either 30 or 32 in number .....7a.

7a. Teeth total 30 in number, six on each side of upper and nine on each side of lower jaw; skull shorter, about 9/16 in (14.5 mm) in length .... EVENING BAT (*Nycticeius humeralis*), page 132.

7b. Teeth total 32 in number; skull longer, more than 9/16 in (14.5 mm) in length ...... 8a.

8a. Upper jaw with two incisors (in front of canine tooth) and four cheek teeth (behind canine tooth) on each side; first premolar (first behind canine) on each side of upper jaw not minute .... BIG BROWN BAT *(Eptesicus fuscus)*, page 116.

8b. Upper jaw with only one incisor and five cheek teeth on each side; first upper premolar on each side of upper jaw minute .... 9a.

9a. Skull larger, more than 11/16 in (17 mm) long and ½ in (11.5 mm) wide .. HOARY BAT *(Lasiurus cinereus)*, page 127.

9b. Skull smaller, less than ⅝ in (15 mm) long and 7/16 in (11 mm) wide .... RED BAT *(Lasiurus borealis)*, page 122.

RELATIONSHIPS OF BATS TO HUMANS

For centuries bats have lived in close proximity to humans. Loosely constructed buildings provide numerous entry ways through roof coverings (shingles or thatching), under eaves, around chimneys, and under warped siding. It is estimated that the Michigan bat population is now many times what it may have been in pre-settlement times. Today, aside from treeholes, there are very few natural retreats such as caves, cracks in rocky ledges and cliffs (Davies, 1955), for this species. However, buildings dot the landscape and provide abundant bat housing, especially in summer. Further, this partiality of bats for human dwellings has allowed many folktales concerning bats and people to develop and endure. Strange indeed is the vampire legend of central Europe, because there are no blood-feeding bats in the Old World, and the legend became part of the lore long before the true vampire bats of the tropical parts of the New World were discovered by Europeans.

Fortunately, house-dwelling bats in Michigan do not often congregate in the thousands as they do in the caves of such places as southern Indiana, Kentucky and the Ozarks. Michigan bats may even earn their keep by feeding on insects (some of which are noxious) and usually do not cause serious problems, as they attempt to avoid humans. Even so, deposits of fecal material (guano) of only a few house-dwellers can exude a musty smell in attics, double walls, churches, and even hotels. In quantity, this guano is valuable as fertilizer and is mined from caves in parts of the world where large colonies of bats live.

Bats are infested with an array of parasites and disease organisms not especially different from other mammals. Aside from fleas, mites, ticks, and lice, bats may carry parasitic flies (Streblidae, Nycteribiidae, and Hippoboscidae) and bedbugs (Cimicidae), most of which do not infest people. On the other hand, bats earn human distrust (often unwarranted) as they can contract and spread rabies. A major rabies problem exists in tropical Latin America where blood-feeding vampire bats infected with the disease may transmit it to cattle on whose blood the bats feed (Greenhall, 1968). Bats preferring insects or other foods can also contract rabies, but their role in rabies transmission to livestock, people, or other mammalian hosts is yet unclear. Conservationists have long shown concern lest "harmless" species of bats are destroyed in the process of eliminating rabies-carriers (Clark, 1981). This is especially true in tropical America where relatively innocuous fruit-eating, nectar-feeding, and insect-eating bats are mistaken for vampire bats.

The presence of rabies in vespertilionid bats in the United States was not positively detected until 1953 in Florida. Since that time, bat rabies has been reported in most parts of North America. Some medical authorities regard species of bats, along with foxes, skunks, coyotes and other mammals, as natural hosts of the rabies virus in the United States. There is also evidence that migratory species, notably the Brazilian free-tailed bat *(Tadarida brasiliensis)* which reaches southern United States, may provide a chain of infection from tropical area bats to temperate areas bats.

In Michigan, accounts of infected bats have appeared in newspaper stories—often unjustly sensationalized. Between 1968 and 1981, out of 1,247 bats examined by the Michigan Department of Public Health only 62 (5%) had rabies, with only 3 to 8 (maximum, 11) positive animals recorded annually (Kurta, 1979 and unpublished data). In contrast, 28% of the 384 Michigan striped skunks *(Mephitis mephitis)* submitted from 1965 through 1978 for examination were rabid. Beginning in 1968, most Michigan bats received by the Michigan Department of Public Health were identified as to species by mammalogists at The Museum at Michigan State University. It is well to point out, however, that the figures of 5% for bats and 28% for skunks are really overestimates, since in both cases sick animals are more likely to be tested than well ones. Seven of the nine species of Michigan bats were among the 1,247 specimens identified as follows (in part, after Kurta, 1979):

Keen's Bat *(Myotis keenii)*—27 (0 rabid)
Little Brown Bat *(Myotis lucifugus)*—75 (0 rabid)
Silver-haired Bat *(Lasionycteris noctivagans)*—12 (2 rabid)
Big Brown Bat *(Eptesicus fuscus)*—1,093 (48 rabid)
Red Bat *(Lasiurus borealis)*—55 (3 rabid)
Hoary Bat *(Lasiurus cinereus)*—12 (2 rabid)
Evening Bat *(Nycticeius humeralis)*—1 (0 rabid)

Naturally, the public is repeatedly warned against handling or touching live bats or other wild mammals. Most communities have animal control officers to assist in removing bats from homes, office buildings, and other human habitation.

Despite some age-old lore to the contrary, bats are remarkable and interesting associates in the human environment and deserve to be better understood. To the outdoorperson, enjoyment of the Michigan scene would be incomplete without the aerial acrobatics of these tiny and gentle creatures. They can be seen at twilight darting for flying insects over gardens, along streams, or even in the glare of city streetlights.

Knowledge of bats has been notably enhanced first by bat-banding programs and second by the post WWII introduction to the United States of Japanese-made mist nets. The same metal band used to ring a small bird's hind leg fits a bat's forearm. Most banding has been accomplished using the easily accessible aggregations of colonial species, especially in wintering cave or mine tunnel sites. Banding programs help document daily and seasonal movements as well as longevity.

Mist nets, made of fine silk or nylon, have been used by the Japanese for generations to catch birds to supplement meager diets. Using mist nets for this purpose is unlawful in the United States; however, scientists with special government permits can use these nets to snare live birds and bats. When stretched between two poles a mist net performs much like a trammel net in catching fish. The bat, perhaps using echolocation, can often avoid a mist net stretched across a small stream, a forest opening, or at the mouth of a mine shaft. However, if the flying mammal does hit the net, it can become enmeshed in the netting. When disengaged, the bats can be identified, sexed, weighed, banded, and released without difficulty. By using mist nets, chiropteran specialists are gaining much more knowledge of the population biology of tree bats and other non-colonial species, rarely captured by other means.

BATS IN BUILDINGS

Bats like the red bat and hoary bat are usually solitary, foliage-dwelling species. These bats spend the summer months in Michigan and then fly southward to warmer areas for the winter. They rarely enter human dwellings. On the other hand, several other Michigan bats such as the big brown bat, little brown bat, and Keen's bat are usually gregarious (colonial) cavity-dwellers. In summer, they use protective roosts as daytime retreats and forage for flying insects at twilight and at night. In winter, they hibernate in caves, tree holes, openings in cliffs, mine tunnels, and buildings. Using human habitation as roosts, however, brings these bats into close contact with people.

Bats gain entry into the double-wall spaces of a building through an opening usually caused by warped siding up near the eaves or roof. In summer, a group of bats may hang quietly during daylight hours (upside down, see Fig. 20), either in the space between the double walls or even in the attic (providing the temperature of the surround-

Figure 20. Big brown bats (*Eptesicus fuscus*) hanging from attic roost in Ingham County, Michigan.
Photograph by L. A. Ryel

ings is not excessively warm). At dusk, the bats arouse, crawl to the opening and fly out to forage (see Fig. 21). They might return or use another retreat for a midnight rest period then forage again before daylight when they resume their daytime quiescence.

In winter, some of these bats, in groups of less than ten to as many as several hundred depending on species, will hibernate, usually without much motion (unless the cool temperature of their immediate surroundings changes) from autumn to spring. In this state of hibernation, bodily processes (heartbeat, respiration, the maintenance of body temperature, and other metabolic activities) decline drastically, allowing each bat to ration stored energy (including body fat) to provide for survival during this extended period.

Generally, efforts are made to eliminate bats from buildings for the following reasons:

(1) Offensive Odors. Bat fecal material and urine deposited below roosts, often in sizeable amounts, can produce musty odors in cottages, houses, hotels, and churches.

(2) Scratching Noises. When bats crawl to and from building entries to resting places, their movements are noisy and annoying to human occupants.

(3) Room Invasion. It is supposed that bats may lose their way occasionally when moving to and from building entries to interior resting areas. They also may move around if the temperature or other environmental conditions change in the vicinity of usual resting areas. During such times, bats might be able to get from attic or double-wall spaces into other rooms. Such an intruder can be swatted down with a tennis racquet or entangled in a filamentous dish towel thrown across its path of flight.

It is not practical to try to control bats living in buildings by use of fumigants such as napthalene flakes or paradichlorobenzene, since the entry ways remain to attract other individuals, even through the gassed animals are killed or dispersed. The only sure way to eliminate these animals is bat-proofing (Anon., n.d.; Barclay *et al.,* 1980), which simply means plugging up likely entries in outer building walls. These openings, perhaps no more than ½ in (13 mm) in diameter, may be difficult to find as they are often near the roof, the eaves or around a chimney. If a building is painted white, a dark stain will often point to an entry. For best results, observers should be stationed at each corner of the building on a cloudless, warm, summer evening, preferably in August. Just at dusk or shortly after dark, bats may emerge, dropping down as they leave the structure, thus identifying their point of entry. All such openings should be nailed shut, filled with caulking compound, or covered with sheet metal or fine-mesh hardware cloth. Naturally these openings must be closed at night while the bats are out foraging; otherwise, the bats are trapped inside the building.

Figure 21. Big brown bat (*Eptesicus fuscus*) departing from opening between brick siding and eaves on house in Ingham County. Photograph by L. A. Ryel

LITERATURE CITED

Anon.
n.d. Controlling bats. U.S.D.I., Bur. Sport Fisheries and Wildlife, Leaflet 402, 2 pp.

Barbour, R. W., and W. H. Davis
1969 Bats of America. The Univ. Press of Kentucky, Lexington. 285 pp.

Barclay, R. M. R., D. W. Thomas, and M. B. Fenton
1980 Comparison of methods used for controlling bats in buildings, Jour. Wildlife Mgt., 44(2):502–506.

Clark, D. R., Jr.
1981 Bats and environmental contaminants: A review. U.S.D.I., Fish and Wildlife Service, Sp. Sci. Rpt.—Wildlife No. 235, 27 pp.

Davies, W. E.
1955 Caves and related features of Michigan. National Speleological Soc., Bull. 17, pp. 23–31.

Findley, J. S., and C. Jones
1964 Seasonal distribution of the hoary bat. Jour. Mammalogy, 45(3):461-470.

Greenhall, A. M.
1968 Bats, rabies and control programs. Oryx, 9(4):263–266.

Greenhall, A. M., and J. L. Paradiso
1968 Bats and bat banding. U.S.D.I., Bur. Sport Fisheries & Wildlife, Resource Publ. 72, iv+48 pp.

Griffin, D. R.
1958 Listening in the dark. Yale University Press, New Haven. 413 pp.

Humphrey, S. R., and J. B. Cope
1976 Population ecology of the little brown bat, *Myotis lucifugus*, in Indiana and north-central Kentucky. American Soc. Mammalogists, Sp. Publ. No. 4, vii+81 pp.

Jepsen, G. L.
1966 Early Eocene bat from Wyoming. Science, 154(3754): 1333–1339.

Kurta, A.
1979 Bat rabies in Michigan. Michigan Acad., 12(2):221–230.

Simmons, J. A., M. B. Fenton, and M. J. O'Farrell
1979 Echolocation and pursuit of prey by bats. Science, 203(4375):16–21.

Slaughter, B. H., and D. W. Walton, Eds.
1970 About bats. A chiropteran biology symposium. Southern Methodist Univ. Press, Dallas. 339 pp.

Struhsaker, T. T.
1961 Morphological factors regulating flight in bats. Jour. Mammalogy, 42(2):152–159.

Vaughan, T. A.
1972 Mammalogy, W. B. Saunders Co., Philadelphia. viii+463 pp.

Wimsatt, W. A., Ed.
1970 Biology of bats. Academic Press, New York. 406 pp.

KEEN'S BAT
Myotis keenii (Merriam)

NAMES. The generic name *Myotis,* proposed by Kaup in 1829, is from the Greek and means "mouse ear." The specific name *keenii,* proposed by Merriam in 1895, refers to J. H. Keen who collected the type specimen in 1894. In Michigan, this species is usually called bat or brown bat. The subspecies, *Myotis keenii septentrionalis* (Trouessart), occurs in Michigan. The taxonomic position of this bat, however, is under discussion (van Zyll de Jong, 1979).

RECOGNITION. Body small, the size of a short, small mouse; head and body approximately 1¾ in (45 mm) long; snout pointed; eyes small and beady; ears rounded and long for a myotis bat, extending when laid (not pulled) forward more than ⅛ in (3 mm) beyond the tip of the nose; tragus in ear long and pointed; forearm short (see Table 14); tail long; hind foot short, approximately one-half as long as tibia; calcar with keel-shaped extension (see Fig. 19D).

Hair with silky texture; light brown with a brassy sheen on upper parts; paler and grayer on underside with slate-colored basal parts of hairs frequently exposed; ears and wing membranes dark brown. The length of the ears, pointed tragus, and keeled calcar all assist in distinguishing this bat from other Michigan species (see Table 14).

MEASUREMENTS AND WEIGHTS. Adult Keen's Bats have the following dimensions: length from tip of nose to end of tail vertebrae, 3 to 3½ in (77 to 92 mm); length of tail vertebrae, 1⅜ to 1¾ in (34 to 45 mm); length of hind foot, 5/16 to 7/16 in (8 to 10 mm); height of ear from notch, ⅝ to ¾ in (15 to 19 mm); length of bone of forearm, 1 5/16 to 1½ in (34 to 38 mm); weight of non-hibernating adults, 0.2 to 0.3 oz (6 to 9 g).

DISTINCTIVE CRANIAL AND DENTAL CHARACTERISTICS. The skull of Keen's bat is of delicate construction. It is long and narrow compared with skulls of other Michigan bats of the genus *Myotis,* being closest to the Indiana bat in these features (see Fig. 19). The skull is approximately ⅝ in (15 mm) long and 5/16 in (9 mm) wide (across zygomatic arches). If mummified remains are found in barns, attics, mine tunnels and other roosting sites, it is probably best to send such specimens to museum for identification, since the delicate nature of the skulls and teeth require special cleaning techniques in order to reveal significant identifying characteristics. The dental formula is:

$$\text{I (incisors)}\frac{2\text{-}2}{3\text{-}3}, \text{C (canines)}\frac{1\text{-}1}{1\text{-}1}, \text{P (premolars)}\frac{3\text{-}3}{3\text{-}3}, \text{M (molars)}\frac{3\text{-}3}{3\text{-}3} = 38$$

DISTRIBUTION. Keen's bat is widespread in central and east-central United States and adjacent parts of southern Canada (Fitch and Shump, 1979). To the west and separated by plains habitat is another part of this species population (considered distinctive by van Zyll de Jong, 1979). In Michigan, Keen's bat has been found widely distributed except in the central part of the Lower Peninsula (see Map 13). There are records from ten counties in the Upper Peninsula, seven in the northern part of the Lower Peninsula and eight in the southern part of the Lower Peninsula.

HABITAT PREFERENCES. The forested country of Michigan intermixed with openings, lakes and ponds, and streamways is preferred summer foraging habitat for Keen's bat. In addition, the animal requires proper daytime roosts in summer and hibernating sites in winter. The lack of caves and mine tunnels in southern Michigan probably helps account for the small winter population of Keen's bat noted in this area (Kurta, 1980a).

Keen's bat uses a variety of summer daytime retreats, including the interior of Bear Cave in Berrien County, behind cabin shutters, under tarpaper covering on a shack, in attics and double walls of houses, under loose bark on an elm tree, in barns and outbuildings, and under shingles on a picnic shelter (Brandon, 1961; Cope and Humphrey, 1972; Doutt *et al.,* 1966; Kurta, 1980a; Mumford, 1969; Mumford and Cope, 1964; Timm, 1975; Turner, 1974). One or more of these sites are found in any Michigan county. On the other hand, insulated winter hibernating sites, except for sewers (Goehring, 1954), in caves and mine tunnels, are much more restricted, chiefly in the western part of the Upper Peninsula. Wintering sites for summering colonies of Keen's bat in southern Michigan are yet to be identified (Kurta, 1980a).

DENSITY AND MOVEMENTS. Keen's bat appears to be less common than some of the other colonial myotis bats. In comparison with the little brown bat, overwintering aggregations of Keen's

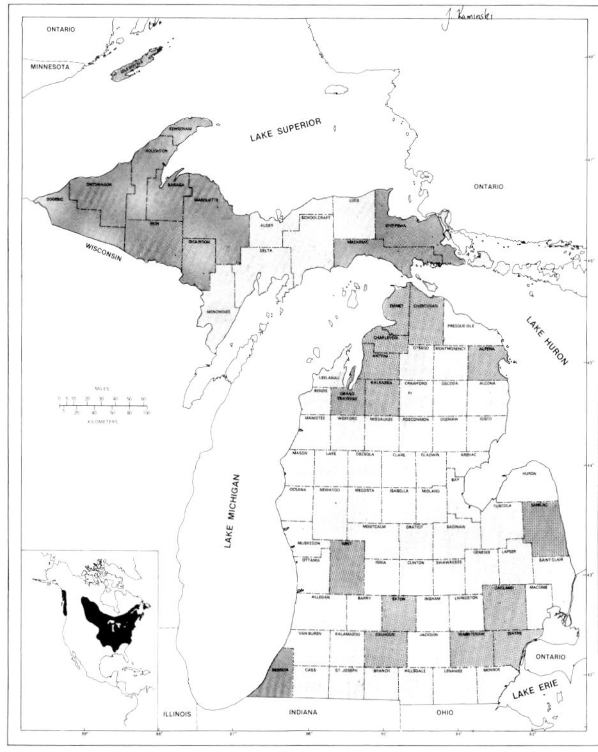

Map 13. Geographic distribution of Keen's bat (*Myotis keenii*). Actual records exist for counties with dark shading.

bat may be outnumbered as much as 16 to one. Hitchcock (1949) counted one winter grouping of 230 individuals in an Ontario corundum mine; Griffin (1940) observed a colony of 304 in a Massachusetts cave; Jackson (1961) reported 100 Keen's bats in an abandoned iron mine shaft in Wisconsin. In Michigan, wintering colonies observed thus far have been small. Groupings of no more than 20 individuals were found in mine tunnels near Iron Mountain on November 23 and near Norway on March 26 in Dickinson County. A group of 87 Keen's bats were counted on seven nights in September in the process of swarming at Bear Cave in Berrien County (Kurta, 1980c). Upper Peninsula mining operations, beginning as early as the 1840s, have produced numerous sites highly suitable as winter quarters for Keen's bat (Stones and Fritz, 1969). These tunnels and shafts may have left the landscape unpleasantly pockmarked but certainly have provided increased hibernacula for several kinds of colonial bats.

Keen's bat makes two basic kinds of movements in its annual life cycle. The first occurs in spring and autumn between winter quarters and summer roosts. In the Upper Peninsula, distances involved probably are not great. However, in extreme southern Michigan, it is likely that summering Keen's bats, like Indiana bats (Kurta, 1980b), spend their winters in caves in the Indiana and Kentucky area.

The second kind of movement is in summer between daytime retreats and foraging grounds. Here again, available information suggests that these foraging flights out from a central shelter and over trees and standing water may not be extensive, especially if a separate night roosting shelter is used between early and late nightly feeding flights. The use of chemical lights attached to individuals has helped trace their flights and determine the extent of the areas covered (LaVal *et al.*, 1977). Keen's bats transported as far as 36 mi (58 km) from their home roosts returned (Stones and Branick, 1968).

BEHAVIOR. We know most about Keen's bats by observing them when they congregate in summer daytime shelters and in winter hibernacula, and when they are captured in mist nets on their nightly, summer flights (Kurta, 1980a). These small, beady-eyed animals are highly nervous, alert and wary, will take to flight hurriedly when disturbed, and attempt to bite when handled. However, this bat's jaws cannot exert sufficient force for the tiny teeth to pierce the handler's skin.

Studies of Keen's bat in Michigan by Stones and Branick (1968) show that loss of vision will reduce homing ability while loss of hearing will prevent this action completely. Like other bats, Keen's bat has developed echolocation. The high frequency utterances which echo back to inform the bat about its surroundings or prey proximity are like those of the little brown bat but are of shorter duration (0.9 milliseconds as compared with 2.3 milliseconds) and higher frequency, averaging 95 kilocycles at the beginning, 57 in the middle and 46 at the end as compared with 78, 48, and 39, respectively, for the little brown bat.

In summer, females in small maternity groupings and males presumably in bachelor groups or as singles move nightly from their daylight shelters to forage for insect foods. There are usually two peaks in this activity, just after dusk and again just prior to daylight. In the middle of the night the bats will often rest between meals in night roosts which may be different from those used by day (Barbour and Davis, 1969). In late summer and early autumn Keen's bat must accumulate a fatty surplus to withstand the winter hibernating period.

After the late summer and autumn swarming, both sexes enter winter quarters together (Griffin, 1940). Stones and Fritz (1969) found Keen's bats in winter retreats in the Upper Peninsula as early as late August, with maximum numbers reached in November. The species often hangs singly wedging itself in cracks and irregular niches in dark, moist, cool places such as walls and ceilings of mine shafts and natural caves. The preferred temperature is 36° to 42° F (2.2° to 5.6° C). The bats remain torpid in winter quarters until the spring exodus begins in March. The earliest record of an active Keen's bat in southern Michigan was for an individual caught by Kurta (1980a) on May 29 in a mist net stretched across the Thornapple River in Eaton County.

ASSOCIATES. Keen's bat's association with other chiropterans has been recorded in both winter quarters and in summer retreats. In one limestone cave at Fourth Chute in Renfrew County, Ontario, Hitchcock (1949) counted a winter congregation consisting of 117 Keen's bats, five little brown bats,

434 small-footed bats (not yet known from Michigan), 751 big brown bats, and 59 eastern pipistrelles. Swanson and Evans (1936) found a similar assemblage, except for the small-footed bat, in a cave in southern Minnesota. In abandoned copper mines in Keweenaw and Ontonagon counties, Stones and Fritz (1969) noted Keen's bats, little brown bats, and big brown bats hanging in the same tunnels. Myotis bats tend to stay apart under these circumstances but occasionally may mix in the same cluster (Griffin, 1940).

In summer, John Hall and Wayne Davis (*in* Barbour and Davis, 1969) found Keen's bats, little brown bats, eastern pipistrelles, and big brown bats all using the same night roost in Kentucky. Jack Mell (pers. comm.) noted both Keen's bats and little brown bats in Bear Cave in Berrien County on August 26. In September of 1978 and 1979, also at Bear Cave, Kurta (1980a) counted a two-year total of 87 Keen's bats, 268 little brown bats, and two red bats. At least the first two species were in the process of autumn swarming. Along the Thornapple River in Eaton County in the summer of 1979, Kurta (1980a) snared Keen's bats, little brown bats, Indiana bats, silver-haired bats, big brown bats, red bats, and hoary bats, in mist nets.

REPRODUCTIVE ACTIVITIES. Most data on reproduction in myotis bats have been obtained for the little brown bat with the assumption that other species in the genus have similar habits. Thus, for Keen's bat it is supposed that courtship and breeding occur in autumn, with swarming as one component of this activity. Following the mating act, viable sperm is retained in the female reproductive tract during the winter with ovulation and fertilization taking place in April and May (delayed fertilization, Barbour and Davis, 1969). During pregnancy, the females abandon their winter quarters for summer nursery colonies in retreats which may be warmer and less well insulated. These colonies, containing as many as 30 pregnant females, can be located in barns, under wooden shingles, or under loose tree bark (Brandon, 1961; Cope and Humphrey, 1972; Mumford and Cope, 1964). The whereabouts of the males at this time is not definitely known.

By early June, about 50 to 60 days following conception, a single young is born (Claire *et al.*, 1979; Easterla, 1968). A female containing a late-term fetus was examined by Kurta (1980a) on May 29. Lactating females have been found from June to August (Turner, 1974), but lactation may be completed when the young is about 32 days of age (Kunz, 1971). The rapidly-developing young are capable of flight as early as late June, or within 21 days after birth.

Considering this bat's slow rate of reproduction, it can be expected to live for several years. Based on banding records, one Keen's bat lived in the wild for more than 18 years (Hall *et al.*, 1957).

FOOD HABITS. Keen's bat obtains insects for food by doggedly pursuing them in flight. Examination of digestive tracts (LaVal and LaVal, 1980) show that this bat savors small moths (Lepidoptera), caddisflies (Trichoptera), stoneflies (Plecoptera), bugs (Homoptera), mayflies (Ephemeroptera), beetles (Coleoptera), and flies (Diptera). Some mosquitoes are ingested, but Keen's bat's influence on populations of these insect pests is probably negligible.

ENEMIES. Although there are no specific reports of Keen's bat being preyed upon, falcons and other hawks are known to feed on other species of myotis bats. Raccoons have also been known to catch bats as they hang from the sides or ceilings of caves.

PARASITES AND DISEASES. Fleas, bedbugs, ticks and chiggers have been identified as infecting Keen's bat (Jones and Genoways, 1967). Whitaker and Mumford (1971) counted 43 intestinal flatworms (trematodes) in a single individual taken in Indiana. Intestinal helminths in bats of this species from Wisconsin are reported by Coggins *et al.* (1981).

MOLT AND COLOR ABERRATIONS. A prolonged summer molt occurs in which the old fur is gradually replaced by the new. No color aberrations are reported for this species.

ECONOMIC IMPORTANCE. Keen's bat appears to have little economic importance. Of 26 individuals examined by the Michigan Department of Public Health between 1968 and 1980, there was no evidence of rabies (in part from Kurta, 1979). Although the species is a potential carrier of this dreaded disease, its weak jaws and tiny teeth do not allow its bite to break the skin of the adult human hand. Summer colonies become nuisances in cottages and other buildings by depositing sizeable amounts of feces (guano) giving off a musty odor. An occasional stray flyer can disrupt the household.

COUNTY RECORDS FROM MICHIGAN. Records of specimens in collections of museums and from the literature show that Keen's bat has been recorded from ten counties in the Upper Peninsula: Baraga, Chippewa, Dickinson, Gogebic, Houghton, Iron, Keweenaw, Mackinac, Marquette, and Ontonagon; and from 15 counties in the Lower Peninsula: Alpena, Antrim, Berrien, Calhoun, Charlevoix, Cheyboygan, Eaton, Emmet, Grand Traverse, Kalkaska, Kent, Oakland, Sanilac, Washtenaw and Wayne. Keen's bats are also known from Isle Royale in Lake Superior, from Big Summer Island in Lake Michigan, and from Mackinac Island (Burt, 1946; Long, 1978)

LITERATURE CITED

Barbour, R. W., and W. H. Davis
1969 Bats of America. Univ. Kentucky Press, Lexington. 286 pp.

Brandon, R. A.
1961 Observations of young Keen bats. Jour. Mammalogy, 42:400–401.

Burt, W. H.
1946 The mammals of Michigan. Univ. Michigan Press, Ann Arbor. xv+288 pp.

Claire, W., R. K LaVal, M. L. LaVal, and R. Clawson
1979 Notes on the ecology of *Myotis keenii* (Chiroptera: Vespertilionidae) in eastern Missouri. American Midl. Nat., 102:404–407.

Coggins, J. R., J. L. Tedesco, and C. Rupprecht
1981 Intestinal helminths of the bat, *Myotis keenii* (Merriam), for southeastern Wisconsin. Proc. Helminthol. Soc. Washington, 48(1):93–96.

Cope, J. B., and S. R. Humphrey
1972 Reproduction of the bats *Myotis keenii* and *Pipistrellus subflavus* in Indiana. Bat. Res. News, 13:9.

Doutt, J. K., C. A. Heppenstall, and J. E. Guilday
1966 Mammals of Pennsylvania. Game Comm., Harrisburg. 273 pp.

Easterla, D. A.
1968 Parturition of Keen's myotis in southwestern Missouri. Jour. Mammalogy, 49(4):770.

Fitch, J. H., and K. A. Shump, Jr.
1979 Myotis keenii. American Soc. Mammalogists, Mammalian Species, No. 121:1–3.

Goehring, H. H.
1954 *Pipistrellus subflavus obscurus, Myotis keeni*, and *Eptesicus fuscus fuscus* hibernating in a storm sewer in central Minnesota. Jour. Mammalogy, 35(3):434–435.

Griffin, D. R.
1940 Notes on the life histories of New England cave bats. Jour. Mammalogy, 21(2):181–187.

Hall, J. S., R. J. Cloutier, and D. R. Driffin
1957 Longevity records and notes on tooth wear of bats. Jour. Mammalogy, 38(3):407–409.

Hitchcock, H. B.
1949 Hibernation of bats in southeastern Ontario and adjacent Quebec. Canadian Field-Nat., 63:47–59.

Jackson, H. H. T.
1961 Mammals of Wisconsin. Univ. Wisconsin Press, Madison. xii+504 pp.

Jones, J. K., Jr., and H. H. Genoways
1967 Annotated checklist of bats of South Dakota. Trans. Kansas Acad. Sci., 70:184–196.

Kunz, T. H.
1971 Reproduction of some vespertilionid bats in central Iowa. American Midl. Nat., 86:477–486.

Kurta, A.
1979 Bat rabies in Michigan. Michigan Acad., 12(2):221–230.
1980a The bats of southern lower Michigan. Michigan State Univ., unpubl. M.S. thesis, ix+147 pp.
1980b Status of the Indiana bat, *Myotis sodalis*, in Michigan. Michigan Acad., 13(1):31–36.
1980c Notes on summer bat activity at Michigan caves. Natl. Speleological Soc., 42(4):68–69.

LaVal, R. K., R. L. Clawson, M. L. LaVal, and W. Claire
1977 Foraging behaviour and nocturnal activity patterns of Missouri bats with emphasis on the endangered species *Myotis grisescens* and *Myotis sodalis*. Jour. Mammalogy, 58(3):592–599.

LaVal, R. K., and M. L. LaVal
1980 Ecological studies and management of Missouri bats, with emphasis on cave-dwelling species. Missouri Dept. Conserv., Terrestral. Ser. No. 8, 53 pp.

Long, C. A.
1978 Mammals of the islands of Green Bay, Lake Michigan. Jack-Pine Warbler, 56(2):59–83.

Mumford, R. E.
1969 Distribution of the mammals in Indiana. Indiana Acad. Sci., Monog. No. 1, vii+114 pp.

Mumford, R. E., and J. E. Cope
1964 Distribution and status of the Chiroptera of Indiana. American Midl. Nat., 72(2):473–489.

Stones, R. C., and L. P. Branick
1968 Use of hearing in homing by two species of Myotis bats. Jour. Mammalogy, 50:157–160.

Stones, R. C., and W. Fritz
1969 Bat studies in upper Michigan's copper mining district. Michigan Acad., 2(1):77–85.

Swanson, G., and C. Evans.
1936 The hibernation of certain bats in southern Minnesota. Jour. Mammalogy, 17(1):39–43.

Timm, R. M.
1975 Distribution, natural history and parasites of mammals of Cook County, Minnesota. Univ. Minnesota, Bell Mus. Nat. Hist., Occas Papers 14, 56 pp.

Turner, R. W.
1974 Mammals of the Black Hills of South Dakota and Wyoming. Univ. Kansas Publ., Mus. Nat. Hist., Misc. Publ. No. 60, 178 pp.

Van Zyll de Jong, C. G.
1979 Distribution and systematic relationships of long-eared *Myotis* in western Canada. Canadian Jour. Zool., 57:987–994.

Whitaker, J. O., Jr., and R. E. Mumford
1971 Notes on a collection of bats taken by mist-netting at an Indiana cave. American Midl. Nat., 85(1):277–279.

SMALL-FOOTED BAT
Myotis leibii (Audubon and Bachman)

NAMES. The generic name *Myotis,* proposed by Kaup in 1829, is from the Greek and means "mouse ear." The specific name *leibii,* proposed by John J. Audubon and John Bachman in 1842, refers to physician George C. Leib who, in Erie County, Ohio, collected the specimen used in the original description. The subspecies, *Myotis leibii leibii* (Audubon and Bachman), occurs in the Upper Great Lakes Region (Glass and Baker, 1965).

RECOGNITION. Body small, the smallest of the myotis bats in the area, head and body less than 1¾ in (45 mm) long; nose pointed; eyes small and beady; ears short, extending when laid (not pulled) forward no more than 1/32 in (1 mm) beyond the tip of the nose; tragus in ear long and straight (see Table 14); forearm short; third metacarpal short, measuring less than length of forearm; hind foot short; calcar with keel-shaped extension (see Fig. 18D).

Color yellowish-brown; darker above with individual hairs having shiny tips giving a burnished appearance; lighter and more buffy below; bases of hairs black except on sides; face from nose to base of ears and including lower lip black, presenting a pronounced masked appearance (a most distinctive characteristic); wings and membranes brownish-black. The small size, yellowish-brown coloring, and black facial mask are distinguishing characters (see Table 14).

MEASUREMENTS AND WEIGHTS. Adult small-footed bats have the following dimensions: length from tip of nose to end of tail vertebrae, 2⅞ to 3 3/16 in (73 to 82 mm); length of tail vertebrae, 1 3/16 to 1⅜ in (30 to 35 mm); length of hind foot, ¼ in (6.5 to 7.0 mm); height of ear from notch, ½ to 9/16 in (12 to 15 mm); length of bone of forearm, 1 3/16 to 1¼ in (30 to 32 mm); weight, 0.12 to 0.28 oz (2 to 8 g).

DISTINCTIVE CRANIAL AND DENTAL CHARACTERISTICS. The skull of the small-footed bat is small and delicate. In profile, it has a distinctly flattened look, gradually rising posteriorly. However, for accurate identification it is almost necessary to visit a museum with a collection of mammals in order to visually compare the skulls of the several myotis bats in the Upper Great Lakes Region. The skull of the small-footed bat is approximately ½ in (13 to 14 mm) long and 5/16 in (8.0 to 8.6 mm) wide (across zygomatic arches). The dental formula is:

$$\text{I (incisors)}\frac{2\text{-}2}{3\text{-}3}, \text{C (canines)}\frac{1\text{-}1}{1\text{-}1}, \text{P (premolars)}\frac{3\text{-}3}{3\text{-}3}, \text{M (molars)}\frac{3\text{-}3}{3\text{-}3} = 38$$

DISTRIBUTION. The small-footed bat occurs in a broad belt across northern United States and southern Canada (Barbour and Davis, 1969). It is not a common species except in the "cave country" in the eastern part of its range. This bat has not been reported in Michigan, but it is included herein because localities in Ontario and Ohio from which the species are known are very close to the Michigan border. Collection sites closest to Michigan in Ontario are: Agawa Bay (Peterson, 1966) and Alona Bay Mine (Fenton, 1972), both on the east shore of Lake Superior about 50 mi (80 km) north and slightly west of Sault Ste. Marie, and Mount Brydges near London (Anderson, 1946), about 69 mi (110 km) east of Port Huron; in Ohio: Erie County (Audubon and Bachman, 1842), about 62 mi (100 km) east-southeast of the Michigan-Ohio line directly north of Toledo. The species is unreported in other states adjacent to Michigan.

HABITAT PREFERENCES. In winter, the small-footed bat inhabits cool, dry caves, fissures, and mine tunnels (Hitchcock, 1949). In summer, the species usually moves to less insulated day and night retreats, such as outbuildings, cottages, behind sliding barn doors, and under loose tree bark (Barbour and Davis, 1969; Hitchcock, 1955; Jones, 1964; Jones *et al.*, 1967). This diminutive black-masked bat should be looked for in any of these roosts, especially in the eastern part of the Upper Peninsula.

DENSITY AND MOVEMENTS. The small-footed bat concentrates in large numbers or hangs alone in winter hibernacula; Hitchcock (1949) counted a record 434 individuals in a cave in Renfrew County, Ontario. These colonies divide and disperse in the warm months to other shelters. Females either singly or in aggregations of 12 to 20 individuals use summer roosts (Barbour and Davis, 1969; Tuttle and Heaney, 1974). Males usually keep to themselves in summer with little known about their movements or gregarious nature. The presence of single individuals, perhaps solitary males, has been noted in summer retreats on several occasions.

BEHAVIOR. Among myotis bats, the small-footed bat is often the last species to enter winter quarters (as late as December, Mohr, 1936) and one of the first to depart (in March). Movements between hibernating sites during the winter months are also reported (Peterson, 1966), demonstrating this species' hardy nature (Banfield, 1974). When awake, this tiny myotis bat seems alive with boundless nervous energy. Despite hibernating in narrow crevices in cave ceilings or in tightly-packed clusters, each bat tends to hold its wings characteristically partly outstretched (Paradiso, 1969), while other myotis bats usually fold their wings parallel to their bodies. In winter quarters the small-footed bat often hangs in the coldest, driest, and draftiest situations. This bat uses echolocation to seek food and avoid objects while in flight.

ASSOCIATES. Little is known about the small-footed bat's association with other species in summer shelters or at feeding sites. In winter hibernating quarters, however, this bat is known to use the same caves and mine tunnels as the little brown bat, Keen's bat, eastern pipistrelle, and the big brown bat (Fenton, 1972).

REPRODUCTIVE ACTIVITIES. The breeding habits of the small-footed bat are very similar to those of other myotis bats. It is presumed that there is the customary mating in autumn prior to entering hibernation with the process of delayed fertilization allowing for early spring conception and single births in late spring (as early as May). Pregnant individuals have been observed as early as April 22 (Cockrum, 1955) and as late as July 12 (Quay, 1948). Details of growth for this species are lacking. Like other myotis, its low productivity is counterbalanced by its ability to live ten years or more (Paradiso and Greenhall, 1967).

FOOD HABITS. In summer, the small-footed bat ingests large quantities of flying insects to sustain itself on a daily basis as well as store food in the form of fat for the long period of winter torpor. Dalquest (1948) found small-footed bats in Washington so crammed with insects that their stomachs were greatly distended. Cockrum (1952) found identifiable remains in bat stomachs of flies (Diptera), bugs (Homoptera and Hemiptera), small beetles (Coleoptera) and ants (Hymenoptera). Individual weight losses between dawn (after a night of feeding) and dusk (after a daytime of fasting in myotis bats, including the small-footed bat, range from 6.4% to 21.9% of initial body weight (Studier *et al.*, 1970).

ENEMIES. There are no published records of predation on this species. Terrestrial carnivores with excellent climbing ability might capture the bats as they hang in torpor in shelters. It is also highly probable that bats fall victim to owls and hawks.

MOLT AND COLOR ABBERATIONS. Specimens taken in June and July in South Dakota were reported to be molting (Jones and Genoways, 1967). Pelage changes and color mutations in this species are poorly understood.

PARASITES AND DISEASES. The small-footed bat is infested by both internal worms and external parasites, including chiggers and bed bugs (Turner, 1974).

ECONOMIC IMPORTANCE. This smallest of bats in the Upper Great Lakes Region is rare and has been observed primarily in winter hibernating quarters. If captured, its tiny jaws are not sufficiently strong to break the skin of the human hand. When the small-footed bat becomes a summer resident of attics and double walls of cottages, its accumulated feces (guano) can give off a musty odor.

LITERATURE CITED

Anderson, R. M.
1946 Catalogue of Canadian recent mammals. Nat. Mus. Canada, Bull. Biol. Ser. No. 31, v+238 pp.

Audubon, J. J., and J. Bachman
1842 Descriptions of new species of quadrupeds inhabiting North America. Jour. Acad. Nat. Sci. Philadelphia, Ser. 1, Vol. 8, pp. 280–323.

Banfield, A. W. F.
1974 The mammals of Canada. Univ. Toronto Press, Toronto. xxiv+438 pp.

Barbour, R. W., and W. H. Davis
1969 Bats of America. The Univ. Kentucky Press, Lexington. 286 pp.

Cockrum, E. L.
1952 Mammals of Kansas. Univ. Kansas Publ., Mus. Nat. Hist., 7:1–303.
1955 Reproduction in North American bats. Trans. Kansas Acad. Sci., 58(4):487–511.

Dalquest, W. W.
1948 Mammals of Washington. Univ. Kansas Publ., Mus. Nat. Hist., 2:1–444.

Fenton, M. B.
1972 Distribution and overwintering of *Myotis leibii* and *Eptesicus fuscus* (Chiroptera: Vespertilionidae) in Ontario. Royal Ontario Mus., Life Sci., Occas. Papers No. 21, 8 pp.

Glass, B. P., and R. J. Baker
1965 *Vespertilio subulatus* Say, 1823: Proposed suppression under the plenary powers (Mammalia, Chiroptera). Bull. Zool. Nomenclature, 22:204–205.

Hitchcock, H. B.
1949 Hibernation of bats in southeastern Ontario and adjacent Quebec. Canadian Field-Nat., 63:47–59.
1955 A summer colony of the least bat, *Myotis subulatus leibii* (Audubon and Bachman). Canadian Field-Nat., 69:31.

Jones, J. K., Jr.
1964 Distribution and taxonomy of mammals of Nebraska. Univ. Kansas Publ., Mus. Nat. Hist., 16:1–356.

Jones, J. K., Jr., E. D. Fleharty, and P. B. Dunnigan
1967 The distributional status of bats in Kansas. Univ. Kansas Publ., Mus. Nat. Hist., Misc. Publ. 46, 33 pp.

Jones, J. K., Jr., and H. H. Genoways
1967 Annotated checklist of bats of South Dakota. Trans. Kansas Acad. Sci., 70:184–196.

Mohr, C. E.
1936 Notes on the least bat, *Myotis subulatus leibii.* Proc. Pennsylvania Acad. Sci., 10:62–65.

Paradiso, J. L.
1969 Mammals of Maryland. U.S.D.I., Bur. Sport Fisheries and Wildlife, N. American Fauna No. 66, iv+193 pp.

Paradiso, J. L., and A. M. Greenhall
1967 Longevity records for American bats. American Midl. Nat., 78(1):251–252.

Peterson, R. L.
1966 The mammals of eastern Canada. Oxford Univ. Press, Toronto. xxxii+463 pp.

Quay, W. B.
1948 Notes on some bats from Nebraska and Wyoming. Jour. Mammalogy., 29:181–182.

Studier, E. H., J. W. Procter, and D. J. Howell
1970 Diurnal body weight loss and tolerance of weight loss in five species of *Myotis.* Jour. Mammalogy, 51(2): 302–309.

Turner, R. W.
1974 Mammals of the Black Hills of South Dakota and Wyoming. Univ. Kansas Publ., Mus. Nat. Hist., Misc. Publ. No. 60, 178 pp.

Tuttle, M. D., and L. R. Heaney
1974 Maternity habits of *Myotis leibii* in South Dakota. Southern California Acad. Sci., Bull., 73(2):80–83.

LITTLE BROWN BAT
Myotis lucifugus (Le Conte)

NAMES. The generic name *Myotis,* proposed by Kaup in 1829, is from the Greek and means "mouse ear." The specific name *lucifugus,* proposed by Le Conte in 1831, is a Latin word meaning "avoiding" or "fleeing light." In Michigan, the name bat is generally applies to all chiropterans by local observers. Mammalogists prefer the name little brown myotis (Jones *et al.,* 1979). The subspecies, *Myotis lucifugus lucifugus* (Le Conte), occurs in both peninsulas of Michigan.

RECOGNITION. Body small, and size of a short, small mouse; head and body approximately 2 in (51 mm) long; nose pointed; eyes small and beady;

ears short, extending when laid (not pulled) forward no more than 1/32 in (1 mm) beyond the tip of the nose; tragus in ear opening long, slightly rounded and asymmetrical; forearm medium in length; hind feet short; hairs on terminal toe joints long, extending beyond claws (Fig. 18C); calcar rarely with keel-shaped extension (see Fig. 19C).

Hair color glossy olive brown on upperparts, sometimes more red or yellow; darker on shoulders; buff colored on underparts; ears and wing membranes dark brown or black. The short ears, rounded and asymmetrical tragus, lack of a keel on the calcar, and olive brown back aid in distinguishing the little brown bat from its close associate in Michigan, Keen's bat.

MEASUREMENTS AND WEIGHTS. Adult little brown bats have the following dimensions: length from tip of nose to end of tail vertebrae, 3⅛ to 3¾ in (80 to 95 mm); length of tail vertebrae, 1 3/16 to 1¾ in (31 to 45 mm); length of hind foot, 5/16 to 7/16 in (8 to 11 mm); height of ear from notch, ½ to ⅝ in (13 to 16 mm); length of bone of forearm, 1 7/16 to 1 9/16 in (36 to 40 mm); weight, 0.2 to 0.4 oz (6 to 12 g).

DISTINCTIVE CRANIAL AND DENTAL CHARACTERISTICS. The skull of the little brown bat is small and delicate (see Fig. 19). However, the only way to positively determine the species of a Michigan myotis using the skull alone is to compare it with skulls of identified bats in a museum collection. In comparison with the skulls of other Michigan myotis bats, the skull is broad with an interorbital width of more than ⅛ in (4 mm). The skull of the little brown bat is approximately 9/16 in (14.7 mm) long and 5/16 in (8.9 mm) wide (across zygomatic arches). The 38 minute teeth are difficult to count, unless magnified. The dental formula is:

$$\text{I (incisors)}\ \frac{2\text{-}2}{3\text{-}3},\ \text{C (canines)}\ \frac{1\text{-}1}{1\text{-}1},\ \text{P (premolars)}\ \frac{3\text{-}3}{3\text{-}3},\ \text{M (molars)}\ \frac{3\text{-}3}{3\text{-}3} = 38$$

DISTRIBUTION. The little brown bat is characteristic of forested areas of central and northern North America, in the east from just north of the Gulf States north to tree line in eastern Canada and westward over a broad region from central Alaska south to central California and into northern Mexico in the mountains. In Michigan, the little brown bat may be found in most parts of the state but occurs most regularly in the Upper Peninsula (see Map 14).

Map 14. Geographic distribution of the little brown bat (*Myotis lucifugus*). Actual records exist for counties with dark shading.

HABITAT PREFERENCES. The little brown bat lives most commonly in forested areas which are in close association with winter roosting quarters in natural caverns, fissures in rocky outcrops, or mine tunnels (Humphrey and Cope, 1976). These requirements are in addition to the variety of less well insulated shelters used as summer retreats (Humphrey, 1975). The number of substantial winter quarters in Upper Peninsula mining districts accounts for a sustained population there of little brown bats, while the lack of such shelters in southern Michigan accounts for the spotty distribution of this species in that area. However, in both peninsulas the little brown bat has acceptable summer foraging areas at forest edges, over woodland, along bushy streams, and at borders of ponds and small cultivated fields.

Summer Shelters for Nursery Colonies—The females leave winter quarters in spring, to associate in small groups and move to summer roosts where they bear and nurse their offspring (Fenton and Barclay, 1980). These sites may be used year after year. Sometimes one or more adult males are found in a nursery colony, but usually they remain apart. A wide choice of harborages are selected for nursery colonies; they may be poorly insulated becoming warm at midday and cold at night, dark or shaded, often poorly ventilated, but usually near a stream or lake (Davis and Hitchcock, 1965). Nursery colonies have been found in Michigan in summer cottages in Houghton and Keweenaw counties (Stones and Fritz, 1969); in frame houses on Mackinac Island (Wesley Maurer, Jr., *in litt.*, February 17, 1974); in buildings on Garden Island (Phillips *et al.*, 1965) and in McCormick Experimental Forest (Robinson and Werner, 1975); behind cabin shutters, between the ridgepole and roofing in a garage, in attics, and in a space in a cabin door at Douglas Lake in Cheboygan County (Hitchcock, 1943); and in a 90 year old barn-like structure near Dowagiac in Cass County (Kurta, 1980b).

Summer Roosts for Males—While nursery colonies containing predominately adult females are fairly well studied, little is known about the whereabouts of males during this warm season period of sexual segregation. Often a solitary male will select one site as a daytime shelter and another to visit at night between feeding forays (Kunz, 1973). Males have been found in summer in various types of buildings (Humphrey and Cope, 1976), beneath a stone in the Huron Mountains in Marquette County (Manville, 1948), and in Bear Cave in Berrien County (Jack Mell, pers. comm.).

Winter Hibernation Quarters—Little brown bats of both sexes gather in autumn and move into the stabilized and moderate environments in natural caverns and mine tunnels. Other wintering sites such as tree hollows are occasionally acceptable (Connor, 1960; Manville, 1948), although their insulating qualities may be marginal for hibernation. While the little brown bat is an acknowledged year-around resident in northern Michigan (Stones and Fritz, 1969), there is little evidence that it winters in the southern part of the state.

DENSITY AND MOVEMENTS. Knowledge of densities of little brown bats is best obtained when these creatures are concentrated in wintering quarters. Since they use these same natural caverns and abandoned mine shafts year after year, annual population estimates can be made. In the Delaware Mine in Michigan's Copper Country on the Keweenaw Peninsula, for example, Stone and Fritz (1969) counted approximately 1,600 little brown bats in hibernation in each of two winters (1966–67 and 1967–68). To illustrate the magnitude of these congregations, Mumford and Cope (1964) counted 1,500 wintering little brown bats in Grotto Cave in Indiana; Griffin (1940) found 792 in a cave near Chester, Massachusetts; Hitchcock (1949) estimated a population of 3,800 in an abandoned corundum mine at Craigmont, Ontario; Fenton (1970) found about 10,000 of these bats in mine shafts in Craigmont and vicinity; Davis and Hitchcock (1964) banded 1,976 little brown bats in one cave in Ulster County, New York; Brenner (1974) noted 6,200 ± 1,500 bats in Laurel Caverns in Pennsylvania, and Barbour and Davis (1969) report a record wintering population of 300,000 in a single cave in Vermont. As yet, the total area used in summer by such enormous congregations of wintering bats is poorly understood.

In most cases, summer nurseries include groupings of far fewer individuals. Even so, when a Michigan family opens the cottage for summer occupancy in late June, the musty odors emerging from the wastes (urine and guano) in a nursery colony in the loosely constructed attic or double walls can be unpleasant. Most summer colonies range in size from perhaps 50 to 2,500 individuals with an average of 400 in Indiana (Mumford and Cope, 1964). A record of 30,000 individuals were counted in one maternity colony in southern Illinois (Cockrum, 1956). Stones and Fritz (1969) estimated that 1,000 individuals used a cottage at Bete Grise on Michigan's Keweenaw Peninsula in the summer of 1966. This maternity colony began to form on May 20, held maximum numbers by mid-June, and had disbanded by July 26. In southern Michigan, Kurta (1980b) counted 11 bats on June 23, 1978, and 13 on June 16, 1979 in a barn-like structure near Dowagiac, Cass County. This was the only summer roost located in the southern three tiers of Michigan counties in the course of Kurta's two-year study. However, he did find about 300 little brown bats in an old farmhouse in La Porte County, Indiana, only 10 mi (16 km) south of Michigan's Berrien County border.

In the western part of the Upper Peninsula,

distances between summer and winter refuge sites may be short. In southern Michigan, on the other hand, summer and winter shelters appear to be much more widely separated. The best data on movements come from studying the activities of tagged bats. Little monitoring of bat movements has been done in Michigan (Hitchcock, 1943; Miller, 1955; Mohrmann, 1943); much more has been done in adjacent Ontario (Fenton, 1970; Hitchcock, 1949) and Indiana (Humphrey and Cope, 1976). These authors recorded movements between shelters separated in distance from 6 to 283 mi (10 to 455 km).

The spotty summer population of the little brown bat in the southern part of Michigan's Lower Peninsula may be caused, in part, by this area's remoteness from adequate winter hibernating shelters. Except for an unusual winter record of these bats in buildings (in Washtenaw County, Wood, 1922) and in Bear Cave (in Berrien County, Kurta, 1980b), they apparently leave the area entirely. The evidence points to north-south direction movements to wintering sites in caves in southern Indiana or perhaps Kentucky. James Cope is cited by Kurta (1980b) as banding two male little brown bats on February 8, 1964, at Grotto Cave in Monroe County, Indiana (see also Humphrey and Cope, 1976). One was recovered seven months later (in August) near Rolling Prairie in La Porte County, Indiana, about 6 mi (10 km) from the Michigan border in Berrien County. The other was recovered eight months later (in September) near Bronson in Branch County, Michigan, an airline flight distance of about 220 mi (350 km) between summer and winter quarters.

There is also the possibility that some summer residents in southern Michigan might winter in mine shafts in the Upper Peninsula. The only evidence of this, however, is the movement of a single little brown bat reported by Miller (1955). He wrote that a bat banded at Douglas Lake in Cheboygan County on May 9 was recaptured at Windsor, Ontario, 23 days later on June 1. This movement may have been from a northern hibernaculum to a southern summering area. Airline distance between these two places of capture is about 240 mi (386 km).

The little brown bat flies at speeds of 6 to 16 mi (10 to 26 km) per hour (Banfield, 1974; Patterson and Hardin, 1969). Based on the recapture of banded individuals, Barbour and Davis (1969) found one bat in Kentucky flew more than 50 mi (85 km) in a single night and another covered a distance of 80 mi (130 km) in three nights. Little brown bats are also capable of homing after being displaced as far as 280 mi (450 km) from point of capture. In Michigan, a bat banded at Douglas Lake in Cheboygan County on August 21 and released near Petoskey in Emmet County returned a distance of about 16 mi (26 km) in two days. Stones and Branick (1969) found that homing may depend on the little brown bat's hearing and echolocation as much as on sight.

BEHAVIOR. The little brown bat is highly nervous, vocal (Barclay *et al.*, 1979), and active (Buchler, 1980). It is an agile and strong flyer, zigzagging about in the summer twilight and night. It spends daylight hours in shelters as previously described and periods between feeding flights in retreats different from those used in the daytime.

The little brown bat qualifies as a warm-blooded mammal; however, its body temperature can be induced to fluctuate from near freezing as low as 20° F (−6.5° C) to as high as 129° F (54° C) without serious effects (Barbour and Davis, 1969; Davis and Reite, 1967). This range of temperature tolerance may explain why summer nursery colonies survive in poorly insulated attics in Michigan cottages, where night temperatures may be at or near freezing and day temperatures hot. While the bats rest in daytime retreats and in night shelters between early and late foraging flights, they save energy by slowing down metabolically and becoming torpid. Stones and Oldenburg (1968) found torpid bats in Keweenaw County to have esophageal temperatures of 41.9° F (5.5° C) in July. When disturbed, these bats took 30 minutes to arouse sufficiently to take flight. However, in summer maternity colonies, mother bats may spend no more than 79% of their time in daylight retreats resting, being active as much as 5% of the time, grooming 14% and actually moving 1% (Burnet and August, 1981).

The metabolic slowdown which occurs during torpor compares with that which takes place during winter hibernation. Bats sleep soundly with reduced heartbeat, respiration, and body heat but can wake up occasionally to drink, to mate and even to change hanging positions within or between wintering quarters. The cool, damp, constant environment of underground shelters aids in the maintenance of hibernation, allowing torpid bats to slowly expend stored nutrients (fat) in order to survive the long winter when no insect food is available.

The little brown bat was used as the model for the original demonstrations by Griffin and Galambos (1940, 1941) that this bat could avoid obstacles in flight (and also detect and pursue flying insects) by means of echolocation. These authors showed that the little brown bat utters a rapid series of calls having vibrations well above the range detectible by the human ear. However, the keen hearing of the bat could receive echoes of these vibrations from nearby objects and thus make successful flight maneuvers. There is also evidence that these utterances are used as a means to communicate between individuals. The calls are sufficiently distinctive as to be recognizable to species (Fenton and Bell, 1981).

In late summer and early autumn, adult females and males and young-of-the-year congregate in swarming associations (Fenton, 1969). In September, mating and hibernation site-selection begin. Using mist nets and a bat trapping device (Tuttle, 1974), Kurta (1980b) caught and examined little brown bats during their swarming activities in September 1978 and 1979 at Bear Cave in Berrien County. Except for records from Bear Cave in October and December, Kurta (1980b) found the last autumn record for this species in the southern part of the Lower Peninsula on September 28 in Eaton County. The earliest spring record is for a male captured in Ingham County on April 12. As mentioned previously, the little brown bat is a year-around resident in northern Michigan (Stones and Fritz, 1969).

ASSOCIATES. On the Keweenaw Peninsula, Stones and Fritz (1969) found the little brown bat using the same mine shafts as Keen's bat, eastern pipistrelle, and the big brown bat. The eastern pipistrelle and the small-footed bat are also known to join this species in common wintering sites in other states (Fenton, 1970; Hitchcock, 1949). In September Kurta (1980b) found the little brown bat, Keen's bat and the red bat together, mostly at the entrance to Bear Cave in Berrien County. The former two species at least were involved in post-breeding swarming behavior. Summer nighttime roosts used between foraging flights can be occupied jointly by the little brown bat, Keen's bat, eastern pipistrelle, and big brown bat (Barbour and Davis, 1969).

Myotis bats and others often forage over the same stream. Along the Thornapple River near Vermontville in Eaton County, Kurta (1980b) snared the little brown bat, Keen's bat, Indiana bat, silver-haired bat, big brown bat, red bat and hoary bat in the same mist nets.

REPRODUCTIVE ACTIVITIES. The reproductive activities of the little brown bat, as well as other gregarious species, are easily appraised by examining individuals concentrated in winter hibernating quarters and in summer nursery shelters (Thomas *et al.,* 1979). Most detailed studies in the Upper Great Lakes have been made in Ontario and Indiana by James Cope, Brock Fenton, Harold Hitchcock, Stephen Humphrey, Russell Mumford and John Whitaker.

It is during swarming behavior in September that courtship and mating are initiated. At this time viable sperm may be stored in the uterine cavity of the female. Sporadic mating also continues in winter quarters right up to departure from hibernation (Humphrey and Cope, 1976). When the female leaves the winter shelter for a summer retreat, ovulation takes place, the stored sperm is available for fertilization and the development of the single young commences. Births in Michigan's Copper Country take place between June 5 and July 8 (Stones and Fritz, 1969), after a gestation period of 50 to 60 days. During delivery, the female hangs head up allowing the newly born young to be supported by the net-like positioning of the interfemoral membrane.

At birth, the young is flesh-colored, possesses erupted deciduous teeth, measures about 1⅞ in (48 mm) in total length, and weighs less than ⅛ oz (2.5 g). It opens its eyes and ears within a day or two. The young quickly becomes sufficiently active to climb up the mother's fur and suckle on one of the two pectoral teats, while clinging with hind feet and clawed thumbs. The young is left hanging in the nursery shelter while the parent forages.

Young bats grow rapidly. On June 16 and 23, Kurta (1980b) found non-volant young in a nursery colony in Cass County; in the Upper Peninsula, bats in a colony under observation by Stones and Fritz (1969) completed nursing activity between July 26 and August 19. At about 21 days of age, the young are able to fly; they attain near adult size in wing spread and body dimension in late summer. Females (but not males) may become sexually mature their first autumn and bear young at the age of one year. Life expectancy in little brown bats is surprisingly long, with individuals known to live in nature 24 to 30 years (Griffin and Hitchcock, 1965; Keen and Hitchcock, 1980).

FOOD HABITS. The little brown bat appears to pick and choose its flying insect prey using echolocation to track its targets (Buchler, 1976). The catch is usually made in the interfemoral membrane held pouch-like in the form of a net. A little brown bat may catch up to 500 insects per hour (Gould, 1959). However, there is evidence that moths can detect the utterances of bats and thereby escape predation (Fenton and Fullard, 1981). Young little brown bats may consume 1.8 g of insects nightly; pregnant females, 2.5 g; and lactating females, 3.7 g (Anthony and Kunz, 1977). This comparatively large intake is important since a weight-loss of as much as 16.2% may be experienced by these bats during the daytime fasting period (Studier *et al.*, 1970), with grooming behavior accounting for more than half of the energy expended (Burnett and August, 1981). The masticated materials found in bat stomachs and the minute insect parts in guano (feces) are often difficult to identify but do show that the diet is a combination of selectivity and variation. Items most often identified are flies (Diptera), mayflies (Ephemeroptera) and caddis flies (Trichoptera) along with some wasps (Hymenoptera), moths (Lepidoptera), beetles (Coeloptera), stoneflies (Plecoptera), and neuropterous insects (Anthony and Kunz, 1977; Barbour and Davis, 1969; Belwood and Fenton, 1976; Buchler, 1976; Freeman, 1981). In captivity, little brown bats will accept a diet of mealworms (Stones, 1965).

ENEMIES. Snakes, voles, raccoons, marten, mink and house cats are known to climb to roosting areas and catch hanging little brown bats (Barbour and Davis, 1969; Fenton, 1970). These bats also have been preyed on by the red-tailed hawk, sharp-skinned hawk, kestrel, merlin, and red-winged blackbird by day, and the barn owl and great-horned owl by night (Doutt *et al.*, 1966; Garber, 1977; Fenton, 1970; Jackson, 1961; Miller, 1962; Wallace, 1948). Little brown bats have died from cave flooding, cave contamination by such materials as pesticides, becoming entangled in vegetation, or being trapped in ceiling crevices (DeBlase *et al.*, 1965; Gillette and Kimbrough, 1970; Kuntz *et al.*, 1977; Kurta, 1970a; Sustare, 1977).

PARASITES AND DISEASES. Fleas, ticks, chiggers, and bedbugs infest the bodies of little brown bats (Benton and Scharoun, 1958; Timm, 1975; Whitaker and Mumford, 1971; Scharf and Stewart, 1980). Bats also harbor internal flatworms and tapeworms (Nickel and Hansen, 1967; Rausch, 1975). Little brown bats are susceptible to rabies, but the incidence is less than 1% (Fenton and Barclay, 1980). To date no rabid little brown bats have been reported in Michigan (Kurta, 1979).

MOLT AND COLOR ABERRATIONS. There is a recognizable summer molt (Turner, 1974). Occasionally little brown bats have blackish color patterns, pied coloring, and all white pelage (Hamilton, 1930; Trapido and Crowe, 1942; Turner, 1974; Walley, 1974).

ECONOMIC IMPORTANCE. Close contact between little brown bats and the human population rarely occurs in Michigan except when nursery colonies occupy houses, especially in northern areas. Odors from urine and feces deposited below roosts can permeate the entire habitation. For example, the musty smell of an attic nursery colony of bats could be detected in all rooms on the second floor of a hotel in Iosco County. In disposing of unwanted bats, great care should be taken not to allow them to bite. Fortunately, myotis bats have a weak bite and usually are incapable of penetrating human skin. The oil extracted from the little brown bat has been used as a medicinal treatment for rheumatic joints (Lyon, 1933).

COUNTY RECORDS FROM MICHIGAN. Reports of specimens preserved in museum collections, documented in the literature, or recorded by knowledgeable observers show that the little brown bat is known from the following counties in the Upper Peninsula: Baraga, Chippewa, Delta, Dickinson, Gogebic, Houghton, Iron, Keweenaw, Luce, Mackinac, Marquette, Menominee, Ontonagon and Schoolcraft; in the Lower Peninsula: Alpena, Antrim, Arenac, Berrien, Branch, Cass, Charlevoix, Cheboygan, Crawford, Eaton, Emmet,

Grand Traverse, Gratiot, Hillsdale, Ingham, Ionia, Iosco, Jackson, Kalamazoo, Kalkaska, Leelanau, Lenawee, Livingston, Macomb, Manistee, Mason, Missaukee, Newaygo, Oakland, Oceana, Ogemaw, Oscoda, Otsego, Presque Isle, St. Joseph, Sanilac, Washtenaw, Wayne, Wexford. The bat is also known on Poverty, Big Summer, Beaver, Garden, North Manitou, Trout, and St. Martin islands in Lake Michigan; on Isle Royale in Lake Superior; and on Mackinaw, Drummond, and Charity Islands in Lake Huron (Burt, 1946; Hatt *et al.,* 1948; Long, 1978; Manville, 1950; Ozoga and Phillips, 1964; Scharf and Jorae, 1980).

LITERATURE CITED

Anthony, E. L. P., and T. H. Kunz
1977 Feeding strategies of the little brown bat, *Myotis lucifugus,* in southern New Hampshire. Ecology, 58:775–786.

Banfield, A. W. F.
1974 The mammals of Canada. Univ. Toronto Press, Toronto. xxiv+438 pp.

Barbour, R. W., and W. H. Davis
1969 Bats of America. Univ. Kentucky Press, Lexington. 286 pp.

Barclay, R. M. R., M. B. Fenton, and D. W. Thomas
1979 Social behaviour of the little brown bat, *Myotis lucifugus.* II. Vocal communication. Behav. Ecol. Sociobiol., 6(2):137–146.

Belwood, J. J., and M. B. Fenton
1976 Variation in the diet of *Myotis lucifugus* (Chiroptera: Vespertilionidae). Canadian Jour. Zool., 54:1674–1678.

Benton, A. H., and J. Scharoun
1958 Notes on a breeding colony of *Myotis.* Jour. Mammalogy, 39(2):293–295.

Brenner, F. J.
1974 A five-year study of a hibernating colony of *Myotis lucifugus.* Ohio Jour. Sci., 74(4):239–244.

Buchler, E. R.
1976 Prey selection by *Myotis lucifugus* (Chiroptera: Vespertilionidae). American Nat., 110:619–628.
1980 The development of flight, foraging, and echolocation in the little brown bat *(Myotis lucifugus).* Behav. Ecol. Sociobiol., 6(3):211–218.

Burnett, C. D., and P. V. August
1981 Time and energy budgets for dayroosting in a maternity colony of *Myotis lucifugus.* Jour. Mammalogy, 62(4):758–766.

Burt, W. H.
1946 The mammals of Michigan, Univ. Michigan Press, Ann Arbor. xv+288 pp.

Cockrum, E. L.
1956 Homing, movements and longevity of bats. Jour. Mammalogy, 37(1):48–57.

Connor, P. F.
1960 The small mammals of Otsego and Schoharie counties, New York. New York State Mus. and Sci. Ser., Bull. No. 382, 84 pp.

Davis, W. H., and H. B. Hitchcock
1964 Notes on sex ratios of hibernating bats. Jour. Mammalogy, 45(3):475–476.
1965 Biology and migration of the bat, *Myotis lucifugus,* in New England. Jour. Mammalogy, 46(2):296–313.

Davis, W. H., and O. B. Reite
1967 Responses of bats from temperate regions to changes in ambient temperature. Biol. Bull., 132-320-328.

DeBlase, A. F., S. R. Humphrey, and K. S. Drury
1965 Cave flooding and mortality in bats in Wind Cave, Kentucky. Jour. Mammalogy, 46(1):96.

Doutt, J. K., C. A. Heppenstall, and J. E. Guilday
1966 Mammals of Pennsylvania. Pennsylvania Game Commission, Harrisburg. 273 pp.

Fenton, M. B.
1969 Summer activity of *Myotis lucifugus* (Chiroptera: Vespertilionidae) at hibernacula in Ontario and Quebec. Canadian Jour. Zool., 47(4):597–602.
1970 Population studies of *Myotis lucifugus* (Chiroptera: Vespertilionidae) in Ontario. Royal Ontario Mus., Life Sci. Contr., No. 77, 34 pp.

Fenton, M. B., and R. M. R. Barclay
1980 *Myotis lucifugus.* American Soc. Mammalogists, Mammalian Species, No. 142, 8 pp.

Fenton, M. B., and G. P. Bell
1981 Recognition of species of insectivorous bats by their echolocation calls. Jour. Mammalogy, 62(2):233–243.

Fenton, M. B., and J. H. Fullard
1981 Moth hearing and the feeding strategies of bats. American Sci., 69(3):266–275.

Freeman, P. W.
1981 Correspondence of food habits and morphology in insectivorous bats. Jour. Mammalogy, 62(1):166–173.

Garber, S. D.
1977 Bat predation by the American kestrel, *Falco sparverius* (Aves: Falconiformes). Bat Res. News, 18(4):37–38.

Gillette, D. D., and J. D. Kimbrough
1970 Chiropteran mortality. Pp. 262–283, *in* B. H. Slaughter and D. W. Walton, eds. About bats. Southern Methodist Univ. Press, Dallas. 339 pp.

Gould, E.
1959 Further studies on the feeding efficiency of bats. Jour. Mammalogy, 40(1):149–150.

Griffin, D. R.
1940 Notes on the life histories of New England bats. Jour. Mammalogy, 21(2):181–187.

Griffin, D. R., and R. Galambos
1940 Obstacle avoidance by flying bats. Anat. Rec., 78:95.
1941 The sensory basis of obstacle avoidance by flying bats. Jour. Exp. Zool., 86:481–506.

Griffin, D. R., and H. B. Hitchcock
1965 Probable 24-year longevity records for *Myotis lucifugus*. Jour. Mammalogy, 46(2):332.

Hamilton, W. J., Jr.
1930 Notes on the mammals of Breathitt County, Kentucky. Jour. Mammalogy, 11(3):306–311.

Hatt, R. T., J. Van Tyne, L. C. Stuart, C. H. Pope, and A. B. Grobman
1948 Island life: A study of the land vertebrates of the islands of eastern Lake Michigan. Cranbrook Inst. Sci., Bull. No. 27, xi+179 pp.

Hitchcock, H. B.
1943 Banding as an aid in studying the activities of the little brown bat, *Myotis lucifugus lucifugus*. Michigan Acad. Sci., Arts and Ltrs., Papers, 29:227–279.
1949 Hibernation of bats in southeastern Ontario and adjacent Quebec. Canadian Field-Nat., 69(1):31.

Humphrey, S. R.
1975 Nursery roosts and community diversity of Nearctic bats. Jour. Mammalogy, 56(2):321–346.

Humphrey, S. R., and J. B. Cope
1976 Population ecology of the little brown bat, *Myotis lucifugus*, in Indiana and north-central Kentucky. American Soc. Mammalogists, Sp. Publ. No. 4, vii+81 pp.

Jackson, H. H. T.
1961 Mammals of Wisconsin. Univ. Wisconsin Press, Madison. xii+504 pp.

Jones, J. K., Jr., D. C. Carter, and H. H. Genoways
1979 Revised checklist of North American mammals north of Mexico, 1979. Texas Tech Univ., Mus., Occas. Papers No. 62, 17 pp.

Keen, R., and H. B. Hitchcock
1980 Survival and longevity of the little brown bat *(Myotis lucifugus)* in southeastern Ontario. Jour. Mammalogy, 61(1):1–7.

Kunz, T. H.
1973 Resource utilization: Temporal and spatial components of bat activity in central Iowa. Jour. Mammalogy, 54(1):14–32.

Kunz, T. H., E. L. P. Anthony, W. T. Rumage III
1977 Mortality of little brown bats following multiple pesticide applications. Jour. Wildlife Mgt., 41(3):476–483.

Kurta, A.
1979 Bat rabies in Michigan. Michigan Acad., 12(2):221–230.
1980a. Accidental death in bats trapped in ceiling crevices. Jack-Pine Warbler, 58(1):34.
1980b. The bats of southern lower Michigan. Michigan State Univ., unpubl. M.S. thesis, ix+147 pp.

Leffler, J. W., L. T. Leffler, and J. S. Hall
1979 Effects of familiar area on the homing ability of the little brown bat, *Myotis lucifugus*. Jour. Mammalogy, 69(1):201–204.

Long, C. A.
1978 Mammals of the islands of Green Bay, Lake Michigan. Jack-Pine Warbler, 56(2):59–83.

Lyon, M. W., Jr.
1933 Bat oil for rheumatism. Jour. Mammalogy, 14:313.

Manville, R. M.
1948 The vertebrate fauna of the Huron Mountains, Michigan. American Midl. Nat., 39(3):615–640.
1950 The mammals of Drummond Island, Michigan. Jour. Mammalogy, 31(3):358–359.

Miller, J. S.
1955 A study of the roosting habits, and of the environmental factors concurrent with the time of evening flight, of little brown bats *(Myotis lucifugus)* in northern Michigan. Univ. Michigan, unpubl. Ph.D. dissertation, vii+75 pp.

Miller, D. H.
1962 Daytime attack on a bat by blackbirds. Jour. Mammalogy, 43(4):546.

Mohrmann, M. E.
1943 Some bat-banding experiences at the University of Michigan Biological Station. Univ. Michigan, unpubl. M.S. thesis, v+27 pp.

Mumford, R. E., and J. B. Cope
1964 Distribution and status of the Chiroptera of Indiana. American Midl. Nat., 72(2):473–489.

Nickel, P. A., and M. F. Hansen
1967 Helminths of bats collected in Kansas, Nebraska, and Oklahoma. American Midl. Nat., 78:481–486.

Ozoga, J. J., and C. J. Phillips
1964 Mammals of Beaver Island, Michigan. Michigan State Univ., Publ. Mus., Biol. Ser., 2(6):305–348.

Patterson, A. P., and J. W. Hardin
1969 Flight speeds of five species of vespertilionid bats. Jour. Mammalogy, 50(1):152–153.

Phillips, C. J., J. T. Ozoga, and L. C. Drew.
1965 The land vertebrates of Garden Island, Michigan. Jack-Pine Warbler, 43(1):20–24.

Rausch, R. L.
1975 Cestodes of the genus *Hymenolepis* Weinland, 1858 *(sensu lato)* from bats in North America and Hawaii. Canadian Jour. Sci., 53:1537–1551.

Robinson, W. L., and J. K. Werner
1975 Vertebrate animal populations of the McCormick Forest. USDA Forest Service, Res. Paper NC-118, 24 pp.

Scharf, W. C., and M. L. Jorae
1980 Birds and land vertebrates of North Manitou Island. Jack-Pine Warbler, 58(1):4–15.

Scharf, W. C., and K. R. Stewart
1980 New records of Siphonaptera from northern Michigan. Great Lakes Ento., 13(3):165–167.

Stones, R. C.
1965 Laboratory care of little brown bats at thermal neutrality. Jour. Mammalogy, 46(4):681–682.

Stones, R. C., and L. P. Branick
1969 Use of hearing in homing by two species of myotis bats. Jour. Mammalogy, 50(1):157–160.

Stones, R. C., and W. Fritz
1969 Bat studies in Upper Michigan's copper mining district, Michigan Acad., 2(1):77–85.

Stones, R. C., and T. Oldenburg
1968 Occurrence of torpid *Myotis lucifugus* in a cold mine in summer. Jour. Mammalogy, 49(1):123.

Stones, R. C., and J. E. Wiebers
1965 Seasonal changes in food consumption of little brown bats held in captivity at a "neutral" temperature of 92° F. Jour. Mammalogy, 46(1):18–22.

Studier, E. H., J. W. Procter, and D. J. Howell
1970 Diurnal body weight loss and tolerance of weight loss in five species of *Myotis*. Jour. Mammalogy, 51(2):302–309.

Sustare, D.
1977 News. Bat Research News, 18(2/3):1.

Thomas, D. W., M. B. Fenton, and R. M. R. Barclay
1979 Social behavior of the little brown bat, *Myotis lucifugus*. I. Mating behavior. Behav. Ecol. Sociobiol., 6(2):129–136.

Timm, R. M.
1975 Distribution, natural history and parasites of mammals of Cook County, Minnesota. Univ. Minnesota, Bell Mus. Nat. Hist., Occas. Papers 14, 56 pp.

Trapido, H., and P. E. Crowe
1942 Color abnormalities in three genera of northeastern cave bats. Jour. Mammalogy, 23(3):303–305.

Turner, R. W.
1974 Mammals of the Black Hills of South Dakota and Wyoming. Univ. Kansas Publ., Mus. Nat. Hist., Misc. Publ. No. 60, 178 pp.

Tuttle, M. D.
1974 An improved trap for bats. Jour. Mammalogy, 55(3): 475–477.

Wallace, G. J.
1948 The barn owl in Michigan. Its distribution, natural history and food habits. Michigan State Coll., Agri. Exp. Sta., Tech. Bull. 208, 61 pp.

Walley, H. D.
1974 Albino little brown bat, *Myotis lucifugus lucifugus*, new record from Wisconsin with remarks on other aberrant bats. Canadian Field Nat., 88(1):80–81.

Whitaker, J. O., Jr., and R. E. Mumford
1971 Notes on a collection of bats taken by mist-netting at an Indiana cave. American Midl. Nat., 85(1):277–279.

Wood, N. A.
1922 The mammals of Washtenaw County, Michigan. Univ. Michigan, Mus. Zool., Occas. Papers No. 123, 23 pp.

INDIANA BAT
Myotis sodalis Miller and Allen

NAMES. The generic name *Myotis,* proposed by Kaup in 1829, is of Greek origin and means "mouse ear." The specific name *sodalis,* proposed by Miller and Allen in 1928, is a Latin word which may be translated as "comrade" or "associate" and refers to the gregarious nature of this bat. In Michigan, the Indiana bat has no special common name; its general appearance is too similar to that of other small myotis bats in the state. This species is monotypic with no subspecies recognized.

RECOGNITION. Body small, having the dimensions of a small, short mouse; head and body approximately 1⅞ in (48 mm) long; nose pointed; eyes small and beady; ears short, extending when laid (not pulled) forward no more than 1/32 in (1 mm) beyond the tip of the nose; tragus in ear opening short and blunt; forearm medium in length; hind foot small; hairs on terminal toe joints short, not extending in length beyond claws (Fig. 18D); calcar with keel-shaped extension (see Fig. 18D).

Upper parts dull light to dark brown, individual hairs black basally, gray medially, and cinnamon-brown on tips, producing a distinctly tricolor pattern; underparts pale gray or buff; bare parts of nose, ears, and wing membranes dark brown to black. The short ears, rounded and blunt tragus, short hairs on terminal toe joints, pronounced keel on the calcar, and dull brown pelage aid in distinguishing the Indiana bat from other Michigan chiropterans.

MEASUREMENTS AND WEIGHTS. Adult Indiana bats have the following dimensions: length from tip of nose to end of tail vertebrae, 2⅞ to 3⅞ in (73 to 100 mm); length of tail vertebrae, 1 to 1¾ in (27 to 44 mm); length of hind foot, ¼ to ⅜ in (7 to 10 mm); height of ear from notch, 7/16 to ⅝ in (11 to 15 mm); length of bone of forearm, 1⅜ to 1⅝ in (36 to 41 mm); weight, 0.15 to 0.33 oz (4.5 to 9.5 g).

DISTINCTIVE CRANIAL AND DENTAL CHARACTERISTICS. The skull of the Indiana bat is small and delicate (see Fig. 19), differing only in fine details from that of the little brown bat. Comparing the skulls of the two species, that of the Indiana bat has a more obvious ridge along the midline (sagittal crest) and a braincase and interorbital space usually less than 7.0 mm and 4.0 mm in width, respectively; in the little brown bat, these are wider. Nevertheless, the only way of positively identifying the skulls of Michigan myotis bats is to make comparisons with identified museum specimens. The skull of the Indiana bat is about 9/16 in (13.7 to 15.2 mm) long and 5/16 in (8.3 to 9.3 mm) wide (across zygomatic arches). The minute teeth are difficult to distinguish without the aid of a magnifying glass. The dental formula is:

$$\text{I (incisors)}\frac{2\text{-}2}{3\text{-}3},\ \text{C (canines)}\frac{1\text{-}1}{1\text{-}1},\ \text{P (premolars)}\frac{3\text{-}3}{3\text{-}3},\ \text{M (molars)}\frac{3\text{-}3}{3\text{-}3} = 38$$

DISTRIBUTION. The Indiana bat is an inhabitant of the cave country of northeastern United States, from Maine southwestward to Oklahoma. In winter, these bats hibernate in large numbers in caverns, diminishing their range drastically compared with summer when small groups, especially of females, disperse over large areas to establish nursery colonies. It is during the warm season that Indiana bats have been found to occupy sites in nine counties in the three southernmost tiers of Michigan counties (see Map 15). Studies by Kurta (1980a, 1980b) in 1978 and 1979 showed that summer residency of this endangered species is more widespread than was previously supposed, with individuals recorded as early as May 17 (Eaton County) and as late as October 11 (Ingham County). He also established, by capturing lactating females and young juveniles in mist nets, that Indiana bats rear their young in Michigan. To date no adult males have been captured in Michigan. Banding studies show that the likely wintering areas for these summer residents are Kentucky caves (W. H. Davis *in* Kurta, 1980b).

HABITAT PREFERENCES. For at least six months of the year, November through April (Mumford and Cope, 1958), the Indiana bat hibernates in large concentrations in caverns. Within these protective, winter shelters the temperature varies little, usually between 39° and 43° F (4° to 6° C), and the humidity is high (Hall, 1962). In summer, females disperse in small groups to bear and rear their young in various kinds of less substantial retreats. The summer habitat of the bachelor males is little known.

Although Kurta (1980b) captured lactating females in mist nets in Michigan, he obtained no information on the sites of nursery colonies. In other states, observers have found females with young using summer shelters under loose tree bark, in hollow trees, under a bridge, and in buildings (Barbour and Davis, 1969; Cope *et al.*, 1974; Engel, 1976, Hall, 1962; Humphrey *et al.*, 1977; Mumford and Cope, 1958; Mumford *et al.*, n.d.). In Michigan, available evidence shows that slow-moving streams lined with trees are favored

Map 15. Geographic distribution of the Indiana bat (*Myotis sodalis*). Actual records exist for counties with dark shading.

foraging areas. Kurta (1980b) snared Indiana bats in mist nets stretched across Mill and Big Swan creeks in St. Joseph County, the Shiawassee River in Livingston County, the East Branch of the St. Joseph River in Hillsdale County, and the Thornapple River in Eaton County.

DENSITY AND MOVEMENTS. The Indiana bat was given endangered status by the U.S. Fish and Wildlife Service mainly because of its tendency to congregate in such enormous numbers in winter quarters, thus making it more vulnerable. Certain caves in Indiana, Kentucky, and Missouri, for example, have been known to harbor winter populations of as many as 100,000 individuals (Hall, 1962; Richter *et al.*, 1978). Wintering concentrations in other caves are smaller, 3,000 individuals or less. In late spring and early summer, on the other hand, nursery colonies are small, numbering no more than 50 (Cope *et al.*, 1974; Humphrey *et al.*, 1977). In southern Michigan, Kurta (1980a) caught a total of 16 Indiana bats in 1978 and 1979. These were taken in the course of 91 nights of mist-netting at 39 streams in fifteen southern Michigan counties and constituted about 4% of his total catch of bats.

BEHAVIOR. With other myotis bats, the Indiana bat shows marked seasonal behavioral differences. The winters are spent concentrated and inactive in well-insulated caves and mine tunnels. In summers the bats disperse and usually segregate according to sex in different and less stable shelters by day and forage in adjacent areas in the nighttime (Brenner, 1974; Henshaw and Folk, 1966; Humphrey *et al.*, 1977). Between feeding flights, the bats may use still other quarters for resting. Indiana bats become torpid while inactive in both winter and summer daytime retreats. The reduction in vital processes, as described in greater detail for the little brown bat, allows this small warm-blooded animal to ration its energy output between periods of activity (Hardin and Hassell, 1970). Huddling in clusters is another means by which the bats conserve body heat and thereby save energy.

Movements between summer and winter quarters can be extensive (Hall, 1962). Band returns indicate that Indiana bats may migrate as far as 312 mi (500 km) from Michigan in order to hibernate in Kentucky caves (Kurta, 1980a). The long-time attachment of these bats to their customary retreats is indicated by their ability to home from as far away as 200 mi (330 km) after being forceably displaced (Davis and Barbour, 1965; Barbour *et al.*, 1966). While on both migratory and foraging flights, Indiana bats, like little brown bats and Keen's bats, depend on echolocation to avoid obstacles and pursue flying insect prey.

ASSOCIATES. In winter quarters in such states as Indiana and Kentucky, the Indiana bat associates with hibernating little brown bats, Keen's bats, small-footed bats, eastern pipistrelles, and big brown bats. However, when several of these species hibernate in a single cave or mine shaft, each species may select different roosting places depending on their tolerance to temperature and humidity and also congregate in different arrangements of clusters (Hall, 1962). The Indiana bat does not appear to join other bats in summer daytime retreats. For example, the Indiana bat appears much less tolerant of high temperatures in roosts than the little brown bat (Brenner, 1974; Henshaw and Folk, 1966).

REPRODUCTIVE ACTIVITIES. Late summer swarming, including nightly forays into caves by large numbers of adults of both sexes and young-of-the-year, marks the beginning of courtship and mating (Cope and Humphrey, 1977; Mumford, 1969). In September and October, the bats begin to enter hibernating retreats, where some continued mating will occur even up until the bats depart in April (Hall, 1962). Ovulation and fertilization take place at or near departure time, with pregnant females, each carrying a single fetus, reported in Missouri and Indiana from May 20 to June 24 (Cope *et al.*, 1974; Easterla and Watkins, 1969; Hall, 1962). Although no pregnant females have been detected in Michigan, Kurta (1980a; 1980b) examined lactating females between June 28 and August 15, with flying young recorded as early as July 31. Like other myotis bats, the Indiana bat may survive for up to 13 years(Humphrey and Cope, 1977; Paradiso and Greenhall, 1967).

FOOD HABITS. The Indiana bat preys on flying insects which it pursues using echolocation and catches with the aid of the interfemoral membrane shaped in the form of a net. Identifying food eaten by the Indiana bat, from stomach contents or from guano accumulated beneath roosts, requires expert assistance from entomologists. Moths (Lepidoptera) may be preferred foods, along with small beetles (Coleoptera), small wasps (Hymenoptera), true bugs (Homoptera), flies (Diptera), caddis-flies (Trichoptera) and other insect groups (LaVal and LaVal, 1980; Whitaker, 1972).

ENEMIES. Although there are no reports, hawks and owls probably capture some Indiana bats. Meat-eaters capable of climbing to bat roosts also are potential predators. There is one reference in the literature of predation by a pilot black snake (Barr and Norton, 1965). A bat impaled on barbed wire was reported by DeBlase and Cope (1967).

PARASITES AND DISEASES. Indiana bats are known to be infested with mites externally and trematodes internally (Whitaker and Mumford, 1971).

MOLT AND COLOR ABERRATIONS. A prolonged molt in early summer allows for the replacement of old hair (Hall, 1962). Individual Indiana bats with white spots have been observed (Barbour and Davis, 1969; Metzger, 1965).

ECONOMIC IMPORTANCE. The Indiana bat has been declared an endangered species in accordance with the provisions of the Federal Endangered Species Act of 1973 (Public Law 93-205)—an appropriate action because these bats congregate in just a few caves in winter. Should sudden environmental changes take place in these shelters, this species would be highly vulnerable. Such changes could occur as a result of pesticides being introduced (either intentionally to kill the bats or inadvisedly by amateur cave explorers); opening the caves as winter tourist attractions; or using the caves for storage or mineral exploration. Natural catastrophes, such as flooding or cave-ins, also endanger these bats (DeBlase *et al.,* 1965). Although Michigan residents can have only minor influence on the status of this species as a summer resident in southern Michigan, conservation groups, private and public, should support bat conservation recommendations resulting from investigations in the cave country of eastern United States (Mumford *et al.,* n.d).

COUNTY RECORDS FROM MICHIGAN. The few records of the Indiana bat in Michigan indicate that the species is a summer resident in the three southern tiers of Michigan counties. Records are from the following nine counties: Barry, Calhoun, Eaton, Hillsdale, Ingham, Livingston, St. Joseph, Washtenaw and Wayne (see Map 15).

LITERATURE CITED

Barbour, R. W., and W. H. Davis
1969 Bats of America. Univ. Kentucky Press, Lexington. 286 pp.

Barbour, R. W., W. H. Davis, and M. D. Hassell
1966 The need for vision in homing by *Myotis sodalis.* Jour. Mammalogy, 47:356–357.

Barr, T. C., Jr., and R. M. Norton
1965 Predation on cave bats by the pilot black snake. Jour. Mammalogy, 46(4):672.

Brenner, F. J.
1974 Body temperature and arousal rates of two species of bats. Ohio Jour. Sci., 74(5):296–300.

Burt, W. H.
1946 The mammals of Michigan. Univ. Michigan Press, Ann Arbor. xv+288 pp.

Cope, J. B., and S. R. Humphrey
1977 Spring and autumn swarming behavior in the Indiana bat, *Myotis sodalis.* Jour. Mammalogy, 58(1):93–95.

Cope, J. B., A. R. Richter, and R. S. Mills
1974 A summer concentration of the Indiana bat, *Myotis sodalis,* in Wayne County, Indiana. Proc. Indiana Acad. Sci., 83:482–484.

Davis, W. H., and R. W. Barbour
1965 The use of vision in flight by the bat, *Myotis sodalis.* American Midl. Nat., 74:497–499.

DeBlase, A. F., and J. B. Cope
1967 An Indiana bat impaled on barbed wire. American Midl. Nat., 77(1):238.

DeBlase, A. F., S. R. Humphrey, and K. S. Drury
1965 Cave flooding and mortality in bats in Wind Cave, Kentucky. Jour. Mammalogy, 46(1):96.

Easterla, D. A., and L. C. Watkins
1969 Pregnant *Myotis sodalis* in northwestern Missouri. Jour. Mammalogy, 50(2):372–373.

Engel, J. M.
1976 The Indiana bat, *Myotis sodalis.* A bibliography. U.S. Fish and Wildlife Ser., Sp. Sci. Rpt., Wildlife No. 196, ii+11 pp.

Fenton, M. B.
1969 Summer activity of *Myotis lucifugus* (Chiroptera: Vespertilionidae) at hibernacula in Ontario and Quebec. Canadian Jour. Zool., 47(4):597–602.

Hall, J. S.
1962 A life history and taxonomic study of the Indiana bat, *Myotis sodalis.* Reading Pub. Mus. and Art Gallery, Sci. Publ. No. 12, 68 pp.

Hardin, J. W., and M. D. Hassell
1970 Observation on waking periods and movements of *Myotis sodalis* during hibernation. Jour. Mammalogy, 51(4):829–831.

Hassell, M. D.
1963 A study of homing in the Indiana bat, *Myotis sodalis*. Trans. Kentucky Acad. Sci., 24:1–4.

Hassell, M. D., and M. J. Harvey
1965 Differential homing in *Myotis sodalis*. American Midl. Nat., 74:501–503.

Henshaw, R. E., and G. E. Folk, Jr.
1966 Relation of thermoregulation to seasonally changing micro-climate in two species of bats (*Myotis lucifugus* and *M. sodalis*). Physiol. Zool., 39:223–236.

Humphrey, S. R., and J. B. Cope
1977 Survival rates of the endangered Indiana bat, *Myotis sodalis*. Jour. Mammalogy, 58(1):32–36.

Humphrey, S. R., and J. B. Cope
1977 Survival rates of the endangered Indiana bat, *Myotis sodalis*. Jour. Mammalogy, 58(1):32–36.

Humphrey, S. R., A. R. Richter and J. B. Cope
1977 Summer habitat and ecology of the endangered Indiana bat, *Myotis sodalis*. Jour. Mammalogy, 58(3): 334–346.

Kurta, A.
1980a The bats of southern lower Michigan. Michigan State Univ., unpubl. M.S. thesis ix+147 pp.
1980b Status of the Indiana bat, *Myotis sodalis*, in Michigan. Michigan Acad., 13(1):31–36.

LaVal, R. K., and M. L. LaVal
1980 Ecological studies and management of Missouri bats, with emphasis on cave-dwelling species. Missouri Dept. Conserv., Terrestrial Ser. No. 8, 53 pp.

Metzger, B.
1956 Partial albinism in *Myotis sodalis*. Jour. Mammalogy, 37(4):546.

Mumford, R. E.
1969 Distribution of the mammals of Indiana. Indiana Acad. Sci., Monog. No. 1, vii+114 pp.

Mumford, R. E., and J. B. Cope
1958 Summer records of *Myotis sodalis* in Indiana. Jour. Mammalogy, 39(4):586–587.

Mumford, R. E., J. S. Hall, and J. O. Whitaker, Jr.
n.d. Distributional studies of the Indiana bat (*Myotis sodalis*) on three national forests in Region Three. U.S. Forest Ser., Region Three, vi+64 pp.

Mumford, R. E., and J. O. Whitaker, Jr.
1975 Seasonal activity of bats at an Indiana cave. Proc. Indiana Acad. Sci., 84:500–507.

Paradiso, J. L. and A. M. Greenhall
1967 Longevity records for American bats. American Midl. Nat., 78(1):251–252.

Richter, A. R., D. A. Seerley, J. B. Cope, and J. H. Keith
1978 A newly discovered concentration of hibernating Indiana bats, *Myotis sodalis*, in southern Indiana. Jour. Mammalogy, 59(1):191.

Whitaker, J. O., Jr.
1972 Food habits of bats from Indiana. Canadian Jour. Zool., 50:877–883.

Whitaker, J. O., Jr., and R. E. Mumford
1971 Notes on a collection of bats taken by mist-netting at an Indiana cave. American Midl. Nat., 85(1):277–279.

SILVER-HAIRED BAT
Lasionycteris noctivagans (Le Conte)

NAMES. The generic name *Lasionycteris*, proposed by Peters in 1866, is of Greek origin and literally means "shaggy bat" with *nycteris* referring to "bat" and *lasio* meaning "shaggy." The specific name *noctivagans*, proposed by Le Conte in 1831, is derived from two Latin words, the stem *nocti* referring to the "night" and *vagans* meaning "wander." This is truly a shaggy bat that wanders around at night. Although the common name, silver-haired bat, is used generally by both professional and amateur field biologists, this distinctive, dark species is also called black bat, silvery bat, and silver-black bat. The silver-haired bat belongs to the monotypic species, *Lasionycteris noctivagans*. This means that geographic variation is insufficiently expressed for proposing subspecific names.

RECOGNITION. Body medium, the size of a small, short mouse; head and body approximately 2⅜ in (60 mm) long; wingspread about 12 in (305 mm); nose pointed; eyes small and beady; ears short, broad and rounded; tragus in ear opening short, broad and symmetrical; tail short, less than 40% of total length; first joints (metacarpals) of flight digits of the third, fourth and fifth fingers almost the same length; calcar without keel-shaped extension (see Fig. 18C).

Long, soft hair sooty brown to nearly black and except on head and neck washed with silver producing a frosted appearance; basal half (sometimes more) of interfemoral membrane (connecting tail with two hind limbs) lightly clothed with white-tipped, black hair; ears black; wing membranes dark brown or black. The frosted black fur, small size, and lack of a keel on the calcar all aid in distinguishing this species from other Michigan bats.

MEASUREMENTS AND WEIGHTS. Adults measure from tip of nose to end of tail vertebrae, 3½ to 4½ in (90 to 115 mm); length of tail vertebrae, 1⅜ to 2 in (35 to 50 mm); length of hind foot, 5/16 to 7/16 in (6 to 12 mm); height of ear from notch, ⅝ in (15 to 17 mm); length of bone of forearm, 1 7/16 to 1⅝ in (37 to 42 mm); weight, 0.18 to 0.35 oz (6 to 12 g).

DISTINCTIVE CRANIAL AND DENTAL CHARACTERISTICS. The skull of the silver-haired bat is rather large (compared with those of myotis bats), broad and flattened (see Fig. 19). In profile, its flatness is shown by the very gradual elevation from the front tip of the nasal bones to the hindmost part of the braincase. The noticeably broad rostrum tapers to the front, being concave in shape on each side and back of the nasal openings. The skull is about ⅝ in (16 to 17 mm) long and ⅜ in (9.5 to 10.3 mm) wide (across zygomatic arches). The silver-haired bat is unique among Michigan bats in having 36 teeth. The dental formula is:

$$\text{I (incisors)}\frac{2\text{-}2}{3\text{-}3}, \text{C (canines)}\frac{1\text{-}1}{1\text{-}1}, \text{P (premolars)}\frac{2\text{-}2}{3\text{-}3}, \text{M (molars)}\frac{3\text{-}3}{3\text{-}3} = 36$$

DISTRIBUTION. The silver-haired bat occurs in a broad band across northern North America, from tree line in the north, southward to central California, the southern Rocky Mountain states, and eastward just north of the Gulf coast (Map 16). In Michigan, this bat is a summer resident in both peninsulas; records in winter are exceptional.

HABITAT PREFERENCES. Many Michigan anglers may be surprised to learn that the dark bats which may attack their trout flies in late afternoon are very likely to be of this species. Indeed, foraging along wooded streams is a major activity of the silver-haired bat in the summer twilight and night (Kurta, 1980; Manville, 1942; Scharf, 1973; Wood, 1914, 1922). Its attraction to both the cold water streams of northern Michigan and the warmer ones in southern Michigan is thought to be emerging aquatic insects.

In summer the silver-haired bat is chiefly solitary and, as a result, is more difficult to study than Michigan bats which congregate in colonies. After nightly foraging, this bat spends the daytime in seclusion in such retreats as tree hollows, bulky bird nests, under loose bark, in hollow snags, sometimes in cabins and other buildings, in woodpecker holes, and even under piles of brush (Banfield, 1974; Barbour and Davis, 1969; Stones and Holyoak, 1970; Timm, 1975).

In winter, silver-haired bats usually join in groups containing both sexes and migrate to hibernation sites in more southern parts of their range. The bats have been found in cold weather retreats such as under loose bark, in tree holes, in caves, in cracks in sandstone ledges, in various kinds of buildings, in mines, in the folds of a bathroom shower curtain, and in the access passage of a root cellar (Fowler, 1959; Frum, 1953; Gosling, 1977; Hamilton, 1943; Izor, 1979; Jackson, 1961; Krutzsch, 1966; Mumford, 1969; Pearson, 1962; Turner, 1974).

DENSITY AND MOVEMENTS. Kurta (1980) regarded the silver-haired bat as a rather uncommon summer resident in southern Michigan. Although there have been occasional reports of small aggregations of nursing females (Barbour and Davis, 1969), the adults are generally dispersed

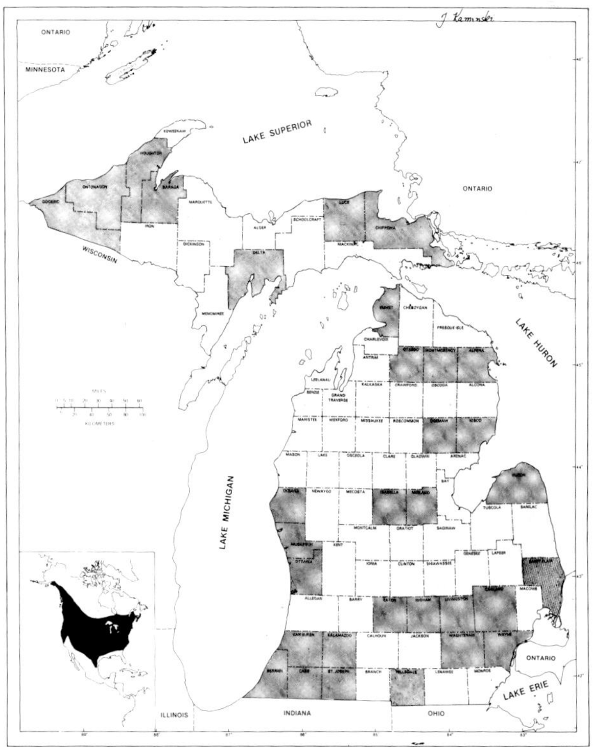

Map 16. Geographic distribution of the silver-haired bat (*Lasionycteris noctivagans*). Actual records exist for counties with dark shading.

and apart during the entire warm season. However in late summer and autumn, silver-haired bats congregate for migrating purposes. Most observations on these gatherings have been along the Atlantic coast (Banfield, 1974; Connor, 1971). Peterson (1966) found groups of these bats assembled on the north shore of Lake Erie for southward movements. Wood (1914) noted large numbers flying over the beach in the Whitefish Point area of Lake Superior on August 10.

BEHAVIOR. Apparently the seasonal movements of the silver-haired bat are not extensive, as indicated by the rather narrow north-south distribution of the species (see Map 16). Even so, it seems likely that few bats spend the winter in Michigan. The only three winter records published, two from Oakland County and one from St. Joseph County (Fowler, 1959; Gosling, 1977), are from January. In spring the earliest record is for May 2 on South Manitou Island (William Scharf *in* Kurta, 1980) with others on May 19 in Ingham and Houghton counties (Stones and Holyoak, 1970). The autumn flights south begin in September, with the last Michigan record on October 24.

In summer, the silver-haired bat is one of the first chiropterans to begin evening flights, often appearing along streams in broad daylight. The flight pattern is slow, fluttering and somewhat feeble. Even so, this bat's zig-zag aerial antics are highly efficient for capturing insects, with the aid of echolocation. During night hours the bat's whereabouts between early and late forage flights are unknown. A hibernating bat, found in a South Dakota limestone cave by Turner (1974), had a rectal temperature of 49.3° F (9.6° C) and was found where the relative humidity was 57% and the ambient temperature was 44.1° F (6.7° C).

ASSOCIATES. The silver-haired bat occupies its solitary summer roosts by day but at night associates with other bats in common feeding and/or drinking areas along streams. Along the Thornapple River in Eaton County, for example, Kurta (1980) snared silver-haired bats in the same mist nets with Keen's bats, little brown bats, Indiana bats, big brown bats, red bats and hoary bats. Although these bats seem concentrated along streams, it is most likely that there is food partitioning, perhaps by separating activity periods and habitats (Kunz, 1973; Reith, 1980).

In winter, silver-haired bats have been found using the same hibernating shelters (caves, cracks in limestone ledges) with Keen's bats, eastern pipistrelles, Indiana bats, and big brown bats (Barbour and Davis, 1969; Frum, 1953).

REPRODUCTIVE ACTIVITIES. Courtship and mating in the silver-haired bat are thought to occur in autumn when both sexes congregate to migrate. Evidence for this includes the increase in size of testes in males in late July followed by a decline in November (Turner, 1974). This suggests that delayed fertilization of the female occurs with the sperm remaining viable to fertilize eggs shed in the spring. Kurta (1980) gathered records to show that silver-haired bats are born and nursed in southern Michigan. Females pregnant with one or two embryos were obtained in late May and June. One caught on June 18 in a mist net across the Thornapple River in Eaton County, gave birth to two full-term offspring (one male and one female) on June 20. This birth date is well within the time (late June and early July) for births in this species in Iowa (Easterla and Watkins, 1967). Delivery takes place with the pregnant female assuming a head-up position. At birth the young are pinkish, have wrinkled skin with no body hair, closed eyes and folded black ears, mottled (black and tan) wing membranes, measure about 1¾ in (46 mm) long and less than 0.09 oz (2.5 g) (Kurta 1980). At about 21 days of age, the young are capable of flight and able to catch insect food for themselves.

FOOD HABITS. Moths (Lepidoptera) and flies (Diptera) are reported as major food choices of the silver-haired bat in western states (Freeman, 1981). It is suspected that along Michigan streams this species may eat insects with aquatic nymphal and larval stages. While the silver-haired bat is generally thought to catch its food while in flight, there are some unexplained records of silver-haired bats being caught in buildings in mouse traps baited with cheese and raisins (Bartsch, 1956; Getz, 1961). On another occasion, young bats were found feeding on fly larvae which had accumulated in the bottom of a woodpecker hole (Novakowski, 1956). In captivity, silver-haired bats have been fed vitamin-fortified hamburger and mealworms. This bat's daily food requirements on the latter diet were 0.50 kcal per gram of body weight (Neuhauser and Brisbin, 1969).

ENEMIES. The solitary nature of the silver-haired bat, at least in summer, presents a very different life pattern to a would-be predator than bats gathered in large conspicuous nursery colonies. Owls are probably the most efficient predators (Jackson, 1961), although Sperry (1933) identified the remains of two silver-haired bats in the stomach contents of a striped skunk in North Carolina. These bats were probably hibernating as the event took place in January, but no explanation was given as to how the skunk managed to catch them.

PARASITES AND DISEASES. Fleas, mites, chiggers, a bat fly, bed bugs, and internal flat worms (trematodes) are reported to infest silver-haired bats (Barbour and Davis, 1969: Jackson, 1961; Turner, 1974).

MOLT AND COLOR ABERRATIONS. There appears to be a single molt in midsummer (Turner, 1974).

ECONOMIC IMPORTANCE. This bat remains fairly apart from most human habitations in Michigan because of its solitary nature and its adaptation to sylvan retreats. Although *L. noctivagans* is known to invade houses and other buildings in winter, it rarely does so in sufficiently large gatherings to cause alarm. Caution, of course, should be taken when handling a live silver-haired bat, as there have been two positive records of this bat harboring rabies in Michigan (Kurta, 1979). The teeth and jaws are powerful enough to produce an effective bite. To the outdoorsperson, the silver-haired bat's aerobatics lend charm to the stream-side vista on warm Michigan evenings.

COUNTY RECORDS FROM MICHIGAN. Reports of specimens preserved in museum collections and/or mentioned in the literature show that the silver-haired bat is known from the following counties in the Upper Peninsula: Baraga, Chippewa, Delta, Gogebic, Houghton, Luce, Ontonagon; in the Lower Peninsula: Alpena, Berrien, Eaton, Emmet, Hillsdale, Huron, Ingham, Iosco, Isabella, Kalamazoo, Livingston, Midland, Montmorency, Muskegon, Oakland, Oceana, Ogemaw, Otsego, Ottawa, St. Clair, St. Joseph, Van Buren, Washtenaw, and Wayne. The species is also known from South Manitou Island in Lake Michigan and Charity Island in Saginaw Bay.

LITERATURE CITED

Banfield, A. W. F.
1974 The mammals of Canada. Univ. Toronto Press, Toronto. xxiv+438 pp.

Barbour, R. W., and W. H. Davis
1969 Bats of America. Univ. Kentucky Press, Lexington, 286 pp.

Bartsch, P.
1956 An interesting catch. Jour. Mammalogy, 37(1):111.

Connor, P. F.
1971 The mammals of Long Island, New York. New York State Mus. and Sci. Serv., Bull. 416, v+78 pp.

Easterla, D. A., and L. C. Watkins
1967 Silver-haired bat in southwestern Iowa. Jour. Mammalogy, 48(2):327.

Fowler, J. A.
1959 Hibernating bats. Cranbrook Inst. Sci., News Letter, 28(9):106.

Freeman, P. W.
1981 Correspondence of food habits and morphology in insectivorous bats. Jour. Mammalogy, 62(1):166–173.

Frum, W. G.
1953 Silver-haired bat, *Lasionycteris noctivagans*, in West Virginia. Jour. Mammalogy, 34(4):499–500.

Getz, L. L.
1961 New locality records of some Kansas mammals. Jour. Mammalogy, 42:282–283.

Gosling, N. M.
1977 Winter record of the silver-haired bat, *Lasionycteris noctivagans* Le Conte, in Michigan. Jour. Mammalogy, 58(4):657.

Hamilton, W. J., Jr.
1943 The mammals of eastern United States. Comstock Publ. Assoc., Ithaca. 432 pp.

Izor, R. J.
1979 Winter range of the silver-haired bat. Jour. Mammalogy, 60(3):641–643.

Jackson, H. H. T.
1961 Mammals of Wisconsin. Univ. Wisconsin Press, Madison. xii+504 pp.

Krutzsch, P. H.
1966 Remarks on silver-haired and Leib's bats in eastern United States. Jour. Mammalogy, 47(1):121.

Kunz, T. H.
1973 Resource utilization: Temporary and spatial components of bat activity in central Iowa. Jour. Mammalogy, 54(1):14–32.

Kurta, A.
1979 Bat rabies in Michigan. Michigan Acad., 12(2):221–230.
1980 The bats of southern lower Michigan. Michigan State Univ., unpubl. M.S. thesis, ix+147 pp.

Manville, R. H.
1942 A report on wildlife studies at the Huron Mountain Club, 1939 to 1942. vii+181 pp.

Mumford, R. E.
1969 Distribution of the mammals of Indiana. Indiana Acad. Sci., Monog. No. 1, vii+114 pp.

Novakowski, N. S.
1956 Additional records of bats in Saskatchewan, Canadian Field-Nat., 70:142.

Neuhauser, H. N., and I. L. Brisbin, Jr.
1969 Energy utilization in a captive silver-haired bat. Bat Research News, 10(3):30–31.

Pearson, E. W.
1962 Bats hibernating in silica mines in southern Illinois. Jour. Mammalogy, 42(1):27–33.

Peterson, R. L.
1966 The mammals of eastern Canada. Oxford Univ. Press, Toronto. xxxii+463 pp.

Reith, C. C.
1980 Shifts in times of activity by *Lasionycteris noctivagans*. Jour. Mammalogy, 61(1):104–108.

Scharf, W. C.
1973 Birds and land vertebrates of South Manitou Island. Jack-Pine Warbler, 51(1):3–19.

Sperry, C. C.
1933 Opossum and skunk eat bats. Jour. Mammalogy, 14(2):152–153.

Stones, R. C., and G. W. Holyoak
1970 Spring occurrence of a silver-haired bat in upper Michigan. Jour. Mammalogy, 51(4):811–812.

Timm, R. M.
1975 Distribution, natural history and parasites of mammals of Cook County, Minnesota. Univ. Minnesota, Bell Mus. Nat. Hist., Occas. Papers 14, 56 pp.

Turner, R. W.
1974 Mammals of the Black Hills of South Dakota and Wyoming. Univ. Kansas Publ., Mus. Nat. Hist., Misc. Publ. No. 60, 178 pp.

Wood, N. A.
1914 Results of the Shiras expeditions to Whitefish Point, Michigan—mammals. Michigan Acad. Sci., 16th Rept., pp. 92–98.
1922 The mammals of Washtenaw County, Michigan. Univ. Michigan, Mus. Zool., Occas. Papers No. 123, 23 pp.

EASTERN PIPISTRELLE
Pipistrellus subflavus (Cuvier)

NAMES. The generic name *Pipistrellus*, proposed by Kaup in 1829, is derived from an Italian word meaning "bat." The specific name *subflavus*, proposed by F. Cuvier in 1832, is a Latin word which refers to the characteristic yellowish color of this bat. Sometimes this species is called the pygmy bat or yellowish-brown bat. The eastern pipistrelle in Michigan belongs to the subspecies, *Pipistrellus subflavus subflavus* (Cuvier).

RECOGNITION. Body small, the size of a short, small mouse; head and body approximately 1¾ in (45 mm) long; nose pointed; eyes small; ears short, tapered and rounded, extending when laid (not pulled) forward to just in front of the nasal openings; tragus in ear opening nearly straight with blunt rounded tip and about half the length of ear pinna; wings short and broad; tail long, almost half of total length; feet large for such a small bat; calcar lacking keel-shaped extension (see Fig. 18C).

Color grizzled yellow-brown to dark brown on upper parts, individual hair with distinctive tricolor pattern, dark basally and on the tip and pale in midsection; pale yellow-buff on underparts; ears and tail membrane light brown; tail membrane sparsely haired at base; wing membranes dark with skin over forearm slightly reddish; hind toes clothed in fine reddish hair.

MEASUREMENTS AND WEIGHTS. Adults measure from tip of nose to end of tail, 2¾ to 3½ in (70 to 88 mm); length of tail vertebrae, 1⅛ to 1¾ in (28 to 45 mm); length of hind foot, ¼ to ⅜ in (7 to 10 mm); height of ear from notch, ½ to ⅝ in (13 to 15 mm); length of bone of forearm, 1¼ to 1⅜ in (32 to 36 mm); weight, 0.11 to 0.25 oz (3.0 to 7.0 g).

DISTINCTIVE CRANIAL AND DENTAL CHARACTERISTICS. The skull of the eastern pipistrelle is small with an inflated braincase and a depressed area in the region of the orbits (See Fig. 19). The nasal area (rostrum) is broad. The hard palate extends only a short distance behind the last molars and has a prominent notch at its anterior end. The skull is approximately ½ in (12.2 to 13.3 mm) long and ¼ in (6.6 to 8.2 mm) wide (across zygomatic arches). The dentition resembles closely that of the myotis bats, except that the eastern pipistrelle has one less premolar tooth (in both upper and lower jaws) giving the species a total of 34 teeth, a number not found in any other Michigan bat. The dental formula is as follows:

$$\text{I (incisors)}\frac{2\text{-}2}{3\text{-}3}, \text{C (canines)}\frac{1\text{-}1}{1\text{-}1}, \text{P (premolars)}\frac{2\text{-}2}{2\text{-}2}, \text{M (molars)}\frac{3\text{-}3}{3\text{-}3} = 34$$

DISTRIBUTION. The eastern pipistrelle occurs in eastern North America from southern Canada southward to the Gulf States and eastern Mexico. At the northern edge of this bat's Michigan range, the species is apparently rare to uncommon and presently known from two counties in the western part of the Upper Peninsula and from only one county in the Lower Peninsula (see Map 17). Although known for a long time from nearby Wisconsin (Greeley and Beer, 1949), Burt (1946) had no early Michigan records. Professor Robert C. Stones and his students of Michigan Technological University found the first state record while studying cave-dwelling bats in the mining district of Michigan's copper country (Stones and Haber, 1965). The single record for the Lower Peninsula is from Grand Mere just west of Stevensville in Berrien County and was taken on November 12, 1964 (Kurta, 1980). It is catalogued as specimen #1,241 in collections at Andrews University.

HABITAT PREFERENCES. In winter, the eastern pipistrelle generally hibernates in substantial, insulated shelters such as caves and mine tunnels. Occasionally in the northern parts of its range this bat may also winter in storm sewers or even hollow stumps (Davis and Mumford, 1962; Goehring, 1954). They are creatures of habit and return to these same wintering retreats year after year.

In summer, eastern pipistrelles move to summer roosts either in small sex-segregated groups or singly. These shelters, used in daytime after nightly forays for food, are generally in the vicinity of the winter quarters but can be as far away as 85 mi (140 km) according to Griffin (1940) and Mumford (1969). Shelters may occasionally include sites in caves and mine tunnels but more usually in attics of buildings, crevices, and even tree foliage (Davis and Mumford, 1962; Findley, 1954). Eastern pipistrelles use caves for swarming purposes midway between late evening and early morning forage flights (Hall and Brenner, 1968). Insect hunting grounds are near treetops and along forested edges of fields and waterways.

There is evidence that eastern pipistrelles are uncommon year-around residents in the western part of Michigan's Upper Peninsula (Kurta, 1980). The sparse population may simply be the result of climatic factors, since this area is at the northern edge of the range of this species. The presence of only a single record of this bat in the Lower Peninsula suggests that southern counties may be

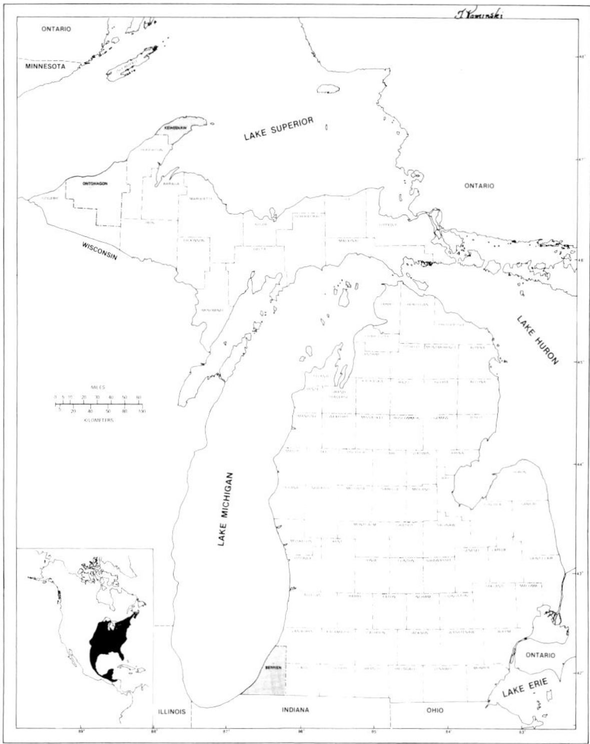

Map 17. Geographic distribution of the eastern pipistrelle (*Pipistrellus subflavus*).

too remote from proper winter harborages in caves and mines for this bat's rather limited flying range.

DENSITY AND MOVEMENTS. As mentioned previously, the Michigan population of eastern pipistrelles is sparse with only a small number counted. Elsewhere, winter concentrations in individual caves may number more than 2,000 (Barbour and Davis, 1969). In Indiana, however, the largest colony reported thus far included only 112 individuals (Mumford and Cope, 1964). Sometimes, single individuals will be found in one winter shelter (Greeley and Beer, 1949; Stones and Haber, 1965). In summer, eastern pipistrelles hang singly or in groups of 30 or more in daytime shelters.

BEHAVIOR. In the field, the eastern pipistrelle is usually easily identified by its moth-like size and its fluttery, weak, erratic flight. Although it has been known to fly at speeds up to 16 mi (22 km) per hour, the bat mostly flutters about no faster than 6 mi (10 km) per hour (Banfield, 1974; Patterson and Hardin, 1969). The eastern pipistrelle is also one of the first bats to appear in flight in late afternoon and on cloudy days and the last one to go to roost after daylight. In the northern part of its range, it appears particularly sensitive to cold weather. Consequently, it enters winter shelters to hibernate at the first sign of autumn frost—by mid-October or even earlier (Doutt *et al.,* 1966)—and the last to emerge in spring. An eight months hibernation in warm, draft-free and humid sites is not unusual. This bat is also noted as a sound sleeper while in hibernation, not waking frequently as do myotis bats (Davis, 1964). When roosting in large groups, the individuals are often spread out rather than forming tight clusters (Hall, 1962). One hibernating bat had an esophageal temperature of about 48° F (9° C), just above ambient temperature (Stones and Haber, 1965). In summer, the eastern pipistrelle characteristically enters a state of torpor when roosting between forage flights (Hall, 1962). As with other bats, it uses echolocation to advantage in avoiding obstacles while in flight and to pursue insects.

ASSOCIATES. In winter colonies, the eastern pipistrelle keeps to itself but will tolerate other species of bats in the same shelter. These winter associations include the little brown bat, Keen's bat, big brown bat, Indiana bat, and silver-haired bat (Hall, 1962; Jones *et al.,* 1967; Pearson, 1962; Swanson and Evans, 1936). In summer, the eastern pipistrelle also uses foraging areas somewhat similar to the species above listed. However, Davis and Mumford (1962) found that it feeds more commonly in association with the big brown bat and the red bat than with the little brown bat. The eastern pipistrelle has also been observed entering caves on summer nights along with the hoary bat, evening bat, silver-haired bat, and red bat (Mumford and Whitaker, 1975).

REPRODUCTIVE ACTIVITIES. Courtship and mating occur much as they do in myotis bats, when the sexes regroup in autumn following summer segregation. Mating apparently takes place in October and November when the bats are ready to enter hibernation (Hahn, 1908). The sperm stored in the female tract remains viable until the bats emerge in spring, when the eggs are shed, fertilization takes place, and embryonic development begins. There may be further matings in the spring as the females become receptive.

Although the exact period of gestation is not known, young are born in the Upper Great Lakes Region from mid-June to early July, somewhat later than myotis bats. Little has been recorded about the early life of the one or two young (sometimes three). For a few days, the mother carries the young on forage flights. Later, she leaves them in shelters. At 30 days of age, the young bats are weaned, capable of flight, and learn to forage for themselves. Records of banded eastern pipistrelles show that individuals are exceedingly long lived for such a small mammal, up to ten years (Davis, 1966).

FOOD HABITS. In spite of its slow, fluttering flight, the eastern pipistrelle manages to catch flying moths (Lepidoptera), true bugs (Homoptera), small beetles (Coleoptera), small wasps and flying ants (Hymenoptera), flies (Diptera) and other insect life (LaVal and LaVal, 1980; Whitaker, 1972). Gould (1953) noted that in 30 minute flights, two of these bats managed to eat insect food equal to 25% of their body weights.

ENEMIES. Agile climbers, including snakes and small four-footed carnivores, might capture eastern pipistrelles as they hang in torpor in summer or winter shelters. Owls probably catch these bats while in flight but no such events have been reported. There are records in the literature, however, to show that this bat has been eaten by a leopard frog, attacked by a hoary bat (Bishop, 1947; Creel, 1963), and involved in natural catastrophies in caves (DeBlase *et al.,* 1965).

PARASITES AND DISEASES. The eastern pipistrelle is reported to be infested with chiggers, three kinds of mites, and intestinal trematodes (Whitaker and Wilson, 1971, 1974).

MOLT AND COLOR ABERRATIONS. In late summer, observers may find young-of-the-year having dorsal coloration much duller than the usual yellowish adult coloring. This immature fur is replaced by the normal adult pelage before the animals enter winter quarters. Adults with conspicuously darkened fur (melanistic) and reddish fur (erythristic) have been reported by several authors (*e.g.,* Barbour and Davis, 1969; Jackson, 1961) while at least one whitish (albinistic) eastern pipistrelle has been described (Blair, 1948). Another specimen, obtained in Ohio, had white wing tips (Goslin, 1947). Annual molt occurs in summer.

ECONOMIC IMPORTANCE. Since Michigan lies at the northern edge of the range of the eastern pipistrelle, its occurrence here may continue to be unusual. Certainly overwintering or nursery groupings of this bat in Michigan should be protected. Although rabies has been reported in *P. subflavus* in Indiana (in two individuals out of 406 examined, Whitaker *et al.,* 1969), this bat's teeth and jaws are not sufficiently strong to inflict a bite on a person's hand.

COUNTY RECORDS FROM MICHIGAN. Reports of specimens preserved in museum collections, documented in the literature, or recorded by knowledgeable observers show that the eastern pipistrelle is known from the following counties in the Upper Peninsula: Keweenaw (Barron, 1981) and Ontonagon; in the Lower Peninsula: Berrien County (see Map 16).

LITERATURE CITED

Banfield, A. W. F.
1974 The mammals of Canada. Univ. Toronto Press, Toronto. xxiv+438 pp.

Barbour, R. W., and W. H. Davis
1969 Bats of America. Univ. Kentucky Press, Lexington. 286 pp.

Barron, R. J.
1981 Second record of the eastern pipistrelle, *Pipistrellus subflavus,* in Michigan. Jack-Pine Warbler, 59(2):68.

Bishop, S. C.
1947 Curious behavior of a hoary bat. Jour. Mammalogy, 28(3):293–294.

Blair, W. F.
1948 A color pattern aberration in *Pipistrellus subflavus subflavus.* Jour. Mammalogy, 29(2):178–179.

Burt, W. H.
1946 The mammals of Michigan. Univ. Michigan Press, Ann Arbor. xv+288 pp.

Creel, G. C.
1963 Bat as a food item of *Rana pipiens.* Texas Jour. Sci., 15(1):104–105.

Davis, W. H.
1964 Winter awakening patterns in the bats *Myotis lucifugus* and *Pipistrellus subflavus.* Jour. Mammalogy, 45(4): 645–647.
1966 Population dynamics in the bat *Pipistrellus subflavus.* Jour. Mammalogy, 47(3):383–396.

Davis, W. H., and R. E. Mumford
1962 Ecological notes on the bat *Pipistrellus subflavus.* American Midl. Nat., 68(2):394–398.

DeBlase, A. F., S. R. Humphrey, and K. S. Drury
1965 Cave flooding and mortality in bats in Wind Cave, Kentucky. Jour. Mammalogy, 46(1):96.

Doutt, J. K., C. A. Heppenstall, and J. E. Guilday
1966 Mammals of Pennsylvania. Pennslyvania Game Commission, Harrisburg. 273 pp.

Findley, J. S.
1954 Tree roosting of the eastern pipistrelle, Jour. Mammalogy, 35(3):433.

Goehring, H. H.
1954 *Pipistrellus subflavus obscurus, Myotis keenii,* and *Eptesicus fuscus fuscus* hibernating in a storm sewer in central Minnesota. Jour. Mammalogy, 35(3):434–435.

Goslin, E.
1947 A bat with white wing tips. Jour. Mammalogy, 28(1):62.

Gould, E.
1953 Feeding efficiency of bats. Biol. Bull., 105(2):364.

Greeley, F., and J. R. Beer
1949 The pipistrelle *(Pipistrellus subflavus)* in northern Wisconsin. Jour. Mammalogy, 30(2):198.

Griffin, D. R.
1940 Notes on the life histories of New England cave bats. Jour. Mammalogy, 21(2):181–187.

Hall, J. S.
1962 A life history and taxonomic study of the Indiana bat, *Myotis sodalis.* Reading Pub. Mus. and Art Gallery, Sci. Publ. No. 12, 68 pp.

Hall, J. S., and F. J. Brenner
1968 Summer netting of bats at a cave in Pennsylvania. Jour. Mammalogy, 49(4):779–781.

Hahn, W. L.
1908 Some habits and sensory adaptations of cave-inhabiting bats. Biol. Bull., 15:135–193.

Jackson, H. H. T.
1961 Mammals of Wisconsin. Univ. Wisconsin Press, Madison. xii+504 pp.

Jones, J. K., Jr., E. D. Fleharty, and P. B. Dunnigan
1967 The distributional status of bats in Kansas. Univ. Kansas Publ., Mus. Nat. Hist., Musc. Publ. 46, 33 pp.

LaVal, R. K., and M. L. LaVal
1980 Ecological studies and management of Missouri bats, with emphasis on cave-dwelling species. Missouri Dept. Conserv., Terrestrial Ser. No. 8, 53 pp.

Kurta, A.
1980 The bats of southern lower Michigan. Michigan State Univ., unpubl. M.S. thesis, ix+147 pp.

Mumford, R. E.
1969 Distribution of the mammals of Indiana. Indiana Acad. Sci., Monog. No. 1, vii+114 pp.

Mumford, R. E., and J. B. Cope
1964 Distribution and status of the Chiroptera of Indiana. American Midl. Nat., 72(2):472–489.

Mumford, R. E., and J. O. Whitaker, Jr.
1975 Seasonal activity of bats at an Indiana cave. Proc. Indiana Acad. Sci., 84:500–507.

Patterson, A. P., and J. W. Hardin
1969 Flight speeds of five species of vespertilionid bats. Jour. Mammalogy, 50(1):152–153.

Pearson, E. W.
1962 Bats hibernating in silica mines in southern Illinois. Jour. Mammalogy, 43(1):27–33.

Stones, R. C., and G. C. Haber
1965 Eastern pipistrelle in Michigan. Jour. Mammalogy, 46(1):688.

Swanson, G., and C. Evans
1936 The hibernation of certain bats in southern Minnesota. Jour. Mammalogy, 17(1):39–43.

Whitaker, J. O., Jr.
1972 Food habits of bats from Indiana. Canadian Jour. Zool., 50:877–883.

Whitaker, J. O., Jr., W. A. Miller, and W. L. Boyko
1969 Rabies in Indiana bats. Proc. Indiana Acad. Sci., 78:447–456.

Whitaker, J. O., Jr. and N. Wilson
1971 Notes on a collection of bats taken by mist-netting at an Indiana Cave. American Midl. Nat., 85(1):277–279.
1974 Host and distribution lists of mites (Acari), parasitic and phoretic, in the hair of wild mammals of North America, north of Mexico. American Midl. Nat., 91(1):1–67.

BIG BROWN BAT
Eptesicus fuscus (Palisot de Beauvois)

NAMES. The generic name *Eptesicus,* proposed by Rafinesque in 1820, is derived from the Greek and literally means "house flyer." The specific name *fuscus,* proposed by Palisot de Beauvois in 1796, is a Latin word meaning dusky or somber. In Michigan, this species is called brown bat or bat. The subspecies *Eptesicus fuscus fuscus* (Palisot de Beauvois) occurs throughout Michigan.

RECOGNITION. Body medium, the size of a plump mouse; head and body approximately 2½ in (65 mm) long; wingspread about 13 in (330 mm); head large; nose broad with sparse vibrissae and fleshy lips; eyes large and bright; ears rounded, extending when laid (not pulled) forward just barely to the nostrils; tragus in ear opening straight and broad with rounded tip; tail less than half total length with tip projecting slightly beyond interfemoral membrane; calcar (extension from heel) with keel-shaped extension (see Fig. 18D).

The long, lax fur of the upperparts appears oily in texture and is rusty brown in color. Below, the coloring is more gray. Naked parts of the face, ears, wings, and tail membrane are black. The large size, big head, large ears, and soft dark brown hair (Fig. 17) all aid in identifying this species from other Michigan bats.

MEASUREMENTS AND WEIGHTS. Adults measure from tip of nose to end of tail vertebrae, 3¾ to 4⅝ in (96 to 118 mm); length of tail vertebrae, 1⅜ to 2 in (34 to 50 mm); length of hind foot, ⅜ to 9/16 in (10 to 14 mm); height of ear from notch, ⅝ to 13/16 in (16 to 20 mm); length of bone of forearm, 1 9/16 to 1 15/16 in (40 to 50 mm); weight, 0.42 to 0.67 oz (12 to 19 g).

DISTINCTIVE CRANIAL AND DENTAL CHARACTERISTICS. The skull of the big brown bat is large and heavily constructed, larger than those of other Michigan bats except the hoary bat which is noticeably wider (see Fig. 19). The skull of the big brown bat is approximately ¾ in (18.1 to 20.6 mm) long and ½ in (11.5 to 13.4 mm) wide (across zygomatic arches). Both the hoary bat and the big brown bat have 32 teeth but they are arranged differently in each species. In both instances, however, they are heavy and sharp, capable of making severe bites on the handler's hand. The dental formula is:

$$\text{I (incisors)}\frac{2\text{-}2}{3\text{-}3},\ \text{C (canines)}\frac{1\text{-}1}{1\text{-}1},\ \text{P (premolars)}\frac{1\text{-}1}{2\text{-}2},\ \text{M (molars)}\frac{3\text{-}3}{3\text{-}3} = 32$$

DISTRIBUTION. The big brown bat occupies much of North and Central America, ranging northward to near treeline and southward into northern South America (Nagorsen, 1980). In Michigan it has the widest and most even distribution of all of the residential, colonial species, being expected to occur in all counties at all seasons (see Map 18).

Map 18. Geographic distribution of the big brown bat (*Eptesicus fuscus*). Actual records exist for counties with dark shading.

HABITAT PREFERENCES. Unlike the myotis bats and the eastern pipistrelle with their apparent need to hibernate in the highly insulated and stable environments characteristic of caves and mine shafts, the big brown bat has a more tolerant constitution allowing it to winter in assorted, less substantial shelters. These include rural and urban human-made structures, such as barns, silos, storm sewers (Goehring, 1954, 1972), churches, expansion joint spaces in concrete athletic stadiums, copper mines (Stones and Fritz, 1969) and the double walls and attics of residences (see Fig. 20, 21).

In presettlement times it is presumed that in most parts of Michigan tree hollows provided both summer and winter quarters in the absence of natural caves or openings in rock ledges (Cahalane, 1932; Christian, 1953; Mills *et al.*, 1975). However, as mentioned above, available buildings now provide abundant shelters. In southern Michigan, Kurta (1980) examined 57 roosts occupied by big brown bats in the summers of 1978 and 1979. These were mostly in barns (22) and houses (27) but also in commercial buildings (3) and churches (4). Only one group of bats was found in a tree cavity, in a large beech in Cass County. The big brown bat needs only to find an entry under loose or warped siding or through narrow openings around chimneys and under eaves to roost in attics or preferably in double walls.

It is reasonable to speculate that Michigan populations of the big brown bat have probably multiplied, perhaps many fold, in the past 100 years because of the increased number of human accomodations. In fact, the big brown bat seems to thrive best where roost site buildings are in recreational lands, farming areas, suburban developments and even inner-city environments. Evening observers in downtown Detroit, for example, have abundant opportunity to watch bat flights. The big brown bat is one native mammal which seems to have benefitted greatly from intensive human occupation.

DENSITY AND MOVEMENTS. Although there are records of large winter concentrations of big brown bats in caves (751 individuals in one cave in Ontario by Hitchcock, 1949, 200 in another in Indiana by Kirkpatrick and Conaway, 1948), most observers conclude that Michigan winter colonies usually are much smaller (Burt, 1946). Stones and Fritz (1969) counted 20–30 wintering in copper mines on the Keweenaw Peninsula. They estimated that the population of this bat in the Delaware Mine was only about 1/50th the size of the associated aggregation of Keen's bat and the little brown bat.

In summer, nursery colonies may also contain up to 400 (Rysgaard, 1942) or even 700 individuals (Mills *et al.,* 1975). Again, however, the numbers are usually much less; Mumford and Cope (1964), for example, found 190 summer breeding colonies in Indiana to average 65 individuals (minimum, 15; maximum, 260). On May 24, Robert Goodsell counted 43 big brown bats (one adult male and 42 pregnant females) behind a shutter of a home in Lansing. In 21 summer nursery roosts examined in southern Michigan, Kurta (1980) found populations to range in size from 5 to 150 bats, with an average of 51.6 individuals.

Although the big brown bat is a strong flyer (see Fig. 22), individuals may have little need to move great distances between winter hibernating shelters and summer roosts. In a study made in Minnesota by Beer (1955), most bats moved no more than

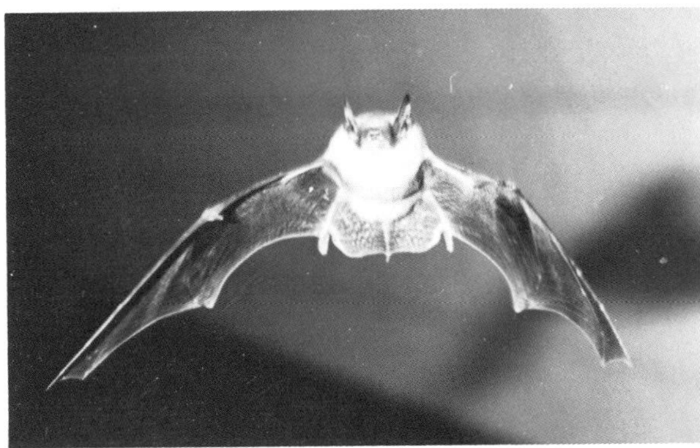

Figure 22. The big brown bat, *Eptesicus fuscus*.
Photograph by William H. Burt

10 mi (16 km), with an extreme of 61 mi (98 km). These data suggest that Michigan big brown bats may spend their entire life spans within areas no larger than a single county. Although both sexes share common winter quarters, the sexes generally segregate in summer. The adult males may even continue to use the winter shelters in summer whereas females will seek nursery roosts elsewhere (Jones *et al.,* 1967; Phillips, 1966).

BEHAVIOR. The big brown bat makes its appearance in spring, emerging from winter quarters occasionally as early as late March. The majority of the population becomes active in southern Michigan by mid-April (Kurta, 1980). Females move rather promptly to traditional summer nursery roosts; Kurta found a perennial summer nursery colony in a church in Livingston County to appear between April 6 and 23. The whereabouts of these particular bats in winter was not determined.

In summer, big brown bats make nightly forays to catch flying insect prey. Feeding generally occurs in one major peak of activity, with a nightly resting period in a shelter which may be different from the usual daytime retreat. In studies in southern Michigan, Kurta (1980, 1981) found big brown bats begin leaving daytime roosts from as early as sunset to as late as 28 minutes after sundown. At 14 different nursery colonies, he found that the average emergence time of the first bat was 17.9 ± 2.4 minutes after sundown. Foraging activity may be reduced or suspended on cold or wet nights; high wind alone may not be a deterrent.

Preparation for winter hibernation requires a buildup in fat, with the big brown bat fattening most rapidly in September (Weber and Findley, 1970). At this time, the sexes rejoin and congregate in caves, mine tunnels, barns, or other open buildings. By September these activities become centered around the winter quarters. Hibernation begins in October with males becoming lethargic earlier than females (Phillips, 1966). By the first week of December, as reported in Minnesota by Beer (1955), the typical wintering population has become generally stabilized in the hibernaculum. Nevertheless, the bats will become restless from time to time throughout the winter (Beer, 1955; Rysgaard, 1942). In fact, big brown bats will arouse, change locations in their shelters, or leave one shelter and go to another on extremely cold days. Some of these movements may be attributable to mid-winter mating urges (Mumford, 1958). Other such activity may be the result of temperature fluctuations within the winter shelter.

Beer and Richards (1956) found the big brown bat hibernating where temperatures averaged 42° F (5.6° C) and the relative humidity 79%. When the ambient temperature drops below 32° F (0° C), the bat wakes and may move to a warmer site (Davis and Reite, 1967). Conversely, when the ambient temperature in the hibernating site rises above the optimum for hibernating, the bat will also become aroused (Banfield, 1974; Wolverton and Edgar, 1969). These occasional winter movements become evident to householders when scratching and sometimes squeaking noises are detected inside walls or rooms. Bird watchers may also be surprised to see a bat flying over a snowy vista (Baker, 1969, Burt, 1946). The large number of reports of big brown bats received from the Lansing area by Michigan State University Museum personnel during the severe 1977–1978 winter seemed correlated with that season's pronounced temperature changes.

The big brown bat is an alert, nervous creature, capable of making highly audible squeaking cries when disturbed. It can fly at speeds up to 24 mi (39 km) per hour (Patterson and Hardin, 1969). The flight is more direct than is characteristic of the myotis bats. Homing seems to be a strong factor in the life of this bat, with records of displaced individuals (banded for positive identification) homing from distances as far away as 450 mi (724 km) by Smith and Goodpaster (1958), even though there are no reported records of normal migrations of this magnitude. In flight, the big brown bat, like other chiropterans, avoids obstacles and pursues prey by using echolocation. The utterances employed in this intriguing skill are usually inaudible to human ears and are at lower frequencies than those employed by the little brown bat. These calls may also be used to communicate between individuals. Keen hearing appears to account for big brown bats being able to locate concentrations of flying insects (Buchler and Childs, 1981).

ASSOCIATES. In winter hibernacula, the big brown bat often uses the same facilities as do other Michigan bats. Hitchcock (1949), for example, counted 751 big brown bats, 434 small-footed bats, 117 Keen's bats, 59 eastern pipistrelles, and five little brown bats sharing a limestone cave in Ontario. In another winter shelter, he counted 3,800 little brown bats, 230 Keen's bats, and five big brown bats. In selecting hibernation sites, Hitchcock noted that the big brown bat was the most tolerant of all the species to cold temperature, low humidity and air movement. On occasion, a male big brown bat will be found within a cluster of little brown bats (Barbour and Davis, 1969). The big brown bat has also been found to associate in winter quarters with the Indiana bat in sites in Kentucky, Indiana, and Iowa (Hall, 1962; Muir and Polder, 1960).

In summer, big brown bats apparently associate less frequently with other colonial bats in daytime roosts but perhaps more commonly in night shelters between feedings (Hall and Brenner, 1968; Humphrey, 1975). Observations show that several species will hunt prey in the same areas (Davis and Mumford, 1962). Kurta (1980) captured the big brown bat, Keen's bat, little brown bat, Indiana bat, silver-haired bat, red bat, and hoary bat in the same mist nets stretched across the Thornapple River in a wooded area near Vermontville in Eaton County.

REPRODUCTIVE ACTIVITIES. According to studies in Maryland (Christian, 1953), spermatozoa are produced beginning in October. As in other hibernating bat species, the big brown bat mates primarily just prior to entering winter quarters, although copulation reportedly can take place at anytime individuals wake during winter hibernacula (Beer, 1955). The viable sperm is then held in the female tract until about the first week in April when ovulation and fertilization take place (Christian, 1953). The birth of one or two (sometimes three; Siamese twinning has been recorded, Peterson and Fenton, 1969) young is usually from late May to late June (Kunz, 1974; Schowalter and Gunson, 1979). Females in late stages of pregnancy have been recorded in Michigan on May 24 (Ingham County), May 25 (Ingham County), and June 3 (Washtenaw County). A female in the process of delivering her two young was observed on May 26 in Lansing, Ingham County. Cahalane (1932) reported a birth on June 19 (Oakland County). In studies of the bats in the southern three tiers of counties in the Lower Peninsula, Kurta (1980, 1981) mist-netted pregnant big brown bats as late as June 18 (Eaton County) and June 24 (Cass County). He obtained lactating females as early as June 5 in 1979 and June 2 in 1980 (Livingston County) and as late as August 6 (St. Joseph County).

At birth, the young are wrinkled, lack body hair, have closed eyes and folded ears, and weigh about 0.09 oz (2.5 g). The mother has two pectoral mammary glands for nursing and the young grow rapidly with eyes opening after about seven days

(Davis *et al.,* 1968). Young are weaned and able to fly as early as the last week in June and first week in July and can reach adult size in August. Reproductive maturity the first year is attained in only part of the young female population (Christian, 1953). Banding returns show that big brown bats may have a life span of 19 years (Goehring, 1972; Paradiso and Greenhall, 1967). There is also evidence that males outlive females (Kurta and Matson, 1980).

FOOD HABITS. With flying insects as its chief prey, the big brown bat has to confine its feeding period to the warm months when these invertebrates are active. The bat must not only support its daily metabolic needs during this time of the year, but also accumulate enough fat to carry over for as long as 194 days of winter hibernation. As much as one-third of the bat's body weight is in the form of stored fat when it enters hibernation in November. In consequence, the big brown bat must be an efficient predator, pursuing insects by echolocation and using the tail and wing membranes as pouch-like devices to assist in capturing them. There are estimates that the bat can catch at least 1.4 grams of insects per hour (Gould, 1955). In captivity, the big brown bat may consume as much as one-third of its weight per day.

Food habit studies have shown that beetles (Coleoptera) are eaten more than any other insect group (Connor, 1971; Hamilton, 1933; Kunz and Fenton, 1973; Whitaker, 1972). As Freeman (1981) has pointed out, the big brown bats has a robust skull and powerful jaws allowing it to feed effectively on hard-shelled beetles. Other insect prey include moths (Lepidoptera), wasps, flying ants, etc. (Hymenoptera), flies (Diptera), stoneflies (Plecoptera), mayflies (Ephemeroptera), true bugs (Hemiptera), caddis flies (Trichoptera), scorpion flies (Mecoptera), lacewing flies (Neuroptera), and dragonflies (Odonata).

ENEMIES. The fact that adult big brown bats often live for several years indicates that they have few determined enemies. Most are probably taken by birds of prey; others are captured by four-footed predators which may climb to their roosting sites or eat grounded individuals. Known predators include barn owl, screech owl, great horned owl, American kestrel, common grackle, Virginia opossum, long-tailed weasel, and striped skunk (Banfield, 1974; Beer, 1953; Black, 1976; Dexter, 1978; Jackson, 1961; Long, 1971; Mumford, 1969; Stupka, 1931). There are also reports of deaths by freezing, severe storms, and even flypaper entanglement (Pine and Duncan, 1975; Rysgaard, 1941).

PARASITES AND DISEASES. The big brown bat is infested with ticks, mites, chiggers, bed bugs, and internal trematode worms and trypanosomes (Jackson, 1961; Whitaker and Mumford, 1971; Whitaker and Wilson, 1974).

MOLT AND COLOR ABERRATIONS. The short, dull pelage of the young bat is shed before the first autumn and replaced by rich, thick adult fur. The annual molt occurs in early summer (Phillips, 1966). Bats with white blotches on the wings as well as albinos are known (Jackson, 1961; Trapido and Crowe, 1942).

ECONOMIC IMPORTANCE. The big brown bat is of major importance in Michigan because (1) it frequently lives in human habitations, (2) it can inflict a severe bite if not handled properly, and (3) it occasionally is infected with rabies. However, statistics obtained from examining 1,093 big brown bats between 1968 and 1981 by Michigan Department of Public Health officials showed that only 55 (5.0%) were rabid (in part from Kurta, 1979). It should also be pointed out that this percentage is an overestimate, since sick bats are much more likely to be submitted for testing than well individuals.

Big brown bats deposit feces (guano) beneath their roosts in attics and other places where the musty odors linger and become offensive. Many of the insects selected by this bat as food are considered noxious.

COUNTY RECORDS FROM MICHIGAN. Reports of specimens preserved in museum collections and/or mentioned in the literature show that the big brown bat is known from the following counties in the Upper Peninsula: Baraga, Chippewa, Delta, Dickinson, Gogebic, Iron, Houghton, Keweenaw, Marquette, Menominee and Ontonagon; in the Lower Peninsula: Allegan, Barry, Bay, Berrien, Branch, Calhoun, Cass, Claire, Clinton, Eaton, Genesee, Gladwin, Gratiot, Hillsdale, Huron, Ingham, Ionia, Iosco, Isabella, Jackson, Kalamazoo, Kent, Lapeer, Lenawee, Leelanau, Livingston, Macomb, Manistee, Mason, Mecosta, Midland, Missaukee, Monroe, Montcalm, Muskegon, Newaygo, Oakland, Ogemaw, Osceola, Oscoda, Otsego, Ottawa, Roscommon, Saginaw, Sanilac, Shiawassee, St. Clair, St. Joseph, Tuscola, Van Buren, Washtenaw, Wayne, and Wexford. The species is also known from South Manitou Island in Lake Michigan and from Isle Royale in Lake Superior.

LITERATURE CITED

Baker, R. H.
1969 Early activity of a big brown bat. Bat Res. News, 10(2):17.

Banfield, A. W. F.
1974 The mammals of Canada. Univ. Toronto Press, Toronto. xxiv+438 pp.

Barbour, R. W., and W. H. Davis
1969 Bats of America. Univ. Kentucky Press, Lexington. 286 pp.

Beer, J. R.
1953 The screech owl as a predator on the big brown bat. Jour. Mammalogy, 34(3):384.
1955 Survival and movements of banded big brown bats. Jour. Mammalogy, 36(2):242–248.

Beer, J. R., and A. G. Richards
1956 Hibernation of the big brown bat. Jour. Mammalogy, 37(1):31–41.

Black, H. L.
1976 American kestrel predation on the bats *Eptesicus fuscus, Euderma maculatum,* and *Tadarida braziliensis.* Southwestern Nat., 21(2):250–251.

Buchler, E. R., and S. B. Childs
1981 Orientation to distant sounds by foraging big brown bats *(Eptesicus fuscus).* Animal Behav., 29(2):428–432.

Burt, W. H.
1946 The mammals of Michigan. Univ. Michigan Press, Ann Arbor. xv+288 pp.

Cahalane, V. H.
1932 Large brown bat in Michigan. Jour. Mammalogy, 13(1):70–71.

Christian, J. J.
1953 The natural history of a summer aggregation of *Eptesicus fuscus fuscus.* Naval Med. Res. Inst., Bethesda, Md., Memo Rep. 53-16:161–193.

Connor, P. F.
1971 The mammals of Long Island, New York. New York State Mus. and Sci. Ser., Bull. 416, v+78 pp.

Davis, W. H., R. W. Barbour, and M. D. Hassell
1968 Colonial behavior of *Eptesicus fuscus.* Jour. Mammalogy, 49(1):44–50.

Davis, W. H., and R. E. Mumford
1962 Ecological notes on the bat *Pipistrellus subflavus.* American Midl. Nat., 68(2):394–398.

Davis, W. H., and O. B. Reite
1967 Responses of bats from temperate regions to changes in ambient temperature. Biol. Bull., 132:320–328.

Dexter, R. W.
1978 Mammals utilized as food by owls in reference to the local fauna of northeastern Ohio. Kirtlandia, No. 24, 6 pp.

Freeman, P. W.
1981 Correspondence of food habits and morphology in insectivorous bats. Jour. Mammalogy, 62(1):166–173.

Goehring, H. H.
1954 *Pipistrellus subflavus obscurus, Myotis keenii,* and *Eptesicus fuscus fuscus* hibernating in a storm sewer in central Minnesota. Jour. Mammalogy, 35(3):434–435.
1972 Twenty-year study of *Eptesicus fuscus* in Minnesota. Jour. Mammalogy, 53(1):201–207.

Gould, E.
1955 The feeding efficiency of insectivorous bats. Jour. Mammalogy, 36(3):399–407.

Hall, J. S.
1962 A life history and taxonomic study of the Indiana bat, *Myotis sodalis.* Reading Pub. Mus. and Art Gallery, Sci. Publ. No. 12, 68 pp.

Hall, J. S., and F. J. Brenner
1968 Summer netting of bats at a cave in Pennsylvania. Jour. Mammalogy, 49(4):779–781.

Hamilton, W. J., Jr.
1933 The insect food of the big brown bat. Jour. Mammalogy, 14(2):155–156.

Hitchcock, H. B.
1949 Hibernation of bats in southeastern Ontario and adjacent Quebec. Canadian Field-Nat., 63:47–59.

Humphrey, S. R.
1975 Nursery roosts and community diversity of Nearctic bats. Jour. Mammalogy, 56(2):321–346.

Jackson, H. H. T.
1961 Mammals of Wisconsin. Univ. Wisconsin Press, Madison. xii+504 pp.

Jones, J. K., Jr., E. D. Fleharty, and P. B. Dunnigan
1967 The distributional status of bats in Kansas. Univ. Kansas, Mus. Nat. Hist. Publ. 46, 33 pp.

Kirkpatrick, C. M., and C. H. Conaway
1948 Some notes on Indiana mammals. American Midl. Nat., 39(1):128–136.

Kunz, T. H.
1974 Reproduction, growth, and mortality of the vespertilionid bat, *Eptesicus fuscus,* in Kansas. Jour. Mammalogy, 55(1):1–13.

Kunz, T. H., and M. B. Fenton
1973 Resource partitioning by *Eptesicus fuscus* and *Lasiurus cinereus.* Bat Res. News, 14(4):55–56.

Kurta, A.
1979 Bat rabies in Michigan. Michigan Acad., 12(2):221–230.
1980 The bats of southern Lower Michigan. Michigan State Univ., unpubl. M.S. thesis, ix+147 pp.

Kurta, A., and J. O. Matson
1980 Disproportionate sex ratio in the big brown bat *(Eptesicus fuscus).* American Midland, Nat., 104(2):367–369.

Long, C. F.
1971 Common grackles prey on big brown bat. Wilson Bull., 83(2):196.

Mills, R. S., G. W. Barrett, and M. P. Farrell
1975 Population dynamics of the big brown bat *(Eptesicus fuscus)* in southwest Ohio. Jour. Mammalogy, 56(3): 591–604.

Muir, T. J., and E. Polder
1960 Notes on hibernating bats in Dubuque County caves. Proc. Iowa Acad. Sci., 67:602–606..

Mumford, R. E.
1958 Population turnover in wintering bats in Indiana. Jour. Mammalogy, 39(2):253–261.
1969 Long-tailed weasel preys on big brown bats. Jour. Mammalogy, 50(2):360.

Mumford, R. E., and J. B. Cope
1964 Distribution and status of the Chiroptera of Indiana. American Midl. Nat., 72(2):473–489.

Nagorsen, D. W.
1980 Records of hibernating big brown bats *(Eptesicus fuscus)* and little brown bats *(Myotis lucifugus)* in northwestern Ontario. Canadian Field-Nat., 94(1):83–85.

Paradiso, J. L., and A. M. Greenhall
1967 Longevity records for American bats. American Midl. Nat., 78(1):251–252.

Patterson, A. P., and J. W. Hardin
1969 Flight speeds of five species of vespertilionid bats. Jour. Mammalogy, 50(1):152–153.

Peterson, R. L., and M. B. Fenton
1969 A record of Siamese twinning in bats. Canadian Jour. Zool., 46(1):154.

Phillips, G. L.
1966 Ecology of the big brown bat (Chiroptera: Vespertilionidae) in northeastern Kansas. American Midl. Nat., 75(1):168–198.

Pine, R. H., and W. B. Duncan
1975 Bat versus flypaper. Bat Res. News, 16(1):9.

Rysgaard, G. N.
1941 Bats killed by severe storm. Jour. Mammalogy, 22(4): 452–453.
1942 A study of the cave bats of Minnesota with special reference to the large brown bay, *Eptesicus fuscus fuscus* (Beauvois). American Midl. Nat., 28:245–267.

Schowalter, D. B., and J. R. Gunson
1979 Reproductive biology of the big brown bat *(Eptesicus fuscus)* in Alberta. Canadian Field-Nat., 93(1):48–54.

Smith, E., and W. Goodpaster
1958 Homing in nonmigratory bats. Science, 127(3299): 644.

Stones, R. C., and W. Fritz
1969 Bat studies in upper Michigan's copper mining district. Michigan Acad., 2(1):77–85.

Stupka, A.
1931 The dietary habits of barn owls. Ohio Dept. Agri., Div. Conserv., Bull. 6, 5 pp.

Trapido, H., and P. E. Crowe
1942 Color abnormalities in three genera of northeastern cave bats. Jour. Mammalogy, 23(3):303–305.

Weber, N. S., and J. S. Findley
1970 Warm-season changes in fat content of *Eptesicus fuscus*. Jour. Mammalogy, 51(1):160–162.

Whitaker, J. O., Jr.
1972 Food habits of bats from Indiana. Canadian Jour. Zool., 50:877–883.

Whitaker, J. O., Jr., and R. E. Mumford
1971 Notes on a collection of bats taken by mist-netting at an Indiana cave. American Midl. Nat., 85(1):277–279.

Whitaker, J. O., Jr., and N. Wilson
1974 Host and distribution lists of mites (Acari), parasitic and phoretic, in the hair of wild mammals of North America, north of Mexico. American Midl. Nat., 91(1):1–67.

Wolverton, C., and A. L. Edgar
1969 Temperature regulation of migrating and hibernating bats in Gratiot County, Michigan. Bios, 40(4):147–153.

RED BAT
Lasiurus borealis (Müller)

NAMES. The generic name *Lasiurus,* proposed by Gray in 1831, is derived from two Greek words, *lasios* meaning "hairy" and *oura* meaning "tail." The specific name *borealis,* proposed by Müller in 1776, is derived from a Latin word meaning "northern." This "northern hairy-tailed" bat is sometimes referred to locally as tree bat, red tree bat, or leaf bat. The subspecies *Lasiurus borealis borealis* (Müller) occurs throughout Michigan.

RECOGNITION. Body medium, the size of a short, fat mouse; head and body approximately 2¼ in (58 mm) long; nose blunt; eyes small and beady; ears short and rounded, when laid (not pulled) forward not reaching as far as nostrils;

tragus in ear broad, short and bent forward slightly; thumbs long; mammary glands four in number.

The sexes of this most attractive of Michigan bats have distinctive colors, with males bright brick-red and females dull yellowish-red. In addition, females have a frosted look caused by cream-colored tips on the dorsal hairs and a more buffy head. The fur in both sexes is long, silky, basally black and possesses none of the oily texture so characteristic of the big brown bat. Furry patches on the shoulder, and at the bases of the thumb, fourth and fifth metacarpals are yellowish-white. The heavily-furred upper tail (interfemoral) membrane (Fig. 18) is colored like the back. The underside of the humerus, radius, and basal parts of the metacarpals are pale colored like the underparts. Ears are densely covered with short orange fur. Wing membranes are black. The reddish coloring distinguishes this species from all other Michigan bats.

MEASUREMENTS AND WEIGHTS. Adults measure from tip of nose to end of tail vertebrae, 3⅝ to 4⅝ in (93 to 117 mm); length of tail vertebrae, 1½ to 2⅛ in (40 to 55 mm); length of hind foot, ¼ to ⅜ in (6 to 11 mm); height of ear from notch, 5/16 to ½ in (8 to 13 mm); length of bone of forearm, 1⅜ to 1¾ in (36 to 46 mm); weight, 0.24 to 0.45 oz (7 to 13 g).

DISTINCTIVE CRANIAL AND DENTAL CHARACTERISTICS. The skull of the red bat is small, heavily constructed, broad, and deep (see Fig. 19). The nasal openings and the rostrum are distinctly wide, the latter almost as wide as the braincase. There is a pronounced ridge directly above the lacrimal bone at the forward edge of each orbit. The skull is approximately ½ in (13.1 to 14.4 mm) long and ⅜ in (9.0 to 10.4 mm) wide (across zygomatic arches). The arrangement of the 32 teeth is distinctive, in common in Michigan only with that of the hoary bat. A tiny, cone-shaped upper premolar nestled at the inner junction of the large upper canine and second premolar is seen best with a magnifying glass. The dental formula is:

$$\text{I (incisors)}\frac{1\text{-}1}{3\text{-}3}, \text{C (canines)}\frac{1\text{-}1}{1\text{-}1}, \text{P (premolars)}\frac{2\text{-}2}{2\text{-}2}, \text{M (molars)}\frac{3\text{-}3}{3\text{-}3} = 32$$

DISTRIBUTION. The red bat enjoys one of the largest distributions of any Michigan bat and occurs from southern Canada southward (avoiding most non-forested plains and deserts) throughout the Americas, across the Equator and as far south as Argentina and Chile. The red bat is widely distributed in Michigan, (see Map 19), and is even found on Isle Royale (Nichols and Stones, 1971). As far as is known, the red bat is a warm-weather resident, migrating southward in the cold months

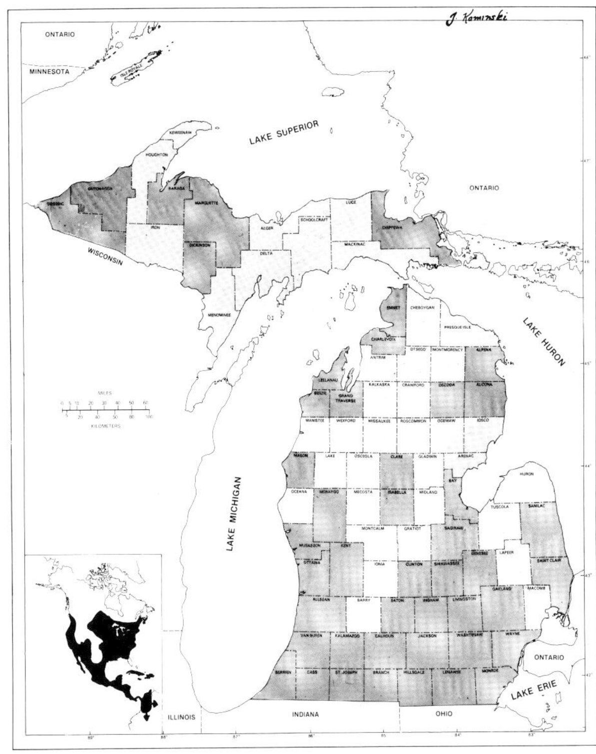

Map 19. Geographic distribution of the red bat (*Lasiurus borealis*). Actual records exist for counties with dark shading.

of the year. According to Kurta (1980), the earliest spring record for Michigan is April 30 and the last autumn record in November.

HABITAT PREFERENCES. The red bat requires at least some tree growth and can be expected in most Michigan locales: forested areas, cities where shade trees abound in parks, cemeteries, and along avenues, and farming areas with scattered woodlots.

DENSITY AND MOVEMENTS. Unlike Michigan's colonial bats, the red bat is mostly solitary, although more than one bat may use a single tree. Even so, the species is abundant; Mumford and Cope (1964) regard it as one of the most numerous bats in Indiana, while McClure (1942) estimated in his study area in Iowa that the roosting density was about one individual per acre. Comparable numbers may occur in some parts of southern Michigan, although no estimates are available. Following the summer young-rearing, red bats tend to swarm in August at and near mouths of caves (Barbour and Davis, 1969; Mumford and Whitaker, 1975). In addition, red bats will forage in close association in summer (LaVal and LaVal, 1979) and seem to migrate in groups to southern wintering places.

Because of their solitary nature, red bats have been difficult to capture, band, and recapture in large enough numbers to determine the extent of their movements. Most dramatic observations have been of bats seen flying across the Great Lakes (Superior, Nichols and Stones, 1971; Erie, Banfield, 1974; Saunders, 1930) or out at sea in the Atlantic (Barbour and Davis, 1969; Connor, 1971). Other evidences of migration include finding an occasional dead red bat at the bases of lighthouses, buildings and television transmitting towers—where both migrating birds and bats accidentally dash themselves against structures or guy wires (Van Gelder, 1956).

BEHAVIOR. The red bat is most conspicuous to the Michigan observer in two summer situations: one, when this vivid reddish flyer darts through the beam of an automobile headlight or into the glow of a streetlight; the other, when the observer happens to see a roosting red bat hanging in the foliage of a tree. The bat's reddish pelage does little to camouflage the creature against the background of green leaves. For would-be red bat watchers, sycamores, oaks, elms, and box elders seem to be preferred trees (Constantine, 1966), where the bats hang at the tree crown edges. The bats leave their roosts in early evening to catch flying insects. Foraging activity may be concentrated in a single peak, but there are records of active red bats at all hours of the night.

Red bats differ markedly from the myotis bats and the big brown bat in being able to find adequate vegetation for summer shelters. Sites used, according to Mumford (1973), have been found as low as 2 ft (0.6 m) to as high as 40 ft (12.2 m). At low levels, red bats may forsake trees and cling to brushes, vines, briar patches, weed stalks, and even the underside of a sunflower leaf (Downes, 1964). In such exposed situations, these durable bats not only withstand extreme temperature changes and direct sunlight, but also must endure the buffeting by wind on their unstable perches. No doubt their abundant fur and hairy tails afford some protection, as the bats are able to curl up into ball-like shapes to gain maximum heat conservation. While in their shelters, red bats conserve energy by going into torpor, dropping their body temperatures and slowing down other metabolic functions (Reite and Davis, 1966).

There are records of red bats being found in caves and trees cavities (Barbour and Davis, 1969; Fassler, 1975; Myers, 1960; Quay and Miller, 1955). In most cases these finds have been in late summer or when the bats are on migration. Although red bats have not yet been recorded in Michigan in the winter months, this could occasionally happen (Davis and Lidicker, 1956). In more moderate Indiana, for example, Mumford (1973) has recorded the red bat for every month except February. Although it is supposed that red bats hibernate in vegetation roosts in the south, Mumford suspects that a few individuals might over-winter in tree holes in northern areas.

Long, narrow wings provide the red bat with a strong flight pattern. In straightaway flight, Jackson (1961) thought that this bat might attain flight speeds of 50 mi (64 km) per hour; in a hallway, Patterson and Hardin (1969) timed a flying red bat at 7.9 mi (18 km) per hour. Although red bats may associate during migrations, they generally maintain their solitary nature except during the mating season (Jones *et al.*, 1966).

ASSOCIATES. The tree habitat of the red bat naturally tends to keep this species fairly segregated from other kinds of bats. In summer, red bats join other kinds of bats in using common foraging and drinking areas (Davis and Mumford, 1962). In mist nets set across the Thornapple River in a wooded sector in Eaton County, for example, Kurta (1980) caught the red bat in company with Keen's bat, little brown bat, Indiana bat, big brown bat, silver-haired bat, and hoary bat. While using a bat trapping device (Tuttle, 1974), Kurta caught two red bats along with 87 Keen's bats and 267 little brown bats in a sample of swarming individuals on seven nights in September at Bear Cave in Berrien County.

REPRODUCTIVE ACTIVITIES. Red bats mate in late summer and early autumn (August and September). Stuewer (1948) found a pair mating at dusk (2021 hours) on August 20, on the bank of Swan Creek Pond in Allegan County. This act may be initiated in flight with the individuals then descending to the ground. The male spermatozoa are stored in the female reproductive tract until spring when ovulation and fertilization occur (Barbour and Davis, 1969).

In April and May, the females journey northward to Michigan, presumably preceding the males by almost two months (*e.g.*, Cockrum, 1955). It has been estimated that the gestation period extends from 80 to 90 days (Jackson, 1961). In Indiana, Mumford (1973) found females containing embryos as early as April 30. In Michigan births have been recorded in June. Females with attached young have been examined as early as June 20 in Washtenaw County (Wood, 1922) and as late as July 15 in Kalamazoo County (Drake, 1957). Normally a litter will contain three or four and occasionally five offspring (Hamilton and Stalling, 1972; Mumford, 1973). Each may weigh only 0.5 grams at birth. While the mother forages for insect foods, she is obliged to leave her young at the roost, except when moving them to other retreats. Stains (1965) reported four young being carried by a female red bat in Illinois and weighing almost twice (23.4 g) as much as the parent (12.9 g). Young bats are probably able to fly in three to four weeks after birth but may not be weaned for five to six weeks (Barbour and Davis, 1969). LaVal and LaVal (1979) recorded flying juveniles as early as July 21 and 25 in Louisiana and Iowa. In Michigan, Kurta (1980) mist-netted a juvenile on July 24 in Eaton County.

FOOD HABITS. The red bat selects a diverse assortment of insect foods during its nightly feeding flights. Representatives of six insect orders (Homoptera, Coleoptera, Hymenoptera, Diptera, Orthoptera, and Lepidoptera) were found in stomach contents of red bats (Freeman, 1981; Ross, 1967). In New York, moths of the family Geometridae and small beetles were noted as foods of this bat (Connor, 1971); moths and beetles were also found to be preferred foods in Indiana (Whitaker, 1972). Red bats have been so aggressive in pursuing insects that they have even been caught in ultraviolet light insect traps (Wilson, 1965). Crickets have also been recorded as red bat food (Jackson, 1961) indicating that some prey may be captured at or near ground level.

ENEMIES. Red bats, because of their solitary, tree-living habits, are subject to different predators than colonial cave bats. One of these is the blue jay, which seems able to locate the tree roosts of red bats. Attacks by this heavy-beaked bird have been reported in Iowa, Texas, Illinois, and Michigan (Kalamazoo County, Drake, 1957). Other successful attackers include the opossum (Sperry, 1933), sharp-shinned hawk (Downing and Baldwin, 1961), merlin (Johnson and Coble, 1967), other hawks and various owls (Lowery, 1974). Accidents befalling red bats include: impalement on barbed wire (Long, 1964); entanglement on the head of a burdock plant (Mumford, 1973); hitting television towers and their guywires (Long, 1976; Van Gelder, 1956); entrapment in road surface oil (Koestner, 1942); flying into lighthouses, buildings (Barbour and Davis, 1969) and the radiator grill of an automobile (Long, 1976); and drowning during a severe Lake Michigan storm (Mumford, 1973).

PARASITES AND DISEASES. The red bat is infested with mites (Whitaker and Wilson, 1974), fleas, bat bugs *(Cimex),* and internal trematodes, tapeworms, and protozoan parasites (Jackson, 1961; Lowery, 1974).

MOLT AND COLOR ABERRATIONS. The difference in color between the two sexes is one remarkable characteristic of the red bat which is not shared with other Michigan bats. Young bats also display similar sexual differences in color. The juvenile males are dark red and the immature females grayish with slight traces of red (Dice, 1927). No albinistic or melanistic individuals have been reported.

ECONOMIC IMPORTANCE. The red bat with its attractive reddish color and solitary nature has a more favorable rapport with the general public than do the myotis bats and the big brown bat. Also, a furry red bat with attached squirming young in an oak tree is an interesting sight. That the red bat rarely invades houses is also a point in its favor and acceptance. Only three rabid red bats have been found in Michigan (Kurta, 1979). There are also records from Indiana (Whitaker *et al.*, 1969) and nationwide (Baer and Adams, 1970). Again, the general public must be cautious in handling bats and visit a physician promptly if bitten. Of course, this same precaution should be taken if bitten by any wild (or tame) mammal.

COUNTY RECORDS FROM MICHIGAN. Reports of specimens preserved in museum collections and/or mentioned in the literature show that the red bat has a state-wide distribution and is known from the following counties in the Upper Peninsula: Baraga, Chippewa, Dickinson, Gogebic, Marquette, Ontonagon; in the Lower Peninsula: Alcona, Allegan, Alpena, Bay, Benzie, Berrien, Branch, Calhoun, Cass, Charlevoix, Clare, Clinton, Eaton, Emmet, Genesee, Grand Traverse, Hillsdale, Ingham, Isabella, Jackson, Kalamazoo, Kent, Leelanau, Lenawee, Livingston, Mason, Monroe, Muskegon, Newaygo, Oakland, Oscoda, Ottawa, Saginaw, Sanilac, Shiawassee, St. Claire, St. Joseph, Van Buren, Washtenaw, and Wayne. The species is also known from Isle Royale and Huron Island in Lake Superior, from North Manitou, South Fox, and South Manitou islands in Lake Michigan, and the Charity Islands in Saginaw Bay (Burt, 1946, Hatt *et al.*, 1948; Kurta, 1980; Scharf and Jorae, 1980).

LITERATURE CITED

Baer, G. M., and D. B. Adams
1970 Rabies in insectivorous bats in the United States, 1953–65. Public Health Rpts., 85(7):637–645.

Banfield, A. W. F.
1974 The mammals of Canada. Univ. Toronto Press, Toronto. xxiv+438 pp.

Barbour, R. W., and W. H. Davis
1969 Bats of America. Univ. Kentucky Press, Lexington. 286 pp.

Burt, W. H.
1946 The mammals of Michigan. Univ. Michigan Press, Ann Arbor. xv+288 pp.

Cockrum, E. L.
1955 Reproduction in North American bats. Trans. Kansas Acad. Sci., 58(4):487–511.

Connor, P. F.
1971 The mammals of Long Island, New York. New York State Mus. and Sci. Ser., Bull. 416, v+78 pp.

Constantine, D. G.
1966 Ecological observations of lasiurine bats in Iowa. Jour. Mammalogy, 47(1):34–41.

Davis, W. H., and W. Z. Lidicker, Jr.
1956 Winter range of the red bat, *Lasiurus borealis.* Jour. Mammalogy, 37(2):280–281.

Davis, W. H., and R. E. Mumford
1962 Ecological notes on the bat *Pipistrellus subflavus.* American Midl. Nat., 68(2):394–398.

Dice, L. R.
1927 Notes on the young of the red bat *(Nycteris borealis borealis).* Jour. Mammalogy, 8(3):243–244.

Downes, W. L., Jr.
1964 Unusual roosting behavior in red bats. Jour. Mammalogy, 45(1):143–144.

Downing, S. C., and D. H. Baldwin
1961 Sharp-shinned hawk preys on red bat. Jour. Mammalogy, 42(4):540–541.

Drake, J. J.
1957 Blue jay attacks nursing red bat. Jack-Pine Warbler, 35(2):79.

Fassler, D. J.
1975 Red bat hibernating in a woodpecker hole. American Midl. Nat., 93(1):254.

Freeman, P. W.
1981 Correspondence of food habits and morphology in insectivorous bats. Jour. Mammalogy, 62(1):166–173.

Hamilton, R. B., and D. T. Stalling
1972 *Lasiurus borealis* with five young. Jour. Mammalogy, 53(1):190.

Hatt, R. T., J. Van Tyne, L. C. Stuart, C. H. Pope, and A. B. Grobman
1948 Island life: A study of the land vertebrates of the islands of eastern Lake Michigan. Cranbrook Inst. Sci., Bull. No. 27, xi+179 pp.

Jackson, H. H. T.
1961 Mammals of Wisconsin. Univ. Wisconsin Press, Madison. xii+504 pp.

Johnson, W. J., and J. A. Coble
1967 Notes on the food habits of pigeon hawks. Jack-Pine Warbler, 45(3):97–98.

Jones, J. K., Jr., E. D. Fleharty, and P. B. Dunnigan
1966 The distributional status of bats in Kansas. Univ. Kansas, Mus. Nat. Hist., Misc. Publ. 46, 33 pp.

Koestner, E. J.
1942 A method of collecting bats. Jour. Tennessee Acad. Sci., 17:301.

Kurta, A.
1979 Bat rabies in Michigan. Michigan Acad., 12(2):221–230.
1980 The bats of southern lower Michigan. Michigan State Univ., unpubl. M.S. thesis, ix+197 pp.

LaVal, R. K., and M. L. LaVal
1979 Notes on reproduction, behavior, and abundance of the red bat, *Lasiurus borealis.* Jour. Mammalogy, 60(1): 209–212.

Long, C. A.
1964 Red bat impaled on barbed wire. Trans. Kansas Acad. Sci., 67(1):201.
1976 The occurrence, status and importance of bats in Wisconsin with a key to the species. Wisconsin Acad. Sci., 64:62–82.

Lowery, G. H., Jr.
1974 The mammals of Louisiana and its adjacent waters. Louisiana State Univ. Press, Baton Rouge. xxiii+565 pp.

McClure, H. E.
1942 Summer activities of bats (genus *Lasiurus*) in Iowa. Jour. Mammalogy, 23(4):430–434.

Mumford, R. E.
1973 Natural history of the red bat *(Lasiurus borealis)* in Indiana. Period. Biol., 75(1):155–158.

Mumford, R. E., and J. B. Cope
1964 Distribution and status of the Chiroptera of Indiana. American Midl. Nat., 72(2):473–489.

Mumford, R. E., and J. O. Whitaker, Jr.
1975 Seasonal activity of bats in an Indiana cave. Proc. Indiana Acad. Sci., 84:500–507.

Myers, R. F.
1960 *Lasiurus* from Missouri caves. Jour. Mammalogy, 41(1):114–117.

Nichols, G. E., and R. C. Stones
1971 Occurrence of red bats near and in Isle Royale. Jack-Pine Warbler, 49(4):130–131.

Patterson, A. P., and J. W. Hardin
1969 Flight speeds of five species of vespertilionid bats. Jour. Mammalogy, 50(1):152–153.

Quay, W. B., and J. S. Miller
1955 Occurrence of the red bat, *Lasiurus borealis,* in caves. Jour. Mammalogy, 36(3):454–455.

Reite, O. B., and W. H. Davis
1966 Thermoregulation in bats exposed to low ambient temperatures. Proc. Soc. Exp. Biol. Med., 121:1212–1215.

Ross, A.
1967 Ecological aspects of the food habits of insectivorous bats. Proc. Western Found. Vert. Zool., 1:205–263.

Saunders, W. E.
1930 Bats in migration. Jour. Mammalogy, 11(2):225.

Scharf, W. C. and M. L. Jorae
1980 Birds and land vertebrates of North Manitou Island. Jack-Pine Warbler, 58(1):4–15.

Sperry, C. C.
1933 Opossum and skunk eat bats. Jour. Mammalogy, 14(2):152–153.

Stains, H. J.
1965 Female red bat carrying four young. Jour. Mammalogy. 46(2):333.

Stuewer, F. W.
1948 A record of red bats mating. Jour. Mammalogy, 29(2):180–181.

Tuttle, M.
1974 An improved trap for bats. Jour. Mammalogy, 55(3): 475–477.

Van Gelder, R. G.
1956 Echo-location failure in migratory bats. Trans. Kansas Acad. Sci., 59(2):220–222.

Whitaker, J. O., Jr.
1972 Food habits of bats from Indiana. Canadian Jour. Zool., 50:877–883.

Whitaker, J. O., Jr., W. A. Miller, and W. L. Boyko
1969 Rabies in Indiana bats. Proc. Indiana Acad. Sci., 78:447–456.

Whitaker, J. O., Jr., and N. Wilson
1974 Host and distribution lists of mites (Acari), parasitic and phoretic, in the hair of wild mammals of North America, north of Mexico. American Midl. Nat., 9(1): 1–67.

Wilson, N.
1965 Red bats attracted to insect light traps. Jour. Mammalogy, 46(4):704–705.

Wood, N. A.
1922 The mammals of Washtenaw County, Michigan. Univ. Michigan, Mus. Zool., Occas. Papers No. 123, 23 pp.

HOARY BAT
Lasiurus cinereus (Palisot de Beauvois)

NAMES. The generic name *Lasiurus,* proposed by Gray in 1831, is derived from two Greek words, *lasios* meaning "hairy" and *oura* meaning "tail." The specific name *cinereus,* proposed by Palisot de Beauvois in 1796, is derived from a Latin word meaning "grayish" or "ash colored." This species is sometimes referred to as the frosty bat or great northern bat. The subspecies *Lasiurus cinereus cinereus* (Palisot de Beauvois) occurs throughout Michigan.

RECOGNITION. Body large, the size of a fat mouse; head and body approximately 3½ in (88 mm) long; wingspread 17 in (432 mm) wide; nose blunt and rounded; eyes small and beady, ears short, thick, broad, and rounded, when laid (not pulled) forward not reaching as far as nostrils;

tragus in ear short and blunt; thumbs long; mammary glands four in number.

The long soft hair provides a thick body covering, extends over the entire dorsal surface and to basal part of the lower surface of the tail (interfemoral) membrane, to the elbow, to the median ventral border of the undersides of the wings adjacent to the body, and to the ventral sides of the long bones making up the upper arm and forearm. The color of the upperparts including the hairy tail membrane are mixed browns and grays with a heavy tinge of white producing a hoary or frosted effect. The individual silky hairs are basally dark, medially yellowish, and distally brownish-black with white tips. The underparts are more yellowish-brown and lack the white tips to the individual hairs. The throat has a distinct yellowish patch. The hair on the elbow, at the base of the clawed thumb, and on the upper arm is yellowish. The ears are yellow with blackish rims. The large size and frosty-brown coloring easily set this species apart from all other Michigan bats.

MEASUREMENTS AND WEIGHTS. Adults measure as follows: length from tip of nose to end of tail vertebrae, 5⅛ to 5⅞ in (130 to 150 mm); length of tail vertebrae, 2⅛ to 2⅜ in (53 to 64 mm); length of hind foot, ⅜ to ½ in (10 to 14 mm); height of ear from notch, ⅝ to ¾ in (17 to 19 mm); length of bone of forearm, 1⅞ to 2¼ in (46 to 56 mm); weight, 0.87 to 1.22 oz (25 to 35 g).

DISTINCTIVE CRANIAL AND DENTAL CHARACTERISTICS. The skull of the hoary bat is large, massive and broad (Fig. 19). The rostrum is noticeably wide with a broad nasal opening. In most features, the skull of the hoary bat resembles closely that of its smaller relative, the red bat. The skull is approximately ⅝ in (17 to 18 mm) long and ½ in (11.8 to 13.9 mm) wide (across zygomatic arches). The arrangement of the teeth with the diminutive first premolar nestled at the inner junction of the large canine and second premolar is like that of the red bat. The dental formula is:

$$\text{I (incisors)}\frac{1\text{-}1}{3\text{-}3},\ \text{C (canines)}\frac{1\text{-}1}{1\text{-}1},\ \text{P (premolars)}\frac{2\text{-}2}{2\text{-}2},\ \text{M (molars)}\frac{3\text{-}3}{3\text{-}3} = 32$$

DISTRIBUTION. The hoary bat ranges from near tree line at the edge of the Canadian arctic southward through Central America into South America as far southward as Argentina and Chile. This species has the most expansive distribution of all American bats, using all environments except for treeless plains and deserts. In Michigan, the hoary bat is classed as rare but can be expected in any part of the state in summer (see Map 20). Except for an occasional hardy individual, the entire population flies to more southern areas in winter.

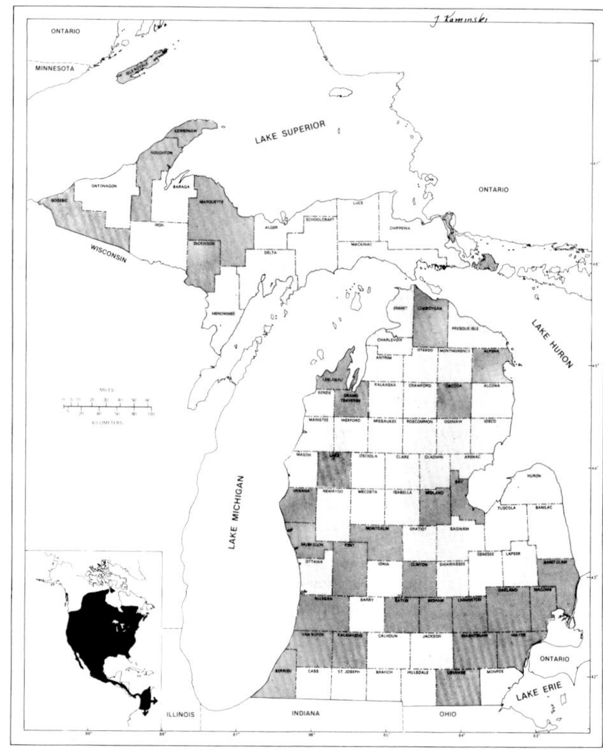

Map 20. Geographic distribution of the hoary bat (*Lasiurus cinereus*). Actual records exist for counties with dark shading.

HABITAT PREFERENCES. The hoary bat, like the red bat, occurs in areas where at least some tree growth is present. This includes heavy forest, open wooded glades, and shade trees along urban streets or in city parks. There is some undocumented evidence that hoary bats prefer coniferous trees (pine, fir, spruce) to the big-leaved deciduous trees, especially in northern Michigan.

DENSITY AND MOVEMENTS. In summer, males and nursing females roost separately in trees or other daytime retreats. This dispersed population allows little chance for observers to obtain density figures. Catches in mist nets stretched across streamways provide data about their presence but no real quantitative information (Mumford, 1969). Apparently, hoary bats congregate for their seasonal movements north and south, with groups noted from as early as February to as late as June and again from August to as late as November (Barbour and Davis, 1969; Findley and Jones, 1964; Mumford, 1963; Zinn and Baker, 1979). For Michigan, which is near the northern edge of this bat's range, the hoary bat seems to be a late arrival, with the earliest active individuals taken on April 10 in Grand Traverse County, on April 21 in Oakland County (Kurta, 1980), and on May 13 on South Manitou Island (Scharf, 1973). As a rule, females are the first to arrive. No positive autumn departure dates of this bat are recorded, although Kurta (1980) caught his last hoary bat of the season on August 24 in Livingston County.

The hoary bat occasionally occurs in Michigan during cold months. For example, individuals have been found behind a log cabin door in Berrien County on October 14 (David Mohrhardt, pers. comm.), on a brick siding under the eaves of a home in East Lansing in Ingham County on November 4 (Bowers *et al.*, 1968), in a barn in Washtenaw County in December (Wood, 1922a), and in a house in Ingham County in early April (William Lovis, pers. comm.). There was some evidence that these bats were in torpor when found.

Obvious difficulties in banding large numbers of these bats have prevented the acquisition of much positive data about migratory pathways and wintering grounds (Constantine, 1966). However, Findley and Jones (1964) suggest that at least part of the population which summers in the United States may actually migrate as far as Latin America.

BEHAVIOR. The hoary bat, like its smaller relative, the red bat, spends summers alone, chiefly using trees as daytime retreats. Roosts are generally between 10 and 15 ft (3.0 to 4.6 m) above ground in the foliage of either evergreen or broadleafed trees. In Iowa, Constantine (1966) found hoary bats in elm, box elder, black cherry, plum and osage orange trees. Hoary bats tend to select sites at the edge of a grove and usually use the periphery of the foliage as hanging places. The frosted-brown coloration offers much better camouflage against the plant growth than does the reddish color of the red bat. At any rate, fewer observations of hoary bats in trees are reported. In addition to trees, hoary bats have been reported in a woodpecker hole (Cowan and Guiguet, 1960), in the leaf nest of a gray squirrel (Neill, 1952), under a piece of driftwood (Connor, 1971), but rarely inside of buildings (Mumford, 1969). Although individuals have been found in caves (Myers, 1960; Mumford, 1953), there is no real evidence that they are used for prolonged shelter.

On a typical summer evening, the hoary bat is usually one of the last chiropterans to emerge from its daytime retreat, with an activity peak reached the fifth hour after sunset (Kunz, 1973). This bat has a strong and rather direct flight, up to at least 13 mi (21 km) per hour. When sighted in the glare of city streetlights, the hoary bat is sometimes mistaken for a common nighthawk because of its size and wing length (Banfield, 1974). Hoary bats forage about treetops, along streams and lake shores, and in urban areas where shade trees abound. Very likely they have rest stops between meals at night, but there is no evidence for this.

The hairy covering of the body and tail membrane offer the hoary bat insulation as it hangs in curled fashion in its tree roost during adverse weather. While inactive by day or between feeding flights at night, hoary bats can become torpid with physiological processes reduced. This is, of course, an adaptation shared by many bats to ration more slowly energy supplies for life support systems. In a resting (non-torpid) pose, a hoary bat was found to have a metabolic rate of 1.19 cc of oxygen per gram of body weight per hour (Bowers *et al.*, 1968). Besides ultrasonic sounds uttered for echolocation purposes while flying, hoary bats make shrill, hissing noises when disturbed.

ASSOCIATES. The only time hoary bats appear to associate with other species is while they are feeding or drinking during summer flights (Mumford, 1969). In mist nets stretched across a wooded sector of the Thornapple River in Eaton County, Kurta (1980) captured hoary bats in company with Keen's bat, the little brown bat, big brown bat, silver-haired bat, and red bat.

REPRODUCTIVE ACTIVITIES. The hoary bat is thought to mate in late summer and early autumn (late August to October). At this time the sexes have assembled for southward movement, but it is not known whether copulation occurs prior to, during, or after the bats have begun their southward flights. There is also the suggestion that courtship activity may proceed even during daylight flights (Hall, 1946). However, it is possible that a major part of this reproductive activity may take place after the bats reach southern areas, since there are some indications that males travelling with southbound females are mostly their offspring and not adult breeders (Findley and Jones, 1964). In fact, the literature suggests that adult males, to a large extent, do not accompany the females to northern summering areas, although Kurta (1980) did catch an adult male on July 12 in Eaton County. Much more study of seasonal movements of hoary bats is needed.

In spring, females return to Michigan in a pregnant condition—a result of delayed fertilization by which the sperm is stored in the female tract all winter and is available to fertilize the eggs when ovulation takes place in spring. The length of the gestation period is not known; females in late pregnancy have been examined in early June in Michigan (Kurta, 1980; Wood, 1922b). Births have been reported in other states in late May and early June (Bogan, 1972; Whitaker and Mumford, 1971). A pregnant female was captured May 19 and a lactating female on May 31 in Indiana (Provost and Kirkpatrick, 1952). Mumford (1969) witnessed the birth of two young in Indiana on May 28. Captive pregnant females in New Mexico gave birth to young from June 4 through June 27 (Bogan, 1972).

Although two usually make up a litter, there can be as many as four, accommodated by the four teats. At birth the young possess fine, silvery-gray hair on the head, back, tail membrane and feet and weigh about 4.5 grams. By the third day the ears are open and erect, and by the 12th day, the eyes are open. By the 22nd day, the pelage has grown sufficiently to resemble that of the adults. By the 33rd day, the young bats are capable of purposeful flight (Bogan, 1972). The young are left hidden in foliage while the mother seeks insect foods. On one occasion a female left her two young in an elm tree at 2050 hours, returning at 2325 hours (Goehring, 1955).

In Michigan, it appears that nursing which began in early June may be completed anywhere from early July to early August. Kurta (1980) snared juveniles in mist nets over the Paw Paw River in Van Buren County on July 14 and over the Thornapple River in Eaton County on July 19. Since testes in adult males may be descended and in reproductive readiness by mid-August, mating activities may begin soon after the young are weaned (Baker and Ward, 1967). Juvenile males are reported capable of mating their first year (Druecker, 1972). The hoary bat is probably long-lived; on one occasion an individual was recaptured 25 months after being banded and released (Cockrum, 1973).

FOOD HABITS. The hoary bat's diet appears to be chiefly moths (Lepidoptera), with smaller numbers of flies (Diptera), beetles (Coleoptera), small wasps and relatives (Hymenoptera), and other insects (Black, 1974; Freeman, 1981; Whitaker, 1962). There are also records of this bat eating leaves and shed snake skin (Whitaker, 1967) and even an eastern pipistrelle (Bishop, 1947). In captivity, hoary bats relish mealworms, other insects and parts of the viscera and muscle tissue of a freshly-killed laboratory mouse (Bowers *et al.*, 1968; Jackson, 1961).

ENEMIES. Hawks and owls are probably the chief predators on hoary bats (Lowery, 1974). Specific predators include the kestrel and a rat snake (Church, 1967; Wiseman, 1963). On several occasions, hoary bats have been found hanging on barbed wire fences (Denys, 1972; Hibbard, 1963; Iwen, 1958; Wisely, 1978).

PARASITES AND DISEASES. The hoary bat is infested with mites (Whitaker and Wilson, 1974) and other external parasities as well as internal trematodes (Whitaker, 1967), nematodes and cestodes (Tromba, 1954).

ECONOMIC IMPORTANCE. Except for occasionally hanging in a menacing pose on the sides of houses, the hoary bat rarely disturbs Michigan homemakers. In fact, because of its size, attractive coloring, the tree-living habits, this species usually enjoys an acceptance by the general public. Caution should be taken in handling these bats because they do have sharp teeth and strong jaws and also because there are a few records of rabid hoary bats in the United States (Baer and

Adams, 1970)—only two, however, from Michigan.

COUNTY RECORDS FROM MICHIGAN. Reports of specimens in museum collections or reported in the literature show that the hoary bat has a statewide distribution and is known from the following counties in the Upper Peninsula: Dickinson, Gogebic, Keweenaw, Houghton, and Marquette; in the Lower Peninsula: Allegan, Alpena, Bay, Berrien, Cheboygan, Clinton, Eaton, Grand Traverse, Ingham, Kalamazoo, Kent, Lake, Leelanau, Lenawee, Livingston, Macomb, Midland, Montcalm, Muskegon, Oakland, Oceana, Oscoda, St. Claire, Van Buren, Washtenaw, and Wayne. The species is also known from Isle Royle in Lake Superior and from South Manitou Island in Lake Michigan.

LITERATURE CITED

Baer, G. M,, and D. B. Adams
1970 Rabies in insectivorous bats in the United States, 1953–65. Public Health Reports, 85(7):637–645.

Baker, R. J., and C. M. Ward
1967 Distribution of bats in southeastern Arkansas. Jour. Mammalogy, 48(1):130–132.

Banfield, A. W. F.
1974 The mammals of Canada. Univ. Toronto Press, Toronto, xxiv+438 pp.

Barbour, R. W., and W. H. Davis
1969 Bats of America. Univ. Kentucky Press, Lexington. 286 pp.

Bishop, S. C.
1947 Curious behavior of a hoary bat. Jour. Mammalogy, 28(3):293–294.

Black, H. L.
1974 A north temperate bat community: Structure and prey populations. Jour. Mammalogy, 55(1):138–157.

Bogan, M. A.
1972 Observation on parturition and development in the hoary bat, *Lasiurus cinereus*. Jour. Mammalogy, 53(3): 611–614.

Bowers, J. R., G. A. Heidt, and R. H. Baker
1968 A late autumn record for the hoary bat in Michigan. Jack-Pine Warbler, 46(1):33.

Church, R. L.
1967 Capture of a hoary bat, *Lasiurus cinereus*, by a sparrow hawk. Condor, 69(4):426.

Cockrum, E. L.
1973 Additional longevity records for American bats. Jour. Arizona Acad. Sci., 8:108–110.

Connor, P. F.
1971 The mammals of Long Island, New York. New York State Mus. and Sci. Ser., Bull. 416, v+78 pp.

Constantine, D. G.
1966 Ecological observations of lasiurine bats in Iowa. Jour. Mammalogy, 47(1):34–41.

Cowan, I. M., and C. J. Guiguet
1960 The mammals of British Columbia. British Columbia Prov. Mus., Handbook No. 11, 413 pp.

Denys, G. A.
1972 Hoary bat impaled on barbed wire. Jack-Pine Warbler, 50(3):63.

Druecker, J. D.
1972 Aspects of reproduction in *Myotis volans, Lasionycteris noctivagans*, and *Lasiurus cinereus*. Univ. New Mexico, unpubl. Ph.D. dissertation, 68 pp.

Findley, J. S., and C. Jones
1964 Seasonal distribution of the hoary bat. Jour. Mammalogy, 45(3):461–470.

Freeman, P. W.
1981 Correspondence of food habits and morphology in insectivorous bats. Jour. Mammalogy, 62(1):166–173.

Goehring, H. H.
1955 Observations on hoary bats in a storm. Jour. Mammalogy, 36(1):130–131.

Hall, E. R.
1946 Mammals of Nevada. Univ. California Press, Berkeley and Los Angeles. xi+710 pp.

Hibbard, E. A.
1963 Another hoary bat found hanging on a fence. Jour. Mammalogy, 44(2):265.

Iwen, F. A.
1958 Hoary bat the victim of a barbed wire fence. Jour. Mammalogy,39(3):438.

Jackson, H. H. T.
1961 Mammals of Wisconsin. Univ. Wisconsin Press, Madison. xii+504 pp.

Kunz, T. H.
1973 Resource utilization: Temporal and spatial components of bat activity in central Iowa. Jour. Mammalogy, 54(1):14–32.

Kurta, A.
1979 Bat rabies in Michigan. Michigan Acad., 12(2):221–230.
1980 The bats of southern lower Michigan. Michigan State Univ., unpubl. M.S. thesis, ix+147 pp.

Lowery, G. H., Jr.
1974 The mammals of Louisiana and its adjacent waters. Louisiana State Univ. Press, Baton Rouge, xxiii+565 pp.

Mumford, R. E.
1953 Hoary bat skull in an Indiana cave. Jour. Mammalogy, 34(1):121.
1963 A concentration of hoary bats in Arizona. Jour. Mammalogy, 44(2):272.
1969 The hoary bat in Indiana. Proc. Indiana Acad. Sci., 78:497–501.

Myers, R. F.
1960 *Lasiurus* from Missouri caves. Jour. Mammalogy, 41(1):114–117.

Neill, W. T.
1952 Hoary bat in a squirrel's nest. Jour. Mammalogy, 33(1):113.

Provost, E. E., and C. M. Kirkpatrick
1952 Observations on the hoary bat in Indiana and Illinois. Jour. Mammalogy, 33(1):110–113.

Scharf, W. C.
1973 Land vertebrates of South Manitou Island. Jack-Pine Warbler, 41(1):2–19.

Tromba, F. G.
1954 Some parasites of the hoary bat, *Lasiurus cinereus* (Beauvois). Jour. Mammalogy, 34(2):253–254.

Whitaker, J. O., Jr.
1967 Hoary bat apparently hibernating in Indiana. Jour. Mammalogy, 48(4):663.

Whitaker, J. O., Jr., and R. E. Mumford
1971 Notes on occurrence and reproduction of bats in Indiana. Proc. Indiana Acad. Sci., 81:376–382.

Whitaker, J. O., Jr., and N. Wilson
1974 Host and distribution lists of mites (Acari), parasitic and phoretic, in the hair of wild mammals of North America, north of Mexico. American Midl. Nat., 91(1):1–67.

Wisely, A. N.
1978 Bat dies on barbed wire fence. Blue Jay, 36(1):53.

Wiseman, J. S.
1963 Predation by the Texas rat snake on the hoary bat. Jour. Mammalogy, 44(4):581.

Wood, N. A.
1922a The mammals of Washtenaw County, Michigan. Univ. Michigan, Mus. Zool., Occas. Papers No. 123, 23 pp.
1922b Notes on the mammals of Berrien County, Michigan. Univ. Michigan, Mus. Zool., Occas. Papers No. 124, 4 pp.

Zinn, T. L., and W. W. Baker
1979 Seasonal migration of the hoary bat, *Lasiurus cinereus*, through Florida. Jour. Mammalogy, 60(3):634–635.

EVENING BAT
Nycticeius humeralis (Rafinesque)

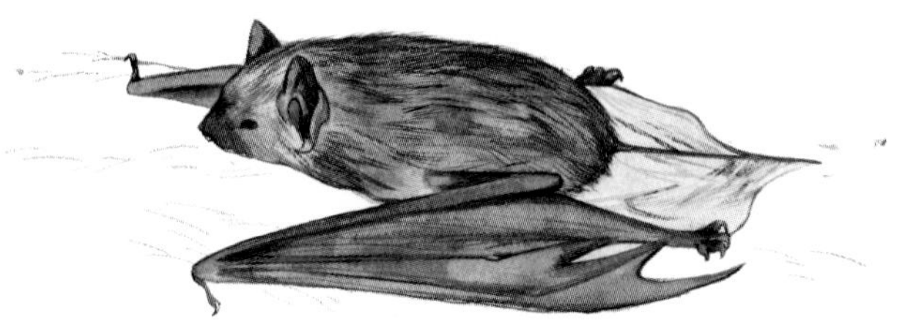

NAMES. The generic name *Nycticeius,* proposed by Rafinesque in 1819, is derived from the Greek and means "night." The specific name *humeralis,* proposed by Rafinesque in 1818, is from the Latin and can be translated "of the shoulder." This species is sometimes referred to as the twilight bat. The subspecies *Nycticeius humeralis humeralis* (Rafinesque) occurs in Michigan.

RECOGNITION. Body medium, the size of a short, fat mouse; head and body approximately 2¼ in (57 mm) long; nose pointed; eyes small and beady; ears rounded, short, extending when laid (not pulled) forward just to or short of the nostrils; tragus in ear short, blunt, and broad; calcar with slight or no keel-shaped extension (Fig. 28C).

The short, thin pelage is drab brown above and pale brown below. The tip of the muzzle, ears and membranes are hairless and blackish. Its small size distinguishes the evening bat from its near look-alike, the much larger big brown bat. It differs from the myotis bats by the presence of a blackish muzzle, rounded ears, and short tragus.

MEASUREMENTS AND WEIGHTS. Adults measure from tip of nose to end of tail vertebrae, 3⅜ to 4 in (86 to 103 mm); length of tail vertebrae, 1 5/16 to 1 9/16 in (33 to 40 mm); length of hind foot, ¼ to ⅜ in (6 to 9 mm); height of ear from notch, ⅜ to 9/16 in (9 to 14 mm); length of bone of forearm, 1 5/16 to 1½ in (33 to 38 mm); weight, 0.21 to 0.42 oz (6 to 12 g).

DISTINCTIVE CRANIAL AND DENTAL CHARACTERISTICS. The skull of the evening bat is short, broad and flattened (see Fig. 29). In profile, there is little elevation of the braincase above the level of the rostrum. The skull is approximately 9/16 in (14.1 to 14.6 mm) long and ⅜ in (9.6 to 10.2 mm) wide (across zygomatic arches).

This species has only 30 teeth, the fewest of any Michigan chiropteran. The upper incisor is separated from the canine by a space less than one-half the diameter of the incisor (Watkins, 1972). The dental formula is:

$$\text{I (incisors)}\frac{1\text{-}1}{3\text{-}3}, \text{C (canines)}\frac{1\text{-}1}{1\text{-}1}, \text{P (premolars)}\frac{1\text{-}1}{2\text{-}2}, \text{M (molars)}\frac{3\text{-}3}{3\text{-}3} = 30$$

DISTRIBUTION. The evening bat is distributed in southeastern North America from the Great Plains in the west and the southern Great Lakes in the north east to the Atlantic coast and southward to the Gulf States and northeastern Mexico. Its presence in Michigan is based on the preservation of three specimens, one from Kalamazoo County (May 23, 1938), one from Washtenaw County (May 21, 1956), and one from Berrien County (June 2, 1969), see Map 21 and Kurta (1980). Because Michigan is at the extreme northern part of this bat's range, it is expected that local populations may be small and scattered. The specimen from Ann Arbor in Washtenaw County is the most northern record for the species (Hall, 1981:227).

Map 21. Geographic distribution of the evening bat (*Nycticeius humeralis*). Actual records exist for counties with dark shading.

HABITAT PREFERENCES. The evening bat is characteristic of the mixed deciduous forests of southeastern United States. It appears to thrive where forest habitat is interspersed with cultivated areas.

DENSITY AND MOVEMENTS. As far as is known, the evening bat is strictly a summer visitor in Michigan, perhaps from early April to as late as early November (Mumford and Cope, 1964). Females gather in nursery colonies (Jones, 1967; Watkins, 1969). The latter author examined 28 such colonies in Missouri and Iowa, and found that 12 had fewer than 100 individuals, five had 200 or more, and one had 950. Although an occasional male may associate with nursing females, the whereabouts of males in summer is virtually unknown. In winter the evening bat, unlike the myotis bats, apparently does not band together in large groupings. Actually, little is known about its cold season habits.

Following the breakup of the summer nursery colonies, there is evidence that bats of both sexes congregate and swarm about cave entrances (Barbour and Davis, 1969). North-south migrations have been determined by the recovery of banded individuals as much as 328 mi (547 km) south of the point of first capture (Humphrey and Cope, 1968; Watkins, 1969). Since there are no winter records for either Michigan or Indiana, the evening bat must move southward, perhaps to concentrate in Gulf States (Mumford, 1969). Less is known of the evening bat, in terms of its annual activity cycle, than of any Michigan bat.

BEHAVIOR. In summer in northern parts of this bat's range, groups of pregnant females select a variety of day-time retreats in which to bear and nurse their young. They arrive by early May to complete these maternity duties and depart by mid-September (Watkins and Shump, 1981). Buildings seem to be preferred sites for these colonies. A female was captured on May 23 in the window of a house in Climax in Kalamazoo County (Burt 1939). Summer colonies in other states have been discovered in commercial and farm buildings, a cistern, college buildings, a belfry, an attic, and other constructed facilities (Baker and Ward, 1967; Cope and Humphrey, 1967; Doutt *et al.*, 1966; Hooper, 1939; Lowery, 1974; Watkins, 1969). Evening bats rarely select caves and mine tunnels as summer roosts, but do use natural hollows in trees and spaces under loose bark (Barbour and Davis, 1969; Easterla, 1965; Watkins, 1972). Summer shelters, as noted above, are in places protected from excessive moisture and probably drafts, but not necessarily from fluctuating temperatures. Watkins (1972) found temperatures in summer roosts to vary from 47° to 113° F (8° to 45° C). Lactating females prefer roosts with higher temperatures than do pregnant or postlactating individuals (Watkins and Shump, 1981). At lower temperatures, evening bats tend to cluster; at higher temperatures, they generally separate. There is also evidence, according to Cope and Humphrey, (1967), that evening bats are strongly attached to their summer roosts with displaced individuals homing from as far away as 96 mi (155 km).

In winter, evening bats depart for winter quarters and probably hibernate, although Lowery (1974) obtained records of active individuals in Louisiana in every month except February and December. There are indications, however, that males may not make northward excursions (Jones *et al.*, 1967). All three of the known Michigan specimens are females.

The evening bat is a strong flyer. As its name implies it begins nightly flights in the late twilight. According to Lowery (1974), this bat first uses airspace above treetop level before descending, as darkness comes, to forage near ground level. He reports catches of evening bats in mist nets stretched over farm ponds between 1830 and 2030 hours.

ASSOCIATES. The lack of social contact with other bat species in roosts may be primarily related to the evening bat's avoidance of caves and mine tunnels. In the more confined roosting spaces in double walls and other parts of buildings, there are records of associations of the evening bat with such species as the big brown bat in an attic in Indiana (Mumford, 1953) and with the big brown bat, the little brown bat, and the Indiana bat in buildings in Missouri (Watkins, 1972). The evening bat joins several other species in using common feeding areas. In Indiana, for example, Davis and Mumford (1962) found the evening bat foraging with the eastern pipistrelle; in Missouri, Easterla (1965) observed the red bat and evening bat flying in the same area. In Iowa, Kunz (1966) found these latter two species along with the big brown bat catching flying insects along a creek.

REPRODUCTIVE ACTIVITIES. Matings apparently take place in autumn, with both sexes congregating prior to moving to winter quarters (Watkins, 1972). Two young (occasionally one, three, or four) are born from mid-May (in more southern sites) to mid-June or even early July. Burt (1939) found two well-developed embryos in a specimen taken on May 23 in Kalamazoo County. Presumably, the birth of these young might have taken place about the first of June in Michigan. Growth, development, sex ratio, and behavior of developing young have been described in nursery colonies in Mississippi (Jones, 1967), Louisiana (Gates, 1941), Alabama (Hooper, 1939), Missouri (Watkins, 1969), and Indiana (Humphrey and Cope, 1970). The newly born grow rapidly, with eyes opening in two days and body hair appearing in five days. The young are capable of sustained flight 20 days after birth. The young mature by late summer; there is evidence that males may breed their first year (Baker and Ward, 1967).

Although a pregnant female was taken on May 23 in Kalamazoo County (Burt, 1939), there is no other evidence of reproduction by this bat in Michigan. Were nursery colonies present, more individuals would probably have been reported. Watkins (1972) found evening bats in Missouri to have an average life span of two years, with some individuals living more than five years.

FOOD HABITS. There are no specific studies analyzing the insect foods selected by evening bats. It can be presumed that they seek similar insects (flies, beetles, moths, etc.) as the other bats with whom they associate in foraging areas. The feeding of evening bats in captivity has been described by Gates (1941) and Jones (1967).

ENEMIES. Raccoons, house cats, and black rat snakes are considered likely predators on nursery colonies of evening bats (Watkins, 1972). Because this bat's sites are frequently inaccessible, only those bats which fall to the ground may become victims of predation.

PARASITES AND DISEASES. Evening bats are parasitized by mites and bat bugs externally and roundworms and tapeworms internally (Watkins, 1972; Whitaker and Wilson, 1974).

MOLT AND COLOR ABERRATIONS. The blackish coloring of the young evening bats is lost in about six weeks after birth; the pelage then resembles that of the adults. There are records of albino or partly albino individuals (Easterla and Watkins, 1969; Watkins, 1969).

ECONOMIC IMPORTANCE. There is evidence in states more to the south that nursery colonies of evening bats in human residences can cause some of the same problems as myotis bats and big brown bat do in Michigan. However, at the present time the summer population of the evening bat in Michigan appears too insignificant to present difficulties for homemakers. The one individual from Michigan examined for rabies was negative (Kurta, 1979).

COUNTY RECORDS FROM MICHIGAN. Reports of specimens preserved in museum collections show that the evening bat is known from the following counties in the Lower Peninsula: Berrien, Kalamazoo, Washtenaw.

LITERATURE CITED

Baker, R. J., and C. M. Ward
1967 Distribution of bats in southeast Arkansas. Jour. Mammalogy, 48(1):130–132.

Barbour, R. W., and W. H. Davis
1969 Bats of America. Univ. Kentucky Press, Lexington. 286 pp.

Burt, W. H.
1939 The Rafinesque bat in Michigan. Jour. Mammalogy, 20(1):103.

Cope, J. B., and S. R. Humphrey
1967 Homing experiments with the evening bat, *Nycticeius humeralis*. Jour. Mammalogy, 48(1):136.

Davis, W. H., and R. E. Mumford
1962 Ecological notes on the bat *Pipistrellus subflavus*. American Midl. Nat., 68(2):394–398.

Doutt, J. K., C. A. Heppenstall, and J. E. Guilday
1966 Mammals of Pennsylvania. Pennsylvania Game Commission, Harrisburg. 273 pp.

Easterla, D. A.
1965 A nursery colony of evening bats in southern Missouri. Jour. Mammalogy, 46(3):498.

Easterla, D. A., and L. C. Watkins
1969 An aberrant evening bat. Southwestern Nat., 13(4): 447–448.

Gates, W. H.
1941 A few notes on the evening bat, *Nycticeius humeralis* Rafinesque. Jour. Mammalogy, 22(1):53–56.

Hall, E. R.
1981 The mammals of North America. Second Edition. John Wiley and Sons, New York. Vol. 1, xv+600+90 pp.

Hooper, E. T.
1939 Notes on the sex ratio in *Nycticeius humeralis*. Jour. Mammalogy, 20(3):369–370.

Humphrey, S. R., and J. B. Cope
1968 Records of migration of the evening bat, *Nycticeius humeralis*. Jour. Mammalogy, 49(3):329.
1970 Population samples of the evening bat, *Nycticeius humeralis*. Jour. Mammalogy, 51(2):399–401.

Jones, C.
1967 Growth, development, and wing loading in the evening bat, *Nycticeius humeralis* (Rafinesque). Jour. Mammalogy, 48(1):1–19.

Jones, J. K., Jr., E. D. Fleharty, and P. B. Dunnigan
1967 The distributional status of bats of Kansas. Univ. Kansas, Mus. Nat. Hist., Misc. Publ. 46, 33 pp.

Kunz, T. H.
1966 Evening bat in Iowa. Jour. Mammalogy, 47(2):341.

Kurta, A.
1979 Bat rabies in Michigan. Michigan Acad., 12(2):221–230.
1980 The bats of southern lower Michigan. Michigan State Univ., unpubl. M.S. thesis, ix+147 pp.

Lowery, G. H., Jr.
1974 The mammals of Louisiana and its adjacent waters. Louisiana State Univ. Press, Baton Rouge. xxiii+565 pp.

Mumford, R. E.
1953 Status of *Nycticeius humeralis* in Indiana. Jour. Mammalogy, 34(1):121–122.
1969 Distribution of the mammals of Indiana. Indiana Acad. Sci., Monog. No. 1, vii+114 pp.

Mumford, R. E., and J. B. Cope
1964 Distribution and status of the Chiroptera of Indiana. American Midl. Nat., 72(2):473–489.

Watkins, L. C.
1969 Observations on the distribution and natural history of the evening bat *(Nycticeius humeralis)* in northwestern Missouri and adjacent Iowa. Trans. Kansas Acad. Sci., 72(3):330–336.
1972 *Nycticeius humeralis*. American Soc. Mammalogists, Mamm. Species, No. 23, 4 pp.

Watkins, L. C., and K. A. Shump, Jr.
1981 Behavior of the evening bat *Nycticeius humeralis* at a nursery roost. American Midl. Nat., 105(2):258–268.

Whitaker, J. O., Jr., and N. Wilson
1974 Host and distribution lists of mites (Acari), parasitic and phoretic, in the hair of wild mammals of North America, north of Mexico. American Midl. Nat., 91(1):1–67.

Rabbits and Hares

ORDER LAGOMORPHA

Rabbits and hares, plus the montane and arctic pikas (which look more like guinea pigs than small cottontails) make up an ancient line of mammals, dating from the late Paleocene of Asia. Because of their many superficial similarities to rodents, lagomorphs were formerly arranged as a suborder of Rodentia. When these seemingly close relationships were shown to be the result of parallel development (convergences) instead of merely divergences of two phylogenetic herbivorous lines from a common ancestoral stock (Gidley, 1912), the two mammalian groups were classified in separate orders. Like the rodents, the lagomorphs have chisel-shaped front teeth. The lagomorphs, however, also have two small, peg-like teeth (behind the large front ones) on each side of the upper jaws.

Unlike rodents, the rabbits, hares and pikas are not a highly diverse group. They include only ten genera and 63 living species, although they occur naturally nearly worldwide except for Australia, New Zealand, and southern South America. Rabbits and hares inhabit both forested and open lands and are perhaps best adapted to the temperate conditions of North America and Eurasia in both numbers of species and abundance of individuals. Populations in these areas have been characterized by their conspicuous multiannual fluctuations, marked by periods of abundance alternating with periods of scarcity (MacLulich, 1937).

Modern lagomorphs are divided into two families, the Ochotonidae (pikas) and the Leporidae (rabbits and hares). Only the family Leporidae is represented in Michigan. A resident of Michigan must travel westward to the Rocky Mountains to observe pikas and hear their distinctive calls.

Many of the distinguishing features of the members of the Lagomorpha relate to their plant-eating habits and in the leporids, at least, to rapid movements with a hopping gait (saltatorial locomotion). As mentioned, one chisel-shaped front incisor occurs on each side of the lower jaw and one on each side of the upper jaw with an additional peg-like upper incisor centered directly behind each upper incisor. This arrangement is in marked contrast to that of other mammalian groups where the teeth are usually in a continuous arc around the upper edges of the jaws. As with rodents, these gnawing front teeth (incisors), unlike those of many other mammalian groups, continue to grow (ever-growing) throughout life. This growth is important to the welfare of the individual since these teeth must withstand a great deal of wear in normal food chewing. The upper front incisors have distinct grooves on their anterior surfaces. In Lagomorpha there are no canines and a wide space (diastema) separates the incisors from the high-crowned (hypsodont) premolars and molars, which are highly adapted for masticating coarse vegetable matter. The motion of the jaws in the chewing process is generally lateral or oblique.

Other features of lagomorphs include: collar bone (clavicle) developed in pikas and rudimentary in rabbits and hares; bones of the lower hind limbs (tibia and fibula) fused distally; soles of the feet covered with hair; forefeet with five toes and hind feet with four; fur dense and short; skin paper-thin, making the pelt of poor quality for the fur trade; tail short; penis bone (baculum) characteristic in rodents, absent in Lagomorpha; testes in a scrotum located in front of penis, rather than behind as in most mammals; bones at the front of the skull (maxillaries) and at the back of skull (occipitals) pitted with small window openings (fenestrated bones); eyes large; ears of leporids elongate, proximal part tubular with actual opening well above skull.

Lagomorphs practice coprophagy (refection or reingestion), whereby the moist, greenish fecal pellets (as opposed to the dry, brownish pellets) which contain much partly-digested food (through caecal fermentation) are re-eaten to pass again through the digestive tract (Geis, 1957; Houpt, 1963; Janis, 1976; Kirkpatrick, 1956). This physiological action functions somewhat like rumination in even-toed ungulates (Artiodactyla) whereby partly masticated food is regurgitated from one of the chambers of their complex stomachs to be chewed and swallowed again (cud-chewing).

FAMILY LEPORIDAE

Three lagomorphs (two native and one introduced) in Michigan belong to this widespread but only moderately diversified family (49 living species in nine genera). Of the two native species, the eastern cottontail (*Sylvilagus floridanus*) has an extensive tropical-temperate distribution in the Western Hemisphere and reaches the northernmost part of its range in the western part of Michigan's Upper Peninsula. In contrast, the snowshoe hare (*Lepus americanus*) has boreal and subarctic affinities and in Michigan is near the southern edge of its northern distribution except for populations in the eastern and western mountains.

Because of their highly edible meat, rabbits and hares have been raised in captivity and domesticated; are widely hunted (for food and sport); and have been captured, transported, and released to successfully propagate in such non-lagomorph regions as southern South America, Australia and New Zealand. The herbivorous diets of rabbits and hares sometimes causes them to interfere with agricultural pursuits, whether it be cottontails invading farm crops; jackrabbits eating irrigated alfalfa or competing with livestock for grass on western ranges; introduced European rabbits (*Oryctolagus cuniculus*) eating forage needed by native marsupials and introduced domestic sheep in Australia; or introduced European hares (*Lepus capensis*) causing similar problems in Argentina or Chile.

Rabbits and hares have some distinctive morphological differences (see Key to Species). They may also be distinguished by their contrasting early development. Rabbit young, including Michigan's eastern cottontail, are born in a fur-lined nest (plucked by the female from her underside) and in an altricial state, meaning the neonates are not highly developed and are virtually helpless; their eyes and ears are closed, and their skins are almost devoid of hair covering. On the other hand, hares, including Michigan's snowshoe hare and the introduced European hare, are born in the open (perhaps in a "form" in the grass but certainly not in a nest) and in a precocial state, meaning the neonates are advanced in their development; their eyes and ears are open, they are able to run about within a few minutes of birth, and their bodies are fully haired.

KEY TO SPECIES OF RABBITS AND HARES OF THE FAMILY LEPORIDAE IN MICHIGAN
Using Characteristics of Adult Animals in the Flesh

1a. Size small (total length of head and body without tail usually less than 15⅝ in or 400 mm); hind foot short (less than 4 in or 100 mm) .... EASTERN COTTONTAIL *(Sylvilagus floridanus)*, page 139.

1b. Size large (total length of head and body without tail usually greater than 15⅝ in or 400 mm); hind foot long (more than 4 in or 100 mm) .... 2a.

2a. Pelage except for ear tips white in winter; ears (measured from tip to notch) less than 3¼ in (85 mm); basal parts of hairs on back slate colored .... SNOWSHOE HARE *(Lepus americanus)*, page 148.

2b. Pelage not white in winter; ears (measured from tip to notch) more than 3¼ in (85 mm); basal parts of hairs on back whitish .... EUROPEAN HARE (*Lepus capensis*), page 156.

KEY TO SPECIES OF RABBITS AND HARES OF THE FAMILY LEPORIDAE IN MICHIGAN
Using Characteristics of the Skull of Adult Animals

1a. Skull short, usually less than 3 in (75 mm); interparietal not fused with parietal bones at the posterior part of the top of the skull; postorbital processes not flared, with tips usually fused with frontals; zygomatic arch, in lateral view, not noticeably broader in middle than at ends . . . . EASTERN COTTONTAIL *(Sylvilagus floridanus)*, page 139.

1b. Skull long, usually more than 3 in (75 mm); interparietal fused with parietal bones at the posterior part of the top of the skull; post-orbital processes flared, with tips not fused with frontals; zygomatic arch in lateral view, noticeably broader in middle than at ends . . . . 2a.

2a. Skull small, less than 3½ in (90 mm) in length; nasal bones usually less than 1⅝ in (42 mm) in length; skull not massive in construction . . . SNOWSHOE HARE *(Lepus americanus)*, page 148.

2b. Skull large, more than 3½ in (90 mm) in length; nasal bones usually more than 1⅝ in (42 mm) in length; skull massive in construction . . . . . .EUROPEAN HARE *(Lepus capensis)*, page 156.

LITERATURE CITED

Geis, A. D.
1957 Coprophagy in the cottontail rabbit. Jour. Mammalogy, 38(1):136.

Gidley, J. W.
1912 The lagomorphs an independent order. Science, N.S., 36:285–286.

Houpt, T. R.
1963 Urea utilization by rabbits fed a low protein diet. American Jour. Physiol., 205:1144–1150.

Janis, C.
1976 The evolutionary strategy of the Equidae and the origins of rumen and cecal digestion. Evolution, 30(4):757–774.

Kirkpatrick, C. M.
1956 Coprophagy in the cottontail. Jour. Mammalogy, 37(3):300.

MacLulich, D. A.
1937 Fluctuations in the numbers of varying hare *(Lepus americanus)*. Univ. Toronto, Biol. Studies No. 43, 136 pp.

EASTERN COTTONTAIL
Sylvilagus floridanus (Allen)

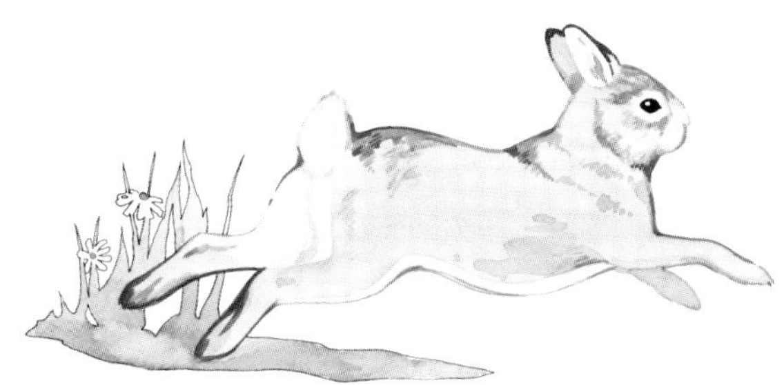

NAMES. The generic name *Sylvilagus,* proposed by Gray in 1867, has both Latin and Greek origins, *sylva* meaning "a wood" and *lagos* meaning "hare." The specific name *floridanus,* proposed by Allen in 1890, refers to the state of Florida, where the describer first recognized and named the species. This animal is often referred to as rabbit or cottontail rabbit. The subspecies *Sylvilagus floridanus mearnsii* (Allen) occurs in Michigan.

RECOGNITION. Body size medium for a leporid (Fig. 23); head and body approximately 15 in (385 mm) long; ears large and elongate (see Measurements and Weights); tail short and fluffy with white under surface exposed when animal is running; eyes large and conspicuous, mammary glands arranged in four pairs.

Color of upperparts dark buffy-brown darkened by grayish and black tips of long, fairly coarse guard hairs with some of the short, slate-colored, dense, silky underfur visible; rump area darker gray; sides of body paler and grayer than upperparts; nape (area between ears and shoulder) prominently rufus; throat and legs dark reddish-brown; most of underparts whitish; tail brownish-gray above and whitish below.

MEASUREMENTS AND WEIGHTS. Adults measure as follows: length from tip of nose to end of

Figure 23. The eastern cottontail (*Sylvilagus floridanus*). Sketches from life by Bonnie Marris

tail vertebrae, 15¾ to 19¼ in (400 to 490 mm); length of tail vertebrae, 1½ to 2½ in (40 to 70 mm); length of hind foot, 3⅛ to 4¼ in (80 to 110 mm); height of ear from notch, 2⅛ to 2½ in (55 to 65 mm); weight, 32 to 64 oz (915 to 1,828 g).

DISTINCTIVE CRANIAL AND DENTAL CHARACTERISTICS. The skull of the eastern cottontail is only slightly smaller than that of Michigan's snowshoe hare. It is further distinguished from the Michigan snowshoe hare's skull by the lack of fusion between the interparietal and parietal bones at the hind end of the skull top, and by the fusion of the unflared postorbital processes (at their posterior ends) with the frontals (Fig. 24). The skull averages 3 in (77 mm) long and 1½ in (38 mm) wide. Conspicuous, of course, are the large orbits to accommodate the prominent eyes of this leporid. The elongated space between the chisel-shaped incisors (with the dimunitive peg-like teeth directly behind the upper ones) is called a diastema and it is a means of distinguishing rabbits and hares from most other Michigan mammals. The dental formula is:

$$\text{I (incisors)}\frac{2\text{-}2}{1\text{-}1},\ \text{C (canines)}\frac{0\text{-}0}{0\text{-}0},\ \text{P (premolars)}\frac{3\text{-}3}{2\text{-}2},\ \text{M (molars)}\frac{3\text{-}3}{3\text{-}3} = 28$$

DISTRIBUTION. The eastern cottontail is widespread in eastern United States, northward to the Canadian border, westward to the Great Plains, and southward through Central America to northern South America (Chapman *et al.,* 1980). In Michigan, this species seems to thrive best in mixed woodlots, cultivated and fallow fields, and fence-row and roadside vegetation in the southern part of the Lower Peninsula (Map 22). It is also found in all counties of the northern part of the Lower Peninsula and in the western and southwestern part of the Upper Peninsula, especially near the Wisconsin border, and eastward as far as Alger County (Ozoga and Verme, 1967). The eastern cottontail is probably a latecomer to the

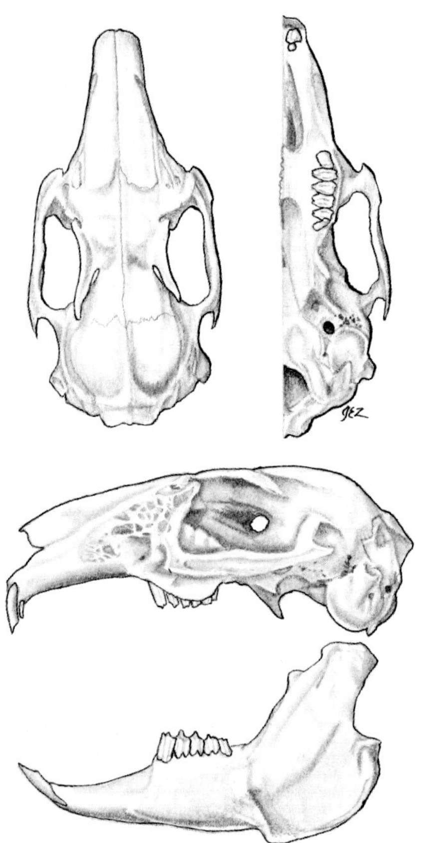

Figure 24. Cranial and dental characteristics of the eastern cottontail (*Sylvilagus floridanus*).

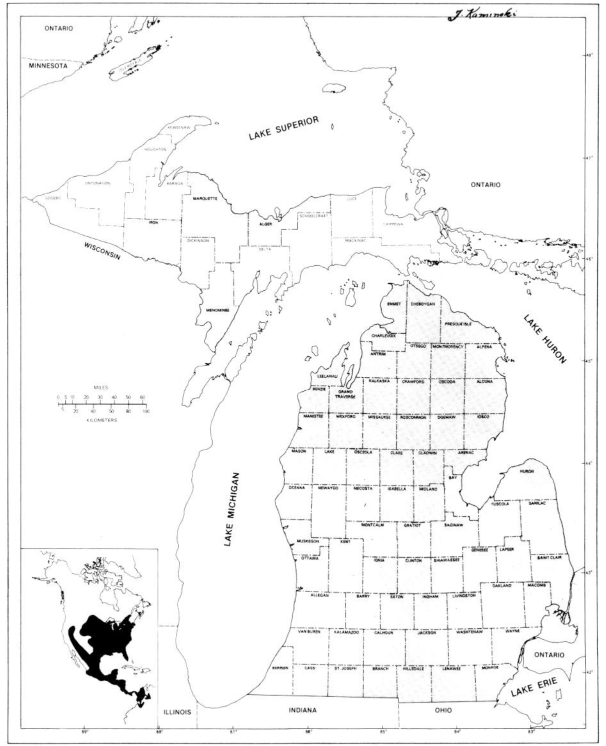

Map 22. Geographic distribution of the eastern cottontail (*Sylvilagus floridanus*). It is to be expected in other counties in the western part of the Upper Peninsula.

Upper Peninsula (perhaps since the turn of the century, DeVos, 1964; Hickie, 1940), but has been a long-time resident in the southern part of the Lower Peninsula (Cleland, 1966).

HABITAT PREFERENCES. The areas often cleared and/or covered with second-growth shrubs, vines, and low trees in what were once the decidous forests of southern Michigan are the preferred habitats of the eastern cottontail. The mosaic pattern of land-use practices with large amounts of the edge environment—between growths of woody vegetation and open pasture and croplands—has produced favorable conditions for a large and widespread cottontail population. Where conifers and related boreal forested vegetation dominate in northern parts of the state, the eastern cottontail disappears and the snowshoe hare becomes the featured leporid.

Although there is limited historic evidence (De Vos, 1964), it is suspected that cutting of southern Michigan's hardwood forests, in the past 80–90 years, which brought about habitat diversification as observed today, has allowed for the northward increase in this southern rabbit (Conger, 1920). The fact that Michigan is at the northern edge of the range of this rabbit may be a factor in eastern cottontail population fluctuations. Winter weather severity may be a depressing influence. The eastern cottontail's winter use of burrows (especially those of the woodchuck, Grizzell, 1955) in the northern part of its range may be an adaptation to cold temperatures (Allen, 1939).

DENSITY AND MOVEMENTS. With three or more litters produced from spring until early autumn, the cottontail population reaches high densities, as many as 100 rabbits in a two-acre (5 ha) orchard (Hickie, 1940), in summer. Following this summer peak and the end of the reproductive season, cottontail populations face the end of the vegetative growing season. Happily, production exceeds winter mortality, and provides sufficient breeding stock to initiate the next reproductive activity. Wildlife scientists in Michigan (notably Allen, 1939; Eberhardt *et al.,* 1963; Geis, 1956; Haugen, 1941, 1942a, 1942b, 1943; Hickie, 1940; Trippensee, 1936) have studied this rabbit's annual cycle.

Eastern cottontail numbers also fluctuate markedly in relation to habitat quality (Trippensee, 1934). In the best mixed environments (clearings, croplands, brushy creek edges and fence rows, small woodlots) of southern Michigan, this rabbit may attain autumn populations of more than three individuals per acre (about 7.5 per ha) in abundant times; in times of scarcity, less than one cottontail per three acres (less than one per ha). In studies using live-traps to capture, mark, release, and recapture individual cottontails at the Allegan State Game Area, Haugen (1942a) calculated that male rabbits moved over areas as large as 100 acres (40 ha) while less mobile females averaged about 14 acres (5.6 ha) in winter and about 22.5 acres (9 ha) in the summer breeding period. In Kalamazoo County farmland, Allen (1938b, 1939) found that cottontails occupied areas from 5 to 10 acres (2 to 4 ha), with a December population amounting to one animal per 2.1 acres (0.84 ha). In studies in Wisconsin (Trent and Rongstad, 1974) where cottontails were radio-tagged (fitted with collars containing radio transmitters), adult males were found to have home ranges of about 10 acres (4 ha) in early summer and only about 4 acres (1.5 ha) in the late summer post-breeding period. Adult females had home ranges averaging 4.3 acres (1.7 ha) in spring and 2.1 acres (0.8 ha) in late summer. Other methods of appraising cottontail densities include the use of telemetry with transmitter-tagged individuals tracked by radio, roadside counts (Lord, 1961), counts of rabbits killed by hunters (Hickie, 1940), counts of rabbits flushed per hour (McCabe, 1943), and track and fecal pellet counts (Lord, 1963).

In general, the individual cottontail resides for its entire lifetime in an area of a few hundred acres at most, even though there are records in southern Michigan, according to Hickie (1940), of tagged animals moving as much as 16 mi (26 km).

During the summer growing season, non-woody cover and widespread food supplies allow eastern cottontails to range into most areas of southern Michigan. In winter, there is a tendency for this rabbit to restrict its movements to heavy woody cover, even moving from upland openings to lowland thickets (Haugen, 1943). During periods of snow cover, holes, woodpiles and other protected retreats are used (Trent and Rongstad, 1974). Fortunately, three active burrowers, the striped skunk, woodchuck and badger, provide underground dens for Michigan cottontails.

BEHAVIOR. Although cottontails can be observed any hour of the day, they are most active in dim light, sometimes on drab days, but mostly at night. They are often flushed in bright sunlight from brushy fencerows and grassy-weedy vegetation. At that time the animals are ordinarily reposing in "forms" fashioned in ground vegetation and flee only when disturbed. Most visits to the vegetable garden in summer or to orchards or cover plantings in winter are nightly activities. These actions begin in late afternoon (about 1700 hrs) and subside the next morning (about 0700 hrs). According to Lord (1961; 1963), Illinois cottontails start this activity later in the evening and complete it later in the morning in summer. There is also some evidence, based on Lord's roadside counts, of activity peaks in early evening and again in the two or three hours prior to daylight. Haugen (1944) noted greatest numbers of cottontails killed by Michigan motorists occur in March and April, at the onset of the breeding activity.

Eastern cottontails normally move about by short hops. Longer leaps allow the animals to attain speeds of 18 mi (29 km) per hour, although they cannot maintain such speeds for very long (Jackson, 1961). After their nocturnal and crepuscular activities, individuals usually repose in surface "forms," resting and sometimes grooming. Tunneling in heavy accumulations of snow has been noted (Fitzsimmons and Weeks, 1981). At the approach of possible danger, the cottontail remains motionless to avoid detection but ready for instant flight if the intruder approaches too closely. When necessary, escape is accomplished by a rapid, often zig-zag, series of bounds. These flights are not far, usually to a burrow, hollow log or woody cover (Fig. 25).

Young nestlings may utter shrill squeals. Adults make grunting sounds, sometimes accompanied by thumping the hind feet. One of the most pathetic calls in nature is the scream of an adult rabbit

Figure 25. The eastern cottontail reposing in a hollow log. Photo by Robert Harrington

caught by a predator. This same shrill and plaintive call may be given by a hand-held rabbit which is struggling to escape (Haugen, 1942b). This penetrating scream has been mimicked by predator-calling devices, because it attracts foxes and coyotes.

In spite of its ability to thrive in large numbers in favorable natural habitats, eastern cottontails are not very tolerant of each other and live fairly solitary existences. In nature, home ranges of adult males overlap those of females and other males. However, adult females maintain home ranges generally separated from those of other females (Haugen, 1942a; Trent and Rongstad, 1974). In captivity, adult cottontails are apt to be antagonistic when caged together.

REPRODUCTIVE ACTIVITIES. Knowledge of the Michigan eastern cottontail's reproductive habits comes chiefly from studies by Allen (1938a), Haugen (1942b), Hickie (1940), and Trippensee (1936). Data from throughout the range of the species are summarized by Wainright (1969). Enlargement of the male testes is the first indication of the onset of the breeding season which begins in Michigan from late February to mid-March. The males may mate until September when the testes reduce in size. Pregnant rabbits reported on November 21 in Wisconsin (Lemke, 1957) and on January 17 in Illinois (Lord, 1959) are exceptions to the regular breeding season. Pre-copulatory courtship includes nocturnal antics by the male, often called a mating dance. Male/female interaction includes a chase, head-to-head encounter, and hops in the air prior to actual mating. The gestation period is 29 or 30 days. Prior to parturition, the pregnant female usually digs or finds a shallow burrow (or depression), which she lines with dry grass and fur plucked from her abdomen (see Fig. 42). Haugen (1942b) found litters of cottontails in southern Michigan to vary in number from four to seven young with an average of 5.4. A record of 12 has been reported on one occasion in Indiana (Kirkpatrick, 1960). Since the parent female may mate and become pregnant again shortly after the birth of a litter, she must finish nursing and otherwise caring for her present young before her next litter arrives. This maternal activity (including weaning) is usually completed within 22 days. In Michigan, one female may have as many as four litters in a single reproductive season.

The newly-born cottontails are virtually helpless (altricial), naked, and blind, often suffering high mortality if rainfall is heavy (Jacobs and Dixon, 1981). Each weighs about 1 oz (30 g). In their fur-insulated nest, the female nurses her young at least twice a day (near dawn and near dusk) and leaves them to feed herself, mostly at night. The offspring develop rapidly; their eyes usually open by the sixth day. They are fully furred and can leave the nest on short forays by the 13th or 16th day, with weaning occurring shortly after. The fecundity of Michigan's eastern cottontail is also demonstrated by the fact that females born early in the season may mature sufficiently to have litters themselves before the end of the same breeding season (Cooley, 1946).

The average cottontail is short-lived. In autumn (following cessation of breeding), the Wisconsin population may contain as few as 17% parent animals and 83% young-of-year (Trent and Rongstad, 1974). Haugen (1942b) concluded from his study that at least 25% of the Michigan cottontail population lived to an age of approximately 21 months. Most cottontails probably do not survive in nature beyond the third summer, although at Rose Lake Wildlife Research Station in Clinton County, Linduska (1947) found that two individuals—out of more than 1,400 live-trapped, marked and released for recapture—survived four years or longer.

FOOD HABITS. Many of the vegetative foodstuffs eaten by the eastern cottontail and other leporids consist of highly complex carbohydrates and cellulose, ordinarily difficult to digest. To maximize digestion of these materials, a digestive process called caecal fermentation has evolved in rabbits and hares (Janis, 1976). Foods which are not digestible in the upper alimentary tract are chemically reduced to absorbable substances by fermentation (with bacterial action) in the large sac-like caecum in the lower part of the tract. Because this area of fermentation is beyond that part of the system where the end-products can be assimilated by the body, these mammals practice coprophagy (refection). To make use of these processed foods, the cottontail must reingest the fecal pellets from this caecal fermentation (Kirkpatrick, 1960; Geis, 1957b), so that the nutrients can be absorbed in the upper tract. These pellets differ from normal cottontail droppings in that they are soft instead of hard and dry; soft pellets may amount to 30% of the total material defecated (Bailey, 1969). This remarkable method of extracting nourishment from coarse foods requires that the cottontail have the opportunity to eat the soft pellets at the time of deposition or shortly thereafter.

In general, the diet of the eastern cottontail varies considerably from season to season depending on food availability. During the summer growing season, green plants are favored; in winter, woody plant parts (twigs, bark, and buds) of shrubs and young trees are major food sources (Allen, 1939; Dice, 1945; Dusi, 1952; Haugen, 1942b; Hickie, 1940). Grasses constitute at least half of the summer diet with herbs making up much of the rest. Preferred Michigan summer foods include bluegrass, orchard grass, wild rye, timothy, crabgrass, foxtail, red top, plantain, golden-rod, wild strawberry, smartweed, sheep sorrel, chickweed, buttercup, various clovers, and, of course, garden vegetables when available.

The most predominant rabbit winter foods from among about 43 woody plants listed as eaten (according to Michigan observers Allen, Dice and Haugen) are the bark, twigs and buds of white oak, flowering dogwood, sassafras, black oak, New Jersey tea, honey locust, dwarf and staghorn sumac. Deep snow drifts in southern Michigan, while covering some winter foods, literally raise the level of the ground surface to allow cottontails to eat parts of woody plants well above their normal reach.

ENEMIES. The eastern cottontail survives despite the many mammalian, avian, and reptilian predators which daily confront these gentle creatures. Fortunately, the high annual turnover rate of about 80% is offset by an equally high reproductive potential. For southern Michigan, Haugen (1942b) and Hickie (1940) list the following predators: long-tailed weasel, feral house cat, domestic dog, red fox (see also Arnold, 1956; Murie, 1936; Scofield, 1960), opossum, broad-winged hawk, red-tailed hawk, marsh hawk, barred owl, great horned owl (Errington *et al.,* 1940), crow, and black snake. Cottontails are also captured by gray wolf (Van Ballenberghe *et al.,* 1975), mink (Sealanders, 1943), barn owl (Wallace, 1950), snowy owl (Chamberlin, 1980), red squirrel (Hamilton, 1934), great blue heron (Peifer, 1979), and American kestrel (Hubard, 1941). There is evidence in Michigan that red fox populations fluctuate somewhat in accordance with, but lagging behind, the ups and downs of numbers of eastern cottontails (Marvin Cooley, pers. comm.).

Although human land-use practices have improved cottontail habitat in Michigan, free-ranging house cats and, to a lesser extent, dogs have sometimes reduced thriving populations. Highway mortality is a major factor (Haugen, 1944). Steven L. Hartley (1975) counted 52 dead rabbits between March 1 and August 4, 1974, on a 19-mi stretch of M-21 between Grand Rapids and Lowell.

PARASITES AND DISEASES. Fleas, ticks, and mites are reported as ectoparasites by Haugen (1942b), Scharf and Stewart (1980), Wassel *et al.,* (1980), and Whitlock (*in* Hickie, 1940) along with warble fly larvae, round worms (nematodes), tapeworms (cestodes) and flat worms (trematodes) as endoparasites. Warbles located just beneath the skin of the throat and chest were noted in a large number of cottontails examined at the Kellogg Biological Station in Kalamazoo County (Geis, 1957a), while Whitelock (1939b) found infections of the larval stage of the dog tapeworm in Michigan cottontails. Papillomas (sometimes causing horn-like growths, Haugen, 1960), staphylococcal infections (Allen, 1939), neoplasms (Lopushinsky and Fay, 1967), and salmonella infections (Youatt and Fay, 1968) are also reported. Perhaps the most well-known of the diseases to inflict the eastern cottontail is tularemia (Jacobson *et al.,* 1978), a bacterial disease known to infect humans and other mammals as well. According to Whitlock (1939a; *in* Hickie, 1940) during the four years from 1935 to 1938, about 50 human cases of tularemia were reported in Michigan. Today, with new medications, this disease is not a great problem with humans. Even so, sluggish and inactive cottontails should be avoided as this behavior is sometimes symptomatic of tularemia. Hunters should use gloves when skinning or eviscerating animals.

MOLT AND COLOR ABERRATIONS. Naked at birth, the growing cottontail develops a soft, gray coat of fur (nestling fur) until about five weeks of age. This is replaced by a darker salt-and-pepper juvenal coat of heavier fur, which is molted after about 20 days and replaced by a subadult pelt, somewhat paler and more buff than true adult pelage. The subadult coat may be retained for as long as 100 days (Negus, 1958). There are also the usual fall and spring molts, allowing for the heavy winter coat and the lighter summer coat.

Abnormal coat colors are sometimes observed, including both light (albinistic) and dark (melanistic) shades (Casteel, 1961; Manville, 1960). Among Michigan specimens of the eastern cottontail in the Michigan State University Museum are a completely white male taken in January from Jackson County and a buff-colored (angora) male taken in December from Wayne County. A black male was found in Tuscola County in December (Charles Shick, pers. comm.).

ECONOMIC IMPORTANCE. The eastern cottontail is abundant, highly edible, and appears skillful at evading hunting beagles and shotgun pellets—all excellent qualities in a sporting animal. As a consequence, throughout its vast range, this species is pursued for its meat and for sport. The Michigan cottontail ranked first in abundance as a game species at Rose Lake Wildlife Research Area (Linduska, 1950). There, the season kill for each year from 1940 to 1942 was 12 rabbits per 100 acres. Paul Hickie, in his informative and highly readable booklet "Cottontails in Michigan," called the cottontail the "big game of the small boy." Hickie also described one Ogemaw County group of five market hunters, each of whom shipped a barrel containing 144 rabbits to either Chicago or Detroit each week in December and January between 1915 and 1920. Hickie estimated that in 1938, there were at least 300,000 hunters who killed more than two million Michigan cottontails.

Estimates by wildlife biologists for Michigan's Department of Natural Resources noted ups and downs in the numbers of registered hunters and their rabbit takes. In 1974, for example, some 409,980 hunters bagged an estimated 2.25 million cottontails; in 1975, the take was judged to be 2.47 million. Although cottontails can be hunted legally from October through February (through March in the northern part of the Lower Peninsula and in the Upper Peninsula), December, when tracking snow is on the ground and hunting is closed on most other game species, is the traditional cottontail month in Michigan.

Long-time field observer Marvin Cooley of the Department of Natural Resources, has commented (*Detroit Free Press,* December 14, 1975) on the difficulty in predicting well in advance the annual population fluctuations of Michigan cottontails. However, there has always been evidence that the annual take by hunters rarely depresses rabbit populations, and that the bag limit of five per day for the long season seems biologically realistic (Schofield, 1957). The only threat which might emerge is from the unwarranted use of pesticides and herbicides (Malecki *et al.,* 1974).

If the meat from an average adult Michigan cottontail is worth a minimum of $1.50, the annual harvest would have a value of between three and four million dollars. However, it is difficult to determine whether this estimated monetary return really offsets cottontail damage to gardens, field crops of grains and fodders, orchards, nurseries, forest reproduction, and yard ornamentals. Cottontails can be pestiferous to the green growth of the gardener and farmer in summer and to the array of woody tissues from shrubs and trees of the forester, orchardist and landscaper in winter (Eadie, 1954). Discussions of such damage in Michigan have been presented by Dice (1945), Geis (1954), Hickie (1940), Mathies and Schneider (1967), and Pirnie (1949). The fall-winter hunting season might be helpful in reducing the population in areas where woody plants can be girdled or damaged by hungry cottontails. However, as Pirnie (1949) points out, heavy winter hunting may not depress cottontail numbers sufficiently to alleviate the ravages of the over-wintering adults and their first-litter offspring.

Rabbit-proof fencing (poultry netting), low-strung electric fences surrounding gardens and nurseries, and wrapping the bases of large shrubs and trees with heavy paper discourage cottontails. Chemical repellents have been tested in Michigan (Hayne and Cardinell, 1958) with some success; an up-to-date listing with instructions may be obtained by requesting the leaflet "Repelling Rabbits" available from the Division of Technical Assistance, U.S. Fish and Wildlife Service, Washington, D.C. 20240.

COUNTY RECORDS FROM MICHIGAN. Records of specimens preserved in collections of museums and from the literature show that the eastern cottontail is known from all counties in the Lower Peninsula and the following counties in the Upper Peninsula: Alger, Iron, Marquette, Menominee. The species is also known from South Manitou Island in Lake Michigan, Marion Island in Grand Traverse Bay, and Charity Island in Saginaw Bay. Eastern cottontails released in 1950 on Beaver Island in Lake Michigan (Ozoga and Phillips, 1964) did not survive.

LITERATURE CITED

Allen, D. L.
1938a Breeding of the cottontail rabbit in southern Michigan. American Midl. Nat., 20(2):464–469.
1938b Ecological studies on the vertebrate fauna of a 500-acre farm in Kalamazoo County, Michigan. Ecol. Monogr., 8(3):348–436.
1939 Michigan cottontails in winter. Jour. Wildlife Mgt., 3(4):307–322.

Arnold, D. A.
1956 Red foxes of Michigan. Michigan Dept. Conservation, Lansing. 48 pp.

Bailey, J. A.
1969 Quantity of soft pellets produced by caged cottontails. Jour. Wildlife Mgt., 33(2):421.

Casteel, D. A.
1961 A white cottontail rabbit. Jour. Mammalogy, 42(4): 541.

Chamberlin, M. L.
1980 Winter hunting behavior of a snowy owl in Michigan. Wilson Bull., 92(1):116–120.

Chapman, J. A., J. G. Hockman, and M. M. Ojeda C.
1980 *Sylvilagus floridanus.* American Society of Mammalogists, Mammalian Species, No. 136, 8 pp.

Cleland, C. E.
1966 The prehistoric animal ecology and ethnozoology of the Upper Great Lakes region. Univ. Michigan, Mus. Anthro., Anthro. Papers, No. 29, x+294 pp.

Conger, A. C.
1920 A key to Michigan vertebrates except birds. Michigan Agri. College, East Lansing. 76 pp.

Cooley, M. E.
1946 Cottontails breeding in their first summer. Jour. Mammalogy, 27(3):273–274.

deVos, A.
1964 Range changes of mammals in the Great Lakes region. American Midl. Nat., 71(1):210–231.

Dice, L. R.
1945 Some winter foods of the cottontail in Michigan. Jour. Mammalogy, 26(1):87–88.

Dusi, J. L.
1952 The food habits of several populations of cottontail rabbits in Ohio. Jour. Wildlife Mgt., 16(2):180–184.

Eadie, W. R.
1954 Animal control in field, farm and forest. Macmillan Co., New York. viii+257 pp.

Eberhardt, L., T. J. Peterle, and R. D. Schofield
1962 Problems in a rabbit population study. Wildlife Monog. No. 10, 51 pp.

Errington, P. L., F. Hamerstrom and F. N. Hamerstrom, Jr.
1940 The great horned owl and its prey in north-central United States. Iowa State Col., Agri. Exp. Sta., Res. Bull. 277:757–850.

Fitzsimmons, M., and H. P. Weeks, Jr.
1981 Observations on snow tunneling by *Sylvilagus floridanus.* Jour. Mammalogy, 62(1):211–212.

Geis, A. D.
1954 Rabbit damage to oak reproduction at the Kellogg Bird Sanctuary. Jour. Wildlife Mgt., 18(3):423–424.
1956 A population study of the cottontail rabbit in southern Michigan. Michigan State Univ., unpubl. Ph.D. dissertation, 184 pp.
1957a Incidence and effect of warbles on southern Michigan cottontails. Jour. Wildlife Mgt., 21(1):94–95.
1957b Coprophagy in the cottontail rabbit. Jour. Mammalogy, 38(1):136.

Grizzell, R. A., Jr.
1955 A study of the southern woodchuck, *Marmota monax monax.* American Midland Nat., 53(2):259–293.

Hamilton, W. J., Jr.
1934 Red squirrel killing young cottontail and young grey squirrel. Jour. Mammalogy, 15(4):322.

Hartley, S. L.
1975 Life's cycle. Michigan Out-of-Doors, 29(3):41.

Haugen, A. O.
1941 Life history and management studies of the cottontail rabbit in southwestern Michigan. Univ. Michigan, unpubl. Ph.D. dissertation, 201 pp.
1942a Home range of the cottontail rabbit. Ecology, 23(3): 354–367.
1942b Life history studies of the cottontail rabbit in southwestern Michigan. American Midl. Nat., 28(1):204–244.
1943 Management studies of the cottontail rabbit in southwestern Michigan. Jour. Wildlife Mgt., 7(1):102–119.
1944 Highway mortality of wildlife in southern Michigan. Jour. Mammalogy, 25(2):177–184.

Hayne, D. W., and H. A. Cardinell
1958 New materials as cottontail repellents. Michigan State Univ., Agri. Exp. Sta., Quart. Bull., 41(1):88–98.

Hickie, P.
1940 Cottontails in Michigan. Michigan Dept. Conserv., Game Div., 109 pp.

Hubbard, D. H.
1941 Sparrow hawk preys on cottontail. Jour. Mammalogy, 22(4):454.

Jackson, H. H. T.
1961 Mammals of Wisconsin. Univ. Wisconsin Press, Madison. xii+504 pp.

Jacobs, D., and K. R. Dixon
1981 Breeding-season precipitation and the harvest of cottontails. Jour. Wildlife Mgt., 45(4):1011–1014.

Jacobson, H. A., R. L. Kirkpatrick, and B. S. McGinnes
1978 Disease and physiologic characteristics of two cottontail populations in Virginia. Wildlife Monogr., 60:1–53.

Janis, C.
1976 The evolutionary strategy of the Equidae and the origins of rumen and cecal digestion. Evolution, 30(4):757–774.

Kirkpatrick, C. M.
1960 Unusual cottontail litter. Jour. Mammalogy, 41(1): 119–120.

Lemke, C. W.
1957 An unusually late pregnancy in a Wisconsin cottontail. Jour. Mammalogy, 38(2):275.

Linduska, J. P.
1947 Longevity of some Michigan farm game mammals. Jour. Mammalogy, 28(2):126–129.
1950 Ecology and land-use relationships of small mammals on a Michigan farm. Michigan Dept. Conserv., Game Div., ix+144 pp.

Lopushinsky, T., and L. D. Fay
1967 Some benign and malignant neoplasms of Michigan cottontail rabbits. Bull. Wildlife Disease Assoc., 3:148–151.

Lord, R. D.
1959 Winter pregnancy of the cottontail. Jour. Mammalogy, 40(3):443.
1961 Seasonal changes in roadside activity of cottontails. Jour. Wildlife Mgt., 25(2):206–209.
1963 The cottontail rabbit in Illinois. Illinois Dept. Conserv., Tech. Bull. No. 3, xi+94 pp.

Malecki, R. A., S. H. Allen, and J. O. Elliston
1974 Cottontail reproduction related to dieldrin exposure. Bur. Sport Fisheries and Wildlife, Sp. Sci. Rep. Wildlife No. 177, iii+61 pp.

Manville, R. H.
1961 Angora cottontail from Georgia. Jour. Mammalogy, 42(2):255.

Mathies, J. B., and G. Schneider
1967 Influence of the eastern cottontail on tree reproduction in sugar maple-beech stands of southern Michigan. Michigan State Univ., Agri. Exp. Sta., Quart. Bull., 50(1):63–73.

McCabe, R. A.
1943 Population trends in Wisconsin cottontails. Jour. Mammalogy, 24(1):18–22.

Murie, A.
1936 Following fox trails. Univ. Michigan, Mus. Zool., Misc. Publ. No. 32, 45 pp.

Negus, N. C.
1958 Pelage stages in the cottontail rabbit. Jour. Mammalogy, 39(2):246–252.

Ozoga, J. J., and C. J. Phillips
1964 Mammals of Beaver Island, Michigan. Michigan State Univ., Publ. Mus., Biol. Ser., 2(6):305–348.

Ozoga, J. J., and L. J. Verme
1966 Noteworthy locality records for some Upper Peninsula Michigan mammals. Jack-Pine Warbler, 44(1): 52.

Peifer, R. W.
1979 Great blue herons foraging for small mammals. Wilson Bull., 91(4):630–631.

Pirnie, M. D.
1949 A test of hunting as cottontail control. Michigan State Col., Agri. Exp. Sta., Quart. Bull., 31(3):304–308.

Scharf, W. C., and K. R. Stewart
1980 New records of Siphonaptera from northern Michigan. Great Lakes Ento., 13(3):165–167.

Schofield, R. D.
1957 A study of rabbit survival on a public hunting area. Michigan State Univ., unpubl. M.S. thesis.
1960 A thousand miles of fox trails in Michigan's ruffed grouse range. Jour. Wildlife Mgt., 24(4):432–434.

Sealander, J. A.
1943 Winter food habits of mink in southern Michigan. Jour. Wildlife Mgt., 7(4):411–417.

Trent, T. T., and O. J. Rongstad
1974 Home range and survival of cottontail rabbits in southwestern Wisconsin. Jour. Wildlife Mgt., 38(3): 459–472.

Trippensee, R. E.
1934 The biology and management of the cottontail rabbit. University of Michigan., unpubl. Ph.D. dissertation, 217 pp.
1936 The reproductive function in the cottontail rabbit (*Sylvilagus floridanus mearnsii* Allen) in southern Michigan. Proc. North American Wildlife Conf., 1936:344–350.

Van Ballenberghe, V., A. W. Erickson, and D. Byman
1975 Ecology of the timber wolf in northeastern Minnesota. Wildlife Monogr., No. 43, 43 pp.

Wainright, L. C.
1969 A literature review on cottontail reproduction. Colorado Div. Game, Fish and Parks, Sp. Rept. No. 19, ii+24 pp.

Wallace, G. J.
1950 In support of the barn owl. Michigan State Col., Agri. Exp. Sta., Quart. Bull., 33(2):96–105.

Wassel, M. E., J. O. Whitaker, Jr., and E. J. Spika
1980 The ectoparasites and other associates of the cottontail rabbit, *Sylvilagus floridanus,* in Indiana. Proc. Indiana Acad. Sci., 89:418–420.

Whitlock, S. C.
1939a Tularemia. Michigan Conserv., Feb., pp. 3–4.
1939b Infection of cottontail rabbits by cycticercus pisiformis (*Taenia pisiformis*). Jour. Wildlife Mgt., 3(3):258–260.
1940 Parasites and diseases. Pp. 54–61 *in* P. Hickie, Cottontails in Michigan. Michigan Dept. Conserv., Game Div., 109 pp.

Youatt, W. G., and L. D. Fay
1968 Survey of salmonella in some wild birds and mammals in Michigan. Michigan Dept. Conserv., Res. and Dev. Rpt. No. 135 (mimeo), 1 p.

SNOWSHOE HARE
Lepus americanus Erxleben

NAMES. The generic name *Lepus,* proposed by Linnaeus in 1758, is of Latin origin and means "hare." The specific name *americanus,* proposed by Erxleben in 1777, refers to America. In Michigan, this species is often called varying hare, snowshoe rabbit, or white rabbit. Snowshoe hares in the Upper and Lower Peninsulas belong to the subspecies *Lepus americanus phaeonotus* Allen. Those on Isle Royale belong to *Lepus americanus americanus* Erxleben.

RECOGNITION. Body size medium for a leporid (see Fig. 26); head and body approximately 15¾ in (405 mm) long; ears small in relation to body size (see Measurements and Weights); eyes large and conspicuous; mammae arranged in four pairs.

In summer, color of upperparts gray-brown to yellowish-brown, with slight darkish mid-dorsal line; individual hairs basically slate-colored, medially buff, and tipped with brownish; flanks buffy-gray; abdomen white; face, throat and legs buffy-brown; ears brown, black-tipped behind, laterally bordered with yellowish-white; top of tail blackish, undersurface white; feet large and whitish to buffy; soles of feet heavily furred. In winter, pelage entirely white, obscuring the short, gray-buff underfur, except for dark eyelids and ear tips; hind feet densely padded with coarse hair, producing the snowshoe effect. In autumn and spring, individuals often observable in partly brown and partly white pelage.

Figure 26. The snowshoe hare (*Lepus americanus*) in winter pelage. Photograph by Robert Harrington

MEASUREMENTS AND WEIGHTS. Adults measure as follows: length from tip of nose to end of tail vertebrae, 15 to 19¾ in (380 to 505 mm); length of tail vertebrae, 1 to 1¾ in (25 to 45 mm); length of hind foot, 4¾ to 5⅞ in (120 to 150 mm); height of ear from notch, 2½ to 2¾ in (60 to 70 mm); weight, 48 to 80 oz (1,370 to 2,285 g).

DISTINCTIVE CRANIAL AND DENTAL CHARACTERISTICS. The large hind feet, which help give the animal an elongated look, contrast sharply with its small skull, which is only slightly larger than the skull of an adult eastern cottontail and dramatically smaller than that of the introduced European hare (Fig. 27). The skull averages 3¼ in (8 mm) long and 1 9/16 in (40 mm) wide. Although almost equal in size to that of the eastern cottontail, the skull of the snowshoe hare lacks the separate interparietal bone in the center of the back part of the top of the skull and the extending postorbital processes ("roofing" the back part of the orbits) are not fused at their posterior ends with the cranium (frontals). The dental formula is:

$$\text{I (incisors)}\ \frac{2\text{-}2}{1\text{-}1},\ \text{C (canines)}\ \frac{0\text{-}0}{0\text{-}0},\ \text{P (premolars)}\ \frac{3\text{-}3}{2\text{-}2},\ \text{M (molars)}\ \frac{3\text{-}3}{3\text{-}3} = 28$$

DISTRIBUTION. The snowshoe hare lives in the broad coniferous forest belt across Canada and Alaska from tree line south in the boreal environments to the northern tier of border states in the United States. The species also thrives in the

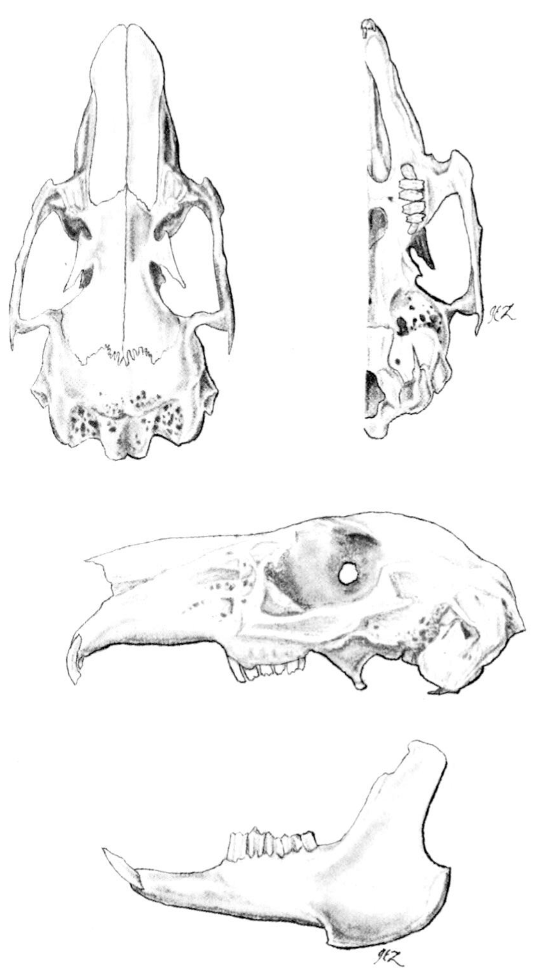

Figure 27. Cranial and dental characteristics of the snowshoe hare (*Lepus americanus*).

montane coniferous forests in the eastern mountains south to North Carolina and in the Rocky Mountains south to New Mexico and California. In Michigan today, the snowshoe hare occurs in most parts of the Upper Peninsula and in the northern half of the Lower Peninsula (see Map 23). In the colonial period the snowshoe rabbit was probably found throughout much of the southern part of Michigan's Lower Peninsula (Bookhout and Chase, 1965; de Vos, 1964; Wood, 1922), with populations chiefly concentrated in the vicinity of swampy areas covered with stands of evergreen trees. Apparently this hare persisted in tamarack swamps in Jackson County as late as 1907 (Wood and Dice, 1923), but authorities seem convinced that the species ranged no further south than the second tier of southernmost counties. As of 1977, according to records kept by the Michigan Department of Natural Resources (Marvin Cooley, pers. comm.), the southernmost limit of the snowshoe hare is the following counties (west to east): Muskegon, Newaygo, Mecosta, Isabella, Midland, and Bay. Natural populations are now gone in Michigan's Thumb area, although they were reported in the mid-1960s in the Sleeper State Park in Huron County (Bookhout, 1965b). In Sanilac County, 32 hares (live-captured in Oscoda County) were released on December 13, 1976, in the Minden City State Game Area. An attempt to reestablish this species in the Gratiot-Saginaw State Game Area was unsuccessful.

HABITAT PREFERENCES. The snowshoe hare is a characteristic mammal of Michigan's Canadian Biotic Province (Dice, 1938). The species seems to thrive best in areas having overhead woody protection, which is partly coniferous. Snowshoe hares are probably attracted mostly to northern swamps shaded by white cedar and black spruce. They also

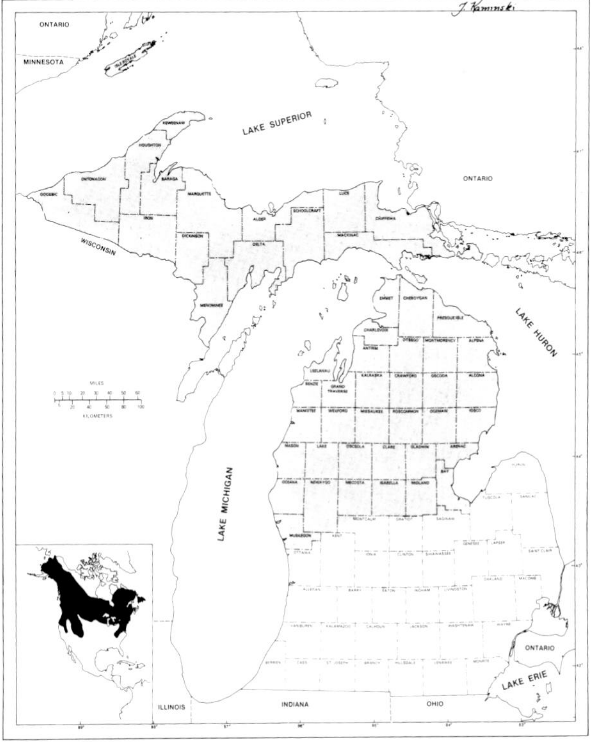

Map. 23. Geographic distribution of the snowshoe hare (*Lepus americanus*). In previous times its range was more extensive in southern Michigan.

use patchy bracken and brushy understory in spruce-fir woodlands and pine plantations (Wolff, 1980). The animals withdraw from areas ecologically changed or radically disturbed by human activities. However, if such areas are allowed to revert or are planted with conifers, snowshoe hares will become re-established (Conroy, 1976). According to Bookhout (1965b), the swamp conifer environment is the preferred habitat for the snowshoe hare in Michigan. To a lesser extent the alder swamp is also used, followed by aspen-balsam fir and paper birch. Bookhout noted that northern hardwood stands and grassy openings are little used by snowshoe hares. The snowshoe hare, like the eastern cottontail, likes the transition (edge) between two of its favored habitats, with well-worn connecting runways. The effects of fire on snowshoe hare range are not clearly known, although fire-induced plant successions probably increase food supplies. A spring fire in Alberta, for example, produced no evident fire-induced mortality among hares. The animals moved back into the area the second summer after the burn, when brushy cover had redeveloped (Keith and Surrendi, 1971), emphasizing again the importance of ground cover, especially low-growing conifers, for the snowshoe hare.

DENSITY AND MOVEMENTS. Snowshoe hare populations are, of course, responsive to the quality of their environments. In Michigan, food, cover, and other life essentials seem most available to hares in swamp conifer habitat with adjacent alder swamp and mixed evergreen-deciduous forest with brushy understory. There are population increases during the growing season as a result of recruitment of young-of-the-year, with a gradual decline in local densities through the winter season. Besides the annual fluctuations due to local environmental factors, snowshoe hares are subject to multiannual population fluctuations. Banfield (1974) noted that in Canada snowshoe hare numbers can fluctuate from as few as one individual to as many as 3,400 individuals per sq mi (from 0.39 per sq km to 1,308 per sq km), with even higher numbers estimated in favorable habitat. These dramatic variations have mystified early explorers, settlers, and wildlife biologists. Long-time observations show that peak densities occur, on the average, every 10 years (Keith and Windberg, 1978), some as often as every six years, others as infrequently as every 13 years. These numbers are reflected in the records of the fur trade kept by the Hudson Bay Company (Elton, 1933; MacLulich, 1937) and even in hunter success. In Michigan, for example, the annual hunting season harvest of snowshoe hares may be as low as 150,000 animals or as high as 750,000 animals. Suggested causes for these fluctuations include suspected movements of animals out of areas of abundance to marginal habitats having low hare densities (Banfield, 1974), disease (Stuht *et al.*, 1978), variations in adult and juvenile survival (Wood and Munroe, 1977), and major changes in the quality and quantity of the food supply especially in winter (Keith, 1974; Windberg and Keith, 1976; Vaughan and Keith, 1981).

Movements by snowshoe hares are not extensive. Bookhout (1965b) found individuals in Alger County living in areas averaging 20 acres (8 ha). Some animals used only five acres (2 ha) while others ranged in areas of 57 acres (22.8 ha). Living areas and daily movements are more extensive in summer than in winter. Females are prone to cover smaller areas when nursing litters. Males involved in courtship will move in larger territories than non-breeders. Methods of estimating populations and individual movements include live-trapping-tagging-releasing-retrapping procedures, counts of tracks in snow or animals flushed, hunter's bag records, fecal pellet counts, and radio telemetry.

BEHAVIOR. During daylight hours the snowshoe hare rests (and grooms) in a secluded place (under a fallen log, in a grassy form or in brush). It is from such resting places that most hares are flushed and observed by passers-by. At night and in the twilight of early morning and late evening, the snowshoe hare forages, with activity reaching a peak at or near 2300 hrs (Keith, 1964). The hare uses an array of criss-crossing runways and tunnels through herbaceous and woody ground cover in summer and along trails in the snow in winter.

Although not often seen in groups, the snowshoe hare does tolerate several neighbors in the same cedar swamp or thicket. Occasionally hares will use the same daytime forms and may share dusting areas in summer. A night observer, using a vehicle's headlights or a hand-held spot light, may see several individuals in forest openings or at the edges of swamps interacting in brief chases. Banfield (1974) and Jackson (1961) summarize some of our knowledge of the social activities of this hare which relies on its running speed of up to 31 mi (52 km) per hour (Terres, 1941) to escape predators. Single leaps may be as much as 10 ft (3 m). The hare's furred hind feet enable it to negotiate deep winter snow without extreme difficulty. In swamps, on lake shores or at stream sides, the snowshoe hare will readily take to water when threatened and swim to safety as observed in Iron County by George Hunt, (1950). Occasionally snowshoe hares will enter underground burrows or holes in snow banks. However this behavior is unusual—certainly not a way of life as it is for the eastern cottontail.

Thumping of the large hind feet seems to be one type of communication used by snowshoe hares. A low chirp has been described (Trapp and Trapp, 1965); also a snort or grunt is detectable, often when an individual is handled. Like the eastern cottontail, the snowshoe hare utters a shrill, bleating cry, when seized by a predator, or when struggling to escape from the grasp of even a gentle handler. The animals must rely largely on acute hearing to detect approaching danger, while they themselves are exceedingly quiet creatures.

ASSOCIATES. The snowshoe hare occurs in habitats with all other northern forest creatures. Its relationships with most, however, may be remote since this leporid rarely uses communal situations, especially burrows or other shelters either constructed or occupied by many of the rodents. Aside from supplying meals for carnivores, the snowshoe hare has its closest interspecific relationships with the resident cervids. This is chiefly because of their overlapping food selections. Feeding coactions between hares and moose have been examined in such places as Newfoundland (Dodds, 1960) and New Brunswick (Telfer, 1974); between hares and wapiti in Alberta (Telfer, *loc. cit.*); and between hares and white-tailed deer in Alberta (Telfer, *loc. cit.*), in Nova Scotia and New Brunswick (Telfer, 1972), in the Upper Great Lakes Region (Krefting, 1975; Krefting and Stoeckeler, 1953), and in Michigan (Bookhout, 1965b).

In lowland coniferous areas in northern Michigan, hares and white-tailed deer often eat the same kinds of foods. However, it was Bookhout's (1965c) opinion that the use of ground cover as food by hares seems not to detract from the quality of the deer yard environment. There is even less association in summer when deer tend to move from the swamps to the uplands. On the other hand, the moose-hare coaction may be a more serious matter (Bookhout, 1965b). In Newfoundland, for example, Dodds (1960) found that 27 of the 30 woody plants browsed by moose were also eaten by hares. He also noted that when moose fed on balsam fir reproduction, ground cover for hare could be seriously affected.

REPRODUCTIVE ACTIVITIES. The Michigan male snowshoe hare's testes begin enlarging in early February, mating occurs from early April through early August, and the testes decline in size in early September. Aldous (1937), Bookhout (1965a, 1965b), Grange (1932), and Meslow and Keith (1968) have made major contributions to our knowledge of the life history of the snowshoe hare. Reproductive receptiveness in the female hare lags behind that of the male, perhaps by as long as six weeks. Mating and conception occur beginning in early April in Alger County, according to Bookhout (1965a, 1965b). Matings are usually promiscuous—one receptive female may be followed in a procession-like dance by several males, with leaping and dodging involved before the actual mating.

The pregnant female prepares no elaborate nest as does the eastern cottontail. Instead, the place usually selected is in a sheltered area of grass, brush or low trees, sometimes under a log or even in the entrance of an old woodchuck burrow. The vegetation is packed down to produce a form. Here, after a gestation period of about 35 days (Bookhout, 1964), a litter of from one to seven (generally two or four in Michigan, Bookhout, 1965b) is produced. The doe hare is receptive to the buck for mating shortly after the birth of her young; in fact, she can have three and sometimes four litters during a single season (Banfield, 1974; Bookhout, 1965a). Litters produced later in the season usually contain more offspring than the first (Rowan and Keith, 1956). There is also evidence that litter sizes may average more than four young per litter during the rising population phase of the multiannual fluctuations and approximately two young per litter during the declining population phase (Meslow and Keith, 1968). The breeding season is generally completed in late August and early September. Although there is an occasional record of a female having a litter her first season (Keith and Meslow, 1967; Vaughan and

Keith, 1980), the young usually do not reproduce until after their first winter.

At birth the young snowshoe hares are precocial; that is, born in an advanced stage of development. They possess soft, dense, brownish fur, grizzled and long. The eyes are open at or soon after birth. On the first day, the young leverets each weigh from 2.2 to 2.8 oz (65 to 80 g) and can crawl and hop feebly (Banfield, 1974). By the ninth day, the young hares have doubled their weight, are fully active, and are beginning to eat green vegetation. At two to three weeks of age the offspring are weaned and leave the nesting site. At this age, however, the leverets "freeze" at the approach of danger rather than attempting to escape, while the mother hare distracts adversaries away from her litter.

Although there is some evidence that snowshoe hares may survive in nature for four or five years (Banfield, 1974), most individuals complete their life spans before or by the end of the second year of life, with marked reductions in survivors during a third season—probably no more than 15% live to produce litters for a second season. Wild-taken yearlings, two-year-olds, and older age classes may be identified by eye lens weight (Keith and Cary, 1979). The average annual productivity for Michigan hares is 6.54 young per female (Bookhout, 1965a).

FOOD HABITS. Green plant growth in summer and buds, twigs, bark, and conifer needles (leaves) in winter are standard fare for the snowshoe hare. Unlike plant-feeders which generally ingest nutrients concentrated in seeds and fruits, hares eat large quantities of the plant's vegetative parts, leaves, and fibrous tissues containing complex carbohydrates. Winter daily requirements for captive hares are about 300 g of woody browse having maximum diameters of 3–4 mm (Pease *et al.*, 1979). Like the eastern cottontail, caecal fermentation in the lower gut breaks down much of this vegetation to assimilable compounds. Reingestion of soft fecal pellets (coprophagy) is practiced so the hare can achieve maximum nutritional use of its food (Bookhout, 1959).

Preferred summer foods of the snowshoe hare are not as well understood as preferred winter foods. Summer forage includes clovers, various grasses (especially brome and bluegrass), vetches, asters, dandelions, strawberry, ferns, horsetails, and the tender parts of such woody vegetation as hazelnut, willow, aspen, alder, and birch. In winter in Alger County, Bookhout (1965b) graded as "very good" hare forage, the edible parts of yellow birch, paper birch, white cedar, sugar maple, trembling aspen, American elm, jack pine, and red pine. White spruce, black spruce, eastern hemlock, white pine, and beaked hazel were graded "good." There is growing evidence (Walski and Mautz, 1977) that despite the fact that such choice winter fare as aspen, white cedar and red maple are eaten in quantity by the snowshoe hare, these plants are actually rather poor nutritionally. Even so, Conroy *et al.*, (1979) found red maple (*Acer rubrum*) along with speckled alder (*Alnus rugosa*) to be the most intensively browsed plant species in lowland habitat in Roscommon County. There is evidence, based on studies in Alaska (Bryant, 1981), that some severely browsed deciduous saplings and shrubs produce adventitious shoots that are unpalatable to snowshoe hares because of higher concentrations of terpene and phenolic resins than the normal mature twigs. Occasionally, snowshoe hares will eat table refuse, insects, sand (Conroy, 1978), and even meat from a dead carcass (including white-tailed deer, Field, 1970).

ENEMIES. Studies of the food habits of predators indicate that every carnivore of sufficient size (even weasels) finds the snowshoe hare fair game. Naturally the snowshoe hare is an important link in the plant-herbivore-carnivore food chain of the north woods (Brand *et al.*, 1976). When populations of the hare are high or low, such important carnivores as the lynx and coyote are noticeably affected (Brand *et al.*, 1976; Brand and Keith, 1979; Todd *et al.*, 1981). Other meat-eating mammals which eat hares include the red fox (Arnold, 1956; Johnson, 1970; Manville, 1948; Schofield, 1960), fisher (de Vos, 1951), mink (Sealander, 1943), domestic house cat (Doucet, 1973; Manville, 1948), gray wolf (Voigt *et al.*, 1975), bobcat (Manville, 1958), coyote and both the ermine and the long-tailed weasels (Jackson, 1961). There is even a record of a short-tailed shrew attacking a two-day-old snowshoe hare (Rongstad, 1965). The great horned owl (Adams, 1959; Errington *et al.*, 1940), goshawk (Pettingill, 1976), snowy owl (Jackson, 1961) and other northern birds of prey also catch snowshoe hares.

PARASITES AND DISEASES. Despite the constant drain on snowshoe hare populations by meat-eaters, it is still generally thought that disease, not predation, is more important in snowshoe hare population regulation (Banfield, 1974). On South Fox Island in Lake Michigan, for example, a sampling of snowshoe hares obtained during a die-off of this species in August of 1974, showed infections of tularemia, coccidiosis, and babesiosis—diseases which Stuht *et al.* (1978), believe contributed to the drastic death rate. However, Bookhout (1965b) concluded from studies made in Alger County (in Michigan's Upper Peninsula) from 1957 to 1962 that parasites did not seriously affect the size of snowshoe hare populations, despite the fact that the number of parasites this hare carries is at times extremely high. He found the rabbit tick (*Haemaphysalis leporis-palustris*) to be the most prevalent ectoparasite of Michigan snowshoe hares. Mites and fleas were also commonplace. Ten species of parasitic worms have been obtained from Michigan hares. The lungworm (*Protostrongylus boughtoni*) and the stomach worm (*Obeliscoides cuniculi*) can produce tissue damage. As in the eastern cottontail, the snowshoe hare has a history of tularemia. As a result, persons handling any leporids should use rubber gloves or take other precautions, especially when skinning and dressing the animals.

MOLT AND COLOR ABERRATIONS. The adult snowshoe hare undergoes an autumn molt and a spring molt (Grange, 1932). This is the usual mammalian pattern, except that in this species the summer fur is brownish and the winter fur is whitish (see Plate 2). The autumn molt begins in September (a hare from Garden Island was in molt on August 28, Phillips *et al.,* 1965) and continues until December. The spring molt commences in March and is complete in May. Although there is still some question as to whether separate hair roots are responsible for the two coat colors (Severaid, 1945), the color change first described by Allen (1894) and later examined in detail by Lyman (1943), is closely correlated with photoperiod. The spring or vernal molt is activated by the increasing interval of daylight per 24 hours, and the autumn molt is activated when the interval of daylight per 24 hours decreases.

There is considerable variation in coat color in snowshoe hares. Black (melanistic) individuals have been reported (Jackson, 1961); one was taken in Houghton County in the winter of 1976 by John Vollner of Hancock.

ECONOMIC IMPORTANCE. For centuries snowshoe hare meat has been standard fare for residents and travelers in the North Country. Despite its dietary importance and the value of its pelt, this hare was generally taken for granted by early writers who extolled the moose, caribou and lynx but rarely mentioned the hare. Bookhout and Chase (1965), in their history of snowshoe hare hunting in Michigan, report that "rabbits" (not identified by species) were mentioned in Michigan laws as early as 1897, but the first specific reference to the snowshoe hare, as a species, was in 1915–16 when it was designated as a legal game species and a hunting season established. In 1922–23, a daily bag limit, a possession limit, and a season harvest limit were declared, much the same as current regulations. Today, a special class of small game hunters and their well-trained dogs hunt the Upper Peninsula and northern part of the Lower Peninsula (Zones 1 and 2), where the season is open from October through March. There has been little commercial interest in snowshoe hare pelts in Michigan, although as many as 10,000 are marketed annually in Canada (Banfield, 1974).

Snowshoe hare kill statistics, maintained by the Michigan Department of Natural Resources, reflect natural hare population oscillations and also the post–World War II increase in licensed hunters. Looking back over the past 40 years these kill estimates (calculated using data from post-season hunter survey cards) have shown ups and downs. Lows in the neighborhood of 150,000 animals were harvested in 1944–45 and 1969–70, while highs of about 550,000 in 1939–40 and 762,000 in 1975–76 were recorded. Despite hunters' annual harvest and other losses, the hare's reproductive potential allows for a sustained population.

Sharp, chisel-shaped front incisors serve the snowshoe hare well in obtaining nutrients. At the same time the forester may take a dim view of the hare's choice of species in a conifer plantation. In fact, Aldous (1947) thought the snowshoe hare was the major animal pest to foresters in the Lake States. Studies, reviewed by Bookhout (1965b), show that snowshoe hares damage jack pine, red pine, white pine, white spruce, white cedar and balsam fir, sometimes affecting 50% of a planting. The hares do much less damage to aspen and paper birch. On occasion the animals actually do a service by thinning dense natural tree reproduction.

The hunt for the snowshoe hare begins after a heavy snowfall and is capped by bagging a brace of these attractive animals. Getting them from their hiding places to the game bag is high adventure. The culmination is the preparation of a delicacy known as hasenpfeffer. The following is a recipe as given by Dreis (1976):

> Put the pieces of meat from the snowshoes (minus rib cages) in a pot and add 1½ cups of water, 1 cup of vinegar, 2 sliced onions, 3 bay leaves, 10 whole cloves, 2 teaspoons of salt and ⅔ teaspoon of pepper. Stir it all up well, put on a tight cover, stick it in the refrigerator and forget about it for a couple of days. You should be good and hungry by then.
>
> Take out the pieces of meat, roll in flour, salt and fry until a golden brown. Now add ¼ of a cup of the mixture (strained) that the meat soaked in, and let it simmer for an hour. Add sour cream just before serving (don't boil the cream).
>
> It tastes great! And with that, the snowshoe hares have fulfilled their destiny.

COUNTY RECORDS FROM MICHIGAN. Records of specimens in collections of museums and from the literature show that the snowshoe hare occurs in all counties in the Upper Peninsula and in those counties in the Lower Peninsula including and to the north of the following, from west to east: Muskegon, Newago, Mecosta, Isabella, Midland and Bay (see Map 23). Early settlement records show that showshoe hares lived in most counties north of the southernmost tier in southern Michigan (Burt, 1946). The species is also known from Isle Royale and the Huron Islands in Lake Superior; Big Summer Island, High Island, Hog Island, Beaver Island, Garden Island, North and South Fox islands, and North (formerly) and South Manitou islands in Lake Michigan; Bois Blanc Island and Drummond Island in Lake Huron; and Charity Island in Saginaw Bay.

LITERATURE CITED

Adams, L.
1959 An analysis of a population of snowshoe hares in northwestern Montana. Ecol. Monogr., 20(2):141–170.

Aldous, C. M.
1937 Notes on the life history of the snowshoe hare. Jour. Mammalogy, 18(1):46–57.

Aldous, S. E.
1947 Some forest-wildlife problems in the Lake States. Lake States Forest Exp. Sta., Paper No. 6, 11 pp.

Allen, J. A.
1894 On the seasonal changes of color in the varying hare (*Lepus americanus* Erx.). Bull. American Mus. Nat. Hist., 6(4):107–128.

Arnold, D. A.
1956 Red Foxes of Michigan. Michigan Dept. Conservation, Lansing. 48 pp.

Banfield, A. W. F.
1974 The mammals of Canada. Univ. Toronto Press, Toronto. xxiv + 438 pp.

Bookhout, T. A.
1959 Reingestion by the snowshoe hare. Jour. Mammalogy, 40(2):250.
1964 Prenatal development of snowshoe hare. Jour. Wildlife Mgt., 28(3):338–345.
1965a Breeding biology of snowshoe hares in Michigan's Upper Peninsula. Jour. Wildlife Mgt., 29(2):296–303.
1965b The snowshoe hare in Upper Michigan its biology and feeding coaction with white-tailed deer. Michigan Dept. Conserv., Res. and Dev. Rept. No. 38, x + 191 pp.
1965c Feeding coactions between snowshoe hares and white-tailed deer in northern Michigan. Trans. 30th North American Wildlife and Nat. Res. Conf., Washington, D.C., pp. 321–335.

Bookhout, T. A., and W. W. Chase
1965 History and status of the snowshoe hare in Michigan. Michigan Acad. Sci., Arts, and Ltrs., Papers, 50:31–37.

Brand, C. J., and L. B. Keith
1979 Lynx demography during a snowshoe hare decline in Alberta. Jour. Wildlife Mgt., 43(4):827–849.

Brand, C. J., L. B. Keith, and C. A. Fischer
1976 Lynx responses to changing snowshoe hare densities in central Alberta. Jour. Wildlife Mgt., 40(3):416–428.

Brand, C. J., R. H. Vowles, and L. B. Keith
1975 Snowshoe hare mortality monitored by telemetry. Jour. Wildlife Mgt., 39(4):741–747.

Bryant, J. R.
1981 Phytochemical deterrence of snowshoe hare browsing by adventitious shoots of four Alaskan trees. Science, 213(4510):889–890.

Burt, W. H.
1946 The mammals of Michigan. Univ. Michigan Press, Ann Arbor. xv + 288 pp.

Conroy, M. J.
1976 Winter habitat structure of the snowshoe hare. Michigan State Univ., unpubl. M.S. thesis, v + 41 pp.
1978 Ingestion of sand by snowshoe hares. Jack-Pine Warbler, 56(3):160.

Conroy, M. J., L. W. Gysel, and G. R. Dudderar
1979 Habitat components of clear-cut areas for snowshoe hares in Michigan. Jour. Wildlife Mgt., 43(3):680–690.

de Vos, A.
1951 Recent findings in fisher and marten ecology and management. Trans. North American Wildlife Conf., 16:498–507.

1964 Range changes of mammals in the Great Lakes region. American Midl. Nat., 71(1):210–231.

Dice, L. R.
1938 The Canadian Biotic Province with special reference to the mammals. Ecology, 19:503–514.

Dodds, D. G.
1960 Food competition and range relationships of moose and snowshoe hare in Newfoundland. Jour. Wildlife Mgt., 24(1):52–60.

Doucet, G. J.
1973 House cat as predator of snowshoe hare. Jour. Wildlife Mgt., 37(4):591.

Dreis, R. E.
1976 Snowshoe hares from hiding place to hasenpfeffer. Wisconsin Nat. Res. Bull., 41(2):8–9.

Elton, C.
1933 The snowshoe rabbit inquiry, 1931–32. Canadian Field-Nat., 47:63–69.

Errington, P. L., F. Hamerstrom, and F. N. Hamerstrom, Jr.
1940 The great horned owl and its prey in north-central United States. Iowa State Col., Agri. Exp. Sta., Res. Bull. 277, pp. 757–850.

Field, R. J.
1970 Scavengers feeding on a Michigan deer carcass. Jack-Pine Warbler, 48(2):73.

Grange, W. B.
1932 The pelages and color changes in the snowshoe hare, *Lepus americanus phaeonotus* Allen. Jour. Mammalogy, 13(2):99–116.

Hunt, G. S.
1950 Aquatic activity of a snowshoe hare. Jour. Mammalogy, 31(2):193–194.

Jackson, H. H. T.
1961 Mammals of Wisconsin. Univ. Wisconsin Press, Madison. xii+504 pp.

Johnson, W. J.
1970 Food habits of the red fox in Isle Royale National Park, Lake Superior. American Midl. Nat., 84(2):568–572.

Keith, L. B.
1964 Daily activity patterns of snowshoe hares. Jour. Mammalogy, 45(4):626–627.
1974 Some features of population dynamics in mammals. Proc. Int. Cong. Game Biol., Stockholm, 11:17–58.

Keith, L. B., and J. R. Cary
1979 Eye lens weights from free-living adult snowshoe hares of known age. Jour. Wildlife Mgt., 43(4):965–969.

Keith, L. B., and E. C. Meslow
1967 Juvenile breeding in the snowshoe hare. Jour. Mammalogy, 48(2):327.

Keith, L. B., and D. C. Surrendi
1971 Effects of fire on a snowshoe hare population. Jour. Wildlife Mgt., 31(1):16–26.

Keith, L. B., and L. A. Windberg
1978 A demographic analysis of the snowshoe hare cycle. Wildlife Monog., No. 58, 70 pp.

Krefting, L. W.
1975 The effect of white-tailed deer and snowshoe hare browsing on trees and shrubs in northern Minnesota. Univ. Minnesota, Agri. Exp. Sta., Tech. Bull. 302-1975, For. Ser. 18, 43 pp.

Krefting, L. W., and J. H. Stoeckeler
1953 Effect of simulated snowshoe hare and deer damage on planted conifers in the Lake States. Jour. Wildlife Mgt., 17(4):487–494.

Lyman, C. P.
1943 Control of coat color in the varying hare *Lepus americanus* Erxleben. Bull. Mus. Comp. Zool., 93:393–461.

MacLulich, D. A.
1937 Fluctuations in the numbers of the varying hare *(Lepus americanus)*. Univ. Toronto Studies, Biol. Ser., 43:1–136.

Manville, R. H.
1948 The vertebrate fauna of the Huron Mountains, Michigan. American Midl. Nat., 39(3):615–640.
1958 Odd items in bobcat stomachs. Jour. Mammalogy, 39(3):439.

Meslow, E. C., and L. B. Keith
1968 Demographic parameters of a snowshoe hare population. Jour. Wildlife Mgt., 32(4):812–834.

Pease, J. L., R. H. Vowles, and L. B. Keith
1979 Interaction of snowshoe hares and woody vegetation. Jour. Wildlife Mgt., 43(1):43–60.

Pettingill, O. S., Jr.
1976 The prey of six species of hawks in northern Lower Michigan. Jack-Pine Warbler, 54(2):70–74.

Phillips, C. J., J. T. Ozoga, and L. C. Drew
1965 The land vertebrates of Garden Island, Michigan. Jack-Pine Warbler, 43(1):20–24.

Rongstad, O. J.
1965 Short-tailed shrew attacks young snowshoe hare. Jour. Mammalogy, 46(2):328–329.

Rowan, W. M., and L. B. Keith
1956 Reproductive potential and sex ratios of snowshoe hares in northern Alberta. Canadian Jour. Zool., 34:273–281.

Schofield, R. D.
1960 A thousand miles of fox trails in Michigan's ruffed grouse range. Jour. Wildlife Mgt., 24(4):432–434.

Sealander, J. A.
1943 Winter food habits of mink in southern Michigan. Jour. Wildlife Mgt., 7(4):411–417.

Severaid, J. H.
1945 Pelage changes in the snowshoe hare (*Lepus americanus struthopus* Bangs). Jour. Mammalogy, 26(1):41–63.

Stuht, J., R. Huff, R. Odom, and C. Stewart
1978 South Fox Island snowshoe hare die-off. Michigan Dept. Nat. Res., Wildlife Div., Rpt., No. 2823, 7 pp.

Telfer, E. S.
1972 Browse selection by deer and hares. Jour. Wildlife Mgt., 36(4):1344–1349.
1974 Vertical distribution of cervid and snowshoe hare browsing. Jour. Wildlife Mgt., 38(4):944–946.

Terres, J. K.
1941 Speed of the varying hare. Jour. Mammalogy, 22(2): 453–454.

Todd, A. W., L. B. Keith, and C. A. Fischer
1981 Population ecology of coyotes during a fluctuation of snowshoe hares. Jour. Wildlife Mgt., 45(3):629–640.

Trapp, G. G., and C. Trapp
1965 Another vocal sound made by snowshoe hares. Jour. Mammalogy, 45(4):705.

Vaughan, M. R., and L. B. Keith
1980 Breeding by juvenile snowshoe hares. Jour. Wildlife Mgt., 44(4):948–951.
1981 Demographic response of experimental snowshoe hare populations to overwinter food shortage. Jour. Wildlife Mgt., 45(2):354–380.

Voigt, D. R., G. B. Kolenosky, and D. H. Pimlott
1975 Changes in summer foods of wolves in central Ontario. Jour. Wildlife Mgt., 40(4):663–668.

Walski, T. W., and W. W. Mautz
1977 Nutritional evaluation of three winter browse species of snowshoe hares. Jour. Wildlife Mgt., 41(1):144–147.

Windberg, L. A., and L. B. Keith
1976 Snowshoe hare population response to artificial high densities. Jour. Mammalogy, 57(3):523–553.

Wolff, J. O.
1980 The role of habitat patchiness in the population dynamics of snowshoe hares. Ecol. Monog., 50(1): 111–130.

Wood, N. A.
1922 The mammals of Washtenaw County, Michigan. Univ. Michigan, Mus. Zool., Occas. Papers No. 123, 23 pp.

Wood, N. A., and L. R. Dice
1923 Records of the distribution of Michigan mammals. Michigan Acad. Sci., Arts and Ltrs., 3:425–469.

Wood, T. J., and S. A. Monroe
1977 Dynamics of snowshoe hare populations in the Maritime Provinces. Canadian Wildlife Ser., Occas. Paper No. 30, 21 pp.

EUROPEAN OR CAPE HARE
Lepus capensis Linnaeus

NAMES. The generic name *Lepus,* proposed by Linnaeus in 1758, is of Latin derivation and means "hare." The specific name *capensis,* also proposed by Linnaeus in 1758, refers to the Cape area of southern Africa. In Michigan, this species is sometimes called introduced hare or jack rabbit, but generally European hare. The subspecies *Lepus capensis hybridus* Desmarest occurs in Michigan (Dean and de Vos, 1965).

RECOGNITION. Body size large for a leporid; head and body approximately 23 in (585 mm) long; eyes and ears large; tail short. Color of upperparts grizzled yellowish (usually in summer) to grayish brown (usually in winter), more grayish on cheeks, flanks, and rump; underfur generally grayish; long guard hairs black-banded with buffy terminal tips; nape area, throat, and forefeet richly buffy; hind feet buffy; ears black-tipped and mostly brown with grayish-white interiors; tail blackish above and whitish below; most of underparts white. It is always possible that a sight record for a European hare may actually prove to be a tame domestic rabbit *(Oryctolagus cuniculus).* Such a rabbit might resemble the European hare in bulk and general coloring, but the shorter, less yellowish ears and rather slow, hopping gait of the rabbit should distinguish these two large leporids in the field.

MEASUREMENTS AND WEIGHTS. Adults measure as follows: length from tip of nose to end of tail vertebrae, 23½ to 29¼ in (600 to 750 mm); length of tail vertebrae, 2¾ to 3⅞ in (70 to 100 mm); length of hind foot, 5¾ to 6¼ in (145 to 160 mm); height of ear from notch, 3⅛ to 3⅞ in (80 to 100 mm); weight, 6 to 12 lbs. (2.7 to 5.4 kg).

DISTINCTIVE CRANIAL AND DENTAL CHARACTERISTICS. This introduced hare has a truly massive skull, much larger than that of the native snowshoe hare (Fig. 28). The skull averages 3⅞ in (98 mm) long and 1⅞ in (48 mm) wide.

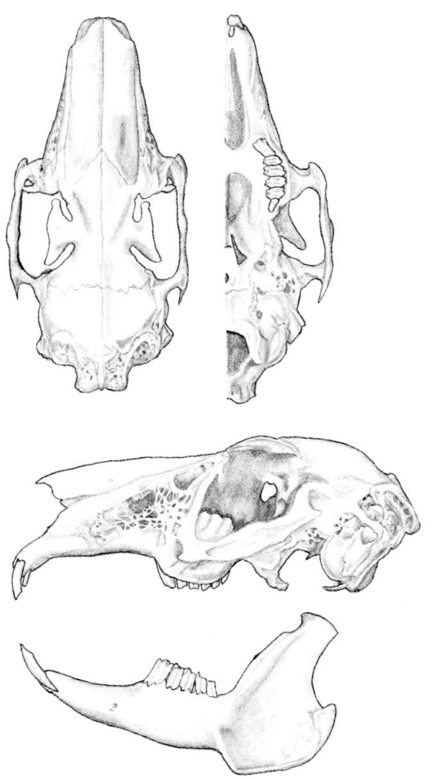

Figure 28. Cranial and dental characteristics of the introduced European hare (*Lepus capensis*).

Major features are the broad nasal bones and large orbits. The dental formula is:

$$\text{I (incisors)}\frac{2\text{-}2}{1\text{-}1}, \text{C (canines)}\frac{0\text{-}0}{0\text{-}0}, \text{P (premolars)}\frac{3\text{-}3}{2\text{-}2}, \text{M (molars)}\frac{3\text{-}3}{3\text{-}3} = 28$$

DISTRIBUTION. The European hare have come to Michigan from neighboring Ontario, perhaps initially across ice on the St. Claire River (Ninth Biennial Report for 1937–38 of the Michigan Department of Conservation). Successful introductions of this Eurasian species were made in the southeastern part of Ontario as early as 1912 (Dean and de Vos, 1965). The Michigan population may also have been enhanced by stock released by Michigan residents, although such releases had no official sanction from the Michigan Department of Natural Resources. Specimens have been recorded from at least two Upper Peninsula and ten Lower Peninsula counties but there is no real evidence to support the contention that the species is completely established in Michigan (Map 24).

HABITAT PREFERENCES. A mixture of fallow and cultivated fields and small woodlots seems to provide the best Michigan habitat for the European hare, while Burt (1946) noted that the hare may prefer open hilly lands. The European hare normally makes its escape across open areas rather than through woody vegetation, as the eastern cottontail and snowshoe hare often do. The European hare does not use burrows, generally selecting unprotected resting places where visibility across open lands is extensive. Much of our knowledge of the habitats used by this species come from European or Canadian sources (Banfield, 1974; Peterson, 1966), since little positive data on the Michigan population are available.

Map 24. Counties from which there are known records for the introduced European hare (*Lepus capensis*).

DENSITY AND MOVEMENTS. According to Banfield (1974), a population density of 25 hares per sq mi (about 10 per sq km) may be average, with double or triple that number occasionally obtained. No distinctive or regular multiannual fluctuations, as noted in Eurasian populations, have been observed in the rather recently-established North American populations.

The European hare, like other leporids, is chiefly nocturnal or crepuscular in activity. The animal crouches in a "form" (de Vos and Dean, 1967) consisting of crushed or parted grasses or low shrubs or a depression in soil during daylight hours. This resting place may be partly or entirely covered with snow in winter. In fact, the hare may huddle under complete snow coverage but remain sufficiently alert to bound off across open country in case of danger. Like the snowshoe hare, the European hare relies almost entirely on its rapid leaping gait to escape enemies and, like the western jack rabbit, can outrun many of its adversaries.

BEHAVIOR. The nocturnal activity pattern of the European hare plus its rarity in Michigan have given observers little opportunity to determine its behavioral responses to a new environment. The hare, like other rabbits and hares, has keen eyesight and hearing. In general, the animals are solitary except during the reproductive season, although contact must occur frequently in nature owing to the large cruising territory of each individual.

REPRODUCTIVE ACTIVITIES. Data on the reproductive cycle of the European hare in Michigan are sparse. Observations in Europe and Ontario show that breeding may occur at any time, but chiefly from late winter to early summer. Buck hares fight one another aggressively for mates. The gestation period is between 30 and 40 days. The young (varying in number from one to seven) are born above ground in a precocious state, fully-haired and able to move about. According to observers, the female may segregate members of her litter, spacing them in different grassy or brushy forms and moving from one to another during daylight hours for nursing. Immature hares, exhibiting characteristic white ear patches, are weaned and capable of self-maintenance within a few weeks after birth. It is possible for one female to have two litters within a single season. Apparently young animals do not breed until the second year.

FOOD HABITS. The European hare's diet is similar to that of the eastern cottontail and the snowshoe hare. Greens are the chief fare in summer and bark, buds and twigs of woody plants in winter. Most feeding activity occurs at night or in the dim light of early morning or late evening. No specific food studies of Michigan animals have been made. Like other leporids, the European hare reingests its soft green feces which contain foods previously reduced to a digestible state by means of caecal fermentation.

ENEMIES. The European hare was introduced into Michigan after the disappearance of such large carnivores as the gray wolf and mountain lion. Very likely the large size of the adult hares saves them from many would-be predators except for red fox, gray fox, coyote, bobcat, lynx and larger hawks and owls. A wider assortment of meat-eaters can, of course, prey on young hares.

ECONOMIC IMPORTANCE. The European hare has not been as serious a pest as the European rabbit *(Oryctolagus cuniculus)* in areas foreign to its native habitat. Nonetheless, introduced populations of the hare have damaged agricultural crops and competed with domestic livestock for pasturage in southern South America, New Zealand and elsewhere (de Vos *et al.*, 1956; Greer, 1965). In North America, especially in Ontario and New York (Banfield, 1974; Dean and de Vos, 1965), the European hare is guilty of raiding gardens and girdling orchard trees. The latter is probably the species' worst offense. Because the European hare is basically an open lands (steppe) animal, its range in much of northeastern North America is probably not optimum. In fact, when and where the country is allowed to revert slowly to shrub lands and forest, this hare should gradually deline in number. On the other hand, in areas where the country is held open through managed pastures for livestock, the European hare should thrive.

The big, meaty European hare has served as a staple food for rural peoples in the Old World for centuries. It is also prized as a sporting animal. Not only will European hares outdistance many hunting hounds, they do not hesitate to cross marshes, streams, and ice to evade the pursuers. With the sporting and eating qualities of this introduced hare, it should have a profitable stay in Ontario and perhaps in Michigan, where its ecological distribution does not seriously overlap those of the eastern cottontail and the snowshoe hare.

COUNTY RECORDS FROM MICHIGAN. Records of specimens in collections and from the literature show that the European hare has been found in the following Upper Peninsula counties: Alger and Ontonagon; in the Lower Peninsula: Alcona, Calhoun, Clinton, Ingham, Ionia, Macomb, Presque Isle, Sanilac, St. Claire, Tuscola, and Washtenaw. There is always the possibility that one or more of these records are for an escaped domestic rabbit instead of for the European hare.

LITERATURE CITED

Banfield, A. W. F.
1974 The mammals of Canada. Univ. Toronto Press, Toronto. xxiv+438 pp.

Burt, W. H.
1946 The mammals of Michigan. Univ. Michigan Press, Ann Arbor. xv+288 pp.

Dean, P. B., and A. de Vos
1965 The spread and present status of the European hare, *Lepus europaeus hybridus* Desmarest, in North America. Canadian Field-Nat., 79(1):38–48.

de Vos, A., and P. B. Dean
1967 The distribution and the use of forms by European hares, *Lepus europeaus hybridus* (Desmarest 1822), in southern Ontario. Säug. Mitt., 15(1):57–61.

de Vos, A., R. H. Manville, and R. G. Van Gelder
1956 Introduced mammals and their influence on native biota. Zoologica, 41(4):163–194.

Greer, J. K.
1965 Mammals of Malleco Province, Chile. Michigan State Univ., Publ. Mus., Biol. Ser., 3(2):49–152.

Peterson, R. L.
1966 The mammals of eastern Canada. Oxford Univ. Press, Toronto. xxxii+463 pp.

Wood, N. A., and L. R. Dice
1923 Records of the distribution of Michigan mammals. Michigan Acad. Sci., Arts and Ltrs., Papers, 3:425–469.

Rodents

ORDER RODENTIA

Of the slightly more than 4,000 species of mammals on earth, almost 1,700 species (42%) belong to that most diverse and cosmopolitan of mammalian orders, the Rodentia. Not only are there more species in this order than in any other, but numbers of individuals (not species), of the prolific rodents are more abundant than all other mammals combined. Further, more than one-third (about 354) of the approximately 1,000 mammalian genera are also in the Rodentia. By means of their spectacular ability through time (since the late Paleocene) to radiate adaptively, rodents have become specialized in an array of terrestrial, subterranean, arboreal, and aquatic environments. Some have even invaded the atmosphere—not as true flyers like bats, but as gliders. In these remarkable adaptations, rodents have adjusted to eat all kinds of foods—seeds, fruits, green plants, bark, roots, fungi, meat, insects, carrion, and fish.

Rodents vary in size from a tiny pocket mouse (*Perognathus flavus* of the American Southwest) weighing less than 1½ oz (5 g) to the ponderous capybara (*Hydrochoerus hydrochoeris* of South America) weighing more than 100 lbs (45 kg). A Pleistocene "beaver" (*Castoroides ohioensis*), which lived in Michigan as late as 11,000 years ago, is supposed to have weighed more than 400 lbs (180 kg). Rodents are characterized by two large front teeth (incisors, in both the upper and lower jaws). Rabbits, hares, and pikas have somewhat similar front teeth except that in the upper jaws there are two small peg-like teeth directly behind the large incisors. These chisel-shaped teeth grow throughout life; sharpened edges are produced by the differential wear of the harder front layer of enamel (often yellow or orange in color) and the softer hind layer of dentine. There are no canines and there is a sizable space (diastema) between the front incisors and the grinding cheek teeth (mostly molars, sometimes premolars). The digestive tract is often highly specialized to digest well-masticated greens and other foods.

Because of their abundance, both in numbers of species and in numbers of individuals, rodents occupy the major intermediate position in the food chain between the vegetation (producers) and the carnivores (secondary consumers). Rodents are almost always a major part of the meat-eater's diet. Hence, their reproductive potential must be great to balance the constant drain on their numbers by both terrestrial and avian predators. Many rodents live in areas remote from human habitations; others have developed a liking for agricultural plantings, forage crops and pasturage, stored foodstuffs, and human dwellings. The annual damage to the human economy wrought by these latter species is estimated in the billions of dollars.

Most rodents can be arranged into three suborders. This classification is based chiefly on the arrangement of the jaw muscle attachment (the zygomasseteric pattern) from the lower jaw to the zygomatic arch and the rostral part of the skull. The most generalized condition is found in the Sciuromorpha, represented in Michigan by the families Sciuridae and Castoridae. A second suborder, the Myomorpha, is represented in Michigan by the families Cricetidae, Zapodidae, and the introduced Muridae. The third suborder, Caviomorpha (also called Hystricomorpha) is represented in Michigan by the family Erethizontidae. In all, there are 21 species of rodents native to Michigan (one-third of the total mammalian population) plus two introduced species, and one species (Franklin's ground squirrel) for which Michigan records should be forthcoming.

KEY TO THE SPECIES OF RODENTS IN MICHIGAN Using Characteristics of Adult Animals in the Flesh

1a. Hair either soft or coarse but not mixed with heavy, sharp quills on back, sides, and tail . . . . 2a.

1b. Hair mixed with heavy, sharp quills on back, sides and tail . . . . . . family Erethizontidae—PORCUPINE (*Erethizon dorsatum*), page 371.

2a. Tail hair bushy, at mid-length of tail hairs longer than diameter of tail . . . . . . family Sciuridae . . . . 3a.

2b. Tail hair sparse, at mid-length of tail hairs shorter than diameter of tail . . . . 12a.

3a. Size large, length of head and body more than 13¾ in (350 mm); tail short, less than one-fourth as long as head and body; feet black . . . . WOODCHUCK (*Marmota monax*), page 179.

3b. Size small, length of head and body less than 13¾ in (350 mm); tail long, more than one-fourth as long as head and body; feet usually not black . . . . 4a.

4a. Large, loose fold of skin along sides between front and hind legs (stretched out for gliding); hairs on tail arranged laterally to give flattened appearance . . . . 5a.

4b. No loose fold of skin along sides between front and hind legs; hairs on tail arranged around tail to give round or oval appearance . . . . 6a.

5a. Hair on underside white from tip to root; length of tail vertebrae usually less than 4½ in (110 mm); length of hind foot including claws usually less than 1¼ in (33 mm). . . . . . . SOUTHERN FLYING SQUIRREL (*Glaucomys volans*), page 243.

5b. Hair on underside white with slate or grayish base; length of tail vertebrae usually more than 4½ in (110 mm); length of hind foot including claws usually more than 1¼ in (33 mm). . . . . NORTHERN FLYING SQUIRREL (*Glaucomys sabrinus*), page 236.

6a. Hair at mid-length of tail short, less than ¾ in (20 mm) long . . . . 7a.

6b. Hair at mid-length of tail long, more than 1⅜ in (35 mm) long . . . . 10a.

7a. Upperparts of body with longitudinal stripes . . . . 8a.

7b. Upperparts of body without longitudinal stripes . . . . FRANKLIN'S GROUND SQUIRREL (*Spermophilus franklinii*), page 189.

8a. Longitudinal stripes on sides of face as well as on body, stripes not broken into series of whitish dots; tail colored differently above than below . . . . 9a.

8b. Longitudinal stripes absent on sides of face; back and sides with 13 narrow stripes, some a series of white dots . . THIRTEEN-LINED GROUND SQUIRREL (*Spermophilus tridecemlineatus*), page 193.

9a. Median dark stripe separating two broad gray stripes on mid-back; all longitudinal stripes not extending to base of tail; rump reddish; tail short, usually less than 40% of total length . . . . . . EASTERN CHIPMUNK (*Tamias striatus*), page 167.

9b. Median dark stripe separating two narrow gray stripes on mid-back; all longitudinal stripes extending to base of tail; rump similar to back in coloring; tail long, usually more than 40% of total length . . . . LEAST CHIPMUNK (*Eutamias minimus*), page 174.

10a. Size small, total length of head, body and tail less than 13¾ in (350 mm); summer pelage with blackish, longitudinal stripe on each side of body, winter pelage with yellowish or reddish dorsal stripe . . . . . RED SQUIRREL (*Tamiasciurus hudsonicus*), page 226.

10b. Size large, total length of head, body and tail more than 13¾ in (350 mm); summer pelage lacking blackish stripes on sides and winter pelage lacking yellowish or reddish dorsal stripe . . . . 11a.

11a. Dorsum grayish or rusty gray in normal condition, blackish or rusty black in melanistic condition; underparts whitish or grayish; tips of tail hairs, tufts behind ears (in winter), and eye-ring (in summer) whitish . . . GRAY SQUIRREL (*Sciurus carolinensis*), page 202.

11b. Dorsum mixed buffy and blackish; underparts and feet reddish orange; tips of tail hairs and tufts behind ear reddish . . . . FOX SQUIRREL (*Sciurus niger*), page 214.

12a. Size large, head and body more than 15¾ in (400 mm) long; tail paddle-shaped, flattened dorsoventrally; second claw on hind foot double . . . . . . family Castoridae—BEAVER (*Castor canadensis*), page 251.

12b. Size small, head and body much less than 15¾ in (400 mm) long; tail closely haired, round or flattened laterally; second claw on hind foot normal .13a.

13a. Hair on belly white from tip to base, never gray or lead-colored; hind feet long (adapted for leaping); tail long, at least 20% longer than length of head and body . . . . family Zapodidae—jumping mice . . . . .25a.

13b. Hair on belly lead-colored at base; hind feet normal length for non-jumping rodents; tail short, equal in length or less than length of head and body . . . . families Cricetidae and Muridae . . . . .14a.

14a. Size large, weight more than 2 lb (0.9 kg); tail flattened laterally; five clawed toes on front feet; hind toes partly webbed with stiff fringe of hairs on edges of feet and toes . . . . MUSKRAT (*Ondatra zibethicus*), page 326.

14b. Size small, weight much less than 2 lb (0.9 kg); tail round; four clawed toes on front feet; hind toes without webbing or edges with stiff hairs . . . . .15a.

15a. Tail short, less than one-third the length of head and body; ears short, round and partly concealed in fur; feet and belly never white . . . . .16a.

15b. Tail long, more than one-third the length of head and body; ears prominent, not concealed in fur of head; feet and belly whitish or sometimes grayish . . . . .20a.

16a. Tail short, less than 1 in (25 mm) in length . . . . .17a.

16b. Tail long, more than 1 in (25 mm) in length . . . . .18a.

17a. Upperparts auburn or chestnut in color; fur plush, lacking long guard hairs; ear inconspicuous, hidden in soft fur . . WOODLAND VOLE (*Microtus pinetorum*), page 318.

17b. Upperparts grayish brown or grizzled gray; fur with long guard hairs; ear conspicuous . . . . SOUTHERN BOG LEMMING (*Synaptomys cooperi*), page 336.

18a. Upperparts with broad reddish median band contrasting with gray sides . . . SOUTHERN RED-BACKED VOLE (*Clethrionomys gapperi*), page 291.

18b. Upperparts and sides brownish or grayish . . . . .19a.

19a. Upperparts brownish or blackish brown; hair of underparts tipped with whitish; feet black; six tubercules on soles of hind feet; tail usually more than 1½ in (38 mm) long . . . . . MEADOW VOLE (*Microtus pennsylvanicus*), page 304.

19b. Upperparts grizzled grayish brown; hair of underparts tipped with warm buff or ochre; feet brown; five tubercules on soles of hind feet; tail usually less than 1½ in (38 mm) long . . . . PRAIRIE VOLE (*Microtus ochrogaster*), page 298.

20a. Size large, total length of head, body and tail more than 8 in (200 mm) . . NORWAY RAT (*Rattus norvegicus*), page 345.

20b. Size small, total length of head, body and tail less than 8 in (200 mm) . . . . .21a.

21a. Upperparts generally grayish or gray brown; upperparts not sharply contrasting with paler underparts; underparts grayish or yellowish; tail scaly in appearance and not distinctly bicolored (dark above and light below) . . . . HOUSE MOUSE (*Mus musculus*), page 351.

21b. Upperparts brownish or grayish (young animals gray or gray black); upperparts sharply contrasting with white underparts; tail not scaly in appearance, clothed in short hairs, and usually distinctly bicolored .22a.

22a. Throat and forearms with hair usually white from base to tip; tail indistinctly bicolored, sparsely furred and lacking pencil of hair at tip . . . . WHITE-FOOTED MOUSE (*Peromyscus leucopus*), page 266.

22b. Throat and forearms with hair usually white with grayish bases; tail conspicuously bicolored, heavily furred with pencil of hair at tip . . . . .23a.

23a. Upperparts slate grayish or slightly reddish; tail short, usually less than 2½ in (65 mm); ears short, usually less than ⅝ in (15 mm) from notch to rim; habitat cultivated and fallow fields in Upper Peninsula (mostly Menominee County) and in Lower Peninsula . . . . DEER MOUSE (*Peromyscus maniculatus bairdii*), page 278.

23b. Upperparts reddish or fulvous; tail long,

usually more than 2½ in (65 mm); ears long, usually more than ⅝ in (15 mm) from notch to rim; habitat woodlands . . . . .24a.

24a. Upper lip whitish; ears long, usually more than ¾ in (19 mm) from notch to rim; habitat woodlands of Upper Peninsula and northern part of Lower Peninsula . . . . DEER MOUSE (*Peromyscus maniculatus gracilis*), page 278.

24b. Upper lip dusky; ears short, usually less than ¾ in (19 mm) from notch to rim; habitat Isle Royale . . . . DEER MOUSE (*Peromyscus maniculatus maniculatus*), page 278.

25a. Upperparts with dark median band and yellowish-olive sides; tail not tipped with white . . . . . . MEADOW JUMPING MOUSE (*Zapus hudsonius*), page 357.

25b. Upperparts with dark median band and bright reddish sides; tail tipped with white . . . WOODLAND JUMPING MOUSE (*Napaeozapus insignis*), page 364.

KEY TO THE SPECIES OF RODENTS IN MICHIGAN
Using Characteristics of the Skulls of Adult Animals

1a. Infraorbital foramen not conspicuously large, sometimes slit-like; zygomatic plate not horizontal, always broader than infraorbital foramen . . . . .2a.

1b. Infraorbital foramen conspicuously large, often oval; zygomatic plate horizontal, less broad and below infraorbital foramen .25a.

2a. Skull small, less than 4 in (100 mm) long . .3a.

2b. Skull massive, much more than 4 in (100 mm) long . . . . family Castoridae—BEAVER (*Castor canadensis*), page 251.

3a. Total number of teeth 20 or 22; postorbital processes on frontals well developed . . . . family Sciuridae . . . . .4a.

3b. Total number of teeth 16; postorbital processes not developed . . . . .14a.

4a. Total number of teeth 20, four cheek teeth on each side of upper jaw . . . . .5a.

4b. Total number of teeth 22, five cheek teeth on each side of upper jaw, first cheek tooth in upper jaw may be minute or sometimes lost . . . . .7a.

5a. Skull large, more than 2 in (50 mm) in total length, bone coloring pinkish . . . . FOX SQUIRREL (*Sciurus niger*), page 214.

5b. Skull small, less than 2 in (50 mm) in total length, bone coloring whitish . . . . .6a.

6a. Skull large, more than 1⅝ in (42 mm) in total length; length of upper cheek tooth row more than 5/16 in (7 mm) . . . . RED SQUIRREL (*Tamiasciurus hudsonicus*), page 226.

6b. Skull small, less than 1⅝ in (42 mm) in total length; length of upper cheek tooth row less than 5/16 in (7 mm) . . . . EASTERN CHIPMUNK (*Tamias striatus*), page 167.

7a. Skull large, more than 2 4/5 in (70 mm) long; zygomatic width at least 1⅝ in (40 mm); interorbital space flattened or concave; anterior surface of incisors pale yellow or whitish . . . WOODCHUCK (*Marmota monax*), page 179.

7b. Skull small, less than 2 4/5 in (70 mm) long; zygomatic width less than 1⅝ in (40 mm); interorbital space not flattened or concave; anterior surface of incisors more orange than yellowish . . . . .8a.

8a. Skull large, more than 2 in (50 mm) in length . . . . .9a.

8b. Skull small, less than 2 in (50 mm) in length . . . . .10a.

9a. Skull length more than 2⅜ in (59 mm); first upper tooth (P^3) in cheek-tooth row peg-like and minute, less than one-half the breadth of the second tooth (P^4); sides of rostrum somewhat concave . . . . GRAY SQUIRREL (*Sciurus carolinensis*), page 202.

9b. Skull length less than 2⅜ in (59 mm); first upper tooth (P^3) in cheek-tooth row peg-like but larger, at least half the breadth of the second tooth (P^4); sides of rostrum not concave . . . . FRANKLIN'S GROUND SQUIRREL (*Spermophilus franklinii*), page 189.

10a. Skull length more than 1⅝ in (42 mm) . . . . . . RED SQUIRREL (*Tamiasciurus hudsonicus*), page 226.

10b. Skull length less than 1⅝ in (42 mm) . . . . .11a.

11a. Skull length more than 1¼ in (33 mm); infraorbital foramen forms canal . . . . .12a.

11b. Skull length less than 1¼ in (33 mm); no infraorbital canal, infraorbital foramen piercing zygomatic plate of maxillary . . . . LEAST CHIPMUNK (*Eutamias minimus*), page 174.

12a. Zygomata not parallel, contracted anteriorly, anterior part twisted toward a horizontal plane; frontal bones just behind postorbital processes more than ⅜ in (10 mm) wide; posterior border of zygomatic plate, when viewed from below, opposite third and fourth teeth (M^{1-2}) in cheek tooth row; no distinct U-shape notch in interorbital constriction anterior to postorbital process . . . . THIRTEEN-LINED GROUND SQUIRREL (*Spermophilus tridecemlineatus*), page 193.

12b. Zygomata almost parallel, slightly contracted anteriorly, anterior part not twisted and nearly vertical; frontal bones just behind postorbital processes less than ⅜ in (10 mm) wide; posterior border of zygomatic plate, when viewed from below, opposite first and second teeth (P^{3-4}) in cheek tooth row; no distinct U-shape notch in interorbital constriction anterior to postorbital processes . . . . 13a.

13a. Skull length usually more than 1 7/16 in (36 mm); skull width usually more than ⅞ in (22 mm); upper cheek tooth row more than 5/16 in (7 mm) in length . . . NORTHERN FLYING SQUIRREL (*Glaucomys sabrinus*), page 236.

13b. Skull length usually less than 1 7/16 in (36 mm); skull width usually less than ⅞ in (22 mm); upper cheek tooth row less than 5/16 in (7 mm) in length . . . . . . SOUTHERN FLYING SQUIRREL (*Glaucomys volans*), page 243.

14a. Occlusal surfaces of cheek teeth flat with pattern of folded enamel loops surrounding exposed islands of dentine . . . . family Cricetidae (part) . . . . 15a.

14b. Occlusal surfaces of cheek teeth irregular with low rounded cusps, often worn flat in older individuals; crown of molars completely covered with enamel . . . . 20a.

15a. Skull large, more than 2 in (50 mm) long . . . . . MUSKRAT (*Ondatra zibethicus*), page 326.

15b. Skull small, less than 2 in (50 mm) long . . . . 16a.

16a. Upper incisors each with single longitudinal groove on front surface . . . . . . SOUTHERN BOG LEMMING (*Synaptomys cooperi*), page 336.

16b. Upper incisors smooth on front surface, no longitudinal grooves . . . . 17a.

17a. Posterior palate terminating as simple shelf, lacking median spinous projection, connecting with palatine bones only at sides of narial cavity; cheek teeth rooted . . . SOUTHERN RED-BACKED VOLE (*Clethrionomys gapperi*), page 291.

17b. Posterior palate terminating with median spinous projection, connecting with palatine bones at center as well as at sides; cheek teeth not rooted . . . . 18a.

18a. Least interorbital width usually less than 3/16 in (4 mm) . . . . 19a.

18b. Least interorbital width usually more than 3/16 in (4 mm) . . . . WOODLAND VOLE (*Microtis pinetorum*) page 318.

19a. Last upper molar (M^3) with four or five inner and outer angular enamel loops; auditory bullae rounded; anterior palatine foramen more than 3/16 in (5 mm) long . . MEADOW VOLE (*Microtus pennsylvanicus*), page 304.

19b. Last upper molar (M^3) with three inner and outer angular enamel loops; auditory bullae slightly flattened; anterior palatine foramen less than 3/16 in (5 mm) long . . . . PRAIRIE VOLE (*Microtus ochrogaster*), page 298.

20a. Palate not extending posteriorly beyond plane of last molars; upper cheek teeth, in unworn condition, with cusps arranged in two longitudinal rows . . . . . family Cricetidae (part) . . . . 21a.

20b. Palate extending posteriorly beyond plane of last molars; upper cheek teeth, in unworn condition, with cusps arranged in three longitudinal rows . . . . . . family Muridae . . . 24a.

21a. Anterior border of zygomatic plate covering infraorbital foramen when viewed from side; incisive foramen constricted anteriorly, forming slight angle near the junction of the premaxillary and maxillary; molars smaller and uniform in width . . . WHITE-FOOTED MOUSE (*Peromyscus leucopus*), page 266.

21b. Anterior border of zygomatic plate not covering infraorbital foramen when viewed from side; incisive foramen with evenly arched or nearly parallel lateral sides, usually not forming an angle near the junction of the premaxillary and maxillary; molars larger and not uniform in width, the first being the widest and the last the narrowest . . . . . 22a.

22a. Size smaller, length of skull usually less than 1 in (25 mm) . . . . DEER MOUSE (*Peromyscus maniculatus bairdii*), page 278.

22b. Size smaller, length of skull usually more than 1 in (25 mm) . . . . .23a.

23a. Found in Upper Peninsula and northern part of Lower Peninsula . . . DEER MOUSE (*Peromyscus maniculatus gracilis*), page 278.

23b. Found on Isle Royale . . . . . . DEER MOUSE (*Peromyscus maniculatus maniculatus*), page 278.

24a. Skull large, more than 1¼ in (30 mm) long; first upper molar (M^1) relatively short, less than the lengths of the second and third molars (M^{2-3} combined . . . . NORWAY RAT (*Rattus norvegicus*), page 345.

24b. Skull small, less than 1¼ in (30 mm) long; first upper molar (M^1) relatively long, longer than the lengths of the second and third molars (M^{2-3} combined . . . HOUSE MOUSE (*Mus musculus*), page 351.

25a. Skull small, less than 1 in (25 mm) in length; upper incisors grooved in front . . . . . family Zapodidae . . . . .26a.

25b. Skull large, more than 3 in (76 mm) in length; upper incisors not grooved in front . . family Erethizontidae . . . . PORCUPINE (*Erethizon dorsatum*), page 371.

26a. Three cheek teeth (M^{1-3}) present on each side
of the upper jaw . . . . . WOODLAND JUMPING MOUSE (*Napaeozapus insignis*), page 364.

26b. Four cheek teeth (P^4, M^{1-3}) present on each side of the upper jaw . . . MEADOW JUMPING MOUSE (*Zapus hudsonius*), page 357.

FAMILY SCIURIDAE

Sciurids are well-distributed in varied environments in all terrestrial areas of the globe except the Australian region, southern South America, Malagasy, and other islands. There are approximately 261 species arranged into about 51 genera (Ellerman, 1940). About 93 species occur in North America. In Michigan, there are nine species representing six genera plus another probable resident, the Franklin's ground squirrel, which has a geographic distribution reaching just to Michigan's southwestern border. Members of the squirrel family are diverse in their adaptations; some are diurnal (tree squirrels), some nocturnal (flying squirrels), some arboreal (tree squirrels), some gliders (flying squirrels), some terrestrial (chipmunks), and some burrowers (marmots and prairie dogs). Sciurids are among the most vocal of rodents. Their diets are varied with green vegetation, nuts, seeds, fruits, insects, and other animal matter as standard fare. Some hibernate while others remain active year around. Larger species are edible, and several are classed as game animals. In fact, gray and fox squirrels are major game species in many parts of eastern United States (Allen, 1943).

Sciurids have rather conservative and similar features which are characterized externally by large eyes, usually stout, well-haired bodies, varied coloration, and often bushy tails. When feeding or on the alert, members of the squirrel family sit up on their haunches, freeing their forefeet for food manipulation. The thumb is reduced; the hind feet have five functional toes; the fourth digit on each foot is the longest. Mammary glands vary from two pairs in some Neotropical tree squirrels to as many as six pairs (seven have also been reported) in Nearctic ground squirrels (Bryant, 1945). Sciurids have rather large, smooth, roundish brain cases. The zygomatic arches are strongly constructed with slanting, flat, anterior surfaces. The supraorbital processes and rounded auditory bullae are prominent. The incisive foramina are usually short and considerably anterior to the cheek tooth-rows. The masseter (jaw) muscle has no part extending forward on the sides of the skull to penetrate the infraorbital canal, a condition pointing to the primitive position of sciurids among rodents. The rooted cheek teeth have prominent cusps or ridges, useful in crushing foods. There are four teeth (the fourth premolar and the three molars) in each side of the lower jaw and either four or five (the reduced third premolar may be present plus the fourth premolar and the three molars) in each side of the upper jaw. Another primitive characteristic is that the bones of the lower hind limb, the tibia and fibula, are not fully fused.

LITERATURE CITED

Allen, D. L.
1943 Michigan fox squirrel management. Michigan Dept. Cons., Game Div., Publ. 100, 404 pp.

Bryant, M. D.
1945 Phylogeny of Nearctic Sciuridae. American Midl. Nat., 33:257–390.

Ellerman, J. R.
1940 The families and genera of living rodents. Vol. 1. Rodents other than Muridae. British Mus. (Nat. Hist.), xxvi+689 pp.

EASTERN CHIPMUNK
Tamias striatus (Linnaeus)

NAMES. The generic name *Tamias,* proposed by Illiger in 1811, is of Greek derivation meaning "steward" or "one who stores and takes care of provisions." The specific name *striatus,* proposed by Linnaeus in 1758, is derived from the Latin and means "striped." In Michigan, most observers call this mammal a chipmunk, sometimes a rock squirrel or ground squirrel. The subspecies *Tamias striatus griseus* Mearns occurs in the Upper Peninsula; the subspecies *T. s. peninsulae* Hooper, in the northern Lower Peninsula (southward at least to Bay County) and on islands in the Beaver, Fox and Manitou groups; the subspecies *T. s. rufescens* Bole and Moulthrop, in the southern half of the Lower Peninsula.

RECOGNITION. Body size small for a sciurid (Fig. 29); head and body approximately 5½ in (140 mm) long; in overall appearance, the short, fine pelage of the upperparts appears striped and reddish-brown; underparts white; head and flanks reddish or yellowish-brown becoming much richer on rump; buffy cheeks crossed by a brownish stripe from nose to under ear; eye-ring buffy; nape and shoulders grayish; back stripes include a central one extending from back of head as reddish brown, changing to blackish behind shoulders, and blending into rich russet coloring on rump; grayish stripes on either side of median stripe extending from shoulder to rump, bordered laterally by blackish, then pale buffy; most lateral brownish stripes bordering reddish or yellowish-brown body sides; upper surface of tail pale brownish fringed with black and white tipped hairs, undersurface tawny; hind feet reddish brown; eyes large and conspicuous. There are two inguinal, four abdominal, and two pectoral mammae. Internal cheek pouches are characteristic; when filled with nuts and seeds, the sides of a chipmunk's head bulge enormously. Where the eastern chipmunk and the least chipmunk occur together in the Upper Peninsula, *T. striatus* can be

Figure 29. The eastern chipmunk, *Tamias striatus.* Sketches from life by Bonnie Marris.

easily distinguished by its larger size and heavy-bodied appearance.

MEASUREMENTS AND WEIGHTS. Adults measure as follows: length from tip of nose to end of tail vertebrae, 9 to 11 in (230 to 280 mm); length of tail vertebrae, 2½ to 4¼ in (65 to 110 mm); length of hind foot, 1¼ to 1⅝ in (32 to 40 mm); height of ear from notch, ½ to ¾ in (12 to 20 mm); weight, 2.4 to 4.0 oz (70 to 115 g).

DISTINCTIVE CRANIAL AND DENTAL CHARACTERISTICS. The skull is long and narrow but has a wide interorbital space (Fig. 30). The overall length is approximately 1⅝ in (40 mm). Like other sciurids, the various cranial bones are smooth and regular in apearance (Fig. 31). In comparison with related Michigan mammals with similar-sized skulls, the presence of only four upper cheek teeth (P^4, M^{1-3}) is most distinctive. Like the least chipmunk, the infraorbital foramen is large and pierces the zygomatic plate. The adult skull averages 1½ in (39 mm) long and ⅞ in (22 mm) wide. The color of the enamel on the front of the least chipmunk's incisors is dark orange; the eastern chipmunk, pale orange. The dental formula is:

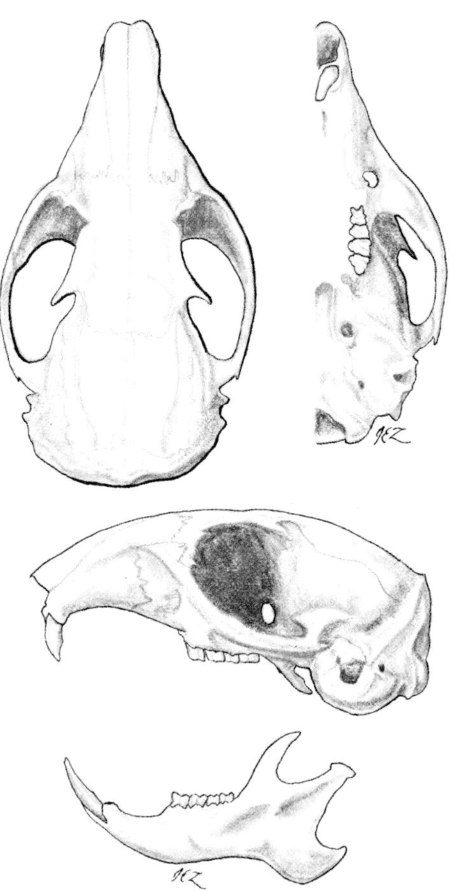

Figure 30. Cranial and dental characteristics of the eastern chipmunk (*Tamias striatus*).

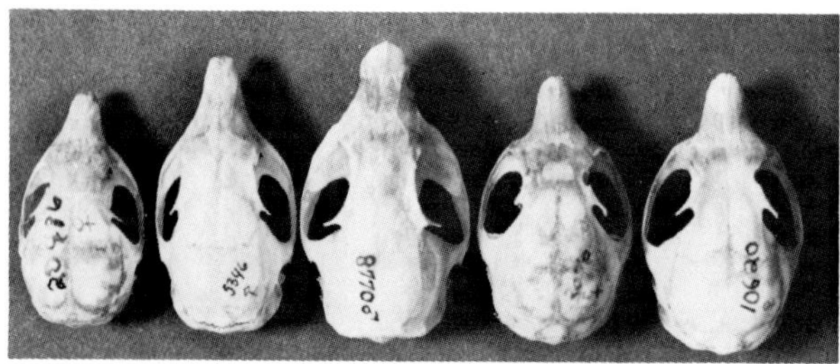

Figure 31. Dorsal aspects of the skulls of the five small Michigan sciurids; from left to right: least chipmunk, eastern chipmunk, thirteen-lined ground squirrel, southern flying squirrel, northern flying squirrel.

$$\text{I (incisors)}\,\frac{1\text{-}1}{1\text{-}1},\ \text{C (canines)}\,\frac{0\text{-}0}{0\text{-}0},\ \text{P (premolars)}\,\frac{1\text{-}1}{1\text{-}1},\ \text{M (molars)}\,\frac{3\text{-}3}{3\text{-}3} = 20$$

DISTRIBUTION. The eastern chipmunk inhabits deciduous hardwood forests of eastern North America, from southeastern Canada southward as far as Georgia and Louisiana and westward to the edge of the Great Plains. The species is at home in both the Upper and Lower Peninsulas, occurring in all counties and on several islands in Lake Michigan (see Map 25).

HABITAT PREFERENCES. The eastern chipmunk is chiefly characteristic of Michigan's Carolinian Biotic Province (Dice, 1943); that is, in deciduous hardwoods of both the Upper and Lower Peninsulas. The species will thrive in all successional stages of this forest type, from shrub growth in cutover to climax hardwood forest (Wrigley, 1969). Mature beech-maple forests may support maximum numbers, while moist swamps are usually avoided. In mixed evergreen-deciduous areas, chipmunks may occupy open situations and forest edges rather than shaded interiors (Gates, 1976). It is probable that the average hiker and outdoor observer associate the eastern chipmunk with wooded glens, rail fences, rock out-

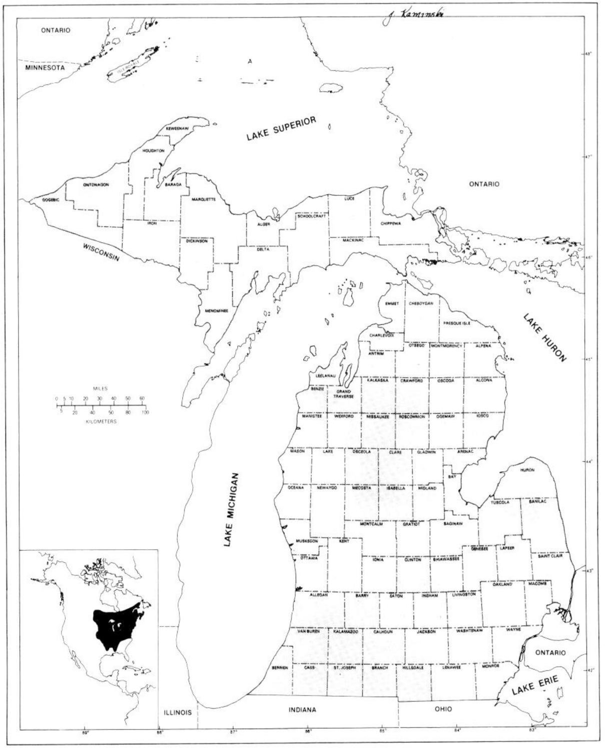

Map 25. Geographic distribution of the eastern chipmunk (*Tamias striatus*).

crops or walls (Burt, 1940). Such areas provide a variety of summer foods, woody cover, and either underground or rocky shelter for this versatile woodland resident. Eastern chipmunks also thrive near human habitations, using outbuildings, brush piles, rocky foundations and even basements as refuges. However, except for an occasional invasion of a rural cottage, this sciurid is not prone to cause serious problems to homemakers.

DENSITY AND MOVEMENTS. Mature beech-maple woodlands, according to Wrigley (1969), may support the highest numbers of eastern chipmunks. The best habitat also contains dense canopy and sparse cover on the forest floor (Svendsen and Yahner, 1979). High populations of 20 or more per acre (50 per ha), as described by Jackson (1961), may result from a typical sciurid characteristic of concentrating around a food source, such as campgrounds where hand-outs are provided by tourists or in small groves of trees when mast yields are high. Rural garbage dumps are also attractive to concentrations of eastern chipmunks (Courtney and Fenton, 1976). More than likely, a population of two or three individuals per acre (about five to seven per ha) is normal in early summer and perhaps four or more per acre (10 per ha) in late summer (Burt, 1946). Although as many as 10 animals have been captured in a area no larger than one-third of an acre (0.12 ha) by Yerger (1953) in New York, it was his opinion that densities of from four to nine per acre (10 to 22 per ha) are usual in good habitat. The highest population density would be expected just after the emergence of weaned young.

In August and September, Blair (1942) calculated that individual eastern chipmunks may use areas in Alger County averaging slightly more than two acres (one ha), with males covering more space than females. He found that home ranges broadly overlapped even though chipmunks are generally antagonistic toward each other. Somewhat similar use of space by eastern chipmunks was found in Marquette County (Manville, 1949) and in Livingston County (Burt, 1940). Nevertheless, eastern chipmunk home ranges vary in extent and are generally correlated with appearance and depletion of resources, with peripheral areas used less frequently than central areas from where movements radiate (Getty, 1981b).

BEHAVIOR. Because the eastern chipmunk is active in the daytime, the behavioral ecologist can study its summer habits fairly easily. In southern Michigan, it becomes a conspicuous member of the wildlife community in mid to late March. By June, the animals, almost always solitary, show major peaks of activity in early morning and late afternoon (Yerger, 1955). This pace is continued through the early growing season with a marked decline in late July and August. Activity increases again in late September with the appearance of weaned offspring from the second litter. In November, the animals disappear into their winter quarters. The chipmunk's decreased daily activity in late summer has not been fully explained (Allen, 1938; Dunford, 1972; Wrazen, 1980).

Unlike the thirteen-lined ground squirrel which hibernates in a deep lethargic state with the reduced metabolic rates sustained for the entire cold period, the eastern chipmunk does not put on fat prior to winter and enters an often brief torpor, allowing the animal to become active periodically in winter (Maclean, 1981; Panuska, 1959; Pivorum,

1976). From its curled posture (rolled in a ball with head tucked between hind legs and tail wrapped over head and shoulders), the eastern chipmunk may awaken in its underground nest to feed on cached food or even emerge and move around in the cold and snow. Elaine and Edward Rybak reported active eastern chipmunks on the campus of Michigan State University in Ingham County on January 27 and March 8. Charles Shick (*in litt.*) watched a chipmunk filling its cheek pouches with cracked corn at a bird feeder in Okemos in Ingham County on a snowy afterooon (1600 hours) on January 11. Another was seen at a bird feeder in Berrien County on March 10, again a cold snowy day. In a letter to Dr. E. A. Mearns dated March 1, 1880, Dr. H. A. Atkins remarks about the mild January and February for that year in Locke Township in northeastern Ingham County. He writes that active eastern chipmunks were seen on February 19, 24, 25, 26, 27, 28, and 29. There is evidence of such activity for all of the winter months (Davis and Beer, 1959; Engels, 1951; Test, 1932), although most sight records are for late winter (Banfield, 1974).

Eastern chipmunks occupy underground dens, sometimes tree hollows (Getty, 1981a). In constructing these home-sites, the animals frequently conceal the entrances (usually only one but sometimes two) under a rock, beside a log, between roots, or in shrubs. Any telltale pile of earth is removed from the vicinity, sometimes by way of an entrance which is later plugged, or by spreading it in nearby vegetation. Burrows are often complicated systems of tunnels extending for as much as 12 ft (about 4 m), with old burrows being extended annually (Panuska and Wade, 1956). The nest chamber is situated as much as three ft (one m) below ground and may be 12 in (300 mm) in diameter. The nest; which may not be below the frost-line, can also serve as the food cache (Allen, 1938; Howell, 1939; Jackson, 1961).

Tamias striatus spends much of its foraging time on the ground. However, chipmunks will join tree squirrels in climbing shrubs and trees to obtain seeds and fruits. An eastern chipmunk was observed 30 ft (9 m) above ground in a fruiting elm tree on the campus of Michigan State University.

The eastern chipmunk is solitary (in contrast to statements by early writers, Seton, 1909) except during the breeding periods and when parental care is needed for offspring. Chipmunks essentially avoid each other even when their home ranges broadly overlap (Elliott, 1978; Getty, 1981a). In Livingston County, Burt (1940) described how an adult female living under a building for three years gave chase and successfully defended her denning area from other chipmunks of both sexes. She defended an area of about 50 yds (40 m) around her home site, but seemed indifferent to the use of space by other individuals beyond this limit. The eastern chipmunk's attachment to a home site is indicated by Layne (1957) who found that individuals removed from their homes as far as 775 ft (about 200 m) returned, while Seidel (1961) found adults to home from as far away as 1,800 ft (about 550 m)

Like most sciurids, the eastern chipmunk is vocal, delivering high-pitched, often "scolding" calls with associated tail-jerks. A "cluck" or "chip" either as a single call or repeated is characteristic. Sometimes a chipmunk will keep up a lively chatter (often more than 100 calls per minute) for as long as a half hour. At the appearance of possible danger, the animal may utter one loud "chip" which is extended into a trill before dashing to safety.

ASSOCIATES. The eastern chipmunk probably comes in contact with all other Michigan mammals because of its use of various woodland habitats and the borders between these and open environments. Even so, its diurnal habits rarely allow it to actually interact with shrews, mice, and other nocturnal neighbors. Blair (1942), for example, live-trapped, marked, released, and retrapped 151 woodland deer mice and 154 eastern chipmunks in about 30 days on 18.18 a (7.2 ha) of hardwood forest in Alger County. However, it is doubtful that these two rodents have much actual contact, except for possibly using the same refuge sites on occasion. Eastern chipmunks are in closest relationship with other diurnal sciurids. These may reside compatibly by keeping their distance. One spring morning in East Lansing, this author observed an eastern chipmunk, a red squirrel, and a fox squirrel feeding without incident in a large Chinese elm, each animal occupying well separated sites. Nevertheless, Allen (1938) reported the killing of an eastern chipmunk by a red squirrel, presumably because of their proximity while feeding in the same hazelnut tree. Eastern chipmunks may occasionally come in contact with thirteen-lined ground squirrels where their preferred habitats join. In the Upper Peninsula, however, eastern chipmunks and least chipmunks have abundant chances of meeting in forested situations. The former's inclination for upland woods and the latter's for swamp habitat may separate them to some extent ecologically (Quinby, 1944), although Reilly (1970) obtained no evidence of competitiveness where both occurred on his study area in the Upper Peninsula.

REPRODUCTIVE ACTIVITIES. Generally there are two sets of litters born each year, the first in late March and early April and the second in late July and early August (Schooley, 1934). Actually, males come into reproductive readiness in late February (even before emerging from winter quarters) and remain so until August (Banfield, 1974; Condrin, 1936). However, females only have two estrous cycles, in early spring and in midsummer. After matings, the young are born following a gestation period of about 31 days. Although the usual litter numbers from three to four, there can be as few as one or as many as eight. The young are born and nutured in the underground nest cavity; usually the one used previously as a wintering site and food storage container. Leaf fragments are arranged as a nest above the other debris. A hole about 10 ft (3 m) above ground in a white oak was used by a female as a nursery in Livingston County (Getty, 1981a).

The young are born naked with their eyes closed. In 10 days the dorsal stripe appears. The reddish unhaired body skin is enveloped in short, silky hair by the 18th day. By 30 days after birth, the young chipmunks have taken on adult appearance and opened their eyes. They leave the nest chamber when 35 to 40 days old. At this time, eastern chipmunks seem most abundant to the average observer. On the campus of Michigan State University the abundance of these playful and conspicuous animals is always a telltale sign of graduation in early June. After dismissing this spring litter at age of five to six weeks, the mother prepares for her second family. Some of her spring-born female offspring (but not males) are also destined to perform parental duties that same year. Autumn litters are above ground in late September and early October with more than a month to prepare their own burrows and stock them with seeds and nuts for the coming winter.

It was Burt's (1940) opinion that Michigan chipmunks might live for three or more years in the wild. Tryon and Synder (1973) noted a mean life expectancy at first capture (in a live trap) for a chipmunk of between 1.0 and 2.25 years, with an average of 1.29 years. June Nelson (*in litt.*) kept a female eastern chipmunk in captivity in Caseville, Michigan, from the time the animal was born on August 26, 1964, until its death on August 8, 1976, a few days short of 12 years. This eastern chipmunk had litters until 1972 (to the age of eight years) and continued to become reproductively responsive (but was not mated) each summer subsequently through 1975, indicating that she might have been capable of having young for 11 years. Her litters varied in number from two to six; the usual number was four or five.

FOOD HABITS. While the eastern chipmunk will eat insects, other small organisms, birds' eggs, and even mice (Torres, 1937) and snakes (Harriot, 1940), the major diet consists of the reproductive parts of plants—fruit, flowers, seeds, and nuts. The species dines less often on green foliage, buds, bark and other vegetative parts of plants (Aldous, 1941). From the time of emergence in spring until November, the eastern chipmunk can select from an array of different fruits, seeds and nuts, as shrubs and trees become productive. Early in spring, the chipmunk gathers buds and reproductive parts from the American elm and subsequently from cherries (Jameson, 1943), maples, other trees, various herbaceous plants, berries, grapes, even mushrooms; in autumn, the seeds and nuts from hazel, hickory, pine, walnut, oak, and beech (Timm, 1975; Verme, 1957).

As winter approaches, eastern chipmunks accelerate their caching of nuts and seeds. Under laboratory conditions, Brenner and Lyle (1975) and Wrazen and Wrazen (1982) showed, for example, that chipmunks store foods as part of their normal activity, but when preparing for winter inactivity individuals store significantly more food, presumably in preparation for seasonal change in habits. The highly variable pattern of winter activity, including short-term torpor followed by periods of arousal and feeding either in underground chambers or occasionally above ground, may account for a likely high energetic cost of hibernation (Edward Rybak, pers. comm.). In times of food abundance, the energetic chipmunk almost exhausts itself caching foods. This has been demonstrated dramatically by captive individuals (Engels, 1947), and the action is probably necessitated by the likelihood that caches will be robbed by other chipmunks (Shaffer, 1980).

ENEMIES. Because of the chipmunk's diurnal activities, it does evade some nighttime predators. Wallace (1948) did not find this mammal in the Michigan barn owl's diet, although Audubon depicts a chipmunk in the owl's clutches in his famous painting. Errington *et al.* (1940) discovered few records of the eastern chipmunk among the items eaten by the great horned owl while Allen (1938) reported the screech owl and barred owl as enemies. Diurnal birds of prey which eat eastern chipmunks include common raven (Harlow *et al.*, 1975), great blue heron (Peifer, 1979), goshawk (Porter, 1951), red-tailed hawk (Connor, 1971), marsh hawk (Burt, 1940), and other large hawks. Four-footed predators include all Michigan carnivores from the small least weasel to the large raccoon, red fox, bobcat, lynx and coyote. Snakes also catch a few eastern chipmunks (Jackson, 1961). House cats (Allen, 1938; Toner, 1956; Wood, 1922) account for part of this chipmunk's mortality. In eastern Canada, Banfield (1974) considers the long-tailed weasel and the ermine the major terrestrial predators of eastern chipmunks; he also cites an instance where a [Norway] rat attacked and killed one of these animals.

PARASITES AND DISEASES. Burt (1940), Ozoga and Phillips (1964), and Dorney (1965) have noted cases of botfly infestation in eastern chipmunks in Michigan. The large larvae, generally appearing in late summer and autumn, live directly under the skin, usually on the underside in the throat or inguinal areas. Fleas, at least one louse, and ticks infest the body surface of chipmunks (Allen, 1938; Scharf and Stewart, 1980; Timm, 1975). Internally, trematodes, cestodes, horny-headed worms, protozoa, and nematodes can be present.

MOLT AND COLOR ABERRATIONS. Young eastern chipmunks wear a juvenile coat for the first few weeks of their lives and then molt to change into adult pelage. This molt occurs in spring litters mostly in July and August and in late summer litters in October and November. Adults undergo two molts, in summer and autumn. Aberrant coloration in this chipmunk, including both albinism and melanism, is recorded (Smith and Smith, 1972). Wood (1922) reported an albino eastern chipmunk from Washtenaw County; Jackson (1961) described a white-spotted individual from Wisconsin. John Haslem (pers. comm.) saw a partial albino on the Michigan State University campus in the summer of 1977. This animal was normally colored except for a white tail and splotches of white on the head and one side. A female from Roscommon County in the collections of The Museum at Michigan State University (MSU 13064) has cream-buff upperparts and appendages, with the back striping obscurely showing in different shades of yellowish, buff and reddish-brown. The tail is whitish above and orange below.

ECONOMIC IMPORTANCE. Because of the eastern chipmunk's affinity for woody cover, rocky outcrops, and stone walls, it will only use open areas, such as grain fields, where brushy cover is nearby. In such situations the animals are known to damage corn. Eastern chipmunks are attractive additions around campsites, park areas, and rural cottages, although they can be nuisances when they take up residence under houses, rock patios or other foundations. They may occasionally gnaw tulip bulbs or cut down lilies and other plants. Generally, however, this species is one of the least harmful of dooryard associates.

COUNTY RECORDS FROM MICHIGAN. Records of specimens in collections of museums and from the literature show that the eastern chipmunk occurs in all counties in both the Upper Peninsula and Lower Peninsula. The species is also known from the Beaver, Fox and Manitou island groups in Lake Michigan (Hatt *et al.*, 1948).

LITERATURE CITED

Aldous, S. E.
1941 Food habits of chipmunks. Jour. Mammalogy, 22(1): 18–24.

Allen, E. G.
1938 The habits and life history of the eastern chipmunk, *Tamias striatus lysteri*. New York State Mus., Bull. No. 314, 22 pp.

Banfield, A. W. F.
1974 The mammals of Canada. Univ. Toronto Press, Toronto. xxiv+438 pp.

Blair, W. F.
1942 Size of home range and notes on the life history of the woodland deer-mouse and eastern chipmunk in northern Michigan. Jour. Mammalogy, 23(1):27–36.

Brenner, F. J., and P. D. Lyle
1975 Effects of previous photoperiodic conditions and visual stimulation on food storage and hibernation in the eastern chipmunk (*Tamias striatus*). American Midl. Nat., 93:227–234.

Burt, W. H.
1940 Territorial behavior and populations of some small mammals in southern Michigan. Univ. Michigan, Mus. Zool., Misc. Publ. No. 45, 58 pp.
1946 The mammals of Michigan. Univ. Michigan Press, Ann Arbor. xv+288 pp.

Condrin, J. M.
1936 Observations on the seasonal and reproductive activities of the eastern chipmunk. Jour. Mammalogy, 17(3):231–234.

Connor, P. F.
1971 The mammals of Long Island, New York. New York State Mus. and Sci. Ser., Bull. No. 416, v+78 pp.

Courtney, P. A., and M. B. Fenton
1976 The effects of a small rural garbage dump on populations of *Peromyscus leucopus* Rafinesque and other small mammals. Jour. Applied Ecology, 13:413–422.

Davis, W. H., and J. R. Beer
1959 Winter activity of the eastern chipmunk in Minnesota. Jour. Mammalogy, 40(3):444–445.

Dice, L. R.
1943 The biotic provinces of North America. Univ. Michigan Press, Ann Arbor. viii+78 pp.

Dorney, R. S.
1965 Incidence of botfly larvae (*Cuterebra emasculator*) in the chipmunk (*Tamias striatus*) and the red squirrel (*Tamiasciurus hudsonicus*) in northern Wisconsin. Jour. Parasitology, 51(5):893–894.

Dunford, C.
1972 Summer activity of eastern chipmunks. Jour. Mammalogy, 53(1):176–180.

Elliott, L.
1978 Social behavior and foraging ecology of the eastern chipmunk (*Tamias striatus*) in the Adirondack Mountains. Smithsonian Contr. Zool., No. 265, vi+107 pp.

Engels, W. L.
1947 Nest building and food storage of a captive chipmunk. Jour. Mammalogy, 28(3):296–297.
1951 Winter inactivity of some captive chipmunks (*Tamias s. striatus*) at Chapel Hill, North Carolina. Ecology, 32: 549–555.

Errington, P. L., F. Hamerstrom, and F. N. Hamerstrom, Jr.
1940 The great horned owl and its prey in north-central United States. Iowa State Col., Agri. Exp. Sta., Res. Bull. No. 277:758–850.

Gates, J. E.
1976 An ecological analysis of forest edge suitability for avian populations. Michigan State Univ., unpubl. Ph.D. dissertation, 57 pp.

Getty, T.
1981a Territorial behavior of eastern chipmunks (*Tamias striatus*): Encounter avoidance and spatial timesharing. Ecology, 62(4):915–921.
1981b Structure and dynamics of chipmunk home range. Jour. Mammalogy, 62(4):726–737.

Harlow, R. F., R. G. Hooper, D. R. Chamberlain, and H. S. Crawford
1975 Some winter and nesting season foods of the common raven in Virginia. Auk, 92(2):298–306.

Harriot, S. C.
1940 Chipmunk eating a red-bellied snake. Jour. Mammalogy, 21(1):92.

Hatt, R. T., J. Van Tyne, L. C. Stuart, C. H. Pope, and A. B. Grobman
1948 Island life: A study of the land vertebrates of the islands in eastern Lake Michigan. Cranbrook Inst. Sci., Bull. No. 27, xi+179 pp.

Howell, A. H.
1929 Revision of the American chipmunks (genera *Tamias* and *Eutamias*). N. American Fauna No. 52, 157 pp.

Jackson, H. H. T.
1961 Mammals of Wisconsin. Univ. Wisconsin Press, Madison. xii+504 pp.

Jameson, E. W., Jr.
1943 Source of food for chipmunks. Jour. Mammalogy, 24(4):500.

Layne, J. N.
1957 Homing behavior of chipmunks in central New York. Jour. Mammalogy, 38(4):519–520.

Maclean, G. S.
1981 Torpor patterns and microenvironment of the eastern chipmunk, *Tamias striatus*. Jour. Mammalogy, 62(1): 64–73.

Manville, R. H.
1949 A study of small mammal populations in northern Michigan. Univ. Michigan, Mus. Zool., Misc. Publ. No. 73, 93 pp.

Ozoga, J. J., and C. J. Phillips
1964 Mammals of Beaver Island, Michigan. Michigan State Univ., Publ. Mus., Biol. Ser., 2(6):305–348.

Panuska, J. A.
1959 Weight patterns and hibernation in *Tamias striatus*. Jour. Mammalogy, 40(4):544–566.

Panuska, J. A., and N. J. Wade
1956 The burrow of *Tamias striatus*. Jour. Mammalogy, 37(1):23–31.

Peifer, R. W.
1979 Great blue herons foraging for small mammals. Wilson Bull., 91(4):630–631.

Pivorum, E. B.
1976 A biotelemetry study of the thermoregulatory patterns of *Tamias striatus* and *Eutamias minimus* during hibernation. Comp. Biochem. Physiol., 53A:265–271.

Porter, T. W.
1951 A second nest of the goshawk near Douglas Lake, Cheboygan County, Michigan. Jack-Pine Warbler, 29(3):89–90.

Quinby, D. C.
1944 A comparison of overwintering populations of small mammals in a northern coniferous forest for two consecutive years. Jour. Mammalogy, 25(1);86–87.

Reilly, R. E.
1970 Factors influencing habitat selection by the least chipmunk in Upper Michigan. Michigan State Univ. unpubl. Ph.D. dissertation, 108 pp.

Scharf, W. C., and K. R. Stewart
1980 New records of Siphonaptera from northern Michigan. Great Lakes Ento., 13(3):165–167.

Schooley, J. P.
1934 A summer breeding season in the eastern chipmunk, *Tamias striatus.* Jour. Mammalogy, 15(3):194–196.

Seidel, D. R.
1961 Homing in the eastern chipmunk. Jour. Mammalogy, 42(2):256–257.

Seton, E. T.
1909 Life histories of northern animals. An account of the mammals of Manitoba. Vol. I. Grass-eaters. Charles Scribner's Sons, New York. xxx+673 pp.

Shaffer, L.
1980 Use of scatterhoards by eastern chipmunks to replace stolen food. Jour. Mammalogy, 61(4);733–734.

Smith, D. A., and L. C. Smith
1972 Aberrant coloration in Canadian eastern chipmunks, *Tamias striatus.* Canadian Field-Nat., 86:253–257.

Svendsen, G. E., and R. H. Yahner
1979 Habitat preference and utilization by the eastern chipmunk (*Tamias striatus*). Kirtlandia, No. 31, 14 pp.

Test, G. C.
1932 Winter activities of the eastern chipmunk. Jour. Mammalogy, 13(3):278.

Timm, R. M.
1975 Distribution, natural history and parasites of mammals of Cook County, Minnesota. Univ. Minnesota, Bell Mus. Nat. Hist., Occas. Papers 14, 56 pp.

Toner, G. C.
1956 House cat predation on small mammals. Jour. Mammalogy, 37(1):119.

Torres, J. K.
1937 A chipmunk captures a mouse. Jour. Mammalogy, 18(1):100.

Tryon, C. A., and D. P. Snyder
1973 Biology of the eastern chipmunk, *Tamias striatus:* Life tables, age distributions, and trends in population numbers. Jour. Mammalogy, 54(1):145–189.

Verme, L. J.
1957 Acorn consumption by chipmunks and white-footed mice. Jour. Mammalogy, 38(1):129–132.

Wallace, G. J.
1948 The barn owl in Michigan. Its distribution, natural history and food habits. Michigan State Col., Agri. Exp. Sta., Tech. Bull. No. 208, 61 pp.

Wood, N. A.
1922 The mammals of Washtenaw County, Michigan. Univ. Michigan, Mus. Zool., Occas. Papers No. 123, 23 pp.

Wrazen, J. A.
1980 Late summer activity changes in populations of eastern chipmunks (*Tamias striatus*). Canadian Field-Nat., 94(3):305–310.

Wrazen, J. A., and L. A. Wrazen
1982 Hoarding, body mass dynamics, and torpor as components of the survival strategy of the eastern chipmunk, Jour. Mammalogy, 63(1):63–72.

Wrigley, R. E.
1969 Ecological notes on the mammals of southern Quebec. Canadian Field-Nat., 83:201–211.

Yerger, R. W.
1953 Home range, territoriality, and populations of the chipmunk in central New York. Jour. Mammalogy, 34(4):448–458.
1955 Life history notes on the eastern chipmunk, *Tamias striatus lysteri* (Richardson), in central New York. American Midl. Nat., 53:312–323.

LEAST CHIPMUNK
Eutamias minimus (Bachman)

NAMES. The generic name *Eutamias,* proposed by Trouessart in 1880, is of Greek origin and means "well" or "true" (*Eu*) and "steward" or "one who stores and takes care of provisions" (*tamias*). There is some evidence that *Eutamias* should hold only a subgenic rank and that this species can be placed in the genus *Tamias.* The specific name *minimus,* proposed by Bachman in 1839, is derived from the Latin and means "least." In Michigan, local residents sometimes call this animal rock squirrel, gopher, ground squirrel, or simply chipmunk. The subspecies *Eutamias minimus jacksoni* Howell occurs in the Upper Peninsula.

RECOGNITION. Body size small, the smallest of Michigan sciurids; head and body approximately 4½ in (115 mm) long; the short, fine pelage of the upperparts traversed by five blackish stripes (edged with brownish) and four light stripes (lateral ones pale, median ones mixed with grayish) extending from near base of tail and rump to shoulder with median blackish stripe to forehead; top of head brownish; nape gray; three brown stripes with white ones in between cross cheeks from nose to ears, with middle blackish one through eye; pale gray spots behind ears; sides orange-brown; underparts whitish; upper surface of tail brownish, fringed with buffy orange; undersurface buffy orange with hairs having submarginal black bands and buff tips; feet grayish. Winter pelage is slightly paler. Eyes are large and conspicuous. There are two inguinal, four abdominal and two pectoral mammae. Internal cheek pouches, often bulging when filled with seeds and nuts, are characteristic. Of all Michigan sciurids, the least chipmunk most closely resembles, in external appearance, the eastern chipmunk. Where the two species occur together in the Upper Peninsula, the least chipmunk is easily distinguished from the eastern chipmunk by its slight build, small size and more "striped" appearance.

MEASUREMENTS AND WEIGHTS. Adults measure as follows: length from tip of nose to end of tail vertebrae, 7¼ to 9 in (185 to 225 mm); length of tail vertebrae, 3¼ to 4 in (80 to 100 mm); length of hind foot, 1⅛ to 1⅜ in (28 to 35 mm); height of ear from notch, ½ to ¾ in (13 to 18 mm); weight, 1.2 to 1.8 oz (35 to 53 g).

DISTINCTIVE CRANIAL AND DENTAL CHARACTERISTICS. The small skull of the least chipmunk has the same long, narrow shape with the wide interorbital space as does that of the larger eastern chipmunk (Fig. 31). The skull is approximately 1¼ in (32 mm) long and 11/16 in (17 mm) wide. The smooth, regular appearance of the bones on the dorsal surface of the cranium is a typical sciurid characteristic. The skulls of the least chipmunk and the eastern chipmunk are similar in the large size of their infraorbital foramina, but the latter species is distinctive because of its larger overall size and in having one less premolar tooth in the upper jaw making a total of only four cheek teeth instead of five as in the least chipmunk (see Fig. 32). The front surface of the upper incisors of the least chipmunk is darker orange than that of the upper incisors of the eastern chipmunk. The dental formula is:

$$\text{I (incisors)}\ \frac{1\text{-}1}{1\text{-}1},\ \text{C (canines)}\ \frac{0\text{-}0}{0\text{-}0},\ \text{P (premolars)}\ \frac{2\text{-}2}{1\text{-}1},\ \text{M (molars)}\ \frac{3\text{-}3}{3\text{-}3} = 22$$

DISTRIBUTION. The least chipmunk enjoys a wide distribution in western North America, from near the Alaskan border in northwestern Canada southward in the Rocky Mountain area to Arizona and New Mexico and eastward to Michigan and Ontario along the Canadian-United States border. Its easternmost distribution in the United States is Upper Peninsula Michigan; in Canada the species extends to western Quebec. The least chipmunk has been reported in every county in Michigan's Upper Peninsula (see Map 26).

HABITAT PREFERENCES. The least chipmunk is remarkably adaptable to various environments

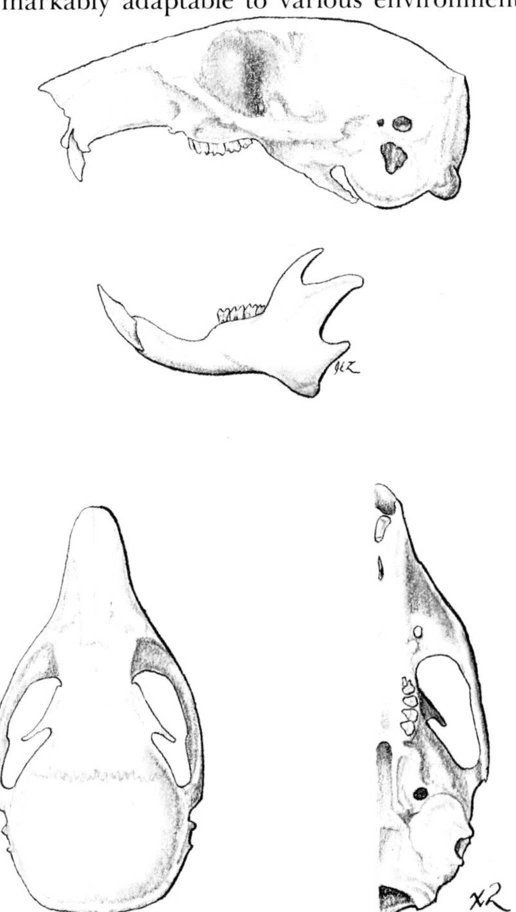

Figure 32. Cranial and dental characteristics of the least chipmunk (*Eutamias minimus*).

Map 26. Geographic distribution of the least chipmunk (*Eutamias minimus*). Michigan counties from where this species is known are shaded.

in its vast range. In western areas the species thrives in eroded badlands, in sagebrush flats, and even in alpine tundra. In Michigan, however, the least chipmunk is most at home in the various coniferous habitats of the Canadian Biotic Province (Dice, 1943). Although similar habitat also occurs in the northern part of Michigan's Lower Peninsula, the least chipmunk has become established only in the Upper Peninsula, perhaps being unable to cross the Straits of Mackinac water barrier.

The least chipmunk is rarely found in the deep, dense boreal forests but prefers the edges, especially surrounding small openings. It also lives along streams and lakeshores; in jack pine, sandy ridges, swamp edges, cutover, burnouts, cleared land, other disturbed areas, lumber camps, and under cabins. Apparently, the ecological successional changes caused by lumbering is a stimulus to population increase. Strip clear-cutting in conifer swamps in Alger County allowed for a higher number of least chipmunks (Verme and Ozoga, 1981). Like the eastern chipmunk, the least chipmunk will occupy log piles, rock walls and brush piles. Stumps and felled logs are favored den sites (Burt, 1946). Reilly (1970) found that Upper Peninsula least chipmunks occurred in largest numbers in sylvan habitats with good horizontal visibility, medium to dense brush piles for cover and open canopies with correspondingly high light intensities. The amount of low, woody ground vegetation appeared insignificant, although dense stands of bracken fern (*Pteridium aquilinum*) seemed to be avoided. Reilly noted that horizontal visibility was important in behavioral interactions and spacing of the population. The animals also occupy deciduous hardwood forest habitat, where they occur most often in edge situations and disturbed areas (Forbes, 1964; Timm, 1975). In hardwood habitats least chipmunks associate frequently with eastern chipmunks (Manville, 1949). Although there are some indications that the least chipmunk is retreating northward in central Wisconsin (Wydeven and Wydeven, 1976), there is no evidence of any such decline in Michigan. Instead, it is thought that land management programs often produce highly favorable habitats for least chipmunks.

DENSITY AND MOVEMENTS. Observers (notably Seton, 1953) usually note the characteristic sciurid tendency of the least chipmunk to congregate at picnic grounds or places where seasonal foods are localized. After food concentration is reduced, however, such groupings of chipmunks are likely to disperse. Local populations are, of course, highest with the emergence of weaned offspring, usually in July in Michigan (Reilly, 1970). In Wisconsin, Jackson (1961) reports populations of 30 or more per acre (75 per ha) in highly favorable habitat, although figures ranging from 5 to 15 per acre (12 to 37 per ha) are more usual. Population estimates made by Manville (1949) and Reilly (1970) in the Upper Peninsula showed densities to vary from less than one chipmunk per acre (1.8 per ha) to 5.6 per acre (14 per ha). Individual chipmunks may range over areas from less than one acre (0.5 ha) to as large as 4.3 acres (1.5 ha) in size (Jackson, 1961; Banfield, 1974; Reilly, 1970).

BEHAVIOR. The least chipmunk, with its quick actions, vigorous tail movements, and loud chirp-ings, is a conspicuous member of the sylvan environment of Michigan's Upper Peninsula. Like most sciurids, it adheres strictly to daytime activity (especially sunny periods), being up and about soon after daylight, perhaps most active at midday, and safely in its den or burrow prior to nightfall. Most movement is terrestrial, but it also perches on or scurries across piles of debris, rock fences, outcroppings, and treefalls (Reilly, 1970). Chipmunks travel expertly through bushy understory or climb to the tops of food trees; Manville (1949) observed these chipmunks at heights of 12 ft (4 m) in tag alders and 30 ft (10 m) in jack pine. In Michigan, its active life is seasonal, from April to as late as early November (Banfield, 1974; Burt, 1946; Manville, 1949).

The winter months are spent in a state of torpor, a condition similar to that of the wintering eastern chipmunk. In the Upper Peninsula, Manville (1948) concluded that winter torpor began on October 16 and continued until March 30. Prior to this, the least chipmunk stocks food caches convenient to the sleeping chamber for feeding during arousal periods; the nesting site is often located on top of these caches for added convenience. The winter nest is usually positioned as much as 3 ft (1 m) below the surface of the ground. Insulating nest materials include mammalian hair, shredded grass and bark, silky fibers from poplar and willow catkins, and feathers (Banfield, 1974; Howell, 1929).

Like the eastern chipmunk, the least chipmunk excavates its burrow through a workhole, which is later plugged. The main entry is an obscure opening bearing no evidence of excavation. The winter burrow is usually situated under a thick layer of insulating surface leaves and other duff of the forest floor. In summer, dens may be in hollow trees or stumps, in felled logs, debris and rock piles, or even woodpecker holes (Jackson, 1961). There are also records of nests of grasses and leaves exposed on tree branches (Banfield, 1974); Such bulky, rounded nests duplicate to some extent those of the gray squirrel. One summer nest was found 18 ft (6 m) above ground in a black spruce (Orr, 1930).

Manville (1949) found least chipmunks to be more tolerant of one another than were eastern chipmunks. However, there is evidence that least chipmunks also maintain spatial separation between themselves, especially adults (Reilly, 1970). Reilly noted that individuals "evaluated" the number and social status of its neighboring relatives by visual means; if the horizontal visibility was restricted, direct confrontations between chipmunks were more common.

Typically sciurid, the least chipmunk is vocal. Frequently in late afteroon in spring and fall, Reilly (1970) noted the ventriloquistic call which he described as *qwip*. The higher pitched *chit-chit* announces danger while *kek-kek*, uttered more lowly, is used to communicate between individuals (Jackson, 1961). Seton (1953) quoted Max M. Peet as noting that the "song" of the least chipmunk resembles that of a junco with a "prolonged twitter." Its various calls are accompanied by characteristic tail-jerks.

ASSOCIATES. Because of its diurnal habits, the least chipmunk rarely comes in contact with deer mice and other small, nocturnal mammals of the northern forests. Aside from predators, its usual daytime associates are other squirrels. Interactions between least chipmunks and gray and red squirrels are unclear; however, the eastern and least chipmunks associate in most woodland types in the Upper Peninsula. Since least chipmunks in the Upper Great Lakes Region may actually prefer swamp habitat while eastern chipmunks seem more at home in upland forested situations, there is some ecological separation (Quinby, 1944). In his study of the least chipmunk in the Upper Peninsula, Reilly (1970) observed both species on his study area but obtained no evidence that the two sciurids were competing actively for food, den sites, or observation posts.

REPRODUCTIVE ACTIVITIES. Both female and male least chipmunks emerge from hibernation in early spring in near readiness for courtship and mating. The mating urge probably becomes dominant once these small sciurids, long inactive in their winter chambers, have feasted on newly emerging vegetation and acorns, nuts, seeds and other edibles left over from the previous growing season. Interplay during courtship includes considerable chatter and chases involving both sexes as well as between rival males. Within 10 or 20 days, usually by the first week in May, mating activties have been completed. Pregnant females choose nursery nests in stumps and logs, and under brush or rock piles, with underground storage caches connected. The nests are often lined with grass and positioned so that rainfall or ground moisture will not affect the offspring (Jackson, 1961).

After a gestation period of approximately 31 days, a litter varying from two to seven (usually five or six) is born. Manville (1949) found lactating females in Michigan's Upper Peninsula from June 9 until August 5. The newly born weigh about 2.3 g and are 2 in (50 mm) long. They are hairless and have closed eyes and ear openings. The eyes open at about 28 days and the body is fully haired at 40 days. In less than two months after birth, the offspring are nearly two-thirds grown and ready to depart from parental care. Field observers note a conspicuous surge of active least chipmunks above ground at this time.

Unlike the eastern chipmunk and some of the other larger sciurids, the least chipmunk seems able to survive admirably by having only a single litter each year. Although a second litter may not be out of the ordinary, most late litters (born in late June or thereafter) result when a first litter is aborted or otherwise lost and the female remates (Banfield, 1974). There is no evidence that young of either sex mature sufficiently in their first summer to carry out reproductive functions. Despite the seemingly low reproductive rate in this small mammal, the least chipmunk, according to Manville (1949), appears to have a shorter natural life span than does the larger eastern chipmunk, which breeds twice each growing season.

FOOD HABITS. The least chipmunk, like other sciurids, is primarily a vegetarian (eating mostly seeds, nuts, acorns and other reproductive parts of plants), although it will ingest insects, birds' eggs, and other animal materials. However, various kinds of seeds and nuts from shrubs and trees seem preferred since these items can also be transported easily in cheek pouches and preserve well in storage caches. This means a great reliance on seasonal food supplies produced at intervals during the growing season but storable for use in winter. Jackson (1961) and Banfield (1974) noted that least chipmunks eat green foods including some grass, weeds and sedges but favorite summer foods include the fruits and seeds of strawberries, blueberries, blackberries, smartweed, and cherry pits. Banfield (1974) reported that the cheek pouches of one individual contained as many as 3,700 blueberry seeds, another held 86 smartweed seeds, and another 800 timothy seeds. He also found a cache containing 487 acorns and 2,734 cherry pits. Both eggs and young of birds, grasshoppers, and various other invertebrates are obtained as the warm season progresses. In essence, the least chipmunk is mostly a selective seed-eater but at the same time remains opportunistic.

ENEMIES. This small sciurid has a host of would-be predators. Being strictly diurnal, it is safe from at least some of the night prowlers. Major threats include hawks, snakes, weasels, mink, red fox and domestic dog and cat (Jackson, 1961; Manville, 1949).

PARASITES AND DISEASES. Manville (1949) found fleas, a tick, and mites on least chipmunks in the Upper Peninsula. A louse was identified from Minnesota chipmunks (Timm, 1975). In Wisconsin, Jackson (1961) also lists such internal parasites as nematode worms. In Wyoming, this chipmunk has been known to be infected with tularemia.

MOLT AND COLOR ABERRATIONS. Adult least chipmunks undergo two molts annually. The worn grayish winter coat is replaced in late June and early July by the brighter and more reddish-orange summer pelage (Howell, 1929). This bright-colored coat is worn for only 60 to 90 days and is replaced by the more drab winter fur in late September and early October (Jackson, 1961).

ECONOMIC IMPORTANCE. The pleasure given the observer by this active and colorful member of the outdoor community far exceeds any mischief attributed to the least chipmunk. To be sure, the species will forage for corn, oats and other domestic grains—Jackson (1961) counted 84 kernels of wheat in the cheek pouches of a female in Wisconsin—or occasionally take eggs and nestling young of birds. But their invasions of cottages or other outbuildings rarely put them at odds with outdoors persons. Their role in forest regeneration may be more important than is provable by the little good data on the success of their "planting" of nuts, acorns and other tree mast. To the field observer, the least chipmunk's reputation as a "friend in the woodlands" is summed up by Seton (1953) who described the small sciurid as "a cross between a striped squirrel and a song bird; and the bird that it most resembles is the chickadee—usually confiding, pert, and noisy, but also quick to escape and disappear when in evident danger."

COUNTY RECORDS FROM MICHIGAN. Records of specimens of the least chipmunk in collections of museums and from the literature show that this species occurs in all counties of the Upper Peninsula.

LITERATURE CITED

Banfield, A. W. F.
1974 The mammals of Canada. Univ. Toronto Press, Toronto. xxiv+438 pp.

Burt, W. H.
1946 The mammals of Michigan. Univ. Michigan Press, Ann Arbor. xv+288 pp.

Dice, L. R.
1943 The biotic provinces of Norh America. Univ. Michigan Press, Ann Arbor. viii+78 pp.

Forbes, R.
1964 Ecological studies of the eastern and least chipmunks. Univ. Minnesota, unpubl. Ph.D. dissertation, 82 pp.

Howell, A. H.
1929 Revision of the American chipmunks (genera *Tamias* and *Eutamias*). N. American Fauna No. 52, 157 pp.

Jackson, H. H. T.
1957 The status of *Eutamias minimus jacksoni*. Jour. Mammalogy, 38(4):518–519.
1961 Mammals of Wisconsin. Univ. Wisconsin Press, Madison. xii+504 pp.

Manville, R. H.
1948 The vertebrate fauna of the Huron Mountains, Michigan. American Midl. Nat., 39(3):615–640.
1949 A study of small mammal populations in northern Michigan. Univ. Michigan, Mus. Zool., Misc. Publ. No. 73, 93 pp.

Orr, L. W.
1930 An unusual chipmunk nest. Jour. Mammalogy, 11(3) 315.

Quinby, D. C.
1944 A comparison of overwintering populations of small mammals in a northern coniferous forest for two consecutive years. Jour. Mammalogy, 25(1):86–87.

Reilly, R. E.
1970 Factors influencing habitat selection by the least chipmunk in Upper Michigan. Michigan State Univ., unpubl. Ph.D. dissertation, 108 pp.

Seton, E. T.
1953 Lives of game animals. Vol. IV, Pt. 1, Charles T. Branford Co., Boston, xxii+440 pp.

Timm, R. M.
1975 Distribution, natural history and parasites of mammals of Cook County, Minnesota. Univ. Minnesota, Bell Mus. Nat. Hist., Occas. Papers No. 14, 56 pp.

Verme, L. J., and J. J. Ozoga
1981 Changes in small mammal populations following clear-cutting in Upper Michigan conifer swamps. Canadian Field-Nat., 95(3):253–256.

Wydeven, A. P., and P. R. Wydeven
1976 The status of the least chipmunk (*Eutamias minimus jacksoni*) in central Wisconsin. Univ. Wisconsin, Stevens Point, Mus. Nat. Hist., Rpts. Fauna and Flora Wisconsin, No. 11:3.

WOODCHUCK
Marmota monax (Linnaeus)

NAMES. The generic name *Marmota,* proposed by Blumenbach in 1779, is of Latin origin and derived from two words, *Mar* meaning "mouse" or "rat" and *mota* meaning "mountain." The specific name *monax,* proposed by Linnaeus in 1758, is derived from early American accounts of this mammal in which it is referred to as *monax,* a name given it by American Indians who admired the animal's ability to dig. In Michigan, local residents sometimes call this animal groundhog, chuck, whistler, or whistlepig. Woodchucks in the Upper Peninsula belong to the subspecies *Marmota monax rufescens* Howell; in the Lower Peninsula to *Marmota monax monax* (Linneaus).

RECOGNITION. Body size massive, the largest of Michigan sciurids (Fig. 33); head and body of an adult averaging approximately 18 in (460 mm) long, with a bushy tail one-fourth again as long; body heavy and compact, supported by short, powerfully built limbs; pelage of back and sides with dense, woolly underfur somewhat obscured by longer, less-dense guard hair; underfur basally gray with yellowish tips; guard hairs banded with alternating dark and light coloring from base to tip, with the darkish subterminal band and buffy tip blending with the buffy underfur to present an

Figure 33. The massive facial feature of the woodchuck (*Marmota monax*). Photograph by Robert Harrington.

overall grizzled yellowish-gray appearance; ventral surface lacking woolly underfur with the sparse, coarse guard hairs basally blackish with tawny or rufus tips, giving a less grizzled appearance to the belly area; cheeks, ears, and throat whitish buff; top of head brownish black; shoulders and forelegs tawny rufus; feet black; small eyes and facial vibrissae blackish. There are two pairs of thoracic, one pair of abdominal, and one pair of inguinal mammary glands with teats. The slightly flattened feet are equipped with heavily-constructed digits terminating in flattened, curved claws, useful in digging. The first digit (pollex) of the forelimb is reduced. The thickset body, short and powerful limbs, and smallish ears and eyes are all adaptations for the semi-fossorial life of the woodchuck.

MEASUREMENTS AND WEIGHTS. Adults measure as follows: length from tip of nose to end of tail vertebrae, 19¾ to 25 3/5 in (500 to 650 mm); length of tail vertebrae, 4⅛ to 6¼ in (105 to 160 mm); length of hind foot, 2½ to 3½ in (65 to 90 mm); height of ear from notch, 1 to 1½ in (25 to 40 mm); weight of adults, 5 to 12 lbs (2.3 to 5.4 kg).

DISTINCTIVE CRANIAL AND DENTAL CHARACTERISTICS. The skull is large, broad and flat (Figs. 34 and 35). There is a pronounced depression in the space between the orbits adding to this flattened effect. The prominent postorbital processes are strongly developed and extend at right angles laterally to the long axis of the skull.

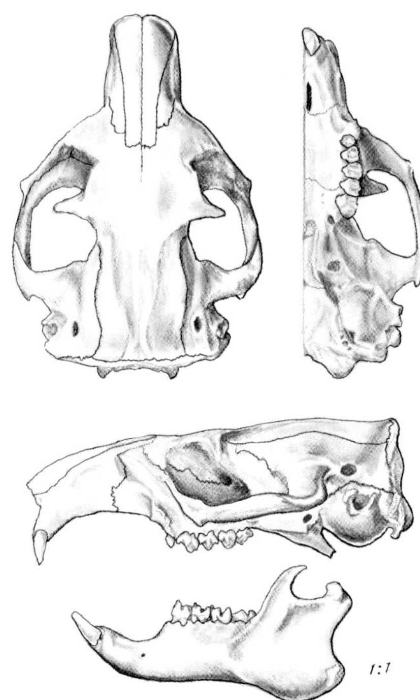

Figure 34. Cranial and dental characteristics of the woodchuck (*Marmota monax*).

The rostrum is large and massive and characteristically depressed anteriorly. The skull of an adult is approximately 3⅜ in (86 mm) long and 2¼ in (57 mm) wide. The incisors are heavily constructed with the anterior faces ivory or yellowish ivory in color. As in other rodents, each incisor is evergrowing. If not worn down properly by occluding with the tooth opposite it in the course of chewing plant fibers, it may continue to grow with often fatal results (see Fig. 36). When such a malocclusion occurs, the growth of the tooth follows a circular pattern. The dental formula is:

$$\text{I (incisors)}\frac{1\text{-}1}{1\text{-}1}, \text{C (canines)}\frac{0\text{-}0}{0\text{-}0}, \text{P (premolars)}\frac{2\text{-}2}{1\text{-}1}, \text{M (molars)}\frac{3\text{-}3}{3\text{-}3} = 22$$

DISTRIBUTION. The woodchuck occurs in most of northeastern North America avoiding the

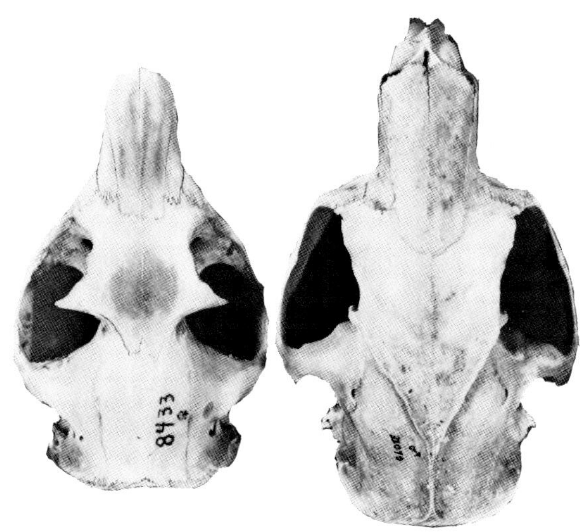

Figure 35. Dorsal aspects of the skulls of the woodchuck (left) and porcupine (right).

coastal areas of the southern Atlantic and Gulf states. It expands its range north at the edge of the Great Plains to spread across southern Canada to the Pacific coastal provinces, extending south to northern Idaho. The animal is widespread in both peninsulas of Michigan (see Map 27) and has been introduced on Beaver Island (Ozoga and Phillips, 1964).

HABITAT PREFERENCES. In presettlement days in eastern North America, the woodchuck was perhaps more restricted in numbers than today because heavy-canopied forest cover was less to this large rodent's liking (Howell, 1915; Banfield, 1974) than a mosaic of openings. The axe, torch and plow produced a vast amount of forest edge, open hillsides, brushy property lines and fencerows—all conducive to a larger woodchuck population (Wood, 1922a). This environmental disturbance by human occupation allowed the woodchuck to thrive by setting back plant successional stages from the climax and subclimax forest conditions to the early stages in the vegetative successions (Hamilton, 1934). In fact, de Vos and Gillespie (1960) found woodchucks more populous on pasturage that was fertilized and seeded with tame grasses than on unmanaged pasture. Earthen constructed dams and the slopes of right-of-ways for railroads and highways have attracted these burrowers. In Pennsylvania, highly concentrated populations in earthern mounds have allowed for detailed studies of the animal's life habits (Bronson, 1962; Davis 1967a).

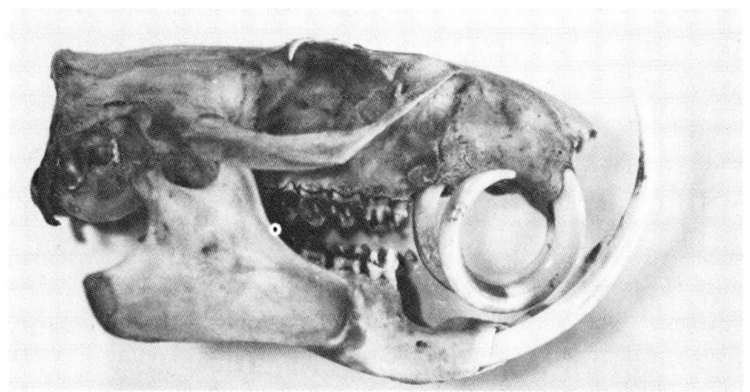

Figure 36. Side view of the skull and lower jaws of a female woodchuck (*Marmota monax*) showing excessive (and almost always fatal) growth of both upper and lower incisors as a result of malocclusion. Specimen obtained by Jim Lloyd, 10 km E and 2 km S Evart, Osceola County, Michigan, on 17 April 1977 (MSU Museum cat. no. 25100).

Map 27. Geographic distribution of the woodchuck (*Marmota monax*).

The uneven topography of the southern Michigan countryside with mixed farmlands, pasturage and scattered woodlots provides excellent habitats, especially on the crests of some of the glacial features. The animal has been reported on spruce barrens (Robinson and Werner, 1975), around cabins (Laundre, 1975), and in jack pine and cutover hardwoods (Manville, 1948). Allen (1938) described burrows excavated in mixed farmlands in Kalamazoo County. According to Grizzell (1955), at Rose Lake Wildlife Research Center in Clinton County woodchucks seemed partial to loam and sandy loam soils for digging. Wood (1922b) found woodchucks in dune habitat near Lake Michigan in Berrien County.

DENSITY AND MOVEMENTS. Woodchuck numbers vary with environmental conditions, although the numbers are always greatest directly after the emergence of the young-of-the-year in late spring. These burrowers may concentrate in favorable areas: Hamilton (1934) noted 30 occupied dens in an area of less than three acres (one ha); while Anthony (1962) counted 15 woodchucks in a clover field in late summer in Illinois. Such population concentrations, possibly local and temporary, may result from an abundance of foods or suitable terrain for den sites. In New York, Manville (1966) calculated that woodchucks occurred at the rate of one per 46 acres (18.4 ha) in the Hudson Valley along the Taconic Parkway. Other counts show highly variable densities. In Missouri, Twichell (1939) found a summer population of one animal per 36 acres (14.4 ha), with greatest densities of one to 11.5 acres (4.6 ha); in Maryland, Grizzell (1955) counted 18 individuals per 100 acres (40 ha); in Ontario, de Vos and Gillespie (1960) found a density of one woodchuck to about two acres (less than one ha) in June; in Wisconsin, Jackson (1961) reported that the state-wide density may be no more than four or five individuals per sq mi (256 ha) but in selected areas as many as 30 to 40. Perhaps the most astonishing record is the taking of 1,700 woodchucks from a 2,000 acre Pennsylvania tract (800 ha) during a two-year period (Snyder *et al.*, 1961).

At Rose Lake Wildlife Research Center in southern Michigan, Linduska (1950) reported overall densities in autumn to be as high as 35 to 40 woodchucks per 640 acres (256 ha); in selected areas as many as 25 to 30 animals per 100 acres (40 ha) were counted (Linduska, 1947). In the Huron Mountains of Michigan's Upper Peninsula, Laundre (1975) suggested that woodchuck populations are as high as 30 to 40 per 640 acres (256 ha). Various observers in eastern United States point to the ups and downs of woodchuck populations; some authors, notably Jackson (1961) in Wisconsin, describe high populations at the turn of the century. Others comment on years of population abundance followed by years of scarcity (Schwartz and Schwartz, 1981). In Michigan, 1974, 1975, and 1976 were characterized by high woodchuck populations, as evidenced by the unusual highway kill (Hartley, 1975) and by the number of telephone calls received by the Michigan Department of Natural Resources asking for methods of discouraging the animals from living under houses and invading gardens.

Adult woodchucks center their daily activities around underground burrows (Bronson, 1962), from which they develop surface trails leading to feeding areas. Evidence of currently used burrows includes freshly-worked dirt piles at den entrances and well-kept trails. Most movements are only short distances from burrow systems, usually well within a radius of 100 yd (100 m) but sometimes as much as three times that distance (de Vos and Gillespie, 1960; Grizzell, 1955; Seton, 1953; Trump, 1950; Twichell, 1939).

BEHAVIOR. Most observers consider adult woodchucks to be solitary, although on occasion two to four individuals occupy a single den (Anthony, 1962; Banfield, 1974; Hamilton, 1934; Twichell, 1939). These multiple occupants may sometimes be a female and her young. There is also evidence that woodchucks associate in groups which an adult male dominates, discouraging other males (Anthony, 1962; Bronson, 1964; Brown, 1948; Grizzell, 1955; Merriam, 1971).

Although the woodchuck is considered diurnal in its aboveground activities (Bronson, 1962), on occasion, especially in early spring, it has been observed feeding in dim light or even at night (Banfield, 1974). As a rule, most spring and autumn activity is near midday, perhaps due to the cooler temperatures early and late in the day (Anthony, 1962). In summer, however, the animal appears more active in morning and afternoon periods. Woodchucks rarely appear during rainy weather, preferring clear skies and sunny days for maximum activity. When not foraging, the woodchuck often squats at the mouth of the den, seemingly surveying the countryside while sunning, scratching, hair-pulling or otherwise preening (Anthony, 1962). Some authors (Jackson, 1961) term this "sentry duty" as the animal peers around possibly looking for potential enemies or for other woodchucks whose presence may need to be discouraged. At times the animal will "freeze" in one position for as long as one minute. The animal may change from a flattened posture to an upright position, which is typical of ground squirrels, prairie dogs and other members of the squirrel family. From this erect posture the woodchuck may slowly drop from sight into the den when possible adversaries are detected.

The woodchuck, like other sciurids, has a repertoire of several sounds, although not so many perhaps as its western relative, the yellow-bellied marmot (*Marmota flaviventris*). The woodchuck is often called "whistle-pig" because of its vocalizations. When alarmed and ready to "dive" to safety, it gives a loud, shrill whistle. When cornered or disturbed, the animal may grind its teeth while chattering. When handled, a muffled bark may be emitted. When fighting with other woodchucks, the animal may squeal, bark, and whistle (Anthony, 1962).

This heavy-bodied marmot retreats into a nearby den instead of running from danger, although when cornered it is a worthy and fierce competitor. When necessary, the woodchuck changes from a rather slow, cumbersome, waddling gait to a loping gallop (Jackson, 1961), despite its low-hung body and widespread feet. Hamilton (1934) reported that the distance between tracks of a running woodchuck was about 12 in (305 mm). Twichell (1939) measured those distances for a Missouri woodchuck as being from 25½ to 34 in (645 to 865 mm); these measurement are from tracks of a startled animal crossing a muddy field. Woodchucks swim with only the nose and top of the head above water (Jackson, 1961). Unlikely as it would seem for such a bulky creature, woodchucks forage without difficulty in trees (Bowdish, 1922; Johnson, 1926; Robb, 1926; Saunders, 1922). In the Huron Mountains of Michigan, Manville (1948) watched a young animal, startled by an automobile, climb to a height of 15 ft (about 5 m) in a jack pine. Wood (1922a) reported sighting woodchucks in Washtenaw County in trees at heights of 30 ft (9 m).

Strong, "stubby" forelimbs equipped with heavy claws provide the woodchuck with efficient digging tools. The large teeth are handy to cut roots and clear debris. Dirt and rocks loosened when excavating may be pushed out of the burrow by the head or by using the hind limbs as a drag. According to Schwartz and Schwartz (1981), the weight of soil removed while digging one burrow can be as much as 716 lbs (325 kg). Dens are situated in well-drained locations, with the main entryways often adjacent to a stump, rock or rocky ledge. When den sites are next to gardens or under houses, woodchuck/human relationships may become strained.

Dens may be single or in clusters when social groupings develop. A single burrow system may have as many as five entrances (Grizzell, 1955). Ordinarily, there is an earthen mound at the main entrance with trails leading away to foraging areas. The well-used appearance is due partly to the weekly addition of soil brought up from the tunnels. A plunge hole is often near the major den entrance; this may have a vertical drop of as much as 2 ft (610 mm) to a main tunnel (Merriam, 1971). The plunge hole is usually hidden and bears no telltale signs of fresh earth. Burrow systems may be short, especially when freshly dug by young individuals. Extensive dens may include up to 45 ft (about 15 m) of tunnels, extending as far underground as 5 ft (almost 2 m), sometimes to almost 14 ft (5 m). Dens found in open country, often on sloping hillsides, are generally used in summer. Dens under stumps, at the edge of rock ledges, and near other protection are generally occupied in winter, although they can be used the year around. Nesting chambers are usually located about 2 ft (610 mm) below ground and are about 15 in (½ m) in diameter; some are lined with leaves, others with grass and weeds, some remain bare. Fecal pellets are disposed of in side passages or in the soft earth of the entry mound.

HIBERNATION. Biologists and other observers have long been interested in this large, bulky animal's ability to retire to its underground retreat for the winter months. This hibernation takes place in a side chamber along one of the underground corridors of the burrow system (Allen and Shapton, 1942; Grizzell, 1955). The woodchuck tucks its head between its hind legs, curls itself into a ball and enters a state of prolonged torpor during which various metabolic processes are reduced—heartbeat slows (from more than 100 per minute down to as few as 15 per minute), body temperature is reduced (from about 35°C to near 8°C), and respiration is down (oxygen consumption reduced from 1,400 ml per kg per h to less than 100 ml per kg per h). All of these lowered life processes (Lyman, 1958) are essential because the fat and other bodily nutrients accumulated just prior to the start of the winter sleep must be rationed over several months (Davis, 1967b) to allow for minimal life-process maintenance. Unlike the eastern chipmunk which stores nuts and other seed staples for winter consumption, the woodchuck relies solely on body fat for winter survival in the hibernaculum and may experience a weight loss of from 30 to 50% (Grizzell, 1955). If the summer food supply is poor, individuals, especially young-of-the-year, may fail to gain sufficient weight for surviving the hibernation period (Davis and Ludwig, 1981). This may account for marked reductions observed in local populations.

Woodchuck populations do not enter hibernation simultaneously; the senior and fatter individuals proceed first and the yearlings and young-of-the-year follow. When emerging at the end of winter, the same order is followed with the older animals (especially males) appearing above ground first. In Michigan, especially in the northern areas, the woodchuck may enter hibernation as early as late September and early October with the latest autumn record in the Huron Mountains on October 11 (Manville, 1948). In southern Michigan, animals may be above ground as late as mid-November. In more southern areas, as in southern Illinois, Anthony (1962) thought that woodchucks were active most of the winter.

The exact causes for this winter sleep are unknown. Contributing factors must include the frosts which eliminate green foods, the increasing length of darkness per 24 hours, the lethargy resulting from fat accumulation, perhaps low oxygen content in the den, and hormonal triggering action termed a biologic clock by some authors (Davis, 1967a).

The awakening process is equally mysterious. Undoubtedly the woodchuck's underground winter chamber varies little in temperature. Yet, the animal seems to know when to emerge—again there is inference in the literature that some kind of biologic clock is the awakening mechanism. Davis (1977) concludes that arousal is controlled by a circannual clock but that actual emergence depends on daily temperature. However, before finally forsaking their winter quarters, Michigan woodchucks may emerge, look around a bit and then resume their sleep. There are records of woodchucks being above ground in all winter months in Michigan; Glen Belyea observed an active woodchuck on the grounds of Okemos High School (Ingham County) at 0930 on a sunny January 21. Holt (1929) reports a Pennsylvania woodchuck caught in a trap on January 1 under blizzard conditions, although Davis (1977) found that first emergence in south-central Pennsylvania was between January 29 and February 13.

In Michigan's Upper Peninsula, Manville (1948) obtained his earliest spring record of emergence on March 26; in southern Michigan this can be earlier, even rarely in late February. Emerging woodchucks are primarily concerned with foraging to fill empty stomachs and regain lost body weight, although reproduction behavior can also be noted. The active season gets into full swing as the spring "green-up" time unfolds.

GROUNDHOG DAY. The woodchuck's emergence after winter sleep has been the subject of much American folklore. The supposition is that if the animal surfaces on February 2 and is able to see its own shadow, there will, unhappily, be frosty weather for six more weeks. Of course, if the day is cloudy, the subsequent weather is supposed to be mild. Punxsutawney, Pennsylvania, holds a vigilant watch to determine whether or not Punxsutawney Phil (the woodchuck) will see his shadow on Groundhog Day. Although Seton (1953) attributed this tradition to early settlers in mid-eastern America, it undoubtedly was derived from a European notion that such Old World creatures as the hedgehog (*Erinaceus europaeus*), but more likely the European badger (*Meles meles*), were weather prophets. When the tradition was brought to North America, the woodchuck replaced the European badger as the long-range weather forecaster. February 2 was also significant for the religious celebration of Candlemas; ancient tradition held that a sunny Candlemas presaged a cold spring.

ASSOCIATES. In biotic communities, one animal may initiate environmental changes used to advantage by other animals. Good examples include the beaver, whose dams offer sanctuary and food for many semiaquatic and riparian species, and the meadow vole, whose surface runways through grassy plots offer means of travel through overhead protective cover to other small mammals. Burrows excavated by woodchucks harbor a host of small mammals and even some birds (see impressive lists in Grizzell, 1955 and Schmeltz and Whitaker, 1977) seeking underground refuges. Species like the eastern cottontail, Virginia opossum, striped skunk (Anthony, 1962), and otter (Liers, 1951) can come and go, especially in abandoned burrows, sometimes making side chambers for their own use. In Michigan, the eastern cottontail used woodchuck burrows, especially in cold weather (Linduska, 1947). These "roomers" use burrows as summer nurseries and winter dens. During the latter season, the woodchuck, hibernating in a plugged side chamber, is oblivious to the intruders. Other species, including red fox, coyote and raccoon (Butterfield, 1954), may be obliged to enlarge the tunnel system; these larger species probably exclude the woodchuck or in some cases use it as food (Grizzell, 1955). The woodchuck's den excavations are important to the Michigan small mammal community, providing numerous abandoned burrows for species which are unable or unwilling to dig their own.

REPRODUCTIVE ACTIVITIES. Although there is conflicting opinion (Jackson, 1961), it is likely that most juveniles (especially males) do not become sexually mature until their second spring (Banfield, 1974; Christian *et al.,* 1972; Grizzell, 1955). Those adults which are sexually mature on leaving the hibernating chambers begin the reproductive process soon after. Males are sexually active almost at once and may joust with neighbors of the same sex, with scars about the head and shoulders often the result. Considerable movement may also occur during the mating season, with manifestations of this shown by the number of dead woodchucks seen on highways in March and April. Sexually mature females may make their appearances above ground as much as three to four weeks after the emergence of the males (Snyder and Christian, 1960). Most females are inseminated by mid-March. In Missouri, Twichell (1939) found matings to occur as early as mid-February. The females are monoestrus (receptive to mating only once each year).

The gestation period is between 31 and 32 days (Hoyt, 1952). Many young are probably born in southern Michigan by mid-April (as early as April 5 in Maryland, Grizzell, 1955), later in the Upper Peninsula. The average litter contains from three to five young (Snyder and Christian, 1960), although as few as one and as many as nine are reported (Howell, 1915; Jackson, 1961). At birth, young are covered with pink and wrinkled skin with no evidence of hair. They are blind and virtually helpless, with forelegs better developed than hind ones (Grizzell, 1955). Newly born individuals weigh 26 to 27 g and are about 4¼ in (104 mm) in length. At one week, the skin becomes pigmented; at two weeks, short, black body hair emerges; at three weeks, the young are capable crawlers; at four weeks, the eyes open, and shortly thereafter green food is taken. The weaning process is completed by six weeks after birth. Just prior to weaning, the female allows her charges romping room in front of the den. She will also nurse them while maintaining her vigilance at the mouth of her retreat. By the time the young are capable of fending for themselves, they weigh at least 1 lb (450 g). By mid-July, the mother is usually rid of her offspring, which disperse to occupy abandoned burrows or excavate ones of their own. However, some females remain with the mother until the following year.

A large rodent like the woodchuck logically would not need to produce as many offspring annually as would small rodents which are more susceptible to predators. Three to five young per breeding female seem to provide the necessary annual recruits to maintain woodchuck numbers. Naturally, individuals must be fairly long-lived. At Rose Lake Wildlife Research Center, Linduska (1947) recorded a probable minimum life span for one male of four years and two months. In trapping-tagging-releasing-retrapping studies he found many individuals to survive to the second summer of life but only four of 196 marked woodchucks lived to the third summer. There are other records of woodchucks living as long as five to six years in nature and up to 10 years in captivity (Schwartz and Schwartz, 1981).

FOOD HABITS. The woodchuck is primarily a grazer, eating the vegetative parts of plants. This habit is unlike that of other Michigan sciurids which are more prone to consume the reproductive parts (seeds, nuts, flowers, etc.). After emergence in early spring, the woodchuck may eat stems, buds, bark, and twigs of such plants as sumac, dogwood, cherry, and other fruit trees (Banfield, 1974), but rapidly turns to greenery as soon as it appears. The woodchuck is almost completely vegetarian, rarely eating insects, snails, or birds' eggs (Schwartz and Schwartz, 1981; Seton, 1953). Although there is seasonal variation in food consumed (Fall, 1971), a woodchuck may eat as much as one and one-half lbs (0.67 kg) daily. Some favorite foods include the leaves of sassafras and other trees (obtained by climbing), wild lettuce, white clover, red clover, sweet clover, bluegrass and other grasses, buttercup, chickweed, dandelions, plantains, asters, goldenrod, and numerous other herbs (Banfield, 1974; Grizzell, 1955; Howell, 1915; Seton, 1953; Twichell, 1939). In addition, the woodchuck relishes farm crops, especially alfalfa, planted clovers, corn (especially when winds blow the ears down to the woodchuck's level), oats, and assorted fruits and vegetables including beets, carrots, beans, peas, celery, lettuce, melons, pumpkin, potato vines, cabbage, turnips, blackberries and strawberries. Such foraging places the woodchuck high on the pest list of farmers and gardeners.

ENEMIES. Although human occupation of the woodchuck's geographic range brought about habitat changes favorable to the species, the human remains the woodchuck's most persistent enemy, primarily because of the animal's depredations on farm crops and its pesky habit of digging immense burrows under foundations and at yard edges. Because of its large size, the adult woodchuck is safe from an array of small predators. Those predators known to eat woodchucks (mostly young and juveniles) include the red-tailed hawk (Craighead and Craighead, 1956), raven (Harlow *et al.*, 1975), domestic dog, and black bear (Grizzell, 1955), gray wolf (Van Ballenberghe *et al.*, 1975), coyote (Laundre, 1975), and bobcat (Schwartz and Schwartz, 1981).

With the gray wolf (Voigt *et al.*, 1976) and mountain lion no longer threats and the bobcat and lynx on the wane, major predators in Michigan are probably the coyote, red fox and perhaps the gray fox and the badger. Most authors agree with Seton (1953) and Jackson (1961) that the red fox is an important adversary (Scott and Klimstra, 1955), and that young woodchucks experience the highest mortality. In Ontario, de Vos and Gillespie (1960) recorded seven such predators. In Michigan, Murie (1936) and Schofield (1960) found little evidence that the woodchuck was a prey species for the red fox as a result of their snow-tracking investigations. Most likely, the woodchucks were hibernating during the time that these authors were conducting their studies.

Although there is no way of predicting accurately the kill of woodchucks by automobiles, the annual number must be astounding. This mortality is probably highest at mating time in spring and after the young animals become independent in midsummer—times when individual animals tend to move around. Some evidence of woodchuck destruction in Michigan has been presented by Manville (1949) for areas in the Upper Peninsula and by Haugen (1944) in the Lower Peninsula. On the 19-mile stretch of roadway between Lowell and Grand Rapids, Hartley (1975) counted 15 dead woodchucks from March 1 to August 30, 1974.

PARASITES AND DISEASES. Grizzell (1955) felt that parasites and diseases were of only negligible importance in the life of the woodchuck. In a study of 91 Indiana woodchucks, Whitaker and Schmeltz (1973) collected seven species of mites, chiggers and ticks, one species of flea, and one louse. In Michigan, Rausch and Tiner (1946) found woodchucks in the area of East Lansing infected with roundworms (*Obeliscoides cuniculi*); a woodchuck heavily parasitized with a mite (*Haemolaelaps glasgowi*) was found wandering aimlessly around in the snow near Ann Arbor on February 19 (Grizzell, 1955); and the flea (*Oropsylla arctomys*) was collected from a woodchuck in Leelanau County (Scharf and Stewart, 1980). Our present knowledge of the array of diseases and parasites which may not only infect but also cripple and kill the woodchuck is rudimentary.

MOLT AND COLOR ABERRATIONS. The pelage, both the coarse guard hairs and the soft, dense underfur, is molted once each year in summer, from late May to early September (Davis, 1966). Albinistic (whitish) and melanistic blackish) color phases are known (Banfield, 1974). The Museum at Michigan State University had a mounted albino individual on display for many years. Wood (1922a) reports an albino taken near Saline about 1885; he (1922b) also noted a dark individual in Berrien County in 1919. Jackson (1961) found a melanistic woodchuck in August 1922 in Wisconsin.

ECONOMIC IMPORTANCE. To the outdoorsperson devoted to animal watching with field glasses, the antics of the woodchuck at and about its burrow on a green or rocky hillside are exciting to watch. On the other hand, farmers, ranchers and truck gardeners often consider this animal the least desirable of all local mammals. In Michigan, this aversion to the woodchuck is illustrated by the bounty paid between August, 1919, and August, 1923, for killing the animals (Hickie, 1940). The bounty was 50¢ per animal for the first two years and 25¢ for the last two, with the county and state each contributing half the cost. The total bounties paid were $281,116.17, with approximately 1,500,000 animals falling victim. Most bounties were paid in Barry, Calhoun, Eaton, Isabella, Jackson, Montcalm and Washtenaw counties. In each of these counties the catch was more than 79 animals per sq mi (256 ha). No unusual decline was noted as a consequence of this campaign and expense. Today, the woodchuck receives no protection by law in the Upper Peninsula, but is protected in the Lower Peninsula except during the hunting season beginning October 1 (in Zone 2) and October 20 (in Zone 3) and closing January 31. However, a landowner may take this animal on his own property at any time.

The large, often conspicuous woodchuck has also been the target of many a hunter, especially the farm boy. With few restrictions placed on the harvest of woodchucks, the animal has become a major summer quarry, especially for the small-caliber rifle. Manufacturers of arms and ammunition have stressed in numerous advertisements the woodchuck's qualities as sporting game for the .22. Although its feeding habits place the woodchuck in a class to be hunted, trapped (Ludwig and Davis, 1975), and poisoned (de Vos and Merrill, 1957) at will, the animal's reproductive potential seems sufficient to prevent local extermination.

On the positive side, it has been shown earlier that the burrowing activities of the woodchuck provide homes for many kinds of wildlife (Grizzell, 1955). In fact, woodchucks were introduced on Beaver Island in hopes they would dig homes for the also-introduced eastern cottontail (Ozoga and Phillips, 1964). Both Merriam and Merriam (1965) and Wallihan (1947) have demonstrated that growth of vegetation, including trees, in the vicinity of burrows is enhanced by fertilization from buried fecal materials. The woodchuck's sporting qualities have been further emphasized by Davis (1962), who studied ways to regulate the harvest. Many writers including Paul Kelsey (1976) describe methods of cooking tasty woodchuck meat dishes.

COUNTY RECORDS FROM MICHIGAN. Records of specimens of woodchucks in collections of museums and from the literature show that this species occurs in every county of Michigan's Upper and Lower Peninsulas.

LITERATURE CITED

Allen, D. L.
1938 Ecological studies on the vertebrate fauna of a 500-acre farm in Kalamazoo County, Michigan. Ecol. Monog., 8(3):347–436.

Allen, D. L., and W. W. Shapton
1942 An ecological study of winter dens, with special reference to the eastern skunk. Ecology, 23:59–68.

Anthony, M.
1962 Activity and behavior of the woodchuck in southern Illinois. Occas. Papers C. C. Adams Center Ecol. Studies, 6:1–24.

Banfield, A. W. F.
1974 The mammals of Canada. Univ. Toronto Press, Toronto. xxiv+438 pp.

Bowdish, B. S.
1922 Tree-climbing woodchucks. Jour. Mammalogy, 3(4): 259.

Bronson, F. H.
1962 Daily and season activity patterns in woodchucks. Jour. Mammalogy, 43(3):425–427.
1964 Agonistic behaviour in woodchucks. Animal Behaviour, 12:470–478.

Brown, C. P.
1948 Woodchucks observed while fighting. Jour. Mammalogy, 29(1):70.

Burt, W. H.
1943 Changes in nomenclature of Michigan mammals. Univ. Michigan, Mus. Zool., Occas. Papers No. 481, 9 pp.

Butterfield, R. T.
1954 Some raccoon and groundhog relationships. Jour. Wildlife Mgt., 18(4):433–437.

Christian, J. J., E. Steinberger, and T. D. McKinney
1972 Annual cycle of spermatogensis and testis morphology in woodchucks. Jour. Mammalogy, 53(4):708–716.

Craighead, J. J., and F. C. Craighead, Jr.
1956 Hawks, owls and wildlife. The Stackpole Co., Harrisburg, Pennsylvania. xix+443 pp.

Davis, D. E.
1962 The potential harvest of woodchucks. Jour. Wildlife Mgt., 26(2):144–149.
1966 The moult of woodchucks (*Marmota monax*). Mammalia, 30(4):640–644.
1967a The role of environmental factors in hibernation of woodchucks (*Marmota monax*). Ecology, 48(4): 683–689.
1967b The annual rhythm of fat deposit in woodchucks (*Marmota monax*). Physiol. Zool., 40(4):391–402.
1977 Role of ambient temperature in emergence of woodchucks (*Marmota monax*) from hibernation. American Midl. Nat., 97(1):224–229.

Davis, D. E., and J. Ludwig
1981 Mechanism for decline in a woodchuck population. Jour. Wildlife Mgt., 45(3):658–668.

de Vos, A., and D. I. Gillespie
1960 A study of woodchucks on an Ontario farm. Canadian Field-Nat., 74(2):130–145.

de Vos, A., and H. A. Merrill
1957 Results of a woodchuck control experiment. Jour. Wildlife Mgt., 21(4):454–456.

Fall, M. W.
1971 Seasonal variations in the food consumption of woodchucks (*Marmota monax*). Jour. Mammalogy, 52(2): 370–375.

Grizzell, R. A., Jr.
1955 A study of the southern woodchuck, *Marmota monax monax*. American Midl. Nat., 53(2):257–293.

Hamilton, W. J., Jr.
1934 The life history of the rufescent woodchuck. Ann. Carnegie Mus., 23:85–178.

Harlow, R. F., R. G. Hooper, R. D. Chamberlain, and H. S. Crawford
1975 Some winter and nesting season foods of the common raven in Virginia. Auk, 92(2):298–306.

Hartley, S. L.
1975 Life's cycle. Michigan Out-of-Doors, 29(3):41.

Haugen, A. O.
1944 Highway mortality of wildlife in southern Michigan. Jour. Mammalogy, 25(2):160–170.

Hickie, P.
1940 Cottontail in Michigan. Michigan Dept. Conservation, Game Div., 107 pp.

Holt, E. G.
1929 Midwinter record of the woodchuck in western Pennsylvania. Jour. Mammalogy, 10(1):80.

Howell, A. H.
1915 Revision of the American marmots. N. American Fauna, No. 37, 80 pp.

Hoyt, S. F.
1952 Additional notes on the gestation period of the woodchuck. Jour. Mammalogy, 33(3):388.

Jackson, H. H. T.
1961 Mammals of Wisconsin. Univ. Wisconsin Press, Madison. xii+504 pp.

Johnson, A. M.
1926 Tree-climbing woodchucks again. Jour. Mammalogy, 7(2):132–133.

Kelsey, P. M.
1976 Chuck hunters can advance sportsman-farmer relations. Michigan Out-of-Doors, 30(5):44.

Laundre, J.
1975 An ecological survey of the mammals of the Huron Mountain area. Huron Mt. Wildlife Foundation, Occas. Papers No. 2, x+69 pp.

Liers, E. E.
1951 Notes on the river otter (*Lutra canadensis*). Jour. Mammalogy, 32(1):1–9.

Linduska, J. P.
1947 Longevity of some Michigan farm game animals. Jour. Mammalogy, 28(2):126–129.
1950 Ecology and land-use relationships of small mammals on a Michigan farm. Michigan Dept. Conservation, Game Div., ix+144 pp.

Ludwig, J., and D. E. Davis
1975 An improved woodchuck trap. Jour. Wildlife Mgt., 39(2):439–442.

Lyman, C. P.
1958 Oxygen consumption, body temperature and heart rate of woodchucks entering hibernation. American Jour. Physiol., 194:83–91.

Manville, R. H.
1948 The vertebrate fauna of the Huron Mountains, Michigan. American Midl. Nat., 39(3):615–640.
1949 Highway mortality in northern Michigan. Jour. Mammalogy, 30(3):311–312.
1966 Roadside abundance of woodchucks. American Midl. Nat., 75(2):537–538.

Merriam, H. G.
1966 Temporal distribution of woodchuck interburrow movements. Jour. Mammalogy, 47(1):103–110.
1971 Woodchuck burrow distribution and related movement patterns. Jour. Mammalogy, 52(4):732–746.

Merriam, H. G., and A. Merriam
1965 Vegetative zones around woodchuck burrows. Canadian Field-Nat., 79:177–180.

Murie, A.
1936 Following fox trails. Univ. Michigan, Mus. Zool., Misc. Publ. No. 32, 45 pp.

Ozoga, J. J., and C. J. Phillips
1964 Mammals of Beaver Island, Michigan. Michigan State Univ., Publ. Mus., Biol. Ser., 2(6):305–348.

Rausch, R., and J. Tiner
1946 *Obeliscoides cuniculi* from the woodchuck in Ohio and Michigan. Jour. Mammalogy, 27(2):177–178.

Robb, W. H.
1926 Another tree-climbing porcupine. Jour. Mammalogy, 7(2):133.

Robinson, W. L., and J. K. Werner
1975 Vertebrate animal populations of the McCormick Forest. USDA Forest Service, Res. Paper NC-118, 25 pp.

Saunders, A. A.
1922 More tree-climbing woodchucks. Jour. Mammalogy, 3(4):161.

Scharf, W. C., and K. R. Stewart
1980 New records of Siphonaptera from northern Michigan. Great Lakes Ento., 13(3):165–167.

Schmeltz, L. L., and J. O. Whitaker, Jr.
1977 Use of woodchuck burrows by woodchucks and other mammals. Trans. Kentucky Acad. Sci., 38(1–2):79–82.

Schofield, R. D.
1960 A thousand miles of fox trails in Michigan's ruffed grouse range. Jour. Wildlife Mgt., 24(4):432–434.

Schwartz, C. W., and E. R. Schwartz
1981 The wild mammals of Missouri. Revised Edition. Univ. Missouri Press and Missouri Dept. Conservation, Columbia. viii+356 pp.

Scott, T. G., and W. D. Klimstra
1955 Red foxes and a declining prey population. South. Illinois Univ., Monog. Ser., No. 1, 123 pp.

Seton, E. T.
1953 Lives of game mammals. Vol. IV, Pt. I. Charles T. Branford Co., Boston. xxii+440 pp.

Snyder, R. L., and J. J. Christian
1960 Reproductive cycle and litter size of the woodchuck. Ecology, 41(4):647–656.

Snyder, R. L., D. E. Davis, and J. J. Christian
1961 Seasonal changes in the weights of woodchucks. Jour. Mammalogy, 42(3):297–312.

Trump, R. F.
1950 Home range of the southern woodchuck. Iowa Acad. Sci., 57:537–540.

Twichell, A. R.
1939 Notes on the southern woodchuck in Missouri. Jour. Mammalogy, 20(1):71–74.

Van Ballenberghe, V., A. W. Erickson, and D. Byman
1975 Ecology of the timber wolf in northeastern Minnesota. Wildlife Monogr., No. 43, 43 pp.

Voigt, D. R., G. B. Kolenosky, and D. H. Pimlott
1976 Changes in summer foods of wolves in central Ontario. Jour. Wildlife Mgt., 40(4):663–668.

Wallihan, E. F.
1947 Growth rate of trees in the vicinity of woodchuck burrows. Jour. Forestry, 45:372–373.

Whitaker, J. G., and L. L. Schmeltz
1973 External parasites of the woodchuck, *Marmota monax*, in Indiana. Ento. News, 84:69–72.

Wood, N. A.
1922a The mammals of Washtenaw County, Michigan. Univ. Michigan. Mus. Zool., Occas. Papers No. 123, 23 pp.
1922b Notes on the mammals of Berrien County, Michigan. Univ. Michigan. Mus. Zool., Occas. Papers No. 124, 4 pp.

FRANKLIN'S GROUND SQUIRREL
Spermophilus franklinii (Sabine)

NAMES. The generic name *Spermophilus,* proposed by Cuvier in 1825, is of Greek origin and is derived from two words, *Spermo* meaning "seed" and *philus* meaning "loving" or "having a fondness or affinity for." The specific name *franklinii,* proposed by Sabine in 1822, refers to British explorer Sir John Franklin, who was lost in the Canadian Arctic in 1846. In many publications on mammals the generic name of this ground squirrel will be listed as *Citellus* instead of *Spermophilus;* however, *Citellus,* proposed by Oken in 1816 and thus older than the first usage of *Spermophilus* in 1825 by Cuvier, has been discarded, because *Citellus* does not qualify as a proper generic designation according to the Regales of the International Commission on Zoological Nomenclature. In the Upper Great Lakes Region, local residents sometimes call this animal "gray gopher." The species *Spermophilus franklinii* (Sabine) is monotypic, meaning that geographic variation is judged to be insufficiently discernible so as to formally distinguish subspecies.

RECOGNITION. Body size large for a ground squirrel; head and body of an adult averaging approximately 9 in (230 mm) long, with a short, sparsely bushy tail almost half again as long; body slender and elongate; pelage short and coarse; color of back and sides brownish gray, speckled with black, with a yellowish cast; underfur thin, especially in summer coat; eye-ring white; head and tail grayish, mostly resulting from alternating bands of black and white on individual hairs; belly grayish to whitish, sometimes slightly tawny; feet pale gray; ears short and ovate, in marked contrast to longer ears of tree squirrels. There are two pairs of thoracic, one pair of abdominal, and two pairs of inguinal mammary glands. There are four functioning toes on front feet and five on hind feet. The legs are short, the feet large, and the claws of forefeet short and heavy. Internal cheek pouches are large. Among Michigan sciurids, the Franklin's ground squirrel resembles most closely the gray squirrel, although the latter is slightly larger and has a longer and more bushy tail.

MEASUREMENTS AND WEIGHTS. Adults measure as follows: length from tip of nose to end of tail vertebrae, 13¾ to 16½ in (350 to 420 mm); length of tail vertebrae, 5⅛ to 6¼ in (130 to 160 mm); length of hind foot, 2 to 2¼ in (50 to 57 mm); height of ear from notch, ¾ in (16 to 18 mm); weight, 13.3 to 17.2 oz (380 to 490 g). Males are slightly heavier than females.

DISTINCTIVE CRANIAL AND DENTAL CHARACTERISTICS. The skull of Franklin's ground squirrel is long and narrow, of a more "streamlined" construction than any other area sciurids (Fig. 37). The nasal (rostral) part of the skull is especially long and broad. In dorsal outline,

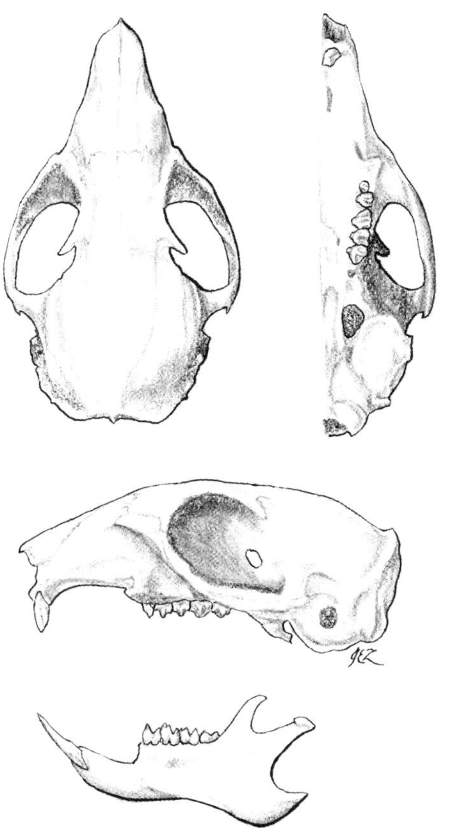

Figure 37. Cranial and dental characteristics of Franklin's ground squirrel (*Spermophilus franklinii*).

the skull also has a more flattened appearance than those of other local sciurids. The auditory meatus (ear opening) is clearly visible from above. In contrast with the width of the interorbital space (between the orbits) of the smaller skull of the thirteen-lined ground squirrel, this space on the skull of Franklin's ground squirrel is broader. The postorbital processes (ridges above and behind the orbits) are prominent and angle backward. The mature skull is approximately 2⅛ in (54 mm) long and 1¼ in (32 mm) wide. The heavy upper incisors are yellowish to deep orange on the front surfaces. The dental formula is:

$$\text{I (incisors)}\frac{1\text{-}1}{1\text{-}1}, \text{C (canines)}\frac{0\text{-}0}{0\text{-}0}, \text{P (premolars)}\frac{2\text{-}2}{1\text{-}1}, \text{M (molars)}\frac{3\text{-}3}{3\text{-}3} = 22$$

DISTRIBUTION. The northern part of the American tallgrass prairie is the home of the Franklin's ground squirrel. In the north, the species occurs from extreme southwest Ontario west to central Manitoba (avoiding shortgrass habitats); southward in the United States from central North Dakota to southcentral Kansas; eastward to westcentral Indiana, and northwestward to the Lake Michigan shore in the Michigan City-Chicago area, southern Wisconsin and central Minnesota. Franklin's ground squirrel has not been collected or observed in Michigan. However, space in this report is given to this species because of its presence in northwestern Indiana in the South Bend area (St. Joseph County) and at Michigan City (LaPorte County) on sand dunes bordering Lake Michigan (Mumford, 1969). This latter record would account for this species as being no more than 1 or 2 mi (1.6 to 3.2 km) from the Indiana-Michigan line or near Lake Michigan at extreme southwestern Berrien County. Numerous local observers have been alerted to watch for evidence of this large and conspicuous ground squirrel with its short salt-and-pepper colored pelage in this county, in sand dune habitat as well as along rights-of-way of north-south highways and on railroad enbankments.

HABITAT PREFERENCES. The Franklin's ground squirrel has an affinity for the tallgrass prairie of northcentral United States and adjacent parts of Canada, especially the edge situation between grassy areas and woody vegetation. This preference allows for its distributional pattern to follow drainage systems along which woody riparian growth extends in dendritic fashion well into otherwise open prairie habitat. With the woody plantings by land operators in shelter belts, along fencerows, and at homesteads plus the natural spread of shrubs and trees as a result of prairie fire control, Franklin's ground squirrel actually may have more extensive living space today within the tallgrass habitat than in presettlement times. On the other hand, forest clearing may have allowed for the eastward spread of Franklin's ground squirrel, perhaps accounting for its presence at the Michigan-Indiana border (de Vos, 1964). However, overgrazing of pasturage and extensive mowing of meadows, cemeteries, parks, and rights-of-way, all conducive to the welfare of the thirteen-lined ground squirrel, probably have discouraged local populations of Franklin's ground squirrel. In Indiana, Lyon (1932) and Mumford (1969) report the species in tall, weedy growth along roadsides and railroad embankments, in open fields and pastures containing brush or herbaceous plants, in a stone pile surrounded by an oat field, in an abandoned gravel pit, and in sand dune areas. Most of these habitats also exist in Berrien County as well as other counties in southwestern Michigan.

DENSITY AND MOVEMENTS. Franklin's ground squirrel does not seem common in any locale as reported by observers from Missouri (Schwartz and Schwartz, 1981) to southern Canada (Banfield, 1974). Some biologists (notably Seton, 1953) suggest that the species was more abundant following settlement. Although Franklin's ground squirrel has often been seen in aggregations, the animal is not considered colonial (Hall, 1955) but prefers to live in loose communities (Gunderson and Beer, 1953). This semigregarious behavior may account for records of as few as four to five animals per acre (10 to 12.5 per ha) to as many as eight per acre (20 per ha) in average habitats (Banfield, 1974; Jackson, 1961). These authors and others also describe multiannual fluctuations in ground squirrel populations. Apparently population peaks occur every four to six years, with populations at these peaks as dense as 30 animals per acre (75 per ha) noted (Schwartz and Schwartz, 1981). Sowls (1948) found large populations of ground squirrels in 1938 at Delta, Manitoba, where on July 2, two hunters shot 14 individuals at one location in less than an hour. In 1943, in the same area, the species was rarely seen.

BEHAVIOR. For a large, diurnal ground squirrel, this species is often inconspicuous, lacking the noisy, "curious" actions of related species. In fact, in recent years, evidence of the Franklin's ground squirrel's presence has come more from individuals run over on highways than by actual field sightings. This may be due in part to the fact that this ground squirrel prefers to skulk in tall ground cover (weeds, grass and shrubs). Conversely, other ground squirrels, like the related prairie dog, thrive in short vegetation which gives them maximum visibility (Hall, 1955). Further, Franklin's ground squirrel does not normally strike the picket-pin upright posture (so characteristic of other ground squirrels) when alarmed; it also rarely stops when the observer imitates its call but instead heads for the burrow (Jackson, 1961).

The classic work on the activities of Franklin's ground squirrel is by Sowls (1948), who was concerned mostly with the influence of this species on nesting success of ducks at the Delta Waterfowl Research Station in Manitoba. Some of his findings are summarized here: Franklin's ground squirrel is most active on bright, sunny days and much less active on drab, cold, and windy days. It was Sowls' estimate that less than 10% of the animal's time was spent above ground. Individuals tend to center their lives in areas usually no more than 300 ft (100 m) in diameter. Ground burrows are the focal points of this activity. Burrows are usually dug by the animal itself and may extend underground as deep as 8 ft (2½ m), with several branches and openings, each of which may be plugged with soil part of the time. Each opening is no more than 3 in (75 mm) in diameter, often with a small mound of soil in front of the entrance. Burrows can be found in well-drained tall grass/weed cover, on rocky slopes, along road rights-of-way, on railroad embankments, and under logs, rock piles, or fences. A nest of herbaceous materials is usually placed in a side branch of the main burrow (Jackson, 1961). If need be, Franklin's ground squirrel can swim. The species also can climb trees; in fact, Hollister (1909) once shot a individual well up in a tamarack tree, mistaking it for a gray squirrel.

In loosely constituted colonies, individual Franklin's ground squirrels appear to maintain some gregarious relationship by means of calls. Kennicott (1857) described its whistle as clear and musical. Jackson (1961) noted that this call is lower in pitch and louder than that of the thirteen-lined ground squirrel and that it is used more when the animals are feeding and less when they are disturbed or on the alert. Although the variety of Franklin's ground squirrel calls has yet to be correlated closely with other activity patterns, Seton (1953) felt that this species' clear whistle makes it the musician of the squirrel family, and points to the common names "whistling gopher" and "musical ground squirrel" given to it.

In late summer, Franklin's ground squirrel begins to take on a heavy layer of fat in preparation for winter sleep. By late September, the species selects an underground cavity for hibernation. As a rule, younger animals will linger above ground longer than older ones. In the latitude of southern Michigan, this ground squirrel emerges in late March or early April. As is the case in other hibernating sciurids, the males appear above ground first with the females following a week or so later. Their distinctive whistling calls may be their first announcements of spring emergence.

REPRODUCTIVE ACTIVITIES. Evidently the first-appearing males waste little time in courting the later-emerging females, with the mating phase of the reproductive process being completed by mid-April. Courtship involves much rivalry among males in their pursuit of females. Musky discharges from anal glands play a role in the sex-attracting process.

The gestation period is approximately 28 days. The young are born in May or June in an underground cavity lined with grasses and forbs. The single annual litter is large, from five to 11 young with an average of seven. This one reproductive effort each year provides sufficient recruits to sustain local populations. At birth, the young are naked and blind. Fuzzy hair appears in about 10 days; the eyes open at 20 days; whistle calls can also be emitted at 20 days; the young venture outside at about 30 days; weaning is completed by 40 days; and the young are almost adult size by hibernating time. There is no evidence of sexual activity by these young until after emergence from hibernation at the end of their first winter.

FOOD HABITS. The ground squirrel's dental characteristics—chisel-like incisors, no canines, and the space between the gnawing front teeth and the chewing and grinding cheek teeth—adapt to a diet of tough vegetable fibers or hard-shelled seeds and fruits which must be reduced to fragments for the digestive process. The species will eat assorted animal foods as well. Most observers report that plant foods comprise two-thirds and animal foods one-third of the spring, summer and autumn diets (Howell, 1936). As plant growth and animal supplies vary through these seasons, so do the food items in the Franklin's ground squirrel's diet. Plant foods include vegetative parts of grasses, clovers, mustard, dandelion, strawberry, thistle and other herbaceous plants; and seeds, fruits, and other reproductive parts of elderberry, blackberry, basswood, needlegrass, cockleburs, nightshade and many other plants plus cultivated crops (corn, oats, wheat, and a variety of garden vegetables). Animal foods identified in one study included beetles, beetle larvae, caterpillars, grasshoppers, crickets, ants, eggs of ants, and a small bird (Bailey, 1893). Seton (1953) noted that this ground squirrel also eats deer mice (*Peromyscus maniculatus*), young eastern cottontails (*Sylvilagus floridanus*), and young domestic fowl. Frogs, toads, birds' eggs, nestling birds, and even other ground squirrels may be eaten. Sowls (1948) provided details of Franklin's ground squirrel's predation on duck eggs and ducklings in Manitoba.

ENEMIES. The red-tailed hawk is a major predator of Franklin's ground squirrel (Seton, 1953). Most authors include in the list of four-footed enemies the red fox, badger, coyote, striped skunk, mink, and long-tailed weasel. The badger is probably most efficient at digging the ground squirrels out of their burrows. Owls may be less important predators than hawks, mainly because the ground squirrel is safely in its burrow when these nocturnal birds of prey are most active.

PARASITES AND DISEASES. Lice and fleas are known ectoparasites of Franklin's ground squirrel. Internal parasites include protozoans, tapeworms, and round worms (Jackson, 1961).

MOLT AND COLOR ABERRATIONS. The short and coarse pelage of the Franklin's ground squirrel is shed and regrown once each year, in early summer. The heavier winter coat is somewhat paler in color than the summer coat. In summer, the underfur is noticeably sparse.

ECONOMIC IMPORTANCE. Franklin's ground squirrel has long been a nuisance to the prairie farmer and gardener. Before the advent of modern control methods, it was often the task of farm boys to snare these "gray gophers" when they lived close by the vegetable garden. In high population years, the animals can be serious competitors for agriculturists' vegetables and grains. Sowls (1948) found that this ground squirrel accounted for the destruction of 19% of the duck nests under study in Manitoba. Young ducklings, especially those which become separated from the brood, also fall victim to ground squirrels.

COUNTY RECORDS FROM MICHIGAN. There are no records for this species but it can be expected to occur in Berrien and possibly Cass Counties in southwestern Michigan. Watch for this large, grayish animal on sand dunes and along highways and railroad embankments on bright days in summer.

LITERATURE CITED

Bailey, V.
1893 The prairie ground squirrels or spermophiles of the Mississippi Valley. U.S. Dept. Agriculture, Div. Ornith. and Mammalogy, Bull. 4, 39 pp.

Banfield, A. W. F.
1974 The mammals of Canada. Univ. Toronto Press, Toronto. xxiv+438 pp.

de Vos, A.
1964 Range changes of mammals in the Great Lakes Region. American Midl. Nat., 71(1):210–231.

Gunderson, H. L., and J. R. Beer
1953 The mammals of Minnesota. Univ. Minnesota Press, Minneapolis. xii+190 pp.

Hall, E. R.
1955 Handbook of mammals of Kansas. Univ. Kansas, Mus. Nat. Hist., Misc. Publ. No. 7, 303 pp.

Hollister, N.
1909 Notes on Wisconsin mammals. Bull. Wisconsin Nat. Hist. Soc., 6(3–4):137–142.

Howell, A. H.
1936 Revision of the North American ground squirrels. U.S. Dept. Agriculture, N. American Fauna No. 56, 256 pp.

Jackson, H. H. T.
1961 Mammals of Wisconsin. Univ. Wisconsin Press, Madison. xii+504 pp.

Kennicott, R.
1857 The quadrupeds of Illinois, injurious and beneficial to the farmer. U.S. Patent Office Rept. (Agri.) for 1856, pp. 52–110.

Lyon, M. W., Jr.
1932 Franklin's ground squirrel and its distribution in Indiana. American Midl. Nat., 13:16–20.

Mumford, R. E.
1969 Distribution of the mammals of Indiana. Indiana Acad. Sci., Monog. No. 1, vii+114 pp.

Schwartz, C. W., and E. Schwartz
1981 The wild animals of Missouri. Revised Edition. Univ. Missouri Press and Missouri Conservation Commission, Columbia. viii+356 pp.

Seton, E. T.
1953 Lives of game animals. Vol. IV, Pt. I. Charles T. Branford Co., Boston. xxii+440 pp.

Sowls, L. K.
1948 The Franklin ground squirrel, *Citellus franklinii* (Sabine), and its relationship to nesting ducks. Jour. Mammalogy, 29(2):113–137.

THIRTEEN-LINED GROUND SQUIRREL
Spermophilus tridecemlineatus (Mitchill)

NAMES. The generic name *Spermophilus,* proposed by Cuvier in 1825, is of Greek origin and is derived from two words, *Spermo* meaning "seed" and *philus* meaning "loving" or "having a fondness or affinity for." The specific name *tridecemlineatus,* proposed by Mitchill in 1821, is of Latin origin and means "thirteen-lined" with reference to the dorsal stripes so characteristic of this species. In many publications, the generic name of this ground squirrel is listed as *Citellus* instead of *Spermophilus;* however, *Citellus,* proposed by Oken in 1816 and thus older than the first usage of *Spermophilus* in 1825 by Cuvier, has been discarded because it does not qualify as a proper generic designation according to the Regales of the International Commission on Zoological Nomenclature. In the Upper Great Lakes Region, local residents commonly call this conspicuous animal "gopher," "striped gopher" or "striped ground squirrel." The subspecies *Sper-*

mophilus tridecemlineatus tridecemlineatus (Mitchill) occurs in Michigan.

RECOGNITION. Body size slight for a ground squirrel; head and body of an adult averaging approximately 6¾ in (175 mm) long, with a short, moderately bushy tail approximately one-third as long as the head and body; body slender and elongate; pelage dense and coarse (especially summer pelage), underfur inconspicuous under longer guard hairs; back and sides from back of head to rump conspicuously marked with thirteen longitudinal stripes, with seven wider dark brown ones (each enclosing a row of buffy, squarish spots) alternating with six narrower buffy stripes; sides of body buff to yellowish; underparts buff to pale cinnamon buff; top of head brown, finely streaked with buff; nose, eye-ring, cheeks, sides of neck, and feet buff; tail yellowish-brown fringed with coarse, blackish hairs with buff tips. The eyes are noticeably large; the ears are small, ovate and set low on the head. The short legs possess rather large feet; the claws on the four toes of the forefeet long and slender; the claws on the five toes of the hindfeet shorter and heavier. There are two pairs of thoracic, one pair of abdominal, and two pairs of inguinal mammary glands with teats. Well-developed cheek pouches open into the sides of the mouth.

Among Michigan sciurids, the thirteen-lined ground squirrel most closely resembles the more colorful eastern chipmunk (*Tamias striatus*) and the least chipmunk (*Eutamias minimus*) in size. However, the longitudinal stripes, although similarly conspicuous in the three mammals, extend onto the face and cheeks only in the two chipmunks. Both chipmunks hold their tails vertically when running, in sharp contrast to the thirteen-lined ground squirrel, which carries its tail straight out behind.

MEASUREMENTS AND WEIGHTS. Adults measure as follows: length from tip of nose to end of tail vertebrae, 8¾ to 12 in (225 to 300 mm); length of tail vertebrae, 3 to 4¼ in (75 to 110 mm); length of hind foot, 1¼ to 1⅝ in (32 to 41 mm); height of ear from notch, ⅜ in (7 to 8 mm); weight of adults, from 4 oz (115 g) in May to near 9 oz (225 g) in September (just prior to hibernation). Males are slightly heavier than females.

DISTINCTIVE CRANIAL AND DENTAL CHARACTERISTICS. The skull of the thirteen-lined ground squirrel is long and delicate with a narrow braincase (Fig. 38). The zygomatic arches are not widely expanded; the postorbital processes (above and behind each orbit) are short and slender; the interorbital space is characteristically

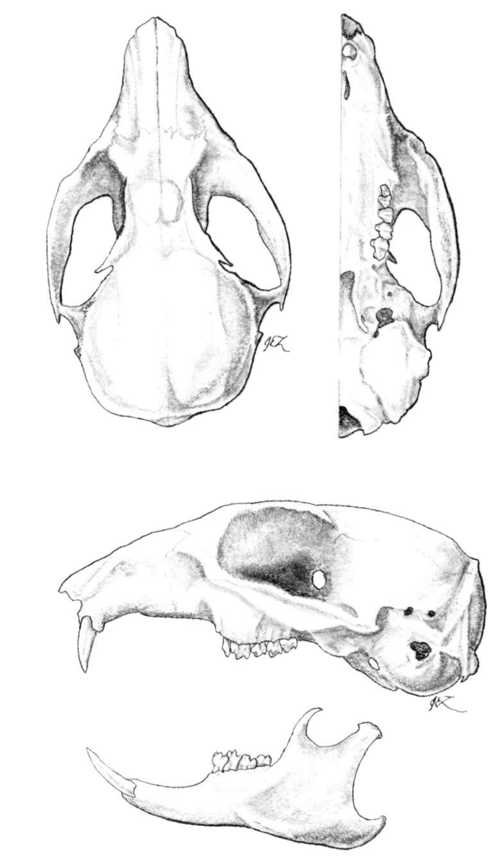

Figure 38. Cranial and dental characteristics of the thirteen-lined ground squirrel (*Spermophilus tridecemlineatus*).

flattened. The snout (rostrum) is long, tapering and depressed anteriorly. The upper maxillary toothrows diverge slightly anteriorly. The upper incisors have yellowish front surfaces. The skull of the thirteen-lined ground squirrel is similar in size to those of the least and eastern chipmunks, and southern and northern flying squirrels (Fig. 31). The thirteen-lined ground squirrel has five cheek teeth (premolars and molars) in each upper jaw whereas the eastern chipmunk (*Tamias striatus*) has only four. The skull of the thirteen-lined ground squirrel is easily distinguished from those of the southern and northern flying squirrels (genus *Glaucomys*), by the absence of deep notches in the interorbital space and by incisors with yellow rather than orange front surfaces. The skull of the thirteen-lined ground squirrel is longer, more than 1⅜ in (34 mm), and wider, more than ¾ in (20 mm) than the skull of the least chipmunk (*Eutamias minimus*). The dental formula for the thirteen-lined ground squirrel is:

$$\text{I (incisors)}\frac{1\text{-}1}{1\text{-}1},\ \text{C (canines)}\frac{0\text{-}0}{0\text{-}0},\ \text{P (premolars)}\frac{2\text{-}2}{1\text{-}1},\ \text{M (molars)}\frac{3\text{-}3}{3\text{-}3} = 22$$

DISTRIBUTION. The thirteen-lined ground squirrel is widespread in both the tallgrass and shortgrass prairies of central United States and southcentral Canada, from southern parts of Manitoba, Saskatchewan, and Alberta south through eastern Montana, to northern New Mexico and southeastward to coastal Texas, and northeastward to western and northern Missouri, western Ohio and Lower Peninsula Michigan, and northwestward through Wisconsin, extreme southwestern Upper Peninsula Michigan, and most of Minnesota (see Map 28). In Lower Peninsula Michigan the species occurs widespread but in the Upper Peninsula is known only from Iron, Dickinson, Menominee and Marquette counties (Burt, 1946; Ozoga and Gaertner, 1963; Ozoga and Verme, 1966; Laundre, 1975).

In presettlement days, the forested regions centered around the western Great Lakes all but excluded prairie plants and animals. After the 1820 Land Law was enacted (Hurley and Franks, 1976), however, settlement of the Michigan area expanded and Lower Peninsula lumbering which began before the mid-nineteenth century reached a high level in the 1870s. The pine forest and much of the beech-maple woodlands were removed, and the resulting cleared lands were burned repeatedly (Kendeigh, 1948). This clearing allowed for the eastern and northern spread of plants (both native and introduced) and animals having open-land affinities (Evans, 1951). The thirteen-lined ground squirrel, which may actually have occurred in early times in sandy openings in the southwestern part of the Lower Peninsula, was one of the species which rapidly moved northward. In 1860, State Zoologist Manly Miles (1861) wrote that this animal "... is very common in the southern counties, but has not been known in the central parts of the state until within a few years past. It is gradually extending its range northward, where the timber has been removed and the land brought under cultivation." By the early part of the twentieth century the species had occupied most of the southern half of the Lower Peninsula (de Vos, 1964). Ultimately, the thirteen-lined ground squirrel moved into cleared and/or cultivated areas in all Lower Peninsula counties (Wood, 1922). It is also reported on Bois Blanc Island in Lake Huron, where it was probably introduced (Barry *in lit.*, December 20, 1973). The animal's entry into the southwestern part of the Upper Peninsula has been traced on the Wisconsin side by Jackson (1961). It was probably the first decade of the twentieth century before it arrived at the Michigan line. However, its spread in the Upper Peninsula has been slow and perhaps more northward than eastward (Ozoga and Verme, 1966).

Map 28. Geographic distribution of the thirteen-lined ground squirrel (*Spermophilus tridecemlineatus*). Michigan counties from where this species is known are shaded.

HABITAT PREFERENCES. The thirteen-lined ground squirrel occurs in a variety of grass/weed environments mostly in sandy soils and in dry, well-drained locations. Although the animal can be found in tall vegetation, it prefers areas where its view is not impaired, such as grassy fields, pastures, hay meadows, and along weedy/grassy fence lines between cultivated fields. The species generally avoids plowed fields (Linduska, 1950). Urban dwellers are most apt to notice these animals scurrying across well-mowed lawns of estates, cemeteries, parks, and on campus grounds. The species may have been encouraged in its rapid spread throughout the Lower Peninsula by the building of railroad embankments and road rights-of-way. Beach habitat (Wood, 1922) and dry pine barrens (Green, 1925) are also acceptable living areas.

DENSITY AND MOVEMENTS. The thirteen-lined ground squirrel has a characteristic spotty distribution because open areas containing short vegetation usually occur in a mosaic pattern, easily seen when one looks out of an airplane while flying across southern Michigan. Of course, scattered woodlots, inhospitable to this ground squirrel, dominate the countryside. Perhaps because of this seemingly disjunct living pattern, most observers (Jackson, 1961) find little evidence of multi-annual fluctuations in thirteen-lined ground squirrel populations. Sometimes a local population will increase noticeably; James Drake (a graduate student at Michigan State University) noted such an increase during the summer of 1956 in the formal gardens surrounding the Kellogg Biological Station headquarters on Gull Lake. In the same area from 1934 to 1937, Allen (1938) noted that the thirteen-lined ground squirrel was the most abundant of larger than mouse-size mammals.

Local densities of this ground squirrel are regulated, according to McCarley (1966), by dispersal of the juveniles in late summer and early autumn. In his three year study in northern Texas, he found that adult populations ranged from one individual per 3 to 4.5 acres (one per 1 to 2 ha) to a peak annual population, after the weaned young and a modest number of transients appeared, of almost two ground squirrels per acre (5 per ha). At the Edwin S. George Reserve in Livingston County, Evans (1951) obtained comparable figures. In the summer of 1950 in a 15-acre field, he found 30–35 thirteen-lined ground squirrels of all ages (2 per acre or 5 per ha). In Ohio, Lishak (1977) estimated a local population to attain as many as 12 adults and juveniles per acre (30 per ha). Other authors (Jackson, 1961; Schwartz and Schwartz, 1981) reported thirteen-lined ground squirrels in densities varying with locality from 1.2 to 20 per acre (3 to 50 per ha).

Thirteen-lined ground squirrels, once established in a home area, are not noted for extensive movements. On the other hand, juveniles are apt to disperse considerably from their birth site to their own established home areas (McCarley, 1966). In Livingston County, Evans (1951) calculated that an individual lived in a space with a radius of no more than 150 ft (50 m) with a total area of 1.6 acres (less than 0.5 ha). Other workers (Gunderson, 1976) report home ranges as smaller for females, up to 3.4 acres (1.4 ha), and larger for males, up to 11.7 acres (4.7 ha).

BEHAVIOR. The thirteen-lined ground squirrel is diurnal and best observed in bright sunshine. It will almost always be secure in its underground burrow on cold, overcast or rainy days. On the campus at Michigan State University, Haigh (1979) found that thirteen-lined ground squirrels orient toward their homesites using sun-compass navigation. Often, the animal will be seen in its characteristic upright "picket-pin" posture in short grass. If the observer begins to approach, the ground squirrel will scurry to a burrow opening, and, if potential danger comes closer, duck into its burrow, uttering a chirping call. If the intruder whistles an imitation of this call, the thirteen-lined ground squirrel may reappear, at least far enough out of its burrow to take a look around. Like Franklin's ground squirrel, this species will also spend only a few hours each day above ground, especially in Michigan where cloudy skies are typical of many summer days. When inclement weather continues for several days, the animal eats stored foods.

The waddling gait of the thirteen-lined ground squirrel is accentuated by its short legs. Despite its low profile when moving, Jackson (1961) notes that it can cover the ground at speeds as high as 8 mi (13 km) per hour by means of a lumbering gallop. It is not a strong swimmer and rarely will enter water voluntarily. Although there are observations of thirteen-lined ground squirrels climbing trees (Morrisey, 1941), this is probably an unusual occurrence.

In general, members of the squirrel family (Sciuridae), except for the prairie dogs (genus *Cynomys*), are rarely considered highly colonial. The thirteen-lined ground squirrel, like the Franklin's ground squirrel, lives in loosely constituted colonies or families in gregarious associations. In fact, McCarley (1966) proposed that the animals tend to congregate because of attraction to a particular habitat rather than to one another. However, he does show that both social and individual behavior have evolved.

Early writers alluded to different kinds of burrows used by this ground squirrel (Howell, 1936; Seton, 1953). Rongstad (1965) concluded that there are basically three types: for nesting, for hiding, and for hibernating. There is overlap in use, especially between the nesting and hibernating burrows. A nesting and/or hibernating burrow is characterized by one entrance (sometimes two, Johnson, 1917), no more than 2 in (50 mm) across. The burrow is designed so that the passageway descends to a level as far below the surface as 6 ft (2 m), shallower in more southerly areas. The passageway may extend for as much as 15 ft (4.6 m) to a nesting chamber, with connected food storage cavities, side tunnels and chambers. A second opening may be dug (usually with a sharp bend) with earth being removed by way of the first opening so that no conspicuous soil is left at the surface of the second entrance. As a rule, the first opening will then be plugged. An extensive burrow like this also contains a special grass-lined side chamber for hibernation.

Nesting and hibernating burrows are elaborate compared to hiding burrows. The latter are shallow, short, and constructed in various parts of the area in which the thirteen-lined ground squirrel is active in summer. In Michigan, Evans (1951) concluded that the average individual excavated three to four of these summer refuges. In the course of a single season, Desha (1966) found one male used a record 15 different burrow entrances. Often these hiding burrows are connected by inconspicuous trails. Each night most ground squirrels plug the entrances to their burrows with soil or grass. Simultaneous use of the same burrow by different squirrels is rare (Desha, 1966).

Like most other sciurids, the thirteen-lined ground squirrel is vocal (Harris, 1967; Matocha, 1977; Schwagmeyer, 1980). Seton (1953) described a variety of whistles and bird-like chirps and suggested that the adults are most vocal when the young first appear above ground in mid-summer. McCarley (1966) was able to discern a high-pitched "alarm" call or trill (given by adults when an intruder approaches), a short, low "peep" call (given by the young when away from the nest and seemingly lost), and a low-pitched "distress" call (usually given by the young when in danger). Lishak (1977) broadcast recorded alarm calls of both young and adult thirteen-lined ground squirrels to elict responses from resident populations in order to census them.

HIBERNATION. Much attention has been given to the hibernating behavior of the thirteen-lined ground squirrel (Hoy, 1875; Johnson and Hanawalt, 1930; Wade, 1930, 1950). As autumn approaches, this ground squirrel takes on a layer of fat just under the skin; it is not unusual for individuals to double their spring weights. Green (1925) found very fat animals in Montmorency County on July 18. Beginning in late September, the fat adults retreat to the hibernating chambers well below ground surface. The less obese adults and the young-of-the-year may remain above ground as much as a month longer (November 1 in Wisconsin; Jackson, 1961). In its grassy cavity, the animal curls into a ball, with the nose turned under to touch the pelvic area of the body. A torpor sets in which greatly reduces life processes.

Observers of hibernating ground squirrels, both in the field and in the laboratory, have shown that body temperature drops from an awake condition of near 100° F (37° C) to a deep-sleep temperature of near 37° F (3° C). The heart beat drops from 200 to 350 per min to as low as 5 per min. Breathing drops from 50 times per min to 4 times per min. When normally active, the thermoregulatory system controlled largely by a biological thermostat in the hypothalmus, a small section of the brain just dorsal to the pituitary gland, keeps the body temperature almost constant. In the hibernation state this biological thermostat readjusts to allow for a reduction in metabolic function; of course this reduction means the stored fat and other body nutrients can be rationed to last throughout the several months of torpor. In fact, Dawe (1960) noted that a hibernating animal survives on about 1/100th of its normal supply of food and oxygen. Within limits, the hibernator is relatively uninfluenced by radiation, sound, light, and slight temperature changes. However, it is highly sensitive to tactile stimuli.

During the period of winter sleep, animals sometimes arouse and even appear above ground (Engels, 1932). Although temperature and other conditions in their deep hibernation chambers must change very little, even during the spring thawing period, the aboveground appearance of the thirteen-lined ground squirrel happens regularly each year at almost the same time. During each of three springs in northern Texas, McCarley (1966) noted these squirrels emerge on March 12. In Nebraska, Wade (1927, 1950) found squirrels appeared one week later; in Minnesota, Beer (1962) first noted them above ground on March 26. Males usually appear earliest (Wade, 1927; McCarley, 1966; Beer, 1962) and by the third week in April, the entire hibernating population has aroused. The "biological clock" which triggers the awakening process has yet to be determined. However, as Wade (1950) commented, the date of ground thaw also determines when it is physically possible for the ground squirrel to dig its way out. At any rate, the awakening process requires considerable energy to accelerate the life processes from the hibernating to the active state—especially the necessity of raising the body temperature. One of the prices paid, of course, is a substantial weight loss (⅓ to ½ of the body weight), no doubt leaving the newly aroused animals in a severe state of hunger.

To the biologist interested in how mammals adapt to their surroundings, hibernation, which is found in only a few mammals in widely unrelated groups, seems one highly successful method for species survival at times of the year when nutrient resources may be low and the need to reproduce is unnecessary. It therefore seems strange that hibernation has not become a way of life in many other mammals.

ASSOCIATES. The thirteen-lined ground squirrel probably has little opportunity to associate with other small mammals. In winter, for example, it hibernates in its underground nest. At the same time, however, its burrow system might very well be invaded by shrews or mice. In summer, the diurnal habits of this squirrel generally separate its period of activity from those of nocturnal mammalian neighbors. Even when such small mammals as the short-tailed shrew and the meadow vole do move around in daytime, they require at least some protective overhead cover whereas the ground squirrel prefers that such vegetation be much shorter. The reason for this preference is the apparent need for the sciurid to have an unobstructed view of its surroundings; hence, the use by this squirrel of mowed lawns.

REPRODUCTIVE ACTIVITIES. Most males are sexually active immediately after emergence from hibernation (McCarley, 1966). Courtship includes aggressive interactions between males in their pursuit of females. When more than one male mates with a receptive female, it is possible that multiple paternity of a single litter may occur (Hanken and Sherman, 1981). The breeding season begins by mid-April (earlier in southern parts of the animal's range) and is over in early May (Rongstad, 1965; for Wisconsin). Generally there is a single annual litter, but there are cases of second matings (McCarley, 1966; Banfield, 1974), perhaps when females fail to conceive during the initial matings or when a litter is lost.

The gestation period is 27 to 28 days in length. Litters are large, numbering as many as 13 but normally about eight (Zimny, 1965). With only 10 teats available for nourishing young, there may be rather rapid mortality when more than 10 young are produced. At the time of birth from mid-May to early June the young are hairless and pink-skinned, have closed eyes, are toothless, weigh about ¼ oz (6–7 g), and are completely helpless in the grass-lined underground nests. At eight days after birth, dark pigment first appears in the skin; at 12 days, the back stripes are distinct, hair growth is noted, and trilling calls are made; at 20 days, upper and lower incisors erupt and the young are able to walk; at 26 days, the eyes begin to open and the young are mostly weaned; and at 31 to 34 days, the young are fully weaned and venture out of their burrows (Wade, 1927; Bridgwater, 1966). At this time, in late June or early July, the young are about one-third grown. Soon they learn to dig their own burrows and thrive on the abundant natural foods. They are more than three-fourths the size of adults by the time of hibernation. On emergence in spring, these yearlings have reached full size and sexual maturity.

FOOD HABITS. The thirteen-lined ground squirrel is a true omnivore, with a liking for both plant and animal foods. Vegetable foods include cultivated plants (oats, wheat, corn, currants, peas, sugar beets, melons, and other garden crops) and roots, herbage, and seeds of such wild species as goosefoot, knotweed, sunflower, ragweed, black locust, dandelion, vetch, and various grasses (Bailey, 1893; Green, 1925; Howell, 1936; Seton, 1953; Whitaker, 1972). Seeds are usually stuffed in the internal cheek pouches for easy transportation to underground caches. In Montmorency County, for example, Green (1925) found a thirteen-lined ground squirrel with cheek pouches bulging with 196 large seed pods of the sleepy catchfly (*Silene antirrhina*).

Animal foods consumed by thirteen-lined ground squirrels consist mostly of insects, including grasshoppers, crickets, ants, caterpillars, beetle larvae and adults, cockroaches, stoneflies, leaf hoppers, and hemipterous insects (Bailey, 1893; Whitaker, 1972). Whitaker also reported that thirteen-lined ground squirrels in Indiana ate earthworms and diplopods. Both Burt (1946) and Yeager (1937) found these squirrels eating white grubs (beetle larvae) in Michigan. The thirteen-lined ground squirrel also consumes, as the opportunity is presented, assorted kinds of vertebrates; these include bird eggs, young birds, lizards, young chickens, small snakes, short-tailed shrews, young eastern cottontails, and small rodents (B. Bailey, 1923; V. Bailey, 1893; Bridgwater and Penny, 1966; Seton, 1953; Whitaker, 1972). Thirteen-lined ground squirrels are also known to eat their own young as well as the bodies of animals crushed on paved roadways.

ENEMIES. Despite producing only a single litter per year, the thirteen-lined ground squirrel does an adequate job of maintaining its numbers, though a variety of predators are known to capture and eat the animals. Even so, individuals manage to survive for several years; McCarley (1966), for example, noted that 40% of the females captured at his Texas study area in 1962 were still present in 1965. Perhaps the most persistent enemy of the thirteen-lined ground squirrel is the badger, which is highly efficient at digging out this prey (Leedy, 1947). Also able to invade this squirrel's burrows effectively in Michigan are the weasels (*Mustela* spp.) and snakes. Owls, which hunt at night, rarely catch the diurnal ground squirrels (Wallace, 1948); however, hawks are active predators. Craighead and Craighead (1956) and Linduska (1950) noted successful catches of Michigan ground squirrels by the marsh hawk, red-tailed hawk and Cooper's hawk. Additional Michigan enemies of the thirteen-lined ground squirrel include red fox (Schofield, 1960), badger (Dearborn, 1932), coyote, gray fox, skunk, domestic cat, great blue heron, kestrel, shrikes, and crow (Banfield, 1974; Fitzpatrick, 1925; Jackson, 1961; Peifer, 1979; Schwartz and Schwartz, 1981).

PARASITES AND DISEASES. The prevalence of disease in the thirteen-lined ground squirrel is poorly documented. However, the animal does carry a high burden of protozoans, cestodes, nematodes, thorny-headed worms, and botfly larvae internally, and mites, ticks, lice, and fleas externally (Jackson, 1961; Scharf and Stewart, 1980; Whitaker, 1972).

MOLT AND COLOR ABERRATIONS. There are two annual molts. The more brownish summer coat grown in June is molted in late summer to allow for the appearance of the grayer winter coat. There have been few reports of albinistic individuals. Hweston (1962) found a dark melanistic animal in North Dakota; others have been found in Ohio (Goslin, 1959).

ECONOMIC IMPORTANCE. A detailed account of some of the agricultural crops relished by the thirteen-lined ground squirrel is given under Food Habits. A most serious charge by farmers is that the animals dig up and eat planted corn kernels (Howell, 1936). However, since seeds and other parts of many noxious weed pests also are parts of the diet, many observers conclude that these small ground squirrels do as much good as harm to the farmer. Even so, the thirteen-lined ground squirrel has been considered a nuisance, and in years past, some states or counties have authorized bounties for ground squirrel tails.

There are numerous ways to capture these animals in traps and snares (see Jackson, 1961; Lishak, 1976; Beer, 1959; Burnett, 1917, and others). In farming areas and on golf courses and cemetery grounds, these small squirrels are easily captured in rat traps baited with rolled oats. Placing poisoned grain in the burrows early in the summer season also effectively removes unwanted animals. The method most farm boys learn is to snare them with a loop of stout cord which can be tightened about the squirrel's neck when it sticks its head out of its burrow. The patience needed to accomplish this task has been described by both Seton (1953) and Garland (1899). Garland remarked, snaring "... gophers was like fishing, an excuse for enjoying the prairie." Whether there is need for snaring the animals or not, watching their antics on a bright, sunny, summer day is as engaging as bird-watching.

COUNTY RECORDS FROM MICHIGAN. Records of specimens of the thirteen-lined ground squirrel in collections of museums and from the literature show that this species lives in all parts of the Lower Peninsula with extant specimens preserved from most counties. In the Upper Peninsula, this ground squirrel has been recorded in only four western counties: Dickinson, Iron, Marquette, and Menominee.

LITERATURE CITED

Allen, D. L.
1938 Ecological studies on the vertebrate fauna of a 500-acre farm in Kalamazoo County, Michigan. Ecol. Monog., 8(3):347–436.

Bailey, B.
1923 Meat-eating propensities of some rodents of Minnesota. Jour. Mammalogy, 4(1):129.

Bailey, V.
1893 The prairie ground squirrels or spermophiles of the Mississippi Valley. U.S. Dept. Agri., Div. Ornith. and Mammalogy, Bull. 4, 39 pp.

Banfield, A. W. F.
1974 The mammals of Canada. Univ. Toronto Press, Toronto. xxiv+438 pp.

Beer, J. R.
1959 A method of trapping ground squirrels. Jour. Mammalogy, 40(3):445.
1962 Emergence of thirteen-lined ground squirrels from hibernation. Jour. Mammalogy, 43(1):109.

Bridgwater, D. D.
1966 Laboratory breeding, early growth, development and behavior of *Citellus tridecemlineatus* (Rodentia). Southwestern Nat., 11(3):325–337.

Bridgwater, D. D., and D. F. Penny
1966 Predation by *Citellus tridecemlineatus* on other vertebrates. Jour. Mammalogy, 47(2):345–346.

Burnett, W. L.
1917 Suggestions for combating prairie dogs, ground squirrels and other rodent pests. Office of State Entomologist, Fort Collins, Colorado. Circular 24, 6 pp.

Burt, W. H.
1946 The mammals of Michigan. Univ. Michigan Press, Ann Arbor. xv+288 pp.

Craighead, J. J., and F. C. Craighead, Jr.
1956 Hawks, owls and wildlife. The Stackpole Co., Harrisburg, Pennsylvania. xix+443 pp.

Dawe, A. R.
1960 Natural mammalian hibernation. Naval Research Rev., Aug., pp. 12–15.

Dearborn, N.
1932 Foods of some predatory, fur-bearing animals in Michigan. Univ. Michigan, Sch. For. and Cons., Bull, No. 1, 52 pp.

Desha, P. G.
1966 Observations on the burrow utilization of the thirteen-lined ground squirrel. Southwestern Nat., 11(3):408–410.

de Vos, A.
1964 Range changes of mammals in the Great Lakes Region. American Midl. Nat., 71(1):210–231.

Engels, W. L.
1932 Mid-winter activity in a striped spermophile. Jour. Mammalogy, 13(2):164–165.

Evans, F. C.
1951 Notes on a population of the striped ground squirrel (*Citellus tridecemlineatus*) in an abandoned field in southeastern Michigan. Jour. Mammalogy, 32(4):437–449.

Fitzpatrick, F. L.
1925 The ecology and economic status of *Citellus tridecemlineatus* (Mitchill). Iowa State Col. Studies, Nat. Hist., 11(1):1–40.

Garland, H.
1899 Boy life on the prairie. Harper and Bros., New York. 389 pp.

Goslin, R. M.
1959 Melanistic ground squirrels from Ohio. Jour. Mammalogy, 40(1):145.

Green, M. M.
1925 Notes on some animals of Montmorency County, Michigan. Jour. Mammalogy, 6(3);173–178.

Gunderson, H. L.
1976 Mammalogy. McGraw-Hill Book Company, New York. viii+483 pp.

Haigh, G. R.
1979 Sun-compass orientation in the thirteen-lined ground squirrel, *Spermophilus tridecemlineatus*. Jour. Mammalogy, 60(3):629–632.

Hanken, J., and P. W. Sherman
1981 Multiple paternity in Belding's ground squirrel litters. Science, 212(4492):351–353.

Harris, J. P.
1967 Voice and associated behavior in *Citellus tridecemlineatus* and other ground squirrels. Univ. Michigan, unpubl. Ph.D. dissertation, 99 pp.

Howell, A. H.
1936 Revision of the North American ground squirrels. N. American Fauna No. 56, 256 pp.

Hoy, P. R.
1875 On hibernation as exhibited in the striped gopher. Proc. Amer. Assoc. Adv. Science, 24:148–150.

Hurley, R. J., and E. C. Franks
1976 Changes in the breeding ranges of two grassland birds. Auk, 93(1):108–115.

Hweston, J.
1962 A melanistic 13-lined ground squirrel. Jour. Mammalogy, 43(1):107.

Jackson, H. H. T.
1961 Mammals of Wisconsin. Univ. Wisconsin Press, Madison. xii+504 pp.

Johnson, G. E.
1917 The habits of the thirteen-lined ground squirrel. Univ. North Dakota, Quart. Rev., 7(3):261–271.

Johnson, G. E., and V. B. Hanawalt
1930 Hibernation of the thirteen-lined ground squirrel, *Citellus tridecemlineatus* (Mitchill). American Nat., 64(296):272–284.

Kendeigh, S. C.
1948 Bird populations and biotic communities in northern lower Michigan. Ecology, 29:101–114.

Laundre, J.
1975 An ecological survey of the mammals of the Huron Mountain area. Huron Mountain Wildlife Foundation, Occas. Papers, No. 2, x+69 pp.

Leedy, D. L.
1947 Spermophiles and badgers move eastward in Ohio. Jour. Mammalogy, 28(3):290–292.

Linduska, J. P.
1950 Ecology and land-use relationships of small mammals on a Michigan farm. Michigan Dept. Conservation, Game Div., ix+144 pp.

Lishak, R. S.
1976 A burrow entrance snare for capturing ground squirrels. Jour. Wildlife Mgt., 40(2):364–365.
1977 Censusing 13-lined ground squirrels with adult and young alarm calls. Jour. Wildlife Mgt., 41(4):755–759.

Matocha, K. G.
1977 The vocal repertoire of *Spermophilus tridecemlineatus*. American Midl. Nat., 98(2):482–487.

McCarley, H.
1966 Annual cycle, population dynamics and adaptive behavior of *Citellus tridecemlineatus*. Jour. Mammalogy, 47(2):294–314.

Miles, M.
1861 A catalogue of the mammals, birds, reptiles, and mollusks of Michigan. Geological Survey of Michigan, First Biennial Rept., pp. 219–241.

Morrisey, T. J.
1941 Ground squirrel in an oak sapling. Jour. Mammalogy, 22(1):88.

Ozoga, J. J., and R. Gaertner
1963 Noteworthy locality records for some Michigan mammals. Jack-Pine Warbler, 41(2):89–90.

Ozoga, J. J., and L. J. Verme
1966 Noteworthy locality records of some Upper Michigan mammals. Jack-Pine Warbler, 44(1):52.

Peifer, R. W.
1979 Great blue herons foraging for small mammals. Wilson Bull., 91(4):630–631.

Rongstad, O. J.
1965 A life history study of thirteen-lined ground squirrels in southern Wisconsin. Jour. Mammalogy, 46(1):76–87.

Scharf, W. C., and K. R. Stewart
1980 New records of Siphonaptera from northern Michigan. Great Lakes Ento., 13(3):165–167.

Schofield, R. D.
1960 A thousand miles of fox trails in Michigan's ruffed grouse range. Jour. Wildlife Mgt., 24(4):432–434.

Schwagmeyer, P. L.
1980 Alarm calling behavior of the thirteen-lined ground squirrel, *Spermophilus tridecemlineatus*. Behav. Ecol. Sociobiol., 7(3):201–205.

Schwartz, C. W., and E. R. Schwartz
1981 The wild animals of Missouri. Revised Edition. Univ. Missouri Press and Missouri Conservation Commission, Columbia. viii+356 pp.

Seton, E. T.
1953 Lives of game animals. Vol. IV, Pt. I. Charles T. Branford Co., Boston. xxii+440 pp.

Wade, O.
1927 Breeding habits and early life of the thirteen-striped ground squirrel, *Citellus tridecemlineatus* (Mitchill). Jour. Mammalogy, 8(4):269–276.
1930 The behavior of certain spermophiles with special reference to aestivation and hibernation. Jour. Mammalogy, 11(2):160–188.
1950 Soil temperatures, weather conditions, and emergence of ground squirrels from hibernation. Jour. Mammalogy, 31(2):158–161.

Wallace, G. J.
1948 The barn owl in Michigan. Its distribution, natural history and food habits. Michigan State Coll., Agri. Exp. Sta., Tech. Bull. No. 208, 61 pp.

Whitaker, J. O., Jr.
1972 Food and external parasites of *Spermophilus tridecemlineatus* in Vigo County, Indiana. Jour. Mammalogy, 53(3):644–648.

Wood, N. A.
1922 The mammals of Washtenaw County, Michigan. Univ. Michigan, Mus. Zool.., Occas. Papers No. 123, 23 pp.

Yeager, L. E.
1937 Thirteen-lined ground squirrel feeds on white grubs. Jour. Mammalogy, 18(2):243.

Zimny, M. L.
1965 Thirteen-lined ground squirrels born in captivity. Jour. Mammalogy, 46(3):521–522.

GRAY SQUIRREL
Sciurus carolinensis Gmelin

NAMES. The generic name *Sciurus,* proposed by Linneaus in 1758, is from the Latin meaning "squirrel" and is derived from the Greek with the first syllable *Sci* from *skia* meaning "shadow" or "shade" and the last syllable *urus* from *oura* meaning "tail." The specific name *carolinensis,* proposed by Gmelin in 1788, means "of Carolina," the place from which the species was originally described. In Michigan, the melanistic phase of the gray squirrel is known as the black squirrel. The term cat squirrel, a popular name in southern states, is rarely heard in Michigan. Gray squirrels in the Upper Peninsula belong to the subspecies *Sciurus carolinensis hypophaeus* Merriam, and show closest affinities to those in adjacent Wisconsin. In the Lower Peninsula, the subspecies is *Sciurus carolinensis pennsylvanicus* Ord, reflecting relationships with populations in more southern states. These two designations point to the effectiveness of the Straits of Mackinac as a barrier to north-south dispersal of gray squirrels (see Table 12).

RECOGNITION. Body size medium for a tree squirrel, about as large as an adult Norway rat (Fig. 39); head and body of an adult averaging approximately 10½ in (270 mm) long, with a bushy, well-haired tail almost half again as long; body slender and elongate, with prominent ears without tufts; pelage soft and dense, with heavy dark gray underfur and much longer, banded guard hairs. Color somewhat variable but generally grizzled gray on the back with more yellowish brown tones on the head, mid-back, sides and upper surfaces of feet; chin and under parts whitish to brownish-yellow with color of upperparts encroaching on each side, somewhat variable depending on season; eye-ring and spot behind ears whitish; tail hairs banded, basally tan, medially blackish, and terminally white, individual hairs long and slightly wavy; sexes colored alike. Young are generally grayer than adults. In winter, upperparts are washed with silvery with ears silvery-tipped, the pelage is long and thick, and the soles of the feet are haired.

Besides the overall grayish and grayish brown coloring of normal individuals, there is the common black (melanistic) phase in Michigan, sometimes a reddish (erythristic) phase, and only rarely a white (albinistic) phase. The gray squirrel has no internal cheek pouches. The first digit of the forefoot is minute; the other digits have short, flattened, curved and sharp claws as aids in tree-climbing. There are eight mammae present in the females. The gray pelage and slender build distinguish the gray squirrel from its reddish and heavier built close relative, the fox squirrel.

MEASUREMENTS AND WEIGHTS. Adults measure as follows: length from tip of nose to end of tail vertebrae, 16½ to 21 in (420 to 535 mm); length of tail vertebrae, 8 to 10 in (200 to 250 mm); length of hind foot, 2⅜ to 3 in (60 to 74 mm); height of ear from notch, 1 to 1⅜ in (25 to 35 mm);

Figure 39. The gray squirrel (*Sciurus carolinensis*). Drawing by Bonnie Marris.

weight of adults, 12 to 24 oz (340 to 680 g). Males are slightly heavier than females.

DISTINCTIVE CRANIAL AND DENTAL CHARACTERISTICS. The skull of the gray squirrel is long and broad, with a high, rounded braincase, especially anteriorly (Fig. 40). It is approximately 2⅜ in (61 mm) long and 1⅜ in (34

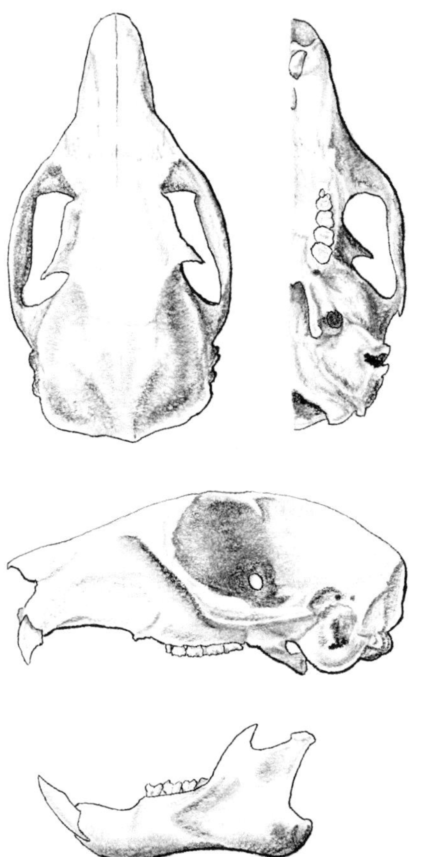

Figure 40. Cranial and dental characteristics of the gray squirrel (*Sciurus carolinensis*).

mm) wide. The rather regular and smooth lines of the dorsal aspect of the skull are typical of sciurids. Skulls of many other groups of rodents are irregular, rugose and sculptured. Characteristic also are the short and narrow rostrum (nasal area) and the wideness between the orbits. Except for the large skull of the woodchuck, those of the gray squirrel and slightly larger fox squirrel are the largest of any Michigan members of the squirrel family. The skulls of the latter two can be distinguished by the slightly smaller size of that of the gray squirrel and the presence of a very small peg-like upper (third) premolar directly in front and to the lingual (tongue) side of the first large tooth on each side of the upper jaw of the gray squirrel. This tiny tooth is present in all but about 1% of the skulls of gray squirrels and is always absent in those of fox squirrels. When exposed to fluorescent light, the pinkish bones of the fox squirrel give off a glow, whereas those of the gray squirrel do not. This distinction results from a skeletal chromatophore called uroporphyrin I in the bones of the fox squirrel (Levin and Flyger, 1971). However, the closeness of these two squirrels is shown by their similar kinds of chromosomes (Nadler and Sutton, 1967). The dental formula is:

$$\text{I (incisors)}\frac{1\text{-}1}{1\text{-}1}, \text{C (canines)}\frac{0\text{-}0}{0\text{-}0}, \text{P (premolars)}\frac{2\text{-}2}{1\text{-}1}, \text{M (molars)}\frac{3\text{-}3}{3\text{-}3} = 22$$

DISTRIBUTION. The gray squirrel occupies most of eastern North America, extending westward to the prairies (in the vicinity of the 100th meridian) and northward into southern Canada (see Map 29). Land-clearing and lumbering activi-

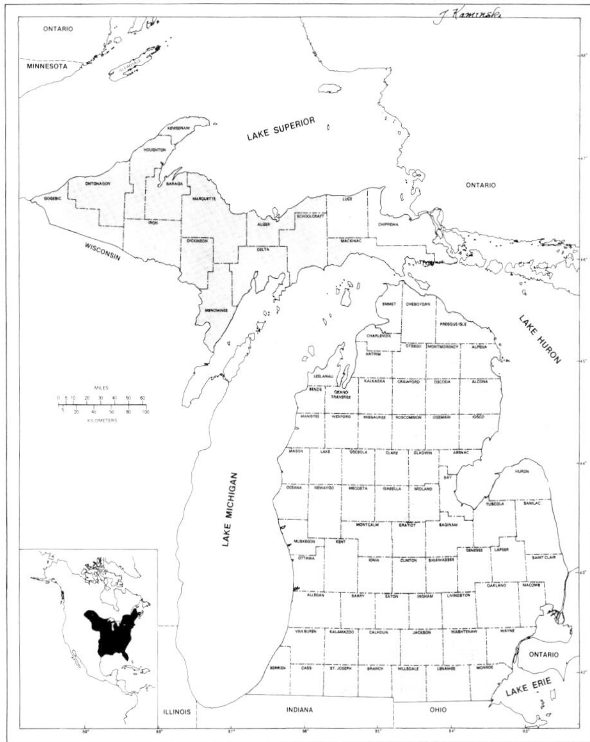

Map 29. Geographic distribution of the gray squirrel (*Sciurus carolinensis*).

ties which began in the east and spread slowly westward reduced, in 150 years, the mixed hardwood forests to isolated woodlots and stream-following groves and severely limited the quality of woodlands for the gray squirrel. Concerning the general decline of the gray squirrel, Baker (1959) commented, "Some comfort may be derived from the following statements: (1) The species still occurs in its pre-settlement range, although its distribution often is discontinuous and spotty; (2) favored habitat is becoming reestablished slowly, especially along streams; (3) the gray squirrel has become resident in groves of nut-bearing shade trees in numerous communities, especially in city parks."

In Michigan, the gray squirrel was the conspicuous arboreal squirrel when the early settlers arrived, even being found in the fossil record in postglacial times (Holman, 1975). Apparently only in southwestern Michigan did it share part of its geographic range with the fox squirrel (Allen, 1943). Gray squirrels were possibly less attracted to extensive pine stands, preferring the hardwoods, and lumbering interests initially focused on white and red pine and other conifers. However, when cutting of the hardwoods began in earnest (mostly from 1890 to 1910, according to Allen, 1943), gray squirrel populations declined drastically.

Today, the gray squirrel still occupies all counties in the state, although its pattern of distribution is a patchwork one, since suitable forested habitat, some of it regrown, is scattered (Marvin Cooley, pers. comm.). Naturally occurring populations have spread from residual stock and have been augmented by the release of live-trapped individuals, always from Michigan but mostly of the black color phase. Successful introductions of both gray and black phases have been made on Beaver and High islands in the Beaver Island Group in Lake Michigan (Hatt *et al.*, 1948, Ozoga and Phillips, 1964) and on Bois Blanc Island in Lake Huron (Berry, 1971). The black phase was introduced at the Kellogg Bird Sanctuary in Kalamazoo County (Johnson, 1973), and in the East Lansing area of Ingham County (Baker, 1973).

HABITAT PREFERENCES. Throughout much of forested eastern United States, the fox squirrel often occupies the "upland" woods, the second growth, and farm woodlots, while the gray squirrel is confined to the "big" timber along streams, especially where there is broad river bottom habitat (Barkalow and Shorten, 1973). One ideal habitat, according to Pack *et al.* (1967), is "overmature" oak-hickory woodlands, with an age of 150 to 200 years. Although second-growth timber is now maturing, much of it seems more capable of supporting fox squirrels than gray squirrels. Fortunately, groves of nut-bearing shade trees established in city parks and along community streets now provide areas where gray squirrels can thrive.

The gray squirrels' attraction to mature groves of nut-bearing (and cavity-filled) oaks, hickories, and beeches has been cited by the numerous field observations of Michigan naturalists. The animals have been noted in forest stands of beech, beech-maple, hemlock-hardwoods, red and white pine, oaks, and mixed hardwoods (Allen, 1943; Laundre, 1975; Johnson, 1973; Ozoga and Phillips, 1964; Manville, 1948; Wood, 1922a, 1922b). Although the gray squirrel is slowly returning to areas where the second-growth forests are maturing, the mosaic patterns will prevail where suitable woodlands are scattered and frequently logged. However, as Beckwith (1954) suggested, the gray squirrel in Michigan is overshadowed by the fox squirrel in oak-hickory stands but comes to the fore when the beech-basswood-maple forest stage is reached.

DENSITY AND LOCAL MOVEMENTS. Early accounts, as summarized by Allen (1943), Baker (1959), Jackson (1961), Seton (1953) and others, call attention to the staggering numbers of gray squirrels which concentrated in the mixed hardwood forests from the Atlantic Seaboard to eastern Iowa in the last century. In Michigan, for example, Seton (1953) quotes W. B. Mershon of Saginaw as saying that in 1860, ". . . the woods around Saginaw were full of gray and black squirrels. One could sit down under a hickory tree or an oak in September, and shoot from that one tree 12 to 15 squirrels." Wenzel (1912) reported that in the early 1900s in Osceola County, hunters could shoot 10 to 20 squirrels in a day. D. L. Allen (1943) writes of remarkable kills of these squirrels in Eaton County beginning in 1866; by the turn of the century, three hunters keeping kill records found the take of grays had diminished and that of fox squirrels had increased. Such high numbers undoubtedly reflected localized concentrations of squirrels attracted by a bountiful food supply (Sharp, 1959) or aggregations involved in cross country movements (see Mass Movements).

As a rule, gray squirrels live most of their lives in and around a single nest tree, perhaps moving no more than a few hundred yards in the course of any season (Brown and Yeager, 1945; Cordes and Barkalow, 1972; Flyger, 1960; Jackson, 1961; Uhlig, 1956). However, since gray squirrels are arboreal, this movement is three dimensional and their in-tree movements are more difficult to ascertain, because most movement data have been obtained by live-trapping, marking, releasing and re-live-trapping on or near the ground (Bouffard and Hein, 1978). Some movement information has also been obtained through the field identification of individuals marked with dyes (Nyanzol A, Flyger, 1960, and Nyanzol D, Pack *et al.*, 1967) and through directional radio signals from gray squirrels wearing small transmitter equipped collars.

Although gray squirrels usually tend to maintain a fairly restricted area of activity, seasonal changes in food supplies or local scarcity of acorns or nuts might cause movements. There is evidence that gray squirrels will cover as much as 5 mi (8 km) in foraging, even moving from one woodlot to another across as much as ½ mi (1 km) of open space (Longley, 1963). The homing instinct also appears to be strong in these squirrels, with marked animals released as far as 3 mi (5 km) away from the places of capture able to return home (Schwartz and Schwartz, 1981).

Today, observers generally agree that gray squirrel densities rarely average two per acre (5 per ha) when viewing total populations in large wooded tracts. In Minnesota, Longley (1963) found densities of one gray squirrel per acre (2.5 per ha) in September and only 0.65 per acre (1.5 per ha) in winter. Thompson (1978) concluded that the behavior of the established adult population toward young individuals is a factor in regulating local population numbers, influencing both recruitment and dispersal of the annual production. Because of the reduction in quality of much gray squirrel habitat and the patchy condition of the currently favorable living places, there now seems little chance for observing the astounding numbers of these prolific squirrels as were recorded in Michigan a century or more ago.

BEHAVIOR. Tree squirrels are among the least social members of the sciurids (Fitzwater and Frank, 1944; Hazard, 1960), and rarely congregate except when mating or when attracted by an abundant food supply. Some authors, however, note gregarious behavior and correlate this with the apparent lack of territoriality (defense of living area) in the species. The exception is the lactating female defending her den tree in which young are located (Sharp, 1959). On the other hand, the general reluctance of gray squirrels to socialize is illustrated by Ackerman and Weigl (1970). These authors found that captive gray squirrels might share a nest box with a red squirrel but not with another gray (for exceptions, see Sharp, 1959). In nature, there are some records of several gray squirrels being found in one tree den (Bakken, 1959), but the tendency for individuals to live alone seems strong. Based on social interactions observed mostly at feeding stations, several authors (Bakken, 1959; Pack *et al.*, 1967) have noted that gray squirrels can be ranked in a social hierarchy. Males are more dominant than females, and older animals are more dominant than younger ones. Dominant males are, according to these authors, responsible for the majority of observed sexual contacts. Also, these authors determined that the size of individuals' minimum home ranges increased somewhat with social rank.

Gray squirrel watching has been a favored occupation of many naturalists because this highly visible species is diurnal in habits and is active in both summer and winter. Detailed studies of the gray squirrel's seasonal activities (Thompson, 1977) show that winter feeding activity is confined to a single midday period, generally the warmest time. In summer, the longer days allow for two feeding periods (morning and afternoon). This bimodal activity in summer also permits the gray squirrel to avoid the midday heat as well as the accompanying extreme light intensities. While females are most conspicuously active during spring and summer, males, according to Bakken (1959) and Thompson (1977), are more conspicuous in winter.

In the course of their activities, gray squirrels are apt to spend as much time on the ground as in trees. The animals walk along slowly when foraging in leaf litter, but they can also move as fast as 15 mi (25 km) per hour by bounding along with leaps of up to 5 ft (almost 2 m). Their agile movements through the trees include leaps between branches of as much as 6 ft (2 m) and drops of up to 15 ft (5 m) if the animal is startled or pursued. In escaping, gray squirrels often seek high branches, flatten out on top of them, and become immobile. On other occasions, gray squirrels effectively elude enemies by scurrying through understory, usually toward their home tree.

Gray squirrels use two types of arboreal refuges: dens in tree cavities, and leaf nests. Tree cavities are usually high in mature nut-bearing trees, as much as 40 ft (13 m) above ground. At least two to three trees—with suitable cavities—per acre (6 to 8 per ha) appear important for optimum gray squirrel habitat (Nixon *et al.*, 1980). The opening is at least 3 in (80 mm) across and the cavity is more than 12 in (0.3 m) deep. This refuge can be either an enlarged woodpecker hole or an opening resulting from rot or a broken branch. Shredded plant fibers and leaves are nest materials. Nests soiled by urine and feces or infested with ectoparasites can be cleaned out or temporarily abandoned. Forest management, in which hollow and overaged trees are weeded out, and the wasteful cutting of a tree to extract a wounded squirrel from a hollow have contributed to the paucity of tree cavities in many woodland habitats. However, the erection of artificial nest boxes (made of rough, untreated boards and measuring about 10 × 10 × 20 in (25 × 25 × 50 cm) on tree trunks 10–25 ft (3–8 m) above ground can be usable substitutes for natural tree cavities (Barkalow and Soots, 1965; and others). For gray squirrels, however, artificial nest boxes do not seem to be used unless forest stands are more than 25 years old (Nixon, 1979).

Nests of twigs and leaves are usually situated at least 20 to 30 ft (7–10 m) above ground in mature oaks, hickories, maples or elms. In more boreal woodlands, especially in northern areas, nests are frequently found in hemlock and white pine. Most of these nests sites are in forks where a substantial branch grows out from the main tree trunk. These refuges are rarely placed on smaller limbs where wind might disturb the nest organization, although grapevine tangles are favored sites (Sanderson *et al.*, 1980). Leaf nests, when viewed from ground level, may appear to be loosely contrived, but there is usually a well-constructed and "weatherproofed" roof over a ball-like nest. Entry to the inner chamber is from the side; at least one observer (Packard, 1956) found most openings facing east. Grasses and finely shredded leaves line this inner chamber. Leaf nests are constructed during the warmer months, usually by midsummer. In Michigan latitudes, leaf nests are normally occupied only in summer and generally fall apart in winter. One way to determine if one of these structures is currently being used is to view it from the ground with field glasses, watching for cobwebs especially over the entry to the inner nest cavity.

There is some correlation between the location of leaf nests and summer food supplies (Brown and Yeager, 1945). Spring-born juveniles construct their own leaf nests by the time they are about 18 weeks old (Uhlig, 1955). Several workers have suggested that young animals are the major nest builders because tree holes are already occupied by the anti-social adults. However, adult squirrels will leave tree holes and use leaf nests when the cavities become heavily infested with fleas (Fitzwater and Frank, 1944; Lowery, 1974). The number of leaf nests present, if censused annually at the same time of year in the same area, may give the observer some indication of fluctuations in population size (Goodrum, 1940; Allen, 1952). Uhlig (1956) thought that the juvenile gray squirrel population could be estimated by assuming that there was one juvenile for every 1.5 leaf nests counted.

Arboreal life has undoubtedly played a major role in special behavioral development in tree squirrels. A good climber, for example, needs excellent senses of sight, balance, and touch to judge distances for jumps and maintain equilibrium and clawhold while negotiating various arboreal pathways or feeding on buds and fruits at the end of delicate and swinging branchlets. Smell is important for food detection and sex recognition. Hearing is also important to the forest dweller, especially in summer in deciduous woodlands where visibility is impaired by thick leaves in the upperstory of the trees and often in brushy foliage of ground cover.

The gray squirrel has its noisy moments (Bakken, 1959; Seton, 1953). There are "buzzing" calls associated with mating chases; "barking," "chucking," and "screaming" calls associated with alarm or warnings; and "mews" (hence the name cat squirrel), "whistles" and "purrs" associated with group interactions. Nestlings utter squeaks, usually as distress calls, but also produce growls, screams, tooth chatter, and lip smacking. Closely related are postures and tail signals (identified by Bakken, 1959). There are rapid stiff or flexible back-and-forth jerks or waves of the tail used to warn or intimidate; and short, back-and-forth flips and circular wavings involved with mating chases and other interactions between sexes. Foot stamping and tail presses over back are postures associated with disturbance and unidentified objects or intruders. To the hunter and outdoorperson, the nervous "barking" fuss made by a gray squirrel, often shaking its tail in a head-down position on a tree trunk, is most characteristic.

The heavily-furred tail of the gray squirrel plays a significant role in behavioral interactions between individuals. This structure is used remarkably as a balancing device in climbing, in running along branches, and in jumping—even as a "parachute" to help break an unintentional fall. The tail also serves as a warm muff in cold weather, as a sunshade on bright days, and even as a rudder in swimming.

MASS MOVEMENTS. There are many records of mass movements of gray squirrels. The attraction of individual squirrels from surrounding parts of a woodland to a concentration of food seems to be a usual occurrence in the annual cycle of the gray squirrel. However, the gathering together of hundreds—even thousands—of these squirrels to move en masse through the countryside and across open areas, streams and other "barriers" is a strange but highly authenticated happening. Seton (1920) recounts many of the early observations of these concentrations, with records dating as far back as the mid-18th century. He quotes from the Bay City Tribune of February 17, 1907, in which a Michigan observer in 1866 counted 1,400 squirrels along a two-mile stretch of road. Another gray squirrel concentration in Michigan occurred in the autumn of 1840, when hundreds of these squirrels crossed the Raisin River near Adrian (Wood, 1922a). Both gray and black phases of the squirrel were involved. Early Michigan naturalist Percy Selous noted similar movements in the Greenville area in the 1890s (Wood and Dice, 1923).

There are accounts of hordes of gray squirrels crossing the Niagara River near Buffalo in the nineteenth century (Banfield, 1974) and the Mississippi River in 1925 from Wisconsin into Iowa (Fryxell, 1926), and again in 1905 and either 1914 or 1915 from Wisconsin into Minnesota (Jackson, 1921). These inhospitable waters apparently caused large numbers to drown. P. R. Hoy (*in* Seton, 1953) reported one great emigration lasting four weeks in Wisconsin in 1842. Schorger (1947; 1949) also records such movements in Wisconsin on 20 occasions between 1842 and 1946.

Since 1900, mass concentrations have been rare (Hoover, 1936; Larson, 1962). The last great movement on record was in October 1968, when gray squirrels in a 250 mi area in southern Appalachia traveled by the tens of thousands into parts of Tennessee, Georgia, and North Carolina (Bioscience, 18(10):985, 1968). Observers reported that thousands were killed by automobiles on highways, by drowning in reservoirs, and by starvation. The movements seemed to have had no special directional patterns.

Biologists have long been puzzled by these widely-spaced antics of gray squirrels. Like many other unusual happenings in nature, the observer gets the chance to carefully document what went on after it started (as in the case of the Appalachia event in 1968), but has little opportunity to make observations in advance and so determine what initiated the event. The gray squirrel's urge to congregate and travel could be due to sudden overpopulation combined with food shortages and weather conditions (Jackson, 1961). Observers at the site of the 1968 movements (Science News, 94(15):359, 1968) reasoned that (a) the previous autumn (1967) had produced a highly-favorable crop of mast (nuts and acorns of the forest trees); (b) this abundant food allowed for over-winter survival of a larger than usual squirrel population which produced a large number of young squirrels in the breeding season of 1968; (c) summer food supplies were sufficient to maintain this higher-than-usual population, but a late spring frost severely damaged blooming mast-bearing trees; (d) by autumn the over-abundant gray squirrels throughout much of southern Appalachia were faced with a shortage of mast to store as winter food; (e) this shortage and the normal autumn urge to gather and store these foods probably caused abnormally high squirrel populations to move out of their home territories in search of storable winter foods. Certainly this explanation is plausible, but perhaps overly simplified.

ASSOCIATES. The squirrels (except for flying squirrels) are the only small Michigan mammals which are active in the daytime. Consequently, there is little close interaction with other species. There may even be, under these circumstances, some multiple den use with diurnal species using refuges at night and nocturnal ones using them by day. The three Michigan tree squirrels have some distinctive habitat preferences but can be close associates in many parts of southern Michigan, especially in urban parks and along tree-lined streets. When these three occur together, as on the campus of Michigan State University, the red squirrel, although smaller, seems the most aggressive. A more complete discussion of these relationships is given in the account for the fox squirrel.

REPRODUCTIVE ACTIVITIES. Gray squirrels have two litters each year. Mating periods are generally confined to January and February (accounting for the first) and to late May, June and early July (accounting for the second), with limited amounts of breeding activity occurring between these peaks (Hoffman and Kirkpatrick, 1959). In Michigan, females are normally in anestrus from August through December, with vaginas imperforate, although there are records of year-around breeding in more southern latitudes (Goodrum, 1940). The start of each breeding season becomes evident by the progressively increasing antagonism between males and by their vigorous courtship tactics toward females. Individual females may maintain a breeding readiness for no more than 10–14 days and will accept several males during the promiscuous breeding pattern. Gray squirrels in rut are noisy (barks, purrs, and sucking sounds) and highly active, with several males often in pursuit of a single female in estrus (Baker, 1944; Banfield, 1974). A dominant male from among the pursuing group may ultimately mate the female; however, because the composition of the male group changes, each of several males becomes a mating partner.

It might seem that this noisy demonstration, combined with the squirrels' preoccupation with mating, would make them easy prey for hawks, bobcats and the like. Nevertheless, tree squirrels seem to hold their own in the wildlife community despite such conspicuous behavioral ritual during which basic caution appears abandoned. Females in estrus during the first (winter) mating period include adults and the yearlings born as a result of the previous winter's breeding; during the second (spring) matings, the yearlings conceived the previous spring also become reproductive.

The gestation period in the gray squirrel is about 44 days. Pregnant females seem to prefer tree holes as nurseries; although artificial nest boxes (Longley, 1963; Burger, 1969) and sometimes leaf nests for second litters (perhaps 10% of the time, Uhlig, 1955) are also used. Most first litters are born in late March and early April and second ones in late July and early August. Litters may contain from one to six young but usually two to four (with a record in the wild of eight, Barkalow, 1967). Since the female is solely responsible for the offspring, she often moves them if the nest area is disturbed (Nixon *et al.*, 1968; Schwartz and Schwartz, 1981), transporting her young mostly by grasping their loose belly skin.

The newly born weigh ½ oz (15 g), are hairless, with eyes and ears closed, but possess developed claws. At birth, the head and feet of the young seem disproportionally large for the rest of the body. The external sexual structures are obvious. The offspring grow rapidly, with short body hair appearing, ears opening, and lower incisors erupting in about 21 days; the eyes begin to open at about 35 days; the hair is long and the tail bushy at about 60 days; and weaning begins to take place at about 50 days, when the young are half-grown. The mother gray squirrel usually tolerates her first litter progeny, even continuing to nurse them, until her second litter is born. This second litter may remain in close association with the mother during the first winter. However, at about 80 days after birth the offspring can be fully independent.

Each adult female produces an average of six young during the two breeding seasons each year. Despite the array of environmental hazards which may befall these offspring, sufficient numbers survive to replace their parental stock and thus maintain the species' local existence. In one study (Longley, 1963), almost half (40% to 48%) of marked young gray squirrels survived from one autumn to the next while 11% to 12% survived to the third autumn. The gray squirrel's population stability is due in part to its longevity. Females may be reproductively active for more than six years and males for more than eight (Barkalow and Soots, 1975). Once a gray squirrel has survived to adulthood (at least one year old), it has an excellent chance of living three to four years (Mosby, 1969; Jackson, 1961; Uhlig, 1955; Lowery, 1974).

FOOD HABITS. The gray squirrel has a varied but basically conservative taste for both plant and animal foods (Korschgen, 1981; Thompson and Thompson, 1980). In summer there is usually a choice of animal materials and the vegetative parts of plants (Montgomery *et al.*, 1975); in winter the gray squirrel relies mostly on plant reproductive parts (seeds, nuts, and acorns). As Bakken (1959) pointed out, the general impression of the gray squirrel is its characteristic pose sitting on its haunches holding an acorn with its forepaws. In late winter and early spring, when stored acorns and other nuts are depleted, gray squirrels, and other arboreal sciurids, eat buds and later blooms of such trees as oaks, elms, maples, willow (Dudderar, 1967; Judd, 1955; Nichols, 1927, 1958; Terres, 1939) and even the sap and inner bark of maple and elm (Hatfield, 1937; Jackson, 1961). As the new growing season progresses, fungi, mushrooms, insects (Nixon, 1970), insect galls, bird eggs, and fruits such as grapes and thorn apples (Dambach, 1942) are eaten. In fact, about 76 different kinds of plants are consumed (Schwartz and Schwartz, 1981).

In autumn, maturation of acorns, walnuts, hickory nuts and other tree fruits brings about a flurry of activity among tree squirrels, as they not only feed on this crop but also store it for winter use. There appears to be a great deal of waste as gray squirrels feed in nut-bearing trees. There is often a constant "rain" of nuts, partly eaten nuts, and shell parts falling to the forest floor below, with no effort by the squirrel to retrieve these edibles. However, these cracked and broken nuts provide food for many other forest creatures. Gray squirrels generally secrete these durable foods stocks in debris covering the forest floor, sometimes in openings in fallen logs and occasionally in tree cavities and in forks and trees (Habeck, 1960). The nuts may be stored in quantities in cavities but usually singly (never more than two or three) in a hole in the ground. Gray squirrels have an uncanny way of relocating these buried caches, probably by means of smell, even after deep snowdrifts or high water obliterate landmarks. One study showed that 85% of these cached nuts are recovered (Thompson and Thompson, 1980); those that are not recovered no doubt result in new tree growth. Gray squirrels may raid cornfields in autumn and winter, although these are indications that corn is not a highly nutritious supplemental food for squirrels (Havera and Nixon, 1980). Like most mammals dependent on the reproductive parts of vegetation for the bulk of their nutrients, the Michigan non-growing season from late autumn to spring can be of serious consequence to squirrels, especially if stored food supplies run short, if crusty snow prevents retrieval, or if competitors rob caches. A surprising array of materials may be eaten at this time, including bird seeds in backyard feeders.

ENEMIES. Gray squirrels spend time on the ground and in trees and are therefore subject to both terrestrial and arboreal predators. There is some evidence that the bobcat may have been a major squirrel predator in much of eastern North America in the presettlement days (Progulske, 1955), but this threat is now of minor importance because of the decline of the bobcat. Snakes, carnivorous mammals from the size of a long-tailed weasel to a gray wolf, and large birds of prey are known to eat gray squirrels. In Michigan, Schofield (1960) counted gray squirrel kills by red foxes tracked in winter snow. Red-shouldered hawk, red-tailed hawk, broad-winged hawk, Cooper's hawk, goshawk, sometimes large owls, and ravens take gray squirrels (Pettingill, 1976; Harlow *et al.*, 1975; Lowery, 1974; Uhlig, 1955; Banfield, 1974). There is even a record of a red squirrel killing a young gray squirrel (Hamilton, 1934). In addition to hunters' kills and human encroachment through land-use changes, gray squirrels suffer considerable mortality from automobiles.

PARASITES AND DISEASES. Gray squirrels are infested both internally and externally with parasites. Many of these host-parasite relationships are maintained by the squirrel's use of tree holes and leaf nests, which also harbor the parasites. An obvious parasitic condition is caused by the mange mite, which burrows in the skin causing scabs, hair loss, and can be fatal if too much of the hair covering is lost. Virus tumors can also be serious (Kilham, 1959; Jackson, 1961). Outbreaks of squirrel pox were reported in this species in the

Harrison area in 1971 and 1973 (Stuht and Harte, 1973).

External parasites include sucking lice, fleas, mange mites, ticks, and botfly warbles (Clark, 1959; Whitaker *et al.*, 1976). Scharf and Stewart (1980) identified the fleas, *Ctenophthalmus pseudagyrtes* and *Orchopeas howardi*, from gray squirrels in Grand Traverse County. Internal parasites include several kinds of Protozoa, roundworms, microfilaria, flukes, tapeworms, and thorny-headed worms (Clark, 1959; Parker and Holliman, 1971).

MOLT AND COLOR ABERRATIONS. In the annual cycle, adult gray squirrels molt in late spring with a complete summer coat appearing by late June, and in early autumn with the heavy, more silvery, winter coat completely grown in early November. In young squirrels, there is also a molt with new pelage-growth just after weaning.

In the Upper Great Lakes Region, both the normal gray and the black (melanistic) phases of the gray squirrel have been commonplace since early days of settlement. Because both black and gray phases are found in close association, the two have often been thought, incorrectly, to belong to different species. In fact, there have been laws enacted (later repealed) to protect black squirrels from hunting. Both color phases were noted in early times in such counties as Alger (as early as 1908), Allegan (1914), Berrien (1919), Cass (1911), Chippewa (1910), Dickinson (1909), Grand Traverse (1908), Kent (1898), Lenawee (1840), Luce (1914), Montmorency (1910), Oscoda (1903), Ottawa (1901), and Washtenaw (1875) as recorded by Wood (1922a, 1922b) and Wood and Dice (1923).

In addition to the black phase and the much less common white (albino) phase (Jackson, 1961; Barkalow, 1967), Michigan gray squirrels are also observed in reddish or mixed reddish, grayish and blackish coats, sometimes with brown on the sides. Local observers often believe that these intermediate color phases may represent successful crosses resulting in hybrids between gray squirrels and fox squirrels. There is no scientific evidence that hybrids between these two species exist in nature (Lowery, 1974). There is also no plausible explanation for the success of the black and gray phases in the same habitats. The black fur allows less heat loss than normal gray pelage in winter (Innes and Lavigne, 1979); however, hair density and hair length are also important in thermoregulation. The black phase is more conspicuous against the forested background than the more camouflaged normal gray coat but it is possible that color has little to do with survival since predators (mostly birds of prey) which hunt by visual means may not make sufficiently serious selective inroads into these populations to cause a reduction in the non-camouflaged black squirrels.

In southern Michigan locales where the gray squirrel was extirpated through habitat removal and human habitation, there have been some efforts to re-establish the species primarily by stocking with black phase individuals. Most notable success has been in the area of Gull Lake and the Kellogg Bird Sanctuary (Johnson, 1973). On Bois Blanc Island in Lake Huron, 41 squirrels, both gray and black color phases, from the vicinity of Houghton Lake, were released in August, 1971 (Barry, 1971). There are also gray squirrels in black phase on Belle Isle and adjacent parts of Detroit (Miller, 1979).

Only red squirrels and fox squirrels were found in the East Lansing area in 1955. Gray squirrels had been present but were reported as rare in Lansing by 1908 (Wood and Dice, 1923) and evidently disappeared shortly thereafter. Officials at Michigan State University live-trapped black squirrels at the Kellogg Biological Station in Kalamazoo County (eight in 1958, 12 in 1962, several more in 1974) and released them in Beal Gardens on the campus. By 1963, a resident population, presumably from these plantings, had become established in the nearby Glencairn district of East Lansing, and by 1968, in parts of Okemos (Baker, 1973). Apparently the original stocks liberated on the Campus left immediately not to return until 1974, when the first one was found on September 25 by Larry Bowdre (pers. comm.). All individuals observed then were in black phase until June 12, 1975, when a normal gray-coated individual was observed in the Whitehills section of East Lansing. After that, gray individuals became more conspicuous, with both color phases frequently seen together on lawns and in the mature shade trees on the campus and along streets in the East Lansing-Okemos area.

ECONOMIC IMPORTANCE. Archaeological studies provide little evidence (Cleland, 1966) that Indian peoples made much use of gray squirrels.

The same might be assumed for early European settlers, who probably preferred to expend scarce powder and lead shot on larger and more meaty targets. Instead, to the early farmers the gray squirrel was a nuisance and raided corn crops and vegetable gardens (Wood, 1922a). Although gray squirrel bounties were paid as early as 1749 in Massachusetts (Baker, 1959), the pestiferous nature of large gray squirrel concentrations in the Midwest is best illustrated by the drastic measures enacted in Ohio. The Ohio Legislature passed a statute in 1807, providing a bounty for gray squirrel scalps; when this proved ineffective in discouraging high squirrel numbers around farm plots, countywide squirrel hunts were also initiated. In 1822, a prize of a barrel of whiskey was offered to the district in Franklin County producing the most scalps. According to the Columbus Gazette for September 12, 1822, at least 19,660 scalps were collected. In Ohio, as in southern Michigan, more and more land was cleared and forest areas dwindled, so by the 1850s the mass squirrel hunts in Ohio ended (Anon., 1976).

In the late nineteenth and early twentieth centuries, gray squirrel populations in southern Michigan declined while fox squirrel populations increased (Allen, 1943). It was at this time that the sporting qualities of tree squirrels became more appreciated, because hunting gray squirrels for meat alone would produce too small a return to justify the cost of ammunition. Even so, dressed gray squirrels were offered for sale in Detroit city markets in the 1880s and 1890s (Wood and Dice, 1923). Squirrel, fried, stewed or even baked, has traditionally been relished fare on the game table. Squirrel tails are also prized by fish-lure makers and fly-tiers.

Between 1897 and 1911, both the gray squirrel (normal and black phases) and the fox squirrel were included in state regulations on open seasons, with no bag-limit restrictions. In 1911, however, Michigan's squirrel populations were considered sufficiently low to have the hunting season closed on all species. Although the season for fox squirrels was reopened in 1919, normal-colored gray squirrels did not become legal game again until 1939. The black phase remained protected until 1956.

Tree replanting and growth, enforced squirrel harvest regulations, den box installations, and various other conservation practices (Allen, 1943; Uhlig, 1959) have favored tree-squirrel production. Estimates of the annual legal kill of all squirrels by Michigan hunters have shown marked upswings since the early 1940s, when less than 50,000 were taken, to a high of over 1,000,000 in 1976. Using Michigan Department of Natural Resources statistics for the annual take in the Upper Peninsula (Region 1), where only gray squirrels are widespread, the annual kill is estimated to be 25,000 to 30,000, with a high of 51,910 reported for 1965. Sustained yields as shown by these annual figures indicate that natural squirrel productivity for both gray and fox squirrels is keeping well ahead of the harvest. It is suspected that gray squirrels will gradually increase in the southern half of the Lower Peninsula (perhaps at the expense of the fox squirrel) as more and more nut-bearing trees mature in city parks, along streams, and in nature sanctuaries.

Crop damage by gray squirrels is part of the historic record for the early settlement days, and the species still causes problems for land managers. Attempts to plant walnuts are sometimes thwarted by foraging gray squirrels. The animals annoy homeowners by invading garages, farm buildings, and attics (Bakken, 1959). In Great Britain, the introduced gray squirrel's destructiveness is second only to that of the Norway rat (Shorten, 1946, 1959; de Vos *et al.*, 1956)—an example of a valued American wildlife resource being almost a total nuisance in an alien situation.

The gray squirrel's antics are, of course, a highly visible attraction for Michigan's nature-watchers. Some people provide separate backyard feeding stations for both squirrels and birds, being careful to squirrel-shield the ones for birds.

COUNTY RECORDS FROM MICHIGAN. Records of gray squirrel specimens in collections of museums and from the literature show that his species occurs in all counties of the Upper and Lower Peninsulas. There are also records of successful introductions on the Beaver Island group in Lake Michigan (Ozoga and Phillips, 1964) and on Bois Blanc Island in Lake Huron (Barry, 1971).

LITERATURE CITED

Ackerman, R., and P. D. Weigl
1970 Dominance relations of red and gray squirrels. Ecology, 51(2):332–334.

Allen, D. L.
1943 Michigan fox squirrel management. Michigan Dept. Cons., Game Div., Publ. 100, 404 pp.

Allen, J. M.
1952 Gray and fox squirrel management in Indiana. Indiana Dept. Cons., Fed. Aid Bull. No. 1., 102 pp.

Anon.
1976 Attack of the squirrel horde. Metro. Park News, 27(2):1.

Baker, R. H.
1944 An ecological study of tree squirrels in eastern Texas. Jour. Mammalogy, 25(1):8–24.
1959 The gray squirrel—past, present and future. Pp. 390–392 *in* V. Flyger, Ed. Symposium on the gray squirrel. Maryland Dept. Res. and Ed., Contr. No. 162, pp. 356–407.
1973 Black phase of the gray squirrel in East Lansing. Jack-Pine Warbler, 51(2):91.

Bakken, A.
1959 Behavior of gray squirrels. Pp. 393–407 *in* V. Flyger, Ed. Symposium on the gray squirrel. Maryland Dept. Res. and Ed., Contr. No. 162, pp. 356–407.

Banfield, A. W. F.
1974 The mammals of Canada. Univ. Toronto Press, Toronto. xxiv+438 pp.

Barkalow, F. S., Jr.
1967 A record gray squirrel litter. Jour. Mammalogy, 48(1): 141.

Barkalow, F. S., Jr., and M. Shorten
1973 The world of the gray squirrel. J. B. Lippincott Co., Philadelphia, Pa. 160 pp.

Barkalow, F. S., Jr., and R. F. Soots, Jr.
1965 An analysis of the effect of artificial nest boxes on a gray squirrel population. Trans. 30th North American Wildlife and Nat. Res. Conf., pp. 349–360.
1975 Life span and reproductive longevity of the gray squirrel, *Sciurus c. carolinensis* Gmelin. Jour. Mammalogy, 56(2):522–524.

Barry, W. J.
1971 The introduction of gray and black squirrels to Bois Blanc Island, Michigan. Unpubl. project report to Michigan Dept. Nat. Res., 11 pp.

Beckwith, S. L.
1954 Ecological succession on abandoned farmlands and its relationship to wildlife management. Ecol. Monog., 24:349–376.

Bouffard, S. H., and D. Hein
1978 Census methods for eastern gray squirrels. Jour. Wildlife Mgt., 42(3):550–557.

Brown, L. G., and L. E. Yeager
1945 Fox and gray squirrels in Illinois. Illinois Nat. Hist. Surv., 23(5):449–532.

Burger, G. V.
1969 Response of gray squirrels to nest boxes at Remington Farms, Maryland. Jour. Wildlife Mgt., 33(4):796–801.

Burt, W. H.
1946 The mammals of Michigan. Univ. Michigan Press, Ann Arbor. xv+288 pp.

Clark, G. M.
1959 Parasites of the gray squirrel. Pp. 368–373 *in* V. Flyger, Ed. Symposium on the gray squirrel. Maryland Dept. Res. and Ed., Contr. No. 162, pp. 356–407.

Cleland, C. E.
1966 The prehistoric animal ecology and ethnozoology of the Upper Great Lakes Region. Univ. Michigan, Mus. Anthro., Anthro. Papers, No. 29, x+294 pp.

Cordes, C. L., and F. S. Barkalow, Jr.
1972 Home range and dispersal in a North Carolina gray squirrel population. Proc. 26th Ann. Conf. Southeastern Assoc. Game and Fish Comm., 1972:124–135.

Dambach, C. A.
1942 Gray squirrel feeding on *Crataegus*. Jour. Mammalogy, 23(3):337.

de Vos, A.
1964 Range changes of mammals in the Great Lakes Region. American Mid. Nat., 71(1):210–231.

de Vos, A., R. H. Manville, and R. G. Van Gelder
1956 Introduced mammals and the influence on native biota. Zoologica, 41(4):163–194.

Dudderar, G. R.
1967 A survey of the food habits of the gray squirrel (*Sciurus carolinensis*) in Montgomery County, Virginia. Virginia Polytechnic Institute and State Univ., unpubl. M.S. thesis, 65 pp.

Fitzwater, W. D., Jr., and W. J. Frank
1944 Leaf nests of gray squirrel in Connecticut. Jour. Mammalogy, 25(2):160–170.

Flyger, V. F.
1960 Movements and home range of the gray squirrel, *Sciurus carolinensis*, in two Maryland woodlots. Ecology, 41(2):365–369.

Fryxell, F. M.
1926 Squirrels migrate from Wisconsin to Iowa. Jour. Mammalogy, 7(1):60.

Goodrum, P. D.
1940 A population study of the gray squirrel in eastern Texas. Texas Agri. Exp. Sta., Bull. 591, 34 pp.

Habeck, J. R.
1960 Tree-caching behavior in the gray squirrel. Jour. Mammalogy, 41(1):125–126.

Hamilton, W. J., Jr.
1934 Red squirrel killing young cottontail and young gray squirrel. Jour. Mammalogy, 15(4):322.

Harlow, R. F., R. G. Hooper, D. R. Chamberlain, and H. S. Crawford
1975 Some winter and nesting season foods of the common raven in Virginia. Auk, 92(2):298–306.

Hatfield, D. M.
1937 Notes on Minnesota squirrels. Jour. Mammalogy, 18(2):242–243.

Hatt, R. T., J. Van Tyne, L. C. Stuart, C. H. Pope, and A. B. Grobman
1948 Island life: A study of the land vertebrates of the islands of eastern Lake Michigan. Cranbrook Inst. Sci., Bull. No. 27, xi+179 pp.

Havera, S. P., and C. M. Nixon
1980 Winter feeding of fox and gray squirrel populations. Jour. Wildlife Mgt., 44(1):41–55.

Hazard, E. B.
1960 A field study of activity among squirrels (Sciuridae) in southern Michigan. Univ. Michigan, unpubl. Ph.D. dissertation, 295 pp.

Hoffman, R. A., and C. M. Kirkpatrick
1959 Current knowledge of tree squirrel reproductive cycles and development. Pp. 363–367 *in* V. Flyger, Ed. Symposium on the gray squirrel. Maryland Dept. Res. and Ed., Contr. No. 162, pp. 356–407.

Holman, J. A.
1975 Michigan's fossil vertebrates. Michigan State Univ., Pub. Mus., Ed. Bull. No. 2, 54 pp.

Hoover, E. F.
1936 Migration of gray squirrels. Science, n.s., 83(2151): 284–285.

Innes, S., and D. M. Lavigne
1979 Comparative energetics of coat colour polymorphs in the eastern gray squirrel, *Sciurus carolinensis*. Canadian Jour. Zool., 57(3):585–592.

Jackson, H. H. T.
1921 A recent migration of the gray squirrel in Wisconsin. Jour. Mammalogy, 2(2):113–114.
1961 Mammals of Wisconsin. Univ. Wisconsin Press, Madison. xii+504 pp.

Johnson, W. C.
1973 Gray squirrels at the Kellogg Bird Sanctuary. Jack-Pine Warbler, 51(2):75–79.

Judd. W. W.
1955 Gray squirrels feeding upon samaras of elm. Jour. Mammalogy, 36(2):296.

Kilham, L.
1959 Virus tumors of gray squirrels. P. 374 *in* V. Flyger, Ed. Symposium on the gray squirrel. Maryland Dept. Res. and Ed., Contr. No. 162, pp. 356–407.

Korschgen, L. J.
1981 Foods of fox and gray squirrels in Missouri. Jour. Wildlife Mgt., 45(1):260–266.

Larson, J. S.
1962 Notes on a recent squirrel emigration in New England. Jour. Mammalogy, 43(2):272–273.

Laundre, J.
1975 An ecological survey of the mammals of the Huron Mountain area. Huron Mt. Wildlife Foundation, Occas. Papers No. 2, x+69 pp.

Levin, E. Y., and V. Flyger
1971 Uroporphyrinogen III cosynthetase activity in the fox squirrel (*Sciurus niger*). Science, 174(4004):59–60.

Longley, W. H.
1963 Minnesota gray and fox squirrels. American Midl. Nat., 69(1):82–98.

Lowery, G. H., Jr.
1974 The mammals of Louisiana and its adjacent waters. Louisiana State Univ. Press, Baton Rouge. xxiii+565 pp.

Manville, R. H.
1948 The vertebrate fauna of the Huron Mountains, Michigan. American Midl. Nat., 39(3):615–640.

Miller, B.
1979 Squirrels. Detroit Free Press, January 21, 148(262): 12–17 (Magazine Section).

Montgomery, S. D., J. B. Whelan, and H. S. Mosby
1975 Bioenergetics of a woodlot gray squirrel population. Jour. Wildlife Mgt., 39(4):709–717.

Mosby, H. S.
1969 The influence of hunting on the population dynamics of a woodlot gray squirrel population. Jour. Wildlife Mgt., 33(1):59–73.

Nadler, C. F., and D. A. Sutton
1967 Chromosomes of some squirrels (Mammalia-Sciuridae) from the genera *Sciurus* and *Glaucomys*. Experentia, 23(4):249–251.

Nichols, J. T.
1927 Notes on the food habits of the gray squirrel. Jour. Mammalogy, 8(1):55–57.
1958 Food habits and behavior of the gray squirrel. Jour. Mammalogy, 39(3):376–380.

Nixon, C. M.
1970 Insects as food for juvenile gray squirrels. American Midl. Nat., 84(1):283.
1979 Squirrel next boxes—are they effective in young hardwood stands? Wildlife Soc. Bull., 7(4):283–284.

Nixon, C. M., R. O. Beal, and R. W. Donohoe
1968 Gray squirrel litter movement. Jour. Mammalogy, 49(3):560.

Nixon, C. M., S. P. Havera, and L. P. Hansen
1980 Initial responses of squirrels to forest changes associated with selection cutting. Wildlife Soc. Bull., 8(4):298–306.

Ozoga, J. J., and C. J. Phillips
1964 Mammals of Beaver Island, Michigan. Michigan State Univ., Publ. Mus., Biol. Ser., 2(6):305–348.

Pack, J. C., H. S. Mosby, and P. D. Siegel
1967 Influence of social hierarchy on gray squirrel behavior. Jour. Wildlife Mgt., 31(4);720–728.

Packard, R. L.
1956 The tree squirrels of Kansas: Ecology and economic importance. Univ. Kansas, Mus. Nat. Hist. and State Biol. Surv., Misc. Publ. No. 11, 67 pp.

Parker, J. C., and R. B. Holliman
1971 Observations on parasites of gray squirrels during the 1968 emigration in North Carolina. Jour. Mammalogy, 52(2):437–441.

Pettingill, O. S., Jr.
1976 The prey of six species of hawks in northern Lower Michigan. Jack-Pine Warbler, 54(2):70–74.

Progulske, D. R.
1955 Game animals utilized as food by the bobcat in the southern Appalachians. Jour. Wildlife Mgt., 19(2): 249–253.

Sanderson, H. R., C. M. Nixon, R. W. Donohoe, and L. P. Hansen
1980 Grapevines—an important component of gray and fox squirrel habitat. Wildlife Soc. Bull., 8(4):307–310.

Scharf, W. C., and K. R. Stewart
1980 New records of Siphonaptera from northern Michigan. Great Lakes Ento., 13(3):165–167.

Schofield, R. D.
1960 A thousand miles of fox trails in Michigan's ruffed grouse range. Jour. Wildlife Mgt., 24(4):432–434.

Schorger, A. W.
1947 A emigration of squirrels in Wisconsin. Jour. Mammalogy, 28(4):401–403.
1949 Squirrels in early Wisconsin. Trans. Wisconsin Acad. Sci., Arts and Ltrs., 39:195–247.

Schwartz, C. W., and E. Schwartz
1981 The wild animals of Missouri. Revised Edition. Univ. Missouri Press and Missouri Conservation Commission, Columbia. viii+356 pp.

Seton, E. T.
1920 Migrations of the gray squirrel (*Sciurus carolinensis*). Jour. Mammalogy, 1(2):53–58.
1953 Lives of game animals. Vol. IV, Pt. I. Charles T. Branford Co., Boston. xii+440 pp.

Sharp, W. M.
1959 A commentary on the behavior of free-running gray squirrels. Pp. 382–387 *in* V. Flyger, Ed. Symposium on the gray squirrel. Maryland Dept. Res. and Ed., Contr. 162, pp. 356–407.

Shorten, M.
1946 American grey and British red squirrels in England and Wales. Jour. Animal Ecol., 15(1):82–92.
1959 Squirrels in Britain. Pp. 375–378 *in* V. Flyger, Ed. Symposium on gray squirrel. Maryland Dept. Res. and Ed., Contr. No. 162, pp. 356–407.

Stuht, J. N., and H. D. Harte
1973 Disease and mortality factors affecting Michigan Wildlife, 1972. Michigan Dept. Nat. Res., Wildlife Div., Rept. No. 2704, 7 pp.

Terres, J. K.
1939 Gray squirrel utilization of elm. Jour. Wildlife Mgt., 3(4):358–359.

Thompson, D. G.
1977 Diurnal and seasonal activity of the grey squirrel (*Sciurus carolinensis*). Canadian Jour. Zool., 55:1185–1189.
1978 Regulation of a northern gray squirrel (*Sciurus carolinensis*) population. Ecology, 59(4):708–715.

Thompson, D. G., and P. S. Thompson
1980 Foodhabits and caching behavior of the urban gray squirrels. Canadian Jour. Zool., 58(5);701–710.

Uhlig, H. G.
1955 The gray squirrel. Its life history, ecology, and population characteristics in West Virginia. West Virginia, Cons. Comm., Final Rep., P–R Proj. 31–5, 175 pp.
1956 A theory on leaf nests built by gray squirrels on Seneca State Forest, West Virginia. Jour. Wildlife Mgt., 20(3): 263–266.
1959 Squirrel management and research. Pp. 387–389 *in* V. Flyger, Ed. Symposium on the gray squirrel. Maryland Dept. Res. and Ed., Contr. No. 162, pp. 356–407.

Wenzel, O. J.
1912 A collection of mammals from Osceola County, Michigan. Michigan Acad. Sci., Arts and Ltrs., 14th Rept., pp. 198–205.

Whitaker, J. O., E. J. Spicka, and L. L. Schmeltz
1976 Ectoparasites of squirrels of the genus *Sciurus* from Indiana. Proc. Indiana Acad. Sci., 85:431–436.

Wood, N. A.
1922a The mammals of Washtenaw County, Michigan. Univ. Michigan, Mus. Zool., Occas. Papers No. 123, 23 pp.
1922b Notes on the mammals of Berrien County Michigan. Univ. Michigan, Mus. Zool., Occas. Papers No. 124, 4 pp.

Wood, N. A., and L. R. Dice
1923 Records of the distribution of Michigan mammals. Michigan Acad. Sci., Arts and Ltrs., Papers, 3:425–469.

FOX SQUIRREL
Sciurus niger Linnaeus

NAMES. The generic name *Sciurus,* proposed by Linnaeus in 1758, is of Latin derivation and means "squirrel," but has an even earlier origin from the Greek in which the first syllable *Sci* refers to "shadow" or "shade" and the last syllable *urus* refers to "tail." The specific name *niger,* also proposed by Linnaeus in 1758, is from the Latin and means "black." A name meaning black is used because Swedish biologist Linnaeus first described this squirrel using a black specimen. Later, of course, it was determined that the name, although valid, was based on the black color phase of the fox squirrel. In Michigan, local residents usually call this species fox squirrel. The name red squirrel is occasionally used although this causes some confusion because the smaller *Tamiasciurus hudsonicus* also has this common name. The subspecies *Sciurus niger rufiventer* Geoffroy-Saint-Hilaire occurs in Michigan.

RECOGNITION. Body size average for a large tree squirrel (Fig. 41); head and body of an adult averaging approximately 11½ in (290 mm) long,

Figure 41. The fox squirrel.
Sketches from life by Bonnie Marris.

with a bushy, well-haired tail almost one-half again as long; body cylindrical, robust, and elongate; head with somewhat square rather than rounded profile as in the gray squirrel; untufted ears comparatively shorter and more rounded than in the gray squirrel; pelage moderately soft and dense, with heavy underfur and long, coarse guard hairs. Color variable but generally reddish to cinnamon (sometimes buffy) on the cheeks, belly, feet, and around the ears; back and sides mixed with grizzled grayish produced by the tricolored guard hairs (basally buffy, medially blackish, terminally grayish); upper side of tail black and buff banded; underside of tail basally reddish, medially blackish and terminally cinnamon; sexes colored alike. Young-of-the-year are generally paler red than adults. Winter and summer coats are very similar. Occasionally individuals have grayish, blackish or white pelages or even mixtures—white tails, black bellies, etc. (see Molt and Color Aberrations). Unlike ground squirrels, fox squirrels have no cheek pouches. Each hind foot has five clawed toes; each forefoot has four clawed toes and a knob-like thumb. The mammary glands are arranged in four pairs, two pectoral, one abdominal, and one inguinal pair of teats. The large size and reddish color distinguish the fox squirrel from the gray squirrel in Michigan.

MEASUREMENTS AND WEIGHTS. Adult fox squirrels measure as follows: length from tip of nose to end of tail vertebrae, 19½ to 22 in (500 to 570 mm); length of tail vertebrae, 8½ to 10½ in (210 to 270 mm); length of hind foot, 2½ to 3¼ in (62 to 80 mm); height of ear from notch, ¾ to 1¼ in (20 to 32 mm). Michigan fox squirrels in autumn (Allen, 1943), average 28 oz (800 g) in weight, with females slightly larger; in late winter the average drops to 26 oz (743 g). The heaviest fox squirrel recorded by Allen in Michigan weighed 43.5 oz (1,243 g).

DISTINCTIVE CRANIAL AND DENTAL CHARACTERISTICS. The skull of the fox squirrel is long and broad (Fig. 42) like that of the gray squirrel but somewhat more massive. It measures 2½ in (65 mm) long and 1½ in (38 mm) wide. The slit-like infraorbital foramen slightly above and in front of the beginning of the cheek tooth-row in the upper jaw is about 3/16 in (5 mm) long (dorsoventrally) while that of the gray squirrel is shorter, about ⅛ in (3 mm) long. However, the skulls can be most quickly distinguished by the presence in the gray squirrel (and the absence in the fox squirrel) of the small, peg-like upper (third) premolar directly in front and to the lingual (tongue) side of the upper premolar-molar cheek

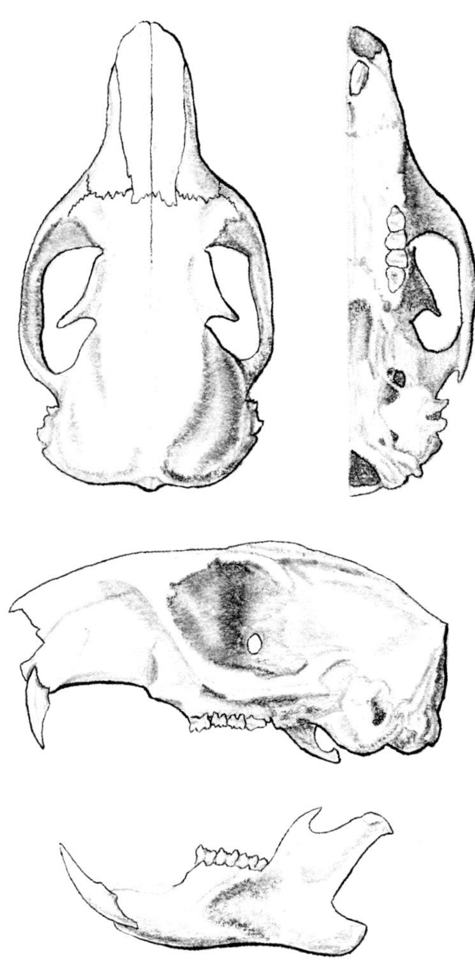

Figure 42. Cranial and dental characteristics of the fox squirrel (*Sciurus niger*).

teeth. Only about 1% of the gray squirrel skulls lack this distinctive peg-like tooth. The whitish bones of the gray squirrel are also a sharp contrast to the pinkish-red ones of the fox squirrel. This distinction in bone color results from the presence in the fox squirrel bones of a skeletal chromatophore called uroporphyrin I (Levin and Flyger, 1971; Flyger and Levin, 1977). The dental formula is:

$$\text{I (incisors)}\frac{1\text{-}1}{1\text{-}1}, \text{C (canines)}\frac{0\text{-}0}{0\text{-}0}, \text{P (premolars)}\frac{1\text{-}1}{1\text{-}1}, \text{M (molars)}\frac{4\text{-}4}{3\text{-}3} = 20$$

DISTRIBUTION. The fox squirrel occupies an almost identical geographic range in North America as the gray squirrel, with these exceptions: It does not extend as far into northeastern United States and not at all into adjacent parts of Eastern Canada; it ranges further westward where woody vegetation occurs on the eastern Great Plains and more southwestward into central Texas and adjacent parts of northeastern Mexico (see Map 30).

Unlike the gray squirrel, the fox squirrel is more at home at the forest edge rather than in the extensive, uncut Michigan woodlands of the earliest pioneer days. Although there are fox squirrel remains from approximately A. D. 1400 (Cleland, 1966) in archaeological sites as far north as what is now southern Ontario (where the species is not known to exist today), fox squirrels apparently lived in Michigan in pioneer times only in the southwestern part of the state where "oak openings" (northeastern vestiges of the western prairies) provided the needed habitat (Kennicott, 1857; Schorger, 1949). As Allen (1943) points out, these openings were scattered in what is now nine of Michigan's southwestern counties (see Veach, 1927), at least as far north as Barry and Eaton counties, but chiefly in Cass, St. Joseph and Kalamazoo counties.

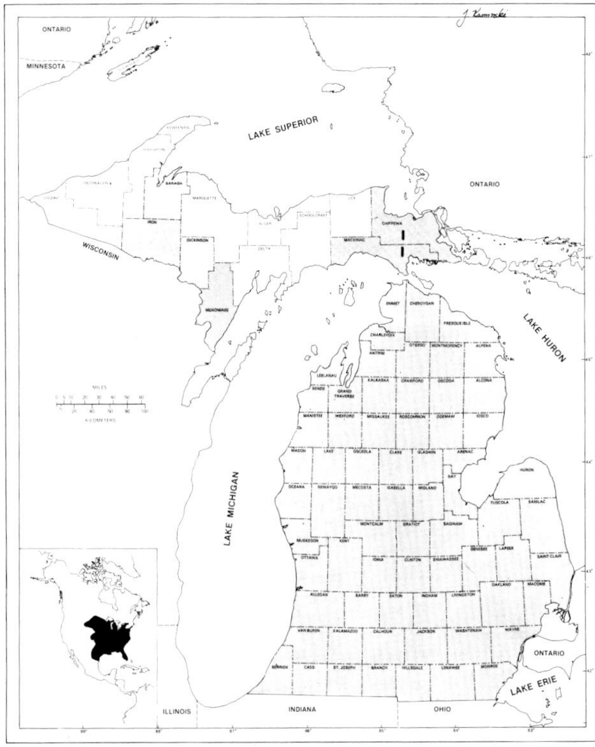

Map 30. Geographic distribution of the fox squirrel (*Sciurus niger*). Populations in counties in the extreme eastern part of the Upper Peninsula are introduced (I).

Logging and farming in Lower Peninsula Michigan produced habitat highly suitable for fox squirrels (de Vos, 1964). This resulted in a steady encroachment by fox squirrels into the areas occupied by the declining gray squirrels, who strictly preferred the mature nut-bearing timber lands. The spread of the fox squirrel throughout Michigan's Lower Peninsula can be gleaned from early observers' accounts (Allen, 1942, 1943; Wood and Dice, 1923). Essentially, the population in southwestern Michigan in the 1840s expanded through the southern half of the Lower Peninsula by the 1870s and finally through all of the counties of the northern half by 1925. In most southern counties, settlers first began to notice fox squirrels by at least the 1850s. Botanist W. J. Beal reported the fox squirrel in Lenawee County in 1858, but others reported the species there in the 1840s. In Ingham County, a local physician and naturalist, H. A. Atkins, first noted the squirrel about 1863. First reports in more northerly counties include Montcalm (1887), Tuscola (1870), Isabella (1880), Grand Traverse (1900), Crawford County (probably about 1910), Leelanau (1895), Cheboygan (1920), and Emmett (1930). Supposedly fox squirrels spread northward more rapidly in lakeshore counties rather than inland ones.

Most of Michigan's Upper Peninsula is still not fox squirrel range (Map 30). This situation may change in years to come as evidenced by the natural spread of fox squirrels from northeastern Wisconsin. In the mid-1950s (Jackson, 1961, see Map 32), the fox squirrel was reported in Brown County, the northern boundary of which is no more than 36 mi (60 km) southward of the Wisconsin-Michigan line. In 1974, Long (1974, see Fig. 20) lists the fox squirrel as resident of both Marinette and Florence counties which border the extreme southwestern counties of Upper Peninsula Michigan. Biologists of Michigan's Department of Natural Resources (Marvin Cooley, pers. comm.) found evidence of the fox squirrel in the early 1970s in the southern quarter of Iron County and the southwestern corner of Dickinson County. Biologist Robert Rafferty found a road-killed fox squirrel in 1977 in Baraga County, directly north of Iron County; a road-kill is also reported from Menominee County by Charles Schneider. Stands of hardwoods in the Upper Peninsula should support these hardy animals.

Introductions in the eastern sector of the Upper Peninsula are apparently responsible for an established population reported there by both Wood (1914) and Burt (*in* Hall and Kelson, 1959:388–389). In the 1970s fox squirrels occurred in Mackinac and Chippewa counties (Robert Strong, pers. comm.). Fox squirrels have been introduced on both North and South Manitou islands in Lake Michigan (Hatt *et al.*, 1948). In the early 1970s Scharf (1973) found the population low on North Manitou Island and restricted to hardwood stands, mostly beech and maple. Fox squirrels introduced to Charity Island in Saginaw Bay later disappeared (Wood, 1914).

Although the gray squirrel has been known to cross formidable rivers during emigrations (Fryxell, 1926), the fox squirrel seems less likely to enter turbulent waters, although it can swim (Jackson, 1961). The St. Clair and Detroit rivers separating southeastern Michigan from Ontario remain ineffective water barriers to the fox squirrel, which is commonplace on the Michigan side but absent on the Ontario side. The gray squirrel occurs on both sides.

HABITAT PREFERENCES. From the oak openings of southwestern Michigan (Allen, 1937), fox squirrels fanned out through the mosaic second-growth shrub and forest lands of Lower Peninsula Michigan. The "edge effect" produced by agricultural practices and lumbering provided the pathways. The fox squirrel's adaptability to shrubby fencerows allowed the species to travel easily between the island-like woodlots now dotting much of the lower Michigan landscape. However, wooded areas of less than five acres (2 ha) may not provide all the basic needs for year-around squirrel survival (Baumgartner, 1943). Allen (1937) noted that grazing of such woodlots by domestic livestock is less harmful to fox squirrels than to other game species.

Much of our knowledge about Michigan's fox squirrel population is derived from Durward L. Allen's 1943 informative volume, "Michigan Fox Squirrel Management." He looked at four major habitats in southern Michigan and concluded that mature oak-hickory woodlands are the best habitats for the fox squirrel. Beech-maple woodlands rank second in providing environmental essentials. In forested stands where beech and maple are well mixed with oaks, ash, elm, basswood, ironwood and other species, the quality is noticeably improved and rivals oak-hickory as a favored fox squirrel habitat. However, when the Michigan forest succession changes from an oak-hickory stage to a beech-basswood-maple stage the fox squirrel tends to decline and the gray squirrel increases (Beckwith, 1954). Scrub oak on sandy soils of southwestern Michigan ranks third as fox squirrel environment in Michigan. The fourth and least productive of the major fox squirrel habitats is streamside woodland. Where the tree species composition is mostly elm and maple, the forest stand may be little used. Where the mix includes ash, elm, red maple, butternut, swamp white oak, sycamore and such bordering species as basswood, red oak, beech, and walnut, the habitat may be as productive for fox squirrels as some of the better upland woodlands. As Allen (1943) noted, forested areas bordering streamways represent the most continuous forested growth in southern Michigan. This habitat is also enhanced as fox squirrel habitat, because it often borders farms where a diversity of land manipulations add to the squirrel's year-around food supply.

DENSITY AND MOVEMENTS. While early day naturalists' accounts describe large numbers of gray squirrels at local concentrations of food supplies or in mass movements, there is little evidence that fox squirrels existed in such densities. Seton (1953) records few early concentrations of fox squirrels in cornfields or high hunter kills. Allen (1942; 1943) cites no unusual early day records of high fox squirrel populations in oak glades of southwestern Michigan.

Fox squirrels are most common in the Michigan woods immediately after the emergence of weaned young in April and May and in September and October. During these peaks, fox squirrels are not only most dense but also most conspicuous (Baker, 1944; Allen, 1943). Both peaks are followed by declines until the next recruitment of young occurs; the winter interval sustains the most drastic losses.

Most information on fox squirrel densities is from early autumn and has usually been evaluated in relation to habitat quality, fluctuations in oak, hickory, and other fall mast crops (Fouch, 1962), and to squirrel harvests by hunters. Allen (1943) estimated Michigan autumn densities in the several basic fox squirrel habitats. He found populations in oak-hickory woodlands (the best of his categories) to vary from about one fox squirrel per two acres (almost 1 per ha) to two animals per acre (5 per ha), with even higher densities in small woodlots in good years. On a year-around basis, Baumgartner (1940) thought that this type of habitat (at least in Ohio) could support one fox squirrel for every 1½ acres (almost 1.5 squirrels per ha) of woodland.

In beech-maple woodlands (the second best environment for Michigan fox squirrels), Allen (1943) estimated autumn populations to range from one animal per two acres (1 per ha) to more than one per acre (perhaps 3 per ha). In Ohio, Baumgartner (1940) estimated that the year-around density in this habitat could be as high as one squirrel to slightly more than two acres (about 1 per ha). In scrub oak country in southwestern Michigan, Allen (1943) found densities lower, with populations of one fox squirrel per two to one per six acres (almost 1 per ha to 1 per more than 2 ha). In the least productive streamside and lowland woodlands, the fox squirrels are less abundant and may occasionally move out of exceedingly wet areas during spring floods. In Michigan, an autumn population of two fox squirrels per acre (5 per ha) of woodlands is high; in most situations densities much lower are to be expected.

Fox squirrels are generally rather sedentary. Michigan animals may satisfy their total requirements for a single season in areas no larger than five to 10 acres (2 to 4 ha), although single squirrels may cover as much as 40 acres (15 ha) during the course of an entire year. In Nebraska, Adams (1976) found radio-tracked males using areas averaging about 19 acres (7.56 ha) while females lived in about one-half as much space, nine acres (3.55 ha).

Major shifts occur during the breeding season when males travel over large areas of woodlands in search of females, and again in autumn when the squirrels move from family groups and from leaf nest shelters in search of winter quarters and food storage areas. Juvenile squirrels account for much of this movement. Allen (1943) noted that one marked squirrel moved approximately 40 mi (65 km) in the space of three months. On another occasion a fox squirrel released in unfamiliar territory traveled approximately 46 mi (75 km) from the point of release. Even so, the fox squirrel appears to lack the dispersal abilities of the gray squirrel and is less likely to concentrate at food sources and to participate in mass movements.

BEHAVIOR. Like gray squirrels, fox squirrels are not highly social creatures, although they tend to nest together in winter and use common environments and food supplies with only modest amounts of friction. The fox squirrel lacks the strong territorial behavior (except nest defense) found in the red squirrel (Smith, 1968), but does demonstrate a social hierarchy which allows for exploitation of a food source by many individuals rather than the restrictive use by only one animal, as is often the case in a territorial system. In a study of social interaction of fox squirrels at a winter feeding station on the Michigan State University campus, Bernard (1972) found a well-defined social structure, perhaps somewhat emphasized by a concentration of the squirrels at a corn-filled feeder. As in the case of gray squirrels (Bakken, 1959; Pack *et al.*, 1967), males dominate females in such instances and older individuals dominate younger ones.

The fox squirrel's daytime activity and use of conspicuous tree holes and leaf nests gives the species high visibility. Light plays a major role in regulating this diurnal activity. Packard (1956) found the fox squirrel to be more tolerant of overall light conditions than the gray squirrel. The fox squirrel, by being more responsive to various light conditions, is usually active longer in the day and in more open, brighter habitats than the gray squirrel. The fox squirrel is most visibly active in December and January and again in April and May, periods correlated somewhat with mating activities (Bakken, 1952; Packard, 1956). The species is sensitive to climatic changes and becomes inactive in rain or snowfall. Activity is also reduced in high winds, unseasonable high or low temperatures, and excessive cloud cover (Allen, 1943; Hicks, 1942, 1949).

Fox squirrels probably spend more time on the ground (foraging or moving between trees) than do gray squirrels. Part of this ground movement is along brushy or tree-lined fencerows and across open areas where gray squirrels rarely venture. The fox squirrel moves with a waddling gait when foraging in leaf litter or for buried nuts. When startled, it runs rapidly with bounding leaps, with its flag-like tail being most conspicuous, and usually heads for leafy understory, ultimately climbing into a tree. Like the gray squirrel, the fox squirrel will disappear into a tree hole, "flatten" on the top side of a horizontal branch, or dodge around the opposite side of a sizable trunk. Most observers agree, however, that the fox squirrel is a less agile climber than the gray squirrel.

The fox squirrel will den in tree hollows and also construct leaf-twig nests (Brown and Yeager, 1945). Tree cavities are the most protective and can be used over many years. Studies by Nixon *et al.* (1980) in Illinois suggest that a fox squirrel may need as many as one to two cavities per acre (3 to 5 per ha). Preferably the entrance should be no more than 5 in (130 mm) wide, with the cavity at least 14 in (360 mm) deep. The best holes result from a mature tree's self-pruning process and a combination of central rotting (plus pecking by woodpeckers and gnawing by squirrels) of the broken branch and surrounding growth of the tree's cambium layer. Unfortunately, such trees are frequently old and may fall or be culled in the course of woodlot management (Allen, 1943). To make a good squirrel den, a tree cavity must be positioned so that water cannot collect in the hole; it must also be small enough to exclude raccoons or other large tree hole users. Allen (1943) found good tree-hole dens for Michigan fox squirrels in mature white oak, soft maple, elm, sycamore, and beech. Of these, he noted that white oak and beech produce the best cavities in the uplands and elm and red maple in the lowlands. Although tree holes provide excellent litter protection, as well as substantial winter quarters, they may be infested by fleas and mites, and they may harbor carcasses or living, den-seeking competitors including red squirrels, flying squirrels, white-footed mice, screech owls, wood ducks, and bees.

In the absence of suitable tree holes, especially in many of Michigan's managed farm woodlots, fox squirrels rely on leaf nests; a combination of leaf nests and cavities provides the best year-around quarters. Although some leaf nests may be no more than 10 ft (3 m) off the ground, most are higher, with grape vines offering favored attachment sites (Sanderson *et al.*, 1980). In a study in Allegan County, Allen (1943) found the average leaf nest was 30 ft (9 m) above ground in a tree 7 in (180 mm) in diameter (dbh). In a study of 120 leaf nests, 55 were in black oak, 42 in white pine, 11 in red maple, 9 in beech, and 3 in white oak. In a 1979 class study on 193 acres (72 ha) of the northern part of the Michigan State University campus, Dale Rezabek examined 30 leaf-twig nests constructed by fox squirrels: 13 were in American elms, 7 in white oaks, and the remaining 10 in 7 other kinds of trees. In Ohio, Baumgartner (1939) found squirrel nests in 22 different tree species. It was Allen's (1943) opinion that the fox squirrel picks the larger trees for nest building, with white pine and black oak often favored.

Most or all of the materials used in leaf nests are cut from the tree in which the nest is built. Summerbuilt nests are bulky and usually have an abundance of green-cut leafy twigs. The framework may be as much as 20 in (500 mm) across, with several layers of greenery packed to form a roundish inner pocket with side entry and a rain-shielding roof. Fox squirrels seem to prefer tree holes in winter but they do use winter-built nests which sometimes have a cross-cross construction of bare twigs and dead leaves. They are compact and roundish in appearance, with many bare twigs projecting. In late summer particularly young squirrels may construct platforms of leafy twigs. There is usually no cavity in this structure and the squirrel merely lies on top. As in the case of the gray squirrel, the number of leaf and twig nests, if counted annually in the same area and at the same season, may give some indication of fluctuations in fox squirrel populations (Goodrum, 1940). Although these nests may remain intact for several months, many disintegrate in a matter of weeks, especially if abandoned. The observer using binoculars can often find signs of use or non-use such as spiderwebs in front of the side entrances and other indications of neglect.

Calls of the fox squirrel resemble those of the gray squirrel, although they are somewhat lower in pitch and more guttural (Jackson, 1961; Zelley, 1971). Seton (1953) regards the fox squirrel as a less noisy woods animal than the gray squirrel. The gray squirrel tends to chatter or bark at first indication of an intruder; the fox squirrel is more inclined to hide or seek the security of a den. Both squirrels call in typical head-down position on tree trunks, while vigorously flicking the tail. Basically, however, these two squirrels vocalize much in the same way and apparently for the same reasons.

ASSOCIATES. It is possible that in southern Michigan the three tree squirrels (gray, fox and red), the southern flying squirrel, the eastern chipmunk, and even the thirteen-lined ground squirrel may interact, perhaps while using similar foods (Allen, 1943; Bakken, 1952; Hazard, 1960). Moving northward to the upper part of the Lower Peninsula, the northern flying squirrel is also a sciurid competitor; in the Upper Peninsula, the least chipmunk is another. Many of these sciurids compete with fox squirrels for tree mast as well as cached acorns, nuts and other seeds. The food supply is probably the major factor governing the associations between these close relatives (Reichard, 1976). In addition, habitat quality and little-documented social interactions may play a role in the relationships between the arboreal fox and the gray and red squirrels.

In general, the fox squirrel favors the mixed southern hardwoods and conifers interspersed with numerous openings and second growth shrubs and trees. The gray squirrel prefers the extensive stands of mature upland or lowland southern or northern hardwoods, shying away from openings and glades (Baker, 1944; Longley, 1963). The red squirrel seems more at home in coniferous or mixed coniferous and hardwood forests than in hardwood (deciduous) stands unless they are in moist lowlands. In Michigan, these species compete for tree holes and artificial den boxes (Allen, 1943). Most observers have noted that fox squirrels and red squirrels rarely associate. For example, on the campus of Michigan State University in late April, a fox squirrel, a red squirrel, and an eastern chipmunk were observed eating blooms in a large elm tree. Each squirrel, however, remained well separated from the others.

In 1958, 1962, and 1974, animals of the gray squirrel black phase captured in Kalamazoo County were released on the Michigan State University campus (Baker, 1973), allowing for the association of this species with existing populations of the fox squirrel and the smaller red squirrel. Although there are as yet no quantitative data, there is some reason to believe that the less aggressive fox squirrel may be declining (in 1981), perhaps because of the intrusion of the more aggressive gray squirrel, with no evident change in the less-common red squirrel. As a rule, the fox squirrel adjusts easily to urban forested areas when gray squirrels are absent, but much less so when both are present. Despite the interest ecologists have in survival strategies and resource uses of animals associated in natural communities, it is difficult to design studies to measure and determine the causes and effects of such associations.

REPRODUCTIVE ACTIVITIES. Like gray squirrels, fox squirrels can have two litters each year

(Allen, 1942; 1943). In Michigan, the two mating periods are in late December to mid-February (peak in mid-January) and in May to early July. The first litter is usually produced in mid-March, with older females and yearlings born the previous spring participating. The second litter is usually produced in July, with older females and yearlings born the previous summer involved. There are occasional exceptions to this reproductive cycle, with young being born from February to November (Moore, 1957).

The fox squirrel's courtship behavior, like the gray squirrel's, is elaborate and noisy (barks, purrs, and sucking sounds). Males who are dominant in the social hierarchy are usually the successful mates for the responsive females. Various aggressive and placative signals, tail flicks, approaches, and grooming are involved before the mating act occurs (McCloskey and Shaw, 1977).

After a gestation period of about 44 days, a litter of three or four (sometimes fewer or as many as six) are born (Allen, 1943; Baumgartner, 1940; Jackson, 1961). In Michigan, female fox squirrels generally give birth to their young in tree holes, but sometimes leaf nests are used as nurseries even in winter (Allen, 1942). At birth, a young fox squirrel weighs 14 to 17 g, is hairless, has closed eyes and ears, noticeable claws and vibrissae, and proportionally large head and feet. At seven days, hair appears on the back and the weight is doubled; at 25 days, the ears open and the lower incisors appear; and at 35 days, the eyes open and the tail hair is growing. At 55 days, the young will venture out of the nest, and at 90 days, they are weaned and able to live independently. The first (spring) litter is generally excluded soon after weaning by the expectant mother, whereas the second (summer) litter may stay with the mother in a family group for several weeks longer.

As in the gray squirrel, the fox squirrel potential production is approximately six young for each adult female annually. In practice, however, Allen (1943) found the age ratio of Michigan fox squirrels in autumn to be approximately 60% young-of-the-year and 40% older individuals. He also concluded that the yearly turnover rate resulted from natural mortality (⅓), hunter kill (⅓), and surviving breeding stock (⅓). Nevertheless, individual fox squirrels are capable of living several years. At Rose Lake Wildlife Research Center in Clinton County, for example, there are records of six tagged squirrels surviving for more than six years in nature (Fouch, 1958; Greene, 1950; Linduska, 1947).

FOOD HABITS. Fox squirrels eat plant and animal foods (Korschgen, 1981). In late April and early May, Michigan fox squirrels respond energetically to emerging tree and shrub growth at a time when winter stores and foods salvaged from under snowdrifts and leaf litter are depleted. Following autumns of low acorn and nut production, food supplies in late winter and early spring can be critically sparse for squirrel populations, especially for females in late pregnancy or with nursing litters (Allen, 1943). As tree buds swell and flowering parts appear, fox squirrels can be seen feeding on this growth on limb ends while attempting to keep their balance on swaying or overweighted branches. Allen (1943) and Reichard (1976) provide most of the Michigan field data on late winter and early spring diets of fox squirrels. Foods in late March, April and May include flower buds and samaras of silver and red maple, buds and flowers of sugar maple and burr oak, pistillate catkins of willow, staminate catkins of cottonwood, flowers of hackberry, elm and beech, and buds of elm and basswood. Some of these may only serve as secondary foods (Reichard, 1976) while more staple items are in short supply. During this period, fox squirrels also probe in forest litter for adult beetles and other insects and for tubers, bulbs and roots; nests of robins, blue jays, and other birds are robbed.

The summer ripening of a variety of fruits offers a bounty to fox squirrels. Included are wild strawberries, serviceberries, haws, plums, raspberries, blackberries, greenbrier, blueberries, grapes, chokeberries, cherries, basswood, box elder, black ash, hard maples, dogwood, elderberries and others (Allen, 1943). Beetles, grasshoppers, grubs and other larval stages of insects, bird eggs, nestling birds, green corn and other agricultural crops are also fair game for fox squirrels. As Packard (1956) remarks, however, the discovery of cached acorns or nuts, even this late, provides the fox squirrel with a meal which is probably preferable to any summer fare.

The appearance of acorns, nuts and other ripening seeds in autumn allows the fox squirrel to "fatten-up" for winter and to gather these non-perishable foods for storage. The next year's breeding stock of squirrels is dependent on this mast production, with lean crops seriously effecting overwintering populations, and, in fact, acorn and nut production is highly variable. In a ten-year study near East Lansing, Gysel (1971) found beechnut production was a failure for two years, low for four years, intermediate for three years, and high for only one year. Allen (1943) lists the major Michigan mast-producing trees as hickories, various oaks (white, swamp-white, bur, and black), butternuts, walnuts, and beech. Squirrel preference for hickory nuts may be due to their high energy content and metabolism efficiency (Havera and Smith, 1979). The only important autumn mast-bearing shrub is the hazelnut. In addition to harvesting acorns and nuts (which are shared with a variety of birds, other squirrels, white-footed mice, white-tailed deer, even raccoons) fox squirrels may feed on corn kernels (preferably the germ portion only, Bernard, 1972) and other agricultural crops, cocklebur seeds (Gates and Gates, 1975), and hardy insects. Normally, however, tree mast is the important autumn food, since corn, for example, is not nutritionally adequate as a supplemental food for squirrels (Havera and Nixon, 1980).

In winter, the fox squirrel subsists on leftovers, having spent considerable time caching a part of this surplus. Whereas red squirrels cache large quantities of acorns and nuts in single deposits, fox squirrels bury them individually in small holes dug no more than 1 in (24 mm) below ground level. Each food item is then covered with soil or litter. This caching activity frequently occupies most of the fox squirrel's daylight hours. Relocating these buried foods requires a keen and specialized sense of smell (Cahalane, 1942; Dice, 1927), although the acorn or nut odor is undoubtedly enhanced by ground moisture. Naturalists like to believe that many a new tree is planted when a squirrel fails to dig up a seed; however, Cahalane has shown that fox squirrels may retrieve up to 99% of their cache. The squirrel's ability to find walnut plantings has hampered efforts of foresters attempting to reestablish walnut groves.

Reduced acorn and nut crops cause some squirrels to move over large areas in quest of foods. They can subsist successfully on buds, fungi, and leftover fruits of bittersweet, grape, haws, hackberry, and rose, and bark of sugar maple, red maple, elm, beech, basswood (Allen, 1943; Fouch, 1962). Their conspicuous stripping of bark from sugar maples in quest of sweetened sap is a common practice in late February and early March. Fortunately, most crop fluctuations are usually localized, caused by such factors as variations in soil quality (Gysel, 1957) and the presence or absence of late frosts affecting flowering success (Gysel, 1956).

ENEMIES. Because in Michigan the fox squirrel is very close to the northern edge of its natural distribution, severe winter weather coupled with periodically short food supplies undoubtedly make serious inroads into local populations. Both young and old fox squirrels are also fair game for many kinds of predators. In Michigan, fox squirrels have been taken by red-tailed hawk, osprey, barred owl, raccoon, and domestic dog (Allen, 1942, 1943); by red fox (Dearborn, 1932; Murie, 1936); by great horned owl, red-shouldered hawk, and Cooper's hawk (Craighead and Craighead (1956); and by marsh hawk (Linduska, 1950). In addition, coyotes, gray foxes, house cats, long-tailed weasels, bobcats, and even great blue herons eat fox squirrels (Korschgen, 1957; Packard, 1956; Peifer, 1979; Progulske, 1955).

Fox squirrels are killed by automobiles on streets and highways, with urban populations perhaps the more frequent victims. In 1941, it was reported that 1,422 fox squirrels were killed on Michigan highways (Haugen, 1944). Highest mortality rates were for June and September—months when fox squirrels seem most active and also when new litters are dispersing to establish their homes.

DISEASES AND PARASITES. Fox squirrels are popular game animals, so their diseases and parasites have probably been given special attention by wildlife pathologists and ecologists. An important parasite, because of its obvious symptoms, is scabies or mange. A minute scabies mite (barely visibly to the naked eye) burrows beneath the skin causing skin irritation, loss of hair and vitality, and even death (probably due to exposure of bare skin). This parasite is an especially serious threat to Michigan fox squirrels in winter (Allen, 1943).

Also infecting Michigan fox squirrels are large warble larvae (of the fly of the genus *Cuterebra*) which form conspicuous elongated lumps just under the skin, usually in the area of the neck or lower abdomen. After completing their larval stage, these worm-like organisms push through their breathing holes and drop in to soil or litter to pupate and ultimately emerge as adult flies. Michigan fox squirrels host an assortment of fleas, ticks, mites, and lice externally and tapeworms, flukes, roundworms and protozoa internally (Baker, 1944; Allen, 1943; Packard, 1956; Scharf and Stewart, 1980; Whitaker *et al.,* 1976). Bacteria of the genus *Salmonella* have been isolated from fox squirrels taken in Michigan (Youatt and Fay, 1968). Squirrel pox (fibromatosis) has also been reported in Michigan fox squirrels (Stuht and Harte, 1973).

MOLT AND COLOR ABERRATIONS. In Michigan, fox squirrels begin a molt of the worn, winter pelage in April and May. This process begins on the head and proceeds posteriorly, often with a distinct molt line, where the old hair is dropping out and the new hair is emerging. These new hairs continue to grow to provide a longer, heavier coat for winter. The first and second litters of young squirrels also molt and grow dense coats for winter, making molting and hair-growth a somewhat continual summer process.

Michigan gray squirrels are better known for color phases, with both the normal gray and the black phases common in many areas. Fox squirrels also occasionally exhibit black coats in Michigan, although this color phase may be more common elsewhere, as in parts of Louisiana (Lowery, 1974). Color mutations for Michigan fox squirrels include albino and black-bellied varieties (Wood, 1922), and an animal with a dark head and a silver tail (Allen, 1943). An albino fox squirrel in Kensington Metropolitan Park in Oakland County attracted the attention of visitors in the summer of 1980. In the collection of the Michigan State University Museum, there are specimens of fox squirrels with mixed patterns of normal pelage and patches of whitish or blackish.

ECONOMIC IMPORTANCE. Fox squirrel meat was a part of the pioneer settler's diet. Allen (1943) used early hunting data from Eaton County to demonstrate the practical use of this species and also show that between 1866 and 1915 the fox squirrels killed by hunters surpassed the number of gray squirrels killed in the same manner. Despite the fact that many hunters feel the fox squirrel lacks the frisky nature and other sporting qualities of the gray squirrel, the fox squirrel has rapidly become a favored target for the small-game hunter in southern Michigan. Certainly the fox squirrel's less wary nature makes it more attractive to some hunters and nature observers.

By the mid-1970s, at least 250,000 to 300,000 Michigan hunters were licensed annually to bag squirrels. Currently, the estimated kill, combining fox and gray squirrels, totals more than 1,000,000 annually. This take, as officials of Michigan's Department of Natural Resources attest, fluctuates largely in response to the abundance of fall acorn and nut production. Records kept over many years at the Rose Lake Wildlife Research Center in Clinton and Shiawassee counties (Black, 1953) show an average annual sustained fox squirrel kill of 5.2 individuals per 100 acres (40 ha).

The fox squirrel's ability to thrive in mixed second-growth woody vegetation allows for the maintenance of a well-distributed population in southern Michigan. The need is obvious for vigorous enforcement of laws regulating the hunter's kill; these laws, based on scientific findings concerning harvesting limits, ensure sustained fox squirrel yields. In addition, Allen (1943) stressed the importance of conserving as well as replanting mast-producing oaks, hickories, walnuts, butternut, elms, maples, beeches, and hazelnut (see Nixon *et al.,* 1980). Restocking and predator control have long been ruled out as important to Michigan squirrel management. Nest boxes (made of nail kegs, old lumber or even of twisted auto tires) are attractive homesites for squirrels when tree cavities are sparse. Backyard feeding stations (ear corn in wire baskets) effectively attract fox squirrels in winter, although bird feeders must be protected by metal shields as fox squirrels will monopolize these as well.

Like other tree squirrels, the fox squirrel becomes a nuisance when it chooses attics, garages and barns as den sites. It also damages ornamentals, lawns, and some flowers; raids fields of green and ripe corn and groves of fruit trees; and gnaws trees in winter. This sciurid will cut bark from almost all types of trees but probably does the most damage to maples (Allen, 1943).

Fox squirrels have sporting qualities and an adequate reproductive rate to rank the species high in popularity as Michigan farm game. The meat, fried or stewed, is esteemed on the dinner table. Fly-tiers and other fishing lure makers prize the tail hairs. The arboreal antics of these diurnal tree squirrels also endear them to nature watchers.

COUNTY RECORDS FROM MICHIGAN. Records of specimens of fox squirrels in collections of museums and from the literature show that this species occurs in each county of the Lower Peninsula. In the Upper Peninsula, where fox squirrels have recently spread from northeastern Wisconsin, they presently occur in Menominee, Dickinson, Iron, and Baraga counties. Fox squirrels have also been successfully introduced in Mackinac and Chippewa counties and on North and South Manitou islands in Lake Michigan.

LITERATURE CITED

Adams, C. E.
1976 Measurement and characteristics of fox squirrel, *Sciurus niger rufiventer*, home ranges. American Midl. Nat., 95(1):211–215.

Allen, D. L.
1937 Ecological studies on the vertebrate fauna of a 500-acre farm in Kalamazoo County, Michigan. Ecol. Monog., 8(3):348–436.
1942 Populations and habits of the fox squirrel in Allegan County, Michigan. American Midl. Nat., 27(2):338–379.
1943 Michigan fox squirrel management. Michigan Dept. Conservation, Game Div., Publ. 100, 404 pp.

Baker, R. H.
1944 An ecological study of tree squirrels in eastern Texas. Jour. Mammalogy, 25(1):8–24.
1973 Black phase of the gray squirrel in East Lansing. Jack-Pine Warbler, 51(2):91.

Bakken, A.
1952 Interrelationships of *Sciurus carolinensis* (Gmelin) and *Sciurus niger* (Linnaeus) in mixed populations. Univ. of Wisconsin, unpubl. Ph.D. dissertation, 188 pp.
1959 Behavior of gray squirrels. Pp. 390–392 *in* V. Flyger, Ed., Symposium on the gray squirrel. Maryland Dept. Res. and Ed., Contr. No. 162, pp. 356–407.

Baumgartner, L. L.
1939 Fox squirrel dens. Jour. Mammalogy, 20(4):456–465.
1940 The fox squirrel: Its life history, habits, and management in Ohio. Ohio State Univ., unpubl. Ph.D. dissertation, 257 pp.
1943 Fox squirrels in Ohio. Jour. Wildlife Mgt., 7(2):193–202.

Beale, D. M.
1962 Growth of the eye lens in relation to age in fox squirrels. Jour. Wildlife Mgt., 26(2):208–211.

Beckwith, S. L.
1954 Ecological succession on abandoned farm lands and its relationship to wildlife management. Ecol. Monog., 24:349–376.

Bernard, R. J.
1972 Social organization of the western fox squirrel. Michigan State Univ., unpubl. M.S. thesis, 41 pp.

Black, C. T.
1953 Fourteen-year game harvest on a 1500-acre Michigan farm. Trans. 18th North American Wildlife Conf., pp. 421–438.

Brown, L. G., and L. E. Yeager
1945 Fox and gray squirrels in Illinois. Illinois Nat. Hist. Surv., 23(5):449–532.

Cahalane, V. H.
1942 Caching and food recovery by the western fox squirrel. Jour. Wildlife Mgt., 6(4):338–352.

Cleland, C. E.
1966 The prehistoric animal ecology and ethnozoology of the Upper Great Lakes Region. Univ. Michigan, Mus. Anthro., Anthro. Papers, No. 29, x+294 pp.

Craighead, J. J., and F. C. Craighead, Jr.
1956 Hawks, owls and wildlife. The Stackpole Co., Harrisburg, Pennsylvania. xix+443 pp.

Dearborn, N.
1932 Foods of some predatory fur-bearing animals in Michigan. Univ. Michigan, School For. and Cons., Bull. No. 1, 52 pp.

de Vos, A.
1964 Range changes of mammals in the Great Lakes Region. American Midl. Nat., 71(1):210–231.

Dice, L. R.
1927 How do squirrels find buried nuts? Jour. Mammalogy, 8(1):55.

Flyger, V., and E. Y. Levin
1977 Congenital erythropoietic porphyria. Animal model: Normal porphyria of fox squirrels (*Sciurus niger*). American Jour. Path., 87(1):269–272.

Fouch, W. R.
1958 Longevity records for the fox squirrel. Jour. Mammalogy, 39(1):154–155.
1962 Mast crops and fox squirrel populations at the Rose Lake Wildlife Experiment Station. Michigan Acad., Sci., Arts and Ltrs., Papers, 47:211–217.

Fryxell, F. M.
1926 Squirrels migrate from Wisconsin to Iowa. Jour. Mammalogy, 7(1):60.

Gates, J. E., and D. M. Gates
1975 Fox squirrel use of cocklebur seeds. Jour. Mammalogy, 56(1):239–240.

Goodrum, P. D.
1940 A population study of the gray squirrel in eastern Texas. Texas Agric. Exp. Sta., Bull. 591, 34 pp.

Greene, H. C.
1950 A record of fox squirrel longevity. Jour. Mammalogy, 31(4):454–455.

Gysel, L. W.
1956 Measurement of acorn crops. For. Sci., 2(4):305–313.
1957 Acorn production on good, medium, and poor oak sites in southern Michigan. Jour. For., 55(8):570–574.
1971 A 10-year analysis of beechnut production and use in Michigan. Jour. Wildlife Mgt., 35(3):516–519.

Hall, E. R., and K. R. Kelson
1959 The mammals of North America. Vol. 1. Ronald Press Co., New York. xxx+546+79 pp.

Hatt, R. T., J. VanTyne, L. C. Stuart, C. H. Pope, and A. B. Grobman
1948 Island life: A study of the land vertebrates of the islands in eastern Lake Michigan. Cranbrook Inst. Sci., Bull. No. 27, xi+179 pp.

Haugen, A. O.
1944 Highway mortality of wildlife in southern Michigan. Jour. Mammalogy, 25(2):160–170.

Havera, S. P., and C. M. Nixon
1980 Winter feeding of fox and gray squirrel populations. Jour. Wildlife Mgt., 44(1):41–55.

Havera, S. P., and K. E. Smith
1979 A nutritional comparison of selected fox squirrel foods. Jour. Wildlife Mgt., 43(3):691–704.

Hazard, E. B.
1960 A field study of activity among squirrels (Sciuridae) in southern Michigan. Univ. Michigan, unpubl. Ph.D. dissertation, 295 pp.

Hicks, E. A.
1942 Some major factors affecting the use of two inventory methods applicable to the western fox squirrel, *Sciurus niger rufiventer* (Geoffroy). Iowa State Col. Jour. Sci., 16:299–305.
1949 Ecological factors affecting the activity of the western fox squirrel, *Sciurus niger rufiventer* (Geoffroy). Ecol. Monog., 19:287–302.

Jackson, H. H. T.
1961 Mammals of Wisconsin. Univ. Wisconsin Press, Madison. xii+504 pp.

Kennicott, R.
1857 The quadrupeds of Illinois, injurious and beneficial to the farmer. U.S. Patent Office Rept. (Agric.) for 1856, pp. 52–110.

Korschgen, L. J.
1957 Food habits of coyotes, foxes, house cats, and bobcats in Missouri. Missouri Cons. Comm., Fish and Game Div., P-R Ser. No. 14, 64 pp.
1981 Foods of fox and gray squirrels in Missouri. Jour. Wildlife Mgt., 45(1):260–266.

Levin, E. Y., and V. Flyger
1971 Uroporphyrinogen III cosynthetase activity in the fox squirrel (*Sciurus niger*). Science, 174(4004):59–60.

Linduska, J. P.
1947 Longevity of some Michigan farm game mammals. Jour. Mammalogy, 28(2):126–129.
1950 Ecology and land-use relationships of small mammals on a Michigan farm. Michigan Dept. Cons., Game Div. ix+144 pp.

Long, C. A.
1974 Environmental status of the Lake Michigan region. Vol. 15. Mammals of the Lake Michigan Drainage basin. Argonne Natl. Lab., 108 pp.

Longley, W. H.
1963 Minnesota gray and fox squirrels. American Midl. Nat., 69(1):82–98.

Lowery, G. H., Jr.
1974 The mammals of Louisiana and its adjacent waters. Louisiana State Univ. Press, Baton Rouge. xxiii+565 pp.

McCloskey, R. J., and K. C. Shaw
1977 Copulatory behavior of the fox squirrel. Jour. Mammalogy, 58(4):663–665.

Moore, J. C.
1957 The natural history of the fox squirrel, *Sciurus niger shermani*. American Mus. Nat. Hist., Bull. 113, 71 pp.

Murie, A.
1936 Following fox trails. Univ. Michigan, Mus. Zool., Misc. Publ. No. 32, 45 pp.

Nixon, C. M., S. P. Havera, and L. P. Hansen
1980 Initial response of squirrels to forest changes associated with selective cutting. Wildlife Soc. Bull., 8 (4):298–306.

Pack, J. C., H. S. Mosby, and P. D. Siegel
1967 Influence of social hierarchy on gray squirrel behavior. Jour. Wildlife Mgt., 31(4):720–728.

Packard, R. L.
1956 The tree squirrels of Kansas: Ecology and economic importance. Univ. Kansas, Mus. Nat. Hist. and State Biol. Surv., Misc. Publ., No. 11, 67 pp.

Peifer, R. W.
1979 Great blue heron foraging for small mammals. Wilson Bull., 91(4):630–631.

Progulske, D. R.
1955 Game animals utilized as food by the bobcat in the southern Appalachians. Jour. Wildlife Mgt., 19(2): 249–253.

Reichard, T. A.
1976 Spring food habits and feeding behavior of fox squirrels and red squirrels. American Midl. Nat., 92(2): 443–450.

Sanderson, H. R., C. M. Nixon, R. W. Donohoe, and L. P. Hansen
1980 Grapevines—an important component of gray and fox squirrel habitat. Wildlife Soc. Bull., 8(4):307–310.

Scharf, W. C.
1973 Birds and land vertebrates of South Manitou Island. Jack-Pine Warbler, 51(1):3–19.

Scharf, W. C., and K. R. Stewart
1980 New records of Siphonaptera from northern Michigan. Great Lakes Ento., 13(3):165–167.

Schorger, A. W.
1949 Squirrels in early Wisconsin. Trans. Wisconsin Acad. Sci., Arts and Ltrs., 39:195–247.

Seton, E. T.
1953 Lives of game animals. Vol. IV, Pt. I. Charles T. Branford Co., Boston. xxiii+440 pp.

Smith, C. C.
1968 The adaptive nature of social organization in the genus of tree squirrels, *Tamiasciurus*. Ecol. Monog., 38:31–63.

Stuht, J. N., and H. D. Harte
1973 Disease and mortality factors affecting Michigan wildlife, 1972. Michigan Dept. Nat. Res., Wildlife Div., Rept. No. 2704, 7 pp.

Veach, J. O.
1927 The dry prairies of Michigan. Michigan Acad. Sci., Papers, 8:269–278.

Whitaker, J. O., and E. J. Spicka
1976 Ectoparasites of squirrels of the genus *Sciurus* from Indiana. Proc. Indiana Acad. Sci., 85:431–436.

Wood, N. A.
1914 An annotated check-list of Michigan mammals. Univ. Michigan, Mus. Zool., Occas. Papers No. 4, 13 pp.
1922 The mammals of Washtenaw County, Michigan. Univ. Michigan, Mus. Zool., Occas. Papers No. 123, 23 pp.

Wood, N. A., and L. R. Dice
1923 Records of the distribution of Michigan mammals. Michigan Acad. Sci., Arts and Ltrs., Papers, 3:425–469.

Youatt, W. G., and L. D. Fay
1968 Survey for Salmonella in some wild birds and mammals in Michigan. Michigan Dept. Cons., Res. and Dev. Rep. No. 135, 1 p.

Zelley, R. A.
1971 The sounds of the fox squirrel, *Sciurus niger rufiventer*. Jour. Mammalogy, 52(4):597–604.

RED SQUIRREL
Tamiasciurus hudsonicus (Erxleben)

NAMES. The generic name *Tamiasciurus,* proposed by the French mammalogist Trouessart in 1880, is of Latin and earlier Greek origin and can be translated as referring to a "chipmunk-like squirrel." Looking at the derivation more closely, however, the first part of the generic name, *Tamias,* means "steward" or "one who stores and takes care of provisions." The second part of the generic name, *sciurus,* means "shade-tail." The specific name *hudsonicus,* proposed by Erxleben in 1777, refers to the Hudson Bay region from where the type specimen used in the original description was derived. In Michigan, local residents sometimes call this animal pine squirrel or chickaree but most commonly red squirrel. Red squirrels living in both the Upper and Lower Peninsulas belong to the subspecies *Tamiasciurus hudsonicus loquax* (Bangs). Those on Isle Royale, as pointed out by Kramm *et al.* (1975), belong to the subspecies *Tamiasciurus hudsonicus regalis* Howell, showing close relationship to the population in Ontario to the north.

Figure 43. The red squirrel (*Tamiasciurus hudsonicus*)
Sketches from life by Bonnie Marris

RECOGNITION. Body size small for a tree squirrel (Fig. 43), about the size of a Norway rat; head and body of an adult averaging approximately 7¾ in (200 mm) long, with a bushy, well-haired tail almost one-half again as long; body slender and elongate; ears prominent, somewhat pointed and tufted (especially in winter pelage); pelage soft and dense. In summer, upperparts and sides are reddish gray; head slightly darker, flecked with blackish; underparts whitish, separated from the coloring of the upperparts by a blackish lateral line extending in the midsection of belly; eye-ring white and conspicuous; upper surface of tail bright rufous medially with individual hairs having subterminal blackish bands and tawny tips; underside of tail grayish mixed with buff; upper surfaces of feet reddish. In winter, pelage is longer, finer, thicker, brighter; upperparts have an orange-red stripe extending along midline from head to tail with sides less rufus and more buffy, flecked with blackish; head generally darker, flecked with blackish; underparts whitish or creamy with black lateral line absent or barely present; eye-ring white; tail coloring similar to condition in summer although slightly darker; ear tufts rufous and distinctive; feet grayish, soles covered with silvery hairs. The forefeet have four prominent and clawed digits, the hind feet five. There are no internal cheek pouches. There are eight mammae present in females (one pair pectoral; two pairs abdominal; one pair inguinal). The small size, conspicuous white eye-ring, tufted ears, black lateral stripe, reddish upperparts and bright reddish upper tail surface help to distinguish the red squirrel from the larger gray and fox squirrels in Michigan.

MEASUREMENTS AND WEIGHTS. Adult red squirrels measure as follows: length from tip of nose to end of tail vertebrae, 11 to 13½ in (280 to 340 mm); length of tail vertebrae, 4 to 5½ in (100 to 145 mm); length of hind foot, 1¾ to 2¼ in (40 to 55 mm); height of ear from notch, ¾ to 1 in (18 to 26 mm); weight, 5 to 9 oz (135 to 250 g). Adult males average slightly heavier than adult, non-pregnant females.

DISTINCTIVE CRANIAL AND DENTAL CHARACTERISTICS. The skull is intermediate in size between the ground squirrels (the largest of Michigan's smaller sciurids) and the larger gray and fox squirrel (Fig. 44). The skull of the red squirrel is notably roundish and broad, averaging 1¾ in (45 mm) long and 1 in (26 mm) wide. Males tend to have slightly larger skulls than females (Nellis, 1969). In profile, the forward part of the dorsal outline is almost flat; the sides of the zygomatic arches are not expanded but rather parallel to the skull's axis; and the rostrum is characteristically short and stubby. The anterior (third) premolar on each side of the upper jaw is usually absent (see dental formula below); when present, it is a tiny, non-functional, peg-like structure almost covered by the larger fourth upper premolar. The dental formula is:

$$\text{I (incisors) } \frac{1\text{-}1}{1\text{-}1}, \text{ C (canines) } \frac{0\text{-}0}{0\text{-}0}, \text{ P (premolars) } \frac{1\text{-}1}{1\text{-}1} \text{ or } \frac{1\text{-}2}{1\text{-}1},$$

$$\text{M (molars) } \frac{3\text{-}3}{3\text{-}3} = 20 \text{ or } 22$$

DISTRIBUTION. The red squirrel inhabits northern cone-bearing evergreen forests, adjacent mixed conifers and hardwoods and, sometimes, pure deciduous hardwood stands, especially in moist situations. This hardy squirrel occurs across North America from the northern edge of tree line in Canada and Alaska southward in the western mountains as far as New Mexico and Arizona and, after avoiding the Great Plains, southeastward in the Great Lakes States and New England and southward in the montane forests of the Appalachians to northern Georgia. In Michigan, the

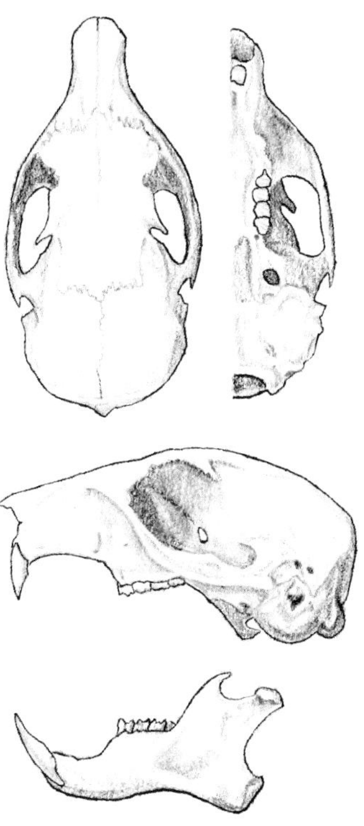

Figure 44. Cranial and dental characteristics of the red squirrel (*Tamiasciurus hudsonicus*).

red squirrel is widespread, thriving in all counties in both the Upper and Lower Peninsulas (see Map 31). The species is also known from North and South Manitou islands in Lake Michigan (Hatt *et al.*, 1948), from Bois Blanc and Drummond islands in Lake Huron (Manville, 1950), and from Isle Royale in Lake Superior (Burt, 1946).

Map 31. Geographic distribution of the red squirrel (*Tamiasciurus hudsonicus*).

HABITAT PREFERENCES. Many observers consider the red squirrel a denizen of damp, cool, shady, boreal, coniferous forest habitats. However, both Hatt (1929) and Layne (1954), major contributors to the natural history of the red squirrel, found the species in assorted woody environments ranging from pure cone-bearing coniferous areas to stands of deciduous trees. It is Layne's contention from his New York studies that red squirrels thrive best in abundant coniferous cover and/or seeds. In Michigan, Beckwith (1954) writes that the red squirrel joins the gray squirrel in taking over the beech-basswood-maple stage in forest succession; whereas, the fox squirrel holds its own in the oak-hickory association. Nevertheless, the species seems to have survived the drastic inroads by axe and plow in Michigan, adjusting compatibly to marginal areas. In non-evergreen areas like northcentral Iowa, red squirrels live in immature hardwood forests near rivers and lakes, while nesting and caching food in mature trees if they are available (Lynch and Folk, 1968).

In Michigan, red squirrels are most prevalent in mixed evergreen and deciduous forests but can be found in older pine plantations, old orchards, city parks, and school grounds. At Rose Lake Wildlife Research Center, Linduska (1950) found red squirrels most common in lowland woodlots, also noting them in other sylvan areas, even in fencerow habitat. Small numbers of red squirrels live in the park-like stands of diverse forest trees on the Michigan State University campus, but are much less in evidence than the fox squirrel. Quality habitat, according to Layne (1954), consists of stands of ungrazed beech, maple and hemlock or of oak-hickory hardwood association with conifers, especially hemlock and white pine.

DENSITY AND MOVEMENTS. The red squirrel is certainly the predominant tree squirrel in most wooded habitats in the northern part of the Lower Peninsula and throughout the Upper Peninsula. Red squirrels are also common in the southern part of the Lower Peninsula (Wood, 1922), although quality woodlands for the animals are scattered. One difficulty in evaluating red squirrel populations is a result of seasonal fluctuations in various habitats. Highest densities are naturally achieved when litters join the active adult populations or when individuals congregate at abundant food sources. Densities, as summarized by Layne (1954) and others, vary as a result of habitat quality from lows of one individual per 5 acres (1 per 2 ha) to highs of almost 2 per acre (5 per ha), with the mixed conifer (especially hemlock) and hardwood habitat having the highest expected numbers. Food concentrations may result in as many as 10 to 12 squirrels per acre (25 to 30 per ha) in limited areas and for short durations (Jackson, 1961). Besides local movements (including spring and autumn dispersals, Rusch and Reeder, 1978) and concentrations at seasonal food supplies, there is some evidence that red squirrels occasionally engage in mass movements (Cahalane, 1947). These are thought to be much less frequent than mass movements of the gray squirrel in earlier times; certainly such red squirrel behavior was probably less conspicuous.

Daily movements for adults of both sexes in the prebreeding period are minimal. Males move most extensively during the mating season while females are most active during the lactation and post-weaning periods of the reproductive cycle (Fancy, 1981). However, red squirrels normally thrive in a home range of about 3 to 6 acres (between 1 and 3 ha), moving in an area perhaps 600 ft (300 m) across, except for occasional moves elsewhere for food or mating. As with other squirrels, the juveniles are involved in much of the movement since they must depart from their birth site to acquire home areas of their own (C. C. Smith, 1968).

BEHAVIOR. Like other tree squirrels, red squirrels are not highly social. Each individual is essentially solitary, but uses a home range which is partly shared with others. Normally, the red squirrel defends its den site and feeding areas, especially from others of the same sex, throughout the year (C. C. Smith, 1968). In practice, however, red squirrels show sufficient tolerance to even share a den at times with another individual, often a younger one. Some studies also show that red squirrels in western parts of the country may be more territorial than those in Michigan. In Alberta, Rusch and Reeder (1978) suggested that territorial behavior regulates red squirrel density around food supply in years of poor pine cone production.

In autumn and winter, males and females often defend contiguous territories while in spring and summer males expand their territories to include those of adjacent females and defend them against intruding males but not females. Red squirrels seeking home territories seem to have two choices (C. C. Smith, 1968): establish themselves in small, undefended spaces between defended areas and then expand them at the expense of the established territory holders, or aggressively oust occupants from established territories.

Like other Michigan tree squirrels (except the flying squirrels), the red squirrel is active by day during all seasons, and occasionally, appears on moonlit, or even dark, rainy, nights (Burt, 1946). Red squirrels are the most conspicuous species in terms of energetic movement and vocalization. Behavior, as classified by Layne (1954), includes alimental (foraging and feeding), social interaction (physical and vocal), resting (motionless in resting posture), and active (defined as movement without obvious objectives). Most summer activity on clear or overcast (but not stormy) days occurs within two hours after sunrise and within two hours of sunset. During the cold months, these two activity peaks may be abandoned for a single one during the middle of the day. Red squirrels seek cover in rain and snowstorms, but often are on the move immediately afterwards. Like other tree squirrels, they are not active during strong winds.

To the average field observer, the red squirrel appears to be constantly on the go and is characterized by its boldness, conspicuousness, curiosity and utterances. It spends more time in trees than either the gray or fox squirrels and, of the three, is perhaps the most agile climber, running along on undersides of branches, leaping from branch to branch, and jumping from trees to the ground. It negotiates ground surfaces, logs, rail fences, rocky outcrops, and rock fences with a rapid, bounding gait. It is also an able swimmer, and is known to cross rivers and small lakes (Banfield, 1974).

The red squirrel faces survival problems in winter which are very different from those in summer. During the cold months the species shows a rapid increase in energy metabolism when the outside temperature drops below 20°C. According to Irving, Krog, and Monson (1955), this temperature is at the lower limit of the thermal neutral zone, below which the red squirrel's use of food to supply body heat must increase drastically. The active life of this sciurid is also energy costly, with oxygen consumption 1.4 to 2.8 times greater while the animal is running (Wunder and Morrison, 1974). When the Michigan winter temperature dips below 20°C, the red squirrel is apt to be less active in trees and more active on the ground, especially in snow tunnels where there is some insulation against the outside cold. To date, however, winter behavioral habits of small Michigan mammals are, at best, little known.

Red squirrels seem much more ubiquitous in selecting den sites than either gray or fox squirrels. The former not only use arboreal refuges—leaf nests and tree cavities—but also den at ground level in logs, rock piles, and underground burrows as well as in barns, houses and other man-made structures (Fancy, 1980; Layne, 1954; Yahner, 1980). Leaf nests are placed on substantial limbs near the main tree trunks, at least 15 ft (5 m) above ground and in trees with dense branching. These structures, as Doutt *et al.* (1966) pointed out, are often "ungainly, basketball-size" tangles of twigs, leaves, shredded plant fibers, bark, grasses, mosses, feathers, and even paper. Sometimes a nest abandoned by a hawk or crow makes an excellent base for a red squirrel nest. Layne (1954) noted three layers of a well-constructed leaf nest: (a) A shell of loosely-woven twigs with attached leaves, (b) a dense intermediate layer of compacted leaves, and (c) an inner layer of finer plant fibers enclosing a chamber about 4 to 6 in (100 to 155 mm) in diameter and lined with dried grasses, conifer needles, vegetable fibers, moss, fur and other fine materials. The chamber is accessible through one or two entrances in the side of the nest. Less complicated nests, perhaps for temporary use only, are also constructed. Leaf nests are usually in larger trees but the squirrels show no real preference for a particular tree species.

Along with the gray and fox squirrels, the red squirrel seems to prefer tree cavities to leaf nests, even though these hollows may be in short supply in evergreen forests (Fancy, 1980). Red squirrels will use natural cavities or enlarge and take over holes excavated by woodpeckers. Dens of this sort are acceptable in either living or dead trees but the dens must be protected from rain or snow. There are records of as many as three or four red squirrels (sometimes a female and her second litter) occupying a tree cavity den in winter.

Red squirrels also occupy ground refuges constructed in log piles, rock fences, beehives, and human habitations. Underground burrow systems often open beside old stumps (Hatt, 1929) and under clumps of woody vegetation, including hedgerows. In winter, red squirrels tunnel under snowdrifts. Near Mio in Oscoda County, Harold Mayfield (1948) found red squirrel young in a nest on the ground at the base of a small jack pine. The nest was set in a depression, constructed of grass, and overlain with a layer of oak leaves. Red squirrels appear to select a wider array of den sites than either fox or gray squirrels.

In the Michigan forest, an observer will probably hear a red squirrel before actually seeing one. When an intruder is detected, the squirrel utters an oft-repeated "scolding" call sounding somewhat like *chick-chick-chick.* If the intruder approaches ever closer, the chatter becomes more nervously intense sounding like *tcher-r-r-r-r-r* (Jackson, 1961). Calls resembling *kak-kak-kak* and *whuck-whuck* have also been described.

Red squirrels apparently use a great deal of vocalization in protecting their territories. According to C. C. Smith (1968, 1978), there are five calls associated with territorial behavior: One informs an intruder that the area is occupied; two calls announce the aggressive intentions of the resident toward the invader; a fourth is an appeasing call by the intruder; a fifth warns of predators.

ASSOCIATES. In much of Michigan's Lower Peninsula, the red squirrel has abundant opportunity to associate with both the fox and gray squirrels. In the Upper Peninsula gray squirrels are its main associates. Despite the diminutive size of the red squirrel, it has a high degree of interaction with the two other species (Hazard, 1960), and because of its aggressiveness, appears to dominate its larger relatives, although the common belief that red squirrels emasculate male gray squirrels has little scientific basis (Hatt, 1929; Doutt *et al.*, 1966). The red squirrel is often the aggressor in a chase, with the fox or gray squirrel shying away (Preston, 1948). On the Michigan State University campus, the author observed a red squirrel chase a fox squirrel for at least 50 yds (46 m) across a mowed lawn, with the latter always holding a 25 ft (8 m) gap. At the end of the chase the red squirrel returned to the starting area and the fox squirrel foraged at the stopping place.

Allen (1943) found the red squirrel competed with the fox squirrel for tree cavities; all three tree squirrels probably interact occasionally in this competition. Even so, Layne (1954) noted evidence of tolerance when he found a red squirrel and a gray squirrel living in the same tree but in different cavities. In studies of red and gray squirrels confined together in large outdoor cages (Ackerman and Weigl, 1970), red squirrels not only shared nest boxes with other red squirrels but also with gray squirrels. These authors concluded that "the much publicized aggressiveness of the red squirrel toward the gray may be a function of territorial behavior . . . rather than an inherent characteristic of the species." Even so, the red squirrel is accused of discouraging both gray and fox squirrels. On the University of Michigan campus, Wood (1922) reported an effort to obliterate the red squirrel in order to encourage the other two species. In May of 1974, a red squirrel, a fox squirrel, and an eastern chipmunk were feeding in the same elm tree at sunup on the Michigan State University campus. However, each animal occupied a different part of the large tree and maintained a respectable distance from the others, at least during the observation. In a woodlot near Lansing, Reichard (1976) observed red squirrels and fox squirrels feeding in the same trees on 11 occasions, with aggressive or avoidance interactions occurring nine times. Twice, the red squirrel chased the fox squirrel out; and on four occasions, the red squirrel avoided interaction by climbing to a higher or lower branch or moving to another tree. Usually the first squirrel in the tree drove the other out.

Numerous studies have been made of interactions between mammals using what appears to be similar resources including food supplies, den sites, and spatial arrangements. However, many questions concerning how organisms have achieved a degree of compatibility in sharing habitats and their nutrients and other attributes without serious conflict are yet to be answered. At the present time in Michigan it is possible to find three species of diurnal tree squirrels, plus two species of nocturnal flying squirrels, obviously using arboreal facilities in the same area. If this interaction was not successful, competition would certainly have caused some species to be eliminated by now. The survival strategies of each sciurid must be sufficiently distinctive to reduce competition to a minimum. These differences are admittedly hard to identify and evaluate. Hopefully, ecologists will closely watch the status of the Michigan populations of the red, gray, and fox squirrels, because there are likely to be future habitat apportionments differing from those of today, at least in the southern half of Michigan's Lower Peninsula with the reemergence of the gray squirrel in rapidly maturing woodlands.

REPRODUCTIVE ACTIVITIES. Red squirrels normally have two litters annually in Michigan (Linduska, 1950), although in more northern areas (northeastern Minnesota, Timm, 1975; Quebec, Ferron and Prescott, 1977) there may be only one. It is suspected that the limited growing season in the northern part of the red squirrel's range is a major factor in determining the number of litters (Ferron and Prescott, 1977). Breeding may also be influenced by weather (Millar, 1970). In Michigan, courtship and mating occur in February or March and again in June or July (Burt, 1946).

Like other tree squirrels, the red squirrel mates promiscuously, often with two or more males involved with one female. According to C. C. Smith (1968), the female's territorial defense is continued until the day when she comes into heat and is receptive to males. On one occasion, Layne (1954) noted seven squirrels involved in a single courtship; even so, he suspects there is some tendency toward monogamy because there is evidence that a breeding male may expand his territory to include one of an adjacent female and then try to keep other males away. The act of mating, following an active chase accompanied by soft coughing calls, takes place either on the ground or in a tree. The red squirrel, like the fox and gray squirrels, is noisy and conspicuous during courtship and almost oblivious to other environmental events.

The gestation period is approximately 40 days (perhaps 35, according to Ferron and Prescott, 1977). Young are born and reared in tree cavities, leaf nests, even on the ground (Mayfield, 1948). A single litter may contain four to seven individuals (five perhaps is usual). At birth, the offspring are hairless with wrinkled, pink skin, closed eyes and ears, and have some claw development. They are 2½ in (70 mm) long and weigh about ¼ oz (6.5 g). At about 10 days of age, hair development is obvious; at 18 days the ears open; between 27 and 35 days the eyes open; at 38 days the young venture from the nest; between 49 and 56 days the young are weaned; and by 70 days they are half-grown (Ferron, 1980; Layne 1954; Svihla, 1930). Ordinarily, the young will remain with the mother no more than 120 days. During this latter period, the offspring explores the nest surroundings, learns to eat adult foods, often through observing a parent (Weigl and Hanson, 1980), participates in sexual play, and becomes acquainted with territorial defense (C. C. Smith, 1968). These young do not breed until one year of age.

Although there is some question as to whether all adult females produce two litters or have a late litter after failing to produce in spring, the potential production may be as many as 10 offspring per adult female per year. At Rose Lake Wildlife Research Station, Linduska (1950) detected a 50% turnover in local red squirrel populations from one year to the next, although he was unsure whether the losses were due to death or dispersal.

He noted the survival of marked squirrels for two years; at least one survived for three. Individuals survive for 10 years in captivity (Banfield, 1974); in the wild they could survive for perhaps seven (Jackson, 1961).

FOOD HABITS. Red squirrels, like gray and fox squirrels, are food opportunists (Weigl and Hanson, 1980) and rely on seasonal foods, in addition to the year-around availability of acorns, conifer seeds, and nuts from the autumn production (Hatt, 1929). C. C. Smith (1968) stressed that red squirrels need high energy foods such as the reproductive products (seeds, nuts, etc.) of conifers, deciduous trees and bushes and the cambium layer of pine trees as providing such needs. It was also his opinion (Smith, 1981) that red squirrels are essentially adapted to exploit conifer seeds and fungi in boreal coniferous forests. He measured the food consumption for an adult male as about 117 kg cal per day; for a lactating female, 323 kg cal per day.

In late winter and early spring, Michigan red squirrels depend on cached foods (nuts, acorns, and pine seeds). As for other tree squirrels (Allen, 1943), these seasons are critical times for the Michigan red squirrel. Nutritious foods may be in short supply, especially if the tree seed crop from the previous autumn was inadequate. In the Lansing area, Reichard (1976) found buds, flowers, and fruits of maples and other trees were indispensable links in the food chain during the crucial period from March through June, with flower buds of red and silver maple and buds and flowers of sugar maple the most important. Buds of oaks, elms and other deciduous tree species, plus sap from sugar maples also supply emergency rations in this interim before "green-up" begins (Klugh, 1927; Hatt, 1929; Hamilton, 1939; Layne, 1954).

In summer, buds of deciduous trees, seeds of elm, and green samaras of the several maples continue as important foods for red squirrels. They are also attracted to fleshy fruits (including raspberries and mulberries), sap from sapsucker holes in jack pines (Hatfield, 1937), gray birches, sugar maples and other trees (Hatt, 1929; Kilham, 1958), and a variety of mushrooms. Summer animal foods include numerous insects, eggs and young of birds, and small mammals (Banfield, 1974; Hatt, 1929; Layne, 1954). By August, red squirrels may show resin smears about the mouth because of their liking for the green cones of conifers. Cones become edible almost as soon as they develop on the various pines, spruce, fir, hemlock, white cedar, and common juniper, with hemlock and white pine perhaps most preferred. Red squirrels also seek cones in Christmas tree plantations and on ornamental plantings in cemeteries and parks.

In late summer and early autumn, red squirrels begin to store a part of the crop of seeds, nuts, and acorns. Some food items are cached individually, similar to the practice of other tree squirrels. However, red squirrels traditionally store foods for winter and early spring use in "middens." Such stockpiles may contain from a dozen to hundreds of food items (Finley, 1969). Acorns from oaks and nuts from hickories, walnut, beech, hawthorn, sumac, hazelnut and butternut are cached especially in non-conifer areas (Lynch and Folk, 1968). However, the most characteristic middens are those filled with conifer cones. The red squirrel spends considerable time in the tops of cone-bearing trees cutting terminal twigs on which clusters of cones are growing, preferably when the cones are green and unopened so that the seeds are not lost prior to caching. After cutting numerous cone-bearing twigs, the squirrel descends to the ground to either bury the cones in underground chambers or merely pile them at the bases of trees or fallen logs. Yeager (1937) found caches of Norway spruce cones covered with moss near Ann Arbor. Again, cones of hemlock and white pine are perhaps most preferred by Michigan red squirrels. The close tie between red squirrels and cone-bearing trees has pointed to the possibility of coevolution of the squirrel (the predator) and the conifer (the prey) with possible evolutionary responses on the part of the trees (Elliott, 1974; Smith, 1970).

In winter, the red squirrel relies chiefly on its caches of conifer seeds, nuts and acorns. Resident red squirrels energetically exclude non-residents from their caches (Rusch and Reeder, 1978). When the autumn crop is short, squirrel populations can be seriously impaired. In Alaska, M. C. Smith (1968) noted a winter decline of 67% in squirrel numbers following crop failures. Spruce buds became the primary food during such a catastrophe. The Michigan red squirrel's ability to scavenge in winter is shown by the species' attraction to a deer carcass in Schoolcraft County

(Field, 1970) and to beef bones in Emmet County (Howard, 1935).

ENEMIES. The red squirrel is Michigan's smallest mammal with both diurnal and arboreal habits, making it fair game for both medium-sized and large daylight predators. However, Layne (1954) felt that predation was not a highly significant factor in red squirrel mortality, even though other authors point to the red squirrel as being an important link in the forest food chain.

The traditional enemy of the red squirrel in Michigan presettlement days was perhaps the marten (Seton, 1953). Although once completely extirpated, this successful red squirrel predator has been reestablished in the Upper Peninsula, along with the fisher, another mustelid which relishes red squirrel. The several carnivores known to feed on Michigan red squirrels include the badger and bobcat (Dearborn, 1932), red fox (Johnson, 1970; Schofield, 1960), gray wolf (Mech, 1966), screech owl and barn owl (Wilson, 1938), great horned owl and red-shouldered hawk (Linduska, 1950), Cooper's hawk (Craighead and Craighead, 1956), and goshawk and broad-winged hawk (Pettingill, 1976). Other red squirrel enemies are coyote, mink, long-tailed weasel, lynx, possibly house cat, sparrow hawk, barred owl, and even large fishes and snakes (Banfield, 1974; Hatt, 1929; Jackson, 1961; Seton, 1953). Highway traffic also takes its toll; Manville (1949) counted nine dead red squirrels over 650 mi of road in the Upper Peninsula in early August, a time when tourist traffic is high.

PARASITES AND DISEASE. Red squirrels are infested with fleas, lice, ticks, mites and chiggers as external parasites and tapeworms, flukes, and roundworms as internal parasites. Coccidiosis is also known in red squirrels (Dorney, 1966) and botfly larvae are carried occasionally (Dorney, 1965). In New York, Layne (1954) recorded 11 species of fleas from red squirrels. Two fleas, *Monopsyllus vison* and *Orchopeas caedens caedens,* are known from Michigan red squirrels (Scharf and Stewart, 1980). Scab mites do not seem to cause as serious an epidermal problem in red squirrels as in gray and fox squirrels.

MOLT AND COLOR ABERRATIONS. The red squirrel molts twice each year. The spring molt beginning in late March and continuing until early July sheds the heavier and brighter winter coat (Nelson, 1945). The autumn molt beginning in late August and continuing until early December sheds the thinner and darker summer coat. The spring molt new growth begins on the front of the head and then on the feet; the autumn molt begins on the tail and progresses forward (Layne, 1954). This is an energy-expensive process but is no doubt of major survival value to a small warm-blooded, active animal.

Although most red squirrels have normal colored pelages, there are the occasional albinistic (all white) and melanistic (all black) individuals (Layne, 1954). There are also animals with intermediate coat colors (Dapson, 1963). In Michigan, Dice (1925) and Wood (1922) report albinos and part albinos. A specimen from Barry County in the collections of The Museum at Michigan State University has white spotting above, some white on the feet, a whitish belly, and basally whitish hairs on the underside of the tail. C. R. Peebles reported a white-tailed individual on the Michigan State University campus in March, 1979. True melanistic individuals are apparently rare (Benton, 1958). A blue-eyed individual was reported in Kalamazoo County in 1979.

ECONOMIC IMPORTANCE. Red squirrels' fondness for conifer seeds has long been a concern of those interested in reforestation. In Minnesota, red squirrels obtained in jack pine slash had the remains of an average of 392 pine seeds in their stomachs (Smith and Aldous, 1947). In New York, Hatt (1929) reported that a pair of captive red squirrels consumed the seeds from 422 white pine cones in one week's time. Red squirrels also damage forests by eating buds of young conifers, girdling small trees, and gnawing away bark. The species may raid fields of cultivated crops and stores of corn and other grains. Red squirrels sometimes kill poultry. In both urban and rural areas the animals may invade homes and other structures, using out-of-the-way spaces in attics and in double-walls as den sites. Ornithologists have never really appreciated the role played by the red squirrel in the woodland community. For example, Ingham County physician and student of bird life H. A. Atkins wrote to ornithologist Edgar A. Mearns on April 3, 1880: "Had I been consulted at the creation or immediately after with my present knowledge of natural history, I should have had this most impudent and "cheeky" nuisance with its mate annihiliated at once. They are so destructive, ruining the nests and young of my precious little birds that I hate them with a perfect hatred and I of late embrace every opportunity that I have of blowing them to perdition."

Despite its small size, the red squirrel has been a source of meat for Indians (Cleland, 1966), for pioneers, and for present-day outdoorspersons. Even so, there are presently no Michigan laws protecting red squirrels. The durable and attractive pelts of the red squirrel have long been used in making clothes and decorations by Indians. Marvin Cooley of Michigan's Department of Natural Resources reports that Michigan red squirrel pelts are not marketed commercially. In Canada, however, more than 25,000 skins are sold annually (Banfield, 1974). Officials of the Ontario Trappers Association buy squirrels skins primarily from western provinces although some are imported from Colorado. In 1978, the price per raw pelt was as high as $2.95, with skins being shipped to Europe for processing and sale. Pelts (especially tail hairs) are also in demand by flytiers and fishing lure manufacturers.

The red squirrel will probably never live down its overrated reputation for discouraging fox and gray squirrels. Even so, the red squirrels' acrobatics are fascinating to the mammal watcher. Biologists also like to believe that red squirrels play a part in reforestation by leaving a few unfound tree seeds in caches.

COUNTY RECORDS FROM MICHIGAN. Records of specimens of red squirrels in collections of museums and from the literature show that this species lives in all counties of the Upper and Lower Peninsulas of Michigan. There are also records from North and South Manitou islands in Lake Michigan (Hatt *et al.,* 1948), from Isle Royale in Lake Superior (Burt, 1946), and from Bois Blanc and Drummond islands in Lake Huron (Barry, 1971; Manville, 1950).

LITERATURE CITED

Ackerman, R., and P. D. Weigl
1970 Dominance relations of red and gray squirrels. Ecology, 51(2):332–334.

Allen, D. L.
1943 Michigan fox squirrel management. Michigan Dept. Conservation, Game Div. Publ. 100, 404 pp.

Banfield, A. W. F.
1974 The mammals of Canada. Univ. Toronto Press, Toronto. xxiv+438 pp.

Barry, W. J.
1971 The introduction of gray and black squirrels to Bois Blanc Island, Michigan. Unpubl. project report to Michigan Dept. Nat. Res., 11 pp.

Beckwith, S. L.
1954 Ecological succession in abandoned farm lands and its relationship to wildlife management. Ecol. Monog., 24:349–376.

Benton, A. H.
1958 Melanistic red squirrels from Cayuga County, New York. Jour. Mammalogy, 39(3):445.

Burt, W. H.
1946 The mammals of Michigan. Univ. Michigan Press, Ann Arbor. xv+288 pp.

Cahalane, V. H.
1947 Mammals of North America. The Macmillan Co., New York. x+682 pp.

Cleland, C. E.
1966 The prehistoric animal ecology and ethnozoology of the Upper Great Lakes Region. Univ. Michigan, Mus. Anthro., Anthro. Papers, No. 29, x+294 pp.

Craighead, J. J., and F. C. Craighead, Jr.
1956 Hawks, owls and wildlife. The Stackpole Co., Harrisburg, Pa. xix+443 pp.

Dapson, R. W.
1963 Color aberrations in the red squirrel. Jour. Mammalogy, 44(1):123.

Dearborn, N.
1932 Foods of some predatory fur-bearing animals in Michigan. Univ. Michigan, School For. and Cons., Bull. No. 1, 52 pp.

Dice, L. R.
1925 A survey of the mammals of Charlevoix County, Michigan, and vicinity. Univ. Michigan., Mus. Zool., Occas. Papers 159, 33 pp.

Dorney, R. S.
1965 Incidence of botfly larvae (*Cuterebra emasculator*) in the chipmunk (*Tamias striatus*) and the red squirrel (*Tamiasciurus hudsonicus*) in northern Wisconsin. Jour. Parasitology, 51(5):893–894.
1966 Quantitative data on four species of *Eimeria* in eastern chipmunks and red squirrels. Jour. Protozool., 13(4): 549–550.

Doutt, J. K., C. A. Heppenstall, and J. E. Guilday
1966 Mammals of Pennsylvania. Pennsylvania Game Comm., Harrisburg. 273 pp.

Elliott, P. F.
1974 Evolutionary responses of plants to seed-eaters: Pine squirrel predation on lodgepole pine. Evolution, 28(2):221–231.

Fancy, S. G.
1980 Nest-tree selection by red squirrels in a boreal forest. Canadian Field-Nat., 94(2):198.

1981 Daily movements of red squirrels, *Tamiasciurus hudsonicus*. Canadian Field-Nat., 95(3):348–350.

Ferron, J.
1980 Ontogénèse du comportement de l'Ecureuil roux (*Tamiasciurus hudsonicus*). Canadian Jour. Zool., 58(6): 1090–1099.

Ferron, J., and J. Prescott
1977 Gestation, litter size, and number of litters of the red squirrel (*Tamiasciurus hudsonicus*) in Quebec. Canadian Field-Nat., 91:83–84.

Field, R. J.
1970 Scavengers feeding on a Michigan deer carcass. Jack-Pine Warbler, 48(2):73.

Finley, R. B., Jr.
1969 Cone caches and middens of *Tamiasciurus* in the Rocky Mountain Region. Univ. Kansas, Mus. Nat. Hist., Misc. Publ. No. 51:233–273.

Hamilton, W. J., Jr.
1939 Observations on the life history of the red squirrel in New York. American Midl. Nat., 22(4):732–745.

Hatfield, D. M.
1937 Notes on Minnesota squirrels. Jour. Mammalogy, 18(2):242–243.

Hatt, R. T.
1929 The red squirrel: Its life history and habits, with special reference to the Adirondacks of New York and Harvard Forest. New York State Col. For. at Syracuse Univ. Roosevelt Wild Life Annals, 2(1):1–146.

Hatt, R. T., J. Van Tyne, L. C. Stuart, C. H. Pope, and A. B. Grobman
1948 Island life: A study of the land vertebrates of the islands of eastern Lake Michigan. Cranbrook Inst. Sci., Bull. No. 27, xi+179 pp.

Hazard, E. B.
1960 A field study of activity among squirrels (Sciuridae) in southern Michigan. Univ. Michigan, unpubl. Ph.D. dissertation, 295 pp.

Howard, W. J.
1935 Apparently neutral relations of weasel and squirrel. Jour. Mammalogy, 16(4):322–323.

Irving, L., H. Krog, and M. Monson
1955 The metabolism of some Alaskan animals in winter and summer. Physiol. Zool., 29:173–185.

Jackson, H. H. T.
1961 Mammals of Wisconsin. Univ. Wisconsin Press, Madison. xii+504 pp.

Johnson, W. J.
1970 Food habits of the red fox in Isle Royale National Park, Lake Superior. American Midl. Nat., 84(2):568–572.

Kilham, L.
1954 Territorial behaviour of red squirrel. Jour. Mammalogy, 35(2):252–253.
1958 Red squirrels feeding at sapsucker holes. Jour. Mammalogy, 39(4):596–597.

Klugh, A. B.
1927 Ecology of the red squirrel. Jour. Mammalogy, 8(1):1–32.

Kramm, K. R., D. E. Maki, and J. M. Glime
1975 Variation within and among populations of red squirrel in the Lake Superior region. Jour. Mammalogy, 56(1):258–262.

Layne, J. N.
1954 The biology of the red squirrel, *Tamiasciurus hudsonicus loquax* (Bangs), in central New York. Ecol. Monog., 24(3):227–267.

Linduska, J. P.
1950 Ecology and land-use relationships of small mammals on a Michigan farm. Michigan Dept. Cons., Game Div. ix+144 pp.

Lynch, G. R., and G. E. Folk, Jr.
1968 Distribution and habitat of the red squirrel, *Tamiasciurus hudsonicus* in the North Central States. Iowa Academy Sci., 75:463–466.

Manville, R. H.
1949 Highway mortality in northern Michigan. Jour. Mammalogy, 30(3):311–312.
1950 The mammals of Drummond Island, Michigan. Jour. Mammalogy, 31(3):358–359.

Mayfield, H.
1948 Red squirrel nesting on the ground. Jour. Mammalogy, 29(1):186.

Mech, L. D.
1966 The wolves of Isle Royale. National Park Service, Fauna Ser. 7, xiv+210 pp.

Millar, J. S.
1970 The breeding season and reproductive cycle of the western red squirrel. Canadian Jour. Zool., 48:471–473.

Nellis, C. H.
1969 Sex and age variation in red squirrel skulls from Missoula County, Montana. Canadian Field-Nat., 83: 324–330.

Nelson, B. A.
1945 The spring molt of the northern red squirrel in Minnesota. Jour. Mammalogy, 26(4):397–400.

Pettingill, O. S., Jr.
1976 The prey of six species of hawks in northern Lower Michigan. Jack-Pine Warbler, 54(2):70–74.

Preston, F. W.
1948 Red squirrels and gray. Jour. Mammalogy, 29(3):297–298.

Reichard, T. A.
1976 Spring food habits and feeding behavior of fox squirrels and red squirrels. American Midl. Nat., 96(2): 443–450.

Rusch, D. A., and W. G. Reeder
1978 Population ecology of Alberta red squirrels. Ecology, 59(2):400–420.

Scharf, W. C., and K. R. Stewart
1980 New records of Siphonaptera from northern Michigan. Great Lakes Ento., 13(3):165–167.

Schofield, R. D.
1960 A thousand miles of fox trails in Michigan's ruffed grouse range. Jour. Wildlife Mgt., 24(4):432–434.

Seton, E. T.
1953 Lives of game animals. Vol. IV, Pt. I. Charles T. Bradford Co., Boston. xiii+440 pp.

Smith, C. C.
1968 The adaptive nature of social organization in the genus of three [*sic*] squirrels *Tamiasciurus*. Ecol. Monog., 38(1):31–63.
1970 The coevolution of pine squirrels (*Tamiasciurus*) and conifers. Ecol. Monog., 40(3):369–371.
1978 Structure and function of the vocalizations of tree squirrels (*Tamiasciurus*). Jour. Mammalogy, 59(4): 793–808.
1981 The indivisible niche of *Tamiasciurus*: an example of nonpartitioning of resources. Ecol. Monog., 51(3): 343–363.

Smith, C. F., and S. E. Aldous
1947 The influence of mammals and birds in retarding artificial and natural reseeding of coniferous forests in the United States. Jour. Forestry, 45:361–369.

Smith, M. C.
1968 Red squirrel responses to spruce cone failure in interior Alaska. Jour. Wildlife Mgt., 32(2):305–317.

Svihla, R. D.
1930 Development of young red squirrels. Jour. Mammalogy, 11(1):79–80.

Timm, R. M.
1975 Distribution, natural history, and parasites of mammals of Cook County, Minnesota. Univ. Minnesota, Bell Mus. Nat. Hist., Occas. Papers No. 14, 56 pp.

Weigl, P. D., and E. V. Hanson
1980 Observational learning and the feeding behavior of the red squirrel *Tamiasciurus hudsonicus:* The ontogeny of optimization. Ecology, 61(2):213–218.

Wilscon, K. A.
1938 Owl studies at Ann Arbor, Michigan. Auk, 55(2):187–197.

Wood, N. A.
1922 The mammals of Washtenaw County, Michigan. Univ. Michigan, Mus. Zool., Occas. Papers No. 123, 23 pp.

Wunder, B. A., and P. R. Morrison
1974 Red squirrel metabolism during incline running. Comp. Biochem. Physiol., 48A:153–161.

Yahner, R. H.
1980 Burrow system used by red squirrels. American Midl. Nat., 103(2):409–411.

Yeager, L. E.
1937 Cone-piling by Michigan red squirrels. Jour. Mammalogy, 18(2):191–194.

NORTHERN FLYING SQUIRREL
Glaucomys sabrinus (Shaw)

NAMES. The generic name *Glaucomys*, proposed by the English mammalogist Oldfield Thomas in 1908, is of Greek derivation and translated as "gray" (from *glaukos*) and "mouse" (from *mys*). The specific name *sabrinus*, proposed by Shaw in 1801, is derived from the Roman name of an English river, later changed to Severn. It is supposed that Shaw chose this name because the specimen on which the original description was based came from the Severn River in northwestern Ontario. In Michigan, local residents generally call this species flying squirrel and also apply this name to the southern flying squirrel because there are no obvious, distinguishable differences between them. The subspecies *Glaucomys sabrinus macrotis* (Mearns) occurs in Michigan.

RECOGNITION. Body size that of a small Norway rat; head and body of an adult averaging approximately 6¼ in (160 mm) long, with fluffy, well-haired tail flattened dorsoventrally and almost one-half as long as the body; body flattened, accentuated by a lateral fold of loose skin between the wrists and the ankles which forms a flattened, fur-covered gliding surface (plagiopatagium, see Gupta, 1966) when the feet are extended; a cartilaginous "spur" (styliform cartilage) extends from the wrist to support the lateral spread at the anterior margin of the gliding membrane; size of head accentuated by large, black-lidded eyes; ears long, somewhat pointed and sparsely haired; pelage silky in texture and dense; vibrissae prominently long. Color of upperparts and sides cinnamon, individual hairs basally slate-colored and terminally light yellowish brown; sides of head smoky gray; underparts whitish and washed with pale buff, individual hairs basally slate colored; tail drab grayish brown above, darker toward the tip, and paler and more buffy-gray below; feet drab gray above, whitish below, with soles well-furred in winter. There are no size or color differences between sexes. The mammary glands are arranged in four pairs, with two pectoral, one abdominal, and one inguinal pair of teats. The four functional digits on the forefoot and the five on the hind foot have sharp, curved claws. In external features, the northern flying squirrel is distinguished from the southern flying squirrel by its slightly larger size, more reddish and brighter coloring on upperparts, and by the presence of basally slate colored but white-tipped hairs on the underparts (instead of pure white as in *G. volans*).

MEASUREMENTS AND WEIGHTS. Adult northern flying squirrels measure as follows: from tip of nose to end of tail vertebrae, 9¾ to 12¼ in (250 to 310 mm); length of tail vertebrae, 4¼ to 5¾ in (110 to 150 mm); length of hind foot, 1⅜ to 1½ in (36 to 40 mm); height of ear from notch, ¾ to 1 in (18 to 25 mm); weight of adults, 2.6 to 4.4 oz (70 to 130 g).

DISTINCTIVE CRANIAL AND DENTAL CHARACTERISTICS. The skulls of both the northern flying squirrel and the southern flying squirrel have a short and roundish appearance, more like the larger red squirrel than the Michigan ground squirrels and chipmunks (Fig. 31). The skull of the northern flying squirrel averages almost 1½ in (37 mm) long and ⅞ in (23 mm) wide. Features include brain case depressed posteriorly, rostrum narrow and flat, space between orbits broad, and postorbital processes broad at base and tapering distally to a point (Fig. 45). The skulls of both species of flying squirrels can be distinguished

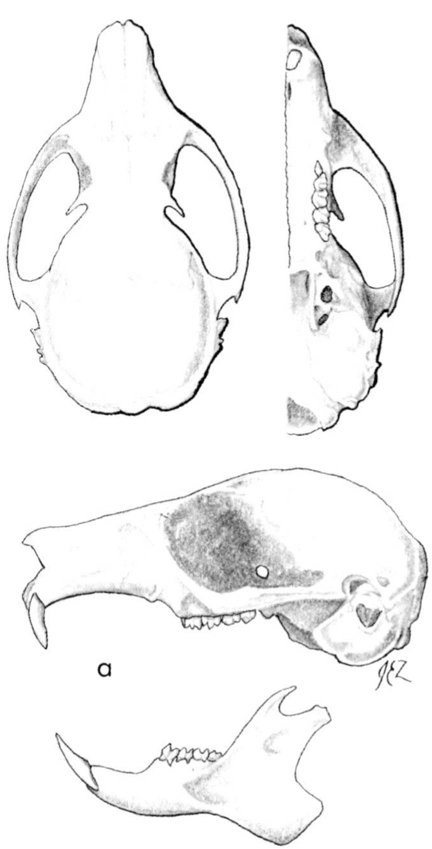

Figure 45a. Cranial and dental characteristics of the northern flying squirrel (*Glaucomys sabrinus*).

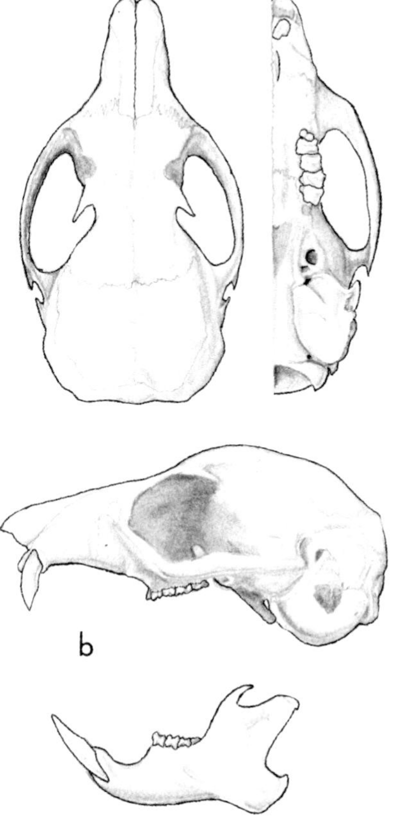

Figure 45b. Cranial and dental characteristics of the southern flying squirrel (*Glaucomys volans*).

from those of other Michigan mammals by a combination of skull length and dental formula. The skull of the northern flying squirrel appears heavier (less fragile), larger, and possesses a longer cheek tooth-row than that of the southern flying squirrel. The dental formula for the northern flying squirrel is:

$$\text{I (incisors)}\frac{1\text{-}1}{1\text{-}1},\ \text{C (canines)}\frac{0\text{-}0}{0\text{-}0},\ \text{P (premolars)}\frac{2\text{-}2}{1\text{-}1},\ \text{M (molars)}\frac{3\text{-}3}{3\text{-}3} = 22$$

DISTRIBUTION. The northern flying squirrel is distributed in the broad belt of boreal and primarily evergreen forests across northern North America from the treeline in Canada and Alaska southward in the west to the montane forests of California and Colorado and in the east to the montane forests of North Carolina and Tennessee. The boreal forests of the Upper Great Lakes Region are also occupied. In Michigan, the northern flying squirrel occurs throughout the Upper Peninsula and in most counties in the northern half of the Lower Peninsula, at least as far south as Gratiot and Montcalm counties (see Map 32). The

Map 32. Geographic distribution of the northern flying squirrel (*Glaucomys sabrinus*). Actual records exist for counties with dark shading.

precise systematic relationships of the isolated Lower Peninsula population to the one in the Upper Peninsula, across the Straits of Mackinac, are not fully known.

HABITAT PREFERENCES. Habitat associations of the nocturnal northern flying squirrel are less documented than those of diurnal members of the squirrel family, although it is agreed that the species thrives in heavily wooded areas containing mixed conifers and northern hardwoods having mature growth (N. M. Gosling, pers. comm.). Preferred environments are moist with a forest floor littered with moss-covered logs. Pure stands of white cedar, spruce, balsam fir or hardwoods are perhaps less productive habitat (Jackson, 1961) than mixtures of hemlock-maple or hemlock-birch. In Michigan, northern flying squirrels have been observed in coniferous swamps in Montmorency County (Green, 1925) and in jack pine and cutover hardwoods and in mixed hardwoods and conifers in the Huron Mountains (Laundre, 1975; Manville, 1948). In Quebec, Wrigley (1960) listed a favorable habitat as a mature mixed forest of yellow birch, sugar maple, hemlock, balsam fir, and white spruce.

DENSITY AND MOVEMENTS. The northern flying squirrel's nocturnal habits make study difficult. Grids of traps set in typical forest habitat seem to be more attractive to other small mammal species. Tapping on tree trunks containing cavities usually causes a resident flying squirrel to show itself or leave the cavity (Jackson, 1961). Tracks in snow also indicate the animal's presence (Laundre, 1975). Studies in North Carolina and Pennsylvania (Weigl and Osgood, 1974) showed that northern flying squirrels (tracked by means of radiotransmitter-bearing individuals) may occupy a home range from 8 to 32 acres (3.2 to 12.6 ha). In the Huron Mountains, Manville (1949) estimated that at least one flying squirrel per 14 acres (5.6 per ha) was present on two of his study areas—one containing jack, white and Norway pines, and a few hardwoods; the other primarily sugar maple and birch with some hemlocks. In Michigan, field observers' estimations vary as to northern flying squirrel populations. Manville (1949) felt the species was abundant, while Laundre (1975) recorded it as being common. On the other hand, Jackson (1961) believed that population densities of three or four individuals per acre (7 to 10 per ha) might exist in favorable habitats in Wisconsin. Populations would be most dense, of course, immediately following the appearance of the weaned offspring.

BEHAVIOR. The northern flying squirrel has many of the behavioral characteristics of other Michigan sciurids but, like the southern flying squirrel, is able to glide and is nocturnal. In diurnal squirrels there is preretinal absorbance of considerable light (the blue and ultraviolet regions of the spectrum), whereas in flying squirrels this absorbance is minimal—an adaptation for night vision (Yolton *et al.*, 1974). By being active at night, the flying squirrel can occupy almost identical environments to those of neighboring diurnal chipmunks and tree squirrels. This habitat sharing allows for utilization of similar resources, including foods, without actual competitive association. Certainly the flying squirrels, especially the northern species, must be hardy and well-protected by dense fur in order to survive active nocturnal life during winter. The species ordinarily remains in its protected nest during stormy and cold spells, although there is abundant evidence that flying squirrels are active on sub-zero nights (Jackson, 1961).

The ability to glide, by means of a flattened body and extended plagiopatagium (which is not flapped like a wing) stretched between fore and hind limbs, has evolved in a few remotely-related mammals. Remarkably similar specializations are also found in the pouched order Marsupialia (the sugar gliders of Australia and New Guinea) and the primitive placental order Dermoptera (the flying lemurs of Southeast Asia and adjacent islands). Each of these mammals is an adept climber and uses gliding only as one means of locomotion from one object to another—in flying squirrels usually from tree to tree, or tree to a fallen log, stump or forest floor. The animal begins the leap from an elevated position, with the glide initially at an angle of about 30°, then levels off, and swings upwards a bit just before landing (Muul and Alley, 1963; Walker, 1947). Gliding leaps have been examined and photographed; the animal can change course or maneuver in flight by altering body shape. Volplaning distances up to 50 yds (46 m) have been measured (Howell, 1918). The derivation of the gliding adaptation in both North American and Old World flying squirrels is obscure. Because these sciurids have been highly successful members of forest communities, gliding must play an important role in the survival of flying squirrels and their welfare as viable woodlands species.

Unlike the less social, diurnal, tree squirrels, the nocturnal flying squirrels are conspicuously compatible with one another—more so in winter than in summer. Groups share dens during the daytime (Wenzel, 1912); Seton (1953) reported finding nine individuals in one stump in December. In the Huron Mountains, Manville (1948) counted three in a cavity in a dead Norway pine on October 15. The animals also associate at feeding stations and during play; the adults usually dominate the young. There is some evidence that female southern flying squirrels aggressively defend home territories while males do not (Madden, 1974). Similar action should be expected of female northern flying squirrels. Interactions between northern and southern flying squirrels, where their ranges overlap (as in parts of Michigan) need more study (Weigl, 1978).

Flying squirrels begin their nightly forays at dusk, probably overlapping occasionally with twilight-active red squirrels. Maximum activity occurs just after dark and again near dawn (Weigl and Osgood, 1974). In normal nightly activity the northern flying squirrel spends considerable time foraging on the ground (Jackson, 1961), as is evident from the large number of captures in ground-set traps and from extensive tracks in snow. These squirrels move around primarily by a hopping gait, jumping from log to log or across the litter-covered forest floor. Jackson (1961) suggested that they attain speeds of eight mi (13 km) per hour. In severe and stormy weather northern flying squirrels remain holed-up, using stored food for sustenance.

Northern flying squirrels den in cavities in living or dead trees, and sometimes in houses and other structures; they probably have to compete with the larger tree squirrels, raccoons, and other arboreal animals for hollows with large openings. A favorite northern flying squirrel site is a woodpecker hole, positioned so that rain or snow does not drain into the denning space. Rotted out areas in snags, topped-off trees, and even cavities in old telephone poles can be enlarged and deepened to provide suitable housing, especially for winter occupancy. In summer, like other tree squirrels, the northern flying squirrel constructs bark and twig nests (Howell, 1918). These are usually located in young evergreen trees well above the ground and on branches adjacent to sturdy trunks; they consist of shredded bark, lichens, mosses, with coarser material to the outside and finer material to the inside as lining to the nest cavity (Cowan, 1936). A side entrance, usually facing the tree trunk, is normal. On occasion, the northern flying squirrel will remodel a nest abandoned by a red squirrel or even by a crow or raven.

Observers have reported that the northern flying squirrel is less vocal than the southern flying squirrel. Even so, the animals appear to communicate by soft, low-pitched "chuckling" chirps. According to Jackson (1961), they probably do not utter an alarm call (a "kuk-kuk-kuk" sound as given by other squirrels), although Goodwin (1924) described a shrill cry emitted when the animals were captured and heard again in the Quebec woods at night. It is suspected, however, that further study of flying squirrel vocalizations will reveal that they have a much more sophisticated repertoire of calls than is generally known.

ASSOCIATES. The northern flying squirrel certainly has an opportunity to interact in terms of den and food selections with other woodland sciurids inhabiting northern Michigan. However, its association may not be close with the diurnal gray, fox, and red squirrels and the eastern and least chipmunks. There must be those occasions, nevertheless, when the nocturnal flying squirrel does interact with these species when attempting to share the same dens or helping themselves to each other's food caches. There is evidence of conflict between the northern and the southern flying squirrels in their zone of distributional overlap in the northern part of the Lower Peninsula and in the western end of the Upper Peninsula. These relationships have been examined by Muul (1968) and Weigl (1978) and summarized herein under Associates in the account of the southern flying squirrel. In areas of the Upper Peninsula where the southern species does not occur, the only nocturnal semi-arboreal rodent which might be a close associate of the northern flying squirrel and perhaps a competitor is the northern subspecies of the deer mouse (*Peromyscus maniculatus gracilis*). It is interesting to contemplate that Michigan's northern forests of cone-bearing and nut-producing trees have so few nocturnal arboreal rodents as residents. On the forest floor and adjacent borders with open areas, the species of nocturnal rodents is comparatively high. In contrast, in the oak forests in the highlands of Chiapas in southern Mexico, the resident southern flying squirrel has indeed a large number of species of nocturnal climbing rodents with which to share the mast crop (del Toro, 1977). One might conclude as far as interspecific competition for foods and refuges is concerned that Michigan flying squirrels may have less complicated life strategies than their close relative in Latin America. In either place, however, tree-climbing predators are diverse in species numbers and always welcome meals consisting of tender flying squirrels.

REPRODUCTIVE ACTIVITIES. Northern flying squirrels normally have a single litter in spring, although some females have a second litter in late summer (Banfield, 1974; Doutt *et al.,* 1966), possibly because the first litter failed to survive. In Ontonagon County, Dice and Sherman (1922) counted five small embryos in a female taken on July 6, indicating the possibility of a late July-early August birth. The courtship and breeding activity begins in March and may continue until late May. Muul (1969) noted a captive male chasing a female and stamping his feet intermittently while uttering high pitched, pulsed, nasal sounds ("widy-widy-widy"). Calls by confined individuals during the mating act include a "churring" sound by the female and a nasal "whine" by the male.

Following a gestation period of about 37 days (Muul, 1969), an average of three young (as few as two and as many as six) are born in a lined nest either in a tree cavity or often in a twig and bark nest. Newborn young are 2¾ in (70 mm) long and weigh about 0.4 oz (5.8 g). They are hairless, pinkish in color, have both eyes and ears closed, and show evidence of the lateral gliding membrane. At six days of age vibrissae begin to grow, at 18 days the body is well-furred, at 31 days the eyes open, at 35 days the flattened tail hair growth is evident, at 40 days the young venture from the nest for short periods, at 47 days solid food is taken and the weaning process begins, at 60 days weaning is complete, and at 90 days gliding is accomplished and parental independence achieved (Banfield, 1974; Jackson, 1961; Muul, 1969).

A rather low reproductive rate infers only a modest annual population turnover for this small sciurid, meaning that individuals are probably long-lived in comparison to similar-sized mammals with higher reproductive rates. In fact, Jackson (1961) suggested that the normal longevity in the wild may be as much as four years.

FOOD HABITS. Like other squirrels, the northern flying squirrel basically prefers seeds but will also feed on assorted animal and plant foods when seasonally available. In late winter and spring, stored seeds of conifers and hardwoods are the major food supply; tree buds, bark, and lichens provide supplementary nourishment. In summer, insects (especially moths and beetles), bird eggs and nestlings provide animal protein, while fleshy fruits of trees and shrubs supply vegetative bulk. Fungi (Wrigley, 1969) and maple sap (Seton, 1953) are eaten in spring. The caching of the autumn seed crop is probably the most important requirement for this species survival. Northern flying squirrels consume quantities of conifer seeds (Brink and Dean, 1966; Smith and Aldous, 1947), which are cached in the cones. Beechnuts, hazelnuts, acorns, and evergreen cones may be piled on the ground or in hollow stumps but are usually stored well above ground in hollow trees. Northern flying squirrels occasionally supplement their own supplies by raiding the stores cached by diurnal tree squirrels.

ENEMIES. Because northern flying squirrels are active both on the ground and in trees, they are taken by a variety of medium-sized and large predators. These nocturnal sciurids escape most diurnal hawks but are fair game for owls. Chief enemies include the barred owl (Seton, 1953), other owls, occasionally the goshawk, red and gray foxes, weasels, marten and other flesh-eating mammals. Gray wolf (Voigt *et al.*, 1976), bobcat (Pollack, 1951), and house cats (Toner, 1956) also eat northern flying squirrels. Seton (1953) even recorded a northern flying squirrel being found in the digestive system of a trout in British Colombia. The domesticated house cat is reportedly very proficient at catching these animals. Forest fires cause high mortality in the northern flying squirrel. Trappers of mink and marten often catch this squirrel in their sets. In Mackinac County, one animal died after becoming entangled in barbwire on a fence (Findley, 1945). Plant material found embedded in subcutaneous tissue indicates a natural-occurring accident which may befall this climber and glider (Wrigley, 1975).

PARASITES AND DISEASE. Manville (1949) reported two fleas, *Opisodasys pseudarctomys* and *Orchopeas wickhami*, from northern flying squirrels taken in Marquette County. Other external parasites include lice, mites, and chiggers. Internal parasites are flatworms and roundworms (Jackson, 1961).

MOLT AND COLOR ABERRATIONS. The northern flying squirrel molts once annually, in early autumn. The dense, new pelage provides insulation for the long, boreal winter. By summer, however, normal wear on hair tips gives the flying squirrel a less than well-groomed appearance and darker color. So far, there have been no published records of albinism or melanism in the northern flying squirrel.

ECONOMIC IMPORTANCE. Northern flying squirrels, being nocturnal, rarely make their presence known to the human population unless they select houses or barns as den sites. This may be a nuisance for householders because the animals can (1) be noisily active at night, and (2) litter areas with nest materials and seed caches. The flying squirrel's use of conifer seeds possibly inhibits forest reproduction (Smith and Aldous, 1947). Northern trappers are apt to dislike northern flying squirrels because of their fondness for meat (Gunderson and Beer, 1953), which leads to their being killed in traps set for martin and mink. Their soft fur has never had commercial value. Because the northern flying squirrel is rarely abundant, Long (1974) felt that the species should be on the threatened list in Wisconsin and Michigan, as it is in such states as North Carolina (Cooper *et al.*, 1977).

COUNTY RECORDS FROM MICHIGAN. Records of specimens of northern flying squirrels in collections of museums and from the literature show that this species lives in all counties of the Upper Peninsula. The species occurs in suitable habitat in the northern half of the Lower Peninsula, including Bay, Charlevoix, Cheboygan, Crawford, Emmet, Grant Traverse, Gratiot, Isabella, Leelanau, Montcalm, Montmorency, Oscoda, Otsego, Presque Isle, Osceola, Roscommon, and Wexford counties. It is also known from Drummond and Sugar islands in the Superior-Huron Waterway Connection (Manville, 1950).

LITERATURE CITED

Banfield, A. W. F.
1974 The mammals of Canada. Univ. Toronto Press, Toronto. xxv+438 pp.

Brink, C. H., and F. C. Dean
1966 Spruce seed as a food of red squirrels and flying squirrels in interior Alaska. Jour. Wildlife Mgt., 30(3): 503–512.

Cooper, J. E., S. S. Robinson, and J. B. Funderburg
1977 Endangered and threatened plants and animals of North Carolina. North Carolina State Mus. Nat. Hist., Raleigh. xvi+444 pp.

Cowan, I. M.
1936 Nesting habits of the flying squirrel *Glaucomys sabrinus*. Jour. Mammalogy, 17(1):58–60.

del Toro, M. A.
1977 Los mamiferos de Chiapas. Univ. Auton. de Chiapas. Tuxtla Gutierrez. 147 pp.

Dice, L. R., and H. B. Sherman
1922 Notes on the mammals of Gogebic and Ontonagon counties, Michigan. 1920. Univ. Michigan, Mus. Zool., Occas. Papers No. 109, 40 pp.

Doutt, J. K., C. A. Heppenstall, and J. E. Guilday
1966 Mammals of Pennsylvania. Pennsylvania Game Comm., Harrisburg. 273 pp.

Findley, J. S.
1945 The interesting fate of a flying squirrel. Jour. Mammalogy, 26(4):437.

Goodwin, G. G.
1924 Mammals of the Gaspé peninsula, Quebec. Jour. Mammalogy, 5(4):246–257.

Green, M. M.
1925 Notes on some mammals of Montmorency County, Michigan. Jour. Mammalogy, 6(3):173–178.

Gunderson, H. L., and J. R. Beer
1953 The mammals of Minnesota. Univ. Minnesota Press, Minneapolis. xii+190 pp.

Gupta, B. B.
1966 Notes on the gliding mechanism in the flying squirrel. Univ. Michigan, Mus. Zool., Occas. Papers No. 645, 7 pp.

Howell, A. H.
1918 Revision of the American flying squirrels. N. American Fauna, No. 44, 64 pp.

Jackson, H. H. T.
1961 Mammals of Wisconsin. Univ. Wisconsin Press, Madison. xii+504 pp.

Laundre, J.
1975 An ecological survey of the mammals of the Huron Mountain area. Huron Mountain Wildlife Foundation, Occas. Papers No. 2, x+69 pp.

Long, C. A.
1974 Mammals of the Lake Michigan drainage basin. Vol. 15, Environmental status of the Lake Michigan region. Argonne Nat. Lab., 108 pp.

Madden, J. R.
1974 Female territoriality in a Suffolk County, Long Island, population of *Glaucomys volans*. Jour. Mammalogy, 55(3):647–652.

Manville, R. H.
1948 The vertebrate fauna of the Huron Mountains, Michigan. American Midl. Nat., 39(3):615–640.
1949 A study of small mammal populations in northern Michigan. Univ. Michigan, Mus. Zool., Misc. Papers, No. 73, 93 pp.
1950 The mammals of Drummond Island, Michigan Jour. Mammalogy, 31(3):358–359.

Muul, I.
1069 Mating behavior, gestation period, and development of *Glaucomys sabrinus*. Jour. Mammalogy, 50(1):121.

Muul, I., and J. W. Alley
1963 Night gliders of the woodlands. Nat. Hist. 72:18–25.

Pollack, E. M.
1951 Food habits of the bobcat in the New England states. Jour. Wildlife Mgt., 15(2):209–213.

Seton, E. T.
1953 Lives of game animals. Vol. IV, Pt. I. Charles T. Branford Co., Boston. xxii+440 pp.

Smith, C. F., and S. E. Aldous
1947 The influence of mammals and birds in retarding artificial and natural reseeding of coniferous forests in the United States. Jour. Forestry, 45:361–369.

Toner, G. C.
1956 House cat predation on small animals. Jour. Mammalogy, 37(1):119.

Voigt, D. R., G. B. Kolenosky, and D. H. Pimlott
1976 Changes in summer foods of wolves in central Ontario. Jour. Wildlife Mgt., 40(4):663–668.

Walker, E. P.
1947 "Flying" squirrels, nature's gliders. Nat. Geogr. Mag., 91(5):662–674.

Weigl, P. D.
1969 The distribution of the flying squirrels, *Glaucomys volans* and *G. sabrinus*: an evaluation of the competitive exclusion idea. Duke Univ., unpubl. Ph.D. dissertation, 246 pp.
1978 Resource overlap, interspecific interactions and the distribution of the flying squirrels, *Glaucomys volans* and *G. sabrinus*. American Midl. Nat., 100(1):83–96.

Weigl, P. D., and D. W. Osgood
1974 Study of the northern flying squirrel, *Glaucomys sabrinus*, by temperature telemetry. American Midl. Nat., 92(2):482–486.

Wenzel, O. J.
1912 A collection of mammals from Osceola County, Michigan. Michigan Acad. Sci. Arts and Ltrs., 14th Rept., pp. 198–205.

Wrigley, R. E.
1969 Ecological notes on the mammals of southern Quebec. Canadian Field-Nat., 83:201–211.
1975 Poplar bud in the subcutaneous tissue of a northern flying squirrel. Canadian Field-Nat., 89:466.

Yolton, R. L., D. P. Yolton, J. Renz, and G. H. Jacobs
1974 Preretinal absorbance in sciurid eyes. Jour. Mammalogy, 55(1):14–20.

SOUTHERN FLYING SQUIRREL
Glaucomys volans (Linnaeus)

NAMES. The generic name *Glaucomys,* proposed by the English mammalogist Oldfield Thomas in 1908, is of Greek origin and translated as "gray" (from *glaukos*) and "mouse" (from *mys*). The specific name *volans,* proposed by Linnaeus in 1758, is from the Latin and means "flying." In Michigan, local residents generally call this species flying squirrel or sometimes eastern or small flying squirrel. The subspecies *glaucomys volans volans* (Linnaeus) occurs in Michigan.

RECOGNITION. Body size small for an arboreal squirrel (Fig. 46), about the size of a small Norway rat; head and body of an adult averaging approximately 5½ in (140 mm) long, with a dorsoventrally flattened and densely-haired tail slightly less than half as long as the body; body with flattened appearance, accentuated by loosely folded and fully haired gliding membrane or plagiopatagium (Gupta, 1966) extending laterally between wrists and ankles; a cartilaginous "spur" (styliform cartilage) leading laterally and posteriorly from each wrist supports the lateral spread of the anterior margin of the gliding membrane; size of head large and rounded, black-lidded eyes, large and conspicuous; ears long, somewhat pointed and sparsely haired; pelage silky in texture, dense and rather uniformly colored; vibrissae long, blackish and prominent. Color of upperparts and sides pale, drab brown resulting from the color of the tips of the body hairs which are basally slate colored; gliding membrane darker because brownish tips of hairs are reduced or absent; sides of head and neck smoky gray; underparts from chin to base of tail with hairs creamy white from base to tip; flattened tail pale buffy-brown above and pale grayish brown below; feet gray above and whitish below. Sexes do not differ in color. Mammary glands are arranged in four pairs, with two pectoral, one abdominal, and one inguinal. The four functional digits on the forefoot and the five on the hind foot have sharp, curved claws. In external features, the southern flying squirrel is easily distinguished from all other sciurids except the northern flying squirrel by the presence of the lateral gliding membranes. The southern flying squirrel is smaller than the northern flying squirrel and the belly fur is pure creamy white from base to tip in the southern flying squirrel but basally slate-colored in the northern flying squirrel.

MEASUREMENTS AND WEIGHTS. Adult southern flying squirrels measure as follows:

Figure 46. The southern flying squirrel (*Glaucomys volans*).
Sketches from life by Bonnie Marris

length from tip of nose to end of tail vertebrae, 8½ to 10 in (220 to 255 mm); length of tail vertebrae, 3⅛ to 4¼ in (80 to 110 mm); length of hind foot, 1 to 1¼ in (26 to 33 mm); height of ear from notch, ½ to ⅝ in (13 to 18 mm); weight, 1.8 to 2.5 oz (50 to 79 g).

DISTINCTIVE CRANIAL AND DENTAL CHARACTERISTICS. The skull of the southern flying squirrel closely resembles that of the northern flying squirrel except in dimension (Fig. 31), with the northern flying squirrel having a longer, wider skull and a longer upper cheek tooth-row (see Key to the Species of Rodentia). The cranium of the southern flying squirrel has a short, roundish appearance with a stubby rostrum (nasal area). The skull of the southern flying squirrel averages 1⅜ in (35 mm) long and ⅞ in (21 mm) wide. Cranial features (Fig. 45b) include delicate construction (compared to those of ground and tree squirrels), nasals short and depressed anteriorly, dorsal profile of skull straight from nasals to area above orbits, space between orbits broad, and postorbital processes broad basally and pointed distally (Dolan and Carter, 1977). The skulls of the southern flying squirrel and the northern flying squirrel differ from those of other Michigan mammals by a combination of skull length and dental formula (Burt, 1946). The dental formula for the southern flying squirrel is:

$$\text{I (incisors)}\,\frac{1\text{-}1}{1\text{-}1},\ \text{C (canines)}\,\frac{0\text{-}0}{0\text{-}0},\ \text{P (premolars)}\,\frac{2\text{-}2}{1\text{-}1},\ \text{M (molars)}\,\frac{3\text{-}3}{3\text{-}3} = 22$$

DISTRIBUTION. The southern flying squirrel occupies temperate forested areas in eastern United States from the Canadian border (into southern Ontario, Banfield, 1974) south to the Gulf of Mexico and west to the edge of the Great Plains. The species also is distributed in scattered montane forested habitats in Mexico, Guatemala and Honduras (Dolan and Carter, 1977). In Michigan, the southern flying squirrel is found primarily in the Lower Peninsula (see Map 33). Northward, its range overlaps that of the northern flying squirrel; only scattered reports of the southern flying squirrel are known from the northern half of the Lower Peninsula (Burt, 1946). In the Upper Peninsula, the southern flying squirrel has been recorded from only four western counties. Although there is little actual evidence, the southern flying squirrel probably made inroads from the south into areas of secondary hardwood forests of the northern part of the Lower Peninsula as well as the western Upper Peninsula, with the northern flying squirrel moving north as a result of lumbering (Gosling, 1980). The difficulties of making exact field identifications of these two Michigan flying squirrels account for the limited knowledge of their interactions and space allocations.

HABITAT PREFERENCES. The southern flying squirrel's inconspicuous, nocturnal habits perhaps account for the spotty records of the species. Certainly, the animals require tree growth and prefer hardwoods which produce seed crops. The tree stands should be parklike with scattered thickets (Jordan, 1948), probably for effective gliding. Mature beech, maples, oaks, and hickories may be preferred, with southern flying squirrels occurring in extensive forests as well as small woodlots, old orchards, city parks and on campuses. In Michigan, southern flying squirrels have been observed in elm groves, oak-hickory

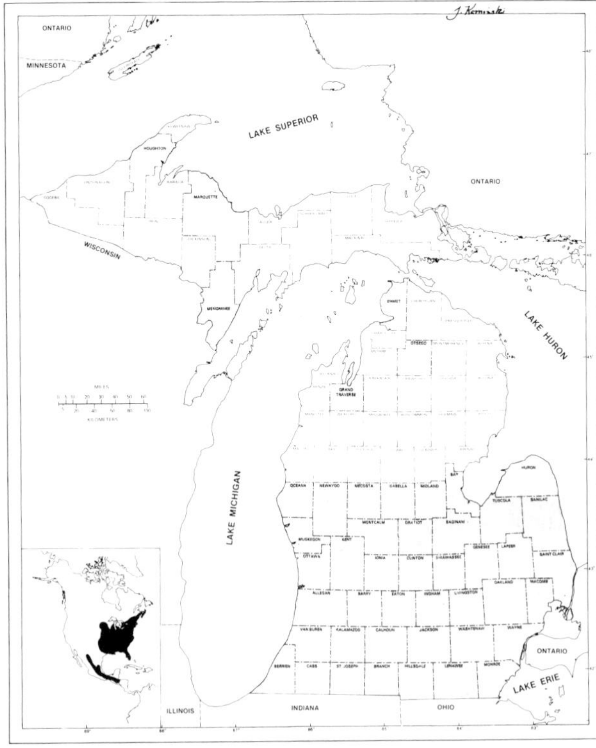

Map 33. Geographic distribution of the southern flying squirrel (*Glaucomys volans*).

association (Jordan, 1948; Muul, 1968), small oak and pine (Hodgson and Brewer, 1975), upland oak-hickory and lowland elm-maple (Linduska, 1950), aspen and scattered oaks (Muul, 1968), virgin hardwoods (Dice, 1925); and narrow ecotone between northern forest and swamp featuring northern white cedar, eastern hemlock, yellow birch and some ash and maple in Emmet County (John F. Douglass, pers. comm., 25 September 1981).

DENSITY AND MOVEMENTS. Southern flying squirrels may be more numerous than general observations can document. Some authors felt that southern flying squirrels may outnumber other tree squirrels in favorable beech-maple or oak-hickory habitat (Banfield, 1974). Sollberger (1940) suggested that the presence of flying squirrels might be detected by trapping with tree sets baited with hickory nuts, listening for the animal's nocturnal bird-like squeaks and feeding noises, knocking on trees with cavities in hope that a flying squirrel will emerge, and looking for elliptical openings in hickory nuts, which are characteristic of flying squirrel gnawings.

In favorable habitat, the species may be as dense as two individuals per acre (5 per ha), and, as Jackson (1961) pointed out, there may be local concentrations of 20 to 30 animals seeking winter protection in one or two cavities in the same tree. Summer and early autumn populations are highest just after the weaned young emerge. In Ohio, Bole (1939) estimated that a population of six individuals per acre (21 per ha) in beech-maple was near maximum density. In Pennsylvania, Sollberger (1943) found almost five animals per acre (12 per ha). In Michigan, Burt (1940) recorded approximately three southern flying squirrels per two acres (4 per ha) in Livingston County; Jordan (1948) counted one individual per acre (2.5 per ha) in Washtenaw County; and in 1940 Linduska (1950) trapped 50 individuals in July and August, 36 in September and October, and 8 in December on 102 acres (41 ha) in Clinton County. In 1941, Linduska caught only 11 individuals in the same area in June and July and none in September and October; in 1942 he caught only six individuals in five trapping periods between January and November. Multi-annual fluctuations in flying squirrel numbers occur but are poorly understood.

Southern flying squirrels usually remain in fairly confined places, moving to other areas because of seasonal food fluctuations, courtship and breeding activities, and the need for adequate winter quarters. In studies of trapped-marked-released-retrapped animals (Jordan, 1948), the usual cruising area may be 60 to 100 yds (55 to 93 m). Burt (1940) noted movements of as much as 120 yds (111 m). Linduska (1950) found flying squirrels at Rose Lake Wildlife Research Station moved as much as 440 yds (406 m) between woodlots separated by open lands. He discovered that 14 of 82 marked animals were live-trapped in both of two such separated woodlots.

BEHAVIOR. The most spectacular characteristic of the southern flying squirrel is its ability to glide. High-speed photography (Walker, 1947) has enabled mammalogists to examine the details of this unusual form of mammalian locomotion. This involves spreading the gliding membrane, flared anteriorly by the wrist "spur," and pointing the vibrissae forward, perhaps to provide first contact with landing sites or obstructions. By manipulating the gliding surfaces, the animal can change course as much as 90° (Jackson, 1961). An upward turn at the end of the glide is often characteristic (Sollberger, 1940). Gliding provides a rapid means of moving from tree to tree, or tree to log or forest floor. Gliding across streams has rarely been reported. However, in Eaton County, Kurta (1979) snared an adult female on September 28 in a mist net stretched across the Thornapple River. The animal struck the net 14.6 ft (4.5 m) above the water and 7.4 ft (2 m) from the nearest wooded bank, at a point where the river is about 42 ft (13 m) across.

Southern flying squirrels spend considerable time on the forest floor and on felled logs, although authorities differ as to just how much time (Hodgson and Brewer, 1975). On the ground, the gait of the flying squirrel is awkward at best (Jackson, 1961) and slow; an observer can easily run one down. Swimming is also difficult. Most collections of flying squirrels have been obtained from traps set on the ground or on logs, rather than in trees.

Sollberger (1940) believes flying squirrels are the most nocturnal of all mammals, being active from sunset to sunrise. Peaks of nocturnal activity, according to King (1883), occur in late evening (near 1045 hours) and again in early morning (near 0330 hours). Flying squirrels sometimes venture out before sunset (Banfield, 1974; Kelker, 1931), perhaps interacting with red squirrels at this time. A combination of strong winds and precipitation, according to Muul (1968), markedly inhibits flying squirrel activity.

The literature often refers to the flying squirrels' sociability; they hole up together to huddle against the winter's cold, eating hoarded foods (Banfield, 1974; Lowery, 1974; Jackson, 1961). Huddling in a well-insulated nest supposedly allows for a sustained temperature approximating thermal neutrality (Neumann, 1967). In Washtenaw County, Wood (1922) counted 20 flying squirrels in one winter den. This grouping behavior appears less common at other seasons. In fact, Madden (1974) has shown that in summer, females actively exclude other flying squirrels (both males and other females) from their home ranges. On the other hand, males seem tolerant of one another, occupying broadly over-lapping living sites.

Southern flying squirrels spend most of their summer and winter daylight hours in tree holes. Most observers have reported that the availability of nest cavities is more important than the species of trees (Jackson, 1961; Muul, 1974). Southern flying squirrels often enlarge woodpecker holes; the squirrels require an opening of 1.5 to 2 in (40 to 50 mm). Banfield (1974) noted that abandoned holes of downy woodpeckers in poplar trees are frequently used. Southern flying squirrels also select a wide array of other inside den sites. They are especially attracted to nest boxes (Goertz *et al.*, 1975; Heidt, 1977). Lowery (1974) found the squirrels nesting in a purple martin house, also being occupied by the birds. In Michigan, flying squirrels have used wooden nest boxes built in trees (Linduska, 1950), spaces in attics (Burt, 1946), a ledge in a car port in Okemos, and double walls in a school house at Onondaga. Nest materials include finely shredded bark, leaves, and other plant fibers, usually obtained from the tree in which the den is located.

Outside leaf nests are also used by flying squirrels but to a lesser extent than by other tree squirrels. These may be reconstructed nests of diurnal tree squirrels or birds and are normally oval-shaped, usually about 1 ft (305 mm) in diameter, with an entry to the inner chamber from near the top (Burt, 1940; Landwer, 1935; Moore, 1947; Sollberger, 1943; Snyder, 1921). Coarsely woven leaves and twigs make up the outer covering with finely-shredded materials lining the living chamber. Outside nests are used more in summer than in winter (Linduska, 1950); they also are used as secondary nests strategically placed within the flying squirrel's area of activity.

Naturalists have long noted the nocturnal vocalizations of flying squirrels describing them as "chuckles," "squeals," "squeaks," and "chirps" (Dolan and Carter, 1977; Gosling, 1978; Seton, 1953). Sollberger (1940) considered three basic sounds: A soft "tssep" call resembling that of a small bird; a similar call only much louder, perhaps when the animal is alarmed, (these two are apt to be heard on cool, autumn nights); and a "squeal" which may be uttered in times of distress. Young emit high-pitched squeaks.

ASSOCIATES. In the Michigan woodlands, the activities of both the southern and northern flying squirrels closely resemble those of the other tree-climbing sciurids, the red squirrel, gray squirrel, fox squirrel, and somewhat less the eastern chipmunk and least chipmunk. However, their paths rarely cross because the flying squirrels are nocturnal and the other species are diurnal. Even so, nest cavity space is probably disputed occasionally, especially between flying squirrels and gray squirrels, and each may raid the other's food hoards, especially in winter and early spring. Recent evidence indicates that gray squirrels may be the successful competitor and actually cause depressed flying squirrel populations, especially in mature woodlands. The only other small nocturnal Michigan mammals, except for would-be predators, to use the dark, leafy woodland upperstory are the white-footed mouse and the northern subspecies of the deer mouse; these might confront flying squirrels at den sites or food caches. Woodpeckers sometimes discourage flying squirrels from using tree cavities (Stickel, 1963).

The ranges of the southern and northern flying squirrels overlap in the northern half of Michigan's Lower Peninsula and in the western end of the Upper Peninsula (N. W. Gosling, 1981). Crider (1979) captured both species in a grid of traps within 200 ft (60 m) of each other in mixed hardwoods in Houghton County. There may be some interspecific interactions (compatible or not compatible) between these species, perhaps involving competitive exclusion (Weigl, 1969). Muul (1968) found no clear-cut dominance of either Michigan species when the two animals lived together under laboratory conditions. In nature, however, he thought the southern flying squirrel might have an advantage over the northern flying squirrel, because female southern flying squirrels tend to breed earlier and might secure the most favorable nest sites for young-rearing. Weigl (1969) also found that a nematode parasite (a roundworm of the genus *Strongyloides*) infests the southern flying squirrel without apparent detrimental effects, but can cause mortality in the northern flying squirrel. Consequently, this parasite's presence in a community of both flying squirrels might depress the population of the "non-immune" northern flying squirrel. In mountainous areas of North Carolina, Weigl (1978) noted evidence that where the northern flying squirrel declined, the southern species increased. Interactions with the diurnal tree squirrels involved competition for denning cavities and perhaps leaf nests. Serious confrontations between flying squirrels and red squirrels probably only occur when three holes are large enough to accomodate both the smaller flying squirrel and the larger red squirrel (Muul, 1968). Osgood (1938) also noted a decrease in southern flying squirrels as a result of the intrusion of gray squirrels in Vermont.

REPRODUCTIVE ACTIVITIES. Southern flying squirrels breed twice annually, although there is some question as to whether all adult females are involved in both the spring and autumn breeding periods (Dolan and Carter, 1977; Sollberger, 1943). Females ordinarily attain at least one year of age before breeding (Jordan, 1948). Courtship and mating for the first litter take place in February and March; for the second litter in June and July (Jackson, 1961). Kelker (1931) observed a mating near Ann Arbor on March 12. Muul (1969) has shown that southern flying squirrels achieve breeding readiness by changes in photoperiod (extension of length of daylight) and not by temperature cues.

After a gestation period of about 40 days, first litters are born from April to early June and second ones in August (Bert, 1946; Muul, 1969). In southern states, litters can appear as late as early October (Goertz, 1965). Litter size may also vary with latitude. In Michigan, Burt (1946) and Muul (1970) found that litters usually number four or five (as few as two and as many as six); in Alabama, Florida and Louisiana, Goertz *et al.* (1975), Linzey and Linzey (1979), and Moore (1947) reported litters of two (as few as one and as many as six). The young are kept in nest cavities lined with finely-shredded materials; when these become fouled or flea-ridden, the mother flying squirrel promptly moves the young to other quarters (Madden, 1974). The mother flying squirrel's devotion to her young has been described frequently (Muul, 1970; Seton, 1953). At East Lansing, Stack (1925) described how a female retrieved her young by climbing the pant leg of a student holding their nest.

Newborn young have closed eyes and ears, are pink in color, lack hair except for some evidence of vibrissae, show distinct evidence of the gliding membranes, weigh 0.2 to 0.3 oz (3 to 5 g), and measure 2½ in (64 mm) long. At six days vibrissae become conspicuous, at seven days the youngs' weight doubles, at 14 days hair covers the body except for the belly, at 18 days they are almost fully furred (see photo in Gosling, 1980), at 28 to 32 days the eyes and ears are open, at 35 days the hair growth on the tail takes on the adult dorso-ventral flattened appearance, at 41 days the young move about actively, at 42 days they become noticeably vocal, at 47 days they take solid food, at 60 days they attain two-thirds the adult weight, at 65 days they are weaned, at 90 days they can glide, and at 120 days the young are completely independent (Banfield, 1974; Hatt, 1931; Jackson, 1961; Linzey and Linzey, 1979; Sollberger, 1943).

The normal life span of the southern flying squirrel may be as much as five years (Jackson, 1961; Sollberger, 1943), although Linduska (1950), in his study in Clinton County, recaptured only four of 78 marked individuals the second year and none the third. Jackson (1961) suggested that captive individuals may live for as long as eight years.

FOOD HABITS. Southern flying squirrels forage in trees and on the adjacent forest floor year-around; stored nuts and acorns are the major dependable staple foods. Moisture is obtained in the non-frigid months from succulent foods, surface pools, and from water in tree holes (Madden,

1974). In spring, when caches of nuts may be depleted, tree buds and bark provide some essential nourishment. Late spring brings a flurry of insects which southern flying squirrels eat readily (Seton, 1953). Nesting birds also lose eggs and young to hungry flying squirrels. Bailey (1923) found a young purple martin's remains in the stomach of one of these sciurids which had taken up residence in a martin house. Stoddard (1920) recorded a flying squirrel eating an adult yellow-bellied sapsucker when both were confined in the same cage. Dolan and Carter (1977) concluded that flying squirrels are among the most carnivorous of sciurids.

In summer, ripening seeds and fruits become part of the southern flying squirrel's diet, while larval and adult insects continue as select foods; Banfield (1974) and Burt (1946) reported flying squirrels feeding on moths and beetles which congregate around lights at night. The flying squirrel energetically begins stockpiling food in autumn. Muul (1968) has shown that this period of active food gathering is also triggered by changes in photoperiod, in this instance the shortening of daylight. This nightly foraging gives field observers an excellent opportunity to detect flying squirrels, as they are noisy both vocally and in rustling nut tree foliage and leafy ground cover, and some of their normal wariness is abandoned. Rituals relating to nut selection and processing have been described by Avenoso (1968) and Muul (1968). Hickory nuts, filberts, beechnuts, and acorns are stored in tree cavities, under loose bark, in forks of branches, and sometimes in holes dug in the forest floor. The southern flying squirrel uses fewer seeds from conifer cones than does the northern flying squirrel. There is some evidence that chestnuts were favorite winter foods in pre-blight times (Howell, 1918). However, hickory nuts appear to be the preferred fare in Michigan. If hickory nuts are in short supply, the flying squirrel will use acorns, but when both nuts are available, hickory nuts are chosen (Muul, 1968). If there is a bumper nut crop, southern flying squirrels can survive the winter with supplies left over to supplement spring and summer foods. When autumn nut production is short, winter survival can be a problem, and competition among all the food-storing sciurids for caches may cut these resources to a minimum. Fortunately for southern flying squirrels, they are able to cache nuts in tree holes too small for the larger red, gray, and fox squirrels to enter. However, tree-climbing deer mice and white-footed mice may be able to raid their stores.

ENEMIES. Although snakes might invade den cavities in daylight (Pearson, 1954), most of the flying squirrels' enemies probably hunt them at night. These predators may include great horned owls (Errington *et al.*, 1940) and other owls, hawks, bobcats, raccoons, weasels, foxes (Dolan and Carter, 1977), and house cats (Jackson, 1961). Normal activities, including gliding, may account for unusual injuries (Barkalow, 1956). Forest fires, lumbering, and the felling of cavity-filled trees by forest managers all contribute to the flying squirrel's survival problems.

PARASITES AND DISEASE. Ectoparasites infesting southern flying squirrels include at least six kinds of fleas as well as mites and lice (Day and Benton, 1980; Dolan and Carter, 1977; Jackson, 1961; Lowery, 1974; Scharf and Stewart, 1980). There are also protozoans, acanthocephalan worms, roundworms, and tapeworms as endoparasites (Dolan and Carter, 1977). This flying squirrel has also been known to contract rabies (Venters, 1962) and a specific kind of malaria (Dasgupta and Chatterjee, 1967).

MOLT AND COLOR ABERRATIONS. Old pelage is molted and new hair takes its place annually in autumn (September to November). The new pelage emerges first on the sides of the body, then spreads over the back, and finally to the head. Climbing and gliding cause considerable pelage wear, giving animals examined in late summer a ragged and darker appearance. An albino individual has been recorded from Wisconsin by Jackson (1961).

ECONOMIC IMPORTANCE. Because of this squirrel's efficiency in harvesting nut crops, high concentrations of southern flying squirrels may be a concern to foresters. However, there is no evidence that their hoarding and eating habits cause a decline in the development of new forest growth. Flying squirrels can be pests when lodging in city houses (Fryxell, 1926), although most authors have classified the flying squirrel as harmless or at least neutral in its relationships with humans. Being an unprotected species in Michigan and most other states, flying squirrels have been exploited as pets. Dealers not only live-trap them but establish breeding colonies in large outdoor cages. This commercialization does not seem to have any bad effects on local flying squirrel populations and these soft-haired, easily tamed creatures no doubt make wholesome and affectionate pets, although owners may find their noctural habits somewhat frustrating.

COUNTY RECORDS FROM MICHIGAN. Records of specimens of southern flying squirrels in collections of museums and from the literature show that this species lives in all counties in the southern half of Michigan's Lower Peninsula but is recorded from only three counties in the northern half: Grand Traverse (W. C. Scharf, *in litt.,* July 17, 1978), Emmet (N. M. Gosling, pers. comm.) and Otsego (Burt, 1946). This paucity of records is due in part to the minimal fieldwork being done in the area on small mammal distribution. In the Upper Peninsula, the species is known from only three western counties: Houghton (Crider, 1979; Stormer and Sloan, 1976), Marquette (Haveman and Robinson, 1976) and Menominee (Burt, 1946).

LITERATURE CITED

Avenoso, A. C.
1968 Selection and processing of nuts by the flying squirrel, *Glaucomys volans.* Univ. Florida, unpubl. Ph.D. dissertation, 126 pp.

Bailey, B.
1923 Meat-eating propensities of some rodents of Minnesota. Jour. Mammalogy, 4(2):129.

Banfield, A. W. F.
1974 The mammals of Canada. Univ. Toronto Press, Toronto. xxiv+438 pp.

Barkalow, F. S., Jr.
1956 A handicapped flying squirrel, *Glaucomys volans.* Jour. Mammalogy, 37(1):122–123.

Bole, B. P.
1939 The quadrat method of studying small mammal populations. Cleveland Mus. Nat. Hist., Sci. Publ., 5:15–77.

Burt, W. H.
1940 Territorial behavior and populations of some small mammals in southern Michigan. Univ. Michigan, Mus. Zool., Misc. Publ., No. 45, 58 pp.
1946 The mammals of Michigan. Univ. Michigan Press, Ann Arbor, xv+288 pp.

Crider, J. E.
1979 A wildlife inventory of the Sturgeon River Wilderness Study Area. Michigan Tech. Univ., unpubl. M.S. thesis, vii+67 pp.

Dasgupta, B., and A. Chatterjee
1967 A new malarial parasite of the flying squirrel. Parasit., 57(3):467–474.

Day, J., and A. H. Benton
1980 Population dynamics and coevolution of adult siphonapteran parasites of the southern flying squirrel (*Glaucomys volans volans*). American Midl. Nat., 103(2): 333–338.

Dice, L. R.
1925 A survey of the mammals of Charlevoix County, Michigan, and vicinity. Univ. Michigan, Mus. Zool., Occas. Papers No. 159, 33 pp.

Dolan, P. G., and D. C. Carter
1977 *Glaucomys volans.* Mammalian Species, No. 78, 6 pp.

Errington, P. L., F. Hamerstrom, and F. N. Hamerstrom, Jr.
1940 The great horned owl and its prey in north-central United States. Iowa State Col., Agri. Exp. Sta., Res. Bull. 277:757–850.

Fryxell, F. M.
1926 Flying squirrels as city nuisances. Jour. Mammalogy, 7(2):133

Goerz, J. W.
1965 Late summer breeding of flying squirrels. Jour. Mammalogy, 46(3):510.

Goertz, J. W., R. M. Dawson, and E. E. Mowbray
1975 Response to nest boxes and reproduction by *Glaucomys volans* in northern Louisiana. Jour. Mammalogy, 56(4):933–939.

Gosling, N. M.
1978 Michigan's night gliders. Michigan Out-of-Doors, 32(3):56, 60–61.
1980 The night gliders. Michigan Nat. Res. Mag., 49(5):44–49.

Gupta, B. B.
1966 Notes on the gliding mechanism in the flying squirrel. Univ. Michigan, Mus. Zool., Occas. Papers No. 645, 7 pp.

Hatt, R. T.
1931 Habits of a young flying squirrel (*Glaucomys volans*). Jour. Mammalogy, 12(3):233–238.

Haveman, J. R., and W. L. Robinson
1976 Northward range extension of the southern flying squirrel. Jack-Pine Warbler, 54(1):40–41.

Heidt, G. A.
1977 Utilization of nest boxes by the southern flying squirrel, *Glaucomys volans,* in central Arkansas. Proc. Arkansas Acad. Sci., 31:55–57.

Hodgson, J. R., and R. Brewer
1975 Mammal populations of oak forests in southwestern Michigan. Jack-Pine Warbler, 53(4):131–140.

Howell, A. H.
1918 Revision of the American flying squirrels. North American Fauna, No. 44, 64 pp.

Jackson, H. H. T.
1961 Mammals of Wisconsin. Univ. Wisconsin Press, Madison. xii+504 pp.

Jodan, J. S.
1948 A midsummer study of the southern flying squirrel. Jour. Mammalogy, 29(1):44–48.

Kelker, G.
1931 The breeding time of the flying squirrel (*Glaucomys volans volans*). Jour. Mammalogy, 12(2):166–167.

King, F. H.
1883 Instinct and memory exhibited by the flying squirrel in confinement, with a thought on the origin of wings in bats. American Nat., 17(1):36–42.

Kurta, A.
1979 Southern flying squirrel caught in mist net. Jack-Pine Warbler, 57(3):170.

Landwer, M. F.
1935 An outside nest of a flying squirrel. Jour. Mammalogy, 16(1):67.

Linduska, J. P.
1950 Ecology and land-use relationships of small mammals on a Michigan farm. Michigan Dept. Conservation, Game Div., ix+144 pp.

Linzey, D. W. and A. V. Linzey
1979 Growth and development of the southern flying squirrel (*Glaucomys volans volans*). Jour. Mammalogy, 60(3):615–620.

Lowery, G. H., Jr.
1974 The mammals of Louisiana and its adjacent waters. Louisiana State Univ. Press, Baton Rouge. xxiii+565 pp.

Madden, J. R.
1974 Female territoriality in a Suffolk County, Long Island, population of *Glaucomys volans*. Jour. Mammalogy, 55(3):647–652.

Moore, J. C.
1947 Nests of the Florida flying squirrel. American Midland Nat., 38:248–253.

Muul, I.
1968 Behavior and physiological influences on the distribution of the flying squirrel, *Glaucomys volans*. Univ. Michigan, Mus. Zool., Misc. Publ., No. 134, 66 pp.
1969 Photoperiod and reproduction in flying squirrels, *Glaucomys volans*. Jour. Mammalogy, 50(3):542–549.
1970 Intra- and inter-familial behaviour of *Glaucomys volans* (Rodentia) following parturition. Anim. Behav., 18:20–25.
1974 Geographic variation in the nesting habits of *Glaucomys volans*. Jour. Mammalogy, 55(4):840–844.

Neumann, R. L.
1967 Metabolism in the eastern chipmunk (*Tamias striatus*) and the southern flying squirrel (*Glaucomys volans*) during the winter and summer. Pp. 64–74 *in* K. C. Fisher, A. R. Dawe, C. P. Lyman, E. Schonbaum, and F. E. South, Jr., Eds. Mammalian Hibernation III. American Elsevier Publ. Co., Inc., New York, xiv+535 pp.

Osgood, F. L., Jr.
1938 The mammals of Vermont. Jour. Mammalogy, 19(4): 435–441.

Pearson, P. G.
1954 Mammals of Gulf Hammock, Levy County, Florida. American Midland Nat., 51:468–480.

Scharf, W. C., and K. R. Stewart
1980 New records of Siphonaptera from northern Michigan. Great Lakes Ento., 13(3):165–167.

Seton, E. T.
1953 Lives of game animals. Vol. IV, Pt. I. Charles T. Branford Co., Boston. xxii+440 pp.

Snyder, L. L.
1921 An outside nest of a flying squirrel. Jour. Mammalogy, 2(3):171.

Sollberger, D. E.
1940 Notes on the life history of the small eastern flying squirrel. Jour. Mammalogy, 21(3):282–293.
1943 Notes on the breeding habits of the eastern flying squirrel (*Glaucomys volans volans*). Jour. Mammalogy, 24(2):163–173.

Stack, J. W.
1925 Courage shown by flying squirrel, *Glaucomys volans*. Jour. Mammalogy, 6(2):128–129.

Stickel, D. W.
1963 Interspecific relations among red-bellied and hairy woodpeckers and a flying squirrel. Wilson Bull., 75(2):203–204.

Stoddard, H. L.
1920 The flying squirrel as a bird killer. Jour. Mammalogy, 1(2):95–96.

Stormer, F. A., and N. Sloan
1976 Evidence of the range extension of the southern flying squirrel in the Upper Peninsula of Michigan. Jack-Pine Warbler, 54(4):176–177.

Venters, H. D.
1962 Epidemiologic note: Rabies in a flying squirrel. Public Health Rept., 77:200.

Walker, E. P.
1947 "Flying" squirrels, nature's gliders. Nat. Geogr. Mag., 91(5):662–674.

Weigl, P. D.
1969 The distribution of the flying squirrels, *Glaucomys volans* and *G. sabrinus:* An evaluation of the competitive exclusion idea. Duke Univ., unpubl. Ph.D. dissertation, 246 pp.
1978 Resource overlap, interspecific interactions and the distribution of the flying squirrels, *Glaucomys volans* and *G. sabrinus*. American Midl. Nat., 100(1):83–96.

Wood, N. A.
1922 The mammals of Washtenaw County, Michigan. Univ. Michigan, Mus. Zool., Occas. Papers No. 123, 23 pp.

FAMILY CASTORIDAE

Castor is the only genus in this family, which consists of one species, *C. canadensis,* in North America and another, *C. fiber,* in Eurasia. These two species of beaver are so alike that some workers regard them as a single species (Freye, 1960), in which case the older name *C. fiber* would have precedence. However, most mammalogists prefer to consider them as two distinct species on the basis of cranial, as well as chromosomal, distinctions (Lavrov and Orlov, 1973).

The beaver is large and heavy-bodied; the only larger living rodent is the capybara of South America. A now-extinct giant beaver (*Castoroides*), the size of a black bear and truly a monster among rodents (see Table 2) lived in post-glacial Michigan, and as late as approximately 10,000 years ago. The chief characteristics of modern members of the Castoridae, other than large size, are broad, blunt head with small eyes; short neck and limbs; scaly, flattened (dorsoventrally), paddle-shaped tail containing flattened caudal vertebrae; small, dextrous forefeet; large, webbed hind feet; digestive, urinary and reproductive tracts opening into a common exit or cloaca; and well-furred lips meeting behind the incisors and blocking water entry to allow the beaver to cut and peel vegetation under water. Many of these characteristics are adaptions which assist the beaver in its semi-aquatic mode of life. The skull is massive in construction, with large, laterally compressed zygomatic arches; each infraorbital foramen forming a small, narrow canal with no entry of the masseter muscle; and each elongate auditory canal extending upward and outward. The massive incisors are chisel-shaped and orange-fronted. There are four jaw teeth (the fourth premolar and three molars) on each side of the upper and lower jaws. These are high-crowned, flat-surfaced, with enamel folds for chewing and grinding.

LITERATURE CITED

Freye, H.
1960 Zur systematik de Castoridae (Rodentia, Mammalia). Mitt. Zool. Mus. Berlin, 36(1):105–122.

Lavrov, L. S., and V. N. Orlov
1973 Karyotypes and taxonomy of modern beavers (Castor, Castoridae, Mammalia). [In Russian with English summary], Zool. Jour., 52(5):734–742.

BEAVER
Castor canadensis Kuhl

NAMES. The generic name *Castor,* proposed by Swedish naturalist Linnaeus in 1758, is derived from *kastor,* a Greek word meaning "beaver." The specific name *canadensis,* proposed by Kuhl in 1820, is a Latinized word meaning "of Canada." The name "beaver" is derived from the Anglo-Saxon word *beofar,* meaning "brown," and perhaps earlier from the Old Aryan *bebhrus* and Persian *baovara.* In Michigan, local residents refer to this species as beaver. The subspecies *Castor canadensis michiganensis* Bailey occurs in Michigan.

RECOGNITION. Body size extremely large for a rodent; head and body averaging 28 in (720 mm) long, with tail slightly less than one-third as long (Fig. 47); body robust; head broad and blunt; eyes and ears small; nostrils valvular; tail broadly flattened horizontally, naked, scaly, leathery; forefeet not massive but equipped with strong, digging claws; hind feet large, broad, fully webbed; claws on three outer hind toes rather flat and blunt; claws on the inner two with horny pads; claws on the second having sharp, serrated edges—specialized and cleft for use in grooming the fur. Long, moderately coarse guard hairs give the upperparts a rich brown color with the underparts somewhat paler; head and shoulders bright brown; underfur dense, wavy, and brown in color; ears, nasal area, feet, and tail blackish.

Sexes differ slightly in size but not color. There are four pectoral (chest) mammary glands, usually obscure except when nipples are enlarged during the young-rearing period. Two types of glands (in both sexes) open in the anal area (Svendsen, 1978). To each side of the anus are large castor glands which excrete musky fluid (castoreum) into the cloaca at the posterior end of the digestive tract. Posterior to the anal opening are anal glands which

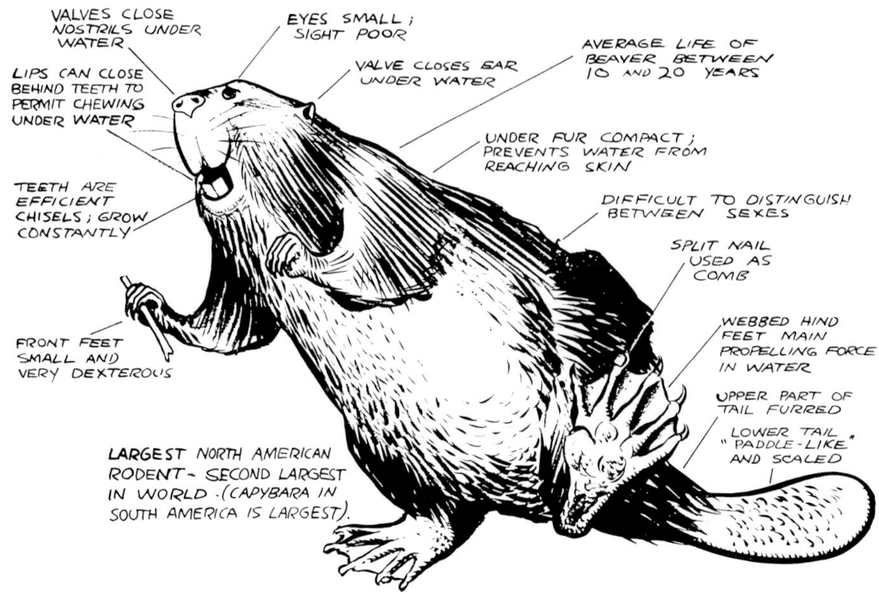

Figure 47. The beaver (*Castor canadensis*) and some of its characteristics. Drawing courtesy of Oscar Warbach and the Michigan Department of Natural Resources.

discharge oil through skin pores onto hair roots. The beaver uses the forefeet and second toes of the hind feet to spead this oil over the body to keep the pelage sleek, oily and water repellent. The beaver has 2N = 40 chromosomes and a fundamental number of 80 chromosomal arms (Lavrov and Orlov, 1973).

MEASUREMENTS AND WEIGHTS. Adult beavers measure as follows: length from tip of nose to end of tail vertebrae, 36 to 43 in (900 to 1100 mm); length of tail vertebrae, 11¾ to 15¾ in (300 to 400 mm); maximum breadth of tail, 4¼ to 6½ in (115 to 165 mm); height of ear from notch, 1¼ in (33 mm); weight of adults, 30 to 60 lbs (13.7 to 27.4 kg), with unusually large individuals weighing upwards to 100 lb (45.5 kg). Males are slightly larger than females.

DISTINCTIVE CRANIAL AND DENTAL CHARACTERISTICS. The massive skull with large spaces for jaw muscle attachments, large incisors, and high-crowned, rootless cheek-teeth with their numerous cross ridges of enamel on the flat grinding surfaces all reflect the beaver's ability to gnaw and masticate coarse vegetation including wood fibers (Fig. 48). The Michigan beaver's skull is approximately 5 in (130 mm) long and 3¾ in (95

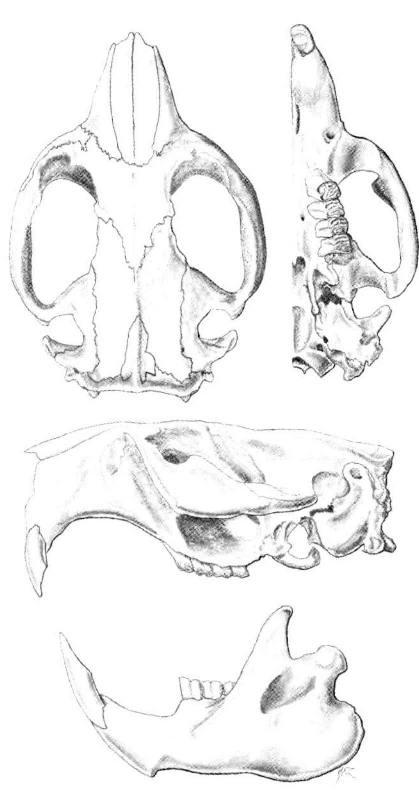

Figure 48. Cranial and dental characteristics of the beaver (*Castor canadensis*).

mm) wide across the zygomatic arches. There is a distinctive ear opening at the termination of an elongated bony auditory canal at each side of the posterior ends of the braincase. Among North American rodents, only woodchuck and porcupine skulls approach the size of that of the beaver. The skull of the woodchuck is smaller, being only about five-sevenths the length of that of the beaver. The skull of the porcupine is larger than the woodchuck's, but is still only about four-fifths the size of an adult beaver skull. In addition, the greatly enlarged infraorbital canal (to permit passage of a large sector of the masseter muscle) on each side of the rostrum of the porcupine is highly distinctive and contrasts sharply with the insignificant infraorbital canal openings of the more massively constructed beaver skull. The dental formula for the beaver is:

$$\text{I (incisors)}\frac{1\text{-}1}{1\text{-}1}, \text{C (canines)}\frac{0\text{-}0}{0\text{-}0}, \text{P (premolars)}\frac{1\text{-}1}{1\text{-}1}, \text{M (molars)}\frac{3\text{-}3}{3\text{-}3} = 20$$

DISTRIBUTION. In presettlement days, the beaver had an extensive North American geographic range from tree line in the American subarctic southward throughout most of the United States and adjacent parts of Mexico except for peninsula Florida and parts of the arid Southwest (Ffolliott *et al.*, 1976; Jenkins and Busher, 1979). The beaver population at that time has been estimated at between 100 and 400 million individuals (Beidleman, 1959). The beaver's valued pelt (especially for European export in pioneer times), its marketable musk glands, and its highly edible flesh, plus the ease of trapping the animal near conspicuous dams or stick lodges, contributed to the marked decline of the species in many parts of this vast area (Hay, 1956). By 1850–1900, the beaver population had been reduced to scattered isolated groups, especially in heavily settled parts of the United States. Since 1930, however, enforced trapping regulations and total protection have allowed these remnant stocks to increase and spread naturally to become reestablished in formerly occupied areas along drainage systems. Live-trapping and subsequent release in areas where they had previously been extirpated have also aided the beaver's reestablishment. In fact, the beaver's recovery has been so successful that it is now considered pestiferous in some localities due to felled fruit trees, damaged ornamentals and cultivated crops, and inundated road ways and streamside farmlands caused by backwater from dams.

Soon after the melt-back of the last glaciation (Holman, 1975), the beaver was found on all major waterways of Michigan as well as in swamps, marshes, and lakes. There is abundant evidence, based on faunal remains at archaeological sites (Cleland, 1966; Higgins, 1980), that Indians used the beaver's pelt, meat, bones, sinews, and other anatomical parts but rarely exploited the species on a widespread basis. With the arrival of the French, however, the value of the fur in trade with European countries provided a new incentive for trapping beaver (Fasquelle, 1950). By the middle of the seventeenth century, the French were exchanging weapons, cloth, tools, decorative objects, and spirits for furs with Michigan's Chippewa, Ottawa, and Huron tribes (Baker, 1899). Some writers have called the beaver "black gold" during this period (Bump and Cook, 1941). Although the political climate changed with British and American take-overs, the fur trade boomed through the eighteenth century, peaking just after the War of 1812 (Petersen, 1960). By this time, however, the Michigan beaver population had been largely depleted (Alcoze, 1981). The settlers who were replacing the French trappers and their Indian associates possibly considered the meat of the beaver more important to their survival than the pelt. By 1950, damsites were the only remaining evidence of beaver in most counties in the southern part of Michigan's Lower Peninsula (Wood and Dice, 1923). In the Detroit area, beaver were present in Cadillac's time in the early 1700s with no traces of the species left by 1877 (Johnson, 1919). In southern counties, the beaver was reported as being extirpated in Monroe by 1833 (Wood and Dice, 1923) and Washtenaw by 1836 (Wood, 1922). Northward, the beaver held out in isolated pockets. These served as dispersal points for the spread and reestablishment of the species in the present century after trapping was closely regulated.

The Michigan season for trapping beaver was closed beginning in 1921. In 1928, officials of the Michigan Department of Conservation (now the Department of Natural Resources) live-trapped and moved surplus or nuisance beaver to unoccupied localities in the upper part of Michigan's Lower Peninsula: Crawford, Gladwin, Iosco, and

Ogemaw counties (Ruhl and Lovejoy, 1930). Animals from Ontario replenished the extirpated population on Beaver Island (Ozoga and Phillips, 1964). The beaver's return to most areas, with the exception of farming districts in southern counties (Burt, 1946), has been remarkable. In 1978, the Wildlife Division of the Michigan Department of Natural Resources (Marvin Cooley, pers. comm.) estimated that the beaver was either present in high or medium densities in most of the counties of Michigan's Upper Peninsula and northern half of the Lower Peninsula. In the southern part of the Lower Peninsula beaver were present in low densities or scarce or absent (see Map 34).

Map 34. Geographic distribution of the beaver (*Castor canadensis*). Although this species at one time occurred throughout the state, its present distribution is within those counties which are shaded.

HABITAT PREFERENCES. The semiaquatic habits of the beaver are well-served by the more than 10,000 mi of slow-flowing streams and rivers, and the inland lake shorelines of Michigan (Ruhl, 1931). Fast water and streams subject to flash floods, both deterrents to beaver settlement, are rare. Essential aquatic areas must be associated with proper vegetative growth to ensure proper harborage for beaver. Mature waterside forests may attract an occasional beaver colony, but stands of young trees, of the early successional stages in forest growth, make much more attractive habitat. In fact, beaver populations have prospered where waterside secondary forests of aspen-birch-conifer stands have followed lumbering, clearing and burning. The replacement of the mature forest by deciduous seral species has produced an overall Michigan environment which is probably more conducive to the beaver's welfare than was the prehistoric. Aspens (trees of the genus *Populus*), along with willows, alders, and water plants, are needed to make the wetland environment preferable for beaver in Michigan.

DENSITY AND MOVEMENTS. Numerous studies have been done concerning the beaver's ecology and economic importance (Denney, 1952; Hay, 1957). Lewis H. Morgan (1868), a director of the Marquette and Ontonagon Railroad, conducted a study from 1855 through 1867 in Marquette County (see Manville, 1949) which was the first of a series of important investigations of Upper Great Lakes beaver populations. Almost a century later the more sophisticated work by G. W. Bradt (1938; 1947) added greatly to our knowledge of the species in Michigan.

Beaver populations are arranged linearly along streams or at pond and lake edges. Thus, numbers in one Michigan county might potentially be much higher than in another, depending on the extent of aquatic habits and food supplies. Well-maintained dams and lodges are conspicuous signs of the beaver's presence and have long provided a gauge for estimating beaver populations (Johnson, 1927). A lodge generally houses a colony consisting of an adult pair, the kits (young-of-the-year), and yearlings of the previous year (Bradt, 1938; Novak, 1977). Bradt noted high variability in the number of lodge occupants, from one to 12 individuals with an average of about five in Michigan. He discounted average estimates of as many as 10 in New York (Johnson, 1927), seven in the Lake Superior region (Morgan, 1868), and trappers' figures of as many as eight and 12. Lodge counts can be made on foot, or from canoes or airplanes (Payne, 1981; Swank and Glover, 1948). Bradt (1947) suggested that counts should be made in mid-autumn (October or November) using evidence of fresh food caches or repairs as criteria for current lodge occupation. The total population, according to Bradt, could be estimated by multiplying the number of lodges by five. Of course, bank-dwelling beavers often leave little occupational evidence and may be missed in such counts. Bradt (1938) also concluded that a colony having no lodge contained fewer individuals, approximately 2.4.

Bradt's pioneer work in Michigan also emphasized the importance of food in determining the size and length of residency of a given colony. He calculated that the average diameter of aspen (and other trees) cut by beavers was about 2 in (50 mm), with each colony member needing to cut, on the average, 216 trees per year. If aspen of this diameter grows at the rate of 1,500 to 3,000 trees per acre (600 to 1,200 per ha), a single acre (0.4 ha) will support one beaver for about seven years or seven beavers for one year.

Beaver density along a river, its small tributaries, and in marshes, ponds and lakes is therefore dependent on stands of aspen and other edible softwoods either bankside or nearby. In St. Croix State Park in Minnesota, Erickson (1939) found that beaver populations along small creeks emptying into the St. Croix River varied, depending on habitat quality, from 1.87 individuals per mi (1.2 per km) to 14.58 per mi (9.1 per km); along the St. Croix River, 2.52 per mi (1.6 per km); and in ponds, from 1.13 per acre (2.83 per ha) to 8.82 per acre (22.05 per ha) of water. In well-watered areas in the Upper Peninsula, beaver populations have been estimated at 2.6 to 3.8 per sq mi (1.0 to 1.5 per sq km) in McCormick Experimental Forest of northwestern Marquette and eastern Baraga counties (Robinson and Werner, 1975), and as dense as 4.7 per sq mi (1.8 per sq km) in the Huron Mountain area of Marquette County (Manville, 1942).

The beaver's ability to consume young stands of softwood faster than the trees can regenerate is a major factor in population stability. The excluding of yearlings from the colony soon after the birth of the second annual litter (Bradt, 1938) is one means of limiting local concentrations and allowing the food supply to last longer. In reality, however, the beaver "logs off" seral stage food trees, chiefly aspen in Michigan, and accelerates the development of "climax" vegetation. Since the beaver is essentially adapted to early successional stages of Michigan streamside forest growth, it rarely stays in a given area for long.

HOME SITES. The beaver's well-being depends on its ability to adapt to a variety of aquatic environments, where it is at home year-around. Although the beaver is a slow swimmer, it can remain underwater, according to Beidleman (1959), for as long as 15 minutes and can swim successfully under ice-covered water for as far as ½ mi (0.8 k). Its digestive system accepts and assimilates nutrients from a variety of vegetable foods including fibrous tissues hewn from trees. Massive chisel-shaped teeth not only gnaw chips to fell trees but are efficient in stripping bark and clipping off small branches and twigs. Felled trees of various softwood species serve as year-around food and as construction materials for dams, lodges, and bank-den roofs.

The beaver's need to manipulate its habitat is largely governed by the site. In large streams, water sufficiently deep at all seasons eliminates the need for the beaver to manipulate the depth by constructing a dam. Consequently, only minor efforts are required to locate adequate food supplies and construct a nearby bank den with an underwater entrance under tree roots or an overhanging ledge. The entry tunnel, often sufficiently large in diameter to provide crawlspace for an adult human, leads to a nest chamber above waterline (Seton, 1953). Often the roof of this cavity is thin to admit air for ventilation. If the roof is not protected overhead by roots or covering bushes, the beavers interlace sticks across the opening to secure the roof as well as provide for an interchange of nest cavity air.

In marshes, swamps, ponds, and lakes, a beaver colony may construct either a bank lodge or a moated lodge. A typical beaver lodge (Bailey, 1926) is massive, as much as 9 ft (3 m) high and from 10 to 20 ft (3.3 to 6.6 m) across at its base. Perhaps the largest recorded beaver lodge was observed by Barber (1919) in Wisconsin. This structure was 16 ft (4.9 m) high and fully 40 ft (12.4 m) long. Wood (1914) described a lodge on Michigan's Whitefish Point as containing ". . . several wagon-loads of peeled sticks." Dice and Sherman (1922) examined lodges in leatherleaf bogs in Gogebic County. Bank lodges and island lodges are normally based on firm ground, but the moated lodge requires underwater support, such as a submerged log, a rock, or a mass of roots, on which an accumulation of sticks and mud is heaped. The lodge may be constructed with a hollow space inside or this nesting cavity may be gnawed out later. The water surrounding the lodge is deepened to accommodate one or more underwater entrances. There is usually a central chamber above water level with a floor covering generally consisting of grasses, twigs, shredded bark and wood. This space can be 5 ft (1.5 m) across and 3 ft (1 m) high. In addition, there may be a lower landing platform onto which the entry passages open. The spacious interior may accommodate as many as eight beaver (Jackson, 1961). The side walls may be 4 ft (1.3 m) thick, with the apex thin enough to permit ventilation. In winter, the moisture in the lodge walls usually freezes, not only adding insulation but making it almost impossible for a would-be predator to dig out the beaver family. Ice covering the surrounding water further decreases the likelihood that the beaver's winter quarters will be invaded. As a result, the river otter may be the major winter predator in Michigan. Despite low and fluctuating winter temperatures, the beaver maintains rather stable lodge temperatures mostly due to body heat (Stephenson, 1969). Snow cover on the roof provides excellent insulation. Lodges are generally constructed and stocked with winter food during the autumn months. They are reinforced from time to time with poles and mud. The very large lodges, of course, have been occupied for many years by a succession of beaver families; they are abandoned only when local food or water supplies wane.

The beaver shows its greatest ability as an "aquatic engineer," in small brooks and creeks. The construction of dams, lodges, and canal systems is considered part of the beaver's natural instinctive capability. However, with experience these individual skills improve. The backwater from the dam is intended to support an impoundment sufficiently deep to provide moated protection to one or more lodges and allow swimming space in winter under an ice layer which might be as thick as 3 ft (1 m). The beaver often selects as the dam site a narrow point in the stream bed where the flow is rapid and the banks are firm (Banfield, 1974). The beaver then cuts branches and drags them into position paralleling the current with the butt ends heading upstream. Branches from a variety of trees and shrubs and even corn stalks (Henderson, 1960) and conifers (Manville, 1948) are used as building materials. The spreading branches, extending downstream, tend to become anchored in the stream bed. Stones, mud, and sod are also used to weigh down and position these branches. Occasionally, a large tree is felled in correct position across the dam site. The first layer of pole branches is plastered with stones, roots, and mud, with the stream current adding silt to fill up spaces between the leaves and twigs. Additional layers of parallel branches and filler produce the proper height to accommodate a pond large enough for the beaver's requirements. An early Michigan observer, L. H. Morgan (1868), identified two kinds of dams: 1) the open stick variety, faced with mud and sod on the upstream side and the top through which overflow water trickles all along; and 2) a more solid and perhaps older dam which permits overflow water to be discharged in only one place. In either case, the dam is in need of almost daily repair, with the beaver colony constantly patching breaks and eroded overflow areas as well as merely adding to the bulk. Although the construction of a dam seems a major accomplishment, beavers build them rapidly. Beidleman (1959) reported that one 40 ft (12.4 m) dam was constructed in about one week. Beaver dams are notable for their length and height. In Missaukee County, one dam was estimated to be at least 100 yds (91 m) long, with moss-covered sod, rounded stones and skinned aspen sticks conspicuous structural materials. The enterprising beaver colony may also construct smaller dams downstream or along feeder water courses (Lawrence, 1952) to stabilize fluctuation of water levels. In the pond produced by the dam, a major lodge and perhaps auxillary lodges are constructed as centers of the colony's activity.

The beaver often must fell trees at some distances from the pond. Instead of dragging these overland, the beaver constructs canals or mud slides (downhill) to carry cuttings from the felled timbers and other materials to the dam or lodge site. Some authors (Seton, 1953) regard the beaver's canal digging ability as equal in "technological expertise" to dam building. Canals are known to be as long as ½ mi (0.8 k), contain sufficient water in which to float and paddle a canoe, and may even contain "locks" or small check dams designed to extend the canal at elevations higher than the beaver pond level (Beidleman, 1959). One downhill slide observed in Missaukee County had been used to get aspen sections down a steep bank (45° slope) almost 50 ft (15 m) above the stream.

BEHAVIOR. The family unit, centered around the breeding female, is the basic social group. The family normally consists of the two parents, their first-year kits, and yearlings, which are excluded when the next litter appears. If the male is lost, the parent female remains with the family until another adult male comes along. If the female is lost and no female offspring are present, the males tend to abandon the locality. These family groupings maintain stablility and compatibility with little obvious antagonism (Tevis, 1950), although adult males sometimes exhibit mild dominance over yearlings. All family members become involved in maintenance of dams, canals, lodges, and food storage, with the adult male often taking the initiative. These family activities have been viewed and even photographed on many occasions in Michigan by such naturalists as Gregory (1931) and Harris and DuCharme (1928).

The family occupies a territory centered about the lodge with no overlap in areas used by neighboring families (Bradt, 1938). Adult males may patrol borders between adjacent territories but overt aggression seems to be avoided. Territoriality is maintained by scent mounds which resemble mud pies and are placed along trails or at territory edges, especially in July (Townsend, 1953). The mounds range in size from ones barely perceptible to piles of mud as large as 2 ft (610 mm) high. On close inspection, the mounds show beaver claw marks and deposits of musky castoreum from the anal castor glands. The castoreum deposit on the scent mounds helps maintain a close family association, communicates with members of adjacent families, and ensures the territorial boundaries from transient outsiders (Aleksiuk, 1968).

Beavers seem to communicate vocally, with the high pitched whines of kits most conspicuous (Tevis, 1950). Yearlings and adults also utter whines and bellows. Loud hissing is one adult reaction to the detection of castoreum of strange, transient beavers (Aleksiuk, 1968). The loud report of the flattened tail being whacked on the surface of still pond water serves as an alarm call.

Beavers spend considerable time grooming. As one beaver frequently grooms another inside the lodge, this serves a social as well as hygienic purpose.

Because of the closeness of the family, females ordinarily do much less moving than males. The yearling beaver, after being excluded from the family unit, may travel considerably (mostly in summer) before becoming established. Movements up to 11.3 mi (18 km) have been recorded (Leege, 1968). As Bradt (1938) has shown with Michigan beavers, the emigration of the two-year-old beavers is the means by which new colonies become established in previously unoccupied areas. The beavers normally follow water courses in these movements, but occasionally undertake overland journeys.

Summer tourists in Michigan can enjoy the view of a serene beaver pond edged with forest. They can also examine the lodge and dam and their intricate constructions. However, the beaver itself is rarely conspicuous during summer daylight hours (Banfield, 1974; Jackson, 1961). In winter, however, beavers may be active night or day, although their movements may be confined to under-ice activities which can be detected by air bubble formation. Within the lodge, the beaver family may be active at any hour, especially when kits are being nursed. Their actions can be monitored by putting one's ear next to a lodge roof or by inserting microphones into living spaces and recording utterances and noises remotely on tape.

The webbed hind feet, the dorsoventrally flattened and rudder-like tail, and the rather fusiform body shape are all adaptations for swimming, although speeds of more than 3 mi (5 km) per hour are unusual. In swimming, the beaver keeps its forefeet pressed to the sides of the body, produces rapid propelling strokes with its hind appendages, and uses its tail as a rudder and for quick dives. Tevis (1950) classified diving behavior as (1) the dive when an undisturbed individual submerges quietly leaving barely a ripple; (2) the dive when a disturbed animal makes a loud warning slap on the water with its tail before submerging; and (3) the dive when a frightened animal notes imminent danger and submerges instantaneously without fanfare.

ASSOCIATES. The beaver's habitat manipulations produce major changes in the local environments by raising water temperatures and by discouraging some plant and animal species and encouraging others. The aquatic life is perhaps most affected, with the creation of quiet-water pond habitats along flowing streams, profoundly altering populations of trout and other fishes (Bradt, 1947). The flooding provides an increased aquatic area but eliminates all terrestrial vegetation. However, trees killed by the flooding become nesting sites for woodpeckers and other birds. The pondside and emergent vegetation is attractive for nesting ducks, geese, and marsh birds. Turtles, aquatic snakes, frogs, and other amphibians flourish. Predacious species such as mink, river otter, great blue heron, and bittern benefit from the impoundment. Beaver lodges may become bird roosting sites or nesting platforms for marsh hawks and Canada geese. The major semiaquatic mammalian associate is usually the muskrat, which commonly occupies beaver ponds. These two species associate on rather friendly terms, sometimes using the same caches of stored food (Tevis, 1950).

However, the beaver pond and lodges are rarely permanent additions to the local environment, although there are exceptional cases where dams are maintained for decades. Nevertheless, beaver logging operations sooner or later deplete the food trees. Removing the seral stage aspen and birch, for example, accelerates the plant succession and the development of climax vegetation. The beaver, primarily a seral stage animal, is obliged to depart. The dam then falls into disrepair and the pond drains, leaving the stream trickling through a bare ground environment. This silt-filled area is destined to pass through successional plant-animal stages, the first being a meadow, then a shrub growth, and finally a wooded environment perhaps once again attractive to beaver. In short, the beaver's engineering activities set into motion a dramatic series of environmental changes attractive to an array of aquatic and terrestrial life which comes and goes in accordance with the handiwork of this remarkable animal.

REPRODUCTIVE ACTIVITIES. Except for conspicuous teats on the pectoral (chest) section of the female abdomen, the beaver shows little external evidence of its sex because both the male and the female systems are obscured in a cloacal opening, where urinary and digestive wastes also empty. Beaver reach breeding maturity during their second year (Brenner, 1964; Larson, 1967), with older females producing, as a rule, larger litters than younger females. Matings probably take place in water (Bradt, 1941) during late winter (January and February). Males appear to be monogamous. The gestation period is about three and one-half months, with the Michigan young (generally called kits) born between late April and late June (Benson, 1936; Bradt, 1947). Embryonic counts vary from one to nine (Brenner, 1964; Hay, 1957), but there may be some embryo resorption before birth that reduces the number born (Banfield, 1974). Studies in New Mexico (Huey, 1956) show that there is a relationship between the kind of food available and productivity, indicating that reproductive performance is correlated with the quality of the mother's diet during pregnancy. In Michigan, Bradt (1938; 1939; 1947) determined that litters contain either three or four kits, each about 5 in (125 mm) long and weighing on the average 16 oz (450 g).

The young are born in either a lodge or bank burrow, from which the female excludes the male parent shortly before the birth. Kits arrive in a somewhat advanced (precocial) stage, being fully haired with richly colored, dark brown fur. Their eyes are either open at birth (Seton, 1953) or shortly thereafter, but their tails are not flattened as is characteristic of adults. The young grow rapidly and are able to eat solid food and venture from the nest area in company with the mother after one month of age. According to Tevis (1950), young beaver rarely go out in daylight, at least not until August. Kits may weigh as much as 10 lbs (4.5 kg) by August or September (Bradt, 1941). During their first summer, kits rarely assist the family group in cutting woody vegetation for food or repairing dams and lodges. However, as juveniles the following spring, summer and autumn, they actively participate in colony work programs. They reach reproductive maturity during their second winter and the adults then exclude them from the family, forcing them to seek other areas in which to establish colonies either with litter mates or other individuals (Svendsen, 1980b).

Since the beaver and its luxurious pelt have great economic importance, numerous studies relating to productivity and population age structure have been made (Henry and Bookout, 1969; Larson, 1967; Leege and Williams, 1967; Novak, 1977). Methods have been developed for aging individuals by examining tooth development (Van Nostrand and Stephenson, 1964), determining the growth of the baculum or bone in the penis (Friley, 1949), and by weighing the eye lens (Larson, 1967; Malcolm and Brooks, 1981). It is suspected that individuals are long-lived, although the maximum age is unknown (Henderson, 1960). In Michigan, Bradt (1938) reported a beaver surviving for 11 years, while Larson (1967) noted that one lived as long as 21 years in Maryland.

FOOD HABITS. Beavers are almost entirely vegetarians with the ability to eat the soft cambium layer of the bark and the buds, leaves and twigs of certain trees as well as pond and bankside herbaceous vegetation. There is evidence that nonwoody vegetation is the beaver's primary summer food, while woody vegetation is the chief winter fare (Svendsen, 1980a). In northeastern North America, beaver will accept assorted plants including cultivated fruit trees and corn stalks (Bradt, 1947). However, the species seems most closely associated with aspens of the genus *Populus,* with this food tree governing to a large extent the beaver's distribution (Stegeman, 1954). As mentioned previously, aspen and beaver are characteristic of early stages in community succession. In northern Michigan, when aspen is in short supply, the beaver population will reflect this.

Although felled trees and piles of chips next to neatly-chewed stumps are easily recognized evidence of hungry beavers, summer fare also includes an array of aquatic and semiaquatic plants (Bradt, 1947; Seton, 1953). Bradt identified some of these foods as eelgrass, duck potato, duckweed, water weed, and the roots of both white and yellow water lilies. In autumn, winter and spring, the beaver's basic diet is derived from woody plants stored mostly in under-ice caches. Bradt (1947) found that, following aspens, Michigan beaver were fond of by maples, then willows. When these favorites become exhausted, Michigan beaver resort to ash (Wood, 1917), elm, pin cherry, tag alder, white and yellow birch, and sometimes even white cedar, hemlock, and Norway pine (Manville, 1948). Supposedly, conifers are cut primarily for dam and lodge construction and maintenance. Fruit trees, red oak, thornapple, blackberry, and basswood are also cut (Tevis, 1950). It has not been fully determined if beaver can survive in northern parts of Michigan without some aspen. Birch trees are certainly not good substitutes. They are often cut down but rarely consumed. The beaver will venture as far from water as 300 to 450 ft (90 to 135 m), according to Ruhl (1931) and Hiner (1938), in search of food trees.

Usually trees of less than 15 in (37.5 cm) in diameter are cut, with beaver apparently making size selections in relation to the distance the cuttings have to be transported. One study showed that there is a decrease in mean size of trees cut with increasing distance from homesites (Jenkins, 1980). However, large trees seem to be no problem for the beaver's powerful incisors. One of the largest trees reported as felled by this animal was a British Columbia cottonwood which measured 67 in (166 mm) in diameter just above the cut and was approximately 110 ft (33 m) high (Hatt, 1944). Usually a single beaver fells a tree, although occasionally two animals are involved. To accomplish this, the beaver squats on its haunches with the tail as a brace (Fig. 49). As it clasps the tree with its forepaws, it gnaws out chips and slowly rotates

Figure 49. The beaver (*Castor canadensis*) in the process of felling a part of its food supply.
Photograph by Robert Harrington

around the trunk biting deeply into the wood. Shadle (1957) measured gnawed chips as short as 2 in (50 mm) to as long as 6 in (150 mm).

Normally the beaver does not cut completely through the gnawed area but allows the tree to topple under its own weight when the central column becomes sufficiently small. Cut timbers and branches are hauled to the pond site where they are used to repair the dam and lodge or cached as winter food. To make winter caches, a raft is usually constructed, the food materials are placed under the raft and when the entire structure becomes waterlogged it sinks below ice-level for easy access in winter (Slough, 1978). Beaver may require as many as 216 trees per beaver per year for food (Bradt, 1947). According to Warren (1940), each beaver requires approximately one ton of green aspen per year in order to survive. Some of this nutrient is stored as fat in the beaver's flattened tail (Aleksuik, 1970).

ENEMIES. The human is and has been for hundreds of years the major factor limiting beaver populations. The earliest human intruders in America probably lived rather harmoniously with the beaver, sparingly using its hides, sinews, teeth, bones, meat, and other vitals for various survival and cultural needs (Cleland, 1966). This situation changed dramatically when the pelt become highly valued in the fur trade and the Indians and the European immigrants joined in depleting the species. The beaver's comeback today is a result of modern efforts to preserve habitat and regulate the harvest.

Young beaver are perhaps most subject to natural predation—especially those yearlings which are forced from parental homes to find new living quarters. The domestic dog (Brooks, 1959) and its wild counterparts, the coyote (Krefting, 1969; Packard, 1940), the gray wolf (Seton, 1953; Timm, 1975), and even the smaller red fox (Johnson, 1970; Payne and Finley, 1975) are known to prey on beaver. Other Michigan beaver predators include eagles, black bear (Banfield, 1974; Rutherford, 1964; Seton, 1953), lynx (Seton, 1953), bobcat (Henderson, 1960), fisher (Banfield, 1974), and even mink (Denney, 1952). Some observers have reported that the river otter, which gains access to the interior of beaver lodges from under water or ice, preys on the beaver; others insist that this association is mostly amicable (Doutt *et al.*, 1966). Bradt (1947) felt that Michigan beaver populations suffer only minor predator mortality. Nevertheless, on Isle Royale, the beaver was second to moose as a summer food item for gray wolf (Peterson, 1970; Shelton, 1966). Felling trees may be hazardous, as beaver are sometimes pinned by a falling tree (Ellarson and Hickey, 1952).

PARASITES AND DISEASES. External parasites infesting beaver include mites, ticks (Lowery, 1974), and small beetles, often called lice by trappers (Rutherford, 1964). These beetles do not seem to cause any serious skin or fur problems. Internally, beaver may have nematodes in the respiratory tract (Goble and Cook, 1942) and in the alimentary tract (Brenner, 1970). Trematodes and protozoans are also recorded (Jackson, 1961; Lowery, 1974). Tularemia is a serious beaver disease, causing countless deaths in Michigan and adjacent areas in 1952–1955 (Jackson, 1961; Lawrence *et al.,* 1956; Stenlund, 1953). During the 1953 trapping season in Michigan alone, approximately 180 beaver sick or dead from this affliction were reported by trappers. Rabies can also be contracted by beaver (Brakhage and Sampson, 1952). Pesticide residues are likely to affect beaver as well (Hill and Lovett, 1976).

MOLT AND COLOR ABERRATIONS. There are two molts per year, one in autumn and one in spring. Occasionally a beaver will sport an exceedingly blackish-brown (melanistic) coat or a silvery or even a white (albinistic) coat, although these mutations are rare (Jackson, 1961; Jones, 1923; Seton, 1953).

ECONOMIC IMPORTANCE. The beaver is of value to human society because of its: (1) dense luxuriant fur, making the pelt important in the manufacture of various kinds of wearing apparel; (2) castor glands, an extract from which is used in the perfume industry; (3) high quality meat, perhaps of more value to humans in pioneer times than today; (4) manipulation of fresh-water and adjacent terrestrial environments, thereby enhancing their recreational and esthetic dimensions for human enjoyment; and (5) ability to build dams which aid in soil and water conservation, especially at drainage headwaters in the uplands. These attributes far outweigh the occasional reports that the backwater from dams floods timber, grazing, or farming lands; that plant-cutting activities (including corn stalks and fruit trees) reduce landowner income; and that the ponds themselves reduce the quality of trout streams (Yeager and Hay, 1955).

The Early Michigan Fur Trade—Nowhere in the United States has the economic importance of a single fur-bearing mammal—the beaver—figured more importantly in a state's early history than in Michigan. According to Bradt (1947), the beaver's status shifted from one of chiefly local Indian use to one of value in the European trade when Nicollet first visited the Straits of Mackinac in 1635. The exploitation of Michigan beaver began in earnest in 1660 when Radisson and Groseiliers, after an 18-month outing in what is now Michigan, reached Montreal with 60 canoes full of valuable pelts. The currency of the land was the beaver skin (Fasquella, 1950). The French and Indian War, which erupted in 1754, was, in part, a struggle between France and England, at least in the New World, to see who would control the lucrative Great Lakes fur trade (Petersen, 1979).

Three factors emphasize Michigan's important role in this trade. The first was the early recognition by enterprising Frenchmen of the value of Michigan beaver pelts, the willingness of the Michigan area Indians to assist in beaver procurement in exchange for trade goods, and the 1638 proclamation by the British Parliament that only beaver fur could be used in the manufacture of certain hats. As Petersen (1979) pointed out, the demand for the fur for hats began in the 1600s at the time of Charles I and lasted until the 1800s in Victorian England. In 1767, for example, 50,938 beaver pelts were shipped by British traders from Michilimackinac. Second, the geographic position of what is now Michigan facilitated the accumulation of the beaver hides by the trading companies and their shipment by lake carriers to the European-bound ships at coastal ports. Third was the vigorous fur trading operation at Michilimackinac with John Jacob Astor's American Fur Company (organized in 1808 at Mackinac Island) assuming leadership soon after the end of the War of 1812 (Johnson, 1919; Petersen, 1979). However, by the time the Americans entered the fur trade, the Michigan and Upper Great Lakes beaver catch was on the decline, as was the demand for beaver hats after 1830 (Beidleman, 1959). The depleted beaver population was allowed to return to much of its original range in Michigan by some restocking (Couch, 1937) and by the protection of closed seasons between 1920 and 1930, and again in 1947. Today, a regulated trapping program permits harvestable surpluses to be removed in selected areas while a sustained breeding population remains. Henry and Bookout (1969) estimated that about one-third of the resident beaver population in Ohio could be trapped each year without reducing the basic breeding stock.

In Michigan, the trapping seasons extend from December through April, beginning earlier in the Upper Peninsula and later in the Lower Peninsula. Carefully-kept statistics by the Michigan Department of Natural Resources (courtesy of Marvin Cooley) show marked fluctuations in the annual catches and the number of persons buying trapping licenses. Some of this reflects the prices received for the pelts as well as the snow depth, weather, and other factors affecting trapping success and trapper mobility. Between 1931 and 1978, the smallest catch was 1,479 animals in 1932–1933; the largest was 27,800 animals in 1972–1973. When pelts can be marketed for as much as $30 to $50 each (only about $21 in 1978, $35 in 1979, $36 in 1980 but as high as $60 in 1981), the annual catch provides a substantial income to the trappers, many of whom are from the rural sectors of the state. Although distasteful to many, trapping animals for their pelts will continue so long as there is a demand in the fashion marketplace for the beaver's fur. With wise management, the annual beaver catch will not damage the breeding stock of this renewable resource. Imitation beaver fur, an increasingly popular substitute for the real article, is manufactured from fossil fuel, a resource which is not renewable.

Castoreum—The paired musk glands opening on each side of the rectum near the anus, useful to the beaver in exuding odor for the recognition of individuals and in courtship, produce a secretion called castoreum. In early times, trappers used these glands as lures in the trapping process. Also, the dried castor glands had commercial value as a base or fixative in the manufacture of perfume and for their alleged medicinal qualities (Henderson and Craig, 1932). These glands are not so salable today. In 1920, the price per pound was between $10 and $20; in 1978, the price was $5 (Marvin Cooley, pers. comm.).

Beaver as a Meat Producer—The importance of beaver meat in the early-day rural economy is often overlooked. In fact, the meat was probably more important to the settlers than the pelt, especially if the animal was taken at a season when the fur was not in the best marketable condition (Baker, 1956). Beaver meat is rich flavored and juicy, and many northern Michigan churches serve beaver dinners to raise funds. Recipes have been concocted for roast, baked, and fried beaver and even for cooking beaver tail (Stains and Baker, 1958; Swank, 1949).

Nuisance Beaver—By 1920, when the closed season on beaver trapping was imposed in Michigan, public sentiment supported the restoration program, including restocking efforts. However, as beaver increased, nuisance animal complaints began and increased in proportion to the spread of the population. The stream and pond engineering projects of the beaver caused flooding of roadways, timber (McNeel, 1964), pasture and cultivated areas, and even railroad tracks. Unpardonable also was the cutting of shade trees in resort areas. Trout fishermen thought that pond formation along their fishing streams was harmful. In Michigan, J. C. Salyer and G. W. Bradt (Bradt, 1935) examined this latter problem; studies were also made in other states (in New York, Cook, 1940). Findings showed that waters impounded by beaver dams were favorable to trout in the first two years after the dam was constructed; trout later found these ponds less favorable. It was suggested that some control be exerted over beaver numbers along such streams (Bradt, 1932). At the same time, the many obvious advantages of beaver occupancy of rural streams should be weighed against the possible reduction of local trout populations (Brooks, 1959; Rutherford, 1964). In most instances, however, except perhaps in farmlands where streamside areas are intensely cultivated, the beaver is welcomed.

COUNTY RECORDS FROM MICHIGAN. Records of specimens of beaver in collections of museums and from the literature show that this species has occurred within historic times in every county of the state. According to Marvin Cooley (pers. comm.), the Michigan Department of Natural Resources had reports of beaver in all counties in 1978 except for some southeast (Macomb, Monroe, Wayne) and southwest (Allegan, Berrien, Cass, Ottawa, Van Buren) counties. The animal also lives on Isle Royale in Lake Superior (Burt, 1946), on Beaver Island (Ozoga and Phillips, 1964) on Garden Island (Phillips *et al.*, 1965), and on North Manitou Island (Scharf and Jorae, 1980) in Lake Michigan, and on Drummond Island in Lake Huron (Manville, 1950).

LITERATURE CITED

Alcoze, T. M.
1981 Pre-settlement beaver population density in the Upper Great Lakes Region. Michigan State Univ., unpubl. Ph.D. dissertation, vii+107 pp.

Aleksiuk, M.
1968 Scent-mound communication, territoriality, and population regulation in beaver (*Castor canadensis* Kuhl). Jour. Mammalogy, 49(4):759–762.
1970 The function of the tail as a fat depot in the beaver (*Castor canadensis*). Jour. Mammalogy, 51(1):145–148.

Bailey, V.
1926 How beavers build their houses. Jour. Mammalogy, 7(1):41–43.

Baker, G. A.
1899 The St. Joseph-Kankakee portage. Northern Indiana Hist. Soc., Publ. No. 1, 48 pp.

Baker, R. H.
1956 Remarks on the former distribution of animals in eastern Texas. Texas Jour. Sci., 8(3):356–359.

Banfield, A. W. F.
1974 The mammals of Canada. Univ. Toronto Press, Toronto. xxiv+438 pp.

Barber, W. E.
1919 A beaver dam of huge proportions. Sci. American, 122(7):167.

Beidleman, R. G.
1959 The American beaver, Thorne Ecol. Res. Sta., Bull. No. 7:1–14.

Benson, S. B.
1936 Notes on the sex ratio and breeding of the beaver in Michigan. Univ. Michigan, Mus. Zool., Occas. Papers No. 325, 6 pp.

Bradt, G. W.
1932 Report on nuisance-beaver control, 1931. Michigan Acad. Sci., Arts and Ltrs., Papers, 17:509–513.
1935 Michigan's beaver-trout management program. American Fish. Soc., Trans., 65:253–257.
1938 A study of beaver colonies in Michigan. Jour. Mammalogy, 19(2):139–162.
1939 Breeding habits of beaver. Jour. Mammalogy, 20(4): 486–489.
1941 Notes on breeding of beavers. Jour. Mammalogy, 22(2):220–221.
1947 Michigan beaver management. Michigan Dept. Conserv., Game Div., 56 pp.

Brakhage, G. K., and F. W. Sampson
1952 Rabies in beaver. Jour. Wildlife Mgt., 16(2):226.

Brenner, F. J.
1964 Reproduction of the beaver in Crawford County, Pennsylvania. Jour. Wildlife Mgt., 28(4):743–747.
1970 Observations of the helminth parasites in beaver. Jour. Mammalogy, 51(1):171–173.

Brooks, D. M.
1959 Fur animals of Indiana. Indiana Dept. Conserv., P–R Bull. No. 4, 195 pp.

Bump, G., and A. H. Cook
1941 Black Gold. The story of the beaver in New York

State. New York State Conserv. Dept., Div. Fish and Game, Mgt. Bull. No. 2, 16 pp.

Burt, W. H.
1946 The mammals of Michigan. Univ. Michigan Press, Ann Arbor. xv+288 pp.

Cleland, C. E.
1966 The prehistoric animal ecology and ethnozoology of the Upper Great Lakes Region. Univ. Michigan, Mus. Anthro., Anthro. Papers, No. 29, x+294 pp.

Cook, D. B.
1940 Beaver-trout relations. Jour. Mammalogy, 21(4):397–401.

Couch, L. K.
1937 Trapping and transplanting live beavers. U.S. Dept. Agri., Farmers' Bull. No. 1768, 18 pp.

Denney, R. N.
1952 A summary of North American beaver management, 1946–1948. Colorado Game and Fish Dept., Current Rept. 28, vii+58 pp.

Dice, L. R., and H. B. Sherman
1922 Notes on the mammals of Gogebic and Ontonagon counties, Michigan, 1920. Univ. Michigan, Mus. Zool., Occas. Papers, No. 109, 40 pp.

Doutt, J. K., C. A. Heppenstall, and J. E. Guilday
1966 Mammals of Pennsylvania. Pennsylvania Game Comm., Harrisburg, 273 pp.

Ellarson, R. S., and J. J. Hickey
1952 Beaver trapped by tree. Jour. Mammalogy, 33(4):482–483.

Erickson, A. B.
1939 Beaver populations in Pine County, Minnesota. Jour. Mammalogy, 20(2):195–201.

Fasquelle, E. R.
1950 When Michigan was young. Wm. B. Eerdmans Publ. Co., Grand Rapids. 156 pp.

Ffolliott, P. F., W. P. Clary, and F. R. Larson
1976 Observations of beaver activity in an extreme environment. Southwestern Nat., 21(1):131–133.

Friley, C. E. Jr.
1949 Use of the baculum in age determination of Michigan beaver. Jour. Mammalogy, 30(3):261–267.

Goble, F. C., and A. H. Cook
1942 Notes on nematodes from the lungs and frontal sinuses of New York fur-beavers. Jour. Parasitol., 28(6):451–455

Gregory, T.
1931 Beavers at work on their house. Jour. Mammalogy, 12(3):242–244.

Harris, W. P., Jr., and H. Du Charme
1928 Notes on set camera work with beavers in northern Michigan. Jour. Mammalogy, 9(1):17–19.

Hatt, R. T.
1944 A large beaver-felled tree. Jour. Mammalogy, 25(3): 313.

Hay, K. G.
1957 Record beaver litter for Colorado. Jour. Mammalogy, 38(2):268–269.

Hay, K. G., and W. H. Rutherford
1956 A guide to beaver trapping and pelting. Colorado Dept. Game and Fish, Tech. Bull., No. 3, 21 pp.

Henderson, F. R.
1960 Beaver in Kansas. Univ. Kansas State Biol. Surv. and Mus. Nat. Hist., Misc. Publ. No. 26, 85 pp.

Henderson, J., and E. L. Craig
1932 Economic mammalogy. Charles C. Thomas, Springfield. x+397 pp.

Henry, D. B., and T. A. Bookout
1969 Productivity of beavers in northeastern Ohio. Jour. Wildlife Mgt., 33(4):927–932.

Higgins, M. J.
1980 An analysis of the faunal remains from the Schwerdt Site, a late prehistoric encampment in Allegan County, Michigan. Western Michigan Univ., unpubl. M.A. thesis, v+77 pp.

Hill, E. P., and J. W. Lovett
1976 Pesticide residues in beaver and river otter from Alabama. Proc. Ann. Conf. Southeastern Assoc. Game and Fish Comm., 29:365–369.

Hiner, L. E.
1938 Observations on the foraging habits of beavers. Jour. Mammalogy, 19(3);317–319.

Holman, J. A.
1975 Michigan's fossil vertebrates. Michigan State Univ., Pub. Mus., Ed. Bull. No. 2, 54 pp.

Huey, W. S.
1956 New Mexico beaver management. New Mexico Dept. Game and Fish, Bull. 4, 49 pp.

Jackson, H. H. T.
1961 Mammals of Wisconsin. Univ. Wisconsin Press, Madison. xii+504 pp.

Jenkins, S. H.
1980 A size-distance relation in food selection by beavers. Ecology, 61(4):740–746.

Jenkins, S. H., and P. E. Busher
1979 Castor canadensis. American Soc. Mammalogists, Mammalian Species, No. 120, 8 pp.

Johnson, C. E.
1927 The beaver in the Adirondacks: Its economics and natural history. Roosevelt Wild Life Bull., 4(4):501–641.

Johnson, I. A.
1919 The Michigan fur trade. Michigan Hist. Comm., Univ. Ser., xii+201 pp.

Johnson, W. J.
1970 Food habits of the red fox in Isle Royale National Park, Lake Superior. American Midl. Nat., 84(2):568–572.

Jones, S. V. H.
1923 Color variations in wild animals. Jour. Mammalogy, 4(3):172–177.

Krefting, L. W.
1969 The rise and fall of the coyote on Isle Royale. Naturalist, 20(4):24–32.

Larson, J. S.
1967 Age structure and sexual maturity within a western Maryland beaver (*Castor canadensis*) population. Jour. Mammalogy, 48(3):408–413.

Lavrov, L. S., and V. N. Orlov
1973 Karyotypes and taxonomy of modern beavers (*Castor*, Castoridae, Mammalia). [In Russian with English summary], Zool. Jour., 52(5):734–742.

Lawrence, W. H.
1952 Evidence of the age of beaver ponds. Jour. Wildlife Mgt., 16(1):69–79.

Lawrence, W. H., L. D. Fay, and S. A. Graham
1956 A report on the beaver die-off in Michigan. Jour. Wildlife Mgt., 20(2):184–187.

Leege, T. A.
1968 Natural movements of beavers in southeastern Idaho. Jour. Wildlife Mgt., 32(4):973–976.

Leege, T. A., and R. M. Williams
1967 Beaver productivity in Idaho. Jour. Wildlife Mgt., 31(2):326–332.

Lowery, G. H., Jr.
1974 The mammals of Louisiana and its adjacent waters. Louisiana State Univ. Press, Baton Rouge. xxiii+565 pp.

Malcolm, J. R., and R. J. Brooks
1981 Eye lens weight and body size as criteria of age in beaver (*Castor canadensis*). Canadian Jour. Zool., 59(6):1189–1192.

Manville, R. H.
1942 A report on wildlife studies at the Huron Mountain Club 1939 to 1942. (mimeo.), 165 pp.
1948 The vertebrate fauna of the Huron Mountains, Michigan. American Midl. Nat., 39(3):615–660.
1949 The fate of Morgan's beaver. Sci. Monthly, 69(3):186–191.
1950 The mammals of Drummond Island, Michigan Jour. Mammalogy, 31(3):358–359.

McNeel, W., Jr.
1964 Beaver cuttings in aspen indirectly detrimental to white pine. Jour. Wildlife Mgt., 28(4):861–863.

Morgan, L. H.
1868 The American beaver and his works. Lippincott, Philadelphia. 330 pp.

Novak, M.
1977 Determining the average size and composition of beaver families. Jour. Wildlife Mgt., 41(4):751–754.

Ozoga, J. J., and C. J. Phillips
1964 Mammals of Beaver Island, Michigan. Michigan State Univ., Publ. Mus., Biol. Ser., 2(6):305–348.

Packard, F. M.
1940 Beaver killed by coyotes. Jour. Mammalogy, 21(3): 359–360.

Payne, N. F.
1981 Accuracy of aerial censusing for beaver colonies in Newfoundland. Jour. Wildlife Mgt., 45(4):1014–1016.

Payne, N. F., and C. Finley
1975 Red fox attack on beaver. Canadian Field-Nat., 89(4): 450–451.

Petersen, E. T.
1960 Wildlife conservation in Michigan. Michigan Hist., 44(2):129–146.
1979 Hunters' heritage. A history of hunting in Michigan. Michigan United Cons. Clubs., Lansing. 55 pp.

Peterson, R. O.
1977 Wolf ecology and prey relationships on Isle Royale. Natl. Park Serv., Sci. Monog. Ser., No. 11, xx+210 pp.

Phillips, C. J., J. T. Ozoga, and L. C. Drew
1965 The land vertebrates of Garden Island, Michigan. Jack-Pine Warbler, 43(1):20–25.

Robinson, W. L. and J. K. Werner
1975 Vertebrate animal populations of the McCormick Forest. U.S. Dept. Agri., For. Serv. Res. Paper NC-118, 25 pp.

Ruhl, H. D.
1931 Fur beavers in Michigan's forest areas. Michigan Acad. Sci., Arts and Ltrs., Papers, 15:261–266.

Ruhl, H. D., and P. S. Lovejoy
1930 Beaver plantings in Michigan. Michigan Acad. Sci., Arts and Ltrs., Papers, 11:465–469.

Rutherford, W. H.
1964 The beaver in Colorado. Its biology, ecology, management and economics. Colorado Game, Fish and Parks Dept., Tech. Publ. No. 17, 49 pp.

Scharf, W. C., and M. L. Jorae
1980 Birds and land vertebrates of North Manitou Island. Jack-Pine Warbler, 58(1):4–15.

Seton, E. T.
1953 Lives of game animals. Vol. IV, Pt. I. Charles T. Branford Co., Boston. xii+440 pp.

Shadle, A. R.
1957 Sizes of beaver chips cut from aspen. Jour. Mammalogy, 38(2):268.
1980a Seasonal change in feeding patterns of beaver in southeastern Ohio. Jour. Wildlife Mgt., 44(1):285–290.
1980b Population parameters and colony composition of beaver (*Castor canadensis*) in southeast Ohio. American Midl. Nat., 104(1):47–56.

Shelton, P. C.
1966 Ecological studies of beavers, wolves, and moose in Isle Royale National Park, Michigan. Purdue Univ., unpubl. Ph.D. dissertation, 308 pp.

Slough, B. G.
1978 Beaver food cache structure and utilization. Jour. Wildlife Mgt., 42(3):644–646.

Stains, H. J., and R. H. Baker
1958 Furbearers in Kansas. A guide to trapping. Univ. Kansas, Mus. Nat. Hist., Misc. Publ. No. 18, 100 pp.

Stegeman, L. C.
1954 The production of aspen and its utilization by beaver on the Huntington Forest. Jour. Wildlife Mgt., 18(3): 348–357.

Stenlund, M. H.
1953 Report of Minnesota beaver die-off, 1951–1952. Jour. Wildlife Mgt., 17(3):376–377.

Stephenson, A. B.
1969 Temperatures within a beaver lodge in winter. Jour. Mammalogy, 50(1):134–136.

Svendsen, G. E.
1978 Castor and anal glands of the beaver (*Castor canadensis*). Jour. Mammalogy, 59(3):618–620.
1980a Seasonal change in feeding patterns of beaver in southeastern Ohio. Jour. Wildlife Mgt., 44(1):285–290.
1980b Population parameters and colony composition of beaver (*Castor canadensis*) in southeast Ohio. American Midl. Nat., 104(1):47–56.

Swank, W. G.
1949 Beaver ecology and management in West Virginia. Conserv. Comm. of West Virginia, Div. Game Mgt., Bull. 1, 65 pp.

Swank, W. G., and F. A. Glover
1948 Beaver censusing by airplane. Jour. Wildlife Mgt., 12(2):214.

Tevis, L., Jr.
1950 Summer behavior of a family of beavers in New York State. Jour. Mammalogy, 31(1):40–65.

Timm, R. M.
1975 Distribution, natural history, and parasites of mammals of Cook County, Minnesota. Univ. Minnesota, Bell Mus. Nat. Hist., Occas. Papers No. 14, 56 pp.

Townsend, J. E.
1953 Beaver ecology in western Montana with special reference to movements. Jour. Mammalogy, 34(4):459–478.

Van Nostrand, F. C., and A. B. Stephenson
1964 Age determination for beavers by tooth development. Jour. Wildlife Mgt., 28(3):430–434.

Warren, E. R.
1940 A beaver's food requirements. Jour. Mammalogy, 21(1):93.

Wood, N. A.
1914 Results of the Shira Expeditions to Whitefish Point, Michigan–Mammals. Michigan Acad. Sci., 16th Rept., pp. 92–98.
1917 Notes on the mammals of Alger County, Michigan. Univ. Michigan, Mus. Zool., Occas. Papers No. 36, 8 pp.
1922 The mammals of Washtenaw County, Michigan. Univ. Michigan, Mus. Zool., Occas. Papers No. 123, 23 pp.

Wood, N. A., and L. R. Dice
1923 Records of the distribution of Michigan mammals. Michigan Acad. Sci., Arts and Ltrs., Papers, 3:425–469.

Yeager, L. E., and K. G. Hay
1955 A contribution toward a bibliography on the beaver. Colorado Dept. Game and Fish, Tech. Bull. No. 1, ii+103 pp.

FAMILY CRICETIDAE

Approximately one-third of the living rodents—almost 570 species in 97 genera—belong to the family Cricetidae. They live in a variety of terrestrial environments; some species are also arboreal, semiaquatic, and fossorial. Most are small, generally weigh less than 7 oz (200 g), and have a total length, including tail, of less than 10 in (250 mm). One of the largest, the muskrat, is a prominent member of the Michigan mammalian fauna. Most cricetids are mouselike or ratlike in shape, with expressive noses, conspicuous vibrissae, beady eyes, round ears, and long, mostly hairless tails. The skull may be variable in shape and size but always has large front (first) incisors and a moderately-developed V-shaped infraorbital foramen through which a part of the jaw musculature (masseter medialis) passes. This muscle arrangement is a distinctive characteristic of the rodent Suborder Myomorpha. Another part of the jaw musculature (masseter lateralis) originates on the enlarged and tilted zygomatic plate at the anterior end of the zygomatic arch. There are no supraorbital processes on the frontal bones. The cheek (jaw) teeth include three molars; there are no premolars.

There are two subfamilies of cricetids in Michigan. One, the Cricetinae, is represented by two species of mice of the genus *Peromyscus*. These and most other members of this large grouping are mouselike or ratlike in appearance. The rather unspecialized molar teeth are generally low-crowned (brachyodont) with chewing surfaces covered with low, rounded cusps or tubercules useful for crushing seeds, insects, and other foods which characterize their generally omnivorous diet.

The other subfamily, the Microtinae, is represented in Michigan by volelike rodents of the genera *Clethrionomys, Microtus, Ondatra,* and *Synaptomys*. These have specialized molar teeth which are high-crowned (hypsodont), have open roots (allowing for continual growth), and possess chewing surfaces with flat prismatic (triangular) enamel patterns useful for masticating coarse grasses and forbs.

WHITE-FOOTED MOUSE
Peromyscus leucopus (Rafinesque)

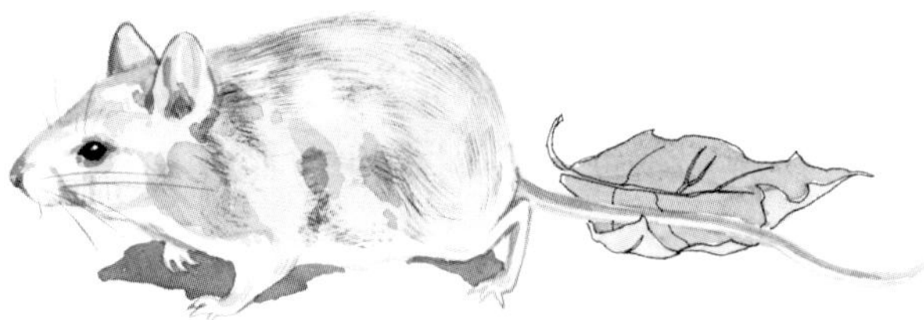

NAMES. The generic name *Peromyscus,* proposed by Gloger in 1841, is of obscure origin. The first part of the word may be from the Latin *pero,* meaning pointed, perhaps referring to the characteristic pointed nose. However, an earlier derivation, from the Greek *pera,* meaning pouched (in reference to internal cheek pouches), is the preferred etymon among mammalogists. The last part of the generic name comes from the Greek *myskos,* meaning little mouse. The specific name *leucopus,* proposed by the American biologist Rafinesque in 1818, comes from two Greek words, *leukon* meaning white and *pous* meaning foot. In Michigan, this species is called the wood mouse, mouse, or white-footed mouse. The subspecies *Peromyscus leucopus noveboracensis* (Fischer) occurs in Michigan.

RECOGNITION. Body size small, no larger than a house mouse; head and body of an adult averaging 3½ in (90 mm) long, with a finely-haired, slim tail, almost as long as the head and body; body roundish and slender, covered with short, soft, and dense pelage; head with pointed nose; large, black, beady eyes; large, conspicuous, membranous ears; long, prominent vibrissae; hind limbs longer than fore limbs. In adults (Plate 3), coloring is: head and back, brown or cinnamon brown, with faintly darker dorsal stripe; underparts and feet white; belly hairs, basally dark, throat hairs white to bases; eyelids and base of vibrissae, blackish; ears, gray faintly edged with white; no white preauricular spot which is often prominent in the deer mouse; tail usually indistinctly bicolored, brownish above and cream colored below. For about the first 40 days of life, juvenile pelage is slate gray above and white below. Subsequently and for about 25 days the subadult coloring is somewhat dull brownish above and white below (Coventry, 1937). The slate colored fur begins molting on the flanks and is replaced with new brown hairs along a molt line moving anteriorly to cover the base of tail and head. The glossy cinnamon-brown upperparts of the adult may not appear until the regular autumn molt. There are no color distinctions between sexes. There are three pairs of mammary glands, one pectoral and two inguinal. Internal cheek pouches are characteristic. The four functional digits on the forefeet and the five on the hind feet have slender, delicate claws.

Adult Michigan white-footed mice may be confused with the introduced house mouse (*Mus musculus*) and the three subspecies of Michigan deer mice (*Peromyscus maniculatus bairdii, P. m. gracilis,* and *P. m. maniculatus*). Adult white-footed mice differ from adult house mice, in that they have rich cinnamon brown upperparts instead of grayish or grayish brown; white underparts instead of grayish or yellowish; fully haired tail instead of scaly tail; bony palate of skull not extending beyond plane of last upper molars; underside of cutting edges of upper incisors (as seen in profile) approximately straight instead of showing angular wear; and cheek teeth (molars) with cusps arranged in two longitudinal rows instead of in three.

Deer mice of the subspecies *Peromyscus maniculatus bairdii* and white-footed mice occur in the same counties: in Michigan's Lower Peninsula and in the Menominee County area of the southwestern part of the Upper Peninsula. The two differ in that the adult white-footed mouse (see Table 15) has a more rounded "roman" facial profile and less bulging eyes; throat and forearms with hairs usually (but not always) white to their bases instead of grayish bases; upper lips usually dusky in color instead of white; ear (measured in a fresh specimen from notch to crown) longer, usually more than 9/16 in (15 mm); tail longer, usually more than 2⅜ in (65 mm); tail usually indistinctly bicolored instead of being conspicuously dark brown above and pure white below; tail less heavily furred and lacking distinct growth of long hairs at tip; base of undertail generally whitish instead of reddish brown; hind feet longer, usually more than 13/16 in (20 mm); hairs on soles of feet usually not extending to plantar tubercules instead of covering their edges; incisive foramina constricted anteriorly and forming a slight angle near the junction of the suture between the premaxillary and maxillary bones instead of being more parallel-sided, more rounded anteriorly and usually lacking the angle; upper (maxillary) tooth

*Table 15. Characteristics distinguishing Michigan white-footed mice (*Peromyscus leucopus*) and deer mice (*Peromyscus maniculatus*).*

	PEROMYSCUS MANICULATUS		
PEROMYSCUS LEUCOPUS	P. m. bairdii	P. m. gracilis	P. m. maniculatus
Biocolored tail usually blends from darker top to paler bottom	Bicolored tail sharply delimited between top and bottom	Biocolored tail sharply delimited between top and bottom	Biocolored tail sharply delimited between top and bottom
Tail medium, 2⅝ to 3⅜ in (65–85 mm)	Tail short, 1¾ to 2¾ in (45–70 mm)	Tail long, 2¾ to 4⅛ in (70–105 mm)	Tail long, 2¾ to 4⅛ in (70–105 mm)
Ear medium, ⅝ to ¾ in (16–18 mm)	Ear short, ½ to 3/5 in (12–16 mm)	Ear long, ¾ to 13/16 in (18–21 mm)	Ear long, ¾ to 13/16 in (18–21 mm)
Hind foot long, ¾ to ⅞ in (18–23 mm)	Hind foot short, ⅝ to 13/16 in (16–20 mm)	Hind foot long, ¾ to ⅞ in (18–23 mm)	Hind foot long, ¾ to ⅞ in (18–23 mm)
Braincase rounded, not parallel-sided	Braincase flattened, parallel-sided	Braincase flattened, parallel-sided	Braincase flattened, parallel-sided
Skull long, 1 in (25 mm) or more	Skull short, 1 in (25 mm) or less	Skull long, 1 in (25 mm) or more	Skull long, 1 in (25 mm) or more
Third upper molar tooth almost as wide as first two	Third upper molar less wide than first two	Third upper molar less wide than first two	Third upper molar less wide than first two

row longer averaging 3.5 mm (in *P. m. bairdii*, 3.2 mm); and individual upper cheek teeth (molars) smaller and relatively uniform in width. Skulls of both species should be compared to appreciate this distinction.

Deer mice of the subspecies *Peromyscus maniculatus gracilis* and white-footed mice occur in the many of the same counties in the northern half of Michigan's Lower Peninsula and in the southwestern part of the Upper Peninsula. The two adult mice differ (see Table 15 and Choate, 1973) in most characteristics mentioned for *P. m. bairdii* except for size. The white-footed mouse displays less "nervous" behavior; softer, more silky pelage; and an ear (measured in a fresh specimen from notch to crown) usually less than ¾ in (18 mm) insead of longer, with a 10% overlap. Since deer mice of the subspecies *P. m. maniculatus* occur in Michigan only on Isle Royale, which is not included in the geographic range of the white-footed mouse, there is no need to distinguish them here.

Some of the above differences may serve to distinguish Michigan mice of the genus *Peromyscus* in the gray juvenile pelage; however, accurate identification using skin characteristics, body dimensions, or cranial features may not always be reliable.

MEASUREMENTS AND WEIGHTS. Adult white-footed mice measure as follows: length from tip of nose to end of tail vertebrae, 5⅝ to 7⅝ in (145 to 195 mm); length of tail vertebrae, 2⅜ to 3¾ in (60 to 95 mm); length of hind foot, 11/16 to 15/16 in (18 to 23 mm); height of ear from notch, 9/16 to ¾ in (15 to 19 mm); weight of adults, 0.5 to 1.0 oz (15 to 30 g).

DISTINCTIVE CRANIAL AND DENTAL CHARACTERISTICS. The skull of the white-footed mouse is delicately formed with an elongated rostrum and broad, rounded cranium. The zygomatic arches are fragile in construction; in side view the slit-like infraorbital foramina are obscured by the zygomatic plates. The skull averages 1 in (25 mm) long and ½ in (13 mm) wide, across the zygomatic arches. The skull of the white-footed mouse is distinguished from that of the deer mouse or the house mouse by characteristics listed in the section entitled Recognition. To distinguish white-footed mice from other Michigan mammals, see Key under Order Rodentia and Fig. 50. The dental formula of the white-footed mouse is:

$$I\,(\text{incisors})\,\frac{1\text{-}1}{1\text{-}1},\ C\,(\text{canines})\,\frac{0\text{-}0}{0\text{-}0},\ P\,(\text{premolars})\,\frac{0\text{-}0}{0\text{-}0},\ M\,(\text{molars})\,\frac{3\text{-}3}{3\text{-}3} = 16$$

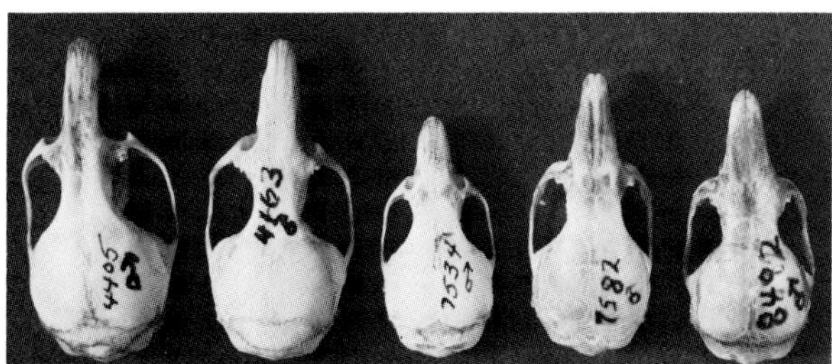

Figure 50. Dorsal aspects of the skulls of five small Michigan rodents; from left to right: white-footed mouse (*Peromyscus leucopus*), deer mouse (*Peromyscus maniculatus*), introduced house mouse (*Mus musculus*), meadow jumping mouse (*Zapus hudsonius*), woodland jumping mouse (*Napaeozapus insignis*).

DISTRIBUTION. The white-footed mouse occupies the eastern United States as far west as the Great Plains, where it follows forested borders along west-east flowing streams. The species extends into parts of southern Canada, including southeastern Ontario, and southward in eastern Mexico along the Gulf Coastal Plain as far as Yucatán. In Michigan (see Map 35), the white-footed mouse occurs throughout the Lower Peninsula, with specimens identified in almost all counties (Burt, 1946). In the Upper Peninsula, specimens have been identified from Menominee and Alger counties (Ozoga and Verme, 1966). It also is to be expected in other counties in the southwestern part of the Upper Peninsula. A specimen obtained near Sault Ste Marie in Chippewa County, recorded by Lawrence L. Master (pers. comm.), seems out of range and is perhaps an escapee.

HABITAT PREFERENCES. The white-footed mouse inhabits woodlands and brushlands (Baker, 1968), and is active on the litter-strewn forest floor, in stumps, and under logs, as well as above ground in trees and shrubs (Pruitt, 1959; Dueser and Shugart, 1978). According to M'Closkey (1975a), shrub understory provides a "vertical complexity" for exploration, predator evasion, and foraging. Distribution is thus influenced by the successional stage of the woody growth and mosaic paterns of the habitats, especially in southern Michigan (M'Closkey and Lajoie, 1975; Tardiff, 1979). White-footed mice can also occasionally be found in grassy and weedy areas (Blair, 1940; Howard, 1949, Pearson, 1959). Such individuals, according to Burt (1940) and Whitaker (1967), are often young or unsettled animals. However, this rodent may inhabit disturbed, early successional stages of Michigan farm lands where it is locally sympatric with the prairie form of the deer mouse (Master, 1977). Heavy herbaceous cover seems less desirable, although seeds from various weeds (forbs) are attractive to the mouse in grass-sedge marsh areas (Getz, 1961). Woodland edges may provide some of the best habitat for this mouse (Nicholson, 1941).

The white-footed mouse's ability to climb efficiently allows it to utilize forested areas in three dimensions (Horner, 1954). Climbing activity declines in winter (Newton *et al.*, 1980). In Michigan,

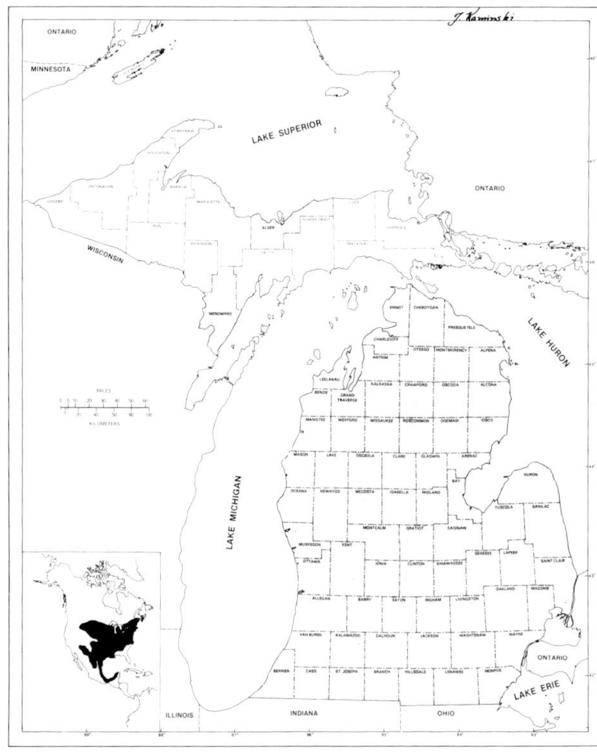

Map 35. Geographic distribution of the white-footed mouse (*Peromyscus leucopus*). Counties from which this species is known are shaded. It is to be expected in other counties in the southwestern part of the Upper Peninsula.

at least half of the mice captured in woodlands may be above ground (Gosling, 1977), in traps set as high as 13 ft (4 m) in trees. Many individuals nest above ground (Nicholson, 1941). Mice are also occasionally caught underground, especially in winter in eastern mole burrows which crisscross the forest substrat (Connor, 1971). According to Getz (1961), white-footed mice generally avoid excessive moisture on the forest floor by living above ground (also see Ruffer, 1961 and Blem and Blem, 1975). Although bodies of water may form barriers to white-footed mice (Waters, 1961; Savidge, 1973), there are records of dispersal by swimming (Sheppe, 1965) and possibly by winter ice crossings (Christianson, 1977). Powerline corridor clearings as wide as 340 ft (104 m) also seem to offer no obstacle to crossing movements (Schreiber and Graves, 1977).

The present mosaic pattern of woody habitats in much of southern Michigan might cause considerable isolation of white-footed mouse populations were it not for the animal's ability to spread from one woody area to another along brushy fence rows, roadsides, and streamways. These habitats provide minimum cover for encounters with mice from other habitats nearby. Even without such avenues, however, the mice are known to cross open areas between woodland patches (Tardiff, 1979).

Supposedly, the best Michigan habitat for white-footed mice is oak-hickory woodlands (Burt, 1940; Getz, 1961). In Leelanau County, Hatt (1923) found the species ". . . in most terrestrial communities in which there is shelter." In Washtenaw County, Wood (1922a) caught mice in forests and corn fields. In Charlevoix County, Dice (1925) trapped white-footed mice in various woody and brushy areas but obtained most of them in virgin and second-growth hardwood forests. In Montmorency County, Green (1925) obtained white-footed mice in pine barrens and coniferous swamps. In Clinton County, Linduska (1950) captured white-footed mice in shrub fencerows, upland woods, lowland woods, and even by isolated trees. His discovery of the species' abundance in lowland woods has been confirmed by other studies (*i.e.*, Batzli, 1977; Blem and Blem, 1975; and Stickel, 1948). In the Allegan area, Hodgson and Brewer (1975) noted that white-footed mice were widespread in wooded areas with no apparent preference for any habitat type. In Kalamazoo County, Allen (1938) concluded that the species was the most common small mammal in brush and woodland. Consequently, white-footed mice can probably be found in southern Michigan in almost any complex, heterogeneous and forested habitat, with an understory of shrubs, rather than grass, being the most acceptable (Van Deusen and Kaufman, 1977).

DENSITY AND MOVEMENTS. White-footed mice produce litters several times a year. Individuals are normally short-lived with a wide array of predators. The interaction between mouse production and negative environmental influences means that a local population may fluctuate widely from month to month or even day to day throughout the year (Burt, 1940; Myton, 1974). Nevertheless, white-footed mouse populations in Michigan are largest in late spring and again in mid autumn (Blair, 1948). Populations dwindle in early spring (April-May) and again (to a lesser extent) in midsummer (July-August). While white-footed mice are experiencing these seasonal fluctuations, their numbers are also greatly influenced by environmental factors, defined by M'Closkey and Lajoie (1975) as habitat structure. Forested areas with abundant and varied shrub understory seem to hold the highest densities in the Michigan area (Getz, 1961; M'Closkey, 1975b). Although it is recognized that populations are dense in such areas (Hodgson and Brewer, 1975), it is not fully known whether this environment provides abundant food, or whether it facilitates escape from predators (McCloskey and Lajoie, 1975; Van Deusen and Kaufman, 1977).

Densities of the white-footed mouse populations are influenced by seasons and existing food supplies (Batzli, 1977 and others). In the hardwood forest habitat in central Illinois (not unlike that found in southern Michigan), Blem and Blem (1975) found population densities of white-footed mice in uplands to range from 0.9 per acre (2.2 per ha) in April to 9.7 per acre (23.2 per ha) in October. In floodplain habitat, these authors found mice to range in numbers from 3.4 per acre (8.6 per ha) in June to 19.5 per acre (48.7 per ha) in October. Specific counts for Michigan white-footed mice show similar low populations in spring. In Livingston County, for example, Burt (1940) counted only about three mice per acre (7.5 per ha) in mid-May. Master (1978) found densities as high as 13 per acre (33–34 per ha) in early successional fields in July through October in southern Michigan. In early October both Burt and Getz (1960) in Washtenaw County, and Linduska (1950) in Clinton County obtained densities of 7.0 to 8.0 individuals per acre (17.5 to 20.0 per ha). In late November, Burt calculated that the density of his study population at the Edwin S. George Reserve in Livingston County peaked at almost 11 per acre (27.5 per ha). These numbers are, of course, profoundly influenced by the spatial needs of the individual mice, whose home ranges may overlap broadly but whose home sites (territory) may be defended from others (Burt, 1943; Stickel, 1946).

BEHAVIOR. In his pioneering study of small mammals in southern Michigan, Burt (1940) determined that the white-footed mouse is not highly social. During the breeding season in the warm months, individuals were less tolerant of one another than in the cold months. Breeding females also seemed much more antagonistic to each other than adult males to each other. Subsequent work by Myton (1974) showed that these mice may live in "family" aggregations, consisting of one adult female, several adult males, and a number of juveniles. Mature females held separate and non-overlapping areas of activity (home ranges), with each surrounded by the activity areas of males (also see Metzgar, 1973). In winter, groups of both sexes might assemble and huddle to reduce individual heat loss (Howard, 1951; Thomsen, 1945).

In the cold months, a combination of freezing temperatures and a shortage of food may cause white-footed mice to succumb to "cold weather starvation" (Howard, 1951). To counteract this, these mice can become torpid (Hill, 1975; Lynch *et al.*, 1978), meaning that the body temperature of an inactive mouse in the shelter of its nest can drop from about 98°F (37°C) to as low as 59°F (15°C). The energy saved by maintaining a lower body temperature has obvious advantages for winter survival. According to Hill (1975), fasting (food shortage) as well as lowered temperatures may be inducements for initiating torpor.

White-footed mice, like other members of the diverse group of species in the genus *Peromyscus*, are creatures of habit and tend to establish home sites, seek mates, and forage in consistent areas. The mice make their presence known to other individuals by urine odor and fecal deposits, as well as by marking places with exudate from anal, clitoridal and preputial glands (Eisenberg, 1968). The animals use set patterns for nocturnal activity (except under snow cover, Behney, 1936), to adapt to arboreal life (Horner, 1954; King, 1968), to find their way back home after being displaced (Cooke and Terman, 1977), to gather seeds and other foodstuffs, to defend against predators, to exclude competitive species or conspecifics, to communicate vocally or visually with associates or intruding conspecifics, to develop social groupings, to groom, to engage in sexual behavior, and to care for young (Eisenberg, 1968). White-footed mice do not establish conspicuous pathways through grass or underbrush (Baker, 1971), although as Osterberg (1962) noted, they do use runs constructed by meadow voles (*Microtus pennsylvanicus*) and bog lemmings (*Synaptomys cooperi*).

The white-footed mouse centers its activity about a nest site, which can be underground, under logs, in stone walls or rocky ledges, in stumps, under brush piles, in hollow trees, in dense growths of branches or vines, in bottles and cans in garbage dumps (Courtney and Fenton, 1976), in remodeled nests of birds, squirrels or even bald-faced hornets (Jackson, 1961), and in special nest boxes (Nicholson, 1941). When mice invade houses, barns or cottages, their nest sites might be in litter, mattresses, drawers or on shelves (Connor, 1971). In Livingston County, Burt (1940; 1965) observed a mouse building a nest and bearing a litter of young in a rubber boot between late afternoon and mid-evening. Near Grand Ledge in Eaton County, Jim Walter found a family of white-footed mice (adult female and three young) using a wren house attached to the limb of a cherry tree; a house wren had raised a brood there earlier that same year. Edwards and Pitts (1952) found nests of white-footed mice with the aid of a trained Labrador retriever.

Nests are composed of soft, shredded materials, including grass, feathers, hair, and cloth (Jackson, 1961). According to Eisenberg (1968), white-footed mice construct their nests by: (1) boring into the gathered pile of material, (2) combing the nest material to the side and above with the forepaws, (3) molding the nest cup by turning movements of the body, (4) pushing movements with the forepaws and the nose, and (5) shredding and stripping the plant material by holding it with the forepaws, nibbling along one edge with the incisors and jerking the head back. The completed nest may be much as 10 in (255 mm) in diameter and 7 in (180 mm) deep. One entrance hole is usually located on the side and near the top. Male and female mice may work together in constructing the nest. White-footed mice move from one nest to another, possibly because they foul their nests with urine and feces (Nicholson, 1951).

ASSOCIATES. Living in wooded and brush lands, along shrub-lined fence rows or field borders, and on edges of fields and marshes, the white-footed mouse comes in contact with almost every terrestrial and arboreal mammal in Michigan. Allen (1938), Burt (1940), Getz (1961), Linduska (1950), and Nicholson (1941) discuss these associations. This distribution subjects the white-footed mouse to the entire gamut of four-footed and winged predators. The white-footed mouse has survived in these varied habitats chiefly because of its high reproductive rate.

Mammalian neighbors which encroach on home sites, shelters, and food supplies are probably as great a threat to the mouse as predators. If competition is great for limited resources, several solutions result: (1) the exclusion of one or more species by another, (2) the interdigitation of species in relation to their preferences for different seral stages of plant communities, and (3) some degree of compatibility (Baker, 1968). More than likely the second option prevails in Michigan communities, where coexisting species differ sufficiently to minimize the struggle for existence. This species separation has been the basis for the formulation of the competitive exclusion (or Gause's), principle (Diamond, 1978).

Documenting small mammal associations in nature depends largely on finding two or more species caught in the same or nearby traps, under the same pile of brush, decaying log or shock of corn, or possibly photographed using the same runways (Osterberg, 1962). Under laboratory conditions, interspecific compatibility can be tested but it is difficult in nature to infer tolerance and close association simply because species are caught in closely-spaced traps in similar habitats. The interrelations of the white-footed mouse and the meadow vole, both common in many areas, show that there is no serious competition and little diet overlap (Getz, 1961; Bowker and Pearson, 1975; M'Closkey and Fieldwick, 1975; Newton *et al.*, 1980). These small mammals live together at woodland/grass borders, along brushy and grassy fence rows, and in marshes. Their association depends partly on the habitat succession (Pearson, 1959). White-footed mice increase as the plant community advances to the brushy/forest stage; but meadow voles decline in this habitat. In some instances, local distributions may not overlap (Morris, 1979); in others, they may use the same nest boxes (Nicholson, 1941). White-footed mice and southern red-backed voles live in similar habitats in northern Michigan, but their different life styles separate them (Beer, 1961). The house mouse, on the other hand, appears more aggressive, often restricting or excluding the white-footed mouse from certain joint habitats (Courtney and Fenton, 1976). Nichols and Conley (1981) also found evidence in Ingham County of meadow jumping mice excluding white-footed mice.

The white-footed mouse occurs in Michigan areas which are also used by the deer mouse. In the Lower Peninsula (mostly the southern counties) and the extreme southwestern part of the Upper Peninsula, the subspecies *P. m. bairdii,* known commonly as the prairie deer mouse lives in plowed and cultivated fields, stubble, and in fallow situations in which the seral successions allow for open grassy and weedy areas (Whitaker, 1967). In these environments and also where such environments border on brushy or woody growth, these two mice often associate and may, in fact, compete for space (Master, 1977). However, Nicholson (1941) found both species sharing an underground nest box.

In the northern Lower Peninsula and also in parts of the southwestern Upper Peninsula, the white-footed mouse may occur with the deer mouse of the subspecies *P. m. gracilis,* known commonly as the woodland deer mouse. Since both of these mice live in woodlands, they associate but may prefer slightly different kinds of environments (Drickamer and Capone, 1977). For example, there is evidence that where only one of these two mice occurs, the preferred nesting sites are above ground. However, where they occur together, the deer mouse has a greater tendency to occupy the above-ground nesting area and the white-footed mouse uses sites on the forest floor (Steh, 1980). Wrigley (1969) found the white-footed mouse in Quebec most often lives in dry and warm pine-hardwood and forest edge situations, and the woodland deer mouse lives in moist and cool depths of the forest. Klein (1960) obtained similar findings in Ontario. In the Charlevoix area of Michigan, Dice (1925) also found the white-footed mouse in open (edge) areas and second growth forest, with the woodland deer mouse more numerous in mature hardwoods. The exact ecologic relationships between these two mice are poorly understood (Smith and Speller, 1970); however, the evidence does suggest that the white-footed mouse may have made its distributional gains in northern conifer/hardwood habitats following logging, replacing (or joining) the woodland deer mouse in much of the second growth and brush lands resulting from timber cutting. The cool, moist, older forest stands probably remain the principal habitat for the woodland deer mouse.

REPRODUCTIVE ACTIVITIES. This short-lived mouse has a high annual reproduction rate. In the Michigan area, major reproductive peaks occur in spring and autumn, with marked reduction in mid-summer and little or no activity in December and January (Burt, 1940). Evidently, the spring-born mice have greater life expectation than those born in autumn (Millar and Gyng, 1981; Rintamaa *et al.*, 1976). Manville (1952) noted that mild weather may augment the autumn reproductive period for Michigan white-footed mice. In Ingham County, he found few pregnant or lactating females in October of 1948 and 1949, but in the milder autumn of 1950, all sample females were either pregnant or lactating in mid-October. Females are seasonally polyestrous, with the ratio of darkness to daylight per 24 hours playing an important role in initiating the annual reproductive cycle (Whitaker, 1936).

Mating begins as early as late February, with the first litter arriving in March after a gestation of about 22 to 25 days (Svihla, 1932). A newly-born female can reach sexual maturity as early as four weeks of age (Clark, 1938), and can bear a litter when she is from seven to 10 weeks old (Jackson, 1961; Lowery, 1974). David Rapp, a student at Michigan State University, caught a pregnant female in immature gray pelage on May 17, 1977. This individual was presumably born no earlier than the previous March. After birth of the young, the parent female undergoes post-partum estrus at which time mating can again take place. However, the process of nursing young tends to delay the development of the next litter, extending the gestation period to as long as 37 days (Hill, 1972; Svihla, 1932). Although there are records of females having as many as 10 litters in 12 months in laboratory environments, female Michigan white-footed mice probably produce only about four litters yearly, with two as a result of late winter and spring breeding activity and two more as a result of late summer and autumn activity (Snyder, 1956). The reproductive cycle in the male correlates with that of the female. The testes enlarge and descend from the abdominal cavity into the scrotum during the breeding season. The few records of winter breeding may result from the availability of a combination of superior shelter and abundant food, as in a grain bin or under corn shocks (Burt, 1946). For example, in Clinton County in mid-January, one of three female white-footed mice under field-shocked corn was pregnant (Linduska, 1942).

A litter of white-footed mice may vary in number from one to eight or nine, with four being average. At birth, each offspring weighs about 0.06 oz (1.5 to 2.0 g), is hairless and has closed eyes. Hair begins to grow and the ear pinnae become erect almost immediately, the lower incisors erupt in less than six days, and the eyes open in 12–14 days. Weaning is completed as early as 19 days but generally in 24 to 28 days (Layne, 1968), whereupon the young learn to fend for themselves. Although mother white-footed mice maintain protective care of their litters, they must leave the young periodically to forage. Hill (1972) found that mothers may be away from the nest as much as 5.0 to 8.4 hours each 24 hours.

White-footed mice thrive in captivity for several years, but their life expectancy in nature is short; the average female is fortunate to survive long enough to complete the nuturing of her first litter, let alone live to see her second year. In his study at the Edwin S. George Reserve in Livingston County, Burt (1940) recorded as his oldest mouse a female that lived in nature about 21 months.

Normally, spring-born individuals produce the autumn litters and autumn-born individuals survive the winter to produce the spring litters. However, in a study of the survival of marked white-footed mice conducted in Ohio (Rintamaa *et al.*, 1975), it was found that of the pregnant females examined in the spring of 1974, five were born in previous spring (one year earlier), six were born the previous autumn (six months earlier), and one was born the same spring. In the autumn of 1974, the pregnant females included four born in the spring of 1973 (one and one-half years previous), two born in the autumn of 1973 (one year previous), and 13 born in the spring of 1974 (six months previous). In Livingston County, Blair (1948) found that mice remained on his study area for an average of 4.6 months, with the longest stay being 12 months. Despite the precarious status of the white-footed mouse in what appears to be a hostile environment, the reproductive potential of this species seems sufficient to sustain local populations.

FOOD HABITS. The white-footed mouse eats primarily seeds and insects. Its low-crowned, cusp-covered teeth and uncomplicated digestive tract facilitate mastication and assimilation of these concentrated foodstuffs. These characteristics are in sharp contrast to those of the meadow vole and other grass-eaters of the genus *Microtus,* which have elaborate dental chewing surfaces, esophageal sacs, complex stomachs, and voluminous caeca to process bulky and fibrous vegetative plant parts (Baker, 1971).

White-footed mice adapt easily to the seasonal food supplies, apparently developing large body fat reserves only when nutrient supplies are uncertain (Millar, 1981). They are inveterate hoarders, caching seeds and other staples in assorted places underground, in logs, above the ground in trees, and even in human habitations and outhouses (Abbott and Quink, 1970; Burt, 1940). However, seed caching is chiefly a late summer and autumn activity, whereas most seeds of earlier-growing woodland herbs are eaten when found (Heithaus, 1981). Hamilton (1941) found that insects provided a major part of the summer and winter diet of the white-footed mouse. Larvae of beetles, moths, and butterflies (Whitaker, 1963) and cocoons of larch sawflies, wasps, and bald-faced hornets (Graham, 1929) are generally seasonal in abundance. Seeds of grasses, weeds, clover, and grains, green stuff and fungal parts, and fruits such as grapes (Gosling, 1977) and cherries (Whitaker, 1967) are favored during the Michigan growing season (Getz, 1961). In autumn, winter, and early spring, acorns, other nuts, and pine seeds are shared with many other forest creatures (Abbott and Quick, 1970; Green, 1925; and Getz, 1961). In the vicinity of East Lansing, Verme (1957) calculated that the average daily acorn consumption by a white-footed mouse was 1.8 for mid-October.

White-footed mice are also attracted to garbage dumps, where they scavenge for various plant and animal food (Courtney and Fenton, 1976). Large doses of insecticides may be ingested when the mice devour insects sprayed with these chemicals (Stehn *et al.,* 1976).

ENEMIES. Besides environmental disasters due to weather changes or food shortages, the mice are fair game for almost all meat-eaters, including cannibalistic members of their own species. Fortunately, some mouse-eating carnivores only hunt in daytime and may not threaten the nocturnal white-footed mouse (for the red-tailed hawk as a predator, see Fitch and Bare, 1978).

The most successful mouse catchers are probably snakes, owls, and weasels. Owl catches are relatively easy to identify because of the partly digested skulls, bones, and hair in their coughed-up pellets. Despite the differences in crania between sympatric white-footed mice (*P. leucopus*) and deer mice (*P. maniculatus*), some predator food-habit analysts do not attempt to distinguish between the two species. Those that have, however, show that white-footed mice are eaten by the barn owl, saw whet owl, screech owl, long-eared owl, barred owl, and great-horned owl (Dexter, 1978; Geis, 1952; Jackson, 1961; Voight and Glenn-Lewin, 1978; Wallace, 1948). According to a Michigan study by Metzgar (1967), the screech owl's ability to catch white-footed mice is enhanced by the mouse being transient rather than having an established home site. Four-footed carnivores preying on mice include weasels, minks, house cats, striped skunks, and raccoons. In Michigan, Murie (1936) found remains of white-footed mice in the fecal droppings of red foxes in March and May. In captivity, a young white-footed mouse was eaten by an adult deer mouse in Wisconsin (Sterling, 1953).

PARASITES AND DISEASES. White-footed mice are infested with assorted parasites. Whitaker (1968) reports that the species can carry at least four kinds of flat worms (trematodes), five tapeworms (cestodes), 15 round worms (nematodes), and one pentastomid internally, and 23 mites, 33 chiggers, six ticks, 28 fleas, three lice (Anoplura), and six botfly larvae (Cuterebridae) externally. At least two protozoan parasites also occur in the species (Doran, 1954). Not all of these parasites occur in or on Michigan white-footed mice. Perhaps the most conspicuous on Michigan specimens are the mites and the botfly larvae (Burt, 1940). Some mites attack the skin, producing inflamed, swollen ears and tail, as well as bare spots on the head and body.

From June to October, large botfly larvae (sometimes called warbles) are obvious on the undersides of many Michigan white-footed mice. The adult fly lays its eggs and the hatched larvae invade the host early in development. Ultimately, the larvae settle just under the skin, most often in the inguinal region between the hind legs. Each larva extracts nourishment from its host, makes a small hole in the mouse's skin for ventilation and excretion, and enlarges to maturity after about four weeks (see summary in Whitaker, 1968). The cylindrical and elongated body of the larva reaches a length of 1 in (25 mm) and a width of ¾ in (18 mm). After the larva has completed its growth, it enlarges the breathing hole to emerge and drop to the ground where it burrows into the soil to go through the pupal stage before emerging as an adult fly. Despite the obvious discomfort and burden of carrying this parasite, there is no evidence of lethal or sterilizing effects on the white-footed mouse (Abbott and Parsons, 1961; Dunaway *et al.*, 1967; Miller and Getz, 1969; Timm and Cook, 1979; Timm and Lee, 1982).

White-footed mice may also be plagued with hair balls (trichobenzoars) in their stomachs (Horner, 1950). This condition does not seem to result from any parasite encroachment but may be fatal.

MOLT AND COLOR ABERRATIONS. Beginning students of mammalogy are often confused by the various color phases of mammals at certain stages of growth. Some stages are standard, and some are a result of mutations. The color variations in white-footed mice are conspicuous as the individual matures (Layne, 1968). Essentially, weaned, young white-footed mice are dorsally grayish while mature adults are brownish to cinnamon-red. The young animals retain their grayish pelage until a molt, beginning when the individual is between 40 and 50 days of age, begins the change to the adult pelage (Gottschang, 1956). Coventry (1937) identified progressive immature pelages as (1) pure gray, (2) gray-brown, (3) brown-gray, and finally (4) adult brown. Adults molt and grow new coats of hair in summer (Lensing, 1977). Color phases include white spotting and yellow (Jackson, 1961).

ECONOMIC IMPORTANCE. In the outdoor community, the white-footed mouse is an important prey species for many birds and furbearing mammals. It is sometimes a pest to farmers when it damages standing or shocked grains or invades storage bins in search of seeds. One serious, though slightly humorous problem was called to this author's attention in 1968. In Van Buren County, a mechanical grape harvester was not only picking grapes but also vine-dwelling white-footed mice. The matter was of such severity that an extra person had to be employed to pick out the mice as they rode down the conveyer belt.

The white-footed mouse can be an unwelcome visitor to lake cottages and other human establishments in Michigan (Wood, 1922b). White-footed mice, along with other seed-eating forest rodents, may exert some influence on conifer regeneration (Pank, 1974; Smith and Aldous, 1947). However, since the white-footed mouse is not widely distributed in the northern coniferous forests, its role as a pine-seed consumer is a minor one.

COUNTY RECORDS FROM MICHIGAN. Records of specimens of the white-footed mouse in collections of museums and from the literature show that this species lives in all counties in Michigan's Lower Peninsula but is reported only in Menominee and Alger counties in the Upper Peninsula's southwestern side (Burt, 1946; Ozoga and Verme, 1966).

LITERATURE CITED

Abbott, H. G., and M. A. Parsons
1961 *Cuterebra* infestation in *Peromyscus*. Jour. Mammalogy, 42(3):383–385.

Abbott, H. G., and T. F. Quink
1970 Ecology of eastern white pine seed caches made by small forest mammals. Ecology, 51(2):271–278.

Allen D. L.
1938 Ecological studies on the vertebrate fauna of a 500-acre farm in Kalamazoo County, Michigan. Ecol. Monog., 8(3):347–436.

Baker, R. H.
1968 Habits and distribution. Pp. 98–126 *In* J. A. King, ed. Biology of *Peromyscus* (Rodentia). American Soc. Mammalogists, Sp. Publ. No. 2, xiii+593 pp.
1971 Nutritional strategies of myomorph rodents in North American grasslands. Jour. Mammalogy, 52(4):800–805.

Batzli, G. O.
1977 Population dynamics of the white-footed mouse in floodplain and upland forests. American Midl. Nat., 97(1):18–32.

Beer, J. R.
1961 Winter home ranges of the red-backed mouse and white-footed mouse. Jour. Mammalogy, 42(2):174–180.

Behney, W. H.
1936 Nocturnal explorations of the forest deer-mouse. Jour. Mammalogy, 17(3):225–230.

Blair, W. F.
1940 A study of the prairie deer-mouse population in southern Michigan. American Midl. Nat., 24(2):273–305.
1948 Population density, life span, and mortality rates of small mammals in the blue-grass meadow and blue-grass field associations of southern Michigan. American Midl. Nat., 40(2):395–419.

Blem, L. B., and C. R. Blem
1975 The effect of flooding on length of residency in the white-footed mouse, *Peromyscus leucopus*. American Midl. Nat., 94(1):232–236.

Bowker, L. S., and P. G. Pearson
1975 Habitat orientation and interspecific interaction of *Microtus pennsylvanicus* and *Peromyscus leucopus*. American Midl. Nat., 94:491–496.

Burt, W. H.
1940 Territorial behavior and populations of some small mammals in southern Michigan. Univ. Michigan, Mus. Zool., Misc. Publ. No. 45, 58 pp.
1943 Territorial behavior and home range concepts as applied to mammals. Jour. Mammalogy, 24:346–352.
1946 The mammals of Michigan. Univ. Michigan Press, Ann Arbor. xv+288 pp.
1965 Michigan's pouched little mouse. Michigan Audubon Newsletter, 13(1).

Choate, J. R.
1973 Identification and recent distribution of white-footed mice (*Peromyscus*) in New England. Jour. Mammalogy, 54(1):41–49.

Christianson, L.
1977 Winter movements of *Peromyscus* across a lake in northern Minnesota. Jour. Mammalogy, 58(2):244.

Clark, F. H.
1938 Age of sexual maturity in mice of the genus *Peromyscus*. Jour. Mammalogy, 19(2):230–234.

Connor, P. F.
1971 The mammals of Long Island, New York. New York State Mus. and Sci. Serv. Bull. 416, 78 pp.

Cooke, J. A., and C. R. Terman
1977 Influence of displacement distance and vision on homing behavior of the white-footed mouse (*Peromyscus leucopus noveboracensis*). Jour. Mammalogy, 58(1):58–66.

Courtney, P. A., and M. B. Fenton
1976 The effects of a small rural garbage dump on populations of *Peromyscus leucopus* Rafinesque and other small mammals. Jour. Appl. Ecol., 13:413–422.

Coventry, A. F.
1937 Notes on the breeding of some Cricetidae in Ontario. Jour. Mammalogy, 18(4):489–496.

Dexter, R. W.
1978 Mammals utilized as food by owls in reference to the local fauna of northeastern Ohio. Kirtlandia, No. 24, 6 pp.

Diamond, J. M.
1978 Niche shifts and the rediscovery of interspecific competition. American Sci., 66(3):322–331.

Dice, L. R.
1925 A survey of the mammals of Charlevoix County, Michigan, and vicinity. Univ. Michigan, Mus. Zool., Occas. Papers No. 159, 33 pp.

Doran, D. J.
1954 A catalogue of the protozoa and helminths of North American rodents. I. Protozoa and Acanthocephala. American Midl. Nat., 52:118–128.

Drickamer, L. C., and M. R. Capone
1977 Weather parameters, trappability and niche separation in two sympatric species of *Peromyscus*. American Midl. Nat., 98(2):376–381.

Dueser, R. D., and H. H. Shugart, Jr.
1978 Microhabitats in a forest-floor small mammal fauna. Ecology, 59(1):89–98.

Dunaway, P. B., J. A. Payne, L. L. Lewis, and J. D. Story
1967 Incidence and effects of *Cuterebra* in *Peromyscus*. Jour. Mammalogy, 48(1):38–51.

Edwards, R. L., and W. H. Pitts
1952 Dog locates winter nests of mammals. Jour. Mammalogy, 33(2):243–244.

Eisenberg, J. F.
1968 Behavior patterns. Pp. 451–495, *in* J. A. King, ed. Biology of *Peromyscus* (Rodentia). American Soc. Mammalogists, Sp. Publ. No. 2, xiii+593 pp.

Fitch, H. S., and R. O. Bare
1978 A field study of the red-tailed hawk in eastern Kansas. Trans. Kansas Acad. Sci., 81(1):1–13.

Geis, A. D.
1952 Winter food habits of a pair of long-eared owls. Jack-Pine Warbler, 30(3):93.

Getz, L. L.
1959 Activity of *Peromyscus leucopus*. Jour. Mammalogy, 40(3):449–450.
1960 Populations of small mammals on an ecological island. Jack-Pine Warbler, 38(1):16–19.
1961 Notes on the local distribution of *Peromyscus leucopus* and *Zapus hudsonius*. American Midl. Nat., 65(2):486–500.

Gosling, N. M.
1977 Observations on the three-dimensional range of *Pero-*

myscus leucopus noveboracensis. Jack-Pine Warbler, 55(1):43–44.

Gottschang, J. L.
1956 Juvenile molt in *Peromyscus leucopus noveboracensis*. Jour. Mammalogy, 37(4):516–520.

Graham, S. A.
1929 The larch sawfly as an indicator of mouse abundance. Jour. Mammalogy, 10(3):189–196.

Green, M. M.
1925 Notes on some mammals of Montmorency County, Michigan. Jour. Mammalogy, 6(3):173–178.

Hamilton, W. J., Jr.
1941 The food of small forest mammals in eastern United States. Jour. Mammalogy, 22(3):250–263.

Hatt, R. T.
1923 The land vertebrate communities of western Leelanau County, Michigan, with an annotated list of the mammals of the county. Michigan Acad. Sci., Arts and Ltrs., Papers, 3:369–402.

Heithaus, E. R.
1981 Seed predation by rodents on three ant-dispersed plants. Ecology, 62(1):136–145.

Hill, R. W.
1972 The amount of maternal care in *Peromyscus leucopus* and its thermal significance for the young. Jour. Mammalogy, 53(4):774–790.
1975 Daily torpor in *Peromyscus leucopus* on an adequate diet. Comp. Biochem. Physiol., 51A:413–423.

Hodgson, J. R., and R. Brewer
1975 Mammals of oak forests in southwestern Michigan. Jack-Pine Warbler, 53(4):131–140.

Horner, B. E.
1950 Trichobenzoars (hair balls) in *Peromyscus*. Jour. Mammalogy, 31(1):94–95.
1954 Arboreal adaptations of *Peromyscus* with special reference to use of the tail. Univ. Michigan, Lab. Vert. Biol., Contr. No. 61, 84 pp.

Howard, W. E.
1949 Dispersal, amount of inbreeding, and longevity in a local population of prairie deermice on the George Reserve, southern Michigan. Univ. Michigan, Lab. Vert. Biol., Contr. No. 43, 52 pp.
1951 Relation between low temperature and available food to survival in small rodents. Jour. Mammalogy, 32(3): 300–312.

Jackson, H. H. T.
1961 Mammals of Wisconsin. Univ. Wisconsin Press, Madison. xii+504 pp.

Jackson, W. B.
1953 Use of nest boxes in wood mouse population studies. Jour. Mammalogy, 34(4):505–507.

King, J. A.
1968 Psychology, Pp. 496–542, *in* J. A. King, ed. Biology of *Peromyscus* (Rodentia). American Soc. Mammalogists, Sp. Publ. No. 2, xiii+593 pp.

Klein, H. G.
1960 Ecological relationships of *Peromyscus leucopus noveboracensis* and *Peromyscus maniculatus gracilis* in central New York. Ecol. Monog., 30:387–407.

Layne, J. N.
1968 Ontogeny. Pp. 148–253, *in* J. A. King, ed. Biology of *Peromyscus* (Rodentia). American Soc. Mammalogists, Sp. Publ. No. 2, xiii+593 pp.

Lensing, B. A.
1977 Seasonal molt in the white-footed mouse, *Peromyscus leucopus*. Trans. Kentucky Acad. Sci., 38(1–2):88–94.

Linduska, J. P.
1942 Winter rodent populations in field-shocked corn. Jour. Wildlife Mgt., 6(4):353–363.
1950 Ecology and land-use relationships of small mammals on a Michigan farm. Michigan Dept. Conservation, Game Div., ix+144 pp.

Lowery, G. H., Jr.
1974 The mammals of Louisiana and its adjacent waters. Louisiana State Univ. Press, Baton Rouge. xxiii+565 pp.

Lynch, G. R., F. D. Vogt, and H. Smith
1978 Seasonal study of spontaneous daily torpor in the white-footed mouse, *Peromyscus leucopus*. Physiol. Zool., 51:389–399.

Manville, R. H.
1952 A late breeding cycle in *Peromyscus*. Jour. Mammalogy, 33(3):389.

Master, L. L.
1977 The effect of interspecific competition on habitat utilization by two species of *Peromyscus*. Univ. Michigan, unpubl. Ph.D. dissertation, xiii+179 pp.

M'Closkey, R. T.
1975a Habitat dimensions of white-footed mice, *Peromyscus leucopus*. American Midl. Nat., 93(1):158–167.
1975b Habitat sucession and rodent distribution. Jour. Mammalogy, 56(4):950–955.

M'Closkey, R. T., and B. Fieldwick
1975 Ecological separation of sympatric rodents (*Peromyscus* and *Microtus*). Jour. Mammalogy, 56(1):119–129.

M'Closkey, R. T., and D. T. Lajoie
1975 Determinants of local distribution and abundance in white-footed mice. Ecology, 56(2):467–472.

Metzgar, L. H.
1967 An experimental comparison of screech owl predation on resident and transient white-footed mice (*Peromyscus leucopus*). Jour. Mammalogy, 48(3):387–391.
1973 Home range shape and activity in *Peromyscus leucopus*. Jour. Mammalogy, 54(2):383–390.

Millar, J. S.
1981 Body composition and energy reserves of northern *Peromyscus leucopus*. Jour. Mammalogy, 62(4):786–794.

Millar, J. S., and L. W. Gyng
1981 Initiation of breeding by northern *Peromyscus* in relation to temperature. Canadian Jour. Zool., 59(6): 1094–1098.

Miller, D. H., and L. L. Getz
1969 Botfly infections in a population of *Peromyscus leucopus*. Jour. Mammalogy, 50(2):277–283.

Morris, D. W.
1979 Microhabitat utilization and species distribution of sympatric small mammals in southwestern Ontario. American Midl. Nat., 101(2):373–384.

Murie, A.
1936 Following fox trails. Univ. Michigan, Mus. Zool., Misc. Publ. no. 32, 45 pp.

Myton, B.
1974 Utilization of space by *Peromyscus leucopus* and other small mammals. Ecology, 55:277–290.

Newton, S. L., T. D. Nudds and J. S. Millar
1980 Importance of arboreality in *Peromyscus leucopus* and *Microtus pennsylvanicus* interactions. Canadian Field-Nat., 94(2):167–170.

Nichols, J. D., and W. Conley
1981 Observations suggesting competition between *Peromyscus* and *Zapus* in southern Michigan. Jack-Pine Warbler, 59(1):3–6.

Nicholson, A. J.
1941 The homes and social habits of the woodmouse *Peromsycus leucopus noveboracensis* in southern Michigan. American Midl. Nat., 25(1):196–223.

Osterberg, D. M.
1962 Activity of small mammals as recorded by a photographic device. Jour. Mammalogy, 43(2):219–229.

Ozoga, J. J., and L. J. Verme
1966 Noteworthy locality records for some upper Michigan mammals. Jack-Pine Warbler, 44(1):52.

Pank, L. F.
1974 A bibliography on seed-eating mammals and birds that affect forest regeneration. U.S. Bur. Sport Fisheries and Wildlife, Sp. Sci. Rep.—Wildlife No. 174, 28 pp.

Pearson, P. R.
1959 Small mammals and old field succession on the Piedmont of New Jersey, Ecology, 40(2):249–255.

Pruitt, W. O., Jr.
1959 Microclimates and local distribution of small mammals on the George Reserve, Michigan. Univ. Michigan, Mus. Zool., Misc. Publ. No. 109, 27 pp.

Rintamaa, D. L., P. A. Mazur, and S. H. Vessey
1976 Reproduction during two annual cycles in the population of *Peromyscus leucopus noveboracensis*. Jour. Mammalogy, 57(3):593–595.

Ruffer, D. G.
1961 Effect of flooding on a population of mice. Jour. Mammalogy, 42(4):494–502.

Savidge, I. R.
1973 A stream as a barrier to homing in *Peromyscus leucopus*. Jour. Mammalogy, 54(4):982–984.

Schreiber, R. K., and J. H. Graves
1977 Powerline corridors as possible barriers to the movements of small mammals. American Midl. Nat., 97(2): 504–508.

Sheppe, W.
1965 Dispersal by swimming in *Peromyscus leucopus*. Jour. Mammalogy, 46(2):336–337.

Smith, C. F., and S. E. Aldous
1947 The influence of mammals and birds in retarding artificial and natural reseeding of coniferous forests in the United States. Jour. Forestry, 45:361–369.

Smith, D. A., and S. W. Speller
1970 The distribution and behavior of *Peromyscus maniculatus gracilis* and *Peromyscus leucopus noveboracensis* (Rodentia: Cricetidae) in a southeastern Ontario woodlot. Canadian Jour. Zool., 48:1187–1199.

Snyder, D. P.
1956 Survival rates, longevity and population fluctuation in the white-footed mouse, *Peromyscus leucopus*, in southeastern Michigan. Univ. Michigan, Mus. Zool., Misc. Publ. 95, 33 pp.

Steh, C. D.
1980 Vertical nesting distribution of two species of *Peromyscus* under experimental conditions. Jour. Mammalogy, 61(1):141–143.

Stehn, R. A., J. A. Stone, and M. E. Richmond
1976 Feeding responses of small mammal scavengers to pesticide-killed arthropod prey. American Midl. Nat., 95:253–256.

Sterling, K. B.
1953 Cannibalism in *Peromyscus*. Jour. Mammalogy, 34(2): 262.

Stickel, L. F.
1946 Experimental analysis of methods for measuring small mammal populations. Jour. Wildlife Mgt., 10(2):150–159.
1948 Observations of the effect of flood on animals. Ecology, 29:505–507.

Svihla, A.
1932 A comparative life history study of the mice of the genus *Peromyscus*. Univ. Michigan, Mus. Zool., Misc. Publ. No. 24, 39 pp.

Tardiff, R. R.
1979 Dispersal of *Peromyscus leucopus* within and between woodlots. Michigan State Univ., unpubl. Ph.D. dissertation, 122 pp.

Thomson, H. P.
1945 The winter habits of the northern white-footed mouse. Jour. Mammalogy, 26(2):138–142.

Timm, R. M., and E. F. Cook
1979 The effect of bot fly larvae on reproduction in white-footed mice, *Peromyscus leucopus*. American Midl. Nat., 101(1):211–217.

Timm, R. M., and R. E., Lee, Jr.
1982 Is host castration an evolutionary strategy of bot flies? Evolution, 36(2):416–417.

Van Deusen, M., and D. W. Kaufman
1977 Habitat distribution of *Peromyscus leucopus* within prairie woods. Trans. Kansas Acad. Sci., 80(3 and 4):151–154.

Verme, L. J.
1957 Acorn consumption by chipmunks and white-footed mice. Jour. Mammalogy, 38(1):129–132.

Voight, J., and D. C. Glenn-Lewin
1978 Prey availability and prey taken by long-eared owls in Iowa. American Midl. Nat., 99(1):162–171.

Wallace, G. J.
1948 The barn owl in Michigan. Its distribution, natural

history and food habits. Michigan State Col., Agri. Exp. Sta., Tech. Bull. 208, 61 pp.

Waters, J. H.
1961 *Clethrionomys* and *Peromyscus* in southeastern Massachusetts. Jour. Mammalogy, 42(2):263–265.

Whitaker, J. O., Jr.
1963 Food of 120 *Peromyscus leucopus* from Ithaca, New York. Jour. Mammalogy, 44(3):418–419.
1967 Habitat and reproduction of some of the small mammals of Vigo County, Indiana, with a list of mammals known to occur there. Western Michigan Univ., C. C. Adams Center Ecol. Studies, Occas. Papers, No. 16, 24 pp.
1968 Parasites. Pp. 254–311 *In* J. A. King, ed. Biology of *Peromyscus* (Rodentia). American Soc. Mammalogists, Sp. Publ. No. 2, xiii+593 pp.

Whitaker, W. L.
1936 Effect of light on reproductive cycle of *Peromyscus leucopus noveboracensis*. Proc. Soc. Exp. Biol. and Medicine, 34:329–330.

Wood, N. A.
1922a The mammals of Washtenaw County, Michigan. Univ. Michigan, Mus. Zool., Occas. Papers No. 123, 23 pp.
1922b Notes on the mammals of Berrien County, Michigan. Univ. Michigan, Mus. Zool., Occas. Papers No. 124, 4 pp.

Wrigley, R. E.
1969 Ecological notes on the mammals of southern Quebec. Canadian Field-Nat., 83:201–211.

DEER MOUSE
Peromyscus maniculatus (Wagner)

NAMES. The generic name *Peromyscus,* proposed by Gloger in 1841, is of obscure origin. The first part of the name could be derived from the Latin *pero,* meaning pointed and perhaps referring to the mouse's nose, but more likely from the Greek *pera,* meaning pouched in reference to internal cheek pouches. The last part of the word is from the Greek *myskos,* meaning little mouse. Possibly the best translation is the pouched little mouse. The specific name *maniculatus,* proposed by Wagner in 1845, is from the Latin and means small handed. In Michigan, local residents call this species wood mouse, forest mouse and merely mouse, with the subspecies *P. m. bairdii* known in much of the literature as the prairie deer mouse, although for Michigan a more appropriate name would be cornfield mouse; for *P. m. gracilis,* woodland deer mouse (Burt, 1946). However, mammalogists wishing to standardize common names recommend the name deer mouse for the entire species. The subspecies *Peromyscus maniculatus bairdii* (Hoy and Kennicott) occurs in the Lower Peninsula and the southwestern part of the Upper Peninsula. The subspecies *Peromyscus maniculatus gracilis* (Le Conte) occurs in the northern part of the Lower Peninsula and in the Upper Peninsula. The subspecies *Peromyscus maniculatus maniculatus* (Wagner) occurs on Isle Royale.

RECOGNITION. In the northern part of Michigan's Lower Peninsula and in the southwestern part of the Upper Peninsula two distinctive subspecies of the deer mouse (Plate 4a) occur in adjacent areas but show no evidence of interbreeding. These two subspecies may remain reproductively isolated because they occupy different habitats (Barbehenn and New, 1957), although under laboratory conditions they will mate and produce fertile offspring (Foster, 1959). The northern subspecies *Peromyscus maniculatus gracilis* thrives in forests while the southern subspecies *Peromyscus maniculatus bairdii* lives in open lands, perferably plowed or cultivated fields, early stages of grasslands and along lakes shores (Dice, 1932, Hooper, 1942). Still another subspecies *Peromyscus maniculatus maniculatus* on Isle Royale in Lake Superior is isolated by water from the other Michigan deer mouse populations. This subspecies shows closest genetic relationships to the deer mice on the adjacent Canadian mainland to the north. Because of the distinctive characteristics and the differences in habitat preference between the subspecies (Harris, 1952; Welker, 1963), *P. m. bairdii* will be described separately from *P. m. gracilis* and *P. m. maniculatus.*

Peromyscus maniculatus bairdii. Body size small, no longer than that of a house mouse; head and body of an adult averaging 5⅝ in (145 mm) long; short, finely-haired tail, 2⅜ in (60 mm) long, no more than 40–45% as long as the head and body; body round and slender, covered with short, soft and dense pelage; head with pointed nose; large, black, beady eyes; large, scantily haired ears, long, prominent whiskers (vibrissae); hind limbs longer than forelimbs. In adults, color of head and back grayish-brown (russet), dark middorsally; underparts and feet, white; belly hair, basally dark, throat hair, white with gray base; eye ring and vibrissae, blackish; ears, dark with narrow, yellow-white margins; tail, distinctly bicolored, white below and gray brown above (most conspicuous in fresh autumn pelage). Immatures are darker and more gray than adults.

Peromyscus maniculatus gracilis and *Peromyscus maniculatus maniculatus.* The former, *P. m. gracilis,* occurring in forested areas in northern Michigan and the latter, *P. m. maniculatus,* isolated from other Michigan deer mice on Isle Royale, are similar in adult characteristics and in habits. They share all of the above features described for *P. m. bairdii* except for distinctive differences as follows: tail longer, usually more than 2¾ in (70 mm), in *P. m. bairdii* shorter; ear from notch conspicuously larger usually more than ¾ in (18.5 mm) long, in *P. m. bairdii* usually less than ⅝ in (15 mm); skull longer, 1 in (25 mm) or more, in *P. m. bairdii* shorter.

For characteristics distinguishing these subspecies from the white-footed mouse and the house mouse, consult the species account for the white-footed mouse (*Peromyscus leucopus*) and Table 15.

MEASUREMENTS AND WEIGHTS. Adult deer mice of the subspecies *P. m. bairdii* measure as follows: length from tip of nose to end of tail vertebrae, 4⅞ to 6¼ in (125 to 160 mm); length of tail vertebrae, 2 to 2¾ in (50 to 70 mm); length of hind foot, ⅝ to ¾ in (16 to 20 mm); height of ear from notch, ½ to ⅝ in (12 to 16 mm); weight of adults, 0.4 to 0.8 oz (12 to 24 g).

Adult deer mice of the subspecies *P. m. gracilis* and *P. m. maniculatus* measure as follows: length from tip of nose to end of tail vertebrae, 6¼ to 8 in (160 to 205 mm); length of tail vertebrae, 2¾ to 4⅜ in (70 to 110 mm); length of hind foot, ¾ to ⅞ in (18 to 23 mm); height of ear from notch, ¾ to 13/16 in (18 to 21 mm); weight of adults, 0.5 to 1.0 oz (15 to 30 g).

DISTINCTIVE CRANIAL AND DENTAL CHARACTERISTICS. The skull of the deer mouse is smooth, rounded and thin-walled (Fig. 50). The zygomatic arches are slender and depressed to the level of the palate; the interparietal is developed; the anterior palatine foramen are slitlike and long. The skull of *P. m. bairdii* averages smaller, ⅞ in (23 mm) long and ½ in (13 mm) wide (across the zygomatic arches); while those of *P. m. gracilis* and *P. m. maniculatus* average slightly larger, 1 in (26 mm) long and ½ in (13 mm) wide. The dental formula for the deer mouse is:

$$\text{I (incisors) } \frac{1\text{-}1}{1\text{-}1}, \text{ C (canines) } \frac{0\text{-}0}{0\text{-}0}, \text{ P (premolars) } \frac{0\text{-}0}{0\text{-}0}, \text{ M (molars) } \frac{3\text{-}3}{3\text{-}3} = 16$$

DISTRIBUTION. The deer mouse occurs in varied environments from northern treeline in Alaska and Canada southward to central Mexico, with the exception of southeastern United States and some coastal areas of Mexico. The ubiquitous and diverse nature of this dimunitive rodent is shown by its division into 67 recognizable subspecies (Hall, 1981). Deer mice adapt to such contrasting habitats as the cool, moist, forests of coastal British Columbia and the arid, bleak shrub growth of the Chihuahuan desert of northern Mexico. Because of the distinct geographical ranges of the three subspecies of the deer mouse in Michigan (Burt, 1946), their specific distributional patterns are described spearately (see Map 36).

Peromyscus maniculatus bairdii. This distinctive population of the deer mouse occurs in non-forested areas in eastern and central parts of the United States. It reaches the extreme southwestern part of Michigan's Upper Peninsula by way of Wisconsin (Long, 1974). In the Lower Peninsula, it may be expected in suitable open habitat in most counties.

Peromyscus maniculatus gracilis. This subspecies is found throughout Michigan's Upper Peninsula (also occurring westward into Wisconsin and eastward into Ontario). In the northern part of the Lower Peninsula, it lives in suitable forest habitat at least as far south as Missaukee County (Burt, 1946) and Lake County (L. D. Caldwell, pers. comm., November 29, 1978). Records of *P. m. gracilis* on islands in lakes Michigan and Huron are listed in the section entitled County Records.

Peromyscus maniculatus maniculatus. This deer mouse, widespread in east-central Canada, apparently gained access to Isle Royale from adjacent Ontario, with which the island holds close biological affinity (Burt, 1946).

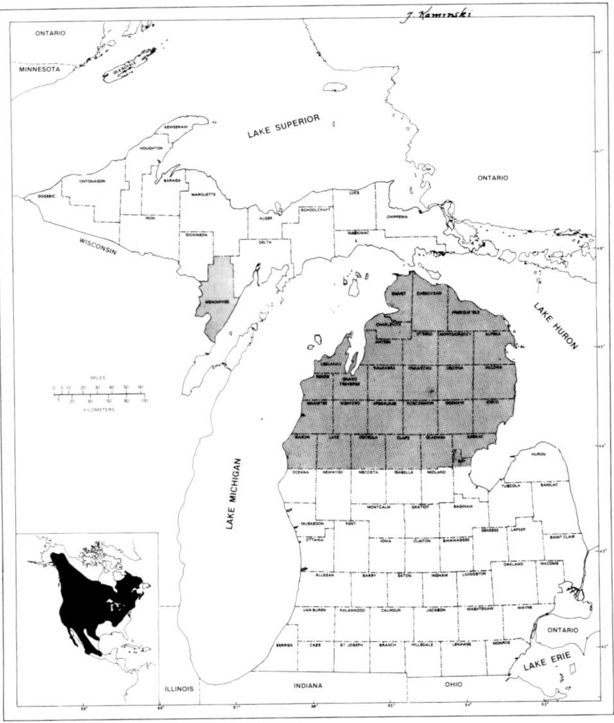

Map 36. Geographic distribution of the deer mouse (*Peromyscus maniculatus*). Pale shading in the Lower Peninsula shows where the subspecies *P. m. bairdii* occurs alone; in the Upper Peninsula where *P. m. gracilis* occurs alone. Dark shading shows where these two species overlap geographically. *P. m. maniculatus* occurs on Isle Royale.

HABITAT PREFERENCES. Specific habitat preferences separate the two Michigan mainland subspecies, *P. m. bairdii* and *P. m. gracilis,* while the third subspecies *P. m. maniculatus,* being insular, is widely segregated by the waters of Lake Superior. The environments used by these deer mice are examined separately as follows:

Peromyscus maniculatus bairdii. This prairie subspecies of deer mouse probably was an unusual mammal in extreme southern Michigan in presettlement days. Its major habitats then were perhaps sandy beaches along lake shores, edges of marshes, openings in oak woodlands in southwestern counties or situations where severe fires had obliterated woody cover (Hooper, 1942). After settlement, this plains-adapted mouse was able to extend its range slowly eastward, thriving in the early vegetation stages produced by deforestation and agriculture (Dice, 1932; Hooper, 1942). Ultimately, this midwestern deer mouse reached the Atlantic coastal states of New York and Maryland (Baker, 1968). It progressed north in cleared lands of Michigan's Lower Peninsula to reach northernmost Emmet County (Hitchcock, 1943). Its entry into the southwestern part of the Upper Peninsula (Burt, 1946) has been less obvious than that of another prairie dweller, the thirteen-lined ground squirrel. This deer mouse's preference for the early successional vegetation produced by agriculture has caused it to be called the farmer's mouse. It is closely associated with crop and recently-fallow lands. Beckwith (1954) and others have shown that the prairie subspecies of the deer mouse is common in fallow agricultural land when vegetation is in the annual-biennial stage of successional growth. When the successional process continues to a strictly perennial grass stage, the deer mouse decreases and the meadow vole becomes common. Because of this rodent's proclivity for disturbed environments in the earliest plant associations from open ground to mature forest, its distribution in Michigan has both a mosaic and changeable pattern. Its preferred environments, if not re-disturbed, will go through a series of plant changes producing habitats progressively unfavorable and causing the decline and ultimate disappearance of the animal. Michigan mammalogists have reported this deer mouse in open grassy fields in Kalamazoo County (Allen, 1938), on gravel and sandy beaches in Charlevoix and Huron counties (Dice, 1925; 1932), in open fields in Washtenaw County (Wood, 1922a), on sand beach under driftwood in Berrien County (Wood, 1922b), in fields near marshes and tamarack swamps in Cass County (Wood and Dice, 1923), in wheat, oats, and alfalfa, and field-shocked corn in Clinton County (Linduska, 1942; 1946; 1950), in bare places and sparse vegetation in Livingston County (Pruit, 1959), in bluegrass association in Livingston County (Blair, 1940), in sparse numbers on dunes, in orchards and in newly-invaded shrub seres in Grand Traverse County (W. C. Scharf, pers. comm., November 13, 1978), and based on studies by classes in mammalogy each autumn on the Michigan State University campus, along disturbed railroad right-of-way and in fields of cultivated plants. This mouse's ability to appear suddenly in cornfields, even when new ground is broken for planting, should earn it the local name of cornfield mouse (John King, pers. comm.; Linduska, 1950).

Peromyscus maniculatus gracilis. The subspecies of the deer mouse occurring in northern Michigan is, in marked contrast to the more southern *P. m. bairdii,* a woodland dweller, widespread in the Upper Peninsula and on many wooded islands in the Great Lakes. In the upper third of the Lower Peninsula this long-eared and long-tailed deer mouse occupies suitable forested areas. In presettlement times, *P. m. gracilis* could conceivably have occurred in forested areas in southern Michigan. However, Hooper (1942) does not believe that was the case because 1) some relict populations should still remain in selected forested areas of southern Michigan, but none have been found, and 2) there are no museum specimens attributable to this subspecies from southern Michigan among the collections preserved from the last half of the nineteenth century. Consequently, it is thought that proper northern hardwood environments, to which this rodent is adapted, exist no further south in Michigan than at a line of latitude drawn through the mouth of Saginaw Bay. In contrast, the prairie-dwelling subspecies *P. m. bairdii* has taken advantage of clearing operations to spread northward to the Straits of Mackinac, with the result that the two ranges remain ecologically separated but geographically interdigitated. As Hooper (1942) stated, in Michigan *P. m. gracilis* thrives best in dense upland forests of beech, maple, yellow birch, and other hardwoods. Other observers have recorded *P. m. gracilis* in virgin and second-growth hardwoods, bogs, brush, jackpine, dune heath, swamp, and marsh edge in Charlevoix County (Dice, 1925), in mature upland forest in Marquette and Baraga counties (Robinson *in* Robinson and Werner, 1975), in a variety of forest and shrub habitats in Gogebic and Ontonagon counties (Dice and Sherman, 1922), in mixed grass and weeds on mine tailings in Houghton County (F. A. Stormer, pers. comm., March 1978), in mature forest in Marquette County (Manville, 1949), in upland hardwoods and conifer swamps in Alger and Schoolcraft counties (Ozoga and Verme, 1968), in various woody habitats, grassy meadows, and ground juniper on Beaver Island and other island environments (Ozoga and Phillips, 1964; Scharf, 1973), and in isolated woodlots and surrounding open lichen-grass associations of the Kingston Plains in Alger County (Fitch, 1979). Forest fires and the varied habitats produced in the prolonged period of vegetation regrowth do not seem to have been major deterents to the survival of woods-dwelling deer mice (Briech *et al.,* 1977). In fact, a Minnesota study shows that this mouse was the most abundant small mammal for the first seven years after a burn (Krefting and Ahlgren, 1974). Further, in Alger County, Verme and Ozoga (1981) noted that deer mice numbers increased in conifer swamps following successional changes resulting from both strip clear-cutting and the burning of the resultant slash.

Peromyscus mainiculatus maniculatus. This subspecies of deer mouse occupies most of the same habitat on Isle Royale as does *P. m. gracilis* on the Michigan mainland and on Beaver Island. In a study made by Johnson *et al.* (1969), this rodent occurred most commonly in boreal (conifer) forest, and in smaller numbers in second growth resulting from forest fires (which had taken place at least 32 years previous) and in mature sugar maple-yellow birch woodland. The authors concluded that the deer mouse could be found on all island habitats except wet sites.

DENSITY AND MOVEMENTS. Deer mice have long been among the more obvious small mammals. Whether liked or considered pestiferous, their abundance periodically amazes observers. For example, deer mice in northern Ontario were so abundant in the autumn of 1949 that individuals were reported swimming or dead in the Severn River, but by the following summer the mice in this same area were rare (de Vos, 1951). In Michigan, O. H. Perry recorded in his diary, while on an October hunting expedition in Tuscola County in 1854, that deer mice (perhaps actually white-footed mice instead, but there is no way to be sure) overran his camp, eating tent strings, shoes, other leather items, and food (Schorger, 1956). In northern Michigan, field observers conducting inventories of mammals in the early years of this century were almost unanimous in reporting that the deer mouse was the most abundant of the readily-trappable small mammals in such counties as Alger (Wood, 1917), Gogebic and Ontonagon (Dice and Sherman, 1922).

Deer mice populations, like those of other small mammals, fluctuate. Linduska (1950) in his studies in Clinton County, noted a dramatic upward trend in deer mouse (*P. m. bairdii,* in this case) populations for a two-month period in each of several summers, from a low of 0.6 mice per trap line in 1940 to a high of 4.3 per trap line in 1946. Density, of course, varies seasonally. Michigan deer mice populations are at an annual low in spring prior to the new plant growth and the emergence of insects (Blair, 1940). Overwinter survivors produce a high population increase in late spring, followed by a mid-summer breeding lull and the resultant drop in numbers; there is another increase in autumn, and again the inevitable winter decline (Howard, 1949). Blair (1940) found summer densities of prairie deer mice to attain 9.2 individuals per acre (23 per ha) in Livingston County. On Isle Royale, Johnson *et al.* (1968) obtained a high of six mice per acre (15 per ha). In northern Michigan, reported densities were as high as 43 woodland deer mice per acre (108 per ha) in mature upland forest in Marquette and Baraga counties (Robinson *in* Robinson and Werner, 1975); 36 per acre (80 per ha) in mature maple forest in Marquette County (Laundre, 1974). Densities were heavier in upland hardwoods than in conifer swamps in Alger and Schoolcraft counties (Ozoga and Verme 1968). In a study on the Keweenaw Peninsula, Nofz (1972) found woodland deer mice in summer to be most dense in sugar maple habitat, less dense in birch-aspen, old field and swamp, with lowest density in spruce-fir. High numbers may represent local extremes, since densities over large areas usually seem no greater than 11 per acre (27 per ha) at peak periods of the year (Blair, 1941; Manville, 1949; Terman, 1968). Food supplies, habitat quality, and spatial needs play major roles in determining local deer mouse abundance.

Deer mice are generally sedentary, spending most of their lives in areas from 0.5 to 2.5 acres (0.2 to 1.0 ha) to as much as 5.6 acres (1.8 ha), with males more mobile than females (Blair, 1940; 1942). In addition, there are indications that the woodland subspecies have larger home ranges than the prairie subspecies (Blair, 1942). Homing ability seems well-developed, with deer mice experimentally removed several hundred yards from their home sites able to return (Murie, 1963).

BEHAVIOR. The deer mouse, perhaps because of its widespread distribution, abundance, and tractability in captivity, is the most studied of North American mammals, at least among non-game species. Much of our knowledge about the deer mouse is summarized in a 593-page treatise edited by a Michigan State University professor, J. A. King (1968). In addition to field studies, deer mice and their habits have been extensively examined under laboratory conditions, where the animals breed regularly, and readily use experimental data-gathering devices such as activity wheels. Behavioral studies of deer mice and related species in the genus *Peromyscus* are grouped, according to King (1968), under three categories: motor patterns (locomotion, swimming, climbing, digging, nest building, food and water consumption, grooming, vocalization); sensory capacities (light, odor, taste, temperature and humidity, gravity); and learning (studies designed to determine habituation, use of mazes, means of avoidance, discrimination, early learning). The reader can best consult King (1968) and subsequent studies including Enders (1977) for details concerning many of the aspects of deer mouse behavior mentioned here.

According to Mihok (1979), the basic social unit consists of a mature male, a few mature females, and several young. The mature male tends to exclude other adult males from his area of activity while the superimposed ranges of females and young overlap extensively. Pregnant females tend to avoid other mice (especially other females) and defend their nests of young from intruders (Blair, 1942; King, 1958). There is also some indication that adult aggressiveness (especially by males) in spring causes dispersal and even death of juveniles (Sadleir, 1965). In winter, groups of ten or more, of mixed ages and sexes, will huddle together in shelters to conserve heat (Howard, 1951). Although burrows are used as refuges (Master, 1977), this huddling practice suffered a major setback in Michigan a few years ago when farmers abandoned the practice of leaving shocked corn in fields in winter. The bases of these shocks offered deer mice insulated refuges as well as ear corn for food. These aggregations often consist of a family group plus unrelated individuals (Howard, 1949). As in the case of the white-footed mouse, the deer mouse conserves energy in winter by entering daily torpor to reduce body temperature (Howard, 1951).

Deer mice are chiefly active at night. In his extensive live-trapping of deer mice in Alger County, Fitch (1979) caught only two individuals during daylight hours. The mice may use runways established by voles but are more apt to move seemingly at random, leaving their footprints and tail marks on sandy soils and snow cover. Deer mice of the subspecies *P. m. gracilis* and *P. m. maniculatus* are equally at home on the ground or in arboreal situations. In Marquette County, Manville (1949) found that mice of the former subspecies, when released from live traps, were often apt to escape by climbing oaks, sugar maple, basswood, and hemlock. Terrestrial *P. m. bairdii* also climb well. The deer mouse's methods of moving around and establishing its presence are much the same as described for the white-footed mouse. Deer mice usually make buzzing sounds and sharp squeaks; however, young individuals have special vocalizations including comfort sounds, pain squeaks, and a distinctive abandoned cry (Eisenberg, 1968; Hart and King, 1966). Tests show that deer mice can detect a broad range of ultrasonic sounds (Dice and Barto, 1952).

Activity patterns of the deer mouse center around a nest site and a food cache, although both can be at the same place. The terrestrial prairie subspecies usually constructs a nest just below ground level in its own burrow or in one abandoned by an other small mammal (Houtcooper, 1971). Nests can also be found under debris, boards or other objects discarded by humans, and in shocked field crops. The nests of the forest-dwelling deer mice are placed near the ground, in stumps, logs, in brush piles, in tree cavities, under tree bark, in reconstructed bird nests, in burrows, and in cottages or outbuildings. Nests are rounded masses of vegetable material, as much as 4 in (100 mm) in diameter, with outer parts of grass, tree leaves, bits of bone, cloth and paper, and an interior lining of dandelion or thistle down, hair, feathers, and finely-shredded plant fibers.

ASSOCIATES. Occurring in both open lands (*P. m. bairdii*) and northern forested areas (*P. m. gracilis* and *P. m. maniculatus*), the deer mouse comes in contact with most other Michigan mammals. Night-hunting predators (some snakes, owls, various mammals) are, of course, a constant threat. In addition, non-predatory mammals similar in size can compete with deer mice for food, shelters, and home sites. Species of shrews, jumping mice, the white-footed mouse, chipmunks, ground squirrels, and feral house mice may share various insects, other invertebrates, seeds, nuts and fruits with deer mice. If these commodities are in short supply, competition could be a major survival factor. Voles of the genera *Microtus, Synaptomys,* and *Clethrionomys,* usually with a food preference of plant foliage rather than seeds and other plant reproductive parts, may compete with deer mice for living space (Grant, 1971).

Association with the White-footed Mouse— P. leucopus. In the southern areas where the prairie deer mouse (*P. m. bairdii*) occurs, these two species are in contact primarily where brush lands or trees occupied by the white-footed mouse are bordered by the cleared or disturbed open lands mostly preferred by the deer mouse. Although this deer mouse rarely enters the brush and forest habitat of the white-footed mouse (Blair, 1940), the latter species does share with the deer mouse the early seral stages of cultivated fields consisting of weeds, crops, and annual grasses and, of course, eventually replaces the deer mouse when these early successional stages give way to shrublands (Master, 1977; Whitaker, 1967). Master even suggests that these two species may compete for space in this habitat. In northern forested areas, there is a close interaction between the white-footed mouse and the woodland deer mouse (Drickamer and Capone, 1977). Although these two rodents have been captured in similar areas and even in adjacent traps where their ranges overlap from Minnesota and Wisconsin to Vermont, Maine and Quebec, there is a tendency, in areas where they occur together, for the more aggressive white-footed mouse to occupy dry, exposed, and more open hardwood forests while the deer mouse prefers the moist, secluded, and cool coniferous forest interiors (Klein, 1960; Smith and Speller, 1970; Wrigley, 1969). Also, in areas where both occur, the woodland deer mouse usually nests in trees and the white-footed mouse on the forest floor (Steh, 1980). However, when only one of these two species occurs in such habitat, it may very well occur in hardwoods, in mixed hardwoods and conifers, and in conifers. To date there has been little evaluation of the interaction between these two forest rodents where they occur together in the northern part of Michigan's Lower Peninsula (for Montmorency County, see Green, 1925).

Association with Other Small Mammals—Knowledge of environmental associations among small, inconspicuous, nocturnal mammals is accumulated mostly from trapping when individuals of different species are captured in adjacent traps, or one after the other in the same traps in comparable environments. In cultivated crop lands, the prairie deer mouse probably associates most closely with feral house mice, according to work done in Indiana by Whitaker (1967). In Clinton County, Linduska (1950) found this deer mouse with such small mammals as the short-tailed shrew, bog lemming, meadow vole, and house mouse. However, in southwestern Ontario, Morris (1979) never captured deer mice and short-tailed shrews together, each perhaps using differing microhabitats. In Livingston County, nevertheless, Blair (1940), obtained the same species, plus meadow jumping mice, in association with deer mice. In Van Buren County, Brewer and Reed (1977) reported somewhat similar species associations.

In northern Michigan, the woodland subspecies of deer mouse is an associate of the woodland jumping mouse; Blair (1942), on one occasion, caught both species simultaneously in the same live trap in Alger County. In the McCormick Experimental Forest (Baraga and Marquette counties), Haveman (1973) captured deer mice in pit traps along with arctic shrews, masked shrews, pygmy shrews, short-tailed shrews, southern red-backed voles, meadow voles, southern bog lemmings, and meadow jumping mice. The deer mouse has also been found in the same habitats as the southern red-backed mouse and the meadow vole in Marquette County (Manville, 1949). Kirkland and Griffin (1974) noted negative interspecific associations between woodland deer mice and the southern red-backed vole. In a study on the Keweenaw Peninsula, Nofz (1972) found the deer mouse to be the most abundant small mammal in six habitat types examined: sugar maple, birch-aspen, spruce-fir, old field, pine, and swamp. Other small mammals present were short-tailed shrew, eastern chipmunk, least chipmunk, southern red-backed vole, meadow vole, meadow jumping mouse, and woodland jumping mouse. On Isle Royale, the deer mouse (*P. m. maniculatus*) has no small, nocturnal rodent associates (Johnson *et al.*, 1969). Likewise, on species-impoverished Beaver Island, Ozoga and Phillips (1964) noted that *P. m. gracilis* had "a rather wide-ranging ecological distribution" perhaps because of the absence of nocturnal small rodent associates. These observations were substantiated as well for deer mice living on Summer Island (Rhoades, 1970).

REPRODUCTIVE ACTIVITIES. In early spring, the overwintering population, diminishing steadily without significant replacements, reaches an annual low. These winter survivors begin a flurry of breeding activity which coincides with a temperature rise (Millar and Gyng, 1981) and the appearance of insects and new plant growth. In southern Michigan, litters of prairie deer mice can appear as early as March (Burt, 1946); first litters of woodland deer mice may appear as late as May or even June in northern Michigan (Manville, 1949). The later start in the north for population recovery is emphasized by the small catch of spring-born individuals (only 20%) among deer mice examined at Cusino in June; by mid-July, however, the young made up 55% of the population sampled and 80% in August and September (Ozoga and Verme, 1968).

The overwintering breeding stock nurtures its offspring to just prior to their sexual maturity and then slowly disappears. Howard (1949) showed that only 13% of the individuals alive in March remained in the population in July. These adults, however, may produce two litters (in spring and early summer) in rather rapid succession, before the offspring begin reproducing. Some breeding females are still in gray juvenile pelage (Zimm, 1975) and as young as 28 days old (Clark, 1938). On Isle Royale, it was observed that these early litters could themselves produce one or two sets of offspring prior to the cessation of breeding in late autumn (Johnson *et al.*, 1970). Deer mice in more southern parts of Michigan—where the growing season is longer—may have even more (Burt, 1946).

The autumn population flourishes because of the large spring/summer production of individuals and the growing season's culmination of seeds, insects and other foods (Blair, 1940; Howard, 1949). These late-born offspring constitute the major segment of hardy individuals needed to survive winter. However, in areas where shelter and food are in abundance (under shocks of grain, in storage bins, etc.) deer mice, especially prairie deer mice, may produce some litters throughout winter (Burt, 1946). Linduska (1942) found eight of 41 females living under field-shocked corn in Clinton County in mid-January were pregnant or nursing litters.

Deer mice breed readily in captivity, therefore

considerable knowledge of their reproductive behavior and life history is available (Layne, 1968). Females are seasonally polyestrous, with reproductive readiness in Michigan occurring usually between March and November. Ovulation is spontaneous in the absence of tactile contact with mates (Bradley and Terman, 1979). Females demonstrate post-partum estrus, which maximizes productivity by allowing a mother of newly-born young to mate and become pregnant again just shortly after giving birth. The non-lactational gestation period lasts for approximately 22 to 23 days with an average delay of five days for lactating females (Lawrence Master, pers. comm.). Litters may contain from one to 11 offspring with the usual number either four, five, or six.

At birth, the young mouse weighs about one-half oz (1.5 g), is hairless, has wrinkled, pink skin, has closed eyes and folded-over ear pinnae. Growth and development is rapid, with juvenile hair beginning to emerge on the second day, the pinnae of the ears unfolding on the third, the ear canal opening by the 10th, the eyes opening about the 15th, (earlier by two or three days in the prairie form of the deer mouse), and weaning taking place between the 25th and the 35th day. By this time, the female is near full term with her next litter and usually departs for another nest site to give birth. The deer mouse mother will move her offspring to new quarters at the slightest disturbance or provocation. Characteristically, the nursing young may cling to their mother's nipples while being transported or she may carry them away one at a time in her mouth.

In captivity, deer mice can live as long as eight years (Dice, 1933). In nature, the struggle of the individual for survival is demonstrated by Hansen and Fleharty (1974) who found the population turn-over for prairie deer mice in Kansas to be 42% complete in six months and 95% complete in 12 months. In Livingston County, Blair (1948) noted that deer mice remained on his study sites (based on recaptures of marked animals in live traps) for an average of five months. Even so, some animals defy the odds and live more than one year; in Marquette County, Manville (1949) noted that 10% of the woodland deer mice first live-trapped, marked, and released in 1940 were recaptured in 1942. One female, his oldest record, was lactating when first captured on July 17, 1940, and was last taken on June 18, 1942, possibly being as old as 33 months.

FOOD HABITS. Food is an important population-limiting factor to small rodents of the genus *Peromyscus.* In experimental studies in South Carolina, supplementary food added to that available in nature for the old-field mouse (*Peromyscus polionotus*) caused an increase in the rodent's density (Smith, 1971). It is likely that the same effect would be obtained in Michigan with the deer mouse. Michigan winters, may cause cold-weather starvation (combining the effects of low temperature and food shortage), in deer mice (Howard, 1951). The recycling of eaten material through the ingestion of fecal deposits (coprophagy) has been observed in deer mice (Karch, 1974) and may be one means of acquiring nutrients in times of food shortages. However, as protection against such eventualities, deer mice spend considerable time during autumn collecting food stuffs (seeds mostly) and storing them in secret granaries (Burt, 1965). Olfactory cues in food getting seem more important than visual cues (Howard *et al.,* 1968). Winter caches may be lost, however, when other mice or grain-eating squirrels raid storage chambers. Deer mice have low-crowned, cuspidate teeth, adapting them to a general diet of insects, other invertebrates, seeds, fruits, flowers, and other plant products (Baker, 1971)—foods produced in the warm, growing period of the Michigan annual cycle. During the cold, non-growing period of late autumn, winter and early spring, caches of staples are necessities, especially when individual deer mice may eat between 2.0 and 3.5 grams of food per day (Drickamer, 1970; Sealander, 1952).

The omnivorous diet of the prairie subspecies of the deer mouse is shown by a study of its food habits in Indiana where Houtcooper (1978), discovered that insects and other animal material accounted for 15% of the diet in winter, 35% in spring, 54% in summer, and 47% in autumn. Plant material constituted 81% of the diet in winter, 65% in spring, 45% in summer and 52% in autumn. At Rose Lake in Clinton County, Linduska (1950) identified food items from three autumn caches of prairie deer mouse: seeds of ragweed were most abundant, followed by those of lamb's quarter, lady's thumb, black bindweed, redroot, tumbling pigweed, yellow foxtail-grass, green foxtail-grass, old-witch grass, catnip, wheat and corn plus hickory nuts, cherry pits and remains of grasshoppers and crickets. Seeds of the ragweed (*Ambrosia elatior*), according to Linduska, were abundantly available in the early 1940s, with about 212 lbs found per acre (238 kilo per ha) in wheat stubble in October (Baumgras, 1943). With the prevalence of herbicides used in modern agricultural pursuits, it is highly probable that ragweed and other such plants in and adjacent to wheat fields are now at a minimum on well-managed farms. In Livingston County, Howard and Evans (1961) also identified food items stored by deer mice in nest-box caches in mid-autumn. They found seeds of ragweed, bush clover, and panic grass, plus acorns of three species of oak, were the major stored food (by volume). One unusual food item was a salamander (genus *Ambystoma*). Illustrating the deer mouse's range of foods is its taste for ground nesting birds' eggs, even those of the spotted sandpiper (Maxson and Oring, 1978).

Woodland deer mice in northern areas also depend on fruits and nuts of trees and shrubs and on forest insects (Martell and Macaulay, 1981). This mammal's intensive use of seeds of pine and other conifers has been a major concern to foresters (Smith and Aldous, 1947). Woodland deer mice prefer seeds of white pine (Abbott and Quink, 1970), and jack pine in autumn and winter. Tests show that deer mice may locate conifer seeds by olfactory cues (Record *et al.,* 1976). In autumn and winter, these deer mice also use a variety of previously harvested nuts and fruits, as well as insects (in over-wintering larval, pupal and adult stages, depending on species). In spring and summer, insects become more prevalent in the diet (Hamilton, 1941). On Isle Royale, for example, Johnson *et al.* (1968) found that summer foods consisted chiefly of insects, with a winter shift back to seeds, nuts, and fruits. When forest areas are cut or burned, diet changes apparently are made without difficulty (Krefting and Ahlgren, 1974). Schloyer (1976) regards deer mice as opportunistic eaters when they favor brush and other second growth.

ENEMIES. Vertebrate associates of deer mice are rarely friendly. If they are not competing for food, shelter, or nest sites, they are deer mice predators. Nocturnal hunters of Michigan deer mice are primarily owls and meat-eating mammals, with snakes of less importance. Hawks, being daytime hunters, rarely catch deer mice, although there are records that the red-tailed hawk has done so (de la Perriere, 1970; Fitch and Bare, 1978). Michigan owls are highly efficient mouse catchers; there are records of deer mice in the pellets of great horned owl, short-eared owl, long-eared owl, screech owl, and barn owl (Geis, 1952; Linduska, 1950; Short and Drew, 1962; Wallace, 1948). The ring-billed gull is also able to catch deer mice (Blokpoel and Haymes, 1979). Various mammalian carnivores, including shrews, felids, mustelids, and canids along with raccoon and probably the black bear, eat deer mice. Michigan records of carnivores feeding on deer mice are for gray wolf, red fox, mink, and domestic house cat (Christian, 1975; Johnson *et al.,* 1970; Manville, 1948).

PARASITES AND DISEASES. Deer mice are afflicted by numerous internal and external parasites: protozoans, flatworms (trematodes), tapeworms (cestodes), round worms (nematodes), thorny-headed worms (Acanthocephala), mites, chiggers, ticks, fleas, biting lice, and botfly larvae (Doran, 1954; Scharf and Stewart, 1980; Whitaker, 1968). In Marquette County, Manville (1949) collected one species of flea (infesting as many as 82% of the mice examined), a chigger and a tick from deer mice. Cuterebrid botfly larvae are common parasites of Michigan deer mice (Smith, 1977; also see account under the white-footed mouse). Wilson and Johnson (1971) identified botfly larvae, one chigger, two ticks, one louse, and five fleas in deer mice on Isle Royale.

MOLT AND COLOR ABERRATIONS. Both the deer mouse and the white-footed mouse have distinctive, sometimes confusing, juvenile pelages. The weaned young deer mouse has lead gray upperparts, which slowly become more gray-brown as brown hairs are added (Coventry, 1937). The subadult deer mouse has a brownish-gray coat. At about 40 days of age, this drab color is molted and the rufous coloring of the adult appears. Adults molt and grow a new fur coat each summer or autumn. There may also be some hair replacement at other times of the year (Layne, 1968). Individuals in both dark (melanistic) and white (albinistic) coats have been reported. In

Marquette County, Manville (1949) caught a deer mouse in partial albino fur.

ECONOMIC IMPORTANCE. Perhaps the most serious charge made against deer mice is their consumption of seeds of valued forest trees (Pank, 1974). According to Smith and Aldous (1947), Lake States deer mice (chiefly *P. m. gracilis*) rank second only to chipmunks in consuming jack pine seeds. On the other hand, deer mice eat forest insects, such as the larch sawfly (Graham, 1929). Campers and cottage dwellers may at least tolerate deer mice, but in large numbers these rodents can be destructive (Schorger, 1956) by raiding stored grains and other foods supplies, accumulating litter, and gnawing. The northern woodland subspecies (*P. m. gracilis* and *P. m. maniculatus*) are more inclined to invade cottages and other human habitation. The southern Michigan prairie subspecies (*P. m. bairdii*), is more apt to occur in cultivated fields, where crop losses can occasionally be severe (Linduska, 1942). Allen (1940) noted that mice reduced the yield of field-shocked grain (left in the field until spring) in a Clinton County corn field by one-half.

To protect seeds in conifer stands and plantations from deer mice, various repellents have been devised (Lindsey *et al.*, 1974). Mouse traps and poisoned grain (commercially prepared) reduce deer mouse numbers in human habitations. Bed clothing (mattresses and pillows), in which mice often burrow for shelter and nesting, should be stored in mouse-proof places or, at least, hung over thin wire lines where they are inaccessible. The prolific deer mice serve a useful purpose in the environment as food for carnivores, including valued fur-bearing mammals.

COUNTY RECORDS FROM MICHIGAN. Records of specimens of the deer mouse in collections of museums and from the literature show that this species lives in all counties of the state. *Peromyscus m. maniculatus* is confined to Isle Royale. *Peromyscus m. gracilis* occurs in suitable habitat throughout the Upper Peninsula and northern one-half of the Lower Peninsula, at least as far south as Lake County. This subspecies is also found on various islands in the Great Lakes including: in Lake Michigan, Summer Island (Rhodes, 1970), St. Martin Island (Long, 1978), Marion Island (Hatt *et al.*, 1948), Fisherman's Island (Hatt *et al.*, 1948), Beaver Island Group including Beaver, Garden, Gull, Squaw, Trout, Whiskey (Hatt *et al.*, 1948; Ozoga and Phillips, 1964; Phillips *et al.*, 1965), North and South Fox islands (Hatt *et al.*, 1948), North and South Manitou islands (Hatt *et al.*, 1948; Scharf, 1973; Scharf and Jorae, 1980); in Lake Superior, Huron Island (Corin, 1976); in Lake Huron, Drummond Island (Manville, 1950), Bois Blanc (Cleland, 1966). *Peromyscus m. bairdii* occurs in suitable habitat throughout the Lower Peninsula and in the extreme southwestern part of the Upper Peninsula, in Menominee County (Burt, 1946).

LITERATURE CITED

Abbott, H. G., and T. F. Quink
1970 Ecology of eastern white pine seed caches made by small forest mammals. Ecology, 51(2):271–278.

Allen, D. L.
1938 Ecological studies on the vertebrate fauna of a 500-acre farm in Kalamazoo County, Michigan. Ecol. Monog., 8(3):347–436.
1940 Millions of little teeth. Michigan Conserv., Sept., pp. 2–6.

Baker, R. H.
1968 Habits and distribution. Pp. 98–126. *In* J. A. King, ed. Biology of *Peromyscus* (Rodentia). American Soc. Mammalogists, Sp. Publ. No. 2, xiii+593 pp.
1971 Nutritional strategies of myomorph rodents in North American grasslands. Jour. Mammalogy, 52(4):800–805.

Barbehenn, K. R., and J. G. New
1957 Possible natural intergradation between prairie deer and forest deer mice. Jour. Mammalogy, 38(2):210–218.

Baumgras, P. S.
1943 Winter food productivity of agricultural land for seed-eating birds and mammals. Jour. Wildlife Mgt., 7(1):13–18.

Beckwith, S. L.
1954 Ecological succession on abandoned farm lands and its relationship to wildlife management. Ecol. Monog., 24:349–376.

Blair, W. F.
1940 A study of prairie deer-mouse populations in southern Michigan. American Midl. Nat., 24(2):273–305.
1941 The small mammal population of a hardwood forest in nothern Michigan. Univ. Michigan, Lab. Vert. Biol., Contr. No. 17, 10 pp.
1942 Size of home range and notes on the life history of the woodland deer-mouse and eastern chipmunk in northern Michigan. Jour. Mammalogy, 23(1):27–36.
1948 Population density, life span, and mortality rates of small mammals in the blue-grass meadow and blue-

grass field associations of southern Michigan. American Midl. Nat., 40(2):395–419.

Blokpoel, H., and G. T. Haymes
1979 Small mammals and birds as food items of ring-billed gulls on the lower Great Lakes. Wilson Bull., 91(4): 623–625.

Bradley, E. L., and C. R. Terman
1979 Ovulation in *Peromyscus maniculatus bairdii* under laboratory conditions. Jour. Mammalogy, 60(30):543–549.

Brewer, R., and M. L. Reed
1977 Vertebrate inventory of wet meadows in Kalamazoo and Van Buren counties. Michigan Dept. Nat. Res., Mimeo. Rpt., 22 pp.

Briech, R. R., K. Siderits, R. E. Radtke, H. L. Sheldon, and D. Elsing
1977 Small mammal populations after a wildfire in northeast Minnesota. USDA Forest Service, Res. Paper NC-151, 8 pp.

Burt, W. H.
1946 The mammals of Michigan. Univ. Michigan Press, Ann Arbor. xv+288 pp.
1965 Michigan's pouched little mouse. Michigan Audubon Newsletter, 13(1).

Christian, D. B.
1975 Vulnerability of meadow voles, *Microtus pennsylvanicus*, to predation by domestic cats. American Midl. Nat., 93(2):498–502.

Cleland, C. L.
1966 The prehistoric animal ecology and ethnozoology of the Upper Great Lakes Region. Univ. Michigan., Mus. Anthro., Anthro. Papers, No. 29, x+294 pp.

Cooke, J. A., and C. R. Terman
1977 Influence of displacement distance and vision of homing behavior of the white-footed mouse (*Peromyscus leucopus noveboracensis*). Jour. Mammalogy, 58(1):58–66.

Corin, C. W.
1976 The land vertebrates of the Huron Islands, Lake Superior. Jack-Pine Warbler, 54(4):138–147.

Coventry, A. F.
1937 Notes on the breeding of some Cricetidae in Ontario. Jour. Mammalogy, 18(4):489–496.

de la Perriere, C. I.
1970 Food habits of red-tailed hawks, Summer Island, Michigan, 1969. Summer Sci. Jour., 2(2):72–73.

de Vos, A.
1951 Peak Populations of *Peromyscus maniculatus gracilis* in northern Ontario. Jour. Mammalogy, 32(4):462.

Dice, L. R.
1925 A survey of the mammals of Charlevoix County, Michigan, and vicinity. Univ. Michigan, Mus. Zool., Occas. Papers No. 159, 33 pp.
1932 The prairie deer-mouse. Cranbrook Inst. Sci., Bull. No. 2, 8 pp.
1933 Longevity in *Peromyscus maniculatus gracilis*, Jour. Mammalogy, 14(2):147–148.

Dice, L. R., and E. Barto
1952 Ability of mice of the genus *Peromyscus* to hear ultrasonic sounds. Science, 116:110–111.

Dice, L. R., and H. B. Sherman
1922 Notes on the mammals of Gogebic and Ontonagon counties, Michigan, 1920. Univ. Michigan, Mus. Zool., Occas. Papers No. 109, 40 pp.

Doran, D. J.
1954 A catalogue of the protozoa and helminths of North American rodents. I. Protozoa and Acanthocephala. American Midl. Nat., 52(1):118–128.

Drickamer, L. C.
1970 Seed preferences in wild caught *Peromyscus maniculatus bairdii* and *Peromyscus leucopus noveboracensis*. Jour. Mammalogy, 51(1):191–194.

Drickamer, L. C., and M. R. Capone
1977 Weather parameters, trapability and niche separation in two sympatric species of *Peromyscus*. American Midl. Nat., 98(2):376–381.

Eisenberg, J. F.
1968 Behavior patterns. Pp. 451–495, *in* J. A. King, ed. Biology of *Peromyscus* (Rodentia). American Soc. Mammalogists, Sp. Publ. No. 2, xiii+593 pp.

Enders, J. E.
1977 Social interactions between confined juvenile and adult *Peromyscus maniculatus bairdii*. Effects of social factors on juvenile settlement and growth. Michigan State Univ., unpubl. Ph.D. dissertation, xii+180 pp.

Fitch, H. S., and R. O. Bare
1978 A field study of the red-tailed hawk in eastern Kansas. Trans. Kansas Acad. Sci., 81(1):1–13.

Fitch, J. H.
1979 Patterns of habitat selection and occurrence in the deermouse, *Peromyscus maniculatus gracilis*. Michigan State Univ., Publ. Mus., Biol. Ser., 5(6):443–484.

Foster, D. D.
1959 Differences in behavior and temperament between two races of the deer mouse. Jour. Mammalogy, 40:496–513.

Geis, A. D.
1952 Winter food habits of a pair of long-eared owls. Jack-Pine Warbler, 30(3):93.

Graham, S. A.
1929 The large sawfly as an indicator of mouse abundance. Jour. Mammalogy, 10(3):189–196.

Grant, P. R.
1971 Experimental studies of competitive interaction in a two-species system. III. *Microtus* and *Peromyscus* species in enclosures. Jour. Animal Ecol., 40(2):323–350.

Green, M. M.
1925 Notes on some mammals of Montmorency County, Michigan. Jour. Mammalogy, 6(3):173–178.

Hall, E. R.
1981 The mammals of North America. Second Edition. John Wiley and Sons, New York. Vol. 2, vi+600–1181+90 pp.

Hamilton, W. J., Jr.
1941 The food of small forest mammals in eastern United States. Jour. Mammalogy, 22(3):250–263.

Hansen, C. M., and E. D. Fleharty
1974 Structural ecological parameters of a population of *Peromyscus maniculatus* in west-central Kansas. Southwestern Nat., 19(3):293–303.

Harris, V. T.
1952 An experimental study of habitat selection by prairie and forest races of the deer mouse, *Peromyscus maniculatus*. Univ. Michigan, Lab. Vert. Biol., Contr. No. 56, 53 pp.

Hart, F. M., and J. A. King
1966 Distress vocalizations of young in two subspecies of *Peromyscus maniculatus*. Jour. Mammalogy, 47(2):287–293.

Hatt, R. T., J. Van Tyne, L. C. Stuart, C. H. Pope, and A. B. Grobman
1948 Island life: A study of the land vertebrates of the islands of Lake Michigan. Cranbrook Inst. Sci., Bull. No. 27, xi+179 pp.

Haveman, J. R.
1973 A study of population densities, habitats and foods of four sympatric species of shrews. Northern Michigan Univ., unpubl. M.S. thesis, vii+70 pp.

Hitchock, H. B.
1943 *Peromyscus maniculatus bairdii* and *Sorex palustris hydrobadistes* in the Lower Peninsula of Michigan. Jour. Mammalogy, 24(3):402–403.

Hooper, E. T.
1942 An effect on the *Peromyscus maniculatus* rassenkreis of land utilization in Michigan. Jour. Mammalogy, 23(2): 193–196.

Houtcooper, W. C.
1971 Rodent seed supply and burrows of *Peromyscus* in cultivated fields. Proc. Indiana Acad. Sci., 81:384–389.
1978 Food habits of rodents in a cultivated ecosystem. Jour. Mammalogy, 59(2):427–430.

Howard, W. E.
1949 Dispersal, amount of inbreeding, and longevity in a local population of prairie deer mice on the George Reserve, southern Michigan. Univ. Michigan, Lab. Vert. Biol., Contr. No. 43, 52 pp.
1951 Relation between low temperature and available food to survival of small rodents. Jour. Mammalogy, 32(3): 300–312.

Howard, W. E., and F. C. Evans
1961 Seeds stored by prairie deer mice. Jour. Mammalogy, 42(2):260–263.

Howard, W. E., R. E. Marsh, and R. E. Cole
1968 Food detection by deer mice using olfactory rather than visual cues. Anim. Behav., 16(1):13–17.

Johnson, W. J., M. L. Wolfe, Jr., and D. L. Allen
1968 Community relationships and population dynamics of terrestrial mammals of Isle Royale, Lake Superior. Second An. Rpt., Isle Royale Studies (mimeo.), 23 pp.
1969 Community relationships and population dynamics of terrestrial mammals of Isle Royale, Lake Superior. Third An. Rpt., Isle Royale Studies (mimeo.), 22 pp.
1970 Community relationships and population dynamics of terrestrial mammals of Isle Royale, Lake Superior. Fourth An. Rpt., Isle Royale Studies (mimeo.), 12 pp.

Karch, P.
1974 The role of experience in the behavioral/physiological adaptation of *Peromyscus maniculatus bairdii* to extended periods of a protein free diet. Michigan State Univ., unpubl. Ph.D. dissertation, 140 pp.

King, J. A.
1958 Maternal behavior and behavioral development in two subspecies of *Peromyscus maniculatus*. Jour. Mammalogy, 39(2):177–190.

King, J. A., ed.
1968 Biology of *Peromyscus* (Rodentia). American Soc. Mammalogists, Sp. Publ. No. 2, xiii+593 pp.

Kirkland, G. L., Jr., and R. J. Griffin
1974 Microdistribution of small mammals at the coniferous-deciduous forest ecotone in northern New York. Jour. Mammalogy, 55(2):417–427.

Klein, H. G.
1960 Ecological relationships of *Peromyscus leucopus noveboracensis* and *Peromyscus maniculatus gracilis* in central New York. Ecol. Monog., 30:387–407.

Krefting, L. W., and C. E. Ahlgren
1974 Small mammals and vegetation changes after fire in a mixed conifer-hardwood forest. Ecology, 55(6):1391–1398.

Laundre, J. W.
1974 An ecological survey of the mammals of the Huron Mountain area. Northern Michigan Univ., unpubl. M.A. thesis, 117 pp.

Layne, J. N.
1968 Ontogeny. Pp. 451–495, *in* J. A. King, ed. Biology of *Peromyscus* (Rodentia). American Soc. Mammalogists, Sp. Publ. No. 2, xiii+593 pp.

Lindsey, G. D., R. M. Anthony, and J. Evans
1974 Mestranol as a repellent to protect Douglas-fir seed from deer mice. Proc. Sixth Vert. Pest Conf., Davis, California, pp. 272–279.

Linduska, J. P.
1942 Winter rodent populations in field-shocked corn. Jour. Wildlife Mgt., 6(4):353–363.
1946 Edge effect as it applies to small mammals on southern Michigan farmland. Trans. Eleventh North American Wildlife Conf., pp. 200–204.
1950 Ecology and land-use relationships of small mammals on a Michigan farm. Michigan Dept. Conser., Game Div., ix+144 pp.

Long, C. A.
1974 Environmental status of the Lake Michigan region. Vol. 15. Mammals of the Lake Michigan drainage. Argonne Nat. Lab., 108 pp.
1978 Mammals of the islands of Green Bay, Lake Michigan. Jack-Pine Warbler, 56(2):59–83.

Manville, R. H.
1948 The vertebrate fauna of the Huron Mountains, Michigan. American Midl. Nat., 39(3):615–640.
1949 A study of small mammal populations in northern Michigan. Univ. Michigan, Mus. Zool., Misc. Publ. No. 73, 83 pp.
1950 The mammals of Drummond Island, Michigan. Jour. Mammalogy, 31(3):358–359.

Martell, A. M., and A. L. Macaulay
1981 Food habits of deer mice (*Peromyscus maniculatus*) in northern Ontario. Canadian Field-Nat., 95(3):319–324.

Master, L. L.
1977 The effect of interspecific competition on habitat utilization by two species of *Peromyscus*. Univ. Michigan, unpubl. Ph.D. dissertation, xii+179 pp.

Maxson, S. J., and L. W. Oring
1978 Mice as a source of egg loss among ground-nesting birds. Auk, 95(3):582–584.

Mihok, S.
1979 Behavioral structure and demography of subarctic *Clethrionomys gapperi* and *Peromyscus maniculatus*. Canadian Jour. Zool., 57:1521–1535.

Millar, J. S., and L. W. Gyng
1981 Initiation of breeding by northern *Peromyscus* in relation to temperature. Canadian Jour. Zool., 69(6): 1094–1098.

Morris, D. W.
1979 Microhabitat utilization and species distribution of sympatric small mammals in southwestern Ontario. American Midl. Nat., 101(2):373–384.

Murie, M.
1963 Homing and orientation of deer mice. Jour. Mammalogy, 44(3):338–349.

Nofz, E. T.
1972 A survey of small mammals in the Keweenaw Peninsula during the summer of 1971. Michigan Tech. Univ., unpubl. M.S. thesis, 31 pp.

Ogoga, J. J., and C. J. Phillips
1964 Mammals of Beaver Island, Michigan. Michigan State Univ., Publ. Mus., Biol. Ser., 2(6):307–347.

Ozoga, J. J., and L. J. Verme
1968 Small mammals of conifer swamp deer yards in northern Michigan. Michigan Acad. Sci., Arts and Ltrs., Papers, 53:37–49.

Pank, L. F.
1974 A bibliography on seed-eating mammals and birds that affect forest regeneration. U.S. Bur. Sport Fisheries and Wildlife, Sp. Sci. Rpt.-Wildlife No. 174, 28 pp.

Phillips, C. J., J. J. Ozoga, and L. C. Drew
1965 The land vertebrates of Garden Island, Michigan. Jack-Pine Warbler, 43(1):20–25.

Pruitt, W. O., Jr.
1959 Microclimates and local distribution of small mammals on the George Reserve, Michigan. Univ. Michigan, Mus. Zool., Misc. Publ., No. 109, 27 pp.

Record, C. R., R. E. Marsh, and W. E. Howard
1976 Olfactory responses of deer mice to Douglas-fir seed volatiles. Proc. Seventh Vert. Pest Conf., Monterrey, California, pp. 291–297.

Rhoades, J. A.
1970 The woodland deer mouse of Summer Island, Michigan. Summer Sci. Jour., 2(2):76–81.

Robinson, W. L., and J. K. Werner
1975 Vertebrate animal populations of the McCormick Forest. USDA Forest Service, Research Paper NC-118, 24 pp.

Sadleir, R. M. F. S.
1965 The relationship between agonistic behaviour and population changes in the deer-mouse, *Peromyscus maniculatus* (Wagner). Jour. Animal Ecol., 34:331–352.

Scharf, W. C.
1973 Birds and land vertebrates of South Manitou Island. Jack-Pine Warbler, 51(1):3–19.

Scharf, W. C., and M. L. Jorae
1980 Birds and land vertebrates of North Manitou Island. Jack-Pine Warbler, 58(1):4–15.

Scharf, W. C., and K. R. Stewart
1980 New records of Siphonaptera from northern Michigan. Great Lakes Ento., 13(3):165–167.

Schloyer, C. R.
1976 Changes in food habits of *Peromyscus maniculatus nubiterrae* Rhoads on clearcuts in West Virginia. Proc. Pennsylvania Acad. Sci., 50(1):78–80.

Schorger, A. W.
1956 Abundance of deer mice in Tuscola County, Michigan, in 1854. Jour. Mammalogy, 37(1):121–122.

Sealander, J. A., Jr.
1952 Food consumption in *Peromyscus* in relation to air temperature and previous thermal experience. Jour. Mammalogy, 33(2):206–218.

Short, H. L., and L. C. Drew
1962 Observations concerning behavior, feeding, and pellets of short-eared owls. American Midl. Nat., 67(2):424–433.

Smith, C. F., and S. E. Aldous
1947 The influence of mammals and birds in retarding artificial and natural reseeding of coniferous forests in the United States. Jour. Forestry, 45:361–369.

Smith, D. A., and S. W. Speller
1970 The distribution and behavior of *Peromyscus maniculatus gracilis* and *Peromyscus leucopus noveboracensis* (Rodentia: Cricetidae) in a southeastern Ontario woodlot. Canadian Jour. Zool., 48:1187–1199.

Smith, D. H.
1977 The natural history and development of *Cuterebra approximata* (Diptera: Cuterebridae) in its natural host, *Peromyscus maniculatus* (Rodentia: Cricetidae) in western Montana. Jour. Med. Entom., 14(2):137–145.

Smith, M. H.
1971 Food as a limiting factor in the population ecology of *Peromyscus polionotus* (Wagner). An. Zool. Fennici, 8:109–112.

Steh, C. D.
1980 Vertical nesting distribution of two species of *Peromyscus* under experimental conditions. Jour. Mammalogy, 61(1):141–143.

Terman, C. R.
1968 Population dynamics. Pp. 412–450, *in* J. A. King, ed., Biology of *Peromyscus* (Rodentia). American Soc. Mammalogists, Sp. Publ. No. 2, xiii+593 pp.

Timm, R. M.
1975 Distribution, natural history, and parasites of mammals of Cook County, Minnesota. Univ. Minnesota, Bell Mus. Nat. Hist., Occas. Papers No. 14, 56 pp.

Verme, L. J., and J. J. Ozoga
1981 Changes in small mammal populations following clear-cutting in upper Michigan conifer swamps. Canadian Field-Nat., 95(3):253–256.

Wallace, G. J.
1948 The barn owl in Michigan. Its distribution, natural history and food habits. Michigan State Coll., Agri. Exp. Sta., Tech. Bull. 208, 61 pp.

Welker, S. C.
1963 The role of early experience in habitat selection by the prairie deer-mouse, *Peromyscus maniculatus bairdii.* Ecol. Monog., 33:307–325.

Whitaker, J. O., Jr.
1967 Habitat and reproduction of some of the small mammals of Vigo County, Indiana, with a list of mammals known to occur there. Western Michigan Univ., C. C. Adams Center Ecol. Studies, Occas. Papers No. 16, 24 pp.
1968 Parasites. Pp. 254–311, *in* J. A. King, ed., Biology of *Peromyscus* (Rodentia). American Soc. Mammalogists, Sp. Publ. No. 2, xiii+593 pp.

Wilson, N., and W. J. Johnson
1971 Ectoparasites of Isle Royale, Michigan. Michigan Ento., 4(4):109–115.

Wood, N. A.
1917 Notes on the mammals of Alger County, Michigan. Univ. Michigan, Mus. Zool., Occas. Papers No. 35, 8 pp.
1922a The mammals of Washtenaw County, Michigan. Univ. Michigan, Mus. Zool., Occas. Papers No. 123, 23 pp.
1922b Notes on the mammals of Berrien County, Michigan. Univ. Michigan, Mus. Zool., Occas. Papers No. 124, 4 pp.

Wood, N. A., and L. R. Dice
1923 Records of the distribution of Michigan mammals. Michigan Acad. Sci., Arts and Ltrs., Papers, 3:425–469.

Wrigley, R. E.
1969 Ecological notes on the mammals of southern Quebec. Canadian Field-Nat., 83:201–211.

SOUTHERN RED-BACKED VOLE
Clethrionomys gapperi (Vigors)

NAMES. The generic name *Clethrionomys,* proposed by Tilesius in 1850, is from the Greek; the first part probably means "alder" and the last part refers to "mouse." The specific name *gapperi,* proposed by Vigors in 1830, refers to a Dr. Gapper of Ontario. In Michigan, local residents generally call this species a vole or a red-backed mouse. The subspecies *Clethrionomys gapperi gapperi* (Vigors) occurs in Michigan.

RECOGNITION. Body size small and slender for a vole; length of head and body of an adult averaging 3⅝ in (95 mm) long; slim, finely-haired tail, 1½ in (40 mm) long, almost equal in length to one-third of the head and body; body roundish and slender, covered with medium-long, soft, dense pelage; head moderately large; eyes small; ears hairy and rounded, extending conspicuously above the fur; vibrissae (whiskers) less prominent than in deer mice; feet slim and medium in length; hind limbs longer than fore limbs. In adults, the color of the upperparts from the forehead to the rump is a distinctive, bright chestnut (as indicative of this vole's common name), intermixed with a few blackish hairs; face, sides of head, body and flanks, buffy gray; underparts and feet, whitish or buffy; ears covered with brown hair; tail, faintly bicolored, brown above and light gray or buffy gray below. Juvenile and subadult coats are darker and less reddish than adult coats. In winter, adult pelage is longer, finer and brighter than in summer. Four pairs of mammary glands are present; two pectoral and two inguinal. There are four functional digits on the forefeet and five on the hind. Males possess small flank glands. The chestnut-colored back stripe serves to distinguish the southern red-backed vole from other small, short-tailed Michigan mammals (Plate 4b).

MEASUREMENTS AND WEIGHTS. Adult red-backed voles measure as follows: length from tip of nose to end of tail vertebrae, 4¾ to 6¼ in (120 to 160 mm); length of tail vertebrae, 1⅛ to 2 in (30 to 50 mm); length of hind foot, 5/16 to ¾ in (17 to 20 mm); height of ear from notch, ½ to ⅝ in (13 to 16 mm); weight of adults, 0.5 to 1.3 oz (15 to 36 g).

DISTINCTIVE CRANIAL AND DENTAL CHARACTERISTICS. The skull of the red-backed vole is delicate in construction; lacks conspicuous ridges or angularity characteristic of other members of the Subfamily Microtinae (see Fig. 51); has a smooth, rounded cranium, inflated auditory bullae and slender zygomatic arches. The lower jaws (mandibles) are also delicate in form. The skull averages 15/16 in (24 mm) long and ½ in (13 mm) wide across the zygomatic arches. The straight posterior border of the palate of the skull of the southern red-backed vole is a characteristic feature and differs from those of other Michigan voles. Distinctive also are the rooted conditions of the molar teeth in adults and the pattern of folded enamel loops surrounding exposed islands of dentine on the flat grinding surfaces of the molars (see Fig. 52). The dental formula of the southern red-backed vole is:

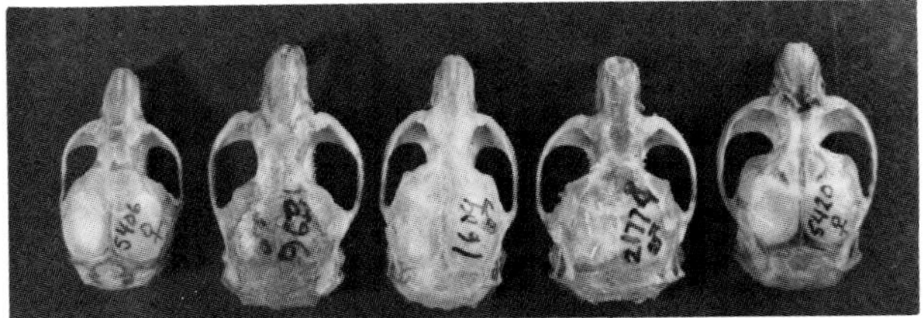

Figure 51. Dorsal aspects of the skulls of five Michigan microtines: From left to right: southern red-backed vole, prairie vole, meadow vole, woodland vole, southern bog lemming.

$$\text{I (incisors)}\ \frac{1\text{-}1}{1\text{-}1},\ \text{C (canines)}\ \frac{0\text{-}0}{0\text{-}0},\ \text{P (premolars)}\ \frac{0\text{-}0}{0\text{-}0},\ \text{M (molars)}\ \frac{3\text{-}3}{3\text{-}3} = 16$$

DISTRIBUTION. The southern red-backed vole occurs in wooded and shrub lands across boreal North America from central Canada south to the United States border, further south in suitable habitats in the eastern mountains to at least Tennessee and North Carolina, and in the western mountains to New Mexico and Arizona. In Michigan (see Map 37), this distinctive vole lives in parts of the upper four tiers of counties in the Lower Peninsula and throughout the Upper Peninsula.

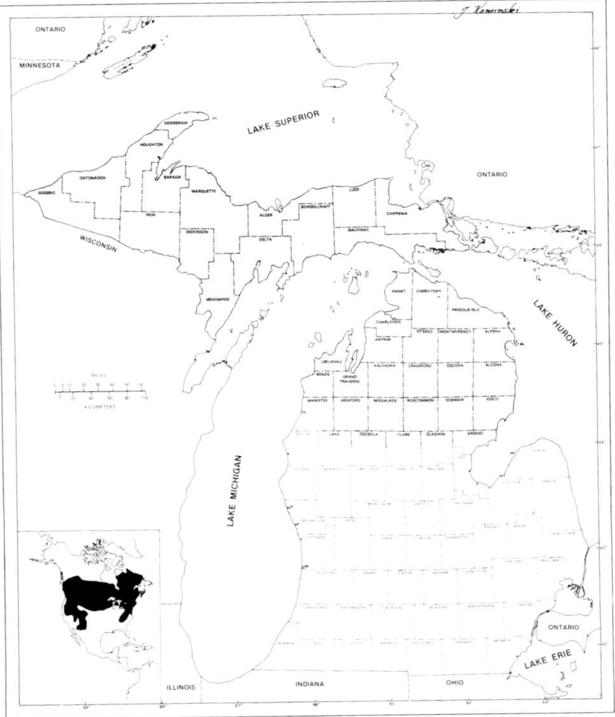

Map 37. Geographic distribution of the southern red-backed vole (*Clethrionomys gapperi*).

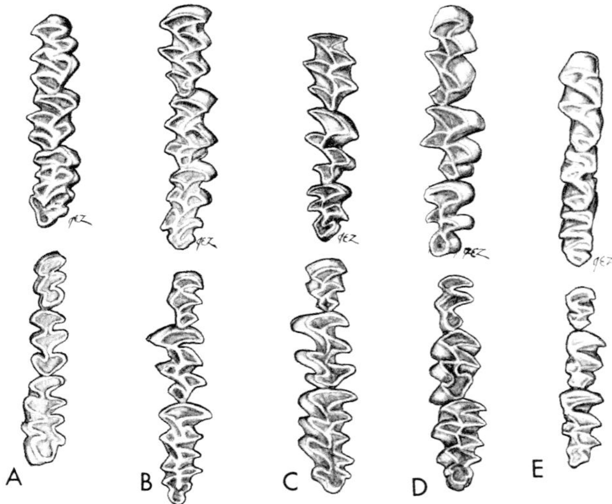

Figure 52. Occlusive faces of the right molar teeth (uppers above) of the five small Michigan microtines: (A) southern red-backed vole, (B) prairie vole, (C) meadow vole, (D) woodland vole, (E) southern bog lemming.

HABITAT PREFERENCES. The southern red-backed vole inhabits northern forests. The naturalist expects to find them most often in cool, shaded, moist, and mature woodland (mixed hardwoods and conifers) containing scattered shrubs, decaying stumps, exposed tree root systems, logs partly buried in mossy growth, and leaf litter (Butsch, 1954). Successional changes induced by clear-cutting reduce habitat quality, especially food supply, and vole populations tend to decrease in these situations (Martell, 1981). According to Getz (1968), these rodents display a preference for areas near water (such as swamp borders), although such environments as dry birch woods are also used (Jackson, 1961). In Michigan, these small mammals have been found in virgin hardwood forests in Alger County (Blair, 1941); cedar swamps in Leelanau County (Hatt, 1923); most forest types, especially conifers, swamps, and bogs, in Marquette County (Manville, 1949); in a root cellar in Alger County (Wood, 1917); a clearing bordered by a highway and a coniferous swamp in Roscommon County (Ozoga and Gaertner, 1963); widespread (except in bogs) in McCormick Forest in Marquette and Baraga counties (Robinson *in* Robinson and Werner, 1975); lowland conifer associations on Beaver Island (Ozoga and Phillips, 1964); forests, especially hardwoods, in Gogebic and Ontonagon counties (Dice and Sherman, 1922); hardwoods in Charlevoix County (Dice, 1925); and forest swamp in Montmorency County (Green, 1925).

DENSITY AND MOVEMENTS. Populations of the southern red-backed vole fluctuate widely during its annual cycle. Its highest populations occur following the appearance of weaned juveniles at the end of major breeding periods. In subalpine forests in Colorado, Merritt and Merritt (1978) noted about 7.2 voles per acre (18 per ha) in winter, declining to a low of 4.0 per acre (10 per ha) in May, to 5.6 per acre (14 per ha) in July, and increasing to a peak of 16.4 per acre (42 per ha) in November. In Michigan, Robinson (*in* Robinson and Werner, 1975) counted 6 to 8 voles per acre (15 to 20 per ha) in Marquette and Baraga counties. In another study in Marquette County, Manville (1949) found a high of 4.4 per acre (11 per ha) in white cedar habitat in September.

Density is linked to daily movements of individuals. Beer (1961) and Gunderson (1962) examined the home ranges of southern red-backed voles in Minnesota and found that individuals used about one-third of an acre (0.12 ha). Manville (1949) recorded almost the same summer ranges for voles in Marquette County. On the other hand, Blair (1941) caught a marked individual 22 times in live-traps set in a grid in Alger County in summer and found this male vole used about 3.6 acres (1.4 ha). As a rule, however, cruising areas are noticeably smaller, with no major difference in movements between adults of either sex.

BEHAVIOR. Richard Manville's study (1949) of small mammals in northern Michigan emphasized the antagonism southern red-backed voles exhibit towards each other. This vole shows little gregarious or colonial behavior (Merritt, 1981). Adult females tend to be more territorial than males, and defend their living spaces from other females more vigorously (Perrin, 1981). Ranges of males overlap broadly; female ranges do not (Mihok, 1979). In captivity, individuals display a nervous temperament; some exhibit a "waltzing condition" (Benton, 1966). Manville even found they enter shock and become unconscious when handled or marked. They may also fight with each other when placed together in cages. In nature, there may be some tendency to form midwinter aggregations for warmth (at least in the northern red-backed vole, *C. rutilus*; West, 1977), although Mihok (1979) obtained few multiple captures in his studies. Watts (1970) found no evidence that male adults killed or drove away juveniles, as does the bank vole (*Clethrionomys glareolus*) of Eurasia (Chitty and Phipps, 1966).

Southern red-backed voles are rather quiet when under observation. Benton (1955) described shrill but almost inaudible squeaks in young within one day after birth. Adults grind their teeth noisily and utter sharp, shrill chattering ("churr-r-r") squeaks (Manville, 1949).

Southern red-backed voles are active night and day, both winter and summer (Merritt and Merritt, 1978). Michigan deer hunters and naturalists (Dice and Sherman, 1922) are often entertained by this vole's antics. Unlike the meadow vole, prairie vole, and southern bog lemming, the southern red-backed vole constructs no special runways, except for tunnels under ground litter, under logs, or in sphagnum moss. In winter, however, they may riddle activity areas with tunnels beneath snow cover (Beer, 1961). The voles apparently stay close to these tunnels and snow trails in winter, in contrast to their greater summer mobility. According to Getz (1968), temperature and humidity have little effect on activity. The vole can climb trees (Muul and Carlson, 1963), jump over obstacles, and is a modestly capable swimmer. Jackson (1961) estimated that this vole can move at a rate of 5 mi (8 km) per hour.

Forest fires are certainly threats to these voles, although a residual population appears to survive in ground cover or in small unburned refuges (Buech *et al.,* 1977; Wrigley, 1975). Southern red-backed voles apparently reappear more slowly than deer mice after fires, perhaps because it may take up to seven years for suitable ground cover to return (Krefting and Ahlgren, 1974).

The focus of the vole's activity is the nest, which can sometimes be detected by following tunnels under logs, leaf litter or moss. Nests are generally global or oval in shape, with diameters as large as 3 to 4 in (76 to 101 mm). Construction materials include grasses, mosses, lichens, and shredded leaves. Nests are usually secluded under logs, in stumps, under brush or in clumps of moss. They are occasionally found in tree holes, or in limb crotches as high as 20 ft (6 m) above ground (Banfield, 1974). In winter, nests can be built at ground level under the protective blanket of snow. Edwards and Pitts (1952) used a trained Labrador retriever to locate vole nests.

ASSOCIATES. Although the southern red-backed vole can climb, its major associates are no doubt the terrestrial species. As the resident woodland vegetation-eater, this vole seems to show little hostility toward the more ubiquitous, seed-eating white-footed mouse (Waters, 1961). In fact, an individual of each species was caught in the same trap in Massachusetts (Muul and Carlson, 1963). However, Kirkland and Griffin (1974) have recorded a negative association between this vole and the deer mouse. According to these authors and Vickery (1979), the vole also maintains some separation from the eastern chipmunk and the woodland jumping mouse. In Minnesota, Timm (1975) frequently caught masked and short-tailed shrews in the same trap line with southern red-backed voles. In the McCormick Experimental Forest (Baraga and Marquette counties), Haveman (1973) captured southern red-backed voles in pit traps along with arctic shrews, masked shrews, pygmy shrews, short-tailed shrews, deer mice, meadow voles, southern bog lemmings, and meadow jumping mice. At grassland/woodland borders, meadow voles often interact with the southern red-backed vole (Grant, 1969); these two apparently coexist in either the grassland or woodland side of this edge when interspecific aggression is low during the non-breeding season; this relationship seems to cease when breeding begins in spring (Iverson and Turner, 1972).

REPRODUCTIVE ACTIVITIES. The southern red-backed vole reproduces during the summer. Merritt and Merritt (1978) found that breeding activities can begin under snow in late March, and conclude by October. As a rule, these animals do not reproduce during winter, perhaps because of reduced body weight. Abundant food supplies in winter, however, may allow for some reproductive activity, as in the case of deer mice (Linduska, 1942). Normally, three or four litters are produced by a single female; the number of weaned juveniles become most conspicuous in the local population in May, July and September.

In Schoolcraft and Alger counties, Ozoga and Verme (1968) found juvenile recruits constituted 30% of the vole population in mid-July and 56% by September. The prolific females can mate immediately after giving birth. Svihla (1930) reported the birth of two litters by a single Michigan female (in captivity) only 18 days apart. There is also some evidence that these voles reproduce more successfully when population numbers are low (Patric, 1962). Evidence of breeding activity has been found for the southern red-backed vole in Michigan from June to August in Gogebic and Ontonagon counties (Dice and Sherman, 1922), July and August in Charlevoix County (Dice, 1925), August and September in Alger County (Blair, 1941), and June and September in Schoolcraft and Alger counties (Ozoga and Verme, 1968).

In his studies in Michigan, Svihla (1930) found the gestation period can vary from 17 to 19 days. A litter may contain from two to eight young, although four to six are usual. The young are born in a grass-lined nest, with eyes closed and body hairless. Their birth weight is about 0.07 oz (2 g). At four days of age, hair begins to appear; at seven, incisors erupt; at eight, some crawling ability is observed; at 10 or 11, the eyes open; and by 17 to 21 days, the process of weaning begins.

Life expectancy in the wild for the southern red-backed vole is short, although some individuals are known to live from 20 to 36 months in nature, and longer in captivity (Banfield, 1974; Jackson, 1961). Normally, young-of-the-year will overwinter and produce the spring breeders; the summer generations then replace the parental stocks. The survival of the species, not the individual, is the basic objective.

FOOD HABITS. The southern red-backed vole, like other microtines, has flat grinding surfaces on the molar teeth. These are kept irregular by a pattern of folding enamel loops which surround exposed, more-wearable, islets of dentine (Fig. 68). This dentition allows these rodents to masticate the coarse and fibrous vegetative parts of plants, which then pass through a complex digestive tract so that nutrients can be extracted and assimilated. In his classic work on the food habits of small mammals in northeastern United States, Hamilton (1941) commented on how finely-chewed the food was in a vole stomach, compared to that in a deer mouse stomach. He found green material predominant in summer, but so well masticated that identification of the plant source was difficult. Paste-like masses of berries, nuts, and seeds were also difficult to identify. Preferred vegetable matter is generally available, especially during the growing season, so that southern red-backed voles may not need to travel as far in search of food as white-footed and deer mice, which have to find widely-scattered seed-producing plants. One observer (Getz, 1968) found that local distribution of this vole is correlated with water availability, rather than with food supplies.

Although best adapted to eating the growing, vegetative parts of weeds and some grasses, the southern red-backed mouse is also opportunistic in food selection, even being classed as omnivorous (Banfield, 1974). Insects (mostly ground beetles, lepidopterous larvae, and flies) make up about 10% of summer foods; centipedes, spiders, and snails are also eaten. Voles in successional stages of deciduous habitat in West Virginia chose succulent plant materials, fungal materials, some woody bark, and roots as summer food (Schloyer, 1977). In autumn and winter, southern red-backed voles may partially rely on cached foods, including beechnuts, hazelnuts, seeds of pine, cherry, alder, and dogwood (Banfield, 1974; Krefting and Ahlgren, 1974; Abbott and Quink, 1970; Vickery, 1979). If green materials and seeds and nut supplies become exhausted, winter and early spring fare can include bark and twigs of various shrubs and small trees. However, this vole does not seem to accumulate the large winter stores of seeds and nuts so characteristic of deer mice. Merritt and Merritt (1978) concluded that the southern red-backed vole has reduced food requirements in the cold season because of lower body weight and lower metabolic rate.

ENEMIES. The southern red-backed vole is exposed to both diurnal and nocturnal predators. Hawks and owls known to catch voles include the red-shouldered, red-tailed, and rough-legged hawks, and long-eared, barred, great-horned, and screech owls (Banfield, 1974; de la Perriere, 1970; Errington *et al.,* 1940). This vole is also pursued by the black bear, raccoon, coyote, gray wolf, red fox, weasel, striped skunk, bobcat and house cat (Banfield, 1974; Jackson, 1961; Pollack, 1951; Toner, 1956; Voigt *et al.,* 1976).

PARASITES AND DISEASES. Internal parasites, including protozoans and tapeworms (cestodes), infest these voles (Jackson, 1961). External parasites include fleas, lice, mites, chiggers, and ticks (Merritt, 1981; Scharf and Stewart, 1980; Timm, 1975). In Marquette County, Manville (1949) found southern red-backed voles carrying one species of flea, one mite, and one tick.

MOLT AND COLOR ABERRATIONS. Adult southern red-backed voles have two annual molts: one in autumn producing the longer, brighter winter coat; and one in spring producing the less dense, darker summer pelage. Occasionally, this vole is found in a grayish brown phase instead of the usual red-backed condition; Green (1925) reported one individual in the grayish brown phase in Montmorency County. Two specimens from Beaver Island had white-spotted flanks (Ozoga and Phillips, 1964).

ECONOMIC IMPORTANCE. The southern red-backed vole may prefer succulent vegetation but also consumes jack pine seeds (Smith and Aldous,

1947) and seeds of other forest trees, although foresters do not believe that it endangers forest reproduction. This vole, however, can be pestiferous in girdling saplings, including fruit trees and young confiers, on plantations. The species has also been known to enter cabins and other outbuildings and feed on stored grains (Banfield, 1974).

COUNTY RECORDS FROM MICHIGAN. Records of specimens of the southern red-backed vole in collections of museums and from the literature show that this species lives in all counties in Michigan's Upper Peninsula and in the Lower Peninsula as far south as Roscommon County (Ozoga and Gaertner, 1963). The species also occur on various islands in the Great Lakes including: in Lake Michigan, Summer Island (Long, 1978; Nellis, 1970); Poverty Island (Long, 1978); North Manitou Island (Scharf and Jorae, 1980); Beaver Island (Ozoga and Phillips, 1964); in Lake Superior, East Huron Island (Corin, 1976); in Lake Huron, Bois Blanc Island (Cleland, 1966); and Drummond Island (Manville, 1950). According to Johnson *et al.* (1969), the southern red-backed vole has been reported on Isle Royale, but no specimens have been preserved.

LITERATURE CITED

Abbott, H. C., and T. F. Quink
1970 Ecology of eastern white pine seed caches made by small forest mammals. Ecology, 51(2):271–278.

Banfield, A. W. F.
1974 The mammals of Canada. Univ. Toronto Press, Toronto. xxiv + 438 pp.

Beer, J. R.
1961 Winter home ranges of the red-backed mouse and white-footed mouse. Jour. Mammalogy, 42(2):174–180.

Benton, A. H.
1955 Notes on the behavioral development of captive red-backed mice. Jour. Mammalogy, 36(4):566–567.
1966 Waltzing in the red-backed mouse. Jour. Mammalogy, 47(2):357.

Blair, W. F.
1941 Some data on the home ranges and general life history of the short-tailed shrew, red-backed mouse, and woodland jumping mouse in northern Michigan. American Midl. Nat., 25(3):681-685.

Buech, R. R., K. Siderits, R. E. Radtke, H. L. Sheldon, and D. Elsing
1977 Small mammal populations after a wildfire in northeast Minnesota. U.S.D.A. Forest Service, Res. Paper NC-151, 8 pp.

Butsch, R. S.
1954 The life history and ecology of the red-backed vole *Clethrionomys gapperi* Vigors in Minnesota. Univ. Michigan, unpubl. Ph.D. dissertation, 161 pp.

Chitty, D., and E. Phipps
1966 Seasonal changes in survival in mixed populations of two species of voles. Jour. Animal Ecol. 35:313.331.

Cleland, C. L.
1956 The prehistoric animal ecology and ethnozoology of the Upper Great Lakes Region. Univ. Michigan, Mus. Anthro., Anthro. Papers, No. 29, x + 294 pp.

Corin, A. F.
1976 The land vertebrates of the Huron Islands, Lake Superior. Jack-Pine Warbler, 54(4):138–147.

de la Perriere, C. I.
1970 Food habits of red-tailed hawks, Summer Island, Michigan, 1969. Summer Sci. Jour., 2(2):72–73.

Dice, L. R.
1925 A survey of the mammals of Charlevoix County, Michigan, and vicinity. Univ. Michigan, Mus. Zool., Occas. Papers No. 159, 33 pp.

Dice, L. R., and H. B. Sherman
1922 Notes on the mammals of Gogebic and Ontonagon counties, Michigan, 1920. Univ. Michigan, Mus. Zool., Occas. Papers No. 109, 40 pp.

Edwards, R. L., and W. H. Pitts
1952 Dog locates winter nests of mammals. Jour. Mammalogy, 33(2):243–244.

Errington, P. L., F. Hammerstrom, and F. N. Hammerstrom, Jr.
1940 The great horned owl and its prey in north-central United States. Iowa State Col., Agri. Exp. Station, Res. Bull. 277:757–850.

Getz, L. L.
1968 Influence of weather on the activity of the red-backed vole. Jour. Mammalogy, 49(3):565–570.

Grant, P. R.
1969 Experimental studies of competitive interaction in a two-species system. I. Microtus and Clethrionomys species in enclosures. Canandian Jour. Zool., 47(5): 1059–1082.

Green, M. M.
1925 Notes on some mammals of Montmorency County, Michigan. Jour. Mammalogy, 6(3):173–178.

Gunderson, H. L.
1962 An eight and one-half year study of the red-backed vole (*Clethrionomys gapperi* Vigors) at Cedar Creek Forest, Anoka and Isanti counties, Minnesota. Univ. Michigan, unpubl. Ph.D. dissertation, 110 pp.

Hamilton, W. J., Jr.
1941 The food of small forest mammals in eastern United States. Jour. Mammalogy, 22(3):250–263.

Hatt, R. T.
1923 The land vertebrate communities of western Leelanau County, Michigan, with an annotated list of the mammals of the county, Michigan Acad. Sci., Arts and Ltrs., Papers, 3:369–402.

Haveman, J. R.
1973 A study of population densities, habitats and foods of four sympatric species of shrews. Northern Michigan University, unpubl. M.S. thesis, vii+70 pp.

Iverson, S. L., and B. N. Turner
1972 Winter coexistence of *Clethrionomys gapperi* and *Microtus pennsylvanicus* in a grassland habitat. American Midl. Nat., 88(2):440–445.

Jackson, H. H. T.
1961 Mammals of Wisconsin. Univ. Wisconsin Press, Madison. xii+504 pp.

Johnson, W. J., M. L. Wolfe, Jr., and D. L. Allen
1969 Community relationships and population dynamics of terrestrial mammals of Isle Royale, Lake Superior. Third Ann. Rept., Isle Royale Studies, (mimeo.), 22 pp.

Kirkland, G. L., Jr., and R. J. Griffin
1974 Microdistribution of small mammals at the coniferous-deciduous forest ecotone in northern New York. Jour. Mammalogy, 55(2):417–427.

Krefting, L. W., and C. E. Ahlgren
1974 Small mammals and vegetation changes after fire in a mixed conifer-hardwood forest. Ecology, 55(6):1391–1398.

Linduska, J. P.
1942 Winter rodent populations in field-shocked corn. Jour. Wildlife Mgt., 6(4):353–363.

Long, C. A.
1978 Mammals of the islands of Green Bay, Lake Michigan. Jack-Pine Warbler, 56(2):59–82.

Manville, R. H.
1949 A study of small mammal populations in northern Michigan. Univ. Michigan, Mus. Zool., Misc. Publ. No. 73, 83 pp.
1950 The mammals of Drummond Island, Michigan. Jour. Mammalogy, 31(3):358–359.

Martell, A. M.
1981 Food habits of southern red-backed voles (*Clethrionomys gapperi*) in northern Ontario. Canadian Field-Nat., 95(3):325–328.

Merritt, J. F.
1981 *Clethrionomys gapperi*. American Soc. Mammalogists, Mammalian Species, No. 146, 9 pp.

Merritt, J. F., and J. M. Merritt
1978 Population ecology and energy relationships of *Clethrionomys gapperi* in a Colorado subalpine forest. Jour. Mammalogy, 59(3):576–598.

Mihok, S.
1979 Behavioral structure and demography of subarctic *Clethrionomys gapperi* and *Peromyscus maniculatus*. Canadian Jour. Zool., 57:1520–1535.

Muul, I., and F. W. Carlson
1963 Red-backed voles in trees. Jour. Mammalogy, 44(3): 415–416.

Nellis, C. H.
1970 Mammals of Summer Island, Michigan. Summer Sci. Jour., 2(2):66–67.

Ozoga, J., and R. Gaertner
1963 Noteworthy locality records for some Michigan mammals. Jack-Pine Warbler, 41(2):89–90.

Ozoga, J. J., and C. J. Phillips
1964 Mammals of Beaver Island, Michigan. Michigan State Univ., Publ. Mus., Biol. Ser., 2(6):305–348.

Ozoga, J. J., and L. J. Verme
1968 Small mammals of conifer swamp deeryards in northern Michigan. Michigan Acad. Sci., Arts and Ltrs., Papers, 53:37–49.

Patric, E. F.
1962 Reproductive characteristics of the red-backed mouse during years of differing population densities. Jour. Mammalogy, 43(2):200–205.

Perrin, M. R.
1981 Seasonal changes in agonistic behavior of *Clethrionomys gapperi* in southeastern Manitoba and its possible relation to population regulation. American Midl. Nat., 106(1):102–110.

Pollack, E. M.
1951 Food habits of the bobcat in the New England states. Jour. Wildlife Mgt., 15(2):209–213.

Robinson, W. L., and J. K. Werner
1975 Vertebrate animal populations of the McCormick Forest. U.S.D.A. Forest Serv., Res. Paper NC-118, 25 pp.

Scharf, W. C., and M. L. Jorae
1980 Birds and land vertebrates of North Manitou Island. Jack-Pine Warbler, 58(1):4–15.

Scharf, W. C., and K. R. Stewart
1980 New records of *Siphonaptera* from northern Michigan. Great Lakes Ento., 13(3):165–167.

Schloyer, C. R.
1977 Food habits of *Clethrionomys gapperi* on clearcuts in West Virginia. Jour. Mammalogy, 58(4):677–679.

Smith, C. F., and S. E. Aldous
1947 The influence of mammals and birds in retarding artificial and natural reseeding of coniferous forests in the United States. Jour. Forestry, 45:361–369.

Svihla, A.
1930 Breeding habits and young of the red-backed mouse, *Evotomys*. Michigan Acad. Sci., Arts and Ltrs., Papers, 11:485–490.

Timm, R. M.
1975 Distribution, natural history, and parasites of mammals of Cook County, Minnesota. Univ. Minnesota, Bell Mus. Nat. Hist., Occas. Papers No. 14, 56 pp.

Toner, G. C.
1956 House cat predation on small mammals. Jour. Mammalogy, 37 (1):117.

Vickery, W. L.
1979 Food consumption and preferences in wild populations of *Clethrionomys gapperi* and *Napaeozapus insignis*. Canadian Jour. Zool., 57(8):1536–1542.

Voigt, D. R., G. B. Kolenosky, and D. H. Pimlott
1976 Changes in summer foods of wolves in central Ontario. Jour. Wildlife Mgt., 40(4):663–668.

Waters, J. H.
1961 *Clethrionomys* and *Peromyscus* in southeastern Massachusetts. Jour. Mammalogy, 42(2):263–265.

Watts, C. H. S.
1970 A field experiment of intraspecific interactions in the red-backed mouse, *Clethrionomys gapperi*. Jour. Mammalogy, 51(2):341–347.

West, S. D.
1977 Midwinter aggregation in the northern red-backed vole, *Clethrionomys rutilus*. Canadian Jour. Zool., 55(9): 1404–1409.

Wood, N. A.
1917 Notes on the mammals of Alger County, Michigan. Univ. Michigan, Mus. Zool., Occas. Papers No. 36, 8 pp.

Wrigley, R. E.
1975 Survival of small mammals in a coniferous burn. Blue Jay, 33(1):56–58.

PRAIRIE VOLE
Microtus ochrogaster (Wagner)

NAMES. The generic name *Microtus,* proposed by Schrank in 1798, is derived from two Greek words, *mikros* meaning small and *otos* referring to the ear. The specific name *ochrogaster,* proposed by Wagner in 1842, is also from two Greek words, *ochra* meaning yellow-ochre and *gaster* referring to the belly. The word vole is of Scandinavian derivation. In Michigan, local residents refer to this rodent as a field mouse. The subspecies *Microtus ochrogaster ochrogaster* (Wagner) occurs in Michigan.

RECOGNITION. Body size medium and stout; head and body of an adult averaging 4½ in (115 mm) long; tail finely-haired and medium in length, averaging 1½ in (40 mm) long, almost equal to one-third of the length of the head and body; body covered with medium-long, soft, dense pelage; head large with blunt nose and no evidence of neck; eyes small, dark and beady; ears short, rounded, hairy, and almost concealed in the pelage at the sides of the head; vibrissae (whiskers) medium-long and not prominent; feet slim and proportionately short; hind limbs slightly longer than fore limbs.

In adults, the upperparts are brownish gray with a grizzled look caused by a mixture of black and buff tips to the long (guard) hairs; body sides are paler; underparts are covered with hair basally black but tipped with light cinnamon to buff (alluding to the specific name *ochrogaster*); tail is well-furred and distinctively bicolored, dark gray-brown above and buff-gray below; feet pale gray. Three pairs of mammary glands characterize the prairie vole, one is pectoral and two are inguinal. There are four functional digits on the forefeet and five on the hind.

The grizzled gray-brown upperparts, the buff-washed underparts, the plump, rounded body, the partly concealed ears, and the short, bicolored tail are, in combination, external characteristics which distinguish the prairie vole from other Michigan mammals living in association with it in southwestern Michigan.

MEASUREMENTS AND WEIGHTS. Adult prairie voles measure as follows: length from tip of nose to end of tail vertebrae, 4⅞ to 6⅛ in (125 to 155 mm); length of tail vertebrae, 1⅛ to 1½ in (30 to 40 mm); length of hind foot, 11/16 to 13/16 in (18 to 21 mm); height of ear from notch, 7/16 to ½ in (11 to 13 mm); weight of adults, 1.1 to 1.8 oz (30 to 50 g).

DISTINCTIVE CRANIAL AND DENTAL CHARACTERISTICS. The skull of the prairie vole, like that of the meadow vole (Fig. 51), is heavy in construction, arched when viewed from the side, somewhat rectangular, and slightly ridged; has a medium projection on the posterior border of palate; possesses incisive foramina less than five mm long and not conspicuously narrow at posterior ends; has auditory bullae less enlarged than in the meadow vole; and shows the premaxillary bones extending further back than the nasal bones. The skull averages 1⅛ in (29 mm) long and 9/16 in (15 mm) wide across the zygomatic arches. Dental characteristics include: molar teeth not rooted and ever-growing; anterior faces of upper incisors not grooved and yellow in color; upper row of the three molar teeth medium long (av., 6.8 mm); all molar teeth with prismatic, flat-crowned grinding surfaces displaying folded, and unusually angular, enamel loops surrounding islands with wide, re-entrant angular loops of equal depth on each side of individual teeth; each last (third) upper molar with grinding surface distinctly characterized (except for being similar to that of the smaller woodland vole) by an anterior crescent-shaped loop, followed by two closed triangular-shaped loops, and a posterior loop with two small inner lobes (see Fig. 52). In overall view the third upper molar has three inner (lingual) and three outer (labial) loops. The dental formula of the prairie vole is:

$$\text{I (incisors)}\,\frac{1\text{-}1}{1\text{-}1},\ \text{C (canines)}\,\frac{0\text{-}0}{0\text{-}0},\ \text{P (premolars)}\,\frac{0\text{-}0}{0\text{-}0},\ \text{M (molars)}\,\frac{3\text{-}3}{3\text{-}3} = 16$$

DISTRIBUTION. The prairie vole occurs in prairie grasslands from southern Alberta, Saskatchewan, and Manitoba, south to eastern Colorado and northern Oklahoma, east to Louisiana, northern Tennessee, extreme western West Virginia, and northwestward through western Ohio, southwestern Michigan, southwestern Wisconsin, and Minnesota. This distribution includes both the true prairie of the American Midwest and areas resembling this habitat to the east and west, into which the prairie vole presumably dispersed after human settlement (Choate and Williams, 1978).

Evidence in both the fossil and archaeological records indicates that the prairie vole has inhabited western Michigan since postglacial times (Cleland, 1966; Handley, 1971; Holman, 1975). This was first documented when remains of the prairie vole (dated by radio-carbon as A.D. 1,220 ± 250) were discovered in the Sleeping Dune area in what is now Leelanau County (Griffin, 1960; Pruitt, 1954). Presumably, beach, dune, and possibly prairie environments which extended northward along Lake Michigan's east shore during the Neo-Atlantic Warm episode, and supported midwestern species such as the prairie vole, were eliminated during the succeeding cool climatic episode (Cleland, 1966).

In the historic period, land clearing which began in southern Michigan in the mid-nineteenth century created a vast habitat of weeds, grasses and shrubs suitable, once again, to prairie species (Kendeigh, 1948); this allowed grassland birds such as the horned lark and the dickcissel to appear, initially in the 1860s and 1870s (Hurley and Franks, 1976). However, it was not until 1918 that the first modern record of the prairie vole was taken in Michigan, in southwesternmost Berrien County (Wood, 1922). Currently (Map 38) it is also known from Cass County (Burt, 1946), Kalamazoo

Map 38. Geographic distribution of the prairie vole (*Microtus ochrogaster*).

County (Brewer, 1970), and Van Buren County (Master, 1978). Although the status of the prairie vole in Michigan has not been closely monitored by field observers, it is suspected that the species may be slowly extending its range northward, and might be observed elsewhere in southwestern Michigan. The species is also reported from Elkhart County, Indiana, directly south of Michigan counties, Cass and St. Joseph (Mumford, 1969).

HABITAT PREFERENCES. Throughout most of its range, the prairie vole may occur in various open habitats. The population is highest in grass and mixed grass and weeds where the vegetation is rank and lush; it is lower where plant growth is short and sparse (Fitch, 1958; Martin, 1956; Mumford, 1969). Prairie voles also thrive in hay fields, in plantings of alfalfa and soybeans, and along field borders where shrubs and other woody vegetation may be interspersed with ground plants. In Tennesee, Severinghaus and Beasley (1973) obtained 65% of their catch in old field habitat and 35% at field edges. Prairie voles use right-of-ways along roads as dispersal routes (Getz *et al.*, 1978), although frequent close mowing of such grassy strips can discourage their movements (Cole, 1978).

In northern parts of the prairie vole's range, it often shares habitats with the meadow vole. These two voles may occur in the same areas (Krebs, 1977), although the prairie vole tends to occupy the dry grassland sites while the meadow vole prefers more moist situations (Findley, 1954; Birney *et al.*, 1976). The prairie vole's habitat preferences in Michigan have not yet been analyzed. Wood (1922) found this rodent ". . . in a small meadow near Birchwood Beach" in Berrien County; Dice (1920) obtained the species in wheat stubble, sedge and rush habitat at Warren Woods in the same county; Brewer (1970) caught the animal in ". . . a strip of weeds and grass on the right-of-way of the Grand Trunk Railroad" in Kalamazoo County; and Master (1978) reported a male taken in farm land near Paw Paw in Van Buren County.

DENSITY AND MOVEMENTS. The prairie vole, like other microtine relatives, can exhibit marked periodic (cyclic) fluctuations, with peak densities occurring at intervals of two to four years (Beasley, 1978; Gaines and Rose, 1976; Krebs *et al.*, 1973). However, there is evidence that noncycling populations exist (Birney *et al.*, 1976). Local population changes are also likely to occur from season to season, since prairie voles can either breed year-round (as in eastern Kansas, Gaines and Rose, 1976) or primarily in the summer months (as in southern Indiana, Krebs *et al.*, 1969). In midwestern prairie vole populations, densities of less than 10 individuals per acre (25 per ha) may be expected, with numbers during peak years increasing to more than 145 per acre (362 per ha). A remarkable high of 429 per acre (1,725 per ha), was estimated for a Missouri population in August (Crawford, 1971).

Despite evidence of emigration as one means of regulating population size and of immigration as a factor in cyclic fluctuations (Abramsky and Tracy, 1979), individual prairie voles are fairly sedentary. Studies of the movements of these voles have been made by trapping, marking, releasing, and re-trapping individuals in grids of live-traps and by tracking radioisotope-tagged animals (Harvey and Barbour, 1965; Martin, 1956). These studies have shown that prairie voles occupy areas varying between 0.05 to 0.18 acre (less than 0.10 ha) and as much as 0.26 acre (0.10 ha) over a period of several months (Fitch, 1958); males have larger home ranges than females (Krebs, 1970).

In Michigan, there are no data on population size for the prairie vole in its limited distribution in the southwestern part of the state. However, its habits do compare closely with those of the meadow vole, and information about the meadow vole in Michigan can generally apply to the prairie vole.

BEHAVIOR. Prairie voles share much of their activity and foraging areas with their neighbors. Groups of these rodents exhibit colonial behavior by using common surface runway systems and underground burrows. At least two individual voles were caught in the same live-trap on several occasions (Fitch, 1957), an indication of intra-specific tolerance. Like meadow voles, several prairie voles may share the same nest, huddling together during cold spells. However, during the breeding season and also at the peak phase of population fluctuations, adults, especially males, behave aggressively (Krebs, 1970; Rose and Gaines 1976) and frequently wound each other (Rose and Heuston, 1978).

Prairie voles can be active at any hour, although Martin (1956) found Kansas populations to be active primarily in dim light beween dawn and 0800 hours, and from sunset until dark. Calhoun (1945), on the other hand, found prairie voles under laboratory conditions concentrating their major activity in the early part of the evening, up to midnight, and then again just before dawn. In winter weather, these rodents are active for a greater number of daylight hours.

The prairie vole is a grazer, and can clear runways at ground level through dense, grassy vegetation. These runways are maintained by clipping plant stems at their bases and clearing debris by constant use. The prairie vole can move around rapidly, and is protected by overhead herbaceous cover within its confined home range. Vegetation adjacent to the runways is the major food supply. Considerable information about the ecology of voles can be gained by examining these runways, as they show evidence of plant cuttings, other residues of feeding activities, resting sites, and fecal deposits.

Prairie voles also spend a great deal of time in underground burrow systems which are used principally for feeding. These burrows are dug by the rodents, although mole tunnels may be used. Underground passages are usually no more than 1½ in (40 mm) in diameter and rarely extend into the ground below 4 in (100 mm). In thin stands of grassy vegetation or in fields of alfalfa or clover, prairie voles may not use well-defined runways (Jameson, 1947). On occasion, prairie voles have been known to climb trees (Crawford, 1971).

Nests constructed by prairie voles are much like those made by meadow voles. In summer, the round, flattened structures, composed mostly of shredded grasses, are usually found in clumps of vegetation. In winter, they may be placed in underground burrows, or sometimes under small hillocks (Fisher, 1945). Nests may measure 5 in (125 mm) wide and 3 in (80 mm) deep.

ASSOCIATES. Several small mammals associate with the prairie vole and use its runway systems. Although there are no studies to show these interspecific relationships in Michigan, investigations in other states indicate that prairie voles are found in the same areas as thirteen-lined ground squirrels, deer mice, white-footed mice, meadow jumping mice, house mice, short-tailed shrews, masked shrews, and least shrews, plus three close relatives: meadow voles, woodland voles, and bog lemmings (DeCoursey, 1957; Fisher, 1945; Martin, 1956; Zimmerman, 1965). The interactions between this variety of grass-eaters, seed-eaters, and insectivores/carnivores in relation to space and nutrient supplies have been the subject of considerable study (Baker, 1971; Krebs, 1977; Rose and Gaines, 1978; Zimmerman, 1965, etc.). The prairie vole seems to be slowly acclimating to certain grassy habitats in Michigan's southwestern counties. It will be interesting to observe how this apparent encroachment affects local small mammal populations.

REPRODUCTIVE ACTIVITIES. Prairie voles are capable of producing large numbers of offspring during the peak years of their population fluctuations (Keller and Krebs, 1970). Prairie voles in Kansas may breed more or less continuously throughout the year (Rose and Gaines, 1978), but in Indiana, breeding occurs mainly in the summer (Krebs *et al.*, 1969). There is, however, some evidence that abundant, highly nutritious food will elicit reproductive activity at any time of the year (Cole and Batzli, 1978). Unlike meadow voles, prairie voles appear to practice monogamous mating (Gavish *et al.*, 1981; Getz *et al.*, 1981; Thomas and Birney, 1979; Wilson, 1982). Females can conceive again shortly after the birth of their litters. Young animals (especially females) may begin breeding at about 35 to 40 days of age even though each individual may be one-half to two-thirds its adult weight. In nature, a female may annually produce as many as five litters, each containing from two to six individuals. After a gestation period of about 21 days, the young are born in a grassy nest. At birth each weighs about three g; has thin, pinkish skin, finely-haired; possesses short vibrissae; and is capable of making faint squeaking sounds. On the second day after birth, the lower incisors erupt; at five days, the young can crawl; at nine days, the eyes open; at 13 days, the dorsal hair averages 8 mm in length; at 17 days, cheek (molar) teeth can be used to masticate plant food and weaning takes place (Fitch, 1957).

Jackson (1961) suggested that prairie voles can live as long as four or five years, although in nature the population turn-over is so rapid that it is doubtful if individuals survive much more than 12 months. Fisher (1945) found that voles live in captivity as long as 16 months, while Getz (1965) held a male in captivity for 35 months.

FOOD HABITS. The prairie vole, like the meadow vole, is equipped with a sac-like caecum in which large quantities of green plant material can be stored; the cellulose is acted on by intestinal microorganisms and reduced to digestible nutrients (Golley, 1960). These voles can also ingest almost any available edibles (seeds, including acorns and grain, fruits, bulbs, tree bark, insects, invertebrates, small vertebrates, each other). In a southern Indiana study of the prairie vole's year-round food habits, Zimmerman (1965) found considerable material (roots and other plant parts) in stomachs which could not be specifically identified. Bluegrass (*Poa compressa*) was determined to be an important food and insect remains (mostly larvae of moths and butterflies and beetles) constituted 4.7% of the diet. At least 15 other identifiable plant remains were obtained. In Kansas, Martin (1956) found that brome grass (*Bromus inermis*) and bluegrass (both occur also in southern Michigan) are commonly eaten by prairie voles. However, recent studies in Illinois show that prairie voles thrive better on a diet of dicotyledon flowering plants, rather than on the grasses (monocotyledons), which meadow voles favor (F. R. Cole and L. Verner, pers. comm.).

Prairie voles excavate underground storage chambers in which seeds, fruits, grass clippings, and tubers can be stored. These chambers can be as much as 36 in (92 cm) in length and 18 in (46 cm) in height and hold sizable quantities of food (Banfield, 1974). In Missouri, Fisher (1945) found a prairie vole cache containing about 1 gal (4 l) of yellow fruits of the horse nettle (*Solanum carolinense*). Prairie voles also will girdle ornamental shrubs and such trees as crab apple, honey locust, and Scotch pine (Martin, 1956; Crawford, 1971).

ENEMIES. The prolific prairie vole is active both day and night, and therefore is fair game for both diurnal and nocturnal mouse eaters. These include a variety of carnivores: large fish, bullfrogs, snakes, hawks, owls, other birds, and various mammals (Martin, 1956). Specific enemies reported in the literature include: Cooper's hawk, sharp-skinned hawk, red-tailed hawk, red-shouldered hawk, long-eared owl, barn owl, barred owl, screech owl, great horned owl, shrike, crow, Virginia opossum, short-tailed shrew, coyote, red fox, gray fox, mink, least weasel, house cat, and bullfrog (Criddle, 1926; Fitch and Bare, 1978; Korschgen, 1952 and 1957; Martin, 1956; Voight and Glenn-Lewin, 1978).

PARASITES AND DISEASES. The prairie vole is infested internally by flatworms (trematodes), tapeworms (cestodes), and round worms (nematodes). In a study of ectoparasites of this mouse in Kansas, Jameson (1947) identified six species of fleas, one species of sucking louse (Anoplura), two species of mites, and three species of ticks.

MOLT AND COLOR ABERRATIONS. Weaned prairie voles have a dark juvenile pelage which molts in about 30 days. A subadult pelage then appears, remaining for 60 to 90 days until replaced by the adult pelage. There is a single molt annually, usually in summer. Coat color mutations reported in prairie voles include albinism (DeBlase and Humphrey, 1965), dark buff (Pinter and Negus, 1971), melanism (Mumford, 1964), xanthochromism (Stalling, 1974), and white spotting (Cockrum, 1953).

ECONOMIC IMORTANCE. It is doubtful that the minor enclave of prairie voles living in southwestern Michigan has much effect on agricultural pursuits. However, in other parts of its range this vole is of economic importance. In cultivated areas, prairie voles cache wheat, rye, and oats (Criddle, 1926). Voles have also been known to girdle fruit trees and pines, and damage ornamental shrubs, bulbs, and vegetables in gardens (Barbour and Davis, 1974; Crawford, 1971).

COUNTY RECORDS FROM MICHIGAN. Records of specimens of the prairie vole in collections of museums and from the literature show that this species has been reported in only four southwestern counties: Berrien, Cass, Kalamazoo, and Van Buren (Brewer, 1970; Burt, 1946; Master, 1978).

LITERATURE CITED

Abramsky, Z., and C. R. Tracy
1979 Population biology of a "noncycling" population of prairie voles and a hypothesis on the role of migration in regulating microtine cycles. Ecology, 60(2):349–361.

Baker, R. H.
1971 Nutritional strategies of myomorph rodents in North American grasslands. Jour. Mammalogy, 52(4):800–805.

Banfield, A. W. F.
1974 The mammals of Canada. Univ. Toronto Press, Toronto. xxiv+438 pp.

Barbour, R. W., and W. H. Davis
1974 Mammals of Kentucky. Univ. Kentucky Press, Lexington. xii+322 pp.

Beasley, L. E.
1978 Demography of southern bog lemmings (*Synaptomys cooperi*) and prairie voles (*Microtus ochrogaster*) in southern Illinois. Univ. Illinois, unpubl. Ph.D. dissertation, iv+95 pp.

Birney, E. C., W. E. Grant, and D. D. Baird
1976 Importance of vegetative cover to cycles of *Microtus* populations. Ecology, 57:1043–1051.

Brewer, R.
1970 The prairie vole in Kalamazoo County, Michigan. Jack-Pine Warbler, 48(2):45.

Burt, W. H.
1946 The mammals of Michigan. Univ. Michigan Press, Ann Arbor. xv+288 pp.

Calhoun, J. B.
1945 Diel activity rhythms of the rodents *Microtus ochrogaster* and *Sigmodon hispidus hispidus*. Ecology, 26:251–273.

Choate, J. R., and S. L. Williams
1978 Biogeographic interpretation of variation within and among populations of the prairie vole, *Microtus ochrogaster*. Texas Tech Univ., Mus., Occas. Papers No. 49, 25 pp.

Cleland, C. L.
1966 The prehistoric animal ecology and ethnozoology of the Upper Great Lakes Region. Univ. Michigan, Mus. Anthro., Anthro. Papers, No. 29, x+294 pp.

Cockrum, E. L.
1953 Aberrations in the color of the prairie vole, *Microtus ochrogaster*. Trans. Kansas Acad. Sci., 56(1):86–88.

Cole, F. R.
1978 A movement barrier useful in population studies of small mammals. American Midl. Nat., 100(2):480–482.

Cole, F. R., and G. O. Batzli
1978 Influence of supplemental feeding on a vole population. Jour. Mammalogy, 59(4):809–819.

Crawford, R. D.
1971 High population density of *Microtus ochrogaster*. Jour. Mammalogy, 52(2):478.

Criddle, S.
1926 The habits of *Microtus minor* in Manitoba. Jour. Mammalogy, 7(3):193–200.

DeBlase, A. F., and S. R. Humphrey
1965 Additional record of an albino prairie vole. Jour. Mammalogy, 46(3):501.

DeCoursey, G. E.
1957 Identification, ecology and reproduction of *Microtus* in Ohio. Jour. Mammalogy, 38(1):44–52.

Dice, L. R.
1920 The mammals of Warren Woods, Berrien County, Michigan. Univ. Michigan, Mus. Zool., Occas. Papers No. 86, 20 pp.

Findley, J. S.
1954 Competition as a possible limiting factor in the distribution of *Microtus*. Ecology, 35:418–420.

Fisher, H. J.
1945 Notes on voles in central Missouri. Jour. Mammalogy, 26(4):435–437.

Fitch, H. S.
1957 Aspects of reproduction and development in the prairie vole (*Microtus ochrogaster*). Univ. Kansas Publ., Mus. Nat. Hist., 19(4):129–161.
1958 Home ranges, territories, and seasonal movements of vertebrates of the Natural History Reservation. Univ. Kansas Publ., Mus. Nat. Hist., 11(3):63–326.

Fitch, H. S., and R. O. Bare
1978 A field study of the red-tailed hawk in eastern Kansas. Trans. Kansas Acad. Sci., 81(1):1–13.

Gaines, M. S., and R. K. Rose
1976 Population dynamics of *Microtus ochrogaster* in eastern Kansas. Ecology, 57(6):1145–1161.

Gavish, L., C. S. Carter, and L. L. Getz
1981 Further evidences for monogamy in the prairie vole. Animal Behav., 29(3):955–957.

Getz, L. L.
1965 Longevity of two captive prairie voles. Jour. Mammalogy, 46(3):514.

Getz, L. L., C. S. Carter, and L. Gavish
1981 The mating system of the prairie vole, *Microtus ochrogaster*: Field and laboratory evidence of pair-bonding. Behav. Ecol. Sociobiol., 8(3):189–194.

Getz, L. L., F. R. Cole, and G. L. Gates
1978 Interstate roadsides as dispersal routes for *Microtus pennsylvanicus*. Jour. Mammalogy, 59(1):208–212.

Golley, F. B.
1960 Anatomy of the digestive tract of *Microtus*. Jour. Mammalogy, 41(1):89–99.

Griffin, J. B.
1960 A hypothesis for the prehistory of the Winnebago. Pp. 809–865 *in* S. Diamond, ed. Culture in history. Essays in honor of Paul Radin. Columbia Univ. Press, New York. xxiv+1014 pp.

Handley, C. O., Jr.
1971 Appalachian mammalian geography—Recent Epoch. Virginia Poly. Inst. and State Univ., Research Monog. 4:263–303.

Harvey, H. J., and R. W. Barbour
1965 Home range of *Microtus ochrogaster* as determined by a modified minimum area method. Jour. Mammalogy, 46(3):398–402.

Holman, J. A.
1975 Michigan's fossil vertebrates. Michigan State Univ., Publ. Mus., Ed. Bul. No. 2, 54 pp.

Hurley, R. J., and E. C. Franks
1976 Changes in the breeding ranges of two grassland birds. Auk, 93(1):108–115.

Jackson, H. H. T.
1961 Mammals of Wisconsin. Univ. Wisconsin Press, Madison. xii+504 pp.

Jameson, E. W., Jr.
1947 Natural history of the prairie vole (mammalian genus *Microtus*). Univ. Kansas Publ., Mus. Nat. Hist., 1(7):-125–151.

Keller, B. L., and C. J. Krebs
1970 *Microtus* population biology III. Reproductive changes in fluctuating populations of *M. ochrogaster* and *M. pennsylvanicus* in southern Indiana, 1965–67. Ecol. Monog., 40(3):263–294.

Kendeigh, S. C.
1948 Bird populations and biotic communities in northern lower Michigan. Ecology, 29:101–114.

Korschgen, L. J.
1952 A general summary of the food of Missouri predatory and game animals. Missouri Conserv. Comm., Pittman-Robertson Prog., 61 pp.
1957 Food habits of coyotes, foxes, house cats and bobcats in Missouri. Missouri Conserv. Comm., P-R Ser. No. 15, 64 pp.

Krebs, C. J.
1970 *Microtus* population biology: Behavioral changes associated with the population cycle in *M. ochrogaster* and *M. pennsylvanicus*. Ecology, 51(1):34–52.
1977 Competition between *Microtus pennsylvanicus* and *Microtus ochrogaster*. American Midl. Nat., 97(1):42–49.

Krebs, C. J., B. L. Keller, and R. H. Tamarin
1969 *Microtus* population biology: Demographic changes in fluctuating populations of *M. ochrogaster* and *M. pennsylvanicus* in southern Indiana. Ecology, 50(4): 587–607.

Krebs, C. J., M. G. Gaines, B. L. Keller, J. H. Myers, and R. H. Tamarin
1973 Population cycles in small rodents. Science, 179:35–41.

Martin, E. P.
1956 A population study of the prairie vole (*Microtus ochrogaster*) in northeastern Kansas. Univ. Kansas Publ., Mus. Nat. Hist., 8(6):361–416.

Master, L. L.
1978 A survey of the current distribution, abundance, and habitat requirements of threatened and potentially threatened species of small mammals in Michigan. Univ. Michigan. (mimeo.), 52 pp.

Mumford, R. E.
1964 A melanistic prairie vole. Jour. Mammalogy, 45(1): 150.
1969 Distribution of the mammals of Indiana. Indiana Acad. Sci., Monog. No. 1, vii+114 pp.

Pinter, A. J., and N. C. Negus
1971 Coat color mutations in two species of voles (*Microtus montanus* and *Microtus ochrogaster*) in the laboratory. Jour. Mammalogy, 52(1):196–199.

Pruitt, W. O., Jr.
1954 Additional animal remains from under Sleeping Bear Dune, Leelanau County, Michigan. Michigan Acad. Sci., Arts and Ltrs., Papers, 39:253–256.

Rose, R. K., and M. S. Gaines
1976 Levels of aggression in fluctuating populations of the prairie vole, *Microtus ochrogaster*, in eastern Kansas. Jour. Mammalogy, 57(1):43–57.
1978 The reproductive cycle of *Microtus ochrogaster* in eastern Kansas. Ecol. Monog., 48(1):21–42.

Rose, R. K., and W. D. Heuston
1978 Wound healing in meadow voles. Jour. Mammalogy, 59(1):186–188.

Severinghaus, W. D., and L. E. Beasley
1973 A survey of the microtine and zapoid rodents of west Tennessee. Jour. Tennessee Acad. Sci., 48(4):129–133.

Stalling, D. T.
1974 A xanthochromic prairie vole and notes on associated literature. Southwest. Nat., 19(1);115–117.

Thomas, J. A., and E. C. Birney
1979 Parental care and mating system of the prairie vole, *Microtus ochrogaster*. Behav. Ecol. Sociobiol., 5(2):171–186.

Voight, J., and D. C. Glenn-Lewin
1978 Prey availability and prey taken by long-eared owls in Iowa. American Midl. Nat., 99(1):162–171.

Wilson, S. C.
1982 Parent-young contact in prairie and meadow voles. Jour. Mammalogy, 63(2):300–305.

Wood, N. A.
1922 Notes on the mammals of Berrien County, Michigan. Univ. Michigan, Mus. Zool., Occas. Papers No. 124, 4 pp.

Zimmerman, E. G.
1965 A comparison of habitat and food of two species of *Microtus*. Jour. Mammalogy, 46(4):605–612.

MEADOW VOLE
Microtus pennsylvanicus (Ord)

NAMES. The generic name *Microtus,* proposed by Schrank in 1798, is derived from two Greek words, *mikros* meaning small and *otos* referring to the ear. The specific name *pennsylvanicus,* proposed by Ord in 1815, refers to the state of Pennsylvania from which this vole was originally described. In Michigan, local residents call this rodent a field mouse or vole. The subspecies *Microtus pennsylvanicus pennsylvanicus* (Ord) occurs in Michigan.

RECOGNITION. Body size medium and stout; head and body of an adult averaging 4¾ in (120 mm) long; tail finely-haired and medium in length, averaging 2 in (50 mm) long, almost equal to one-third of the length of the head and body; body covered with medium-long, soft, dense pelage; head large with blunt nose and no evidence of neck; eyes small, dark, beady; ears short, rounded, hairy, and almost concealed in the pelage at the sides of the head; vibrissae (whiskers) medium-long and not prominent; feet slim and proportionally short; hind limbs slightly longer than fore limbs. In adults, upperparts are uniformly dark brown produced by a dense, slate gray underfur and long, buff-banded or black-tipped guard hairs; body sides paler; underparts dusky gray produced by the silver-tipped dark belly fur; feet and tail dark brown with underside of tail paler, producing a slight bicolored effect. The summer coat is shorter and browner while the winter coat is longer, silkier and grayer. The pelage of weanling young is short and dark, with feet almost black. Subadults have fur much like adults. Four pairs of mammary glands are present, two pectoral and two inguinal. There are four functional digits on the forefeet and five on the hind.

The dark brown upperparts, the plump, rounded body, the partly concealed ears, the absence of a conspicious neck, and the short tail are a combination of external features serving to distinguish the meadow vole from other Michigan mammals (Plate 4b).

MEASUREMENTS AND WEIGHTS. Adult meadow voles measure as follows: length from tip of nose to end of tail vertebrae, 5⅞ to 7¼ in (150 to 185 mm); length of tail vertebrae, 1½ to 2⅜ in (40 to 60 mm); length of hind foot, ¾ to 1 in (18 to 25 mm); height of ear from notch, ⅜ to ⅝ in (10 to 16 mm); weight of adults, 1.4 to 2.1 oz (40 to 60 g).

DISTINCTIVE CRANIAL AND DENTAL CHARACTERISTICS. The skull of the meadow vole is heavy in construction (Fig. 51), somewhat rectangular, and slightly ridged; possesses a median projection posteriorly on the bony palate, which does not end abruptly as a thin bony shelf; has long anterior palatine foramina (more than 5 mm); features moderately large, slightly elongated, and rounded auditory bullae; and is characterized by a least interorbital width (between orbits) of less than 4 mm. The skull averages 1 in (27 mm) long and 9/16 in (15 mm) wide across the zygomatic arches. Dental characteristics include molar teeth not rooted and ever-growing; anterior faces of upper incisors not grooved and deep yellow in color; upper row of the three molar teeth long (7.2 mm); all molar teeth with flat-crowned grinding surfaces displaying folded and unusually angular enamel loops surrounding islands of softer dentine with re-entrant angular loops of equal depth on each side of individual teeth; last (third) upper molar with flat-crowned grinding surface distinctively characterized by an anterior crescent-shaped loop, following by three closed triangular-shaped loops, and a posterior loop with two small, inner lobes (see Fig. 52), in overall view this third upper molar has four (sometimes five) inner (lingual) and three outer (labial) angular loops. The skull and dentition of the meadow vole most closely resemble those of the prairie vole found only in extreme southwestern Michigan, but is distinguished clearly from the latter by the condition of the angular loops on the third upper molar; in the meadow vole there are four (or five) inner and three outer angular loops while in the prairie vole there are only three inner and three outer angular loops. The dental formula of the meadow vole is:

$$\text{I (incisors)}\,\frac{1\text{-}1}{1\text{-}1},\ \text{C (canines)}\,\frac{0\text{-}0}{0\text{-}0},\ \text{P (premolars)}\,\frac{0\text{-}0}{0\text{-}0},\ \text{M (molars)}\,\frac{3\text{-}3}{3\text{-}3} = 16$$

DISTRIBUTION. The meadow vole occurs in moist grassland habitats across boreal North America from northern Canada and Alaska south (except west of the Rocky Mountains) to north-central United States, extending southward in the eastern mountains and in the western mountains, with one isolated population occurring in the Mexican state of Chihuahua (Reich, 1981). In Michigan, the meadow vole thrives in suitable open environments throughout the state (see Map 39).

HABITAT PREFERENCES. The meadow vole is a grassland species, preferring moist, open grass-herb vegetation (Dice, 1922; Getz, 1961a). In fallow agricultural areas in Michigan, Beckwith (1954) found the meadow vole became the prominent small mammal in the community when the vegetative succession moved from the annual-biennial stage to the perennial grass stage. The rodent will also live in some woody cover (Getz, 1970; Grant, 1971). Nevertheless, overhead grassy cover seems essential for optimum habitat, serving to provide both seclusion and food. In Michigan, meadow voles have been studied in a deer yard containing even-age saplings with rank herbaceous ground cover in Alger and Schoolcraft counties

Map 39. Geographic distribution of the meadow vole (*Microtus pennsylvanicus*).

(Ozoga and Verme, 1968); in grassy bogs and beaver meadows in Marquette and Baraga counties (Robinson *in* Robinson and Werner, 1975); in meadow and white cedar swamp in Marquette county (Manville, 1948); in meadow, sedge, black ash swamp, white cedar swamp, leather leaf and sphagnum bogs, and black spruce-tamarack bog in Gogebic and Ontonagon counties (Dice and Sherman, 1922); in sedge, rush, grass, sphagnum, tamarack, black spruce, and white cedar bogs, in aspen and second-growth hardwoods, and in cultivated fields in Charlevoix County (Dice, 1925); in sedges bordering a brook and a beaver meadow in Montmorency County (Green, 1925); in timothy and other grasses in Washtenaw County (Greffenius, 1939); in grass and cultivated areas and in old beaver meadows in Washtenaw County (Wood, 1922); in old field and grass sedge marsh in Washtenaw County (Getz, 1961a); in grass, clover, and alfalfa, and in winter under deep snow into upland brush or coniferous plantings in Kalamazoo County (Allen, 1938); in bluegrass association in Livingston County (Blair, 1940b); in grassy areas along roads and adjacent to swamps in Missaukee County (R. H. Baker, unpubl. field notes); in cultivated fields in winter, grass-herb areas and infrequently in woods in Clinton County (Linduska, 1950); in old field habitat in Ingham County (Golley, 1961); along railroad tracks in Ingham County (field trips by Michigan State University class in mammalogy); and in orchards, conifer plantations, nurseries, parks, landscaped areas, and cultivated fields containing field-shocked grain (Allen, 1940; Linduska, 1942; Shick, 1965).

In winter, deep snow provides the meadow voles with a protective cover for tunneling. The snow's insulative effect also ensures a rather stable microclimate (Pruitt, 1957; 1960). Although meadow voles have a wide degree of freedom to tunnel in the subnivean habitat, periodic thaws, at least in southern Michigan, might have serious effects if individuals become stranded under islands of melting snow. In northern parts of Michigan, where deep snow cover remains longer, meadow voles probably have use of more kinds of habitats during winter than in any other season of the year.

Distribution in Correlation with Vegetative Succession.— Much of presettlement Michigan was forested. Rodents chiefly adapted for open lands were restricted to streamside growth, swamp and marsh edges, beach, beaver meadows, forest openings, including natural ones like those found in southwestern Michigan, and other areas in various stages of second growth after forest disturbance caused by fires, wind storms, and clearing. Although the meadow vole appears to have resided in Michigan since postglacial times (Holman, 1979), this author suspects (without adequate data) that the southern bog lemming, which today plays only a minor role as a grass-eating rodent, may originally have been the dominant microtine in suitable Michigan habitat. During settlement, pronounced changes resulted from forest clearing, intensive farming, and the spread and introduction of numerous herbaceous plants including bluegrass (*Poa*), quackgrass (*Agropyron*), and brome (*Bromus*). These changes provided abundant and attractive habitat and food for grass-eating microtines (Thompson, 1965). The meadow vole then presumably increased in these environments. In discussing the meadow vole in Washtenaw County, Wood (1922) wrote that the species "was formerly found in beaver meadows, but with the clearing of the forests it has extended its range to the fields of grass and grain and has become the most numerous of all the mammals of the county." In short, the meadow vole thrived as a result of these environmental changes, possibly to the detriment of the southern bog lemming. The success of the meadow vole in adapting to human land-use is well illustrated by its highly successful occupation of isolated strips of grassy vegetation, some no more than five meters wide, on the right-of-ways bordering roads, railroad tracks and, more recently, pipe lines and power lines (Getz *et al.,* 1978). The meadow vole's ability to respond to agricultural operations was documented in a study done in a Wisconsin alfalfa field (LoBue and Darnell, 1959) where the animal's major habitat was shown to be the edge environment around the field. However, as the alfalfa increased in height, the vole spread into the field; when the crop was mowed, the animals used only the edge. The meadow vole's dominant position in Michigan's open lands today depends on the presence of proper grassy seral stages in the ecological succession of plant growth (M'Closkey, 1975).

DENSITY AND MOVEMENTS. The meadow vole (like its near relative, the lemming) can produce unusually high densities, followed by marked declines, in 3 to 4 year fluctuations (Elton, 1942; Hamilton 1937a; Krebs and Myers, 1974). In Michigan, "mouse years," when voles are extremely abundant, have long been recognized by agriculturists. These dense populations are often local. Hayne (1950) noted that high vole numbers caused severe damage to orchards in southern Michigan in the autumn and winter of 1948–49. In 1940, on the other hand, Linduska (1950) found the species almost nonexistent in favorable habitat in the Rose Lake area of Clinton County. Fortunately, devastating vole irruptions similar to those in Oregon which increased populations of montane voles (*M. montanus*) to an estimated 2,000 to 3,000 individuals per acre (5,000 to 7,500 per ha) in November 1957 (Spencer, n.d.) have not been recorded in Michigan. High populations in Michigan, as reported by Aumann (1965), were 50 to 60 meadow voles per acre (125 to 150 per ha).

The causes of these multiannual (cyclic) fluctuations in vole populations have puzzled biologists for years. Mechanisms involved, according to Krebs and Myers (1974) Rosenzweig and Abransky (1980), and others, include such environmental factors as food shortages, habitat heterogeneity (patchiness), unusual predation, immigration, inclement weather, disease outbreaks, and the manifestation of genetically-controlled inherited behavioral factors. The latter appears to play a major role as vole populations on the increase have different behavioral characteristics from those on the decrease.

In the normal annual cycle, the fecund meadow vole usually attains its highest numbers no later than early winter (Hamilton, 1937a), although Michigan studies by both Golley (1961) and Linduska (1950) showed population highs in January and February. The decline in breeding activity in late winter allows for a gradual depression of the numbers until the appearance of the first crop of young mice at the onset of vegetation regrowth in mid-spring or later. Although there have been few population studies of meadow voles in Michigan's Upper Peninsula, there is some evidence that densities are generally lower than in the Lower Peninsula, perhaps because a greater percentage of the land is wooded. However, successional changes resulting from clear-cutting do stimulate the increase of local meadow vole populations, even in conifer swamps (Verme and Ozoga, 1981). In McCormick Forest in Marquette and Baraga counties, Robinson (*in* Robinson and Werner, 1975) suggested that vole populations do not exceed 10 individuals per acre (25 per ha) and obtained a count of eight per acre (20 per ha) in bog habitat. In his classic work in the Huron Mountain region of Marquette County, Manville (1949a) counted approximately only one individual per acre (2.5 per ha). This was in late summer, when populations had not yet reached their annual peaks.

In contrast, moist grass-herb areas adjacent to swamps or in old field swales in the Lower Peninsula have produced local populations of more than 50 individuals per acre (125 per ha). In viewing the population fluctuations throughout a single year in an old field bluegrass community in Ingham County, Golley (1960b) reported a low of two meadow voles per acre (5.2 per ha) in late May, which increased steadily (as a result of recruitment) to a peak population of 56 meadow voles per acre (139 per ha) in late February. In terms of total biomass, the combined weight of these mice expanded from a low of 2.2 oz per acre (151 grams per ha) in May to a high of 57.6 oz per acre (4,034 grams per ha) in February, a remarkable 25-fold increase in a single season. In other studies in the Lower Peninsula, Blair (1948) found about 30 meadow voles per acre (75 per ha) in November in a bluegrass meadow association in Livingston County; Linduska (1950) counted more than 300 meadow voles under 315 field corn shocks spaced over 28 acres (11 ha) in January in Clinton County; Greffenius (1939) estimated that 40 to 45 voles per acre (100 to 112 per ha) occurred in timothy and other grasses in the vicinity of Ann Arbor; Getz (1960) found vole densities as high as 25 individuals per acre (60 per ha) in grass-sedge marsh in Washtenaw County.

Despite pronounced multiannual and annual population fluctuations, individual meadow voles occupy the same area for months at a time (Getz, 1961a), and have the ability to return to a locality when displaced (Robinson and Falls, 1965). This activity area may be shared with other individuals, with areas used by females being superimposed on the larger ones occupied by males. However, within each home range, the vole has its own territory, such as a nest or feeding place, which is defended by the resident adult vole from intrusion by other members of the species of the same sex (Ambrose, 1973; Burt, 1943). As a rule, females are more vigorous in excluding other females from their areas of activities than are males (Madison, 1980; Webster and Brooks, 1981). In Michigan, meadow vole home ranges have been calculated mostly by setting live-traps at regular intervals on field grids and then capturing, marking, releasing, and recapturing resident animals. This technique has proved somewhat unreliable (Hayne, 1950a); nevertheless, home ranges as small as 0.15 acre (0.08 ha) to as large as 0.75 acre (0.30 ha) have been determined by such workers as Blair (1940b), Getz (1961a), and Manville (1949). These home ranges may expand or contract from one season to the next, depending on population numbers and habitat conditions.

BEHAVIOR. Field evidence shows that meadow voles are promiscuous and live in close association with each other in nature (even in communal nests in winter, Webster and Brooks, 1981) but are socially organized into territorial maternal-offspring units during the breeding season (Madison, 1980). Getz (1972) found little antagonism between adults and/or immature individuals of either sex when captured together in multiple-capture live-traps, although fighting between males in breeding condition has been documented (Christian, 1971; Rose and Heuston, 1978). This behavior is often intensified when animals are kept in confinement (Getz, 1962; Turner and Iverson, 1973).

The meadow vole can be active at any time, although Hamilton (1937c) concluded that it is primarily diurnal (active by day) and crepuscular (active at dawn and dusk in dim light). There is some evidence that voles are more nocturnal and crepuscular in summer and more diurnal and less crepuscular in winter. In Ingham County, Osterberg (1962) photographed animals (day and night) moving through runways in a grassy habitat and found that meadow voles were most active from sunset to five hours after sunset in October and November. He also found that vole activity was only slightly hampered by fluctuations in temperature or humidity and during rain (see also Getz, 1961c). The vole's preference for moist habitats is also shown by its swimming ability (Blair, 1939) and its association with vegetation bordering streams, swamps and marshes.

Perhaps the most interesting behavioral characteristic of the meadow vole is its ability to construct surface pathways through heavy grassy vegetation. This behavior is shared with other members of the genus *Microtus* and with other grass-eating rodents, including the southern bog lemming. As a grassland dweller, the meadow vole prefers moving at ground level by pushing its way through the basal parts of plants rather than exposing itself by climbing over the plant growth. Where the stem count per unit area is low, such as in Michigan stands of brome grass, runways are less obvious. Where the stem count per unit area is high, the meadow vole must clip vegetation to open its trail and then keep the narrow corridor clear by constant growth-cropping and back-and-forth travel. Well-used runways, no more than 1¼ in (30 mm) wide, are conspicuous when the grass is parted and are often reminiscent of the deep-cut cattle trails approaching a water hole on a western ranch. This trail-making ability is closely correlated with the graminivorous habits of the meadow vole, which eats (but not exclusively) the vegetative growth through which its trails extend. Other small rodents in the community (deer mice, white-footed mice, meadow jumping mice, house mice), feeding mostly on plant products (seeds, fruits, and other reproductive parts rather than the plants themselves), sometimes use vole trails as well. What puzzles mammalogists is, that with other creatures using this elaborate system of undergrass runways, the vole may benefit little from developing them (Osterburg, 1962). Perhaps it is much like the beaver whose dam and impounded water provide facilities for other species.

Mammalogy students at Michigan State University have been required to survey a small, staked grid (16 by 16 feet, divided by string into 16 four-foot squares) for evidence of vole activities and map all surface and underground runways in their study plots. In one such study Hugh R. Wygmans (field report for October 30, 1965) found 146 ft (37 m) of runways in his plot. Projecting his figures to an area the size of an acre would show a runway system 4.6 mi long (to an area the size of 1 ha, 18.6 km long). Such statistics dramatize the influence meadow voles and their runways have on the ground-level ecology. Vole movements may be even more impressive in winter when they tunnel under snow cover into areas, including residential lawns, where grass cover is insufficient at other times of the year. The ultrasonic calls used by voles in their runway systems decrease rapidly above $20kH_2$, while the calls of small rodents living in woodlands, or where ground cover is less dense, have a greater range of high frequency sounds (Smith, 1979).

Meadow voles construct nests in clumps of grass, along suface runways, under boards or rocks, underground in shallow burrows, in bird nests (Low, 1944) or in winter in surface locations protected by a blanket of snow. Weilert and Shump (1977) examined 20 vole nests in a grassy, old field community in Ingham County; all consisted of grass clippings woven into somewhat globular shapes and situated in slight depressions on the ground surface. These nests averaged 5¾ in (148 mm) across and 2¾ in (72 mm) high; each contained a distinct inner cavity with one side entry way. There is sometimes more than one entry; occasionally a nest will consist of merely a grassy platform with no inner cavity. Nests in use are kept in good repair and free of debris and fecal matter.

ASSOCIATES. The grass-herb communities of Michigan contain several resource-using small mammals. Generally (but not always) meadow voles are considered the representative graminivores or grazers (consumers of the vegetative parts of plants) while deer mice, white-footed mice, jumping mice, and the introduced house mouse are the granivores and/or frugivores (consumers of seeds and fruits), and shrews are the principal small insectivores/carnivores (consumers of insects, other invertebrates and small vertebrates). In Michigan, the diversity of small mammal species in grass-herb communities varies. For example, in Alger County, Fitch (1979) observed small numbers of meadow voles in company with deer mice on the bleak and sparsely-vegetated Kingston Plains; in Alger and Schoolcraft counties, Ozoga and Verme (1968) found the meadow vole associated with the masked shrew, arctic shrew, short-tailed shrew, deer mouse, southern red-backed vole, meadow and woodland jumping mice in deer yards; in Marquette County, Manville (1949a) captured the meadow vole with the masked shrew, short-tailed shrew, pygmy shrew, deer mouse, and southern red-backed vole in white cedar swamp; in Baraga and Marquette counties, Haveman (1973) caught the meadow vole in pit traps in which the arctic shrew, masked shrew, pygmy shrew, short-tailed shrew, deer mouse, southern red-backed vole, southern bog lemming, and meadow jumping mouse were also caught; in Roscommon County, Rabe (1981) caught the meadow vole with the masked shrew, pygmy shrew, water shrew, and meadow jumping mouse in peatland dominated by sedge-willow and leatherleaf-bogbirch; in Ingham County, Michigan State University mammalogy students found the meadow vole along with the masked shrew, short-tailed shrew, white-footed mouse, deer mouse, meadow jumping mouse, and house mouse in grass-herb, old field habitat; in Clinton County, Linduska (1942, 1950) obtained the meadow vole, deer mouse and house mouse in croplands, and these species, plus the short-tailed shrew and meadow jumping mouse, were most numerous in moist places in the bluegrass community.

Several ecologists have studied the interaction between meadow voles and other small mammalian competitors in their common environments (Grant, 1971). This competition can be for food resources and/or living space. There is evidence that the grass-eating vole may restrict the activities of granivorous deer mice and white-footed mice (Grant, 1971; M'Closkey, 1975). In southwestern Ontario, for example, Morris (1979) failed to catch white-footed mice and meadow voles in the same trapping stations. There seems to be diet overlap between meadow voles and forest deer mice, plus some habitat separation, as the deer mouse is capable of using vertical (arboreal) habitat and the meadow vole is restricted to the horizontal (terrestrial). However, white-footed mice are no more arboreal when meadow voles are present (Newton *et al.,* 1980). There may be some diet overlap between sympatric meadow voles and prairie deer mice but less habitat separation because both occur in the horizontal (terrestrial) habitat. However, the meadow vole normally occurs in dense, successionally more mature, grassy habitat than does the prairie deer mouse, which thrives best in grain fields and in mixed annual weeds and grasses characteristic of the earliest successional stages in Michigan old field growth (Linduska, 1950). In this latter type of habitat in southern Michigan, deer mice outnumber meadow voles as much as 12 to one (Brewer and Reed, 1977).

Meadow voles and short-tailed shrews are commonly associated in open grassy areas. They use the same surface runways to move about in dense grass (Osterburg, 1962), with the voles apparently attempting to avoid the shrews (Fulk, 1972). Short-tailed shrews have been reported to prey on meadow voles, especially in winter (Eadie, 1952), although Getz (1961a) found little evidence in a Washtenaw County study of any serious predation by the shrews.

In northern Michigan, meadow voles and southern red-backed voles may share a common food supply in forest-grassland edges but their habitats overlap little because of the former's preference for grassland and the latter's preference for woodland (Iverson and Turner, 1972). Apparently, these two species share this edge compatibly although interspecific aggression among breeding males can occur (Grant, 1969). Forest-dwelling woodland voles, although little observed in Michigan, can be found in association with meadow voles in orchards; when confined together, the smaller woodland vole is the more aggressive (Novak and Getz, 1969).

Meadow voles also associate with other grass-eating microtines. In Michigan, these include the prairie vole, which is found in only the extreme southwestern part, and the southern bog lemming, which has a widespread but spotty distribution. Interactions between prairie voles and meadow voles have been studied in Indiana by Krebs (1977) and associates and by Miller (1969) and Zimmerman (1965). Although prairie voles prefer drier areas than do meadow voles, there is some evidence of co-existence by the two species, despite the general indication that grassland habitats are usually inhabited by only one dominant grass-eating small rodent but normally can support several species of seed-eaters (Baker, 1971). In meadow vole and southern bog lemming associations, Getz (1961b) and Linzey (1981) found the meadow vole most attracted to pure strands of grasses whereas the southern bog lemming seems at home in a mixture of woody plants as long as some grass cover is also present. As mentioned earlier, this author has a strong suspicion that the habitat preferred by the southern bog lemming has deteriorated considerably in Michigan, while the preferred meadow vole habitat has improved because this vole seems more adaptable to habitat changes resulting from human occupation.

REPRODUCTIVE ACTIVITIES. The meadow vole is often used (along with the cod fish and the house fly) as a classic example of fecundity. Under laboratory conditions, it is possible for a single female vole to produce as many as 17 litters (totaling as many as 100 offspring) in 12 months. In the field, four litters are normal, with more (perhaps as many as 10) in years of high populations (Bailey, 1924; Hamilton, 1941). Golley (1961), examining an annual cycle in Ingham County, found meadow vole density in old field habitat low in May; it then increased rapidly with the recruitment of young through the summer, autumn, and early winter. Although (in his study) breeding ceased in January, the population continued to rise locally in February as a result of immigration before a general decline in late winter and early spring. Golley found mortality by age was highest in postnestling juveniles and young adults; seasonally, it was highest in summer and lowest in winter. Although Golley did not note any breeding in February and March, there is abundant evidence that meadow voles can produce young in every month (Keller and Krebs, 1970), with births occurring under field-shocked corn (Linduska, 1942) and under heavy snow cover (Beer and MacLeod, 1961). However, because multiannual fluctuations reach population peaks every three to four years (Elton, 1942), records of meadow vole productivity from one year to the next may be highly variable, even in the same habitat (Blair, 1948).

Meadow voles, in marked contrast to prairie voles (Getz *et al.*, 1981), are highly promiscuous in their breeding activity. A female in estrous will accept several males as mates (Madison, 1980). Males may not reach sexual maturity until 45 days of age (Bailey, 1924; Hamilton, 1937) or may not even reach breeding condition their first year if born after mid-June (Barbehenn, 1955). However, females can mate when 25 days old and when their weight is no more than 19 to 24 g (Linduska, 1942; Timm, 1975). The pregnant female's body weight begins to increase markedly from the eighth day of gestation (Madison, 1978b). The gestation period is 20 to 21 days. Following birth of an average of six young (as few as one; as many as 11), the mother undergoes post-partum estrous immediately and again readily accepts mates, meaning she must have her present litter weaned and on their way in less than 20 days.

At birth, the meadow vole weighs 3 g (Whitmoyer, 1956); young born earlier in the season may be heavier than those born later (Kott and Robinson, 1963). The newly-born have closed eyes, folded ear pinnae, and are toothless, hairless, and cling to their mother's teats much of the time, permitting themselves to be dragged about, even to a new nest site. Growth is rapid; weight gain is as much as 1 g per day for the first few days after birth. On the fifth day after birth, the incisor teeth erupt; on the sixth day, the body is fully clothed in velvety hair; on the seventh day the molar (cheek) teeth erupt; on or about the eighth day the eyes and ears open and the young begin to move about; On or about the 12th day, the young, with body weights averaging 14 g, are weaned, although individuals weighing no more than 10 g have been live-trapped in Michigan (Golley 1960b). The mother then may have a short respite prior to the birth of her next brood.

Life expectancy in the wild for an individual vole is less than one year, although individuals may live for as long as 65 weeks for a male and 86 weeks for a female (Rose and Dueser, 1980). In captivity, the animals can survive for several years. Meadow vole age can be determined by the weight of the eye lens (Thomas and Bellis, 1980).

FOOD HABITS. Meadow voles primarily eat herbaceous vegetation, the leaves and other growing parts of grasses, sedges, forbs, and other plants. They also will eat seeds, fruits, insects, snails and other invertebrates, and various small vertebrates including, on occasion, each other. As a grazer, meadow voles must eat large amounts of bulky green vegetation (up to 60% of their own weight per day, Jackson, 1961) to obtain needed energy and nutrients. On the other hand, deer mice and other seed-eaters ingest concentrated and easily digested foods. However, seed-eaters may have to spend considerable time hunting for their foods, while forage for the grazing meadow voles includes grass which usually conveniently borders their surface runways. The efficiency of the meadow vole digestive tract in reducing and assimilating complicated carbohydrates and cellulose in plant vegetation results from the large caecum in which large quantities of the green forage can be stored and acted on (cecal fermentation) by intestinal flora and fauna (Carleton, 1981; Golley, 1960a). Reingestion of feces (coprophagy) to salvage digested foods is practiced (Ouellette and Heisinger, 1980).

In spring, meadow voles thrive on the new green shoots of grasses and sedges; in summer and autumn they clip basal parts of plants (leaving small piles of cuttings along runways) to obtain the succulent tips and fruiting heads; and in winter and early spring they eat the basal green parts of grasses and forbs preserved under snow cover. Although meadow voles' stomachs and intestines contain finely chewed materials, microscopic examination can distinguish grasses, herbs, roots, bark and other woody plant parts, seeds, fruits, mosses, fungi, insect parts, and tissues from other animals (Golley, 1960b). In Golley's study (conducted in Ingham County), he found grasses (with herbs less prominent) dominant in food samples from vole stomachs at every season. Fruits appeared most often in the spring diet, while some evidence of such foods as wood, seeds, and mosses was primarily noted in autumn. Fungi were principally summer foods. In a food habits study conducted in Indiana, Zimmerman (1965) identified bluegrass (*Poa compressa*) as the major food with lesser amounts of other grasses of the genera *Panicum* and *Muhlenbergia*. The herb most commonly eaten was lance-leaved plantation (*Plantago lanceolata*). Zimmerman also found that voles ate the dandelion (*Taraxacum officinale*); the basal parts of this plant are relished especially in winter by southern Michigan meadow voles. Recent work in Illinois has shown that the meadow vole thrives better on a diet of grasses than does the prairie vole (F. R. Cole and L. Verner, pers. comm.). Caterpillars of moths and butterflies and various species of beetles are the main insect foods. Eggs of ground-nesting birds may also be eaten (Maxson and Oring, 1978). Winter food caches of basal parts and root stocks of plants belonging to the buckwheat, pulse, morning glory, legume and composite families of plants are reported (Connor, 1971; Gates and Gates, 1980). Cultivated grains may also be stored.

One reason the meadow vole has done so well in the formerly-forested and presently grassy open lands in Michigan is its ability to thrive on introduced grasses and forbs. In a food-preference study, Thompson (1965) found that captive meadow voles preferred eating forage crops and non-native, introduced plants to native grasses. Hence, the meadow vole survives easily where the human population has disturbed the natural environment and introduced plants into open lands which were previously woodlands.

ENEMIES. The prolific meadow vole, being active day and night, is a source of food for both diurnal and nocturnal predators. It also is a source of energy in the ecosystem of Michigan's grassy community. In his investigation in Ingham County, Golley (1960b) traced the food chain energy flow from the producer (vegetation) through the primary consumer (meadow vole) to the secondary consumer (least weasel). The long list of mammalian meadow vole predators also includes the Virginia opossum, short-tailed shrew, eastern chipmunk, gray fox, red fox, coyote, gray wolf, black bear, raccoon, short-tailed weasel, long-tailed weasel, marten, mink, striped skunk, badger, bobcat, and house cat (Banfield, 1974; Christian, 1975; Dearborn, 1932; Eadie, 1952; Hatt, 1930; Koehler and Hornocker, 1977; Linduska, 1950; Murie, 1936; Progulske, 1955; Schofield, 1960; Toner, 1956; Torres, 1937; Voight *et al.,* 1976). In his study of the food habits of Michigan fur-bearing mammals, Dearborn (1932) listed the meadow mouse (without specific designation) as included in the diets of the long-tailed weasel, mink, bobcat, and badger. Snakes find meadow voles easily obtained along runways or in their nests, while large predatory fish and snapping turtles sometimes catch swimming voles (Banfield, 1974; Hatt, 1930; Madison, 1978a).

Hawks which prey on meadow voles include the harrier, red-shouldered hawk, red-tailed hawk, American kestrel, American rough-legged hawk, Cooper's hawk, and osprey (Craighead and Craighead, 1956; de la Pierriere, 1970; Fitch and Bare, 1978; Hamerstrom, 1979; Hatt, 1930; Linduska, 1950; Proctor, 1977). Owls are among the most efficient meadow vole predators (see Fig. 53); some of the species involved are the great horned owl, barn owl, screech owl, short-eared owl, snowy owl, great gray owl, and long-eared owl (Allan, 1977; Armstrong, 1958; Chamberlin, 1980; Craighead and Craighead, 1956; Dexter, 1978; Geis, 1952; Master, 1979; Short and Drew, 1962; Spiker, 1933; VanCamp and Henny, 1975; Wallace, 1948; Warthin and Van Tyne, 1922; Wilson, 1938). Ravens, crows, jays, shrikes, herons, cranes, and

Figure 53. Skulls of the meadow vole (*Microtus pennsylvanicus*) extracted from the pellets of the barn owl (*Tyto alba*) from a roost on the campus of Michigan State University.
Photo courtesy of Michigan Agricultural Experiment Station (Wallace, 1950).

gulls also eat meadow voles (Banfield, 1974; Blokpoel and Haymes, 1979; Getz, 1970; Harlow *et al.*, 1975; Hatt, 1930). Automobiles are also responsible for an occasional meadow vole death (Manville, 1949b).

PARASITES AND DISEASES. Meadow voles are infested with both internal and external parasites. Most conspicuous are tapeworms (cestodes) found in the digestive tract, but parasitic protozoans, round worms (nematodes), flat worms (trematodes), and spiny-headed worms (acanthocephalans) also live in meadow voles (Doran, 1954; Hatt, 1930; Jackson, 1961). The blood fluke, *Schistosomatium douthitti,* is known to live in Michigan meadow voles (Price, 1931; Zajac and Williams, 1980, 1981). Scharf and Stewart (1980) obtained fleas from meadow voles in Benzie and Grand Traverse counties. Manville (1949) collected mites and fleas from meadow voles in Marquette County. Botfly larvae (*Cuterebra* sp.) also parasitize these rodents (Hatt, 1930; Mauer and Skaley, 1968).

MOLT AND COLOR ABERRATIONS. When first weaned, young meadow voles have short, dark pelage. This hair is lost when subadult and ultimately adult hair appears. For those long-lived individuals (who survive for as long as 12 months), there are two molts followed by new hair growth, one in spring, producing the normally bright chestnut-brown dorsal coloration and one in autumn, producing a duller, more gray dorsal appearance.

Various color phases have been reported in the literature (Bloom, 1942; Heidt and Bowers, 1968; Owen and Shackelford, 1942; Stalling, 1974) including white (albinistic), dark (melanistic), various shades of yellow and brown, and spotting. In the Michigan State University Museum collections there are study specimen of meadow voles with dorsal pelage light yellow-brown (Ingham County), light yellow (Saginaw and Wayne counties), and dusky white (Ingham County). Individuals of normal color except for spots of white on the dorsum have been kept for several years as a breeding colony for bioassay studies at Michigan State University (Elliott, 1963).

ECONOMIC IMPORTANCE. The meadow vole can be a major pest because of its attraction to cultivated areas in the vicinity of human habitations and its high reproductive potential (Allen, 1940; Hatt, 1930). In the early part of this century, Conger (1919) noted some of the problems which meadow voles can cause Michigan agriculturists and pointed out that many of the vole's predators were on the decline and therefore did not provide a substantial degree of control.

Grain fields, hay meadows, alfalfa plantings and formal lawns can be seriously damaged by this vole. Hine (1950) estimated that 11 tons (10 M.T.) of grass or 5.5 tons (5 M.T.) of hay can be eaten annually by voles having densities of 10 animals per acre (25 per ha). Linduska (1942, 1950) noted damage which meadow voles can do to field-shocked grains. Fertilized lawns and garden plots attract meadow voles, especially in suburbia where such developments border vacant grasslands. Scores of these mice, usually under snow cover in winter, invade grasses, ornamentals, bulbs, and other garden plants.

Meadow voles in Michigan seriously damage woody ornamental shrubs, fruit trees in orchards, and young trees being propagated in forest nurseries (Hayne, 1950b). Meadow voles find the bark and roots of many shrubs and trees an important source of nutrients, especially in late winter and early spring and their gnawing almost invariably girdles and destroys these plants. Trapping, poison baits, and wrapping tree and shrub bases with tar paper or hardware cloth diminish some of this destruction (Byers, 1976; Hayne, 1951; Libby and Abrams, 1966; and Shick, 1965). In Michigan, advice for controlling meadow voles can be obtained from the county offices of the Michigan Cooperative Extension Service or from the Wildlife Extension Specialist, Department of Fisheries and Wildlife at Michigan State University.

Despite its destructive activities, the meadow vole holds an important position in Michigan's grassland ecosystems as a provider of animal protein for almost all meat-eating creatures, from the shrew to the bear (Golley, 1960b). If the meadow vole was not so abundantly available, carnivores might prey on a larger number of the more visible game animals and song birds. Meadow voles are also important in the laboratory in medical research and as a bioassay test organism for individual forage plants (Elliott, 1963; Poiley, 1949; Reich, 1981). There is also evidence that the meadow vole is an indicator of the concentration of lead and other heavy metals along highways, since these elements from automobile discharges are picked up by plants and subsequently ingested by voles (Blair *et al.*, 1977; Getz *et al.*, 1977).

COUNTY RECORDS FROM MICHIGAN. Records of specimens of the meadow vole in collections of museums and from the literature show that this species lives in all counties in Michigan. The species also occurs on various islands in the Great Lakes including: in Lake Michigan, Summer Island (Nellis, 1970), Marion Island in West Arm of Grand Traverse Bay (Hatt *et al.*, 1948); in Lake Superior, Huron Islands (Corin, 1976); in Lake Huron, Drummond Island (Manville, 1950).

LITERATURE CITED

Allan, T. A.
1977 Winter food of the snowy owl in northwestern lower Michigan. Jack-Pine Warbler, 55(1):42.

Allen, D. L.
1938 Ecological studies on the vertebrate fauna of a 500-acre farm in Kalamazoo County, Michigan. Ecol. Monog., 8(3):347–436.
1940 Millions of little teeth. Michigan Conserv., Sept., pp. 1–6.

Ambrose, H. W., III
1973 An experimental study of some factors affecting the spatial and temporal activity of *Microtus pennsylvanicus*. Jour. Mammalogy, 54(1):79–110.

Armstrong, W. L.
1958 Nesting and food habits of the long-eared owl in Michigan. Michigan State Univ., Publ. Mus., Biol. Ser., 1(2):61–96.

Aumann, G. D.
1965 Microtine abundance and soil sodium levels. Jour. Mammalogy, 46(4):594–604.

Bailey, V.
1924 Breeding, feeding, and other life habits of meadow mice (*Microtus*). Jour. Agric. Res., 27(8):523–538.

Baker, R. H.
1971 Nutritional strategies of myomorph rodents in North American grasslands. Jour. Mammalogy, 52(4):800–805.

Banfield, A. W. F.
1974 The mammals of Canada. Univ. Toronto Press, Toronto. xxiv+438 pp.

Barbehenn, K. R.
1955 A field study of growth in *Microtus pennsylvanicus*. Jour. Mammalogy, 36(4):533–543.

Beckwith, S. L.
1954 Ecological succession on abandoned farm lands and its relationship to wildlife management. Ecol. Monog., 24:349–376.

Beer, J. R., and C. F. MacLeod
1961 Seasonal reproduction in the meadow vole. Jour. Mammalogy, 42(4):483–489.

Blair, C. W., A. L. Hiller, and P. F. Scanlon
1977 Heavy metal concentrations in mammals associated with highways of different traffic densities. Virginia Jour. Sci., 28(2):61.

Blair, W. F.
1939 A swimming and diving meadow vole. Jour. Mammalogy, 20(3):375.
1940a Notes on home ranges and populations of the short-tailed shrew. Ecology, 21(2):284–288.
1940b Home ranges and populations of the meadow vole in southern Michigan. Jour. Wildlife Mgt., 4(2):149–161.
1948 Population density, life span, and mortality rates of small mammals in the blue-grass meadow and blue-grass field associations of southern Michigan. American Midl. Nat., 40(2):395–419.

Blokpoel, H., and G. T. Haymes
1979 Small mammals and birds as food items of ring-billed gulls on the lower Great Lakes. Wilson Bull., 91(4):623–625.

Bloom, P. M.
1942 Total melanism in *Microtus* from Michigan. Jour. Mammalogy, 23(2):214.

Brewer, R., and M. L. Reed
1977 Vertebrate inventory of wet meadows in Kalamazoo and Van Buren counties. Michigan Dept. Nat. Res., Rept., (mimeo), 20 pp.

Burt, W. H.
1943 Territoriality and home range concepts as applied to mammals. Jour. Mammalogy, 24(3):346–352.

Byers, R. E.
1976 New compound (RH 787) for use in control of orchard voles. Jour. Wildlife Mgt., 40(1):169–171.

Carleton, M. D.
1981 A survey of gross stomach morphology in Microtinae (Rodentia: Muroidea). Zeitschrift für Saügetierkunde, 46(2):93–108.

Chamberlin, M. L.
1980 Winter hunting behavior of a snowy owl in Michigan. Wilson Bull., 92(1):116–120.

Christian, D. P.
1975 Vulnerability of meadow voles, *Microtus pennsylvanicus*, to predation by domestic cats. American Midl. Nat., 93(2):498–502.

Christian, J. J.
1971 Fighting, maturity, and population density in *Microtus pennsylvanicus*. Jour. Mammalogy, 52(3):556–567.

Conger, A. C.
1919 Rodent control—the field mouse. Michigan Agric. College, Agric. Expt. Sta., Quart. Rept., 2(1):50–52.

Connor, P. F.
1971 The mammals of Long Island, New York. New York State Mus. and Sci. Serv., Bull. 416, 78 pp.

Corin, C. W.
1976 The land vertebrates of the Huron Islands, Lake Superior. Jack-Pine Warbler, 54(4):138–147.

Craighead, J. J., and F. C. Craighead, Jr.
1956 Hawks, owls and wildlife. Stackpole Co., Harrisburg, Pa. xix+443 pp.

Dearborn, N.
1932 Foods of some predatory fur-bearing animals in Michigan. Univ. Michigan, School Forestry and Conserv., Bull. No. 1, 52 pp.

de la Pierriere, C. I.
1970 Food habits of red-tailed hawks, Summer Island, Michigan, 1969. Summer Sci. Jour., 2(2):72–73.

Dexter, R. W.
1978 Mammals utilized as food by owls in reference to the local fauna of northeastern Ohio. Kirtlandia, 24:1–8.

Dice, L. R.
1922 Some factors affecting the distribution of the prairie vole, forest deer mouse, and prairie deer mouse. Ecology, 3:29–47.
1925 A survey of the mammals of Charlevoix County, Michigan, and vicinity. Univ. Michigan, Mus. Zool., Occas. Papers No. 159, 33 pp.

Dice, L. R., and H. B. Sherman
1922 Notes on the mammals of Gogebic and Ontonagon counties, Michigan. Univ. Michigan, Mus. Zool., Occas. Papers No. 109, 40 pp.

Doran, D. J.
1954 A catalogue of the protozoa and helminiths of North American rodents. I. Protozoa and Acanthocephala. American Midl. Nat., 52:118–128.

Eadie, W. R.
1952 Shrew predation and vole populations. Jour. Mammalogy, 33:185–189.

Elliott, F. C.
1963 The meadow vole (*Microtus pennsylvanicus*) as a bioassay test organism for individual forage plants. Michigan State Univ., Agric. Exp. Sta., Quart. Bull., 46(1):58–72.

Elton, C. S.
1942 Voles, mice and lemmings. Clarendon Press, Oxford. 496 pp.

Fitch, H. S., and R. O. Bare
1978 A field study of the red-tailed hawk in eastern Kansas. Trans. Kansas Acad. Sci., 81(1):1–13.

Fitch, J. H.
1979 Patterns of habitat selection and occurrence in the deermouse *Peromyscus maniculatus gracilis*. Michigan State Univ., Publ. Mus., Biol. Ser., 5(6):443–484.

Fulk, G. W.
1972 The effect of shrews on the space utilization of voles. Jour. Mammalogy, 53(3):461–478.

Gates, J. E., and D. M. Gates
1980 A winter food cache of *Microtus pennsylvanicus*. American Midl. Nat., 103(2):407–408.

Geis, A. D.
1952 Winter food habits of a pair of long-eared owls. Jack-Pine Warbler, 30(3):93.

Getz, L. L.
1960 A population study of the vole, *Microtus pennsylvanicus*. American Midl. Nat., 64:392–405.
1961a Home ranges, territoriality, and movement of the meadow vole. Jour. Mammalogy, 42(1):24–36.
1961b Factors influencing the local distribution of *Microtus* and *Synaptomys* in southern Michigan. Ecology, 42(1): 110–118.
1961c Responses of small mammals to live-traps and weather conditions. American Midl. Nat., 66(1):160–170.
1962 Aggressive behavior of the meadow and prairie voles. Jour. Mammalogy, 43(3):351–358.
1970 Influence of vegetation on local distribution of the meadow vole in southern Wisconsin. Univ. Connecticut, Occas. Papers, Biol. Sci., 1(4):213–241.
1972 Social structure and aggressive behavior in a population of *Microtus pennsylvanicus*. Jour. Mammalogy, 53(2):310–317.

Getz, L. L., C. S. Carter, and L. Gavish
1981 The mating system of the prairie vole, *Microtus ochrogaster:* Field and laboratory evidence of pair-bonding. Behav. Ecol. Sociobiol., 8(3):189–194.

Getz, L. L., F. R. Cole, and D. L. Gates
1978 Interstate roadsides as dispersal routes for *Microtus pennsylvanicus*. Jour. Mammalogy, 59(1):208–212.

Getz, L. L., L. Verner, and M. Prather
1977 Lead concentrations of small mammals living near highways. Enviro. Pollution, 13:151–157.

Golley, F. B.
1960a Anatomy of the digestive tract of *Microtus*. Jour. Mammalogy, 41(1):89–99.
1960b Energy dynamics of a food chain of an old-field community. Ecol. Monog., 30(2):187–206.
1961 Interaction of natality, mortality and movement during one annual cycle in a *Microtus* population. American Midl. Nat., 66:152–159.

Grant, P. R.
1969 Experimental studies of competitive interaction in a two-species system. I. *Microtus* and *Clethrionomys* species in enclosures. Canadian Jour. Zool., 47(5): 1059–1082.
1971 Experimental studies of competitive interaction in a two-species system. III. *Microtus* and *Peromyscus* species in enclosures. Jour. Animal Ecol., 40:323–350.

Green, M. M.
1925 Notes on some mammals of Montmorency County, Michigan. Jour. Mammalogy, 6(3):173–178.

Greffenius, R. J.
1939 A method for determining the relative abundance of *Microtus pennsylvanicus*. Jour. Wildlife Mgt., 3(3):199–200.

Hamerstrom, F.
1979 Effect of prey on predator: Voles and harriers. Auk, 96(2):370–374.

Hamilton, W. J., Jr.
1937a The biology of microtine cycles. Jour. Agric. Res., 54(10):779–790.
1937b Growth and life span of the field mouse. American Nat., 71(736):500–507.
1937c Activity and home range of the field mouse, *Microtus pennsylvanicus pennsylvanicus* (Ord). Ecology, 18:255–263.

1941 Reproduction of the field mouse *Microtus pennsylvanicus* Ord. Cornell Univ., Agric. Exp. Sta., Mem. 237, 23 pp.

Harlow, R. F., R. G. Hooper, D. R. Chamberlain, and H. S. Crawford
1975 Some winter and nesting season foods of the common raven in Virginia. Auk, 92:298–306.

Hatt, R. T.
1930 The biology of the voles of New York. Roosevelt Wildlife Bull. 5(4):513–623.

Hatt, R. T., J. Van Tyne, L. C. Stuart, C. H. Pope, and A. B. Grobman
1948 Island Life: A study of the land vertebrates of the islands of eastern Lake Michigan. Cranbrook Inst. Sci., Bull. No. 27, xi+179 pp.

Haveman, J. R.
1973 A study of population densities, habitats and foods of four sympatric species of shrews. Northern Michigan Univ., unpubl. M.S. thesis, vii+70 pp.

Hayne, D. W.
1950a Apparent home range of *Microtus* in relation to distance between traps. Jour. Mammalogy, 31(1):26–39.
1950b Mouse populations in orchards and a new method of control. Michigan State Col., Agric. Exp. Sta., Quart. Bull., 33(2):160–168.
1951 Zinc phosphide: Its toxicity to pheasants and effect of weathering upon its toxicity to mice. Michigan State Col., Agric. Exp. Sta., Quart. Bull., 33(4):412–425.

Heidt, G. A., and J. R. Bowers
1968 An albino meadow vole in Michigan. Jack-Pine Warbler, 46(1):33.

Hine, R.
1950 The field mouse: Characteristics and control. Wisconsin Conserv. Bull., 15(11):16–19.

Holman, J. A.
1959 New fossil vertebrate remains from Michigan. Michigan Acad., 11(4):391–397.

Iverson, S. L., and B. N. Turner
1972 Winter coexistence of *Clethrionomys gapperi* and *Microtus pennsylvanicus* in a grassland habitat. American Midl. Nat., 88(2):440–445.

Jackson, H. H. T.
1961 Mammals of Wisconsin. Univ. Wisconsin Pess, Madison. xii+504 pp.

Keller, B. L., and C. J. Krebs
1970 *Microtus* population biology. III. Reproductive changes in fluctuating populations of *M. ochrogaster* and *M. pennsylvanicus* in southern Indiana, 1965–1967. Ecol. Monog., 40:263–294.

Koehler, G. M., and M. G. Hornocker
1977 Fire effects on marten habitat in the Selway-Bitterroot Wilderness. Jour. Wildlife Mgt., 41(3):500–505.

Kott, E., and W. L. Robinson
1963 Seasonal variation in litter size of the meadow vole in southern Ontario. Jour. Mammalogy, 44(4):467–470.

Krebs, C. J.
1977 Competition between *Microtus pennsylvanicus* and *Microtus ochrogaster*. American Midl. Nat., 97(1):42–49.

Krebs, C. J., and J. H. Myers
1974 Population cycles in small mammals. Adv. Ecol. Res., 8:267–399.

Libby, J. L., and J. I. Abrams
1966 Anticoagulant rodenticide in paper tubes for control of meadow mice. Jour. Wildlife Mgt., 39(3):512–518.

Linduska, J. P.
1942 Winter rodent populations in field-shocked corn. Jour. Wildlife Mgt., 6(4):353–363.
1950 Ecology and land-use relationships of small mammals on a Michigan farm. Michigan Dept. Conserv., Game Div., ix+144 pp.

Linzey, A. V.
1981 Patterns of coexistence in *Microtus pennsylvanicus* and *Synaptomys cooperi*. Virginia Poly. Inst. and State Univ., unpubl. Ph.D. dissertation, 97 pp.

LoBue, J., and R. M. Darnell
1959 Effect of habitat disturbance on a small mammal population. Jour. Mammalogy, 40(3):425–437.

Low, J. B.
1944 Meadow mice use wren's nests. Jour. Mammalogy, 25(3):308.

Madison, D. M.
1978a Behavioral and sociochemical susceptibility of meadow voles (*Microtus pennsylvanicus*) to snake predators. American Midl. Nat., 100(1):23–28.
1978b Movement indicators of reproductive events among female meadow voles as revealed by radiotelemetry. Jour. Mammalogy, 59(4):835–843.
1980 Space use and social structure in meadow voles, *Microtus pennsylvanicus*. Beh. Ecol. Sociobiol., 7:65–71.

Manville, R. H.
1948 The vertebrate fauna of the Huron Mountains, Michigan. American Midl. Nat., 39(3):615–640.
1949a A study of small mammal populations in northern Michigan. Univ. Michigan, Mus. Zool., Misc. Publ. No. 73, 83 pp.
1949b Highway mortality in northern Michigan. Jour. Mammalogy, 39(3):311–312.
1950 The mammals of Drummond Island, Michigan. Jour. Mammalogy, 31(3):358–359.

Master, L.
1979 Some observations on great gray owls and their prey in Michigan. Jack-Pine Warbler, 57(4):215–217.

Maurer, F. W., Jr., and J. E. Skaley
1968 Cuterebrid infestation of *Microtus* in eastern North Dakota, Pennsylvania, and New York. Jour. Mammalogy, 49(4):773–774.

Maxson, S. J., and L. W. Oring
1978 Mice as a source of egg loss among ground-nesting birds. Auk, 95(3):582–584.

M'Closkey, R. T.
1975 Habitat succession and rodent distribution. Jour. Mammalogy, 56(4):950–955.

Miller, W. C.
1969 Ecological and ethological isolating mechanisms between *Microtus pennsylvanicus* and *Microtus ochrogaster* at Terre Haute, Indiana. American Midl. Nat., 82: 140–148.

Morris, D. W.
1979 Microhabitat utilization and species distribution of sympatric small mammals in southwestern Ontario. American Midl. Nat., 101(2):373–384.

Murie, A.
1936 Following fox trails. Univ. Michigan, Mus. Zool., Misc. Publ. No. 32, 45 pp.

Nellis, C. H.
1970 Mammals of Summer Island, Michigan. Summer Sci. Jour., 2(2):66–67.

Newton, S. L., T. D. Nudds, and J. S. Millar
1980 Importance of arboreality in *Peromyscus leucopus* and *Microtus pennsylvanicus* interactions. Canadian Field-Nat., 94(2):167–170.

Novak, M. A., and L. L. Getz
1969 Aggressive behavior of meadow voles and pine voles. Jour. Mammalogy, 50(3):637–639.

Osterburg, D. M.
1962 Activity of small mammals as recorded by a photographic device. Jour. Mammalogy, 43(2):219–229.

Ouellette, D. E., and J. F. Heisinger
1980 Reingestion of feces by *Microtus pennsylvanicus*. Jour. Mammalogy, 61(2):366–368.

Owen, R. D., and R. M. Shackelford
1942 Color aberrations in *Microtus* and *Pitymys*. Jour. Mammalogy, 23(3):306–314.

Ozoga, J. J., and L. J. Verme
1968 Small mammals of conifer swamp deeryards in northern Michigan. Michigan Acad. Sci., Arts and Ltrs., Papers, 53:37–49.

Poiley, S. M.
1949 Raising captive meadow voles (*Microtus p. pennsylvanicus*). Jour. Mammalogy, 39(3):317–318.

Price, H. F.
1931 Life history of Schistosomatium douthitti. American Jour. Hyg., 13:685–727.

Proctor, N. S.
1977 Osprey catches vole. Wilson Bull., 89(4):625.

Progulske, D. R.
1955 Game animals utilized as food by the bobcat in the southern Applachians. Jour. Wildlife Mgt., 19(2):249–253.

Pruitt, W. C., Jr.
1957 Observations on the bioclimate of some taiga mammals. Arctic, 10:130–138.
1960 Animals in the snow. Sci. American, 202:60–68.

Rabe, M. L.
1981 New locations for pygmy (*Sorex hoyi*) and water (*Sorex palustris*) shrews in Michigan. Jack-Pine Warbler, 59(1):16–17.

Reich, L. M.
1981 *Microtus pennsylvanicas*. American Soc. Mammalogists, Mammalian Species, No. 159, pp. 1–8.

Robinson, W. L., and J. B. Falls
1965 A study of homing in meadow mice. American Midl. Nat., 73(1):188–224.

Robinson, W. L., and J. K. Werner
1975 Vertebrate animal populations of the McCormick Forest. U.S.D.A. Forest Serv., Res. Paper NC-118, 25 pp.

Rose, R. K., and R. D. Dueser
1980 Lifespan of Virginia meadow voles. Jour. Mammalogy, 61(4):760–763.

Rose, R. K., and W. D. Heuston
1978 Wound healing in meadow voles. Jour. Mammalogy, 59(1):186–188.

Rosenzweig, M. L., and Z. Abramsky
1980 Microtine cycles: the role of habitat heterogeneity. Oilos, 34:141–146.

Scharf, W. C., and K. R. Stewart
1980 New records of Siphonaptera from northern Michigan. Great Lakes Ento., 13(3):165–167.

Schofield, R. D.
1960 A thousand miles of fox trails in Michigan's ruffed grouse range. Jour. Wildlife Mgt., 24(4):432–434.

Shick, C.
1965 Controlling meadow mice in residential areas, parks, orchards, forests and Christmas tree plantations. Michigan State Univ., Coop. Ext. Serv., Ext. Bull. E-430, 4 pp.

Short, H. L., and L. C. Drew
1962 Observations concerning behavior, feeding, and pellets of short-eared owls. American Midl. Nat., 67(2): 424–433.

Smith, J. C.
1979 Factors affecting the transmission of rodent ultrasounds in natural environments. American Zool., 19(2):432–442.

Spencer, D. A.
n.d. Biological and control aspects. Pp. 15–25 *in* J. R. Beck, S. B. Osgood, M. D. Smith, eds. The Oregon meadow mouse irruption of 1957–1958. Oregon State Coll., Fed. Coop. Ext. Serv., 88 pp.

Spiker, C. J.
1933 Analysis of two hundred long-eared owl pellets. Wilson Bull., 45(4):198.

Stalling, D. T.
1974 A xanthochromic prairie vole and notes on associated literature. Southwest. Nat., 19(1):115–117.

Thomas, R. E., and E. D. Bellis
1980 An eye-lens weight curve for determining age in *Microtus pennsylvanicus*. Jour. Mammalogy, 61(3):561–563.

Thompson, D. Q.
1965 Food preferences of the meadow vole (*Microtus pennsylvanicus*) in relation to habitat affinities. American Midl. Nat., 74(1):76–86.

Timm, R. M.
1975 Distribution, natural history, and parasites of mammals of Cook County, Minnesota. Univ. Minnesota, Bell Mus. Nat. Hist., Occas. Papers No. 14, 56 pp.

Toner, G. C.
1956 House cat predation on small animals. Jour. Mammalogy, 37(1):119.

Torres, J. K.
1937 A chipmunk captures a mouse. Jour. Mammalogy, 18(1):100.

Turner, B. N., and S. L. Iverson
1973 The annual cycle of aggression in male *Microtus pennsylvanicus*, and its relation to population parameters. Ecol., 54(5):967–981.

VanCamp, L. F., and C. J. Henny
1975 The screech owl: Its life history and population ecology in northern Ohio. U.S.D.A. Fish and Wildlife Serv., North American Fauna No. 71, iv+65 pp.

Verme, L. J., and J. J. Ozoga
1981 Changes in small mammal populations following clear-cutting in upper Michigan conifer swamps. Canadian Field-Nat., 95(3):253–256.

Voight, D. R., G. B. Kolenosky, and D. H. Pimlott
1976 Changes in summer foods of wolves in central Ontario. Jour. Wildlife Mgt., 40(4):663–668.

Wallace, G. J.
1948 The barn owl in Michigan. Its distribution, natural history and food habits. Michigan State Coll., Agric. Exp. Sta., Techn. Bull. 208, 61 pp.
1950 In support of the barn owl. Michigan Agri. Exp. Sta., Quart. Bull., 33(2):96–105.

Warthin, A. S., Jr., and J. Van Tyne
1922 The food of long-eared owls. Auk, 39(3):417.

Webster, A. B., and R. J. Brooks
1981 Social behavior of *Microtus pennsylvanicus* in relation to seasonal changes in demography. Jour. Mammalogy, 62(4):738–751.

Weilert, N. G., and K. A. Shump, Jr.
1977 Physical parameters of *Microtus* nest construction. Trans. Kansas Acad. Sci., 79(3-4):161–164.

Whitmoyer, T. F.
1956 A laboratory study of the growth rate in young *Microtus pennsylvanicus*. Michigan State Univ., unpubl. M.S. thesis, 62 pp.

Wilson, K. A.
1938 Owl studies at Ann Arbor, Michigan, Auk., 55(2):187–197.

Wood, N. A.
1922 The mammals of Washtenaw County, Michigan. Univ. Michigan, Mus. Zool., Occas. Papers No. 123, 23 pp.

Zajac, A. M., and J. F. Williams
1980 Infection with *Schistosomatium douthitii* (Fam. Schistosomatidae) in the meadow vole (*Microtus pennsylvanicus*) in Michigan. Jour. Parasitol., 66(2):366–367.
1981 The pathology of infection with *Schistosomatium douthitii* in the laboratory mouse and the meadow vole, *Microtus pennsylvanicus*. Jour. Comp. Path., 91:1–10.

Zimmerman, E. G.
1965 A comparison of habitat and food of two species of *Microtus*. Jour. Mammalogy, 46(4):605–612.

WOODLAND VOLE
Microtus pinetorum (Le Conte)

NAMES. The generic name *Microtus,* proposed by Schrank in 1798, is derived from two Greek words, *mikros* meaning small and *otos* referring to the ear. The specific name *pinetorum,* proposed by Le Conte in 1830, is from Latin and can be translated as "of the pines." In Michigan, local residents refer to this rodent as a field mouse. The name pine vole was also once used; however, this was misleading since the species actually prefers areas of deciduous woods. The subspecies *Microtus pinetorum scalopsoides* (Audubon and Bachman) occurs in Michigan.

RECOGNITION. Body size small and stout; head and body of an adult averaging 3⅞ in (100 mm) long, tail finely-haired and short, averaging ¾ in (20 mm) long, only slightly longer than length of hind foot and no more than one-fifth the length of the head and body; body covered with short, velvety, and molelike pelage; head large and blunt with no evidence of neck; eyes small, dark and beady; ears short, rounded, hairy, and almost concealed in the pelage at the sides of the head; vibrissae (whiskers) medium-long; feet short; hind limbs longer than fore limbs. In adults, the upperparts and sides are dull reddish brown, individual hairs are soft, not grizzled, basally dark, and dusky-tipped; underparts are essentially lead gray with pale hair tips providing a buff wash; tail is distinctly bicolored, brown above and gray below; feet are grayish brown. The winter fur is slightly darker than the summer pelage. Only two pairs of mammary glands are characteristic of the woodland vole. There are four functional toes on the forefeet and five on the hind.

The reddish brown and non-grizzled coloring of the upperparts and sides, the short tail, and the thick-set body serve, in combination, to distinguish the woodland vole from other Michigan microtines and mammals of small size.

MEASUREMENTS AND WEIGHTS. Adult woodland voles measure as follows: length from tip of nose to end of tail vertebrae, 4¼ to 5⅛ in (110 to 130 mm); length of tail vertebrae, ¾ to ⅞ in (18 to 24 mm); length of hind foot, ⅝ to 11/16 in (16 to 18 mm); height of ear from notch, ¼ to ⅜ in (7 to 10 mm); weight of adults, 0.7 to 1.2 oz (20 to 25 g).

DISTINCTIVE CRANIAL AND DENTAL CHARACTERISTICS. The skull of the woodland vole, like those of most other microtines (Fig. 51), is heavily constructed, rather flat and arched when viewed from the side, rectangular and short in appearance in dorsal aspect, and slightly ridged; posterior border of the palate with medium projection not ending in a thin bony shelf; least interorbital width more than 4 mm; incisive foramina short, less than the length of the upper molar tooth-row; and auditory bullae small. The skull averages 15/16 in (24 mm) long and 9/16 in (14 mm) wide across the zygomatic arches. Dental characteristics include: molar teeth not rooted and ever-growing; anterior faces of upper incisors yellow in color and not grooved (one specimen with grooved upper incisors reported in Indiana by Fish and Whitaker, 1971); upper row of the three molar teeth short (6.2 mm); all molar teeth with prismatic, flat-crowned grinding surfaces displaying folded and unusually-angular enamel loops surrounding islands with wide re-entrant angular loops of somewhat similar depth on each side of individual teeth; third (last) upper molar with grinding surface distinctly characterized (except for being similar to that of the larger prairie vole) by an anterior crescent-shaped loop, followed by two somewhat-closed triangular-shaped loops, and a posterior loop with two small and sometimes indistinct inner lobes (see Fig. 52).

The skull of the woodland vole can be distinguished from those of other small Michigan mammals by the following combination of features: the shortness and flatness of the rectangular-shaped skull; the posterior border of the bony palate with median projection not ending in a thin bony shelf; the shortness of the incisive foramina, less than the length of the upper molar tooth-row; the flat-crowned grinding surfaces displaying folded and unusually-angular enamel loops surrounding islands; interorbital space wide, more than 4 mm; and "neck" between anterior crescent-shaped loop on first lower molar and first triangular loop narrow, less than 0.2 mm wide.

The dental formula of the woodland vole is:

$$\text{I (incisors)}\frac{1\text{-}1}{1\text{-}1}, \text{C (canines)}\frac{0\text{-}0}{0\text{-}0}, \text{P (premolars)}\frac{0\text{-}0}{0\text{-}0}, \text{M (molars)}\frac{3\text{-}3}{3\text{-}3} = 16$$

DISTRIBUTION. The woodland vole is found throughout the eastern one-third of the United States, except for parts of upper New England, the Florida peninsula, the southern part of the Gulf Coastal Plain, and northern parts of the Upper Great Lakes states. It extends westward to the Great Plains and southward to central Texas. The first modern record of this inconspicuous and semi-fossorial species preserved as a museum specimen in Michigan was in 1896 near East Lansing in Ingham County (Wood, 1913). Among the early records is one obtained in 1910 in Emmet County by the celebrated naturalist and artist Ernest Thompson Seton (Wood, 1914). The occurrence of the woodland vole in the postglacial fossil record, and at the Schultz Archaeological Site of Late Woodland culture (dated 900 to 1,400 years B.P.) in what is now Saginaw County, indicates that the species has been a long-time resident of Michigan (Holman, 1975; Luxenberg, 1972). The woodland vole is now known from 19 counties in Lower Peninsula Michigan (Map 40). Although the woodland vole is presently unreported in the Upper Peninsula, it should be watched for in the extreme southwestern counties, since the species occurs in Brown County, Wisconsin, no more than 40 mi (66 k) south of the Michigan border (Long, 1974).

HABITAT PREFERENCES. The woodland vole is commonly associated with mature hardwood forests where there are loose, sandy soils and deep humus suitable for burrowing (Doutt *et al.*, 1966; Jameson, 1949). It is also fond of grassy areas in parklike stands of orchard trees, on hillsides, along fence-rows, and in road right-of-ways. Of the less than 100 specimens captured in Michigan, most were taken in open or closed wooded areas, characterized by oak-maple, oak-hickory, oak-hickory-maple-beech, maple-ash, maple-beech, fir-spruce, floodplain growth, and orchards (summarized by Master, 1978). For these habitats to be

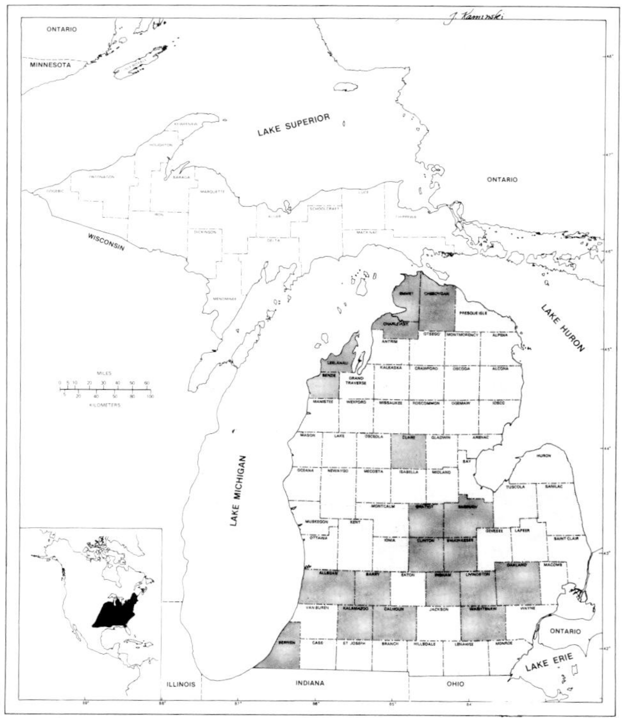

Map 40. Geographic distribution of the woodland vole (*Microtus pinetorum*). Counties from which specimens have been recorded are darkly shaded.

suitable for woodland voles, there needs to be a thick layer of loose soil and humus with fallen leaves or rank grass cover (Butsch, 1958; noted in Livingston County by Burt, 1940). In Clinton County, Linduska (1950) found woodland voles in an ungrazed woodlot and under field-shocked corn in winter. A repetition of this trapping program almost 40 years later by Shier (1981) failed to reveal any of these rodents. In Charlevoix County, Dice (1925) caught one animal in wet, second-growth fir and spruce forest. In Berrien County, Dice (1920) captured woodland voles under leaf litter in mature beech-maple and floodplain woods. In Kalamazoo County, Allen (1938) found two woodland voles in a basement.

In other parts of its range, the woodland vole shows remarkable ability to thrive in diverse woodland habitats (Hamilton, 1938). In New Jersey, Pearson (1959) found the species to be associated with perennial forbs in early successional stages of vegetative growth and also with shrub tree cover in late stages. These voles have been found in pine plantations, damp sphagnum, and cranberry bogs (Mumford, 1969); in upland woods (Whitaker, 1967); on rocky hills (Hahn, 1908); in open sandy field and apple orchards (Benton, 1955); in pine-oak woods (Conner, 1971); in well-drained uplands but also in swamps (Miller and Getz, 1969); in pine-hardwood forest and woodland edges (Lowery, 1974); and within the root systems of large trees (Barbour and Davis, 1974).

DENSITY AND MOVEMENTS. The secretive and partly underground (fossorial) habits of the woodland vole make studies of its density and population fluctuations difficult. Although there is some evidence for the lack of periodical (cyclic) fluctuations, the preponderance of the data supports the occurrence of these fluctuations (Benton, 1955; Gottschang, 1965; Hamilton, 1938). At Rose Lake Wildlife Experiment Station in Clinton County, Linduska (1950) found woodland voles in an ungrazed woodland, in home gardens and in winter under field-shocked corn in 1940 but no evidence of the species in the same areas in either 1941 or 1942; he had no explanation for this. Similarly, Burt (1940) caught 30 individuals in oak-hickory woods at the Edwin S. George Reserve in Livingston County in 1935 and 1936 but only a single individual there in 1937. That high local densities can occur is shown by the fact that 11 woodland voles were captured by hand in two hours in Franklin County, Indiana (Quick and Butler, 1885). In a five-year study in Connecticut, Miller and Getz (1969) found population densities for woodland voles to fluctuate from 0 to 5.8 per acre (14.6 per ha). In Adams County, Pennsylvania, Gettle (1975) found lowest numbers in summer and highest in autumn, with populations ranging from 7.2 to 49.2 per acre (18 to 123 per ha). In New York, Hamilton (1938) estimated high populations in orchards as between 200 and 300 individuals per acre (500 to 750 per ha).

Woodland voles appear to confine themselves to established systems of runways and burrows. Using the trap-mark-release-retrap method with live-trap stations arranged in grids in eastern Kansas (Fitch, 1958), most woodland voles were recaptured no further than 30 ft (9 m) from original trapping sites, with maximum distances between capture sites recorded of more than 200 ft (60 m). Other studies (Benton, 1955; Gettle, 1975; Miller and Getz, 1969; Paul, 1970; Stickel, 1954; and Stickel and Warbach, 1960), show that the range-diameter of home ranges for woodland voles varies between 16 and 115 ft (5 to 35 m). In Michigan, Burt (1940) found that 17 woodland voles were recaptured in an average range-diameter of 114 ft (32 m). He concluded that the size of the home range of individual woodland voles at his study area in Livingston County was approximately one-fourth acre (0.10 ha). The woodland vole is classified as a rare species in Michigan because of its local and highly variable occurrence.

BEHAVIOR. Woodland voles appear to be sedentary, ordinarily staying within a confined system of surface runways and underground tunnels. These passageways are shared by a small group of adults and young of both sexes (Smolen, 1981). Although woodland voles can be aggressive towards one another and even eat their own kind, intraspecific tolerance is shown by the fact that several individuals were caught in the same live-trap on numerous occasions (Miller and Getz, 1969) and three females, each with a litter, were found occupying a single nest (Raynor, 1960). Although Fitch (1958) found no hostility between woodland voles in live-traps or in cages, Wertheim and Giles (1971) recorded fighting and cannibalism in captive groups and concluded that the voles were sensitive to crowding. In experimental behavioral studies with captive woodland voles and meadow voles, Novak and Getz (1969) found the woodland voles to be the more aggressive in both intraspecific and interspecific test situations. The woodland vole also appears to have a higher metabolic rate, with greater daily oxygen consumption than the meadow vole (Pearson, 1947). When aggressive, the rodents chatter noisily and utter birdlike alarm calls. Woodland voles can swim but are not adept climbers. They can move through their trails at the rate of 3.8 mi per hour (6.1 km/hr), according to Layne and Benton (1954).

Woodland voles' underground burrow systems contain tunnels 1⅛ in (30 to 35 mm) wide and 1 in (25 to 30 mm) high. Above ground the voles follow their well-used trails through grassy stands and under forest leaf litter. Both burrows and surface runways closely resemble those constructed by meadow voles. Although there is some evidence that woodland voles are most active at twilight and at night (Benton, 1955), the species can very likely be found moving around in its cruising area any time of day or night. Miller and Getz (1969) felt that activity in surface runways is most intense at mid-day. Surface trails are obscured from view above by roof coverings of grass matting or leaf litter and may gradually extend deeper into the surface soil until they become tunnels 3 in (80 mm) or more below ground. Abandoned eastern mole burrows are frequently incorporated into the woodland vole burrow system. Because of the differential use of surface runways and underground burrows, traps set above ground may yield smaller catches than those set in subsurface tunnels (Stickel and Warbach, 1960; Connor, 1960). However, mammalogists consider woodland voles rather "trap shy" and neither above nor below ground catches reveal true numbers. According to Myton (1974), woodland voles may tunnel under and around traps to avoid capture.

Woodland voles usually construct their nests below ground at depths up to 12 in (300 mm). On occasion, nests can be found at locations only slightly below the surface, under logs, stumps, and debris. In orchards, nests are often placed just below ground among roots or next to tree trunks. Conspicuous runways and tunnels lead to these home sites which are round or oval in shape, about 6 or 7 in (155 to 180 mm) in diameter, and usually constructed of shredded grasses and leaves. A woodland vole nest in Wisconsin was constructed of maple leaves (Schmidt, 1931). Benton (1955) found that voles living in an orchard had taken cotton from live-traps and incorporated it in their nests.

ASSOCIATES. Although the woodland vole appropriates tunnels excavated by the eastern mole, it is not known if these two species are compatible. There is a suspected negative interaction between the woodland vole and the short-tailed shrew, since both may use the same underground burrows (Benton, 1955; Connor, 1971). Other small mammals of the woodlands and adjacent open areas may have some effect on the woodland vole survival (Hamilton, 1938). In Connecticut, Miller and Getz (1969) studied the woodland vole in association with the white-footed mouse and the southern red-backed vole. In New York, Connor (1971) found the woodland vole and the meadow vole together in fields, but pointed out that the differences in habitat preferences probably kept this association to a minimum. The woodland vole and the prairie vole are also known to associate (Jameson, 1947). In Michigan, Linduska (1942) noted woodland voles in winter under field-shocked corn where meadow voles, deer mice, house mice, white-footed mice, southern bog lemmings, short-tailed shrews, and Norway rats were also found.

REPRODUCTIVE ACTIVITIES. Compared to the breeding style of the meadow vole, the woodland vole is far less prolific, chiefly because litters are smaller and, as a rule, fewer litters per breeding season (Schadler and Butterstein, 1979). In years when more animals are obviously present (Linduska, 1950), it can be assumed that more young per litter and more litters per year must be a factor. However, woodland vole population fluctuations have not been studied sufficiently for complete understanding. Most workers have found that the woodland vole does have a long breeding season, from January to November (Benton, 1955; Hamilton, 1938; Miller and Getz, 1969). There is even evidence that some woodland vole populations may breed year around (Horsfall, 1963; Paul, 1970) if there is a highly nutritious winter food supply, such as a managed grassy turf in apple orchards (Cengel *et al.*, 1978). Even so, peak numbers of pregnant females usually occur in July and August (Valentine and Kirkpatrick, 1970).

Female woodland voles tend to take the initiative in courtship (Benton, 1955). Following mating and a gestation period of 20 to 24 days (Kirkpatrick and Valentine, 1970; Smolen, 1981), litters of two to four are born. At birth the young weigh about 0.07 oz (2 g) and are hairless and pale skinned, with eyes and ears closed. At three days, according to Hamilton (1938), the back of each young takes on a dull gray color; at seven days, fine hair covers the head and back, and the body weight has doubled; at nine days, the undersides are well-furred, crawling is accomplished with vigor, and the eyes begin to open; at 12 days, muscular coordination is highly developed and eyes are mostly open; at 16 days, solid food is eaten and weaning takes place; at 24 days, the young are able to provide for themselves, are well-furred (although the rich brown coat of the adult has not appeared), and weigh at least 0.5 oz (15 g)—more than one-half of the adult weight.

Woodland voles become sexually mature later than other Michigan voles; male puberty occurs between 48 and 56 days after birth; in females it occurs between 70 and 84 days (Schadler and Butterstein, 1979). Breeding females usually have several litters during each long reproductive season; there is evidence of a post-partum estrus. In the wild, it is unlikely that woodland voles live for more than a few months. Miller and Getz (1969) found that only about 19% of the tagged individuals from one trapping period were recovered in the next period two months later. These authors and Burt (1940) reported some tagged woodland voles survived for 12 months in the field. Woodland voles can be successfully aged by the weights of the eye lens (Gourley and Jannett, 1975).

FOOD HABITS. As a consequence of its preference for living in underground burrows, the woodland vole relies on succulent tubers and roots as a major part of its food supply. Observers have reported underground caches of tubers and fruits of such plants as wild violet, white clover, red haw, Dutchman's britches, wild morning glory, quack grass, broad-leafed dock, wild onion, acorns, hickory nuts, and hazelnuts (Hamilton, 1938). Stores are often accumulated in autumn, presumably for winter use. As a result, according to Jackson (1961), woodland voles do not subsist on green vegetation to the extent that meadow voles and prairie voles do. Like these other voles, however, the woodland vole also eats a variety of ground-living insects, their larvae and other invertebrates; it is also known to be cannibalistic (Barbour and Davis, 1974). Woodland voles like the bark from roots and bases of trees and shrubs, and roots and bulbs of garden plantings.

ENEMIES. Because the woodland vole's activity is chiefly underground, many predators are probably discouraged (Schadler and Butterstein, 1979). Jackson (1961) regarded snakes and shrews (chiefly the short-tailed shrew) as the major enemies of woodland voles. Predators recorded as eating woodland voles include house cat, gray fox, red fox, coyote, raccoon, domestic dog, Virginia opossum, mink, least weasel, and striped skunk (Barbour and Davis, 1974; Hamilton, 1938; Jackson, 1961; Korschgen, 1957; Lowery, 1974; Schofield, 1960). Long-eared owls and barn owls are known avian predators (Korschgen, 1952; Stupka, 1931). The only published Michigan record of predation on the woodland vole was obtained by Raymond Schofield (1960), who identified the remains of this mouse in the feces of red fox.

PARASITES AND DISEASES. Woodland voles are infested with numerous parasites. Internally, a cestode (genus *Catenotaenia*), a nematode (genus *Trichuris*), and a spiny-headed worm (genus *Moniliformis*) have been identified (Benton, 1955; Doran, 1954; Lowery, 1974). Ectoparasites include fleas of the genera *Ceratophyllus* and *Ctenophthalmus*, lice of the genus *Hoplopleura*, and mites of the genera *Haemolaelaps* and *Laelaps* (Benton, 1955; Hamilton, 1938; Jameson, 1947; Lowery, 1971). An affliction which causes patches of fur and epidermis to be sloughed off and forms encrustations about the eyes has been described by Hamilton (1938).

MOLT AND COLOR ABERRATIONS. The dark juvenile pelage of the weaned woodland vole is completely exchanged for the rich chestnut fur of the adult by 40 to 70 days of age. A winter coat of darker fur replaces the paler summer pelage in late autumn (probably November and early December). The winter coat is molted and exchanged for the summer coat in late spring (May and June). Abnormal pelage colors are occasionally reported, including white, white spotted, buff, yellow, cream, and light brown (Benton, 1955; Handley, 1953; Jackson, 1961; Owen and Shackelford, 1942; Paul, 1964).

ECONOMIC IMPORTANCE. The woodland vole is reputed to be highly destructive to orchards and various other plantings in eastern United States. Farm and garden products eaten by this vole (Henderson and Craig, 1932) include white and sweet potatoes, various kinds of bulbs, strawberries, blackberries, a variety of planted seeds, stored vegetables, and nursery stock. Orchard depredations, which cause up to $50 million annually in damage to commercial apple orchards in northeastern states, are perhaps the most serious charges against this rodent. Since it attacks the bark and cambium of the roots and lower parts of the trunks of fruit trees and shrubs from subsurface runways, the woodland vole's damage, including tree girdling, goes undetected until it is too late to save the vegetation (Hamilton, 1935, 1938).

Control methods using ground sprays (endrin and dieldrin) or poisoned baits (zinc phosphide and strychnine) have been generally ineffective or have been shown to damage other aspects of the local ecosystem (Anthony and Fisher, 1977; Byers, 1976; Gettle, 1975). The use of DMCT, an antibiotic, coated on apple slices may be an effective control (Stehn *et al.*, 1980). Better orchard management by keeping ground cover under and between trees clear of debris and mowed regularly (Davis, 1977), is considered one of the most efficient ways of controlling populations of woodland voles.

In Michigan, however, woodland voles have not yet been identified as orchard pests. In 1974 and 1978, apple-growers in western Michigan (Allegan, Muskegon, Oceana and Ottawa counties) reported tree damage typical of woodland voles, but it was finally decided that meadow voles were responsible. Nevertheless, Shick (1965) described control measures for both the woodland vole and the meadow vole in a Michigan State University Extension Service pamphlet. The only published evidence describing destructive habits of the woodland vole in Michigan is that of Linduska (1950), who found these rodents under field shocks of corn (and presumably feeding on the kernels) and eating root crops in home gardens in Clinton County in 1940.

COUNTY RECORDS FROM MICHIGAN. Records of specimens of the woodland vole in collections of museums and from the literature show that this species is reported from 19 counties in Lower Peninsula Michigan: Allegan, Benzie, Berrien, Berry, Calhoun, Charlevoix, Cheboygan, Clare, Clinton, Emmet, Gratiot, Ingham, Kalamazoo, Leelanau, Livingston, Oakland, Saginaw, Shiawassee, and Washtenaw.

LITERATURE CITED

Allen, D. L.
1938 Ecological studies on the vertebrate fauna of a 500-acre farm in Kalamazoo County, Michigan. Ecol. Monog., 8(3):348–436.

Anthony, R. G., and A. R. Fisher
1977 Wildlife damage in orchards—a need for better management. Wildlife Society Bull., 5(3):107–112.

Barbour, R. W., and W. H. Davis
1974 Mammals of Kentucky. Univ. Press Kentucky, Lexington. x+322 pp.

Benton, A. H.
1955 Observations on the life history of the northern pine mouse. Jour. Mammalogy, 36(1):52–62.

Burt, W. H.
1940 Territorial behavior and populations of some small mammals in southern Michigan. Univ. Michigan, Mus. Zool., Misc. Publ., No. 45, 58 pp.

Butsch, R. S.
1958 Michigan's short-tailed mice. Jack-Pine Warbler, 36(2):74–78.

Byers, R. E.
1976 New compound (RH 787) for use in control of orchard voles. Jour. Wildlife Mgt., 40(1):169–171.

Cengel, D. J., J. E. Estep, and R. L. Kirkpatrick
1978 Pine vole reproduction in relation to food habits and body fat. Jour. Wildlife Mgt., 42(4):822–833.

Connor, P. F.
1960 The small mammals of Otsego and Schokarie counties, New York. New York State Mus. and Sci. Serv., Bull. 382, 84 pp.
1971 The mammals of Long Island, New York. New York State Mus. and Sci. Serv., Bull. 416, 78 pp.

Davis, D. E.
1977 Advances in rodent control. Pp. 193–211 *in* Zeitscrift für Angewandte Zoologie. Duncker and Humbolt, Berlin.

Dice, L. R.
1920 The mammals of Warren Woods, Berrien County, Michigan. Univ. Michigan, Mus. Zool., Occas. Papers No. 86, 20 pp.
1925 A survey of the mammals of Charlevoix County, Michigan, and vicinity. Univ. Michigan, Mus. Zool., Occas. Papers No. 159, 33 pp.

Doran D. J.
1954 A catalogue of the protozoa and helminths of North American rodents. II. Cestoda. American Midl. Nat., 52:469–480.

Doutt, J. K., C. A. Heppenstall, and J. E. Guilday
1966 Mammals of Pennsylvania. Pennsylvania Game Comm., Harrisburg. 273 pp.

Fish, P. G., and J. O. Whitaker, Jr.
1971 *Microtus pinetorum* with grooved incisors. Jour. Mammalogy, 52(4):827.

Fitch, H. S.
1958 Home ranges, territories, and seasonal movements of vertebrates of the Natural History Reservation. Univ. Kansas Publ., Mus. Nat. Hist., 11(3):63–326.

Gettle, A. S.
1975 Densities, movements, and activities of pine voles (*Microtus pinetorum*) in Pennsylvania. Pennsylvania State Univ., unpubl. M.S. thesis, 66 pp.

Gottschang, J. L.
1965 Winter populations of small mammals in old fields of southwestern Ohio. Jour. Mammalogy, 46(1):44–52.

Gourley, R. S., and F. J. Jannett, Jr.
1975 Pine and montane vole age estimates from eye lens weights. Jour. Wildlife Mgt., 39(3):550–556.

Hahn, W. L.
1908 Notes on the mammals and cold-blooded vertebrates of Indiana University farm, Mitchell, Indiana. Proc. United States Nat. Mus., 35:545–581.

Hamilton, W. J., Jr.
1935 Field mouse and rabbit control in New York orchards. Cornell Ext. Bull. No. 338.
1938 Life history notes on the northern pine mouse. Jour. Mammalogy, 19(2):163–170.

Handley, C. O., Jr.
1953 Abnormal coloration in the pine mouse (*Pitymys pinetorum*). Jour. Mammalogy, 34(2):262–263.

Henderson, J., and E. L. Craig
1932 Economic mammalogy. Charles C. Thomas, Springfield, Illinois. x+397 pp.

Holman, J. A.
1975 Michigan's fossil vertebrates. Michigan State Univ., Publ. Mus., Ed. Bull. No. 2, 54 pp.

Horsfall, F., Jr.
1963 Observations of fluctuating pregnancy rate of pine mice and mouse feed potential in Virginia orchards. Proc. American Soc. Hort. Sci., 83:276–279.

Jackson, H. H. T.
1961 Mammals of Wisconsin. Univ. Wisconsin Press, Madison. xii+504 pp.

Jameson, E. W., Jr.
1947 Natural history of the prairie vole (mammalian genus Microtus). Univ. Kansas Publ., Mus. Nat. Hist., 1(7): 125–151.
1949 Some factors influencing the local distribution and abundance of woodland small mammals in central New York. Jour. Mammalogy, 30(2):221–235.

Kirkpatrick, R. L., and G. L. Valentine
1970 Reproduction in captive pine voles, *Microtus pinetorum*. Jour. Mammalogy, 51(4):779–785.

Korschgen, L. J.
1952 A general summary of the food of Missouri predatory and game animals. Missouri Conserv. Comm., Pittman-Robertson Prog., 61 pp.
1957 Food habits of coyotes, foxes, house cats and bobcats in Misouri. Missouri Conserv. Comm., P-R Ser. No. 15, 64 pp.

Layne, J. N., and A. H. Benton
1954 Speeds of some small mammals. Jour. Mammalogy, 35(1):103–104.

Linduska, J. P.
1942 Winter rodent populations in field-shocked corn. Jour. Wildlife Mgt., 6(4):353–363.
1950 Ecology and land-use relationships of small mammals on a Michigan farm. Michigan Dept. Conserv., Game Div., ix+144 pp.

Long, C. A.
1974 Environmental status of the Lake Michigan region. Volume 15. Mammals of the Lake Michigan drainage. Argonne Nat. Lab., ANL/ES-40, 108 pp.

Lowery, G. H., Jr.
1974 The mammals of Louisiana and adjacent waters. Louisiana State Univ. Press, Baton Rouge. xxiii+565 pp.

Luxenberg, B.
1972 Faunal remains. Pp. 91–115 *in* J. E. Fitting, Ed. The Schultz Site at Green Point. A stratified occupation area in the Saginaw Valley of Michigan. Univ. Michigan, Mus. Anthro., Memoir No. 4, xii+317 pp.

Master, L. L.
1978 A survey of the current distribution, abundance, and habitat requirements of threatened and potentially threatened species of small mammals in Michigan. Univ. Michigan. (mimeo.), 52 pp.

Miller, D. H., and L. L. Getz
1969 Life-history notes on *Microtus pinetorum* in central Connecticut. Jour. Mammalogy, 50(4):777–784.

Mumford, R. E.
1969 Distribution of the mammals of Indiana. Indiana Acad. Sci., Monog. No. 1, vii+114 pp.

Myton, B.
1974 Utilization of space by *Peromyscus leucopus* and other small mammals. Ecology, 55:277–290.

Novak, M. A., and L. L. Getz
1969 Aggressive behavior of meadow voles and pine voles. Jour. Mammalogy, 50(3):637–639.

Owen, R. D., and R. M. Shackelford
1942 Color aberrations in *Microtus* and *Pitymys*. Jour. Mammalogy, 23(3):306–314.

Paul, J. R.
1964 Second record of an albino pine vole. Jour. Mammalogy, 45(3):485.
1970 Observations on the ecology, populations and reproductive biology of the pine vole, *Microtus pinetorum*, in North Carolina. Illinois State Mus., Rept. Invest. No. 20, vii+28 pp.

Pearson, O. P.
1947 The rate of metabolism of some small mammals. Ecology, 28(2):127–145.

Pearson, P. G.
1959 Small mammals and old field succession on the Piedmont of New Jersey. Ecology, 40(2):249–255.

Quick, E. R., and A. W. Butler
1885 The habits of some Arvicolinae. American Nat., 19: 113–118.

Raynor, G. S.
1960 Three litters in a pine mouse nest. Jour. Mammalogy, 41(2):275.

Schadler, M. H., and G. M. Butterstein
1979 Reproduction in the pine vole, *Microtus pinetorum*. Jour. Mammalogy, 60(4):841–844.

Schmidt, F. J. W.
1931 Mammals of western Clark County, Wisconsin. Jour. Mammalogy, 12(2):99–117.

Schofield, R. D.
1960 A thousand miles of fox trails in Michigan's ruffed grouse range. Jour. Wildlife Mgt., 24(4):432–434.

Shick, C.
1965 Controlling meadow mice in residential areas, parks, orchards, forests and Christmas tree plantations. Michigan State Univ., Coop. Ext. Serv., Extension Bull. E-430, 4 pp.

Shier, J. L.
1981 Habitats of threatened small mammals on the Rose Lake Wildlife Research Area, Clinton County, Michigan—early forties and late seventies. Michigan State Univ., unpubl. M.S. thesis, vi+106 pp.

Smolen, M. J.
1981 Microtus pinetorum. American Soc. Mammalogists, Mammalian Species, No. 147, 7 pp.

Stehn, R. A., E. A. Johnson, and M. E. Richmond
1980 An antibiotic rodenticide for pine voles in orchards. Jour. Wildlife Mgt., 44(1):275–280.

Stickel, L. F.
1954 A comparison of certain methods of measuring ranges of small mammals. Jour. Mammalogy, 35(1):1–15.

Stickel, L. F., and O. Warbach
1960 Small mammal populations of a Maryland woodlot, 1949–1954. Ecology, 41(2):269–286.

Stupka, A.
1931 The dietary habits of barn owls. Ohio Dept. Agric., Div. Conserv., Bull. 6, 5 pp.

Valentine, G. L., and R. L. Kirkpatrick
1970 Seasonal changes in reproductive and related organs in the pine vole, *Microtus pinetorum*, in southwestern Virginia. Jour. Mammalogy, 51(3):543–560.

Wertheim, R. F., and R. H. Giles, Jr.
1971 Effects of bacterial endotoxin and crowding on pine voles and white mice. Jour. Mammalogy, 52(1):238–242.

Whitaker, J. O., Jr.
1967 Habitat and reproduction of some of the small mammals of Vigo County, Indiana, with a list of mammals known to occur there. C. C. Adams Center Ecol. Studies, No. 16, 24 pp.

Wood, N. A.
1913 Two additions to the mammalian fauna of Michigan. Science, N.S. 37(953):522–523.
1914 On the occurrence of *Neosorex palustris* (Rich.), *Sorex richardsonii* Bach. and *Pitymys pinetorum scalopsoides* (Aud. and Bach.) in Michigan. Univ. Michigan, Mus. Zool., Occas. Papers No. 6, 2 pp.

MUSKRAT
Ondatra zibethicus (Linnaeus)

NAMES. The generic name *Ondatra*, proposed by Link in 1795, is derived from a French Canadian word of Huron Indian origin. The specific name *zibethicus*, proposed by Linnaeus in 1766, is derived from the Latin and refers to the musky odor characteristic of this mammal. In Michigan, local residents call this animal a muskrat, or sometimes a marsh rat or mushrat. There is also a rich legacy of Indian and fur-trade names which were given to this important fur-bearer. Michigan muskrats are classified as belonging to the subspecies *Ondatra zibethicus zibethicus* (Linnaeus).

RECOGNITION. Body size large and robust; head and body of an adult averaging 12½ in (320 mm) long; tail laterally compressed, thinly haired, scaly, averaging 9½ in (240 mm) long; fur dense, luxuriant, waterproof, with air trapped in the underfur for buoyancy and insulation (Johansen, 1962); head large with blunt nose and little evidence of neck; eyes small, beady; ears hairy, small, barely extending beyond the confines of the heavy pelage on the sides of the head; vibrissae (whiskers) medium long, prominent; legs short; feet large, modified for swimming; feet and toes of both front and hind feet fringed by short, stiff hairs; hind feet partly webbed; prominent perineal glands (enlarged during breeding season) in anal area capable of secreting a penetrating musk.

In adults, upperparts (especially midback) glossy dark brown (darker in winter; paler in summer); underparts paler, grayer with dense under fur (slate with chalk gray tips) showing through; small spot on chin, nasal pads, soles of feet, and tail blackish; tops of feet dark brown. Three pairs of mammary glands are characteristic; one pair is pectoral and two are inguinal. The short thumb (pollex) on the forefeet has a well-developed claw; and the five hind toes have strongly-developed claws.

This largest of Michigan's cricetid rodents, with its laterally-flattened tail and semi-aquatic habits, is a distinctive-looking creature; however, at a distance and while swimming, the muskrat might be mistaken for a young beaver, otter or even a mink.

MEASUREMENTS AND WEIGHTS. Adult muskrats measure as follows: length from tip of nose to end of tail vertebrae, 18½ to 24¾ in (470 to 630 mm); length of tail vertebrae, 7⅞ to 10⅝ in (200 to 270 mm); length of hind foot, 2¾ to 3½ in (70 to 90 mm); height of ear from notch, ¾ to 1 in (19 to 25 mm); weight of adults, 28 to 53 oz (800 to 1,500 g).

DISTINCTIVE CRANIAL AND DENTAL CHARACTERISTICS. The skull of the muskrat resembles that of an oversized meadow vole, having the same heavy, rectangular and ridged construction (Fig. 54). Other features include a long rostrum (nasal area); massive, flaring zygomatic arches; a narrow and dorsally-keeled interorbital region; prominent ridges on the squamosal bones; and large auditory bullae. The skull averages 2½ in (65 mm) long and 1⅝ in (41 mm) wide across the zygomatic arches. The smooth-fronted upper incisors are yellow-orange; the molar (cheek) teeth are rooted and possess flat-crowned grinding surfaces with folded and angular enamel loops. The presence on the third upper molar of anterior and posterior enamel loops with two or three closed triangles in between is distinctive. The large size and typical microtine construction of the skull and dentition are distinguishing features and readily separate this species from all other Michigan mammals. The dental formula of the muskrat is:

$$\text{I (incisors)}\frac{1\text{-}1}{1\text{-}1}, \text{C (canines)}\frac{0\text{-}0}{0\text{-}0}, \text{P (premolars)}\frac{0\text{-}0}{0\text{-}0}, \text{M (molars)}\frac{3\text{-}3}{3\text{-}3} = 16$$

DISTRIBUTION. The muskrat, according to Hollister (1911), occurs in suitable swamps, marshes, and waterways in northern North America from treeline in the Arctic southward to the Gulf Coast on the east and to near the Mexican border on the watersheds of the Rio Grande and the Colorado in Texas and Arizona (see Map 41).

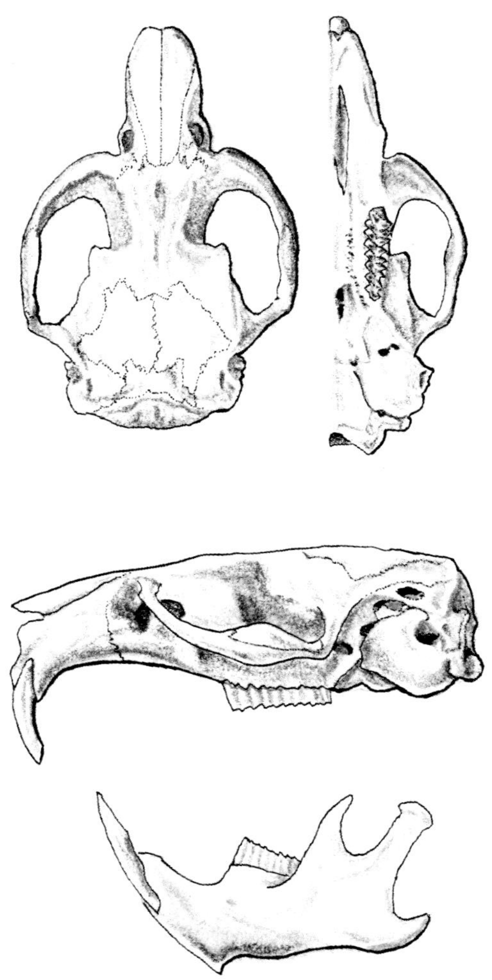

Figure 54. Cranium and lower jaw of muskrat (*Ondatra zibethicus*).

Introduced in northern Eurasia early in this century, the muskrat (usually unwelcome) has become well established (de Vos *et al.*, 1956). In Michigan, the muskrat is known from both the postglacial fossil record and early archaeological sites (Cleland, 1966; Higgins, 1980; Holman, 1975); from the peak of the fur trade (Johnson, 1919; Petersen, 1979); and presently in all counties and many of the islands in Michigan's sector of the Upper Great Lakes.

HABITAT PREFERENCES. The muskrat is a semi-aquatic rodent which spends most of its time in wetland environments. Year-around water is important, preferably 4 to 6 ft (1.2 to 1.8 m) deep so that it will not freeze to the bottom. If water is more than 12 ft (3.7 m) deep, aquatic vegetation (as potential food) is often absent. Muskrat living areas, according to Johnson (1925), can be classified as: (1) ponds, lakes, streams, canals, and reservoirs with or without marshy borders and related emergent vegetation, (2) swamps, and (3) marshes. The first category provides the least favorable muskrat habitat for it often contains high, sheer banks; muddy, unvegetated shorelines; deep water, fast currents; turbid water and irregular bottom topography resulting from disturbance by watering livestock; borders of cultivated fields, sod, pasture, or woods with little transitional vegetation; and fluctuating waterlevels, especially in streams and dammed farm ponds. In these situations, muskrats primarily use bank burrows as dens and may forage considerable distances for scattered food supplies.

Swamps, as defined here, consist of pools of still water and sluggish streams bordered by dense thickets of water-tolerant growth, such as buttonbush, alder, willow, red-osier dogwood, arrow-

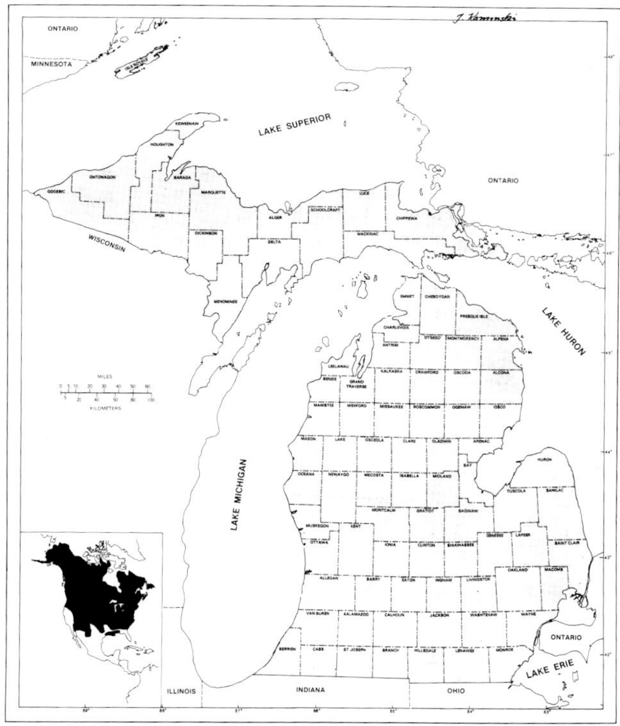

Map 41. Geographic distribution of the muskrat (*Ondatra zibethicus*).

wood, and sweet gale. Although swamp water levels may drop considerably during summer, there are usually patches of open water sometimes covered by dense mattings of duckweeds, *Wolffia,* and various algae bordered by patches of cattails, rushes, grasses, sedges, duck potato, and other low-growing and non-woody plants. Muskrats construct trails under the dense canopy of emergent swamp vegetation, much of which serves as food. The overhead cover also provides protection from flying predators. The muskrat dens in hollow stumps, fallen tree trunks, under exposed root systems of standing timber, in banks of hummocks or other high ground, or in constructed houses.

The third habitat category, marshes, is the most favorable environment in Michigan for muskrat. These consist of wetlands with fairly constant water levels, little or no water currents, and vast stands of cattails and other emergent vegetation among scattered areas of open water. Recent glaciation in Michigan caused a vast array of lakes and ponds. As these shallow due to filling by natural plant succession, ideal marsh situations develop. Some of these also occur along rivers where sluggish water overflows (perhaps with the help of dams). Extensive muskrat marsh habitats, such as along the Maple River in Gratiot County, the Kalamazoo River in Allegan County, in the Erie Marsh in Monroe County, and the Munuscong Marsh in Chippewa County, are also noted concentration areas for waterfowl. Muskrats use marsh vegetation, such as sedges, rushes, grasses and cattails, as food and to construct their picturesque houses (Bishop *et al.,* 1979).

DENSITY AND MOVEMENTS. Muskrats have both annual and multiannual fluctuations in population numbers (Errington, 1963). Dramatic long-term fluctuations appear to some authorities to follow 10 year cycles (Errington, 1951), although there is evidence that regular cycles are non-existent (Jackson, 1961). Habitat conditions, severe winter weather, flooding, drought, disease (epizootics), and overtrapping also affect population size (Errington, 1939a; Mathiak, 1966; Smith, 1938). Although the causes of multiannual fluctuations are obscure, environmental and behavioral factors may be involved (see Krebs and Myers, 1974). For example, there is some evidence that during population highs muskrats develop greater tolerance for crowding which allows for increased numbers of precocial breeders and for higher survival of young (Sather, 1958).

Annually, local populations are first enhanced by the production of young in the warm months; these populations decline in the cold months, when mortality continues without replacements. In Louisiana and other southern areas, however, year-around breeding masks these fluctuations (Lowery, 1974). During the breeding season in northern areas, there can be three peaks of litter production—in late April and early May, in June, and in late August and early Sepember (Sather, 1958).

In the less favorable habitats along streams and in lakes or impoundments, is is difficult to estimate population size because bank dens and other signs of the animals are obscure. There may be no more than three muskrats per acre (7.5 per ha), or perhaps no more than 15 or 16 animals per mi (9 or 10 per km) of flowing stream or lakefront. In open marsh, however, muskrat numbers are estimated with greater accuracy because of the conspicuous houses dotting the landscape. Individual muskrat houses may harbor anywhere from three to 15 or 20 animals (Lay, 1945); however, Dozier (1948a) considers the average number to be five. It is fairly easy to obtain accurate estimates of muskrat houses either by sampling along transects or by rather complete counts from aerial photographs. High populations in favorable habitat can reach 35 muskrats per acre (87 per ha), according to Banfield (1974) and Lay (1945).

Although in autumn and spring there may be population shifts (involving the dispersal of young), established muskrats appear to lead a rather restricted life. Movements have been documented by the recapture of animals tagged with metal markers (Dozier, 1948) or by radiotracking individuals to which FM transmitters are attached or implanted (MacArthur, 1978). In the warmer period of the year, muskrats rarely move far from centers of activity around their houses, resting areas or feeding shelters. Maximum distances are no more than 585 ft (180 m) for summer activity (Takos, 1944; Wragg, 1955). Winter tracking of individuals is more difficult because their actions are obscured by ice cover. However, winter movements seem no more extensive than in summer, with most foraging done within a radius of 36 ft (10 m) of the home site. When populations are at peak numbers, muskrats probably are obliged to use even smaller home ranges. Even so, the muskrat is capable of moving up to 20 mi (33 km) across country (Errington, 1939a). Michigan motorists see muskrats killed on highways, especially in late spring and autumn, in places distant from the nearest water. It is not unusual to find the animals in cornfields or even occupying woodchuck burrows (Errington, 1961). Movements away from the home site appear to be motivated principally by (1) overcrowding, (2) dispersal of young, (3) reproductive activities, (4) severe cold or drought, and (5) food shortages.

Major movements occur in spring, late summer and autumn, and winter. The spring shuffle (Sprugel, 1951) is closely correlated with high water and the break-up of ice; it primarily involves adult males and young. Females remain in the home territory except for surplus females which must search for unoccupied living quarters. In late summer and autumn, sparse surface water supplies may result in conspicuous movement. If summer weather has caused muskrat habitats to dry up completely or become too shallow for winter occupancy, muskrats will often move cross-country. In this highly vulnerable state, many are taken by predators or run over by automobiles. Severely cold winters can cause ice to become so deep in marshes and swamps that it seals off major food resources; the muskrats are again forced to move outside their protective aquatic environments (Mathiak, 1966; Sather, 1958).

BEHAVIOR. Muskrats tend to live in family groups and occupy pronounced territories—especially females with young. There is evidence that over-crowding where resources are in short supply causes females to eject weaned offspring, often by force (Errington, 1940). Most observers (Errington, 1939a; Mathiak, 1966) report fighting and cannibalism among muskrats. Errington wrote that there is ". . . a tremendous amount of natural friction in muskrat populations, this ranging from mere displays of ill temper to lethal attacks and involving both sexes of old or young at all seasons of the year." Nevertheless, muskrats will often live in crowded quarters; 10 or more were found in a single bank burrow in winter (Bailey, 1937).

The muskrat may be active at any time during the 24-hour day, especially in late spring and early summer. Normally, however, they tend to remain in their houses or burrows from daylight to mid-afternoon, and forage most actively in late afternoon and between sunset and shortly after dark (Smith, 1938). Diurnal activity increases in dark, cloudy or rainy weather (Stewart and Bider, 1977).

The muskrat swims efficiently, using its hind feet as paddles and its tail as a rudder and to generate thrust; backward swimming has also been observed (Peterson, 1950). On the surface, the muskrat holds its front feet against its chin with its head and upper dorsal surface being the only parts of the body exposed. The muskrat may attain the speed of at least 2 mi (3.0 k) per hour and can remain under water for as long as 12 to 17 minutes. In winter, air bubbles trapped under the ice are thought to be used for breathing (Johnson, 1925; Smith, 1938). On land, the muskrat has an ambling gait, sometimes a slow hop.

The chief method of communication between a family group of muskrats is the odor of musk deposited in strategic places. This also serves to notify would-be intruders of the presence of an established population. Muskrats vocalize using squeaking notes when in the company of associates, loud squeals when alarmed or irritated, and hissing sounds when cornered or attacked. Breeding females utter birdlike calls. While in the water, a muskrat will dive suddenly with a noisy splash of the tail when alarmed (Seton, 1953). Its senses of sight, hearing, and smell seem poorly developed. Muskrats tend to defecate regularly at places called "posts." These may be on logs, high ground in the marsh or on floating material. Musk is also deposited at these sites.

Living both on land and in the water, muskrats are subjected to a variety of environmental conditions (MacArthur, 1979). They are most vulnerable to sudden changes in temperature and suffer when exposed to severe cold or freezing winds for long periods. Dry, hot weather also seriously affects their survival. Muskrat houses and bank burrows afford protection from weather, especially severe cold in winter (MacArthur and Aleksiuk, 1979). Body insulation (fat layering and non-wettable, air-entrapping fur) is helpful in preventing major loss or gain of heat while the animal is in water. A special muskrat adaptation is regional heterothermia (Fish, 1979), by which the flow of blood to the feet and tail is regulated; this, in turn, controls the amount of body heat which is lost or gained. This adaptation allows the muskrat to survive extreme temperature fluctuations without seriously draining its nutrient supply.

HOME SITES. When not in water, muskrats spend much of their time in their tunnels, houses, burrows, and feeding shelters (Willner *et al.*, 1980). Home sites are of two basic types: (1) the bank burrow and (2) the house; the local topography dictates which type is used (Johnson, 1925). If there is elevated, firm ground at the edge of the pond, marsh or stream, a bank den is constructed. This burrow is dug into the bank, from an underwater tunnel which leads upwards, for distances as much as 35 ft (11 m), to a dry, above-waterline nest chamber. A burrow in use for many years may have several entrances and two or more nesting chambers.

Muskrat houses (also called beds or rat-lodges) are conspicuous in Michigan marsh and swamp habitats. In order to construct a proper house, the water should be between 15 and 40 in (38 to 100 cm) deep. The house is usually based on a firm bottom condition, a stump, or the base of a clump of brush or root stalks. House materials include plant parts from the immediate vicinity—roots, peaty remains of plants, emergent and submergent plants (Smith, 1938). Broad-leafed cattails (*Typha*) or rushes (*Scirpus*), which may be plastered down with mud and pond weeds, are preferred. Houses may be as large as 7 or 8 ft (2 to 3 m) in diameter and sometimes more than 4 ft (122 cm) high. A house may contain one to three nests, each of which is entered from a plunge hole in the floor. The plunge hole joins an underground tunnel that opens several feet underwater from the house. House construction usually begins in August and is completed in October. Spring floods or high water often severely damage or destroy houses. Salvageable structures may be remodeled for another year's use. Muskrats also construct feeding houses, often two or more for every large house (Sather, 1958); these platforms are low in profile and constructed so that the muskrats sit above water while feeding.

Not only are muskrats proficient in excavating tunnels leading to bank dens or plunge-holes underneath marsh houses, they also construct water-filled canals and tunnels in shallow water, or even in mud if their excavations can extend down to the water table and fill with water. These canals and tunnels generally radiate from houses and lead to food sources; they may not be visible in high water but muskrats can follow them by swimming underwater or under-ice. In dry parts of marshes and swamps, muskrats may also follow surface runways, which resemble closely the much smaller ones maintained by the meadow vole and other small microtine relatives.

Because muskrats in northern parts of their range have severe problems in winter with thickening surface ice, they frequently prepare domes of vegetation (push-ups) which become solidly frozen and snow-covered (Dozier, 1948) and act as an insulating cover for a plunge-hole below. These are constructed just after the water begins to freeze since the ice is needed for basal support. The ice opening is made by gnawing, often at the site of a trapped air bubble, followed by pulling up submergent vegetation from below (bulky upper parts of cattails are not used) to construct a lumpy, ball-shaped mass. The muskrat makes an inside cavity at surface-ice level to serve as a resting place, eating station, and breather. Frequent visitations keep the plunge-hole open. When the ice melts, the structure collapses and submerges. MacArthur (1977) estimated that the winter temperature in well-insulated houses may be as much as 67°F (35°C) higher than the outside temperature; temperatures in "push-ups" seldom drop below 20°F (−7°C) even when outside temperatures are below −21°F (−33°C).

ASSOCIATES. Mink and river otter are the two semi-aquatic mammals which share the muskrat's Michigan environment, and can be predators as well. Red foxes, coyotes and especially raccoons also disturb the tops of muskrat houses, which they reach easily in winter by ice travel. The muskrat house can be a convenient nesting or roosting place for such birds as black terns, several kinds of ducks and geese, and harriers. Water snakes of the genus *Natrix* can inhabit muskrat houses, and several species of turtles have been known to lay their eggs in the protective and decaying vegetation used in muskrat house construction (Johnson, 1925). Bank burrows are also shared on occasion with small rodents whose burrows may intersect those of the muskrat. Beaver ponds, of course, are attractive to muskrats. These two semiaquatic rodents appear compatible, sometimes using the same caches of foods (Tevis, 1950).

REPRODUCTIVE ACTIVITIES. Muskrats in the southern part of their range (*e.g.*, Louisiana, Lowery, 1974) breed the year around; females undergo postpartum estrus, allowing them to accept mates shortly after the birth of their litters. In Michigan, reproductive activity is confined to the warmer months (normally March through August), with litters spaced further apart than in southern populations. Muskrats have an active courtship with considerable antagonism between individuals of both sexes, although they seem to generally be monogamous (Beer, 1950). The gestation period is 29 to 30 days (estimates of 25 to 35 days have also been made). Litters may contain from one to 11 offspring, the average being six; the larger litters are born in the north. In Michigan, a mature female may be expected to produce as many as three litters (Burt, 1946); in Louisiana, five to six (Lowery, 1974). The earliest litter reported from Wisconsin's Horicon Marsh was on April 16 (Mathiak, 1966). Mathiak also recorded a pregnant female taken by a Horicon trapper as late as November 3.

The young are born in a grass-lined nest either in a bank burrow or marsh house. At birth, according to Smith (1938), young muskrats have short, fine and dark fur, pink feet and tail, closed eyes, are about 4 in (120 mm) long, and weigh about 0.8 oz (22 g). At five days of age, they are able to cling to their mother's nipples even while she is swimming, vocalize with squeals, are covered with coarse black hair, and can move about in the nest; at 10 days, they are able to swim; at 14 to 16 days, the young have their eyes open and their fur takes on a soft, wooly texture and gray color; at 21 days, the young can accept green food; at 30 days, weaning is complete and the young achieve independence; and at 200 days, the young reach near-adult size (Dorney and Rusch, 1953; Errington, 1939c). Schofield (1955) classified Michigan juvenile animals as weighing about 33 oz (953 g) and exhibiting on the skin side in the dorsal area lyre-shaped unprime areas of their pelts; adults were classified as weighing 44 oz (1,267 g) and showing no unprime patterns on their pelts.

Although muskrats might live in captivity as long as 10 years (Johnson, 1925), life expectancy in nature may be no more than three or four years (Jackson, 1961; Mathiak, 1966). Mathiak (1966) estimated that 87% of the muskrat population succumbs during the first year and 98% the second.

FOOD HABITS. Muskrats are primarily vegetarians but will eat animal matter as well. They have healthy appetites and consume as much as one-third of their weight per day (O'Neil, 1949). According to Errington (1941), muskrats can be very opportunistic in feeding. Green vegetation is a basic food, and the muskrat's digestive tract is adapted for bulky amounts of this diet. For north-central United States, Willner *et al.* (1975) listed the important foods as cattail (*Typha latifolia*); bulrush (*Scirpus validus, S. fluviatilis, S. acutus*); arrowhead (*Sagittaria latifolia*); water-lily (*Nympaea tuberosa*); grasses of the genera *Eragrostis, Panicum, Echinochloa, Bromus, Muhlenbergia, Agropyron, Elymus;* corn (*Zea maize*); reed (*Phragmites communis*); duckweed (*Lemna minor*); and such animal matter as crayfish, frogs, small turtles, mollusks, and fish. Numerous other plant materials also are eaten.

In summer, muskrats thrive primarily on roots and stems of cattail, bulrush, and other emergent aquatic vegetation. In winter, however, submerged vegetation is most easily available, since muskrats rarely are known to cache food. Under the ice, the muskrats have a choice of pondweeds, coontail, water lily tubers, water milfoil, water weed, bladderwort, and bur reed (Banfield, 1974). In Michigan, there are records of muskrats eating pondweeds (*Potomogeton*) in Gogebic and Ontonagon counties (Dice and Sherman, 1922); mussels (*Onadonta marginata*) in Marquette County (Manville, 1948); and mussels, fish, and young poultry in Osceola County (Wenzel, 1912). Besides a decided preference for ear corn (Errington, 1941), muskrats are known to pilfer clover, alfalfa, soybeans, carrots, apples and other fruits from cultivated fields, gardens and orchards.

ENEMIES. Errington (1939) pointed to intraspecific strife, movements away from home sites, and drought as factors conducive to accelerating muskrat losses from predatory animals. Young muskrats, capable of swimming at an early age (10 days following birth), become fair game for snapping turtles, bass, pickerel, and great northern pike. Adult muskrats attract larger predators. Food habit studies show that muskrats are eaten by herons (Mathiak, 1966), harriers and bald eagles (Smith, 1938), great horned owl, red-shouldered hawk, and red-tailed hawk (in Michigan, Craighead and Craighead, 1956); raven (Harlow *et al.*, 1975); gray wolf (Voigt *et al.*, 1976), coyote (Korschgen, 1957), gray fox (Errington, 1935), otter (Wilson, 1954), and bobcat (Pollack, 1951). The major four-footed predators in Michigan include red fox (Johnson, 1970; Murie, 1936; Schofield, 1960) and mink (Dearborn, 1932; Sealander, 1943). The mink is undoubtedly the leading muskrat predator, as the lives of these two are effectively interwoven in the marsh ecosystem (Errington, 1943).

Drought and reduced water availability also constitute major environmental hazards for the semiaquatic muskrat (Errington, 1961). Ironically, drowning takes its toll of these animals (Errington, 1937). Muskrats are frequently killed on Michigan highways, usually in the months of April, August and September, in both the Upper Peninsula (Manville, 1949) and the Lower Peninsula (Haugen, 1944; Kasul, 1976).

PARASITES AND DISEASES. Muskrats are infested with external and internal parasites. Most external parasites are mites (Smith, 1938). Internally, there is an array of parasites—protozoans (Lowery, 1974), cestodes, trematodes, and nematodes (Doran, 1954; Smith, 1938). The muskrat is one of the primary hosts of the blood fluke (*Schistosomatium douthitti*) in Michigan (A. C. Carmichael, pers. comm.; Penner, 1938).

Primarily due to the work on Iowa muskrats done by Paul L. Errington, a great deal is known about diseases in muskrats. Septicemia, coccidiosis, eye inflamation, abcesses, kidney trouble and leukemia are noted afflictions (Smith, 1938). A fungus skin disease produces a dandruff-like scruff and removes hair (Errington, 1942; Sather, 1958). Since tularemia has also been reported from muskrats, persons are cautioned to use care in handling the animals, especially sick ones (Parker *et al.*, 1951). This disease may be responsible for occasional die-offs. Midwestern muskrats are also subjected to a serious infectious, hemorrhagic epizootic, known in the literature as Errington's disease. This fatal malady causes hemorrhagic intestines and lungs as well as liver lesions. There seems to be some correlation between this disease and over-population (Sather, 1958) but this is not yet established.

MOLT AND COLOR ABERRATIONS. The young muskrat undergoes two molts and new hair growths in the first year, and the adult pelage is apparently molted more or less continuously throughout the year. Growing hair, which gives the pelt an unprime appearance, can be detected from the skin-side of the muskrat hide. It shows a bluish to blackish color because of the concentration of pigment granules in the roots of the growing hairs. As the hair lengthens, it is supplied with this dark pigment, leaving the skin creamy white in appearance on completion of the growth, denoting the prime condition (Linde, 1963). At this stage, which occurs during late winter (Stains, 1979), the pelts are the most valuable.

Color mutations are occasionally reported in nature. Animals with coats of fawn color, maltese gray, yellow, black and white are noted in the literature (Benton, 1953; Dozier, 1948b; Long and Hays, 1962). In Michigan, two albino kits were captured by officials of the Michigan Department of Natural Resources at Indian River in Cheboygan County on July 25, 1978.

ECONOMIC IMPORTANCE. Muskrat rapidly became part of the export trade after the arrival of the Europeans, although beaver received the most attention at the time. British traders at Fort Michlimackinac, for example, exported 514 pelts of muskrats (as compared with 50,938 pelts of beaver) in the summer of 1767 (Petersen, 1979). As trappable populations of beaver waned in the Upper Great Lakes Region after the beginning of the nineteenth century, muskrats were more widely sought. Johnson (1969) noted that in 1835–1839 the American Fur Company received 282,736 muskrat pelts from their Detroit Department (furs collected in Illinois, Indiana, Ohio, and Lower Peninsula Michigan) and 190,368 from their Northern Department (furs collected in Upper Peninsula Michigan and the area adjacent to Lake Michigan). Relatively few beaver hides were produced from these areas at that time—1,142 and 1,206 respectively. The muskrat eventually became the most important fur-bearing mammal in Michigan (Ruhl and Baumgartner, 1942)—a position it maintains today.

Annual estimates by the Michigan Department of Natural Resources show that rarely since 1940 has the legal take of muskrat pelts been fewer than 200,000. A high of 995,443 pelts was obtained in the 1943–1944 season. The price per pelt often governs the take. In 1822, fur buyers working for John Jacob Astor paid trappers 25¢ to 27½¢ per pelt (Anon, 1962). During the depression of the 1930s, trappers did little better; they were paid an average of 57¢ per pelt in Ingham County in the

1937–38 season (Hayne, 1941). According to newspaper accounts, prices rose steadily in the 1970s, when these furs became fashionable once again and the export market (mostly to Europe) flourished. In the early part of that decade, Michigan muskrat pelts brought an average of $1.40. In 1973, pelts were selling for an average of $2.75; in 1976, the high was $4.75; in 1977, $7.15; in 1978, $6.00; in 1979, $6.70; and in 1981, $7.30.

Hayne (1941) found that most licensed muskrat trappers in Ingham County in 1937–1938 were between the ages of 15 to 19 years. The opportunity to trap muskrats at a local marsh, slough, or stream has enabled youngsters to gain an appreciation for the value of wildlife and the need to conserve "seed" stock of the animals for the next year, while acquiring patience and technical skills in trapping and pelt preparation (see Petosky, 1969).

The idea of taking a muskrat's life for its hide may have an unpleasant connotation. However, if this renewable source is cropped so that the yield is sustained, the prolific muskrat will continue to be a valued source of furs and at the same time provide an additional cash crop for the land operator. Beginning trappers should seek advice from the officials of the Michigan Department of Natural Resources as to regulations, license requirements and techniques. Local or regional fur buyers or organizations like the Southern Michigan Trappers Association and the National Trappers Association can provide suggestions on good trapping methods, ethics, and the marketing of pelts.

The flesh of the muskrat is also esteemed as a delicacy. Lantz (1910) reported that near the turn of the century the Marsh Club in Monroe served an annual muskrat banquet. Today, according to DuFresne (1982), muskrats may be on the bill of fare at restaurants from November to mid-April in such Detroit River communities as Ecorse, Grosse Ile, Trenton, and Wyandotte. Earlier, the meat was sometimes listed as "aquaba." Now, it is fashionable to simply use the term rat. Recipes for preparing muskrat dishes appear in many cookbooks and outdoor publications (Johnson, 1925; Strains and Baker, 1958). One of the best is Susie Blaskiewicz' muskrat recipe as reported by DuFresne (1982) as follows:

4 muskrats
2 onions, sliced
1 carrot (optional)
celery tops
1 T salt
1 T pickling spice (available in supermarket spice sections)
Salt and pepper

Soak muskrats in cold, salted water for one hour. In a pot, place muskrats, one onion, carrot, celery tops, one tablespoon salt and pickling spice; cover with water. Cover tightly and parboil (simmer) for about one hour, or until fork tender. Remove fat from the meat, sprinkle with salt and pepper, and place in refrigerator at least overnight. Check again for fat, and then pan-fry in butter and remaining onions for five to 10 minutes. Serve with mashed potatoes and sauerkraut.

The meat is sold in Detroit area markets; Du Fresne reported that at one market packets of three rats, weighing in total between 4 and 5 lbs (1.8 to 2.2 kg), cost about $3.50. The musk (perineal) glands at the base of the tail and also the small lymphatic glands in the thigh muscles must be removed before cooking. Also, as is the case with any wild meat, the flesh may contain accumulations of heavy metals (Everett and Anthony, 1977) or be contaminated with polychlorinated biphenyl (PCB), as were muskrats examined from the drainage of the Shiawassee River in Livingston and Shiawassee counties in 1978.

Despite the food and fur value of muskrats, the animals are capable of being pests to land operators. Muskrats can plug drain tiles on farms, undermine road grades by tunneling, and molest farm crops when fields are adjacent to their aquatic homes. Muskrats like corn, especially the tender July stalks in Michigan. However, as Errington (1939b) explained, damage to garden and field crops is usually less than the value of the pelts. Dikes to control flooding and earthen dams constructed across small streams or drains are attractive home sites for muskrats. They can cause damage, leakage or even breakage of these structures when tunneling to construct bank dens (Cook, 1957). The possible use of muskrats as laboratory animals for behavioral research has been considered (Nagel and Kemble, 1974).

COUNTY RECORDS FROM MICHIGAN. Records of specimens of the muskrat in collections of museums and from the literature show that this species lives in all counties in Michigan. The animal also occurs on various islands in the Great Lakes including: in Lake Michigan, South Manitou (Scharf, 1973), Beaver (Ozoga and Phillips, 1964), Garden (Phillips *et al.*, 1965); in Lake Superior, Isle Royale (Burt, 1946); in Lake Huron, Drummond (Manville, 1950).

LITERATURE CITED

Anon.
1962 Fur and hide values, 1822. Michigan Conserv., 31:36–37.

Bailey, V.
1937 The Maryland muskrat marshes. Jour. Mammalogy, 18(4):350–354.

Banfield, A. W. F.
1974 The mammals of Canada. Univ. Toronto Press, Toronto. xxiv+438 pp.

Beer, J. R.
1950 The reproductive cycle of the muskrat in Wisconsin. Jour. Wildlife Mgt., 14(2):151–156.

Benton, A. H.
1953 An unusual concentration of albino muskrats. Jour. Mammalogy, 34(2):262.

Bishop, R. A., R. D. Andrews, and R. J. Bridges
1979 Marsh management and its relationship to vegetation, waterfowl and muskrats. Proc. Iowa Acad. Sci., 86(2): 50–56.

Burt, W. H.
1946 The mammals of Michigan. Univ. Michigan Press, Ann Arbor. xv+288 pp.

Cleland, C. E.
1966 The prehistoric animal ecology and ethnozoology of the Upper Great Lakes Region. Univ. Michigan, Mus. Anthro., Anthro. Papers, No. 29, x+294 pp.

Craighead, J. J., and F. C. Craighead, Jr.
1956 Hawks, owls and wildlife. Stackpole Co., Harrisburg, Pennylvania. xix+443 pp.

Cook, A. H.
1957 Control of muskrat burrow damage in earthen dikes. New York Fish and Game Jour., 4(2):213–218.

Dearborn, N.
1932 Foods of some predatory fur-bearing animals in Michigan. Univ. Michigan, Sch. For. and Conserv., Bull. No. 1, 52 pp.

Dice, L. R., and H. B. Sherman
1922 Notes on the mammals of Gogebic and Ontonagon counties, Michigan, 1920. Univ. Michigan, Mus. Zoology, Occas. Papers No. 109, 40 pp.

Doran, D. J.
1954 A catalogue of the Protozoa and helminths of North American rodents. II. Cestoda. American Midl. Nat., 52(2):469–480.

Dorney, R. S., and A. J. Rusch
1953 Muskrat growth and litter production. Wisconsin Conserv. Dept., Tech. Wildlife Bull. No. 8, 32 pp.

deVos, A., R. H. Manville, and R. G. Van Gelder
1956 Introduced mammals and their influence on native biota. Zoologica, 41(4):163–194.

Dozier, H. L.
1948a Estimating muskrat populations by house counts. U.S. Fish and Wildlife Service, Wildlife Leaflet 306, 17 pp.
1948b Color mutations in the muskrat (*Ondatra z. macrodon*) and their inheritance. Jour. Mammalogy, 29(4):393–405.

DuFresne, J.
1982 M-m-m-m-m-m muskrat love. *Detroit Free Press* (*Detroit* magazine) 151(314), March 14, pp. 19–20.

Errington, P. L.
1935 Food habits of mid-west foxes. Jour. Mammalogy, 16(3):192–200.
1937 Drowning as a cause of mortality in muskrats. Jour. Mammalogy, 18(4):497–500.
1939a Reaction of muskrat populations to drought. Ecology, 20(2):168–186.
1939b Observations on muskrat damage to corn and other crops in central Iowa. Jour. Agric. Res., 57(6):415–421.
1939c Observations on young muskrats in Iowa. Jour. Mammalogy, 20(4):465–478.
1940 Natural restocking of muskrat-vacant habitat. Jour. Wildlife Mgt., 4(2):173–185.
1941 Versatility in feeding and population maintenance of the muskrat. Jour. Wildlife Mgt., 5(1):68–89.
1942 Observations on a fungus skin disease of Iowa muskrats. American Jour. Vet. Res., 3(7):195–201.
1943 An analysis of mink predation upon muskrats in northcentral United States. Iowa State Coll., Agric. Exp. Sta., Res. Bull. 320, pp. 798–924.
1951 Concerning fluctuations in populations of the prolific and widely distributed muskrat. American Nat., 85(824):273–292.
1961 Muskrat and marsh management. The Stackpole Co., Harrisburg, PA. 183 pp.
1963 Muskrat populations. Iowa State Univ. Press, Ames, 665 pp.

Everett, J., and R. G. Anthony
1977 Heavy metal accumulation in muskrats in relation to water quality. Trans. Northeast Sect. Wildlife Soc., 33:105–118.

Fish, F. E.
1979 Thermoregulation in the muskrat (*Ondatra zibethicus*): The use of regional heterothermia. Comp. Biochem. Physiol., 64A(3):391–397.

Harlow, R. F., R. G. Hooper, D. R. Chamberlain, and H. S. Crawford
1975 Some winter and nesting season foods of the common raven in Virginia. Auk, 92(2);298–306.

Haugen, A. O.
1944 Highway mortality of wildlife in southern Michigan. Jour. Mammalogy, 25(2):160–170.

Hayne, D. W.
1941 Michigan trappers. Michigan State Coll., Agric. Exp. Sta., Sp. Bull. 307, 34 pp.

Higgins, M. J.
1980 An analysis of the faunal remains from the Schwerdt Site, a late prehistoric encampment in Allegan County, Michigan. Western Michigan Univ., unpubl. M.A. thesis. v+77 pp.

Hollister, N.
1911 A systematic synopsis of the muskrats. N. American Fauna No. 32, 47 pp.

Holman, J. A.
1975 Michigan's fossil vertebrates, Michigan State Univ., Publ. Mus., Ed. Bull. No. 2, 54 pp.

Jackson, H. H. T.
1961 Mammals of Wisconsin. Univ. Wisconsin Press, Madison. xii+504 pp.

Johansen, K.
1962 Buoyancy and insulation in the muskrat. Jour. Mammalogy, 43(1):64–68.

Johnson, C. E.
1925 The muskrat in New York: Its natural history and economics. Roosevelt Wildlife Bull., 3(2):205–320.

Johnson, D. R.
1969 Returns of the American Fur Company, 1835–1839. Jour. Mammalogy, 50(4):836–839.

Johnson, I. A.
1919 The Michigan fur trade. Michigan Hist. Comm., Lansing, xii+201 pp.

Johnson, W. L.
1970 Food habits of the red fox in Isle Royale National Park, Lake Superior. American Midl. Nat., 84(2):568–572.

Kasul, R. L.
1976 Mortality and movement of mammals and birds on a Michigan interstate highway. Michigan State Univ., unpubl. M.S. thesis, ix+39 pp.

Korschgen, L. J.
1957 Food habits of coyotes, foxes, house cats and bobcats in Missouri. Missouri Conserv. Comm., Fish and Game Div., P-R Ser. No. 15, 64 pp.

Krebs, C. J., and J. H. Myers
1974 Population cycles in small mammals. Adv. Ecol. Res., 8:267–399.

Lantz, D.
1910 The muskrat. U.S.D.A., Farmers Bull. 396, 38 pp.

Lay, D. W.
1945 Muskrat investigations in Texas. Jour. Wildlife Mgt., 9(1):56–76.

Linde, A. F.
1963 Muskrat pelt patterns and primeness. Wisconsin Conserv. Dept., Tech. Bull. No. 29, 66 pp.

Long, C. A., and H. A. Hays
1962 Yellow mutant in the muskrat, *Ondatra zibethicus*. Jour. Mammalogy, 43(1):104

Lowery, G. H., Jr.
1974 The mammals of Louisiana and its adjacent waters. Louisiana State Univ. Press, Baton Rouge. xxiii+565 pp.

MacArthur, R. A.
1977 Muskrats and cold. A study of survival. Manitoba Nature, Winter, pp. 20–25.
1978 Winter movements and home range of the muskrat. Canadian Field-Nat., 92(4):345–349.
1979 Seasonal patterns of body temperature and activity in free-ranging muskrats (*Ondatra zibethicus*). Canadian Jour. Zool., 57(1):25–33.

MacArthur, R. A., and M. Aleksiuk
1979 Seasonal microenvironments of the muskrat (*Ondatra zibethicus*) in a northern marsh. Jour. Mammalogy, 60(1):146–154.

Manville, R. H.
1948 The vertebrate fauna of the Huron Mountains, Michigan. American Midl. Nat., 39(3):615–640.
1949 Highway mortality in northern Michigan. Jour. Mammalogy, 30(3):311–312.
1950 The mammals of Drummond Island, Michigan. Jour. Mammalogy, 31(3):358–359.

Mathiak, H. A.
1966 Muskrat population studies at Horicon Marsh. Wisconsin Conserv. Dept., Tech. Bull. No. 35, 56 pp.

Murie, A.
1936 Following fox trails. Univ. Michigan, Mus. Zool., Misc. Publ. No. 32, 45 pp.

Nagel, J. A., and E. D. Kemble
1974 The muskrat (*Ondatra zibethicus*) as a laboratory animal. Behavior. Res. Meth. and Instru., 6(1):37–39.

O'Neil, T.
1949 The muskrat in the Louisiana coastal marshes. Louisiana Dept. Wildlife and Fish., 152 pp.

Ozoga, J. J., and C. J. Phillips
1964 Mammals of Beaver Island, Michigan. Michigan State Univ., Publ. Mus., Biol. Ser., 2(6):305–348.

Parker, R. R., E. A. Steinhaus, M. Kohls, and W. L. Ellison
1951 Contamination of natural waters and mud with *Pasteurella tularensis* and *tularemia* in beavers and muskrats in the northwestern United States. Nat. Inst. Hlth., Bull. No. 193, 61 pp.

Penner, L. R.
1938 *Schistosomatium* from the muskrat *Ondatra zibethica* in Minnesota and Michigan. Jour. Parasitol., 24:26.

Petersen, E. T.
1979 Hunters' heritage. A history of hunting in Michigan. Michigan United Conserv. Clubs, Lansing. 55 pp.

Peterson, A. W.
1950 Backward swimming of the muskrat. Jour. Mammalogy, 31(4):453.

Petoskey, M. L.
1969 To trap a muskrat or mink. Michigan Nat. Res., 38(1):19–22.

Phillips, C. J., J. J. Ozoga, and L. C. Drew
1965 The land vertebrates of Garden Island, Michigan. Jack-Pine Warbler, 43(1):20–25.

Pollack, E. M.
1951 Food habits of the bobcat in the New England states. Jour. Wildlife Mgt., 15(2):209–213.

Ruhl, H. D., and L. L. Baumgartner
1942 Michigan's million dollar muskrats. Michigan Conserv., 11(8):6–7, 11.

Sather, J. H.
1958 Biology of the Great Plains muskrat in Nebraska. Wildlife Monog., No. 2, 35 pp.

Scharf, W. C.
1973 Birds and land vertebrates of South Manitou Island. Jack-Pine Warbler, 51(1):3–19.

Schofield, R. D.
1955 Analysis of muskrat age determination methods and

their application in Michigan. Jour. Wildlife Mgt., 19(4):463–466.

1960 A thousand miles of fox trails in Michigan's ruffed grouse range. Jour. Wildlife Mgt., 24(4):432–434.

Sealander, J. A.
1943 Winter food habits of mink in southern Michigan. Jour. Wildlife Mgt., 7(4):411–417.

Seton, E. T.
1953 Lives of game animals. Charles T. Branford Co., Boston. 4(2):441–949.

Smith, F. R.
1938 Muskrat investigations in Dorchester County, MD, 1930–34, U.S.D.A. Cir. No. 474, 24 pp.

Sprugel, G., Jr.
1951 Spring dispersal and settling activities of central Iowa muskrats. Iowa State Coll., Jour. Sci., 26(1):71–84.

Stains, H. J.
1979 Primeness in North American furbearers. Wildlife Soc. Bull., 7(2):120–124.

Stains, H. J., and R. H. Baker
1958 Furbearers in Kansas: A guide to trapping. Univ. Kansas, State Biol. Serv. and Mus. Nat. Hist., Misc. Publ. No. 18, 100 pp.

Stewart, R. W., and J. R. Bider
1977 Summer activity of muskrats in relation to weather. Jour. Wildlife Mgt., 41(3):487–499.

Takos, M. J.
1944 Summer movements of banded muskrats. Jour. Wildlife Mgt., 8(4):307–311.

Tevis, L., Jr.
1950 Summer behavior of a family of beavers in New York state. Jour. Mammalogy, 31(1):40–65.

Voigt, D. R., G. B. Kolenosky, and D. H. Pimlott
1976 Changes in summer foods of wolves in central Ontario. Jour. Wildlife Mgt., 40(4):663–668.

Wenzel, O. J.
1912 A collection of mammals from Osceola County, Michigan. Michigan Acad. Sci., 14th Rept., pp. 198–205.

Willner, G. R., J. A. Chapman, and J. R. Goldsberry
1975 A study and review of muskrat food habits with special reference to Maryland. Maryland Wildlife Adm., Publ. Wildlife Ecol., No. 1, v+25 pp.

Willner, G. R., G. A. Feldhamer, E. E. Zucker, and J. A. Chapman
1980 *Ondatra zibethicus*. American Society of Mammalogists, Mammalian Species, No. 141, 8 pp.

Wilson, K. A.
1954 The role of mink and otter as muskrat predators in northeastern North Carolina. Jour. Wildlife Mgt., 18(2):199–207.

Wragg, L. E.
1955 Notes on movements of banded muskrats. Canadian Field-Nat., 69(1):9–11.

SOUTHERN BOG LEMMING
Synaptomys cooperi Baird

NAMES. The generic name *Synaptomys,* proposed by Baird in 1858, is derived from two Greek words, *synapt* meaning joined together and *omys* meaning mouse. This word combination relates to the idea that this rodent serves as a connecting link between two groups of microtines, the lemmings and the voles. The specific name, also proposed by Baird, refers to the donor of the specimen used in the original description, William Cooper of Hoboken, New Jersey. In Michigan, local residents call this rather unusual rodent a field mouse. The subspecies *Synaptomys cooperi cooperi* Baird occurs in Michigan.

RECOGNITION. Body size small and robust (Fig. 55); head and body of an adult averaging 3⅞ in (100 mm) long; tail finely-haired and characteristically short, averaging ¾ in (20 mm) long, about the same length as the hind foot; body covered

Figure 55. The southern bog lemming, like other Michigan voles, has a nondescript, furry, and cylindrical body, rounded at both ends. Anteriorly, the snubbed nose is adorned on each side by sparse whiskers, small beady eyes, and rounded ears almost hidden in fur. Posteriorly, the tail is short and inconspicuous. Photograph taken in Livingston County on July 7, 1981, by Dr. Lawrence L. Master.

with soft, long, thick pelage; head large, blunt with no evidence of neck; eyes small, dark, beady; ears short, rounded, and almost concealed by the hair on the sides of the head; vibrissae (whiskers) medium long, inconspicuous; feet short; hind limbs longer than fore limbs. In adults, the upperparts and sides have a grizzled gray, or gray-brown look caused by a mixture of black-based hairs tipped with gray, brown, and cinnamon; underparts are lead gray resulting from silvery tips on the basally-dark hairs; tail is indistinctly bicolored, gray above and silver-gray below; feet are gray; hip glands of males sometimes support tufts of dingy white hairs. The winter pelage is grayer and longer than summer pelage. Three pairs of mammary glands are characteristic of the southern bog lemming; one pair is pectoral and two are inguinal. There are four functional toes on the small forefeet (the vestigial pollex posesses a small claw) and five on the hind feet. The karyotype of the southern bog lemming is 2n = 50; NF = 52 (Hoffmann and Nadler, 1976).

The grizzled gray to gray-brown dorsal pelage, the short tail (no longer than the length of the hind feet), the short, plump appearance, and the presence of only four functional toes on the forefeet serve, in combination, as external characteristics to distinguish the southern bog lemmings from all other small mammals in Michigan (Plate 4b).

MEASUREMENTS AND WEIGHTS. Adult southern bog lemmings measure as follows: length from tip of nose to end of tail vertebrae, 4¼ to 5⅛ in (110 to 130 mm); length of tail vertebrae, 9/16 to ⅞ in (15 to 23 mm); length of hind foot, ⅝ to 13/16 in (16 to 20 mm); height of ear from notch, ⅜ to ½ in (9 to 13 mm); weight, 0.7 to 1.4 oz (20 to 40 g).

DISTINCTIVE CRANIAL AND DENTAL CHARACTERISTICS. The southern bog lemming has the massive, flattened, angular skull characteristic of other microtines in Michigan, except for the southern red-backed vole (Fig. 51). The braincase is long; the rostrum is depressed and short; the outline of the postorbital areas is almost at right angles to the long axis of the skull; and the zygomatic processes are heavily constructed. The skull averages 1 in (26 mm) long and ⅝ in (15.5 mm) wide. Dental characteristics include: upper incisors with distinctive longitudinal grooves on outer edges and yellow color on front surfaces; molar (cheek) teeth not rooted and evergrowing; upper row of molar teeth noticeably long, more so than the length of the incisive foramina; molar teeth also with prismatic, flat-crowned grinding surfaces displaying folded and distinctively deep reentrant angles on the inner (lingual) side of the lower teeth and on the buccal (cheek) side of the upper teeth (see Fig. 52).

The skull of the southern bog lemming can be distinguished from those of all other small mammals in Michigan by the presence of conspicuous longitudinal grooves on the outer edges (not centered) of the upper incisors, the extreme length of the maxillary cheek tooth-row, and the prismatic, flat-crowned grinding surfaces on the molar teeth with deep reentrant angles on the inner side of these teeth in the lower jaw and on the outer side of those in the upper jaw. The dental formula for the southern bog lemming is:

$$\text{I (incisors)}\frac{1\text{-}1}{1\text{-}1}, \text{C (canines)}\frac{0\text{-}0}{0\text{-}0}, \text{P (premolars)}\frac{0\text{-}0}{0\text{-}0}, \text{M (molars)}\frac{3\text{-}3}{3\text{-}3} = 16$$

DISTRIBUTION. The southern bog lemming is found throughout northeastern United States and adjacent southern Canada. Its concentration ranges from southern Quebec west to southeastern Manitoba, south in the eastern sector of the Great Plains as far as western Kansas, and east to the Atlantic coast of North Carolina. In Michigan, the southern bog lemming occurs in both peninsulas (see Map 42). Although its modern distribution is known from specimens acquired for preservation in museums beginning in the latter part of the last century (Master, 1978; Wood, 1914), archaeological evidence also exists. The southern bog lemming's presence in Michigan is thought to date back to at least 1,700 years B.P. (Middle Woodland Culture) at the Schultz Site in what is now Saginaw County (Luxenberg, 1972) and to 600 to 1,200 years B.P. at the Juntenen Site on Bois Blanc Island (Cleland, 1966). These records lead to the conclusion that the southern bog lemming was part of the late Pleistocene/Recent postglacial faunal assemblage of the area (Parmalee, 1968; Wetzel, 1955). Today, despite the belief that the southern bog lemming exists in most, if not all, Michigan counties, its distribution is spotty and often unpredictable, causing mammalogists to consider classifying the species as threatened.

Of about 300 museum specimens (excluding those obtained from the examination of owl pellets)

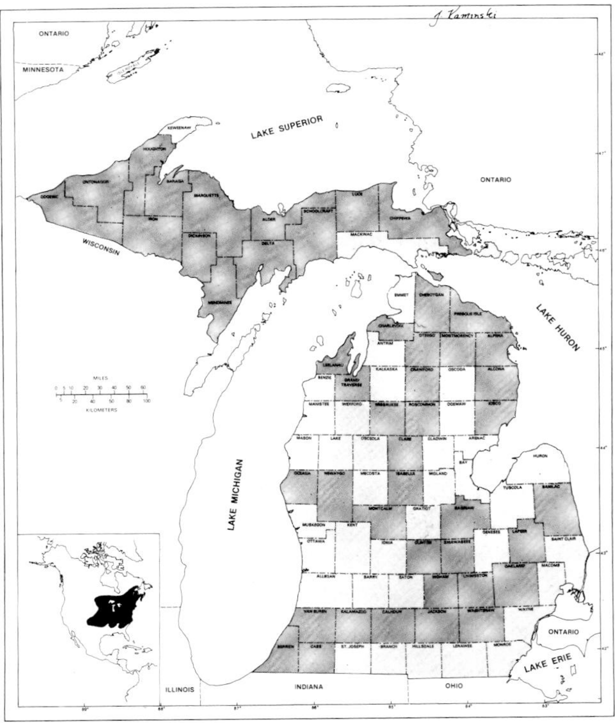

Map 42. Geographic distribution of the southern bog lemming (*Synaptomys cooperi*). Counties from which the species has actually been recorded are dark shaded.

collected in Michigan from the 1880s to the 1970s (records mostly from Master, 1978), 78% were secured in the 1930s, 1940s and 1950s, with only 12% preserved in the 1960s and 1970s. There may have been less emphasis on collecting and preserving specimens in the past 20 years than in the previous three decades. Even so, it is suspected, without substantial data, that the southern bog lemming has declined generally in the southern part of the state, with the meadow vole perhaps usurping some of its habitat. It is also possible that grassy areas, forest edges, and marsh environments conducive to the southern bog lemming's survival have been altered by human land-use practices, while many exotic grasses and forbs to which the meadow vole may be better adapted have increased (Brewer and Reed, 1977; Connor, 1959; Master, 1978).

HABITAT PREFERENCES. The southern bog lemming, contrary to its name, does not inhabit only bogs and other damp areas but also dry and well-drained uplands covered with grass or forests (Connor, 1959; Hatt, 1930; Howell, 1927). Its rather discontinuous distribution and appearance in a variety of open and forested habitats have perplexed naturalists; however, an improved knowledge of pronounced population fluctuations in this species is now providing some basis for many of the conclusions drawn by the often brief and casual field observations in the past (Gaines *et al.,* 1977). Southern bog lemmings have been trapped, for example, in recent clearcuts of northern deciduous and coniferous forests in West Virginia (Kirkland, 1977); in wet, mixed forest in Minnesota (Timm, 1975); in dense upland hardwoods in Wisconsin (Clark, 1972); in dry, lush ground cover of unmown bluegrass in Indiana (Mumford, 1969); in habitats containing green, succulent, monocotyledonous plants, primarily sedges and grasses, in New Jersey (Connor, 1959); in sphagnum-sedge bog in Quebec (Wrigley, 1969); on hillside meadows overgrown with blackberry vines and matted crabgrass in Ohio (Oehler, 1942); in heavy beech-hemlock woods in New York (Hamilton, 1941a); in abandoned fields dominated by brome grass and invaded to varying degrees by woody vegetation in Kansas (Gaines *et al.,* 1977).

In Michigan, the southern bog lemming demonstrates its broad tolerance of a wide variety of land types, including intermediate non-woody/woody plant stages in ecological successions (Master, 1978). Michigan habitats include blue-grass association in Livingston County (Blair, 1948); tall sedge, black spruce bog, tamarack bog, and both wet and dry hardwood forest in Ontonagon and Gogebic counties (Dice and Sherman, 1922); sphagnum bog under spruce trees in Montmorency County (Green, 1925); sphagnum bogs near Burt Lake and along Carp Creek in Cheboygan Couny (Wenzel, 1911); grassy areas containing woody plants in Washtenaw County (Getz, 1961); field-shocked corn in winter in Cass, Clinton, and Washtenaw counties (Linduska, 1942; Wood, 1922; Wood and Dice, 1923); bog, spruce-blueberry barrens, and hardwood-pine in Baraga and Marquette counties (Robinson, 1975); wet grassy depressions and tamarack swamp in Washtenaw County (Pruitt, 1959); cedar swamp in Leelanau County (Hatt, 1923); second-growth hardwood and conifer forests in Charlevoix County (Dice, 1925); and a brush pile in Crawford County (Philip Myers, pers. comm.).

There is abundant evidence (Blair, 1948; Burt, 1940; Linduska, 1950) that southern bog lemmings in Michigan generally occupy habitats in which shrub or tree growth is intermixed with herbaceous vegetation. Less woody and more grassy sites may be used by the species only when its local populations are high. Comparable studies of small mammals were made by Linduska (1950) in 1940–1943 and by Shier (1981) in 1978–1980 at the Rose Lake Wildlife Research Center in Clinton County. In the 1940s, Linduska captured 41 southern bog lemmings in such places as old hedgerows, brushy old fields, winter corn shocks, and open grassy woodlots. Southern bog lemmings were abundant during these years, at least in 1942 and 1943. Repeating this study almost 40 years later during a period when southern bog lemmings were less numerous, Shier failed to obtain this vole in the exact habitats in which they were found by Linduska. Instead, the 20 individuals which she did examine were captured exclusively in pine hedgerows and plantations, where the ground cover consisted of pine needles interspersed with grasses (mostly *Bromus*). The southern bog lemmings had surface runways in the grassy patches, and underground tunnels where grass was sparse and pine needles blanketed the ground. Shier's results allow for the speculation that the southern bog lemming may have flourished in similar habitat in the uncut, park-like stands of mixed conifers and hardwoods which are reported to have dominated the scene in presettlement days.

DENSITY AND MOVEMENTS. Modern biologists are now aware that the southern bog lemming's presence in a single habitat one season or year and its absence in another season or year may result from dramatic cyclic fluctuations in local populations (at three to four-year intervals, two years in a study in southern Illinois, Beasley, 1978). An alternative explanation is that the southern bog lemming is trap-shy (Master, 1978); however, studies by Gaines *et al.* (1977) in Kansas, as well as Linduska's (1950) investigations at Rose Lake in Clinton County in the 1940s, suggest that the southern bog lemming, at least on occasion, is as easily trapped as some other small mammals in their own communities.

In Michigan, a series of trapping studies were conducted at the Edwin S. George Reserve of the University of Michigan in Livingston County from 1936 to 1942 (Blair, 1948; Burt, 1940) and at the Rose Lake Wildlife Research Station of the Michigan Department of Natural Resources in Clinton County from 1940 to 1943 and again in 1946 (Linduska, 1950). These observations showed southern bog lemmings to be abundant in 1936, 1938, 1942–43, and 1946 but either scarce or absent (at least not readily trappable) in 1937, 1939, 1940, and 1941. Not only are the reasons for these density fluctuations obscure but so are the means by which the residual animals survive when populations are reduced.

In these three studies, trapping records showed local populations of five individuals on a 7.6 acre (3 ha) study plot in 1938 (Blair, 1948) and 18 on a one-half acre (0.2 ha) study plot in 1943 (Linduska, 1950). An estimate of 14 animals per acre (5.6 per ha) was obtained by Stegeman (1930). Banfield's (1974) report of populations varying between about two to 14 individuals per acre (0.8 to 5.6 per ha) probably reflects the population sizes to be expected in Michigan. The high of 35 southern bog lemmings per acre (14 per ha) reported in Illinois by Hoffmeister and Warnock (1947) illustrates the magnitude of population size during a peak in the multi-annual cycle. As is the case with other microtines, population densities are apparently controlled by predation and other mortality at the living sites, as well as by the rodents' tendency to disperse (Gaines *et al.*, 1979).

Because of the dramatic fluctuations in local populations of the southern bog lemming, it is difficult to assess the significance of various population figures obtained from the literature. As Burt (1940) reported, when the species is abundant, it can be found in low, moist places supporting matlike growths of grass as well as in various habitats on high ground and in woodland; when at a population low, the animals seem untrappable in any of these places.

Burt live-trapped, marked, released, and retrapped one adult female seven times in 29 days in August and September, during which time she moved within an area with a diameter of 40 yd (37 m). Eighteen days later (in October), she had moved away about 115 yd (106 m). Burt estimated that her home range might be 0.5 acre (0.2 ha). In New Jersey, Connor (1959) found southern bog lemmings confined themselves to home ranges no greater than 1/10 to 1/4 acre (0.04 to 0.1 ha); in Manitoba, Buckner (1957) calculated the home range of a subadult female as 0.8 to 1.0 acre (0.3 to 0.4 ha), a cruising range of about 105 to 118 ft (32 to 36 m). Getz (1960) obtained an average home range size of 0.11 acre (0.04 ha) for males and 0.14 acre (0.06 ha) for females and concluded that southern bog lemmings and meadow voles have home ranges of almost the same size.

BEHAVIOR. The often disjunct distribution of populations of the southern bog lemming leads field observers to believe that this rodent lives in small colonies, perhaps consisting of family groups (Stegeman, 1930). Although these rodents can be active both day and night, Oehler (1942) thought they were most active in late afternoon and at night, especially after rains or during periods of thawing snow. In contrast, Linduska (1950) observed southern bog lemmings moving conspicuously at mid-day across snow surfaces in between tunnels.

Southern bog lemmings seem to spend much of their time in their runway systems. These trails may be cut through rank bluegrass, beds of sphagnum, or pushed-up leaf mold in forested areas (Connor, 1959). Surface trails are usually well concealed from above by grassy cover. Runways average 1.5 in (38 mm) wide and are normally kept cleared of debris, except for vegetation cuttings (scattered and in characteristic piles) and latrine-like deposits of feces. In Washtenaw County, Stegeman (1930) mapped a system of surface runways, noting side trails which he thought were used for resting, feeding and depositing wastes. Descriptions of the runway systems of the meadow vole and the prairie vole fit the southern bog lemming runway systems as well; all are basically graminivorous animals and use vegetation along trail borders as food. All three of these microtines make piles of the basal parts of grasses and sedge stems and similarly-shaped feces (Connor, 1959). However, recent studies showed that clippings (especially of sedges) and color of fecal droppings are distinguishable in most cases between the southern bog lemming and the two voles (Linzey, 1981; Master, 1978; Wilkinson, 1980). Fresh bright green (almost iridescent) droppings in sizable piles found in close association with even-length cuttings of sedges (*Carex*) are almost surely indicative of the presence of southern bog lemmings. In contrast, fresh dull green or dark green droppings in scattered piles and not closely associated with small accumulations of uneven length cuttings of grasses and sedges are generally characteristic of meadow and prairie voles.

Southern bog lemmings are docile when handled and seem tolerant of each other. On one occasion, Connor (1959) found two subadults in the same trap. In captivity, the rodents can be quarrelsome; Burt (1928) even noted cannibalism among captives. Workers agree that the southern bog lemming is difficult to keep in captivity; to date, only Connor (1959) has induced breeding in captivity. The southern bog lemming can become lethargic in confinement, resulting in a high mortality rate.

Southern bog lemmings are among the most vocal of Michigan microtines. Connor (1959) categorized vocalizations as: quarrelsome-sounding notes variously expressing threats, anger, or fear; courting notes; and maternal calls. Captive voles also spend considerable time grooming themselves. They have an ambling gait, and move rapidly through their runway systems, less rapidly in unfamiliar territory. According to Connor (1959), the species can climb and swim without difficulty. Wenzel (1911) observed a bog lemming swimming across Carp Creek in Cheboygan County.

Nests of the southern bog lemming are located on the ground surface, below ground in side chambers of tunnels, and in hollow logs or in stumps (see summary of observations in Connor, 1959). The nests are similar to those of meadow voles, being somewhat round and constructed mostly of dried grasses or sedges with moss, woody material, feathers, fur, and other items. Nests may measure from 3½ to 6 in (90 to 165 mm) in diameter. There may be one to four entrances. Burt (1928) was of the opinion that underground nests were used more in winter and surface nests in summer. Connor (1959) noted that the interior and vicinity of nests are kept clear of droppings and other debris.

ASSOCIATES. The southern bog lemming's runways offer other small animals—including snakes and small weasels—the opportunity to move rapidly under a canopy of grass or matting. In Michigan and adjacent states, environmental associates reported include: (essentially seed-eaters) white-footed mouse, deer mouse, house mouse, meadow jumping mouse, and woodland jumping mouse; (essentially insectivores) arctic shrew, masked shrew, pygmy shrew, short-tailed shrew, least shrew; and (essentially vegetation-eaters and grazers) southern red-backed vole, prairie vole, meadow vole, and woodland vole (Blair, 1940; Fisher, 1945; Getz, 1961; Haveman, 1973; Linduska, 1942, 1950; Linzey, 1981; Ryan, 1982; Shier, 1981; and Timm, 1975).

In competing for space and for nutritional resources, the southern bog lemming probably interacts mostly with its microtine relatives, especially the grass-eating prairie vole (only in southwestern Lower Peninsula Michigan) and the meadow vole. Getz's (1961) study in Washtenaw County showed that although both the meadow vole and the southern bog lemming are dependent on graninoids as food sources, normally these two species will isolate themselves spatially, with the meadow vole avoiding grassy areas containing woody plants and the southern bog lemming using these areas. In Kansas, Rose and Spevak (1978) also found the southern bog lemming used shrubby and more moist parts of old fields while the prairie vole occupied drier, upland sites, but the two interspersed in old fields lacking woody vegetation. These two species have also been captured in the same runways in Kansas (Burt, 1928) and in Nebraska (Fichter and Hansen, 1947). Even so, there is some evidence that if two of these grass-eating species occur in the same area, one is usually more abundant (Baker, 1971). Beasley (1978) found prairie vole and southern bog lemming populations cycling concurrently in Illinois. Recent work shows that the southern bog lemming has a broader tolerance for habitats than the meadow vole. When the two occur in the same area, although the latter species may occupy the more open grassy areas, the less aggressive southern bog lemming can still survive in the brushy fringes, being able to thrive in sparse cover and on foods low in nutritional value (Linzey, 1981).

REPRODUCTIVE ACTIVITIES. Most of the information concerning the breeding habits of the southern bog lemming has been obtained from Connor's (1959) summary for animals in New Jersey. Normally the breeding season corresponds with the warmer months of the year, with most births occurring in spring and autumn (Schwartz and Schwartz, 1981), although the breeding season is extended during peaks in local populations (Beasley, 1978). However, year-round reproductive activity can be expected in Michigan, if shelter and food conditions are ideal (Burt, 1928). Normally, male testes are scrotal from April to December; females are seasonally polyestrous and undergo post-partum estrus within a few hours after the delivery of litters. There is no evidence that lactation prolongs the gestation period.

Following conception the gestation period lasts for 23 days (Connor, 1959). Litters may contain as few as one or as many as seven young with an average of three (Coventry, 1937). The young are born in the grassy interior of the nest. At birth, they weigh about 3.9 g; have closed eyes and ear pinnae; have semi-transparent, wrinkled, and pink skin; are devoid of conspicuous hair except vibrissae (whiskers); have claws; and can emit squeaks or cheeps. At three days (Connor, 1959), the back darkens as hair begins to emerge; at four days, the incisors are visible through the jaw tissue and calls are more audible; at seven days, the young are well-furred and upper and lower incisors have erupted; at eight to nine days, the ear canals and the eyes open, and the young can move vigorously, calling in adult-like chatter; at 12 to 13 days, the young begin to eat solid food; at 14 or more days, they become independently active and the dark, gray-brown dorsal pelage appears; at 16 to 21 days, the young are completely weaned; at about 28 days, the juvenal pelage is replaced by a subadult growth of gray-brown hair; at about 60 days, the young weigh 30 g; and at about 95 days, the young animals are in adult pelage. If the southern bog lemmings have a reproductive life resembling those of the related meadow and prairie voles, the young are probably sexually mature at 60 days; however, Connor (1959) suggested that males may breed at 35 days. He reported that a single female in captivity produced six litters in 26 weeks for a total of 22 young. It is unlikely that similar production is achieved very often in nature.

FOOD HABITS. In his study of southern bog lemmings in New Jersey, Connor (1959) concluded that major food sources are green, succulent, monocotyledonous plants, primarily grasses and sedges. He also listed, from his own studies and those reported in the literature, a broad array of plant materials including fungi, mosses, liverworts, ferns, several species of grasses and sedges, various forbs, woody plant parts, fruits of raspberry, blackberry, huckleberry, and cranberry, plant roots and leaves, and bark; beetles, slugs, snails, and other invertebrates (see also Hamilton, 1941b). Although Connor reported no food caches in his study, there are references to winter stores containing stems of bluegrass, white clover, and tuberous roots of the wild artichoke (*Helianthus doronicoides*) accumulated by the southern bog lemming in Indiana (Quick and Butler, 1885). Linduska (1950) found this vole under field-shocked corn in Clinton County, presumably for both cover and a ready food supply.

Like other microtines, the southern bog lemming has a digestive system especially adapted for utilizing bulky amounts of grass, usually high in quantity and low in quality (Baker, 1971). Grasses and sedges are basic foods and normally obtainable along runway systems. The animals usually squat under this foliage and snip off sections of the basal parts of the grass or sedge, pulling down the top parts to eat and leaving a deposit of the stem sections. These stems are usually 1 to 3 in (25 to 75 mm) long and scattered or in large piles in runways or side chambers. According to Connor (1959), the southern bog lemming makes larger piles of such cuttings (usually of equal length) than does the meadow vole. In Michigan, cover/forage plants known to be palatable to the southern bog lemming include June-grass (*Koeleria*), bluegrasses (*Poa*), brome (*Bromus*), white clover, rushes (*Juncus*), and sedges (*Carex*, especially *C. oligosperma*, Master, 1978 and Wilkinson, 1980). Shier (1981) found this rodent to feed from April to November almost exclusively on brome in pine plantations in Clinton County.

ENEMIES. Being active both day and night, the southern bog lemming is exposed to both diurnal and nocturnal predators. Numerous instances have been cited of meat-eaters preying on this rodent; the number of references is partially due to the ease which which the southern bog lemming's distinctive molars can be accurately identified in stomach contents, pellets, and fecal droppings of predators.

Michigan southern bog lemming predators include representatives of at least three vertebrate classes: Reptilia (snakes), Aves (principally hawks and owls), and Mammalia (probably Insectivora, certainly Marsupialia and Carnivora). Major avian predators are the screech owl, barn owl, long-eared owl, and the great horned owl (Armstrong, 1958; Craighead and Craighead, 1956; Errington *et al.*, 1940; Linduska, 1950; Voight and Glenn-Lewin, 1978; Wallace, 1948; Wilson, 1938). Four-footed predators include gray wolf, coyote, red fox, gray fox, badger, house cat, and bobcat (Dearborn, 1932; Geis, 1952; Korschgen, 1957; Murie, 1936; Spiker, 1933; Voigt *et al.*, 1976; Worthin and Van Tyne, 1922). Although there are no published records, short-tailed shrews, least weasels, ermine, long-tailed weasels, and mink probably also eat southern bog lemmings.

PARASITES AND DISEASES. External parasites obtained from southern bog lemmings include mites, ticks, lice (Anoplura), and fleas (Connor, 1959; Timm, 1975); internal parasites include tapeworms (Doran, 1954).

MOLT AND COLOR ABERRATIONS. As mentioned under Reproductive Activities, weanling southern bog lemmings have a fuzzy, gray-brown juvenal coat which is molted about 28 days after birth and replaced by a darker and duller subadult pelage. About 95 days after birth, the adult pelage is attained. There are two molts annually with a grayer and longer winter coat appearing in late autumn (October to December) and a browner and shorter coat in late spring and early summer (June to August). An albino individual was captured in Clinton County (Manville, 1955).

ECONOMIC IMPORTANCE. The southern bog lemming has the ability to gnaw roots and trunks of economically important trees, but there are no reports that it does so. In most of its living places, it is segregated from agricultural crops and managed orchards. However, legumes (clovers and alfalfa) may be attractive foods and field-shocked corn has been used as winter havens. The southern bog lemming seems of little economic consequence in land management practices.

COUNTY RECORDS FROM MICHIGAN. Records of specimens of the southern bog lemming in collections of museums and from the literature show that this species is reported from 13 of the 15 counties in the Upper Peninsula and from 33 of the 68 counties in the Lower Peninsula, plus a record from an archaeological site (Cleland, 1966) from Bois Blanc Island (see Map 42).

LITERATURE CITED

Armstrong, W. L.
1958 Nesting and food habits of the long-eared owl in Michigan. Michigan State Univ., Publ. Mus., Biol. Ser., 1(2):61–96.

Baker, R. H.
1971 Nutritional strategies of myomorph rodents in North American grasslands. Jour. Mammalogy, 52(4):800–805.

Banfield, A. W. F.
1974 The mammals of Canada. Univ. Toronto Press, Toronto. xxiv+438 pp.

Beasley, L.
1978 Demography of southern bog lemmings (*Synaptomys cooperi*) and prairie voles (*Microtus ochrogaster*) in southern Illinois. Univ. Illinois, unpubl. Ph.D. dissertation, iv+95 pp.

Blair, W. F.
1940 Notes on home ranges and populations of the short-tailed shrew. Ecology, 21(2):284–288.
1948 Population density, life span, and mortality of small mammals in the blue-grass meadow and blue-grass field associations of southern Michigan. American Midl. Nat., 40(2):395–419.

Brewer, R., and M. L. Reed
1977 Vertebrate inventory of wet meadows in Kalamazoo and Van Buren counties. Michigan Dept. Nat. Res., (mimeo). 20 pp.

Buckner, C. H.
1957 Home range of *Synaptomys cooperi*. Jour. Mammalogy, 38(1):132.

Burt, W. H.
1928 Additional notes on the life history of the Goss lemming mouse. Jour. Mammalogy, 9(3):212–216.
1940 Territorial behavior and populations of some small mammals in southern Michigan. Univ. Michigan, Mus. Zool., Misc. Publ., No. 45, 58 pp.

Clark, T. W.
1972 An ecological survey of the mammals of north central Wisconsin. Stevens Point Mus. Nat. Hist., Rept. No. 8, 37 pp.

Cleland, C. E.
1966 The prehistoric animal ecology and ethnozoology of the Upper Great Lakes Region. Univ. Michigan, Mus. Anthro., Anthro. Papers, No. 29, x+294 pp.

Connor, P. F.
1959 The bog lemming *Synaptomys cooperi* in southern New Jersey. Michigan State Univ., Publ. Mus., 1(5):161–248.

Coventry, A. F.
1937 Notes on the breeding of some Cricetidae in Ontario. Jour. Mammalogy, 18(4):489–496.

Craighead, J. J., and F. C. Craighead, Jr.
1956 Hawks, owls and wildlife. Stackpole Co., Harrisburg, Pennsylvania. xix+443 pp.

Dearborn, N.
1932 Foods of some predatory fur-bearing animals in Michigan. Univ. Michigan, Sch. For. and Conserv., Bull. No. 1, 52 pp.

Dice, L. R.
1925 A survey of the mammals of Charlevoix County, Michigan, and vicinity. Univ. Michigan, Mus. Zool., Occas. Papers No. 159, 33 pp.

Dice, L. R., and H. B. Sherman
1922 Notes on the mammals of Gogebic and Ontonagon counties, Michigan, 1920. Univ. Michigan, Mus. Zool., Occas. Papers No. 109, 40 pp.

Doran, D. J.
1954 A catalogue of the Protozoa and helminths of North American rodents. II. Cestoda. American Midl. Nat., 52(2):469–480.

Errington, P. L., F. Hamerstrom, and F. N. Hamerstrom, Jr.
1940 The great horned owl and its prey in north-central United States. Iowa State Col., Agric. Exp. Sta., Res. Bull. No. 277, pp. 759–850.

Fichter, E., and M. F. Hansen
1947 The Goss lemming mouse, *Synaptomys cooperi gossii* (Coues), in Nebraska. Nebraska State Mus., Bull. 3(2):1–8.

Fisher, H. J.
1945 Notes on voles in central Missouri. Jour. Mammalogy, 26(4):435–437.

Gaines, M. S., C. L. Baker, and A. M. Vivas
1979 Demographic attributes of dispersing southern bog lemmings (*Synaptomys cooperi*) in eastern Kansas. Oecologia, 40:91–101.

Gaines, M. S., R. K. Rose, and L. R. McClenaghan
1977 The demography of *Synaptomys cooperi* populations in eastern Kansas. Canadian Jour. Zool., 55(10):1584–1594.

Geis, A. D.
1952 Winter-food habits of a pair of long-eared owls. Jack-Pine Warbler, 30(3):93.

Getz, L. L.
1960 Home ranges of the bog lemming. Jour. Mammalogy, 41(3):404–405.
1961 Factors influencing the local distribution of *Microtus* and *Synaptomys* in southern Michigan. Ecology, 42(1):110–119.

Green, M. M.
1925 Notes on some mammals of Montmorency County, Michigan. Jour. Mammalogy, 6(3);173–178.

Hamilton, W. J., Jr.
1941a On the occurrence of *Synaptomys cooperi* in forested regions. Jour. Mammalogy, 22(2):195.
1941b The food of small forest mammals in eastern United States. Jour. Mammalogy, 22(3):240–263.

Hatt, R. T.
1923 The land vertebrate communities of western Leelanau County, Michigan, with an annotated list of the mammals of the county. Michigan Acad. Sci., Arts and Ltrs., Papers, 3:369–402.
1930 The biology of the voles of New York. Roosevelt Wild Life Bull., 5(4):513–623.

Haveman, J. R.
1973 A study of population densities, habitats and foods of

four sympatric species of shrews. Northern Michigan Univ., unpubl. M.S. thesis, vii+70 pp.

Hoffman, R. S., and C. F. Nadler
1976 The karyotype of the southern bog lemming, *Synaptomys cooperi* (Rodentia: Cricetidae). Mammalia, 40(1):79–82.

Hoffmeister, D. F., and J. E. Warnock
1947 A concentration of lemming mice (*Synaptomys cooperi*) in central Illinois. Illinois Acad. Sci., Trans., 40:190–193.

Howell, A. B.
1927 Revision of the American lemming mice (genus *Synaptomys*). N. American Fauna, No. 50, 37 pp.

Kirkland, G. L., Jr.
1977 Responses of small mammals to the clearcutting of northern Appalachian forests. Jour. Mammalogy, 58(4):600–609.

Korschgen, L. J.
1957 Food habits of coyotes, foxes, house cats and bobcats in Missouri. Missouri Conserv. Comm., Fish and Game Div., P-R Ser. No. 15, 64 pp.

Linduska, J. P.
1942 Winter rodent populations in field-shocked corn. Jour. Wildlife Mgt., 6(4):353–363.
1950 Ecology and land-use relationships of small mammals on a Michigan farm. Michigan Dept. Conserv., Game Div., ix+144 pp.

Linzey, A. V.
1981 Patterns of coexistence in *Microtus pennsylvanicus* and *Synaptomys cooperi*. Virginia Poly. Inst. and State Univ., unpubl. Ph.D. dissertation, 97 pp.

Luxenberg, B.
1972 Faunal remains. Pp. 91–115 *in* J. F. Fitting, ed. The Schultz Site at Green Point. A stratified occupation area in the Saginaw Valley of Michigan. Univ. Michigan, Mus. Anthro., Memoirs No. 4, xii+317 pp.

Manville, R. H.
1955 Dichromatism in Michigan rodents. Jour. Mammalogy, 36(2):293.

Master, L. L.
1978 A survey of the current distribution, abundance, and habitat requirements of threatened and potentially threatened species of small mammals in Michigan. Univ. Michigan, Mus. Zool., iv+52 pp.

Mumford, R. E.
1969 Distribution of the mammals of Indiana. Indiana Acad. Sci., Monog. No. 1, vii+114 pp.

Murie, A.
1936 Following fox trails. Univ. Michigan, Mus. Zool., Misc. Publ. No. 32, 45 pp.

Oehler, C.
1942 Notes on lemming mice in Cincinnati, Ohio. Jour. Mammalogy, 23(3):341–342.

Parmalee, P. W.
1968 Cave and archaeological faunal deposits as indicators of post-Pleistocene animal populations and distribution in Illinois. Pp. 104–113 *in* R. E. Bergstrom, ed. The Quaternary of Illinois. Univ. Illinois, Col. Agric. Sp. Publ. No. 14, 179 pp.

Pruitt, W. O. Jr.
1959 Microclimates and local distribution of small mammals on the George Reserve, Michigan. Univ. Michigan, Mus. Zool., Misc. Publ., No. 109, 27 pp.

Quick, E. R., and A. W. Butler
1885 The habits of some Arvicolinae. American Nat., 19: 113–118.

Robinson, W. L.
1975 Birds and mammals. Pp. 1–17 *in* W. L. Robinson and J. K. Werner. Vertebrate animal populations of the McCormick Forest, USDA Forest Serv., Res. Paper NC-118, 25 pp.

Rose, R. K., and A. M. Spevak
1978 Aggressive behavior in two sympatric microtine rodents. Jour. Mammalogy, 59(1):213–216.

Ryan, J. M.
1982 Seven new locations for the pygmy shrew (*Sorex hoyi*) in the Lower Peninsula, Michigan. Jack-Pine Warbler, 60 (in press).

Schwartz, C. W., and E. R. Schwartz
1981 The wild mammals of Missouri. Revised Edition. Univ. Missouri Press and Missouri Dept. of Conserv. Columbia. viii+356 pp.

Shier, J. L.
1981 Habitats of threatened small mammals on the Rose Lake Wildlife Research Area, Clinton County, Michigan—early forties and late seventies. Michigan State Univ., unpubl. M.S. thesis, vi+106 pp.

Spiker, C. J.
1933 Analysis of two hundred long-eared owl pellets. Wilson Bull., 45(4):198.

Stegeman, L. C.
1930 Notes on *Synaptomys cooperi cooperi* in Washtenaw County, Michigan. Jour. Mammalogy, 11(4):460–466.

Timm, R. M.
1975 Distribution, natural history and parasites of mammals of Cook County, Minnesota. Univ. Minnesota, Bell Mus. Nat. Hist., Occas. Papers No. 14, 56 pp.

Voight, J., and D. C. Glenn-Lewin
1978 Prey availability and prey taken by long-eared owls in Iowa. American Midl. Nat., 99(1):162–171.

Voigt, D. R., G. B. Kolenosky, and D. H. Pimlott
1976 Changes in summer foods of wolves in central Ontario. Jour. Wildlife Mgt., 40(4):663–668.

Wallace, G. J.
1948 The barn owl in Michigan. Its distribution, natural history and food habits. Michigan State Col., Agric. Exp. Sta., Tech. Bull. 208, 61 pp.

Warthin, A. S., Jr., and J. Van Tyne
1922 The food of long-eared owls. Auk, 39(3):417.

Wenzel, O. J.
1911 Observations on the mammals of the Douglas Lake region, Cheboygan County, Michigan. Michigan Acad. Sci., Thirteenth Rept., 136–143.

Wetzel, R. M.
1955 Speciation and dispersal of the southern bog lemming, *Synaptomys cooperi* (Baird). Jour. Mammalogy, 36(1):1–20.

Wilkinson, A. M.
1980 Status and distribution of threatened and rare mammals in certain wetland ecosystems in Upper Michigan. Michigan Tech. Univ., unpubl. M.S. thesis, vii+85 pp.

Wilson, K. A.
1938 Owl studies at Ann Arbor, Michigan. Auk, 55(2):187–197.

Wood, N. A.
1914 An annotated check-list of Michigan mammals. Univ. Michigan, Mus. Zool., Occas. Papers No. 4, 13 pp.
1922 The mammals of Washtenaw County, Michigan. Univ. Michigan, Mus. Zool., Occas. Papers No. 123, 23 pp.

Wood, N. A., and L. R. Dice
1923 Records of the distribution of Michigan mammals. Michigan Acad. Sci., Art and Ltrs., Papers, 3:425–469.

Wrigley, R. E.
1969 Ecological notes on the mammals of southern Quebec. Canadian Field-Nat., 83:201–211.

FAMILY MURIDAE

Members of this large rodent family number approximately 98 genera and 457 species. All are natives of the Eastern Hemisphere. However, at least three pest species, the Norway rat (*Rattus norvegicus*), the black rat (*Rattus rattus*), and the house mouse (*Mus musculus*), because of their close association with human society and commerce, have ended up in the United States and adapted to successful life here. Two of these, the Norway rat and the house mouse, are well-established in Michigan and can be expected in human settlements in all counties.

Murids display considerable diversity in size (mouse-size to near beaver-size) and in habitat selection (semi-aquatic, terrestrial, fossorial, and arboreal). Naked tails characterize some murids, like the American imports, but others have heavily furred or bushy tails. Skulls are mostly rat-like in appearance with crown patterns of the molar teeth normally simplified with rounded cusps (as shown in unworn dentition) arranged in three longitudinal rows. Usually all three molar teeth are present on each side of the upper and lower jaws; in one genus (*Mayermys* of New Guinea), the number of molars is reduced to one on each side of the upper and lower jaws.

The introduced murids which live closely with human habitation damage foodstuffs and property, and carry diseases. They also have become sufficiently adapted to Michigan environments to be found (mostly in summer) in cultivated and fallow lands, along streams, and in other habitats remote from human dwellings and outbuildings.

NORWAY RAT
Rattus norvegicus (Berkenhout)

NAMES. The generic name *Rattus,* proposed by Fischer in 1803, is probably derived from the Anglo-Saxon word *raet* meaning rat and latinized to *Rattus.* The specific name *norvegicus,* proposed by Berkenhout in 1769, is a Latin word meaning "of Norway," referring to the country from which the specimen used in the description was derived. In Michigan, the Norway rat has a host of names: barn rat, brown rat, domestic rat, sewer rat, and white rat (for the albinistic form used in laboratory studies).

RECOGNITION. Body size large and robust, of a size most persons consider typical when defining a rat-sized rodent (Fig. 56); head and body averaging 8⅝ in (222 mm) long; sparsely haired and scaly tail shorter than length of head and body, averaging 6¾ in (173 mm) long; fur coarse and short; head with prominent snout; eyes small; ears sparsely haired, almost naked in appearance; vibrissae long and prominent; legs short; front feet with four clawed toes and small thumb; hind feet with five clawed toes; soles of hind feet naked, each with six tubercules; six pairs of mammary glands; paired glands opening just inside of anus producing musky odor.

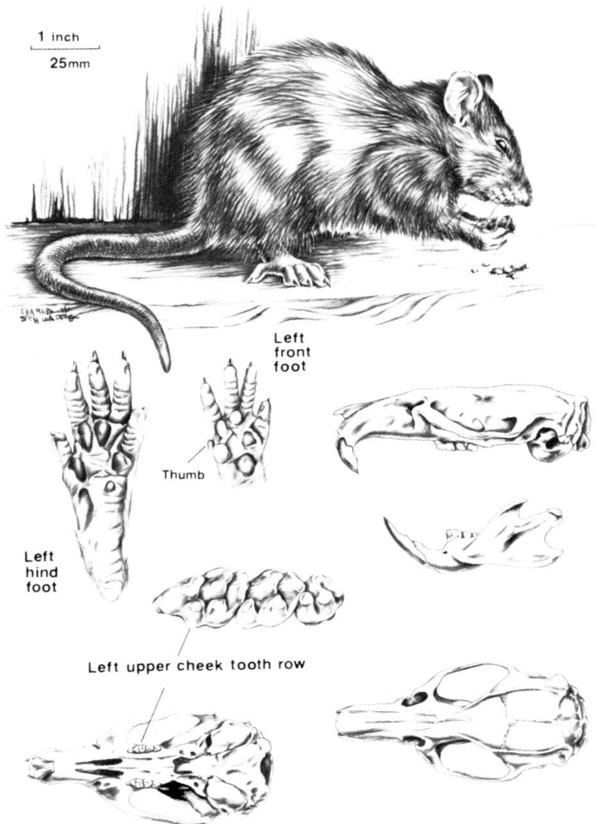

Figure 56. The introduced Norway Rat (*Rattus norvegicus*). Reprinted from *The Wild Mammals of Missouri* by Charles W. and Elizabeth R. Schwartz by permission of the University of Missouri Press and the State Historical Society of Missouri. Copyright 1959, 1981 by the Curators of the University of Missouri.

In adults, upperparts (especially midback) grizzled gray-brown to brown; underparts mostly yellow-gray or pale gray; tail dark gray above, paler below.

This successfully adapted rodent is sufficiently well-known that no detailed description is needed. Its size (that of a small squirrel), easily distinguishes it from other small rodents of Michigan. A close relative, the black rat (*Rattus rattus* and varieties), rarely survives in the north temperate climate of Michigan, but may be expected in Detroit and other lakeshore cities, where they are carried in on ships from more tropical areas. The black rat (often light gray in color) has features similar to the Norway rat, including parchment-like ears and naked, scaly tail, but is readily distinguished by its long, slim build and very long tail (always longer than the head and body).

MEASUREMENTS AND WEIGHTS. Adult Norway rats measure as follows: length from tip of nose to end of tail vertebrae, 12⅛ to 17½ in (320 to 440 mm); length of tail vertebrae, 4¾ to 7½ in (120 to 190 mm); length of hind foot, 1⅛ to 1¾ in (30 to 45 mm); height of ear from notch, ⅝ to ¾ in (16 to 20 mm); weight, 7 to 17 oz (200 to 500 g). Occasionally male rats become exceedingly large, up to 20 oz (581 g), according to Salmon *et al.* (1977). City rats may also grow larger than farm rats, but this may be the result of dietary differences (Davis, 1949).

DISTINCTIVE CRANIAL AND DENTAL CHARACTERISTICS. The skull of the Norway rat is long and rather slender, resembling somewhat a very large skull of a deer mouse (Fig. 56). Major features are the prominent brow (supraorbital) ridges which begin at the interorbital constriction and extend backward in parallel fashion along the cranial sides (as temporal ridges) to the back of the skull. The skull averages 1⅝ in (43 mm) long and 13/16 in (22 mm) wide across the zygomatic arches.

The molar teeth have the distinctive murid pattern on the grinding surfaces of three lengthwise rows of simple, rounded cusps. In worn teeth, these row patterns are obscured because the cusps are somewhat eroded by the mastication of coarse foods. Unlike the house mouse, the upper incisors of the Norway rat are not notched as viewed from the side. In Michigan, a skull of a large Norway rat might be confused with that of a small muskrat. However, the Norway rat's skull is easily distinguished by rows of cusps (instead of a flat triangular-shaped pattern) on the grinding surfaces of the molars; by the distinctive brow and temporal ridges; and by the length of the hard palate which in the Norway rat extends well behind a line drawn between the posterior ends of the last teeth in the upper molar row. The dental formula for the Norway rat is:

$$\text{I (incisors)}\,\frac{1\text{-}1}{1\text{-}1},\ \text{C (canines)}\,\frac{0\text{-}0}{0\text{-}0},\ \text{P (premolars)}\,\frac{0\text{-}0}{0\text{-}0},\ \text{M (molars)}\,\frac{3\text{-}3}{3\text{-}3} = 16$$

DISTRIBUTION. This persistent rodent (probably south Asiatic in origin) is thought to have been introduced into what is now the United States from England around the time of the American Revolution (Guilday, 1970). The species arrived in Michigan at about the same time—its spread presumably facilitated by the lively shipping trade in the Upper Great Lakes—and soon became established in inland settlement areas. The rodent became common in Washtenaw County soon after communities were founded beginning in the 1820s (Wood, 1922). Norway rats were not found in Ingham County until after the country was settled, probably just prior to the Civil War, according to H. A. Atkins, a local naturalist (Wood and Dice, 1923). Burt (1946) listed actual records from 21 Michigan counties in the mid-1940s. Today, the rodent can be expected all over the state with records also reported from several islands in the Upper Great Lakes. It is apt to occur in small and large communities, suburban homes, manufacturing establishments, farm homes, farm storage sites, urban and rural dumps, boat docking areas, recreational campsites, and anywhere human food and shelter are present.

HABITAT PREFERENCES. The Norway rat is at home in backyards and alleyways in the residential, business, and manufacturing districts of cities and towns (Calhoun, 1962). Populations concentrate where food supplies (generally garbage or grain stores) and adjacent cover sites are abundant. Urban and rural garbage dumps, grain storage facilities, strawstacks, dairy barns, and farm outbuildings provide seasonal or year-round living places. Norway rats move into cultivated fields in summer. Favorite harborages are tunnels under banks of streams which are adjacent to human habitation. For example, the Norway rat can be observed at all hours moving under the cut banks of the Red Cedar River where it flows through the campus of Michigan State University.

Although the Norway rat is persistent, it very likely suffers during Michigan winters and usually moves into warmer protective quarters in human establishments during cold weather. Boat dockings provided a ready haven for Norway rats at the Huron Mountain Club but when boat dockings were discontinued in 1906, these rodents disappeared (Manville, 1948). Laundre (1975) felt they left because of the harsh winters. At any rate, the Norway rat thrives in human presence in Michigan despite all efforts to discourage it.

GENERAL HABITS. The Norway rat is rarely found in nonhuman environments. There is some evidence that there is approximately one rat per person in cities and towns; in farming areas the number of rats per capita would be higher (Burt, 1946). Much of our basic information about *Rattus norvegicus* comes from intensive studies of the albino form used in biomedical studies and from the many papers published by participants in the Rodent Ecology project in Baltimore (Calhoun, 1962), where the biology of the Norway rat was examined in a "natural" urban community.

The Norway rat lives a highly organized colonial life. Large males are dominant and hold territories (as in city blocks) and exclude other males from their homesites and from their harems of females. Dominance also plays an important role in habitat selection, with the aggressive, large males and their retinues occupying the most favorable den sites and food sources. The subservient individuals are obliged to accept marginal living areas, and can secure food only when the dominant animals are absent, often resorting to daytime activity (Calhoun, 1962). Young rats appear to be tolerated, although threatened, by adults. Weaker adults, on the other hand, may be forced to leave, starve, and are sometimes killed by dominant animals. Other individuals seem to spend their entire lifetimes dissociated from organized groups.

Studies show that Norway rats rarely increase in numbers in excess of available food and cover. There is evidence that when the carrying capacity of the local habitat is reached, the animals self-limit their numbers by (a) a reduction in births, (b) an increase in death rate, and (c) emigration. On the other hand, if habitat conditions improve markedly, the high fertility rate allows for the local population to recuperate quickly, again under the "supervision" of the dominant males of the colonies. The studies in the Baltimore area showed that as many as 25 to 150 Norway rats might live in a single city block, while farm populations could be up to 300 animals at a single farmstead and related outbuildings.

Norway rats, despite aggressive social interactions, are generally sedentary and do not venture great distances from their birth places. Their normal cruising range (Davis *et al.*, 1948) depends on how far food sources are from homesites but appears to cover an area with a dimaeter of no more than 150 ft (46 m). In rural situations, where individuals move back and forth from farmsteads to fields, haystacks, or grain storage containers, seasonal movements, perhaps as much as ½ mi (0.8 km), may prevent, or at least disrupt, the highly organized social hierarchy characteristic of the less mobile urban rat colony (Fitch, 1958; Taylor and Quy, 1978).

The Norway rat's physical capabilities have rarely been exaggerated. Some of their feats in making maximum use of human habitation and food, according to Howard and Marsh (1976), include: (a) gnawing through a wide variety of substances such as lead sheeting, cinder block, and aluminum siding; (b) swimming as far as ½ mi (0.8 km) in open water or through sewer lines against substantial currents; (c) climbing vines and woody vegetation, horizontal and vertical wires, ropes and hawsers between ship and shore, inside vertical pipes as small in diameter as 1½ in (40 mm), and outside vertical pipes of any size if the pipes are within 3 in (75 mm) of a wall or other support; (d) jumping horizontally as far as 48 in (1.2 m) on a flat surface or at least 8 ft (2.4 m) from an elevation of 15 ft (4.5 m); or (e) squeezing through an opening as small as ½ in (14 mm).

Although Norway rats are usually active at night, they can be observed any time of day. They appear to see poorly, using their keen senses of smell, taste, touch, and hearing to locate foods and nest sites, and avoid enemies. Taste perception appears well-developed and seems useful, along with smell, in detecting toxic compounds placed in foods. Rats are presumed to be color blind, responding chiefly to degrees of light and dark.

Norway rats build bulky nests of twigs, leaves, and assorted scrap from alleys or dumps. Nests may be placed in burrows (rarely) or on the surface under debris, in packing boxes, in buildings, in dumps, and under strawstacks and woodpiles. Underground burrows and surface trails lead from nest sites to feeding areas.

REPRODUCTIVE ACTIVITIES. The Norway rat's ability to literally turn on or turn off its prolific breeding activities seems dependent on the quality of its homesite environment. In colonies, breeding is also regulated by the dominant male whose presence apparently restrains productivity; when this male is absent or killed, promiscuous breeding can result in a marked increase in pregnancies. Females are capable of having young at 16 weeks of age, and can mate again within a day or two of littering. The number of litters annually can be six or eight, although 12 are known. Each litter may contain six to eight young; a record of 22 has been recorded (Burt, 1946). The resorption of embryos is also reported (Storer and Davis, 1953).

The young are born in the nest after a gestation period of 21 to 22 days. At birth the young are helpless, hairless, and have closed eyes. At seven days of age, hair covers the body; at nine to 15 days of age, the eyes open and the young are active; at about 21 days of age, the young begin to eat solid food; at about 30 days, the young weigh 1½ oz (45 g), are completely weaned, and can be on their own with the mother producing another litter. This high reproductive potential is offset by high mortality; Davis (1948) determined that probably only about 5% of the offspring live as long as 12 months. In Michigan, cold winters and dry summers may be partly responsible for major breeding peaks in spring and autumn.

FOOD HABITS. Like most rodents, the Norway rat will eat just about anything (Howard and Marsh, 1976; Jackson, 1965; Whitaker, 1977), including cereals, nuts, fruits, various invertebrates, fish, poultry, vegetables, flesh or other small mammals, assorted garbage, and even manure. In Michigan, Linduska (1942) found the Norway rat in field-shocked corn. Although this rat eats a variety of foods, it is what rodent control specialists term a "cautious feeder" (Howard and Marsh, 1974). This hesitant approach to a food source has no doubt often saved it from poisoned bait.

ENEMIES. Because of its aggressive nature, the Norway rat may not be easily caught prey, even by its most persistent adversary, the human species. Younger animals are probably fair game for the smaller owls—long-eared, short-eared and screech (Dexter, 1978; Linduska, 1950); larger rats are captured by barn, snowy, barred, and great horned owls (Allan, 1977; Armstrong, 1958; Lantz, 1910; Linduska, 1950; Wallace, 1948; Warthin and Van Tyne, 1922). Lantz (1910) listed six species of hawks that feed on rats. Four-footed predators include house cats, various weasels, mink, striped skunks, dogs, coyotes, and gray and red foxes (Errington, 1935; Korschgen, 1957; Lantz, 1910). Norway rats are also killed on Michigan highways (Manville, 1949).

PARASITES AND DISEASES. The biomedical importance of the Norway rat has given impetus to major studies of its several parasites and diseases and their transmission to humans and domesticated animals. The long list of ectoparasites and endoparasites includes fleas, lice, mites, protozoans, cestodes, trematodes, nematodes, and acanthocephalans (Banks, 1910; Doran, 1954; Stiles and Crane, 1910; Whitaker, 1977).

The Norway rat has been responsible for major outbreaks of disease in human populations since the dawn of history (Zinsser, 1935), with bubonic plague epidemics (causative agent: *Pasturella pestis*; vector: flea) killing vast numbers of humans. Other diseases of this rat that are transmittable to humans include rat-bite fever (causative agent: *Streptobacillus moniliformis*), leptospirosis, including Weil's disease and spirochetal jaundice (causative agent: species of *Leptospira*), tularemia (causative agent: *Pasteurella tularensis*), salmonellosis or food poisoning (causative agent: species of *Salmonella*), trichinosis (causative agent: *Trichinella spiralis*, a nematode worm infecting man, swine, and other mammals), and murine typhus fever (causative agent: *Rickettsia typhi*; vectors: oriental rat flea, and biting louse).

MOLT AND COLOR ABERRATIONS. Young Norway rats have a gray and woolly pelage, but are easily distinguished from other Michigan rodents by the naked, scaly tail and long feet. Adult pelages of different colors have frequently been observed. The white (albinistic) mutation is characteristic of the common laboratory form of this rodent. Variants of this coloring are black-belted animals; black (melanistic) and mottled individuals are also known.

ECONOMIC IMPORTANCE. The Norway rat is responsible for more destruction of human foods and related materials than any other animal in Michigan or in the United States (Henderson and Craig, 1932). Recent estimates show that there may be 100 million introduced rats (including the more southern black rat, *R. rattus*) in this country and that each individual damages food supplies and other materials valued from one to ten dollars, and contaminates five to ten times more. Thus, in the United States, these rats may be responsible for between 500 million and one billion dollars annually in direct economic losses (Bjornson *et al.*, 1969). In addition, crippling diseases mentioned previously must account for additional millions in medical costs.

Some of the Norway rat's destructive activities are: eating and contaminating corn, wheat, rice and other cultivated crops in the field, in storage, or after processing as human food; chewing electric wiring (may be responsible for ¼ of all house fires of undetermined origin); chewing telephone cables; chewing through lead pipes damaging sewer systems; chewing and eating leather goods; gnawing holes in building walls, boxes, packages, furniture; and biting babies and sleeping children and adults (25 cases of rat bite were reported in Detroit in 1975).

Local, state, and federal agencies have been conducting rodent control programs for decades. Research has concentrated on the basic ecology and behavior of this rat, means of constructing rodent-proof buildings and food storage areas, methods of discouraging the transportation of Norway rats from one place to another, and rodent-borne disease prevention. The results of such studies are widely published and are summarized in various rodent control manuals (Bjornson *et al.*, 1969; Howard and Marsh, 1974, 1976; Silver and Garlough, 1941). Hayne *et al.* (1945) have published a control manual specifically for Michigan. Public health projects in many Michigan cities include special allotments of funds for rodent control. Constructing rat-proof buildings and food storage areas is the most effective way of eliminating rodent damage but rat-proofing is expensive for the average community or farmstead.

The house cat, the baited, wooden-based rat trap, lethal gas, and poisoned bait are traditional ways to discourage Norway rats in buildings, barns, elevators, silos, and other habitations. Manufacturing rodenticides for baits is a major business, with new products appearing frequently. Some rodenticides used presently or in the recent past include warfarin, antu, arsenic, fluoroacetamide (1081), yellow phosphorus, red squill, sodium fluoroacetate (1080), strychnine, and zinc phosphide (Bjornson *et al.*, 1969). Chemosterilants, primarily designed to sterilize the dominant, breeding males, are also on the market (Field, 1971; Marsh and Howard, 1977).

For local rodent control in Michigan, citizens should consult the County Extension Director for the Michigan State University Cooperative Extension Service or public health officials for the latest information and bulletins.

This most destructive of mammals does serve

one useful purpose. In its highly bred albinistic form, the white laboratory rat has served science and human welfare in studies of diseases and their transmission, nutrition, behavior, and even geriatics.

COUNTY RECORDS FROM MICHIGAN. Records of specimens in collections of museums are from less than 30 of Michigan's 83 counties. However, public health authorities report at least some problems with Norway rats in all counties since 1970 (A. A. Therrien, Insect and Rodent Control Section, Michigan Dept. Public Health, pers. comm.). Although the Norway rat has undoubtedly reached most islands in Michigan's part of the Upper Great Lakes at one time or another, published records show its occurrence in Lake Michigan on Beaver, North Fox, South Fox, North Manitou, and South Manitou islands (Hatt, 1923; Hatt *et al.,* 1948; Ozoga and Phillips, 1864), and in Lake Huron on Drummond Island (Manville, 1950).

LITERATURE CITED

Allan, T. A.
1977 Winter food of the snowy owl in northwestern Lower Michigan. Jack-Pine Warbler, 55(1):42.

Armstrong, W. H.
1958 Nesting and food habits of the long-eared owl in Michigan. Michigan State Univ., Publ. Mus., Biol. Ser., 1(2):61–96.

Banks, N.
1910 Ectoparasites of the rat. Pp. 67–85 *in* The rat and its relation to the public health. U.S. Treas. Dept., Publ. Hlth. and Marine-Hosp. Serv., 254 pp.

Bjornson, B. F., H. D. Pratt, and K. S. Littig
1969 Control of domestic rats and mice. U.S. Publ. Hlth. Serv., PHS Publ. No. 563, 41 pp.

Burt, W. H.
1946 The mammals of Michigan. Univ. Michigan Press, Ann Arbor. xv+288 pp.

Calhoun, J. B.
1962 The ecology and sociology of the Norway rat. U.S. Publ. Hlth. Serv., PHS Publ. No. 1008, viii+288 pp.

Davis, D. E.
1948 The survival of wild brown rats on a Maryland farm. Ecology, 29:437–448.
1949 A phenotypical difference in growth of wild rats. Growth, 13:1–6.

Davis, D. E., J. Emlen, and A. W. Stokes
1948 Studies on home range in the brown rat. Jour. Mammalogy, 29(3):207–225.

Dexter, R. W.
1978 Mammals utilized as food by owls in reference to the local fauna of northeastern Ohio. Kirtlandia, No. 24, 6 pp.

Doran, D. J.
1954 A catalogue of the Protozoa and helminths of North American rodents. II. Cestoda. American Midl. Nat., 52(2):469–480.

Errington, P. L.
1935 Food habits of mid-west foxes. Jour. Mammalogy, 16(3):192–200.

Field, R. J.
1971 The use of sexual attractant pheromones to increase the acceptability of the male chemosterilant 3-chloro-1,2-propanediol by wild Norway rats (*Rattus norvegicus*). Michigan State Univ., unpubl. Ph.D. dissertation, 67 pp.

Fitch, H. S.
1958 Home ranges, territories, and seasonal movements of vertebrates of the Natural History Reservation. Univ. Kansas Publ., Mus. Nat. Hist., 11(3):63–326.

Guilday, J. E.
1970 Animal remains from archaeological excavations at Fort Ligonier. Carnegie Mus., Ann., 42:177–186.

Hatt, R. T.
1923 The land vertebrate communities of western Leelanau County, Michigan, with an annotated list of the mammals of the county. Michigan Acad. Sci., Arts and Ltrs., Papers, 3:369–402.

Hatt, R. T., J. Van Tyne, L. C. Stuart, C. H. Pope, and A. B. Grobman
1948 Island life: A study of the land vertebrates of the island of eastern Lake Michigan. Cranbrook Inst. Sci., Bull. No. 27, xi+179 pp.

Hayne, D. W., M. D. Pirnie, and C. H. Jefferson
1945 Controlling rats and house mice. Michigan Agric. Exp. Sta., Cir. Bull. 167, 40 pp.

Henderson, J., and E. L. Craig
1932 Economic mammalogy. Charles C. Thomas, Springfield, Illinois. x+397 pp.

Howard, W. E., and R. E. Marsh
1974 Rodent control manual. Pest Control, 42(8):D–U.
1976 The rat: Its biology and control. Univ. California, Coop. Ext. Serv., Leaflet 2896, 22 pp.

Jackson, W. B.
1965 Feeding patterns in domestic rodents. Pest Control, August, 4 pp.

Korschgen, L. J.
1957 Food habits of coyotes, foxes, house cats and bobcats in Missouri. Missouri Conserv. Comm., Fish and Game Div., P–R Ser. No. 15, 64 pp.

Lantz, D. E.
1910 Natural enemies of the rat. Pp. 163–169 *in* The rat and its relation to the public health. U.S. Treas. Dept., Publ. Hlth. and Marine-Hosp. Serv., 254 pp.

Laundre, J.
1975 An ecological survey of the mammals of the Huron Mountain area. Huron Mt. Wildlife Found., Occas. Papers, No. 2, x+69 pp.

Linduska, J. P.
1942 Winter rodent populations in field-shocked corn. Jour. Wildlife Mgt., 6(4):353–363.
1950 Ecology and land-use relationships of small mammals on a Michigan farm. Michigan Dept. Conserv., Game Div., ix+144 pp.

Manville, R. H.
1948 The vertebrate fauna of the Huron Mountains, Michigan. American Midl. Nat., 39(3):615–640.
1949 Highway mortality in northern Michigan. Jour. Mammalogy, 30(3):311–312.
1950 The mammals of Drummond Island, Michigan. Jour. Mammalogy, 31(3):358–359.

Marsh, R. E., and W. E. Howard
1977 Testing rodent chemosterilants. EPPO Bull., 7(2): 485–493.

Ozoga, J. J., and C. J. Phillips
1964 Mammals of Beaver Island, Michigan. Michigan State Univ., Publ. Mus., Biol. Ser., 2(6):305–348.

Salmon, T. P., R. E. Marsh, and K. White
1977 Record weight wild Norway rats. California Vector Views, 24(1–2):6–9.

Silver, J., and F. E. Garlough
1941 Rat control. U.S.D.I., Fish and Wildlife Serv., Conserv. Bull. No. 8, 27 pp.

Stiles, C. W., and C. G. Crane
1910 The internal parasites of rats and mice in their relation to diseases of man. Pp. 87–110 *in* The rat and its relation to the public health. U.S. Treas. Dept., Publ. Hlth. and Marine-Hosp. Serv., 254 pp.

Storer, T. I., and D. E. Davis
1953 Studies on rat reproduction in San Francisco. Jour. Mammalogy, 34(3):365–373.

Taylor, K. D., and R. J. Quy
1978 Long distance movements of a common rat (*Rattus norvegicus*) revealed by radio tracking. Mammalia, 42(1):63–71.

Wallace, G. J.
1948 The barn owl in Michigan. Its distribution, natural history and food habits. Michigan Agri. Exp. Sta., Tech. Bull. 208, 61 pp.

Warthin, A. S., Jr., and J. Van Tyne
1922 The food of long-eared owls. Auk, 39(3):417.

Whitaker, J. O., Jr.
1977 Food and external parasites of the Norway rat, *Rattus norvegicus*, in Indiana. Indiana Acad. Sci., Proc., 86:193–198.

Wood, N. A.
1922 The mammals of Washtenaw County, Michigan. Univ. Michigan, Mus. Zool., Occas. Papers No. 123, 23 pp.

Wood, N. A., and L. R. Dice
1923 Records of the distribution of Michigan mammals. Michigan Acad. Sci., Arts and Ltrs., Papers, 3:425–469.

Zinsser, H.
1935 Rats, lice and history. Little Brown and Co., Boston, Massachusetts. 228 pp.

HOUSE MOUSE
Mus musculus Linnaeus

NAMES. Both the generic name *Mus* and the specific name *musculus* were proposed by the Swedish naturalist Linnaeus in the tenth edition of *Systema Naturae* in 1758. The generic name *Mus* is the Latin word for mouse; the specific name *musculus* is also from the Latin and means little or diminutive. In Michigan, this introduced species is generally called house mouse, but also barn mouse, pantry mouse, brown mouse, and white mouse (for the albinistic form used in biomedical studies). The subspecies *Mus musculus domesticus* Rutty occurs in Michigan.

RECOGNITION. Body size small and slender (Fig. 57); head and body averaging 3¼ in (85 mm) long; tail thin, tapering, sparsely haired, scaly, and clearly showing annulations averaging 3¼ in (85 mm) long; fur slightly coarse and short; head with prominent pointed snout; eyes small, black, and somewhat protruding; ears large, naked, and parchment-like in texture; vibrissae (whiskers) conspicuous; legs short; front feet with four clawed toes and reduced thumb (pollex) with small nail; hind feet with five clawed toes and six tubercules (pads) on the naked soles; five pairs of mammary glands.

In adults, upperparts variable, yellow-brown to gray-brown; underparts paler, yellow-brown to gray-brown; tail dusky above and paler below; feet light brown with whitish tips on toes.

The house mouse is apt to associate in natural habitats in Michigan with all species of native mammals. It is easily distinguished, however, by its short body hair; naked, parchment-like ears; tapering, scaly tail; yellow tinged belly fur; and five pairs of mammary glands.

MEASUREMENTS AND WEIGHTS. Adult house mice measure as follows: length from tip of nose to end of tail vertebrae, 5⅞ to 7½ in (150 to 190 mm); length of tail vertebrae, 2¾ to 3¾ in (70 to 95 mm);

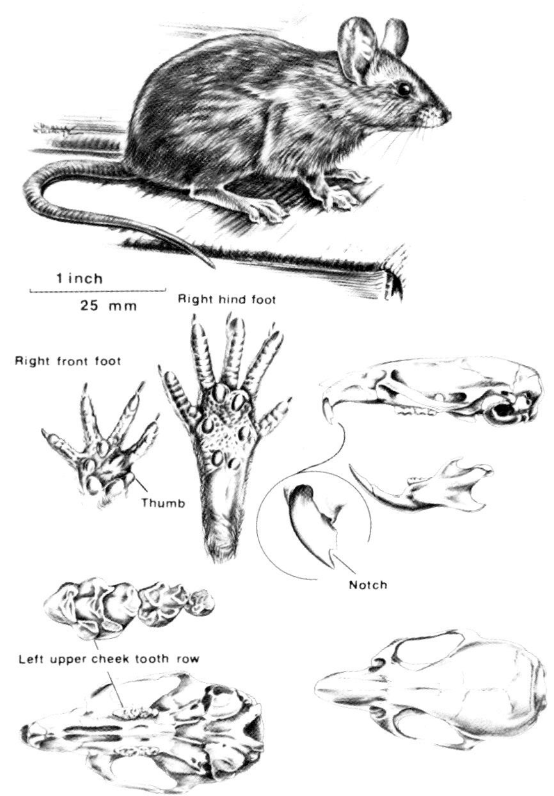

Figure 57. The introduced house mouse (*Mus musculus*). Reprinted from *The Wild Mammals of Missouri* by Charles W. and Elizabeth R. Schwartz by permission of the University of Missouri Press and the State Historical Society of Missouri. Copyright 1959, 1981 by the Curators of the University of Missouri.

length of hind foot, 7/16 to 13/16 in (17 to 29 mm); height of ear from notch, ⅜ to 11/16 in (11 to 18 mm); weight, 0.5 to 0.8 oz (15 to 24 g).

DISTINCTIVE CRANIAL AND DENTAL CHARACTERISTICS. The skull of the house mouse is small and similar to that of the meadow jumping mouse, but is easily distinguished by its slit-like infraorbital foramina as compared with large rounded ones in the latter species (Figs. 50 and 57). The skull averages 13/16 in (21.5 mm) long and 7/16 in (11 mm) wide (across zygomatic arches). The dentition is characterized by three rows of longitudinal cusps on unworn, grinding surfaces of molars; first molars larger than second and third molars combined; and a terminal notch in each upper incisor when viewed from the side. This combination of dental characteristics serves to identify the skull of the house mouse from those of other small rodents in Michigan. The dental formula for the house mouse is:

$$\text{I (incisors)}\frac{1\text{-}1}{1\text{-}1}, \text{C (canines)}\frac{0\text{-}0}{0\text{-}0}, \text{P (premolars)}\frac{0\text{-}0}{0\text{-}0}, \text{M (molars)}\frac{3\text{-}3}{3\text{-}3} = 16$$

DISTRIBUTION. The house mouse occurs in most Michigan environments with the exception of dense woodlands, marshes, and northern swamps. Its chief living areas, however, are cities, towns, farmsteads, outbuildings, resort establishments, cottages, lumber camps, mining facilities, and other human habitations. From its native home in Asia, the species has accompanied human movements and commerce to every part of the world, surviving in ocean and lake going vessels, and overland in wagons. The house mouse probably reached what is now the United States long before the time of the American Revolution (Jackson, 1961). It appears to have arrived in Michigan shortly after permanent settlement (Wood, 1922, for Washtenaw County). No doubt, commercial shipping on the Upper Great Lakes expedited the spread of this persistent mouse. In contrast to the introduced Norway rat, the house mouse seems more successful in establishing permanent residence in cultivated crops and fallow farm areas; winter weather and the absence of surface water may be less limiting to the house mouse than to the rat. The house mouse must be reckoned with as a permanent member of many small mammal ecosystems in Michigan.

HABITAT PREFERENCES. As mentioned previously, the house mouse is at home in human habitations where scrap food, stored grains, and poorly cached garbage provide nourishment, and buildings provide harborage. In Indiana, Whitaker (1967) found house mice in eight different human land-use environments. In order of decreasing mouse abundance, he listed fields of winter wheat, fields of corn, wheat stubble, weedy (early fallow) fields, corn stubble, fields of soybeans, and grassy (late fallow) fields. Whitaker also found that the house mouse was mainly attracted to agricultural areas being actively cultivated. In the Michigan outdoors, as well, the favored habitats for this introduced species are cultivated and recently fallow fields. In Clinton County, Linduska (1950) noted house mice in croplands, a situation it shared chiefly with the prairie form of the deer mouse (*Peromyscus maniculatus bairdii*). In terms of distribution (not abundance), Linduska found that house mice frequented 18 of 45 grainfields, 8 of 23 alfalfa and clover fields, and 3 of 19 fallow (idle) fields. Some of this summer population appeared to depart in autumn and winter, as a result of either the cold weather or the harvesting of crops, although house mice did use adjacent roadside and fencerow cover after crop removal. Linduska (1942) also observed house mice in autumn and winter under shocked corn left in fields. This movement to grainfields in summer and to human habitations in winter may be more pronounced in northern areas (as in Quebec, Wrigley, 1969) than in southern areas, such as Lower Peninsula Michigan.

DENSITY AND MOVEMENTS. House mice populations have been known to fluctuate dramatically—from near zero to as many as 82,800 individuals per acre (33,120 per ha); this latter, staggering figure was for Kern County, California, in 1926 and 1927. Hall (1927), who reported this population explosion, counted 3,520 mice in one granary at one time; 4,000 lbs (1,818 kg) of mice were killed in another granary. Studies in Louisiana record population highs of 500 house mice per acre (200 per ha) according to Lowery (1974). Figures of this magnitude have not been reported for house mice in Michigan.

Normally house mice are most common in outdoor habitats in late summer and autumn (September through December) and least common in early spring (Whitaker, 1967). In fields of wheat, house mice populations may reach 116 per acre (53 per ha) in October with a low in corn stubble of seven per acre (3.2 per ha) in February and March (L. F. Stickel, pers. comm.).

Although there are some obvious shifts of Michigan house mouse populations from cultivated fields in summer to human habitation in winter (Linduska, 1950), most observers consider house mice to be generally sedentary, rarely moving more than 50 ft (15 m) from homesites (Jackson, 1961). However, the mice may also move from one field to another, perhaps attracted by new plantings. In one study, according to L. F. Stickel (pers. comm.), movement of this sort averaged 500 ft (152 m). House mice remaining all winter in field situations appear to range over more territory than do those in the same places in summer. In Kansas, Fitch (1958) found house mice in field situations normally foraged in areas with diameters of 20 to 100 ft (6 to 30 m), with a maximum of 300 ft (90 m). He found that home ranges of these mice averaged 1.5 acres (0.6 ha). Perhaps as part of autumn movements, Linduska (1950) caught two house mice in October almost ½ mi (almost 1 km) from a cultivated field where they had been live-trapped, marked, and released a short time previously.

BEHAVIOR. House mice live in small groups consisting of a male and one or more females and young. The dominant male aggressively discourages other males from joining the family association. During the winter, however, larger assemblages may huddle together in nests to share warmth. There is evidence that population growth in house mice may be arrested in times of food shortages, primarily because of reduced fertility in socially subordinate individuals. Social and pheromonal influences are factors in this population regulatory process (DeLong, 1967; Drickamer, 1979; Southwick, 1955).

House mice are chiefly nocturnal in activity. They construct no obvious runway systems but will follow those of other species such as the meadow vole. In outdoor environments, the mice often excavate their own burrow systems in which nests are constructed. These nests are generally made of grasses with inner chambers lined with soft or shredded material. In human habitations, nests can be in buildings, under debris, or in basements. Mattresses and the interiors of overstuffed furniture make favored nesting sites. House mice climb, jump, and swim without difficulty. They also have the ability to home; Fitch (1958) once moved an individual 860 ft (290 m) from its homesite; it was re-trapped the next day at the place of first capture.

ASSOCIATES. House mice have been found in association with most small mammals living in Michigan's old fields, cultivated croplands, and forest edge situations. In Quebec, Wrigley (1969) caught house mice with white-footed mice, meadow voles, and meadow jumping mice in grainfields and their edges; in Maryland, Myton (1974) found house mice with the species listed above as well as woodland voles, short-tailed shrews, and masked shrews in second-growth woodland and adjacent cultivated fields; in Connecticut, Hirth (1959) found white-footed mice, short-tailed shrews, and southern red-backed voles with house mice in mixed old field and woodland habitat; in New Jersey, Pearson (1959) noted masked shrews, meadow voles, and house mice in early grassland successional stages. Both Linduska (1950) in Michigan and Whitaker (1967) in Indiana emphasized the close association of house mice and the prairie form of the deer mouse (*Peromyscus maniculatus bairdii*). They occur together in cultivated fields, bare stubble after harvest, and weedy and grassy situations in fallow fields in early vegetative successional stages following cultivation. Whitaker noted that the deer mouse was most common in these habitats in spring and early summer, while the house mouse flourished most abundantly in autumn and winter. In Michigan, Linduska (1950) was impressed by the similarity of the habitat requirements of the two species. However, he detected intolerance between them when he found that house mice and deer mice did not share the same shocks of corn in winter, although other species of small mammals associated with them did. It was later shown by King (1957) that these two species seem to be antagonistic, with the house mouse the more aggressive. The same relationship has been noted between the house mouse and the white-footed mouse (*Peromyscus leucopus*), again with the former the aggressor when these species colonize a rural garbage dump (Courtney and Fenton, 1976).

In cultivated croplands and in early successional stages (annual grasses and forbs) of fallow areas in Lower Peninsula Michigan, the population of small rodents will generally consist of the deer mouse, the meadow jumping mouse, and the house mouse.

REPRODUCTIVE ACTIVITIES. In the human environment and under ideal field conditions, the house mouse probably breeds the year around in Michigan. Generally, however, major reproductive effort occurs between early spring and late autumn. Females attain sexual maturity at about 45 days of age, may have as many as eight litters (the record is 14) per year, average six young (as many as 13) per litter, and can potentially produce 100 offspring in a single year, although the number is more likely to be 30. The gestation period is 19 to 21 days. Young are born blind and hairless, weigh no more than 0.03 oz (1 g), and are helpless. At 10 days of age, the young are highly active, and fine fur covers the body; at 14 days, the eyes open; at 21 days, weaning takes place. By this time, the mother may have another litter, as she can be receptive to mating shortly after giving birth. Studies show that the house mouse is short-lived in nature, perhaps surviving, on the average, no longer than five months. In human habitation, these mice may live for more than 12 months. In captivity, a house mouse has been known to live for five years and 11 months.

FOOD HABITS. House mice are omnivorous, although they seem to prefer cereal grains. Their daily requirements include about 0.1 oz (3 g) of dry food and 0.3 oz (0.1 g) of water. In human environments, these commensal rodents will eat almost any kind of material containing nourishment, even cockroaches (Wourms, 1981). In a field study made in Indiana, house mice ate fox-tail grass seeds, larvae of beetles and moths, cockroaches, soybeans, corn, unidentified flesh, and assorted vegetation (Houtcooper, 1978). House mice also eat pine seeds (Smith and Aldous, 1947). In Michigan, Linduska (1950) found house mice living under field-shocked corn which provided convenient food and shelter for the winter. The custom of leaving shocked grain in the fields has been largely abandoned in recent years, no doubt seriously hampering winter survival of small mammals in corn-stubble habitat.

ENEMIES. Feral house mice are fair game for most flying and four-footed meat-eaters, as well as snakes. Birds known to eat house mice include the screech owl, long-eared owl, barn owl, red-tailed hawk, shrike, and common raven (Armstrong, 1958; Fitch and Bare, 1978; Harlow *et al.*, 1975; Jackson, 1961; Linduska, 1950; Wallace, 1948; Wilson, 1938). Coyotes, red foxes, gray foxes, dogs, and house cats also are among the major enemies of these rodents in outdoor situations (Christian, 1975; Korschgen, 1957; Murie, 1936). House mice are known to ingest and accumulate heavy metals such as lead from food materials along highways (Getz *et al.*, 1977).

PARASITES AND DISEASES. House mice harbor a large array of internal parasites belonging to the following major groups: protozoans, trematodes, cestodes, nematodes, acanthocephalans (Doran, 1954; Stiles and Crane, 1910). Ticks, fleas, lice, and mites also infest the house mouse. One mite is a carrier of the causitive agent for rickettsialpox (*Rickettsia akari*). House mice have also been known carriers of murine typhus, leptospirosis, rat-bite fever, ray fungus, and favus (Marsh and Howard, 1977), but have been generally less implicated in human disease problems than the Norway rat.

MOLT AND COLOR ABERRATIONS. House mice in pale and dark pelage colors are occasionally found in nature. In the laboratory, the albinistic form, the white mouse, is a standard animal for biomedical research work. Other color mutations have also been used in research projects.

ECONOMIC IMPORTANCE. The house mouse is regarded as the second most destructive mammal in Michigan, following the Norway rat (Burt, 1946). The amount of damage caused by the mouse alone runs into millions of dollars annually in the United States. Its abundance and general annoyance to humans may have increased in recent years as campaigns to obliterate Norway rats without regard for house mice have intensified in larger communities (Bjornson *et al.*, 1969). Attributes which have helped the house mouse survive in spite of all human discouragement include its ability to: jump 12 in (305 mm) to an elevated surface; scale vertical surfaces of wood, brick, weathered sheet metal, wire mesh, and cables; run horizontally or almost vertically along insulated electrical wires, small ropes, ship-to-shore lines, etc.; squeeze through openings only slightly larger than ¼ in (6 mm) in diameter; travel hanging upside down on such surfaces as hardware mesh; swim adequately (but not underwater as does the Norway rat); negotiate ledges or wall cracks too small for use by Norway rats; jump from heights of 8 ft (2.5 m) without injury; survive and even reproduce in situations where temperatures are as low as 24°F (−3°C); and thrive in such unlikely environments as the lower levels of coal mines (Marsh and Howard, 1977).

Homemakers frequently discourage mice with the small, wooden-based trap baited with cheese. In barns and outbuildings, cats apparently do the best job of catching mice. Modern rodent control specialists also use acute (single-dose) anticoagulant toxicants as poison in baits, including zinc phosphide, strychnine, sodium fluoroacetate (1080), arsenic trioxide, phosphorus, thallium sulfate, and red squill, and chronic (multiple-dose) toxicants including warfarin (Marsh and Howard, 1976). Officials at county cooperative extension or public health offices should be consulted regarding new products (Hayne *et al.*, 1945). Professionals may use gas as a fumigant or lace baits with newly-developed chemosterilants to sterilize breeding males; however, the best control method remains the construction of mouse (and rat) proof habitations and food storage facilities.

The destructiveness of the house mouse to foods (eaten and contaminated), to gnawed materials, and as carriers of diseases demands the attention of all citizens anxious to reduce environmental pollution. Yet, the albinistic form—the white mouse of the laboratory—is an important tool in biomedical studies.

COUNTY RECORDS FROM MICHIGAN. Records of specimens in collections of museums are from less than 30 of Michigan's 83 counties. Nevertheless, public health authorities can vouch for the fact that the ubiquitous house mouse has made a nuisance of itself in all counties of the state. This distribution also includes islands in Michigan's part of the Upper Great Lakes, although published records disclose populations only on North Manitou and Beaver islands in Lake Michigan and on Drummond Island in Lake Huron (Hatt *et al.*, 1948; Manville, 1950; Ozoga and Phillips, 1964).

LITERATURE CITED

Armstrong, W. H.
1958 Nesting and food habits of the long-eared owl in Michigan. Michigan State Univ., Publ. Mus., Biol. Ser., 1(2):61–96.

Bjornson, B. F., H. D. Pratt, and K. S. Littig
1969 Control of domestic rats and mice. U.S. Publ. Hlth. Serv., PHS Publ. 563, 41 pp.

Burt, W. H.
1946 The mammals of Michigan. Univ. Michigan Press, Ann Arbor. xv+288 pp.

Christian, D. P.
1975 Vulnerability of meadow voles, *Microtus pennsylvanicus*, to predation by domestic cats. American Midl. Nat., 93(2):498–502.

Courtney, P. A., and M. B. Fenton
1976 The effects of a small rural garbage dump on populations of *Peromyscus leucopus* Rafinesque and other small mammals. Jour. Applied Ecology, 13:413–422.

DeLong, K. T.
1967 Population ecology of feral house mice. Ecology, 48(4):611–634.

Doran, D. J.
1954 A catalogue of the Protozoa and helminths of North American rodents. II. Cestoda. American Midl. Nat., 52(2):469–480.

Drickamer, L. C.
1979 Acceleration and delay of first estrus in wild *Mus musculus*. Jour. Mammalogy, 60(1):215–216.

Fitch, H. S.
1958 Home ranges, territories, and seasonal movements of vertebrates of the Natural History Reservation. Univ. Kansas Publ., Mus. Nat. Hist., 11(3):63–326.

Fitch, H. S., and R. O. Bare
1978 A field study of the red-tailed hawk in eastern Kansas. Trans. Kansas Acad. Sci., 81(1):1–13.

Getz, L. L., L. Verner, and M. Prather
1977 Lead concentrations in small mammals living near highways. Environ. Pollut., 13:151–157.

Hall, E. R.
1927 An outbreak of house mice in Kern County, California. Univ. California, Publ. Zool., 30:189–203.

Harlow, R. F., R. G. Hooper, D. R. Chamberlain, and H. S. Crawford
1975 Some winter and nesting season foods of the common raven in Virginia. Auk, 92:299–306.

Hatt, R. T., J. Van Tyne, L. C. Stuart, C. H. Pope, and A. B. Grobman
1948 Island life: A study of the land vertebrates of the islands of eastern Lake Michigan. Cranbrook Inst. Sci., Bull. No. 27, xi+179 pp.

Hayne, D. W., M. D. Pirnie, and C. H. Jefferson
1945 Controlling rats and house mice. Michigan Agric. Exp. Sta., Cir. Bull. 167, 40 pp.

Hirth, H. F.
1959 Small mammals in old field succession. Ecology, 40(3): 417–425.

Houtcooper, W. C.
1978 Food habits of rodents in a cultivated ecosystem. Jour. Mammalogy, 59(2):425–427.

Jackson, H. H. T.
1961 Mammals of Wisconsin. Univ. Wisconsin Press, Madison. xii+504 pp.

King, J. A.
1957 Intra- and interspecific conflict of *Mus* and *Peromyscus*. Ecology, 38:355–357.

Korschgen, L. J.
1957 Food habits of coyotes, foxes, house cats and bobcats in Missouri. Missouri Conserv. Comm., Fish and Game Div., P-R Ser. No. 15, 64 pp.

Linduska, J. P.
1942 Winter rodent population in field-shocked corn. Jour. Wildlife Mgt., 6(4):353–363.
1950 Ecology and land-use relationships of small mammals on a Michigan farm. Michigan Dept. Conserv., Game Div., ix+144 pp.

Lowery, G. H., Jr.
1974 The mammals of Louisiana and its adjacent waters. Louisiana State Univ. Press, Baton Rouge. xxiii+565 pp.

Manville, R. H.
1950 The mammals of Drummond Island, Michigan. Jour. Mammalogy, 31(3):358–359.

Marsh, R. E., and W. E. Howard
1976 House mouse control manual. Pest Control, August-November, 27 pp.
1977 The house mouse: Its biology and control. Univ. California, Coop. Ext. Serv., Leaflet 2945, 28 pp.

Murie, A.
1936 Following fox trails. Univ. Michigan, Mus. Zool., Misc. Publ. No. 32, 45 pp.

Myton, B.
1974 Utilization of space by *Peromyscus* and other small mammals. Ecology, 55:277–290.

Ozoga, J. J., and C. J. Phillips
1964 Mammals of Beaver Island, Michigan. Michigan State Univ., Publ. Mus., Biol. Ser., 2(6):305–348.

Pearson, P. G.
1959 Small mammals and old field succession on the Piedmont of New Jersey. Ecology, 40(2):249–255.

Smith, C. F., and S. E. Aldous
1947 The influence of mammals and birds in retarding artificial and natural reseeding of coniferous forests in the United States. Jour. Forestry, 45:361–369.

Southwick, C. H.
1955 The population dynamics of confined house-mice supplied with unlimited food. Ecology, 36:212–225.

Stiles, C. W., and C. G. Case
1910 The internal parasites of rats and mice in their relation to diseases of man. Pp. 87–110 *in* The rat and its relation to the public health. U.S. Treas. Dept., Publ. Hlth. and Marine-Hosp. Serv., 254 pp.

Wallace G. J.
1948 The barn owl in Michigan. Its distribution, natural history and food habits. Michigan State Col., Agric. Exp. Sta., Tech. Bull. 208, 61 pp.

Whitaker, J. O., Jr.
1967 Habitat and reproduction of some of the small mammals of Vigo County, Indiana, with a list of mammals known to occur there. C. C. Adams Center Ecol. Stud., Occas. Papers No. 16, 24 pp.

Wilson, K. A.
1938 Owl studies at Ann Arbor, Michigan. Auk, 55(2):187–197.

Wood, N. A.
1922 The mammals of Washtenaw County, Michigan. Univ. Michigan, Mus. Zool., Occas. Papers No. 123, 23 pp.

Wourms, M. K.
1981 House mouse predation on cockroaches. Jour. Mammalogy, 62(4):853–854.

Wrigley, R. E.
1969 Ecological notes on the mammals of southern Quebec. Canadian Field-Nat., 83:201–211.

FAMILY ZAPODIDAE

In contrast to large and diverse rodent families like Cricetidae and Muridae, the family Zapodidae contains only four living genera. These are divisible into the subfamily Sicistinae (the birch mice of Eurasia) and the subfamily Zapodinae (the jumping mice of western Asia and North America). The American jumping mice, consisting of two genera and four species, are slender and graceful mice, weighing no more than 25 g and sporting elongated hind limbs and long tails. The long appendages give these animals outstanding jumping ability. Their attractiveness is enhanced by bright yellow to rufous upperparts complimenting the white underparts. The back fur is slightly stiff or hispid. The fragile skulls possess enlarged infraorbital foramina through which pass parts of the jaw musculature (much of the medial masseter). Distinctive also are the conspicuous longitudinal grooves on the anterior faces of the upper incisors.

Jumping mice inhabit northern forests and adjacent grasslands, preferring moist situations with ground cover of either shrubs or grasses and forbs. The molar teeth are somewhat high-crowned (hypsodont) with crown patterns consisting of small tubercules (quadritubercular) adapted for masticating seeds, other plant materials, insects and other invertebrates. Species of both North American genera, *Zapus* and *Napaeozapus*, occur in Michigan.

MEADOW JUMPING MOUSE
Zapus hudsonius (Zimmermann)

NAMES. The generic name *Zapus*, proposed by Coues in 1876, is of Greek origin and may be translated as "much" or "very" (*za*, an intensive prefix) and "foot" (*pus*), referring to the well-developed hind limbs for leaping. The specific name *hudsonius*, proposed by Zimmermann in 1780, refers to Hudson Bay, the area from which the specimen used in the original description was obtained. In Michigan, this small rodent is usually called jumping mouse; in Wisconsin, the name jumping jack is sometimes used (Jackson, 1961). The subspecies *Zapus hudsonius hudsonius* (Zimmermann) occurs in the Upper Peninsula; the subspecies *Zapus hudsonius americanus* (Barton) occurs in the Lower Peninsula (Krutsch, 1954).

RECOGNITION. Body size small and slender; head and body averaging 3¼ in (82 mm) long; tail long, thin, tapering, sparsely-haired, wire-like, scaly, averaging 4⅞ in (123 mm) long; head small; ears rounded, conspicuous, well-haired; forelegs and feet short, delicate; hind legs and feet long, wiry; hind feet almost one-third the length of the head and body; forefeet with four well-developed toes, hind feet with five long toes; soles of feet naked. The pelage is thick, short, and somewhat coarse or hispid (not velvety or soft). In adults, upperparts characterized by broad dorsal band from nose to rump consisting of olive to buff hairs mixed with blackish ones to present an olive-brown appearance; sides paler, more yellow-olive; underparts whitish, sometimes buffy; tail sharply bicolored, brown above and white below; ears dark; feet whitish. There are no size or color differences between sexes. Young animals are dorsally paler than adults. The mammary glands are arranged in four pairs, one pectoral, one abdominal and two inguinal. The meadow jumping mouse is easily distinguished from all small Michigan mammals except the woodland jumping mouse by its long tail, long hind limbs and feet, and dark back stripe. The meadow jumping mouse is distinguished externally from the woodland jumping mouse by its generally duller color, somewhat smaller size, and tail lacking white tips.

MEASUREMENTS AND WEIGHTS. Adult meadow jumping mice measure as follows: length from tip of nose to end of tail vertebrae, 7 to 8⅞ in (180 to 225 mm); length of tail vertebrae, 4⅜ to 5½ in (110 to 140 mm); length of hind foot, 1⅛ to 1¼ in (28 to 32 mm); height of ear from notch, ½ to ⅝ in (12 to 16 mm); weight, 0.4 to 0.5 oz (12 to 15 g) in spring and early summer and up to 1.0 oz (28 g) just prior to hibernation in autumn.

DISTINCTIVE CRANIAL AND DENTAL CHARACTERISTICS. The skull of the meadow jumping mouse is small, narrow, and somewhat high-crowned. The rostrum (nasal area) is short and pointed. The zygomatic arches are broadly expanded anteriorly, with a small internal shelf on the anterior ventral surface. The infraorbital foramina are characteristically large, at least one mm in diameter. The skull of an adult averages ⅞ in (22 mm) long and 7/16 in (11 mm) wide (across zygomatic arches). The dentition is characterized by the presence of longitudinal grooves on the anterior faces of the short upper incisors and by small cheek teeth with the upper jaw having a diminutive peg-like premolar (occasionally absent, Klingener, 1964) in front of the molar series. This combination of dental characteristics and the presence of the large infraorbital opening, serve to distinguish this rodent from all other Michigan species (Fig. 50). In addition, no other Michigan mammal has 18 teeth. The dental formula for the meadow jumping mouse is:

$$\text{I (incisors)}\,\frac{1\text{-}1}{1\text{-}1},\ \text{C (canines)}\,\frac{0\text{-}0}{0\text{-}0},\ \text{P (premolars)}\,\frac{1\text{-}1}{0\text{-}0},\ \text{M (molars)}\,\frac{3\text{-}3}{3\text{-}3} = 18$$

DISTRIBUTION. The meadow jumping mouse is found throughout the northeastern and north-central parts of the United States, westward to the Great Plains and northward to the limit of tree growth in Canada and Alaska (Krutsch, 1954). In Michigan, the species occurs throughout both peninsulas, generally in open country and along woodland edges (see Map 43).

HABITAT PREFERENCES. The meadow jumping mouse is apt to be found in all Michigan habitats except extremely heavy woodlands. Preferred habitats are open lands covered with mixed grasses and forbs, scattered shrubs, grassy willow-alder thickets, and early stages of forest growth (both deciduous and coniferous), as well as the edges between these plant types (Quimby, 1951; Whitaker, 1963). The meadow jumping mouse also responds favorably to agriculture, especially recently fallowed fields covered with annual weeds and grasses. Most observers, however, find the meadow jumping mouse occurs in greater numbers when the above habitats are located in poorly drained situations or adjacent to marshes, streams, or ponds. In Michigan (Burt, 1946), most studies tend to correlate abundance with moist environments, especially wet meadow and old field habitats, in Clinton County (Linduska, 1950), Kalamazoo and Van Buren counties (Brewer and Reed, 1977), Livingston County (Blair, 1940a), Marquette County (Manville, 1948), and Montmorency County (Green, 1925). However, Whitaker (1963) in his classic New York study of this species concluded that standing water or high concentrations of soil moisture only affect local distribution indirectly, as these conditions would certainly influence the growth of favored ground cover of grasses and herbs. Nevertheless, there is some evidence (Getz, 1961) that the meadow jumping mouse requires habitats with high humidities.

Some Michigan habitats in which meadow jumping mice have been taken include grassy meadows, swampy land, and white birch-aspen-hardwood saplings (Manville, 1948); bluegrass association (Blair, 1940a); reed-canary grass habitat (Linduska, 1950); open grasses, sedges, and shrubs (Dice and Sherman, 1922); grassy marsh (Dice, 1925); and edge of sedge marsh (Getz, 1961).

Map 43. Geographic distribution of the meadow jumping mouse (*Zapus hudsonius*). Counties from which the species has actually been reported are dark shaded.

DENSITY AND MOVEMENTS. Although there are several studies pointing to annual population fluctuations (Blair, 1940a; Quimby, 1951), there is other evidence showing that local populations of the meadow jumping mouse have considerable year-to-year stability. Kirkland and Kirkland (1979) suggested that hibernation by this species may bring about this near stabilization in annual population size. Compared to nonhibernating mammalian associates of similar size, the meadow jumping mouse appears to survive with fewer litters annually and, on the average, are longer-lived individuals—attributes which are conducive to noneruptive population trends. The annual population of Michigan meadow jumping mice, reaches maximum size just after the emergence of weaned young of the second litter in late summer (Rybak *et al.*, 1975).

Meadow jumping mice can reach densities as high as 12 individuals per acre (30 per ha), according to Quimby (1951), in favorable Minnesota habitat; most of his field studies, however, showed numbers to be closer to three animals per acre (7.5 per ha). Population densities obtained by Michigan workers include up to 0.38 animals per acre (0.95 per ha) in August and September in Marquette County (Manville, 1949a); five animals per acre (12.5 per ha) in late June in Livingston County (Blair, 1940a); 10.8 animals per acre (27 per ha) in late summer in Ingham County (Muchlinski, 1979); and less than one individual per acre (2.5 per ha) at localities in the McCormick Forest in Baraga and Marquette counties (Robinson and Warner, 1975).

Quimby's (1951) Minnesota studies showed that meadow jumping mice restrict their individual movements to small areas. He found that males occupied home ranges averaging 2.7 acres (one ha) while females lived in areas of 1.6 acres (0.6 ha). In Michigan, home ranges of less than one acre (0.4 ha) were found in Livingston County (Blair, 1940a) and Marquette County (Manville, 1949a). However, the meadow jumping mouse's ability to disperse is demonstrated by two individuals (one male, one female) which were live-trapped in July on a grassy hillside in Clinton County. The animals were infected with the blood fluke (*Schistosomatium douthitti*), a parasite which could only have been contracted from pond water no nearer than 1,320 ft (406 m) (Carmichael, 1979).

BEHAVIOR. Fieldworkers sometimes consider the meadow jumping mouse less a "member" of the Michigan small mammal communities than short-tailed shrews, meadow voles and white-footed mice. This is due to the fact that the former actively participates in the environment for only about half of each year and because its special jumping adaptation appears to set it apart in terms of mobility. Certainly its sudden emergence from hibernation in mid-spring could strain existing nutrient supplies for the rest of the small mammal community, although this same hibernation leaves one less species to compete for diminishing winter resources.

As a rule, the meadow jumping mouse is mostly active at night (Muchlinski, 1981). During daylight hours, especially when skies are overcast (and possibly when startled by humans), these mice may suddenly hurtle through the air above the grass. Jumps of 10 ft (3 m) have been measured (Krutsch, 1954). On the ground, this rodent uses short hops or creeps under grass cover along runways constructed by meadow voles (Sheldon, 1938). One individual was timed at 8.1 ft (2.5 m) per second (Layne and Benton, 1954).

Most observers regard the meadow jumping mouse as rather solitary; multiple catches of this species in live traps are unusual (Quimby, 1951). Animals associate compatibly when confined in cages (Whitaker, 1963). In captivity, meadow jumping mice spend considerable time grooming. They are good swimmers, holding their heads high and using their hind limbs for propulsion (Hamilton, 1935); they can also swim underwater. These mice have been known to climb low shrubs (Sheldon, 1938).

Meadow jumping mice are less noisy than other Michigan rodents; however, they are vociferous when young, producing squeaking and suckling sounds (Sheldon, 1938). Adults make clucking calls, especially prior to entering hibernation. Sheldon also describes a drumming noise made by the tail as it vibrates against dry leaves.

Meadow jumping mice are excellent diggers and construct ground burrows (Goodwin, 1935) although they have also been taken in burrows excavated by other mammals (Grizzell, 1949; Sheldon, 1938). The average nest is constructed of grasses and leaves formed into a round, baseball-size structure with a side entrance (Preble, 1899). Nests in underground burrows occur as far as 10 in (0.3 m) below the surface and are usually in well-drained areas. Nests employed in summer often are located at ground level in grass clumps, under logs or protected by roots.

HIBERNATION. Much of the published literature concerning the natural history of the meadow jumping mouse refers to its interesting habit of hibernating, as this is unusual behavior for small mammals. Certainly it is a successful method of winter survival, with fat reserves for energy being accumulated in early autumn (Muchlinski and Rybak, 1978). It is not known why other small mammals have not adopted this means of survival, although degrees of torpor are now known in several rodents, notably members of the genus *Peromyscus* (Hill, 1975).

Most information about hibernation in the meadow jumping mouse comes from field observations by Quimby (1951) in Minnesota and by Whitaker (1963) in New York, while Muchlinski (1979) in Michigan combined field studies with those carried out in the laboratory. The annual cycle of this rodent in southern Michigan, as abstracted from Muchlinski, includes: (1) spring and summer reproduction, triggered by long day-lengths; (2) preparation for the entering into hibernation from late August through mid-October, induced by decreasing day-lengths; and (3) arousal from hibernation from mid-April to early May, possibly triggered by an increase in soil temperature at the site of the animal's hibernaculum.

The obvious mode of preparation for autumn hibernation is weight gain in the form of body fat, from a normal mid-summer weight of 16–18 g to as much as 28 g just prior to the beginning of winter sleep (Gaertner, 1963; Muchlinski, 1979). Muchlinski found adult male meadow jumping mice in southern Michigan enter hibernation first—in late August—and are not caught in standard traplines after early Sepember. Adult females and first-litter juveniles follow by middle to late September, with second-litter juveniles the last to enter hibernation, no later than mid-October. October 20 is the last date recorded for above-ground activity. The meadow jumping mouse survives winter sleep with no nutrients other than those ingested just prior to hibernation. There are also energy demands to meet when entering and arousing from hibernation (Muchlinski and Rybak, 1978). During the winter period, however, the metabolic rate of the sleeping mouse is correlated with body temperature which remains near ambient temperature. Consequently, the hibernating mouse must employ a sleeping chamber sufficiently far underground that the adjacent soil is not likely to freeze. Several of these hibernacula have been described; traditionally, the sleeping meadow jumping mouse is encased in a ball-shaped grassy nest. The mouse also rolls into a ball with its head between its hind legs, its nose against its lower abdomen, its forefeet curled on its breast, and its long tail coiled around its head and body, with its head and hip regions on the bottom (Eadie, 1949; Grizzell, 1949; Nicholson, 1937; Sheldon, 1938).

The males arouse from hibernation, usually in late April and early May, about two weeks earlier than the females. (Muchlinski, 1979); although one active Michigan male was inexplicably found aboveground on February 12 in Barry County (Manville, 1956). All aspects of the arousal have been examined, including the dramatic speeding-up of bodily functions (heartbeat, respiration), the increase in body temperature, and the energy input required. One such awakening was described for a meadow jumping mouse found near Ann Arbor (Clough, 1955).

ASSOCIATES. As mentioned earlier, the meadow jumping mouse appears to lead a rather solitary existence, living apart from other members of the small mammal community. This may be due in part to its winter sleeping habits. It would be interesting to have detailed studies on the impact of the meadow jumping mouse's appearance in mid-spring and disappearance in autumn on the populations of small mammals which are active overwinter. In Ingham County, for example, Nichols and Conley (1981) noted an increase in white-footed and deer mice at a study site in October following the disappearance of meadow jumping mice. These authors concluded that active meadow jumping mice may competitively exclude the white-footed mouse (from its immediate association and probably the deer mouse as well).

Studies by Quimby (1951) in Minnesota and by Sheldon (1938) in Vermont show that the meadow jumping mouse may be found living in association with almost all of the small mammals in Michigan's outdoor communities. Although the species can use trails constructed by microtines (Sheldon,

1938), it is more likely to move about uninhibited either above or below ground cover foraging for grass seeds, insects, and other scattered foods. In Livingston County, Blair (1940b) found the meadow jumping mouse in blue grass association with the short-tailed shrew, eastern mole, thirteen-lined ground squirrel, white-footed mouse, prairie deer mouse, meadow vole, and southern bog lemming. In Marquette County, Manville (1949a) caught meadow jumping mice in the same traplines with the short-tailed shrew, star-nosed mole, least chipmunk, eastern chipmunk, woodland deer mouse, and the woodland jumping mouse. At Cusino near the boundary of Alger and Schoolcraft counties, Ozoga and Verme (1968) found a single meadow jumping mouse in a sedge opening bordering a mixed conifer swamp in company with the arctic shrew, masked shrew, least chipmunk, eastern chipmunk, deer mouse, southern red-backed vole, meadow vole, and woodland jumping mouse. In the McCormick Experimental Forest of Baraga and Marquette counties, Haveman (1973) captured meadow jumping mice in pit traps with arctic shrews, masked shrews, pygmy shrews, short-tailed shrews, deer mice, southern red-backed voles, meadow voles, and southern bog lemmings. In Roscommon County, Rabe (1981) caught meadow jumping mice in sedge-willow and leatherleaf-bog birch peatland in company with the masked shrew, pygmy shrew, water shrew, short-tailed shrew, star-nosed mole, and meadow vole.

It is not altogether clear whether social interactions between the meadow jumping mouse and the woodland jumping mouse are competitive (Brower and Cade, 1966; Getz, 1961; Whitaker, 1963). As Connor (1966) has mentioned, these two species have been caught side by side, especially at meadow-forest boundaries where the meadow jumping mouse seems to be more successful in entering habitats of the woodland jumping mouse than vice versa (Wrigley, 1972). Some observers (Quimby, 1951) place the weasels, especially the ermine, as close associates of the meadow jumping mouse, even though the relationship is strictly a prey-predator one.

REPRODUCTIVE ACTIVITIES. The male meadow jumping mouse, upon emergence from hibernation in mid-spring, reaches a state of reproductive readiness by the time females begin to emerge shortly thereafter (Muchlinski, 1979). As a result, within two weeks following their emergence, most females become pregnant. The first litters appear in June, with weanlings entering the population in early July. The second litters arrive, on the average, in late July and early August, with weanlings entering the active population in August. A single female may produce three litters in one season (Whitaker, 1963), but two are most likely in Michigan. There is also evidence that females born early in the season bear young in the same year (Quimby, 1951; Wrigley, 1969).

The gestation period of the meadow jumping mouse is about 18 days and may be prolonged if the female happens to be in lactation at the time of mating. Two to eight (usually five) young are produced. At birth, the neonate is small (weighing 0.8 g), hairless except for minute vibrissae, pink in color, clawless, has closed eyelids and folded ear pinnae, and emits audible squeaks. In the first seven days, hair develops on the dorsum, the tail becomes bicolored, ear pinnae unfold, and claws appear on toes. In the second week, the dorsum becomes covered with yellow-brown hair, the vibrissae develop, the incisors erupt, and crawling is accomplished. In the third week, the young begin to react to sound, can creep and hop awkwardly, begin to develop elongated hind limbs, and complete the growth of the juvenile pelage. In the fourth week, the juvenile pelage is replaced by the adult coat, teeth are developed, the eyes open, weaning is completed, and the young have the general appearance of the adult.

Hibernation and the less-than-prolific annual breeding cycle make it essential that individual meadow jumping mice live longer than the more fecund deer mice and meadow voles. Quimby (1951) kept records of individuals living more than two years under natural conditions. A study by Muchlinski (1979) showed that if a juvenile survived its first winter it would have a 28.5% chance of living through its second hibernation, and a 8.5% chance of surviving through a third hibernation period. There is also a report of the meadow jumping mouse surviving for five years in captivity (Jones, 1979).

FOOD HABITS. The meadow jumping mouse eats a variety of foods, primarily seasonal seeds, fruits and animal materials (Quimby, 1951; Whitaker, 1963; Whitaker and Mumford, 1971). After emergence from hibernation in spring, the species may obtain as much as half of its food from animal sources (larvae of moths, butterflies, and beetles). Seeds are utilized but are often in short supply early in the growing season. By midsummer, however, seeds and seasonal fleshy fruits are prominent in the diet. Subterranean fungi are eaten, evidently in large amounts. In Michigan, Getz (1961) noted that green vegetation along with seeds were eaten. The array of seeds and other foods identified in the diet of the meadow jumping mouse by Whitaker (1963) and Whitaker and Mumford (1971) indicates that the meadow jumping mouse broadly utilizes the foods most available at a given time in the warm season. High caloric food supplies must certainly be necessary in late summer and early autumn for fat accumulation prior to entering hibernation.

ENEMIES. The meadow jumping mouse probably experiences minimal predation during hibernation. In summer, however, this mouse has to contend with the same meat-eaters as do deer and white-footed mice. Night-flyers recorded as eating meadow jumping mice include the great-horned owl (Errington *et al.*, 1940), screech owl (Wilson, 1938), long-eared owl (Linduska, 1950), and barn owl (Wallace, 1950). Hawks known to catch these rodents include the marsh hawk (Errington, 1933), broadwinged hawk (Linduska, 1950), and red-tailed hawk (Fitch and Bare, 1978). The raven is also a predator (Harlow *et al.*, 1975). Meat-eating mammals found to relish the meadow jumping mouse include the red fox (Murie, 1936), gray fox (Latham, 1950), gray wolf (Voigt *et al.*, 1976), mink (Dearborn, 1932), long-tailed weasel (Blair, 1940a), and house cat (Toner, 1956). This mouse has also been reported from the digestive tracts of a green frog (*Rana clamitans*) by Vergeer (1948) and of a northern pike by Quimby (1951). Meadow jumping mice have also been killed on Michigan highways (Manville, 1949b) and trapped in holes dug for fence and telephone poles (Linduska, 1950).

PARASITES AND DISEASES. Much attention has been given to parasites which infest meadow jumping mice (Quimby, 1951; Whitaker, 1963; Whitaker and Mumford, 1971). Internally, there are Protozoa (trypanosomes, flagellates, coccidians), Trematoda (flukes including schistosomes), Cestoda (larval stages and adult tapeworms), and Nematoda (roundworms). Carmichael and Muchlinski (1980) found that the schistosome, *Schistosomatium douthitti*, survives in the meadow jumping mouse during hibernation. Externally, meadow jumping mice harbor ticks, mites, fleas, and biting lice. Botfly larva may also live in these rodents. As reported by Whitaker (1963), a western relative of the meadow jumping mice (*Z. princeps*) has been found to harbor the tularemia organism. An underground nest of a hibernating meadow jumping mouse uncovered in January in Indiana by Jones and Whitaker (1976) contained species of Collembola, larvae of beetles, Hymenoptera, Pauropoda, mites, and nematodes.

MOLT AND COLOR ABERRATIONS. After developing a juvenile pelage, the young meadow jumping mouse grows the adult coat in the fourth week after birth. Young-of-the-year molt and new pelage appears in August and thereafter they molt and grow new coats once a year beginning in late June (Quimby, 1951; Whitaker, 1963). Animals in pale pelage, sometimes with spotting, are recorded.

ECONOMIC IMPORTANCE. The presence of a meadow jumping mouse under a shock of corn in Washtenaw County (Wood, 1922) appears to indicate that this mouse eats grain. However, as Allen (1938) reported, this animal has little economic significance.

COUNTY RECORDS FROM MICHIGAN. Records of specimens of the meadow jumping mouse in collections of museums and from the literature show that the species is reported from all 15 counties in the Upper Peninsula and 32 in the Lower Peninsula (see Map 43).

LITERATURE CITED

Allen, D. L.
1938 Ecological studies on the vertebrate fauna of a 500-acre farm in Kalamazoo County, Michigan. Ecol. Monogr., 8(3):347–436.

Blair, W. F.
1940a Home ranges and populations of the jumping mouse. American Midl. Nat., 23(1):244–250.
1940b Notes on home ranges and populations of the short-tailed shrew. Ecology, 21(2):284–288.

Brewer, R., and M. L. Reed
1977 Vertebrate inventory of wet meadows in Kalamazoo and Van Buren counties. Michigan Dept. Nat. Res., (mimeo). 20 pp.

Brower, J. E., and T. J. Cade
1966 Ecology and physiology of *Napaeozapus insignis* (Miller) and other woodland mice. Ecology, 47(1):46–63.

Burt, W. H.
1946 The mammals of Michigan. Univ. Michigan Press, Ann Arbor. xv+288 pp.

Carmichael, A. C.
1979 Ecological aspects of a mammalian host-parasite system. Michigan State Univ., unpubl. M.S. thesis, 73 pp.

Carmichael, A. C., and A. E. Muchlinski
1980 Survival of Schistosomatium douthitti during hibernation in the natural host. *Zapus hudsonius*. Jour. Parasitol., 66(2):365–366.

Clough, G. C.
1955 Repeated hibernation in a captive meadow jumping mouse. Jour. Mammalogy, 36(2):301–302.

Connor, P. F.
1966 The mammals of the Tug Hill Plateau, New York. New York State Mus. and Sci. Serv., Bull. No. 406, 82 pp.

Dearborn, N.
1932 Foods of some predatory fur-bearing animals in Michigan. Univ. Michigan, Sch. For. and Conserv., Bull. No. 1, 52 pp.

Dice, L. R.
1925 A survey of the mammals of Charlevoix County, Michigan, and vicinity. Univ. Michigan, Mus. Zool., Occas. Papers No. 159, 33 pp.

Dice, L. R., and H. B. Sherman
1922 Notes on the mammals of Gogebic and Ontonagon counties, Michigan, 1920. Univ. Michigan, Mus. Zool., Occas. Papers No. 109, 40 pp.

Eadie, W. R.
1949 Hibernating meadow jumping mouse. Jour. Mammalogy, 30(3):307–308.

Errington, P. L.
1933 Food habits of southern Wisconsin raptors. II. Hawks. Condor, 34:176–186.

Errington, P. L., F. Hamerstrom, and F. N. Hamerstrom, Jr.
1940 The great horned owl and its prey in north-central United States. Iowa State Col., Agric. Exp. Sta., Res. Bull. No. 277:757–850.

Fitch, H. S., and R. O. Bare
1978 A field study of the red-tailed hawk in eastern Kansas. Trans. Kansas Acad. Sci., 81(1):1–13.

Gaertner, R. A.
1963 Aspects of the physiology of the meadow jumping mouse, *Zapus hudsonius*, in the non-hibernating period. Michigan State Univ., unpubl. M.S. thesis, vi+40 pp.

Getz, L. L.
1961 Notes on the local distribution of *Peromyscus leucopus* and *Zapus hudsonius*. American Midl. Nat., 65(2):486–500.

Goodwin, G. G.
1935 The mammals of Connecticut. Connecticut State Geol. and Nat. Hist. Surv., Bull. 53, 221 pp.

Green, M. M.
1925 Notes on some mammals of Montmorency County, Michigan. Jour. Mammalogy, 6(3):173–178.

Grizzell, R. A., Jr.
1949 Hibernating jumping mice in woodchuck dens. Jour. Mammalogy, 30(1):74–75.

Hamilton, W. J., Jr.
1935 Habits of the jumping mice. American Midl. Nat., 16:187–200.

Harlow, R. F., R. G. Hooper, D. R. Chamberlain, and H. S. Crawford
1975 Some winter and nesting season foods of the common raven in Virginia. Auk, 92:298–306.

Haveman, J. R.
1973 A study of population densities, habitats and foods of four sympatric species of shrews. Northern Michigan Univ., unpubl. M.S. thesis, vii+70 pp.

Hill, R. W.
1975 Daily torpor in *Peromyscus leucopus* on an adequate diet. Comp. Biochem. Physiol., 51A:413–423.

Jackson, H. H. T.
1961 Mammals of Wisconsin. Univ. Wisconsin Press, Madison, xii+504 pp.

Jones, G. S., and J. O. Whitaker, Jr.
1976 The fauna of a hibernation nest of a meadow jumping mouse, *Zapus hudsonius*. Canadian Field-Nat., 90:169–170.

Jones, M. L.
1979 Longevity of mammals in captivity. Intern. Zoo News, 26(3)16–26.

Kirkland, G. L., Jr., and C. J. Kirkland
1979 Are small mammal hibernators K-selected. Jour. Mammalogy, 60(1):164–168.

Klingener, D.
1964 Notes on the range of *Napaeozapus* in Michigan and Indiana. Jour. Mammalogy, 45(4):644–645.

Krutsch, P. H.
1954 North American jumping mice (genus *Zapus*). Univ. Kansas Publ., Mus. Nat. Hist., 7(4):349–472.

Latham, R.
1950 The food of predaceous animals in northeastern United States. Pennsylvania Game Comm., 69 pp.

Layne, J. N., and A. H. Benton
1954 Speeds of some small mammals. Jour. Mammalogy, 35(1):103–104.

Linduska, J. P.
1950 Ecology and land-use relationships of small mammals on a Michigan farm. Michigan Dept. Conserv., Game Div., ix+144 pp.

Manville, R. H.
1948 The vertebrate fauna of the Huron Mountains, Michigan. American Midl. Nat., 39(3):615–640.
1949a A study of small mammal populations in northern Michigan. Univ. Michigan, Mus. Zool., Misc. Publ., No. 73, 83 pp.

1949b Highway mortality in northern Michigan. Jour. Mammalogy, 30(3):311–312.
1956 Hibernation of meadow jumping mouse. Jour. Mammalogy, 37(1):122.

Muchlinski, A. E.
1979 The effects of daylength and temperature on the hibernating rhythm of the meadow jumping mouse (*Zapus hudsonius*). Michigan State Univ., unpubl. Ph.D. dissertation, vii+78 pp.
1981 Activity of *Zapus hudsonius:* A laboratory study. Jack-Pine Warbler, 59(1):13–14.

Muchlinski, A. E., and E. N. Rybak
1978 Energy consumption of resting and hibernating meadow jumping mice. Jour. Mammalogy, 59(2):435–437.

Murie, A.
1936 Following fox trails. Univ. Michigan, Mus. Zool. Misc. Publ. No. 32, 45 pp.

Nichols, J. D., and W. Conley
1981 Observations suggesting competition between *Peromyscus* and *Zapus* in southern Michigan. Jack-Pine Warbler, 59(1):3–6.

Nicholson, A. J.
1937 A hibernating jumping mouse. Jour. Mammalogy, 18(1):103.

Ozoga, J. J., and L. J. Verme
1968 Small mammals of conifer swamp deeryards in northern Michigan. Michigan Acad., Sci., Arts and Ltrs., Papers, 53:37–49.

Preble, E. A.
1899 Revision of the jumping mice of the genus *Zapus*. N. American Fauna No. 15, 42 pp.

Quimby, D. C.
1951 The life history and ecology of the jumping mouse, *Zapus hudsonius*. Ecol. Monogr., 21(1):61–95.

Rabe, M. L.
1981 New locations for pygmy (*Sorex hoyi*) and water (*Sorex palustris*) shrews in Michigan. Jack-Pine Warbler, 59(1):16–17.

Robinson, W. L., and J. K. Werner
1975 Vertebrate animal populations of the McCormick Forest. U.S.D.A. Forest Serv., Res. Paper NC-118, 25 pp.

Rybak, E. J., E. J. Neufarth, and S. H. Vessey
1975 Distribution of the jumping mouse, *Zapus hudsonius*, in Ohio: A twenty-year update. Ohio Jour. Sci., 75(4):184–187.

Sheldon, C.
1938 Vermont jumping mice of the genus *Zapus*. Jour. Mammalogy, 19(4):324–332.

Toner, G. C.
1956 House cat predation on small mammals. Jour. Mammalogy, 37(1):119.

Vergeer, T.
1948 Frog catches mouse in natural environment. Turtox News, 26(3):91.

Voigt, D. R., G. B. Kolenosky, and D. H. Pimlott
1976 Changes in summer foods of wolves in central Ontario. Jour. Wildlife Mgt., 40(4):663–668.

Wallace, G. J.
1950 In support of the barn owl. Michigan Agric. Sta., Quart. Bull., 33(2):96–105.

Whitaker, J. O., Jr.
1963 A study of the meadow jumping mouse, *Zapus hudsonius* (Zimmermann), in central New York. Ecol. Monogr., 33(3):215–254.

Whitaker, J. O., Jr., and R. E. Mumford
1971 Jumping mice (Zapodidae) in Indiana. Proc. Indiana Acad. Sci., 80:201–209.

Wilson, K. A.
1938 Owl studies at Ann Arbor, Michigan Auk, 55(2):187–197.

Wood, N. A.
1922 The mammals of Washtenaw County, Michigan. Univ. Michigan, Mus. Zool., Occas. Papers No. 123, 23 pp.

Wrigley, R. E.
1969 Ecological notes on the mammals of southern Quebec. Canadian Field-Nat., 83:201–211.
1972 Systematics and biology of the woodland jumping mouse, *Napaeozapus insignis.* Illinois Biol. Monog. 47, 117 pp.

WOODLAND JUMPING MOUSE
Napaeozapus insignis (Miller)

NAMES. The generic name *Napaeozapus,* proposed by Preble in 1899, is of Greek origin and may be translated as "belonging to a wooded vale or dell" (*Napaeo*), "much" or "very" (*za*), and "foot" (*pus*), referring to the rodent's habitat and well-developed limbs for jumping. The specific name *insignis,* proposed by Miller in 1891, may be translated as "distinguished by a mark." In the field, the woodland jumping mouse and its near look-alike, the meadow jumping mouse, are commonly referred to as jumping mice. The subspecies *Napaeozapus insignis frutectanus* Jackson occurs in Michigan.

RECOGNITION. Body size medium and stout, larger than the meadow jumping mouse; head and body averaging 3¾ in (95 mm) long; tail long, thin, tapering, sparsely-haired, wire-like, scaly, averaging 5⅝ in (145 mm) long, more than one-half of the total length; head small with rounded, well-haired ears; eyes dark, beady; forelegs and feet short, delicate; hind legs and feet long, wiry, extending laterally at a 45° angle when crouched; pollex on forelimbs reduced, all five digits well developed on hind limbs; soles of feet naked. The pelage is thick, short, and coarse. In adults, upperparts with broad buffy-brown to dark brown dorsal stripe from nose to rump; sides yellow to orange lightly streaked with black hairs; underparts creamy white; ears brown edged with buff; cheeks golden yellow; tail distinctly bicolored, brown above, creamy white below, almost always white-tipped; tops of feet white. There are no color or size differences between sexes. In worn pelage, the back stripe tends to darken and the sides become yellowish gray. The mammary glands are arranged in four pairs, one pectoral, one abdominal, and two inguinal. There are no internal cheek pouches. This colorful mouse with its tricolor pattern, and long tail and hind limbs (Plate 4a) is easily distinguished from all other Michigan mammals except its close relative, the meadow jumping mouse. From the latter, it differs in being larger and having brighter dorsal coloration and a white-tipped tail, which is present in almost all woodland jumping mice (Wrigley, 1972).

MEASUREMENTS AND WEIGHTS. Adult woodland jumping mice measure as follows: length from tip of nose to end of tail vertebrae, 8 to 9¾ in (205 to 250 mm); length of tail vertebrae, 4⅝ to 6 in (119 to 155 mm); length of hind foot, 1 1/16 to 1 5/16 in (28 to 33 mm); height of ear from notch, ⅝ to 11/16 in (16 to 18 mm); weight, 0.7 to 0.9 oz (19 to 26 g) in spring and early summer and up to 1.1 oz (35 g) just prior to hibernation or in pregnant females.

DISTINCTIVE CRANIAL AND DENTAL CHARACTERISTICS. The skull of the woodland jumping mouse is small, narrow, and somewhat high-crowned as is that of the meadow jumping mouse (Fig. 50). The infraorbital foramina are large and obliquely oval, the zygomatic arches narrow and depressed, the palate short and broad, and the auditory bullae small. The skull of an adult averages almost 15/16 in (23.5 mm) long and almost ½ in (1.2 mm) wide (across zygomatic arches). The teeth are characterized by yellow or orange incisors, short upper incisors with longitudinal grooves on the anterior faces, flatcrowned cheek (molar) teeth, and the absence of a peg-like premolar in front of the molar series (Whitaker and Wrigley, 1972). The large infraorbital openings and grooved upper incisors distinguish the skull of the woodland jumping mouse from those of all other Michigan mammals of similar size except its close relative, the meadow jumping mouse, from which it differs by not having the peg-like upper premolars. The dental formula for the woodland jumping mouse is:

$$\text{I (incisors)}\,\frac{1\text{-}1}{1\text{-}1},\ \text{C (canines)}\,\frac{0\text{-}0}{0\text{-}0},\ \text{P (premolars)}\,\frac{0\text{-}0}{0\text{-}0},\ \text{M (molars)}\,\frac{3\text{-}3}{3\text{-}3} = 16$$

DISTRIBUTION. The woodland jumping mouse occurs in boreal forest habitats of northeastern United States and southern Canada from southeastern Manitoba and adjacent Minnesota eastward through the Great Lakes Region to coastal Canada and New England and south in the Appalachians to extreme northern Georgia (Wrigley, 1972). In Michigan, the species occurs throughout the Upper Peninsula and in the northern part of the Lower Peninsula, at least in the northern three tiers of counties (see Map 44). Apparently, this species does reach southwestern Ontario (Banfield, 1974), but does not occur in adjacent parts of Michigan across the Huron-Erie waterway connection (see Table 13).

HABITAT PREFERENCES. The distribution of the woodland jumping mouse is closely related to distributions of the spruce-fir and hemlock-hardwood forest types of northeastern North America (Wrigley, 1972). As Wrigley suggests, however, the near similar distribution of this mouse and these forest associations may be merely coincidental. Instead, herbaceous growth and other ground cover may be more important as distributional factors than specific regimes of trees, abundant moisture, and cool temperatures in summer when the mouse is active above ground (Vickery, 1981). Its major habitat preference seems to be woodland edges near water sources with undergrowth of ferns, grasses, and shrubs and with rocks, logs, and brush piles as cover (Brower and Cade, 1966). Apparently, the woodland jumping mouse rarely ventures far into open meadows, grassy swales,

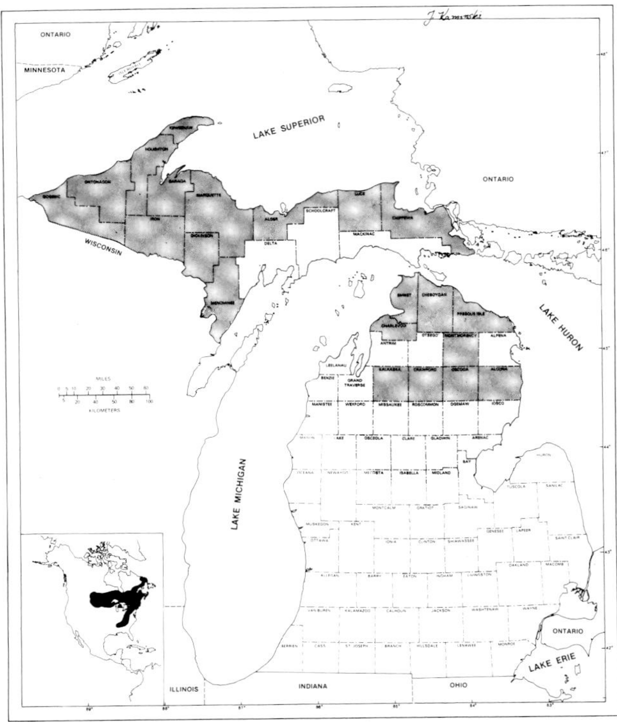

Map 44. Geographic distribution of the woodland jumping mouse (*Napaeozapus insignis*). Counties from which the species has actually been recorded are heavily shaded.

bogs lacking woody cover and cultivated or recently fallow fields (Banfield, 1974; Connor, 1966; Jackson, 1961; Wrigley, 1972).

In Michigan, the woodland jumping mouse has been found living in virgin hardwood forest in Alger County (Blair, 1941); in both dry and wet hardwood forest in Gogebic and Ontonagon counties (Dice and Sherman, 1922); in dense hardwod saplings in Marquette County (Manville, 1948); in a sedge opening in a mixed conifer swamp at the border of Alger and Schoolcraft counties (Ozoga and Verme, 1968); in second-growth hardwood, second-growth fir and spruce, and tamarack and black spruce bog in Charlevoix County (Dice, 1925); along a small trout stream bordered by aspen, red maple, and shrubs in Kalkaska County (Dalby, 1968); and in grassy areas shaded by second-growth ash in Montmorency County (Green, 1925).

DENSITY AND MOVEMENTS. Maximum densities occur when the juveniles join the adults in the active population, reaching a major peak beginning in late June and a minor one in early September, correlated with the two major breeding periods (Wrigley, 1972). Reported estimates of number of individuals in a unit area vary, partly because of the period during the warm months when the studies were made, the array of habitats examined, and likely fluctuations in numbers from year to year (Kirkland and Kirkland, 1979; Sheldon, 1938). Populations of 24 animals per acre (60 per ha) found in moist woodlands in New York by Townsend (1935) seem high, although he calculated populations in other habitats (dry woods) as six to 11 per acre (15 to 27 per ha). Also in New York, Brower and Cade (1966) determined the woodland jumping mouse population in shrubby habitat as 5.2 individuals per acre (13 per ha) from mid-August to mid-September.

In Michigan, Burt (1946) considered autumn populations averaged three mice per acre (7.5 per ha). However, in mostly second-growth yellow birch, sugar maple, and hemlock in Marquette County, Manville (1949) calculated only 0.25 individuals per acre (0.6 per ha). In August and September in Alger County, Blair (1941) determined that the local population in virgin hardwood forest was about 2.7 mice per acre (6.8 per ha).

In contrast to meadow voles, for example, the woodland jumping mouse, like its close relative the meadow jumping mouse, enjoys a high degree of mobility in order to forage for scattered plants seeds and insects. Although Manville (1949) calculated the home range in second-growth woodlands in Marquette County to be only 1.5 acres (0.6 ha), in Alger County, Blair (1941) found females (captured, marked, released, and recaptured on trapping grids) occuping areas from 1.0 to 6.5 acre (0.4 to 2.2 ha) and males moving around from 1.0 to 9.0 acres (0.4 to 3.6 ha). Mobility and concentrations of meadow jumping mice and woodland jumping mice are generally very similar.

BEHAVIOR. The woodland jumping mouse is chiefly nocturnal in activity, perhaps more so than the meadow jumping mouse (Pearson, 1947). Even so, individuals may also be observed in daytime, especially on drab or rainy days (Becker and Christian, 1979; Whitaker and Wrigley, 1972). There is also evidence that this rodent's activity may increase on rainy, cloudy, and chilly nights. As in the case of the meadow jumping mouse, the woodland jumping mouse's habit of hibernation seems to set the species somewhat apart from the active population of small woodland mammals.

Woodland jumping mice move in and around cover using a creeping gait of speeds up to 7.8 ft (2.4 m) per second (Layne and Benton, 1954). They also take short quadrupedal hops (Wrigley, 1972). When disturbed, the animals utilize their long hind limbs in making longer leaps. Most jumps cover no more than 4 ft (1.2 m) and attain a height of 2 ft (0.6 m). Leaps of more than 6 ft (1.8 m), as reported by some observers, are unusual. Despite the long hind limbs, the woodland jumping mouse climbs easily and can forage for seeds and fruits in low shrubs (Hamilton, 1935). Hamilton also described this mouse's swimming ability. Although the animal tires rather quickly while swimming, it can cross small streams without difficulty (Wrigley, 1972). The woodland jumping mouse escapes from potential enemies by jumping, which may be followed by an abrupt stop as the animal remains motionless under cover. In captivity, woodland jumping mice are apt to be nervous and excitable, making sporadic jumps. Tail drumming, also practiced by meadow jumping mice, may be another sign of this species' excitability. However, after several weeks of confinement, woodland jumping mice become docile and are easily handled (Wrigley, 1972).

As a rule, woodland jumping mice carry out their nightly routine without vocalization. However, the young are noisy almost from the day of birth, producing squeaks and suckling sounds (Sheldon, 1938). Adults have a soft, clucking call, used mostly just before hibernation or when the animal is disturbed during sleep.

Like other seed/insect-eating small rodents (deer and meadow jumping mice), the woodland jumping mouse seems to be less a creature of habit than runway-using meadow voles. Woodland jumping mice will use trails constructed by the voles as well as those of muskrats, beaver, and even white-tailed deer. However, its need to forage is more dependent on food-plant distributions and accumulations of dropped seeds than on pathways developed by other mammals.

The woodland jumping mouse is an excellent digger and can excavate its own burrow. It also readily uses cavities established by other species. Burrows may be as long as 5 ft (1.5 m) and plugged during daylight hours (Snyder, 1924). Burrow systems of females are more extensive than those of males (Becker and Christian, 1979). A globular nest as much as 6 in (154 mm) in diameter and constructed of leaves and/or dry grasses often occupies an enlargement in the shallow tunnel. Nests may also be placed in brush piles on the surface of the ground.

HIBERNATION. The normal body temperature for the woodland jumping mouse is just under 100°F (37°C). However, when inactive, it can enter daily torpor with a drop in this body temperature, especially during cold summer nights (Klein, 1957). In preparing for cold-weather hibernation, the woodland jumping mouse, like its close relative the meadow jumping mouse, takes on large amounts of fat, often comprising one-third of the body weight. This nutrient supply is the only nourishment for this rodent's life system during the several months of winter inactivity. Fat accumulation begins as daylengths decrease in autumn (Neumann and Cade, 1964), with adults beginning to accumulate fat earlier than young-of-the-year. Most adults enter underground hibernacula by late September (Wrigley, 1972), with younger animals remaining active until late October and even early November. Hibernation lasts for almost six months (from November until late April). Like the meadow jumping mouse, the woodland jumping mouse coils up in a ball inside of its underground nest during this long winter inactivity. Most males emerge in late April and early May; females do not become a noticeable part of the active population until the latter half of May. The woodland jumping mouse's sudden emergence to share living space and food with the rest of the small mammal community must strain the modest supply of nutrients just beginning to be replenished during the warming season. This effect has yet to be measured and assessed.

ASSOCIATES. The life of the woodland jumping mouse is closely associated with those of other small mammals of northern Michigan's boreal forests. Very likely the woodland deer mouse and the white-footed mouse most closely share the summer living space and food supplies with the woodland jumping mouse. Although it is not altogether clear that active competition between these mice has been demonstrated (Brower and

Cade, 1966; Whitaker and Wrigley, 1972), there is evidence of complementary relative abundance (Kirkland and Griffin, 1974). Wrigley (1972) reported that a captive woodland jumping mouse attacked a deer mouse which came close to the former's nest. On the other hand, Blair (1941) caught an adult male woodland jumping mouse and a young adult male deer mouse together in a single-catch livetrap in Alger County. These two individuals presumably were lured to the bait at the same time and, when caught, showed no evidence of fighting.

Brower and Cade (1966) also observed that the local distribution of the southern red-backed vole (*Clethrionomys gapperi*) complements that of the woodland jumping mouse; when numbers of the former are high, the latter may be scarce or absent. Wrigley (1972) felt that this latter interaction may be the result of aggression by the vole rather than competition between the two for limited resources (see also Vickery, 1979). Ecological preferences seem primarily to segregate the woodland jumping mouse from the meadow jumping mouse, with the two commonly found together only at boundaries between open environments (preferred by the latter) and forest-shrub (preferred by the former). Hamilton (1935) and other observers have shown that the meadow jumping mouse is the more aggressive of the two and much more apt to invade the forest-shrub habitats of the woodland jumping mouse than vice versa. In southern Michigan where the woodland jumping mouse is absent, the meadow jumping mouse appears to utilize woody cover to a greater extent, suggesting that there may be some competitive reason for the rather obvious segregation where the two species occur together in the same general areas.

Small mammals found in similar northern Michigan environments with the woodland jumping mouse include the short-tailed shrew, least chipmunk, eastern chipmunk, deer mouse, and meadow jumping mouse in Marquette County (Manville, 1949); the arctic shrew, masked shrew, short-tailed shrew, least chipmunk, deer mouse, southern red-backed vole, and meadow jumping mouse at the border of Alger and Schoolcraft counties (Ozoga and Verme, 1968); and the white-footed mouse, southern red-backed vole, meadow vole and southern bog lemming in Montmorency County (Green, 1925). A study of species interactions and resource apportionments between the deer mouse, white-footed mouse, meadow jumping mouse, and woodland jumping mouse in habitats where these four seed/insect-eaters may occur in close association in the northern part of Lower Peninsula Michigan would be of interest.

REPRODUCTIVE ACTIVITIES. Male woodland jumping mice emerge from hibernation in late April and early May, approximately two weeks prior to the females. By the time females appear, males are prepared for mating and the females are receptive shortly after emergence. Pregnant females have been noted as early as May 8 and 9 in Pennsylvania, with most females producing litters in June (Wrigley, 1972). Breeding proceeds slowly in July with a second breeding peak in August. As in the case of the meadow jumping mouse, the woodland jumping mouse may also have three litters in a single season.

Rather complete data on postnatal development has been obtained by Layne and Hamilton (1954). Although many litters born to caged females are eaten or deserted by their mothers (Hamilton, 1935; Sheldon, 1938). It is thought that gestation takes at least 23 days, longer than the 18 days noted for the meadow jumping mouse (Whitaker and Wrigley, 1972). Two to seven (generally five) young are produced per litter. At birth, the young woodland jumping mouse weighs less than 1 g (0.87 g); is pink and hairless except for emerging facial vibrissae; has closed ear pinnae; eyes visible merely as dark rings; and blunt claws on all digits. By 10 days, pigment spots appear; ear pinnae unfold; and eyes bulge slightly. By 20 days, the body is covered with fine hair; plantar tubercules on hind feet blacken; claws become well-formed; lower incisors protrude; and the mid-dorsal dark band becomes conspicuous. By 26 days, the body is well-furred; the upper incisors protrude; the eyes and ears open and become functional; and locomotion by short hops is accomplished. By 34 days, weaning is underway. Between 63 and 80 days, according to Wrigley (1972), molt into adult pelage is completed. Gestation is generally briefer and postnatal development in the mother's care longer in jumping mice than in deer and white-footed mice.

Like the meadow jumping mouse, the woodland jumping mouse is long-lived in comparison to other small Michigan rodents. This is attributed to, in part, its ability to hibernate. Even so, there is a rather high annual turnover rate with the September (prehibernating) population consisting of only 30% adults and 70% young-of-the-year (Whitaker and Wrigley, 1972). Individuals may survive in nature for three or four years (Wrigley, 1972). Young-of-the-year presumably do not breed until after their first winter in hibernation (Banfield, 1975), another factor contributing to longevity.

FOOD HABITS. Woodland jumping mice may be apt to eat any edible plant or animal food available; however, their fare, depending on seasonality, contains about 70% seeds and other plant foods and 30% insects and other soft-bodied invertebrates (Connor, 1966; Hamilton, 1935; Sheldon, 1938; Whitaker, 1963; Wrigley, 1972). Insects consist of larval Lepidoptera (moths and butterflies), adult Coeloptera (beetles), larval Diptera (flies), some spiders, centipedes, worms, grasshoppers, and even dragonflies. Plant foods are chiefly seeds of various grasses, forbs, and shrubs. Seasonal fruits, such as blueberries and raspberries, may attract these mice and cause temporary concentrations, perhaps giving observers the impression (thought to be incorrect) that woodland jumping mice tend to live in aggregations (Vickery, 1979; Wrigley, 1972). Various kinds of fungi, notably subterranean species of the genus *Endogone,* appear to be highly important food sources for jumping mice (Whitaker, 1962). Leaves and other green plant materials are eaten, mostly in minor amounts.

ENEMIES. Woodland jumping mice probably are safe from predators in their underground winter hibernacula. In the warm months, however, these rodents are as much fair game for hungry carnivores as are deer mice, meadow voles, and other small mammalian associates. The few records of specific predation on the woodland jumping mouse include those by the screech owl, gray wolf, mink, weasel (unidentified to species), striped skunk, bobcat, and house cat (Blair, 1941; Hamilton, 1935; Linzey and Linzey, 1968; Platt, 1968; Toner, 1956; Voigt *et al.,* 1976). Like the meadow jumping mouse, the woodland jumping mouse may become trapped by falling into holes or containers such as a water-filled bucket (Laundre, 1975).

PARASITES AND DISEASES. The internal parasites in the woodland jumping mouse include protozoans, tapeworms (Cestoda), and roundworms (Nematoda). It is not known how parasites survive in the digestive tract during hibernation when there is no food intake. External parasites, as summarized by Wrigley (1972), are a variety of fleas and mites, a tick, and botflies (genus *Cuterebra*).

MOLT AND COLOR ABERRATIONS. There seems to be no distinctive subadult molt in the woodland jumping mouse (Wrigley, 1972). Normally, animals born early in the warm season molt prior to entering hibernation, whereas young from second litters may not molt and do not grow fresh pelage until the following spring. No color aberrations have been reported.

ECONOMIC IMPORTANCE. There is no evidence that the woodland jumping mouse has any economic importance to foresters or other land operators. Although the woodland jumping mouse is a seed-eater, its diet appears to include few seeds from trees of marketable value.

COUNTY RECORDS FROM MICHIGAN. Records of specimens of the woodland jumping mouse in collections in museums and from the literature show that this species is reported from 12 of the 15 counties in the Upper Peninsula and nine counties in the northern three tiers of counties in the Lower Peninsula (see Map 44).

LITERATURE CITED

Banfield, A. W. F.
1974 The mammals of Canada. Univ. Toronto Press, Toronto. xxiv+438 pp.

Becker, L. R., and J. J. Christian
1979 Nest structure and nest utilization patterns in the woodland jumping mouse (*Napaeozapus insignis*). American Zool., 19(3):934.

Blair, W. F.
1941 Some data on the home ranges and general life history of the short-tailed shrew, red-backed vole, and woodland jumping mouse in northern Michigan. American Midl. Nat., 25(3):681–685.

Brower, J. E., and T. J. Cade
1966 Ecology and physiology of *Napaeozapus insignis* (Miller) and other woodland mice. Ecology, 47(1):46–63.

Burt, W. H.
1946 The mammals of Michigan. Univ. Michigan Press, Ann Arbor. xv+288 pp.

Connor, P. F.
1966 The mammals of the Tug Hill Plateau, New York. New York State Mus. and Sci. Serv., Bull. No. 406, 82 pp.

Dalby, P. L.
1968 Kalkaska County record for woodland jumping mouse. Jack-Pine Warbler, 46(3):95.

Dice, L. R.
1925 A survey of the mammals of Charlevoix County, Michigan, and vicinity. Univ. Michigan, Mus. Zool., Occas. Papers No. 159, 33 pp.

Dice, L. R., and H. B. Sherman
1922 Notes on the mammals of Gogebic and Ontonagon counties, Michigan, 1920. Univ. Michigan, Mus. Zool., Occas. Papers No. 109, 40 pp.

Green, M. M.'
1925 Notes on some mammals of Montmorency County, Michigan. Jour. Mammalogy, 6(3):173–178.

Hamilton, W. J., Jr.
1935 Habits of the jumping mice. American Midl. Nat., 16(2):187–200.

Haveman, J. R.
1973 A study of population densities, habitats and foods of four sympatric species of shrews. Northern Michigan University. unpubl. M.S. thesis, vii+70 pp.

Jackson, H. H. T.
1961 Mammals of Wisconsin. Univ. Wisconsin Press, Madison. xii+504 pp.

Kirkland, G. L., Jr., and R. J. Griffin
1974 Microdistribution of small mammals at the coniferous-deciduous forest ecotone in northern New York. Jour. Mammalogy, 55(2):417–427.

Kirkland, G. L., Jr., and C. J. Kirkland
1979 Are small mammal hibernators K-selected? Jour. Mammalogy, 60(1):164–168.

Klein, H. G.
1957 Inducement of torpidity in the woodland jumping mouse. Jour. Mammalogy, 38(2):272–274.

Laundre, J.
1975 An ecological survey of the mammals of the Huron Mountain area. Huron Mt. Wildlife Found., Occas. Papers No. 2, x+69 pp.

Layne, J. N., and A. H. Benton
1954 Speeds of some small mammals. Jour. Mammalogy, 35(1):103–104.

Layne, J. N., and W. J. Hamilton, Jr.
1954 The young of the woodland jumping mouse, *Napaeozapus insignis insignis* (Miller). American Midl. Nat., 52(1):242–247.

Linzey, D. W., and A. V. Linzey
1968 Mammals of the Great Smoky Mountains National Park. Jour. Elisha Mitchell Soc., 84:384–414.

Manville, R. H.
1948 The vertebrate fauna of the Huron Mountains, Michigan. American Midl. Nat., 39(3):615–640.
1949 A study of small mammals populations in northern Michigan. Univ. Michigan, Mus. Zool., Misc. Publ. No. 73, 83 pp.

Neumann, R., and T. J. Cade
1964 Photoperiodic influence on the hibernation of jumping mice. Ecology, 45:382–384.

Ozoga, J. J., and L. J. Verme
1968 Small mammals of conifer swamp deeryards in northern Michigan. Michigan Acad. Sci., Arts and Ltrs., Papers, 53:37–49.

Pearson, O. P.
1947 The rate of metabolism of some small mammals. Ecology, 28(2):127–145.

Platt, A. P.
1968 Selected predation by a mink on woodland jumping mice confined in live traps. American Midl. Nat., 79:539–540.

Sheldon, C.
1938 Vermont jumping mice of the genus *Napaeozapus*. Jour. Mammalogy, 19(4):444–453.

Snyder, L. L.
1924 Some details on the life history and behavior of *Napaeozapus insignis abietorum* (Preble). Jour. Mammalogy, 5(4):233–237.

Toner, G. C.
1956 House cat predation on small mammals. Jour. Mammalogy, 37(1):119.

Townsend, M. T.
1935 Studies on some of the small mammals of central New York. Roosevelt Wildlife Ann., 4, 120 pp.

Vickery, W. L.
1979 Food consumption and preferences in wild populations of *Clethrionomys gapperi* and *Napaeozapus insignis*. Canadian Jour. Zool., 57(8):1536–1542.
1981 Habitat use by northeastern forest rodents. American Midl. Nat., 106(1):111–118.

Voigt, D. R., G. B. Kolenosky, and D. H. Pimlott
1976 Changes in summer foods of wolves in central Ontario. Jour. Wildlife Mgt., 40(4):663–668.

Whitaker, J. O., Jr.
1962 *Endogone, Hymenogaster,* and *Melanogaster* as small mammal foods. American Midl. Nat., 67(2):152–156.
1963 Food, habitat and parasites of the woodland jumping mouse in central New York. Jour. Mammalogy, 44(3): 316–321.

Whitaker, J. O., Jr., and R. E. Wrigley
1972 *Napaeozapus insignis*. American Soc. Mammalogists, Mammalian Species, No. 14, 6 pp.

Wrigley, R. E.
1972 Systematics and biology of the woodland jumping mouse, *Napaeozapus insignis*. Illinois Biol. Monog. 47, 117 pp.

FAMILY ERETHIZONTIDAE

The family Erethizontidae belongs to the suborder of rodents, the Caviomorpha, which is almost entirely confined to Central and South America. Paleontologists have concluded that ancestral stocks of these rodents entered South America in the early Tertiary period. This group owes its distinctive characteristics and dramatic diversity to long isolation on this southern continent which, throughout much of the early Cenozoic period, was prevented from major species interchange with North America by the obliteration of the land connection at what is now the Isthmus of Panamá. When this land bridge was finally restored, representatives of the Erethizontidae ventured northward to Middle America, with the porcupine (genus *Erethizon*) making the most extensive northward move to occupy vast areas of boreal North America.

Porcupines and their caviomorph relatives possess distinctively enlarged infraorbital foramina. These openings on either side of the anterior part of the skull not only provide a passage for blood vessels and nerves but also for a large portion of the jaw musculature (specifically, the anterior division of the masseter medialis), which originates in large, concave areas on each side of the rostrum. Other unusual familial characteristics include the conspicuous outward flaring of the angular process of each side of the lower jaw, and nearly horizontal and inferior positioning of the zygomatic plate, the lack of prominent postorbital processes, and the absence of typical rodent narrowness of the interorbital area. Porcupines are noted for their stocky bodies and the stiff-barbed quills which adorn the upperparts and tail.

PORCUPINE
Erethizon dorsatum (Linnaeus)

NAMES. The generic name *Erethizon,* proposed by Cuvier in 1822, is of Greek origin and alludes to the porcupine's ability "to excite, to irritate." The specific name, *dorsatum,* proposed by Linnaeus in 1758, is from the Latin and may be translated as "pertaining to the back," no doubt in reference to the spines. The animal is sometimes called quill pig, porky, and porky hog. In Michigan, this heavy-bodied and spiny creature is known everywhere as a porcupine. The subspecies *Erethizon dorsatum dorsatum* (Linnaeus) occurs in Michigan.

RECOGNITION. Body size massive for a rodent (Fig. 58); length of head and body averaging 22 in (560 mm) long; tail short and thick, averaging 8⅜ in (215 mm) long; head small when compared with grossness of body, no obvious connecting neck; muzzle flattened and hairy; eyes and ears small, the latter rounded; legs short and powerfully built; feet flat (plantigrade) with broad soles having a tubercular pattern to increase traction; four functional toes on forefeet and five on hind, each terminating in a strong, curved claw.

The pelage is composed of dense, wooly underfur (drab brown in color); long, coarse, guard hairs (tipped with yellow-white); and stiff, sharp, barbed, and loosely-attached quills (with brown tips and hollow shafts). The quills are located on the head, neck, back, rump, and tail; they are longest on the midback, up to 2½ in (65 mm) long. Quills pulled out are replaced in a period of two weeks to six months. The quills are often obscured by the mass of dark underfur and guard hairs; nevertheless, the number arranged in rows is estimated to be about 30,000 per animal (Banfield, 1974). The ventral surface is clothed in coarse, dark hair containing no quills. In adults, the coat is generally glossy dark brown to black above, with individual hairs often tipped with white or yellow-white. The underparts are dark brown. In winter, the color of the pelage is somewhat darker and fuller. There are two pairs of pectoral mammary glands. The woodchuck and the beaver are the only Michigan rodents to approach the porcupine in size; however, the quill-filled pelage gives the porcupine a distinctive appearance.

MEASUREMENTS AND WEIGHTS. Adult porcupines measure as follows: from the tip of nose to the end of tail vertebrae, 25½ to 35¼ in (650 to 900 mm); length of hind foot, 3⅛ to 4¼ in (80 to 110 mm); height of ear from notch, 1 to 1½ in (25 to 40 mm); weight, 10 to 30 lbs (4.5 to 13.7 kg). In Iron County, Brander (1973) found that 15 adult males averaged 14.4 lbs (5.8 kg) and 36 adult females averaged 14.8 lbs (5.9 kg).

DISTINCTIVE CRANIAL AND DENTAL CHARACTERISTICS. The skull of the porcupine (Fig. 59) is massive in form; zygomatic arches, slender and depressed; nasal bones, short and broad; infraorbital foramina, greatly enlarged, with a diameter greater than that of the foramen magnum; auditory bullae, large and rotund; upper incisors and anterior extensions of the premaxillaries project well forward of the anterior tips of the nasals; and the lower edge of the angular process of each ramus of the mandible is outwardly and strongly flared. The skull of an adult porcupine averages 3⅞ in (100 mm) in greatest length and 2¾ in (70 mm) in greatest width across the zygomatic arches.

The teeth are characterized by heavy incisors which project forward and are yellow-orange in

Figure 58. The porcupine (*Erethizon dorsatum*). Photograph by Robert Harrington and courtesy of the Michigan Department of Natural Resources.

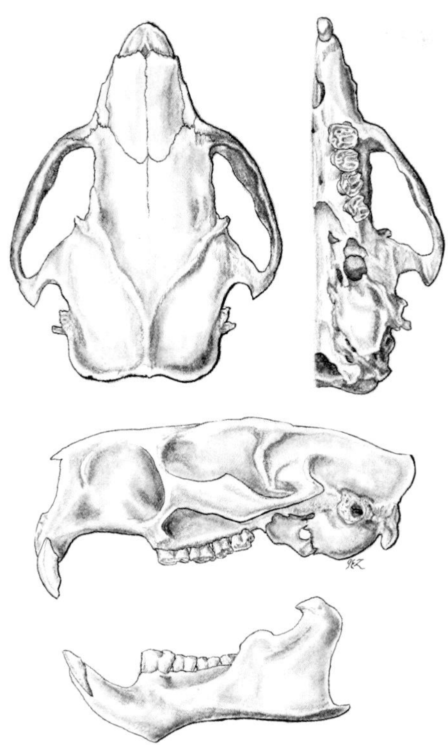

Figure 59. Cranial and dental characteristics of the porcupine (*Erethizon dorsatum*).

color; four cheek teeth on each side of the upper and lower jaws are rooted, molariform in structure, and uniform in size; and occlusal surface of each cheek tooth is flattened with an intricately-folded enamel pattern. The enlarged infraorbital foramina and the projecting upper incisors distinguish the skull of the porcupine from that of the beaver or the woodchuck (Fig. 35). The dental formula for the porcupine is:

$$\text{I (incisors)}\frac{1\text{-}1}{1\text{-}1},\ \text{C (canines)}\frac{0\text{-}0}{0\text{-}0},\ \text{P (premolars)}\frac{1\text{-}1}{1\text{-}1},\ \text{M (molars)}\frac{3\text{-}3}{3\text{-}3} = 20$$

DISTRIBUTION. The porcupine is found throughout most of boreal North America from the arctic prairies near tree line southward, in northern coniferous forest habitats in northeastern United States, and in western montane and prairie environments southward to northwestern Mexico. The original geographic range of this species has been reduced since settlement, especially in the midwestern and eastern United States (Barkalow, 1961). However, there is also evidence that the porcupine has occasionally increased its numbers by spreading into cutover lands and agricultural areas (Spencer, 1946). In Michigan, the porcupine has now vanished from its former range in the Lower Peninsula's five tiers of southernmost counties, except for a small enclave in northern Kent and Montcalm counties (see Map 45 and Michigan Nat. Res. Mag., 46(5):14, 1977). Archaeological sites containing porcupine bones were excavated in the following southern Michigan counties: Berrien (Cleland, 1966) and Ottawa and Allegan (Martin, 1976). Early writers list records for the porcupine in Wayne County in the 1870s (Tibbits, 1877); in Sanilac County in 1852 (Perry, 1899); in Saginaw County in 1860 (Edwards, 1893); and in Washtenaw County as late as 1868 (Wood, 1922).

The porcupine and the opossum are the sole representatives in Michigan of the rodent suborder Caviomorpha and the order Marsupialia, respectively, both of which have their origins in Central and South America (Neotropica). Relatives

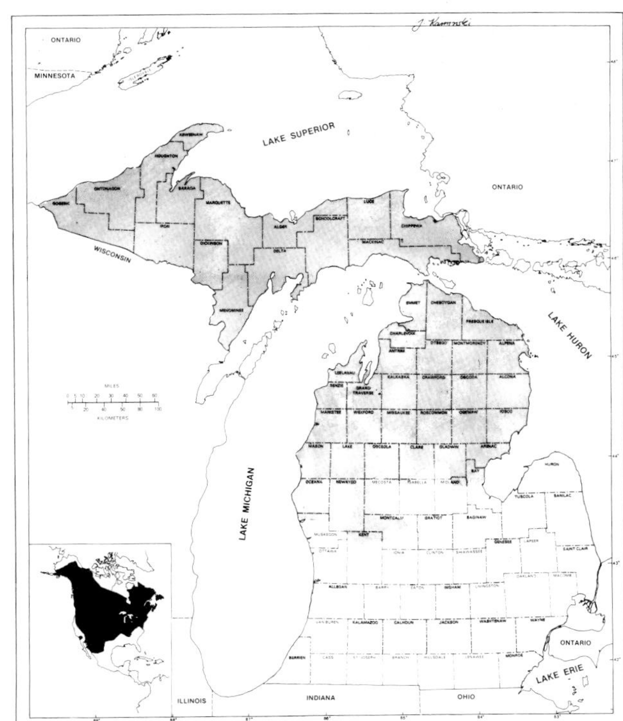

Map 45. Geographic distribution of the porcupine (*Erethizon dorsatum*). Records from past times in southern Michigan are from counties with pale shading.

of these two mammals flourish in these warmer climates, while the porcupine and the opossum have adapted effectively in north-temperate Michigan.

HABITAT PREFERENCES. Most observers in northeastern North America describe the porcupine as a woodland species (Seton, 1953). It frequents both deciduous and coniferous forest areas and prefers hemlock and pine habitats (Earle, 1978; Kelly, 1973). Its food preferences tend to dictate local habitat use. In winter, the animal remains in the heavy cover of tree boughs to feed generally on the phloem layer of trees; in summer, it searches for leaves of herbaceous plants in forest openings, conifer and aspen swamps, and adjacent croplands (Spencer, 1946). In Gogebic, Ontonagon, and Iron counties, Golley (1957) found second-growth maple-hemlock forest was a favored winter range. White-tailed deer hunters in the Upper Peninsula in mid-November most commonly see porcupines in hardwood forest, less in swamp forests, and rarely in pine forests. In Gogebic and Ontonagon counties, Dice and Sherman (1922) most often observed porcupines in hardwood forests in summer and autumn; in Wexford County, Wenzel (1912) reported the animals in hardwood and hemlock forests.

DENSITY AND MOVEMENTS. According to Spencer (1946), the porcupine in natural conditions is rarely found in excessive numbers. However, populations increase locally when orchards and cultivated fields provide extra food supplies, and natural predators are reduced. Field observers have reported densities of 32.5 animals per sq mi (12.3 per sq km) in coniferous forest in Oregon (Smith, 1977); 25.2 animals per sq mi (9.3 sq km) in mixed hardwoods and hemlock in Maine (Curtis, 1944); and a high of 96.7 animals per sq mi (37.2 per sq km) in mixed hardwood-hemlock stands in Wisconsin (Stoeckeler, 1950).

In Michigan, Golley (1957) estimated that the porcupine density over the entire Upper Peninsula was at least two animals per sq mi (0.8 per sq km). Localized studies, however, have revealed densities of 40 to 58 porcupines per sq mi (15.4 to 22.3 per sq km) in second-growth hardwoods and 30 porcupines per sq mi (12.5 per sq km) in old-growth hardwood-hemlock forest in the Ottawa National Forest in western Upper Peninsula (Irving, as quoted in Brander, 1973). In hardwood-hemlock forest in Iron County, Brander (1973) estimated porcupine numbers (based on winter counts) at 24 per sq mi (9.2 per sq km) in areas logged in 1961 shortly before his study the same year; at 32 animals per sq mi (12.3 per sq km) in heavily logged areas in 1957; and at 43 animals per sq mi (16.6 per sq km) in areas lightly logged many years prior to his studies. Brander's study points to the porcupine's preference for older hardwood-hemlock stands as winter habitat.

Data on porcupine movements have been obtained primarily by the catching-tagging-releasing-recatching method (Brander, 1973) or by radio-tracking individuals fitted with radio transmitters (Dodge and Barnes, 1975; Marshall *et al.*, 1962). Mobility appears greatly influenced by habitat quality and season. Winter and summer ranges are often separated (Shapiro, 1949; Taylor, 1935) unless summer foods (mostly herbaceous plants and fruits) and winter foods (mostly phloem layers of favored trees and shrubs) are sufficiently interspersed to make such travel unnecessary (Shapiro, 1949). Some movements between winter dens and summer ranges may be as much as 4,900 ft (1,500 m) in Massachusetts habitats, which are comparable to habitats in Michigan (Dodge, 1967). Average daily movements of two radio-tagged, female porcupines in Minnesota (Marshall *et al.*, 1962) were 393 ft (120 m) in daylight and 264 ft (80 m) at night. Banfield (1974) reported that in a 30 day period, porcupines may range in areas of 32 to 36 acres (13 to 14 ha). In Iron County, Michigan, Brander (1973) found that individuals had an average summer cruising radius of 489 ft (150 m). In winter, this cruising radius declined to an average of only 26 ft (8 m). Brander thought these winter feeding runs were unusually short. In comparable country in New York, Shapiro (1949) found the mean daily radius in winter to be almost as great as the summer radius found by Brander in Iron County.

BEHAVIOR. The ponderous, deliberate porcupine shuffles along the ground and climbs leisurely, appearing oblivious to the surroundings. It rarely seems in a hurry—traveling at 2 mi (3.3 km) per hour at most (Jackson, 1961). As a result, this animal is a victim of automobiles in northern Michigan (Manville, 1949), and is certainly the least sporting target for the hunter. Because the animal is easily caught and edible, it has served for centuries as a ready source of bush meat to hungry travelers in the north woods.

When approached by a potential enemy, the porcupine usually lowers its head between its powerful front legs and exposes its rear to the impending danger in an almost skunk-like manner. The tail is poised and ready to lash out sideways at the intruder. The quickness of this move may have given rise to the incorrect notion that the animals shoot their quills. Such a reaction is triggered more by vibration, touch, and perhaps smell than by sight, as the porcupine's vision is rather poor (Batchelder, 1948).

The porcupine is active by day but is more active by night. It ascends trees with a slow arm-over-arm movement; it descends by reversing this action. There is some evidence that porcupines are most adept at climbing trees with a diameter between 6 and 10 in (150 to 250 mm); trees of massive girth are too large to be readily ascended (Harder, 1979). In winter, porcupines may excavate long snow tunnels leading from their dens. They make no effort to avoid water and can swim vigorously with a dog-paddle stroke. When wading, the tail is held up as if the animal were trying to keep it dry.

Porcupines are social during at least part of the year; they pair during the mating season and may den together as family groups during winter. Burt (1946) reported as many as 20 individuals living together in a single shelter. Brander (1973) described nocturnal gathering sites, usually vacant logging or hunting camps, in Iron County. Individuals joining these associations on summer nights would enter the area on trails from different directions. At other times of the year, porcupines can be solitary and seemingly ill-tempered (Taylor, 1935). They appear to communicate by a variety of vocalizations (Seton, 1953). Manville (1948) described baby-like whines and cries. Squeals, grunts and meows are often emitted during the mating season. When disturbed, a porcupine may hiss and chatter its teeth. Sometimes when searching for food, the animal will sniff the air, producing a snorting sound.

The porcupine rarely digs in the ground for shelter. Instead, it uses hollow trees, rock ledges, logs, windfalls, and even garbage dumps as cover. According to Shapiro (1949), dens may be selected more for protection from winds and snow than for insulation against cold. An individual porcupine may also have several temporary shelters as far as ½ mi (1 km) away from the main den. Dens are used more in winter than in summer. The odor of urine in the vicinity of dens may play an important role in communication (especially in the case of males). Den locations are often discovered because of great accumulations of fecal pellets (Fig. 60).

Figure 60. Fecal droppings of porcupine at den opening at base of hollow tree.
Photograph by Robert P. Carr in Upper Peninsula, Michigan.

Porcupines also use summer "station trees" as roosts, with no protected dens nearby. These roosts are in large trees, usually hemlocks, with large branches often gnawed and pruned (Curtis and Kozicky, 1944). There is the odor of urine, as well as piles of fecal pellets, on the forest floor below. These tree roosts are possibly used in late spring and early summer to avoid mosquitoes (Marshall *et al.*, 1962).

ASSOCIATES. There is a significant relationship between the porcupine, and the snowshoe hare and the white-tailed deer. It is primarily a one-sided association because the latter two species profit from the porcupine's "wasteful" eating habits. When feeding on white cedar, poplar, and white spruce (Curtis and Kozicky, 1944; Ferguson and Merrian, 1978), porcupines are prone to drop numerous unbrowsed and partly-browsed branches. These unretrieved parts can be a winter food bonus for snowshoe hare and white-tailed deer when snow is deep and when the understory in a mature forest is sparse.

REPRODUCTIVE ACTIVITIES. The reproductive habits and related behavior of the porcupine have been documented, primarily by studying confined individuals, in the classic work of Shadle (1944, 1946, 1948, 1951) and in an earlier study of Struthers (1928). Courtship and mating begin in early November and continue into December. Males follow females around, associating with them on the ground and in trees while carrying on various rituals such as rearing up on the hind feet and vocalizing. Michigan hunters become well acquainted with normally inconspicuous porcupines because these activities are in progress during open shooting season in autumn (Golley, 1957). In the courtship process, one female may mate with more than one male and vice versa. If conception does not occur, a female may again become receptive to mating in January or even at monthly intervals thereafter.

According to Shadle (1948), after an unusually long gestation period of about 30 weeks (209 to 217 days), a single offspring is born as early as late April but usually between mid-May and late July (depending on time of mating). At birth, the young's development is well advanced; it weighs about 1.2 lb (530 g); has a body covering of long black hair with the quills just beginning to grow, about 1 in (25 mm) long, soft at first but hardening through exposure to the air; possesses functioning senses of vision and hearing; is able to move about under its own power; has erupted teeth, and can make whining sounds (Dodge, 1967; Shadle and Ploss, 1943; Tyron, 1947).

The porcupine is born at a more developed (precocial) state than other Michigan rodents. At birth it has good control of its basic faculties, and within two weeks is weaned and can thrive independently on solid foods. Its antics as a youngster, in contrast to its sober nature as an adult, have been described by Shadle (1944). Even though the young porcupine may be divorced from maternal care (except for tagging after the mother) shortly after birth, its development to sexual maturity and adult weight is gradual—over a period of one and one-half to two and one-half years (Tryon, 1947). Because of its low breeding potential (one young per female per year), its slow maturity rate, and its apparently low infant mortality, especially after the first year of life, the individual porcupine must survive several seasons to maintain the population. Porcupines have lived as long as 10 years in captivity (Shadle, 1951). In the wild, there are records of the animal surviving six years (Taylor, 1935) with one Michigan porcupine known to have lived almost 11 years in Iron County (Brander, 1971). Age determination in porcupines is best accomplished by sectioning molariform teeth and counting cementum annuli (Earle and Kramm, 1980).

FOOD HABITS. The porcupine is a more complete vegetarian than any other Michigan rodent except the beaver. Its choice of food is directly correlated with the availability of green stuff in summer and inner bark of favored trees in winter. In Iron County, Brander (1973) noted that porcupines fed mostly during daylight hours in summer, with foraging occurring for as little as 15 minutes to as long as one hour. Between feedings, porcupines were inactive or asleep. In forest situations, they seem to prefer the leaves of basswood, elm, and yellow birch, but will also eat the leaves and buds of sugar maple, hemlock, white cedar, aspen, pin cherry, red maple, and red oak (Manville, 1948). Porcupines often move away from woodlands to feed on plant foliage at the meadow-forest edge, in riparian growth along waterways, and in cultivated fields, orchards, and gardens. Eating the leaves of water lilies and other emergent aquatic plant life has also been documented (Schoonmaker, 1930). Dice and Sherman (1922) watched young animals in northern Michigan foraging for yellow water lily leaves by reaching for them from dry perches on submerged logs.

In winter, the porcupine becomes a much less selective feeder (Shapiro, 1949; Speer and Dilworth, 1978), as deep snow and other winter conditions restrict foraging, especially on the ground. Porcupines under surveillance by Brander (1973) in Iron County switched to a diet of inner bark after leaf fall in mid-October. In winter, instead of feeding by day as in summer, porcupines journeyed from their dens to food trees after dark and returned prior to dawn, establishing trails through the snow which were to be used over and over again. The winter diet consists almost entirely of the succulent (sugar and starch-filled) inner bark (cambium and phloem layers) of favored food trees. The outer corky layers of bark are removed by the porcupines powerful incisors. The trunk, major branches, and smaller twigs are stripped and severely damaged by this feeding activity, causing forest managers to initiate strict control measures on porcupine populations. Brander (1973) found Upper Peninsula porcupines feeding on the inner bark of hemlock, sugar maple, basswood, elm, and yellow birch. Golley (1957) added tamarack; jack pine and beech are also used. These authors agree with observers in other areas that second-growth stands are preferred over older trees in mature forests.

In summer and especially in late winter and spring the porcupine has a strong yearning for sodium, an ingredient in table salt (Campbell and LaVoie, 1967; Rogers, 1981). In its quest for this element, the animal is guilty of entering cabins, sheds and logging buildings to graw ax handles, oars, canoe paddles, hand utensils, bedposts, ladders, kitchen tables—any item having acquired a bit of salt from the kitchen or from perspiring humans.

ENEMIES. Today, the greatest enemy of the porcupine is the human, including: the forest manager who wishes to protect valuable timber; the northern Michigan cottage owner who seems unable to construct a porcupine-proof summer shelter or winter camp, and the hunter who finds the porcupine an easy target. It also has other enemies and there are many accounts in the literature of how the fisher and other clever meat-eaters have allegedly learned to turn over a spine-protected porcupine and subdue it by attacking its nonspiny underside (Seton, 1953). Recent studies (Powell and Brander, 1977; Powell and Earle, 1981), however, demonstrate that the fisher is much more successful when its attack is directed at the porcupine's face. Hence, the porcupine is usually safe if it can protect its head by facing outward from the trunk while on a tree branch, facing a tree trunk when on the ground, or occupying a den with only one entrance. Certainly, the rapid swish of the porcupine's tail with its loosely-held barbed spines has discouraged vast numbers of less-skilled molesters, including inquisitive dogs and even an occasional cow.

Porcupine quills and hair in a predator's intestinal tract are evidence that the meat-eater has consumed at least part of this animal. However, it is possible that porcupine bits recorded as ingested by mink, ermine, and marten (Quick, 1963) and great horned owl (Curtis and Kozicky, 1944) may have been carrion. Larger species known to overpower and eat porcupines include gray wolf (Voigt *et al.*, 1976), coyote (Keller, 1935; Manville, 1948), red fox (MacGregor, 1942); gray fox (Weaver, 1939), mountain lion (Young and Goldman, 1946), lynx (Quick, 1953), and bobcat (in Montmorency County, Manville, 1958). In the case of the fisher, studies by Cook and Hamilton (1957) in New York, Dodge (1967) in Massachusetts, and deVos (1951) in Ontario, as well as the investigations in Michigan by Earle (1978), Powell and Brander (1976), and Powell and Earle (1981) show that fisher predation acts as a control to porcupine populations. Continued study of the reestablished fisher population (extirpated but restored recently to the Ottawa National Forest in the Upper Peninsula) should provide further data on this intriguing prey-predator relationship. Increased porcupine road kills from April through June seem to coincide with peak sodium appetites resulting from winter cold (Rogers, 1981) and are probably related to pavement-licking to obtain salt (Earle, 1978).

PARASITES AND DISEASES. The great quantity of parasitic worms found in porcupines' digestive tracts is impressive. Curtis and Kozicky (1944) counted 1,528 tapeworms and 5,184 roundworms in one female porcupine examined in Maine. A tapeworm (*Taenia michiganensis*) found as a cyst in the livers of Michigan porcupines was described as new to science (Gower, 1939). Protozoans and flatworms also infest the inner organs of porcupines. Externally, a tick (Ixodidae), a biting louse (Mallophaga), and a sucking louse (Anoplura) have been identified (Curtis and Kozicky, 1944; Doran, 1954; Jackson, 1961). The long life span of the porcupine probably allows large infestations to accumulate. Mange is also a common affliction.

MOLT AND COLOR ABERRATIONS. The porcupine molts its pelage once a year, in spring to mid-summer. This elimination of old, worn hair and growth of new includes the long and short coat hairs, the long and short coarse hairs, and the quills (Po-Chedley and Shadle, 1955). Replacement generally begins with the hair on the nose and head and progresses to the tail. Albino porcupines have frequently been reported in the literature (Shiras, 1911). A mounted albino individual has been on exhibit at the Michigan State University Museum for 60 years. Black (melanistic) individuals also are reported (Dodge, 1967).

ECONOMIC IMPORTANCE. It appears obvious that humans have little affection for the porcupine. Nevertheless, the animal can be fairly easily caught and its meat is considered highly palatable by some. Certainly the Indians found the porcupine an important source of food. Samuel E. Edwards (1893) gave the following account of a visit to an Indian camp in the area of the Shiawassee and Saginaw rivers (probably Saginaw County) during a hunting trip in the autumn of 1860:

> I went away for a morning hunt, promising to return and take dinner with them. When we came back, there was dangling by a string before the fire, a large porcupine with quills singed off, and two coons suspended in the same manner. The porcupine was evidently stuffed with something, and I was anxious to learn all that was new in the theory of dressing game, accordingly I watched the proceedings.
>
> When roasted sufficiently, the old squaw took it down and commenced dissecting it. When opened, out rolled the feet of both coons, none of them having been dressed at all! The heads of the coons were not skinned below the eyes. They said this was the way to preserve the sweetness of the meat, and the feet, they affirmed, were the best part of the coon, and in distributing the delicate meat, the old squaw was very careful to see that each had a portion of the feet.

Porcupine quills were, and still are, important Indian ornaments. The quills were fastened to garments, robes, pouches, quivers, and moccasins, and combined with feathers on head bonnets. They were attached to pipes and other ceremonial objects, and were also used in the construction of baskets, mats, canoes, and other items.

Even so, foresters dislike the porcupine because it damages trees. If the animal were attracted to the so-called "weed trees" of the northern Michigan forest, it would not be so much in disfavor. However, the porcupine likes to feed on valuable timber species—sugar maple, hemlock, beech, yellow birch, elm, basswood, and others (Golley, 1957; Shapiro, 1949). In their winter feedings on the inner bark, porcupines may girdle trees, which kills them outright or causes the top to die, resulting in a condition known as "staghead." Young trees, including those in plantations, suffer from porcupine damage more than mature trees. Forest damage may be less severe than one is led to believe (Curtis, 1941; 1944; Gensch, 1946; Jackson, 1961); nevertheless, destruction to trees estimated as 36¢ per acre (90¢ per ha) was reported in northeastern forests and 21¢ per acre (62¢ per ha) in Wisconsin and Michigan (Krefting *et al.*, 1962). As a result, the porcupine is not protected by law, and bounties have even been offered as incentives to control populations. In New York, for example, Connor (1966) reported that 6,960 porcupines were bountied in Lewis County in 1964, for one dollar per animal.

In summer, porcupines annoy farmers by eating corn (especially in the milk stage), other grains, and alfalfa (perhaps trampling these as much as eating them), vegetables, truck crops, and fruits. Injuries to a dog or cow resulting from the slap of a porcupine's tail are another potentially serious problem when porcupines invade farm areas and feedlots. The porcupine's attraction to salty objects has been mentioned previously. On one occasion, a porcupine climbed through an open automobile window and chewed away the "salty" steering wheel.

Nuisance porcupines are most effectively controlled by hunting, preferably with a .22 caliber rifle. Trapping (certainly not recommended except in winter because traps are not selective); wire mesh or electric fencing for garden plots (Spencer, 1948); and poison baits (a mixture of strychnine with salt and apple as lure has been recommended but use by untrained persons should be discouraged because this toxic material is nonspecific) have all been suggested, but are not as effective as hunting. Several bulletins describing control methods have been issued (Anon., 1962; Dodge, 1967; Faulkner and Dodge, 1962; Gabrielson and Horn, 1930).

Because there is evidence that porcupine populations are gradually being reduced in areas where human occupation is high, conservationists should be concerned that this exotic-looking creature may be lost in many parts of northern Michigan. Special efforts should be made in areas set aside to support timber and wildlife to protect porcupines from extermination. On the other hand, necessary controls may have to be continued on tree plantations and on commercial forest lands.

COUNTY RECORDS FROM MICHIGAN. Records of specimens of porcupines in collections of museums, from the literature, and provided by personnel of the Michigan Department of Natural Resources show that this species currently occurs throughout the Upper Peninsula and the northern half of the Lower Peninsula and, up to the days of early settlement, occurred in most counties in southern Michigan (Map 45). The species is not known on Isle Royale.

LITERATURE CITED

Anon.
1962 The porcupine, its economic status and control. U.S. Fish & Wildlife Serv., Wildlife Leaflet 328 (rev.), 8 pp.

Banfield, A. W. F.
1974 The mammals of Canada. Univ. Toronto Press, Toronto. xxiv+438 pp.

Barkalow, F. S., Jr.
1961 The Porcupine and Fisher in Alabama Archaeological Sites. Jour. Mammalogy, 42(4):544–545.

Batchelder, C. F.
1948 Notes on the Canada porcupine. Jour. Mammalogy, 29(3):260–268.

Brander, R. B.
1971 Longevity of wild porcupines. Jour. Mammalogy, 52(4):835.
1973 Life-history notes on the porcupine in the hardwood-hemlock forest in upper Michigan. Michigan Acad., 5(4):425–433.

Burt, W. H.
1946 The mammals of Michigan. Univ. Michigan Press, Ann Arbor. xv+288 pp.

Campbell, D. L., and G. K. LaVoie
1967 Chemicals for porcupine control. Denver Wildlife Research Center, Ann. Prog. Rept., Wildlife Res. Work Unit F-45.2, 7 pp.

Cleland, C. E.
1966 The prehistoric animal ecology and ethnozoology of the Upper Great Lakes Region. Univ. Michigan, Mus. Anthro., Anthro. Papers, No. 29, x+294 pp.

Connor, P. F.
1966 The mammals of the Tug Hill Plateau, New York. New York State Mus. and Sci. Serv., Bull. No. 406, 82 pp.

Cook, D. B., and W. J. Hamilton, Jr.
1957 The forest, the fisher, and porcupine. Jour. Forestry, 55(10):719–722.

Curtis, J. D.
1941 The silvicultural significance of the porcupine. Jour. Forestry, 39:583–594.
1944 Appraisal of porcupine damage. Jour. Wildlife Mgt., 8(2):88–91.

Curtis, J. D., and E. L. Kozicky
1944 Observation on the eastern porcupine. Jour. Mammalogy, 25(2):137–146.

deVos, A.
1951 Recent findings in fisher and marten ecology and management. Trans. 16th North Amer. Wildlife Conf., pp. 498–507.

Dice, L. R., and H. B. Sherman
1922 Notes on the mammals of Gogebic and Ontonagon counties, Michigan, 1920. Univ. Michigan, Mus. Zool., Occas. Papers No. 109, 46 pp.

Dodge, W. E.
1967 The biology and life history of the porcupine (*Erethizon dorsatum*) in western Massachusetts. Univ. Massachusetts, unpubl. Ph.D. dissertation, ix+173 pp.

Dodge, W. E., and V. G. Barnes, Jr.
1975 Movements, home range, and control of porcupines in western Washington. U.S. Fish and Wildlife Serv., Wildlife Leaflet 507, 7 pp.

Doran, D. J.
1954 A catalogue of the Protozoa and helminths of North American rodents. II. Cestoda. American Midl. Nat., 52(2):469–480.

Earle, R. D.
1978 The fisher-porcupine relationship in upper Michigan. Michigan Tech. Univ., unpubl. M.S. thesis, viii+126 pp.

Earle, R. D., and K. R. Kramm
1980 Techniques for age determination in the Canadian porcupine. Jour. Wildlife Mgt., 44(2):413–419.

Edwards, S. E.
1893 The Ohio hunter or a brief sketch of the frontier life of Samuel E. Edwards the great bear and deer hunter of the state of Ohio. Review and Herald Publ. Co., Battle Creek, Michigan. 240 pp.

Faulkner, C. E., and W. E. Dodge
1962 Control of the porcupine in New England. Jour. Forestry, 60(1):36–37.

Ferguson, M. A. D., and H. G. Merriam
1978 A winter feeding relationship between snowshoe hares and porcupines. Jour. Mammalogy, 59(4):878–880.

Gabrielson, I. N., and E. E. Horn
1930 Porcupine control in the western states. U.S. Dept. Agric., Leaflet No. 60, 8 pp.

Gensch, R. H.
1946 Observations on the porcupine in northern Wisconsin and northern Michigan from April 16–26, 1946. U.S. Forest Serv., St. Paul, Minnesota, Prog. Rept. (mimeo.), 16 pp.

Golley, F. B.
1957 Distribution of porcupine in upper Michigan. Jour. Mammalogy, 38(4):526–527.

Gower, W. C.
1939 An unusual cestode record from the porcupine in Michigan. Michigan Acad. Sci., Arts and Ltrs., 24(2): 149–151.

Harder, L. D.
1979 Winter feeding by porcupines in montane forests of southwestern Alberta. Canadian Field-Nat., 93(4): 405–410.

Jackson, H. H. T.
1961 Mammals of Wisconsin. Univ. Wisconsin Press, Madison. xii+504 pp.

Keller, L. F.
1935 Porcupines killed and eaten by a coyote. Jour. Mammalogy, 16(4):232.

Kelly, G. M.
1973 The biology of an isolated porcupine population. Univ. Massachusetts, unpubl. M.S. thesis, vi+48 pp.

Krefting, L. W., J. H. Stoeckeler, B. J. Bradle, and W. D. Fitzwater
1962 Porcupine-timber relationships in the lakes states. Jour. Forestry, 60(5):325–330.

MacGregor, A. E.
1942 Late fall and winter food of foxes in central Massachusetts. Jour. Wildlife Mgt., 6(4):221–224.

Manville, R. H.
1948 The vertebrate fauna of the Huron Mountain, Michigan. American Midl. Nat., 39(3):615–640.
1949 Highway mortality in northern Michigan. Jour. Mammalogy, 30(3):311–312.
1958 Odd items in bobcat stomachs. Jour. Mammalogy, 39(3):439.

Marshall, W. H., G. W. Gullion, and R. G. Schwab
1962 Early summer activities of porcupines as determined by radio-positioning techniques. Jour. Wildlife Mgt., 62(1):75–79.

Martin, T. J.
1976 A faunal analysis of five Woodland Period archaeological sites in southwestern Michigan. West. Michigan Univ., unpubl. M.A. thesis, viii+200 pp.

Perry, O. H.
1899 Hunting expeditions of Oliver Hazard Perry of Cleveland verbatim from his diaries. Privately printed, Cleveland, Ohio, viii+246 pp.

Po-Chedley, D. S., and A. R. Shadle
1955 Pelage of the porcupine, *Erethizon dorsatum dorsatum*. Jour. Mammalogy, 36(1):84–95.

Powell, R. A., and R. B. Brander
1976 Adaptations of fishers and porcupines in relation to their predator-prey system. Pp. 45–53. *in* Phillips, R. L., and C. J. Jonkel, eds. Proc. 1975 Predator Symp., Montana Forest and Conserv. Exp. Sta., Univ. Montana, ix+268 pp.

Powell, R. A., and R. Earle
1981 Fisher and porcupine. An odd couple. Michigan Out-of-Doors, 35(12):42–45.

Quick, H. F.
1953 Occurrence of porcupine quills in carnivorous mammals. Jour. Mammalogy, 34(2):256–259.

Rogers, S. P.
1981 The effect of cold exposure on the sodium and water balance on the porcupine, *Erethizon dorsatum*. Michigan State Univ. unpubl. M.S. thesis, v+58 pp.

Schoonmaker, W. J.
1930 Porcupine eats water lily pads. Jour. Mammalogy, 11(2):84.

Seton, E. T.
1953 Lives of game animals. Vol. IV, Pt. II. Charles T. Branford Co., Boston. pp. 441–949.

Shadle, A. R.
1944 The play of American porcupines (*Erethizon d. dorsatum* and *E. epizanthum*). Jour. Comp. Physiol., 37(3): 145–150.
1946 Copulation in the porcupine. Jour. Wildlife Mgt., 10(2):159–162.
1948 Gestation period in the porcupine. *Erethizon dorsatum dorsatum*. Jour. Mammalogy, 29(2):162–164.
1951 Laboratory copulations and gestation of porcupine. *Erethizon dorsatum*. Jour. Mammalogy, 32(2):219–221.

Shadle, A. R., and W. R. Ploss
1943 An unusual porcupine parturition and development of the young. Jour. Mammalogy, 24(4):492–496.

Shapiro, J.
1949 Ecological and life history notes on the porcupine in the Adirondacks. Jour. Mammalogy, 30(3):247–257.

Shiras, G., III
1911 A flashlight story of an albino porcupine and of a cunning but unfortunate coon. Nat. Geogr. Mag., 22(6):572–596.

Smith, G. W.
1977 Population characteristics of the porcupine in northeastern Oregon. Jour. Mammalogy, 58(4):674–676.

Speer, R. J., and T. G. Dilworth
1978 Porcupine winter foods and utilization in central New Brunswick. Canadian Field-Nat., 92(3):271–274.

Spencer, D. A.
1946 A forest mammal moves to the farm—the porcupine. Trans. 11th North American Wildlife Conf., 11:195–199.
1948 An electric fence for use in checking porcupine and other mammalian crop depredations. Jour. Wildlife Mgt., 12(1):110–111.

Stoeckeler, J. H.
1950 Porcupine damage in a northern hardwood-hemlock forest of northeastern Wisconsin. U.S. Forest Serv., Lake States Forest Exp. St., Tech. Note No. 326, 1 p.

Struthers, P. H.
1928 Breeding habits of the Canadian porcupine (*Erethizon dorsatum*). Jour. Mammalogy, 9(4):300–309.

Taylor, W. P.
1935 Ecology and life history of the porcupine (*Erethizon epixanthum*) as related to the forests of Arizona and the southwestern United States. Univ. Arizona, Biol. Sci. Bull. No. 3, 177 pp.

Tibbits, J. S.
1877 Wild animals of Wayne County. Michigan Pioneer Coll., 1:403–406.

Tryon, C. A.
1947 Behavior and post-natal development of a porcupine. Jour. Wildlife Mgt., 11(3):282–283.

Voigt, D. R., G. B. Kolenosky, and D. H. Pimlott
1976 Changes in summer foods of wolves in central Ontario. Jour. Wildlife Mgt., 40(4):663–668.

Weaver, R. L.
1939 Attacks on porcupine by gray fox and wild cats. Jour. Mammalogy, 20(3):379.

Wenzel, O. J.
1912 A collection of mammals from Osceola County, Michigan. Michigan Acad. Sci., 14th Rept., pp. 198–205.

Wood, N. A.
1922 The mammals of Washtenaw County, Michigan. Univ. Michigan, Mus. Zool., Occas. Papers No. 123, 23 pp.

Young, S. P., and E. A. Goldman
1946 The puma, mysterious American cat. Amer. Wildlife Inst., Washington, D.C. xvi+358 pp.

Carnivores or Flesh-Eaters

ORDER CARNIVORA

Although members of this prominent mammalian order are commonly referred to as flesh-eaters, many of them also eat plant materials. Animals bearing dental and skeletal features characteristic of this order diversified early in the evolution of mammals, presumably in response to the development of a great variety of prey species—the herbivorous mammals. Today, the array of carnivores includes seven families, about 96 genera, and approximately 253 species worldwide, excluding the three families of the Pinnipedia (seals, sea lions, walruses), which are sometimes arranged in the same order with the terrestrial carnivores.

Carnivores have dentition adapted to capturing and killing prey and rendering their fleshy parts into bite-sized pieces. Normally, the incisors are small and seldom used in the ingestion process but the canines are long, conical, and recurved to facilate stabbing or seizing prey. The premolars and molars may be highly modified, with the premolars often pointed, or laterally compressed and blade-like, and the molars rotated. The enlarged fourth upper premolar and the first lower molar are the most distinctive; they are designed to shear against each other. A dog chewing a bone is an excellent demonstration of how these teeth are used as shears. These are the carnassial pair of teeth and they appear in all modern carnivores in a highly developed form, except in the raccoons and bears which possess secondarily reduced dentitions fitted for their omnivorous food habits.

Carnivores with long muzzles (Canidae and Ursidae) may have as many as 4/4 premolars and 3/2 molars; with short muzzles (Felidae), as few as 2/2 premolars and 1/1 molars. The braincase is large, ruggedly constructed, and the brain is highly convoluted. The turbinal bones (scroll-shaped structures in the nasal cavity) and the auditory bullae are large. The turbinal bones are correlated with the keen sense of smell in members of this order. Hearing and vision are also efficient, as would be expected in animals which must hunt wary prey. The zygomatic arch is always complete. The lower jaw articulates with the cranium by means of a condyle which is transversely elongated to fit in a similarly expanded glenoid fossa on the squamosal bone of the skull. This snug fit (observe, for example, the skull and lower jaw of a badger) allows up and down but only limited sideways motion in chewing. The limbs of Carnivora are long and have distinctive lower leg bones (radius and ulna in the forelimb and fibula and tibia in the hind limb). There are usually five toes on each foot (four on the hind feet in Michigan Canidae and Felidae). Each digit terminates in a well-developed claw. Some animals (Ursidae and Procyonidae) are flat-footed (plantigrade); others (Canidae and Felidae) walk on the pads of their terminal digits (digitigrade). The latter condition allows for great running speeds (cursorial gait).

Carnivores are found in most terrestrial areas of the world, with the exception of Antarctica and Australia, although the dingo, thought to have been introduced in Australia by humans, has become established there in the wild. Carnivores are as small as the least weasel (*Mustela nivalis*), a Michigan resident weighing as little as 1.4 oz (40 g), and as large as the brown bear (*Ursus arctos*) of western North America and Eurasia, and the polar bear (*Ursus maritimus*) of the North Polar regions, both weighing as much as 1,540 lbs (700 kg). In Michigan, the black bear (*Ursus americanus*) is the largest member of this order.

KEY TO THE SPECIES OF CARNIVORA IN MICHIGAN Using Characteristics of Adult Animals in the Flesh

1a. Five clawed digits on both the front and back feet . . . . .2a.

1b. Forefeet with five clawed digits, the first reduced and high on the foot; hind feet with four clawed toes . . . . 16a.

2a. Weight more than 100 lbs (45 kg); tail short and obscured in thick fur; color mostly brownish black with brown muzzle and occasionally a white spot on chest . . . . . family Ursidae—BLACK BEAR (*Ursus americanus*), page 430.

2b. Weight less than 100 lbs (45 kg); tail sufficiently long to be conspicuous; color not mostly brownish black with a brown muzzle . . . . .3a.

3a. Tail ringed with alternating black and tan bands terminating in a black tip; forehead and eyes crossed with a black mask . . . family Procyonidae—RACCOON (*Procyon lotor*), page 443.

3b. Tail without series of black and tan bands; forehead and eyes not crossed with a black mask . . . .family Mustelidae—mustelids . 4a.

4a. Size medium to large; median white stripe on forehead; claws of forefeet not partly concealed by fur . . . . 5a.

4b. Size small to large; no median stripe on forehead; claws of forefeet partly concealed by fur . . . . 6a.

5a. Weight less than 14 lbs (6.4 kg); overall color mostly black with white on top of head and two white stripes diverging over back; front claws not longer than 1 in (25 mm) and not enlarged for digging . . . . STRIPED SKUNK (*Mephitis mephitis*), page 516.

5b. Weight more than 14 lbs (6.4 kg); overall color yellowish brown; front claws longer than 1 in (25 mm) and greatly enlarged for digging . . . . BADGER (*Taxidea taxus*), page 510.

6a. Size small, weight less than 10 oz (285 g); underparts entirely white or yellowish white . . . . 7a.

6b. Size small to large; underparts not entirely white or yellowish white . . . . 11a.

7a. Tail conspicuously tipped with black; tail length more than 1½ in (40 mm) . . . . . 8a.

7b. Tail not conspicuously tipped with black (occasionally a few black hairs at most), tail length less than 1½ in (40 mm) . . . . LEAST WEASEL (*Mustela nivalis*), page 488.

8a. Sex, male . . . . 9a.

8b. Sex, female . . . . 10a.

9a. Total length (tip of nose to tip of tail) more than 13 in (320 mm); length of tail more than 4 in (100 mm) . . . . LONG-TAILED WEASEL (*Mustela frenata*), page 480.

9b. Total length (tip of nose to tip of tail) less than 13 in (320 mm); length of tail less than 4 in (100 mm) . . . . ERMINE (*Mustela erminea*), page 472.

10a. Total length (tip of nose to tip of tail) more than 10⅞ in (275 mm); length of tail more than 3 in (75 mm) . . . . . . LONG-TAILED WEASEL (*Mustela frenata*), page 480.

10b. Total length (tip of nose to tip of tail) less than 10⅞ in (275 mm); length of tail less than 3 in (75 mm) . . . . ERMINE (*Mustela erminea*), page 472.

11a. Ears brown with narrow border of buff on edges . . . . 12a.

11b. Ears uniformly brown lacking paler color on edges . . . . 14a.

12a. Dorsal color entirely light brown; irregular patch of yellow on throat and chest; longest claw on front digits less than ½ in (13 mm) long . . . . MARTEN (*Martes americana*), page 455.

12b. Dorsal color not entirely light brown; no irregular patch of yellow on throat and chest; longest claw on front digits more than ½ in (13 mm) long . . . . 13a.

13a. Weight more than 15 lbs (6.8 kg); broad light yellowish stripes extending from each shoulder along sides and joining across rump; longest claw on front digits more than ¾ in (20 mm) . . . . WOLVERINE (*Gulo gulo*), page 503.

13b. Weight less than 15 lbs (6.8 kg); no broad light yellowish stripes extending from each shoulder along sides and joining across rump;

longest claw on front digits less than ¾ in (20 mm) . . . FISHER (*Martes pennanti*), page 464.

14a. Total length (tip of nose to tip of tail) more than 31½ in (800 mm); length of tail more than 9⅞ in (250 mm), thickest at base and tapering to a blunt point; digits webbed . . . . RIVER OTTER (*Lutra canadensis*), page 526.

14b. Total length (tip of nose to tip of tail) less than 31½ in (800 mm); length of tail less than 9⅞ in (250 mm), not noticeably thickened at base; digits not webbed . . . . . . 15a.

15a. Weight more than 2.0 oz (60 g); length of tail more than 2 in (50 mm); pelage not white in winter . . . . MINK (*Mustela vison*), page 495.

15b. Weight less than 2.0 oz (60 g); length of tail less than 2 in (50 mm); pelage white in winter . . . . LEAST WEASEL *Mustela nivalis*), page 488.

16a. Claws not retractile and not completely concealed in fur; claws long and blunt; tail bushy; back with mane of stiff, black-tipped hairs . . . . family Canidae—dogs or canids . . . . 17a.

16b. Claws retractile and mostly concealed in fur; claws recurved and sharp; tail not bushy; mane absent . . . . family Felidae—cats or felids . . . . 20a.

17a. Tail vertebrae less than half as long as head and body; total length (head, body, and tail) less than 42 in (1,070 mm); length of hind foot less than 6¼ in (160 mm); width of nose pad less than ¾ in (20 mm); eye in life with pupil vertically elliptical . . . . 18a.

17b. Tail vertebrae more than half as long as head and body; total length (head, body, and tail) more than 42 in (1,070 mm); length of hind foot more than 6¼ in (160 mm); width of nose pad more than ¾ in (20 mm); eye in life with pupil round . . . . 19a.

18a. Body reddish yellow; tail similar in color but mixed with black and tipped with white; legs and feet black . . . . RED FOX (*Vulpes vulpes*), page 412.

18b. Body mixed gray and black with reddish brown on sides of neck and behind ears; tail gray above with mane of stiff hairs and tipped with black; legs and feet reddish brown . . . . . . GRAY FOX (*Urocyon cinereoargenteus*), page 423.

19a. Size large, more than 55 in (1,400 mm) in total length; length of tail more than 15½ in (400 mm); greatest depth of claws on front digits more than 5/16 in (8 mm); width of nose pad more than 1 in (25 mm) . . . . . . GRAY WOLF (*Canis lupus*), page 401.

19b. Size medium, less than 55 in (1,400 mm) in total length; length of tail less than 15½ in (400 mm); greatest depth of claws on front digits less than 5/16 in (8 mm); width of nose pad less than 1 in (25 mm) . . . . . . COYOTE (*Canis latrans*), page 390.

20a. Tail long, more than 19¾ (500 mm) in length and more than one-quarter as long as the length of head and body; tail nearly as thick at tip as at base . . . . MOUNTAIN LION (*Felis concolor*), page 536.

20b. Tail short, less than 8 in (200 mm) in length and much less than one-quarter as long as the length of head and body; tail not as thick at tip as at base . . . . 21a.

21a. Ears with tufts of hair more than 1⅜ in (35 mm) long; legs long; hind feet very large, more than 8 in (200 mm) long; tip of tail entirely black . . . . LYNX (*Felis lynx*), page 544.

21b. Ears with tufts of hair less than 1⅜ in (35 mm) long; legs of normal length; hind feet of normal size, less than 8 in (200 mm) long; tip of tail with black spot on dorsal side and whitish below . . . . BOBCAT (*Felis rufus*), page 552.

KEY TO THE SPECIES OF CARNIVORA IN MICHIGAN Using Characteristics of the Skulls of Adult Animals

1a. Total number of permanent (not deciduous) teeth 34 or more, 16 (eight on either side) or more in the upper jaw and 18 (nine on either side) or more in the lower jaw; last upper molar in each jaw may be small but always more than twice as large as the outer incisor . . . . 2a.

1b. Total number of permanent (not deciduous) teeth 30 or less, 16 (eight on either side) or less in the upper jaw and 14 (seven on either side) or more in the lower jaw; last upper molar in each jaw may be small or smaller than the outer incisor . . . . family Felidae—cats or felids . . . . 17a.

2a. Length of skull more than 10 in (250 mm); last upper molar with front-back (anterior-posterior) diameter about one and one-half times the lateral diameter, distinctly larger than the tooth just in front of it . . . . . family Ursidae—BLACK BEAR (*Ursus americanus*), page 430.

2b. Length of skull less than 10 in (250 mm); last upper molar with front-back (anterior-posterior) diameter about equal to or less than the lateral diameter, about the same size (except in *Mephitis mephitis*) or smaller than the tooth just in front of it . . . . 3a.

3a. Total number of permanent (not deciduous) teeth 40, 20 (10 on either side) in the upper jaw and 20 (10 on each side) in the lower jaw; skull medium in size; 4⅜ in (112 mm) long . . family Procyonidae—RACCOON (*Procyon lotor*), page 443.

3b. Total number of permanent (not deciduous) teeth never 40, either 42 (Canidae), 38 or less (Mustelidae) . . . . 4a.

4a. Total number of permanent (not deciduous) teeth 38 or less, never more than 18 (nine on each side) in the upper jaw . . . . family Mustelidae—mustelids . . . . 5a.

4b. Total number of permanent (not deciduous) teeth 42, 20 (10 on each side) in the upper jaw and 22 (11 on each side) in the lower jaw . . . . family Canidae—dogs or canids . . . . . . 14a.

5a. Skull medium in size, about 2¾ in (70 mm) in length; posterior edge of bony palate ends at or slightly anterior to a line drawn between the posterior borders of the last upper molar teeth; last upper molar tooth almost square in shape and distinctly larger than the preceding tooth . . . . STRIPED SKUNK (*Mephitis mephitis*), page 516.

5b. Skull small to large in size; posterior edge of bony palate extends noticeably beyond (posterior to) a line drawn beween the posterior borders of the last upper molar teeth; last molar tooth not almost square in shape and about the same size or smaller than the preceding . . . . 6a.

6a. Five permanent cheek teeth (four premolars, one molar) on each side of upper jaw behind canines . . . . 7a.

6b. Four permanent cheek teeth (three premolars, one molar) on each side of upper jaw behind canines . . . . 10a.

7a. Skull massive and broad; total length more than 5½ in (140 mm); breadth more than 3⅜ in (85 mm) . . . . . . WOLVERINE (*Gulo gulo*), page 503.

7b. Skull less massive and broad; total length less than 5½ in (140 mm); breadth less than 3⅜ in (85 mm) . . . . 8a.

8a. Skull rather slender; forward part (rostrum) longer than broad; width of braincase across mastoid processes less than 2¼ in (58 mm); auditory bullae rounded . . . . 9a.

8b. Skull rather flat and broad; forward part (rostrum) broader than long; width of braincase across mastoid processes more than 2¼ in (58 mm); auditory bullae flattened . . . . RIVER OTTER (*Lutra canadensis*), page 526.

9a. Skull larger, more than 3¾ in (95 mm) long; length of fourth upper premolar (measured from the outside) more than ⅜ in (9.5 mm); length of lower first molar more than 7/16 in (11 mm) . . . . FISHER (*Martes pennanti*), page 464.

9b. Skull smaller, less than 3¾ in (95 mm) long; length of fourth upper premolar (measured from the outside) less than ⅜ in (9.5 mm); length of lower first molar less than 7/16 in (11 mm) . . . . . MARTEN (*Martes americana*), page 455.

10a. Skull massive and broad; total length more than 3⅞ in (100 mm); breadth more than 2¾ in (70 mm); last upper molar triangular in shape . . . BADGER (*Taxidea taxus*), page 510.

10b. Skull not massive; total length less than 3⅞ in (100 mm); breadth less than 2¾ in (70 mm); last upper molar narrow and dumbbell-shaped . . . . 11a.

11a. Total length of skull less than 1 5/16 in (33 mm); width across mastoid processes usually more than across braincase . . . . LEAST WEASEL (*Mustela nivalis*), page 488.

11b. Total length of skull more than 1 5/16 in (33 mm); width across mastoid processes usually less than across braincase . . . . 12a.

12a. Length of skull more than 2⅛ in (55 mm); width more than 1 3/16 in (30 mm) . . . . MINK (*Mustela vison*), page 495.

12b. Length of skull less than 2⅛ in (55 mm); width less than 1 3/16 in (30 mm) . . . 13a.

13a. Rostrum short and tapering continuously

with the zygomatic arches; postorbital processes blunt and weakly developed, with little postorbital constriction; auditory bullae flattened anteriorly; post-glenoid length usually more than 48% of condylobasal length in males and 50% or more in females; greatest width of skull of males more than ¾ in (19 mm), of females, less . . . . ERMINE (*Mustela erminae*), page 472.

13b. Rostrum long with nearly parallel sides; postorbital processes sharp and strongly developed, with distinct postorbital constriction; auditory bullae bulbous anteriorly instead of flattened; post-glenoid length 48% or less of condylobasal length in males and less than 50% in females; greatest width of skull of males more than ⅞ in (23 mm), of females, less . . . . LONG-TAILED WEASEL (*Mustela frenata*), page 480.

14a. Dorsal surface of skull with prominent paired ridges (parietal or temporal ridges) which begin just behind eye socket (orbit) and enclose a U-shaped area but do not approach each other at the midline (suture between frontals and parietals) by more than ⅜ in (10 mm) in width and not forming a sagittal crest; upper incisors not lobed; lower jaw with prominent notch near rear of bottom edge . . . . GRAY FOX (*Urocyon cinereoargenteus*), page 423.

14b. Dorsal surface of skull with less prominent paired ridges (if present) and separated at the midline (suture between frontals and parietals) by a space less than ⅜ in (10 mm) in width and often coming together to form a conspicuous sagittal crest at the midline, upper incisors lobed; lower jaw lacking prominent notch near rear of bottom edge . . . . 15a.

15a. Total length of skull less than 6¼ in (160 mm); postorbital processes thin and concave dorsally (forming shallow pit); dorsal surface of skull smooth or possessing paired ridges which if they converge do so at rear of cranium and do not form a prominent sagittal crest at the midline . . . RED FOX (*Vulpes vulpes*), page 412.

15b. Total length of skull more than 6¼ in (160 mm); postorbital processes heavy and convex dorsally; dorsal surface of skull having a prominent sagittal crest formed by paired ridges which converge at the midline . 16a.

16a. Total length of skull less than 9 in (230 mm) . . . . COYOTE (*Canis latrans*), page 390.

16b. Total length of skull more than 9 in (230 mm) . . . . GRAY WOLF (*Canis lupus*), page 401.

17a. Four permanent cheek teeth (three premolars, one molar) on each side of upper jaw behind canines; total length of skull more than 6¼ in (160 mm); width more than 4⅜ in (110 mm) . . . . MOUNTAIN LION (*Felis concolor*), page 536.

17b. Three permanent cheek teeth (two premolars, one molar) on each side of upper jaw behind canines; total length of skull less than 6¼ in (160 mm); width less than 4⅜ in (110 mm) . . . . 18a.

18a. Skull with posterior part of presphenoid bone widely flaring, greatest width more than ⅜ in (5 mm); anterior condyloid foramen separate and outside of the jugular foramen; auditory bullae divided externally into two bulbous portions . . . . LYNX (*Felis lynx*), page 544.

18b. Skull with presphenoid bone narrow, greatest width less than ⅜ in (5 mm); anterior condyloid foramen not separate, located within rim of jugular foramen; auditory bullae not divided externally into two bulbous portions . . . . BOBCAT (*Felis rufus*), page 552.

FAMILY CANIDAE

Modern members of the dog family include about 41 species arranged in 15 genera. They thrive in a variety of arctic, temperate and tropical environments on all continents and most major islands with the exception of Australia, as the dingo is considered introduced there by human effort. Believed to have originated during the late Eocene period, canids have retained some rather conservative characteristics including: an elongated rostrum housing complex turbinal bones associated with their remarkable sense of smell; a nearly complete set of teeth (characteristic of ancestral placental mammals), lacking only a single molar on each side in the upper jaw; long, powerfully-built canines; well-developed shearing blades on the carnassials (fourth upper premolar and first lower molar); crushing surfaces on post-carnassial teeth; limbs usually long and adapted for cursorial gait; feet digitigrade; claws blunt and nonretractile; forefeet usually with five toes, hindfeet with four. The clavicle is absent; an intestinal caecum is present.

Canids are chiefly carnivorous, eating all kinds of vertebrates and invertebrates including mollusks, crustaceans, and insects. Carrion and a variety of vegetable material (fruit especially) are also consumed. Members of this family usually capture prey by lengthy pursuit rather than by stealth. Individuals are reported to hunt in groups to tire and subdue prey species. Impressive endurance as well as sustained speed characterize the predatory habits of members of this family. The intelligence and social nature of the canids are well-demonstrated by the domesticated dog.

HYBRIDIZATION BETWEEN MICHIGAN CANIDS. Species of the genus *Canis*, including the coyote (*C. latrans*), the gray wolf (*C. lupus*), and the domestic dog (*C. familiaris*), possess a diploid number (78) of chromosomes. It is not surprising that fertile offspring have been produced from crosses between these three canids (Nowak *in* Bekoff, 1978). In the Great Lakes Region of eastern North America, the coyote slowly began to replace the gray wolf in the latter part of the nineteenth century (Nowak, 1979). Evidently, the rather close association between the adapted coyote, the gray wolf (mostly now north in Ontario), and the domestic dog has produced a somewhat variable wild canid population, especially in New England and adjacent Ontario (Silver and Silver, 1969). These animals are classed as coyotes but apparently also descend from both the gray wolf and domestic dog (Hilton *in* Bekoff, 1978; Kolenosky, 1971; Lawrence and Bossert, 1969). When the domestic dog has played a major role in the ancestry, the term coydog has been used to designate the cross (Mahan *et al.*, 1978; Mengel, 1971). These crosses have been found in Michigan (Whitlock, 1948); one family of Michigan coyote-domestic dog hybrids shared attributes of each parent, a female coyote and a male black and tan hound (Dice, 1942). Although physical features and configurations of the skulls of Michigan coyote and gray wolf differ noticeably (Bailey, 1909; Whitlock, 1948), it can still be difficult to dis-

*Table 16. Characteristics distinguishing the gray wolf, the coyote, various breeds of domestic dogs, and hybrids in Michigan.**

FEATURE	GRAY WOLF**	COYOTE	DOMESTIC DOG	HYBRID
Weight of adult	rarely less than 60 lbs (27 kg)	rarely more than 50 lbs (23 kg)	variable	variable
Total length of adult (nose to tail tip)	rarely less than 53 in (1,325 mm)	rarely more than 63 in (1,570 mm)	variable	variable
Tail shape	bottle-shaped	bottle-shaped	brush-like	variable
Tail carry when running	high	low	high	variable
Toes on hind feet	four	four	occasionally five	unknown
Position of tips of ears	erect and straight	erect and straight	often lopped	variable
Width of nose pad	at least 1¼ in (33 mm)	no more than 1 in (25 mm)	variable	variable
Width of heel pad	at least 1½ in (38 mm)	no more than ⅞ in (22 mm)	variable	variable
Condition of underfur	thick	thick	sparse	variable
Forehead in profile	brow fairly prominent	brow barely evident	well developed***	variable
Diameter of upper canine tooth at base	at least ½ in (13 mm)	less than ½ in (13 mm)	variable	variable

*Canids not readily definable as coyote or gray wolf should be presented for examination to a wildlife biologist of the Michigan Deparment of Natural Resources or to a mammalogist at one of the universities in the Upper or Lower Peninsulas. Both the skin and skull (with lower jaw) will be helpful in identification, especially in distinguishing hybrids.

**With the present rarity of the gray wolf in Upper Peninsula Michigan, there is little chance that recently taken canids will be of this species.

***Purebred collies lack a well-defined forehead brow.

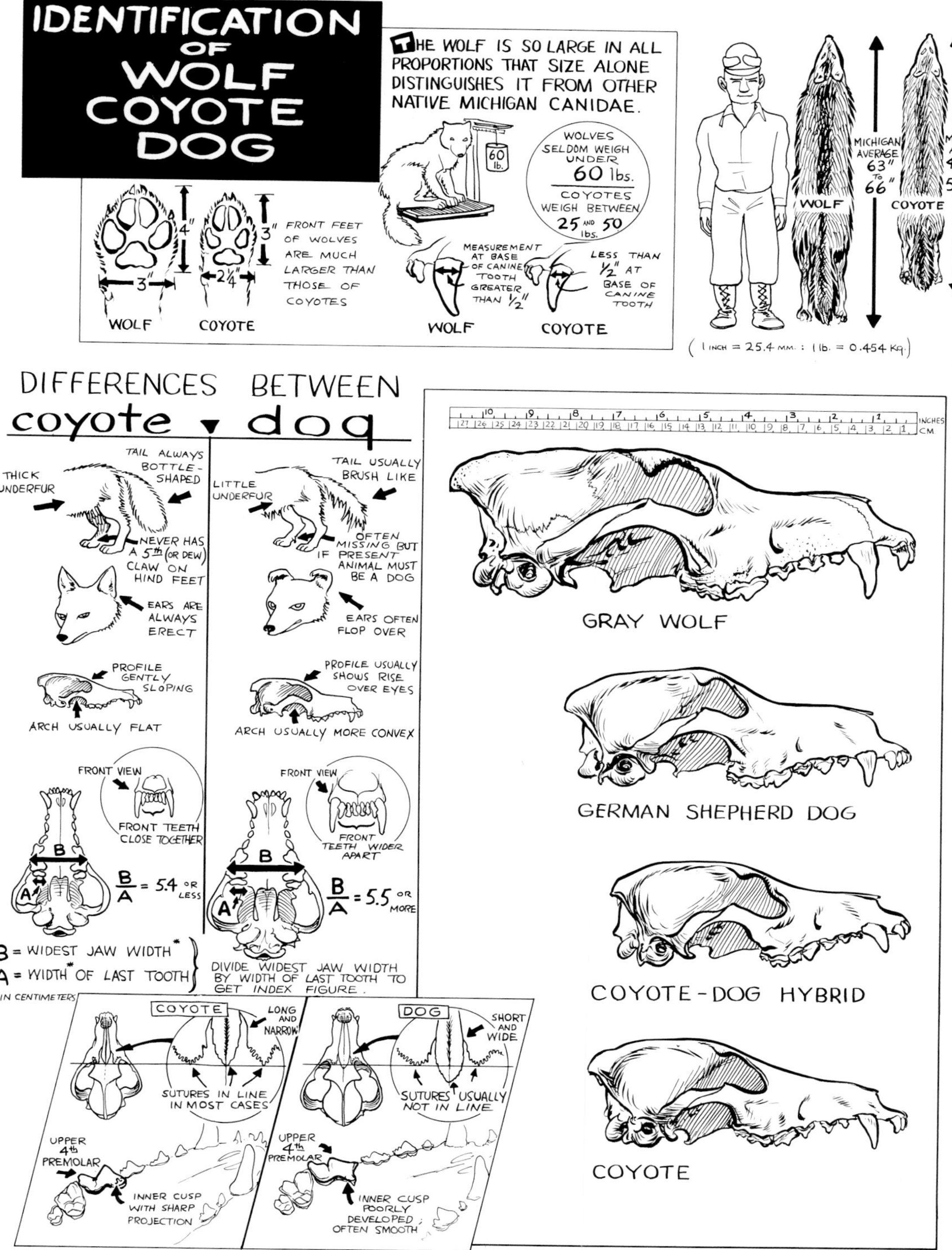

Figure 61. Some characteristics to identify gray wolf, coyote, domestic dog and hybrids. Courtesy of Oscar Warbach and the Michigan Department of Natural Resources. From S. C. Whitlock (1948), Identification of wolf, coyote, dog. Michigan Conserv., 17(10):6–7, 15.

Table 17. Distinguishing features of the pups of Michigan canids (after Bailey, 1909).

Gray Wolf pups:
- Muzzle blackish at birth, fading in a month or six weeks to grayish.
- Head grayish in decided contrast to black of back, nose, and ears.
- Ears black at tips, fading to grayish in a month or six weeks.
- Tail black, fading to gray with black tip.

Coyote pups:
- Muzzle tawny, or yellowish brown, becoming more yellowish with age.
- Head yellowish gray, not strongly contrasted with rest of body.
- Ears dark brown at tips and back, soon fading to yellowish brown.
- Tail black, fading to gray with black tip.

Red Fox pups:
- Muzzle blackish.
- Head dusky with sides of face light yellowish.
- Ears large, nearly the whole back of ears bright black at all ages.
- Eyes and ears relatively larger and nose pad smaller than in coyote or wolf.
- Tail dusky, tip white at all ages.

Gray Fox pups:
- Muzzle blackish.
- Head grayish, face back of eyes sharply pepper and salt gray.
- Ears large, back of ears dusky at tip, fulvous at base.
- Eyes and nose pad small.
- Tail with tip black at all ages.

tinguish these two species from hybrids, as well as from some breeds of domestic dogs (Howard, 1949). It is almost essential for a professional mammalogist to compare the individual specimen in question with specimens (skins, skulls, and lower jaws) of known identity preserved in museum collections (see Table 16 and Fig. 61). Table 17 aids in identifying pups of the four wild Michigan canids.

LITERATURE CITED

Bailey, V.
1909 Key to animals on which wolf and coyote bounties are often paid. U.S.D.A., Bur. Biol. Survey. Circular No. 69, 3 pp.

Bekoff, M., Ed.
1978 Coyotes. Biology, behavior, and management. Academic Press, New York. xx+384 pp.

Dice, L. R.
1942 A family of dog-coyote hybrids. Jour. Mammalogy, 23(2):186–192.

Howard, W. E.
1949 A means to distinguish skulls of coyotes and domestic dogs. Jour. Mammalogy, 30(2):169–171.

Kolenosky, G. B.
1971 Hybridization between wolf and coyote. Jour. Mammalogy, 52(2):446–449.

Lawrence, B., and W. H. Bossert
1969 The cranial evidence for hybridization in New England Canis. Breviora, 330:1–13.

Mahan, B. R., P. S. Gipson, and R. M. Case
1978 Characteristics and distribution of coyote X dog hybrids collected in Nebraska. American Midl. Nat., 100(2):408–415.

Mengel, R. M.
1971 A study of dog-coyote hybrids and implications concerning hybridization in *Canis*. Jour. Mammalogy, 52(2):316–336.

Nowak, R. M.
1979 North American Quaternary *Canis*. Univ. Kansas, Mus. Nat. Hist., Monogr. No. 6, 154 pp.

Silver, H., and W. T. Silver
1969 Growth and behavior of the coyote-like canid of northern New England with observations on canid hybrids. Wildlife Monog., No. 17, 41 pp.

Whitlock, S. C.
1948 Identification of wolf, coyote, dog. Michigan Conserv., 17(10):6–7, 10.

COYOTE
Canis latrans Say

NAMES. The generic name *Canis,* proposed by Linnaeus in 1758, is of Latin origin and means "dog." The specific name *latrans,* proposed by Thomas Say in 1823, is also derived from the Latin and means "barking." In Michigan, this animal is sometimes called a wolf or brush wolf. The name coyote is the Spanish adaptation of the Nahuatl Indian word *coyotl.* The subspecies *Canis latrans thamnos* Jackson occurs in Michigan.

RECOGNITION. Body size large with dimensions of a half-grown German shepherd; head and body averaging 34 in (895 mm) long; tail slightly less than half the body length and well-haired, averaging 13 in (330 mm) in vertebral length; head with slender and somewhat pointed muzzle (Fig. 62); nose pad small, usually less than 1 in (25 mm) wide; ears large and pointed; eyes medium

Figure 62. The coyote (*Canis latrans*) showing its slender head and somewhat pointed muzzle. Sketch by Bonnie Marris.

with yellow iris and round pupil; legs long and slender; feet relatively small for the size of the body; digits terminating in blunt, heavy, non-retractile claws; front feet with five digits, the first digit small; hind feet with four digits; scent gland on back at upper base of tail prominent; mammary glands consist of four pairs; males slightly larger than females in body and cranial dimensions.

The pelage is moderately long with the guard hair rather coarse while the buff colored underfur is thick and soft. The overall color of the coyote is highly variable but generally is grizzled buffy and

Figure 63. The coyote (*Canis latrans*) with its "dog-like" stance. Photograph by Robert Harrington and courtesy of the Michigan Department of Natural Resources.

yellowish gray with a whitish throat and belly. The muzzle, sides of head, forelegs, and feet are reddish brown. The entire back from the neck to the tail is mixed blackish and buffy-gray, with the tip of the tail black. The black-tipped guard hairs form a blackish dorsal stripe and a dark cross on the shoulders.

At a distance, the coyote may be difficult to distinguish from a medium-sized German shepherd or similar domestic dog, although its pointed, erect ears, drooping tail, and rather shuffling trot aid in identification (Fig. 63). The coyote is easily differentiated from the red and gray fox by coat color and larger size, but a casual observer could mistake it (at a distance) for a half-grown gray wolf. In fact, some of the early literature about the coyote and the gray wolf is suspect because of possible mistaken identification.

MEASUREMENTS AND WEIGHTS. Adult coyotes (with males being slightly larger than females) measure as follows: length from tip of nose to end of tail vertebrae, 43 to 52 in (1,075 to 1,300 mm); length of tail vertebrae, 11½ to 15½ in (290 to 390 mm); length of hind foot, 7 to 8⅝ in (180 to 220 mm); height of ear from notch, 3¾ to 4⅞ in (95 to 125 mm); weight, 25 to 45 lbs (11.4 to 20.5 kg), rarely more than 50 lbs (22.7 kg).

DISTINCTIVE CRANIAL AND DENTAL CHARACTERISTICS. The skull of the coyote has a rather low profile; is slender and elongate; shows wide-flaring zygomatic arches; and possesses a prominent crest (especially in mature males) at midline on the braincase (Fig. 64). The conspicuous postorbital processes have convex dorsal surfaces; the interorbital and frontal regions are flat in appearance and slope slightly anteriorily. The auditory bullae are long and fairly narrow. The skull of an adult averages 7⅜ in (187 mm) long and 4 in (101 mm) wide (across zygomatic arches).

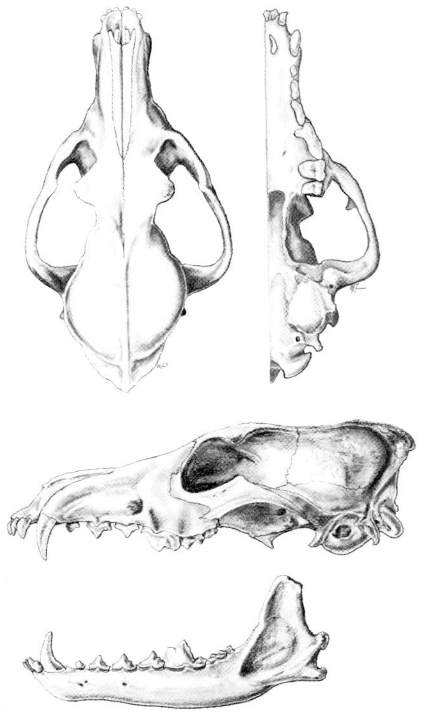

Figure 64. Cranial and dental features of the coyote (*Canis latrans*).

The teeth are moderate in size; the canines are rather long and slender, being no more than 7/16 in (11 mm) in anterior-posterior diameter at the base and about 1½ in (38 mm) long. When the lower jaw is correctly attached and closed, the tips of the canines protrude (when viewed from the front) to below the openings of the anterior mental foramina on the mandible (see Fig. 65). The skull of the adult coyote is larger than those of the gray and red fox but is smaller than that of the gray wolf (other distinctive characteristics are listed in the key). The skull of the coyote is compared with those of various breeds of domestic dogs under a separate heading. The dental formula for the coyote is:

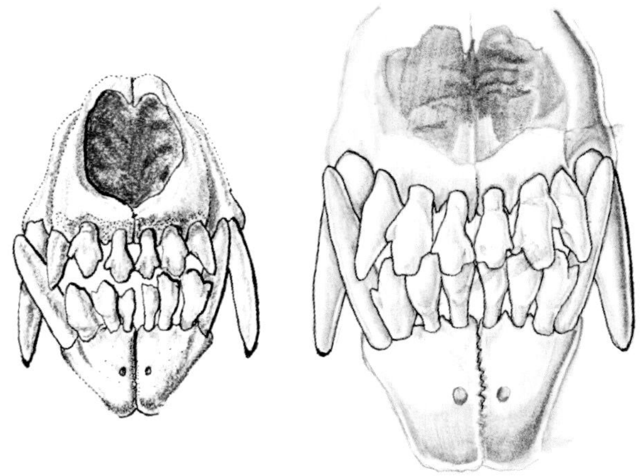

Figure 65. Anterior views of skulls with occluded lower jaws of coyote (left) and gray wolf (right). Note in the upper canines of the coyote that the tips extend below a line drawn through the anterior mental foramina. In the gray wolf, the shorter and heavier upper canines do not reach a line drawn between these foramina.

$$\text{I (incisors)}\frac{3\text{-}3}{3\text{-}3}, \text{C (canines)}\frac{1\text{-}1}{1\text{-}1}, \text{P (premolars)}\frac{4\text{-}4}{4\text{-}4}, \text{M (molars)}\frac{2\text{-}2}{3\text{-}3} = 42$$

DISTRIBUTION. The coyote in pre-Columbian times is thought to have been restricted primarily to the open plains of western North America, perhaps from the southern part of the Mexican Plateau, north through the short-grass and desert regions of the United States and into the Prairie Provinces of Canada (Nowak *in* Bekoff, 1978; Young and Jackson, 1951). With the advent of the Spanish settlements and environmental alterations in Mexico and Central America plus the introduction of livestock, the coyote spread southward. It was also able to extend its distribution both northward and eastward into what is now the United States and Canada as a result of settlement, land-clearing, introductions of livestock (including sheep and poultry), and perhaps elimination of other large and less compatible carnivores. Today (see Map 46), this persistent mammal occurs over most of North America from near Point Barrow, Alaska, south to the Central American republics of Costa Rica and Panama and eastward to Hudson Bay on the north and New Brunswick, Quebec, and Maine (Hilton *in* Bekoff, 1978; Richens and Hugie, 1974) on the east. Although some reports

Map 46. Geographic distribution of the coyote (*Canis latrans*). Although records of occurrence show that the species may be found statewide, it is most common in the Upper Peninsula, less common in the northern part of the Lower Peninsula (dark shading), and rare in the southern part of the Lower Peninsula.

from southeastern United States result from the capture of released individuals (Young and Jackson, 1951), there is evidence of a gradual extension of the coyote range in that direction (Bekoff, 1977; Golley, 1962).

In Lower Peninsula Michigan, the coyote was undoubtedly part of the mammalian fauna in presettlement days (Stebler, 1951; Wood and Dice, 1924) and moved northward and eastward from natural prairie habitat following clearing and settlement. Although early reports often fail to distinguish coyotes from gray wolves, Wood (1922) placed the coyote in Washtenaw County prior to 1881. Apparently the coyote spread throughout the Lower Peninsula; some stock may have spread southward across the Straits of Mackinac from the Upper Peninsula (Stebler, 1951). However, according to the Fourth Biennial Report of the Michigan Conservation Department for 1927–1928, the species disappeared over much of the southern half of the Lower Peninsula, while continuing to thrive in the northern half. Today, coyotes are reported sporadically in southern Michigan counties—Berrien (1955), Clinton (1961), Eaton (1973), Ingham (1963 and 1980), Kent (1981), Livingston (1977), Montcalm (1961), Shiawassee (1977), and St. Clair (1961).

The coyote presumably arrived in Upper Peninsula Michigan after 1900 (deVos, 1964); according to Goldman (1930), the date was about 1906. Less than 10 years following that date, however, the coyote had been reported in Alger, Dickinson, Gogebic, Marquette, and Menominee counties (Wood, 1914). Today, the coyote occurs in all counties in the Upper Peninsula. Its success in living in varied habitats is illustrated by its invasion of islands (see section on County Records in Michigan). On the other hand, coyotes which invaded Isle Royale about 1906 (probably by way of the Sibley Peninsula of Ontario) survived only until 1957 or 1958. According to Krefting (1969), forest growth protected from fire, a reduced food supply, and competition with the gray wolf were some of the factors which led to the coyote's demise.

HABITAT PREFERENCES. Extensive and dense woodlands may discourage the coyote, but lumbering and other clearing have reduced much of Michigan's "trackless" forest environments to patchwork, providing numerous openings and brushy glades conducive to the coyote's survival. Studies in the Upper Great Lakes Region have shown how the coyote has effectively adapted to the mixed forest and cleared environments. Berg and Chesness (*in* Bekoff, 1978) found Minnesota coyotes frequenting woodland cover of aspen, aspen-fir, and lowland conifers (mostly white cedar) in winter. These cover-types were also being used by favored prey: snowshoe hare and white-tailed deer. In summer and autumn, Minnesota coyotes made abundant use of open grassy fields, where small rodents were common. In Michigan, Ozoga and Harger (1966b) found the coyote mostly in aspen-conifer, swamp conifers, lowland brush, and occasionally in upland hardwoods. In open country, coyotes tend to use, and hunt on, the woodland edges. Windswept shorelines and frozen streamways generally serve as travel routes between habitats. The combination of farmlands, fallow fields, orchards, woodlots, and forested strips along streams in the southern part of Michigan's Lower Peninsula would seem to be favorable habitat for the coyote. Yet today in this area, the coyote is rare or, at best, uncommon—a blessing for local stock raisers but puzzling to ecologists.

DENSITY AND MOVEMENTS. The movements and distribution of the coyote are highly correlated with topography, vegetative cover, and food supply (Linhart and Knowlton, 1975). In addition, population numers fluctuate due to such factors as mortality of the annual pup crop, ups and downs in abundance of the food supply (notably microtine rodents and the snowshoe hare), outbreaks of mange, and kills by animal control specialists (Bekoff, 1977; Nellis and Keith, 1976; Todd *et al.*, 1981). Over a large portion of the coyote's western range, density has been estimated by Knowlton (1972) as one individual per 2 sq mi (5.2 sq km) and by Bowen (1982) as one to 5.3 sq mi (14 sq km). In Michigan, Ozoga and Harger (1966b) concluded that the autumn population of coyotes on Beaver Island amounted to one individual per 2 sq mi (5.2 sq km) and in the Shingleton area of Alger and Schoolcraft counties, one individual per 4 sq mi (10.4 sq km).

Data on coyote movements, densities, and home ranges have been obtained primarily by snow-tracking (Ozoga and Harger, 1966b; Stebler, 1951), by scent-station visitations (Roughton and Sweeny, 1982), by mark and recapture (Young *in* Young and Jackson, 1951), and by monitoring individuals and/or groups which had previously been marked with colored dyes or collared with radio transmitters (Andelt and Gipson, 1979; Bowen, 1982; Gipson and Sealander, 1972). Adult animals generally stay in prescribed areas, perhaps for life (Bowen, 1982), with evidence that females may exclude other females. Young-of-the-year may move considerable distances in their first autumn and winter dispersals—up to 100 mi (165 km) or more. Seasonally, adult coyotes may be more sedentary; females are especially so during the reproductive period but become more mobile in the prebreeding and mating seasons (Andelt and Gipson, 1979). In Minnesota, Berg and Chesness (*in* Bekoff, 1978) found radio-tagged coyotes to move 2½ mi (4 km) per day in winter—the same distance Ozoga and Harger (1966b) obtained in Michigan by means of snow-tracking. These latter authors also estimated winter areas of activity for Michigan coyotes as 20 to 25 sq mi (52 to 50 sq km). Tagged coyotes were found to move in a straight line as far as 36 mi (58 km) from Beaver Island, across at least 15 mi (24 km) of Lake Michigan water (or ice) to Emmet County in the Lower Peninsula and to Mackinac County in the Upper Peninsula (Ozoga and Phillips, 1964).

BEHAVIOR. The general behavior of the coyote is similar in many ways to that of the domestic dog (*Canis familaris*). Coyotes viewed in the field (but more commonly in captivity) have been the subject of numerous studies, including behavioral ontogeny, communication, courtship activity, play, and hunting methods (Bekoff, 1978; Gier, 1975). Laurence Frank, who reviewed Bekoff's previously cited work (Jour. Mammalogy, 60(3):658–659, 1979), lamented the fact that most of our information on the coyote's social behavior comes from unnaturally-disrupted and heavily persecuted populations. He noted that when coyote populations are subjected to a 40% to 50% annual human-caused mortality, the social organization may not even remotely resemble the one which would prevail if the human factor were removed.

Although coyotes are most active from shortly before sunset to shortly after sunrise (Ozoga, 1963; Smith *et al.*, 1981), the animals, especially pups, can be seen at any time during daylight hours (Gipson and Sealander, 1972). Motorists have unexpectedly seen them cross the road at midday; nature observers have watched the coyote's daytime antics as it stalks and pounces on meadow voles.

The animals are social, with mated pairs and pups constituting the important unit, at least until autumn and early winter, when the young disperse (Stebler, 1951). Both parents are involved in rearing the offspring (Andelt *et al.*, 1979), supplying food, and occasionally joining the romps and scuffles in which the pups, like young dogs, are so often engaged. Coyotes are less apt than gray wolves to form packs (groups of three to seven are reported). Instead, the animals usually hunt individually, in pairs, or in family units. They are regarded as solitary-social hunters by Fox (1975). There is evidence that social behavior within coyote groups is strongly dependent on prey availability, with certain individuals assuming dominant behavioral roles while others remain submissive (Bekoff, 1978; Bekoff and Wells, 1980). Following the departure of the pups, the parents may stay together over the winter or separate. Many coyotes operate alone though this season.

Coyotes communicate by visual, auditory, olfactory, and tactile signals. There is some evidence that sight is less developed than other senses. Social interactions within family groups often rely on touch. Coyotes also use an elaborate assortment of postures, tail positions, and facial expressions when relating to each other (Bekoff, 1978). Vocalizations are varied and may even differ seasonally (Laundre, 1981), although Alcorn (1946) condensed them to three categories: squeaks, howls, and distress calls. Yelps, followed by a falsetto howl, create an eerie atmosphere for campers. The squeaking whines are like those of small dogs. Coyotes deposit scent (from urine and feces) at marking places in their living areas (Bowen and Cowan, 1980). It is not known whether such acts are specifically for territory identification and demarcation. However, there is evidence that scent-marking may be more important in orienting individuals than in serving as barriers to movements (Wells and Bekoff, 1980). The habit of investigating these scent posts has led many a coyote into a waiting steel trap. In fact, chemical attractants and repellents for coyotes have been developed from urine volatiles (Fagre *et al.*, 1981; Teranishi *et al.*, 1981).

Coyotes are timid and suspicious in the presence of humans. Their gaits include a walk, trot, and bound, with maximum speeds of 35 to 43 mi (56 to 68 km) per hour over long distances (Cottam, 1945; Young and Jackson, 1951). The animal can swim well but is not a good climber. Ground dens are often constructed by enlarging the burrow of a woodchuck or badger, although the coyote is also perfectly capable of digging its own burrow as long as 30 ft (9 m) and as deep as 4 ft (1.3 m). Burrows can be used year after year. There may be as many as three entrances and merely an enlargement, without nest materials, to serve as the lair. Coyotes leave their dens to defecate and urinate. Surface resting beds may be on elevated places, in dense conifers, or by tree falls (Ozoga and Harger, 1966b). Winter coyote beds are often located on snowdrifts or on top of snow-packed brush.

ASSOCIATES. In early times in the Upper Great Lakes Region, the coyote probably held an "intermediate " canid position between the red and gray foxes and the gray wolf. As the gray wolf slowly gave ground to human encroachment, the coyote probably assumed more and more the role of the top canid, and with the demise of the mountain lion, became the top carnivore (assuming that the black bear is an omnivore). There is evidence that the coyote-gray wolf relationship has never been a compatible one (Mech, 1974), with records of gray wolves killing coyotes (Stenlund, 1955). On Isle Royale, the coyote's decline coincided with the build-up of the gray wolf (Krefting, 1969). In northern Minnesota, coyotes and gray wolves still survive in the same general areas, but each reaches highest densities in places where the other is uncommon. According to Chesness (1973) and Young and Jackson (1951), coyotes can be dependent on gray wolves, even feeding on deer carcasses abandoned by the wolves. Coyotes are known to associate with badgers but, according to Young (*in* Young and Jackson, 1951), the coyote is antagonistic toward foxes and bobcats, killing individuals of these species when they are caught in steel traps. Certainly the carnivore's life is almost devoid of interspecific associations, except in eat-or-be-eaten situations.

REPRODUCTIVE ACTIVITIES. Although coyotes rarely mate for life, pairs may remain together for several years. Courtship begins as early as December in Michigan and lasts for around 60 days, with mating occurring from late January to late March. Unlike the domestic dog, the female coyote is monoestrus, having only one annual period of heat lasting for two to five days (Whiteman, 1940; Bekoff, 1977; Kennelly *in* Bekoff, 1978; Kennelly and Johns, 1976). Pregnancy rate varies considerably, and can be as low as 10% to 14% in young females (Nellis and Keith, 1976). Embryos are aborted under stress (Gipson, 1977) as well as reabsorbed, perhaps as a result of food shortages (Gier, 1968; Young *in* Young and Jackson, 1951). After a gestation period of about 63 days, litters averaging six individuals (1 to 19) are born. There is ample evidence that in years when food is scarce, litter size may be as low as four; in years of abundant food, as high as seven (Bekoff, 1977). There is little chance for all individuals of large litters to survive because the female coyote only has eight nipples.

The young are born in a helpless state; they weigh about 9 oz (250 g) and their eyes are closed. These pug-nosed whelps are covered with short, wooly, grayish brown hair with evidence of a black dorsal stripe. By 10 days after birth, the eyes open, the ears begin to erect in true coyote fashion, and they weigh 1¼ lbs (600 g). By 21 to 28 days after birth, the active young may emerge from the den (if underground). By 28 days after birth, the pups begin to resemble their parents and are eating regurgitated and semi-solid food, although they are not totally weaned until at least 35 days. The young do not reach adult size until the ninth month of life.

The coyote in the wild may rarely achieve its maximum life span because older and less fit individuals have difficulty obtaining food, and because of the human population's constant control measures (see Connolly, 1978). There are records of individuals living six to eight years in the wild (Bekoff, 1977), and in captivity, animals have lived as long as 18 years (Young *in* Young and Jackson, 1951). A Crawford County coyote was maintained in captivity in East Lansing for 15 years (Manville, 1953). Age in the coyote can be determined by sectioning the permanent canine teeth and counting the cementum layers. The permanent canines erupt at four to five months after birth, the root canals close during the eighth or ninth month, the first cementum layer is formed on the root at about the 20th month, and additional identifiable layers are formed each year thereafter (Lindhart and Knowlton, 1967; Roberts, 1978).

FOOD HABITS. The foods eaten by the coyote have been documented primarily from identifiable kills found along coyote snow trails and by the examination of residues of materials found in stomachs, intestinal tracts, and feces. Food identified by these means shows what the coyote ate in a given area but fails to provide evidence as to what else was available which the species did not eat. Consequently, the ecologist can never be sure what prey species constitute the coyote's first choice if a selection is available. It has generally been assumed, however, that coyotes have a rather broad diet and are opportunistic eaters most of the time (Wells and Bekoff, 1982). This theory led Krefting ((969) to infer that the coyote is basically a rather poor hunter. According to Bekoff and Wells (1980), three important factors which influence coyote social organization are: the size of the available prey, its spatial distribution, and its temporal or seasonal distribution.

Sperry (1941) found the foods in stomachs of coyotes from the western United States (including upper Michigan) included rabbits and hares (33%), carrion (25%), rodents (18%), sheep, goats, and other domestic stock (13.5%), deer (3.5%), birds (3%), insects (1%), other animal matter (1%), and plants (2%). Korschgen (1957) listed 56 species of animals and 28 species of plants in the diet of Missouri coyotes. Coyotes are occasionally guilty of killing prey in excess of needs. This has especially been true with domestic stock like turkeys, which lack the wild species' ability to avoid predators (Andelt *et al.*, 1980).

Table 18 lists some of the mammalian foods reported as eaten by Michigan coyotes. The white-tailed deer is a major source of food (Ozoga and Harger, 1966b), although, despite popular opinion, the coyote rarely succeeds in downing an adult, healthy deer. Fawns and sickly or undernourished individuals, however, are fair game. Ozoga and Harger described coyotes chasing white-tailed deer for distances as far as 2.8 mi (4.5 km). The size of the deer herd in the Upper Peninsula indicates that this predation is not serious. The majority of deer meat eaten by Michigan coyotes is in the form of carrion (Field, 1970; Ozoga and Harger, 1966b). In late autumn, winter, and early spring, this carrion is available as a result of deaths caused by nutritional and other natural causes, and by hunters who are unable to retrieve felled animals. To distinguish the bodies of deer actually killed by coyotes from those fed on as carrion, Ogle (1971) noted that coyote kills show: (1) large patches of hide leading to the carcass; (2) separation of the vertebral column in the thoracolumbar (end-of-rib) area in adults, and at the atlas (where the head joins the neck) in fawns;

(3) nasal and maxillary bones chewed away; (4) ribs, vertebrae, and scapulae chewed; and (5) limbs widely scattered. Studies in Michigan by Ozoga and Harger (1966b) pointed out that coyote attacks on white-tailed deer are generally aimed at the head or neck.

Coyotes are known to harass calves of moose (Krefting, 1969) and wapiti (Moran and Ozoga, 1965). Krefting found that before the coyote disappeared from Isle Royale, a major item in its diet was moose carrion. Wapiti carrion, when available in Michigan, is undoubtedly eaten by coyotes, as it is in western states (Murie, 1951; Young and Jackson, 1951).

In addition to the assortment of mammalian foods (Table 18) eaten by Michigan coyotes, there is abundant evidence that these animals will eat just about anything. Birds reported as eaten include great blue heron, black duck, American goldeneye, red-breasted merganser, greater prairie chicken, ruffed grouse, herring gull, and barred owl; cold-blooded vertebrates eaten include sturgeon, green frog, leopard frog, snapping turtle, and common garter snake; plants eaten include leaves of balsam fir and white cedar, sarsaparilla, strawberry, beechnut, thimbleberry seeds, and apple; plus numerous unidentified plant materials, insects, and other invertebrates (Dearborn, 1932; Krefting, 1969; Miller, 1935; Ozoga and Harger, 1966b; Ozoga and Phillips, 1964; Stabler, 1951). Coyotes are even known to catch fish (Springer, 1980).

*Table 18. Partial list of mammalian foods eaten by Michigan coyotes, as reported in the literature.**

MAMMALS:
Arctic Shrew (*Sorex arcticus*)
Masked Shrew (*Sorex cinereus*)
Eastern Chipmunk (*Tamias striatus*)
Gray Squirrel (*Sciurus carolinensis*)
Red Squirrel (*Tamiasciurus hudsonicus*)
Beaver (*Castor canadensis*)
Deer Mouse (*Peromyscus maniculatus*)
Southern Red-Backed Vole (*Clethrionomys gapperi*)
Meadow Vole (*Microtus pennsylvanicus*)
Muskrat (*Ondatra zibethicus*)
Meadow Jumping Mouse (*Zapus hudsonius*)
Porcupine (*Erethizon dorsatum*)
Snowshoe Hare (*Lepus americanus*)
Eastern Cottontail (*Sylvilagus floridanus*)
Wapiti (*Cervus elaphus*)**
White-Tailed Deer (*Odocoileus virginianus*)**
Moose (*Alces alces*)**
Domestic Sheep (*Ovis aries*)**
Domestic Cow (*Bos taurus*)**

*Dearborn (1932), Krefting (1969), Manville (1950), Miller (1935), Moran (1973), Ozoga and Harger (1966b), Ozoga and Phillips (1964), Schofield (1959), Stebler (1951).

**Chiefly as carrion.

ENEMIES. In presettlement days, the coyote lived in association with the larger carnivores—the gray wolf and the puma—which are physically capable of dispatching adult coyotes (Stenlund, 1955; Young *in* Young and Jackson, 1951). Young coyotes are sufficiently small to be preyed upon by golden eagles, and perhaps by lynx and bobcat. Today, the coyote is the dominant terrestrial carnivore on most of the continent, with the human as its greatest enemy.

PARASITES AND DISEASES. Coyotes are infested with a diverse assortment of internal and external parasites (Bekoff, 1977, 1978; Gier and Ameel, 1959; Young and Jackson, 1951), many of which are common to domestic dogs. Internally, there are roundworms (nematodes) such as filaria, hookworms, whipworms, heartworms, and pinworms; several tapeworms (cestodes); flukes (trematodes); thorny-headed worms (acanthocephalans); and coccidia (protozoans). External parasites include fleas, lice, ticks, and mites. Michigan coyotes also suffer from distemper (Monson and Stone, 1976), rabies (Youatt *et al.*, 1971), and mange. The mange-mite (*Sarcoptes scabei*) can cause severe infections. Hair-loss, a common manifestation of this parasite's activity, was noted in large numbers of coyotes in Delta County in the late winter of 1975; in Alger, Delta, Chippewa, and Marquette counties in the late winter and early spring of 1979; and in Dickinson County in the winter of 1981–1982. Mange was also thought to have been responsible for the demise of coyotes in a large area of the northern part of the Lower Peninsula in the winter of 1974–1975. The effects of heartworm (*Dirofilaria immitus*), transmitted from animal to animal by a mosquito vector, are not known for the Michigan coyote, although filariasis infection is reported in the species in other states (Kazacos, 1977). There is only one authenticated record of a Michigan coyote being infected with rabies (Kurta, 1979).

MOLT AND COLOR ABERRATIONS. Coyotes undergo a rather gradual molt during the warm months (late spring to early autumn). In summer, a coat of short, harsh guard hairs develops. In early autumn, these guard hairs, as well as the underfur, lengthen and become more luxuriant in preparation for winter (Whiteman, 1940). Time of primeness is in December (Stains, 1979). White (albinistic) and black (melanistic) coyotes are reported (Green, 1947; Young *in* Young and Jackson, 1951). In Michigan, albinos have been captured in Iron and Ontonogan counties, and a black animal was shot in Marquette County (Ozoga and Harger, 1966a). There are also records of reddish coyotes being captured in Upper Peninsula Michigan. The skin of one such animal, taken in Schoolcraft County on January 30, 1963, by Lawrence Sanders, is preserved in the collections of the Museum at Michigan State University. The unusual coloration is principally the result of an inherited condition in which the guard hair is absent. An animal in this pelage is called a Sampson. Two others, very possibly with the same type of coat, were reported by the *Iron Mountain News* as being taken in the early 1960s in the Ralph area of Dickinson County by trapper Douglas Rasmussen of Channing.

ECONOMIC IMPORTANCE. No less than 4,181 publications relating to this animal's economic importance and biology are listed by Dolnick *et al.* (1976). The "voice" of the coyote (Dobie, 1961) adds a distinctive dimension to the outdoor world. The animal's predation on crop-and-forage-eating rodents is also appreciated. But to the livestock and poultry producer, the coyote is just another menace whose depredation can cause profound economic losses (Shelton, 1973). To the hunter, the coyote is one of the conspicuous meat-eaters constantly preying on the wildlife species which the hunter has gone to some expense to hunt. Consequently, the coyote has been persecuted in Michigan and elsewhere by means of bait poisoning (using strychnine and other lethal substances), trapping, trailing with hounds (Rickey, 1981), bountying, hunting from the ground or air, and other destructive means of human design.

Despite the long-time, open season on the coyote, this wily, cunning, and adaptive creature persists. In essence, it seems clear, according to Connally (*in* Bekoff, 1978), that ". . . indiscriminate killing of coyotes, with the techniques presently available, is not a very feasible means of reducing populations over large areas." Coyote depredations on sheep and other livestock can be reduced or controlled only *locally* by such practices as (a) properly disposing of dead livestock and (b) selectively eliminating the problem coyotes (Andrew and Boggess, *in* Beckoff, 1978). In northern Michigan, the coyote seems destined to remain as a top wild carnivore in the ecosystem. The gradual increase in livestock raising in the Upper Peninsula will probably be another reason for the animal's continued unpopularity.

For hundreds of years (in England beginning as early as 957 A.D.; in New England in 1659), governmental officials have considered bounties an efficient means of reducing populations of supposedly noxious animals. This solution has appealed to many sectors of human society. By the time modern environmental science began to demonstrate the bounty system's ineffectiveness at alleviating the problems of domestic livestock producers, the custom of bountying animals was widely accepted and difficult to eradicate. In Michigan, bounties were first paid on gray wolves from 1838 through 1920. Coyotes were undoubtedly included as part of the wolf catch, at least after 1910. The bounty system for wild canids and other top carnivores was abandoned from 1921 through 1934 in favor of a state-paid trapper system supervised, in part, by the U.S. Bureau of the Biological Survey (now the U.S. Fish and Wildlife Service). Major reasons for this change were the strong suspicion of fraud and the fact that the number of animals bountied each year did not seem to change appreciably, indicating that expensive bounties were causing no decrease in statewide populations. The Biennial Report of the Michigan Department of Conservation for 1921–1922 recounted some of the exorbitant amounts paid to particular bounty hunters, who allegedly imported pelts from other states for the higher Michigan bounty payments. The Biennial Report for 1927–1928 mentioned that 30 state-paid trappers were assigned to trap and/or poison predatory animals (Stebler, 1944). Catches of 714 coyotes in 1926–1927 and 751 in 1927–1928 were reported; the pelts were sold and the proceeds were placed in the Game Protection Fund.

In 1935, a Michigan legislative act again placed a bounty on coyotes—$15.00 for males and $22.00

for females. Although many stockmen and hunters held fast to their belief that coyotes could be controlled by the bounty system, increased pressure was placed on the Michigan legislature in the 1960s and 1970s to do away with this ineffective and expensive practice. Citizen groups joined forces with governmental conservation agencies to back this repeal move, as more and more ecologic data were accumulated to demonstrate the futility of the bounty program (Arnold, 1971).

In 1978, when raw fur prices were rising, the Michigan Department of Natural Resources began retaining the bountied coyote pelts. In 1978, coyote pelts brought a high of $95.00 at the Amasa Fur Auction near Crystal Falls; in 1981, a high of $75.00 per pelt was reported. As a result, coyote trappers became less interested in bountying their pelts, and more interested in selling them to fur buyers; the number bountied fell from 3,075 in 1976 to 709 in 1979. In March of 1980, the Michigan bounty on coyotes was finally removed by legislative action. From the resumption of the bounty in 1935 to its termination in early 1980, 133,496 coyotes were bountied, costing the state $2,288,610 in payments. The average annual catch was 2,967; the maximum of 4,328 was bountied in 1960 and the minimum of 709 in 1979. The economic importance of the coyote as a member of the Michigan outdoor community is illustrated by reviewing the history of the bounty as it was influenced by such issues as prey-predator ecologic studies, public education, citizen opinion, the fluctuating deer herd, the livestock industry, hunters and trappers, the fur market, and, of course, politics.

COUNTY RECORDS FROM MICHIGAN. Records of specimens of the coyote in collections of museums, from the literature, and reported by personnel of the Michigan Department of Natural Resources show that this species now occurs in all counties in the Upper Peninsula and the northern half of the Lower Peninsula, and previously (but rarely today), in the southern tiers of counties in the Lower Peninsula (see Map 46). This ubiquitous canid is also known from various islands in the Great Lakes including: in Lake Michigan, Summer (Nellis, 1970), Beaver (Ozoga and Phillips, 1964), and Garden (Phillips *et al.,* 1965); in Lake Superior, Isle Royale from 1906 until 1958 (Krefting, 1969), and Huron (Corin, 1976); in Lake Huron, Drummond (Manville, 1950) and Bois Blanc (Cleland, 1966).

LITERATURE CITED

Alcorn, J. R.
1946 On the decoying of coyotes. Jour. Mammalogy, 27(2): 122–126.

Andelt, D. F., D. P. Althoff, and P. S. Gipson
1979 Movements of breeding coyotes with emphasis on den site relationships. Jour. Mammalogy, 60(3):568–575.

Andelt, W. F., D. P. Althoff, R. M. Case, and P. S. Gipson
1980 Surplus-killing by coyotes. Jour. Mammalogy, 61(2): 377–378.

Andelt, W. F., and P. S. Gipson
1979 Home range, activity, and daily movement of coyotes. Jour. Wildlife Mgt., 43(4):944–951.

Arnold, D. A.
1971 The Michigan bounty system. Michigan Dept. Nat. Res., Wildlife Div., Info. Circ. No. 162, (rev.) (mimeo.) 3 pp.

Bekoff, M.
1977 Canis latrans. American Soc. Mammalogists, Mammalian Species, No. 79, 9 pp.
1978 Coyotes. Biology, behavior, and management. Academic Press, New York. xx+384 pp.

Bekoff, M., and M. C. Wells
1980 The social ecology of coyotes. Sci. American, 242(4): 130–148.

Bowen, W. D.
1982 Home range and spatial organization of coyotes in Jasper National Park, Alberta. Jour. Wildlife Mgt., 46(1):201–216.

Bowen, W. D., and I. M. Cowan
1980 Scent marking in coyotes. Canadian Jour. Zool., 58(4): 473–480.

Chesness, R. A.
1973 Ecology of the coyote (*Canis latrans*) in northern Minnesota. Minnesota Dept. Nat. Res., Game Res. Proj., Quart. Prog. Rept., 33(3):151–157.

Cleland, C. E.
1966 The prehistoric animal ecology and ethnozoology of the Upper Great Lakes Region. Univ. Michigan, Mus. Anthro., Anthro. Papers, No. 29, x+294 pp.

Connolly, G. E.
1978 Predator control and coyote populations: A review of simulation models. Pp. 327–345 *in* M. Bekoff, ed. Coyotes. Biology, behavior, and management. Academic Press, New York. xx+384 pp.

Corin, C. W.
1976 The land vertebrates of the Huron Island, Lake Superior. Jack-Pine Warbler, 54(4):138–147.

Cottam, C.
1945 Speed and endurance of the coyote. Jour. Mammalogy, 26(1):94.

Dearborn, N.
1932 Foods of some predatory fur-bearing animals in Michigan. Univ. Michigan, Sch. For. and Conserv., Bull. No. 1, 52 pp.

deVos, A.
1964 Range changes of mammals in the Great Lakes Region. American Midl, Nat., 71(1):210–231.

Dobie, J. F.
1961 The voice of the coyote. Univ. Nebraska Press, Lincoln. 386 pp.

Dolnick, E. H., R. L. Medford, and R. J. Schied
1976 Bibliography on the control and management of the coyote and related canids with selected references on animal physiology, behaviour, control methods and reproduction. Agric. Res. Serv., Beltsville, Maryland, 248 pp.

Fagre, D. B., B. A. Butler, W. E. Howard, and R. Teranishi
1981 Behavioral responses of coyotes to selected odors and tastes. World Furbearer Conf. Proc., 2:966–983.

Field, R. J.
1970 Scavengers feeding on a Michigan deer carcass. Jack-Pine Warbler, 48(2):73.

Fox, M. W.
1975 Evolution of social behavior in canids. Pp. 429–460 *in* M. W. Fox, ed. The wild canids. Their systematics, behavioral ecology and evolution. Van Nostrand and Reinhold Co., New York. xvi+508 pp.

Gier, H. T.
1957 Coyotes in Kansas. Kansas State Col., Agric. Exp. Sta., Bull. 393, 97 pp.
1968 Coyotes in Kansas. Revised. Kansas State Col., Agric. Exp. Sta., Bull. 393, 118 pp.
1975 Ecology and behavior of the coyote (*Canis latrans*). Pp. 247–262 *in* M. W. Fox, ed. The wild canids. Their systematics, behavioral ecology and evolution. Van Nostrand and Reinhold Co., New York. xvi+508 pp.

Gier, H. T., and D. J. Ameel
1959 Parasites and diseases of Kansas coyotes. Kansas State Col., Agric. Exp. Sta., Tech. Bull. 91, 34 pp.

Gipson, P. G.
1977 Abortion and consumption of fetuses by coyotes following abnormal stress. Southwestern Nat., 21(4): 558–559.

Gipson, P. S., and J. A. Sealander
1972 Home range and activity of the coyote (*Canis latrans frustror*) in Arkansas. Proc. 26th Ann. Conf. Southeastern Assoc. Game and Fish Comm., pp. 82–95.

Goldman, E. A.
1930 The coyote—archpredator. Jour. Mammalogy, 11(3): 325–334.

Golley, F. B.
1962 Mammals of Georgia. Univ. Georgia Press, Athens. xii+218 pp.

Green, D. D.
1947 Albino coyotes are rare. Jour. Mammalogy, 28(1):63.

Kazacos, K. R.
1977 *Dirofilaria immitis* in wild Canidae in Indiana. Proc. Helm. Soc., Washington, 44(2):233–234.

Kennelly, J. J., and B. E. Johns
1976 The estrous cycle of coyotes. Jour. Wildlife Mgt., 40(2):272–277.

Knowlton, F. F.
1972 Preliminary interpretations of coyote population mechanics with some management implications. Jour. Wildlife Mgt., 36(3):369–382.

Korschgen, L. J.
1952 A general summary of the food of Missouri predatory and game animals. Missouri Conserv. Comm., 61 pp.

Krefting, L. W.
1969 The rise and fall of the coyote on Isle Royale. Naturalist, 20(4):25–31.

Kurta, A.
1979 Bat rabies in Michigan. Michigan Acad., 12(2):221–230.

Laundre, J. W.
1981 Temporal variation in coyote vocalization rates. Jour. Wildlife Mgt., 45(3):767–769.

Linhart, S. B., and F. F. Knowlton
1967 Determining age of coyotes by tooth cementum layers. Jour. Wildlife Mgt., 31(2):362–365.
1975 Determining the relative abundance of coyotes by scent station lines. Wildlife Soc. Bull., 3(3):119–124.

Manville, R. H.
1950 The mammals of Drummond Island, Michigan. Jour. Mammalogy, 31(3):358–359.
1953 Longevity of the coyote. Jour. Mammalogy, 34(3):390.

Mech, L. D.
1974 *Canis lupus*. American Soc. Mammalogists, Mammalian Species, No. 37, 6 pp.

Miller, H. J.
1935 Work on coyote routes of travel, food, etc., Muniscong State Park and vicinity (winter of 1934–35). Michigan Dept. Conserv., Game Div., unpubl. rept., 18 pp.

Monson, R. A., and W. B. Stone
1976 Canine distemper in wild carnivores in New York. New York Fish and Game Jour., 23(2):149–154.

Moran, R. J.
1973 The Rocky Mountain elk in Michigan. Michigan Dept. Nat. Res., Wildlife Div., Res. and Dev. Rept. No. 267, x+93 pp.

Moran, R. J., and J. J. Ozoga
1965 Elk calf pursued by coyotes in Michigan. Jour. Mammalogy, 46(3):498.

Murie, O. J.
1951 The elk of North America. Stackpole Co., Harrisburg, Pennsylvania. 376 pp.

Nellis, C. H.
1970 Mammals of Summer Island, Michigan. Summer Sci. Jour., 2(2):66–67.

Nellis, C. H. and L. B. Keith
1976 Population dynamics of coyotes in central Alberta, 1964–68. Jour. Wildlife Mgt., 40(3):389–399.

Ogle, T. F.
1971 Predator-prey relationships between coyotes and white-tailed deer. Northwest Sci., 45:213–218.

Ozoga, J. J.
1963 An ecological study of the coyote on Beaver Island, Lake Michigan, Michigan. Michigan State Univ., unpubl. M.S. thesis, 64 pp.

Ozoga, J. J., and E. M. Harger
1966a Occurrence of albino and melanistic coyotes in Michigan. Jour. Mammalogy, 47(2):339–340.
1966b Winter activities and feeding habits of northern Michigan coyotes. Jour. Wildlife Mgt., 30(4):809–818.

Ozoga, J. J., and C. J. Phillips
1964 Mammals of Beaver Island, Michigan. Michigan State Univ., Publ. Mus., Biol. Ser., 2(6):305–348.

Phillips, C. J., J. J. Ozoga, and L. C. Drew
1965 The land vertebrates of Garden Island, Michigan. Jack-Pine Warbler, 43(1):20–25.

Richens, V. B., and R. D. Hugie
1974 Distribution, taxonomic status, and characteristics of coyotes in Maine. Jour. Wildlife Mgt., 38(8):447–454.

Richey, D.
1981 Hunting Michigan's coyotes. Michigan Out-of-Doors, 35(12):40–41, 54.

Roberts, J. D.
1978 Variation in coyote age determination from annuli in different teeth. Jour. Wildlife Mgt., 42(2):454–456.

Roughton, R. D., and M. W. Sweeny
1982 Refinements in scent-station methodology for assessing trends in carnivore populations. Jour. Wildlife Mgt., 46(1):217–229.

Schofield, R. D.
1959 Summary report on coyote trailing studies in Michigan. Michigan Dept. Conserv., Game Div., Rept. No. 2229 (mimeo), 4 pp.

Shelton, M.
1973 Some myths concerning the coyote as a livestock predator. Bioscience, 23(12):719–720.

Smith, G. J., J. R. Cary, and O. J. Rongstad
1981 Sampling strategies for radio-tracking coyotes. Wildlife Soc. Bull., 9(2):88–93.

Sperry, C. C.
1941 Food habits of the coyote. U.S.D.I., Fish and Wildlife Serv., Wildlife Res. Bull. 4, iv+70 pp.

Springer, J. T.
1980 Fishing behavior of coyotes on the Columbia River, southcentral Washington. Jour. Mammalogy, 61(2): 373–374.

Stains, H. J.
1979 Primeness in North American furbearers. Wildlife Soc. Bull., 7(2):120–123.

Stebler, A. M.
1944 Fox and coyote trapping simplified. Michigan Dept. Conserv., 15 pp.
1951 The ecology of Michigan coyotes and wolves. Univ. Michigan, unpubl. Ph.D. dissertation, x+198 pp.

Stenlund, M. H.
1955 A field study of the timber wolf (*Canis lupus*) on the Superior National Forest, Minnesota. Minnesota Dept. Conserv., Tech. Bull. No. 4, 55 pp.

Teranishi, R., E. L. Murphy, D. J. Stern, W. E. Howard, and D. F. Fagre
1981 Chemicals useful as attractants and repellants for coyotes. World Furbearer Conf. Proc., 2:1839–1851.

Todd, A. W., L. B. Keith, and C. A. Fischer
1981 Population ecology of coyotes during a fluctuation of snowshoe hares. Jour. Wildlife Mgt., 45(3):629–640.

Wells, M. C., and M. Bekoff
1981 An observational study of scent-making in coyotes, *Canis latrans*. Animal Behav., 29(2):332–350.
1982 Predation by wild coyotes: Behavioral and ecological analyses. Jour. Mammalogy, 63(1):118–127.

Whiteman, E. E.
1940 Habits and pelage changes in captive coyotes. Jour. Mammalogy, 21(4):435–438.

Wood, N. A.
1914 An annotated check-list of Michigan mammals. Univ. Michigan, Mus. Zool., Occas. Papers No. 4, 13 pp.
1922 The mammals of Washtenaw County, Michigan. Univ. Michigan, Mus. Zool., Occas. Papers No. 123, 23 pp.

Wood, N. A., and L. R. Dice
1923 Records of the distribution of Michigan mammals. Michigan Acad. Science, Arts and Ltrs., 3:425–469.

Youatt, W. G., J. N. Stuht, and H. D. Harte
1971 Wildlife rabies in Michigan 1958–1970. Michigan Dept. Nat. Res., Res. and Dev. Rept. No. 229, (mimeo) 6 pp.

Young, S. P., and H. H. T. Jackson
1951 The clever coyote. Stackpole Co., Harrisburg, Pennsylvania. vx+411 pp.

GRAY WOLF
Canis lupus Linnaeus

NAMES. The generic name *Canis*, proposed by Linnaeus in 1758, is of Latin origin and means "dog." The specific name *lupus*, also proposed by Linnaeus in 1758, is the Latin word for "wolf." In

Michigan, this canid is usually called wolf or sometimes timber wolf. The subspecies *Canis lupus lycaon* Schreber occurs in Michigan.

RECOGNITION. Body size large and massive with dimensions of a full-grown German shepherd; head and body averaging 45 in (1,150 mm) long; tail less than one-third of the total length and bushy, averaging 18 in (450 mm) in vertebral length; head large with muzzle broad, more so than in coyote (Fig. 66), nose pad usually more than 1¼ in (33 mm) wide; ears erect, slightly less pointed and conspicuous than those of the coyote;

Figure 66. The gray wolf (*Canis lupus*) showing its large head and broad muzzle. Sketch by Bonnie Marris.

body lanky with narrow chest; limbs long and slender; feet large and broad; heel pad of front foot at least 1½ in (38 mm) wide; digits terminating in blunt, heavy, and nonretractile claws; front feet with five digits, with the small pollex high on the inside of the forelegs; hind feet with four digits; scent gland at upper base of tail prominent and covered with bristly hair; mammary glands consist of five pairs; males average about 10% larger than females in both external and cranial dimensions.

The short, soft underfur is covered by long, coarse guard hairs, which usually are black-tipped with variable amounts of cinnamon or buff on the subterminal shafts. In overall view, wolves vary considerably in color with gray tones being most prevalent, usually lacking are yellow to reddish hues which are often common in coyotes. On close inspection, the head and ears of the gray wolf show some buff to cinnamon tones; a mixture of black covers the back from neck to rump; white to buffy coloring is on the underparts, being darker on the chin; the gray of the dorsal side of the tail is mixed with black which is more prominent on the tip; the ventral surface of the tail is more buff in color; and cinnamon coloring adorns the lateral sides of the legs.

The heavy build and general appearance easily distinguish the gray wolf from the smaller and more slightly built coyote. Although the gray wolf normally runs with its tail carried high (as do domestic dogs), instead of low like the coyote (Fig. 67), the two can be confused at a distance. (See the account of the coyote for more distinctive characteristics.)

MEASUREMENTS AND WEIGHTS. Adult male and female gray wolves measure (on the average for animals from Minnesota, Van Ballenberghe, 1977), respectively, as follows: length from tip of nose to end of tail vertebrae, 61 and 59 in (1,545 and 1,486 mm); length of tail vertebrae, 17 and 16 in (433 and 409 mm); length of hind foot, 10½ and 10¼ in (268 and 260 mm); height of ear from notch, 4¾ and 4½ in (121 and 115 mm); weight, 67.3 and 57.9 lbs (30.6 and 26.3 kg). A mature male (MSU No. 10596) taken in 1965 in Chippewa County had the following measurements: 65-18-10.5-5.4 (1,645-453-268-135). When taken on August 31, this wolf weighed 67.5 lbs (30.7 kg) with its viscera removed. An individual weighing 90 lbs (41 kg) was taken in Alger County in 1907 (Wood, 1917); one weighing 112 lbs (46 kg) was captured in Minnesota (Stenlund, 1955).

DISTINCTIVE CRANIAL AND DENTAL CHARACTERISTICS. The skull of the gray wolf is large, moderately elongate, heavily constructed, and has massive teeth, including canines short and thick in comparison with those of the coyote (Fig. 68). The front end (rostrum) is heavily formed and long. The sagittal crest is prominent and developed posteriorly, particularily in mature males. The auditory bullae are large and elongate. View-

Figure 67. The gray wolf (*Canis lupus*).
Photograph by Robert Harrington and courtesy of the Michigan Department of Natural Resources.

ing the skull with attached lower jaw from the front, the ends of the upper canines do not extend (as in the case of the coyote) below a line drawn between the anterior mental foramina on either side of the lower jaw (see Fig. 65). In both jaws, the premolars are noticeably widely spaced. The skull is conspicuously larger than that of the coyote. The skull of the gray wolf averages 9¾ in (245 mm) long and 5¼ in (135 mm) wide (across zygomatic arches). Again, the males average about 10% larger in cranial dimensions than females. The dental formula for the gray wolf is:

$$\text{I (incisors)}\frac{3\text{-}3}{3\text{-}3}, \text{C (canines)}\frac{1\text{-}1}{1\text{-}1}, \text{P (premolars)}\frac{4\text{-}4}{4\text{-}4}, \text{M (molars)}\frac{2\text{-}2}{3\text{-}3} = 42$$

DISTRIBUTION. The gray wolf has a circumpolar distribution, which means a range encompassing most lands in the northern part of the globe. In Eurasia, the gray wolf occurs from the arctic regions to central Europe and Arabia, and east to India and southern China. In the New World, this canid also occurs throughout the arctic and formerly occurred southward throughout the United States (except for the southeast) and into central Mexico. Because it is unwelcome near human settlement and in livestock-raising areas, the species has been extirpated from almost all of the southern part of its North American range (Mech, 1970; Mech and Rausch, 1975; Young and Goldman, 1944).

Immediately following the last glacial melt-back (Holman, 1975), the gray wolf thrived in all parts of the Upper Great Lakes Region and has been successful in persisting to the present in the northern parts of this area. Sparse human settlements, limited range for livestock, and areas devoted chiefly to timber, wildlife, and recreation have allowed the gray wolf to remain. In Michigan, however, its fortunes have steadily declined (Wood and Dice, 1924). At the time of first settlement, the gray wolf occurred throughout the state (Stebler, 1951); in the early years of the eighteenth century, the animal was reported in the Detroit area by Cadillac (Johnson, 1919) and in the vicinity of Fort St. Joseph in what is now Berrien County (Baker, 1899).

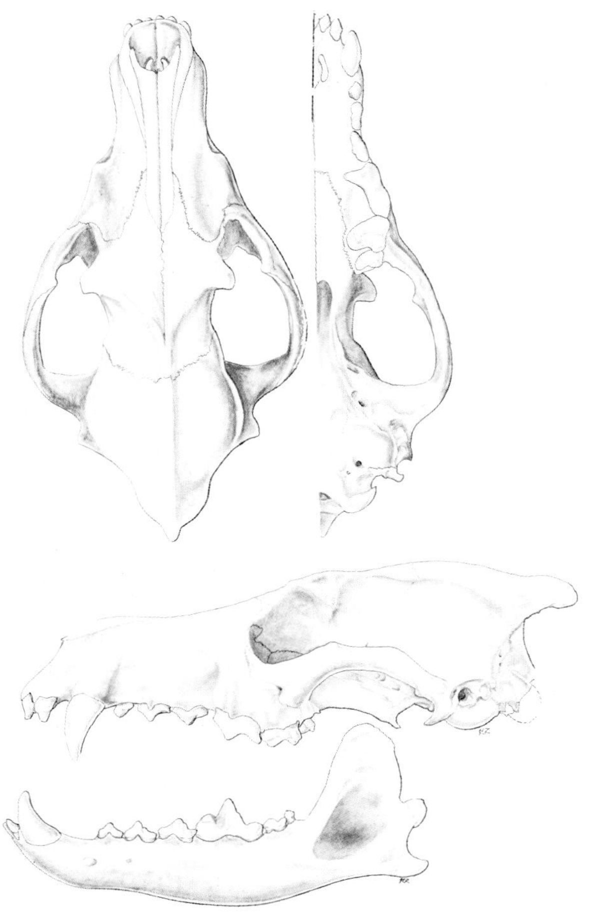

Figure 68. Cranial and dental features of the gray wolf (*Canis lupus*).

By the time a bounty was established in 1838, the gray wolf was rapidly being eliminated in the southern part of the Lower Peninsula (Stebler, 1944). Although many of the reports may have been for the coyote, gray wolves lingered in remote parts of the northern Lower Peninsula until at least 1907 or 1910 (Raymond Schofield *in* Stewart and Negus, 1961). In Upper Peninsula Michigan, however, the gray wolf has survived into the fourth quarter of the twentieth century, with its range and numbers gradually shrinking. In 1955, the total population in Upper Peninsula Michigan was estimated at no more than 100 individuals (Arnold, 1955; Arnold and Schofield, 1956). The number of gray wolves bountied began to dwindle markedly: in 1956, the total was 30; in 1957, 7; in 1958, 7; in 1959, 1; and in 1960, the bounty was repealed. In 1965 the species was given full protection under the law for the first time, but perhaps too late (Douglass, 1970). In 1964 and 1965, published stories suggested that the total population could be as high as 25 or as low as 12, with sightings in 1965 in at least nine counties: Alger, Baraga, Chippewa, Delta, Gogebic, Iron, Luce, Ontonagon, and Schoolcraft. In 1973, Hendrickson *et al.* (1975) estimated the total population as approximately six.

In March 1974, four wolves from northwestern Minnesota, equipped with collars containing radio transmitters, were released in Marquette County in an attempt to revive the waning Michigan stock (Weise *et al.*, 1975). This release had failed by November: two were reported shot, one was trapped, and the fourth was killed by an automobile (Robinson and Smith, 1977). Subsequent to that year, the slim reports of Upper Peninsula gray wolves included the following killings: one in 1974 and one in 1975, both in Menominee County; one in 1976 in Dickinson County; and one in 1980, north of Amasa in Iron County. Sightings and/or tracks of animals thought to be gray wolves were reported in the summer of 1974 in Schoolcraft County; in November 1977 in Iron County; in January 1978 in Delta County; in early 1980 in Mackinac County; and in the winter of 1980–81 in Iron County. Crossings from Canada undoubtedly supply some of the recent records, at least in the eastern Upper Peninsula. In March, 1981, for example, William Jensen found tracks of two gray wolves on Drummond Island. These apparently crossed False Detour Channel from Cockburn Island, Ontario, where a pack of gray wolves occurs (William Robinson, pers. comm.). Gray wolves may also presently frequent other islands in the Superior-Huron Waterway Connection, including Sugar Island (Lawrence Master, pers. comm.). Even with recent records, it seems likely that by the late 1960s the native population of Upper Peninsula wolves, without adequate recruits crossing from Canada, northeastern Wisconsin (Mech and Nowak, 1981), or from introduced individuals, may have been reduced to so few individuals that the ability to propagate has been lost (see Map 47).

Although there is some evidence that wolves may have occasionally visited Isle Royale in the early part of the twentieth century, the species did not become established there until sometime in 1948 or early 1949 (Allen, 1979; Mech, 1966b),

Map 47. Geographic distribution of the gray wolf (*Canis lupus*). Formerly found statewide (pale shading); in the mid-1970s the gray wolf had only patchy distribution in the Upper Peninsula and on Isle Royale (dark shading).

evidently by crossing the ice of Lake Superior from the Ontario mainland 14 mi (22.5 km) to the north.

HABITAT PREFERENCES. Folklore generally places the stealthy, large-fanged gray wolf in the deep forest. However, early travelers in the American West found the species following herds of bison far out on the prairie, while explorers in the Arctic attracted curious wolves on the bleak and wind-swept tundra (Munthe and Hutchinson, 1978). Nevertheless, in presettlement Michigan the gray wolf probably was a forest creature which also frequented openings, lakeshores, and riparian growth along streams. Winter lake ice must have been a means of passage, just as it is today at Isle Royale (Mech, 1966b). Another basic habitat requirement seems to be the need for extensive environment encroached on, but to a very minimum, by the human species and by large game species and/or free-ranging domestic livestock (Stebler, 1944). Except for sizable Isle Royale, such wilderness areas have all but disappeared in Michigan. As shown by long-term studies (Allen, 1979; Mech, 1966b; Peterson, 1977), the gray wolf on Isle Royale frequents all habitats, traveling game trails along forest edges and at lake shorelines. As a top carnivore, its special environmental preferences may be overshadowed by its need to pursue moose and other highly mobile prey. This is clearly demonstrated by accounts of the variety of environments in which Michigan gray wolves were sighted or killed in pioneer days (Perry, 1899, etc.).

DENSITY AND MOVEMENTS. Gray wolves need large amounts of animal protein—10 to 14 lbs (4.5 to 5.9 kg) per day (Mech, 1966b). To be assured of a continual supply of food (primarily from kills of white-tailed deer and other large herbivores), gray wolves, whether they operate singly, in family groups, or in packs, require great expanses which can support prey populations sufficiently large so as not to be markedly decimated by the continual predation. In more than 20 years of study of the gray wolf and its prey species on Isle Royale, Professor Durward Allen and his students at Purdue University (Allen, 1979; Mech, 1966b; Peterson, 1977; plus annual reports) found that the total population ranged from just below 20 to just above 40 individuals. This indicates a density of one individual per 5.2 to 10.5 sq mi (13.6 to 27.2 per sq km). Similar data were obtained by deVos (1950) on the Sibley Peninsula in Ontario just north of Isle Royale. These figures may also closely approximate densities in early Michigan days (Jackson, 1961).

The movement of individuals and groups of gray wolves are best followed in winter when snow cover allows for tracking from the ground and actual observation from the air. Radio collars have also been used (Kalenosky and Johnston, 1967). In the Upper Great Lakes Region, there are movement data for animals in Wisconsin (Thompson, 1952), in Minnesota (Mech, 1977b; Mech *et al.*, 1971; Stenlund, 1955; Van Ballenberghe *et al.*, 1975), in Ontario (deVos, 1950; Joslin, 1967; Kalenosky and Johnston, 1967), and in Michigan (Allen, 1979; Mech, 1966b; Peterson, 1977; Stebler, 1944, 1951). In the Cusino Wildlife Area in Alger and Schoolcraft counties (Stebler, 1944), a family of gray wolves (two adults and two young) cruised over a winter range of approximately 260 sq mi (673 sq km). On Isle Royale (Mech, 1966b), a pack of 15 to 21 animals used 210 sq mi (544 sq km). In marked contrast, winter territories of gray wolf packs varied in size from 20 to 56 sq mi (52 to 145 sq km) in northeastern Minnesota (Mech *et al.,* 1971; Van Ballenberghe *et al.,* 1975). In summer, movements may be less; one radio-tagged pack used only 8.1 sq mi (21 sq km) in Ontario (Kalenosky and Johnston, 1967).

Groups of wolves may travel as much as 45 mi (72 km) in a single day (Stenlund, 1955). However, there is considerable variation in this mobility; in studies in northeastern Minnesota, gray wolves moved no more than 12.8 mi (20 km) per day (Mech *et al.,* 1971). In winter on Isle Royale, a pack moved (between kills) an average of 31 mi (50 km) per day at a speed of 5 mi (8 km) per hour (Mech, 1966b).

BEHAVIOR. The gray wolf lives, hunts, and reproduces in distinct family units or packs. Fox (1975) classed the species as a social hunter (pack type). No other Michigan canid has this trait (Mech, 1970). The pack usually consists of a pair of wolves, their yearling or two-year-old offspring, and possibly members of other family groups or bachelors (Mech, 1966b; Seton, 1953; Stenlund, 1955; Young and Goldman, 1944). The size of the pack is thought to be limited by the availability of food (Murie, 1944). In the Upper Great Lakes Region, a pack, efficient as a hunting unit, may include five to seven animals, although in Minnesota, groups as large as 15 or even 30 have been reported (Mech, 1966b; Wolfe and Allen, 1973). The complex hierarchy and the elaborate behavioral patterns which evolve in this close association have been examined in both captive and wild-living animals (Klinghammer, 1978; Rabb *et al.,* 1967; Zimen, 1975).

Social rank is all-important in pack activities, especially courtship and mating. Leadership is controlled by a large, strong male (the alpha male) with submissive members of this sex following. The leader's mate (the alpha female) dominates submissive females, with the pups at the bottom of the social scale. Although subordinate males and females may not be involved with reproductive functions, they appear to have responsibility for helping the high-ranking female care for her offspring. Intrapack aggression consists of rather ritualized behavior which promotes the social order and subdues direct conflicts which might disturb group stability (Marhenke, 1971). In hunting, food gathering, traveling, and even at play, the alpha animals exert tight control (Mech, 1966a). Mutual respect and affection are displayed among members; included are friendly brawls, romps and other play, grooming and caressing, tail wagging, and grimaces. The pack organization not only helps maintain group integrity but acts to defend hunting territory, with the use of scent-marking (Peters and Mech, 1975; Rothman and Mech, 1979), howling (Harrington and Mech, 1978), and agonistic behavior during occasional confrontations between neighboring packs. In winter, however, two or more packs may join forces to hunt.

The gray wolf is mostly active at night or in dim light, especially in summer (Jackson, 1961). From December through April, between 0900 and 1800 hours each day, Mech *et al.* (1971) noted that radio-tagged gray wolves, on the average, rested 62% of the time, traveled 28%, and fed 10%. The gray wolf can trot or run with a lumbering, bounding gait, with the tail held slightly elevated (Fig. 67). The animal can trot at a speed of 4 mi (6.4 km) per hour (Pruitt, 1960) and run at speeds of between 18 and 28 mi (29 to 45 km) per hour. The gray wolf is not as fast as the smaller coyote but seems to have greater endurance (Jackson, 1961). Like other canids, the gray wolf is not averse to water and is a good swimmer (Young and Goldman, 1944).

The gray wolf's distinctive howl is a long, guttural, quavering wail. As it can be to north woods travelers today, this was a dismal and fearsome cry to Michigan's pioneering families. In a two-day trip by ox-drawn wagon from Vermontville to Bellevue in the 1840s, the E. W. Barber family discovered that even their whispered conversations in sylvan campsites elicited howls from gray wolves prowling in the dark nearby. No attacks or harm resulted, although nerves were frayed. Modern biologists have distinguished two types of vocal expressions; one when the animals are together and another when they are widely separated (Theberge, 1971; Theberge and Falls, 1967). In groups, the animals utter low-volume squeaks, whines, barks, and growls. These sounds are almost always connected with other signals such as body and tail posture, facial expressions, emission of odors, and tactile stimulation. When separated, howling seems to be the principal means of pack communication. Vocalizations uttered during the chase and at the time of a kill have also been identified (Young and Goldman, 1944).

Gray wolves use dens for rearing young in the breeding season and also as shelters in winter. The nursery den is often a ground cavity excavated by the animals themselves or an enlarged burrow of a woodchuck, badger, or fox (Seton, 1953; Young and Goldman, 1944). Captive animals dug dens with entrance tunnels up to 72 in (183 cm) long and 16 in (41 cm) in diameter with a slope of about 45° to a depth of almost 20 in (50 cm); horizontal tunnels at that depth were as extensive as 123 in (312 cm) long and 24 in (61 cm) in diameter; and terminal chambers were 50 in (127 cm) long, 45 in (117 cm) wide, and 14 in (36 cm) high (Ryon, 1977). Other den sites for the rearing of young are located in hollow logs, under ledges, and beneath piles of debris. Most den sites are in well-drained situations, often with commanding views of the surroundings, and in the vicinity of travel routes (Jackson, 1961; Joslin, 1967; Young and Goldman, 1944). In winter, various kinds of shelters may be used—snow tunnels, under roots of fallen trees, beneath ledges, and in beaver lodges.

The gray wolf has the disposition and temperament of a wild, untrained domestic dog; a wolf raised in captivity and tamed has many of the characteristics of a dog of similar size. The gray wolf has keen senses of smell and hearing. Vision is less well developed; the animal may not be able to distinguish objects at a distance but can detect movement. In pioneer times and even in the remote arctic wastes today, gray wolves have regarded human intrusion with inquisitiveness, presumably as a result of their curious nature (Mech, 1970). Although there are popular accounts describing gray wolf attacks on humans (Peterson, 1947), most authors investigating such charges found few cases where the aggressive animal was not plagued with rabies. In modern times, the animals tend to shy away from humans (Jackson, 1961; Mech, 1970; Young and Goldman, 1944).

ASSOCIATES. The gray wolf may consider most smaller creatures in the Upper Great Lakes Region to be fair game. This canid has also been known to show marked aggression toward other large carnivores—black bear, coyote, lynx, and the now-extirpated mountain lion. Serious conflicts between these animals occur primarily over the possession of kills, irrespective of which species was involved in the killing. Stebler (1951) noted gray wolves, coyotes and red foxes feeding on a deer carcass in the Cusino area in the Upper Peninsula in early February. An array of wild creatures undoubtedly trails packs of gray wolves to share in feeding on their kills. Marten, wolverine, and bobcat are noted for this, while the common raven, gray jay, bald eagle, and black-capped chickadee also profit from this relationship (Mech, 1970; Peterson, 1977; Stenlund, 1955).

REPRODUCTIVE ACTIVITIES. It has already been pointed out that, under the rather rigid social hierarchy of the pack, pups are usually born only to the highest-ranking (alpha) female and fathered only by the dominant (alpha) male. Pack members assist in some of the rearing duties. The alpha male aggressively discourages breeding by subordinate males. Should a subordinate animal mate, it may actively be rejected from the pack (Peterson, 1979). The reproductive habits of the gray wolf in Michigan have been examined both on Isle Royale (Mech, 1966b; Peterson, 1977) and in the Cusino area of the Upper Peninsula (Stebler, 1951).

The female is receptive to mating for only a short period (less than seven days) in late winter, usually between late February and mid-March. Courtship and related behavior closely resemble that of the domestic dog (Mech, 1966b). Paired wolves evidently maintain this relationship throughout their lives. After a gestation period of approximately 63 days, a litter averaging seven (as few as four and as many as 10 or more) is born in April or May in a burrow or other protected den site prepared by the female as long as three weeks prior to the birth.

At birth, the pups have rounded heads; pug-noses; small, floppy ears; closed eyes and ear openings; and brownish gray heads with dark sooty body coloring (Jackson, 1961; Mech, 1970). From a weight of about one lb (454 g) at birth, the gray wolf may reach adult size in 12 months (Van Ballenberghe and Mech, 1975). By about 10 days of age, the eyes open and the ears take on their characteristic erect shape. At about 25 days of age, the young appear above ground. By 30 days of age, the wooly pup fur molts and an adult pelage appears. At 50 days of age, they are weaned. Following weaning, adult members of the pack feed the litter by disgorging meat from their kills. At 60 days of age, the pups may be moved to another den site. By late summer, the maturing pups are ready to join the pack in hunts. There is evidence that, in times of food shortages, wolf pack numbers decline first because of pup starvation, followed by lower pup production, and ultimately by intraspecific strife among adults (Mech, 1977b). In captivity, gray wolves may live as long as 18 years.

FOOD HABITS. The white-tailed deer is the primary food of the gray wolf in the Upper Great Lakes Region. Some observers go so far as to suggest that this predator-prey relationship is so highly developed that, in presettlement times at least, neither animal, as a species, was severely reduced (Bailey, 1907; Mech, 1966a, 1975; Mech and Frenzel, 1971; Stebler, 1944; Stenlund, 1955; Wood, 1914). In southeastern Alaska, it was estimated that each gray wolf at 7.4 deer in winter (Holleman and Stephenson, 1981). Wolf packs make a kill, consume the edible parts, and then proceed with the next hunt. On Isle Royale, a pack of 16 gray wolves hunting moose traveled an average of 23 mi (37 km) between consecutive kills (Mech, 1966b). Much has been written about wolves' hunting behavior. Mech (1970) noted that the wolves locate their potential prey by direct scenting, chance encounters, or tracking. The pack will chase prey as far as 5 mi (8 km), with Mech and Korb (1978) recording an exceptionally long chase of about 13 mi (20.8 km). Of course, many white-tailed deer and moose are able to elude the pack; in winter, however, deep snow hampers the prey, especially weakened animals, more so than the hungry canids (Mech and Frenzel, 1971). Forturnately, white-tailed deer inhabiting overlapping edges of wolf-pack territories are usually less hunted since wolf packs appear to avoid boundary encounters with each other (Mech, 1977a; Nelson and Mech, 1981).

Wolves down large game by hamstringing, attacking the hams and rump, and seizing the nose or throat. The viscera of the victim appears to be a choice food item, with the rumen cast aside. The alpha male probably gets first choice. If the kill is sufficiently large, the pack may gorge excessively, move away from the prey, take cover, and sleep off the effects of overindulgence (Young and Goldman, 1944).

It is commonly believed, although incorrectly, that the gray wolf selects fat, healthy prey animals. Instead, the very young, weak, sickly, or aged individuals are more likely to be killed (Fuller and Keith, 1980). If the gray wolf could capture the prime, healthy prey, it undoubtedly would. This prey-predator system has been functioning successfully for eons. The predator would surely suffer serious nutritional problems if the prey, always larger in number, declined (Mech and Karns, 1977).

Information about the food habits of the gray wolf has come from observing packs in action, mostly in winter, by surface or aerial tracking (Mech, 1966b). Collecting fecal materials (scats) also enables ecologists to determine what foods are eaten. Elongated and cylindrical droppings of the gray wolf resemble those of coyotes, domestic dogs, and other canids. In general, a canid scat at least 1¼ in (30 mm) in diameter is most likely that of a gray wolf as the fecal matter of a coyote is normally smaller (Weaver and Fritts, 1979). The diet of the gray wolf, based mostly on the analysis of scats, shows a variety of animal and some plant foods. In Upper Peninsula Michigan, gray wolves eat, besides white-tailed deer, shrews of the genus *Sorex,* snowshoe hare, red squirrel, deer mice, meadow voles, meadow jumping mice, ruffed grouse, crawfish and grass (Manville, 1948; Stebler, 1944; 1951). On Isle Royale, where there are no white-tailed deer, Mech (1966b), Peterson (1977) and Skelton (1966) found preferences to be moose, followed by beaver (especially in summer), and then snowshoe hare, red squirrel, deer mice, red fox, unidentified birds, and grass. Scats from other areas of the Upper Great Lakes Region revealed evidence of the eastern cottontail, woodchuck, eastern chipmunk, flying squirrel (species undetermined), southern red-backed vole, southern bog lemming, muskrat, woodland jumping mouse, porcupine, coyote, black bear, raccoon, wolverine, June beetle, and fruits and seeds of assorted plants (Boles, 1977; Quick, 1953; Rogers and Mech, 1981; Stenlund, 1955; Thompson, 1952; Voigt *et al.,* 1976; Van Ballenberghe *et al.,* 1975). The wolf is known to prey on the lynx in northern Europe (Pulliainen, 1965). Records of wolves killing domestic animals (cattle, sheep, horses, swine, etc.) in Michigan are not well documented in the early literature (Wood, 1922). However, there is no doubt that such depredations in pioneer days caused economic loss.

ENEMIES. If pups are unattended, eagles, perhaps hungry black bear, or even bobcat or lynx may kill and eat them. However, in nature such opportunities are unusual, at least in the Upper Great Lakes Region. It has been demonstrated (see section on Reproductive Activities) that packs of gray wolves under wilderness conditions tend to regulate their own numbers in accordance with the social hierarchy and the population fluctuations of major prey species.

The human constitutes the greatest threat to the gray wolf. This species' decline in Michigan has been correlated with the advance of human settlement (Stebler, 1944; 1951), and was accelerated by the bounty system which was in operation almost continuously (except from 1921 through 1934) from 1838 until 1960 (Arnold, 1971). The gray wolf is much more easily trapped than the coyote and numerous devices (Young and Goldman, 1944) have been developed to catch or kill gray wolves. Poisoning, den-hunting, trail dogs, and tracking by airplane have proven effective.

PARASITES AND DISEASES. The gray wolf shares many decimating diseases with the domestic dog—rabies (see Chapman, 1978), sarcoptic mange, distemper, tularemia, numerous internal and external parasites, and infections caused by viruses and other microorganisms (Mech, 1970; Stenlund, 1955). On Isle Royale, Mech (1966b) found the intermediate stages of two tapeworms (*Taenia hydatigena* and *Echinococcus granulosus*) in moose, the adults of which occur in gray wolves. Other tapeworms (Cestoda), flukes (Trematoda), roundworms (Nematoda), and thorny-headed worms (Acanthocephala) are known from the internal organs while fleas, mites, ticks, and lice—especially the mange mite (*Sarcoptes scabei*)—infest the body surface of this canid (Mech, 1970).

MOLT AND COLOR ABERRATIONS. The gray wolf molts once annually from late spring through summer, a process which may take several weeks (Young and Goldman, 1944). The new and rather short summer pelage of guard hairs continues to grow into the long-haired winter coat, with primeness occurring from late November to early February (Stains, 1979). The underfur, important for winter insulation, develops in autumn. Stiff tufts of hair grow between foot and toe pads in winter.

The pelage color of the gray wolf is highly variable. Individual animals may have coats ranging from almost pure white through buff, reddish, gray to black (Mech, 1970). Gray is usual for animals of the Upper Great Lakes Region, although some black individuals, but no white ones, have been reported (Van Ballenberghe, 1977). In pioneer times, a black wolf was reported in Washtenaw County (Wood, 1922). Perry (1899) wrote of seeing three black wolves on the headwaters of the Cass River in Sanilac County on September 28, 1852; they had been attracted to the carcass of a wapiti which he had shot.

ECONOMIC IMPORTANCE. The incompatibility of the gray wolf with high densities of humans and their domestic livestock has already been stressed (Llewellyn, 1978; Rausch and Hinman, 1977); however, there is also evidence that the animals live successfully in northwestern Minnesota, where there is considerable human activity (Fritts and Mech, 1981). Fortunately, the animal thrives well in zoological parks but to glimpse a gray wolf in the wild or even to hear the wolf's howls, a person must travel to areas remote from human population centers. Isle Royale retains a token population and National Park Service officials afford visitors an opportunity to hike in wolf country. Although it is regrettable that the gray wolf has, by necessity, receded with the steady encroachment of human occupation and land use, there probably will always be gray wolves in the arctic regions of North America and Eurasia. However, if market prices for wolf pelts remain high (in 1976, the best pelts brought as much as $97.00 in the Upper Great Lakes Region) and hunters can determine economical and legal ways to catch gray wolves in the frigid wastelands, arctic packs could also be overharvested.

COUNTY RECORDS FROM MICHIGAN. Records of specimens of gray wolves in collections of museums, from pioneer documents, from the literature (Stebler, 1951), and reported by personnel of the Michigan Department of Natural Resources show that the species occurred at one time in all counties in Michigan. Except for Isle Royale (Mech, 1966b), the status of the gray wolf in Michigan is bleak, with remaining residual populations perhaps too reduced and/or scattered to increase successfully by natural reproductive means.

LITERATURE CITED

Allen, D. L.
1979 Wolves of Minong. Their vital role in a wild community. Houghton Mifflin Co., Boston. xxv+499 pp.

Arnold, D. A.
1955 Status of Michigan timber wolves, 1955. Michigan Dept. Conserv., Game Div. Rept. No. 2062 (mimeo.), 2 pp.
1971 The Michigan bounty system. Michigan Dept. Nat. Res., Info. Circ. No. 162 (rev.) (mimeo.), 3 pp.

Arnold, D. A., and R. D. Schofield
1956 Status of Michigan timber wolves. Michigan Dept. Conserv., Game Div. Rept. No. 2097 (mimeo.), 2 pp.

Bailey, V.
1907 Destruction of deer by the northern timber wolf. U.S.D.A., Bur. Biol. Surv., Circ. No. 58, 2 pp.

Baker, G. A.
1899 The St. Joseph-Kankakee portage. Its location and use by Marquette, La Salle and other French voyageurs. Northern Indiana Hist. Soc., Publ. No. 1, 44 pp.

Boles, B. K.
1977 Predation by wolves on wolverines. Canadian Field-Nat., 91(1):68–69.

Chapman, R. C.
1978 Rabies: Decimation of a wolf pack in arctic Alaska. Science, 201:365–367.

deVos, A.
1950 Timber wolf movements on Sibley Peninsula, Ontario. Jour. Mammalogy, 31(2):169–175.

Douglass, D. W.
1970 History and status of the wolf in Michigan. Pp. 6–8 *in* S. E. Jorgensen, C. E. Faulkner, and L. D. Mech, eds. Proc. Symposium on Wolf Management in Selected Areas of North America, U.S.D.I., Bur. Sport Fish. and Wildlife, Region 3, 50 pp.

Fox, M. W.
1975 Evolution of social behavior in canids. Pp. 429–460 *in* M. W. Fox, ed. The wild canids. Their systematics, behavioral ecology and evolution. Van Nostrand and Reinhard Co., New York. xvi+508 pp.

Fritts, S. H., and L. D. Mech
1981 Dynamics, movements, and feeding ecology of a newly protected wolf population in northwestern Minnesota. Wildlife Monogr. No. 80, 79 pp.

Fuller, T. K., and L. B. Keith
1980 Wolf population dynamics and prey relationships in northeastern Alberta. Jour. Wildlife Mgt., 44(3):583–602.

Harrington, F. H., and L. D. Mech
1978 Howling at two Minnesota wolf pack summer homesites. Canadian Jour. Zool., 56(9):2024–2028.

Hendrickson, J., W. L. Robinson, and L. D. Mech
1975 Status of the wolf in Michigan, 1973. American Midl. Nat., 94(1):226–231.

Holleman, D. F., and R. O. Stephenson
1981 Prey selection and consumption by Alaskan wolves in winter. Jour. Wildlife Mgt., 45(3):620–628.

Holman, J. A.
1975 Michigan's fossil vertebrates. Michigan State Univ., Publ. Mus., Ed. Bull. No. 2, 54 pp.

Jackson, H. H. T.
1961 Mammals of Wisconsin. Univ. Wisconsin Press, Madison. xii+504 pp.

Johnson, I. A.
1919 The Michigan fur trade. Michigan Hist. Comm., Univ. Ser., Vol. 5, xii+201 pp.

Joslin, P. W. B.
1967 Movements and home sites of timber wolves in Algonquin Park. American Zool., 7:279–288.

Kalenosky, C. B., and D. H. Johnston
1967 Radio-tracking timber wolves in Ontario. American Zool., 7:289–303.

Klinghammer, E., ed.
1978 The behavior and ecology of wolves. Garland STPM Press, New York. xvii+588 pp.

Llewellyn, L. G.
1978 Who speaks for the timber wolf? Trans. 43rd N. American Wildlife and Nat. Res. Conference, pp. 442–452.

Manville, R. H.
1948 The vertebrate fauna of the Huron Mountains, Michigan. American Midl. Nat., 39(3):615–640.

Marhenke, P., III
1971 An observation of four wolves killing another wolf. Jour. Mammalogy, 52(3):630–631.

Mech, L. D.
1966a Hunting behavior of timber wolves in Minnesota. Jour. Mammalogy, 47(2):347–348.
1966b The wolves of Isle Royale. National Park Serv., Fauna Ser. 7, xiv+210 pp.
1970 The wolf. The ecology and behavior of an endangered species. Natural History Press, Garden City, N.Y. xx+384 pp.
1974 *Canis lupus*. American Soc. Mammalogists, Mammalian Species, 37:1–6.
1975 Population trend and winter deer consumption in a Minnesota wolf pack. Pp. 55–83 *in* R. L. Phillips and C. Jonkel, eds. Proc. 1975 Predator Symposium. Montana Forest and Conserv. Exp. Sta., Missoula, ix+268 pp.
1977a Wolf-pack buffer zones as prey reservoirs. Science, 198 (4314):320–321.
1977b Productivity, mortality, and population trends of wolves in northeastern Minnesota. Jour. Mammalogy, 58(4):559–574.

Mech, L. D., and L. D. Frenzel, Jr.
1971 An analysis of the age, sex, and condition of deer killed by wolves in northeastern Minnesota. Pp. 35–51 *in* L. D. Mech, and L. D. Frenzel, Jr., eds. Ecological studies of the timber wolf in northeastern Minnesota. U.S.D.A., N. Central Forest Exp. Sta., Res. Paper NC-52, 62 pp.

Mech, L. D., L. D. Frenzel, Jr., R. R. Ream, and J. W. Winship
1971 Movements, behavior and ecology of timber wolves in northeastern Minnesota. Pp. 1–35 *in* L. D. Mech, and

L. D. Frenzel, Jr., eds., Biological studies of the timber wolf in northeastern Minnesota. U.S.D.A., N. Central Forest Exp. Sta., Res. Paper NC-52, 62 pp.

Mech, L. D., and P. D. Karns
1977 Role of the wolf in a deer decline in the Superior National Forest. U.S.D.A., N. Central Forest Exp. Sta., Res. Paper NC-148, 23 pp.

Mech, L. D., and M. Korb
1978 An unusually long pursuit of a deer by a wolf. Jour. Mammalogy, 59(4):860–861.

Mech, L. D., and R. M. Nowak
1981 Return of the gray wolf to Wisconsin. American Midl. Nat., 105(2):408–409.

Mech, L. D., and R. A. Rausch
1975 The status of the wolf in the United States–1973. Pp. 83–88 *in* D. H. Pimlot, ed. Wolves. Proc. First Working Meeting of Wolf Specialists and of the First International Conference on the Conservation of the Wolf. IUCN Publ., New Ser., Suppl. Paper No. 43, 145 pp.

Munthe, K., and J. H. Hutchison
1978 A wolf-human encounter on Ellesmere Island, Canada. Jour. Mammalogy, 59(4):876–878.

Murie, A.
1944 The wolves of Mount McKinley. National Park Serv., Fauna Ser. 5, 238 pp.

Nelson, M. E., and L. D. Mech
1981 Deer social organization and wolf predation in northeastern Minnesota. Wildlife Monogr., No. 77, 53 pp.

Perry, O. H.
1899 Hunting expeditions of Oliver Hazard Perry of Cleveland verbatim from his diaries. Cleveland. viii+246 pp.

Peters, R. P., and L. D. Mech
1975 Scent-marking in wolves. American Sci., 63(6):628–637.

Peterson, R. O.
1977 Wolf ecology and prey relationships on Isle Royale. National Park Serv., Sci. Monog. Serv., No. 11, xx+210 pp.
1979 Social rejection following mating of a subordinate wolf. Jour. Mammalogy, 60(1):219–221.

Pruitt, W. O.
1960 Locomotor speeds of some large northern mammals. Jour. Mammalogy, 41(1):112.

Pulliainen E.
1965 Studies on the wolf (*Canis lupus L.*) in Finland. Annales Zoologici Fennici, 2(4):215–259.

Quick, H. F.
1953 Occurrence of porcupine quills in carnivorous mammals. Jour. Mammalogy, 34(2):256–259.

Rabb, G. B., J. H. Woolpy, and B. E. Ginsberg
1967 Social relations in a group of captive wolves. American Zool., 7:305–311.

Rausch, R. A., and R. A. Hinman
1977 Wolf management in Alaska—an exercise in futility? Pp. 147–156 *in* R. L. Phillips, and C. Jonkel, eds. Proc. 1975 Predator Symposium. Montana For. and Conserv. Exp. Sta., Missoula, ix+268 pp.

Robinson, W. L., and G. J. Smith
1977 Observations on recently killed wolves in Upper Michigan. Wildlife Soc. Bull., 5(1):25–26.

Rogers, L. L., and L. D. Mech
1981 Interactions of wolves and black bears in northeastern Minnesota. Jour. Mammalogy, 62(2):434–436.

Rothman, R. J., and L. D. Mech
1979 Scent-marking in lone wolves and newly formed pairs. Anim. Behav., 27:750–760.

Ryon, C. J.
1977 Den digging and related behavior in a captive timber wolf pack. Jour. Mammalogy, 58(12):87–89.

Seton, E. T.
1953 Lives of game animals. Vol. I, Pt. I., Charles T. Branford Co., Boston. xxxix+337 pp.

Shelton, P. C.
1966 Ecological studies of beavers, wolves, and moose in Isle Royale National Park, Michigan. Purdue Univ., unpubl. Ph.D. dissertation, 308 pp.

Stains, H. J.
1979 Primeness in North American furbearers. Wildlife Soc. Bull., 7(2):120–123.

Stebler, A. M.
1944 The status of the wolf in Michigan. Jour. Mammalogy, 25(1):37–43.
1951 The ecology of Michigan coyotes and wolves. Univ. Michigan, unpubl. Ph.D. dissertation, x+198 pp.

Stenlund, M. H.
1955 A field study of the timber wolf (*Canis lupus*) on the Superior National Forest, Minnesota. Minnesota Dept. Conserv., Tech. Bull. No. 4, 55 pp.

Stewart, P. A., and N. C. Negus
1961 Recent record of wolf in Ohio. Jour. Mammalogy, 42(3):420–421.

Theberge, J. B.
1971 Wolf music, Nat. Hist., 80:36–43.

Theberge, J. B., and J. B. Falls
1967 Howling as a means of communication in timber wolves. American Zool., 7(2):331–338.

Thompson, D. Q.
1952 Travel, range, and food habits of timber wolves in Wisconsin. Jour. Mammalogy, 33(4):429–442.

Van Ballenberghe, V.
1977 Physical characteristics of timber wolves in Minnesota. Pp. 213–219 *in* R. L. Phillips and C. Jonkel, eds. Proc. 1975 Predator Symposium. Montana For. and Conserv. Exp. Sta., Missoula, ix+268 pp.

Van Ballenberghe, V., A. W. Erickson, and D. Byman
1975 Ecology of the timber wolf of northeastern Minnesota. Wildlife Monogr., No. 43, 43 pp.

Van Ballenberghe, V., and L. D. Mech
1975 Weights, growth, and survival of timber wolf pups in Minnesota. Jour. Mammalogy, 56(1):44–63.

Voigt, D. R., G. B. Kolenosky, and D. H. Pimlott
1976 Changes in summer foods of wolves in central Ontario. Jour. Wildlife Mgt., 40(4):663–668.

Weaver, J. L., and S. H. Fritts
1979 Comparison of coyote and wolf scat diameters. Jour. Wildlife Mgt., 43(3):786–788.

Weise, T. F., W. L. Robinson, R. A. Hook, and L. D. Mech
1975 An experimental translocation of the eastern timber wolf. Audubon Cons. Rept. 5, U.S. Fish and Wildlife Serv., Twin Cities, Minnesota. 28 pp.

Wolfe, M. L., and D. L. Allen
1973 Continued studies of the status, socialization, and relationships of Isle Royale wolves, 1967 to 1970. Jour. Mammalogy, 54(3):611–633.

Wood, N. A.
1914 Results of the Shiras Expedition to Whitefish Point, Michigan—Mammals. Michigan Acad. Sci., 16th Rept., pp. 92–98.
1917 Notes on the mammals of Alger County, Michigan. Univ. Michigan, Mus. Zool., Occas. Papers No. 36, 8 pp.
1922 The mammals of Washtenaw County, Michigan. Univ. Michigan, Mus. Zool., Occas. Papers No. 123, 23 pp.

Wood, N. A., and L. R. Dice
1923 Records of the distribution of Michigan mammals. Michigan Acad. Sci., Arts and Ltrs., Papers, 3:425–469.

Young, S. P., and E. A. Goldman
1944 The wolves of North America. American Wildlife Inst., Washington, DC. xx+636 pp.

Zimen, E.
1975 Social dynamics of the wolf pack. Pp. 336–362 *in* M. W. Fox ed. The wild canids. Their systematics, behavioral ecology and evolution. Van Nostrand and Reinhold Co., New York. xvi+508 pp.

RED FOX
Vulpes vulpes (Linnaeus)

NAMES. The generic name *Vulpes,* proposed by Oken in 1816, is of Latin origin and means "fox." The specific name *vulpes,* proposed by Linnaeus in 1758, has the same origin. In Michigan, this bright-colored canid is generally called red fox. The subspecies *Vulpes vulpes fulva* (Desmarest) occurs in Michigan.

RECOGNITION. Body size resembling that of a slender, small dog; head and body averaging 24 in (610 mm) long; tail long and bushy, averaging 15 in (380 mm) in vertebral length; head medium with pointed nose (Plate 5 and Fig. 69); eyes moderate in size, pupils vertically elliptical; ears erect and pointed; body lanky; limbs rather long and slender; feet small with digits terminating in long, sharp, non-retractile claws; foot pads furred; front feet with five digits, with the small pollex high on the inside of the foreleg; hind feet with four digits; mammary glands consist of four pairs; scent gland at base of tail on the dorsal side elliptical in shape; males only slightly larger than females.

Figure 69. The red fox (*Vulpes vulpes*).
Sketches by Bonnie Marris.

The overall pelage of the red fox is soft and silky due to the long, thick underfur (gray at base and buff colored at tip) and long, banded guard hair. The vibrissae are long and black. The normal red coloration of this species predominates (perhaps 99% of the time) in Michigan; other color phases (black, silver, cross) are described in the section under Molt and Color Aberrations. In the characteristically red coloration, the face, top of head, and nape are yellow or rusty red (Plate 0); the back is bright yellowish red or fulvous, darker on midline; the tail reddish mixed with black, always with white tip; the outer sides of ears, legs, and feet blackish; the inside of ears, lips, chest, and belly creamy white.

The striking reddish color of the upperparts and the long, heavily-furred and round tail distinguish this slightly built canid from all other Michigan mammals (Fig. 70).

Figure 70. The red fox (*Vulpes vulpes*). Note heavily furred tail. Photograph by Robert Harrington and courtesy of the Michigan Department of Natural Resources.

MEASUREMENTS AND WEIGHTS. Adult red foxes measure as follows: length from tip of nose to end of tail vertebrae, 38 to 41 in (965 to 1,050 mm); length of tail vertebrae, 14 to 15½ in (355 to 395 mm); length of hind foot, 6½ to 7 in (165 to 178 mm); height of ear from notch, 3 to 3½ in (77 to 89 mm); weight, 8 to 15 lbs (3.6 to 6.8 kg).

DISTINCTIVE CRANIAL AND DENTAL CHARACTERISTICS. The skull of the red fox is slender in construction with a rather long rostrum (Fig. 71). The interorbital area and forehead are flattened. The postorbital processes are weakly constructed and slightly concave above. The parietal ridges extend posterior from the supraorbital processes and often unite medially to form a single sagittal ridge. At the suture between the frontal and the parietal bones, these two parietal ridges are close together, no more than ⅜ in (10 mm) apart. The auditory bullae are prominent and elongate. The mid part of the bony palate terminates posteriorly just anterior to a line drawn between the posterior borders of the last molar tooth on each side of the upper jaw. The teeth are small for a canid; the upper incisors are lobe-shaped; the canines are long and slender. The skull of the red fox averages 5⅝ in (145 mm) long and 3 in (77 mm) wide (across zygomatic arches). Continual growth of the canines allows for progressive deposition of cementum layers, useful in estimating age (Allen, 1974; Churcher, 1960).

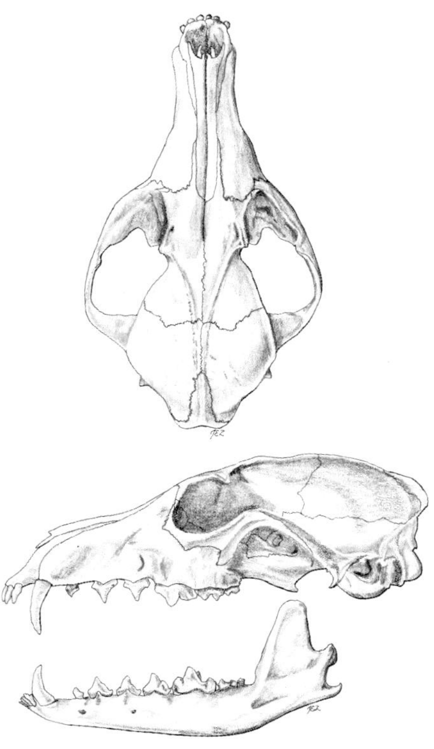

Figure 71. Cranial and dental features of the red fox (*Vulpes vulpes*).

The skull of the red fox most closely resembles in size and shape that of the gray fox, being conspicuously smaller and less massive than the skulls of the coyote or the gray wolf. From the skull of the gray fox, that of the red fox differs by: (1) the space between the parietal ridges at the junction between the frontal and parietal bones being ⅜ in (10 mm) or less, instead of more; (2) the ratio of the length to the width of the auditory bullae being about 10:7, instead of 8:7; (3) the upper incisors being lobed in shape when viewed from the front, instead of not lobed; and (4) having the lower margin on each side of the lower jaw (mandible) gently curving, instead of conspicuously notched near the posterior end. The dental formula for the red fox is:

$$\text{I (incisors)}\frac{3\text{-}3}{3\text{-}3}, \text{C (canines)}\frac{1\text{-}1}{1\text{-}1}, \text{P (premolars)}\frac{4\text{-}4}{4\text{-}4}, \text{M (molars)}\frac{2\text{-}2}{3\text{-}3} = 42$$

DISTRIBUTION. The red fox enjoys an extremely wide range. In Eurasia, the species is found across the entire Paleoarctic land mass from the Japanese islands in the east to the British Isles in the west, north to the Arctic Ocean and south into North Africa and the Arabian Peninsula, and eastward into southern China. (There is a wealth of literature concerning British and European red foxes in Lloyd, 1980 and Zimen, 1980). In North America, the species occurs from the Arctic Ocean and some of the islands in the Canadian Archipelago south to the Gulf coast and west to California. Parts of the Pacific slope and peninsular Florida are not included in this canid's present range. There is also evidence that timbered parts of what is now eastern United States may not have been red fox habitat until logged and/or cleared by settlers. In Michigan, the red fox lives in every county of the state and on several of the larger islands in lakes Michigan, Superior, and Huron (see Map 48). Although it avoids major urban areas, the animal occurs in suburban communities, where housing projects are interspersed with old fields, woodlots, and riparian vegetation.

HABITAT PREFERENCES. Fallow and cultivated fields, meadows, bushy fence lines, woody stream borders, and low shrub cover along the fringes of woods and along beaches bordering larger lakes (Dice, 1925) provide denning and hunting areas for the red fox. Although there is evidence in both the postglacial fossil and the archaeological records that the red fox has been a longtime Michigan resident (Cleland, 1966; Holman, 1975), the species has probably flourished since presettlement days as a result of the pioneers opening woodlands which blanketed much of the state prior to 1800. Lumbering and cultivation provided a major habitat change conducive to an abundant supply of small rodents, insects, fruit-bearing shrubs and a mixed second-growth environment—all to the red fox's liking.

DENSITY AND MOVEMENTS. The red fox is a highly mobile species and needs to forage in an extensive area to secure sufficient food (Ables, 1975). Yet, each individual is confined to an area which is surrounded by areas of other foxes. Biologists have learned most about the density and movements of Michigan red foxes from catching, releasing, and recapturing individuals, by tracking them in winter snow (Arnold and Schofield, 1956; Murie, 1936; Schofield, 1960b), and even by making scent post surveys. Radio-tracking techniques also provide important data on movements (Ables, 1969a; Storm, 1965; Storm *et al.,* 1976).

Much individual red fox movement occurs in autumn—described by Arnold (1956) as the "fall shuffle." At this time, the juveniles disperse from parental den sites to find their own living places. The young males are the first to depart in late

Map 48. Geographic distribution of the red fox (*Vulpes vulpes*).

September and early October with the juvenile females leaving a month later. Studies in Iowa and Illinois (Phillips *et al.,* 1972) showed that males may disperse an average of 18.4 mi (30 km) from the homesite while females average 6.2 mi (10 km). Record movements by juveniles tagged at the den and captured later are 126 mi (194 km) for a female in Minnesota (Longley, 1962) and 245 mi (394 km) for a male from Wisconsin to Indiana (Ables, 1965). In Michigan, a male red fox tagged in Clinton county was later taken in Leelanau County, 166 mi (266 km) away (Arnold, 1956).

By early spring, fox families have again spaced themselves at intervals throughout suitable habitat. Population estimates at this season have primarily used the fox family as the unit, although the number of individuals might vary from six to ten per family. In typical midwestern environment, including a mixture of cultivated fields, fallow and pasture land, scattered woodlots, bushy fencerows, creek banks, and hillsides, it is estimated that there is about one fox family (averaging about 7.4 individuals) per 2,471 acres (1,000 ha). This density estimate comes from studies done in Wisconsin by Pils and Martin (1978), in Iowa by Scott and Klimstra (1955), and in Michigan by Shick (1952).

Red fox home ranges show considerable variation, probably as a result of topography, vegetative cover, and food availability. In Wisconsin, Ables (1969b) reported home ranges from as small as 142 acres (57 ha) to as large as 1,280 acres (512 ha); while another radio-tracking study by Pils and Martin (1978) showed an average winter-spring home range for three vixens as 1,495 acres (598 ha). In Illinois, Storm (1965) found that two transmitter-collared red foxes ranged over an average of 955 acres (385 ha). In Michigan, Murie (1936) concluded that one pair of foxes probably stayed almost entirely within the confines of the 1,200 acre (480 ha) Edwin S. George Reserve of the University of Michigan. In other studies, Arnold (1956) and Arnold and Schofield (1956) reported that the daily winter range of the average Michigan red fox was 900 acres (380 ha); they also determined (by tracking in snow) that in an average night a red fox traversed 5 mi (8 km) of terrain.

BEHAVIOR. Red foxes are usually confined in their behavioral activities in nature (Macdonald *in* Zimen, 1980). One home range may be occupied by a mated pair of adult foxes (sometimes an extra adult or two) and seasonally by their offspring. This home range borders those of other fox families. If nonresidents infringe on these areas, the resident male animals (rarely the females) aggressively harass the intruders (Preston, 1975). The boundaries of these home ranges are also scent-marked (Jorgenson *et al.,* 1978) but are not guarded in the physical sense. An adult red fox displaced from a home territory was found to return from a location 35 mi (56 km) away (Phillips and Mech, 1970). Although the bond between adult males and females is strong when the two share a common range, there is little evidence that red foxes hunt together as do gray wolves. It is more usual for each of the adults to move about as solitary hunters (Fox, 1975).

The red fox is primarily nocturnal in action, using the dim light in early morning and evening as well. Red foxes are also seen in full daylight, especially by motorists in farm country. Typically, a red fox is active for eight to 10 hours each day, with more than 80% of this time spent traveling. The red fox can gallop at speeds as high as 32 mi (51 km) per hour (Jackson, 1961). However, its normal gait is a spirited walk. The red fox swims without difficulty. In tracking small mammals as prey, the red fox trots back and forth across a likely habitat with nose held to the ground, probably also using its keen sense of hearing (Isley and Gysel, 1975). In addition to acute hearing, the red fox also has a superior sense of smell (Jorgenson *et al.,* 1978). On detecting prey, the fox pauses, with ears erect and body tense, and then springs as much as 4 ft (1.2 m) for the capture (Smith, 1944).

Various characteristics allow for close communication between individuals of this species (Montgomery, 1974). Although less vocal than the coyotes or the gray wolf, the red fox makes a variety of sounds, including a high-pitched bark, often consisting of a few short yaps. Whines and screeches are heard in family groups. Odor posts, typical signals of other canids, are placed at strategic sites.

The red fox's antics resemble those of a small dog in many ways. Family groups engage in playful chases and romps, although the species lacks the gregarious social behavior of the gray wolf. Despite some dog-like characteristics, the red fox is shy and extremely nervous in the presence of humans and can rarely be tamed as a pet.

Red foxes generally use ground burrows for

dens which serve as both refuges and for rearing the young. Two fox families occasionally share the same den. Most dens are located in well-drained places. They can be found in fencerows, in the middle of fields, at woodland edges, on ridges, in stone fences, and even in old clogged drain tile or on an island in a river (Arnold, 1956; Layne and McKeon, 1956b; Stanley, 1963; Storm *et al.*, 1976). Dens usually have two or more openings. On a hillside, the excavated tunnel may extend as much as 30 ft (9 m) into the ground. On rather flat terrain, dens can be as deep as four to 10 ft (1.2 to 3 m) below ground level. The nest may be merely a wide place in the burrow with some grass lining. Red foxes are capable of digging their own dens but seem to prefer the excavations of a woodchuck and sometimes of a badger (Arnold, 1956; Grizzell, 1955; Schmeltz and Whitaker, 1977). Generally dens are refurbished in late winter in preparation for the breeding season. A well-used den usually has a conspicuous bare mound of earth surrounding the main entrance.

ASSOCIATES. The red fox appears to have few associates. Although preyed upon by the gray wolf, there is some evidence that these two species coexist, with the red fox trailing a wolf pack to feed on kills. On Isle Royale, Peterson (1977) noted that gray wolves would often chase red foxes, but were unsuccessful in catching them, primarily because the fox is faster and can also run over crusted ice on which the heavier wolf breaks through. Peterson, as well as Johnson (1970), watched red foxes feeding on moose carcasses within sight of resting gray wolves. Interactions between red foxes and gray foxes are not well understood, although studies conducted in Indiana show that the two species tend to select different habitats (with some overlap) when living in the same general areas. The red fox uses uplands with open pasture and woodland edges, while the gray fox remains mostly in lowland woods and woodland edges. Like the gray wolf's large kills of white-tailed deer or moose, the smaller kills by the red fox provide food for other carnivores. At the Edwin S. George Reserve, Murie (1936) found fox-cached prey being eaten by marsh hawk, crow, owls, other birds, and even a striped skunk.

REPRODUCTIVE ACTIVITIES. Artist Oscar Warbach joined with author David A. Arnold to depict some of the intimate details of red fox family life in their prize-winning bulletin entitled "Red Foxes of Michigan" (Arnold, 1956). Their findings are summarized here with those of Pils and Martin (1978) in Wisconsin; Storm *et al.* (1976) in the upper Midwest; Layne and McKeon (1956a) and Sheldon (1949) in New York; and Stanley (1963) in Kansas. Courtship begins as early as late December. The female may be followed by one or more courting males but usually selects a single mate with whom a firm pair bond is formed. The female accepts the male for breeding from late December to perhaps mid-March (Hoffman and Kirkpatrick, 1954). During this period, the mated pair works together in preparing a nursery den. After a gestation period of between 51 and 53 days, the litter is born in March or April in a grass-lined nest in the underground den (Sargeant *et al.*, 1981). In Michigan, the litter averages five pups, although both Schofield (1958) and Switzenburg (1950) found that litters born to red fox vixens in the Lower Peninsula average about one more pup than do litters born in the Upper Peninsula. Sometimes extremely large litters occur; one litter in Calhoun County contained 17 pups (Holcomb, 1965).

At birth, pups weigh 3.5 oz (100 g); have closed eyes; are covered with gray-brown, wooly pelage; and possess the distinctive adult white-tail tip. At 10 days of age, the eyes open. At 20 days of age, the pups venture from the home den and engage in typical puppy romps. Red foxes estimated to be 21 days of age (in the collections of The Museum at Michigan State University) possess gray-brown, wooly body pelage with new growth appearing on the sides of the head. At 60 days, the pups are weaned and brought foods, parts of which are often strewn around the entrance to the den. At 120 days, the pups are three-quarters grown, begin to replace their milk teeth with permanent dentition, engage in hunting on their own, and shortly begin to disperse (from October to January, Tullar and Berchielli, 1980), with males leaving the homesite first. By their first winter, these growing animals will be sexually mature and begin families.

There seems to be a tendency for red fox populations to fluctuate somewhat regularly with high numbers appearing at 10-year intervals, in years ending in five (Arnold, 1956; Cross, 1940; Jackson, 1961). One estimate of annual productivity was about 214% per adult (Layne and

Burt, W. H.
1946 The mammals of Michigan. Univ. Michigan Press, Ann Arbor. xv+288 pp.

Churcher, C. S.
1960 Cranial variation in the North American red fox. Jour. Mammalogy, 41(3):349–360.

Cleland, C. E.
1966 The prehistoric animal ecology and ethnozoology of the Upper Great Lakes Region. Univ. Michigan, Mus. Anthro., Anthro. Papers, No. 29, x+294 pp.

Cross, E. C.
1940 Periodic fluctuations in numbers of the red fox in Ontario. Jour. Mammalogy, 21(3):294–306.
1941 Colour phases of the red fox (*Vulpes fulva*) in Ontario. Jour. Mammalogy, 22(1):24–39.

Dearborn, N.
1932 Foods of some predatory fur-bearing animals in Michigan. Univ. Michigan, Sch. For. and Conserv., Bull. No. 1, 52 pp.

deVos, A.
1951 Recent findings in fisher and marten ecology and management. Trans. 16th N. American Wildlife Conf., pp. 498–507.

Dice, L. R.
1925 A survey of the mammals of Charlevoix County, Michigan, and vicinity. Univ. Michigan, Mus. Zool., Occas. Papers No. 159, 33 pp.

Douglass, D. W., and G. W. Brandt
1945 The red fox—friend or foe? Michigan Dept. Conserv., Game Div., 23 pp.

Eadie, W. R.
1949 Red fox transports play or training items to new den. Jour. Mammalogy, 30(3):308.

Erickson, A. B.
1944 Helminths of Minnesota Canidae in relation to food habits and a host list and key to the species reported from North America. American Midl. Nat., 32(3): 358–372.

Errington, P. L.
1935 Food habits of mid-west foxes. Jour. Mammalogy, 16(3):192–200.

Field, R. J.
1970 Scavengers feeding on a Michigan deer carcass. Jack-Pine Warbler, 48(2):73.

Fox, M. W.
1975 Evolution of social behavior in Canids, Pp. 429–460 *in* M. W. Fox, ed. The wild canids. Their systematics, behavioral ecology and evolution. Van Nostrand and Reinhold Co., New York. xvi+508 pp.

Goble, F. C., and A. H. Cook
1942 Notes on nematodes from the lungs and frontal sinuses of New York fur-bearers. Jour. Parasit., 18(6): 451–455.

Grizzell, R. A., Jr.
1955 A study of the southern woodchuck, *Marmota monax monax*. American Midl. Nat., 53(2):257–293.

Haas, G. E.
1970 Rodent fleas in a red fox den in Wisconsin. Jour. Mammalogy, 51(4):796–798.

Hamilton, W. J., Jr.
1935 Notes on food of red foxes in New York and New England. Jour. Mammalogy, 16(1):16–21.

Hatt, R. T., J. Van Tyne, L. C. Stuart, C. H. Pope, and A. B. Grobman
1948 Island life: A study of the land vertebrates of the islands of eastern Lake Michigan. Cranbrook Inst. Sci., Bull. No. 27, xi+179 pp.

Heit, W. S.
1944 Food habits of red foxes of the Maryland marshes. Jour. Mammalogy, 25(1):55–63.

Hickie, P.
1940 Cottontails in Michigan. Michigan Dept. Conserv., Game Div., 109 pp.

Hoffman, R. A., and C. M. Kirkpatrick
1954 Red fox weights and reproduction in Tippecanoe County, Indiana. Jour. Mammalogy, 35(4):504–509.

Holcomb, L. C.
1965 Large litter size of red fox. Jour. Mammalogy, 46(3): 530.

Holman, J. A.
1974 Michigan's fossil vertebrates. Michigan State Univ., Publ. Mus., Ed. Bull. No. 2, 54 pp.

Isley, T. E., and L. W. Gysel
1975 Sound-source localization by the red fox. Jour. Mammalogy, 56(2):397–404.

Jackson, H. H. T.
1961 Mammals of Wisconsin. Univ. Wisconsin Press, Madison. xii+504 pp.

Johnson, D. R.
1969 Returns of the American Fur Company, 1835–1839. Jour. Mammalogy, 50(4):836–839.

Johnson, W. J.
1970 Food habits of the red fox in Isle Royale National Park, Lake Superior. American Midl. Nat., 84(2):568–572.

Jones, R. E., and R. J. Aulerich
1981 Bibliography of foxes. Michigan State Univ., Agric. Exp. Sta., Jour. Art. No. 9871, 141 pp.

Jorgenson, J. W., M. Novotny, M. Carmack, G. B. Copland, S. R. Wilson, S. Katona, and W. K. Whitten
1978 Chemical scent constituents in the urine of the red fox (*Vulpes vulpes L.*) during the winter season. Science, 199:796–798.

Karpuleon, F.
1958 Food habits of Wisconsin foxes. Jour. Mammalogy, 39(4):591–593.

Kniss, D. H.
1949 Instructions for trapping foxes and other fur-bearers. Michigan Dept. Conserv., 12 pp.

Korschgen, L. J.
1957 Food habits of coyotes, foxes, house cats and bobcats in Missouri. Missouri Conserv. Comm., Fish and Game Div., P-R Ser. No. 15, 64 pp.
1959 Food habits of the red fox in Missouri. Jour. Wildlife Mgt., 23(2):169–176.

Kurta, A.
1979 Bat rabies in Michigan. Michigan Acad., 12(2):221–230.

Latham, R. M.
1952 The fox as a factor in the control of weasel populations. Jour. Wildlife Mgt., 16(4):516–517.

Layne, J. N., and W. H. McKeon
1956a Some aspects of red fox and gray fox reproduction in New York. New York Fish and Game Jour., 3(1):44–74.
1956b Notes on red fox and gray fox den sites in New York. New York Fish and Game Jour., 3:248–249.

Linduska, J. P.
1950 Ecology and land-use relationships of small mammals on a Michigan farm. Michigan Dept. Conserv., Game Div., ix+144 pp.

Lloyd, H. G.
1980 The red fox. B. T. Batsford Ltd., London. 320 pp.

Long, C. A.
1978 Mammals of the islands of Green Bay, Lake Michigan. Jack-Pine Warbler, 56(2):59–82.

Longley, W. H.
1962 Movements of red fox. Jour. Mammalogy, 43(1):107.

MacGregor, A. E.
1942 Late fall and winter food of foxes in central Massachusetts. Jour. Wildlife Mgt., 6(3):221–224.

MacMullan, R. A.
1954 The life and times of Michigan pheasants. Michigan Dept. Conserv., Game Div., 63 pp.

Manville, R. H.
1948 The vertebrate fauna of the Huron Mountains, Michigan. American Midl. Nat., 39(3):615–640.
1950 The mammals of Drummond Island, Michigan. Jour. Mammalogy, 31(3):358–359.

Mech, L. D.
1966 The wolves of Isle Royale. National Park Serv., Fauna Ser. 7, xiv+210 pp.

Montgomery, G. G.
1974 Communication in red fox dyads: A computer simulation study. Smithsonian Contr. Zool., No. 187, iii+30 pp.

Morgan, B. B.
1944 Host list of the genus *Trichomonas* (Protozoa: Flagellata). Part II. Host-parasite list. Trans. Wisconsin Acad. Sci., Arts, and Ltrs., Papers, 35:235–245.

Murie, A.
1936 Following fox trails. Univ. Michigan, Mus. Zool., Misc. Publ. No. 32, 45 pp.

Ozoga, J. J., C. S. Bienz, and L. J. Verme
1982 Red fox feeding habits in relation to fawn mortality. Jour. Wildlife Mgt., 46(1):242–243.

Ozoga, J. J., and C. J. Phillips
1964 Mammals of Beaver Island, Michigan. Michigan State Univ., Publ. Mus., Biol. Ser., 2(6):305–348.

Payne, N. F., and C. Finlay
1975 Red fox attack on beaver. Canadian Field-Nat., 89(4): 450–451.

Petersen, E. T.
1979 Hunters heritage. A history of hunting in Michigan. Michigan United Conserv. Clubs, Lansing. 55 pp.

Peterson, R. O.
1977 Wolf ecology and prey relationships on Isle Royale. National Park Serv., Sci. Monog., No. 11, xx+210 pp.

Phillips, C. J., J. J. Ozoga, and L. C. Drew
1965 The land vertebrates of Garden Island, Michigan. Jack-Pine Warbler, 43(1):20–25.

Phillips, R. L., R. D. Andrews, G. L. Storm, and R. A. Bishop
1972 Dispersal and mortality of red foxes. Jour. Wildlife Mgt., 36(2):237–247.

Phillips, R. L., and L. D. Mech
1970 Homing behavior of a red fox. Jour. Mammalogy, 51(3):621.

Pils, C. M., and M. A. Martin
1978 Population dynamics, predatory-prey relationships and management of the red fox in Wisconsin. Wisconsin Dept. Nat. Res., Tech. Bull. No. 105, 56 pp.

Pils, C. M., M. A. Martin, and E. L. Lange
1981 Harvest, age structure, survivorship, and productivity of red foxes in Wisconsin, 1975–1978. Wisconsin Dept. Nat. Res., Tech. Bull. No. 125, 21 pp.

Preston, E. M.
1975 Home range defense in the red fox, *Vulpes vulpes L.* Jour. Mammalogy, 56(3):645–652.

Pryor, L. B.
1956 Sarcoptic mange in wild foxes in Pennsylvania. Jour. Mammalogy, 37(1):90–93.

Sargeant, A. B., S. H. Allen, and D. H. Johnson
1981 Determination of age and whelping dates of live red fox pups. Jour. Wildlife Mgt., 45(3):760–765.

Scharf, W. C., and K. R. Stewart
1980 New records of Siphonaptera from northern Michigan. Great Lakes Ento., 13(3):165–167.

Schmeltz, L. L., and J. O. Whitaker, Jr.
1977 Use of woodchuck burrows by woodchucks and other mammals. Trans. Kentucky Acad. Sci., 38(1–2):79–82.

Schofield, R. D.
1958 Litter size and age ratios in Michigan red foxes. Jour. Wildlife Mgt., 22(3):313–315.
1960a Determining hunting waste of deer by following fox trails. Jour. Wildlife Mgt., 24(3):342–244.
1960b A thousand miles of fox trails in Michigan's ruffed grouse range. Jour. Wildlife Mgt., 24(4):432–434.

Scott, T. G.
1943 Some food coactions of the northern plains red fox. Ecol. Monog., 13:427–479.

Scott, T. G., and W. D. Klimstra
1955 Red foxes and a declining prey population. Southern Illinois Univ., Monog. ser. No. 1, 123 pp.

Sheldon, W. G.
1949 Reproductive behavior of foxes in New York state. Jour. Mammalogy, 30(3):236–246.

Shelton, P. C.
1966 Ecological studies of beavers, wolves, and moose in Isle Royale National Park, Michigan. Purdue Univ., unpubl. Ph.D. dissertation, 308 pp.

Shick, C. A.
1952 A study of pheasants on a 9,000 acre prairie farm,

McKeon, 1956a) Even so, the annual mortality must be high, especially on the juvenile population as it disperses in autumn and winter.

Hunting and trapping have been shown to account for up to 80% of the mortality in the Midwest (Phillips *et al.*, 1972; Storm *et al.*, 1976). Although red foxes survive in nature for as long as three years (for European red foxes up to six years, according to Stubble *in* Zimen, 1980), there probably is a near total annual turn-over in the population, with the juveniles rapidly replacing the adults (Arnold and Schofield, 1956); in Wisconsin, for example, 74% of 2,153 red foxes taken mostly from November through January in 1975–1978 were juveniles (Pils, *et al.*, 1981).

FOOD HABITS. The general conception of the red fox is of a ravished carnivore in constant search of delectable game birds, poultry, and cottontail rabbits. In truth, however, this canid, usually a solitary hunter as an adult (Fox, 1975), has a varied appetite for both animal and plant foods and eats whatever is available and easy to acquire—which certainly does not describe wary ring-necked pheasants or ruffed grouse. As Arnold (1956) suggested, the Michigan red fox is an "everything eater."

Studies of the food habits of Michigan red foxes have included examining remains of materials in fecal deposits (scats) or in digestive tracts (Manville, 1948); fox kills along their trails in snow (Murie, 1936; Schofield, 1960b); tracks and egg punctures in nesting colonies of herring gulls (Shugart and Scharf, 1977); and entire or partial carcasses of prey found in caches or at entrances to dens (Murie, 1936). The list of red fox prey encompasses virtually all Michigan small mammals (see Table 19), up to the size of woodchuck, porcupine, and even beaver (Payne and Finlay, 1975). The red fox also eats carrion from larger mammals—white-tailed deer, wapiti, moose, and domestic animals. The variety of other foods eaten by the red fox is impressive; the list includes birds (*e.g.*, herring gull, ruffed grouse, bobwhite quail, ring-necked pheasant, woodcock, great horned owl, sandhill crane, crow, ducks, songbirds, and poultry), cold-blooded vertebrates (*e.g.*, snapping turtle, painted turtle, blue racer, garter snake, leopard frog, various fish), invertebrates (*e.g.*, maybeetles and grasshoppers), and numerous fruits, seeds, and other plant parts (*e.g.*, service berry, raspberry, strawberry, blackberry, blueberry, cherry, sarsaparilla, and acorns).

Obviously, Michigan flora and fauna are most available during the warmer months of the year. In

Table 19. Partial list of Michigan mammals known to be eaten by the red fox, as reported in the literature.

Virginia Opossum (*Didelphis virginiana*)*
Masked Shrew (*Sorex cinereus*)*
Short-Tailed Shrew (*Blarina brevicauda*)
Least Shrew (*Cryptotis parva*)*
Eastern Mole (*Scalopus aquaticus*)
Star-Nosed Mole (*Condylura cristata*)
Eastern Cottontail (*Sylvilagus floridanus*)
Snowshoe Hare (*Lepus americanus*)
Eastern Chipmunk (*Tamias striatus*)
Least Chipmunk (*Eutamias minimus*)*
Woodchuck (*Marmota monax*)
Thirteen-Lined Ground Squirrel (*Spermophilus tridecemlineatus*)
Gray Squirrel (*Sciurus carolinensis*)
Fox Squirrel (*Sciurus niger*)
Red Squirrel (*Tamiasciurus hudsonicus*)
Southern Flying Squirrel (*Glaucomys volans*)*
Beaver (*Castor canadensis*)
White-Footed Mouse (*Peromyscus leucopus*)*
Deer Mouse (*Peromyscus maniculatus*)
Southern Red-Backed Vole (*Clethrionomys gapperi*)*
Prairie Vole (*Microtus ochrogaster*)*
Meadow Vole (*Microtus pennsylvanicus*)
Woodland Vole (*Microtus pinetorum*)*
Muskrat (*Ondatra zibethicus*)
Southern Bog Lemming (*Synaptomys cooperi*)*
Norway Rat (*Rattus norvegicus*)*
House Mouse (*Mus musculus*)*
Meadow Jumping Mouse (*Zapus hudsonius*)*
Porcupine (*Erethizon dorsatum*)
Ermine (*Mustela erminea*)*
Long-Tailed Weasel (*Mustela frenata*)*
Least Weasal (*Mustela nivalis*)*
Mink (*Mustela vison*)*
Striped Skunk (*Mephitis mephitis*)
House Cat (*Felis catus*)*
White-Tailed Deer (*Odocoileus virginianus*)**
Wapiti or American Elk (*Cervus canadensis*)**
Moose (*Alces alces*)**

Reported in studies made in Michigan: Arnold (1952, 1956); Dearborn (1932); Field (1970); Johnson (1970); Hickie (1940); Linduska (1950); MacMullan (1954); Manville (1948); Murie (1936); Ozoga *et al.* (1982); Ozoga and Phillips (1964); Schofield (1960a, 1960b); Shelton (1966); Shick (1952); Shugart and Scharf (1977).

*Unreported from the Michigan literature but from other areas: Brickner (1953); Eadie (1949); Errington (1935); Hamilton (1935); Heit (1944); Jackson (1961); Karpuleon (1958); Korschgen (1957, 1959); Latham (1952); MacGregor (1942); Scott (1943); Scott and Klimstra (1955); Timm (1975).

**Eaten as carrion.

winter, there is evidence, especially in northern Michigan, that the nourishment available to the red fox includes large amounts of carrion—including the carcasses of white-tailed deer wasted by hunters during the autumn shooting season (Schofield, 1960a), although fawn carcassses are found in summer as well (Ozoga *et al.,* 1982).

Red foxes apparently prey successfully on weasels despite their elusiveness (Allen, 1979; Latham, 1952). Near Rockford in Kent County at about 1500 hours on December 15, 1977, artist-naturalist Bonnie Marris watched a trio of red foxes circling a group of weasels (judged to be long-tailed weasels). When the foxes became aware of an intruder, they abruptly stopped their aggressive actions and departed, with one fox carrying off a weasel in its mouth. Competition among carnivores is a highly interesting but little-known aspect of prey-predator ecology.

ENEMIES. In many parts of southern Michigan, there are few meat-eating enemies sufficiently large to subdue an adult-sized red fox. Larger hawks and owls might possibly swoop down and carry off a pup from a den entrance. In northern Michigan, however, the coyote, bobcat, lynx, and reestablished fisher may overpower a red fox (Arnold, 1956; Banfield, 1974; deVos, 1951; Jackson, 1961). On Isle Royale, Mech (1966) found only a trace of red fox in fecal droppings (scats) of the gray wolf. Undoubtedly in earlier times in Michigan the wolverine and mountain lion were also potential enemies.

PARASITES AND DISEASES. By far the most noxious of external parasites to infest the skin of the red fox is the mange mite (*Sarcoptes scabei*), which causes hair loss and other serious complications (Pryor, 1956). In the winter of 1974–1975, a heavy mange infestation was thought to be the reason for declines in both the red fox and coyote populations in Michigan. Other problem ectoparasites include the severe ear mites (*Otodectes cynotis*), at least two species of ticks, one louse, and six kinds of fleas (Haas, 1970; Jackson, 1961; Scharf and Stewart, 1980).

Internal parasites found in the red fox often have intermediate stages in the prey species which the canid consumes, thereby allowing entry of the parasite to finish its adult development. Arnold (1956) described the life cycle for a tapeworm which completes its growth by passing from the eastern cottontail, where it exists in the abdominal cavity in a larval (bladder worm) stage, to the red fox where the parasite becomes an elongated tapeworm thriving in the digestive tract. Another internal parasite, the heartworm (*Dirofilaria immitis*), is a serious and lethal agent in all Michigan canids, especially domestic dogs. This microfilarial worm is spread from one host to another by means of mosquito vectors. In the winter of 1970–1971, Stuht and Youatt (1972) found heartworms in 11 of 39 red foxes in an area of Saginaw County known to have a high rate of these worms in domestic dogs. Four of these red foxes also had lungworms (*Paragonimus kellicotti*). At least 11 kinds of flukes (*Trematoda*), six tapeworms (*Cestoda*), 17 roundworms (*Nematoda*), and a flagellate protozoan are recorded in red foxes (Erickson, 1944; Goble and Cook, 1942; Jackson, 1961; Morgan, 1944).

The red fox has problems with tularemia, distemper, and other ailments, perhaps contracted on occasion from domestic dogs. However, rabies is the most dreaded disease for the red fox, domestic stock, and humans (Zimen, 1980). Although the "fall shuffle" by juveniles might serve to spread this virus disease, there is little evidence that this behavior is responsible for the transmission of rabies (Phillips *et al.,* 1972). The first occurrence reported for rabies in Michigan red foxes was in 1958 (Kurta, 1980; Youatt *et al.,* 1971), with local outbreaks again in 1958–1959 and in the spring of 1970 in Chippewa County and adjacent parts of Ontario, across the St. Marys River. According to Dr. Donald Cahoon (pers. comm., October 3, 1973), of 262 red foxes examined by the Michigan Department of Public Health between 1962 and 1972, 32 (12.2%) had rabies infections. However, this percentage was low compared with the 52% (182 out of 337 examined) infection rate for striped skunk in the same period. No gray foxes were examined during that period.

MOLT AND COLOR ABERRATIONS. Red foxes molt once annually. The shedding of the old hair begins in mid-spring and continues until mid-autumn, with new fur not completely grown until November. As a result, foxes often have patchy hair and a scruffy look in summer.

Occasionally, one or more pups in a litter may have color aberrations. White individuals (albinos) do not seem to survive well (Jackson, 1961). At least two distinctive color phases which vary from the typical red to black have become well known, especially to fur fanciers (Cross, 1941). One of these, the cross fox, retains much of the reddish coloring of the normal red fox but has an overall grayish brown cast with a greater amount of black on the legs and underfur. In addition, blackish guard hairs on the back form a conspicuous "cross" between the shoulders. This mutation has occasionally been reported in northern Michigan (Dearborn, 1932; Wood, 1914); an animal, taken in the winter of 1978 on Drummond Island, was probably a cross fox, according to the description published by Smith (1978).

The second distinctive color phase is all black, except for the white-tipped tail. An important variation of the black fox is the popular silver fox, in which the guard hairs on the black body and tail are tipped with white. This mutation has been reported on occasion both in the Upper and Lower Peninsula (Dearborn, 1932; Wood, 1914, 1922; Wood and Dice, 1923) and even on Isle Royale (Allen, 1979). A platinum color phase is also known in captive animals propagated in fur farms. Sometimes a red fox fails to develop part or all of its normal guard hair and has only wooly, pale, yellowish red underfur. This is called a Sampson fox.

ECONOMIC IMPORTANCE. The combined qualities of the long, silky guard hair and soft underfur have made the red fox pelt very important in the fur industry (Jones and Aulerich, 1981). Even in colonial days, the red fox, although overshadowed in the marketplace by the popular beaver, was prominent in trade. For example, Petersen (1979) reported that 1,340 fox pelts were shipped by British traders from Michilimackinac in 1767. Between 1835 and 1839, the American Fur Company, according to Johnson (1969), received 2,401 red fox skins at its Detroit Department (including southern Michigan) and 1,303 at its Northern Department (including northern Michigan).

Fashion, of course, plays a major role in dictating fur prices. In the decades prior to World War I, long-haired furs were in demand. With silver fox pelts bringing as much as $246 apiece in 1919–1920, production of pelts in this color phase was increased by raising more and more mutant foxes on fur farms, which had their start in Canada before the turn of the century. In the post World War I days, the demand for long-haired furs waned; fur farms went out of business, and prices dropped. Many Michigan trappers and hunters were forced to quit the business after World War II when raw pelt prices were less than $5. Prices began to rise in 1970 as long-haired furs once again became fashionable. By the winter of 1973–1974, pelts, depending on size and condition, brought between $25 and $49 at auctions. By 1976–1977, prices of $62 per pelt were paid; in 1977–1978, prices ranged from $69 to $84 per pelt with a high of $97; and in 1978–1979, from $70 to $100 with a high of $150. However, at an auction in Newberry in March, 1980, fox skins brought an average of only $45. Although there have been strong objections raised about methods of capturing red foxes and other fur animals (Kniss, 1949; Stebler, 1944), as long as furs or fur-trimmed apparel remain in demand, the monetary return will insure the continuance of fur trade.

Like the coyote and the gray wolf, Michigan's red foxes came under severe criticism from the sporting public because of alleged depredations on wildlife. As Arnold (1971) pointed out, the bounty on foxes was inaugurated in 1947 when: (1) populations of red foxes appeared high, possibly because of reduced trapping as a result of the low prices paid for pelts by the fur trade; and (2) populations of ring-necked pheasants were low. The argument centered around the important need, as perceived by the public, to restore the status of this important game bird by controlling the foxes (Switzenburg, 1951). Table 20 summarizes the results of the $5.00 bounty beginning in 1947 until its repeal in 1965. A total of 494,635 red foxes were bountied for a total expenditure of $2,473,175 by the Michigan Game and Fish Protection Fund (acquired through the sale of licenses). Arnold (1971) pointed out the irony of this by noting that in the first five years of this experiment about 20,000 red foxes were bountied annually, with the estimated ring-necked pheasant harvest in those years averaging slightly more than 800,000. In the second five years, more red foxes were bountied, an average of 25,000 annually. At the same time, the pheasant harvest increased about 50% to an annual take of 1,200,000 birds. Such statistics indicated that there was little correlation indeed between the ups and downs of the ring-

*Table 20. The Michigan Red Fox Bounty, 1947–1965.**

YEAR	NUMBER OF FOXES BOUNTIED	COST OF BOUNTIES
1947	11,877	$ 59,385
1948	20,968	104,840
1949	24,621	123,105
1950	21,124	105,620
1951	18,681	93,405
1952	16,461	82,305
1953	19,532	97,660
1954	26,964	134,820
1955	25,157	125,785
1956	28,476	142,380
1957	27,629	138,145
1958	31,248	156,240
1959	26,778	133,890
1960	29,241	146,205
1961	34,856	174,280
1962	30,341	151,705
1963	34,461	172,305
1964	26,917	184,585
1965	29,303	146,515
	TOTALS 494,635	$2,473,175

*After Arnold (1971).

necked pheasant and the livelihood of the Michigan red fox. The red fox, however, remained in low esteem until finally, in 1980, legislative action gave this species full protection except for a hunting and trapping season from October 15 to March 1 in the Upper Peninsula and northern Lower Peninsula and from November 1 to March 1 in the southern Lower Peninsula.

Although ecological studies in Michigan have demonstrated that the red fox may have more favorable than unfavorable attributes (Douglass and Brandt, 1945), the average citizen still retains the storybook image of a sly, sometimes-rabid, cunning animal with salivating jowls and big teeth raiding henhouses, eating grapes, and threatening other wild creatures, large and small. Hunters of upland game still are very sure that the red fox relentlessly hunts down favored wildlife. On the other hand, fox hunters have bitterly opposed the destruction of foxes, which they track, capture and release so that the sport may be practiced another day. So the short life of the red fox is not an easy one, primarily due to the bias of a well-meaning but often uninformed public. As Arnold (1956) put it, "May he never vanish from our scene."

COUNTY RECORDS FROM MICHIGAN. Records of specimens of red foxes in collections of museums, from the literature, and reported by personnel of the Michigan Department of Natural Resources show that this canid occurs in all counties in both the Upper and Lower Peninsula. The red fox has also been reported on several islands in the Great Lakes including: in Lake Michigan, Beaver, North Fox, South Fox, Garden, Gull, North Manitou, South Manitou, Squaw, and Whiskey; in Green Bay, St. Martin and Big Summer; in Lake Superior, Isle Royale; in Lake Huron, Bois Blanc and Drummond; and in Saginaw Bay, Charity (Burt, 1946; Cleland, 1966; Hatt *et al.*, 1948; Long, 1978; Manville, 1950; Phillips *et al.*, 1965; Strang, 1855).

LITERATURE CITED

Ables, E. D.
1965 An exceptional fox movement. Jour. Mammalogy, 46(1):102
1969a Activity studies of red foxes in southern Wisconsin. Jour. Wildlife Mgt., 33(1):145–153.
1969b Home-range studies of red foxes (*Vulpes vulpes*). Jour. Mammalogy, 50(1):108–120.
1975 Ecology of the red fox in North America. Pp. 216–236 *in* M. W. Fox, ed. The wild canids. Their systematics, behavioral ecology, and evolution. Van Nostrand Reinhold Co., New York. xvi+508 pp.

Allen, D. L.
1979 Lives and loves of some sly foxes on Isle Royale. Smithsonian Mag., 10(2):91–100.

Allen, S. H.
1974 Modified techniques for aging red fox using canine teeth. Jour. Wildlife Mgt., 38(1):152–154.

Arnold, D. A.
1952 The relationship between ringnecked pheasant and red fox population trends. Michigan Acad. Sci., Arts and Ltrs., Papers, 37:121–127.
1956 Red foxes of Michigan. Michigan Dept. Conserv., 48 pp.
1971 The Michigan bounty system. Michigan Dept. Nat. Res., Wildlife Div., Info. Circ. No. 162 (rev.), (mimeo.) 3 pp.

Arnold, D. A., and R. D. Schofield
1956 Home range and dispersal of Michigan red foxes. Michigan Acad. Sci., Arts and Ltrs., Papers, 41:91–97.

Banfield, A. W. F.
1974 The mammals of Canada. Univ. Toronto Press, Toronto. xxiv+438 pp.

Brickner, J.
1953 Red fox preying on muskrats. Jour. Mammalogy, 34(3):389.

Burt, W. H.
1946 The mammals of Michigan. Univ. Michigan Press, Ann Arbor. xv+288 pp.

Churcher, C. S.
1960 Cranial variation in the North American red fox. Jour. Mammalogy, 41(3):349–360.

Cleland, C. E.
1966 The prehistoric animal ecology and ethnozoology of the Upper Great Lakes Region. Univ. Michigan, Mus. Anthro., Anthro. Papers, No. 29, x+294 pp.

Cross, E. C.
1940 Periodic fluctuations in numbers of the red fox in Ontario. Jour. Mammalogy, 21(3):294–306.
1941 Colour phases of the red fox (*Vulpes fulva*) in Ontario. Jour. Mammalogy, 22(1):24–39.

Dearborn, N.
1932 Foods of some predatory fur-bearing animals in Michigan. Univ. Michigan, Sch. For. and Conserv., Bull. No. 1, 52 pp.

deVos, A.
1951 Recent findings in fisher and marten ecology and management. Trans. 16th N. American Wildlife Conf., pp. 498–507.

Dice, L. R.
1925 A survey of the mammals of Charlevoix County, Michigan, and vicinity. Univ. Michigan, Mus. Zool., Occas. Papers No. 159, 33 pp.

Douglass, D. W., and G. W. Brandt
1945 The red fox—friend or foe? Michigan Dept. Conserv., Game Div., 23 pp.

Eadie, W. R.
1949 Red fox transports play or training items to new den. Jour. Mammalogy, 30(3):308.

Erickson, A. B.
1944 Helminths of Minnesota Canidae in relation to food habits and a host list and key to the species reported from North America. American Midl. Nat., 32(3): 358–372.

Errington, P. L.
1935 Food habits of mid-west foxes. Jour. Mammalogy, 16(3):192–200.

Field, R. J.
1970 Scavengers feeding on a Michigan deer carcass. Jack-Pine Warbler, 48(2):73.

Fox, M. W.
1975 Evolution of social behavior in Canids, Pp. 429–460 *in* M. W. Fox, ed. The wild canids. Their systematics, behavioral ecology and evolution. Van Nostrand and Reinhold Co., New York. xvi+508 pp.

Goble, F. C., and A. H. Cook
1942 Notes on nematodes from the lungs and frontal sinuses of New York fur-bearers. Jour. Parasit., 18(6): 451–455.

Grizzell, R. A., Jr.
1955 A study of the southern woodchuck, *Marmota monax monax*. American Midl. Nat., 53(2):257–293.

Haas, G. E.
1970 Rodent fleas in a red fox den in Wisconsin. Jour. Mammalogy, 51(4):796–798.

Hamilton, W. J., Jr.
1935 Notes on food of red foxes in New York and New England. Jour. Mammalogy, 16(1):16–21.

Hatt, R. T., J. Van Tyne, L. C. Stuart, C. H. Pope, and A. B. Grobman
1948 Island life: A study of the land vertebrates of the islands of eastern Lake Michigan. Cranbrook Inst. Sci., Bull. No. 27, xi+179 pp.

Heit, W. S.
1944 Food habits of red foxes of the Maryland marshes. Jour. Mammalogy, 25(1):55–63.

Hickie, P.
1940 Cottontails in Michigan. Michigan Dept. Conserv., Game Div., 109 pp.

Hoffman, R. A., and C. M. Kirkpatrick
1954 Red fox weights and reproduction in Tippecanoe County, Indiana. Jour. Mammalogy, 35(4):504–509.

Holcomb, L. C.
1965 Large litter size of red fox. Jour. Mammalogy, 46(3): 530.

Holman, J. A.
1974 Michigan's fossil vertebrates. Michigan State Univ., Publ. Mus., Ed. Bull. No. 2, 54 pp.

Isley, T. E., and L. W. Gysel
1975 Sound-source localization by the red fox. Jour. Mammalogy, 56(2):397–404.

Jackson, H. H. T.
1961 Mammals of Wisconsin. Univ. Wisconsin Press, Madison. xii+504 pp.

Johnson, D. R.
1969 Returns of the American Fur Company, 1835–1839. Jour. Mammalogy, 50(4):836–839.

Johnson, W. J.
1970 Food habits of the red fox in Isle Royale National Park, Lake Superior. American Midl. Nat., 84(2):568–572.

Jones, R. E., and R. J. Aulerich
1981 Bibliography of foxes. Michigan State Univ., Agric. Exp. Sta., Jour. Art. No. 9871, 141 pp.

Jorgenson, J. W., M. Novotny, M. Carmack, G. B. Copland, S. R. Wilson, S. Katona, and W. K. Whitten
1978 Chemical scent constituents in the urine of the red fox (*Vulpes vulpes L.*) during the winter season. Science, 199:796–798.

Karpuleon, F.
1958 Food habits of Wisconsin foxes. Jour. Mammalogy, 39(4):591–593.

Kniss, D. H.
1949 Instructions for trapping foxes and other fur-bearers. Michigan Dept. Conserv., 12 pp.

Korschgen, L. J.
1957 Food habits of coyotes, foxes, house cats and bobcats in Missouri. Missouri Conserv. Comm., Fish and Game Div., P-R Ser. No. 15, 64 pp.
1959 Food habits of the red fox in Missouri. Jour. Wildlife Mgt., 23(2):169–176.

Kurta, A.
1979 Bat rabies in Michigan. Michigan Acad., 12(2):221–230.

Latham, R. M.
1952 The fox as a factor in the control of weasel populations. Jour. Wildlife Mgt., 16(4):516–517.

Layne, J. N., and W. H. McKeon
1956a Some aspects of red fox and gray fox reproduction in New York. New York Fish and Game Jour., 3(1):44–74.
1956b Notes on red fox and gray fox den sites in New York. New York Fish and Game Jour., 3:248–249.

Linduska, J. P.
1950 Ecology and land-use relationships of small mammals on a Michigan farm. Michigan Dept. Conserv., Game Div., ix+144 pp.

Lloyd, H. G.
1980 The red fox. B. T. Batsford Ltd., London. 320 pp.

Long, C. A.
1978 Mammals of the islands of Green Bay, Lake Michigan. Jack-Pine Warbler, 56(2):59–82.

Longley, W. H.
1962 Movements of red fox. Jour. Mammalogy, 43(1):107.

MacGregor, A. E.
1942 Late fall and winter food of foxes in central Massachusetts. Jour. Wildlife Mgt., 6(3):221–224.

MacMullan, R. A.
1954 The life and times of Michigan pheasants. Michigan Dept. Conserv., Game Div., 63 pp.

Manville, R. H.
1948 The vertebrate fauna of the Huron Mountains, Michigan. American Midl. Nat., 39(3):615–640.
1950 The mammals of Drummond Island, Michigan. Jour. Mammalogy, 31(3):358–359.

Mech, L. D.
1966 The wolves of Isle Royale. National Park Serv., Fauna Ser. 7, xiv+210 pp.

Montgomery, G. G.
1974 Communication in red fox dyads: A computer simulation study. Smithsonian Contr. Zool., No. 187, iii+30 pp.

Morgan, B. B.
1944 Host list of the genus *Trichomonas* (Protozoa: Flagellata). Part II. Host-parasite list. Trans. Wisconsin Acad. Sci., Arts, and Ltrs., Papers, 35:235–245.

Murie, A.
1936 Following fox trails. Univ. Michigan, Mus. Zool., Misc. Publ. No. 32, 45 pp.

Ozoga, J. J., C. S. Bienz, and L. J. Verme
1982 Red fox feeding habits in relation to fawn mortality. Jour. Wildlife Mgt., 46(1):242–243.

Ozoga, J. J., and C. J. Phillips
1964 Mammals of Beaver Island, Michigan. Michigan State Univ., Publ. Mus., Biol. Ser., 2(6):305–348.

Payne, N. F., and C. Finlay
1975 Red fox attack on beaver. Canadian Field-Nat., 89(4): 450–451.

Petersen, E. T.
1979 Hunters heritage. A history of hunting in Michigan. Michigan United Conserv. Clubs, Lansing. 55 pp.

Peterson, R. O.
1977 Wolf ecology and prey relationships on Isle Royale. National Park Serv., Sci. Monog., No. 11, xx+210 pp.

Phillips, C. J., J. J. Ozoga, and L. C. Drew
1965 The land vertebrates of Garden Island, Michigan. Jack-Pine Warbler, 43(1):20–25.

Phillips, R. L., R. D. Andrews, G. L. Storm, and R. A. Bishop
1972 Dispersal and mortality of red foxes. Jour. Wildlife Mgt., 36(2):237–247.

Phillips, R. L., and L. D. Mech
1970 Homing behavior of a red fox. Jour. Mammalogy, 51(3):621.

Pils, C. M., and M. A. Martin
1978 Population dynamics, predatory-prey relationships and management of the red fox in Wisconsin. Wisconsin Dept. Nat. Res., Tech. Bull. No. 105, 56 pp.

Pils, C. M., M. A. Martin, and E. L. Lange
1981 Harvest, age structure, survivorship, and productivity of red foxes in Wisconsin, 1975–1978. Wisconsin Dept. Nat. Res., Tech. Bull. No. 125, 21 pp.

Preston, E. M.
1975 Home range defense in the red fox, *Vulpes vulpes L.* Jour. Mammalogy, 56(3):645–652.

Pryor, L. B.
1956 Sarcoptic mange in wild foxes in Pennsylvania. Jour. Mammalogy, 37(1):90–93.

Sargeant, A. B., S. H. Allen, and D. H. Johnson
1981 Determination of age and whelping dates of live red fox pups. Jour. Wildlife Mgt., 45(3):760–765.

Scharf, W. C., and K. R. Stewart
1980 New records of Siphonaptera from northern Michigan. Great Lakes Ento., 13(3):165–167.

Schmeltz, L. L., and J. O. Whitaker, Jr.
1977 Use of woodchuck burrows by woodchucks and other mammals. Trans. Kentucky Acad. Sci., 38(1–2):79–82.

Schofield, R. D.
1958 Litter size and age ratios in Michigan red foxes. Jour. Wildlife Mgt., 22(3):313–315.
1960a Determining hunting waste of deer by following fox trails. Jour. Wildlife Mgt., 24(3):342–244.
1960b A thousand miles of fox trails in Michigan's ruffed grouse range. Jour. Wildlife Mgt., 24(4):432–434.

Scott, T. G.
1943 Some food coactions of the northern plains red fox. Ecol. Monog., 13:427–479.

Scott, T. G., and W. D. Klimstra
1955 Red foxes and a declining prey population. Southern Illinois Univ., Monog. ser. No. 1, 123 pp.

Sheldon, W. G.
1949 Reproductive behavior of foxes in New York state. Jour. Mammalogy, 30(3):236–246.

Shelton, P. C.
1966 Ecological studies of beavers, wolves, and moose in Isle Royale National Park, Michigan. Purdue Univ., unpubl. Ph.D. dissertation, 308 pp.

Shick, C. A.
1952 A study of pheasants on a 9,000 acre prairie farm,

Saginaw County, Michigan. Michigan Dept. Conserv., 134 pp.

Shugart, G. W., and W. C. Scharf
1977 Predation and dispersal of herring gull nests. Wilson Bull., 89(3):472–473.

Smith, R. P.
1978 Regional report. Upper Peninsula. Michigan Out-of-Doors, 32(6):29.

Smith, W. P.
1944 Red fox's method of hunting mice. Jour. Mammalogy, 25(1):90–91.

Stanley, W. C.
1963 Habits of the red fox in northeastern Kansas. Univ. Kansas, Mus. Nat. Hist., Misc. Publ. No. 34, 31 pp.

Stebler, A. M.
1944 Fox and coyote trapping simplified. Michigan Dept. Conserv., 15 pp.

Storm, G. L.
1965 Movements and activities of foxes as determined by radio-tracking. Jour. Wildlife Mgt., 29(1):1–13.

Storm, G. L., R. D. Andrews, P. L. Phillips, R. A. Bishop, D. B. Siniff, and J. R. Tester
1976 Morphology, reproduction, dispersal, and mortality of midwestern red fox populations. Wildlife Monog., No. 49, 82 pp.

Strang, J. J.
1855 Some remarks on the natural history of Beaver Island, Michigan. Smithsonian Inst., 9th Ann. Rept. for 1854, pp. 282–288.

Stuht, J. N., and W. G. Youatt
1972 Heartworms and lung flukes from red foxes in Michigan. Jour. Wildlife Mgt., 36(1):166–170.

Switzenburg, D. R.
1950 Breeding productivity in Michigan red foxes. Jour. Mammalogy, 31(2):194–195.
1951 Examination of a state fox bounty. Jour. Wildlife Mgt., 15(3):288–299.

Timm, R. M.
1975 Distribution, natural history, and parasites of mammals of Cook County, Minnesota. Univ. Minnesota, Bell Mus. Nat. Hist., Occas. Papers No. 14, 56 pp.

Tullar, B. F., and L. T. Berchielli, Jr.
1980 Movement of the red fox in Central New York. New York Fish and Game Jour., 27(2):177–204.

Wood, N. A.
1914 Results of the Shiras Expedition to Whitefish Point, Michigan—Mammals. Michigan Acad. Sci., 16th Rept., pp. 92–98.
1922 The mammals of Washtenaw County, Michigan. Univ. Michigan, Mus. Zool., Occas. Papers No. 123, 23 pp.

Wood, N. A., and L. R. Dice
1923 Records of the distribution of Michigan mammals. Michigan Acad. Sci., Arts and Ltrs., Papers, 3:425–469.

Youatt, W. G., J. N. Stuht, and H. D. Harte
1971 Wildlife rabies in Michigan 1958–1970. Michigan Dept. Nat. Res., Res. and Dev. Rept. No. 229, 6 pp.

Zimen, E., Ed.
1980 The red fox. Symposium on behaviour and ecology. Biogeographica, vol. 18, vi+285 pp.

GRAY FOX
Urocyon cinereoargenteus (Schreber)

NAMES. The generic name *Urocyon,* proposed by Baird in 1858, is of Greek origin, is translated as "tailed dog" and is derived from *oura* meaning "tail" and *kyon* meaning "dog." The specific name *cinereoargenteus,* proposed by Schreber in 1775, is of Latin origin and is translated as "silvery gray" and is derived from *cinereus* meaning "ash colored" or "gray" and *argenteus* meaning "silvery." Studies are currently being made which may show that the gray fox could be reclassified in the genus *Canis* (Van Gelder, 1978). In Michigan, this small canid is usually called gray fox, although it is sometimes referred to as tree fox or cat fox. The subspecies *Urocyon cinereoargenteus ocythous* Bangs, with affinity with gray foxes to the west in Wisconsin (Long, 1974), occurs in the Upper Peninsula. The subspecies *U. c. cinereoargenteus* (Schreber), with affinity with gray foxes in the eastern states, occurs in the Lower Peninsula.

RECOGNITION. Body size resembles that of a slender, small dog (Plate 5 and Fig. 72); head and body averaging 24 in (610 mm) long; tail long and bushy, averaging 15 in (381 mm) in vertebrae length; head broad with pointed nose; eyes mod-

Figure 72. The gray fox (*Urocyon cinereoargenteus*). Sketches by Bonnie Marris.

erate, pupils elliptical; ears prominent and erect; body somewhat robust; limbs medium and slender, feet small, large toe pads, digits terminating in long, curved claws; forefeet with five digits, small pollex high on the inside of the foreleg; hind feet with four digits; mammary glands consist of four pairs (possibly all are not functional), one on the chest, one on the belly, and two posterior between the hind limbs; scent gland at base of tail on dorsal side elongated in shape; males slightly larger (as much as 10%) than females.

The pelage of the gray fox is dense but coarse in texture in comparison with that of the red fox. However, the coloring is noticeably different. The buffy underfur of the gray fox is covered by grizzled gray guard hairs from the forehead to the rump. These long, harsh guard hairs have subterminal white bands and black tips. The nose is black with a muzzle patch forming a line of black fur extending to and around each eye (Plate 00). The upper sides of the ears are ferruginous; the lower sides are white. The upper lip, throat, and belly are white. The cheeks, chest, and legs are a buffy ochre. The tail is grizzled much like the back with a prominent dorsal mane terminating in a black-tipped tail.

The two Michigan foxes are much the same size. The gray fox is only slightly smaller, more robust in build, has shorter limbs and feet, has a shorter and broader head, has distinctive coloring as described, and has a black-tipped tail instead of a white-tipped one.

MEASUREMENTS AND WEIGHTS. Adult gray foxes measure as follows: length from tip of nose to end of tail vertebrae, 37½ to 41 in (950 to 1,050 mm); length of tail vertebrae, 13 to 15½ in (330 to 390 mm); length of hind foot, 5⅜ to 5¾ in (135 to 145 mm); height of ear from 2¾ to 3¼ in (70 to 80 mm); weight, 8 to 15 lbs (3.6 to 6.8 kg).

DISTINCTIVE CRANIAL AND DENTAL CHARACTERISTICS. The skull of the gray fox is low in profile and rather slender. The postorbital processes are short and heavy, with a depressed area extending longitudinally on the interior and dorsal surfaces. The rostrum is short; the ends of the maxillary bones extend no further posteriorly than do those of the nasal bones. Conspicuous temporal ridges extend from the supraorbital processes posteriorly in a "U-shaped" configuration to the occipital crest without uniting to form a median sagittal crest (see Fig. 73). The auditory bullae are prominent, short, and broad. The teeth are rather small for a canid. The upper incisors are not lobed. The lower margin on each side of the lower jaw possesses a distinct notch (step) near the posterior end. The skull of the gray fox averages 4⅞ in (124 mm) long and 2¾ in (70 mm) wide across the zygomatic arches.

The skull of the gray fox most closely resembles in size and shape that of the red fox, being conspicuously smaller and less massive than the skulls of the coyote or the gray wolf. From the skull of the red fox, that of the gray fox differs by (1) the space between the parietal ridges at the junction of the frontal and parietal bones being more than ⅜ in (10 mm), instead of less; (2) the

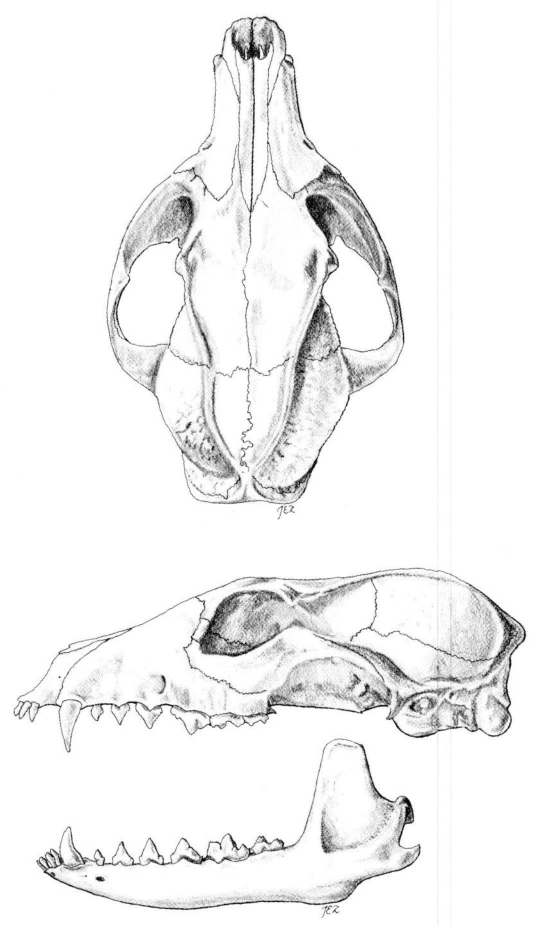

Figure 73. Cranial and dental features of the gray fox (*Urocyon cinereoargenteus*).

ratio of the length to the width of the auditory bullae being about 8:7, instead of 10:7; (3) the upper incisors being evenly rounded when viewed from the front, instead of lobed; and (4) having the lower margin on each side of the lower jaw conspicuously notched (a distinct step) near the posterior end, instead of gently curving. The dental formula for the gray fox is:

$$\text{I (incisors)}\,\frac{3\text{-}3}{3\text{-}3},\ \text{C (canines)}\,\frac{1\text{-}1}{1\text{-}1},\ \text{P (premolars)}\,\frac{4\text{-}4}{4\text{-}4},\ \text{M (molars)}\,\frac{2\text{-}2}{3\text{-}3} = 42$$

DISTRIBUTION. The gray fox, with the eastern cottontail, raccoon, long-tailed weasel, extirpated mountain lion, and white-tailed deer, enjoys an extensive geographic distribution throughout most of the southern half of North America and also southward into northern South America. In Michigan and adjacent parts of southern Ontario (just north of lakes Erie and Huron), the gray fox is at the northern edge of its extensive tropical and temperate range. The problems of survival in this peripheral habitat seem reflected in the observed ups and downs which this species has experienced in the Upper Great Lakes Region (deVos, 1964). Yet, the gray fox is certainly not a new arrival; it is known from archaeological sites of the Woodland Period (1,090 ± 75 years B.P.) in southwestern Michigan (Biesele *in* Fitting, 1968) and from historic sites from as far back as 300 years B.P. (Downing, 1946; Peterson *et al.*, 1953). There are periods of time when no records were obtained for the Michigan area (Allen, 1940; Michigan Dept. Conserv., Ninth Biennial Rept., 1937–1938), but it is likely the species has persisted in low numbers, especially in the temperate deciduous forest habitat.

Although the gray fox has not been officially recorded from every Michigan county, it can be expected to occur in all of them (see Map 49).

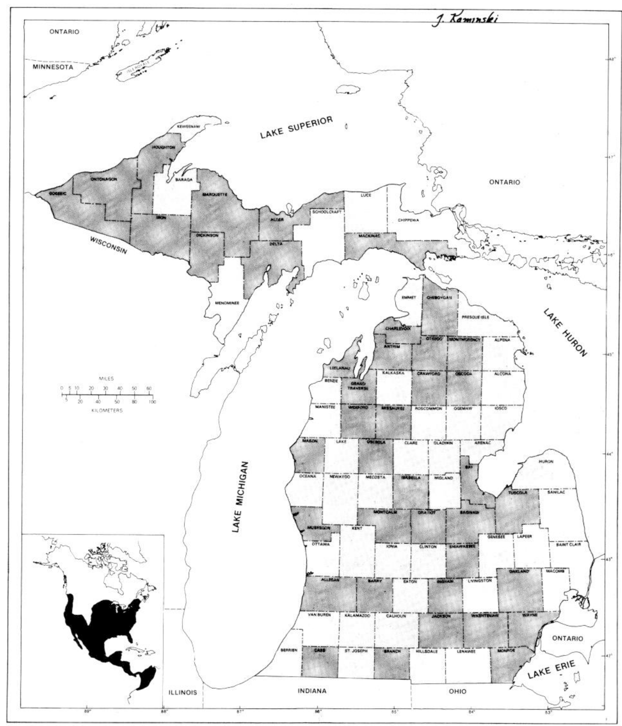

Map 49. Geographic distribution of the gray fox (*Urocyon cinereoargenteus*). Counties from which there are actual records for this species are heavily shaded.

Records of the species in the Upper Peninsula are from a mosaic of counties all the way east to Mackinac County (Smith, 1980) and throughout the Lower Peninsula north to Charlevoix County (Michigan Dept. Conserv., Ninth Biennial Rept., 1937–1938). Apparently the gray fox reached southeastern Ontario by crossing the Detroit or St. Clair rivers, but there is no evidence this fox has made its way into Canada from the Upper Peninsula by crossing St. Marys River or adjacent waterways between lakes Superior and Huron (Banfield, 1974).

HABITAT PREFERENCES. The Michigan gray fox, in marked contrast to the red fox, is primarily a woodland creature, although their ecological ranges overlap considerably. However, the gray fox's preferred Michigan habitat is poorly known because so few field observations have been reported. In Osceola County, Wenzel (1912) found the gray fox using ". . . large boreal islands (tamarack and cedar swamps), which furnish it the best protection." In Wisconsin, Jackson (1961) stressed the gray fox's use of hardwoods or mixed hardwood and conifer forest, brushlands, bottomland woods, and rough, hilly terrain. In Indiana, Mumford (1969) noted that the gray fox uses brushy and wooded habitats in preference to the more open areas which attract red foxes.

DENSITY AND MOVEMENTS. All available information concerning Michigan gray foxes points to a continual low population, probably because the state is peripheral to this animal's extensive southern distribution. Nevertheless, gray fox populations seem to have multiannular fluctuations, reaching peak densities about every 10 years. This tendency may be responsible for reports that the species was increasing in the Jackson area of southern Michigan in 1979 (D. A. Arnold, pers. comm., February 1980). Even so, with such generally low populations in Michigan, it is likely that home-range sizes (normally occupied by family groups) are unpredictably large and/or irregular in comparison with those of the populations in more southern areas. Habitat quality in terms of food, water, and shelter as well as possible interspecific competition affect gray fox populations (Trapp and Hallberg, 1975), but it is not known which aspect is limiting to Michigan gray foxes.

In southeastern United States, Lord (1961a) found densities of three or four gray foxes per sq mi (260 ha), based on data obtained by capture, mark, and recapture studies. Richards and Hine (1953) observed about the same density for gray foxes in Wisconsin, with some ranges as small as 145 acres (59 ha) and others as extensive as 2,155 acres (872 ha). In California, female gray foxes fitted with radio-collars showed considerable seasonal variation in home-range size. Fuller (1978) found areas used in winter (January 18 to February 20) were larger, 457 acres (185 ha), perhaps because of the considerable moving around as a result of courtship activity. In late spring and early summer (May 6 to July 10) home areas were smaller, 74 acres (30 ha), reduced perhaps because of maternal duties. Gray fox home ranges are generally smaller than those of the red fox.

BEHAVIOR. Most of the comments on the behavior of the red fox in the preceding species account apply to the gray fox also. The major behavioral difference is the gray fox's interesting ability to climb trees (Seton, 1953; Yeager, 1938), causing it to be called "cat fox" and "swamp cat" in southern states (Heinold, 1975). The gray fox's claws are more hooked than those of the red fox, an adaptation which helps in negotiating steep trunks. However, the gray fox climbs or descends primarily by jumping from branch to branch. Treeing allows the gray fox to elude dog packs and perhaps other potential enemies.

The gray fox is active mostly at night or in the dim light of early morning or late afternoon. Its usual gait is a trot, but speeds of up to 26 and 28 mi (42 to 45 km) per hour have been recorded (Cottam, 1937; Jackson, 1961). The gray fox swims in a clumsy, dog-like fashion.

There is evidence that the gray fox is less noisy than the red fox. Its barks and yaps are perhaps coarser, lower in pitch, and less loud. In typical canid fashion, the gray fox communicates by urinating on various objects (fence posts, rocks, logs, trunks, ant hills, etc.) at strategic places in its home range (Richards and Hine, 1953; Sullivan, 1956). Its senses are keen. Its antics and behavior toward other family members are much like those of the red fox. The family unit, including the male, female, and young, maintains a close relationship during much of the year. Both mates assist with pup-rearing. In autumn, the young disperse in the usual "fall shuffle" to be relocated at distances as far away as 50 mi (80 km). The male parent may also abandon the female at this time, whereas mated pairs of red fox often stay together the year around.

Gray foxes select a variety of ground dens including abandoned woodchuck and badger burrows (Layne and McKeon, 1956b). Although underground burrows may be better insulated, the animals also use hollow trees, rocky ledges, stumps, and rock piles. The gray fox is more likely to use dens in winter than is the red fox (Seton, 1953); such shelters are often used chiefly as daytime retreats.

ASSOCIATES. Like other carnivores, the gray fox probably has few associates except its own kind. It may follow coyotes or other larger meat-eaters in hopes of feeding on their kills and probably joins other species in eating such carrion. Its relationships to the red fox are little known. Certainly when occurring together, as least in eastern North America, these two canids show different habitat preferences. The gray fox lives in forested areas while the red fox prefers more open habitat, resulting mostly from human land-use. Studies in Indiana have shown that these two foxes tend to overlap at forest-field edges in areas grown to weeds and brush. Presumably, the gray fox, at the periphery of its range in Michigan, may be at a disadvantage in competing for living space with the more northern-adapted red fox.

REPRODUCTIVE ACTIVITIES. The courtship period occurs from late winter to early spring with actual mating activites culminating in early March (Layne and McKeon, 1956a; Richards and Hine, 1953; Sheldon, 1949). At this time, a vixen may be pursued by several males but ultimately selects a single mate and together they prepare a den for the offspring. Two vixens will occasionally use the same den simultaneously. After a gestation period of about 53 days, a litter averaging four (as few as one and as many as seven) is born, probably in May in Michigan. At birth, pups weigh 3.5 oz (100 g), have closed eyes, are almost black in color, and have sparse, wooly fur. At nine to 12 days of age, the eyes open and the pelage thickens. At approximately 20 days, the pups begin to eat solid food, usually brought to the den by the male parent. At 30 days, the wooly fur begins to shed and a more adult-like pelage develops. At 80 days, the pups are generally weaned and can actively follow their parents on hunting forays. At about 120 days, the young foxes have their permanent teeth and are able to forage for themselves, although the family remains together until autumn when the young reach adult size and disperse. Sexual maturity is reached during the first year, with about 54% of the annual young produced by these first-year females, according to a study in southeastern United States (Wood, 1958). Although most of the gray fox population examined by Wood (also see Lord, 1961b) was less than one year old (61.1%), there is evidence based on tagged individuals, that the life span in nature may be as much as six to 10 years (Schwartz and Schwartz, 1981).

FOOD HABITS. Little is known of the foods eaten by the gray fox in Michigan. However, based on studies conducted in other states, this canid, generally a solitary hunter, has been shown to have varied tastes similar to those of the red fox. Although there is evidence that the gray fox eats a larger amount of vegetable matter (fruits) and invertebrates (anthropods) than does the red fox, availability and ease of obtaining food seem to be of major importance and diet selections change from season to season. This is especially true of fruits and other plant foods. The opportunistic eating behavior of the gray fox was noted by Mumford (1969) who observed one of these canids more than ¼ mi (450 m) offshore on the lake ice on December 30 at Michigan City, Indiana, "evidently searching for sick and dead water birds."

Most studies have shown that the eastern cottontail may be the most important year-around food for the gray fox. Mice of the genus *Peromyscus* (usually not identified to species but probably including both Michigan's white-footed and deer mice) and voles of the genus *Microtus* (including Michigan's prairie, meadow, and woodland voles) are also fair game for the gray fox. Other food items occurring in Michigan and reported as being eaten by gray foxes in other states include: short-tailed shrew, least shrew, eastern chipmunk, woodchuck, muskrat, southern bog lemming, Norway rat, house mouse, porcupine, ermine, long-tailed weasel, least weasel, striped skunk, ruffed grouse, bobwhite, ring-necked pheasant, domestic chicken, yellow-shafted flicker, blue jay, catbird, red-winged blackbird, other birds, a variety of fish, reptiles and amphibians, grasshoppers, beetles, spiders, millipedes, other invertebrates, blueberry, grape, haws, apple, corn, acorns, other plant materials, and as carrion, domestic animals and white-tailed deer (Errington, 1935; Hatfield, 1939;

Korschgen, 1952; Latham, 1952; MacGregor, 1942; Richards and Hine, 1953; Weaver, 1939).

ENEMIES. The gray fox has few enemies in Michigan with the exception of the human. Meat-eaters such as the bobcat, coyote, and great horned owl very likely find the young to be more easily subdued than the adults (Jackson, 1961; Lowery, 1974; Progulske, 1955). Occasionally, a gray fox will be run over by an automobile. Certainly most of the enemies of the red fox would find the gray fox equally useful as food.

PARASITES AND DISEASES. External parasites include lice (Anoplura), fleas (Siphonaptera), ticks, and mites (Arachnida). The mange mite (*Sarcoptes scabei*) seems to be a less serious problem to the gray fox than it is to the red fox or coyote (Pryor, 1956). Internally, parasites include species of flukes (Trematoda), tapeworms (Cestoda), roundworms (Nematoda), and one thorny-headed worm (Acanthocephala). Heartworm (*Dirofilaria immites*) has been reported in one Michigan gray fox (Stuht, 1978). Rabies is, of course, always a threat to gray fox populations.

MOLT AND COLOR ABERRATIONS. In typical canid fashion, the gray fox undergoes a single and rather prolonged annual molt in summer. Unlike the red fox, the gray fox has no color phases of economic importance in the fur trade. Black (melanistic) variants are known (Jones, 1923).

ECONOMIC IMPORTANCE. The gray fox has never been sufficiently common in Michigan to be a serious threat to poultry, or to be considered a menace to upland game birds and eastern cottontails. It was not even included in the bounty which was offered for the red fox between 1947 and 1965. It was, however, an unprotected species until 1980, when hunting was limited to the same period as for the red fox—from October 1 to March 1 in the Upper Peninsula and northern Lower Peninsula and from November 1 or March 1 in the southern Lower Peninsula. Nevertheless, many fox hunters find the tree-climbing gray fox has less sporting qualities than the red fox (Brown and Yeager, 1943).

Like the red fox, the gray fox occasionally eats a game bird, eastern cottontail, song bird, eggs of ground-nesting birds, and poultry. If this does happen in Michigan, the more common red fox will probably be blamed. All in all, as Jackson (1961) remarked about Wisconsin gray foxes, this canid is not "a bad citizen."

Gray fox fur is full and thick but lacks the highly marketable soft, luxurious qualities—or the color phases—of that of the red fox (Jones and Aulerich, 1981). It has, however, figured in the Michigan fur trade, even as far back as 1835–1839 (Johnson, 1969), when the American Fur Company received 66 pelts from their Northern Department (including Upper Michigan) and 4,099 from their Detroit Department (including Lower Michigan). Michigan gray fox pelts brought a top price of no more than $15 in the trapping season of 1973–1974. The rising popularity of "long-haired" fur brought the price of a gray fox pelt up to as high as $40 in the 1976–1977 season; red fox pelts marketed that same season were at least $20 higher.

COUNTY RECORDS FROM MICHIGAN. Records of specimens of gray foxes in collections of museums, from the literature, and reported by personnel of the Michigan Department of Natural Resources show that this canid is likely to occur in all counties in both the Upper and Lower Peninsula. Actual reports are from the following counties in the Upper Peninsula: Alger, Delta, Dickinson, Gogebic, Houghton, Iron, Mackinac, Marquette, and Ontonagon; in the Lower Peninsula: Allegan, Antrim, Barry, Bay, Branch, Cass, Charlevoix, Cheboygan, Crawford, Grand Traverse, Gratiot, Ingham, Isabella, Jackson, Leelanau, Mason, Missaukee, Monroe, Montcalm, Montmorency, Muskegon, Oakland, Osceola, Oscoda, Otsego, Saginaw, Shiawassee, Tuscola, Washtenaw, Wayne, and Wexford (Allen, 1940; Biesele *in* Fitting, 1968; Burt, 1946; Dice and Sherman, 1922; Ozoga and Verme, 1966; Perry, 1899; N. F. Sloan, pers. comm., September 19, 1972 for Houghton County; Smith, 1980; Stuht, 1978; Wenzel, 1912; Wood, 1922; Wood and Dice, 1923).

LITERATURE CITED

Allen, D. L.
1940 Two recent mammal records from Allegan County, Michigan. Jour. Mammalogy, 21(4):459–460.

Banfield, A. W. F.
1974 The mammals of Canada. Univ. Toronto Press, Toronto. xxiv+438 pp.

Brown, L. G., and L. E. Yeager
1943 Survey of the Illinois fur resource. Illinois Nat. Hist. Surv., Bull., Vol. 22(6):429–504.

Burt, W. H.
1946 The mammals of Michigan. Univ. Michigan Press, Ann Arbor. xv+288 pp.

Cottam, C.
1937 Speed of the gray fox. Jour. Mammalogy, 18(2):240–241.

deVos, A.
1964 Range changes of mammals in the Great Lakes Region. American Midl. Nat., 71(1):210–231.

Dice, L. R., and H. B. Sherman
1922 Notes on the mammals of Gogebic and Ontonagon counties, Michigan, 1920. Univ. Michigan, Mus. Zool., Occas. Papers, No. 109, 40 pp.

Downing, S. C.
1946 The history of the gray fox in Ontario. Canadian Field-Nat., 60(2):45–46.

Errington, P. L.
1935 Food habits of mid-west foxes. Jour. Mammalogy, 16(3):192–200.

Fitting, J. E.
1968 The Spring Creek Site, 20 MU 3, Muskegon County, Michigan. Pp. 1–80 *in* J. E. Fitting, J. R. Halsey, and M. Wobst, eds. Contributions to Michigan Archaeology. Univ. Michigan, Mus. Anthro., Anthro Papers, No. 32, v+275 pp.

Fuller, T. K.
1978 Variable home-range sizes of female gray foxes. Jour. Mammalogy, 59(2):446–449.

Hatfield, D. M.
1939 Winter food habits of foxes in Minnesota. Jour. Mammalogy, 20(2):202–206.

Heinold, G.
1975 The cat-fox. Ford Times, 68(3):56–58.

Jackson, H. H. T.
1961 Mammals of Wisconsin. Univ. Wisconsin Press, Madison. xii+504 pp.

Johnson, D. R.
1969 Returns of the American Fur Company, 1835–1839. Jour. Mammalogy, 50(4):836–839.

Jones, S. V. H.
1923 Color variations in wild mammals. Jour. Mammalogy, 4(3):172–177.

Jones, R. E., and R. J. Aulerich
1981 Bibliography of foxes. Michigan State Univ., Agric. Exp. Sta., Jour. Art. No. 9871, 141 pp.

Korschgen, L. J.
1952 A general summary of the food of Missouri predatory and game animals. Missouri Conserv. Comm., 61 pp.

Latham, R. M.
1952 The fox as a factor in the control of weasel populations. Jour. Wildlife Mgt., 16(4):516–517.

Layne, J. N., and W. H. McKeon
1956a Some aspects of red fox and gray fox reproduction in New York. New York Fish and Game Jour., 3(1):44–74.
1956b Notes on red fox and gray fox den sites in New York. New York Fish and Game Jour., 3:248–249.

Long, C. A.
1974 Environmental status of the Lake Michigan region. Volume 15. Mammals of the Lake Michigan drainage basin. Argonne Natl. Lab., Enviro. and Earth Sci. Studies, ANL/ES-40, Vol. 15, 108 pp.

Lord, R. D., Jr.
1961a A population study of the gray fox. American Midl. Nat., 66(1):87–109.
1961b The lens as an indicator of age in the gray fox. Jour. Mammalogy, 42(1):109–111.

Lowery, G. H., Jr.
1974 The mammals of Louisiana and its adjacent waters. Louisiana State Univ. Press, Baton Rouge. xxiii+565 pp.

MacGregor, A. E.
1942 Late fall and winter food of foxes in central Massachusetts. Jour. Wildlife Mgt., 6(3):221–224.

Mumford, R. E.
1969 Distribution of the mammals of Indiana. Indiana Acad. Sci., Monog. No. 1, vii+114 pp.

Ozoga, J. J., and L. J. Verme
1966 Noteworthy locality records for some upper Michigan mammals. Jack-Pine Warbler, 44(1):52.

Perry, O. H.
1899 Hunting expeditions of Oliver Hazard Perry of Cleveland verbatim from his diaries. Cleveland. viii+246 pp.

Peterson, R. L., R. O. Standfield, E. H. McEwen, and A. C. Brooks
1953 Early records of the red fox and the gray fox in Ontario. Jour. Mammalogy, 34(1):126–127.

Progulske, D. R.
1955 Game animals utilized as food by the bobcat in the southern Appalachians. Jour. Wildlife Mgt., 19(2): 249–253.

Pryor, L. B.
1956 Sarcoptic mange in wild foxes in Pennsylvania. Jour. Mammalogy, 37(1):90–93.

Richards, S. H., and R. L. Hine
1953 Wisconsin fox populations. Wisconsin Conserv. Dept., Game Mgt. Div., Tech. Wildlife Bull. No. 6, 78 pp.

Schwartz, C. W., and E. R. Schwartz
1981 The wild mammals of Missouri. Revised Edition. Univ. Missouri Press and Missouri Dept. of Conserv., Columbia. viii+356 pp.

Seton, E. T.
1953 Lives of game animals. Vol. I, Part II. Cats, wolves and foxes. Charles T. Branford Co., Boston. pp. 340–640.

Sheldon, W. G.
1949 Reproductive behavior of foxes in New York state. Jour. Mammalogy, 30(3):236–246.

Smith, R. P.
1980 Regional report. Upper Peninsula. Michigan Out-of-Doors, 34(1):71.

Stuht, J. N.
1978 Dirofilariasis in a gray fox. Michigan Dept. Nat. Res., Wildlife Div., Rept. No. 2821, 2 pp.

Sullivan, E. O.
1956 Gray fox reproduction, denning, range, and weights in Alabama. Jour. Mammalogy, 37(3):346–351.

Trapp, G. R., and D. L. Hallberg
1975 Ecology of the gray fox (*Urocyon cinereoargenteus*): A review. Pp. 164–178 *in* M. W. Fox, ed. The wild canids: Their systematics, behavioral ecology, and evolution. Van Nostrand and Reinhold, New York. xvi+508 pp.

Van Gelder, R. G.
1978 A review of canid classification. American Mus. Novit., No. 2646, 10 pp.

Weaver, R. L.
1939 Attacks on porcupine by gray fox and wild cats. Jour. Mammalogy, 20(3):379.

Wenzel, O. J.
1912 A collection of mammals from Osceola County, Michigan. Michigan Acad. Sci., 14th Rept., pp. 198–205.

Wood, J. E.
1958 Age structure and productivity of a gray fox population. Jour. Mammalogy, 39(1):74–86.

Wood, N. A.
1922 The mammals of Washtenaw County, Michigan. Univ. Michigan, Mus. Zool., Occas. Papers No. 123, 23 pp.

Wood, N. A., and L. R. Dice
1923 Records of the distribution of Michigan mammals. Michigan Acad. Sci., Arts and Ltrs., Papers, 3:425–469.

Yeager, L. E.
1938 Tree-climbing by a gray fox. Jour. Mammalogy, 19(3): 376.

FAMILY URSIDAE

The bear family originated in mid-Tertiary geologic times (middle Miocene) in Europe and first appears in the fossil record of North America in the late Pliocene and of South America in the Pleistocene. Today, bears occur in most regions of the world with the exception of Sub-Sahara Africa and the Australian region. Bears have never been highly diversified; modern forms are classified into only six genera and eight species. Most species are noted for their large size, although some are as small as a large dog. The brown bear (*Ursus arctos*), with subspecies found throughout most of the Northern Hemisphere, is commonly called grizzly, Kodiak bear, or Peninsula bear depending on its distribution in North America. This large animal and the polar bear (*Ursus maritimus*) are the most massive terrestrial carnivores, some individuals weighing more than 1,600 lbs (726 kg).

Although bears are true carnivores in terms of lineage and basic characteristics, these often-ponderous creatures (a) have retained the long rostrum typical of the canids; (b) have carnassial teeth (4th upper premolars and 1st lower molars) which lack the shearing configurations typical of most carnivores and instead, resembling other jaw teeth, are flattened and low-crowned (bunodont), crushing structures useful for masticating a variety of fibrous plants as well as animal foods; (c) possess heavy and strongly built limbs; (d) have feet with five toes, each with a long, non-retractile claw; (e) place the entire foot flat on the ground (plantigrade); and (f) have extremely short tails, small ears, heavy and often shaggy pelage, and a generally lumbering gait. Only one species, the black bear *(Ursus americanus),* is a native of Michigan.

BLACK BEAR
Ursus americanus Pallas

NAMES. The generic name *Ursus,* proposed by Linnaeus in 1758, is derived from the Greek word for "bear." The specific name *americanus,* proposed by Pallas in 1780, is the Latinized form of the word American, *americ,* plus the suffix *anus* which means "belonging to" when added to the noun stem. In much of the older literature, the scientific name, *Euarctos americanus,* is used. In Michigan, this large animal is either called black bear or bear. The subspecies *Ursus americanus americanus* Pallas occurs in Michigan.

RECOGNITION. Body size massive (Fig. 74), resembling a very large domestic hog, although more elongated in appearance; head and body averaging 63 in (1,600 mm); tail short and inconspicuous, averaging 4 in (102 mm) in vertebral length; head large with long tapering nose; nose pad broad; openings of nostrils large; facial profile rather straight; eyes small and black; ears round and prominent; head, short neck, shoulders, and back form almost a straight line when the animal's body outline is viewed from the side; legs heavily constructed; feet large, broad and flat (plantigrade); most of sole on each hind foot bare; five digits on both front and hind feet; each digit equipped with a short, curved, non-retractile claw; mammary glands consist of three pairs, two on the chest and one on the lower belly; males are about 10% larger than females.

Figure 74. The black bear (*Ursus americanus*).
Sketches by Bonnie Marris

The pelage of the black bear is long, dense, coarse, and glossy. Except for a cinnamon-brown muzzle and a white blotch (often V-shaped) on the lower throat, the normal body coloring is black or very dark brown. Color phases are well-known and discussed in the section on Molt and Color Aberrations.

The large size, long, shiny-black pelage, and lumbering gait distinguish this Michigan carnivore from all other mammals in the state.

MEASUREMENTS AND WEIGHTS. Adult specimens of the larger male and the smaller female black bear measure, respectively, as follows: length from tip of nose to end of tail vertebrae, 54 to 70 in (1,370 to 1,800 mm) and 47 to 59 in (1,200 to 1,500 mm); length of tail vertebrae, 3½ to 5 in (90 to 125 mm) and 3 to 4½ in (80 to 115 mm); length of hind foot, 8¾ to 11 in (220 to 280 mm) and 7½ to 9½ in (190 to 240 mm); height of ear from notch, 4¼ to 5¼ in (110 to 135 mm) and 4 to 4¾ in (100 to 120 mm); weight, 250 to 500 lbs (113 to 227 kg) and 225 to 450 lbs (102 to 204 kg). Michigan black bear will occasionally weigh more than 600 lbs (272 kg), although determining the exact weight of such a creature prior to its being field-dressed (about 15% of the weight is viscera and fluids) is rarely possible.

DISTINCTIVE CRANIAL AND DENTAL CHARACTERISTICS. The skull of the black bear is large, with an impressive cranium, a conspicuous sagittal crest, and heavy zygomatic arches (Fig. 75). The upper profile is gently rounded with a flattened forehead; auditory bullae flattened and rather inconspicuous; nasal openings large, exposing the well-developed turbinate bones; the posterior edge of the bony palate extending well behind the end of the molar tooth-row; the lower jaw short and heavy. The upper incisors are approximately the same length; the canines long, pointed, and heavily built; the premolars (except for the 4th in both upper and lower jaws) small, weak and often absent in older individuals; and the molars broad with flat crowns and the last upper molar the most elongate. Black bears can be aged by counting the number of cementum layers on the canines or first premolars (Hildebrandt, 1976; Marks and Erickson, 1966; Willey, 1974). Sex can

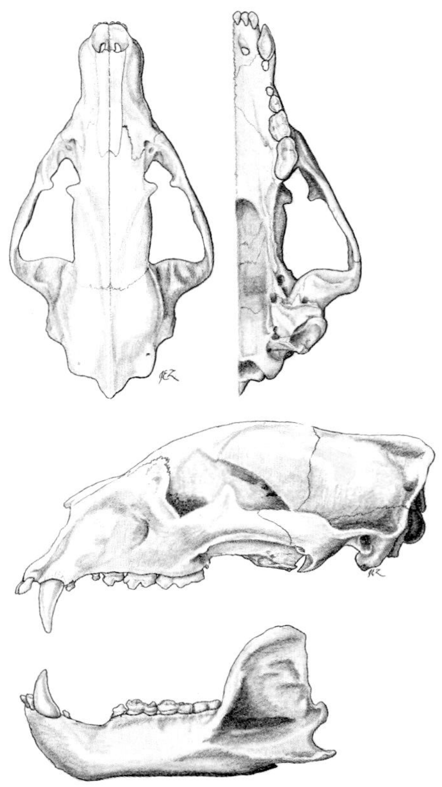

Figure 75. Cranial and dental characteristics of the black bear (*Ursus americanus*).

be determined by the size of the lower canine teeth (Sauer, 1966). The skull continues to grow until the animal is at least six years old. By this age the forehead has terminated its expansion (bulging) and becomes more flattened and the sagittal crest has completed its enlargement (Jackson, 1961).

The skull of the black bear is readily distinguished from those of other Michigan carnivores by its massive size, length and especially its breadth across the zygomatic arches. The skull of an adult male and female black bear, respectively, average 11¼ in (284 mm) and 10½ in (270 mm) in greatest length and 6¾ in (171 mm) and 6¼ in (160 mm) in width across the zygomatic arches. The dental formula is as follows:

$$\text{I (incisors)}\frac{3\text{-}3}{3\text{-}3}, \text{C (canines)}\frac{1\text{-}1}{1\text{-}1}, \text{P (premolars)}\frac{4\text{-}4}{4\text{-}4}, \text{M (molars)}\frac{2\text{-}2}{3\text{-}3} = 42$$

DISTRIBUTION. The black bear occupies most of temperate and boreal North America from tree line limits at the edge of the arctic prairies in northern Alaska and Canada south to central Mexico. Large and conspicuous as the black bear is, this creature has managed to maintain a scattered distribution (except for the Great Plains) through this extensive range despite modern human settlement; the gray wolf, mountain lion, wapiti, and bison have been much less fortunate.

In Michigan, the black bear has a long history of occupation, entering the area soon after melt-back of the last glaciation. Fossil remains from Kent and Oakland counties represent individuals that lived perhaps 8,000 years ago (Eshelman, 1974; Frankforter, 1966; Holman, 1975). Parts of a skeleton of a black bear obtained from silt deposits of Morrison Lake in Branch County may also represent an animal which lived in these early postglacial times (Edward Nofz, *in litt.,* October 8, 1979). In prehistoric times, black bear remains have been unearthed from Indian sites of the Woodland cultural level in Allegan, Berrien, Lapeer, Mackinac, and Saginaw counties (Cleland, 1966; Higgins, 1980; Martin, 1979). In earliest European times, Cadillac reported black bear in the Detroit area about 1700, and Father Hennepin's journal described the populations of this species in what is now Berrien County in 1679 (Engels, 1933; Johnson, 1919). In the eighteenth century, the progressively larger human settlement in the southern half of the Lower Peninsula resulted in the clearing of the black bear's forested habitat and the species' decline. By the mid-nineteenth century, both in-state and out-of-state hunters engaged in Michigan black bear hunts. Sportsman Oliver Hazard Perry of Cleveland described black bear hunting adventures in Monroe, Sanilac and Tuscola counties in the 1840s and 1850s (Burroughs, 1958; Perry, 1899).

In Michigan, black bear first faded from the scene in the southernmost counties. The distinguished Michigan naturalist, Norman Wood, provided some estimates of when black bear disappeared from southern Michigan counties (arranged from south to north): Monroe, 1843; Washtenaw, 1875; Jackson, 1842; Kalamazoo, 1885; Oakland, 1842; Livingston, 1867; Ingham, 1883; Allegan, 1845; Genesee, 1868; Shiawassee, 1897; Montcalm, 1896; (Wood, 1922 and Wood and Dice, 1923). In recent years, an occasional black bear has ventured south of the 44th parallel in southern Michigan—with animals reported in

Saginaw and Bay counties in the spring of 1976 and summer of 1979. A report from Van Buren County in 1979 represents an extreme southern range extension. However, as of 1982, the black bear is restricted to the counties of the Upper Peninsula and to all or parts of those counties in the northern one-half of the Lower Peninsula (see Map 50).

Map 50. Geographic distribution of the black bear (*Ursus americanus*). The darker shaded area depicts its present range. The counties in the extreme southern part of this area in the central Lower Peninsula are only infrequently inhabited by this mammal.

HABITAT PREFERENCES. Two basic black bear environments occur in Michigan: the northern hardwood and conifer forests of the Canadian biotic province and the southeastern hardwood forests of the Carolinian biotic province (Dice, 1943). These broadly overlap in the northcentral part of the Lower Peninsula, with clearing, lumbering, livestock raising, farming, and other human occupations generating a variety of changes in plant diversity and successional stages. Today, the black bear is almost gone from Michigan's southern Carolinian forests, except perhaps in areas of Antrim and Charlevoix counties. The black bear's major habitat today is the boreal forest in northern Michigan. Upland forests, marshes, swamps, and thickets in several plant successional stages provide forage and cover for this mammal (Dice and Sherman, 1922; Shapton, 1956; Wenzel, 1912). The best black bear habitat in Michigan combines a series of forested ridges with an interdigitation of marshes and swamps, which give the animal sufficient ranging room and food, and nearby thickets for seclusion.

DENSITY AND MOVEMENTS. The Michigan Department of Natural Resources reports the largest populations of black bear in the northwestern part of the Upper Peninsula, especially in the counties of Ontonagon, Gogebic, Marquette, and Houghton. In the Lower Peninsula, the northernmost counties, Presque Isle, Cheboygan, Alpena, Montmorency and Alcona, have fairly high black bear densities. The Dead Stream area at the head-waters of the Muskegon River in Missaukee and Roscommon counties also contains some of the best black bear habitat, and was selected as the site for studies in 1977–1980 of the movements of radio-tagged animals in relation to human occupation (Manville, 1981).

Knowledge of density and movements of the black bear has been obtained by live-trapping, tagging (often with radio-collars), releasing, and either retrapping or radio-tracking individuals. Some of the pioneering work on techniques for catching and handling these large, powerful animals was done in Michigan (Erickson, 1957; Erickson *et al.*, 1964). Eight-foot sections of 36-inch steel culverts make excellent catching devices when equipped with heavy screening at one end and a door at the other designed to close when the black bear enters in quest of bait. Less heavy devices consisting of two 55-gallon barrels have been designed to hold even large adults. In a ten-year study beginning in 1952 in the Upper Peninsula at the Cusino Wildlife Research Station, 159 different black bears were handled a total of 182 times (Erickson, 1964). In the Dead Stream Swamp of the Lower Peninsula, 35 black bears (22 males and 13 females) were captured. Of these, 25 were radio-collared allowing for 4,224 air and ground telemetric observations to be made between September 1977 and March 1980 (Manville, 1980, 1981).

Modern quantitative studies show densities of black bears to be highly variable from one habitat to another. In California, Piekielek and Burton (1975) estimated black bear density in one study area at 1 animal per 315 acres (130 ha); in Alberta, Kemp (1972) estimated 1 per 640 acres (255 ha); in Montana, Jonkel and Cowan (1971) estimated 1 per 502 to 1,102 acres (201 to 441 ha); and in Idaho, Amstrup and Beecham (1976) estimated 1 to 4,150 to 13,030 acres (1,660 to 5,212 ha). In an area of 256,000 acres (102,500 ha) in Alger and Schoolcraft counties in Michigan's Upper Peninsula, Erickson (1964) calculated a density of 1 black bear per 2,200 acres (880 ha). In the Lower Peninsula, Harger (1978) estimated 1 per 4,125 to 7,100 acres (1,650 to 3,240 ha) while Manville (1981) found higher populations amounting to 1 per 1,950 to 2,250 acres (780 to 900 ha). It was Banfield's (1974) conclusion that density in most good black bear range may be approximately 1 per 3,560 acres (1,425 ha).

Black bears require large expanses in which to roam—probably as extensive as 50,000 acres (20,000 ha), and as far as 15 mi (24 km) from a home base (Banfield, 1974). Males generally travel over larger areas (perhaps twice as great) than females. Movements of both sexes are regulated by age, dominance, territorial restrictions, breeding activities, and sources of food. On an island off southwestern Washington, Lindzey and Meslow (1977b) estimated average home ranges of radio-collared male black bears at 1,262 acres (505 ha) and females as 587 acres (235 ha). These estimates are smaller than those obtained by workers in mainland areas. In Montana, Jonkel and Cowan (1971) determined that males use an average of 7,707 acres (3,083 ha) and females, 1,295 acres (518 ha); in Great Smoky Mountains National Park, Garshelis and Pelton (1981) found adult males between July and December occupied, on the average, 10,500 acres (4,200 ha) and adult females, 3,750 acres (1,500 ha).

In Michigan's Upper Peninsula, movements of individual black bears are comparable to those living in similar habitats in Minnesota and Pennsylvania (Alt *et al.*, 1980; Rogers, 1977). In his classic work, Erickson (1964) found males moved around in areas averaging 12,955 acres (2,591 ha) and females in 6,477 acres (2,591 ha). In the Lower Peninsula, movements of radio-collared black bears in the Dead Stream Swamp area were considerably more extensive. Manville's (1981) study in this area showed that males used, on the average, 32,400 acres (12,760 ha) and females, 16,550 acres (6,620 ha).

Observations by Manville (1980, 1981) also showed that three male black bears, radio-collared in the Dead Stream Swamp, moved during the summer of 1979, respectively, 65 air mi (104 km) to western Lake County; 95 air mi (253 km) to Emmet and Cheboygan counties; and 35 air mi (56 km) to west-central Claire County. After these unusually long excursions, the animals eventually returned to their home areas by early November of the same year.

The black bear will be attracted by food concentrations even from considerable distances. In 1679 during the construction of Fort Miasmis at the mouth of the St. Joseph River in what is now Berrien County, Father Hennepin wrote that the workers threatened mutiny because of a bear meat diet. It seems that the black bear was the most abundant large animal available to the camp. According to Father Hennepin, the bears assembled in large numbers to feed on a bumper crop of ripe grapes (Engels, 1933). Aggregations of black bears at present-day garbage dumps are comparable (Rogers, 1976).

Michigan black bears are capable of returning to their home areas after being forcibly displaced. Some individuals—live-trapped, marked, transported to other sites, and released—were able to return to the vicinity of the trapping sites from as far away as 96 mi (154 km). A displaced adult female in the Upper Peninsula, Harger (1970) noted, homed from a distance of 142.5 mi (229 km). In many cases, however, displaced black bears became reestablished in their new surroundings (Erickson, 1964; Harger, 1967).

BEHAVIOR. Although the black bear is frequently depicted as a habitue of the rural garbage dump and a freeloader along the roadways in national parks, in the backcountry this animal is wary, and almost timid in actions, quickly departing if danger (perhaps the approach of a human) threatens. Although capable of moving quietly and inconspicuously, the animal usually leaves abundant sign of its presence. Its barefooted-boy tracks are easily followed by an experienced woodsperson with or without snow cover to assist. Trampled berry bushes, excavated old stumps, overturned logs, and gnawings and claw marks on trees all indicate the black bear's presence. Claw marks on trees may result from climbing; some, however, are made by the animal rearing up on its hind limbs and reaching as high as possible to dig its claws into the tree trunk. Marking trees in this fashion may be one way in which the black bear designates its territorial or breeding claims. These claw marks stand out most conspicuously on aspen, birch, and beech trees. The presence of a black bear can also be suspected when a mud wallow is found along a creek. In fact, black bears have been found taking naps in the heat of a summer day half-mired in the cooling mixture of water and mud (Shapton, 1956).

Black bears have been generally considered creatures of the night; however, studies now show that they are most active at twilight (Garshelis and Pelton, 1980), or possibly in broad daylight (Lindzey and Meslow, 1977b). This activity in response to light may vary depending on breeding activity and food supply. According to the above authors, black bears are generally crepuscular (active at twilight) in spring (following winter denning), diurnal in summer, and nocturnal in autumn (prior to winter denning). Females with cubs may not follow this routine. All activity by the black bear may be reduced or curtailed by rain or snowfall and by ambient temperatures either above 77° F (25° C) or below 32° F (0° C).

The black bear's highly sensitive ears and nose compensate for its poor sense of vision. Although the black bear is large and lumbering in gait, it is a good tree-climber, gaining this ability by three months of age. Trees serve not only as refuges, especially for cubs, but also as sources of fruits and even honey from beehives in hollows. Hunters have, of course, used the black bear's climbing ability to advantage; when pursued by hounds the animal will very likely tree and be highly approachable while at bay.

The black bear ambles along with considerable side-swinging of the head and body. Yet, when pursued this huge creature can gallop at a speed of more than 30 mi (48 km) per hour. It is also a good swimmer. Black bears can growl, bellow and even bawl when, alarmed or defending themselves, but usually utter only "woof woof" calls or jaw-snapping sounds. Sow bears may also use "woof woof" calls to answer the whimpering sounds of cubs.

Most early descriptions of the black bear's social life depict the animal as a loner, pairing up briefly only during mating seasons. However, recent studies, aided by such remote data-gathering devices as radio-telemetry, point to a rather sophisticated social organization, which, according to Rogers (1977), varies with the distribution and abundance of the food supply. The social order tends to allow a spatial arrangement to maximize food-foraging efficiency. The major social unit is the female and her cubs. This family group (including yearlings and first-year cubs) remains intact for as long as 16 months (Lindzey and Meslow, 1977b). The adult female maintains a territory and allows her offspring free range. There is usually very little overlap between territories of adult females (Jonkel and Cowan, 1971). Adult males also have territories, but they are larger than those of adult females and generally overlap those of two or more females, but mutual avoidance is usual (Garshelis and Pelton, 1981). Transient young males are excluded since they undoubtedly compete for food with the family groups "maintained" by the dominant adult males. The young males, as a result, are probably obliged to wander considerably in search of an "undefended" area in which to become established. Even so, food acquisition remains a rather solitary endeavor, except when food supplies may be abundantly concentrated at garbage dumps or berry patches. In such instances, aggregations occur rather compatibly and the social system adjusts accordingly.

Black bears use dens for winter dormancy, preparing such quarters in autumn (Novick *et al.*, 1981). Erickson (1964) recorded 229 winter dens in the Upper Peninsula: 46% were in cavities (usually excavated) under stumps and logs, 21% in holes excavated in knolls or hillsides, and 11% under brush or slash. Others were in hollow standing trees, hollow felled logs, rock crevices, unsheltered depressions, a beaver lodge, and even under a deserted dwelling. In Schoolcraft County, Switzenberg (1955) reported a black bear den in a hollow of a standing pine tree, with the opening 53 ft (16 m) above ground. Johnson and Pelton (1981) found black bears in Tennessee prefer tree cavities above ground for winter dens. As a rule, females are more likely than males to line their dens with leaves, grass or ferns. Shelters used in summer are less elaborate, generally consisting of "forms" under brush or leafy boughs.

WINTER DORMANCY. As autumn approaches the black bear accumulates excess fat—up to 40% of the normal body weight—and appears less alert and more lethargic than normal just before entering the winter den. The adult males are last to enter their dens (Amstrup and Beecham, 1976). According to Erickson (1964), the earliest denning date recorded for the Upper Peninsula black bear population was October 13, for the Lower Peninsula, October 27. In the milder winters of more southern areas the length of this inactivity may be shorter. Black bears occupy winter dens, again depending on weather and their physical condition, until mid-April, although records show black bears being active as early as February. In mild years, winter sleep may be shortened or highly irregular (Manville, 1980). Den occupancy may last as long as 160 days; although a study in Washington showed the average denning period to be 126 days (Lindzey and Meslow, 1976).

Why and how black bears spend their winters in protective shelter has been a subject of investigation for many years (Aldous, 1937; Howard, 1935; Matson, 1946; Morse, 1936; Seton, 1953; Svihla and Bowman, 1954; Tietje and Ruff, 1980). The black bear's life processes do not decline as drastically during winter sleep as do those of the true-hibernating woodchuck, thirteen-lined ground squirrel, and meadow jumping mouse. Although there is a major drop in heartbeat to as few as eight to ten beats per minute and a similar reduction in respiration, unlike the true hibernators the body temperature drops less, perhaps 11° F (6° C) from normal. According to Manville (1981), black bears at his study area in the Lower Peninsula had an average rectal temperature of about 100° F (38.2° C) in summer and autumn and about 90° F (34.7° C) in winter. The combination of stored fat as an energy source and a slight depression in body temperature allows for a winter dormancy called carnivorean lethargy, certainly not so profound as true hibernation.

In contrast to typical hibernators which rouse occasionally in winter to eat, urinate, and defecate, the black bear usually does not awaken except when disturbed. It was suggested by Folk *et al.* (1972) that perhaps the digestive and urinary tracts of the black bear are better adjusted for winter sleep than those of the true hibernators. At arousal time in spring, the black bear defecates an extremely dry and hard fecal "plug" prior to beginning the new year's normal elimination of ingested foods. At this time, the newly-aroused animals may still possess some of their fat layer from the previous autumn. It may be two or three weeks after awakening before severe weight loss is noted.

ASSOCIATES. Except for temporary associations with its own kind, the black bear appears intolerant of other animals in the forest community. It naturally gets first chance at such food concentrations as vines and shrubs heavily laden with berries, carrion, or a new batch of garbage just delivered to the dump.

REPRODUCTIVE ACTIVITES. The black bear produces offspring slowly compared with other Michigan carnivores. The young do not become sexually mature until at least three and one-half years old (Erickson *et al.*, 1964). Also, adult females normally have cubs only every second year, because the sow is not receptive to breeding males during those summer mating seasons (mid-June to mid-July) when she is nursing young. Courtship behavior is reported to be noisy but brief, with male-female bonds lasting just a few days (Lindzey and Meslow, 1977a). The female may mate with more than one male, although males other than the dominant one in her territory are usually excluded. The gestation period is long—about 210 days—due, in part, to dormancy of the fertilized egg until as late as October or November when impantation in the uterine wall occurs and normal cleavage and embryo formation begins. Because of this delayed implantation, plus embryonic development of about 70 days, the young are born in winter—mid-January to early February, a time when the sow is undergoing winter dormancy in a sheltered den.

The usual number of young born per litter is two; as many as five cubs in a litter have been reported in Michigan (Matson, 1952). At birth the young are small and underdeveloped (altricial), are about 7 in (178 mm) long, weigh about 9 oz (255 g), have closed ear pinnae and eyes; are just beginning to be clothed in soft, grayish hair; have weak hind limbs and soft but conspicuous claws; and can utter whimpering sounds (Butterworth, 1969). At seven days, blackish fur on the back is discernable. At 25 days, the eyes begin to open. At 46 days, the ear pinnae unfold, the weight is 5⅓ lbs

(2.4 kg), the upper incisors erupt, and a white chest marking, if present, is visible. By the time the female leads her cubs from the den in late April, a full set of milk teeth is present. Their replacement by permanent teeth begins at approximately 90 days of age, with the full set completely grown by the time the young animal is two years old. The young are weaned at about five months and may be self-sufficient shortly thereafter (Erickson, 1959). The cubs may remain with the mother using her den the following winter but disperse prior to the second summer when she will mate again. However, in response to severe environmental pressures, especially food shortages, cubs may either be nursed longer than normal or be abandoned by the parent black bear (Fair, 1978).

Mortality among black bears is greatest during their first two years. In a study made in Alberta, Kemp (1972) found that mortality was 26.7% in cubs, 36.7% in yearlings, and 37.5% in two-year-olds. Black bears are long-lived, surviving perhaps as long as 30 years in captivity; a Michigan black bear shot in 1978 was found to be 13½ years old (as determined by the condition of the teeth).

FOOD HABITS. Black bears not only have a reputation for eating all kinds of animal and plant food but also for being continuously and ravenously hungry. When natural foods are abundant, black bears seem content to remain at a distance from human establishments, agricultural lands, and livestock. Recent work by Rogers (1976) and others show that major economic damage seems correlated with shortages of natural nutrients, notably the seasonal berry and mast crops—blueberries and acorns. Grain fields, orchards, beehives, and livestock suffer molestations (Manville, 1981). Visitations to rural garbage dumps, however, seems to become a fixation with many black bear individuals, with foraging for natural foods perhaps becoming secondary.

Eating plant foods, according to Rogers (1976), may be more selective than previously thought because the black bear's digestive tract contains no special caecum in which microflora and microfauna are stored to aid digestion of complex plant materials. As a result, the black bear can only obtain sustenance from simple carbohydrates in berries, other fruits, nuts, some crops, buds, catkins, certain herbs, tubers, and other roots. When black bears are forced to eat twigs, leaves or grass, they are apt to lose weight and strength (Jonkel and Cowan, 1971). It is Roger's (1976) suspicion that seasonal food shortages (especially plant materials) may be a serious limiting factor to black bear populations in the Upper Great Lakes Region. In times of good mast and berry crops, however, black bears fare well. Apparently beechnuts are a great favorite. These may be obtained from the ground or by climbing the tree where the animal often prepares a "bear nest" consisting of a ball of brush balanced in a substantial crotch. After settling in this position, according to Shapton (1956), the black bear pulls every branch within reach toward his perch in order to strip off the beechnuts into its mouth. This noisy business can also be fatal since a knowledgable hunter with an acute sense of hearing can recognize the sound from a considerable distance (Baker, 1956).

Although its choice of plant foods is somewhat restricted by digestive limitations, the black bear can eat most animal foods and obtain maximum nourishment from their digestible parts. Perhaps the greatest novelty in the diet of this omnivore is the variety of invertebrate animals eaten, with ants and other insects heading the list. The black bear's sizable claws mutilate decayed logs looking for beetles, larval grubs, earthworms, and an occasional unlucky mouse. Beehives in tree hollows, and ground or arboreal nests of paper wasps and bald-faced hornets also provide choice food items (Gilbert and Roy, 1975). Other animal foods include fish (caught by jumping into the water and grasping the prey in its mouth), frogs, other amphibia, snakes, other reptiles including snapping turtles, turtle eggs (excavated), birds and their eggs, hares and rabbits, mice, woodchucks, squirrels, even other bears, moose calves, wapiti, and white-tailed deer, especially fawns (Barmore and Stradley, 1971; Behrend and Sage, 1974; Bennett *et al.*, 1979; Jackson, 1961; Juniper, 1978; Landers *et al.*, 1979; Smith, 1981). The large black bear can easily overtake and slap down most domestic livestock, although it seems to favor piglets (Manville, 1948). Carrion (including garbage, hunter-killed or winter-killed game, and dead domestic livestock) rounds out this animal's usual bill of fare. On an annual basis, Banfield (1974) lists the overall diet as consisting of vegetable matter (76.7%), carrion (15.2%), insects (7.4%), and small mammals (0.7%).

ENEMIES. As determined in a Minnesota study, Rogers (1976) found that more than 90% of the mortality among cubs and yearlings was from natural causes, mostly nutrition related; whereas more than 90% of the mortality among black bears two years of age or older was from human-related causes. In Michigan, human-inflicted mortality includes such unusual fatalities as the electrocution of a sow and one cub who climbed a power pole in Houghton County in November, 1954, and in a car-bear collision in Roscommon County in October, 1974. Natural enemies in the Michigan area are few; the near-extirpated gray wolf, which might attack young bears, is perhaps the only one around in modern times (Rogers and Mech, 1981; Voigt *et al.,* 1976).

PARASITES AND DISEASES. The parasites of black bears in the Upper Great Lakes Region (Michigan, Wisconsin, Minnesota, and Ontario) have been examined by such workers as Addison *et al.,* (1978), Manville (1978, 1981), and Rogers (1975). External parasites identified as infesting this animal include a flea (*Oropsylla arctomys*), louse (*Trichodectes pinguis euarctidos*), dog tick (*Dermacentor variabilis*), winter tick (*Dermacentor albipictus*), black-legged tick (*Ixodes scapularis*), and a mite (*Demodex* sp.). The latter may be the cause of mange observed in black bears in Wisconsin. Internal parasites include a flatworm (*Alaria americana*), ascarid round worm (*Baylisascaris transfuga*), filarial worm (*Dirofilaria ursi*), unidentified hookworm larva, and a roundworm (*Physaloptera rara*). The filarial worm, although harmless to humans, sometimes causes concern among Michigan black bear hunters, because this parasite occurs just under the skin and is conspicuous when an animal is being skinned (J. N. Stuht, pers. comm., 1980). According to Rogers (1975), the broad fish tapeworm (*Diphyllobothrium latum*), formerly found frequently in both black bear and human populations in the Upper Great Lakes Region, is now most uncommon.

Infection rates of the dreaded trichina worm (*Trichinella spiralis*), which causes trichinosis in black bear, swine, humans and other hosts, are extremely low (less than 4%) in black bears in Wisconsin and Ontario, with no evidence of this worm in 23 animals examined in Michigan (Harbottle *et al.,* 1971; Zimmerman, 1977). Even so, it is suspected that ursine trichinosis may be present at low infection levels in Michigan, which makes black bear meat sufficiently suspect to encourage all camp chefs and others to either cook the meat well-done or maintain preserved portions under frozen conditions for at least 20 days to kill any worm cysts which might be present in the diaphragm or other muscle tissue.

In the Dead Stream area of Crawford, Kalkaska, Missaukee, and Roscommon counties, Manville (1981) found peridontal disease in 11 and dental caries (cavities) in 5 of 33 black bears examined.

MOLT AND COLOR ABERRATIONS. Black bear have a single annual molt in spring (April or May) following emergence from winter dens. During winter dormancy, foot pads are shed (Rogers, 1974). In the Upper Great Lakes Region, most black bears have black or extremely dark brown pelages. An occasional light brown individual has been reported in Michigan. The well-known color phases, brown and cinnamon, are mostly found in animals in western North America (Rogers, 1980). Populations characterized by blue or white coats are located on the northwestern Pacific Coast. Wood (1922) reported that an albino was taken in Bay County. In 1979 and 1980, a white bear was seen by residents near Elo in Baraga County (Elsworth Harger, pers. comm.). This same animal may have been shot in this general area in October, 1980. A bear described as having a coat the color of honey was bagged in Baraga County in September, 1980; one having a coat of reddish-brown was seen in Marquette County in 1981.

ECONOMIC IMPORTANCE. Folk stories describe the black bear as big, slow-witted, and friendly, but somewhat cantankerous and mischievous. This deceptive gentleness can quickly become savage strength and hostility. Residents of Michigan's black bear country have witnessed both kinds of behavior, although generally the animals hold their distance. News stories describe incidents when a black bear has (1) broken into a travel trailer and cleaned out the refrigerator, (2) entered a home and eaten apples stored in the cellar, and (3) demolished the door at a club to get to some dog food. All of these events took place in Marquette County in 1978. Black bears can become so attracted to food sources such as garbage dumps, poorly-policed camp grounds, crop lands, and stock farms that it is often necessary to live-trap and relocate persistent individuals (Baptiste *et al.*, 1979; Erickson, 1964; Manville, 1981; Merrill, 1978; Payne, 1978; and Rogers, 1976). Stock killers, according to Davenport (1953), are usually old males. Even in pioneer times in the Vermontville area of Eaton County, the local black bear population was troublesome. According to reports in the press in 1839, one black bear not only raided pig pens at frequent intervals but also wandered into town and looked into windows and doors. Dispatching a black bear in those times required a posse of townspeople, who surrounded the animal where it denned in a swamp only ½ mi (less than 1 km) from the community. In 1955 and 1956, 13 investigated complaints of damage to livestock by Michigan black bears, (Schofield, 1957), showed losses of sheep (6 cases), cattle (4 cases), pigs (2 cases), and horse (1 case).

Meeting a massive black bear in the deep woods has often been a terrifying experience for a hiker. Usually, both participants in such an encounter quickly retrace their steps. However, a black bear has both the speed to overtake and the strength to severely maul or kill a human. Although extremely rare, two human fatalities were reported in Montana's Glacier National Park as a result of attacks by grizzly bears (*Ursus arctos*) in 1967, evoking widespread concern among backpackers and other outdoorspersons (McCullough, 1982). There have been several unprovoked attacks made on humans by Michigan black bear but only three fatalities have been documented, two in the Upper Peninsula and one in the Lower Peninsula (East, 1979). In one case, a backpacker fell to his death from a tree which he had climbed apparently to escape an attacking black bear (June, 1978, Porcupine Mountains State Park in Ontonogan County). Perhaps the most tragic incident occurred on the afternoon of July 7, 1948, at a forest ranger cabin in Hiawatha National Forest near Brimley in Chippewa County. A three-year old girl was attacked on her doorstep by a rather small male black bear (weight about 150 lbs or 68 kg), killed, dragged into the woods, and partially consumed (Whitlock, 1950). There seemed to be no provocation for the act except perhaps hunger.

The long, glossy coat of the black bear has a coarseness which restricts its use to fireplace rugs (mostly dust catchers for housekeepers, Burroughs, 1964); lap robes; trim for hats, helmets, and military great coats; and capes for early-day trappers. Nevertheless, in the summer of 1767, British fur dealers at Michilimackinac exported 1,142 black bear pelts (Petersen, 1979). In 1822, the Astor Trading Post on Mackinac Island bought black bear skins for $2.50 to $4.50, depending on size and quality (Anon., 1962). From 1835–1839, the American Fur Company's Detroit Department (including southern Michigan) accumulated 2,645 black bear pelts and its Northern Department (including northern Michigan) accumulated 2,591 pelts (Johnson, 1969). Today, only a few Michigan black bears hides are sold at fur auctions. In 1975, pelts brought as much as $160; in 1978, $50 to $60; in 1979, $75 to $100.

The black bear also provided needed protein for the frontier family. The flesh, especially the hams, was highly prized. The meat could be only marginal as tablefare, however, when strong and tough (Burroughs, 1964). Bear grease, on the other hand, was a most useful household product serving as cooking oil, leather dressing to waterproof boots, and as a liniment to ease sores, sprains, bruises or inflammation. The teeth and claws, used in Indian ritual and decoration, were fashioned into ornaments or even childrens' toys.

The black bear has always been a sought-after trophy of the Michigan hunter. However, most animals are bagged as by-products of the autumn white-tailed deer hunting season. With the woods fairly saturated with hunters, wary and often secretive black bears will unluckily cross their paths. Hunters are also apt to discover their denning sites; about 37% of the black bears taken in the deer season in the Upper Peninsula are killed in dens (Erickson *et al.*, 1964). Of the black bear kill of 1,185 reported in 1962, deerhunters accounted for 950.

Since 1925, Michigan's black bear has been designated as a game species with open hunting seasons prescribed in both the Upper Peninsula (Region I, Michigan Department of Natural Resources) and the northern Lower Peninsula (Region II). There have been several types of hunting seasons; a special bear stamp was required from 1959 through 1963; drawn permits and special licenses have also been in effect. In 1979, for example, the black bear season in Region I was open from September 10 through October 31 for licensed hunters using hounds, bows and arrows, or guns, and in Region II from September 21 through September 30; for licensed hunters using bows and arrows in Region I from October 1 through November 14; and for licensed white-tailed deer hunters using guns in Region I only from November 15 through November 30. In 1980, the black bear was finally dignified by being given full status as a Michigan big game animal.

Beginning with the first authorized hunting season in 1925 when the estimated legal kill was 150 black bears, the populations of this big game animal have been carefully sustained in northern Michigan under the watchful eyes of wildlife biologists of the Michigan Department of Natural Resources. Since 1955, the estimated annual kill has fluctuated from lows of 430 in 1973 and 524 in 1965 to highs of 1,238 in 1960 and 1,185 in 1962. After 1966, the annual legal kill fluctuated between 430 and 990; in 1979, the take was 907. Cooperation from citizens must also be acknowledged for maintaining the Michigan black bear population. The concerned support given by the Michigan Bear Hunters Association, established in 1946 under the longtime leadership of Cadillac sportsman Carl T. Johnson, is noteworthy. Matters such as disputes between black bear hunters who use hounds and those preferring to hunt the animals by stalking and still hunting have been diplomatically examined by both conservation-minded citizens groups and the Michigan Department of Natural Resources (Smith, 1979).

The black bear appears to have a secure future in Michigan so long as its northern Michigan habitat continues to be partly devoted to timber/wildlife resources uses. However, as an observer of black bear country in the Southeast once pointed out, the major threat to this population might come from the continually growing livestock industry. The demise of the black bear in much of the swamp and riverbottom lands of the Gulf States resulted in part from its depredations on free-ranging pigs, as the latter were considered more economically important.

COUNTY RECORDS FROM MICHIGAN. Records of specimens of black bears in collections of museums, from the literature, and provided by personnel of the Michigan Department of Natural Resources show that this mammal once occurred throughout both the Upper and Lower Peninsulas. The black bear is unknown from Isle Royale or the Beaver group but has been reported from the Huron Islands in Lake Superior and Drummond Island in Lake Huron (Corin, 1976; Manville, 1950).

LITERATURE CITED

Addison, E. M., M. J. Pybus, and J. J. Rietveld
1978 Helminth and arthropod parasites of black bear, *Ursus americanus,* in central Ontario. Canadian Jour. Zool., 56(1):2122-2126.

Aldous, S. E.
1937 A hibernating black bear with cubs. Jour. Mammalogy, 18(4):466–468.

Alt, G. L., G. T. Matula, F. W. Alt, and J. S. Lindzey
1980 Dynamics of home range and movements of adult black bears in northeastern Pennsylvania. Pp. 131–136 *in* C. Martinka and K. L. McArthur, eds. Bears—their biology and management. Bear Biol. Assoc. Conf. Ser., No. 3, 375 pp.

Amstrup, S. C., and J. Beecham
1976 Activity patterns of radio-collared black bears in Idaho. Jour. Wildlife Mgt., 40(2):340–348.

Anon.
1962 Fur and hide values, 1822. Michigan Conserv., 31(6): 36–37.

Baker, R. H.
1956 Remarks on the former distribution of animals in eastern Texas. Texas Jour. Sci., 8(3):356–359.

Banfield, A. W. F.
1974 The mammals of Canada. Univ. Toronto Press, Toronto. xxiv+438 pp.

Baptiste, M. E., J. B. Whelan, and R. B. Frary
1979 Visitor perception of black bear problems at Shenandoah National Park. Wildlife Soc. Bull., 7(1):25–29.

Barmore, W. J., and D. Stradley
1971 Predation by black bear on mature male elk. Jour. Mammalogy, 52(1):199–202.

Behrend, D. F., and R. W. Sage, Jr.
1974 Unusual feeding behavior by black bears. Jour. Wildlife Mgt., 38(3):570.

Bennett, L. J., P. F. English, and R. L. Watts
1943 The food habits of the black bear in Pennsylvania. Jour. Mammalogy, 24(1):25–31.

Butterworth, B. B.
1969 Postnatal growth and development of *Ursus americanus.* Jour. Mammalogy, 50(3):615–616.

Burroughs, R. D.
1958 Perry's deer hunting in Michigan, 1838–1855. Michigan Hist., 42(1):35–58.
1964 Goose grease and bear oil. Michigan Conserv., 33(3): 37.

Cleland, C. E.
1966 The prehistoric animal ecology and ethnozoology of the Upper Great Lakes Region. Univ. Michigan, Mus. Anthro., Anthro. Papers, No. 29, x+294 pp.

Corin, C. W.
1976 The land vertebrates of the Huron Islands, Lake Superior. Jack-Pine Warbler, 54(4):138–147.

Davenport, L. B., Jr.
1953 Agricultural depredation by the black bear in Virginia. Jour. Wildlife Mgt., 17(3):331–340.

Dice, L. R.
1943 The biotic provinces of North America. Univ. Michigan Press, Ann Arbor. viii+78 pp.

Dice, L. R., and H. B. Sherman
1922 Notes on the mammals of Gogebic and Ontonagon counties, Michigan. Univ. Michigan, Mus. Zool., Occas. Papers No. 109, 46 pp.

East, B.
1979 The truth about black bears. Michigan Out-of-Doors, 33(1):44–61.

Engels, W. L.
1933 Notes on the mammals of St. Joseph County, Indiana. American Midl. Nat., 14(1):1–15.

Erickson, A. W.
1957 Techniques for live-trapping and handling black bears. Trans. 22nd North American Wildlife Conf., pp. 520–543.
1959 The age of self-sufficiency in the black bear. Jour. Wildlife Mgt., 23(4):401–405.
1964 Breeding biology and ecology of the black bear in Michigan. Michigan State Univ., unpubl. Ph.D. dissertation. xxxiv+274 pp.

Erickson, A. W., J. Nellor, and G. A. Petrides
1964 The black bear in Michigan. Michigan State Univ., Agric. Exp. Sta., Res. Bull. 4, 102 pp.

Eshelman, R. E.
1974 Black bear from Quaternary deposits in Michigan. Michigan Acad., 6(3):291–298.

Fair, J. S.
1978 Unusual dispersal of black bear cubs in Utah. Jour. Wildlife Mgt., 42(3):642–644.

Folk, G. E., Jr., M. A. Folk, and J. J. Minor
1972 Physiological condition of three species of bears in winter dens. Pp. 107–124 *in* S. Herrero, ed. Bears—their biology and management. IUCN Publ. new ser., No. 23, 371 pp.

Frankforter, W. D.
1966 Some recent discoveries of late Pleistocene fossils in western Michigan. Michigan Acad. Sci., Arts and Ltrs., 51:209–220.

Garshelis, D. L., and M. R. Pelton
1980 Activity of black bears in the Great Smoky Mountains National Park. Jour. Mammalogy, 61(1):8–19.
1981 Movements of black bears in the Great Smoky Mountains National Park. Jour. Wildlife Mgt., 45(4):912–925.

Gilbert, B. K., and L. D. Roy
1977 Prevention of black bear damage to beeyards using adversive conditioning. Pp. 93–102 *in* R. L. Phillips, and C. J. Jonkel, eds. Proc. 1975 Predator Symposium. Montana Forest and Conserv. Exp. Sta., Univ. Montana, ix+268 pp.

Harbottle, J. E., D. K. English, and M. G. Schultz
1971 Trichinosis in bears in northeastern United States. U.S. Public Health Serv., Mental Health Admin., Health Rept. 86:473–476.

Harger, E. M.
1967 Homing behavior of black bears. Michigan Dept. Conserv., Res. Dev. Rept. No. 118, (mimeo) 13 pp.
1970 A study of homing behavior of black bears. Northern Michigan Univ., unpubl. M.A. thesis, 66 pp.
1978 Hunting methods and their effect on the black bear. Pp. 200–205 *in* R. Hugie, ed. Fourth eastern black bear workshop, 409 pp.

Higgins, M. J.
1980 An analysis of the faunal remains from the Schwerdt Site, a late prehistory encampment in Allegan County, Michigan. Western Michigan Univ., unpubl. M.A. thesis, v+77 pp.

Hildebrandt, T. D.
1976 Age determination in Michigan black bear by means of cementum annuli. Michigan Dept. Conserv., Wildlife Div., Rept. No. 2758 (mimeo), 9 pp.

Holman, J. A.
1975 Michigan's fossil vertebrates. Michigan State Univ., Publ. Mus., Ed. Bull., No. 2, 54 pp.

Howard, W. J.
1935 Notes on the hibernation of a captive black bear. Jour. Mammalogy, 16(4):321.

Jackson, H. H. T.
1961 Mammals of Wisconsin. Univ. Wisconsin Press, Madison. xii+504 pp.

Johnson, D. R.
1969 Returns of the American Fur Company, 1835–1839. Jour. Mammalogy, 50(4):836–839.

Johnson, K. G., and M. R. Pelton
1981 Selection and availability of dens for black bears in Tennessee. Jour. Wildlife Mgt., 45(1):111–119.

Jonkel, C. J., and I. McT. Cowan
1971 The black bear in the spruce-fir forest. Wildlife Monog., 27, 57 pp.

Juniper, I.
1978 Morphology, diet, and parasitism in Quebec black bears. Canadian Field-Nat., 92(2):186–189.

Kemp, G. A.
1972 Black bear population dynamics at Cold Lake, Alberta, 1968–70. Pp. 26–31 *in* S. Herrero, ed. Bears—their biology and management. IUCN Publ. new ser., No. 23, 371 pp.

Landers, J. L., R. J. Hamilton, A. S. Johnson, and R. L. Marchinton
1979 Foods and habitat of black bears in southeastern North Carolina. Jour. Wildlife Mgt., 43(1):143–153.

Lindzey, F. G., and E. C. Meslow
1976 Winter dormancy in black bears in southwestern Washington. Jour. Wildlife Mgt., 40(3):408–415.
1977a Population characteristics of black bears on an island in Washington. Jour. Wildlife Mgt., 41(3):408–412.
1977b Home range and habitat use by black bears in southwestern Washington. Jour. Wildlife Mgt., 41(3):413–425.

Manville, A. M.
1978 Ecto- and endoparasites of the black bear in Wisconsin. Jour. Wildlife Dis., 14(1):97–101.

1980 Human impact on the black bear in Michigan's Lower Peninsula. Michigan State Univ., Dept. Fisheries and Wildlife, Prog. Rept. (mimeo.), Winter 1980, 5 pp.

1981 Human impact on the black bear in Michigan's Lower Peninsula. Michigan State Univ., unpubl. Ph.D. dissertation, xii+153 pp.

Manville, R. H.

1948 The vertebrate fauna of the Huron Mountains, Michigan. American Midl. Nat., 39(3):615–640.

1950 The mammals of Drummond Island, Michigan. Jour. Mammalogy, 31(3):358–359.

Marks, S. A., and A. W. Erickson

1966 Age determination in the black bear. Jour. Wildlife Mgt., 30(2):389–410.

Martin, T. J.

1979 Unmodified faunal remains from 20 LP 98. Pp. 65–66 *in* W. A. Lovis, ed., The archaeology and physical anthropology of 20 LP 98: A Woodland burial locale in Lapeer County, Michigan. Michigan Arch., 25(1–2), 69 pp.

Matson, J. R.

1946 Notes on dormancy in the black bear. Jour. 27(3):203–212.

1952 Litter size in the black bear. Jour. Mammalogy, 33(2): 246–247.

McCullough, D. R.

1982 Behavior, bears, and humans. Wildlife Soc. Bull., 10(1):27–33.

Merrill, E. H.

1978 Bear depredations at backcountry campgrounds in Glacier National Park. Wildlife Soc. Bull., 6(3):123–127.

Morse, M. A.

1937 Hibernation and breeding of the black bear. Jour. Mammalogy, 18(4):460–465.

Novick, H. J., J. M. Siperek, and G. R. Stewart

1981 Denning characteristics of black bears, *Ursus americanus,* in the San Bernardino Mountains of southern California. California Fish and Game, 67(1):52–61.

Payne, N. F.

1978 Hunting and management of the Newfoundland black bear. Wildlife Soc. Bull., 6(4):206–211.

Perry, O. L.

1899 Hunting experiences of Oliver Hazard Perry of Cleveland verbatim from his diaries. Cleveland. viii+246 pp.

Petersen, E. T.

1979 Hunter's heritage. A history of hunting in Michigan. Michigan United Conserv. Clubs, Lansing. 55 pp.

Piekielek, W., and T. S. Burton

1975 A black bear population study in northern California. California Fish and Game, 61(1):4–25.

Rogers, L. L.

1974 Shedding of foot pads by black bears during denning. Jour. Mammalogy, 55(3):672–674.

1975 Parasites of black bears of the Lake Superior region. Jour. Wildlife Dis., 11(2):189–192.

1976 Effects of mast and berry crop failures on survival, growth, and reproductive success of black bears. Trans. 42nd North American Wildlife and Nat. Res. Conf., pp. 431–438.

1977 Social relationships, movements, and population dynamics of black bears in northeastern Minnesota. Univ. Minnesota, unpubl. Ph.D. dissertation. 203 pp.

1980 Inheritance of coat color and changes in pelage coloration in black bears in northeastern Minnesota. Jour. Mammalogy, 61(2):324–327.

Rogers, L. L., and L. D. Mech

1981 Interactions of wolves and black bears in northeastern Minnesota. Jour. Mammalogy, 62(2):434–436.

Sauer, P. R.

1966 Determining sex of black bears from the size of the lower canine teeth. New York Fish and Game Jour., 13(2):140–145.

Schofield, R. D.

1957 Livestock and poultry losses caused by wild animals in Michigan. Michigan Dept. Conserv., Game Div., Rept. No. 2118 (mimeo), 6 pp.

Seton, E. T.

1953 Lives of game animals. Vol. II, Pt. I. Bears, coons, badgers, skunks, and weasels. Charles T. Branford Co., Boston. xvii+367 pp.

Shapton, W. W.

1956 Bear in mind. Michigan Conserv., 25(6):18–21.

Smith, R. P.

1979 Big challenge facing bear hunters. Michigan Out-of-Doors, 33(12):36–37.

1981 Regional reports. Michigan Out-of-Doors, 35(8):42.

Svihla, A., and H. S. Bowman

1954 Hibernation in the American black bear. American Midl. Nat., 52(1):248–252.

Switzenberg, D. F.

1955 Black bear denning in tree. Jour. Mammalogy, 36(3): 459.

Tietje, W. D., and R. L. Ruff

1980 Denning behavior of black bears in boreal forest of Alberta. Jour. Wildlife Mgt., 44(4):858–870.

Voigt, D. R., G. B. Kolenosky, and D. H. Pimlott

1976 Changes in summer foods of wolves in central Ontario. Jour. Wildlife Mgt., 40(4):663–668.

Wenzel, O. J.

1912 A collection of mammals from Osceola County, Michigan. Michigan Acad. Sci., 14th Rept., pp. 198–205.

Whitlock, S. C.

1950 The black bear as a predator of man. Jour. Mammalogy, 31(2):135–138.

Willey, C. H.

1974 Aging black bears from first premolar tooth sections. Jour. Wildlife Mgt., 38(1):97–100.

Wood, N. A.

1922 The mammals of Washtenaw County, Michigan. Univ. Michigan, Mus. Zool., Occas. Papers No. 123, 23 pp.

Wood, N. A., and L. R. Dice

1923 Records of the distribution of Michigan mammals. Michigan Acad. Sci., Arts and Ltrs., 3:425–469.

Zimmerman, W. J.
1977 Trichinosis in bears in western and northcentral United States. American Jour. Epidemiol., 106(2): 167–171.

FAMILY PROCYONIDAE

The raccoon and its close relatives are strictly New World in distribution (in North, Central and South America). If the lesser panda of central Asia is considered a procyonid, there are 7 genera and 18 species in this small family. The raccoon and the coati have overall similarities but look unrelated to the ringtail and the tropical olingo and kinkajou. Even so, this seemingly catch-all family does have common characteristics. Its derivation dates from fossil ancestors living in mid-Tertiary times in both North America and Eurasia. The greatest modern diversity has been in tropical America, with the founding stocks reaching South America from the north. Adaptations common to family members include omnivorous feeding habits; climbing ability (prehensile tails present in some); the shearing action of the carnassial teeth (fourth upper premolar and first lower molar) generally lost with the teeth adapted for crushing instead; premolars not reduced, as in the case of the bears; molar teeth low-crowned, lower ones usually reduced in number, uppers generally broader than long; feet usually flat (plantigrade) with all five toes on each foot retained; and weight of family members from as little as 2 lbs (1 kg) to as much as 40 lbs (18 kg).

Although two other procyonids reach southwestern United States—the ringtail and the coati—only the ubiquitious raccoon makes its home in most parts of this vast temperate area, extending into Michigan, and beyond the Canadian border as well.

RACCOON
Procyon lotor (Linnaeus)

NAMES. The generic name *Procyon*, proposed by Storr in 1780, is derived from the Greek *pro* meaning "before" and *kyon* meaning "dog." It is supposed that the author thought the raccoon was close to the ancestral stock which gave rise to dogs as well. The specific name *lotor*, proposed by Linnaeus in 1758, is from a New Latin word meaning "washer" referring no doubt to the raccoon's habit of often wetting its food prior to eating. In Michigan, this black-masked and ring-tailed mammal is called raccoon, sometimes just coon. The name raccoon is from the Algonquian linguistic family and from *arathcone*, which means (depending on the source) either "he scratches with his hands" or "beast like a fox." The subspecies *Procyon lotor lotor* (Linnaeus) occurs in Michigan.

RECOGNITION. Body size resembling that of a small, heavy dog (Fig. 76); head and body averaging 23½ in (600 mm); tail short and moderately bushy, averaging 9½ in (240 mm) in vertebrae length; head broad with pointed muzzle; eyes conspicuous, beady, blackish; ears prominent, rounded, fully-haired, erect; neck inconspicuous; body heavy-set, fairly squat; legs short; feet long, narrow, flat (plantigrade), with naked soles; five toes on each foot, each with strong, recurved claws; mammary glands consist of four pairs; males larger than females.

The body fur is long, fine, thick, and grizzled in appearance. The overall color varies from almost black to shades of dark brown, washed with a yellowish tinge. The underfur is grayish brown and wooly; the long guard hairs are tipped with black on the back and white or cream color on flanks and belly. The face is whitish with a conspicuous black band (mask) across the eyes and

Figure 76. The raccoon (*Procyon lotor*).
Sketches by Bonnie Marris

cheeks and framed by a grayish head and white-edged ears. The sides of the body are grayer than the back; the underparts are browner with a whitish wash. The cylindrical, bushy tail is characteristically annulated with five to seven rings, alternating bands of black and buff brown with a black tip. The legs are grizzled gray with whitish feet.

This stoutly-built mammal is easily distinguished from similar-sized Michigan mammals such as the woodchuck, porcupine and badger, by its dark and glossy coat, black mask, and short ringed tail.

MEASUREMENTS AND WEIGHTS. Adult raccoons (males are generally 10% larger than females) measure as follows: length from tip of nose to end of tail vertebrae, 28 to 38 in (700 to 950 mm); length of tail vertebrae, 8¾ to 10¼ in (220 to 260 mm); length of hind foot, 4¼ to 6 in (110 to 125 mm); height of ear from notch, 2 to 2¼ in (50 to 60 mm); and weight, 12 to 30 lbs (5.4 to 13.6 kg). Animals weighing as much as 56 lbs (25 kg) have been recorded in Michigan (Wood, 1922).

DISTINCTIVE CRANIAL AND DENTAL CHARACTERISTICS. The skull of the raccoon is heavily-constructed, broad, with a high and rounded cranium (Fig. 77). The rostrum is broad and massive. Auditory bullae are laterally compressed. The hard palate extends posteriorly well beyond a line drawn between the posterior ends of the last molar teeth. The mandible is heavy with the coronoid process high and curving posteriorly over the condyle. The skull of an average adult measures 4¾ in (120 mm) in greatest length and 3 in (77 mm) in greatest width across the zygomatic arches.

Dentition is heavy. The crowns of the incisors are slightly grooved. The canines are oval in cross section and grooved. The carnassials (fourth upper premolars and first lower molars) are rounded; the first premolars beyond are peg-like; the molars are large and flat-crowned, with weak,

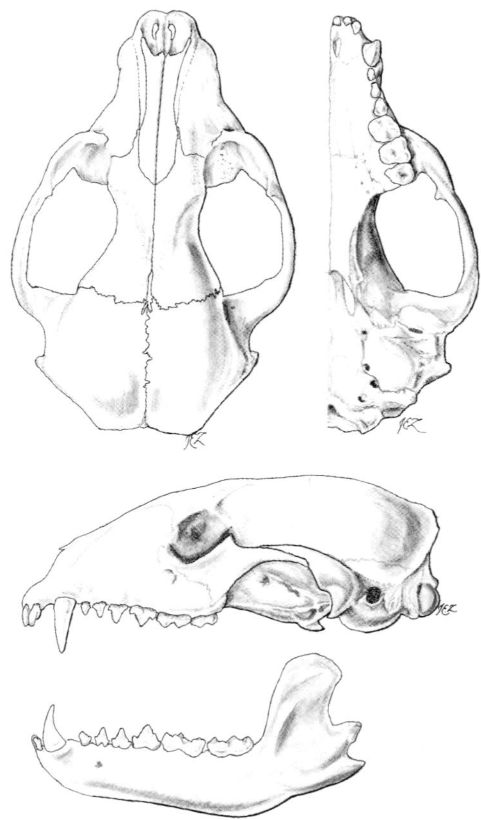

Figure 77. Cranial and dental characteristics of the raccoon (*Procyon lotor*).

rounded cusps. The heavy, flattened jaw teeth are most suitable for masticating the varied diet of an omnivore.

The skull of the raccoon has an overall resemblance to that of a large mustelid, but is readily distinguished by two molars on each side of the upper jaws rather than one as in the Mustelidae. The dental formula is as follows:

$$\text{I (incisors)}\,\frac{3\text{-}3}{3\text{-}3},\ \text{C (canines)}\,\frac{1\text{-}1}{1\text{-}1},\ \text{P (premolars)}\,\frac{4\text{-}4}{4\text{-}4},\ \text{M (molars)}\,\frac{2\text{-}2}{2\text{-}2} = 40$$

DISTRIBUTION. The raccoon lives in a variety of wooded and shrub habitats in North America from north of the United States and Canadian border southward throughout Central America at least to Panama (see Map 51). In grasslands and open desert areas, it may chiefly occur in riparian growth along waterways. The species has also reached offshore islands including parts of the Bahamas (Lotze and Anderson, 1979).

In Michigan, the raccoon is known from all counties in both the Upper and Lower Peninsulas, including some of the islands in the Great Lakes (see County Records). Its presence in Michigan dates back to the early postglacial period (Holman, 1975). Raccoon remains are also known from Indian sites of the Woodland and early European contact periods in the Upper Peninsula in Baraga, Delta, and Mackinac counties (Martin, 1979, 1980; Rhead and Martin, 1978); and in the Lower Peninsula in Allegan, Muskegon, Ottawa, and Saginaw counties (Cleland, 1966; Higgins, 1980; Martin, 1976). With the opening of forest areas by settlers and loggers, especially in the Upper Peninsula, the living space for Michigan raccoons improved with population increases and a more even distribution resulted (de Vos, 1964). Nevertheless, as late as the early 1930s (7th Biennial Rept., Michigan Dept. Conserv.) raccoon populations in many parts of the Upper Peninsula were considered sparse or even non-existent.

Map 51. Geographic distribution of the raccoon (*Procyon lotor*).

HABITAT PREFERENCES. Although the raccoon may be found throughout the state from lakeshores to inland forests, the species shows preference for hardwoods, especially along waterways. The scattered woodlots of southern Michigan provide mixed mature and second-growth woody vegetation with cultivated fields and pasture adjacent for year-around occupancy by these semiarboreal mammals (Fisher, 1977). Hollow trees improve the quality of raccoon habitat, although ground dens are also used (Berner and Gysel, 1967). The best description of the raccoon's environment in southern Michigan is found in the classic work of F. W. Stuewer (1943b). Other observers note raccoon use of both primary and second-growth hardwoods in Osceola County (Wenzel, 1912); creek-bottom woodlands in Clinton County (Linduska, 1950); woody swales in Kalamazoo County (Allen, 1938); garbage dumps and lakeshores in Marquette County (Laundre, 1975); in woodlands and sedge marsh in Charlevoix County (Dice 1925); and along waterways bordered by hardwoods in Chippewa County (Wood, 1914a). Certainly the raccoon has been one Michigan mammal which has prospered as a result of many of the land-use practices instigated by humans. Urban dwellers today are not surprised to see raccoons wandering down back alleys or across streets (Hoffmann and Gottschang, 1977).

DENSITY AND MOVEMENTS. A wide assortment of food preferences may make it unnecessary for the raccoon to be as mobile as other similar-sized carnivores. In some of the best habitat for raccoon, densities may vary from 1 animal per 10 acres (4 ha) to 1 per 16 acres (6 ha) (Dorney, 1954; Johnson, 1970; Urban, 1970; and Yeager and Rennels, 1943). In his pioneer studies of Michigan raccoons, Steuwer (1943b) calculated somewhat similar population densities for raccoons in Allegan County, 1 raccoon per 16 acres (6.5 ha). At Rose Lake Wildlife Research Station in Clinton County, Linduska (1950) thought the raccoon population varied between 1 individual per 43 acres (17 ha) to 1 per 55 acres (23 ha). On a township basis in Washtenaw County, Craighead and Craighead (1956) judged the raccoon population to be sparse with 1 animal per 523 to 640 acres (209 to 256 ha). An overall estimate for the annual raccoon population for the better habitat in the southern part of Michigan's Lower Peninsula is 1 individual per 64 acres (25.6 ha). It must be assumed, however, that the local population expands in mid-summer when the young-of-the-year enter the community, followed by a gradual decline until the next year. Except for highly favorable environments in marsh and swamp lands, populations of raccoon in northern Michigan will be lower and perhaps spotty.

Raccoons maintain fairly well-defined areas of activity (home ranges), as determined both by live-capture, mark, release, and recapture and by radio tracking individuals that have been captured, fitted with radio transmitter collars, and released. When food supplies and other habitat requirements are optimum, raccoons tend to have smaller home ranges than when such environmental factors are limited (Ellis, 1964; Lehman, 1977). Raccoons tend to remain spatially segregated (Fritzell, 1978b). Although raccoons are known to converge on concentrated food supplies, as in corn-fields or at garbage dumps (Johnson, 1970), Urban (1970) found no evidence of an influx of these animals into marsh areas from surrounding habitat during the waterfowl nesting season.

As a rule, males move around in larger areas (often twice as large) than females, although some studies have shown the reverse to be true (Lotze, 1979). By the trap-tag-release-retrap method and by toe-clipping to allow track identification, Stuewer (1943b) in Allegan County found that adult males moved about in areas as small as 45 acres (18 ha) to as large as 2,012 acres (805 ha), with an average of 503 acres (201 ha). Adult females occupied territories as small as 13 acres (5.2 ha) and as large as 930 acres (372 ha), with an average of 268 acres (107 ha), about one-half that of adult males. In a study of shorter duration (May through August) in upland woodlots in Ingham County, Fisher (1977) found home ranges for raccoons to average 137 acres (55 ha), with little difference between the sexes. Fisher's findings are comparable to those obtained for raccoon home ranges in other midwestern states; Mech *et al.* (1966) calculated home range figures of 154 acres (61 ha) in Minnesota; Ellis (1964), 89 acres (36 ha) in Illinois; Lehman (1977), 142 acres (56 ha) for adult males and 27 acres (11 ha) for females in Indiana; and Urban (1970), 120 acres (48 ha) in Ohio.

On an average night, Fisher (1977) found raccoons in Ingham County to move distances of 1,720 ft (524 m). There is a record of a young, tagged male raccoon moving a distance of 165 mi (265 km) in Minnesota (Priewert, 1961). Raccoons released at localities distant from the place of capture show little tendency to return to their former haunts. Rather, these animals become settled at or near the points of release (Johnson, 1970; Stuewer, 1943b).

Hoffmann and Gottschang (1977) found an exceedingly high raccoon population in a suburban community near Cincinnati, Ohio. The density of these animals in a study area (64% residential, 1% business, and 35% fields and woods) of 585 acres (234 ha) was approximately 1 individual per 3.7 acres (1.5 ha); the average minimum home range (calculated by radio-tracking) for each animal was 12.8 acres (5.1 ha); and the home ranges were characteristically 5½ times as long as wide. The authors suggest that the long, narrow home ranges (animals were using street routes going to and from feeding and denning areas) may be a response to the high density, abundance of food, and linear components of an urban habitat.

BEHAVIOR. Although raccoons will gather together to feed when foods are concentrated (Johnson, 1970; Sharp and Sharp, 1956) or will huddle in winter shelters (Mech and Turkowski, 1966), the animals are generally spatially separated (except during courtship) in nature into solitary males and females with litters (which may stay together all winter). Males are considered much more territorial than females, tending to exclude other males from their areas and from access to breeding females living there. According to Fritzell (1978b), adult males seldom approach each other, keeping a distance of at least 2 mi (3 km) apart but ranges overlap slightly (Fritzell (1978a). The home territories of adult females overlap extensively. Although the neighbors are rarely found together, there does seem to be some neighbor recognition among individuals. Antagonism may be more marked between established animals and transients (Barash, 1973).

Raccoons are essentially nocturnal animals. During daylight hours, they use a variety of resting sites: tree cavities, ground dens and, in good weather, exposed situations such as the top of squirrel leaf nests or tree limbs (Mech *et al.*, 1966; Shirer and Fitch, 1970). Fisher (1977) monitored the nocturnal activity of raccoons in spring and summer (May through August) in Ingham County by radio tracking. She found that movements began after sunset and ended within an hour after sunrise. The average duration of activity of radio-tagged raccoons was 8.6 hours. Animals moved an average of 1,720 ft (524 m) at an average rate of 380 ft (115 m) per hour, with peak activity occurring between 0100 and 0400 hours.

Raccoons are known to be curious and playful. They make maximum use of their ability to manipulate food and other items. The so-called "food-washing" behavior, although probably less commonly practiced than generally thought, is one example of this dexterity. The senses of touch, olfaction, hearing, and sight are well developed. Raccoons are natural climbers (except perhaps for smooth-barked beech trees), run rather slowly with a rolling gait, and are easily outdistanced by a human. They excel as swimmers of necessity, since favored habitats are stream-banks, swamps, and marshes. Raccoons utter various audible calls, growls, screeches, and snarls when on the defensive or fighting; muffled growls of recognition when non-combatant; and low twittering purrs by the female when dealing with her young (Stuewer, 1943b; Tevis, 1947). Although seemingly mild in temperament, the raccoon, as described by Stuewer (1943b), can fight savagely, often besting an attacking dog of similar size. The animals are also known to jump from trees as high as 40 ft (12 m) above ground without sustaining injury. Raccoons defecate in specific places in their areas of activities. These scat stations may also have use in territory establishment (Jenkins and Starkey, 1982).

In winter, raccoons may spend considerable time in dens, becoming inactive during spells of severe Michigan weather, especially in the Upper Peninsula. Having stored fat as an energy source and with a slight depression in body temperature, raccoons can survive without eating for long periods in a state of winter dormancy (carnivorean lethargy), a condition less profound than and not to be confused with true hibernation. Occasionally, raccoons are found using the same shelters as striped skunks (Shirer and Fitch, 1970) and with each other, with a record of 23 raccoons found huddling together in a single den in Minnesota (Mech and Turkowski, 1966). Dens can be located in a variety of places. Although hollows in tree trunks at ground level are used (Berner and Gysel, 1967), Michigan raccoons, according to Stuewer (1943b), prefer hollows at an average height above ground of 27.5 ft (8.3 m). One hollow was as low as 7 ft (2 m) and the highest was 65 ft (20 m). In Allegan County, Stuewer found occupied raccoon dens in red maple (22 dens), American elm (4 dens), white oak (3 dens), black oak (1 den), sugar maple (1 den), white ash (1 den), butternut (1 den), and sycamore (1 den). These trees averaged 55 ft (16.8 m) high and 28 in (711 mm) in diameter (DBH). The average distance of these den trees from surface water was 409 ft (125 m).

The quality of much raccoon habitat has probably been reduced by the wanton felling of den trees, either by hunters interested in retrieving game or by foresters concerned with culling less-marketable timber. Artificial dens, constructed as box-shaped structures out of rough planks or nail kegs, have been attractive to raccoons (and other arboreal animals) in woodlots lacking natural cavities (Baker and Newman, 1942; Stuewer, 1948).

Raccoons often use underground burrows as dens, taking over and usually enlarging the living places of woodchucks, badgers, red and gray foxes, and striped skunks. Other shelters include caves and mine tunnels, hollow logs, culverts, buildings, muskrat houses and burrows in irrigation dikes (where trees are absent or cleared), rock ledges, and brushpiles (Mech *et al.*, 1966; Shirer and Fitch, 1970; Stuewer, 1943b; Urban, 1970).

ASSOCIATES. With the gradual reduction of the larger carnivores in Michigan, the raccoon, even though an omnivore in food selection, has become one of the primary meat-eaters in many biotic communities in southern Michigan. As a result, its associates in the environment generally keep their distance. No doubt interspecific actions occur most often at den sites. Raccoons use tree squirrel leaf nests as resting platforms and probably rout other animals from tree hollows. Raccoons may shy away from ground burrows occupied by foxes and badgers, but take over those of the woodchuck (Butterfield, 1954; Grizzell, 1955; Schmeltz and Whitaker, 1977). Shirer and Fitch (1970) found a raccoon and a striped skunk occupying the same burrow in Kansas. In Michigan, Stuewer (1943b) found a close environmental association between the raccoon and the Virginia opossum. Although he found these two animals using the same areas and even food supplies, he found no real evidence of competition between them.

REPRODUCTIVE ACTIVITIES. Courtship activities begin in Michigan in mid-winter with most breeding occurring from the first week in February through at least the first week in March (Stuewer, 1943a, 1943b; George and Stitt, 1951). Both males and females may breed in the first year, but sometimes not until the second year. The actual mating has been described by Stains (1956) as a prolonged, noisy, and vigorous activity. After a gestation period of 63 days, litters averaging three or four individuals (the maximum is seven) are born, usually in late March, April, and May in Michigan. Litters can appear outside of these months in northern states, even as late as August and September (Berard, 1952; Lehman, 1968). Such late births may be the result of a late first conception or a rare second litter in one year (Lotze and Anderson, 1979).

At one day of age, the young raccoon has blackish skin covered with fine, yellow-gray fur, an indistinct black mask and tail rings, closed ears and eyes, and weighs about 3 oz (85 g). At about three weeks of age, the young raccoon has open eyes and ears, a well-furred body, a conspicuous eye mask and a ringed tail, and erupted lower incisors and canines. At six weeks of age, the young moves about actively. At nine weeks of age, the young eats solid food. At 12 to 16 weeks of age, the young is weaned and weighs 3 to 4 lbs (1.4 to 1.8 kg). At 16 to 20 weeks of age, permanent teeth begin replacing the deciduous (milk) teeth. Observations on growth have been made by Hamilton (1936), Montgomery (1964, 1968), Stains (1956), and Stuewer (1943a). During the nursing period, the female raccoon is most attentive to her offspring, moving them to new quarters from time to time (Montgomery *et al.,* 1970). On one occasion, a female raccoon moved her litter (eyes yet unopened) from a nail-keg den affixed to the side of a dead tree in water some 20 ft (6 m) from shore (Baker and Newman, 1942).

The female and her litter become active as a foraging group by late summer. Occasionally, young will depart in the first autumn although most families remain together until the next breeding season and no later than July (Fritzell, 1978b; Stuewer, 1943b). Yearlings, especially the males, are obliged to disperse, often considerable distances, in search of unoccupied homesites. Most fieldworkers find that juvenile mortality is small until the first winter. Losses at this season are much greater, chiefly due to food scarcity, disease, and human factors. According to Johnson (1970), mortality is also great during the second year of life. This is probably closely related to the hazards faced as the young raccoons move away from family groups and attempt to become established in an environment where older animals maintain most of the suitable home territories. Apparently, adult raccoons hold on to their living spaces successfully for several seasons, since the three-year to six-year age classes appear to have few threats to survival, except for human hunting and trapping (Johnson, 1970). Sanderson (1951) calculated that it took about 7½ years for all the raccoons in a Missouri population to be replaced by recruits.

The age of a raccoon can be determined by the extent of cranial suture closure; degree of coossification of diaphyses and epiphyses in the distal ends of the radius and ulna; number of cementum layers on teeth; pigmentation on the nipples; increase in body weight; and changes in characteristics of the fur (Grau *et al.,* 1970; Johnson, 1970; Junge and Hoffmeister, 1980; Petrides, 1959; Sanderson, 1950, 1961; Stains, 1956; Stuewer, 1943b). In the wild, Michigan raccoons often live more than four years; one individual is known to have survived in the area of the Swan Creek Wildlife Research Station in Allegan County for an estimated 12 years and 7 months (Haugen, 1954;

Linduska, 1947). In captivity this hardy animal may live as long as 22 years and 5 months (Jones, 1979).

FOOD HABITS. Raccoons are commonly regarded as being almost constantly hungry and seeking assorted foods in trees, on and in the ground, at the water's edge, and in garbage dumps. House-reared pets have been known to open refrigerator and pantry doors with their dexterous forepaws. Such behavior supports the findings of Dalgish and Anderson (1979) who suggest that learning is involved in the methods which raccoons use in foraging.

Because of the ease in collecting raccoon feces (scats) from raccoon latrines (Stains, 1956), there are numerous published studies of the food habits of this animal. Most show how opportunistic the raccoon is in acquiring the most available food items. A study of the contents of feces obtained at intervals through the year from scat stations in eastern Texas by Baker *et al.* (1945) shows clearly how this omnivore switches foods as one type declines in availability and another appears. Acorns appeared to be a basic year-round staple, often in surplus in autumn and progressively scarcer in subsequent seasons. However, as floodwaters in river bottoms receded in spring, new stocks of this mast crop became available. In spring, summer, and early autumn, crayfish was a favorite as were various fruits, seeds, and invertebrates as they emerged through the growing part of the year. Studies showing comparable food habits have been reported in New York (Hamilton, 1936, 1940, 1951), Kansas (Stains, 1956), Illinois (Yeager and Elder, 1945; Yeager and Rennels, 1943), Missouri (Korschgen, 1952), Minnesota (Schoonover and Marshall, 1951), and Wisconsin (Dorney, 1954).

Michigan studies, mostly by Alexander (1977), Dearborn (1932), Fisher (1977), and Stuewer (1943b), show the variety of nutrients ingested by raccoons include such animals as eastern cottontails, tree squirrels, mice of the genus *Peromyscus*, voles of the genus *Microtus*, various birds including poultry, hen and duck eggs, ring-necked pheasant, snakes, turtles and their eggs, frogs, fish (brook trout, sculpin, creek chub, sucker, fathead minnow), crayfish, grasshoppers, beetles and their larvae, moths and their pupae, earthworms, clams and snails. Among remains of plants found in fecal matter were seeds of dogwood, ragweed, elderberry, cherry, grape, blackberry, acorns, grass, buds, and such cultivated crops as oats, corn, and soybeans. Despite the number of mammals and birds in lists of raccoon foods, the major ones are plant materials and invertebrates.

ENEMIES. Natural enemies, except for the human, are decidedly few in modern times because of the demise of many of the large carnivores. Middle west records of other animals eating the raccoon and its young (in some cases perhaps as carrion) include gray wolf (Voigt *et al.*, 1976), coyote (Korschgen; 1952; Stains, 1956), red fox, (Schofield, 1960), fisher (de Vos, 1951; Seton, 1953), bobcat (Jackson, 1961), raven (Harlow *et al.*, 1975), and great horned owl (Jackson, 1961).

Besides the hazards of surviving hunting and trapping seasons, raccoons are constantly threatened by automobiles when crossing Michigan highways (Haugen, 1944; Kasul, 1975). Haugen found that raccoon highway fatalities were especially high during July. Forty deaths were tabulated by Hartley (1975) as he travelled Michigan highway M-21 to Kent County twice daily for five days a week between March 1 and August 30, 1974. Kasul (1976) counted 64 dead raccoons on 15.5 mi of Interstate Highway I-96 in Ingham County between August 20, 1975 and March 15, 1976. Modern technology can also be a threat to the raccoon and other wildlife through environmental deterioration from additives, mostly as byproducts. For example, lead accumulations in the liver (Sanderson and Thomas, 1961) and the ingestion of such chemicals as PCB (polychlorinated biphenyl) as reported in the Michigan press in January 1978, may have a serious effect on the raccoon population.

PARASITES AND DISEASES. Raccoons are probably among the most parasitized of Michigan carnivores. Stains (1956) compiled a list of 10 species of fleas (Siphonaptera), 1 sucking louse (Anoplura), 6 chewing lice (Mallophaga), and 15 mites and ticks (Acarina) as external parasites and 25 tapeworms and flukes (Platyhelminthes), 31 roundworms (Nematoda), and 3 spiny-headed worms (Acanthocephala) as internal parasites. Stuewer (1943b) recorded 1 flea, 1 louse, ticks, 2 tapeworms, 1 fluke, and 1 roundworm living on or in Michigan raccoons.

Diseases reported in Michigan raccoons include: rabies (Kurta, 1979), bacteria of the genus *Salmonella* (Youatt and Fay, 1968), and a distemper-like illness (Stuht and Harte, 1973). Stains (1956) also lists feline laryngitis and enteritis, gastroenteritis, encephalitis, dog and cat distemper, tuberculosis, pseudotuberculosis, purulent fibrinous pleuritis and pericarditis, listeriosis, and shock disease. Two fungal organisms, *Histoplasma capsulatum* and *Haplosporangium parvum,* were the causitive agents for a serious outbreak of a distemper-like disease throughout the midwest states in 1952 and 1953. This malady was perhaps responsible, at least in part, for the general decline in raccoons at that time in parts of this large area (Menges *et al.,* 1955; Stains, 1956).

MOLT AND COLOR ABERRATIONS. A single annual molt occurs, with shedding and regrowth taking place over several weeks from mid-April to late June in Michigan (Stuewer, 1942). Several color phases have been described in the raccoon (Jones, 1923). Both black (melanistic) and white (albinistic) individuals occasionally occur in Michigan (Wood, 1922). Animals with an over-all orange-red (erythristic) coloring are also reported (Dalby and Dawson, 1973; Neill, 1953). Michigan specimens of both albinistic (from Ingham County) and erythristic (from Washtenaw County) coloring are preserved in the collections of The Museum at Michigan State University.

ECONOMIC IMPORTANCE. The black-masked, ring-tailed raccoon has a temperament and intelligence which has endeared it to generations of American authors of children's stories (Rue, 1964). Food-washing and the antics of raccoons when tamed as house pets are well-known subjects. In general, the citizenry tolerates the animal's pranks and occasional mischievous ways. As a Michigan mammal, the raccoon rates high on the popularity scale, even when one has dumped over the garbage can or shorted out electrical power—an occurrence which blacked-out the Michigan State University campus for several hours on April 13, 1980. This animal has become well adapted to human habitations and land-use practices (Hoffmann and Gottschang, 1977), and seems here to stay as a close associate. As Stuewer (1941) pointed out, there is no need either for restocking in Michigan or for establishing raccoon ranches for the purpose of raising this species.

The raccoon, nevertheless, does incur the farmer's wrath on occasion. In Michigan, Stuewer (1943b) cites the raccoon's destruction of agricultural crops and poultry, although in these times of modern poultry husbandry, chickens are less available. However, grain fields and apple and cherry orchards are not enclosed in raccoon-proof fences. The great attraction is to corn in the milk stage; in 1963, for example, the Michigan Department of Natural Resources received 719 citizen complaints about raccoon depredations, mostly in fields of sweet corn. In Ingham County, radio-tagged raccoons were repeatedly tracked by Fisher (1977) from den sites to food sources. She found raccoons using insects, various seeds, and grasses as major food items (83.4% of scat contents) in May but these items were present only in a minor way (21.3%) in August. At the same time, corn increased from 5.5% in May, to 16.2% in July, to a high of 67.2% in August.

The rich, thick pelage of the raccoon has long been an attraction, first to the Indians and early settlers as items of clothing and later to the lucrative fur trade. In the summer of 1767, for example, British traders at Michilimackinac shipped to Europe 23,005 raccoon pelts, second in number to the popular beaver hides which totaled 50,938 (Petersen, 1979). As the beaver supply declined, the raccoon and the muskrat moved to the forefront in the fur trade. From 1835 to 1839, the American Fur Company received 200,656 raccoon hides at its Detroit Department (including southern Michigan) and 21,023 from its Northern Department (including northern Michigan), according to Johnson (1969). The annual take of raccoon in Michigan has, of course, fluctuated with the ups and downs of the prices offered for pelts by fur buyers. Nevertheless, the average farm boy probably traps a few each winter despite the varying market price for pelts.

Stuewer (1943b) noted with concern that $3.50 was the price for Michigan raccoon pelts in the years just prior to World War II. For several years after the war, fur fashion showed disdain for long-haired fur. Even as late as the early 1970s, an average price per pelt of just $5.50 was typical. At the same time, there were complaints of the over-abundance of raccoons, because trappers and hunters were not removing sufficient surpluses. Finally, the demand for long-haired fur increased, partly as a result of the popularity of fur-trimmed ski clothing. Accordingly, prime pelts brought as much as $10 in 1973, $17 in 1974, $38 in 1977, and $45 in both 1979 and 1980, but back down to an average of $17 in 1981. A high of $81 for a raccoon pelt was paid in November 1978, according to the Michigan press. The estimated harvest of raccoons in Michigan was 232,460 during the 1970–71 season and 376,000 during the 1974–75 season. Less than 10% of these catches were from the Upper Peninsula.

Hunters using trained hounds account for more than 90% of the annual raccoon harvest. To accommodate this active group, the raccoon hunting season in 1979–1980, for example, was open for residents from October 1 through January 31 (for non-residents, November 1 through January 31). For trappers the season was shortened, from October 25 through December 31 (in Region I—the Upper Peninsula), from November 1 through January 15 (in Region II—the northern part of the Lower Peninsula), and from November 18 through January 31 (in Region III—the southern part of the Lower Peninsula). In 1979, there were 26 local hunting clubs affiliated with the Michigan United Coon Hunters Association. Each May, this organization sponsors field trials attracting hundreds of hunters from Michigan, other parts of the United States, and Canada. Various breeds of hounds including bluetick, black and tan, redbone, and treeing walker are entered and awards are given for best performances in trailing and treeing raccoons. No killing or mauling of the raccoon is allowed. It is reported that Michigan had at least 42,000 raccoon hunters (as of 1977) with as many as 50 club hunts staged each year. Some hunters enjoy the hunt for the harvesting of pelts; others live-trap the animals and remove them to places remote from urban or suburban areas. The "woods music" of the hounds on the trail and around the base of the tree in which a raccoon has sought refuge has a sweet and compelling charm (Lowe, 1978).

The dark meat of the raccoon has long been a delicacy to rural Americans, as well as those who just enjoy wild foods (Edwards, 1893). The best-tasting dishes result from selecting animals no more than three-quarters grown. The lymphatic glands are removed from inner parts of the muscles of both the hind and fore limbs to decrease the strong flavor. Raccoon meat can then be baked, chicken-fried, or stewed, although today barbecued raccoon is increasingly popular (Stains and Baker, 1958). It is legal to sell raccoon meat in Michigan and meat dealers at Eastern Market in Detroit often buy and sell raccoon carcasses. Dressed animals, purchased from trappers and hunters, retailed for 59¢ a pound in 1977 and about 89¢ a pound in 1979. In 1979, a Durand fur buyer was selling raccoon carcasses for $2.50 to $3.50 each.

COUNTY RECORDS FROM MICHIGAN. Records of specimens of raccoons in collections of museums, from the literature, and reported by personnel of the Michigan Department of Natural Resources show that raccoons occur in all counties in the state. The animal also lives on Beaver, North Fox, South Fox, North Manitou, and South Manitou islands in Lake Michigan (Hatt *et al.*, 1948; Scharf and Jorae, 1980); on the Huron Islands in Lake Superior (Corin, 1976); on Bois Blanc in Lake Huron (Cleland, 1966), and on Charity Island (introduced) in Saginaw Bay (Wood, 1914b).

LITERATURE CITED

Alexander, G. R.
1977 Food of vertebrate predators on trout waters in north central Lower Michigan. Michigan Acad., 10(2):181–195.

Allen, D. L.
1938 Ecological studies on the vertebrate fauna of a 500-acre farm in Kalamazoo County, Michigan. Ecol. Monog., 8(3):348–436.

Baker, R. H., and C. C. Newman
1942 Note on the den site of a raccoon family. Jour. Mammalogy, 23(2):214–215.

Baker, R. H., C. C. Newman, and F. Wilke
1945 Food habits of the raccoon in eastern Texas. Jour. Wildlife Mgt., 9(1):44–48.

Barash, D.
1973 Neighbor recognition in two solitary carnivores; the raccoon and the red fox. Science, 185:794–796.

Berard, E. V.
1952 Evidence of a late birth for the raccoon. Jour. Mammalogy, 33(2):247–248.

Berner, A., and L. W. Gysel
1967 Raccoon use of large tree cavities and ground burrows. Jour. Wildlife Mgt., 31(4):706–714.

Butterfield, R. T.
1954 Some raccoon and groundhog relationships. Jour. Wildlife Mgt., 18(4):433–437.

Cleland, C. E.
1966 The prehistoric animal ecology and ethnozoology of the Upper Great Lakes Region. Univ. Michigan, Mus. Anthro., Anthro. Papers, No. 29, x+294 pp.

Corin, C. W.
1976 The land vertebrates of the Huron Islands, Lake Superior. Jack Pine Warbler, 54(4):138–147.

Craighead, J. J., and F. C. Craighead, Jr.
1956 Hawks, owls and wildlife. Wildlife Mgt. Inst., Washington, D.C., xix+443 pp.

Dalby, P. L., and G. A. Dawson
1973 Two pink-eyed yellow raccoons, *Procyon lotor*. Carnivore genetics Newsletter, 2(5):104–105.

Dalgish, J., and S. Anderson
1979 A field experiment on learning by raccoons. Jour. Mammalogy, 60(3):620–622.

Dearborn, N.
1932 Foods of some predatory fur-bearing animals in Michigan. Univ. Michigan, Sch. Forestry and Conserv., Bull. No. 1, 52 pp.

de Vos, A.
1951 Recent findings in fisher and marten ecology and management. Trans. 16th North American Wildlife Conf., pp. 498–507.
1964 Range changes of mammals in the Great Lakes Region. American Midl. Nat., 71(1):210–231.

Dice, L. R.
1925 A survey of the mammals of Charlevoix County, Michigan, and vicinity. Univ. Michigan, Mus. Zool., Occas. Papers No. 159, 33 pp.

Dorney, R. S.
1954 Ecology of marsh raccoons. Jour. Wildlife Mgt., 18(2): 217–225.

Edwards, S. E.
1893 The Ohio hunter or a brief sketch of the frontier life of Samuel E. Edwards the great bear and deer hunter of the state of Ohio. Review and Herald Publ. Co., Battle Creek, Michigan. 240 pp.

Ellis, R. J.
1964 Tracking raccoons by radio. Jour. Wildlife Mgt., 28(2):363–368.

Fisher, L. E.
1977 Movements of raccoons in small upland woodlots devoid of water. Michigan State Univ., unpubl. M.S. thesis, v+42 pp.

Fritzell, E. K.
1978a Habitat use by prairie raccoons during the waterfowl breeding season. Jour. Wildlife Mgt., 42(1):118–127.
1978b Aspects of raccoon *(Procyon lotor)* social organization. Canadian Jour. Sci., 56:260–271.

George, J. L., and M. Stitt
1951 March litters of raccoons *(Procyon lotor)* in Michigan. Jour. Mammalogy, 32(1):21.

Grau, G. A., G. C. Sanderson, and J. P. Rogers
1970 Age determination of raccoons. Jour. Mammalogy, 34(2):364–372.

Grizzell, R. A., Jr.
1955 A study of the southern woodchuck, *Marmota monax monax*. American Midl. Nat., 53(2):257–293.

Hamilton, W. J., Jr.
1936 The food and breeding habits of the raccoon. Ohio Jour. Sci., 36(3):131–140.
1940 The summer foods of minks and raccoons on the Montezuma Marsh, New York. Jour. Wildlife Mgt., 4(1):80–84.
1951 Warm weather foods of the raccoon in New York State. Jour. Mammalogy, 32(3):341–344.

Harlow, R. F., R. G. Hooper, D. R. Chamberlain, and H. S. Crawford
1975 Some winter and nesting season foods of the common raven in Virginia. Auk, 92:298–306.

Hartley, S. L.
1975 Life's cycle. Michigan Out-of-Doors, 29(3):41.

Hatt, R. T., J. Van Tyne, L. C. Stuart, C. H. Pope, and A. B. Grobman
1948 Island life: a study of the land vertebrates of the islands of eastern Lake Michigan. Cranbrook Inst. Sci., Bull. No. 27, xi+179 pp.

Haugen, A. O.
1944 Highway mortality of wildlife in southern Michigan. Jour. Mammalogy, 25(2):160–170.
1954 Longevity of the raccoon in the wild. Jour. Mammalogy, 35(3):439.

Higgins, M. J.
1980 An analysis of the faunal remains from the Schwerdt Site, a late prehistoric encampment in Allegan County, Michigan. Western Michigan Univ., unpubl. M.A. thesis, v+77 pp.

Hoffmann, C. O., and J. L. Gottschang
1977 Numbers, distribution, and movements of a raccoon population in a suburban residential community. Jour. Mammalogy, 58(4):623–636.

Holman, J. A.
1975 Michigan's fossil vertebrates. Michigan State Univ., Publ. Mus., Educ. Bull. No. 2, 54 pp.

Jackson, H. H. T.
1961 Mammals of Wisconsin. Univ. Wisconsin Press, Madison. xii+504 pp.

Jenkins, K. J., and E. E. Starkey
1982 Scent marking by captive raccoons. Jour. Mammalogy, 63(2):318–319.

Johnson, A. S.
1970 Biology of the raccoon (*Procyon lotor varius* Nelson and Goldman) in Alabama. Auburn Univ., Agric. Exp. Sta., Bull. No. 402, vi+148 pp.

Johnson, D. R.
1969 Returns of the American Fur Company, 1835–1839. Jour. Mammalogy, 50(4):836–839.

Jones, M. L.
1979 Longevity of mammals in captivity. Internat. Zoo News, 26(3):16–26.

Jones, S. V. H.
1923 Color variations in wild animals. Jour. Mammalogy, 4(3):172–177.

Junge, R., and D. F. Hoffmeister
1980 Age determination in raccoons from cranial suture obliteration. Jour. Wildlife Mgt., 44(3):725–729.

Kasul, R. L.
1976 Mortality and movement of mammals and birds on a Michigan interstate highway. Michigan State Univ., unpubl. M.S. thesis, ix+39 pp.

Korschgen, L. J.
1952 A general summary of the food of Missouri predatory and game animals. Missouri Conserv. Comm., 61 pp.

Kurta, A.
1979 Bat rabies in Michigan. Michigan Acad., 12(2):221–230.

Laundre, J.
1975 An ecological survey of the mammals of the Huron Mountain area. Huron Wildlife Found., Occas. Papers No. 2, x+69 pp.

Lehman, L. E.
1968 September births of raccoons in Indiana. Jour. Mammalogy, 49(1):126–127.
1977 Population ecology of the raccoon on the Jasper-Pulaski Wildlife Study Area. Indiana Dept. Nat. Res., Div. Fish and Wildlife, Pittman-Robertson Bull. No. 9, viii+97 pp.

Linduska, J. P.
1947 Longevity of some Michigan farm game mammals. Jour. Mammalogy, 28(2):126–129.
1950 Ecology and land-use relationships of small mammals on a Michigan farm. Michigan Dept. Conser., Game Div., ix+144 pp.

Lotze, J.-H.
1979 The raccoon (*Procyon lotor*) on St. Catherines Island, Georgia. 4. Comparisons of home ranges determined by livetrapping and radiotracking. American Mus. Novit., No. 2664, 25 pp.

Lotze, J.-H., and S. Anderson
1979 Procyon lotor. American Soc. Mammalogists, Mammalian species, No. 119, 8 pp.

Lowe, K. S.
1978 Chasing coons. Michigan Out-of-Doors., 32(3):44–45, 63.

Martin, T. J.
1976 A faunal analysis of five Woodland Period archaeological sites in southwestern Michigan. Western Michigan Univ., unpubl. M.A. thesis, viii+200 pp.
1979 Faunal remains from the Gros Cap Site (20MK6/7), Mackinac County, Michigan. Appendix E, pp. 133–148 *in* S. R. Martin. Final report: Phase II. Archaeological site examination of the Gros Cap Cemetery area, Mackinac County, Michigan. Michigan Tech. Univ., Cult. Res. Mgt. Rept. No. 6, ix+166 pp.
1980 Animal remains from the Winter Site, a Middle Woodland occupation in Delta County, Michigan. Michigan State Univ., Mus., Div. Anthro., unpubl. Rept., 11 pp.

Mech, L. D., J. R. Tester, and D. W. Warner
1966 Fall daytime resting habits of raccoons as determined by telemetry. Jour. Mammalogy, 47(3):450–466.

Mech, L. D., and F. J. Turkowski
1966 Twenty-three raccoons in one winter den. Jour. Mammalogy, 47(3):529–530.

Menges, R. W., R. T. Habermann, and H. J. Stains
1955 A distemper-like disease in raccoons and isolation of Histoplasma capsulatum and Haplosporangium parvum. Trans. Kansas Acad. Sci., 58(1):58–67.

Montgomery, G. G.
1964 Tooth eruption in preweaned raccoons. Jour. Wildlife Mgt., 28(4):582–584.
1968 Pelage development of young raccoons. Jour. Mammalogy, 49(1):142–145.

Montgomery, G. G., J. W. Lang, and M. E. Sunquist
1970 A raccoon moves her young. Jour. Mammalogy, 51(1):202–203.

Neill, W. T.
1953 Two erythristic raccoons from Florida. Jour. Mammalogy, 34(4):500.

Ozoga, J. J., and C. J. Phillips
1964 Mammals of Beaver Island, Michigan. Michigan State Univ., Publ. Mus., Biol. Ser., 2(6):305–348.

Petersen, E. T.
1979 Hunters' heritage. A history of hunting in Michigan. Michigan United Conserv. Clubs, Lansing. 55 pp.

Petrides, G. A.
1959 Age ratios in raccoon. Jour. Mammalogy, 40(2):249.

Priewert, F. W.
1961 Record of an extensive movement by a raccoon. Jour. Mammalogy, 42(1):113.

Rhead, D., and T. J. Martin
1978 Environment and subsistence at Sand Point. Paper presented Central States Anthro. Soc., South Bend, 8 pp. (mineo)

Rue, L. L., III
1964 The world of the raccoon. J. B. Lippincott Co., Philadelphia and New York. 145 pp.

Sanderson, G. C.
1950 Methods of measuring productivity in raccoons. Jour. Wildlife Mgt., 14(3):389–402.
1951 Breeding habits and a history of the Missouri raccoon population from 1941 to 1948. Trans. 16th North American Wildlife Conf., pp. 481–485.
1961 The lens as an indicator of age in the raccoon. American Midl. Nat., 65(3):481–485.

Sanderson, G. C., and R. M. Thomas
1961 Incidence of lead in livers of Illinois raccoons. Jour. Wildlife Mgt., 25(2):160–168.

Scharf, W. C., and M. L. Jorae
1980 Birds and land vertebrates of North Manitou Island. Jack-Pine Warbler, 58(1):4–15.

Schmeltz, L. L., and J. O. Whitaker, Jr.
1977 Use of woodchuck burrows by woodchucks and other mammals. Trans. Kentucky Acad. Sci., 38(1–2):79–82.

Schofield, R. D.
1960 A thousand miles of fox trails in Michigan's ruffed grouse range. Jour. Wildlife Mgt., 24(4):432–434.

Schoonover, L. J., and W. H. Marshall
1951 Food habits of the raccoon *(Procyon lotor hirtus)* in

northcentral Minnesota. Jour. Mammalogy, 32(4): 422–428.

Seton, E. T.
1953 Lives of game animals. Vol. II, Pt. I. Bears, coons, badgers, skunks and weasels. Charles T. Branford Co., Boston. xvii+367 pp.

Sharp, W. M., and L. H. Sharp
1956 Nocturnal movements and behavior of wild raccoons at a winter feeding station. Jour. Mammalogy, 37(1): 170–177.

Shirer, H. W., and H. S. Fitch
1970 Comparison from radiotracking of movements and denning habits of the raccoon, striped skunk, and opossum in northeastern Kansas. Jour. Mammalogy, 51(3):491–503.

Stains, H. J.
1956 The raccoon in Kansas, natural history, management, and economic importance. Univ. Kansas, Mus. Nat. Hist. and State Biol. Surv., Misc. Publ. No. 10, 76 pp.

Stains, H. J., and R. H. Baker
1958 Furbearers in Kansas: A guide to trapping. Univ. Kansas, Mus. Nat. Hist. and State Biol. Surv., Misc. Publ. No. 18, 100 pp.

Stuewer, F. W.
1941 'Coon stocking not for Michigan. Michigan Conserv., 10(8):3, 11.
1942 Studies of molting and priming of fur of the eastern raccoon. Jour. Mammalogy, 23(4):399–404.
1943a Reproduction of raccoons in Michigan. Jour. Wildlife Mgt., 7(1):60–73.
1943b Raccoons: Their habits and management in Michigan. Ecol. Monog., 13(2):203–258.
1948 Artificial dens for raccoons. Jour. Wildlife Mgt., 12(3): 296–301.

Stuht, J. N., and H. D. Harte
1973 Disease and mortality factors affecting Michigan wildlife, 1972. Michigan Dept. Nat. Res., Wildlife Div., Rept. No. 2704 (mineo.), 7 pp.

Tevis, L., Jr.
1947 Summer activities of California raccoons. Jour. Mammalogy, 28(4):323–332.

Urban, D.
1970 Raccoon populations, movement patterns, and predation on a managed waterfowl marsh. Jour. Wildlife Mgt., 34(2):372–382.

Voigt, D. R., G. B. Kolenosky, and D. H. Pimlott
1976 Changes in summer foods of wolves in central Ontario. Jour. Wildlife Mgt., 40(4):663–668.

Wenzel, O. J.
1912 A collection of mammals from Osceola County, Michigan. Michigan Acad. Sci., 14th Rept., pp. 198–205.

Wood, N. A.
1914a Results of the Shiras expeditions to Whitefish Point, Michigan—Mammals. Michigan Acad. Sci., 16th Ann. Rept., pp. 92–98.
1914b An annotated check-list of Michigan mammals. Univ. Michigan, Mus. Zool., Occas. Papers No. 4, 13 pp.
1922 The mammals of Washtenaw County, Michigan. Univ. Michigan, Mus. Zool., Occas. Papers No. 123, 23 pp.

Yeager, L. E., and W. H. Elder
1945 Pre- and post-hunting season foods of raccoons on an Illinois goose refuge. Jour. Wildlife Mgt., 9(1):48–56.

Yeager, L. E., and R. G. Rennels
1943 Fur yield and autumn foods of the raccoon in Illinois river bottom lands. Jour. Wildlife Mgt., 7(1):45–60.

Youatt, W. G., and L. D. Fay
1968 Survey for Salmonella in some wild birds and mammals in Michigan. Michigan Dept. Conserv., Res. and Dev. Rept. No. 135 (mimeo), 1 p.

FAMILY MUSTELIDAE

The Mustelidae is a large family of 25 genera and approximately 70 species found in all major land surfaces of the world except for the Australian area, where some introductions have thrived. This diverse group of small to medium-sized meat-eaters is characterized, in most cases, by squat, elongated bodies with short limbs, non-retractible claws, short, broad heads, medium to long tails, often rich and glossy pelts valued in the fur trade, and perhaps best known for the enlarged anal glands, the emission of odor from which is a trademark especially for the striped skunk. The typical mustelid skull features an elongated and rather flattened brain case with a short rostrum. The hard palate extends posteriorly well beyond the end of the upper tooth-row. Often the lower jaw is so tightly wedged by its anticular condyles that it may not be disengaged from the upper jaw's glenoid fossa. The molar teeth are reduced to only one on each side in the upper jaw and one or two on each side of the lower. The incisors are weakly constructed, but the canines and carnassial teeth (fourth upper premolar and first lower molar) are strongly developed: the canines for stabbing victims and the carnassial for shearing off chunks of muscle tissue.

Mustelids range in size from the dainty least weasel, weighing no more than 1½ oz (42 g), to the sea otter, weighing as much as 50 lbs (22.7 kg). Members of this family include some of the most well adapted killers of prey found among the carnivores. Small weasels may subdue victims several times larger than themselves, biting effectively about the base of the head and neck. Mustelids are diverse in their environmental adaptations;

some are excellent climbers (marten and tayra), diggers (badger), and some are swimmers (in fresh water, mink and river otter, and in salt water, sea otter). Most smaller species, at least, are noted for their graceful, swift movements. Some, like the weasels, are almost totally meat-eaters; others, like the striped skunk, may be chiefly insectivorous. Michigan is well-represented by this family with ten species belonging to five genera. Of the three which have been extirpated, two have been returned through the release of breeding stock from Canada.

MARTEN
Martes americana (Turton)

NAMES. The generic name *Martes,* proposed by Pinel in 1792, is taken directly from the Latin word for this animal. The specific name *americana,* proposed by Turton in 1806, is the Latinized form of the word American with *americ* referring to America and *anus* meaning "belonging to" when added as a suffix to the noun stem. In Michigan, the marten is sometimes called the American sable or pine marten. Marten living previously in both Upper and Lower Peninsula Michigan as well as those stocks from Ontario recently liberated in the state are classified as belonging to the subspecies *Martes americana americana* (Turton).

RECOGNITION. Body size resembles that of a slender, small house cat or full-bodied mink (Fig. 78); length of head and body averaging 16½ in (420 mm) for males and 14¾ in (375 mm) for the smaller females; tail about one-third the length of animal's total length, bushy and cylindrical in shape, averaging 8 in (200 mm) in the males and 7 in (180 mm) in the females in vertebral length; head broad, narrowing forward to a short, pointed nose; eyes bright, black, beady; ears conspicuous, rounded, erect; neck prominent, slender; body slender, weasel-like; limbs short; feet broad, each with five toes terminating in sharp, flattened, recurved claws (highly suited for climbing); soles of feet furred; stance on terminal digits (digitigrade); mammary glands consist of three pairs, one on the chest, two on the lower belly; anal scent glands prominent especially in female; males about 12–15% larger than females.

The body fur is dense, silky, long, and lustrous. The light-brown coloring of the short, thick, soft underfur is covered over by long, glossy guard

Figure 78. The marten, newly re-introduced in Michigan's Upper Peninsula.
This photograph, copyrighted by Dikran Kashkashian, was taken in boreal forest habitat in 1980.

hairs to produce a dorsal appearance of rich yellow-brown to dark brown. The head is usually lighter in color; the ears are whitish edged; the tail and legs are darker than the mid-back; underparts are dark except for irregular cream or light yellow-brown splotches on the throat and chest.

The marten resembles the mink in size but is distinguished by size from all other Michigan mustelids—the smaller weasels and the larger fisher, otter, badger, striped skunk, and wolverine. The marten is easily differentiated from the mink by its yellow-brown instead of dark chocolate brown coloring, its longer pelage, its bushier tail, and its larger, white-trimmed ears.

MEASUREMENTS AND WEIGHTS. Adult specimens of the larger male and the smaller female marten measure, respectively, as follows: length from tip of nose to end of tail vertebrae, 23⅜ to 25¼ in (600 to 640 mm) and 21¼ to 22½ in (540 to 570 mm); length of tail vertebrae, 7 to 8¾ in (180 to 220 mm) and 6¼ to 8 in (160 to 200 mm); length of hind foot, 3¼ to 3¾ in (85 to 95 mm) and 3 to 3¼ in (24 to 38 mm); height of ear from notch, 1⅜ to 1¾ in (35 to 45 mm) and 1⅜ to 1½ in (34 to 38 mm); weight, 35¼ to 51¼ oz (999 to 1,453 g) and 25½ to 35¼ oz (723 to 999 g).

DISTINCTIVE CRANIAL AND DENTAL CHARACTERISTICS. The skull of the marten is long, slender, smooth, and rounded dorsally (Fig. 79). The frontal region is broad and flat; the rostrum slender and elongate; the brain case moderately flattened; the zygomatic arches slender, delicate; the auditory bullae rounded, somewhat inflated; and the palate elongate with the shelf extending posteriorly beyond the last molar about one-fourth the distance of the entire tooth-row. Older males develop a slight sagittal crest. The broad molars reflect the typical mustelid condition. The presence of four premolars on each side of both the upper and lower jaws clearly distinguishes the skulls of the marten and the larger fisher from those of the weasels and mink (genus *Mustela*). The skulls of an adult male and female marten, respectively, average 3½ in (85 mm) and 3 in (73 mm) in greatest length and 2 in (50 mm) and 1¾ in (43 mm) in greatest width across the zygomatic arches. The dental formula is as follows:

$$\text{I (incisors)}\,\frac{3\text{-}3}{3\text{-}3},\ \text{C (canines)}\,\frac{1\text{-}1}{1\text{-}1},\ \text{P (premolars)}\,\frac{4\text{-}4}{4\text{-}4},\ \text{M (molars)}\,\frac{1\text{-}1}{2\text{-}2} = 38$$

DISTRIBUTION. The marten is at home in the boreal coniferous forests which form a broad belt in northern North America from Newfoundland and New England on the east, westward to Alaska and south to northern California. Its range includes southward extensions into suitable boreal forests in the higher elevations of the Appalachians in the east and in mountainous areas west of the Great Plains. The Upper Great Lakes Region, including Michigan, is a part of this animal's historic distribution; however, as is the case in many other sectors, populations of this species have been depleted or extirpated (de Vos, 1963; Schorger,

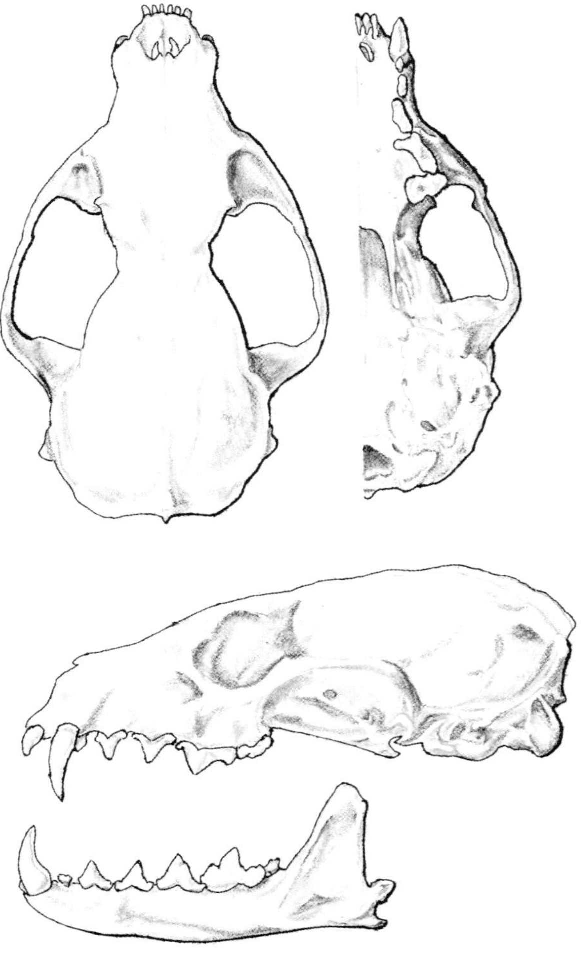

Figure 79. Cranial and dental characteristics of the marten (*Martes americana*).

1942; Schupbach, 1977). A closely related species, the beech or stone marten *(Martes foina)*, occupies comparable habitat in northern Eurasia.

The marten was long a Michigan resident (see Map 52), probably from early post-glacial times until its reported demise in the 1920s and 1930s.

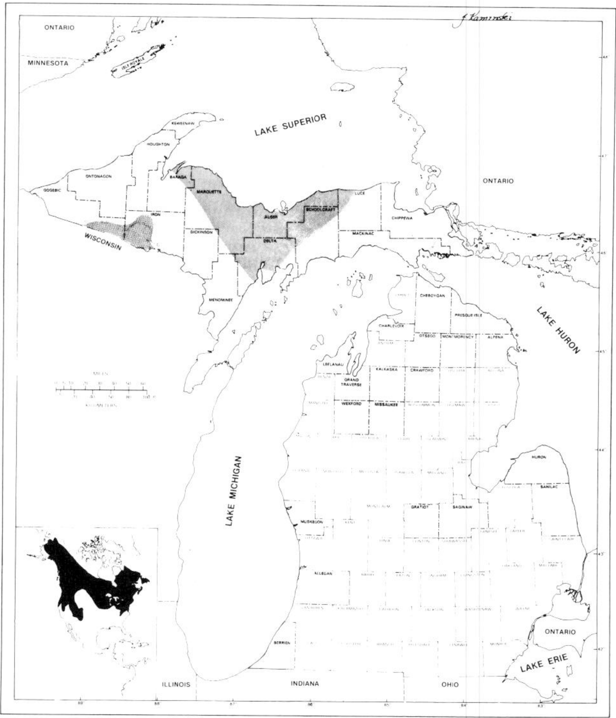

Map 52. Geographic distribution of the marten (*Martes americana*). Areas of known previous occurrence are marked with pale shading. Areas where populations have been recently reestablished are marked with dark shading.

Remains of the animal have been excavated from Woodland Indian sites in both the Upper Peninsula (Mackinac County; Cleland, 1966) and Lower Peninsula (Charlevoix County; Lovis, 1973; Saginaw County; Luxenberg, 1972). Its downfall began in the earliest settlement days when marten fur became an important export for the European trade (Peterson, 1979) and later for the eastern American trade (Johnson, 1969). Schupbach (1977) vividly traced this decline in the Upper Peninsula from 1880 to 1920 and the final extirpation from 1920 to 1940. The Fourth Biennial Report of the Michigan Department of Conservation for 1927–1928 noted that marten numbers had declined at that time to the point where there appeared to be no chance for them to survive. One of the last published reports of the natural occurrence of the marten in the Upper Peninsula was in 1939 in the Huron Mountains of Marquette County (Manville, 1948). The occasional marten occurring on Sugar Island, Chippewa County, in the Superior-Huron waterway connector, may be a transient, crossing to the island from Ontario, perhaps on winter ice (Thomas F. Weise, *in litt.*). One was captured there on November 5, 1981 by Wilfred Eikey.

In Michigan's Lower Peninsula, the marten may have lived in all counties in the earliest of times, although the northern half was better habitat for this animal than the southern half. In their classic report on the distribution of Michigan mammals, Wood and Dice (1924), later quoted by Burt (1946), found evidence of the species in Allegan, Alpena, Crawford, Grand Traverse, Gratiot, Leelanau, Missaukee, Montmorency, Presque Isle, and Wexford counties. Other records are from Berrien (Baker, 1899), Cheboygan, Huron, Muskegon, Kalkaska, Sanilac, Saginaw (Wood, 1914b), Charlevoix, and Otsego counties (Dice, 1925), making at least nineteen Lower Peninsula counties for which there are positive records of occurrence (see Map 52). However, in the Lower Peninsula the marten very likely disappeared before World War I. The last definite report is from Montmorency County in 1911 (Wood and Dice, 1924).

RE-ESTABLISHING THE MARTEN IN MICHIGAN. Marten began increasing in New York in 1946 (Hamilton, 1958); in Minnesota, marten began a return in the northeastern sector in the 1950s (Gunderson, 1965; Mech and Rogers, 1977; Timm, 1975); and Michigan also began a program to bring back the animal (Harger and Switzenberg, 1958; Smith, 1981c). Officials of the Michigan Department of Conservation (now Natural Resources) in cooperation with the Ontario Department of Lands and Forests obtained stock through the live-trapping efforts of department personnel from both units, from commercial trappers, and from breeders. Eight animals were released in the winter of 1955 and another 21 in the winter of 1957 in Porcupine Mountains State Park in Ontonagon County. The status of these tagged individuals was carefully monitored. Apparently the stocked marten dispersed; one animal was taken in Delta County (near Masonville) about 145 mi (230 km) from the place of release. In the winter of 1958–1959, two sets of fresh snow tracks were found in the Porcupine Mountains. Nevertheless, this plant failed with no reliable marten reports obtained after 1962.

Again cooperating with the Ontario officials, the Michigan Department of Natural Resources joined with the U.S. Forest Service to purchase more animals for release. Stock, from the Port Arthur district of Ontario, was obtained at $37.50 per animal from licensed trappers. Of a total of 99 marten (62 males and 37 females), 44 were released in the winter of 1968–1969 and 55 in the winter of 1969–1970 in the Whitefish River area of Hiawatha National Forest. This release site was about 10 mi (16 km) northeast of Rapid River in Delta County. To reduce losses of the stock through normal, legal trapping, the 12 townships in the area were closed to all dry-land trapping beginning in the autumn of 1969.

This large stock of marten (all tagged for subsequent recognition) began dispersal movements as had the stock placed earlier in Ontonagon County. Animals were picked up in 1973 at localities in Luce and Alger counties as well as in Delta; in 1974, eight more observations showed movements as far away as 90 mi (145 km). In an investigation made in 1975 and 1976, Schupbach (1977) detailed these reported movements and was pessimistic at the time about the success of this introduction. In the summer of 1978, a marten was observed at Imp Lake southwest of Watersmeet in Gogebic County (A. M. Wilkerson, pers. comm., September 20, 1978). It is suspected that this sighting, plus similar ones in southern Iron County, were of animals derived from plantings made in Wisconsin. On October 30, 1978, a young male (untagged and evidently the progeny of the planted animals) was captured in a muskrat trap near Chatham in Alger County (N. Sloan, pers. comm., November 28, 1978; Smith, 1979). Another was seen in late summer of 1980 in Schoolcraft County along Forest Highway 13 near the Alger County line (Smith, 1980); another was trapped in October 1980, near Lake Michigamme in Marquette County (Smith, 1981a).

During 1978–1981, the Michigan Marten Reintroduction Program was initiated with the Michigan Department of Natural Resources, the Ontario Ministry of Natural Resources, and the Endangered Species Program of the U.S. Fish and Wildlife Service cooperating (Churchill *et al.*, 1981). Professors Peggy Herman and Norman Sloan of Michigan Technological University supervised the live-trapping, transporting and releasing of the animals from October, 1978, through June, 1980. The Ecological Research Services, Inc. of Iron River under the direction of Dr. James P. Ludwig continued the project from June 28, 1980, through its completion in March, 1981. The 148 marten (77 males and 71 females) were live-trapped in late autumn and winter in Algonquin Provincial Park in Ontario. They were released at the following Upper Peninsula sites: the Huron Mountain Club in northern Marquette County (30 males and 8 females in 1979–1980, 17 males and 23 females in 1980–1981); the McCormick Experimental Forest Tract, a part of the Ottawa National Forest but separated and located in northwestern Marquette and eastern Baraga counties (9 males and 13 females in December, 1980); and in the Iron River District of the Ottawa National Forest in Iron County (21 males and 27 females in 1980–1981). In order to determine movements of some of the animals released at the Huron Mountain Club in October and November, 1980, 15 marten (7 males and 8 females) were fitted with radio transmitter collars. Their movements were monitored until February, 1981. As a result of these releases, plus those made in adjacent parts of Wisconsin (Churchill *et al.*, 1981; Davis, 1978), the prospects in the summer of 1981 looked excellent for the reestablishment of a sustained marten population on the American side of the Upper Great Lakes Region.

HABITAT PREFERENCES. Biologists have noted that marten in the western parts of its North American range appear to thrive in stands of spruce and fir, consisting chiefly of mature trees having a canopy coverage of at least 30% (Koehler and Hornocker, 1977; Marshall, 1951). In contrast, marten in eastern parts of its range not only utilize coniferous forests (including cedar swamps) but also mixed stands of both conifers and hardwoods (Clem, 1975; DeBlaay, 1980; de Vos, 1952; de Vos and Guenther, 1952; Francis and Stephenson, 1972; Mech and Rogers, 1977; Soutiere, 1979). In Michigan, Wood (1914a) noted that beech-maple forests in the vicinity of Whitefish Point in Chippewa County sustained marten populations. Clear-cutting practices and severe fires apparently reduce forest habitat quality drastically for marten (de Vos, 1951). Selective cutting and scattered fires provide some habitat diversity and can increase marten food supplies in the form of small mammals, birds, and fruits (Koehler and Hornocker, 1977; Soutiere, 1979; Steventon and Major, 1982; Yeager, 1950). Ideally, marten are best adapted to mature northern forests, probably mostly coniferous, in which the extensive canopy maintains a shaded and moist interior with a substrate well-strewn with windfalls and other forest debris.

DENSITY AND MOVEMENTS. Studies made in other parts of the marten's range seem to indicate that populations in undisturbed boreal forest in early Michigan days might have averaged, depending on habitat, from 0.5 to 6.2 individuals per sq mi or 1.3 to 1.6 per sq km (Schupbach, 1977). Density studies have been carried out by following marten trails in snow; live-trapping, tagging, releasing, and re-trapping; and by radio-tracking animals collared with small radio transmitters in such areas as Idaho (Marshall, 1942), Montana (Hawley and Newby, 1957; Weckwerth and Hawley, 1962), Ontario (Francis and Stephenson, 1972), and Maine (Soutiere, 1979). Soutiere's investigations in Maine suggest it might be reasonable to assume that in the generally-disturbed situations in which the Ontario animals have recently been released in Delta and Marquette counties, densities of approximately one per sq mi (0.4 per sq km) might be expected.

Home range of the marten, based on studies in other states and provinces, appears varied, perhaps, as Soutiere (1979) points out, because of habitat diversity. Home ranges in poorer (disturbed) forested areas, according to his investigations, would be much larger than those in mature boreal woodlands. The living space for males is also found to be as much as twice that of females. In Minnesota, Mech and Rogers (1977) calculated home ranges from 4.2 to 8.0 sq mi (10.5 to 19.9 sq km) for males and 1.7 sq mi (4.3 sq km) for females; in Maine (Soutiere, 1979; Steventon and Major, 1982), up to 3.8 sq mi (10.0 sq km) for males and 0.9 sq mi (2.5 sq km) for females. In Ontario, Francis and Stephenson (1972) found home ranges smaller for marten living in Algonquin Provincial Park, 1.4 sq mi (3.6 sq km) for males and 0.4 sq mi (1.1 sq km) for females.

Marten forage within their home ranges, moving on a daily basis as much as 3,200 ft (975 m) for males and 1,584 ft (488 km) for females. There is also a tendency for displaced marten to return to their home areas (de Vos and Guenther, 1952). Evidence as to whether marten populations fluctuate multiannually (in cycles), perhaps every 10 years, is not conclusive (Schorger, 1942; Seton, 1953).

BEHAVIOR. Marten are generally solitary animals. Apparently males are intolerant of each other and only associate with females during the brief courtship and mating period. The only individuals that appear to socialize are offspring with their female parent. Adult males maintain territories which overlap little with those of adjacent males. Key places within their territories appear to be "marked" by exudate from their anal glands. The smaller home ranges of the females may fit within those of a single male or perhaps overlap those of more than one male. When the offspring leave the mother's care in late summer or early autumn of their first year, they move distances as far as 25 mi (40 km) to find home places of their own (Hawley and Newby, 1957; Francis and Stephenson, 1972).

The marten is a graceful and skilled climber, as films showing its pursuit of a red squirrel have demonstrated. However, the animal is equally adept at using a bounding gait on the forest floor. The habit of placing their hind feet directly where their forefeet have been leaves distinctive twin tracks in the snow (Seton, 1953). The animals climb and dodge through thick ground cover, often following well-used trails along fallen logs and other debris, especially in summer (Francis and Stephenson, 1972). The marten is active chiefly at night and in the dim light of early morning and dusk. It can also be observed on drab days. There is no recession of activity in winter except in extremely bad weather. The marten can swim (de Vos, 1952; Mech and Rogers, 1977) and, although it is strictly a forest animal, occasionally crosses openings or bare rocky areas. Its attraction for items strange to its environment is well-known to trappers (de Vos *et al.*, 1959; Halvorson, 1961; Seton, 1953) and the animal will even explore camp sites (Hawbecker, 1945).

The marten has the excellent senses of vision, hearing, and smell characteristic of all mustelids. Observers have noted only a few noises uttered by the marten in the wild. Under restraint, the animal makes purr-like grunts or cooing calls to its young. In defense or in anger, martens growl, chatter, snarl, hiss, and even screech (Belan *et al.*, 1978; Jackson, 1961; Seton, 1953).

The young are born and reared in a cavity (lined with vegetation) in a hollow tree, an opening in a fallen log or stump, or in a rock pile. The selection of this site is the responsibility of the pregnant female. Otherwise, martens have no fixed or permanent dens. Temporary resting places, often in exposed locations, are used through the year, mostly as daytime retreats. These sites are located in live tree hollows, or top of branches, in logs, in stumps, in ground burrows, and in rock piles (Francis and Stephenson, 1972; Marshall, 1951; Masters, 1980; Mech and Rogers, 1977; Steventon and Major, 1982). Tree resting sites include hemlock, yellow birch, and balsam fir. The marten often defecates at specific places close to den entrances.

ASSOCIATES. Due to the solitary nature of the marten, it has few associates. Perhaps the closest is the fisher, since both use somewhat similar habitats in northern forests. There is also considerable evidence that the fisher preys on the marten (Seton, 1953; de Vos, 1952). However, in territory where the two occur together in Ontario, Clem (1977) found the marten mostly confined to conifers while the fisher used the edges of these stands as well as forests of mixed conifers and hardwoods. This suggests an apportionment of the environment by these two mustelids. Even so, trappers consider that the two use the same areas, because both are caught on the same trap lines. As de Vos (1951) noted, the fisher may be better adapted to the early successional stages of forest growth than the marten. Nevertheless, there is much about this interesting relationship yet to be determined.

REPRODUCTIVE ACTIVITIES. Courtship and mating activities occur in July and August. By this time the young-of-the-year have become self sufficient, leaving the female parent free for another reproductive cycle. Normally a female accepts only one male, generally the one whose home territory encompasses hers. However, males may mate with more than one female. The mating act, which may involve a period as long as 75 minutes, evokes purring or growling utterances, and sometimes causes damage to the female's neck skin as a result of biting action by the male (Markley and Bassett, 1942). The time between conception and birth may vary from 220 to 275 days—between August and April. This unusually long period results from delayed implantation: after mating and egg fertilization, cell division is arrested at the blastocyst stage for 28 to 32 weeks. Sometime in February or early March, the process of development commences again, although the mechanism which triggers this process remains obscure. Uterine implantation then takes place and normal fetal development is completed in about 28 days.

The young are born in late March or April and maintained by the female parent in a lined nest in a tree cavity, hollow log or stump, or rock pile. Litters may range in size from one to five individuals with two or three average (Brassard and Bernard, 1939). At one day of age, young marten are covered with sparse yellow hair, are physically helpless, have closed eyes and ears, and weigh about 1 oz (28 g) each. At seven days of age, the pelage on the back darkens. At 21 days, a brown fur coat begins to grow and the size difference between the sexes becomes noticeable. At 26 days of age the ears open. At 39 days of age, the eyes open, the brown subadult pelage is well developed, and coordinated movement is good. At about 46 days of age, the young can climb out of the den. Between 42 and 49 days of age, the young are weaned and weigh at least 1 lb (454 g). At about 80 days of age, the young are near adult size and disperse. At about 125 days of age, the growth of the permanent dentition is complete.

Young marten do not breed until their second summer, sometimes waiting even until their third (Strickland, 1975). This slow sexual maturity plus the production of only two or three young per year results in a rather low rate of recruitment. As a consequence, there should necessarily be a rather slow population turnover, seemingly unusual for such a small mammal. Actually, adults can live for at least five or six years in nature; in captivity for as long as 18 (Jones, 1979). This would normally result in successful survival in the north woods were it not for the fur trapper's quest of the marten's fine coat. Its annual low production, the ease by which the marten can be trapped, and habitat destruction by logging helps explain why the animal has been extirpated in so many areas (Quick, 1956). Marten can be aged by the number of cementum layers on the canine teeth and by the development of the suprafabellar tubercle on the femur (Leach *et al.*, 1982).

FOOD HABITS. The marten is a highly efficient predator, catching the red squirrel in a treetop chase or a southern red-backed vole as it scurries across needle-strewn ground litter. Although there are no actual data concerning the food choices of the now-extirpated Michigan marten, very likely the introduced successors are finding the same fare as in Ontario and in other northern habitats. Based on studies from other states and provinces, it seems likely that Michigan marten had a diet consisting mostly of voles, including the meadow vole, the southern red-backed vole and to a lesser extent the bog lemming. Other food selections probably included the masked shrew, pygmy shrew, short-tailed shrew, snowshoe hare, red squirrel, eastern chipmunk, northern flying squirrel, deer mouse, meadow jumping mouse, and woodland jumping mouse. A variety of birds, snakes, frogs, fish, beetles, grasshoppers, wasps, mayflies, snails and assorted fruits in season (blueberry, mountain ash, red raspberry, serviceberry, etc.) round out known food selections of martens studied in Maine, Ontario, and western states (Francis and Stephenson, 1972; Koehler and Hornocker, 1977; Marshall, 1946; Murie, 1961; Soutiere, 1979; Steventon and Major, 1952). In captivity, marten are raised successfully on a variety of table scraps and prepared mink and dog foods.

ENEMIES. As mentioned under Reproductive Activities, apparently the marten's rather low annual reproductive performance is still sufficient to compensate for natural predation and other environmental calamities. However, fur trapping, if not regulated, seems to tip the scales drastically, causing marked population decline. Hence, the human, both as a trapper and as a logger, is certainly the marten's worst enemy (de Vos, 1951; Schupbach, 1977). Major natural enemies of the marten include the bobcat, lynx, fisher, coyote, gray wolf, and the great horned owl (Banfield, 1974; Jackson, 1961; Seton, 1953).

PARASITES AND DISEASES. External parasites on the marten include nine kinds of fleas, one louse, and at least two ticks. These infestations often are heavier in autumn than in spring (J. P. Ludwig, pers. comm.). Internally, four kinds of nematodes (roundworms) are found; probably tapeworms and flukes also live in this mammalian host (de Vos, 1952, 1957; Erickson, 1946; Jackson, 1961).

MOLT AND COLOR ABERRATIONS. The paler winter pelage of the marten is fully grown by mid-October and undergoes a slow replacement by darker hair begun in late April and completed by mid-June (Soutiere and Steventon, 1981). The winter guard hairs are shed initially, giving the summer fur a shorter, patchy appearance. The underfur is shed in summer. A series of pelts taken by trappers will show much dorsal color difference—from almost black to almost pale buff (see photo for Ontario animals in de Vos, 1952). Some individuals are noticeably reddish; almost white individuals are also described (Jones, 1923).

ECONOMIC IMPORTANCE. In the early exploration of the Upper Great Lakes Region, the pelts of fur-bearing animals were almost immediately recognized as major resources for trade with the homelands in Europe. The hides of marten were third behind those of beaver and raccoon for export at Michilmackinac in 1767, when 9,556 were shipped (Petersen, 1979). In 1822, John Jacob Astor's fur trading post on Mackinac Island paid 60¢ for each marten pelt taken in proper season (Anon., 1962). The marten catch continued high from 1835 to 1839, when the American Fur Company's Northern Department (including the Upper Peninsula) and the Detroit Department (including the Lower Peninsula) received catches totalling, respectively, 13,433 and 9,491 pelts (Johnson, 1969). After 1850 until the early years of the twentieth century, marten pelts became a slowly diminishing money crop for Michigan's trappers. Whether recent efforts to reestablish this delightful animal will provide a harvestable surplus or even a token population in Michigan's north woods remains to be seen.

COUNTY RECORDS FROM MICHIGAN. Records of specimens in collections of museums, from the literature, and reported by personnel of the Michigan Department of Natural Resources show that the marten formerly occurred in all counties of the Upper Peninsula and probably in most of the Lower Peninsula as well. However, in the latter area, actual reports are from only 19 counties (see Map 52). The marten formerly lived on Isle Royale, but presumably was trapped out shortly after 1905 (Mech, 1966).

LITERATURE CITED

Anon.
1962 Fur and hide values, 1822. Michigan Conser., 31(6): 36–37.

Baker, G. A.
1899 The St. Joseph-Kankakee portage. Northern Indiana Hist. Soc., South Bend, Publ. No. 1, 48 pp.

Banfield, A. W. F.
1974 The mammals of Canada. Univ. Toronto Press, Toronto. xxiv+438 pp.

Belan, I., P. N. Lehner, and T. Clark
1978 Vocalizations of the American pine marten, *Martes americana*. Jour. Mammalogy, 59(4):871–874.

Brassard, J. A., and R. Bernard
1939 Observations on breeding and development of marten, *Martes a. americana* (Kerr). Canadian Field-Nat., 53(2):15–21.

Burt, W. H.
1946 The mammals of Michigan. Univ. Michigan Press, Ann Arbor. xv+288 pp.

Churchill, S. J., L. A. Herman, M. F. Herman, and J. P. Ludwig
1981 Final report on the completion of the Michigan Marten Reintroduction Program. Ecol. Res. Serv., Inc., Iron River, Michigan. viii+118 pp. (mimeo).

Cleland, C. E.
1966 The prehistoric animal ecology and ethnozoology of the Upper Great Lakes Region. Univ. Michigan, Mus. Anthro., Anthro. Papers No. 29, x+294 pp.

Clem, M. K.
1977 Interspecific relationship of fishers and martens in Ontario during winter. Pp. 165–182 *in* R. L. Phillips, and C. Jonkel, eds. Proc. 1975 Predator Symp. Univ. Montana, Montana Forest and Conserv. Exp. Sta., ix+268 pp.

Davis, M. H.
1978 Reintroduction of the pine marten into the Nicolet National Forest, Forest County, Wisconsin. Univ. Wisconsin, Stevens Point, unpubl. M.S. thesis, 64 pp.

DeBlaay, T. J.
1980 A survey of marten *(Martes americana)* habitat and prey availability in Michigan's Upper Peninsula. Michigan Technological Univ., unpubl. M.S. thesis, 101 pp.

de Vos, A.
1951 Recent findings in fisher and marten ecology and management. Trans. 16th North Amer. Wildlife Conf., pp. 498–507.
1952 The ecology and management of fisher and marten in Ontario. Ontario Dept. Lands and Forests, 90 pp.
1957 Pregnancy and parasites of marten. Jour. Mammalogy, 38(3):412.
1964 Range changes of mammals in the Great Lakes region. American Midl. Nat., 71(1):210–231.

de Vos, A., A. T. Cringan, J. K. Reynolds, and H. G. Lumsden
1959 Biological investigations of traplines in northern Ontario, 1951–56. Ontario Dept. Lands and Forests, Wildlife Ser. No. 8, 62 pp.

de Vos, A., and S. E. Guenther
1952 Preliminary live-trapping studies of marten. Jour. Wildlife Mgt., 16(2):207–214.

Dice, L. R.
1925 A survey of the mammals of Charlevoix County, Michigan, and vicinity. Univ. Michigan, Mus. Zool., Occas. Papers No. 159, 33 pp.

Erickson, A. B.
1946 Incidence of worm parasites in Minnesota Mustelidae and host lists and keys to North American species. American Midl. Nat., 36(2):494–509.

Francis, G. R., and A. B. Stephenson
1972 Marten ranges and food habits in Algonquin Provincial Park, Ontario. Ontario Ministry Nat. Res., Res. Rept. (Wildlife) No. 91, 53 pp.

Gunderson, H. L.
1965 Marten records for Minnesota, Jour. Mammalogy, 46(4):688.

Halvorson, C. H.
1961 Curiosity of a marten. Jour. Mammalogy, 42(1):111–112.

Hamilton, W. J., Jr.
1958 Past and present distribution of marten in New York. Jour. Mammalogy, 39(4):589–591.

Harger, E. M., and D. F. Switzenberg
1958 Returning the pine marten in Michigan. Michigan Dept. Conserv., Game Div. Rept. No. 2199 (mimeo), 7 pp.

Hawbecker, A. C.
1945 Activity of the Sierra pine marten. Jour. Mammalogy, 26(4):435.

Hawley, V. D., and F. E. Newby
1957 Marten home range and population fluctuations. Jour. Mammalogy, 38(2):174–184.

Jackson, H. H. T.
1961 Mammals of Wisconsin. Univ. Wisconsin Press, Madison. xii+504 pp.

Johnson, D. R.
1969 Returns of the American Fur Company, 1835–1839. Jour. Mammalogy, 50(4):836–839.

Jones, M. L.
1979 Longevity of mammals in captivity. Internat. Zoo News, 26(3):16–26.

Jones, S. V. H.
1923 Color variations in wild animals. Jour. Mammalogy, 4(3):172–177.

aging 7¾ in (200 mm) for the larger male and 7 ⅛ in (180 mm) for the smaller female; tail about 35% as long as this mustelid's head and body, well-furred and slender; length of tail vertebrae averaging 3 in (75 mm) in males and 2¼ in (55 mm) in females; head small, narrow, rather blunt-nosed, elongate, no larger than the slender, snake-like neck; ears short, oval, thinly haired; eyes small, intense, beady; vibrissae long, prominent; body cylindrical, slender; legs short with front ones set well back; each foot with five claws; mammary glands consist of five pairs, anal scent glands prominent, a strong musky odor being discharged when the animal becomes disturbed or frightened; males about ⅓ larger in external dimensions and as much as 50% heavier than females. External features presented in greater detail by Hall (1951).

The fur of the ermine is short, moderately fine, not luxuriously dense. In summer, the upperparts and the outer surfaces of the legs are uniformly dark brown. The brown tail is characteristically black-tipped. The underparts are white usually tinged with pale yellowish color, with this color extending on to the insides of the legs, feet, toes (including outsides of the latter) and also anteriorly on to the throat, chin, and upper lip. In preparation for the Michigan winter, the brown coat is molted and replaced by a white one except for retention of the black tail-tip. (See Molt and Color Aberrations.)

In external features and dimensions, the ermine differs from all other Michigan mammals except for the larger long-tailed and smaller least weasels. The ermine is most readily distinguished by tail length and amount of black at the terminal end of the tail. The ermine's tail is not more than ⅓ as long as the length of the head and body. The long-tailed weasel's tail is at least ½ as long as the length of the head and body. In the ermine the black tip amounts to almost ½ the entire length of the tail vertebrae; in the long-tailed weasel the length of the black on the tip of the tail is not more than 40% of the entire length of the tail vertebrae. In summer coat, the ermine generally has white fur on the inner sides of the hind feet, while the long-tailed weasel has brownish fur there. The least weasel is not only distinguished from the ermine by its conspicuously smaller size but also by its shorter tail (no more than ¼ the length of the head and body) and the lack of any black (except for two or three hairs) on the tip of the tail.

MEASUREMENTS AND WEIGHTS. Adult specimens of the larger male and the smaller female ermine measure, respectively, as follows: length from tip of nose to end of tail vertebrae, 10¾ to 13 in (275 to 330 mm) and 9 to 10¼ in (230 to 260 mm); length of tail vertebrae, 2¾ to 3¾ in (70 to 95 mm) and 1½ to 2½ in (40 to 65 mm); length of hind foot, 1⅜ to 1⅝ in (35 to 40 mm) and 1 to 1¼ in (28 to 33 mm); height of ear from notch (almost the same in both sexes), ⅝ to ⅞ in (15 to 22 mm); weight, 3⅜ to 5 oz (90 to 140 g) and 1½ to 2⅝ oz (45 to 75 g).

DISTINCTIVE CRANIAL AND DENTAL CHARACTERISTICS. The skull of the ermine is elongate, narrow, flattened, and smoothly rounded dorsally (see Fig. 82). The rostrum is notice-

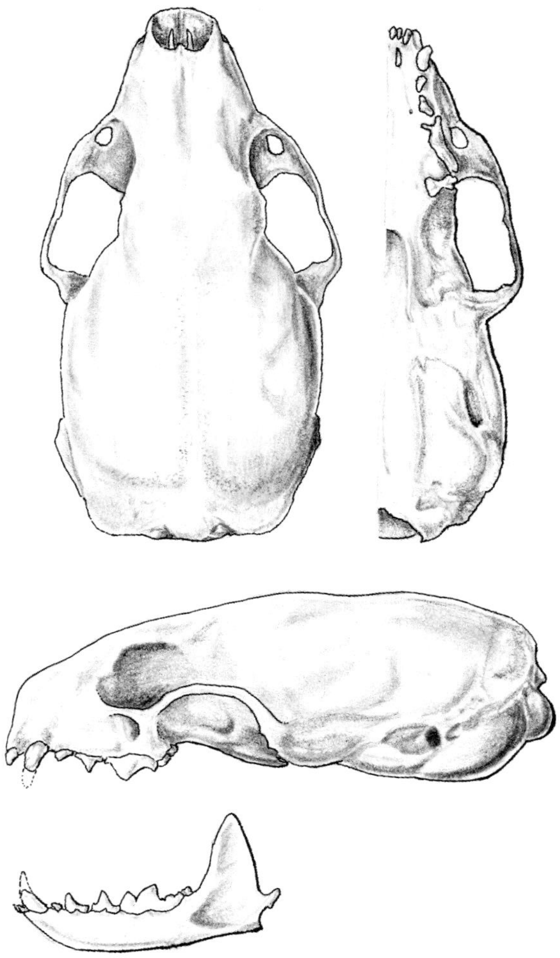

Figure 82. Cranial and dental characteristics of the ermine (*Mustela erminea*).

ably shortened in comparison with the cranial length, which makes the mouth opening appear small; the frontal region flattened and not inflated; the sagittal crest weakly developed at most; the postorbital processes blunt; the zygomatic arches slender, weakly constructed, and somewhat arched; and the auditory bullae inflated and noticeably elongate, with inner sides nearly paralleling each other. The teeth are moderate in size and especially sharp. The premolars are reduced to three in each tooth-row instead of four as found in the genus *Martes* (marten and fisher). The most useful means of distinguishing weasel skulls, if the sex is known, is by size; the large skull is that of the long-tailed weasel; the medium-sized skull, the ermine; and the diminitive skull, the least weasel. If sex is not known, it may be difficult to be certain of the species if the skulls being compared are those of a female long-tailed weasel and a male ermine, because both are so nearly alike in size and configuration. One help in differentiating is the blunt postorbital processes in the ermine and sharply-pointed ones in the long-tailed weasel. Other cranial details are presented in Hall (1951). Holes and other signs of deterioration occasionally observed in the ermine's forehead area are the result of a nematode sinus parasite, *Skrajabingylus nasicola*.

Skulls of adult male and female ermine, respectively, measure 1½ to 1⅞ in (38 to 47 mm) and 1⅜ to 1½ in (34 to 38 mm) in greatest length and ¾ to 15/16 in (19 to 24 mm) and ⅝ to ¾ in (16 to 19 mm) in greatest width across the zygomatic arches. The dental formula for the ermine is:

$$\text{I (incisors)}\frac{3\text{-}3}{3\text{-}3}, \text{C (canines)}\frac{1\text{-}1}{1\text{-}1}, \text{P (premolars)}\frac{3\text{-}3}{3\text{-}3}, \text{M (molars)}\frac{1\text{-}1}{2\text{-}2} = 34$$

DISTRIBUTION. The ermine has circumpolar distribution, being found in suitable habitats in the north temperate and frigid sectors of Eurasia and North America. In the New World, it ranges from east to west in a broad belt from the Arctic Ocean and adjacent islands of the Canadian Archipelago southward into northern United States, avoiding only the Great Plains (see Map 54).

In Michigan, the ermine may be found throughout the Upper Peninsula and on Isle Royale, Sugar, and Marion islands. However, in Michigan's Lower Peninsula this species thrives chiefly in the northern counties, appearing only sporadically in the southern ones, with southernmost specimens obtained in Eaton, Ingham, Kalamazoo, and Washtenaw (see Map 54). Its rarity in southern Michigan is also emphasized by the lack of records of ermine in states immediately to the south and southwest: Illinois, Indiana and Ohio (Hall, 1981; Long, 1974; Mumford, 1969). A possible major contributing factor to the scarcity of ermine in southern Michigan is the lack of continuous winter snow cover. Against a dark substrate, the ermine's white winter coat provides no camouflage protection from predators. In addition, a possible interference interaction between the ermine and the more southern acclimated long-tailed weasel may limit the former's southern distribution (Simms, 1979a).

HABITAT PREFERENCES. The ermine is at home in a variety of arctic and boreal North American environments. Although the animal may occupy tundra (especially rocky areas) in the more southern parts of its range, open country adjacent to forest or shrub borders seems preferred. In

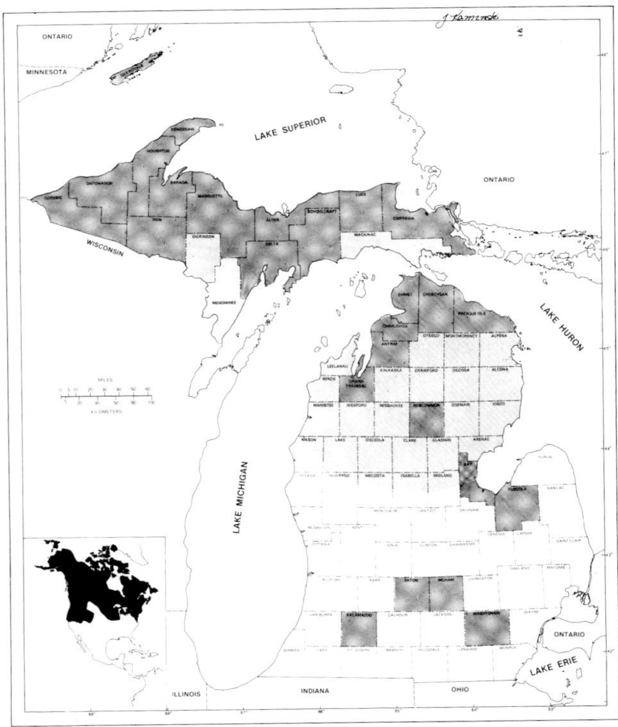

Map 54. Geographic distribution of the ermine (*Mustela erminea*). Counties from which there are actual records are marked with dark shading.

Michigan, the ermine occurs rarely in mature forest areas and more commonly in early successional forest communities. The animal can also be found in woody cover along streams, marshes, and stone walls, in brushy fence rows, and on cleared land reverting to forest. In Gogebic and Ontonagon counties, Dice and Sherman (1922) caught individuals in black spruce and tamarack bog and in dry hardwood forest. Manville (1948) found the ermine to be the most common weasel in the Huron Mountains of Marquette County. In Delta county, the author found the species in woody cover adjacent to the shore of Lake Michigan. The ermine's habitat preference at the southern edge of its range in the Lower Penninsula is not definitely known. However, the gradual southward diminution of the northern hardwoods and boreal conifers, as the temperate hardwoods become dominant, and the lack of winter-long snow cover probably reduce the attractiveness of this environment for the ermine.

DENSITY AND MOVEMENTS. Based on trappers' reports, snow-track counts, and capture-mark-release-recapture studies, the density of this small carnivore has been estimated by various workers including King and Edgar (1977). There have been densities reported as high as one individual per 32 acres (13 ha) by Jackson (1961) and one per 64 acres (26 ha) by Soper (1919). Studies by Simms (1979b) in Ontario habitat comparable to some of that in northern Michigan showed that the ermine had an average monthly density of one animal per 278 acres (111 ha). However, when Simms excluded the "non habitat" in his study area and considered only the early successional areas favorable to ermines, his figure was higher, about one per 160 acres (64 ha). Such population estimates naturally vary seasonally, with highest populations in summer when the young-of-the-year begin to join the adult animals.

The spatial arrangement of ermines in nature is much like that of other related mustelids. Adult males and adult females maintain individual home ranges and discourage juveniles and other adults from encroaching on their living spaces, especially, according to Simms (1979b), when the local prey populations (voles and other rodents) are high. Simms determined that territories occupied by males are often twice the size of female home ranges and usually encompass or overlap those of several females. As a result, pair relationships are non- existent, with males breeding with more than one female. In typical mustelid fashion, female ermine do not associate with males except during the brief courtship period. Based on Simms' (1979b) calculations for Ontario ermine, home ranges for males may cover an area of 63.2 acres (21.3 ha) and for females, 29.2 acres (11.7 ha). He also found that ermine move about actively in their territories, with males more travel-prone than females. Based on live-trapping records, one Ontario ermine was found to move more than 1,800 ft (600 m) in a 24-hour period.

Early literature reports drastic fluctuations in local populations of ermine, especially in accordance with ups and downs of rodent prey species. However, current findings based on this weasel's general intolerance of conspecifics and on its now-recognized spatial arrangements indicate that populations are probably considerably more stable than previously thought. Nevertheless, predation on voles may be an important factor influencing the density of these microtines (Fitzgerald, 1977).

BEHAVIOR. Ermine are generally solitary creatures. Adult males, as mentioned above, maintain bachelor quarters and tend to exclude other males, including juveniles, from their home and foraging territories (Powell, 1979). They tolerate but seem to avoid associating with the females whose home areas may be encompassed or overlapped by theirs, except during summer courtship and mating activity. Likewise, the females remain apart from conspecifics except for offspring which are diligently attended. Apparently, home ranges of adults are so well protected that juveniles leaving the females' care can have difficulty in becoming established and consequently are undoubtedly more susceptible to predation than established individuals. Simms (1979b) noted that these younger ermine are discouraged either by active hostility or by avoidance behavior. Scent marking by resident animals may also discourage encroachment.

In typical weasel fashion, the ermine is a quick, agile, alert, creature with a lithe, snake-like, muscular body. The average person may get only short glimpses of the highly inquisitive ermine as it darts in and out of foliage, other ground cover, snowdrifts, or hollow logs. In a straightaway, the ermine moves in a undulating fashion, arching its body by drawing its forelimbs and hindlimbs somewhat together as it moves in a "humping" or "measuring worm" fashion. Tracks show that its hind feet are placed on the ground very close behind the front feet (King and Edgar, 1977). The ermine climbs efficiently and has been observed pursuing squirrels and chipmunks in trees. Likewise, the ermine swims well, crossing small streams and other waterways without difficulty.

The ermine is classed as a night creature but can be found active anytime during daylight hours. As a result, observers have had the opportunity to watch this remarkably efficient predator (Powell, 1978). The ermine hunts in a solitary fashion employing its senses of hearing, sight, and smell (Seton, 1953). It follows a zig-zag course, changing direction frequently, moving in and out of small rodents' burrows either in the ground or under surface litter. The smaller female naturally has an easier time fitting into rodent burrows than the larger male (Simms, 1979a). The direction-changing hunting pattern allows the ermine not only wide prey selection but maximum predator avoidance. Once the prey is detected, the ermine apparently approaches as close as possible to the victim. Movement by the prey (perhaps from a "frozen" posture) motivates the ermine to make its rush. Once the prey has been caught, the ermine moves with incredible speed, grasping the back of the head and neck with its sharp teeth while literally wrapping its muscular body and feet around the victim (Allen, 1938; Powell, 1978). Uneaten material is cached for a subsequent meal.

The ermine, like other weasels, has a shrill bark and hissing call. There are also low chattering sounds and growls uttered when the animals are disturbed. Purring calls are used around young.

The ermine may use either underground or aboveground dens. Burrows of such prey species as voles, ground squirrels, chipmunks and other rodents are probably confiscated by the ermine. The basic shelter is a lined nest (often containing fur and feathers from prey species) with side cavities for food caches and as latrines. Other den sites include hollow logs, woodpiles, stone walls, and even buildings. Fitzgerald (1977) noted a preference for surface nests of voles as winter dens. Sometimes den entries can be distinguished by food scraps and fecal material deposited nearby.

ASSOCIATES. The small, mostly solitary ermine normally does not associate with conspecifics or other carnivores. This weasel probably does not follow larger carnivores to obtain a few scraps from their kills, although it probably would take carrion if available. Because the ermine is such an efficient prey-catching carnivore, it may need no real assistance in food-getting, either in summer or winter.

In parts of central Michigan, all three weasels as well as the larger, related mink exist in the same general habitat. It might initially seem that competition for similar prey-species would complicate life for these four creatures, similar in mode of life although ranging in size (for males) from 1.5 oz (42 g) for the least weasel, 4.5 oz (130 g) for the ermine, 8.2 oz (230 g) for the long-tailed weasel, and up to 27 oz (756 g) for the mink. Unquestionably, the smaller species, at least, use some of the same foods. The larger species might also eat the smaller ones on occasion. Nevertheless, these predators seem to have achieved coexistence, in part, by specializing in different prey species (Rosenweig, 1966). Studies on the apportionment of prey species between the least weasel and the ermine (stoat) in Great Britain (King and Moors, 1979) and between the ermine and the long-tailed weasel in Ontario (Simms, 1979a) show that each species (as well as the larger males and smaller females) may have different foraging strategies. The smaller species and/or sexes rely more on voles and other small rodents while the larger species and/or sexes may specialize on lagomorphs (*e.g.*, eastern cottontail) or other prey types. The interrelationships between closely related carnivores and their predator/prey activities need additional study by ecologists.

REPRODUCTIVE ACTIVITIES. Courtship and breeding are carried on in early summer. Not only are the parent females who are attending their offspring receptive to courting males at this time, but their female young, perhaps no more than 60 or 70 days old, are sufficiently sexually mature to mate, thus maximizing female reproductive potential (Simms, 1979a). In typical mustelid fashion, females have an extended gestation period of about 300 days because of delayed implantation. The blastocyst remains quiescent in the uterine cavity until the following March, when endocrine action, not fully understood, triggers implantation and development (Wright, 1942).

Between mid-April and early May, a rather large litter averaging six (as many as nine are recorded) is born in a lined nest either underground or in a protected surface refuge. At birth, the young have fine white hair, closed eyes and ears, vibrissae, and the small head, long neck, and fusiform body typical of weasels; are capable of uncoordinated movements and weigh much less than ⅛ oz (3.5 g). At seven days of age, according to Hamilton (1933), they have long hair, darkening on the back, and dark claws on the forefeet; make bird-like squeaks; and weigh as much as ⅛ oz (3.5 g). At 14 days of age, sex of the young can be determined by size; they possess a rather conspicuous brown mane; have obvious claws on all four feet; a darkening nose patch; and weigh ⅓ oz (10 g). At 21 days of age, the young have a well-furred dorsum and fine white belly fur; show erupted deciduous canines and pre-molars; and weigh ⅝ oz (16.5 g). At 35 days of age, the young open their eyes and ears; possess erupted incisors; display uniformly brown coloring above and white below; eat meat from kills made by the female parent; and weigh about 1¼ oz (32.5 g), with males noticeably larger. At 45 days of age, the young are active and follow their female parent on forays; are playful much in the manner of young kittens; and shortly will be able to hunt for food on their own.

As pointed out previously, females mature rapidly sexually and are receptive to mating at two to three months old. On the other hand, their male litter mates do not mature sexually until the following spring. Females in nature may survive for at least two breeding seasons; Simms (1979b) found females in residence on his Ontario study area for as long as 22 months; one male was live-trapped over a period of 13 months.

FOOD HABITS. The diet of the larger male and the smaller female ermine differ very little, with the selection of food items highly dependent on prey availability (Simms, 1979a). When vole numbers are high, the diet of this mustelid reflects this abundance (Fitzgerald, 1977). Adequate daily meals are essential to meet the exorbitant energy and heat production demands of this small carnivore. Food caching is one way in which the ermine seems to have helped solve this problem of sustaining itself (Simms, 1979a).

Studies by Aldous and Manweiler (1942), Fitzgerald (1977), Hall (1951), Hamilton (1933), Osgood (1936), Simms (1978) and others show that one-half of the ermine's diet consists of mice and voles, followed by shrews (22%), hares and rabbits (9%), and lesser amounts of squirrels, birds (including poultry), fish, reptiles, amphibians, earthworms, and carrion (*e.g.,* porcupine). Although no thorough examinations of the food habits of Michigan ermine have been reported, studies in other states show that this weasel is partial to such Michigan species as shrews of the genus *Sorex,* short-tailed shrew, snowshoe hare, eastern cottontail, red squirrel, eastern chipmunk, least chipmunk, mice of the genus *Peromyscus,* southern red-backed voles, meadow voles, and perhaps an occasional house mouse and Norway rat.

ENEMIES. Despite its slight build, agility, quickness, and ferocious—seemingly fearless—nature, the ermine is fair game for several large meat-eaters. Hamilton (1933) listed the blacksnake (*Elaphe obsoleta*), rough-legged hawk, goshawk, barred owl, great horned owl, and house cat as ermine predators. Other enemies include the snowy owl, bald eagle, coyote, red fox, gray fox, marten, fisher, badger, and wolverine (Banfield, 1974; Hall, 1951; Jackson, 1961). Allen (1978) described a red fox's successful catch of an ermine on Isle Royale. Latham (1952) concluded that red and gray foxes in Pennsylvania may be sufficiently important ermine predators as to influence their overall population numbers.

PARASITES AND DISEASES. Internal parasites identified as living in the ermine include flukes, *Alaria mustelae, Alaria taxideae,* and *Euparyphium melis;* tapeworm, *Taenia taeniaformis;* roundworms, *Dracunculus* sp., *Molineus patens, Physaloptera maxillaris,* and *Skrjabingylus nasicola;* and thorny-headed worm, *Moniliformis* sp. (Erickson, 1946; Jackson, 1961). The parasitic sinus nematode, *Skrjabingylus nasicola,* does major damage to the forehead region of the ermine's skull. According to Timm (1975), 11 of 12 skulls examined in Cook county, Minnesota, showed cranial deterioration from this worm. External parasites found on ermine include fleas, *Megabothris asio, Megabothris quirini, Ceratophyllus vison, Nearctopsylla genalis genalis,* and *Nearctopsylla brooksi;* mites and ticks, *Dermancentor variabilis, Laelaps alaskensis,* and *Laelaps kochi;* and the louse,

Trichodectes sp., (Jackson, 1961; Lawrence *et al.*, 1965; Scharf and Stewart, 1980). On ermine examined on Isle Royale, Wilson and Johnson (1971) found the fleas, *Epitedia wenmanni wenmanni*, *Rhadinopsylla difficilis*, and *Orchopeas caedens caedens*.

MOLT AND COLOR ABERRATIONS. Unlike the fisher and the marten, the ermine has two molts annually, one in early spring (late March and April) and one in mid-autumn (late October and November). During the spring pelage change, the old, worn hairs fall out and new growth appears first on the mid-back and then down each side, with the belly fur changing last. In contrast, the autumn molt begins on the belly, reversing the spring shedding and re-growth pattern.

Michigan ermine change coat color to brown dorsal in summer and white in winter. Only the black tip of the tail remains the same. Experiments have shown that the timing of the molt in both spring and autumn is controlled primarily by the duration of daylight per 24 hours. In spring, the increasing amount of daylight initiates the beginning of the pelage change, with temperature of only minor importance (Rust, 1962). In autumn, decreasing amount of daylight per 24 hours apparently governs the change from brown to white. In brief, the ermine's eyes monitor differences in the ratio of night to day; perception is transmitted to the pituitary gland which alters production of gonadotrophic hormones, ultimately causing the pelage change.

Although difficult to measure, a white coat against a snow background in winter and a brown one against a ground cover background in summer must enhance the ermine's chances for survival as well as its ability to approach likely prey species. The field observer, of course, wonders why only the ermine and other Michigan weasels have acquired, through natural selective processes, white winter coats. There appears to be no good explanation for this. According to Simms (1979a), all ermine in eastern North America have white winter coats, which may be a factor limiting southward distribution of the species into largely snow-free areas. The species, for example, has not been reported south of Michigan in Indiana (Mumford, 1969). On the other hand, in western North American areas where winter snow cover is sparse, ermines may remain in their brown coats throughout the winter. Albino animals have been reported occasionally.

ECONOMIC IMPORTANCE. In the Old World, ermine white furs with their contrasting black-tipped tails have for centuries been fashioned into capes and ceremonial costumes for royalty. In the New World fur trade, however, no distinction was made between the white pelts of the ermine and those of the long-tailed weasel. The same confusion has prevailed to this day, with Michigan furs of these two species lumped together merely as weasel. For example, in 1968 and 1969, Michigan trappers took 1,630 and 2,230 weasel pelts, respectively, each having an average value of 75¢ on the fur market.

Although ermine fur, as a marketable product for the trapper, may not be as important today as in the past, the animal in the wild is valued by many agriculturists. Aside from an occasional raid on poultry (chicken and geese, Schofield, 1957), the ermine and other weasels are tolerated because they are excellent mousers. Hamilton (1933) once calculated that a single weasel killed about 400 mice per year and ate 300 of them. Seton (1953) wrote of the admirable mouse control exerted by the ermine when making its home in field-shocked crops, haystacks, or grain bins.

COUNTY RECORDS FROM MICHIGAN. Records of specimens of ermine in collections of museums, from the literature, and reported by personnel of the Michigan Department of Natural Resources show that this species occurs throughout the Upper Peninsula and the northern half of the Lower Peninsula but only in scattered localities in the southern half (see Map 54). The ermine also occurs on Isle Royale (Burt, 1946), Marion Island in Grand Traverse Bay, and Sugar Island in waters separating the mainland of Ontario from the Upper Peninsula.

LITERATURE CITED

Aldous, S. E., and J. Manweiler
1942 The winter food habits of the short-tailed weasel in northern Minnesota. Jour. Mammalogy, 23(3):250–255.

Allen, D. L.
1938 Notes on the killing technique of the New York Weasel. Jour. Mammalogy, 19(2):225–229.

1978 Lives and loves of some sly foxes on Isle Royale. Smithsonian Mag., 10(2):91–100.

Banfield, A. W. F.

1974 The mammals of Canada. Univ. Toronto Press, Toronto. xxiv+438 pp.

Burt, W. H.

1946 The mammals of Michigan. Univ. Michigan Press, Ann Arbor. xv+288 pp.

Dice, L. R., and H. B. Sherman

1922 Notes on the mammals of Gogebic and Ontonagon counties, Michigan. Univ. Michigan, Mus. Zool., Occas. Papers No. 109, 46 pp.

Erickson, A. B.

1946 Incidence of worm parasites in Minnesota Mustelidae and host lists and keys to North American species. American Midl. Nat., 36(2):494–509.

Fitzgerald, B. M.

1977 Weasel predation on a cyclic population of the montane vole *(Microtus montanus)* in California. Jour. Animal Ecol., 46:367–397.

Hall, E. R.

1945 A revised classification of the American ermines with description of a new subspecies from the western Great Lakes Region. Jour. Mammalogy, 26(2):175–182.

1951 American weasels. Univ. Kansas Publ., Mus. Nat. Hist., 4:1–466.

1981 The mammals of North America. Second Edition. John Wiley and Sons, New York. Vol. 2, vi+601-1181+90 pp.

Hamilton, W. J. Jr.

1933 The weasels of New York. Their natural history and economic status. American Midl. Nat., 14(4):289–344.

Jackson, H. H. T.

1961 Mammals of Wisconsin. Univ. Wisconsin Press, Madison. xii+504 pp.

King, C. M., and R. L. Edgar

1977 Techniques for trapping and tracking stoats *(Mustela erminea)*; review, and a new system. New Zealand Jour. Zool., 4(2):193–212.

King, C. M., and P. J. Moors

1979 On co-existence, foraging strategy and biogeography of weasels and stoats (*Mustela nivalis* and *M. erminea*) in Britain Oecologia, 39(2):129–150.

Latham, R. M.

1952 The fox as a factor in the control of weasel populations. Jour. Wildlife Mgt., 16(4):516–517.

Lawrence, W. H., K. L. Hays, and S. A. Graham

1965 Anthropodous ectoparasites from some northern Michigan mammals. Univ. Michigan, Mus. Zool., Occas. Papers No. 639, 7 pp.

Long, C. A.

1974 Environmental status of the Lake Michigan region. Vol. 15. Mammals of the Lake Michigan drainage basin. Argonne Natl. Lab., ANL/ES-40, 108 pp.

Manville, R. H.

1948 The vertebrate fauna of the Huron Mountains, Michigan. American Midl. Nat., 39(3):615–640.

Mumford, R. E.

1969 Distribution of the mammals of Indiana. Indiana Acad. Sci., Monog. No. 1, vii+114 pp.

Osgood, F. L.

1936 Earthworms as a supplementary food for weasels. Jour. Mammalogy, 17(1):64.

Powell, R. A.

1978 A comparison of fisher and weasel hunting behavior. Carnivore, 1(1):28–34.

1979 Mustelid spacing patterns: variations on a theme by *Mustela*. Zeitschrift für Tierpsychologie, 50(2):153–165.

Rosenzweig, M. L.

1966 Community structure in sympatric Carnivora. Jour. Mammalogy, 47(4):602–612.

Rust, C. C.

1962 Temperature as a modifying factor in the spring pelage change of short-tailed weasels. Jour. Mammalogy, 53(3):323–328.

Scharf, W. C., and K. R. Stewart

1980 New records of Siphonaptera from northern Michigan. Great Lakes Ento., 13(3):165–167.

Schofield, R. D.

1957 Livestock and poultry losses caused by wild animals in Michigan. Michigan Dept. Conserv., Game Div., Rept. No. 2118 (mimeo.), 6 pp.

Seton, E. T.

1953 Lives of game animals. Vol. II, Pt. II. Bears, coons, badgers, skunks and weasels. Charles T. Branford Co., Boston. Pp. 369–746.

Simms, D. A.

1978 Spring and summer food habits of an ermine (*Mustela erminea*) in the central Arctic. Canadian Field-Nat., 92(2):192–193.

1979a North American weasels: resource utilization and distribution. Canadian Jour. Zool., 57(3):504–520.

1979b Studies of an ermine population in southern Ontario. Canadian Jour. Zool., 57(4):824–832.

Soper, J. D.

1919 Notes on Canadian weasels. Canadian Field-Nat., 33(1):43–47.

Timm, R. M.

1975 Distribution, natural history, and parasites of mammals of Cook County, Minnesota. Univ. Minnesota, Bell Mus. Nat. Hist., Occas. Papers No. 14, 56 pp.

Wilson, N., and W. J. Johnson

1971 Ectoparasites of Isle Royale, Michigan. Michigan Ento., 4(4):109–115.

Wright, P. L.

1942 Delayed implantation in the long-tailed weasel (*Mustela frenata*), the short-tailed weasel (*Mustela cicognani*) and the marten (*Martes americana*). Anat. Rec., 83(3): 341–353.

LONG-TAILED WEASEL
Mustela frenata Lichtenstein

NAMES. The generic name *Mustela,* proposed by Linnaeus in 1758, is from the Latin and means weasel. The specific name *frenata,* proposed by Lichtenstein in 1831, is from the Latin word *frenum* meaning bridle. This descriptive term refers to the bridle-like black mask characteristic of weasels of this species from more southern parts of its range. In fact, the long-tailed weasel living in southcentral Mexico was the basis for Lichtenstein's original description. The common name weasel is derived from the Anglo-Saxon word *wesle.* In much of the early literature, the Michigan long-tailed weasel was referred to as the species *Mustela noveboracensis* and was commonly called the New York weasel (Dice, 1927). However, Hall (1936) concluded from his systematic studies that this weasel was in reality conspecific with a widespread species to which the older name, *M. frenata,* rightfully applied. In Michigan, this species is usually called weasel, ermine (when in winter white), or sometimes long-tailed weasel. The subspecies *Mustela frenata noveboracensis* (Emmons) occurs in Michigan.

RECOGNITION. Body size resembles that of an elongate, slender rat; length of head and body averaging 10¼ in (260 mm) for the larger males and 8¼ in (210 mm) for the smaller females; tail long, about 50% the length of this weasel's head and body, slender; length of tail vertebrae averaging 5¼ in (134 mm) in males and 4 in (103 mm) in females; head small, narrow, rather blunt-nosed, elongate, no larger in diameter than the slender, snake-like, muscular neck; ears short, oval, covered with short hairs; eyes small, intense, beady; vibrissae (whiskers) prominent; body long, slender, sinuous; tail well-furred, slightly bushy; legs short with front legs set well back; each foot with five clawed toes; mammary glands consist of four pairs of abdominal teats; anal scent glands developed, discharging a strong, irritating, musky odor when the animal is alarmed; males are about 25% larger in external dimensions and as much as 50% heavier than females (Hall, 1951).

The pelage of the long-tailed weasel consists of short, soft underfur and long, glistening guard hairs. In summer, the upper parts, basal part of the tail, legs, and face are cinnamon brown, much like the ermine. The terminal part of the tail (no more than 40% of the length) is black. The under parts (including the chin and upper lip) are white usually tinged with yellow; this color may, although not always, extend to the feet and toes. In winter, the pelage is pure white, sometimes stained with yellow, except for the black terminal part of the tail (see the section entitled Molt and Color Aberrations).

In external dimensions and features, the long-tailed weasel differs from all other Michigan mammals except the ermine. The long-tailed weasel, however, has a tail at least 50% as long as the length of its head and body; the ermine has a shorter tail, no more than one-third the length of the head and body. Further, the amount of black on the terminal part of the tail amounts to no more than 40% of the length in the long-tailed weasel; in the ermine, it amounts to about 50%. In summer pelage, the long-tailed weasel rarely has whitish fur on the inner sides of the hind feet, as does the ermine. If the sexes of the individuals being compared are known, the large size of the long-tailed weasel is clearly recognized. Otherwise, female long-tailed weasels and male ermine can be confused.

MEASUREMENTS AND WEIGHTS. Adult specimens of the larger male and the smaller female long-tailed weasel measure, respectively, as follows: length from tip of nose to end of tail vertebrae, 13¾ to 17 in (350 to 430 mm) and 11 to 13¼ in (280 to 340 mm); length of tail vertebrae, 4 to 5½ in (105 to 140 mm) and 3½ to 4¾ in (85 to 120 mm); length of hind foot, 1½ to 2 in (36 to 50 mm) and 1¼ to 1½ in (30 to 40 mm); height of ear from notch (almost the same in both sexes), ⅝ to ⅞ in (15 to 23 mm); weight, 6 to 8¾ oz (170 to 245 g) and 3 to 4¼ oz (85 to 125 g).

DISTINCTIVE CRANIAL AND DENTAL CHARACTERISTICS. The skull of the long-tailed weasel is elongate, narrow, flattened and ruggedly angular (see Fig. 83). The rostrum is short, as is characteristic of weasels, making the mouth opening appear small. The rostrum is flattened; the sagittal crest prominent in older males; the postorbital space constricted; the postorbital processes pointed; the zygomatic arches slender, weakly constructed, somewhat arched; the auditory bullae noticably longer than wide, inflated, rounded anteriorly; and the hard palate, as in other weasels, extends posteriorly well beyond the end of the tooth-row. The teeth are sharp; the middle lower incisor on each side is larger than those on either side; the premolars are reduced to three; the last upper tooth on each side is a transversely, dumbbell-shaped molar. If the sex of the specimen is known, the large size of the skull of the long-tailed weasel sets it apart from that of the ermine of either sex. If the sex is not known, the skull of a female long-tailed weasel compares favorably in size with that of a male ermine. One distinguishing feature is the pointed condition of the postorbital processes on the skull of the long-tailed weasel as contrasted with the blunt condition of these structures on the skull of the ermine. The frontal area of some skulls is abnormally swollen or pierced by action of the nematode sinus parasite, *Skrajabingylus nasiocola*.

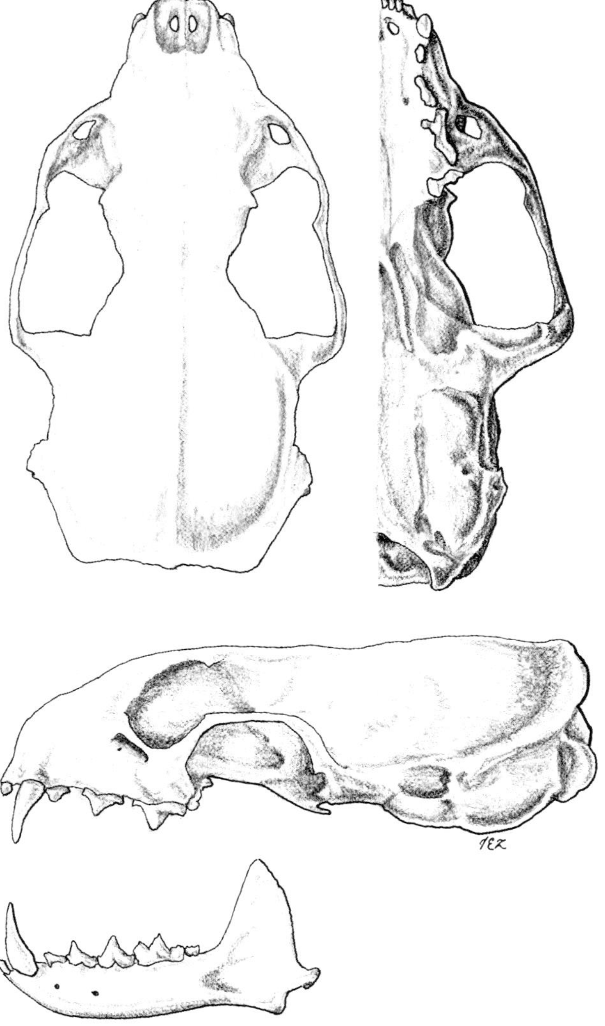

Figure 83. Cranial and dental characteristics of the long-tailed weasel (*Mustela frenata*).

The skulls of adult male and female long-tailed weasels, respectively, measure 1¾ to 2 in (44 to 51 mm) and 1½ to 1¾ in (37 to 45 mm) in greatest length and ⅞ to 1⅛ in (23 to 28 mm) and ¾ to ⅞ in (19 to 23 mm) in greatest width across the zygomatic arches. The dental formula for the long-tailed weasel is:

$$\text{I (incisors)}\,\frac{3\text{-}3}{3\text{-}3},\ \text{C (canines)}\,\frac{1\text{-}1}{1\text{-}1},\ \text{P (premolars)}\,\frac{3\text{-}3}{3\text{-}3},\ \text{M (molars)}\,\frac{1\text{-}1}{2\text{-}2} = 34$$

DISTRIBUTION. The long-tailed weasel, in marked constract to the ermine, has its major distribution in temperate and tropical America; it is found from just north of the Canadian border southward into northern South America (Hall, 1951). In Michigan, this species is near the northern edge of its extensive range. Here it overlaps the southern edge of the range of the ermine (Simms, 1979a). Actual records are from at least 50 of the 79 counties (see Map 55). Although weasels have been reported on both Isle Royale and Beaver Island (Burt, 1946; Ozoga and Phillips, 1964), it is questionable whether these records are for the long-tailed species.

HABITAT PREFERENCES. The long-tailed weasel can be expected in most nonaquatic habitats in Michigan (except perhaps for dense forest), having responded positively to clearing and other land-use practices which the human population has employed in the past 150 years. Today, the long-tailed weasel is certainly at home in the mosaic pattern of crop lands, fallow fields, bushy areas reverting to timber, brushy fence rows, small woodlots, and even suburban residential areas. Perhaps overall vegetation changes in southern Michigan have been more attractive to the long-

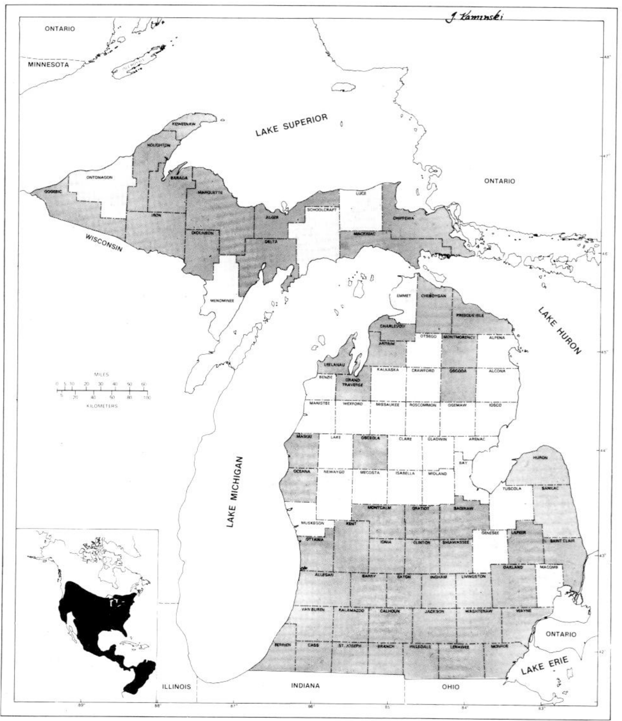

Map 55. Geographic distribution of the long-tailed weasel (*Mustela frenata*). Counties from where there are actual records are heavily shaded.

tailed weasel than those in northern Michigan. Where long-tailed weasels and ermine occur together, their ranges overlap, but the former species, in at least one study area in southern Ontario (Simms, 1979a), showed preference for more advanced successional stages in plant growth. In Michigan, observers reported the occurrence of long-tailed weasels in brushy areas in Osceola County (Wenzel, 1912); on sand dunes in Berrien County (Wood, 1922b); in swamp brush, tamarack and black spruce bog, and in runways under sphagnum in Charlevoix County (Dice, 1925); in woods, oak-hickory brush, and buttonwillow swale in Kalamazoo County (Allen, 1938b); in bluegrass association in Washtenaw County (Blair, 1940b); and concentrated in woodlots where eastern chipmunks were common in Clinton County (Linduska, 1951).

DENSITY AND MOVEMENTS. Several authors, notably Hamilton (1933), have pointed out that much of the early published data on the abundance of the long-tailed weasel is based on the number of pelts marketed. More recent estimations of abundance and spatial arrangement of long-tailed weasels have, however, been based on track counts made in the snow cover of winter and early spring. Animals have usually been tracked from their dens on circuitous routes, with actual distance traversed defining extent of movements and the straight-line distance traveled from the den site defining home range. Winter densities fluctuate widely. In open lands in Iowa, Polderboer *et al.* (1941) found one long-tailed weasel per 40 acres (16 ha). In wooded sectors of Pennsylvania, Glover (1943) found densities as high as one weasel per 6.5 acres (1.6 ha) in areas of forest cutting and slash and one per 13.3 to 38.2 acres (5.3 to 15 ha) in open forest. In Washtenaw Co., Quick (1944) calculated winter numbers no higher than one weasel per 300 acres (120 ha). In Clinton County, Linduska (1950) found local concentrations of long-tailed weasels as high as one animal per 14 acres (6 ha), with autumn population density in more extensive areas to be one individual per 91 acres (36 ha). These varied figures are more or less comparable to those obtained for densities of the ermine.

In spatial distribution in nature, the long-tailed weasel and ermine have similar movement and associations patterns. Males occupy larger home ranges than females and overlap two or more of the latters' living areas. Apparently, both males and females exclude other members of their sexes from their home territories; the two sexes avoid each other except during the summer courtship and mating period. Based primarily on cruising radii calculations ascertained by snow tracking, Polderboer *et al.* (1941) estimated the home range of the long-tailed weasel (sex not ascertained) in open lands in Iowa to be approximately 12 acres (5 ha). In wooded areas of Pennsylvania, Glover (1943b) found weasels to use winter ranges of 3.7 acres (1.5 ha) where slash and other ground cover was heavy and 12.3 acres (5 ha) where woodlands were open. DeVan (1975) found home ranges from 23 to 58 acres (9 to 14 ha) in northern Kentucky. Jackson (1951) thought that the home ranges of Wisconsin long-tailed weasels were between 30 and 40 acres (12 to 16 ha). In Michigan the only estimate of winter home range was 181 acres (73 ha) obtained in Washtenaw County by Quick (1944). He tracked the animals to determine an average daily travel of 1.7 mi (2 km). Populations of long-tailed weasels are probably fairly stable from year to year, even though prey populations may vary.

BEHAVIOR. The long-tailed weasel, despite claims in the early literature, is a rather solitary animal. The sexes remain apart except during courtship and mating in mid-summer, although the larger home ranges of the males generally overlap those of two or more females. Home sites and foraging areas are not shared with other weasels of the same sex. As suggested by Simms (1979b) in the case of the ermine, established long-tailed weasels probably resort to active hostility, avoidance behavior, and scent marking to discourage non-residents, including juveniles. The female has full custody of the offspring. Some of the group activities reported in the early literature may have been accounts of female parents on hunting forays with their offspring or male-female associations during the mating period.

The long-tailed weasel has the attributes of an aggressive, able carnivore. Like the ermine, it is remarkably quick, agile, alert, and persistent in defending its food cache or young against all adversaries. It moves in a loping, humping fashion with its muscular, slender body moving in and out of understory and other cover, rarely allowing the field observer more than a glimpse of its actions. Many characteristics of the ermine are also those of the long-tailed weasel. The larger weasel, however, (especially the male) is less efficient than the ermine in searching mouse or mole burrows for food or foraging under snow cover (Simms, 1979a). The long-tailed weasel is an excellent climber. Jeanne (1965) recorded this weasel's pursuit of an eastern chipmunk in an oak tree and other woody cover in Cheboygan County. The eastern chipmunk did not escape. The weasel is also a successful swimmer but probably does not forage for food in aquatic areas.

The long-tailed weasel, like the ermine, uses a zig-zag course in foraging, as shown by a major study of hunting behavior conducted by Powell (1978) in the Upper Peninsula's Ottawa National Forest. The prey is either sighted or detected by sound or odor, stalked, and generally caught by a quick rush. A wary prey may dash for safety in which case the weasel pursues. In snow cover, Powell measured chases only as far as a few feet (5 m).

Once the capture is made, the long-tailed weasel moves rapidly, wrapping its sinuous body and limbs around its prey and making the kill by a powerful bite at the base of the skull. Allen (1938a), Glover (1943a), and Powell (1978) have described the deadly efficiency of this mammalian predator. When tackling a large prey species, such as an adult eastern cottontail, the long-tailed weasel bites at anyplace on the victim's body while waiting the chance to get at the back of the neck.

The long-tailed weasel, like the ermine, is active in the daytime but is probably more nocturnal than diurnal. Weasels can be noisy. Most of their calls are shrill barks and hisses, reactions to being disturbed. When under restraint they growl and purr. Despite their ferocious reputation, these weasels are successfully kept in captivity, although rarely hand-held with impunity (Svihla, 1931).

The long-tailed weasel may excavate its own burrow or fashion a nest in a stone wall, dead stump, hollow log, rock pile or under a barn. However, more than likely it will usurp an underground dwelling excavated by another mammal—often a victim. Sometimes a larger facility, such as an old woodchuck den, is used. On taking over the burrow of an eastern mole, a weasel enlarged a site in the runway to 9 in (230 mm) in diameter and prepared a grass-lined nest (Polderboer *et al.*, 1941). Although a radiating tunnel was being used as a latrine, the weasel apparently defecated in the nest as well as left bones and other food scraps there. This refuse was then covered by another layer of grasses. When the nest was again fouled, the process was repeated. The result was a multilayered nesting surface. The female long-tailed weasel can probably squeeze into burrows of mice, chipmunks, and ground squirrels; males would have to either enlarge the tunnels or seek more expansive quarters. Fitzgerald (1977) reported this weasel's use of voles' surface nests as winter quarters. These nests provide layers of grasses as well as snow cover as insulation. Weasel dens in strawstacks and shocked grain (Linduska, 1942), and under farm buildings (Seton, 1953) are often welcomed by farmers who recognize the weasel's value as a mouser and ratter.

ASSOCIATES. The long-tailed weasel, like the ermine, lives most of the time by itself, except for the courtship and mating period, or, if a female, with her litter. This weasel rarely follows larger carnivores to eat scraps from kills, although carrion can be part of the diet. One observer (Howard, 1935) in Emmet County described a long-tailed weasel eating meat off beef bones stored on a porch in January in Wilderness State Park. He also noted a red squirrel eating this same meat; it gave way when the weasel approached but was not afraid to return. The long-tailed weasel apparently took little notice of the squirrel with beef scraps more easily available. On the other hand, a long-tailed weasel's association with small mammals such as deer mice, white-footed mice, and meadow voles under field-shocked corn as in Clinton County (Linduska, 1942), can be strictly a predator-prey relationship.

As noted in the chapter on the ermine, a central Michigan habitat might well include all three weasels: the small least weasel, the medium-sized ermine, and the large long-tailed weasel. Although these three live in the same areas, with the larger ones even eating the smaller ones on occasion, their prey preferences may be sufficiently different to prevent excessive competition (Rosenzweig, 1966). In areas where both ermine and long-tailed weasels live, the latter may prefer woodlands and advanced successional stages of shrub vegetation while the former uses more open and disturbed environments (Fitzgerald, 1977; Simms, 1979a); however, there will be some overlap in hunting areas. The animals also tend to partition food resources, with the smaller ermine eating mice and voles and the larger long-tailed weasel eating larger prey. However, different foraging habitats or prey species do not seem sufficient reason to account for successful populations of each of the three weasels in the same area.

Perhaps, as Simms (1979a) pointed out, Michigan is in the transition area where all of these species just happen to coincide. His theory of coexistence included the following points: (a) the opportunity for partitioning prey species and living space is actually small; (b) the northward distribution of the long-tailed weasel is probably limited by snow cover (a difficult medium for the large male weasel because it is too big to use rodents' subnivean tunnels and not big enough to move rapidly on top of the snow); (c) the distribution and sizes of the weasel species are governed largely by prey size; (d) the southward distribution of the smaller weasel species is limited by interference interactions with long-tailed weasels; and (e) the greater reproductive potential of the least weasel (several litters per year as opposed to one litter per year for the larger species) allows the smallest species to range further southward than would otherwise be possible. It seems worth repeating that further investigation of this interplay between three lively Michigan predators should intrigue vertebrate ecologists.

REPRODUCTIVE ACTIVITIES. Long-tailed weasels associate in midsummer for purposes of courtship and mating. Males may become sexually active as early as March, (Wright, 1947), but females are not receptive until July and August. In typical mustelid fashion, implantation of the embryos is delayed and normal growth does not begin until the next March. The entire gestation period averages about 280 days (Wright, 1942).

Michigan young are born in a grass-lined, protected nest, usually in late April or early May (Burt, 1946), with an average litter size of six (maximum, nine). At birth, according to Hamilton (1933), the young are elongate with large heads; long necks; pink, wrinkled skin; closed eyes; folded ear pinnae; facial vibrissae; fine, white body hair; body weight of less than ⅛ oz (3 g); and the capacity to utter squalling sounds. At seven days of age, the young's pink skin coloring is overshadowed by white, fuzzy hair; they have light brown foreclaws and nose pads, can roll effectively but not crawl, and weigh about ¼ oz (7 g). At fourteen days, the young have heavy white hair covering but lack the mane noted in the ermine; sex is easily determined by size; they crawl well, and weigh about ½ oz (17 g for males and 13.5 g for females). At 21 days of age, the young show a grayish cast to the dorsal pelage in sharp contrast with the bare, flesh-colored belly; have black hairs beginning to show on the tail tip; exhibit erupted canine and carnassial teeth; crawl with agility; weigh one oz (27 g) for males and ¾ oz (21 g) for females; and continue making squealing noises. At 36 days, the young are weaned and eating meat brought to the den by the female parent; have open eyes; can move around well; develop their dark brown summer coat on the upper parts; begin to develop characteristic weasel odor; and finally grow their white belly fur. Shortly after, the young can go on foraging parties with their female parent; by 56 days of age they can make their own kills. The female young are sexually mature between 85 and 110 days of age and will mate at the same time, as will their mother. The males in the litter do not mature sexually until the following late spring. This early sexual development in newly-born females maximizes the annual reproductive output—with every female, despite age class, capable of producing offspring each year.

It is probable that individuals of both sexes have life spans of four or more years. In Clinton County, Linduska (1947) recaptured an ear-tagged long-tailed weasel one year after first capture.

FOOD HABITS. The elongated shape of the long-tailed weasel and its relatives is certainly advantageous in pursuing such prey species as small rodents into their burrows. However, the weasel's snake-like body makes it difficult to maintain a constant body temperature or to attain a spherical resting posture (Brown and Lasiewski, 1972). These factors probably contribute to the voracious appetites of these mustelids. DeVan (1977), noted that the smaller female long-tailed weasel consumes 67.5% of her body weight in food per day; the larger male, 43.5%.

Although most accounts of the food habits of the long-tailed weasel do not identify differences in prey selections of the male and female, Simms (1978a) felt that the smaller female has less difficulty capturing small rodents (in their tunnels and under snow) than the male. On the other hand, the male is more apt to catch eastern cottontails (especially young ones), which live above the snow surface and use burrows and trails sufficiently large to accomodate the weasel as well. Simms infers that the male long-tailed weasel may just be the wrong size to negotiate heavy snow. This may be one reason the long-tailed weasel is restricted to northern distribution in the snow belt.

Dearborn (1932) found that long-tailed weasels in Michigan eat about 83% mammalian foods, 10% bird life, and 7% insects. Mammalian prey items in this survey included voles of the genus *Microtus*, 31%; mice of the genus *Peromyscus*, 24%; eastern cottontails, 14%; shrews of the genus *Sorex*, 7%; short-tailed shrews, 5%; and eastern moles, 2%. In Washtenaw County, Quick (1944) found the winter food selection of this weasel to include 65 to 70% mice of the genus *Peromyscus* and 23 to 33% voles of the genus *Microtus*. He also noted that long-tailed weasels relish poultry. Some of the other Michigan prey species known to be eaten by the long-tailed weasel are young eastern cottontails in Saginaw County (Burroughs, 1939), thirteen-lined ground squirrels in Kalamazoo County (Allen 1938b), eastern chipmunks in Clinton County (Linduska, 1950), meadow voles in Clinton County (Linduska, 1942), and meadow jumping mice in Livingston County (Blair, 1940a). Further demonstrating the broad taste range of this weasel (Hall, 1951), the species has been known to eat big brown bats (Mumford, 1969), masked shrews (Nichols and Nichols, 1935), least weasels (Polderboer *et al.*, 1941), reptiles, amphibians, and, in summer months, fruits and berries.

ENEMIES. The remarkable agility and keen senses of the long-tailed weasel undoubtedly assist it in eluding numerous enemies. Even so, most larger four-footed predators—gray fox, red fox, coyote, gray wolf, bobcat, and lynx—catch and eat the long-tailed weasel (Hall, 1951). Latham (1952) felt that the two foxes constituted an important control on weasel populations. In Kent County, Bonnie Marris (pers. comm., December, 1977) noted a red fox successfully attack a weasel on a snowy December mid-afternoon. It is likely, however, that some of the larger birds of prey—the rough-legged hawk, goshawk, great horned owl and snowy owl (Boxall, 1979; Jackson, 1961; Linduska, 1950)—are more effective long-tailed weasel predators. Of course, the winter fur trapper takes a substantial number of these weasels, especially when pelt prices are high. Long-tailed weasels also suffer mortality on Michigan's highways (Manville, 1949; Kasul, 1976).

PARASITES AND DISEASES. Internal parasites reported from the long-tailed weasel include fluke, *Alaria taxideae*; tapeworms of the genus *Taenia*; and roundworms, *Filaroides martis, Filaroides bronchialis, Capillaria mustelorum Physaloptera maxillaris,* and *Skrjabingylus nasicola* (Erickson, 1946; Goble and Cook, 1942; Jackson, 1961). The latter species, the nematode sinus parasite, inflicts considerable damage to the frontal area of the skull of the long-tailed weasel, with obvious dilations and lesions in specimens which have had this infection. External parasites include lice, *Trichodectes retusus* and *Trichodectes kingi*; ticks, *Dermacentor variabilis* and *Ixodes cookei*; and fleas, *Monopsyllus vison, Nearctopsylla genalis genalis, Orchopeas caedens caedens, Ceratophyllus vison, Ceratophyllus fasciatus* and *Neo-*

trichodectes mephitidis (Jackson, 1961; Scharf and Stewart, 1980).

MOLT AND COLOR ABERRATIONS. The long-tailed weasel molts its old fur coat and grows a new one twice a year. The spring molt, taking about 25 days, begins as early as late February and is completed by late April; the autumn molt takes a similar length of time beginning as early as October and is completed by early December. The new spring hair appears first on the neck and shoulders, then on the back and finally on the flanks. The reverse occurs in autumn with the new hair growing first on the belly, feet and tail and spreading gradually over the back and onto the face. As with the ermine, the causative agent for the molt is primarily the length of daylight per 24 hours, with the eyes the sensitive receptors for detecting this seasonal phenomenon.

In the northern parts of the range of this species, the animal grows a white coat (except for the black tail) for winter and a dorsally brown coat (except for the black tail) for summer. However, in the vast southern part of its range, the long-tailed weasel has a brown back all year around. The white winter coat prevails for those animals living far enough north to have snow-covered ground the entire winter. The southern half of Michigan's Lower Peninsula, with its usually fluctuating winter snow cover, is in an intermediate position (Hall, 1951). According to Dearborn (1932), long-tailed weasels from the Upper Peninsula rarely, if ever, stay brown in winter. In the northern tiers of counties in the Lower Peninsula, as far south as Osceola County, between 95 and 100% of the long-tailed weasels turn white. From Bay County south, the weasels' winter coats may be brown or white, with animals from the more southern counties more apt to remain brown. For example, Dearborn found that 75% of the weasels from Montcalm County turn white in winter, 60% in Eaton County, and only 10% in Berrien and St. Joseph counties. Just across the Indiana line in Porter County, Lyon (1923) reported that only 2 of 200 animals trapped over three winters in the Michigan dunes and near Chesterton were white in winter.

ECONOMIC IMPORTANCE. With ermine and long-tailed weasel pelts lumped together in the fur marketing process, there is no sure way of knowing how many of the 2,230 weasel pelts harvested in Michigan in 1969 were long-tailed weasel pelts. Even so, white winter pelts, bringing less than one dollar to the trapper, make up only a minor part of the Michigan fur trade. It is likely that many long-tailed weasels are discarded by trappers who catch them accidentally.

The long-tailed weasel has been known to raid Michigan poultry flocks (Manville, 1948; Schofield, 1957; Wood, 1922a), but like the ermine, is an excellent mouser. Seton (1953) and Jackson (1961) described the efficiency with which this weasel kills Norway rats and house mice in barns or grain bins, as well as meadow voles which may infest cultivated fields or stored grains (Linduska, 1942).

COUNTY RECORDS FROM MICHIGAN. Records of specimens of long-tailed weasels in collections of museums, from the literature, from trapper's reports, and provided by personnel of the Michigan Department of Natural Resources show that the long-tailed weasel occurs throughout both the Upper and Lower Peninsula (Map 55).

LITERATURE CITED

Allen, D. L.
1938a Notes on the killing technique of the New York weasel. Jour. Mammalogy, 19(2):225–229.
1938b Ecological studies on the vertebrate fauna of a 500-acre farm in Kalamazoo County, Michigan. Ecol. Monog., 8(3):348–436.

Banfield, A. W. F.
1974 The mammals of Canada. Univ. Toronto Press, Toronto. xxiv+438 pp.

Blair, W. F.
1940a Home ranges and populations of the jumping mouse. American Midl. Nat., 23(1):244–250.
1940b Notes on home ranges and populations of the short-tailed shrew. Ecology, 21(2):284–288.

Boxall, P. C.
1979 Interaction between a long-tailed weasel and a snowy owl. Canadian Field-Nat., 93(1):67–68.

Brown, J. H., and R. C. Lasiewski
1972 Metabolism of weasels: The cost of being long and thin. Ecology, 53(5):939–943.

Burroughs, R. D.
1939 New York weasel preying on cottontail rabbit. Jour. Mammalogy, 20(2):253.

Burt, W. H.
1946 The mammals of Michigan. Univ. Michigan Press, Ann Arbor. xv+288 pp.

Dearborn, N.
1932 Foods of some predatory fur-bearing animals in Michigan. Univ. Michigan, School Forestry and Conserv., Bull. No. 1, 52 pp.

DeVan, R.
1975 Home range and population density of the long-tailed weasel (*Mustela frenata noveboracensis*) in northern Kentucky. Abstract, Annual Meeting American Soc. Mammalogists, 10 pp.
1977 Natural nutritional requirements of the long-tailed weasel (*Mustela frenata*). Abstract, Annual Meeting American Soc. Mammalogists, 7 pp.

Dice, L. R.
1925 A survey of the mammals of Charlevoix County, Michigan, and vicinity. Univ. Michigan, Mus. Zool., Occas. Papers No. 159, 33 pp.
1927 A manual of the Recent wild mammals of Michigan. Univ. Michigan, Univ. Mus., Michigan Handbook Ser. No. 2, 62 pp.

Erickson, A. B.
1946 Incidence of worm parasites in Minnesota Mustelidae and host lists and keys to North American species. American Midl. Nat., 36(2):494–509.

Fitzgerald, B. M.
1977 Weasel predation on a cyclic population of the montane vole (*Microtus montanus*) in California. Jour. Animal Ecology, 46:367–397.

Glover, F. A.
1943a Killing techniques of the New York weasel. Pennsylvania Game News, 13(10):11 and 23.
1943b A study of the winter activities of the New York weasel. Pennsylvania Game News, 14(6):8–9.

Goble, F. C., and A. H. Cook
1942 Notes on nematodes from the lungs and frontal sinuses of New York fur-bearers. Jour. Parasitol., 28(6):451–455.

Hall, E. R.
1936 Mustelid mammals from the Pleistocene of North America with systematic notes on some of the Recent members of the genera Mustela, Taxidea, and Mephitis. Carnegie Inst. Washington, D.C., Publ. No. 473, pp. 41–119.
1951 American weasels. Univ. Kansas Publ., Mus. Nat. Hist., 4:1–466.

Hamilton, W. J., Jr.
1933 The weasels of New York. Their natural history and economic status. American Midl. Nat., 14(4):289–344.

Howard, W. J.
1935 Apparently neutral relations of weasel and squirrel. Jour. Mammalogy, 16(4):322–323.

Jackson, H. H. T.
1961 Mammals of Wisconsin, Univ. Wisconsin Press, Madison. xi+504 pp.

Jeanne, R. L.
1965 A case of a weasel climbing trees. Jour. Mammalogy, 46(2):344–345.

Kasul, R. L.
1976 Mortality and movement of mammals and birds on a Michigan interstate highway. Michigan State Univ., unpubl. M.S. thesis, ix+39 pp.

Latham, R. M.
1952 The fox as a factor in the control of weasel populations. Jour. Wildlife Mgt., 16(4):516–517.

Linduska, J. P.
1942 Winter rodent populations in field-shocked corn. Jour. Wildlife Mgt., 6(4):353–363.
1947 Longevity of some Michigan farm game mammals. Jour. Mammalogy, 28(2):126–129.
1950 Ecology and land-use relationships of small mammals on a Michigan farm. Michigan Dept. Conserv., Game Div., ix+144 pp.

Lyon, M. W., Jr.
1923 Notes on the mammals of the dune region of Porter County, Indiana. Proc. Indiana Acad. Sci., 32:209–221.

Manville, R. H.
1948 The vertebrate fauna of the Huron Mountains, Michigan. American Midl. Nat., 39(3):615–640.
1949 Highway mortality in northern Michigan. Jour. Mammalogy, 30(3):311–312.

Mumford, R. E.
1969 Long-tailed weasel preys on big brown bats. Jour. Mammalogy, 50(2):360.

Nichols, D. G., and J. T. Nichols
1935 Notes on the New York weasel (*Mustela noveboracensus*). Jour. Mammalogy, 16(4):297–299.

Ozoga, J. J., and C. J. Phillips
1964 Mammals of Beaver Island, Michigan. Michigan State Univ., Publ. Mus., Biol. Ser., 2(6):305–348.

Polderboer, E. B., L. W. Kurh, and G. O. Hendrickson
1941 Winter and spring habits of weasels in central Iowa. Jour. Wildlife Mgt., 5(1):115–119.

Powell, R. A.
1978 A comparison of fisher and weasel hunting behavior. Carnivore, 1(1):28–34.

Quick, H. F.
1944 Habits and economics of the New York weasel in Michigan. Jour. Wildlife Mgt., 8(1):71–78.

Rosenzweig, M. L.
1966 Community structure in sympatric Carnivora. Jour. Mammalogy, 47(4):602–612.

Scharf, W. C., and K. R. Stewart
1980 New records of Siphonaptera from northern Michigan. Great Lakes Ento., 13(3):165–167.

Schofield, R. D.
1957 Livestock and poultry losses caused by wild animals in Michigan. Michigan Dept. Conserv., Game Div., Rept. No. 2118 (mimeo.), 6 pp.

Seton, E. T.
1953 Lives of game animals. Vol. II, Pt. II. Bears, coons, badgers, skunks, and weasels. Charles T. Branford Co., Boston, pp. 369–746.

Simms, D. A.
1979a North American weasels: Resource utilization and distribution. Canadian Jour. Zool., 57(3):504–520.
1979b Studies of an ermine population in southern Ontario. Canadian Jour. Zool., 57(4):824–832.

Svihla, A.
1931 Habits of New York weasel in captivity. Jour. Mammalogy, 12(1);67–68.

Wenzel, O. J.
1912 A collection of mammals from Osceola County, Michigan. Michigan Acad. Sci., 14th Rept., pp. 198–205.

Wood, N. A.
1922a The mammals of Washtenaw County, Michigan. Univ. Michigan, Mus. Zool., Occas. Papers No. 123, 23 pp.
1922b Notes on the mammals of Berrien County, Michigan. Univ. Michigan, Mus. Zool., Occas. Papers No. 124, 4 pp.

Wright, P. L.
1942 Delayed implantation in the long-tailed weasel (*Mustela frenata*), the short-tailed weasel (*Mustela cicognani*) and the marten (*Martes americana*). Anat. Rec., 83(3):341–353.
1947 The sexual cycle of the male long-tailed weasel (*Mustela frenata*). Jour. Mammalogy, 28(4):343–352.

LEAST WEASEL
Mustela nivalis Linnaeus

NAMES. The generic name *Mustela,* proposed by Linnaeus in 1758, is from the Latin and means weasel. The specific name *nivalis,* also proposed by Linnaeus in 1758, is from the Latin and means snow. In much of the earlier literature the North American least weasel has been referred to by the name *Mustela rixosa* (Burt, 1946). However, because it has been shown that this New World population is not specifically distinct from that found in Eurasia (von Reichstein, 1958), the name of the Old World species *Mustela nivalis* (predating *M. rixosa*) now applies. In Michigan, this species is usually called the least weasel. The subspecies *Mustela nivalis allegheniensis* (Rhoads) occurs in Michigan.

RECOGNITION. Body shape resembles that of a long, slender mouse (Fig. 84); length of head and body averaging 6¼ in (160 mm) for males and 5⅞ in (150 mm) for females; tail short, less than 20% as long as this weasel's head and body; length of tail vertebrae averaging 1⅜ in (35 mm) in males and 1 in (25 mm) in females; head small, narrow, blunt-nosed, elongate, no larger than the slender, snake-like, muscular neck; ears short, rounded, inconspicuous; eyes small, beady; vibrissae (whiskers) prominent; body long, slender, sinuous, less than ⅞ in (20 mm) in diameter; tail stubby, well-haired; legs short; each foot with five clawed toes, well-furred soles; three pairs of mammary glands, one abdominal and two inguinal; anal scent glands well-developed, as in other mustelids, producing an irritating odor; males about 10% larger than females in external dimensions and weight (Hall, 1951).

Figure 84. The least weasel (*Mustela nivalis*) stretched out. Photograph by Robert Brown

The fur of the least weasel includes short, fine underfur and longer, shiny, guard hairs. According to Latham (1953), this pelage fluoresces, producing a vivid lavender color under ultra-violet light, whereas the pelages of the ermine and the long-tailed weasel do not. In summer, the upperparts are rich, chocolate brown. This coloring extends down the sides of the body variable distances, meeting midventrally in some individuals. The underparts are white (sometimes splotched with brown spots) from the upper lip to the base of the tail. The brown tail characteristically does not have a black tip (except for an occasional black hair). In winter, the pelage is completely white (except for just a few inconspicuous black hairs at the tip of the tail) in northern counties of Michigan. In southern counties where the ground is less likely to be snow-covered all winter, least weasels, like long-tailed weasels, may retain brown coats all winter. For an explanation of this seasonal color difference, see Molt and Color Aberrations.

In external dimensions and coloring, the least weasel, the smallest of Michigan carnivores, is distinguished from all other Michigan mammals except the long-tailed weasel and the ermine. The least weasel, however, is much smaller than even the female ermine and has a noticeably shorter tail which lacks the striking black tip of the other weasels.

MEASUREMENTS AND WEIGHTS. Adult specimens of the larger male and the smaller female least weasel measure, respectively, as follows: length from tip of nose to end of tail vertebrae, 7¼ to 8 in (185 to 205 mm) and 6¾ to 7⅛ in (170 to 180 mm); length of tail vertebrae, 1 to 1⅝ in (25 to 40 mm) and ⅞ to 1⅛ in (22 to 29 mm); length of hind foot, ⅞ to 15/16 in (21 to 23 mm) and ¾ to 13/16 in (19 to 21 mm); height of ear from notch (almost the same in both sexes), ½ to ⅝ in (12 to 15 mm); weight, 1⅜ to 2⅛ oz (40 to 60 g) and 1 to 1¾ oz (30 to 50 g).

DISTINCTIVE CRANIAL AND DENTAL CHARACTERISTICS. The least weasel has an elongate, narrowly rounded, and smooth skull (Fig. 85). The short rostrum (nasal area) is in marked contrast to the elongated braincase. The mouth appears abnormally small in relation to the long skull. The skull is narrow through the mastoidal area; has a distinct sagittal crest; and possesses sharply pointed, post-orbital processes.

The skull of the least weasel is almost a minature of that of the ermine. Normally, the skulls of the two species are easily distinguished by size alone, however, some skulls of large male least weasels may approach the size of those of small female ermines. The nematode sinus parasite (*Skrajabingylus nasiocola*) is apparently less prevalent in the least weasel than in the ermine and long-tailed weasel. Even so, cranial lesions caused by this roundworm occasionally distort the configuration of the frontal region of the brain case (King, 1977).

The skulls of adult male and female least weasels, respectively, measure approximately 1¼ in

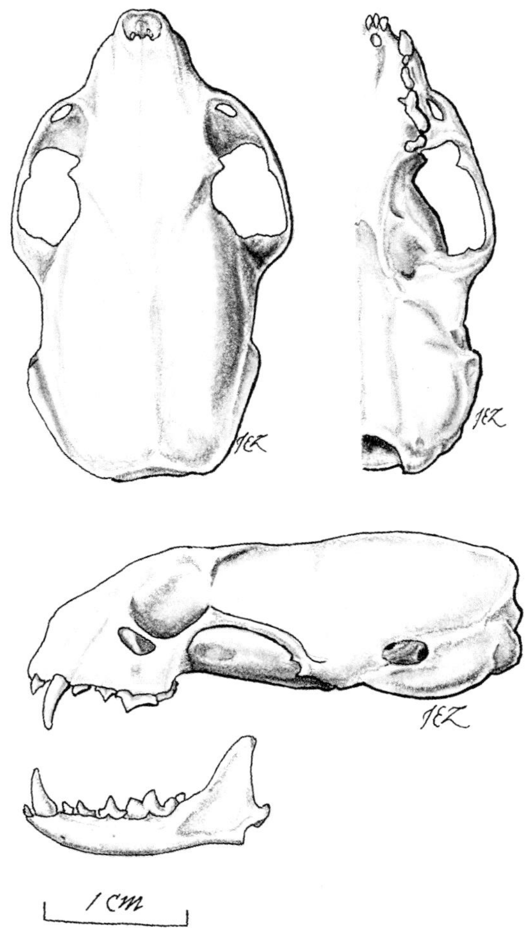

Figure 85. Cranial and dental characteristics of the least weasel (*Mustela nivalis*).

(31.5 to 32.8 mm) and 1 3/16 in (30.2 to 30.5 mm) in greatest length and ⅝ in (15.8 to 17.5 mm) and 9/16 in (14.7 to 15.9 mm) in greatest width across the zygomatic arches. The dental formula is:

$$\text{I (incisors)}\frac{3\text{-}3}{3\text{-}3}, \text{C (canines)}\frac{1\text{-}1}{1\text{-}1}, \text{P (premolars)}\frac{3\text{-}3}{3\text{-}3}, \text{M (molars)}\frac{1\text{-}1}{2\text{-}2} = 34$$

DISTRIBUTION. The least weasel, like the ermine, enjoys a vast circumpolar distribution, including Eurasia (south to northern Africa) and the Japanese Islands. In North America, it is distributed from northcentral and much of eastern United States northward to arctic prairies in Canada and Alaska. Its absence from parts of the eastern arctic, southeastern Canada, and New England is unexplained.

In Michigan (Map 56), its presence was suspected earlier (Dice, 1927), but the first Michigan documentation of the least weasel was not until 1932 in Oakland County (Hatt, 1940). Subsequently, least weasels have been found in most counties in the southern part of the Lower Peninsula (Burt, 1946), with records sparse in more northern localities (Marvin Cooley, pers. comm.). This scarcity occurs in those parts of Michigan where ermine are most common, suggesting that these two species are not compatible (King and Moors, 1979; Simms, 1979b). On the other hand, the least weasel and the long-tailed weasel may have a more compatible existence, although this is also questioned by Simms (1979a).

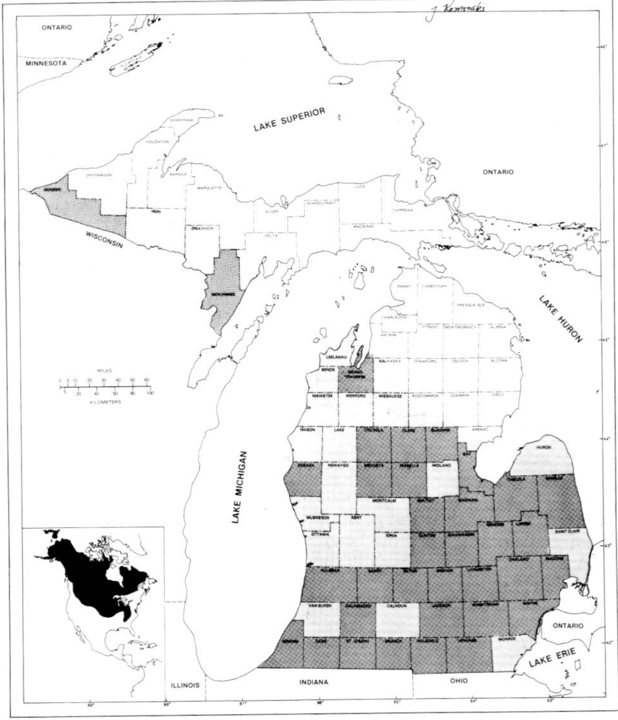

Map 56. Geographic distribution of the least weasel (*Mustela nivalis*). Counties from which there are actual records of occurrence are heavily shaded.

HABITAT PREFERENCES. The least weasel occurs in a variety of environments in southern Michigan, especially in open situations. Apparently the combination of small woodlots, cultivated fields, fallow land, weedy and brushy fence rows, marsh edge, and streamside vegetation is the preferred habitat, with little evidence available that the least weasel uses woodlands extensively. In Ingham County, Heidt (1970) captured numerous least weasels in live-traps placed in mixed grass, weeds, and shrubs along railroad tracks on the campus of Michigan State University; Golley (1960) found this weasel to be an important link in the food chain of an old field community of perennial grasses and herbs; and Osterberg (1962) photographed a least weasel in a small mammal runway in grass cover. Hatt (1940) found least weasels in Oakland County by a small creek, along a drainage ditch, and in a fence row. In residential areas least weasels sometimes become trapped in basement window wells.

DENSITY AND MOVEMENTS. Prior to 1930, the least weasel's presence was suspected but undocumented in Michigan (Hall, 1940). Extensive studies of small mammals at the Rose Lake Wildlife Research Station in Clinton County in the 1940s revealed no evidence of the least weasel (Linduska, 1950). However, when some of these field studies were duplicated in 1979 and 1980, this mustelid was among the species captured (Shier, 1981). In the period 1945–1947, the skull of a least weasel was found in a barn owl pellet in Ingham County (Wallace, 1948). The animal has been a prominent member of the small mammal community in southern Michigan since at least 1955.

Golley (1960) estimated the late autumn population of least weasels in his old-field habitat study area in Ingham County to be one adult per 3 acres (1.2 ha). Very likely, as suggested by Erlinge (1974), least weasel populations fluctuate in relation to the abundance of prey species of small rodents, with populations becoming as dense as one individual per acre (0.4 ha). On one occasion on the Michigan State University campus in Ingham County, skeletons of more than 20 least weasels were found in an abandoned bird trap. These animals apparently had been attracted to a baited but unattended trap and succumbed in a period estimated at less than three months.

The rather meager available knowledge does allow for the conclusion that individual least weasels occupy separate areas in nature, with these spacings variable in size and dependent on the abundance and distribution of prey species (King, 1975). Home ranges of different weasels may overlap but central areas are exclusively controlled by a single individual. As in the case of the ermine and long-tailed weasel, the home range of the male least weasel is much larger than that of the female. King calculated home ranges for males as ranging from 17.5 to 37.5 acres (7 to 15 ha) with a maximum of 67.5 acres (26.2 ha) and for females, 2.5 to 10 acres (1 to 4 ha), with these areas being held by individuals for an average of seven months. In Iowa, Polderboer (1942) found least weasels to occupy areas of about 2 acres (0.8 ha). Although there is some evidence that populations of prey species or other environmental factors cause ups and downs in least weasel numbers (King and Moors, 1979; Swanson and Fryklund, 1935), there appear to be no profound multi-annual fluctuations.

BEHAVIOR. Although the least weasel practices some of the solitary habits of the long-tailed weasel and ermine, the two larger mustelids have only a single annual courtship and mating period. Instead, the least weasel produces several litters each year, indicating a periodic breakdown in the home range system (Heidt *et al.,* 1968). The sexes, otherwise, generally remain apart with individuals defending at least an exclusive central home area from others of their sex (King, 1975; Powell, 1979). The small size of the female may be an advantage, according to Moors (1980), allowing her to use less energy for personal maintenance and more for rearing offspring.

The least weasel is a remarkable animal to watch; it moves with unbelievable speed when seeking refuge, avoiding predators, and attacking its prey—mouse or vole. The least weasel, like the long-tailed weasel and ermine, uses the same quick rush to catch its prey, wraps its body and limbs around the victim, and delivers the lethal bite at the nape of the victim's neck (Heidt, 1972). Males and females have body diameters no larger than those of mice and voles, so they can follow their prey into underground burrows, through snow tunnels, and along runways under matted grass and leaf litter (Osterberg, 1962; Simms, 1979a). The smaller female probably has a slight advantage over the male in negotiating rodent pathways. The male may also be more apt to switch to other food sources in times of rodent scarcity (King and Moors, 1979; Moors, 1980).

The least weasel has acute senses of vision, hearing, and smell. Herman (1973) concluded that the least weasel can detect and find potential prey by olfactory clues alone. The exudate from anal glands may ward off potential enemies and be used to mark home range boundaries and communicate with members of the opposite sex. The lively least weasel remains active the year-round. It seems to be equally adjusted to day or night action. In fact, King (1975), in her studies of British least weasels, caught more individuals in live traps by day than by night. These small, warm-blooded creatures undoubtedly hunt during a major part of the 24-hour cycle to catch sufficient food for their physiological needs (Brown and Lasiewski, 1972; Golley, 1960; Iversen, 1972). In captivity, least weasels are noisy (Heidt and Huff, 1970; Huff and Price, 1968). Adults chirp and hiss when threatened. The female parent utters a trilling sound when calling her young. The newly-born are capable of making high-pitched squeaks, which become chirps after their eyes open.

Least weasel nests are placed in a variety of protective sites either above or below ground. Jackson (1961) found dens in creek bank holes and under corn shocks. More than likely, however, burrows are confiscated from meadow voles, eastern moles, thirteen-lined ground squirrels, or other burrowing creature's. The least weasel will widen a passageway to make a nest and then line it with fine grass or hair and feathers from victims (Banfield, 1974). In Sanilac County a nest containing five young was discovered in March in an abandoned bee hive (James P. Ludwig, pers. comm.).

ASSOCIATES. The least weasel occupies typical small mammal environment in the open lands of southern Michigan. As Golley's (1960) classic study in Ingham County demonstrated, it is an important member of the food chain which transfers energy through the old field ecosystem. Perhaps of greatest interest to the ecologist are the distributions of the long-tailed weasel, the ermine, and the least weasel in Michigan. Although information is fragmentary, it seems likely that all three seldom exist together. Perhaps the ermine and the least weasel are the least compatible (Simms, 1979a). More information is needed concerning (a) the relationship of weasel size to prey selection, (b) predation by the larger weasels on the smaller weasels, and (c) the effects of white coat color in winter and variable snow cover on weasel distribution, especially in the southern counties of the Lower Peninsula.

REPRODUCTIVE ACTIVITIES. Unlike most other mustelids, the least weasel does not have delayed implantation and a prolonged gestation period (Heidt *et al.*, 1968). This briefer period of fetal development allows for two or more litters annually—perhaps offsetting the intense predation to which this small carnivore is subjected. Males are sexually active in all months but December and January, and litters have been reported for most months of the year (Banfield, 1974).

Much of what is known about the American least weasel's breeding activities was obtained from investigations made by Gary Heidt (1970), while a graduate student at Michigan State University. Following mating and conception, the gestation period lasts 35 days and produces a litter of from one to six young. At birth, according to Heidt, the young weigh almost 1½ grams; are wrinkled, pink, and devoid of hair; have closed eyes and folded ear pinnae; utter squeaking sounds; and can move about with fair control of their forelimbs. At six days of age, the young have grayish pigment on the back and possess fine white body hair. At 11 days of age, they have their deciduous canine teeth. At 18 days of age, the young have conspicuous brown hair on their backs and white on their bellies; possess white ear tips; are growing their deciduous premolars; and are eating their first solid food (meat). By 30 days of age, they have opened their eyes, crawl without difficulty, are growing permanent canines, and begin to wean. By 40 days of age, the young can kill prey. By 95 days of age, the young have attained adult weight. At four months of age, the female young reach sexual maturity. At eight months of age, the male young reach sexual maturity. Females born in the early months probably produce offspring later that same year. Individuals can be aged by examination of the wear on their carnassial teeth (King, 1980).

FOOD HABITS. The least weasel is remarkably adapted by size and sensory perception for predation on small rodents. Although data are far from complete, there is some evidence that the least weasel can respond explosively through its "rapid" rate of reproduction to sudden peaks in rodent populations as well as face near extinction locally when these prey animals decline (King and Moors, 1979). The least weasel's appetite is responsive to the vast amounts of energy needed to maintain its physiological requirements for day-to-day survival (Iverson, 1972). There is evidence that the female's smallness gives her an advantage over the male in pursuing rodents into their burrows, thus enabling her to fulfill her energy needs during litter production (Moor, 1980). In Ingham County, Short (1961) studied the feeding habits of a captive least weasel and found that the animal had to ingest proper food at the rate of one gram per hour to survive. Thus the diminutive weasel must each day consume food equal to one-half of its own body weight. This amounts to about 1.6 deer mice or 0.75 meadow vole per 24-hour period.

Besides deer mice and meadow voles, almost any Michigan rodent weighing less than 3½ oz (100 g) is fair game for the least weasel (Hall, 1951). The red-backed vole may be its most important prey species in northern Michigan (Simms, 1979a). Other food species reported include white-footed mice, meadow jumping mice, house mice, and southern bog lemming (Banfield, 1974). Eastern moles, masked shrews, and short-tailed shrews probably are also eaten, although Erlinge (1975) thought that insectivores were possibly distasteful to European least weasels. Birds, amphibians, and insects may also be taken (Jackson, 1961). When the least weasel kills more than it can immediately consume, the surplus is stored in side chambers of underground dens for future meals. Least weasels have been known to thrive on a variety of meats in captivity (Heidt, 1972).

ENEMIES. The least weasel, despite its diminutive size, probably can protect itself against some of its numerous enemies. Even so, it falls victim to its near relative, the long-tailed weasel, as well as such other terrestrial predators as house cats, gray and red foxes, and various snakes (Banfield, 1974; Hall, 1951; Jackson, 1961; Latham, 1952). However, hawks and owls may be the most effective predators. Wallace (1948) found skulls of least weasels in the pellets of barn owls in Ingham County. The great horned owl and rough-legged hawk catch and eat least weasels (Jackson, 1961). Automobiles also take an unknown toll. Kasul (1976) counted two least weasels dead on the highway in his survey in southern Michigan.

PARASITES AND DISEASES. Least weasels are infested with both external and internal parasites. Internally, the roundworms, *Molinens patens*, *Physaloptera maxillaris*, and *Skrjabingylus nasicola*, are reported (Erickson, 1946; King, 1977). Ticks, mites, and lice also infest the least weasel but few positive identifications have been made. Fleas include *Ceratophyllus lunatus*, *Nearctophyllus brooksi*, and *Megabothris atrox*. Apparently fleas from some of the rodents on which the least weasel preys may be found on this small carnivore (King, 1976).

MOLT AND COLOR ABERRATIONS. The least weasel sheds its fur coat and grows a new one twice each year. The shedding and regrowth begins in spring, during March and April on the dorsum and proceeds down each side. In autumn, during November and sometimes into December, the process begins on the ventral surface and extends dorsally. Molting and regrowth are triggered twice annually, as with the long-tailed weasel and the ermine, by the changing percentage of daylight. In the northern part of its range where snow cover for the entire winter is assured, the autumn fur growth is white (except for a few inconspicuous black hairs on the tip of the tail). In the southern part of Michigan's Lower Peninsula, where winter snow cover is normally irregular, a large portion of the least weasel population may not become white.

ECONOMIC IMPORTANCE. Michigan least weasels may occasionally be marketed for their fur (Hatt, 1940) although the skins are usually considered too small for commercial use. The feisty least weasel is able to subdue young poultry, but its value as a mouser probably overshadows any farmyard depredations. In Ingham County, for example, the least weasel is as much at home in vacant lots in urban areas as along weedy fencerows at the edge of cornfields.

COUNTY RECORDS FROM MICHIGAN. Records of specimens of least weasels in collections of museums, from the literature, from trappers' reports, and provided by personnel of the Michigan Department of Natural Resources (Marvin Cooley, pers. comm.) show that this species commonly occurs in the southern half of the Lower Peninsula but is less known from the northern half of the Lower Peninsula and from the Upper Peninsula. County records from the Upper Peninsula include: Gogebic (Laundre, 1975) and Menominee (Gregg Stoll, pers. comm.); from the Lower Peninsula: Allegan, Barry, Bay, Berrien, Branch, Cass, Clare, Clinton, Eaton, Genessee, Gladwin, Grand Traverse, Gratiot, Hillsdale, Ingham, Isabella (L. D. Caldwell *in litt.*, November 24, 1980), Jackson, Kalamazoo, Lapeer, Lenawee, Livingston, Macomb, Mecosta, Oakland, Oceana, Osceola, Saginaw, Sanilac, Shiawassee, St. Joseph, Tuscola, Washtenaw and Wayne.

LITERATURE CITED

Banfield, A. W. F.
1974 The mammals of Canada. Univ. Toronto Press, Toronto. xxiv+438 pp.

Brown, J. H., and R. C. Lasiewski
1972 Metabolism of weasels: The cost of being long and thin. Ecology, 53(5):939–943.

Burt, W. H.
1946 The mammals of Michigan. Univ. Michigan Press, Ann Arbor. xv+288 pp.

Dice, L. R.
1927 A manual of the Recent wild mammals of Michigan. Univ. Michigan, Univ. Mus., Michigan Handbook Ser., no. 2, 63 pp.

Erickson, A. B.
1946 Incidence of worm parasites in Minnesota Mustelidae and host lists and keys to North American species. American Midl. Nat., 36(2):494–509.

Erlinge, S.
1974 Distribution, territoriality, and numbers of the weasel *Mustela nivalis* in relation to prey abundance. Oikos, 25(3):308–314.
1975 Feeding habits of the weasel *Mustela nivalis* in relation to prey abundance. Oikos, 26(3):378–384.

Golley, F. B.
1960 Energy dynamics of a food chain of an old-field comunity. Ecol. Monog., 30(2):187–206.

Hall, E. R.
1951 American weasels. Univ. Kansas Publ., Mus. Nat. Hist., 4:1–466.

Hatt, R. T.
1940 The least weasel in Michigan. Jour. Mammalogy, 21(4):412–416.

Heidt, G. A.
1970 The least weasel *Mustela nivalis* Linnaeus. Developmental biology in comparison with other North American *Mustela.* Michigan State Univ., Publ. Mus., Biol. Ser., 4(7):227–282.
1972 Anatomical and behavioral aspects of killing and feeding by the least weasel, *Mustela nivalis* L. Proc. Arkansas Acad. Sci., 26:53–54.

Heidt, G. A., and J. N. Huff
1970 Ontogeny of vocalization in the least weasel. Jour. Mammalogy, 51(2):385–386.

Heidt, G. A., M. K. Petersen, and G. L. Kirkland, Jr.
1968 Mating behavior and development of least weasels *(Mustela nivalis)* in captivity. Jour. Mammalogy, 49(3): 413–419.

Herman, D. G.
1973 Olfaction as a possible mechanism for prey selection in the least weasel, *Mustela nivalis.* Michigan State Univ., unpubl. M.S. thesis, v+39 pp.

Huff, J. N., and E. O. Price
1968 Vocalizations of the least weasel, *Mustela nivalis.* Jour. Mammalogy, 49(3):548–550.

Iversen, J. A.
1972 Basal energy metabolism of mustelids. Jour. Comp. Physiol., 81(4):341–344.

Jackson, H. H. T.
1961 Mammals of Wisconsin. Univ. Wisconsin Press, Madison. xii+504 pp.

Kasul, R. L.
1976 Mortality and movement of mammals and birds on a Michigan interstate highway. Michigan State Univ., unpubl. M.S. thesis, ix+39 pp.

King, C. M.
1975 The home range of the weasel *(Mustela nivalis)* in an English woodland. Jour. Animal Ecology, 44:636–668.
1976 The fleas of a population of weasels in Wytham Woods, Oxford, Jour. Zool., London, 180:525–535.
1977 The effects of the nematode parasite *Skrjabingylus nasicola* on British weasels (Mustela nivalis). Jour. Zool., London, 182:225–249.
1980 Age determination in the weasel (*Mustela nivalis*) in relation to development of the skull. Zeitschrift für Säugetierkunde, 45(3):153–173.

King, C. M., and P. J. Moors
1979 On co-existence, foraging strategy and biogeography of weasels and stoats (*Mustela nivalis* and *M. erminea*) in Britain. Oecologia, 39(2):129–150.

Latham, R. M.
1952 The fox as a factor in the control of weasel populations. Jour. Wildlife Mgt., 16(4):516–517.
1953 Simple method for identification of least weasel. Jour. Mammalogy, 34(3):385.

Laundre, J.
1975 An ecological survey of the mammals of the Huron Mountain area. Huron Mt. Wildlife Foundation, Occas. Papers, No. 2, x+69 pp.

Linduska, J. P.
1950 Ecology and land-use relationships of small mammals on a Michigan farm. Michigan Dept. Conserv., Game Div., ix+144 pp.

Moors, P. J.
1980 Sexual dimorphism in the body size of mustelids (Carnivora): The roles of food habits and breeding systems. Oikos, 34:147–158.

Osterberg, D. M.
1962 Activity of small mammals as recorded by a photographic device. Jour. Mammalogy, 43(2):219–229.

Polderboer, E. B.
1942 Habits of the least weasel (*Mustela rixosa*) in northeastern Iowa. Jour. Mammalogy, 23(2):145–147.

Powell, R. A.
1979 Mustelid spacing patterns: Variations on the theme by *Mustela.* Zeitschrift für Tierpsychologie, 50(2):153–165.

Shier, J. L.
1981 Habitats of threatened small mammals on the Rose Lake Wildlife Research Area, Clinton County, Michigan. Michigan State Univ., unpubl. M.S. thesis, vi+108 pp.

Short, H. L.
1961 Food habits of a captive least weasel. Jour. Mammalogy, 42(2):273–274.

Simms, D. A.
1979a North American weasels: Resource utilization and distribution. Canadian Jour. Zool., 57(3):504–520.
1979b Studies of an ermine population in southern Ontario. Canadian Jour. Zool., 57(4):824–832.

Swanson, G., and P. O. Fryklund
1935 The least weasel in Minnesota and its fluctuation in numbers. American Midl. Nat., 16(1):120–126.

von Reichstein, H.
1958 Schädelvariabilitat europaischer Mausweisel (*Mustela nivalis* L.) und Hermeline (*Mustela erminea* L.) in Beziehung zu Verbreitung und Geschlecht. Zeitschrift für Säugetierkunde, 22(3–4):151–182.

Wallace, G. J.
1948 The barn owl in Michigan. Its distribution, natural history and food habits. Michigan State Coll., Agric. Exp. Sta., Tech. Bull. 208, 61 pp.

MINK
Mustela vison Schreber

NAMES. The generic name *Mustela*, proposed by Linnaeus in 1758, is from the Latin and means weasel. The specific name *vison*, proposed by Schreber in 1777, has an obscure origin, from either the Icelandic or Swedish word *vison*, which means a kind of weasel or possibly from the Latin word *visor*, which means scout. In Michigan, this species is universally called mink. Mink living in the Upper Peninsula belong to the subspecies *Mustela vison vison* Schreber, intergrading at the Wisconsin border with *Mustela vison letifera* Hollister (Jackson, 1961; Long, 1974). Mink in Michigan's Lower Peninsula are classified as *Mustela vison mink* Peale and Palisot de Beauvois.

RECOGNITION. Body size resembles that of an oversized, heavy-bodied weasel having the characteristic slender form of other members of the genus *Mustela* (Fig. 86); length of head and body averaging 14¾ in (375 mm) for the male and 13 in (330 mm) for the female; tail almost one-half the length of the head and body; length of tail vertebrae averaging 7¼ in (195 mm) for the male and 6 in (155 mm) for the female; head small, flat, no longer around than the long, muscular neck (Fig. 87); nose pointed; ears short, rounded; eyes dark, beady, reflecting green in the glare of a spotlight; vibrissae (whiskers) prominent; body long, slender, sinuous although more heavily constructed than that of the long-tailed weasel; tail long, much more bushy than those of the weasels; legs short with front ones set well back; each foot with five clawed and semiwebbed toes; three pairs of mammary glands, one abdominal and two inguinal; anal scent glands well-developed, emitting a strong, musky odor; males about 10% larger than females.

Figure 86. Museum study skins of adult mink. Note size difference between larger male (left) and smaller female (right). Specimens obtained in Gratiot County.

Figure 87. Portrait of a mink (*Mustela vison*). Photograph by Robert Harrington and courtesy of the Michigan Department of Natural Resources

The pelage of the mink consists of thick, soft underfur and abundant, long, glossy, and somewhat oily guard hairs, which render the coat water-repellent. The grayish-brown underfur is obscured by the uniformly dark-brown coloring toward the tip. This coloring extends to the mink's undersides except for white blotches on the chin, throat, and sometimes on the chest and belly. For a description of color phases, see Molt and Color Aberrations.

In external measurements and coloring, the mink is rarely confused with any other Michigan mammal. The rich brown coloring and larger size (of even the smaller female) readily distingush the mink from the ermine and long-tailed weasel. The otter is somewhat similar in color and shape but conspicuously larger than the mink. A blurred view of a dark animal in the water might be confusing because of a slight similarity between the actions of a mink and those of a young otter or even of a muskrat.

MEASUREMENTS AND WEIGHTS. Adult specimens of the larger male and smaller female mink measure, respectively: length from tip of nose to end of tail vertebrae, 20½ to 25½ in (520 to 650 m) and 16½ to 21½ in (420 to 550 mm); length of tail vertebrae, 7⅛ to 8⅜ in (180 to 210 mm) and 5 to 7 in (130 to 180 mm); length of hind foot, 2¼ to 3 in (56 to 75 mm) and 2 to 2¾ in (50 to 70 mm); height of ear from notch (almost the same in both sexes) ⅝ to 1 in (15 to 25 mm); weight, 20 to 44 oz (565 to 1,250 g) and 17 to 33½ oz (500 to 950 g).

DISTINCTIVE CRANIAL AND DENTAL CHARACTERISTICS. The skull of the mink is somewhat flattened, sturdily constructed, elongate in typical weasel fashion, and rather smooth (Fig. 88). The rostrum is short, flat, and broad. The mouth appears abnormally small in relation to the elongated brain case. The zygomatic arches are narrow and evenly speading. The sagittal crest and lambdoidal ridge are developed in skulls of older individuals. The auditory bullae, only moderately inflated, are one and one-half times as long as wide. The bony palate extends well posterior to the end of the tooth-rows. The last tooth (molar) in the upper tooth-row is transversely dumbell-shaped. The skull of the male is fully 10% larger than that of the female.

The mink's skull resembles that of the long-tailed weasel but is so much larger that confusion is minimal. Compared with other Michigan *Mustela,* the mink's auditory bullae are more flattened and shorter in relation to skull length. The skull of the mink is, of course, much smaller than that of the otter. The skull of the striped skunk is similar in length to that of the mink but its heavier construction, greater height, stout zygomatic arches, and short palate (ending just posterior to the end of the upper tooth-row instead of much further posteriorly as in the mink) are distinguishing features. The skulls of adult male and adult female mink, respectively, measure 2⅜ to 2¾ in (61 to 69 mm) and 2¼ to 2½ in (58 to 65 mm) in greatest length and 1⅜ to 1¾ in (34 to 43 mm) and 1¼ to 1⅝ in (32 to 41 m) in greatest width across the zygomatic arches. The dental formula for the mink is:

$$\text{I (incisors)}\ \frac{3\text{-}3}{3\text{-}3},\ \text{C (canines)}\ \frac{1\text{-}1}{1\text{-}1},\ \text{P (premolars)}\ \frac{3\text{-}3}{3\text{-}3},\ \text{M (molars)}\ \frac{1\text{-}1}{2\text{-}2} = 34$$

DISTRIBUTION. The mink occurs over a vast part of temperate and boreal North America except in parts of the arctic in northcentral Canada and adjacent islands and in the arid American Southwest. In Michigan, the species lives throughout both the Upper and Lower Peninsulas, with actual records of occurrence from almost all counties (see Map 57). The species is also known from archaeological sites in Mackinac County

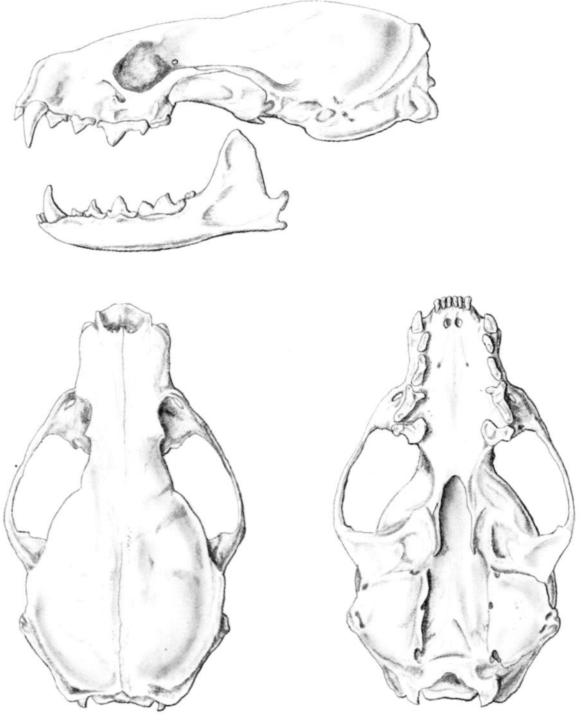

Figure 88. Cranial and dental characteristics of the mink (*Mustela vison*).

Map 57. Geographic distribution of the mink (*Mustela vision*).

(Martin, 1979) and in Saginaw County (Cleland, 1966).

HABITAT PREFERENCES. Michigan has appropriately been called the Water Wonderland State. It is therefore a favored residence for an animal like the mink which is equally at home on land and in streams, rivers, lakes, ponds, and marshes. Although most accounts of the favored environments of Michigan mink place the animal in or along the edge of water, this lively, ubiquitous mustelid can be expected some distance from water, even using a backyard woodpile as shelter on occasion. Wood (1922), for example, tracked one mink through the snow on a circular course cross country in Washtenaw County. He also found other individuals on hunting forays as much as ½ mi (0.8 km) from water. The only habitat which the mink does not use on a regular basis is probably heavy upland woods. When away from aquatic areas, the mink seems most at home in second-growth cover of mixed shrubs, weeds and grasses, and the edges of cultivated fields and pasture land.

DENSITY AND MOVEMENTS. In most places, mink populations rarely attain saturation because of concerted efforts to capture them for their valuable pelts. Banfield (1974) noted that in good habitat prior to the hunting and trapping season mink populations may be as high as 8.5 to 22 animals per sq mi (3.3 to 8.5 per sq km). After the trapping season a decline to 3 to 4 animals per sq mi (1.2 to 1.5 per sq km) can be expected. Calculations of densities and movements have been estimated largely by inventorying catches by trappers in such areas as large marshes, by snow tracking, by live-trapping, and by attaching transmitting collars to individuals and monitoring their movements by means of radio signals (Gerell, 1970; Marshall, 1936; McCabe, 1949; Sealander, 1944).

Mink populations may be highest in large marshes which contain cattails and muskrat beds (Errington, 1943; Hamilton, 1940), such as those at the Seney National Wildlife Refuge in Schoolcraft County or the Maple River Flats in Gratiot County. In these situations mink populations can be estimated in terms of area (two-dimensionally), whereas along stream courses the populations occur linearly and are best estimated in terms of distance. In marsh areas, populations of 9 to 15 per sq mi (3.5 to 5.8 per sq km) can be expected (Errington, 1943). In Washtenaw County, Marshall (1936) estimated one population at 1.5 individuals per sq mi (0.6 per sq km). Another estimate of 0.2 to 0.3 animals per sq mi (0.09 to 0.1 per sq km) was made on a township basis by Craighead and Craighead (1956). The latter two figures probably reflect expected numbers on a large area basis rather than in prime mink habitat alone. In Clinton County, Linduska (1950) estimated mink populations to be as high as 4 animals per sq mi (1.5 per sq km).

The mink's travels are governed in part by intraspecific living space interactions. In southern Michigan, Wood (1922) tracked a mink moving in a circular fashion for about 5 mi (8 km) in the snow—a cruising area as large as 1,300 acres (520 ha). As in the case of weasels, the males move over larger distances than females (McCabe, 1949). Whereas females may occupy space of less than 20 to as much as 50 acres (8 to 20 ha), adult males operate in areas of 1,900 acres (760 ha) or even more (Banfield, 1974; Schwartz and Schwartz, 1981). Along flowing waterways, animals tracked by radio collars foraged up and down streams as far as 1.6 mi (2.6 km) in the case of males and 1.1 mi (1.8 km) in the case of females (Gerell, 1970).

BEHAVIOR. The mink, like its close relatives the weasels, lives a solitary life except during courtship and mating. Reports that the male may assist the female in caring for young are probably incorrect. This appears to be the sole responsibility of the female, with the only groupings being those of the female and her young on forays in early summer. The males occupy larger areas which overlap those of more than one female. Individuals of each sex tend to exclude others of the same sex from defended parts of their home sites. Juvenile males are discouraged from living in areas already occupied by adult males; females are not discouraged (Gerell, 1970). Mink employ scent from their anal glands to mark strategic places within their areas of activity.

The mink is a highly efficient carnivore, employing vision, hearing, and smell in perceiving its environment, prey, and enemies. According to Errington (1943), the mink shows less tendency than the weasels to use zig-zag movement tactics in searching for prey. It does, however, use the same loping, humpbacked gait, and wraps its elongated, sinuous body around its prey while administering a lethal bite to the nape of the neck. The mink will often slide down snowy or muddy inclines in otter-like fashion. The mink can climb but rarely reaches great heights unless treed (Seton, 1953; Wood, 1922). It is an excellent swimmer, moving rapidly through water. The mink may watch for likely fish prey from a stream bank or other out-of-water vantage points, although, according to Dunstone (1978), underwater search may also be employed. Mink can capture muskrats in their beds or bank dens but they also pursue them in open water. Seton (1953) quotes Allan Brooks' observation of a mink's underwater chase of a muskrat, where the mink successfully dispatched the rodent. The mink can descend to a depth of 18 ft (5.5 m) and swim at least 100 ft (30.5 m) underwater at speeds up to 2 mi (3.2 km) per hour (Jackson, 1961). Most of the mink's normal activities are conducted at night, although the animals are seen in daylight especially when the sky is overcast. The mink is active at all seasons but may be less so during extremely cold winter weather (Marshall, 1936).

Behavioral information about mink is abundant because of observations on captives bred for their valued pelts. Gilbert (1969) noted chuckling utterances by males, defensive screams by males and females, warning squeaks, and hissing sounds. Some of the softer calls are made by the female in the presence of her litter or during courtship and mating. Harsher calls are generally associated with fighting or defensive maneuvers.

Mink generally den near water. The animal may excavate its own ground burrow under the protection of tree roots, beneath felled logs, in rock ledges, logjams, drain tiles, brush piles, and stream banks; and even under buildings in rural settings. Mink also take over marsh houses of beaver or muskrat, and more frequently the excavated burrows of the muskrat, woodchuck or ground squirrel. Some burrows, especially those in stream banks, have underwater entrances. Usurped muskrats dens may have as many as five entrances and extend into the ground no deeper than 1 ft (0.3 m), in order to avoid ground water. Entry tunnels range in diameter from 4 to 6 in (102 to 152 mm). A male mink, in particular, may have several dens at strategic places in its home range (Schladweiler and Storm, 1969). Nests in underground burrows are 10 to 12 in (254 to 305 mm) in diameter and lined with plant fibers, grasses, fur, and feathers (Jackson, 1961). Nests are rarely fouled by fecal materials and urine. In contrast to weasels, mink rarely cache food for future use and apparently do not eat much carrion.

ASSOCIATES. The mink is generally a loner. There is probably some degree of interaction between the mink and the large male long-tailed weasel, because their choices of food species as well as den sites overlap. This relationship might be most important in edge habitats such as marsh and old field. Not much is known about mink-otter associations in the aquatic medium. It has been suggested that both size difference and the agility of the smaller mink allow these two mustelids to occupy similar aquatic areas without conflict. Some of the predator/prey relationships between mink and muskrat, as determined by Errington (1943) in his classic work in Iowa, should be extended to include these two species plus beaver and otter where all four live in association in Michigan wetlands.

REPRODUCTIVE ACTIVITIES. Because they are raised in captivity, more is known about the reproductive processes of the mink than about other carnivores (Shump *et al.,* 1976). Much data on the mink's reproductive behavior, physiology, and nutrition have been obtained by Professor Richard Aulerich, his colleagues, and advanced students at the Fur Animal Project at Michigan State University. In nature, males are ready to breed in February and March. Females are receptive to mating from February to early April. As in other mustelids, the fertilized egg undergoes partial development (to the blastula stage) and then becomes inactive—the condition of delayed implantation (Hansson, 1947). The length of this period of embryonic quiescence varies, affected in part by the amount of daylight per 24 hours. The hours of light can, or course, be varied to advantage on mink farms (Holcomb, 1963; Holcomb *et al.,* 1962). Consequently, the total gestation period may be as short as 39 days or as long as 76 days. The average time for this developmental period is 50 days, of which only about 30 to 32 days are actually required to develop the full-term fetus (Enders, 1952).

The single annual litter is usually born in April or May, with an average of five young (maximum ten). At birth, the young seem naked but are covered with fine white hair through which the pink, wrinkled skin shows; they have closed eyes, closed ear pinnae, and weigh about 1/5 oz (6 g). At 14 days of age, the young have reddish-gray hair, and the male members of the litter are obviously larger. At 35 days of age, the young have open eyes and are in the process of being weaned as more solid foods are eaten. Between 42 and 56 days of age, the young are able to catch their own prey and join their female parent in forays for food. The females reach adult weight by four months of age; males take as long as nine to 11 months. The young of both sexes depart from maternal care beginning in August. As Errington (1943) has described, this is a crucial time for mink survival as these dispersing young are exposed to assorted predators while trying to locate homesites for themselves.

In captivity, mink can live as long as 10 years (Altman and Dittner, 1972). In nature, probable life expectancy is no more than three or possibly four years, with the entire local population being replaced during that time period.

FOOD HABITS. Providing the proper nutrients necessary to healthily maintain thousands of farm mink in small cages has become a highly successful commercial endeavor, as attested to by the many publications on the subject (see Shump *et al.,* 1976). Captive mink are fed prepared, balanced rations to maintain maximum reproductive success and proper sheen to their fur. According to Richard Aulerich (*in litt.,* January 27, 1981), a typical ranch mink diet includes 35–45% protein, 22–28% fat, 20–36% carbohydrate, and 6–7% ash. In contrast, the wild mink has to depend on an array of foods, many often seasonal at best, to acquire a comparable diet. Most information about mink food habits has been derived from examination of the contents of digestive tracts and scats (fecal material). Such investigations show that the species is opportunistic in food selection, even eating some vegetable material (Iversen, 1972). Major studies examining food habit diversity have been made in New York (Hamilton, 1940), Missouri (Korschgen, 1958), Iowa (Errington, 1943; 1954), North Dakota (Eberhardt and Sargeant, 1975), and in Michigan (Alexander, 1977; Dearborn, 1932; Manville, 1948; Sealander, 1942, 1943b). Dearborn found Michigan mink eat mostly crayfish in summer (68% by volume) followed by mammals (19%), and frogs (3%); in winter, mammals (56%) followed by fish (18%), frogs (9%), and birds (6%). In another study, Sealander showed that in winter mink preferred mammalian foods (54 to 55% by volume) followed by frogs (up to 20%), crayfish (up to 17%), birds (up to 15%), and fish (up to 13%).

Both Dearborn and Sealander noted that in winter Michigan mink prey chiefly on muskrat and to a lesser extent on eastern cottontail and meadow vole. Other small mammals eaten include shrews of the genus *Sorex,* short-tailed shrew, eastern mole, star-nosed mole, snowshoe hare, red squirrel, northern flying squirrel (Manville, 1948), mice of the genus *Peromyscus,* woodland vole, house mouse, Norway rat, and meadow jumping mouse. Other Michigan mammals likely to be eaten by mink but reported from food habit studies made in other states include bats (Goodpaster and Hoffmeister, 1950), bog lemmings (Korschgen, 1958), and woodland jumping mice (Platt, 1968).

Bird remains found in digestive tracts of Michigan mink taken in winter included American coot, ring-necked pheasant, ruffed grouse, and domestic chicken (Sealander, 1942). The mink also catches and eats other birds, including species of ducks (Eberhardt and Sargeant, 1977). Frogs of several species, snakes, fish, insects and, of course, crayfish, plus some plant material, round out the mink's food selections. Fish of at least six families taken by the mink attest to its swimming ability, and its success with meadow voles, eastern cottontails, snowshoe hares, and other terrestrial species demonstrate the mink's versatility in capturing prey on land. The mink's climbing abilities are demonstrated by the fact that it catches red squirrels or northern flying squirrels in arboreal situations.

ENEMIES. The mink's major enemies in Michigan are probably the hunter and trapper who want the valued pelt. Of importance also are the large birds of prey and major four-footed carnivores. There are reports of mink being captured and eaten by great horned owl (Errington *et al.*, 1940), red fox (Schofield, 1960), bobcat (Lowery, 1974), coyote, and domestic dog (Schwartz and Schwartz, 1981). Automobiles kill mink on roadways (Jackson, 1961). In Ingham County, Kasul (1976) counted six dead individuals between August 20, 1975, and March 15, 1976, along 15.5 mi (24.5 km) of highway (I-96) between the I-496 and M-52 interchanges.

PARASITES AND DISEASES. Parasites and diseases affecting farm mink enterprises have been given considerable study; Shump *et al.* (1976) list 645 publications on these subjects in their bibliography. In the wild, Michigan mink have been found infested with such external parasites as ticks, mites, and fleas including *Ixodes cookei, Laelaps multispinosus, Monopsyllus vison, Nearctopsylla genalis,* and *Orchopeas caedens* (Lawrence *et al.*, 1965). Internal parasites from Michigan mink include lungflukes, *Paragonimus kellicotti;* gall bladder flukes, *Metorchis conjunctus;* lungworms, *Filaroides bronchialis;* round worms, *Physaloptera* sp.; giant kidney worms, *Dioctophyma renale;* Guinea worms, *Dracunculus insignis;* sinus worms, *Skrjabingylus nasicola;* tapeworms, *Mesocestoides litteratus;* and spiny-headed worms, *Pomphorhynchus bulbocalli* and *Gnathostoma spinigerum* (Erickson, 1946; Sealander, 1942; 1943a). Sealander considered the giant kidney worm and the sinus worm the most serious internal parasites to Michigan mink. Wild populations can also be infected with tularemia (Henson *et al.*, 1978).

MOLT AND COLOR ABERRATIONS. In nature, mink sometimes are found with all white (albinistic) pelage (Jones, 1923; Wood, 1922). In addition, in the course of breeding these animals in captivity, several color mutations have appeared. Since these pelt colors often are valued highly in the fur trade, these mutants have been sustained through selective breeding. In the industry, known color phases include pastel, sapphire, violet, heinen, hope, pearl, ambergold, blue iris, gunmetal, platinum, steelblue, and white.

ECONOMIC IMPORTANCE. Commercial mink production in parts of the northern United States and Canada, Europe, Argentina, and elsewhere, is of obvious economic importance. Research carried on for many years by the Fur Animal Project at Michigan State University has helped answer fundamental questions about some of the intricacies of mink raising (Aulerich, 1973). Projects have often been funded by such agencies as the Mink Farmer's Research Foundation (Shump *et al.*, 1976). In Michigan in 1978, there were 45 mink farms, mostly located in the Upper Peninsula. The 106,000 pelts produced that year brought about $39.30 per pelt for a total of $4,200,000 (data from the Michigan Agricultural Reporting Service, July 16, 1979). Several color phases were marketed, with the standard dark brown coat color making up 30% of pelts sold, followed by pastel, 25%, and demi-buff, 15%. Although pelt prices fluctuate—a marked decline in the late 1960s forced some mink farmers out of business—the assured production of quality mink fur for the fashion industry has been achieved. Despite the lack of data, it is also suspected that wild mink populations are now trapped less intensely in many parts of the species' range because of the success of this farming venture.

Ups and downs in the fur market have depended on: (a) fashion trends, (b) high excise taxes on wearing apparel of fur or with fur trim, (c) imports of furs from Australia and the U.S.S.R., and (d) the introduction of synthetic fibers which simulate fur. Even so, mink fur has consistently maintained high appeal, whereas long-hair pelts of red fox and raccoon have not.

In the early days of the fur trade in the Upper Great Lakes Region, mink fur was poorly appreciated compared with beaver fur. For example, in 1767, British traders shipped 50,938 beaver pelts but only 807 mink from Michilimackinac (Petersen, 1979). At the Astor Fur Trading Post on

Mackinac Island in 1822, beaver pelts brought $3 to $4 while mink pelts brought 25¢ (Anon., 1962). With the sharp decline in the beaver catch before the middle of the nineteenth century, muskrat and raccoon took the lead, with mink not far behind. From 1835 to 1838, the American Fur Company, according to Johnson (1969), purchased 22,810 mink pelts through their Detroit Department (including southern Michigan) and 7,705 through their Northern Department (including northern Michigan).

In recent years, the annual Michigan wild mink harvest has varied from 16,971 (in 1957) to 7,180 (in 1967). Annual estimates by the Michigan Department of Natural Resources were discontinued after 1971, due to a change in licensing methods which make identification of hunters and trappers impractical. However, the catch probably increased in the 1970s as a result of the spectacular rise in prices of raw pelts on the fur market. From a rather dismal average of $12 per pelt in 1969, prices for Michigan mink furs rose to about $18 in 1974, $20 in 1976, $26 in 1978, $30 in 1980 and $22 in 1981. Prices for the pelts of the smaller females might be no more than one-half these figures. There apparently will always be a place in the fur market for pelts of wild-taken animals. To the farm youth, winter trapping for muskrat and opossum, the catch of a mink or two is always an occasion for celebration. The rarity of such captures is not so much due to the scarcity of mink, but rather to the mink's wariness (Petoskey, 1969).

In 1979–1980, the legal open season for trapping mink in the Upper Peninsula (Zone I) was from October 25 through December 31; in the northern part of the Lower Peninsula (Zone II), from November 1 through January 15; and in the southern part of the Lower Peninsula (Zone III) for both hunting and trapping, from November 18 through January 31. These dates are, of course, correlated with primeness of the mink pelts in these areas.

The mink's adaptability to forage in both aquatic and terrestrial areas and its opportunistic approach to obtaining prey may conflict with human interests—for example, when mink raid fish hatcheries, feed on favored game fish (Alexander, 1977) or eat prized domestic ducks (Dearborn, 1932). The mink can also be guilty of killing barnyard poultry (Dearborn, 1932; Wood, 1922). However, such depredations are usually minor compared with those of red foxes and raccoons (Schofield, 1957).

COUNTY RECORDS FROM MICHIGAN. Records of specimens of mink in collections of museums, from the literature, from trappers' reports, and provided by personnel of the Michigan Department of Natural Resources show that mink live in all counties of the Upper and Lower Peninsulas. Mink are also recorded from Isle Royale (Johnson, 1970) in Lake Superior and from Bois Blanc (Cleland, 1966) in Lake Huron.

LITERATURE CITED

Alexander, G. R.
1977 Food of vertebrate predators on trout waters in north central Lower Michigan. Michigan Acad., 10(2):181–195.

Altman, P. L., and D. S. Dittner, eds
1972 The biology data book. 2nd ed. Federation of American Soc. Exp. Biol., Bethesda, Md., Vol. I, xvii+606 pp.; Vol. II, xix+607–1432 pp.; Vol. III, xvii+1433–2123 pp.

Anon.
1962 Fur and hide values, 1822. Michigan Conserv., 31(6): 36–37.

Aulerich, R. J.
1973 Michigan's fur bearing animal industry—now and in 1985. Michigan State Univ., Agri. Exp. Sta. and Coop. Ext. Serv., Research Rept. 186, pp. 7–11.

Banfield, A. W. F.
1974 The mammals of Canada. Univ. Toronto Press, Toronto. xxiv+438 pp.

Cleland, C. E.
1966 The prehistoric animal ecology and ethnozoology of the Upper Great Lakes Region. Univ. Michigan, Mus. Anthro., Anthro. Papers, No. 219, x+294 pp.

Craighead, J. J., and F. C. Craighead, Jr.
1956 Hawks, owls and wildlife. The Stackpole Co., Harrisburg, Pennsylvania. xix+443 pp.

Dearborn, N.
1932 Foods of some predatory fur-bearing animals in Michigan. Univ. Michigan, Sch. Forestry and Conserv., Bull. No. 1, 52 pp.

Dunstone, N.
1978 The fishing strategy of the mink (*Mustela vison*): Time-budgeting of hunting effort? Behaviour, 67(3–4):157–177.

Eberhardt, L. E., and A. B. Sargeant
1977 Mink predation on prairie marshes during the waterfowl breeding season. Pp. 33–43 *in* R. L. Phillips, and C. J. Jonkel, eds. Proc. 1975 Predator Sym. Univ. Montana, Forest and Conserv. Exp. Sta. ix+268 pp.

Enders, R. K.
1952 Reproduction in the mink (*Mustela vision*). Proc. American Philos. Soc., 96:691–755.

Erickson, A. B.
1946 Incidence of worm parasites in Minnesota Mustelidae and host lists and keys to North American species. American Midl. Nat., 36(2):494–509.

Errington, P. L.
1943 An analysis of mink predation upon muskrats in northcentral United States. Iowa State Coll., Agric. Exp. Sta., Res. Bull. 320, pp. 797–924.
1954 The special responsiveness of minks to epizootics in muskrat populations. Ecol. Monog., 24(4):377–393.

Errington, P. L., F. Hamerstrom, and F. N. Hamerstrom, Jr.
1940 The great horned owl and its prey in north-central United States. Iowa State Coll., Agric. Exp. Sta. Res. Bull. 277, pp. 759–850.

Gerell, R.
1970 Home ranges and movements of the mink, *Mustela vison*, in southern Sweden. Oikos, 21(2):160–173.

Gilbert, F. F.
1969 Analysis of basic vocalizations of the ranch mink. Jour. Mammalogy, 50(3):625–627.

Goodpaster, W., and D. F. Hoffmeister
1950 Bats as prey for mink in Kentucky cave. Jour. Mammalogy, 31(4):457.

Hamilton, W. J., Jr.
1940 The summer food of minks and raccoons on the Montezuma Marsh, New York. Jour. Wildlife Mgt., 4(1):80–84.

Hansson, A.
1947 The physiology of reproduction in mink (*Mustela vison* Schreb.) with special reference to delayed implantation. Acta Zool., 28:1–136.

Henson, J. B., J. R. Gorham, and D. T. Shen
1978 An outbreak of tularemia in mink. Cornell Vet., 68(1):78–83.

Holcomb, L. C.
1963 Reproductive physiology in mink *(Mustela vison)*. Michigan State Univ., unpubl. Ph.D. dissertation, iv+74 pp.

Holcomb, L. C., P. J. Schaible, and R. K. Ringer
1962 The effects of varied lighting regimes on reproduction in mink. Michigan State Univ., Agri. Exp. Sta., Quart. Bull., 44(4):666–678.

Iversen, J. A.
1972 Basal energy metabolism of mustelids. Jour. Comp. Physiol., 81(4):341–344.

Jackson, H. H. T.
1961 Mammals of Wisconsin. Univ. Wisconsin Press, Madison. xii+504 pp.

Johnson, D. R.
1969 Returns of the American Fur Company, 1835–1839. Jour. Mammalogy, 50(4):836–839.

Johnson, W. L.
1970 Food habits of the red fox in Isle Royale National Park, Lake Superior. American Midl. Nat., 84(2):568–572.

Jones, S. V. H.
1923 Color variations in wild animals. Jour. Mammalogy, 4(3):172–177.

Kasul, R. L.
1976 Mortality and movement of mammals and birds on a Michigan interstate highway. Michigan State Univ., unpubl. M.S. thesis, ix+39 pp.

Korschgen, L. J.
1958 December food habits of mink in Missouri. Jour. Mammalogy, 39(4):521–527.

Lawrence, W. H., K. L. Hays, and S. A. Graham
1965 Arthropodous ectoparasites from some northern Michigan mammals. Univ. Michigan, Mus. Zool., Occas. Papers No. 639, 7 pp.

Linduska, J. P.
1950 Ecology and land-use relationships of small mammals on a Michigan farm. Michigan Dept. Conserv., Game Div., ix+144 pp.

Long, C. A.
1074 Environmental status of the Lake Michigan region. Vol. 15. Mammals of the Lake Michigan drainage basin. Argonne Nat. Lab., ANL/ES-40, 108 pp.

Lowery, G. H., Jr.
1974 The mammals of Louisiana and its adjacent waters. Louisiana State Univ. Press, Baton Rouge. xxiii+565 pp.

McCabe, R. A.
1949 Notes on live-trapping mink. Jour. Mammalogy, 30(4):416–423.

Manville, R. H.
1948 The vertebrate fauna of the Huron Mountains, Michigan. American Midl. Nat., 39(3):615–640.

Marshall, W. H.
1936 A study of the winter activities of the mink. Jour. Mammalogy, 17(4):382–392.

Martin, T. J.
1979 Faunal remains from the Gros Cap Site (20MK6/7), Mackinac County, Michigan. Appendix E, pp. 133–148 *in* S. R. Martin, Final report: Phase II. Archaeological site examination of the Gros Cap Cemetery area, Mackinac County, Michigan. Michigan Tech. Univ., Cult. Res. Mgt. Rept. No. 6, ix+166 pp.

Petersen, E. T.
1979 Hunters' heritage. A history of hunting in Michigan. Michigan United Conserv. Clubs, Lansing. 55 pp.

Poetoskey, M. L.
1969 To trap a muskrat or a mink. Michigan Nat. Res., 38(1):19–22.

Platt, A. P.
1968 Selective predation by a mink on woodland jumping mice confined in live traps. American Midl. Nat., 79:539–540.

Schladweiler, J. L., and G. L. Storm
1969 Den-use by mink. Jour. Wildlife Mgt., 33(4):1025–1026.

Schofield, R. D.
1957 Livestock and poultry losses caused by wild animals in Michigan. Michigan Dept. Conserv., Game Div., Rept. No. 2118 (mineo.), 6 pp.

1960 A thousand miles of fox trails in Michigan's ruffed grouse range. Jour. Wildlife Mgt., 24(4):432–434.

Schwartz, C. W., and E. R. Schwartz

1981 The wild mammals of Missouri. Rev. Edition. Univ. Missouri Press and Missouri Dept. Conserv., Columbia. vii+356 pp.

Sealander, J. A.

1942 Studies on trapped mink *(Mustela vison mink)* in southern Michigan. Michigan State Coll., unpubl. M.S. thesis, iii+94 pp.

1943a Notes on some parasites of the mink in southern Michigan. Jour. Parasitol., 29:361–362.

1943b Winter food habits of mink of southern Michigan. Jour. Mammalogy, 7(4):411–417.

1944 A criticism of Marshall's method for censusing mink. Jour. Mammalogy, 25(1):84–86.

Seton, E. T.

1953 Lives of game animals. Vol. II, Pt. II. Bears, coons, badgers, skunks, and weasels. Charles T. Branford Co., Boston. pp. 369–746.

Shump, A. U., K. A. Shump., Jr., C. A. Heidt, and R. J. Aulerich

1976 A bibliography of mustelids. Part II: Mink. Michigan State Univ., Agri. Exp. Sta., Jour. Art. No. 7390, 156 pp.

Wood, N. A.

1922 The mammals of Washtenaw County, Michigan. Univ. Michigan, Mus. Zool., Occas. Papers No. 123, 23 pp.

WOLVERINE
Gulo gulo (Linnaeus)

NAMES. The generic name *Gulo*, proposed by Pallas in 1780, and the specific name *gulo*, proposed by Linnaeus in 1758, are from the same Latin world meaning "glutton." In Michigan, this species, widely known as wolverine, is the official state animal. The names carcajou and skunk bear have also been used. The subspecies *Gulo gulo luscus* Linnaeus formerly lived in Michigan.

RECOGNITION. Body size resembles that of an adult English bulldog with heavy, muscular body and bearlike proportions (Fig. 89); length of head and body averaging 30 in (760 mm) for the male

Figure 89. The only extant specimen of a Michigan wolverine. This mount represents an individual taken in the Gaylord area in the northern part of the Lower Peninsula in the late 1860s. It was obtained and prepared by naturalist Charles J. Davis (1845–1924) of Lansing and is part of the 600–700 mounted Michigan mammals from the Davis collection and now in the possession of Dr. Gary F. Kaberle of Traverse City.

and 26 in (660 mm) for the female; tail short, no more than 1/5 as long as total length; length of tail vertebrae averaging 9 in (230 mm) for the male and 8¾ in (223 mm) for the female; head broad, heavy, somewhat rounded; nose and muzzle broad, blunt; ears short, rather rounded; eyes small, forward, beady; neck thick, muscular; body massive, arched; tail short, exceedingly bushy; limbs short, stout; feet large, almost flat (plantigrade); each of five toes on the front and back feet with strong, curved, semiretractile claws; four pairs of mammary glands, two abdominal and two inguinal; scent glands developed as in other mustelids with exudate having a strong, musky odor; males about 10% larger than females.

The pelage of the wolverine consists of dense, soft underfur covered by long, coarse guard hairs. General body coloring may vary from dark yellow-brown to almost black, darkest on the midback, feet and tail, and palest on cheeks and forehead, setting off the black muzzle. Two prominent stripes extend from the upper shoulders laterally and posteriorly to meet on the rump and basal part of the tail. These stripes vary in color from whitish buff to light brown. The dark midback area is thus set off as a conspicuous saddle-like patch. The underparts are mostly dark brown often with splotches of whitish buff on the throat and chest. The wolverine is distinguished from other mustelids by its large size, bear cub-like appearance, long, coarse pelage, and conspicuous back stripes.

MEASUREMENTS AND WEIGHTS. Adult specimens of the larger male and the smaller female measure, respectively, as follows: length from tip of nose to end of tail vertebrae, 37¾ to 42 in (950 to 1,070 mm) and 30½ to 37½ in (775 to 940 mm); length of tail vertebrae, 8 to 10¾ in (200 to 260 mm) and 7¼ to 9 in (160 to 230 mm); length of hind foot, 7 to 8 in (175 to 200 mm) and 6¼ to 7 in (160 to 175 mm); height of ear from notch (almost the same in both sexes), 1½ to 2¼ in (45 to 55 mm); and weight, 25 to 36 lbs (11.3 to 16.3 kg) and 22 to 33 lbs (10 to 15 kg). Large adult males may weigh as much as 42 lbs (19 kg).

DISTINCTIVE CRANIAL AND DENTAL CHARACTERISTICS. The skull of the wolverine is large, wide, heavily constructed and, in upper profile, strongly arched in the postfrontal area and depressed in the prefrontal space (Fig. 90). The zygomatic arches are heavily built and laterally spreading. The cranium has a conspicuous sagittal crest, which in mature males may greatly extend posteriorly to the occiput (as much as ½ in or 12 mm). The auditory bullae are small, relative to the size of the skull, but have elongated auditory canals. The lower jaw is short, heavy, and usually so firmly hinged to the cranium that it can be separated only with difficulty. The skull of the mature male is fully 10% larger than that of the female and is also more angular with greater developed sagittal crest and lambdoidal ridges. The teeth are notably massive and, being somewhat hyena-like, ideally suited for crushing large bones. The third upper incisors are proportionately larger than the rather insignificant first and second ones.

The distinctively large wolverine skull may be recognized by the following features: large size, more than 5 in (130 mm) long with heavy lower jaw firmly hinged to cranium; bony palate extending well posterior to the last molars in the tooth-row; and the presence of four premolars on each side of both the upper and lower jaws (as in *Martes*, but not *Mustela, Taxidea, Mephitis,* and *Lutra*). Field observers in Michigan should keep these skull characteristics in mind in the hopes of finding wolverine remains, which could be among the first preserved in the state. Likely locations are post-glacial soil deposits, gravel pit excavations, archaeological sites, or residues from strip mining for peat or marl.

The skulls of adult male and female wolverines, respectively, measure 6½ to 6⅞ in (166 to 175 mm)

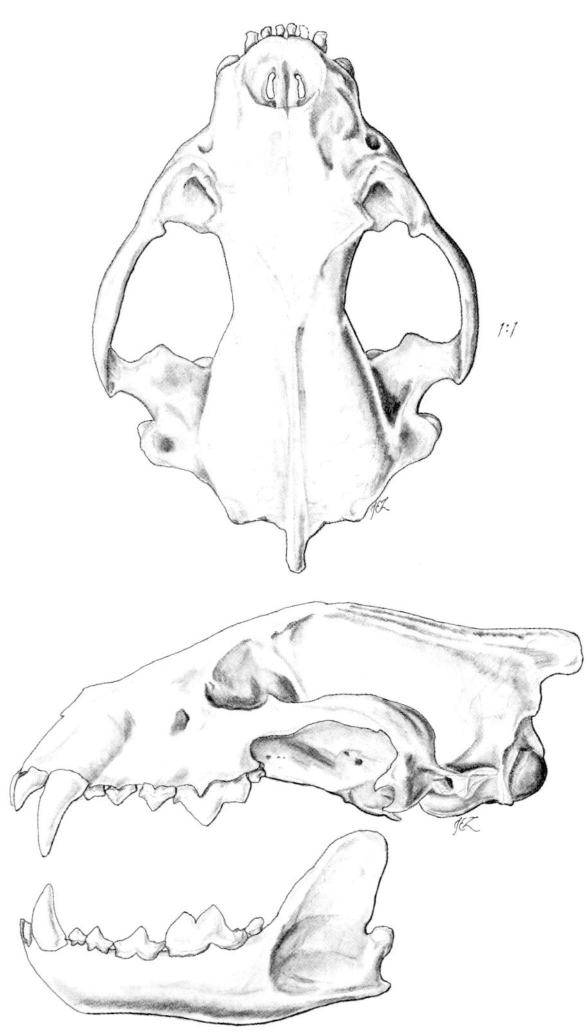

Figure 90. Cranial and dental characteristics of the wolverine (*Gulo gulo*).

and 5¼ to 5⅜ in (134 to 150 mm) in greatest length and 3⅞ to 4⅛ in (99 to 104 mm) and 3½ to 3¾ in (87 to 96 mm) in greatest width across the zygomatic arches. The dental formula is:

$$\text{I (incisors)}\ \frac{3\text{-}3}{3\text{-}3},\ \text{C (canines)}\ \frac{1\text{-}1}{1\text{-}1},\ \text{P (premolars)}\ \frac{4\text{-}4}{4\text{-}4},\ \text{M (molars)}\ \frac{1\text{-}1}{2\text{-}2} = 38$$

DISTRIBUTION. The wolverine has a circumpolar distribution; that is, it occurs on land surfaces in northern latitudes all around the north polar region in North America, Eurasia, and many adjacent islands in the Arctic (Halfpenny *et al.*, 1979). In post-glacial times, the wolverine occurred in most boreal environments of the states bordering the Canadian provinces and into the western mountains at least as far south as latitude 37° in California and Colorado (Nowak, 1973). These southernmost populations, especially in the eastern border states, disappeared early in the colonial period. Remnant populations in the mountainous west held on and, in recent years, have been making some comebacks—generally with legal protection—in California, Colorado, Idaho, Montana, Washington, and perhaps Wyoming (Field and Feltner, 1974; Newby and Wright, 1955; Nowak, 1973). In the Upper Great Lakes Region, wolverines were not only extirpated from the border states, probably prior to 1900 (perhaps later in Minnesota, Birney, 1974; Johnson, 1923), but also from all of southern Ontario. This has eliminated the possibility of any natural restocking of the Great Lakes states from the north, as has apparently happened with the lynx (Peterson, 1966; Schorger, 1942; van Zyll de Jong, 1975). The likelihood that wolverine populations in eastern North America were less dense than those in more western range has contributed to their rather rapid decline in this region (van Zyll de Jong, 1975).

The wolverine once inhabited northern forest habitats of the Upper Peninsula and the northern part of the Lower Peninsula (Map 58). Strangely, however, no actual specimens have been preserved in museum collections as positive records(Dice, 1927), nor have bone fragments been found among materials excavated from Michigan archaeological sites, either prehistoric or historic (Charles Cleland, pers. comm.). Nevertheless, rather substantial evidence has been accumulated to demonstrate that the wolverine was indeed a member of our local fauna in the nineteenth century. In 1860, for example, State Zoologist Manly Miles (1861) noted that the animal ". . . is seldom found in the Lower Peninsula, having been nearly exterminated."

The wolverine apparently occurred in at least 16 counties in the Upper and Lower Peninsulas until the 1880s (Wood and Dice, 1924). Documentation consists mostly of newspaper accounts and hearsay. In Marquette County, for example, a wolverine was reported as trapped "at Marquette, near Lake Superior" on February 15, 1860 (Schorger, 1940), and another near Champion in the 1880s (Cory, 1912). An account by trapper E. N. Woodcock (*in* Harding, 1941; also see Schorger, 1948) put the wolverine in Schoolcraft County about October, 1879. He noted, "We were bothered some by having a wolverine follow a line of deadfalls, tear down the bait pen and take the bait, but we did not allow him to do his cussedness long before we would put a trap in his way." In Menominee County, George Primo reported seeing a wolverine cross the Menominee River from

Map 58. Geographic distribution of the wolverine (*Gulo gulo*) within historic times. Dark shading indicates those counties from where actual records of this extirpated species have been obtained.

Wisconsin on October 8, 1871, just prior to the destructive Peshtigo fire in neighboring Wisconsin (Ford Kellum, *in litt.*, August 18, 1980). One hearsay reference from the Lower Peninsula placed the wolverine in Montmorency County about 1880 (Dice, 1925). A mounted specimen (see Fig. 89) of a wolverine in the collection of Gary F. Kaberle of Traverse City, was obtained in the late 1860s from the Gaylord area by naturalist Charles J. Davis (1845–1924) of Lansing. The last report in the early years of the twentieth century was obtained by Manville (1948) who was told by an old settler that wolverine lived in the Huron Mountains of Marquette County until after 1903.

In the early days of the fur trade, trappers or their agents sold wolverine pelts to buyers at trading posts at Sault Ste. Marie, Mackinac Island, and Detroit (Manville, 1950). It is, of course, entirely possible that most or even all of these hides were trapped in areas north and/or west of Michigan. Because Michigan is near the southern edge of the wolverine's natural distribution, there is evidence the animal also lived in Indiana (Mumford, 1969); its population there must have been sparse at best (de Vos, 1964).

HABITAT PREFERENCES. Field observations of the wolverine in various parts of its range show the animal is at home in a variety of northern environments. It lives on the treeless tundra and along the coast of the Arctic Ocean, and among the rocky outcrops and steep canyon sides of mountainous areas, even to above timber line. It also finds food and refuge in boreal forests. The latter habitat probably appealed to the wolverine of early Michigan and adjacent parts of the Upper Great Lakes Region.

DENSITY AND MOVEMENTS. Studies of wolverine populations indicate these animals normally exist at rather low densities (Krott, 1959; 1960). It is van Zyll de Jong's (1975) contention that this species occurs at lower densities than the gray wolf in northern Canada. He cites a Scandinavian report estimating wolverine density as varying from one wolverine per 77 sq mi (200 sq km) to one per 193 sq mi (500 sq km). In northeastern British Columbia, Quick (1953) calculated the wolverine population, based on track counts and trapping records, at about one individual per 81 sq mi (209 sq km); in Montana Hornocker and Hash (1981) found one individual per 25 sq mi (65 sq km). If these rather low-density estimates can be assumed to be indicative of wolverine numbers in the somewhat limited habitat available in presettlement Michigan, the animal obviously could have been extirpated fairly rapidly by local trapping activity.

Contrary to reports in the early literature, studies in Sweden by Peter Krott (1949, 1960) demonstrate that the wolverine is certainly not nomadic. It has a definite, defended territory and specific strips of terrain in which most movements are made. Hornocker and Hash (1981) and Quick (1953) estimated that wolverines move in territories as extensive as 77 sq mi (200 sq km) to 163 sq mi (422 sq km), with Banfield (1974) reporting that individual animals have been trailed through snow for 60 to 80 mi (96 to 130 km). As is typical of other mustelids, the males move in larger areas than females. One study has shown that males traveled a circuit approximately 60 mi (96 km) long, while females traveled over an area only 25 mi (40 km) long.

BEHAVIOR. The wolverine leads a solitary existence except during courtship and mating. The basic social unit consists of the female parent and her offspring. Like other mustelids, wolverines defend their territories against others of the same sex (Hornocker and Hash, 1981; Krott, 1959; 1960). One or more of the smaller home ranges of females either fit or overlap compatibly within the single larger cruising area of a male. These animals maintain their territories by placing fecal deposits and marking scent posts with urine and exudate of the anal glands at strategic localities on tree trunks and on the ground (Koehler *et al.*, 1980).

Studies of Swedish wolverines by Krott (1959) show that, unlike the weasel, the wolverine is not a highly efficient predator. It lacks the speed, agility, and perhaps the keen eyesight of other mustelids, although its hearing and sense of smell are acute. The wolverine does not stalk or ambush victims, as do weasels and cats, and its slow but persistent pursuit proves ineffectual in summer. In winter, however, the splay-toed wolverine negotiates snow cover more rapidly than most of its would-be prey. Its gait is a clumsy, loping gallop or a slow, deliberate walk, but with either mode of locomotion, it has remarkable endurance. On occasion, a wolverine sits on its haunches rabbit-style or even rears up on its hind legs like a bear. The wolverine is not afraid to take to water and swims well. It cannot jump from tree to tree in marten-like fashion, but climbs readily, even descending head down after caching food in limb crotches (Grinnell, 1921). The wolverine has no fixed time of major activity and is apt to be seen at night or in the daytime. Even severe winter blizzards apparently do not discourage the animal (Banfield, 1974).

The wolverine is, of course, known in fact and fiction as a scoundrel (Seton, 1953). Its strength and viciousness when cornered and its destructive nature in damaging human habitations and raiding traps for bait or trapped animals has given the species a most undesirable reputation. Perhaps the wolverine's most irritating feature is its ability to avoid traps. On the rare occasions when it does get caught, according to Krott (1959), it often frees itself by biting off the trapped limb. The animal also survives normal doses of such poisons as strychnine. Modern field observers have pointed out that the wolverine does not deserve such a reputation. Krott (1959, 1960) describes its ability to "play" and its capacity to alter its behavior as the local environmental situation requires.

Because of its size the wolverine rarely uses dens excavated underground (Seton, 1953). Instead, grass or leaf-lined nests are fashioned under felled logs or roots, in rocky crevices, or under low, leafy cover. Winter dens may be merely holes tunneled in snowdrifts.

Not only is the wolverine noisy as it shuffles over leaf litter, it also utters various calls, perhaps more guttural and deeper but not too different from those produced by weasels. The wolverine hisses when cornered; it also growls and barks. The female parent and her litter exchange purrs and whimpers.

ASSOCIATES. The wolverine has little to do with other animals except as prey or enemies. As a scavenger, however, the wolverine may follow gray wolves, bears, lynx, and even humans (Quick, 1953) in the hopes of feasting on abandoned or cached carcasses of deer, moose, caribou, or wapiti. Its powerful jaw teeth are capable of crushing large bones to obtain nutrients not always available to the canids. Studies in Norway suggest wolverine and lynx coexist because of differences in hunting techniques and usage of prey (Myhre and Myrberget, 1975).

REPRODUCTIVE ACTIVITIES. The courtship and mating season appears prolonged—from April until early September but probably primarily from May to July (Rausch and Pearson, 1972). In typical mustelid fashion, the fertilized egg develops into a blastocyst and remains viable but undeveloped for several months until implantation (Wright and Rausch, 1955). Embryonic development then proceeds normally with litters appearing from February to April. The gestation period, as a result, can vary from 215 to 273 days (Mehrer, 1976). Litters contain two or three (sometimes as many as five) kits.

At birth, the young are almost naked, have closed eyes, and weigh 3½ oz (100 g). The first pelage is pale buff, with a slightly darker face mask and feet. At about 28 days of age, the eyes open. At 70 days of age, the process of weaning is underway. The first solid foods are generally pieces of carrion which the female parent brings to the young from a cache. By autumn, the young follow the mother and forage on their own. The family group will remain together for as long as two years, after which the female will accept a mate, as will her two-year-old female offspring. When the family breaks up, the female young are obliged to depart from their mother's home area. At this time, the young males may be harassed for the first time by the resident male and must disperse in search of unoccupied territories (Krott, 1960). The wolverine probably lives several years in nature; in captivity, life expectancy may be as long as 14 years (Woods, 1944).

FOOD HABITS. The wolverine has perhaps the most varied diet of the mustelids. As Seton (1953) noted, the animal will ingest just about any plant or animal material from which nutrients can be derived. Powerful jaws and heavily constructed teeth allow the wolverine to crack even the largest bones with ease. It was Krott's (1960) contention that this species can rarely catch large, living prey in summer, owing to its slow, noisy gait. Instead, in the warm season it relies mostly on carrion, small mammals, bird eggs, newly hatched young of ground-nesting birds, fish, larvae from nests of wasps and bees, roots, and berries. In winter, the wolverine is able to move through snow cover much faster than many of the prey species which elude it in summer. Larger animals including beaver, porcupine, white-tailed deer, caribou, and even moose have been killed and eaten by the wolverine (Banfield, 1974; Jackson, 1961; Krott, 1959, 1960; Myhre and Myrberget, 1975; Seton, 1953; van Zyll de Jong, 1975).

ENEMIES. The human is the major enemy of the wolverine and the principal factor in eliminating the species from much of its most southern range in Eurasia and North America. Natural enemies are few; perhaps only large bears and the gray wolf have the physical ability to subdue the powerful wolverine (Boles, 1977; Burkholder, 1962).

PARASITES AND DISEASES. Internal parasites known to infect the wolverine include trematodes (flukes), *Opisthorchis felineus* and *Alaria mustelae*; cestodes (tapeworms), *Taenia twitchelli, Taenia martis, Diphyllobothrium* sp., *Bothriocephalus* sp.; and nematodes (roundworms), *Dioctophyme renale, Soboliphyme baturini, Physaloptera torquata, Trichinella spiralis, Bayliscaris devosi* (Erickson, 1946; Haugen, 1961; Addison and Boles, 1978). Haugen (1961) also reported the flea, *Dermacentor variabilis,* from the wolverine.

MOLT AND COLOR ABERRATIONS. There are probably two molts with pelage regrowth each year. Individuals are known with dark (melanistic) and white (albinistic) coats (Jones, 1923).

ECONOMIC IMPORTANCE. The wolverine pelt has been marketed since the early days of the fur trade (Seton, 1953), although this rather coarse fur has never been considered important in fashion. It is used as trim for arctic and ski clothing, because wolverine hair resists the accumulation of frost or, if frost does form, it can be brushed off easily, which is not true of the fur of gray wolf or coyote (Quick, 1952).

Age-old stories of the aggressiveness, strength, cunning, and mischievious ways of this species are legion. This array of "common knowledge" is steeped in folklore, even mythology (Vontobel, 1979). The modern ecologist, intent on determining the animal's true nature, hopes to ultimately disprove some of the exaggerated descriptions of the creature's behavior. Studies by arctic workers such as Peter Krott (1959) have publicized (in both scientific and popular literature) the basic behavioral pattern of this northern animal. Although the species is in no real danger of extirpation because of its occurrence in remote arctic areas, its more southern populations have been seriously exploited and diminished.

Meanwhile, to the majority of people, the wolverine remains Michigan's state animal. The History Division of Michigan's Department of State can find no actual documentation it was ever officially given that title although the custom goes back at least to 1837. There are skeptics (Manville, 1950) as to the actual past occurrence of this animal in Michigan, but there is an array of documentation, mostly of a literary sort. The discovery of wolverine bones or teeth at an archaeological site or a building foundation or in a deposit of marl or river gravel would substantiate these reports.

COUNTY RECORDS FROM MICHIGAN. Records of wolverines from early trapper's reports and from the literature show this animal lived throughout Michigan. As shown in Map 58, there are county records of wolverines in Chippewa, Gogebic, Keweenaw, Marquette, Menominee, Ontonagon, and Schoolcraft counties in the Upper Peninsula and in Alcona, Huron, Iosco, Kent, Leelanau, Montmorency, Oscoda, Otsego, Sanilac, and Tuscola counties in the Lower Peninsula (Burt, 1946; Manville, 1950; Wood, 1914).

LITERATURE CITED

Addison, E. M., and B. Boles
1978 Helminth parasites of wolverine, *Gulo gulo,* from the District of Mackenzie, Northwest Territories. Canadian Jour. Zool., 56(1):2241–2242.

Banfield, A. W. F.
1974 The mammals of Canada. Univ. Toronto Press, Toronto. xxiv+438 pp.

Birney, E. C.
1974 Twentieth century records of wolverine in Minnesota. Loon, 46:78–81.

Boles, B. K.
1977 Predation by wolves on wolverines. Canadian Field-Nat., 91(1):68–69.

Burkholder, B. L.
1962 Observations concerning wolverines. Jour. Mammalogy, 43(2):263–264.

Burt, W. H.
1946 The mammals of Michigan. Univ. Michigan Press, Ann Arbor. xv+288 pp.

Cory, C. B.
1912 The mammals of Illinois and Wisconsin. Field Mus. Nat. Hist., Publ. 153, Zool. Ser. 11, 505 pp.

de Vos, A.
1964 Range changes of mammals in the Great Lakes region. American Midl. Nat., 71(1):210–231.

Dice, L. R.
1925 A survey of the mammals of Charlevoix County, Michigan, and vicinity. Univ. Michigan, Mus. Zool., Occas. Papers No. 159, 33 pp.
1927 A manual of the recent wild mammals of Michigan. Univ. Michigan, Univ. Mus., Michigan Handbook Ser. No. 2, 63 pp.

Erickson, A. B.
1946 Incidence of worm parasites in Minnesota Mustelidae and host lists and keys to North American species. American Midl. Nat., 36(2):494–509.

Field, R. J., and G. Feltner
1974 Wolverine. A bit of history about the carcajou, an endangered species in Colorado. Colorado Outdoors, 23(2):1–6.

Grinnell, G. B.
1921 The tree-climbing wolverine. Jour. Mammalogy, 2(1):36–37.

Halfpenny, J. C., D. Nead, S. J. Bissell, and R. J. Aulerich
1979 A bibliography of mustelids. Part IV: Wolverine. Michigan State Univ., Agric. Exp. Sta., Jour. Art. No. 9214, 121 pp.

Harding, A. R., ed.
1941 Fifty years a hunter and trapper. Experiences of E. N. Woodcock the noted hunter and trapper, as written by himself and published in H-T-T from 1903 to 1913. A. R. Harding, Publisher. Columbus, Ohio. 318 pp.

Haugen, A. O.
1961 Wolverine in Iowa. Jour. Mammalogy, 42(4):546–547.

Hornocker, M. G., and H. S. Hash
1981 Ecology of the wolverine in northwestern Montana. Canadian Jour. Zool., 59(7):1286–1301.

Jackson, H. H. T.
1961 Mammals of Wisconsin. Univ. Wisconsin Press, Madison. xii+504 pp.

Johnson, C. E.
1923 A recent report of the wolverine in Minnesota. Jour. Mammalogy, 4(1):54–55.

Jones, S. V. H.
1923 Color variations in wild animals. Jour. Mammalogy, 4(3):172–177.

Koehler, G. M., M. G. Hornocker, and H. S. Hash
1980 Wolverine marking behavior. Canadian Field-Nat., 94(3):339–341.

Krott, P.
1959 Demon of the north. Alfred A. Knopf, New York. xiii+260 pp.
1960 Ways of the wolverine. Nat. Hist., 69(2):16–29.

Manville, R. H.
1948 The vertebrate fauna of the Huron Mountains, Michigan. American Midl. Nat., 39(3):615–640.
1950 The wolverine in Michigan. Jack-Pine Warbler, 28(4): 127–129.

Mehrer, C. F.
1976 Gestation period in the wolverine, *Gulo gulo*. Jour. Mammalogy, 57(3):570.

Miles, M.
1861 A catalogue of the mammals, birds, reptiles, and mollusks of Michigan. Geological Survey of Michigan, First Biennial Rept., pp. 219–241.

Mumford, R. E.
1969 Distribution of the mammals of Indiana. Indiana Acad. Sci., Monog. No. 1, vii+114 pp.

Myhre, R., and S. Myrberget
1975 Diet of wolverines (*Gulo gulo*) in Norway. Jour. Mammalogy, 56(4):752–757.

Newby, F. E., and P. L. Wright
1955 Distribution and status of the wolverine in Montana. Jour. Mammalogy, 36(2):248–253.

Nowak, R. M.
1973 Return of the wolverine. National Parks and Conserv. Mag., 47(2):20–23.

Peterson, R. L.
1966 The mammals of eastern Canada. Oxford Univ. Press, Toronto. xxxii+465 pp.

Quick, H. F.
1952 Some characteristics of wolverine fur. Jour. Mammalogy, 33(4):492–493.
1953 Wolverine, fisher, and marten studies in a wilderness region. Trans. 18th North American Wildlife Conf., pp. 513–532.

Rausch, R. A., and A. M. Pearson
1972 Notes on the wolverine in Alaska and the Yukon Territory. Jour. Wildlife Mgt., 36(2):249–268.

Schorger, A. W.
1940 Wolverine in Michigan. Jour. Mammalogy, 20(4):503.
1942 Extinct and endangered mammals and birds of the Upper Great Lakes Region. Trans. Wisconsin Acad. Sci., Arts and Ltrs., 34:23–44.
1948 Further records of the wolverine for Wisconsin and Michigan. Jour. Mammalogy, 29(3):295.

Seton, E. T.
1953 Lives of game animals. Vol. II, Pt. II. Bears, coons, badgers, skunks, and weasels. Charles T. Branford Co., Boston, pp. 369–646.

van Zyll de Jong, C. G.
1975 The distribution and abundance of the wolverine (*Gulo gulo*) in Canada. Canadian Field-Nat., 89(4): 431–437.

Vontobel, R.
1979 Reluctant villain. One of Canada's rarest mammals fights a poor public image. Nature Canada, 8(1):38–43.

Wood, N. A.
1914 An annotated check-list of Michigan mammals. Univ. Michigan, Mus. Zool., Occas. Papers No. 4, 13 pp.

Wood, N. A., and L. R. Dice
1924 Records of the distribution of Michigan mammals. Michigan Acad. Sci., Arts and Ltrs., 3:425–469.

Woods, G. T.
1944 Longevity of captive wolverines. American Midl. Nat., 31(2):505.

Wright, P. L., and R. Rausch
1955 Reproduction in the wolverine, *Gulo gulo*. Jour. Mammalogy, 36(3):346–355.

BADGER
Taxidea taxus (Schreber)

NAMES. The generic name *Taxidea,* proposed by Waterhouse in 1839, is derived from the Greek and means badger-like because of its general resemblance to the common badger (*Meles*) of Eurasia. The specific name *taxus,* proposed by Schreber in 1778, is from Middle Latin and means badger. In Michigan, this mammal is known universally as the badger. The subspecies *Taxidea taxus jacksoni* Schantz occurs in Michigan (Long, 1972).

RECOGNITION. Body size resembles that of squat, medium-sized dog with a low, robust, flattened, short-legged body somewhat triangular in shape; length of head and body averaging 24½ in (620 mm); tail short, only about one-sixth of the total length; length of tail vertebrae averaging 5¼ in (135 mm); head tapering anteriorly, slightly flattened, broad between the eyes; eyes small with large nictitating membranes; ears rounded, low, upright, with large ear opening; upper canines protruding beyond lips; neck thick, muscular; tail bushy; legs short, stout; stance subdigitigrade, not completely flat-footed; forelimbs especially robust; forefeet large; foreclaws elongate, as much as 1½ in (38 mm) long, recurved with bony flanges extending dorsally at the base of each claw; four pairs of mammary glands, one pectoral, two abdominal, and one inguinal; anal scent glands well developed as in other mustelids; males only about 5% larger than females (Long, 1972, 1973).

The badger's pelage consists of long, coarse, shaggy fur, especially on the dorsum where the guard hairs may be as long as 3¼ in (85 mm). The grizzled effect of the yellowish gray upper coloring is produced in part by the dorsal hairs being basally buff, subterminally black, and tipped with white. The short hair of the head has a distinct pattern (Plate 3), the muzzle, crown and nape being dark brown or black and divided at the center line by a white, broad line extending from the muzzle to the shoulders. The cheeks and ears are white with the latter trmmed with black. Posterior to each eye is a black, half-moonshaped design. The tail has stiff light brown hairs. The underparts are generally buffy with some midline whitish blotches. The legs are brownish black, and feet blackish.

The badger is distinguished from other Michigan mammals by its flattened and almost triangular body and its long, crisp, yellowish gray pelage. At a distance it might be mistaken for the usually darker woodchuck.

MEASUREMENTS AND WEIGHTS. Adult badgers measure as follows: length from tip of nose to end of tail vertebrae, 28½ to 31½ in (720 to 800 mm); length of tail vertebrae, 4 to 6 in (120 to 150 mm); length of hind foot, 3½ to 5½ in (90 to 140 mm); height of ear from notch, 1¾ to 2¼ in (45 to 60 mm); weight, 14 to 25 lbs (6.4 to 11.3 kg).

DISTINCTIVE CRANIAL AND DENTAL CHARACTERISTICS. The skull of the badger is massive, somewhat triangular in shape, and more flattened than skulls of other Michigan mustelids (Fig. 91). The rostrum (muzzle) is short and tapered. The zygomatic arches are heavily constructed but no wider than the posterior part of the brain case, which is also characteristically truncate. The skull has enlarged infraorbital foramina, large auditory bullae, and a bony palate which extends well posterior to the end of the tooth row. The lower jaw is also massive; in older adults the articular (condyloid) condyles often are locked and not removable from within the glenoid fossae of the cranium. The last upper molar tooth is triangular in shape, becoming narrow posteriorly.

Characteristics distinguishing the badger skull from that of another Michigan carnivore include

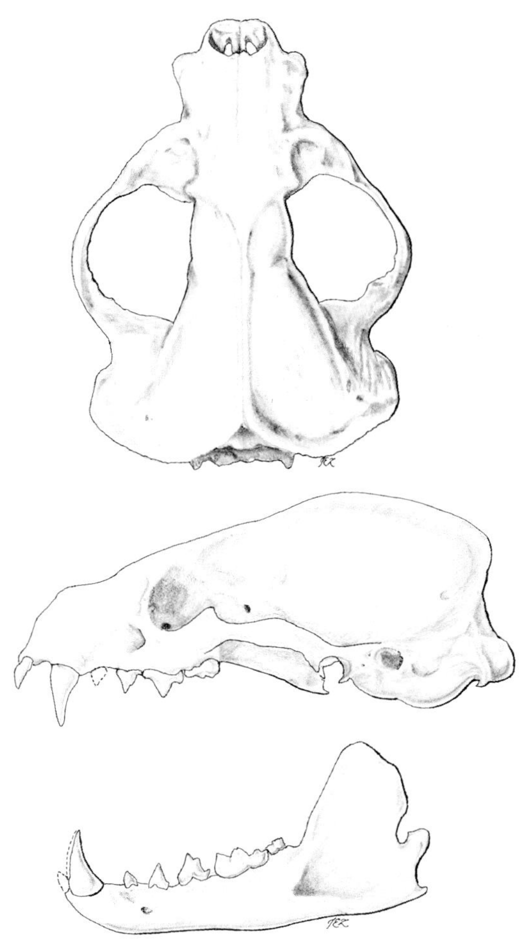

Figure 91. Cranial and dental characteristics of the badger (*Taxidea taxus*).

four upper teeth and five lower teeth behind the canines on each side of the jaws, the triangular-shaped last upper molars, and the extension of the bony palate posteriorly to the upper tooth row. The skulls of adult badgers measure 4¼ to 5 in (110 to 130 mm) in greatest length and 3 to 3½ in (77 to 90 mm) in greatest width across the zygomatic arches. Skulls of adult females are about 5% smaller than those of adult males. The dental formula is:

$$\text{I (incisors)}\,\frac{3\text{-}3}{3\text{-}3},\ \text{C (canines)}\,\frac{1\text{-}1}{1\text{-}1},\ \text{P (premolars)}\,\frac{3\text{-}3}{3\text{-}3},\ \text{M (molars)}\,\frac{1\text{-}1}{2\text{-}2} = 34$$

DISTRIBUTION. There are records of perhaps two separate occurrences of badger in the eastern United States in the recent geologic past (Long, 1972). However, by the time of European settlement, it is most likely that the animal's major distribution was in the open prairie lands of the American West, where it occurred from the Mexican Plateau in the south to the Prairie Provinces of Canada in the north (see Map 59). Its presettlement range also closely approached those of several species of ground squirrels, including the thirteen-lined ground squirrel on the northeastern prairies. The clearing of forest lands eastward from the western prairies and the introduction of exotic grasses and forbs invited the entry of many open-land species. Birds, such as the horned lark, probably moved eastward soon after the beginning of the nineteenth century (Hurley and Franks, 1976). Apparently, the badger also moved eastward shortly thereafter into Ohio and in recent years into New York and even Connecticut (Leedy, 1947; Moseley, 1934; Nugent and Choate, 1970).

While the thirteen-lined ground squirrel and the badger may have been long-time co-residents in the prairie openings in extreme southwestern

Map 59. Geographic distribution of the badger (*Taxidea taxus*). Actual records are from those counties darkly shaded.

Michigan, it seems evident that as soon as human settlement began in earnest after 1840, the species spread in suitable cleared lands throughout the Lower Peninsula (Wood and Dice, 1924). Presumably, both species also entered the Upper Peninsula by way of Wisconsin at least by the early years of the twentieth century. While the thirteen-lined ground squirrel has so far occupied only the western counties, the badger has spread in suitable areas throughout the Upper Peninsula. Records were obtained from Whitefish Point in Chippewa County as early as 1912 (Wood, 1914).

Since the close of the legal badger trapping season in Michigan in 1973, records of these animals have been obtained less frequently, except through highway kills and individuals mistakingly caught by Michigan fox/coyote trappers. So far, the waterway separating Michigan's Upper Peninsula from Ontario has been an effective barrier to the dispersal of the badger in an eastward direction. However, while the watergap between Michigan's Lower Peninsula and Ontario has effectively blocked the eastward spread of such species as the fox squirrel and the least weasel, the badger seems to have had no difficulty crossing the St. Claire and Detroit rivers and now thrives in southwestern Ontario (Peterson, 1966).

HABITAT PREFERENCES. The badger's natural home area is open land (grass and shrub) where visibility of the surroundings may be a necessity and, perhaps more important, the clay or sandy soils are suitable for excavating burrows. In Michigan, old fields, cleared pastures, sparse brushland reverting slowly to timber, and even open woodlots may be homes for the badger. Heavily wooded areas, found principally in northern Michigan, are the least likely habitats for badgers.

DENSITY AND MOVEMENTS. To the field observer, the badger is not the most conspicuous of the larger Michigan mammals, nor are its burrows, with their characteristic deep-drop entrances. Even so, biologists consider these burrowing animals as uncommon and scattered in distribution. In the best Upper Great Lakes badger range, the animals may be expected to be as dense as one individual per 2 to 3 sq mi (5 to 7.8 sq km), according to Jackson (1961) for Wisconsin and Craighead and Craighead (1956) for Washtenaw County.

The spatial arrangement of the badger seems highly dependent on intraspecific interactions, also typical of other mustelids. As a rule, members of one sex will exclude others of that sex from their areas of activity. Further, studies in Idaho of movements of badgers equipped with radio collars (Lindzey, 1978; Messick and Hornocker, 1981) and in Minnesota (Sargeant and Warner, 1972) show that both sexes have seasonal variation in movement patterns and territory use—smaller in autumn and winter (Lampe, 1980; Lampe and Sovada, 1981). In Idaho, Lindzey calculated the average home range of males to be about 1,440 acres (583 ha) and of females, 585 acres (237 ha). In winter, one female used a movement area of only 237 acres (96 ha). The larger ranges of the males overlap those of as many as three to four females. In Minnesota, a radio-tagged female badger moved in a 17 mi (27 km) circuit while using an area of 1880 acres (761 ha) from July to September, an area of 130 acres (52 ha) from September to November, and a greatly reduced area of only five acres (2 ha) in December and January.

BEHAVIOR. Badgers, especially males, are generally solitary during most of the year except for the rather brief courtship and mating period in late summer. While males tolerate (but seem to avoid) females, they definitely discourage other males from using their home sites and areas of activity. Scent marking and agonistic behavior probably are involved (Messick and Hornocker, 1981). On the other hand, females may be more compatible with others of their sex; on one occasion, two radio-tagged females were found in the same den (Lindzey, 1978).

Armed with powerful, digging foreclaws and a squat, streamlined body, the badger has remarkable ability as a fossorial animal. Although it probably digs out thirteen-lined ground squirrels and other prey from underground burrows, it also pursues victims above ground. The badger usually waddles slowly with its body low to the terrain and legs spread sideways. It can, however, run remarkably well. Speeds up to 6 mi (10 km) per hour are reported by Jackson (1961) when the badger raises itself high on its short legs to gallop. In burrowing, it uses both its strong teeth and long front claws to loosen the soil, passing it back with the forelegs so that the hind feet can kick it well to the rear. There is no evidence that badgers can climb trees, although they are efficient swimmers (Wood, 1921). The sense of vision may be less acute in the badger than in other mustelids. Hearing, smell, and touch, however, seem developed to a high order.

The badger is apt to be active either in the daytime or at night. It is probably most active at night (Lindzey, 1978), although there is some evidence that badgers may be more diurnal during summer. In colder months, the amount of time spent above ground at night declines; Harlow (1979; 1981) found badgers to be active outside of their dens an average of five hours per night in November and less than 0.4 hours per night in February. He found that animals spend as long as 70 days underground in cold weather, and that the badger's winter survival seems to depend on energy stores of fat correlated with a slight depression in body temperature. This allows for a winter dormancy called carnivorean lethargy, much less profound than, and certainly not to be confused with, true hibernation.

Badgers utter growls and snarls when on the defensive. In typical mustelid fashion, the badger can also hiss in an effective, aggressive way. These snort-like sounds are the result of heavy breathing through upturned nostrils (Jackson, 1961). In a fight, squealing may be a characteristic utterance. Soft purring is used by the female parent to communicate with offspring.

The den is the focal point for the active life of the badger. Dens vary in length; often have only one, usually large, elliptical entrance surrounded by a mound of fresh soil; and measure 12 to 15 in (305 to 385 mm) in width at the opening. There also may be a down-drop of at least 2 ft (600 mm) before the tunnel curves in a lateral direction. Burrows may extend for 30 ft (9 m) and descend as deep as 10 ft (3 m). An enlarged grass-lined area at the end of a tunnel may serve as a nursery. Excavating dens appears to be almost a daily chore for the badger (Lindzey, 1978; Snead and Hendrickson, 1942). Studies of movements by radio-tagged animals have shown that there will be numerous dens within an individual animal's cruising range. Lindzey (1978) found that an animal rarely uses the same den two days in a row, either digging a new one (sometimes in search of prey) or reusing an older one. These over-day retreats usually lack nesting materials. Dens are frequently in open areas but also are dug in woodlots (Sargeant and Warner, 1972). According to Lindzey (1978), badgers of both sexes may use the same dens, not necessarily at the same time, when the den is within their overlapping home ranges. When caring for her offspring, the female is less likely to change dens frequently. Lindzey (1978) monitored the activities of a female badger, which gave birth on March 27, stayed with her litter for seven days before going in search of food, and did not move them to another den until 25 days after birth. In Barry County, a female with a litter of two lived in a den dug under the front walk and steps of a farmhouse in late June (Joe Johnson, pers. comm.).

ASSOCIATES. The solitary nature of the badger as well as its meat-eating food habits probably keep it socially apart from other animals in its community. A mutually-advantageous interspecific relationship is the much-quoted account by Seton (1953) of the joint hunting foray by a badger and a coyote. The former excavated ground squirrel burrows and the coyote pursued and captured any squirrels escaping the badger (also see Young and Jackson, 1951). Hibbard (1963), however, found considerable rivalry between a badger and a red fox as each attempted to feed on a sheep carcass. In Michigan, the badger undoubtedly usurps burrows originally excavated by woodchucks, but, unlike the latter, apparently allows no den associates such as the striped skunk or Virginia opossum.

REPRODUCTIVE ACTIVITIES. Badger courtship and mating take place in August and September. Delayed implantation occurs (Wright, 1966; 1969), with development of the fertilized egg suspended in the blastula stage until the next January or February. Following the resumption of normal embryonic development, a litter averaging three (two to five) is born in a grass-lined underground den, perhaps as early as late March and as late as May. This means a gestation period of six to nine months. At birth, young badgers are scantily covered with fur and have closed eyes. At 28 days of age, the young open their eyes and are well-furred with characteristic facial markings conspicuous. At 56 days of age, the young are eating solid foods and being weaned. Badgers may be most conspicuous in early and mid-summer when the young can be observed romping around the den entrance. Although adults rarely foul burrows with fecal matter, the female badger appears obliged to move her growing offspring periodically because of their fecal accumulations (Lindzey, 1976). Lindzey (1978) found Idaho badger families hunting together on June 10 with the first evidence of juveniles hunting alone about July 10. In late summer, the female forces her litter to depart, allowing her to prepare for the courtship and mating season. Female young-of-the-year may also be receptive to mating at this time, but juvenile males do not attain sexual maturity until their second summer (Messick and Hornocker, 1981).

In nature, most badgers may have a life expectancy of four or five years, although individuals may live as long as 14 years (Jackson, 1961; Messick and Hornocker, 1981). In captivity, badgers have been known to live for 23 years and 8 months (Jones, 1979). The badger can successfully be aged by examination of the annular deposition in the cementum layer of the teeth (Crowe and Strickland, 1975).

FOOD HABITS. Badgers ingest little vegetable material, preferring fresh animal matter, although carrion is also eaten. As mentioned earlier, the geographic distribution of the badger coincides closely with that of one of its major foods, the various species of ground squirrels. Although the thirteen-lined ground squirrel is most often captured in its burrow, the badger may also wait at the burrow entrance (enlarged to provide crouching room) for the ground squirrel's approach (Balph, 1961). Aside from the female's need for increased nutrients when caring for offspring in late spring and early summer, the badger requires large amounts of food in autumn to accumulate heavy fat stores for winter dormancy (Harlow, 1979).

Information on badgers' food habits has been obtained primarily from the analysis of materials found in fecal deposits. In the only major study in Michigan, Dearborn (1932) identified food items in the badger's diet: voles of the genus *Microtus* (55.8%), eastern cottontail (17.3%), thirteen-lined ground squirrel (13.8%), southern bog lemming (6.1%), mice of the genus *Peromyscus* (3.8%), red squirrel (1.4%), beetles and grasshoppers (0.5%), turtle eggshells (1.1%), and hen eggshells (0.2%). In addition, the woodchuck appears to be a major food, according to Wenzel's (1912) observations in Osceola County. He reported that the badger eats all parts of the woodchuck except for its skin, which is left near the entrance to the victim's burrow. Additional foods eaten by the badger, as reported from studies made in other Midwest states, include birds and bird eggs, snakes, beetles, fish, bumblebees, crickets, butterfly and moth larvae, and some vegetable matter (Drake and Presnall, 1950; Errington, 1937; Potter, 1924; Snead and Hendrickson, 1942). Badgers are known to eat carrion (dead sheep) and raid apiaries and chicken houses (Banfield, 1974; Hibbard, 1963; Schofield, 1957).

ENEMIES. The badger is a powerful and dangerous adversary for any would-be predator. Stories of its successful defense against dogs and other large animals fill the literature (Seton, 1953). Nevertheless, such species as coyotes and golden eagles are reported to have subdued badgers (Grinnell, 1929; Rathbun *et al.*, 1980). Highway accidents account for many badger fatalities (Long, 1973); several specimens in the collections of the Michigan State University Museum were salvaged after the animals had been run down by automobiles. Human predation probably accounts for many badger deaths (Messick and Hornocker, 1981). In Michigan, it is suspected that the legally protected badger may often be captured inadvertently by fox trappers.

PARASITES AND DISEASES. External parasites living on badgers include lice, *Trichodectes interrupto-fasciatus*; ticks, *Ixodes kingi*, *Dermacentor variabilis*, and *Dermacentor andersoni;* and fleas, *Pulex irritans*, *Cediopsylla simplex*, and *Oropsylla arctomys* (Jackson, 1961; Scharf and Stewart, 1980). Internal parasites include flukes, *Alaria taxideae;* tapeworms, *Fossor angertrudae* and *Taenia taxidiensis;* and roundworms, *Ascaris columnaris*, *Filaria martis*, *Molineus patens*, *Monopetalonema eremita*, *Physaloptera maxillaris*, *Physaloptera torquata*, and *Trichinella spiralis* (Erickson, 1946; Worley, 1961). Badgers are also susceptible to such diseases as tularemia and rabies.

MOLT AND COLOR ABERRATIONS. There appears to be a single annual molt in the badger, usually occurring in late spring and early summer in the Upper Great Lakes Region (Long, 1975). Partially albino animals have been reported (Roest, 1961).

ECONOMIC IMPORTANCE. The pelt of the badger has never been highly regarded in the fur trade. The back fur has occasionally been used for trim on coats and parkas. The long guard hairs of the Eurasian badger (*Meles*) are preferred to those of the American species for the manufacture of shaving brushes. In consequence, badger pelts have rarely brought much more than $4 on the fur market in recent years. In a few of the seasons prior to the closing of the badger trapping season in Michigan in 1973, estimated annual catches were low: 130 animals in 1967–68, 100 in 1968–69, and 400 in 1969–70 and 1970–71.

Although badgers occasionally raid chicken houses and beehives (Banfield, 1974; Schofield, 1957), the species is probably disliked most by stock raisers whose livestock can seriously injure themselves by stepping into the badger's pit-like den entrances. The early-day apprehension that badgers were prone to dig into freshly-dug human graves may have been an exaggeration of a few inaccurate observations as noted by Wenzel (1912) in Osceola County. Even so, in most of Michigan, the badger's presence in the outdoor community is more a blessing than a threat. It constructs dens useful for other wildlife and often eats pestiferous rodents.

COUNTY RECORDS FROM MICHIGAN. Records of specimens of badgers in collections of museums, from the literature, and provided by personnel of the Michigan Department of Natural Resources show that this species lives in all counties of the Upper Peninsula. In the Lower Peninsula, actual records of occurrence have been obtained in all but 14 of the 68 counties (see Map 59).

LITERATURE CITED

Balph, D. F.
1961 Underground concealment as a method of predation. Jour. Mammalogy, 42(3):423–424.

Banfield, A. W. F.
1974 The mammals of Canada. Univ. Toronto Press, Toronto. xxiv+438 pp.

Craighead, J. J., and F. C. Craighead, Jr.
1956 Hawks, owls and wildlife. The Stackpole Co., Harrisburg, Pennsylvania. xix+443 pp.

Crowe, D. M., and M. D. Strickland
1975 Dental annulation in the American badger. Jour. Mammalogy, 56(1):269–272.

Dearborn, N.
1932 Foods of some predatory fur-bearing animals in Michigan. Univ. Michigan, Sch. Forestry and Conserv., Bull. No. 1, 52 pp.

Drake, G. E., and C. C. Presnall
1950 A badger preying upon carp. Jour. Mammalogy, 31(3):355–356.

Erickson, A. B.
1946 Incidence of worm parasites in Minnesota Mustelidae and host lists and keys to North American species. American Midl. Nat., 36(2):494–509.

Errington, P. L.
1937 Summer food habits of the badger in northwestern Iowa. Jour. Mammalogy, 18(2):213–216.

Grinnell, G. B.
1929 Eagle's prey. Jour. Mammalogy, 10(1):83.

Harlow, H. J.
1979 A photocell monitor to measure winter activity of confined badgers. Jour. Wildlife Mgt., 43(4):997–1001.
1981 Torpor and other physiological adaptations of the badger (*Taxidae taxus*) to cold environments. Physiol. Zool., 54(3):267–275.

Hibbard, E. A.
1963 A badger-fox episode. Jour. Mammalogy, 44(2):265.

Hurley, R. J., and E. C. Franks
1976 Changes in the breeding ranges of two grassland birds. Auk, 93(1):108–115.

Jackson, H. H. T.
1961 Mammals of Wisconsin. Univ. Wisconsin Press, Madison. xii+504 pp.

Jones, M. L.
1979 Longevity of mammals in captivity. Internat. Zoo News, 26(3):16–26.

Lampe, R. P.
1980 Home ranges of the North American badger in northwestern Iowa. Proc. Iowa Acad. Sci., 87(1):1.

Lampe, R. P., and M. A. Sovada
1981 Seasonal variation in the home range of a female badger (*Taxidae taxus*). Prairie Nat., 13(2):55–58.

Leedy, D. L.
1957 Spermophiles and badgers move eastward in Ohio. Jour. Mammalogy, 28(3):290–292.

Lindzey, F. G.
1976 Characteristics of the natal den of the badger. Northwest Sci., 50(3):178–180.
1978 Movement patterns of badgers in northwestern Utah. Jour. Wildlife Mgt., 42(2):418–422.

Long, C. A.
1972 Taxonomic revision of the North American badger, *Taxidea taxus*. Jour. Mammalogy, 53(4):725–759.
1973 Taxidea taxus. American Society of Mammalogists, Mammalian Species, No. 26, pp. 1–4.
1975 Molt in the North American badger. Jour. Mammalogy, 56(4):921–924.

Messick, J. P., and M. G. Hornocker
1981 Ecology of the badger in southwestern Idaho. Wildlife Monogr., No. 76, 53 pp.

Moseley, E. L.
1934 Increase of badgers in northwestern Ohio. Jour. Mammalogy, 15(2):156–158.

Nugent, R. F., and J. R. Choate
1970 Eastward dispersal of the badger, *Taxidea taxus,* into the northeastern United States. Jour. Mammalogy, 51(3):626–627.

Peterson, R. L.
1966 The mammals of eastern Canada. Oxford Univ. Press, Toronto. xxxii+465 pp.

Potter, L. B.
1924 Badger digs for bank swallows. Condor, 26:191.

Rathbun, A. P., M. C. Wells, and M. Bekoff
1980 Cooperative predation by coyotes on badgers. Jour. Mammalogy, 61(2):373–376.

Roest, A. I.
1961 Partially albino badger from California. Jour. Mammalogy, 42(2):272–276.

Sargeant, A. B., and D. W. Warner
1972 Movements and denning habits of a badger. Jour. Mammalogy, 53(1):207–210.

Scharf, W. C., and K. R. Stewart
1980 New records of Siphonaptera from northern Michigan. Great Lakes Ento., 13(3):165–167.

Schofield, R. D.
1957 Livestock and poultry losses caused by wild animals in Michigan. Michigan Dept. Conserv., Game Div., Rpt. No. 2188 (mimeo.), 6 pp.

Seton, E. T.
1953 Lives of game animals. Vol. II, Pt. I. Bears, coons, badgers, skunks, and weasels. Charles T. Branford Co., Boston. xvii+367 pp.

Snead, E., and G. O. Hendrickson
1942 Food habits of the badger in Iowa. Jour. Mammalogy, 23(4):380–391.

Wenzel, O. J.
1912 A collection of mammals from Osceola County, Michigan. Michigan Acad. Sci., 14th Ann. Rept., pp. 198–205.

Wood, N. A.
1921 The badger as a swimmer. Jour. Mammalogy, 2(3): 170.

Wood, N. A., and L. R. Dice
1924 Records of the distribution of Michigan mammals. Michigan Acad. Sci., Arts and Ltrs., Papers, 3:425–269.

Worley, D. E.
1961 The occurrence of *Filaria martis* Gmelin, 1790, in the striped skunk and badger in Kansas. Jour. Parasitol., 47(1):9–11.

Wright, P. L.
1966 Observations of the reproductive cycles of the American badger *(Taxidea taxus).* Comp. Biol. Reprod. Mammals, Sympos., Zool. Soc. London, 15:37–45.
1969 The reproductive cycle of the male American badger *(Taxidea taxus).* Jour. Reprod. Fert., Suppl., 6:435–445.

Young, S. P., and H. H. T. Jackson
1951 The clever coyote. The Stackpole Co., Harrisburg, Pennsylvania. xv+411 pp.

STRIPED SKUNK
Mephitis mephitis (Schreber)

NAMES. The generic name *Mephitis,* proposed by É. Geoffroy-Saint-Hilaire and G. Cuvier in 1795, and the specific name *mephitis,* proposed by Schreber in 1776, are both derived from the Latin word *mephit,* which means foul odor. In Michigan, the name skunk is universally applied to this common mustelid. Striped skunks living in the Upper Peninsula are classified as belonging to the subspecies *Mephitis mephitis hudsonica* Richardson. Those living in the Lower Peninsula belong to the subspecies *Mephitis mephitis nigra* (Peale and Palisot de Beauvois). This subspecific distribution shows that Upper Peninsula striped skunks have been derived from western populations moving there from Wisconsin. On the other hand, striped skunks in the Lower Peninsula show affinity with populations to the south and east. The presence of these two distinctive taxa indicates that the Straits of Mackinac, separating the two peninsulas, consitutes an effective barrier preventing mixing (and crossbreeding) between the populations on each side of the Michigan-Huron waterway connector (Table 12). Likewise, populations on each side of the Superior-Huron waterway connector (see Table 11) are subspecifically distinctive, with the northern Ontario striped skunks classified *Mephitis mephitis mephitis* (Schreber).

RECOGNITION. Body size resembles that of a fat, short-legged house cat with a stout body and small head (Fig. 92); length of head and body averaging 15¾ in (400 mm) for the larger male and 15¼ in (385 mm) for the smaller female; tail

Figure 92. The striped skunk (*Mephitis mephitis*). Photograph by Robert Harrington and courtesy of the Michigan Department of Natural Resources.

long, about 35% of the animal's total length; length of tail vertebrae averaging 9 in (230 mm) in the male and 8¾ in (224 mm) in the female; head small, somewhat triangular in shape tapering anteriorly to a bulbous nose pad with anteriorly-placed nostrils; ear small, rounded, well-haired; eyes small, black, beady; neck thick, muscular; body stout, heavy, having a posteriorly-placed humped appearance in lateral profile; tail long, bushy; legs short, with hind pair longer; each foot having five slightly-webbed toes; toes tipped with claws, longer and more curved on foretoes with obvious digging adaptations; feet noticeably palmate with body weight supported by most of the foot (plantigrade); mammary glands variable in number (up to 15) but usually seven pairs, two pectoral, three abdominal and two inguinal; anal scent glands highly muscular, emitting the well-known, repelling, musky odor so characteristic of of this species; males almost 15% larger than females (Verts, 1967).

The striped skunk possesses soft, wavy, underfur (grayish with black tips in black areas and all white in white areas) and long, glossy guard hairs beginning on the back of the head and continuing over the back to the tail. The basic lustrous black coloring covers most of the upperparts except for a narrow white stripe between the eyes and a triangular patch on top of the head extending posteriorly as two stripes enclosing a mid-back patch of black, usually extending to the tail which takes on a black-and-white plumed effect. The undersides, legs, and feet are blackish, although in typical mustelid fashion, there can be white blotches on the chest. The amount of dorsal white striping and white tail plume may vary from one individual to another (see section on Molt and Color Aberrations). The glossy black and glaring white coloring of the striped skunk is so distinctive there is really no way it can be mistaken for any other Michigan mammal.

MEASUREMENTS AND WEIGHTS. Adult specimens of the larger male and the smaller female measure, respectively, as follows: length from tip of nose to end of tail vertebrae, 22¾ to 26¾ in (550 to 680 mm) and 20½ to 25½ in (520 to 650 mm); length of tail vertebrae, 7¾ to 11 in (180 to 280 mm) and 7 to 10¾ in (175 to 270 mm); length of hind foot, 2½ to 3½ in (64 to 80 mm) and 2¼ to 3 in (57 to 76 mm); height of ear from notch (almost the same in both sexes), 1 to 1⅜ in (25 to 35 mm); weight, 4.0 to 10.0 lb (1.8 to 4.5 kg) and 3.0 to 8.5 lb (1.4 to 3.8 kg). Excessively fat males (especially in September and October) may weigh as much as 12.0 lb (5.4 kg). Weights of striped skunks taken in late winter and early spring will be lower; in Kalamazoo County, for example, Allen (1939) found males at this season to weigh an average of 3.7 lb (1.66 kg) and females, 2.8 lb (1.29 kg). At this time of the year, of course, the animals are either not eating at all during winter dormancy or, mostly in the case of males, feeding only sporadically.

DISTINCTIVE CRANIAL AND DENTAL CHARACTERISTICS. The skull of the striped skunk is rather short and sturdy in construction, with a snubbed, truncated anterior end (Fig. 93). In lateral profile, the convex outline is highest in

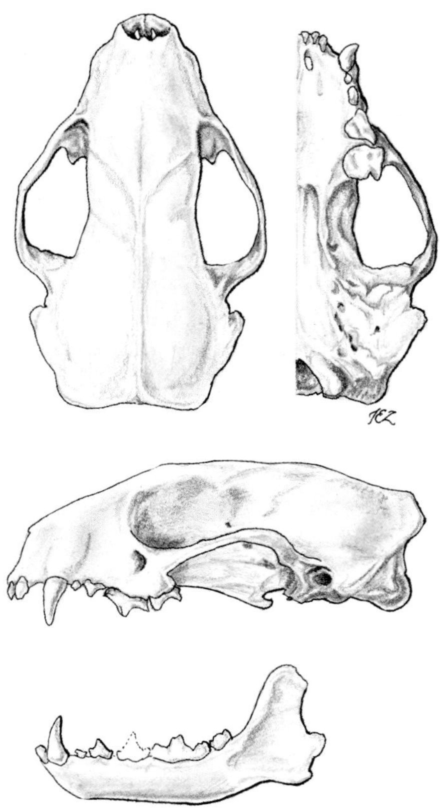

Figure 93. Cranial and dental characteristics of the striped skunk (*Mephitis mephitis*).

the frontal region. The zygomatic arches are widest posteriorly and taper anteriorly. The sagittal crest and lambdoidal ridges are prominent, especially in older animals. The auditory bullae are only slightly inflated. The bony palate is short, extending barely posterior to the ends of the upper tooth-rows. The teeth are heavily built. The last tooth (molar) in the upper tooth-row is transversely somewhat dumbbell-shaped. Skulls of adult males are as much as 8% larger than those of females.

The striped skunk skull can be distinguished from the skulls of other Michigan carnivores by (1) the short bony palate barely extending posterior to the end of the upper tooth-row; (2) the presence of only one molar on each side of the upper jaws and two molars on each side of the lower jaws; and (3) the square shape of the last upper tooth (molar). The skull of a large male mink may be almost as large as that of a female striped skunk; however, the skull of the latter is conspicuously higher in the frontal area, has wider and heavier zygomatic arches, and has a shorter bony palate extending just beyond the end of the upper tooth-row. The skulls of adult male and adult female striped skunks, respectively, measure 2⅞ to 3½ in (72 to 88 mm) and 2¾ to 3 in (69 to 77 mm) in greatest length and 1¾ to 2¼ in (45 to 54 mm) and 1⅝ to 1⅞ in (42 to 47 mm) in greatest width across the zygomatic arches. The dental formula is as follows:

$$\text{I (incisors)}\ \frac{3\text{-}3}{3\text{-}3},\ \text{C (canines)}\ \frac{1\text{-}1}{1\text{-}1},\ \text{P (premolars)}\ \frac{3\text{-}3}{3\text{-}3},\ \text{M (molars)}\ \frac{1\text{-}1}{2\text{-}2} = 34$$

DISTRIBUTION. The striped skunk is widespread in boreal and temperate North America, from near treeline in central Canada (except for the northwest coast) and southward to northern Mexico (Wade-Smith and Verts, 1982). This species is found throughout Michigan (see Map 60). Its presence in presettlement Michigan is also

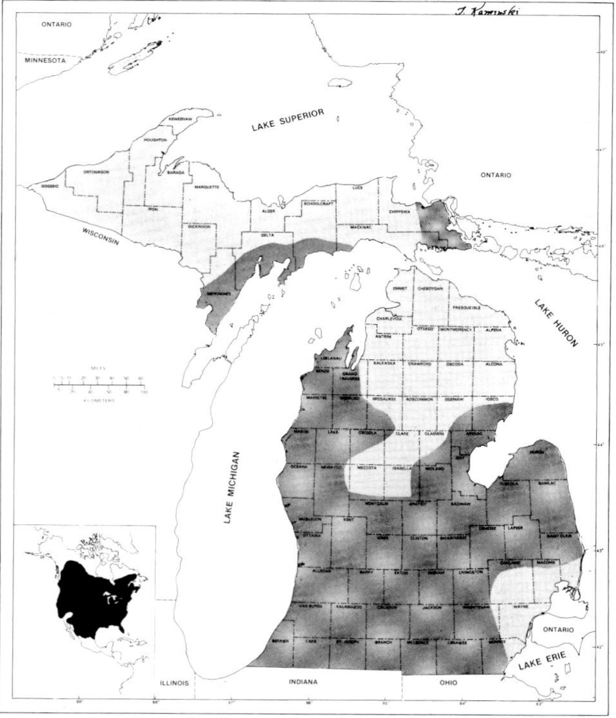

Map 60. Geographic distribution of the striped skunk (*Mephitis mephitis*). Dark shading indicates, according to the Michigan Department of Natural Resources, where populations are at medium to high densities.

reported from archaeological sites in Baraga County in the Upper Peninsula (Rhead and Martin, 1978) and in Saginaw County in the Lower Peninsula (Cleland, 1966). Hatt *et al.* (1948) listed the striped skunk as probably introduced on North Fox Island in Lake Michigan.

HABITAT PREFERENCE. The mosaic pattern of mixed agricultural and tree-cleared land and confined forest tracts, small woodlots, and riparian woody cover in close approximation to ponds, lakes, marshes, and streams makes most of Michigan ideal habitat for the striped skunk (Allen, 1938). The animal thrives also in suburban areas. Human land-use practices, especially agricultural pursuits, have provided early successional stages in old field and shrub areas bordering woodlands—environments which favor higher striped skunk populations today than in presettlement days. That the species is less common in extensive forest stands of today (Dice and Sherman, 1922) is evidence that forest removal has been a favorable influence.

DENSITY AND MOVEMENTS. Striped skunk populations have been estimated by interpreting track counts made in winter snow; recapturing individuals previously caught, tagged, and released; and monitoring the movements of animals equipped with radio collars (Verts, 1963). In prime habitat of mixed farmland, striped skunks may be as uncommon as one animal per 100 acres (40 ha) in Clinton County (Linduska, 1950) and as dense as one animal per 10.4 acres (4.2 ha) in Kalamazoo County (Allen, 1938). Similar figures have been obtained in Alberta (Bjorge *et al.,* 1981), Illinois (Verts, 1967), Maine (Dean, 1965), Manitoba (Lynch, 1972), and Pennsylvania (Jones, 1939).

Mobility studies show that striped skunks travel over larger areas in summer and autumn than in winter (a season when males move around more than do females and juveniles) and that females have smaller ranges than do males (Allen, 1939; Dean, 1965; Storm, 1972). Tracking studies and movements obtained by monitoring animals with transmitters attached to collars show striped skunks use areas varying in expanse from 503 acres (204 ha) to 2,011 acres (814 ha) in Illinois (Storm and Verts, 1966). Females, according to Dean (1965), move about in areas as small as 40 acres (16.2 ha). A female (later determined to have rabies) moved in one night in an area of about 180 acres (73 ha) in August and about 480 acres (194 ha) in September (Storm and Verts, 1966). Nightly movements from one daytime retreat to another are generally no more than 1 mi (1.6 km) and perhaps less in winter (Allen, 1939); however, according to Edmonds (1974), males may cover as much as 4 to 5 mi (6.4 to 8 km) during the height of the breeding season.

BEHAVIOR. In typical mustelid fashion, adults of the two sexes generally remain apart except for the brief courtship and mating period. The exception occurs when a male occasionally occupies a winter den with one or more females (Gunson and Bjorge, 1979). Males, including juveniles, usually occupy winter dens alone (Powell, 1979). There is evidence that in the spring, summer and autumn, males occupy much larger home ranges than do females, with those of two or more females well within the area used by a single adult male. The latter makes an effort to exclude other males from his domain. In winter, at least, females appear much more compatible with each other, even congregating in underground dens. Aboveground activity is essentially a solitary action, except when young are foraging with the female parent.

Field observers have described the rather independent mannerisms which characterize the striped skunk's attitude toward intruders. On the defense, the animal will raise its tail, sometimes stomp its feet or charge a few steps, and if need be, twist its rump toward its adversary and eject its musk with great accuracy. Despite numerous published suggestions, there is no sure way to restrain an irate striped skunk without some musk being ejected.

Normally, the striped skunk's gait is a shuffling walk or pace (Verts, 1967) but the animal can cover ground hurriedly with a bounding lope or canter. Speeds of up to 10 mi (16 km) per hour are attained with a clumsy gallop. These various gaits are used securing food; the prey (perhaps a ground beetle or a small mammal) is located by a cruising-type search pattern, then the skunk approaches closely enough to take one or two bounds and pounce in cat-like fashion on the victim. The striped skunk ambles along somewhat oblivious of its surroundings, giving the impression that its vision is not as acute as is the weasel's. Certainly, its senses of hearing and smell are good (Edmonds,

1974), with hearing especially important in the detection of moving prey (Langley, 1978).

The striped skunk, using a dog-paddle stroke, is a fair swimmer (Wood, 1922). Although Verts (1967) reports striped skunks can climb over woven-wire fences, apparently the animal does not climb trees. Striped skunks can occasionally be seen above ground during daylight hours, but the species is primarily nocturnal and crepuscular (active at night and at twilight).

Striped skunks' utterances are varied. The nursing young screech and twitter, with the female parent answering with low purrs. Adults are normally silent but on occasion can utter growls, churrings, hisses, snarls, and pig-like squeals. The animals can also sniff loudly, make spitting sounds, and click their teeth, depending upon the situation (Laun, 1962; Verts, 1967).

The striped skunk's odor is perhaps its best known feature. Large, muscular anal glands exude a highly conspicuous scent. This scent is used to mark boundaries of home ranges and to communicate with the opposite sex during courtship. In addition, of course, this odoriferous discharge has a decided defense function. As described in the classic study on striped skunks by B. J. Verts (1967), the musk is discharged from the glands through a pair of tubes terminating in projecting nipples as (1) an atomized spray in which droplets are almost microscopic in size, and (2) a short liquid stream which separates into sizeable rainsize drops while in flight. The irritated animal can make a direct hit on its target from as far away as 10 ft (3 m), further if the victim is downwind. The musk, an oily fluid creamy to greenish in color, is highly repugnant to would-be attackers. Ernest Thompson Seton (1953) effectively described the odor as ". . . a mixture of strong ammonia, essence of garlic, burning sulphur, a volume of sewer gas, a vitriol spray, a dash of perfume musk, all mixed together and intensified a thousand times." The scent will sometimes cause nausea and, if spray hits the eyes, temporary blindness may result.

Striped skunks center their activities around their den sites. Most often dens are excavations in the ground (Allen, 1938), but some are refuges under debris, buildings, stumps, or in woodpiles. Even though the animal has strong digging claws, it does not seem to excavate many of its own burrows. In a study in Iowa, for example, Selko (1938) noted that of 27 ground dens used by striped skunks, 14 had been dug originally by woodchucks, four by badgers, three by red foxes, and only four by striped skunks. In Michigan's Clinton County, Allen and Shapton (1942) decided that most of the 36 striped skunk dens which they examined had originally been dug by woodchucks. In Illinois, Verts (1967) found a similar pattern, with excavations made by muskrats added to the list. With woodchucks as well as the less common badgers providing burrows in Michigan's Upper and Lower Peninsulas, the striped skunk will not lack potential "housing."

In their intensive examination of striped skunk dens in Clinton County, Allen and Shapton (1942) found that each had an average of 1.19 entrances and a depth underground of 39.4 in (1.9 m), with a maximum of 78 in (2.4 m). The longest tunnel measured 56 ft 4 in (18.7 m) and had five entrances. The maximum number of nests in a single burrow was three; nests were generally bulky and consisted of dry grasses and leaves. Most were located on slopes above marshes, kettle-holes, and stream bottoms, in patches of brush or at edges of woodlots. Only six nests were in open situations, all in hayfields. In an Illinois study of den use by striped skunks, Storm (1972) found that more than one-half (64%) of 333 underground retreats examined were associated with fencerows. In tracking 43 radio-tagged striped skunks between May 1 and November 15, he found that, on the average, each spent only about 1.6 consecutive days in a single den. In contrast, in winter individuals began staying in a single ground den for weeks at a time. There is some evidence that striped skunks prefer above-ground retreats (under buildings and debris, in logs, etc.) in lowland situations in summer and underground dens in well-drained upland places in winter (Houseknecht and Tester, 1978). In extensive areas of marsh, however, poorly-drained den sites may be acceptable in winter (Mutch, 1977).

The striped skunk's aboveground activity is strongly influenced by seasonal temperatures. Beginning in March and continuing at least through September, Michigan skunks are constantly active. Between October and the middle of December, this activity decreases rapidly, with virtually no surface movements in severe winter weather between late December and late March. In the Upper Peninsula, for example, Manville (1948) found no sign of striped skunk aboveground activity in his study area in the Huron Mountains of Marquette County between October 3, 1939, and March 26, 1940. On the other hand, in the milder winter climate of Washtenaw County, Wood (1922) reported striped skunk tracks in snow in every month throughout the winter (see also Allen, 1939). In northwestern Illinois (a latitude not much different from that of southern Michigan) Storm (1972) found radio-tagged striped skunks spent little more than 30 days at a time in underground retreats in winter. As mentioned earlier, both adult and juvenile males usually occupy winter dens singly. On the other hand, both adult and first-year females join in communal dens which sometimes include single adult males (Gunson and Bjorge, 1979; Mutch and Aleksiuk, 1977). As a general rule, no more than ten skunks occur in a single den, the record is reported to be 19. Commune sizes are thought to increase with latitude. In Minnesota, densharing occurred between the third week in October and the first week in April (Houseknecht and Tester, 1978).

Studies show that in early autumn, striped skunks take on considerable body fat (Aleksiuk and Stewart, 1977). In October, adults were found to have a body fat content totalling as much as 32% of the individual's live weight. By late April, this amount may decline to as little as 10% (Mutch and Aleksiuk, 1977). Females may lose more body weight in winter than males which are more often prone to venture out to feed during warm spells (Banfield, 1974). Winter survival is enhanced not only by the "huddling" effect of communal denning and the use of stored fat as an energy source but also by the striped skunk's ability to conserve some energy output by depressing its body temperature slightly. This action is termed winter dormancy (carnivorean lethargy) and should not be confused with true hibernation, which is characterized by a more profound drop in body temperature.

ASSOCIATES. Normally the striped skunk leads a solitary life, apparently not associating with other species except when jointly occupying the same den on occasion. Based on radiotracking studies in Kansas, Shirer and Fitch (1970) found that striped skunks and raccoons associate in the same dens and apparently are highly tolerant of one another. These authors also found one instance where a striped skunk and a Virginia opossum lived in the same den for several days. Some dens containing hibernating woodchucks, safely sealed in their side chambers, also are reportedly occupied by wintering striped skunks.

REPRODUCTIVE ACTIVITIES. Courtship and mating in Michigan striped skunks take place for a rather brief period in late February and early March (Allen, 1939; Burt, 1946). Males will usually mate with more than one female, probably ones that live within or adjacent to their expansive home ranges. As a result of the increased mobility of males in search of mates (Edmunds, 1974), motorists may find striped skunk mortality on roadways unusually high in late winter. Careful observations on captive animals reveal variation in the length of the gestation period from 59 to 77 days. This suggests some delay in implantation, a condition more pronounced in other mustelids (Wade-Smith and Richmond, 1978). Michigan litters appear in May and as late as early June and contain about six (maximum 10) individuals. On rare occasions, females may mate again in May and deliver a second litter in July (Seton, 1953; Shadle, 1953; Verts, 1967).

Litters are born in underground nests well padded with grasses and leaves (Jackson, 1961; Shadle, 1956). At birth, young striped skunks have wrinkled, pink skin with the characteristic black and white pigmentation apparent, closed eyes and folded ear pinnae, weigh about 1⅛ oz (33 g), and can utter twittering sounds (Verts, 1967; Wade-Smith and Richmond, 1975). At eight days of age, the young can emit musk from the anal glands, have unfolded and upright ear pinnae, and exhibit hardened and dark-pigmented claws. At 22 days of age, the young are fully furred, open their eyes, crawl around, weigh 4 oz (110 g), and appear to be less vociferous. At 32 days of age, the young can assume the offensive position with the raised tail and discharge an effective amount of musk. At 40 days of age, the young possess their first teeth, eat solid food, weigh 7½ oz (210 g), and are in the process of being weaned. At about 46 days of age, the young may be fully weaned and able to follow their mothers on foraging excursions. Although the female parent is capable of moving her litter from one den to another shortly after parturition and during lactation, she usually remains in one den until the young are able to travel, then the family group may change den locations every one or two days (Houseknecht and Tester, 1978). During this time, observers may get a glimpse of this picturesque group, with the female leading her offspring single file down a trail (see Plate LIX in Seton, 1953).

The family group may break up as early as July and August (Bjorge *et al.*, 1981), when the young have gained adult size and must begin to take on necessary fat deposits for winter survival. The juvenile males usually depart and winter alone. If the juvenile females disperse, they are most apt to winter in communal dens with others of their age class (Mutch and Aleksiuk, 1977). Striped skunks may live as long as six and possibly ten years in captivity (Wade-Smith and Richmond, 1975). Verts (1967) reported an Illinois animal that survived in nature for at least 42 months. Linduska (1947) recorded a tagged female surviving in Clinton County for at least 30 months.

FOOD HABITS. The striped skunk is perhaps the most omnivorous mustelid, liking a wide assortment of animal and plant foods. Major studies of food preferences of this species in Michigan include those of Allen (1938) in Kalamazoo County, Kelker (1937) in southeastern Michigan, and Dearborn (1932). Examinations producing somewhat similar findings in other states have been summarized by Verts (1967). In digestive tracts and fecal droppings remains of Michigan striped skunks, (80% of which were taken in summer) Dearborn (1932) found the following classes of items: Insects, 57.32%; fruit, 17.66%; grain, 12.14%; mammals, 10.17%; birds, 2.35%; turtle eggs, 0.21%; bird eggs, 0.11%; reptiles, 0.01%; and nuts, 0.01%. Included were remains of many of the resident small mammals: mice of the genus *Peromyscus,* voles of the genus *Microtus,* meadow jumping mice, house mice, Norway rats, thirteen-lined ground squirrels, eastern chipmunks, red squirrels, fox squirrels, short-tailed shrews, eastern moles, eastern cottontails, woodchucks, and raccoons. Dearborn also listed striped skunk since it is well known that mature males will eat young of their own species. Allen (1938) found that 12 of 61 striped skunk stomachs (taken in autumn, winter and spring) contained remains of the meadow vole. In winter, small mammals are probably a major item in the diet (Lantz, 1923), although Michigan striped skunks actually forage very little in severe winter weather. Other vertebrates eaten include minnows, other fishes, frogs, lizards, snakes, and turtles.

In summer, a major portion of the striped skunk's diet consists of insects and other invertebrates. As in the case of small mammal prey, arthropods may be detected by the striped skunk's sense of hearing (Langley, 1978) or, perhaps more efficiently, by rooting in the ground. Small holes and displaced soil are telltale evidence of striped skunk probings. In examining striped skunk fecal droppings collected in spring, summer, and autumn in southeastern Michigan, Kelker (1937) identified insects of the orders Trichoptera, Hemiptera, Orthoptera, Coleoptera, and Hymenoptera. In Coleoptera, 11 beetle families were represented. In addition, Dearborn (1932) found Lepidoptera (in the form of larval cutworms and caterpillars). Ground beetles (Carabidae) and May beetles (Scarabaeidae) apparently are eaten in the early part of the season while crickets (Gryllidae) and grasshoppers (Acrididae) are the major insect foods later. Clams, earthworms, crayfish, carrion, and garbage are also consumed (Hamilton, 1936).

According to Dearborn (1932), Michigan striped skunks relish such fruits as dogwood, mulberry, raspberry, blackberry, plum, cherry, currant, gooseberry, grape, blueberry, strawberry, nightshade, sarsaparilla, serviceberry, and apple. Other destructive eating habits include raiding beehives (Storer and Vansell, 1935), eating of corn crops and, especially, poultry and their eggs (Schofield, 1957).

ENEMIES. Striped skunks often are able to discourage would-be predators by discharging their musk. Nevertheless, the flesh of this mustelid is an attraction to a variety of large meat-eaters. Michigan mammalian predators include gray wolf, coyote, red fox, gray fox, fisher, badger, bobcat, and lynx (Aleksiuk and Stewart, 1977; Banfield, 1974; Brown and Will, 1979; Murie, 1936; Schofield, 1960; Verts, 1967). Among birds of prey, the great horned owl, raven, and golden eagle eat skunks (Errington *et al.*, 1940; Harlow *et al.*, 1975; Verts, 1967). Even so, the human is the striped skunk's greatest enemy in Michigan. The trapper takes a modest annual toll, usually depending on the market price for fur. But highway deaths are a major mortality factor. In Michigan, Dice (1925) noted highway kills in Charlevoix County as early as 1923. According to Haugen (1944) striped skunks in southern Michigan died on highways most frequently in April and August. Other statistics on highway kills of striped skunks have been reported in the Upper Peninsula by Manville (1949), in Kent County by Hartley (1973), and in Ingham County by Kasul (1976).

PARASITES AND DISEASES. Michigan striped skunks have not been examined thoroughly for ectoparasites or endoparasites. A report by Lawrence *et al.* (1965) listed one flea, *Oropsylla arctomys,* one biting louse, *Trichodectes mephitidis,* one tick, *Ixodes cookei,* and one mite, *Haemolaelaps glasgowi.* Ectoparasites recorded by Verts (1967) from striped skunks in Illinois may also be found on this species in Michigan. He listed five additional fleas, one mite, and one tick. He also discovered infestations of botfly larvae (probably of the genus *Cuterebra*) in a few animals. Verts also listed as internal parasites seven species of Trematoda (flukes), eight Cestoda (cestodes), 19 Nematoda (roundworms), four Acanthocephala (thorny-headed worms), and one Pentastomida. The roundworm (genus *Skrjabingylus*) infests the frontal sinus region of a large percentage of wild-taken striped skunks, as well as other mustelids. Verts (1967) found 65.9% (108 of 164) of the striped skunk skulls examined from northwestern Illinois harbored this worm.

Several diseases have been isolated from striped skunks, including leptospirosis, tularemia, histoplasmosis, and canine distemper (Verts, 1967). Perhaps the most serious disease in terms of relationship to human populations is rabies. Records kept by the Michigan Department of Public Health show that, between 1962 and 1972, 182 (54%) of the 337 striped skunks examined were found to be rabid. Since 1973, however, the number of rabid skunks reported in Michigan has declined noticeably (Kurta, 1979). Persons professionally interested in rabies in striped skunks and other wildlife should consult the excellent presentation in Verts (1967, Chapter XII).

MOLT AND COLOR ABERRATIONS. Striped skunks molt the underfur beginning in April and the longer guard hairs beginning in July (Verts, 1967). Replacement of both coats begins by September. The animals have wide variation in the amount of white striping on the back and upper sides. In the black (or star) condition the amount of white on the back is a minimal U-shaped blob of white at the nape and the ever-present line at the midline of the face. A stripeless skunk with a white-tipped tail and a white spot on its head was seen in Dickinson County in October, 1980 (R. P. Smith, 1980). From this condition (the most valuable in the fur trade), a gradation of the amount of white may increase all the way to the so-called broad stripe condition in which a considerable part of the back is white and the tail has a conspicuous white plume (see drawings in Verts, (1967).

All black (melanistic) individuals are recorded. Other mutations produce all white (albino), seal brown, and yellowish pelages. An occasional albino striped skunk is reported in Michigan.

ECONOMIC IMPORTANCE. The market for striped skunk fur has its ups and downs. When longhaired furs are stylish, the price per pelt may be worth the effort and unpleasantness of catching and skinning the animals. Before World War I, farm boys were especially eager to catch striped skunks. In 1912, Wood (1914) reported that in the Whitefish Point area of Chippewa County trappers would catch these animals in early winter and keep them alive until the fur primed. In recent years, fur market prices for striped skunk pelts have usually been less than $2.00 (about $1.85 in the late winter of 1976). However, in March 1979, Michigan pelts brought as much as $6.50. Michigan's estimated harvest of striped skunk pelts has been 1,370 (in 1965–66), 2,510 (in 1966–67), 1,480 (in 1967–68), 2,810 (in 1968–69), 3,780 (in 1969–70), and 6,350 (in 1970–71).

In pioneer days, fat rendered from the striped skunk had a variety of uses. The animal's use as food by Indians and pioneers is well known. Allen (1939) cleaned and cooked a striped skunk in Kalamazoo County, and commented that "I have tried it and found it very good. An old male, after being steamed and well browned in a skillet, was not at all tough and had a very pleasing flavor. The main objection to the creature for eating purposes is that, as compared with a rabbit, there is little meat on it."

The striped skunk can be a pest to the farmer and orchardist. The animal will eat corn, various berry fruits, apples, grapes, cherries, plums, and currants. It will raid poultry houses, particularly relishing eggs (Dearborn, 1932; Schofield, 1957), and damage hives in search of adult bees and their larvae, as well as honey. In Washtenaw County, Dice (1926) wrote that a striped skunk even ate two halfgrown house cats. Biologists discourage striped skunks in waterfowl nesting areas since they are attracted to the eggs of ducks and geese (Teer, 1964). On the other hand, the striped skunk eats noxious insects and pestiferous small rodents. To Durward L. Allen (1937), who carried on pioneer studies of the striped skunk is southern Michigan, the animal was part "boon" and part "pest."

Many basic biological studies of the striped skunk have been conducted in captivity. The animals have been successfully propagated, with housing needs determined, nutrient requirements understood, and procedures developed for descenting the animals (Wade-Smith and Richmond, 1975). However, as a rabies control measure, Michigan law forbids the possession and selling of animals captured in the state. A deodorized, pet striped skunk may be obtained from another state where the keeping of the species as a pet is legal. However, a Michigan possession permit is required.

Although the striped skunk may become as friendly a house pet as a kitten, the animal's long history of susceptibility to rabies must be kept in mind. It should also receive cat and dog distemper vaccinations. As for dispelling the odor of the musk, Edmonds (1974) listed no less than 15 different home and commercial remedies. R. M. Smith (1980) reports an additional commercial product, "Skunk-Off", as being effective. As home remedies, just soaking clothing, washing one's hands in tomato juice, or applying household ammonia or vinegar are worth trying. Inflammable organic solvents (dry cleaning fluid, gasoline, benzine, cleaning napthas, etc.) may be used with care to dissolve away the musk (consisting of mercaptans which do not dissolve well in water), followed by several rinses.

COUNTY RECORDS FROM MICHIGAN. Records of specimens of striped skunks in collections of museums, from the literature, from trappers' reports, and provided by personnel of the Michigan Department of Natural Resources show that the animal occurs in all counties of the Upper and Lower Peninsulas (see Map 60).

LITERATURE CITED

Aleksiuk, M., and A. P. Stewart
1977 Food intake, weight changes and activity of confined striped skunks (*Mephitis mephitis*) in winter. American Midl. Nat., 98(2):331–342.

Allen, D. L.
1937 The skunk: A boon or pest? Michigan Conserv., 7(4):3,9.
1938 Ecological studies on the vertebrate fauna of a 500-acre farm in Kalamazoo County, Michigan. Ecol. Monog., 8(3):348–436.
1939 Winter habits of Michigan skunks. Jour. Wildlife Mgt., 3(3):212–228.

Allen, D. L., and W. W. Shapton
1942 An ecological study of winter dens, with special reference to the eastern skunk. Ecology, 23(1):59–68.

Banfield, A. W. F.
1974 The mammals of Canada. Univ. Toronto Press, Toronto. xxiv+438 pp.

Bjorge, R. R., J. R. Gunson, and W. M. Samuel
1981 Population characteristics and movements of striped skunks (*Mephitis mephitis*) in central Alberta. Canadian Field-Nat., 95(2):149–155.

Brown, M. K., and G. Will
1979 Food habits of the fisher in northern New York. New York Fish and Game Jour., 26(1):87–92.

Burt, W. H.
1946 The mammals of Michigan. Univ. Michigan Press, Ann Arbor. xv+288 pp.

Cleland, C. E.
1966 The prehistoric animal ecology and ethnozoology of the Upper Great Lakes Region. Univ. Michigan, Mus. Anthro., Anthro. Papers No. 29, x+294 pp.

Dean, F. C.
1965 Winter and spring habits and density of Maine skunks. Jour. Mammalogy, 64(4):673–675.

Dearborn, N.
1932 Foods of some predatory fur-bearing animals in Michigan. Univ. Michigan, School Forestry and Conserv., Bull. No. 1, 52 pp.

Dice, L. R.
1925 A survey of the mammals of Charlevoix County, Michigan, and vicinity. Univ. Michigan, Mus. Zool., Occas. Papers No. 159, 33 pp.
1926 Skunk eats kittens. Jour. Mammalogy, 7(2):131.

Dice, L. R., and H. B. Sherman
1922 Notes on the mammals of Gogebic and Ontonagon counties, Michigan, 1920. Univ. Michigan, Mus. Zool., Occas. Papers No. 109, 46 pp.

Edmonds, W. T., Jr.
1974 Skunks in Kansas: A review of their natural history and methods of control. Univ. Kansas, State Biol. Surv., Bull. No. 3, 32 pp.

Errington, P. L., F. Hamerstrom, and F. N. Hamerstrom, Jr.
1940 The great horned owl and its prey in north-central United States. Iowa Agric. Exp. Sta., Res. Bull. No. 277, pp. 758–850.

Gunson, J. R., and R. R. Bjorge
1979 Winter denning of the striped skunk in Alberta. Canadian Field-Nat., 93(3):252–258.

Hamilton, W. J., Jr.
1936 Seasonal food of skunks in New York. Jour. Mammalogy, 17(3):240–246.

Harlow, R. F., R. G. Hooper, D. R. Chamberlain, and H. S. Crawford
1975 Some winter and nesting season foods of the common raven in Virginia. Auk, 92(1):298–306.

Hartley, S. L.
1975 Life's cycle. Michigan Out-of-Doors, 29(3):41.

Hatt, R. T., J. Van Tyne, L. C. Stuart, C. H. Pope, and A. B. Grobman
1948 Island life: A study of the land vertebrates of the islands of eastern Lake Michigan. Cranbrook Inst. Sci., Bull. No. 27, xi+179 pp.

Haugen, A. O.
1944 Highway mortality of wildlife in southern Michigan. Jour. Mammalogy, 25(2):160–170.

Houseknecht, C. R., and J. R. Tester
1978 Denning habits of striped skunks (*Mephitis mephitis*). American Midl. Nat., 100(2):424–430.

Jackson, H. H. T.
1961 Mammals of Wisconsin. Univ. Wisconsin Press, Madison. xii+504 pp.

Jones, H. W., Jr.
1939 Winter studies of skunks in Pennsylvania. Jour. Mammalogy, 20(2):254–256.

Kasul, R. L.
1976 Mortality and movement of mammals and birds on a Michigan interstate highway. Michigan State Univ., unpubl. M. S. thesis, ix+39 pp.

Kelker, G. H.
1937 Insect food of skunks. Jour. Mammalogy, 18(2):164–170.

Kurta, A.
1979 Bat rabies in Michigan. Michigan Acad., 12(2):221–230.

Langley, W. M.
1978 Senses used in finding moving prey by skunks and opossums. Trans. Kansas Acad. Sci., 81(1):91.

Lantz, D. E.
1923 Economic value of North American skunks. U.S. Dept. Agric., Farmers' Bull. 587, 24 pp.

Laun, H. C.
1962 Loud vocal sounds produced by striped skunk. Jour. Mammalogy, 43(3):432–433.

Lawrence, W. H., K. L. Hays, and S. A. Graham
1965 Arthropodous ectoparasites from some northern Michigan mammals. Univ. Michigan, Mus. Zool., Occas. Papers No. 639, 7 pp.

Linduska, J. P.
1947 Longevity of some Michigan farm game mammals. Jour. Mammalogy, 28(2):126–129.
1950 Ecology and land-use relationships of small mammals on a Michigan farm. Michigan Dept. Conserv., Game Div., ix+144 pp.

Lynch, G. M.
1972 Effects of strychnine control on nest predators of dabbling ducks. Jour. Wildlife Mgt., 36(2):436–440.

Manville, R. H.
1948 The vertebrate fauna of the Huron Mountains, Michigan. American Midl. Nat., 39(3):615–640.
1949 Highway mortality in northern Michigan. Jour. Mammalogy, 30(3):311–312.

Murie, A.
1936 Following fox trails. Univ. Michigan, Mus. Zool., Misc. Publ. No. 32, 45 pp.

Mutch, G. R. P.
1977 Locations of winter dens utilized by striped skunks in Delta Marsh, Manitoba. Canadian Field-Nat., 9(3):289–291.

Mutch, G. R. P., and M. Aleksiuk
1977 Ecological aspects of winter dormancy in the striped skunk (*Mephitis mephitis*). Canadian Jour. Zoology, 55(3):607–615.

Powell, R. A.
1979 Mustelid spacing patterns: Variations on a theme by *Mustela*. Zeitschrift für Tierpsychologie, 50(2):153–165.

Rhead, D., and T. J. Martin
1978 Environment and subsistence at Sand Point. Paper presented Central States Anthro. Soc., South Bend (mimeo), 8 pp.

Schofield, R. D.
1957 Livestock and poultry losses caused by wild animals in Michigan. Michigan Dept. Conserv., Game Div., Rept. No. 2188 (mimeo), 6 pp.
1960 A thousand miles of fox trails in Michigan's ruffed grouse range. Jour. Wildlife Mgt., 24(4):432–434.

Selko, L. F.
1938 Notes on the den ecology of the striped skunk in Iowa. American Midl. Nat., 20(2):455–463.

Seton, E. T.
1953 Lives of game animals. Vol. II, Pt. I. Bears, coons, badgers, skunks, and weasels. Charles T. Branford Co., Boston. xvii+367 pp.

Shadle, A. R.
1953 Captive striped skunk produces two litters. Jour. Mammalogy, 17(3):388–389.
1956 Parurition in a skunk, *Mephitis mephitis*. Jour. Mammalogy, 37(1):112–113.

Shirer, H. W., and H. S. Fitch
1970 Comparison from radiotracking of movements and denning habits of the raccoon, striped skunk, and opossum in northeastern Kansas. Jour. Mammalogy, 51(3):491–503.

Smith, R. M.
1980 Reader offers skunk solution. Michigan Nat. Res., 49(3):7.

Smith, R. P.
1980 Regional report. Michigan Out-of-Doors, 35(1):52.

Storer, T. I., and G. H. Vansell
1935 Bee-eating proclivities of the striped skunk. Jour. Mammalogy, 16(2):118–121.

Storm, G. L.
1972 Daytime retreats and movements of skunks on farmlands in Illinois. Jour. Wildlife Mgt., 36(1):31–45.

Storm, G. L., and B. J. Verts
1966 Movements of a striped skunk infected with rabies. Jour. Mammalogy, 47(4):705–708.

Teer, J. G.
1964 Predation by long-tailed weasels on eggs of blue-winged teal. Jour. Wildlife Mgt., 28(2):404–406.

Verts, B. J.
1963 Equipment and techniques for radio-tracking striped skunks. Jour. Wildlife Mgt., 27(3):325–339.
1967 The biology of the striped skunk. Univ. Illinois Press, Urbana. vii+218 pp.

Wade-Smith, J., and M. E. Richmond
1975 Care, management, and biology of captive striped skunks (*Mephitis mephitis*). Laboratory Animal Sci., 25(5):575–584.
1978 Reproduction in captive stripted skunks (*Mephitis mephitis*). American Midl. Nat., 100(2):452–455.

Wade-Smith, J., and B. J. Verts
1982 Mephitis mephitis. American Soc. Mammalogists, Mammalian Species, No. 173, 7 pp.

Wood, N. A.
1914 Results of the Shiras Expeditions to Whitefish Point, Michigan—Mammals. Michigan Acad. Sci., 16th Ann. Rept., pp. 92–98.
1922 The mammals of Washtenaw County, Michigan. Univ. Michigan, Mus. Zool., Occas. Papers No. 123, 23 pp.

RIVER OTTER
Lutra canadensis (Schreber)

NAMES. The generic name *Lutra,* proposed by Brisson in 1762, is the Latin name for otter. The specific name *canadensis,* proposed by Schreber in 1776, refers to Canada, with the Latin suffix, *ensis,* meaning "belonging to" because the species was first described from Canada. The common name otter has a northern European origin: in Old English, *otor,* in Middle English, *oter,* in Swedish, *utter,* in Danish, *odder,* in German, *otter.* In Michigan, the name otter is universally used when referring to this species, although mammalogists advocate the name river otter to distinguish it from the sea otter. The subspecies *Lutra canadensis canadensis* (Schreber) occurs in Michigan. In some of the recent literature the generic name *Lontra* is used in place of *Lutra* (van Zyll de Jong, 1972).

RECOGNITION. Body large for a mustelid with the diameter of the midsection resembling in size that of a small dog but elongated and streamlined (Fig. 94): length of head and body averaging 27½ in (700 mm) for the larger male and 26¾ in (680 mm) for the smaller female; tail long, at least one-third the total length of the animal; length of tail vertebrae averaging 16¾ in (425 mm) for the male and 16⅜ in (415 mm) for the female; body lithe, muscular, fusiform; head small, broad, distinctly flattened; nose pad prominent; vibrissae (whiskers) long, coarse; eyes small, anteriorly placed; ears small, rounded, meatus closable underwater; neck thick, muscular; tail basally stout, muscular, tapering gradually to the tip; limbs short, powerfully constructed; feet large, broad, each with five toes; toes fully webbed each tipped with a short claw; mammary glands consist of four pairs, all inguinal; anal scent glands less well developed than in other mustelids; males about 5% larger than females.

Figure 94. The river otter (*Lutra canadensis*). Photograph by Robert Harrington and courtesy of the Michigan Department of Natural Resources.

The pelage of the river otter is highly adapted for this animal's semiaquatic life. A layer of fat just under the skin adds to the insulative quality of the body covering. The thick, short, dark underfur has an oily texture, giving it efficient waterproofing. Glossy, dark-brown guard hairs, silky and longer than the underfur, present a rich blackish-brown overall appearance, darker when the outer surface is wet. The underparts are paler, light grayish brown; the muzzle and throat silvery gray.

The otter's muscular "snake-like" body, short oily pelage, powerful tail, short limbs, and webbed feet plus other modifications make it one of the best adapted of amphibious mammals. Its large, fusiform body is distinctive among Michigan mammals.

MEASUREMENTS AND WEIGHTS. Adult specimens of the male and the female river otter measure, respectively, as follows: length from tip of nose to end of tail vertebrae, 37½ to 51½ in (950 to 1,300 mm) and 35½ to 45½ in (900 to 1,150 mm); length of tail vertebrae, 13½ to 20 in (340 to 510 mm) and 12¼ to 17½ in (320 to 445 mm); length of hind foot, 4¼ to 5¼ in (110 to 135 mm) and 4¼ to 5 in (110 to 125 mm); height of ear from notch (almost the same in both sexes), ⅜ to 1 in (10 to 25 mm); weight, 15 to 30 lbs (6.8 to 13.6 kg) and 10 to 25 lbs (4.5 to 11.3 kg).

DISTINCTIVE CRANIAL AND DENTAL CHARACTERISTICS. The skull of the river otter is broad, with a flattened, swollen cranium and a depressed rostrum, the latter shorter than wide (Fig. 95). Like the rest of the skull, the zygomatic arches are sturdy in construction. The interorbital space is noticeably constricted. Auditory bullae are characteristically flattened, and the hard palate extends well posterior to the end of the upper tooth-row. The infraorbital foramina are large and almost oval, in contrast to the triangular shape of these openings in the skull of the badger. The teeth are massive in construction. Incisors and premolars are sturdy but small while the upper molars are large and rectangular. In older individuals, the teeth may be badly worn with some tooth loss. The skull of the mature male river otter may be more than 4% larger than that of the female (Hooper and Ostenson, 1949).

The skull of the river otter with its heavy and flattened configuration and almost "hour-glass" shape is distinctive from other Michigan mammal skulls. It is also the only Michigan carnivore with 36 teeth, having five cheek teeth on each side above and below posterior to the canines. Occasionally, there may be an extra lower premolar, making six teeth behind the canines in the lower jaw (Dearden, 1954). The dental formula is:

$$\text{I (incisors)}\ \frac{3\text{-}3}{3\text{-}3},\ \text{C (canines)}\ \frac{1\text{-}1}{1\text{-}1},\ \text{P (premolars)}\ \frac{4\text{-}4}{3\text{-}3},\ \text{M (molars)}\ \frac{1\text{-}1}{2\text{-}2} = 36$$

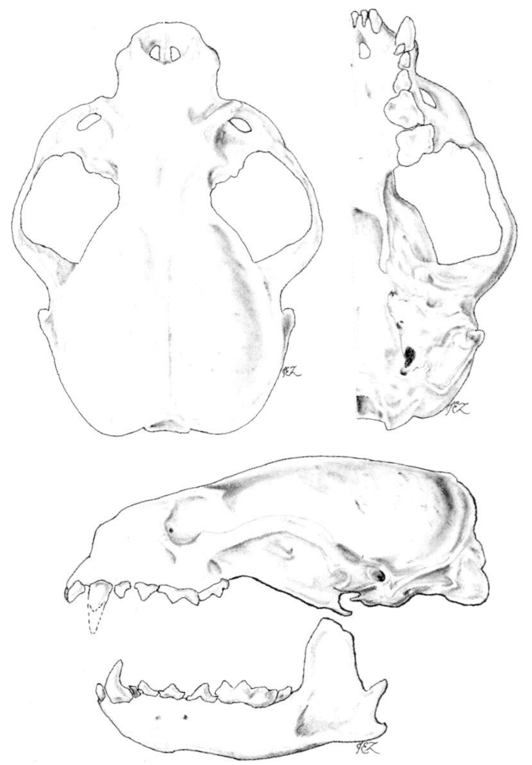

Figure 95. Cranial and dental characteristics of the river otter *(Lutra canadensis)*.

DISTRIBUTION. In presettlement days the otter occurred widespread in arctic, boreal, and temperate North America (Harris, 1968). It ranged from as far north as the Arctic Ocean, except for the open tundra and the islands of the Canadian Archipelago, southward to the Gulf coastal states and westward to California, with arid parts of the Southwest unoccupied (van Zyll de Jong, 1972). Following extirpation or severe range reduction as a result of settlement and overtrapping, especially in the southern and southeastern parts of its range, it is now making a comeback in most states through legalized protection and/or controlled harvest. In Michigan, the river otter probably became resident following glacial melt-back, finding all parts of the state well-watered and inviting. Its bones have been found in early archaelogical sites (Cleland, 1966). As settlement progressed in southern Michigan in the nineteenth century, the river otter was trapped increasingly for its valued pelt, accounting for its gradual disappearance in the southern tiers of counties. It was last known in northern Indiana (St. Joseph County), directly south of Michigan's Berrien and Cass counties, no later than 1905 (Engles, 1933). Wood (1922) wrote that the otter disappeared from the Huron, Raisin and Saline rivers in the Washtenaw County area by 1910. While otters in the Upper Peninsula and northern parts of the Lower Peninsula maintained residual populations in the districts remote from settlements, the animals in the first five or six tiers of counties in the southern Lower Peninsula were apparently extirpated.

A turnaround in the river otter's fortunes resulted from the enactment of legislation giving the species total legal protection from 1925 through 1939 (Lagler and Ostenson, 1942). A map of the distribution of the river otter published in the Ninth Biennial Report of the Michigan Department of Conservation for 1937–1938 shows that south of the 44th parallel the animal was believed to be extinct except in Osceola, Clare, and Gladwin counties, all bisected by this parallel, and Midland county directly south of it. Since 1940, conservative legal trapping seasons and bag limits have allowed the otter to make a decided comeback in many central counties of the Lower Peninsula (see Map 61). Records of sightings by wildlife biologists of the Michigan Department of Natural Resources place the otter on rivers and in lakes in such southern counties as Ingham (1974), Ionia (1979), Kent (1978–1979), Livingston (1978), and Montcalm (1979). As a result of this southward reestablishment, one can hope that otters may again inhabit all major stream systems in southern Michigan.

HABITAT PREFERENCES. The river otter's major habitat requirements are substantial amounts of water (Mowbray *et al.,* 1976) and abundant aquatic life therein. Michigan surely provides these environmental essentials. There seems little direct relationship between otter populations and the types of bankside (riparian), emergent or submergent aquatic vegetation. Apparently the quantity of plant life is more important than the quality in providing shelter and foraging areas—whether in cold-water environments of the Upper Peninsula or warm-water environments of the southern part of the Lower Peninsula. According to Schofield (1961) and J. E. Vogt (*in litt.,* January 8, 1981), high otter populations occur along the watersheds of the Carp,

Map 61. Geographic distribution of the otter (*Lutra canadensis*). In former times this species occurred throughout the Lower Peninsula.

Cedar, Manistique, Menominee, Michigamme, Ontonagon, Presque Isle, Sturgeon, Tahquamenon, Two-hearted, and Whitefish rivers in the Upper Peninsula and the Au Gres, Au Sable, Cheboygan, Manistee, Rifle, Tawas, and Thunder Bay rivers in the Lower Peninsula. Most authors emphasize that the otter requires fairly deep water in streams and lakes. In Michigan, however, the animal appears equally at home in rather shallow and plant-filled marshes. In fact, Field (1970) noted that winter foods were more abundant for otter in extensive marshland (shallow ponds surrounded by earthen dikes) than along streams in the Upper Peninsula's Seney National Wildlife Refuge. Along streams, otter seem to prefer sites with prominent banks.

DENSITY AND MOVEMENTS. Most information on populations and mobility of the otter has been derived from trapping records, tracking of individuals in winter snow, and by radio-telemetry (Melquist and Hornocker, 1979). Many observers feel that the otter does not exist in high densities, perhaps because of its alleged high mobility, its spatial needs for food getting, and its intraspecific relationships. Studies in Sweden by Erlinge (1968) of the closely-related common otter (*Lutra lutra*) showed densities in lake habitat of one animal to as much as 250 acres (100 ha) of water or one per 1.2 to 1.9 mi (2 to 3 km) of lakeshore. Along streamways, his estimate was one otter per each 3 mi (5 km) of bank. In Idaho, otter were found to exist in densities of one animal per 2.2 mi (3.6 km) of waterway. Densities in most northern Michigan favorable habitat might be expected to be comparable. A survey in northern Michigan made in 1949 by wildlife biologists from the Michigan Department of Conservation showed that there were between 900 and 1,500 otter in the Upper and Lower (northern part) Peninsulas combined. Along approximately 16,000 mi (25,755 km) of stream, this would allow for one otter for every 10.7 to 17.7 mi (17 to 28 km) of waterway, representing an estimated population of one otter to every 14 to 25 sq mi (36 to 62 sq km) of potential habitat (Jenkins, 1949).

Highest populations would be anticipated just prior to the trapping season. Annual survival rates, according to Tabor and Wight (1977), appear best for adults from two to as many as eleven years of age (73%), next best was for juveniles less than one year old (68%), and poorest for one-year-old animals (46%). Ostenson and Gross (1940) examined 212 of the total catch of 266 otters in the 1940 season (March 20 to April 5 in the Upper Peninsula and April 1 to April 15 in the Lower Peninsula). One-third of this sample was judged to be yearlings (46 of 103 males and 25 of 38 females). No explanation was given for the capture of a larger number of males than females (2.4 to 1.0).

Much early literature mentions that the otter travels great distances, moving 50 or even 100 mi (80 to 160 km) while foraging or finding living space (Jackson, 1961; Liers, 1951; Seton, 1953). These observations may actually have been mostly of younger animals in search of home territories; modern observations tend to show that an animal, once established in an area possessing necessary year-around life requirements, has a restricted home range. As in the case of most other mustelids, mature males occupy large home areas from which other males are excluded; adult females do much the same but maintain much smaller areas which can be situated within those of males. Careful observations by Erlinge (1967) in Sweden showed that mature males of the common otter of Eurasia have winter home ranges 9 mi (15 km) in diameter, or an overall range of about 68 sq mi (177 sq km), with animals moving distances of 5.6 to 6.2 mi (9 to 10 km) in a single winter night. In contrast, females restrict their home ranges to areas of no more than 2.5 mi (4 km), meaning an overall range of about 4.8 sq mi (12.6 sq km) in diameter. A female with young may cover a larger area with a diameter of as much as 4.3 mi (7 km), having a total area of about 14.5 sq mi (38 sq km). At the Seney National Wildlife Refuge in Alger and Schoolcraft counties, Field (1970) tracked otter in the snow on stream banks or cross-country for as far as 3 mi (5 km). However, recent literature suggests that the extent of the movements and stability of otter is dependent on whether the individuals are adults with established home territories or immature animals seeking such accommodations or transient animals lacking homesites.

BEHAVIOR. As mentioned, otters display territorial behavior. Nevertheless, only about one-third of the population are territory holders, at least in the colder seasons as indicated mostly by studies of the Eurasian common otter (Erlinge, 1978). Another third of the local population consists of young-of-the-year, while the final third includes transients or temporary residents. If these three groups cannot be sufficiently identified, an impression of population instability is obtained.

Otters mark their territories by depositing exudate from their anal scent glands at strategic places. This appears to be a means by which other individuals of the same sex are warned of the presence of the established animal and by which individuals of the opposite sex may communicate. Apparently, members of the opposite sex will either avoid each other, or congregate occasionally, except during courtship and mating. Two individuals of the same sex can exhibit aggression towards each other. In general, however, mature males keep to themselves except at breeding time and are highly mobile in their large territories, especially in winter. Females with offspring occupy smaller areas but apparently always at locations where the food supply is abundant.

The otter displays impressive energy in its active life both on shore and in the water. In winter, neither temperature nor snow depth appears to affect its travel on land in the Upper Peninsula (Field, 1970). On snow, the humping gait usually consists of two to four (often three) loping bounds, each covering 15 to 28 in (0.4 to 0.7 m), and then a glide, with belly dragging and feet trailing backward, for five to 15 ft (1.5 to 4.6 m). On ice they glide as much as 25 ft (7.6 m), according to Field (1970) and Severinghaus and Tanck, (1948). An ability to slide in snow, on ice, or down muddy and grassy banks has always been a prominent characteristic of the otter. The animal also accomplishes a partial sliding action in snow or vegetation by folding back its front legs and pushing forward on its belly by a churning action of the hind feet. On land, the otter has been clocked at speeds up to 18 mi (29 km) per hour.

On land, the otter, despite its lanky bulk, can successfully stalk small mammals. The gait, according to Field (1970), consists of short steps of no more than 5 in (127 mm). From a distance of about 2 ft (0.6 m), the otter makes its quick, weasel-like jump to seize its prey. The otter also digs in muddy areas on land for crayfish, follows grass or snow tunnels in quest of small rodents, and chases snowshoe hares, catching them on occasion (Field, 1970). Most land movements are at night, although otters can be observed at twilight or on drab days. Overland travel between lakes or watersheds is not difficult for the otter, allowing young individuals to find homesites and males to find mates.

The otter is a skillful, graceful swimmer, employing an undulating body motion, both underwater and on the surface, and reportedly can swim at speeds of 7 mi (11 km) per hour. It can tread water rapidly and extend its head and long neck high above the water to look around, can remain submerged for as long as two minutes, and can swim long distances under the ice, using trapped air bubbles or open holes for breathing. Dives to depths of 60 ft (19.8 m) have been reported (Scheffer, 1953). The otter has also been observed riding downstream on a block of river ice (Cook, 1940). It pursues fish by rapid swimming or by ambush (Peterle, 1954). From a vantage point on an overhanging bank or above a hole in the ice, the animal crouches in wait to make a quick dive for an edible fish (Dearborn, 1968; Field, 1970; Smith, 1939). Anglers should be comforted by the fact that the fish taken by otter is usually a slow-swimming forage fish rather than a quicker trout (Ryder, 1955).

Otter, although primarily nocturnal, leave conspicuous signs of their presence. The most obvious are the well-used slides. Other field evidence of this animal, as abstracted from Mowbray *et al.*, (1976), includes: (1) haul-outs, worn trails at water's edge scattered with fish bones, scales, and droppings and used for feeding and resting; (2) bedding sites, concentric impressions on the ground edged with matted leaves or other plant materials; (3) rolling sites, characterized by flattened bankside vegetation covering as much as 20 sq ft (2 sq m); (4) scrapes, areas completely bare of vegetation used by resting otters but free of food scraps or fecal matter; (5) den openings, which may be detected from the animal's distinctive wide track prints; (6) diggings, holes with loose soil thrown around where otters have been probing for clams, turtle eggs, crayfish; and (7) scent posts where musk is deposited to communicate with the opposite sex or warn away members of the same sex and which may often be identified by soil disturbance resulting from digging and ground scratching.

The otter has excellent vision, hearing and smell. Its teeth and jaws, agility, and elongated muscular body give this animal remarkable defensive skills. As Jackson (1961) stated, a dog is no match for river otter in a fair fight. Family groups (female and offspring) seem constantly at play, rolling and tumbling in kitten fashion (Liers, 1951; Seton, 1953). Young animals often make chirping sounds and soft chuckles. Adults utter a snarling growl or a hissing bark when irritated or cornered. Most characteristic perhaps is a shrill whistle, which may be uttered during courtship.

Otters use dens between tree roots, in hollow stumps, beneath decayed logs, under brush piles, and even within woodchuck burrows (Liers, 1951). Some of these may be remote from water. In most instances, however, refuges are in bank dens excavated by a muskrat or beaver, or in log jams, beaver lodges or muskrat beds. Apparently, the otter rarely digs a den of its own, but it will enlarge or remodel one usurped from another animal. The otter chooses bank dens with the entrances either well-hidden or underwater, well below the ice line. A nest, padded with wood chips, grasses, leaves, and aquatic vegetation, is located in an enlarged site in the subterranean tunnel, above the high water line. The same den may be used for many years, although the otter may also have several "over-day" resting areas at strategic places in its home range.

ASSOCIATES. The solitary life of the adult male otter does not seem conducive to interspecific associations. The female's associations are almost entirely related to the rearing of litters, although she may occasionally produce young only every other year. Perhaps the three closest mammalian associates of the Michigan river otter are muskrat, beaver and mink. Apparently, all four thrive along the same streamways. This seems a fairly compatible arrangement except for the mink's predation on the muskrat and the river otter's occasional predation on all three (Greer, 1955). In a study of otter-muskrat relationships in Lake St. Helen in Roscommon County, Miller (1937) found that the muskrat is not a major food item in the otter's diet (see also Wilson, 1954). When the otter usurps the muskrat's dens, the muskrat, if not eaten, simply avoids a conflict by moving out, returning to repair and use the lodge once again after the otter departs. Beaver probably also move out when otters invade their lodges or burrow systems, although there are reports that occasionally the two species are found together in a single lodge. In one case, four otters and three beavers apparently shared the same accommodation. Mink and otter may prey on the same kinds of aquatic life; it has been observed, however, that the two mustelids tend to avoid each other in normal activity.

REPRODUCTIVE ACTIVITIES. The female otter undergoes post-partum estrus, meaning that shortly after the birth of her litter in late March and early April (Ostenson and Gross, 1940), she is receptive to mating. The courtship and mating may occur over an extended period, terminating of course, when successful fertilization takes place. The male otter probably will mate with the several females which occupy home ranges within or adjacent to his extensive area of activity. In contrast, the female may be courted by only the single male in her area. As in most mustelids, delayed implantation causes extension of the gestation period eight or more months beyond the normal time for fetal growth. During this delay, development, which has proceeded only as far as the blastocyst stage, is totally arrested. Ultimately, growth is continued. When the young are finally born, the female will have carried the progeny for as long as 12 months (Hamilton and Eadie, 1964).

Two or three (as many as four or even six) young are born in late winter or early spring in a lined nest in a protected underground den. At birth, the young are fully furred, have closed eyes, partly closed ear pinnae, pinkish-white claws, conspicuous whiskers, and weigh about 9½ oz (275 g). At about 35 days of age, the young open their eyes. At about 40 days of age, the young are active and excel in playing. At about 75 days of age, the young can accompany their female parent on foraging ventures. Subsequently, the family can remain intact through the autumn and winter until such time as the female is obliged to dismiss her litter in order to prepare for another. The offspring are also apt to depart in the first autumn. The young do not reach full adult size until at least one year old. Although there has been a report of a year-old female mating and having young in her second year of life (Liers, 1958), females normally mate initially in their second year and have their first litters in their third year. On the other hand, males may not become successful breeders until their fifth year.

The combination of rather small litters and the length of time required for the female to reach sexual maturity indicates a low recruitment rate. One Oregon study showed that there were 1.14 female young per adult female at the beginning of the trapping season (Tabor and Wight, 1977). However, otters are prone to live long lives, as determined by examining the cementum layering on canine teeth (Stephenson, 1977), and females can be expected to produce litters for several years. In nature, individuals may live as long as eight to 13 years; in captivity, for as long as 14 to 20 years (Jones, 1979; Scheffer, 1958).

FOOD HABITS. Fish probably constitute the bulk of the otter's diet. Because many of the fish which tempt this worthy swimmer are game and human food species, anglers and conservation agencies have supported studies of the otter's diet to determine the extent of its predation on favored species. Reports on these studies include ones from Oregon (Toweill, 1974), Massachusetts (Sheldon and Toll, 1964), Minnesota (Liers, 1951), Montana (Greer, 1955), New York (Hamilton, 1961), and North Carolina (Wilson, 1954). In Michigan, data on otter foods have been accumulated by Alexander (1977), Field (1970), Knudsen and Hale (1968), Lagler and Ostenson (1942), Manville (1948), Miller (1937), Ryder (1955), and Wenzel (1912). In their classic examination of food residues in digestive tracts of otter taken in early spring from Michigan's trout waters, Lagler and Ostenson (1942) found 35.9% of the contents were composed of forage fishes (suckers, minnows, madtoms, mudminnows, darters, muddlers, sticklebacks); 25.3% amphibians (frogs and mudpuppies); 22.7% game and pan fish (trout, bullheads, northern pike, perch, bass, and sunfish); 3.7% crayfish; and 2.9% aquatic insects (mostly in the orders Coleoptera and Hemiptera). Early spring foods of Michigan otter from nontrout waters included 63.3% game and pan fish (bullheads, northern pike, perch, bass, and sunfish); 14.4% amphibians (frogs and mudpuppies); 11.2% forage fishes (suckers, minnows, mudminnows, darters, muddlers, and sticklebacks); 3.7% crayfish; and 2.9% aquatic insects. This study and the others made on Michigan otter show that the animal seems fully capable of catching most species of fish found in the inland waters, but many of the slow-swimming forage fishes may be easier for them to obtain (Ryder, 1955). Other aquatic life eaten include turtles, ducks, clams, mink, muskrat, and perhaps young beaver.

The otter's ability to catch active prey on land is perhaps overlooked by some biologists. What prompts the animal to leave the stream or lake and forage in the riparian growth nearby is not understood. Perhaps is is because aquatic foods are in short supply. In winter, for example, Field (1970) tracked otter footprints in the snow of the Upper Peninsula and found evidence that the animals had successfully stalked, caught, and eaten meadow voles, southern red-backed voles, and even a snowshoe hare, but disdained white-tailed deer carcasses placed on stream banks specifically to tempt them. Hamilton (1961) reported that a river otter in New York successfully caught and ate a ruffed grouse; Lagler and Ostenson (1942) identified the remains of a red-shouldered hawk (possibly eaten as carrion) in the digestive tract of a Michigan river otter. Berry seeds were found in otter fecal droppings in Roscommon County (Miller, 1937).

ENEMIES. The adult otter probably has few real enemies with which to contend in its Michigan aquatic environment, although young animals could conceivably be eaten by large game fish. On overland excursions, however, the animals, especially immature individuals, might fall victim to such meat-eaters as gray wolf, coyote, bobcat, lynx, bald eagle, and even great horned owl (Jackson, 1961). The trapper, especially if his annual catch is not carefully regulated, is the greatest threat to the otter. Underwater entrapment in fishnets (Scott, 1939) must take some toll, although no statistics are available.

PARASITES AND DISEASES. Despite the otter's aquatic habits which should discourage some ectoparasites, three ticks, *Ixodes banksi, Ixodes hexagonus,* and *Ixodes uriae,* have been collected from this animal (Eley, 1977; Jackson, 1961; Lawrence *et al.,* 1965). Internally, there are reports of two trematodes (flukes), *Euparyphium melis* and *Paragonimus kellicotti,* and at least four nematodes (roundworms), *Physaloptera* sp., *Eustrongylides* sp., *Skrjabingylus lutrae,* and *Drancunculus lutrae,* as parasitizing river otters (Cooley *et al.,* 1982; Erickson, 1946; Shump *et al.,* 1976; Stuht, 1978). There are no positive data that Michigan otters have been affected by accumulations of pesticides, heavy metals or other environmental contaminating residues in the fishes which they eat (Cumbie, 1975).

MOLT AND COLOR ABERRATIONS. There may be two molts annually, according to Jackson (1961), but there is little qualitative difference between winter and summer coats. With age, there is a gradual tendency for the otter's coat to take on a frosted look. Variation in pelage coloration includes an occasional white (albino), black, slate, and mottled condition (Jones, 1923).

ECONOMIC IMPORTANCE. In the seventeenth and eighteenth centuries, otter fur did not share the glamorous reputation and popularity of Upper Great Lakes beaver fur with the fashionable set in England and on the continent. However, American otter fur, as well as fur of the common otter of Eurasia, was sought to adorn the greatcoats of many high-ranking officers, courtesans, and diplomatic officials. In 1767, 5,798 otter pelts were exported by British fur buyers at Fort Michilimackinac (Petersen, 1979), only about 10% of the 50,938 beaver pelts handled at the same time. In 1822, the fur trading post of John Jacob Astor on Mackinac Island was buying otter pelts for $3 to $4 (Anon., 1962). The American Fur Company in 1835–1839 (Johnson, 1969) received 2,222 otter pelts at its Detroit Department (including furs from southern Michigan) and 2,163 at its Northern Department (including furs from northern Michigan). In contrast, marketed beaver pelts received by these two departments totalled only 2,348, a dismal decline from the heyday of the eighteenth century.

At the turn of the present century, the otter was disappearing. However, with the legal closing of otter trapping between 1925 and 1939, the animal made a remarkable comeback. Except for closed seasons in 1942 and 1947, a yield to trappers of an average of 533 otter pelts per year (1940 through 1978, with a record high of 1,157 in 1979–80) demonstrates how a scientifically regulated harvest, compatible with the reproductive potential, has allowed for a sustained population, and even the otter's reestablishment in parts of its former range in southern counties. Raymond D. Schofield, Joseph E. Vogt, and other officials of the Michigan Department of Natural Resources, have kept a close surveillance on each year's catch, requiring that pelts be tagged.

Although otter pelts brought a disappointing $10 to $15 in 1940, prices have risen steadily in recent years, from an average price of $20 for a top quality pelt in 1970–71 to $60 to $90 in 1978–79. The otter trapping season is set to coincide with that for beaver, as both are apt to be caught in the same type of set. In the Upper Peninsula (Zone 1 as classified by the Michigan Department of Natural Resources), the 1979–80 season began on December 2 and extended through April 1, 8, or 15 depending on area. In the northern part of the Lower Peninsula (Zone 2), the season in Area D began on December 2 and extended to March 18; in Areas E, F, and G, the season opened February 10 and extended through March 18, April 8, and March 25, respectively. Trapping will continue to be a source of income for Michigan citizens as long as otter pelts have a fair market value and the annual harvest is not detrimental to the overall sustaining population of these interesting animals.

Anglers have always held the otter in contempt because of the animal's like for fish. As noted in the section of Food Habits, this aquatic mustelid does eat a share of Michigan's game and pan fish, although many of the nongame forage fish are more easily caught (Ryder, 1955). According to Bradt (1944), trout fishermen are the most vociferous critics, sometimes calling for the extermination of otter along favored fishing streams. Complaints also come from persons operating fish hatcheries who have witnessed the dire results of an otter invasion of their stock ponds (Alexander, 1977). However, the ecologist takes the attitude that there should be fish enough for human and otter alike. There certainly was an abundance of these prized fish in the early days when otter were probably more numerous along streams than they are today. Trappers have tended to reduce fish-hungry otters along many northern trout streams, while the Michigan Department of Natural Resources has released hatchery-reared fish to supplement natural production, thus tending to keep the balance in favor of the fish.

The naturalist, tourist, and outdoor person are grateful for the river otter's presence in Michigan. There is always a chance that the observer may glimpse an otter sliding down a muddy incline or jumping out of the water and onto a bank, fish in mouth. Emil Liers (1951) has described the delightful experience of having semi-tamed individuals using streamways in one's backyard. Confining Michigan river otters under captive conditions is not legal, except for zoos and educational institutions. Nevertheless, these animals are easily tamed.

COUNTY RECORDS FROM MICHIGAN. Records of otters in collections of museums, from the literature, from trapper's reports, and provided by personnel of the Michigan Department of Natural Resources show that this species occurs in all counties in the Upper Peninsula and in most counties in the upper two-thirds of the Lower Peninsula (see Map 61). The species also is reported from Beaver Island in Lake Michigan (Hatt, *et al.,* 1948; Ozoga and Phillips, 1964; Strang, 1855), Isle Royale in Lake Superior (Peterson, 1977), and Bois Blanc and Drummond Island in Lake Huron (Cleland, 1966; Manville, 1950).

LITERATURE CITED

Alexander, G. R.
1977 Food of vertebrate predators on trout waters in north central Lower Michigan. Michigan Acad., 10(2):181–195.

Anon.
1962 Fur and hide values. 1822. Michigan Conserv., 31(6): 36–37.

Bradt, G. W.
1944 What about the otter? Michigan Dept. Conserv., Game Div. (mimeo.), 4 pp.

Cleland, C. E.
1966 The prehistoric animal ecology and ethnozoology of the Upper Great Lakes Region. Univ. Michigan, Mus. Anthro., Anthro. Papers, No. 29, x+294 pp.

Cook, D. B.
1940 An otter takes a ride. Jour. Mammalogy, 21(2):216.

Cooley, T. M., S. M. Schmitt and P. D. Friedrich
1982 River otter survey—1981–82. Michigan Dept. Nat. Res., Wildlife Div., Rept. No. 2915 (Mimeo.), 7 pp.

Cumbie, P. M.
1975 Mercury levels in Georgia otter, mink and freshwater fish. Bull. Environ. Conserv., 14(2):193–196.

Dearborn, J. H.
1939 Notes on an otter fishing. Jour. Mammalogy, 20(3): 370.

Dearden, L. C.
1954 Extra premolars in the river otter. Jour. Mammalogy, 35(1):125–126.

Eley, T. J., Jr.
1977 *Ixodes uriae* (Acari:Ixodidae) from a river otter. Jour. Med. Entomol., 13(3–4):506.

Engels, W. L.
1933 Notes on the mammals of St. Joseph County, Indiana. American Midl. Nat., 14(1):1–16.

Erickson, A. B.
1946 Incidence of worm parasites in Minnesota Mustelidae and host lists and keys to North American species. American Midl. Nat., 36(2):494–509.

Erlinge, S.
1967 Home range of the otter, *Lutra lutra* L., in southern Sweden. Oikos, 18(2):186–209.
1968 Territoriality of the otter, *Lutra lutra* L. Oikos, 19(1): 81–98.

Field, R. J.
1970 Winter habits of the river otter (*Lutra canadensis*) in Michigan. Michigan Acad., 3(1):49–58.

Greer, K. R.
1955 Yearly food habits of the river otter in the Thompson Lakes region, northwestern Montana, as indicated by scat analyses. American Midl. Nat., 54(2):299–313.

Hamilton, W. J., Jr.
1961 Late fall, winter and early spring foods of 141 otters from New York. New York Fish and Game Jour., 8(2):106–109.

Hamilton, W. J., Jr., and W. R. Eadie
1964 Reproduction in the otter, *Lutra canadensis*. Jour. Mammalogy, 45(2):242–252.

Harris, C. J.
1968 Otters. A study of the Recent Lutrinae. Weidenfeld and Nicolson, London. xiv+397 pp.

Hatt, R. T., J. Van Tyne, L. C. Stuart, C. H. Pope, and A. B. Grobman
1948 Island life: A study of the land vertebrates of the islands of eastern Lake Michigan. Cranbrook Inst. Sci., Bull. No. 27, xi+179 pp.

Hooper, E. T., and B. T. Ostenson
1949 Age groups in Michigan otter. Univ. Michigan, Mus. Zool., Occas. Papers No. 518, 22 pp.

Jackson, H. H. T.
1961 Mammals of Wisconsin. Univ. Wisconsin Press, Madison. xii+504 pp.

Jenkins, D. H.
1949 A report on the otter in Michigan. Michigan Dept. Conserv., Game Div., Rept. No. 1037 (mimeo.), 7 pp.

Johnson, D. R.
1969 Returns of the American Fur Company, 1835–1839. Jour. Mammalogy, 50(4):836–839.

Jones, M. L.
1979 Longevity of mammals in captivity. Internat. Zoo News, 26(3):16–26.

Jones, S. V. H.
1923 Color variations in wild animals. Jour. Mammalogy, 4(3):172–177.

Knudsen, G. J., and J. B. Hale
1968 Food habits of otters in the Great Lakes Region. Jour. Wildlife Mgt., 32(1):89–93.

Lagler, K. F., and B. T. Ostenson
1942 Early spring food of the otter in Michigan. Jour. Wildlife Mgt., 6(3):244–254.

Lawrence, W. H., K. L. Hays, and S. A. Graham
1965 Arthropodous ectoparasites from some northern Michigan mammals. Univ. Michigan, Mus. Zool., Occas. Papers No. 639, 7 pp.

Liers, E. E.
1951 Notes on the river otter (*Lutra canadensis*). Jour. Mammalogy, 32(1):1–9.
1958 Early breeding in the river otter. Jour. Mammalogy, 39(3):438–439.

Manville, R. H.
1948 The vertebrate fauna of the Huron Mountains, Michigan. American Midl. Nat. 39(3):615–640.
1950 The mammals of Drummond Island, Michigan. Jour. Mammalogy, 31(3):358–359.

Melquist, W. E., and M. G. Hornocker
1979 Methods and techniques for studying and censusing river otter populations. Univ. Idaho, Forest Wildlife and Range Exp. Sta., Tech. Rept. 8, 17 pp.

Miller, H. J.
1937 Field work on otter inhabiting Lake St. Helen and vicinity. Michigan Dept. Conserv., Game Div., Rept. no. 339 (mimeo.), 10 pp.

Mowbray, E. E., Jr., J. A. Chapman, and J. R. Goldsberry
1976 Preliminary observations on otter distribution and habitat preferences in Maryland with descriptions of otter field sign. Trans. 33rd Northeastern Sect., The Wildlife Soc., pp. 124–131.

Ostenson, B. T., and J. W. Gross
1940 Size and sex groups, breeding condition of otter taken in Michigan during March and April, 1940. Michigan Dept. Conserv., Game Div., unpubl. Rept., (mimeo)., 8 pp.

Ozoga, J. J., and C. J. Phillips
1964 Mammals of Beaver Island, Michigan. Michigan State Univ., Publ. Mus., Biol. Ser., 2(6):305–348.

Peterle, T. J.
1954 An observation on otter feeding. Jour. Wildlife Mgt., 18(1):141–142.

Petersen, E. T.
1979 Hunters' heritage. A history of hunting in Michigan. Michigan United Conserv. Clubs, Lansing. 55 pp.

Peterson, R. O.
1977 Wolf ecology and prey relationships on Isle Royale. National Park Serv., Sci. Monogr. Ser., No. 11, xx+210 pp.

Ryder, R. A.
1955 Fish predation by the otter in Michigan. Jour. Wildlife Mgt., 19(4):497–498.

Scheffer, V. B.
1953 Otters diving to a depth of sixty feet. Jour. Mammalogy, 34(2):255.
1958 Long life of a river otter. Jour. Mammalogy, 39(4): 591.

Schofield, R. D.
1961 Survey of the distribution of the otter in Michigan. Michigan Dept. Nat. Res., Game Div., Rept. No. 2331 (mimeo.), 5 pp.

Scott, W. E.
1939 Swimming power of the Canadian otter. Jour. Mammalogy, 20(3):371.

Seton, E. T.
1953 Lives of game animals. Vol. II, Pt. II. Bears, coons, badgers, skunks, and weasels. Charles T. Branford Co., Boston. pp. 369–746.

Severinghaus, C. W., and J. E. Tanck
1948 Speed and gait of an otter. Jour. Mammalogy, 29(1): 71.

Sheldon, W. G., and W. G. Toll
1964 Feeding habits of the river otter in a reservoir in central Massachusetts. Jour. Mammalogy, 45(3):449–455.

Shump, K. A., Jr., A. U. Shump, R. A. Aulerich, and G. A. Heidt
1976 A bibliography of Mustelids. Part V: Otters. Michigan State Univ., Agri. Exp. Sta., Jour. Art. No. 7759, 32 pp.

Smith, L. H.
1939 Notes on an otter fishing. Jour. Mammalogy, 20(3): 370

Stephenson, A. B.
1977 Age determination and morphological variation of Ontario otters. Canadian Jour. Zoology, 55(10):1577–1583.

Strang, J. J.
1855 Some remarks on the natural history of Beaver Islands, Michigan. Smithsonian Inst., 9th Ann. Rept., pp. 282–288.

Stuht, J. N.
1978 Paragonimiasis in a river otter. Michigan Dept. Nat. Res., Wildlife Div., Rept. No. 2824 (mimeo.), 2 pp.

Tabor, J. E., and H. W. Wight
1977 Population status of river otter in western Oregon. Jour. Wildlife Mgt., 41(4):692–699.

Toweill, D. E.
1974 Winter food habits of river otters in western Oregon. Jour. Wildlife Mgt., 38(1):107–111.

van Zyll de Jong, C. G.
1972 A systematic review of the Nearctic and Neotropical river otters (genus *Lutra*, Mustelidae, Carnivora). Royal Ontario Mus., Life Sci. Contr., 80, 104 pp.

Wenzel, O. J.
1912 A collection of mammals from Osceola County, Michigan. Michigan Acad. Sci., 14th Ann. Rept., pp. 198–205.

Wilson, K. A.
1954 The role of mink and otter as muskrat predators in northeastern North Carolina. Jour. Wildlife Mgt., 18(2):199–207.

Wood, N. A.
1922 The mammals of Washtenaw County, Michigan. Univ. Michigan, Mus. Zool., Occas. Papers No. 123, 23 pp.

FAMILY FELIDAE

The felids or cats, the most specialized of carnivores, have a rather uniform group of distinctive characteristics. Most are meat-eaters and proficient predators. Exceptions are the fruit-eating Indonesian cat (*Felis planiceps*) and the fish-catching and clam-digging cat (*Felis viverrina*) of southeastern Asia and adjacent islands. The cats, numbering approximately four or five genera and 37 species, are represented on all major land masses except Australia and associated islands, Antarctica, Malagasy, and other offshore islands.

The felids range in size from the small European wild cat (*Felis silvestris*) to the massive tiger (*Panthera tigris*) and the lion (*Panthera leo*). A distinctive feature of this family is the marked reduction in the number of premolar and molar teeth. Michigan felids have 28 teeth (bobcat and lynx) or 30 teeth (extirpated mountain lion). These teeth include large canines (to stab and kill) and well-developed carnassials (to gnaw or help shear off pieces of meat for bolting). This reduction in the number of teeth correlates with the presence of a short, broad rostrum, a singular feature of the cats. This shortness accentuates the powerful bite of these animals. The orbits, conspicuously large in the skulls of most cats, are directed forward. The muscular, wiry body often seems too large for the smallish heads, except in the large roaring species (genus *Panthera*). The felids have a digitigrade stance (walking on tiptoes). The forefeet have five toes, with the first (pollex) high on the wrist; the hind feet have four toes. The sharp claws are compressed, curved and retractable except in the case of the cheetah (*Acinonyx jubatus*). The forelimbs are strongly built and effective in clutching and grappling with prey. Some species are striped or spotted. Although all senses are keenly developed, most cats are traditionally sight hunters who watch for their prey, silently stalk it, and capture the victim with a quick burst of speed.

MOUNTAIN LION
Felis concolor Linnaeus

NAMES. The generic name, *Felis,* proposed by Linnaeus in 1758, is the Latin name for cat. The specific name, *concolor,* proposed by Linnaeus in 1771, consists of the Latin prefix *con,* which means the same, and the Latin noun *color,* which means hue or tint. The name mountain lion is most generally used in western states as the common name for this large cat and is preferred by most mammalogists. However, the species is also called cougar, panther, painter, puma, and catamount. In Michigan, people generally call this large cat either cougar (Burt, 1946) or puma. Mountain lions living formerly in the Upper Peninsula are assumed to have had closest affiliations with populations once occurring westward to Wisconsin (Jackson, 1961) and are classified as belonging to the subspecies *Felis concolor schorgeri* Jackson. In the Lower Peninsula, Burt (1946) placed the now extirpated population in the subspecies *Felis concolor couguar* Kerr.

RECOGNITION. Body size large, rangy, and somewhat similarly proportioned to a rather thin, lanky housecat, although several times larger (Fig. 96); length of head and body averaging 57 in (1,440 mm) for the larger male and 50 in (1,240 mm) for the smaller female; tail long; length of tail vertebrae averaging 30¾ in (780 mm) for the larger male and 26¾ in (675 mm) for the smaller female; head comparatively small in contrast to elongate, muscular body; nose with typical cat-like bluntness; nose pad large; eyes large; ears conspicuous, rounded, erect; vibrissae (whiskers) long, coarse; neck thick, muscular; body standing higher in the rear because forelimbs are shorter than hind limbs; tail round, well-haired, and neither bushy nor tapered; legs rather short, muscular; feet large, broad, with five digits on forefeet (pollex high on wrist) and four on hind feet; claws sharp, curved, sturdy; mammary glands consist of three abdominal pairs; males 15 to 20% larger than females.

Figure 96. The mountain lion (*Felis concolor*).
Sketches by Bonnie Marris.

The pelage of the mountain lion is rather short and coarse. The general coloration ranges from light brown or tawny to grayish-brown and is darker on the back; paler, almost buffy, on the belly; and near white on throat and chest. The nose pad is pinkish bordered by blackish which extends to the lips. The muzzle stripes, back of ears, and tip of tail are black. Individuals less than six months of age have a general pale buff coloration marked with conspicuous dark brown or black blotches, with the tail banded with the same coloring.

FIELD IDENTIFICATION. This large, rangy cat should never be mistaken for any other Michigan mammal, wild or domesticated. Nevertheless, from time to time, experienced field observers report an animal thought to be a mountain lion, especially in remote parts of northern Michigan. Neither actual specimens nor clear photographs have ever been submitted to authorities for verification. On occasion, photographs or plaster castings of tracks allegedly made by mountain lions have been brought to knowledgeable field biologists. To date, these tracks are thought to have been made by such large, big-footed dogs as Great Danes, or St. Bernards. Some of these animals have tracks as wide as 4 in (102 mm). In contrast, mountain lion tracks are rarely more than 3 to 3½ in (76 to 89 mm) wide. Experienced trackers report that: (1) marks of the "thin" toenails of the mountain lion and the "broad" toenails of a dog may not be helpful in distinguishing tracks although claw marks are more apt to be seen in the case of the big cat; (2) the prints of the individual toes tend to be small and teardrop shaped in the mountain lion and larger and less egg-shaped in the dog; in both cases the toes may be spread—in a walking cat and a running dog; (3) the prints of the heel pad are perhaps most distinctive; for the mountain lion the front edge is broad and rather squared off and the rear edge usually has a three-lobed shape; for the dog, the front of the pad is narrow and well rounded and the rear usually straight across or concave. When large and questionable tracks are observed, it is always best to make casts using plaster of Paris from which mammalogists can make a specific determination.

MEASUREMENTS AND WEIGHTS. Adult specimens of the male and the female measure, respectively, as follows: length from tip of nose to end of tail vertebrae, 69 to 109 in (1,710 to 2,740 mm) and 60 to 93 in (1,500 to 2,330 mm); length of tail vertebrae, 27 to 47 in (660 to 900 mm) and 21 to 28½ in (530 to 820 mm); length of hind foot, 9½ to 11½ in (240 to 290 mm) and 8¾ to 10¾ in (220 to 270 mm); height of ear from notch (almost the same in both sexes), 3 to 4 in (75 to 100 mm); weight, 147 to 226 lbs (67 to 103 kg) and 80 to 132 lbs (36 to 60 kg).

DISTINCTIVE CRANIAL AND DENTAL CHARACTERISTICS. The skull of the mountain lion, although seemingly small for its body size, is massive in construction and noticeably short and broad. The width is almost two-thirds of the greatest length (Fig. 97). The region of the forehead is high and arched, the nasal bones are broad as is the entire rostrum. The sagittal crest and the lambdoidal ridges are well developed, the latter overhanging the foramen magnum. The short mandible is deep and powerfully constructed. The incisors are small and straight across. The canines are heavy, long, and compressed. The carnassial teeth (fourth upper premolar and first lower molar) are elongate and massive. The post-canine teeth number four on each side of the upper jaws and three on each side of the lower jaws. The number of teeth is identical to that of the domesticated cat but includes one more small premolar on each side of the upper jaw than in the lynx and bobcat. The dental formula is:

$$\text{I (incisors)}\frac{3\text{-}3}{3\text{-}3}, \text{C (canines)}\frac{1\text{-}1}{1\text{-}1}, \text{P (premolars)}\frac{3\text{-}3}{2\text{-}2}, \text{M (molars)}\frac{1\text{-}1}{1\text{-}1} = 30$$

DISTRIBUTION. The mountain lion has one of the most extensive distributions of all American terrestrial mammals, occurring in the diverse frigid, temperate, and torrid environments of the New World from southern Canada southward to the bleak Patagonian plains of southern South America (Goldman *in* Young and Goldman, 1946). With the encroachment of civilization, especially in the past 200 years, this large cat has lost much of its primeval range in the vicinity of population centers and in areas where its presence has been detrimental to the interests of livestock owners. In eastern United States, the mountain lion had all but disappeared by the mid-eighteenth century, lingering longest in remote mountainous areas. Small enclaves remained chiefly in peninsula Florida (Eaton, 1973) and in the New Brunswick area of eastern Canada (Wright, 1972), and the occasional documented records of mountain lion

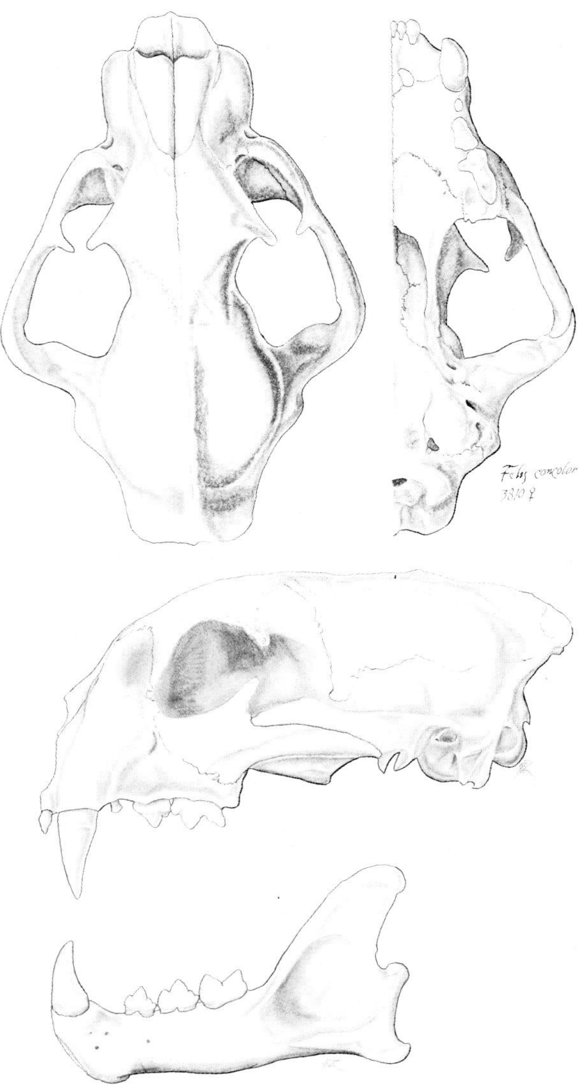

Figure 97. Cranial and dental characteristics of the mountain lion (*Felis concolor*).

in northeastern and southeastern United States probably resulted from cats dispersing from these two foci. In the western Great Lakes sector, however, the mountain lion almost entirely disappeared prior to the beginning of the twentieth century (Schorger, 1942). Today, this part of the border region between the United States and Canada remains remote from extant mountain lion populations in New Brunswick and Florida, although less so from the nearest populations in Manitoba and extreme southwestern Ontario (Banfield, 1974; Nero and Wrigley, 1977). To reach Michigan from any of these places, a mountain lion would have to traverse vast expanses of thickly settled and open countryside where its presence would surely be detected. Nevertheless, it is entirely possible that recent records in Minnesota (Bue and Stenland, 1953) represent this animal moving in from more northwesterly localities.

In Michigan, the earliest mountain lion record is a nearly complete skull with lower jaw (Fig. 98) found in a burial containing four children uncovered in a 1966 archaeological excavation in the southwest quarter of Section 8 just north of the Flint River in Taymouth Township (T 10N,R 5E)

Figure 98. The only extant specimen of a Michigan mountain lion (*Felis concolor*). The partial skull with lower jaw (MSU Museum cat. no. 13178) is from an archaeological site in Taymouth Township in Saginaw County (Foster and Hagge, 1975).
This frontal view showing the complete right upper canine was photographed by Bruce R. Baker.

in Saginaw County (Foster and Hagge, 1975). This specimen may have been placed in association with the burial as early as pre-Columbian times (no datable artifacts were present) and is the only extant specimen of a Michigan mountain lion (MSU Museum, cat. no. 13,178).

In early settlement days, mountain lions were apparently distributed statewide (Map 62), becom-

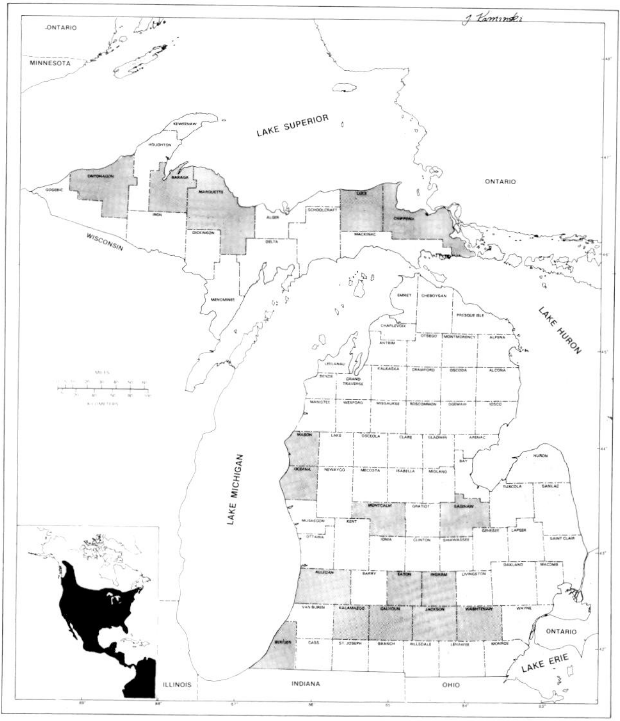

Map 62. Presumed presettlement geographic distribution of the mountain lion (*Felis concolor*). Dark shading indicates those counties from which historic reports of the presence of this large cat are known.

ing rare by the 1830s (Goodrich, 1940). Although many early references do not clearly distinguish between mountain lion, bobcat and lynx, presumably the catamount referred to by Hoyt (1889) as living in Oakland County between 1825 and 1830 was a mountain lion. At that time, mountain lions still roamed the Washtenaw County area, with the last one recorded in the vicinity of Manchester in 1870. Late nineteenth century records included the report of a mountain lion taken in 1875 at Pleasant Lake in Ingham County and of another treed by dogs near Stanton in Montcalm County in 1885 (Wood and Dice, 1924). Wood (1914) and

Burt (1946) listed records of early occurrence from such Lower Peninsula counties as Allegan, Berrien, Calhoun, Eaton, Ingham, Jackson, Kalamazoo, Mason, Montcalm, Oceana, and Washtenaw.

In the Upper Peninsula, the mountain lion apparently lingered longer, although it was Schorger's (1942) opinion that the last records were for 1850. However, Seton (1953) cited an article in the *Sault Ste. Marie News* for January 3, 1907, stating that a mountain lion, 5½ ft long, was captured in a wolf trap at Deerfoot Lodge near the Taquamenow [=Tahquamenon] River, presumably in Chippewa or Luce counties. He also noted that the *Minneapolis Journal* reported a mountain lion in the vicinity of Silver River about 7 mi from L'Anse in Baraga County. Wood (1914) also recorded the animal from Ontonagon County. Although Manville (1948) supposed all Michigan mountain lions were extirpated by about 1900, he does list a "documented record" from the Huron Mountains of Marquette County in 1937.

Professional wildlife biologists do not believe mountain lions have lived naturally in Michigan for many decades. However, seldom a year passes without one or more "records" of these large cats coming to light, primarily in the Upper Peninsula. Between 1976 and 1981, reports have been publicized of mountain lion sightings in such Upper Peninsula counties as Alger, Baraga, Chippewa, Delta, Gogebic, Houghton, Marquette, Menominee, and Ontonagon (Richey, 1981). There is a convincing article by La Pointe (1978) about a big cat in the Porcupine Mountains of Ontonagon County. Despite all of this, mammalogists have yet to examine a specimen, an authentic footprint, or even view a photograph of the animal. It should be remembered that although the mountain lion, in spite of its large size, is most secretive in its actions, it will (1) leave numerous tracks and other characteristic signs, and (2) periodically need to make a kill, either a white-tailed deer or a head of domestic stock, using identifiable feline techniques including the caching of the uneaten portions. Consequently, its presence can not escape detection for long. As Elsworth Harger, felid expert for the Michigan Department of Natural Resources reported in an article by Smith (1978), one of these animals would certainly have been shot, treed by hounds, or trapped, if actually present.

HABITAT PREFERENCES. The mountain lion is a highly adaptative animal as demonstrated by its ability to thrive in a variety of habitats, in both warm and cold climates, and in both lowlands and high mountains. Some woody cover and an abundant supply of prey animals, especially members of the deer family, are two basic environmental requirements. Historic records indicate the mountain lion found all parts of Michigan suitable for its lifestyle, with stream valleys and adjacent uplands perhaps most favorable.

DENSITY AND MOVEMENTS. Population numbers and the mobility of mountain lions which once lived in the Upper Great Lakes Region can only be estimated using data from recent studies made in western states and provinces. Although some data has been based on track counts, work done chiefly in Idaho on mountain lions tracked by means of radio telemetry has produced much new information about this cat's habitats (Seidensticker *et al.*, 1970). Findings summarized in such reports as those by Hornocker (1969) and Seidensticker *et al.* (1973) show that mountain lion populations consists of: (1) resident individuals (male and females) with established home ranges, and (2) transient animals (often juveniles) in search of living space. Normally, resident males have large, nonoverlapping areas while females use smaller, more restricted ranges, one or more of which may be enclosed by that of a single, resident male. Where man is not a constant predator, there is evidence that fairly stable populations are maintained in suitable habitat year after year. Under such circumstances, the animals may occur in densities as low as one individual per 33 sq mi (85 sq km) as estimated for an area in California (Koford, 1978) to perhaps as high as one per 5 to 21 sq mi (13 to 54 sq km) in a Colorado study site (Currier *et al.*, 1977).

Experienced field observers report that mountain lions are highly mobile, covering distances of 25 mi (40 km) or even more (Seton, 1953, and Young and Goldman, 1946). But these high figures may actually be attributable to transient individuals, because monitoring of mountain lions equipped with radio collars show that resident females use relatively small areas from 5 to 20 sq mi (13 to 52 sq km), with resident males using ranges no more than twice as large (Hornocker,

1969). Summer ranges are larger than those used in winter (Seidensticker *et al.,* 1973). Radio studies also show that actual movements of resident individual cats, tracked on a 24-hour basis, may be no more than 7.7 mi (12.9 km) for males and 4.5 mi (7.6 km) for females.

BEHAVIOR. Recent investigations using radio telemetry and other refined data-gathering instruments provide additional evidence that many mammals live in a fairly orderly world, spacing themselves so that local food supplies and other habitat essentials are not overly abused. The mountain lion uses this arrangement for survival. Females with dependent kittens have living spaces within the larger expanses used by the resident males. Apparently the only real struggle develops when dispersing young animals must look for unoccupied places in which to settle. When a resident mountain lion dies, transients probably interact considerably to determine which will gain this place of residence. As Hornocker (1969) points out, territorial defense is developed through mutual avoidance. Residents mark their areas by making scrapes and tree scratch marks (often as far up a trunk as the animal can reach) where urine or fecal materials are deposited, providing both a visual and olfactory signal to others. The solitary existence of the mountain lion is interrupted only during the breeding season and, of course, during the period of juvenile dependency when females and young are together.

The mountain lion is primarily nocturnal in its activity. It relies on keen vision, smell, and hearing to acquaint itself with its environment and food sources. Its movements closely resemble those of the domestic cat. It can travel almost noiselessly, hunting primarily by stalking until close enough to spring on its prey. In the case of a white-tailed deer, the shoulder is the basic point of attack. A deep bite in the neck subdues the victim while the mountain lion's talons rip open the head and flanks. Partly eaten prey may be cached under leaves or other brush. The big cats are rarely attracted to carrion, although their cached meat may become ripe before it is consumed. The mountain lion is an excellent climber and swims when necessary. Many of the mountain lion's behavioral characteristics have been recorded in such treatises as those by Barnes (1960), Seton (1953), Young and Goldman (1946), and Wright (1972).

The mountain lion vocalizes with cat-like hisses, growls, cries and purrs. Barnes (1960) categorized some of these utterances as: (a) courting sounds—yeowings and squalls; (b) warning sounds—hisses and growls; (c) eating sounds—whirring purrs; (d) fighting sounds—snarls, hisses, and whistles; and (e) sounds of contentment—soft, shrill whistles and clucks. Barnes described the so-called scream as a special vocalization about which there is considerable controversy (Young and Goldman, 1946). Many outdoor authorities (McCabe, 1949) will attest that there is such a high piercing yell or "caterwauling." There is, however, some evidence that only the female utters this call. The mountain lion is, a purring cat in contrast to the jaguar, tiger, leopard, and lion which roar. This difference in vocalization appears correlated with the presence of an elastic (cartilagineous) hyoid apparatus supporting the tongue in the "roarers" and a bony (ossified) hyoid apparatus in the "purrers."

Supposedly the mountain lion denned in early Michigan under the cover of a tree with branches low to the ground, beneath an overhanging bank, in a tangle of tree roots, or within a brush pile (Jackson, 1961).

ASSOCIATES. Mountain lions are either solitary males or females with dependent kittens. Opposite sexes appear to associate only during the courtship and mating season. Smaller carnivores may be attracted to mountain lions kills, especially when the food is cached and not under surveillance.

REPRODUCTIVE ACTIVITIES. Observers indicate that the mountain lion passes through a kitten stage, a transient juvenile stage, and finally to the resident adult stage. In the kitten stage, the young animal is dependent on the female parent with its movements restricted to her home range. The transient juvenile is independent and moves around with no parental attachments, seeking a home site. The resident adult is established in a defined area of activity. Apparently, females do not breed successfully until situated in a home area. Transient males rarely become involved in mating.

Courtship and mating generally occur from December to March. After gestation periods of between 90 to 96 days, litters are born from June to September, with a peak in July (Robinette *et al.,* 1961). There is also evidence that births may occur in any month of the year (Banfield, 1974). A litter usually consists of two to four kittens (as many as five or six). At birth, the young have a wooly, spotted coat and a striped tail the same coloring as the body, have folded ear pinnae and closed eyes, chirp vociferously, and weigh about 14 oz (400 g). At 10 days of age, the young have open eyes and unfolded ear pinnae; their first teeth have erupted, and they begin play (Eaton and Velander, 1977). At about 40 days of age, the young undergo weaning, eat solid food, and can join their female parent on foraging trips. At about 180 days of age, the young molt, lose their spots, and grow a juvenile coat of tan. The female parent and her kittens will remain together for as long as 12 months. As a rule, the female has a litter every other year. Females reach sexual maturity at about 2½ years of age; males at three.

Because of the extended maturation time for these large cats and the uncertainty of locating a suitable home range within a short time, individuals may often be more than three years old before engaging in reproductive activities. Individuals could live as long as eight or more years in the wild; in captivity, there are records of mountain lions living for 19 years (James, 1977).

FOOD HABITS. According to Hornocker (1969), in a natural situation, territoriality plays an important role in spacing mountain lions to insure individual cats greater success in obtaining large prey animals. Hornocker (1970) further pointed out that the mountain lion, as a predator on large hoofed animals, exerts ". . . a powerful force acting to dampen and protract severe prey oscillations and to distribute ungulates on restricted, critical range." Hornocker also noted that these big cats remove less fit individuals from the prey population with the ultimate result of ecologic stability in the environment. Although Hornocker's remarks related to studies in Idaho, the mountain lion probably played a similar role in presettlement Michigan.

Because the mountain lion in eastern United States was almost extirpated by the time field ecologists began examining its food requirements (Wright, 1972), most data on this cat's diet are derived from studies made in western states and provinces (Barnes, 1960; Young and Goldman, 1946), including British Columbia (Spalding and Lesowski, 1971), Idaho (Hornocker, 1970), Oregon (Toweill and Meslow, 1977), Arizona (Shaw, 1977), and Nevada and Utah (Robinette *et al.*, 1959). The mountain lion apparently makes a kill of a sizable prey species periodically but between kills catches assorted smaller creatures. Judging from western studies, early Michigan populations of wapiti, white-tailed deer, moose, caribou, and perhaps bison were undoubtedly the major prey species sought by mountain lions. Most resident smaller creatures were also fair game: voles, mice, squirrels, beaver, muskrat, porcupine, snowshoe hare, raccoon, coyote, striped skunk, birds, even fish and snails. With the coming of settlers and their domesticated stock, the mountain lion quickly developed a taste for poultry, pigs, sheep, calves, and colts. Young and Goldman (1946) vividly described the great cat's downfall in the United States as a result of such depredations. The ecologist recognizes that the mountain lion may not be compatible with livestock raising, especially where herds and flocks are unattended on open range. Nevertheless, areas of sufficient magnitude, devoid of domestic livestock, should be set aside as range for this spectacular carnivore.

ENEMIES. The incompatibility of the mountain lion with human interests has, since the time of the earliest settlement, spelled doom for this cat in most of the United States. Aside from the human's traps, trail hounds and guns, the adult mountain lion has little to fear from its associated animal life (Young and Goldman, 1946). Kittens, when away from the female parent's watchful care, could be overpowered by any of the larger meat-eating mammals, including hungry male mountain lions. Eagles and large hawks could also kill young individuals.

PARASITES AND DISEASES. Ectoparasites and endoparasites infest the mountain lion (Jackson, 1961; Young and Goldman, 1946). External parasites include a louse (*Trichodectes felis*), three ticks (*Dermacentor variabilis, Ixodes ricinus, Ixodes cookei*), and a flea (*Arctopsylla setosa*). Internal parasites reported include a roundworm (*Physaloptera praeputialis*), and three tapeworms (*Taenia taeniaformis, Taenia lyncis, Echinococcus granulosis*). Rabies also infects the mountain lion.

MOLT AND COLOR ABERRATIONS. Molt in the mountain lion is apparently a gradual process beginning in late spring. The summer coat tends to be paler and the winter pelage darker. Males are sometimes grayer; females tend to be more reddish. Individuals in the dark (melanistic) color phase have been reported from South America (Young and Goldman, 1946). In North America, reports of blackish animals have been received from Louisiana (Lowery, 1974), Wisconsin (Jackson, 1961), and New Brunswick (Wright, 1972).

ECONOMIC IMPORTANCE. The mountain lion, as mentioned previously, made itself most unwelcome in the early days of settlement. It not only preyed on domestic stock but was also considered a potential danger to children and adults. Wood (1922), for example, cites an 1830 report that the mother of Miss Julia Dexter Stannard was chased by one of these big cats when she was returning at dusk on horseback to her home in Webster Township in Washtenaw County. Such incidents helped seal the fate of the large cats except in the most remote and rugged terrain. In modern times, luridly illustrated sports stories often exaggerate the ferocity of the mountain lion (O'Connor, 1959). In most scientific treaties (Barnes, 1960; Seton, 1953; Young and Goldman, 1946) there is ample evidence presented to show that the mountain lion is normally very shy in the presence of humans and even is treed by small domestic dogs. However, there have been some records of attacks, some fatal.

The mountain lion's pelt has been of little value on the fur market; however, it has considerable trophy value, as a life-like mount or as a flat skin attached to a wall or fashioned into a rug. Several western states have limited open seasons, so it is still possible to sport hunt (usually with trail hounds) this cat. The flesh of the mountain lion, as with other cats, was highly prized as food by Indians and early settlers. Although the mountain lion has passed from the Michigan scene and remains only as a part of the folklore in this state, the species will always persist in some remote areas in North and South America. Because it thrives in captivity, zoos will continue to provide close range inspection of this handsome beast of prey.

COUNTY RECORDS FROM MICHIGAN. Records of mountain lion gleaned mostly from historic documents show that the species probably lived in all parts of both the Upper and Lower Peninsulas, being reduced severely in the nineteenth century and ultimately extirpated in the early years of the twentieth century. The only extant Michigan specimen is a partial skull with lower jaw from a prehistoric Indian grave in Saginaw County (Foster and Hagge, 1975). County records are shown on Map 62.

LITERATURE CITED

Banfield, A. W. F.
1974 The mammals of Canada. Univ. Toronto Press, Toronto. xxiv+438 pp.

Barnes, C. T.
1960 The cougar or mountain lion. The Ralton Co., Salt Lake City, Utah. 176 pp.

Bue, G. T., and M. H. Stenlund
1953 Recent records of the mountain lion, *Felis concolor*, in Minnesota. Jour. Mammalogy, 34(3):390–391.

Burt, W. H.
1946 The mammals of Michigan. Univ. Michigan Press, Ann Arbor. xv+288 pp.

Currier, M. J. P., S. L. Sheriff, and K. R. Russell
1977 Mountain lion population and harvest near Canon City, Colorado, 1974–1977. Colorado Div. Wildlife, Sp. Rept. No. 42, 12 pp.

Eaton, R. L.
1973 The status, management and conservation of the cougar in the United States. The World's Cats, 1:68–86.

Eaton, R. L., and K. A. Velander
1977 Reproduction in the puma: Biology, behavior and ontogeny. The World's Cats, 3(3):45–70.

Foster, D. W., and D. R. Hagge
1975 A unique secondary burial of four children found in Taymouth Township, Saginaw County, Michigan. Michigan Archaeologist, 21(2):63–70.

Goodrich, C.
1940 The first Michigan frontier. Univ. Michigan Press, Ann Arbor. viii+344 pp.

Hornocker, M. G.
1969 Winter territoriality in mountain lions. Jour. Wildlife Mgt., 33(3):457–464.
1970 An analysis of mountain lion predation upon mule deer and elk in the Idaho Primitive Area. Wildlife Monog., No. 21, 39 pp.

Hoyt, J. M.
1889 History of the town of Commerce. Michigan Pioneer Coll., 14:421–430.

Jackson, H. H. T.
1955. The Wisconsin puma. Proc. Biol. Soc. Washington, 68:149–150.
1961 Mammals of Wisconsin. Univ. Wisconsin Press, Madison. xii+504 pp.

James, M. L.
1977 Record keeping and longevity of felids in captivity. The World's Cats, 3(3):132–138.

Koford, C. B.
1978 The welfare of the puma in California, 1976. Carnivore, 1(1):92–96.

LaPointe, D.
1978 The cat that isn't. Michigan Nat. Res. Mag., 47(1):28–30.

Lowery, G. H., Jr.
1974 The mammals of Louisiana and its adjacent waters. Louisiana State Univ. Press, Baton Rouge. xxiii+565 pp.

Manville, R. H.
1948 The vertebrate fauna of the Huron Mountains, Michigan. American Midl. Nat., 39(3):615–640.

McCabe, R. A.
1949 The scream of the mountain lion. Jour. Mammalogy, 30(3):305–306.

Nero, R. W., and R. E. Wrigley
1977 Status and habits of the cougar in Manitoba. Canadian Field-Nat., 91(1):28–40.

O'Connor, J.
1959 The mountain lion. Outdoor Life, 124(5):48–51, 88–89.

Richey, D.
1981 Tall tales? or cougars? The Detroit News, 108(177): 10D, 15 February.

Robinette, W. L., J. S. Gashwiler, and O. W. Morris
1959 Food habits of the cougar in Utah and nevada. Jour. Wildlife Mgt., 23(2);261–273.
1961 Notes on cougar productivity and life history. Jour. Mammalogy, 42(2);204–217.

Schorger, A. W.
1942 Extinct and endangered mammals and birds in the Upper Great Lakes Region. Trans. Wisconsin Acad. Sci., Arts and Ltrs., 34:23–44.

Seidensticker, J. C., IV, M. G. Hornocker, R. R. Knight, and S. L. Judd
1970 Equipment and techniques for radiotracking lions and elk. Univ. Idaho, Forest, Wildlife and Range Exp. Sta., Bull. No. 6, 20 pp.

Seidensticker, J. C., IV, M. G. Hornocker, W. V. Wiles, and J. P. Messick
1973 Mountain lion social organization in the Idaho Primitive Area. Wildlife Monogr., No. 35, 60 pp.

Seton, E. T.
1953 Lives of game animals. Vol. I, Pt. I. Cats, wolves, and foxes. x+337 pp.

Shaw, H. G.
1977 Impact of mountain lion on mule deer and cattle in northwestern Arizona. Proc. 1975 Predator Symp., Univ. Montana, Forest and Conserv. Exp. Sta., pp. 17–32.

Smith, R. P.
1978 Mountain lions in U.P.? Not likely. Michigan Out-of-Doors, 32(7):41.

Spalding, D. J., and J. Leesowski
1971 Winter food of the cougar in south-central British Columbia. Jour. Wildlife Mgt., 35(2);378–381.

Toweill, D. E., and E. C. Meslow
1977 Food habits of cougars in Oregon. Jour. Wildlife Mgt., 41(3):576–578.

Wood, N. A.
1914 An annotated check-list of Michigan mammals. Univ. Michigan, Mus. Zool., Occas. Papers No. 4, 13 pp.
1922 The mammals of Washtenaw County, Michigan. Univ. Michigan, Mus. Zool., Occas. Papers No. 123, 23 pp.

Wood, N. A., and L. R. Dice
1924 Records of the distribution of Michigan mammals. Michigan Acad. Sci., Arts and Ltrs., Papers, 3:425–469.

Wright, B. S.
1972 The eastern panther. A question of survival. Clarke, Irwin & Company Ltd., Toronto. x+180 pp.

Young, S. P., and E. A. Goldman
1946 The puma, mysterious American cat. American Wildlife Inst., Washington, D.C. xvi+358 pp.

LYNX
Felis lynx Linnaeus

NAMES. The generic name *Felis,* proposed by Linnaeus in 1758, is the Latin name for cat. The specific name *lynx,* also proposed by Linnaeus in 1758, is derived from Greek words meaning "lamp" "to see" used apparently to describe the animal's bright eyes and keen sight. In much of the literature, this northern cat is given the scientific name *Lynx canadensis.* Modern taxonomic view is that *Lynx* is correctly ranked as a subgenus within the genus *Felis* and that *canadensis* has only the rank of subspecies, since animals in both North America and Eurasia are now considered conspecific, with the older specific name, *Lynx* for the Old World lynx, having priority. In Michigan, this species is usually called lynx, Canada lynx, or sometimes lynx cat. The subspecies *Felis lynx canadensis* (Kerr) occurs in Michigan.

RECOGNITION. Body size medium for a field, more than twice as large as a large male house cat, rather short, robust; length of head and body averaging 31 in (786 mm) for the larger male and 29½ in (748 mm) for the smaller female; tail characteristically bobbed; length of tail vertebrae averaging 4⅛ in (105 mm) for the male and 3⅞ in (98 mm) for the female; head comparatively small for the size of the body; nose broad, blunt in typical cat configuation; nose pad large; eyes large; ears erect, pointed, distinguished by long black tufts; vibrissae (whiskers) long, coarse; neck short, thick, muscular; legs relatively long compared to body size, hind limbs somewhat longer causing body to stand slightly higher in the rear; feet large, broad, padded; tail stubby, rounded, well-haired; forefeet with five digits (pollex high on wrist); hindfeet with four digits; claws curved, sharp, sturdy; two pairs of mammary glands, one abdominal and one inguinal; adult males approximately 5% larger than adult females.

The pelage of the lynx is thick, long, silky, and luxuriant. Underfur is buffy brown; longer guard hair is gray-banded with black tips. The overall appearance is a frosted gray above and slightly more buffy below. The face (Plate 6) is characterized by a prominent, black-striped ruff on each cheek extending down on each side to below the lower jaw. The general grayish head is also marked with black stripes on the forehead. The long, black ear tufts are set off by the black-edged, grayish ears. Rather indistinct body spotting is more prominent on the belly and the insides of the legs. The short tail is grayish with a completely black tip.

This large, bob-tailed, long-legged cat should not be confused with any other Michigan mammal except its close relative the bobcat. The lynx differs from the bobcat in having an overall grayish rather than a slightly reddish color, distinctive long, black ear tufts, a tail which is completely black-tipped all around, and broad feet.

MEASUREMENTS AND WEIGHTS. Adult specimens of the male and the female lynx measure, respectively, as follows: length from tip of nose to end of tail vertebrae, 31 to 42 in (780 to 1,065 mm) and 30 to 38 in (765 to 965 mm); length of tail vertebrae, 3 to 5½ in (75 to 138 mm) and 3 to 4¾ in (76 to 122 mm); length of hind foot, 8 to 12¾ in (205 to 325 mm) and 7⅛ to 10 in (180 to 250 mm); height of ear from notch (almost the same in both sexes), 2¾ to 3⅛ in (70 to 80 mm); weight, 15 to 39 lbs (6.7 to 17.2 kg) and 11 to 25.5 lbs (5.1 to 11.6 kg).

DISTINCTIVE CRANIAL AND DENTAL CHARACTERISTICS. The skull of the lynx is strongly built, low, and broad (Fig. 99). The short, broad rostrum and the widely-flaring zygomatic arches are cat-like. The sagittal crest is weakly expressed at the midline of the rounded brain

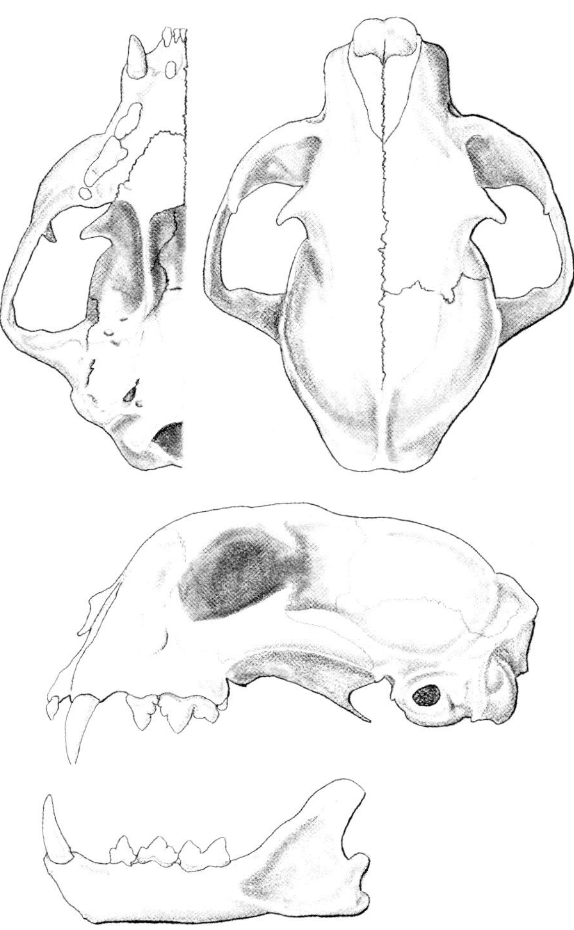

Figure 99. Cranial and dental characteristics of the lynx *(Felis lynx)*.

case. The toothrow of the lynx is characteristically short. Not only is the skull much smaller than that of the mountain lion, less than 6¼ in (160 mm) in breadth, but the lynx, as well as the bobcat, has only three cheek teeth posterior to the canines on each side of the jaw both above and below. The mountain lion and also the domestic housecat have four cheek teeth behind the upper canines.

The skulls of the lynx and the bobcat resemble one another closely. However, the skull of the lynx is usually larger; has a wider interorbital space, usually more than 1 3/16 in (30 mm) instead of less; and possesses smaller auditory bullae. Most conspicuous is the difference in spacing of two small openings at the inner, posterior end of each auditory bulla. In the lynx, these two cranial openings, the anterior condyloid foramen and the foramen lacerum posterius, are separate (see Fig. 100). In the bobcat, these two foramina are confluent, occurring close together in the same cavity. The dental formula for the lynx is:

$$\text{I (incisors)}\frac{3\text{-}3}{3\text{-}3}, \text{C (canines)}\frac{1\text{-}1}{1\text{-}1}, \text{P (premolars)}\frac{2\text{-}2}{2\text{-}2}, \text{M (molars)}\frac{1\text{-}1}{1\text{-}1} = 28$$

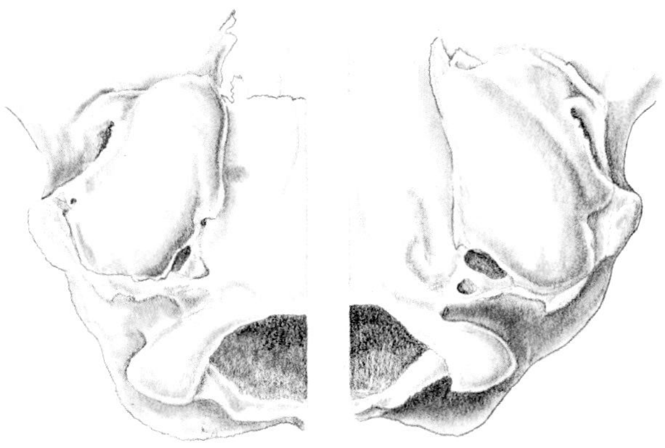

Figure 100. Ventral views of half of the posterior parts of the skulls of the bobcat (left) and lynx (right). Note that at the inner edge of the posterior part of the bullae the condyloid foramen is confluent (in the same cavity) with the anterior foramen lacerum posterius in the bobcat while these two openings are noticeably separated in the lynx.

DISTRIBUTION. The lynx enjoys a circumboreal distribution, completely around the North Polar area in Eurasia and North America. This large northern cat occurs in suitable habitat throughout most of Arctic America southward to the Canadian border of the United States and into suitable northern forests in New England, the Great Lakes Region, and in western mountains. In presettlement times in the Midwest it ranged into northern sectors of eastern deciduous hardwood environments, as far south as central Illinois, Indiana, and Ohio (de Vos, 1964). With the com-

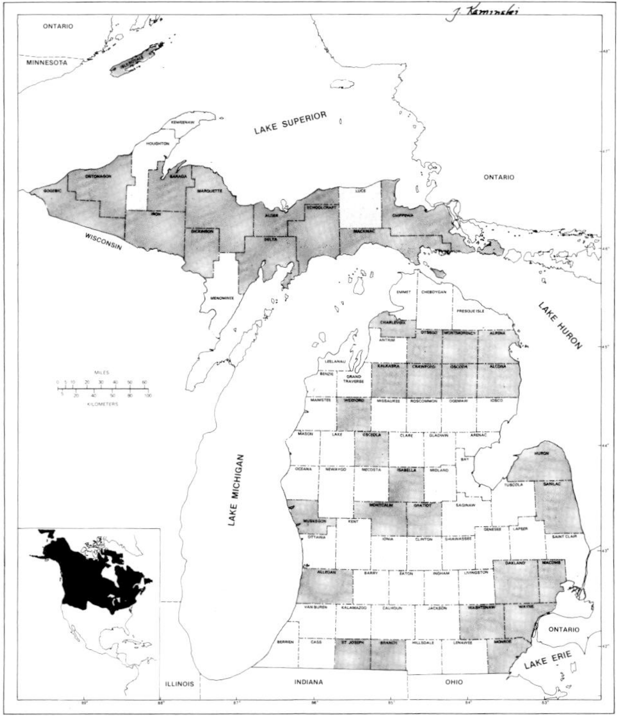

Map 63. Geographic distribution of the lynx (*Felis lynx*). Counties with light shading are those from which records of early occurrence of lynx are known or suspected. Counties with dark shading are those in which the species has been reported beginning in 1940.

ing of the Europeans, the lynx slowly gave way in these southern areas; the last actual record for Indiana, for example, was in 1832 (Mumford, 1969).

In pioneer records, the lynx is reported in much of southern Michigan (Map 63). Cadillac noted the animal's presence in the Detroit area about 1700 (Johnson, 1919). Although there are positive historic records for lynx from only 23 Lower Peninsula counties, most likely this cat was widespread there at the beginning of the nineteenth century (Burt, 1946; Wood, 1914; Wood and Dice, 1924). Selected counties and dates of lynx occurrence include Branch and St. Joseph counties in 1876 (Babcock *et al.,* 1976), Allegan County in 1845 and Montcalm County in 1910 (Wood and Dice, 1924), Charlevoix County in 1922 (Dice, 1925), Washtenaw County in 1842 (Wood, 1922), and Oscoda County in 1917 (Harger, 1965).

In the Upper Peninsula, this valued fur animal declined less noticeably. Dice and Sherman (1922) found the animal rare in Gogebic County by 1911; Wood (1917) reported the animal in Alger County in 1916; Gibson Butler (pers. comm., 1978) trapped one in Ontonagon County in 1928. In its fourth Biennial Report (1927–1928), the Michigan Department of Conservation (now Michigan Department of Natural Resources) maintained that the lynx was evidently gone from the Lower Peninsula and only a residual number was left in the Upper Peninsula, with extinction imminent. In the seventh Biennial Report (1933–1934), it was stated that an occasional lynx was reported in the Upper Peninsula. In the ninth Biennial Report (1937–1938), however, the authorities announced that the lynx was rare or extinct. Through these years the lynx persisted on Isle Royale (Peterson, 1977), with the latest sighting on that island occurring in 1979. Its earliest record from this island was Late Woodland times, based on finds from the Indian Point archaeological site (Cleland, 1968).

The only logical explanation for this valued, rather easily-trapped, fur animal's presence in Michigan in the first half of the twentieth century is the periodic entry of new animals from Ontario. New stock, possibly on the move during ten year peaks, crossed into Chippewa County via the narrow island-dotted waterway (for Sugar Island record, see Pruitt, 1951) between Lake Superior and Lake Huron (Harger, 1965). Unlike the wolverine, which had no chance to repopulate northern Michigan by way of the Canadian Sault because stock had disappeared long before from the immediate area of the water gap, Ontario lynx populations continue to range on the north side of this waterway (Banfield, 1974). Consequently, these Canadian lynx populations have an almost continuous opportunity to cross this narrow water gap into the Upper Peninsula in at least token numbers, so that an occasional animal is trapped or reported on the Michigan side. This also explains the reasoning of experienced field observers, like wildlife biologist Ralph E. Bailey of the Michigan Department of Natural Resources, who has never regarded the Michigan lynx as completely extinct (*in litt.,* October 4, 1978).

The status of the lynx in the Upper Peninsula has been well documented by personnel of the Michigan Department of Natural Resources, especially by Elsworth Harger (1965), who for many years has monitored both lynx and bobcat populations. Modern records for lynx in Michigan first list an animal captured on Bois Blanc Island (Mackinac County) in 1940. Authentic county records with dates as compiled by Harger (also see Erickson, 1954) are: Marquette County (1946, 1955, 1957), Mackinac Island (1949, 1962), Chippewa County (1953, 1958, 1960, 1961, 1962), Schoolcraft County (1961, 1962), Alger County (1961), Dickinson County (1962), and Ontonagon County (1962). Between 1940 and 1962, at least 21 lynxes were actually captured. In 1962, 12 were apprehended, causing a rash of news stories about the return of this "phantom cat." These numbers may be a reflection in northern Michigan of a widespread explosion in lynx populations between 1961–1963, also reported in other border states such as Minnesota and North Dakota (Adams, 1963; Gunderson, 1978). Records of the lynx in the Upper Peninsula subsequent to those of Harger (1965) include animals or recognizable tracks on Drummond Island (1966), in Ontonagon County (1966), Schoolcraft County (1966), Delta County (two records in 1968), Gogebic County (1969), Marquette County (1971, 1972), Iron County (1972), and Schoolcraft County (1973, 1975). Through infiltration, the lynx has returned by natural spread. By being granted full legal protection from hunting and trapping, its status is assured as a sustaining member of the outdoor community of the Upper Peninsula, presumably also spreading westward into northeastern Wisconsin (Doll *et al.,* 1957).

HABITAT PREFERENCES. The lynx is primarily a denizen of the boreal evergreen forests but does range into mountains and to the edge of the arctic prairie. In Michigan, it is generally pictured as a stealthy, quiet animal sneaking along on a forest floor well littered with moss-covered tree falls, bracken fern, alder thickets, and brush piles, and closeting itself in dense white cedar swamps, avoiding as much as possible the open uplands and areas of clearcut.

DENSITY AND MOVEMENTS. The lynx has long been known to have a characteristic multiannular population fluctuation with peaks occurring every nine to ten years (Elton and Nicholson, 1942). This cyclic behavior closely corresponds to the ups and downs of this northern cat's chief prey species, the snowshoe hare (Nellis *et al.*, 1972). These dramatic periodic trends have been found to occur very regularly by scientists examining the fluctuations in fur production statistics of the Hudson's Bay Company since the early part of the nineteenth century (1820). There is evidence that this cyclic condition can be affected (but probably only locally) by overtrapping and habitat deterioration (van Zyll de Jong, 1971).

Most basic information on these fluctuations has been obtained from publications summarizing the findings of a series of long-range studies by Professor Lloyd B. Keith and his University of Wisconsin students on the lynx and associated prey species in the Canadian province of Alberta. Rapid population change was documented in a report by Brand *et al.* (1976) in which the lynx density in the study area was calculated to be one cat per 16 sq mi (43 sq km) in a year of low populations (1966–1967) as compared to a density of one per 3.8 sq mi (10 sq km) in a peak year (1971–1972). According to Mech (1980), a marked increase occurred at the same time in the southern "edge" populations of lynx in Minnesota. It is likely that this same population peak manifested itself in Upper Peninsula Michigan and accounted for the pronounced upswing in numbers, and, as in Minnesota, was possibly due in part to the influx from Canadian population centers. This may also have been true in 1961–1962, when numerous records of lynx in northern Michigan were obtained.

In spite of these observable fluctuations, the lynx population in nature, especially where it is not harried by trappers, has a discrete spatial arrangement. Adults establish home ranges, sharing at least parts of their cruising areas with other cats (Mech, 1980). In these areas, movements on a nightly basis may average 2 mi (3.2 km). Young animals in search of living space may wander exceptional distances; extremes include a young female traveling 64 mi (103 km) in Newfoundland (Saunders, 1963b) and a record journey of a young female moving some 300 mi (483 km) from Minnesota to Ontario (Mech, 1977). Established adults move about in sizable home ranges: up to 8 sq mi (21 sq km) in Newfoundland (Saunders, 1963b); 14 sq mi (36 sq km) in Montana (Koehler *et al.*, 1979); and from 19 to 46 sq mi (51 to 122 sq km) for the less mobile females and from 55 to 92 sq mi (145 to 243 sq km) for males in Minnesota (Mech, 1980).

BEHAVIOR. Much like its larger relative the mountain lion, the lynx is chiefly a solitary creature, associating with the opposite sex only during courtship and mating. Otherwise, individual adult males usually keep to themselves, occupying large home ranges where the essentials of life for each season are available. Established males probably discourage other members of their sex (adult or juvenile) from encroaching on their areas. Scent posts marked with urine may be used to alert transients of the presence of residents in established home areas as well as to communicate with the opposite sex. Adult females and associated kittens occupy home ranges perhaps half the size of those of adult males. There seems to be overlap of ranges between males and females with resulting compatible use of common areas, although additional studies are needed to verify the extent of this overlap (Mech, 1980). At any rate, it is suspected that the juvenile lynx, just dispersing from the female parent's care, faces the difficult task of becoming established, a period during which environmental influences can be critical.

The lynx is essentially nocturnal, although field observers occasionally glimpse the animal just before dusk or shortly after daylight. Its inconspicuous movements at night have been little studied except by winter snow tracking and radio telemetry. The average woodsperson may only be aware of the lynx's presence by observing the tracks of its large furry feet.

Smell seems less developed in the lynx than vision and hearing. The latter senses are used in combination to hunt for food. According to Brand *et al.* (1976), the lynx uses three hunting techniques, especially in pursuing snowshoe hares: (a) following well-used snowshoe hare trails, (b) concentrating activities within small areas of snowshoe hare activity, especially during periods when prey populations are low, and (c) crouching in ambush using "waiting beds" which are alongside trails used by snowshoe hares. According to Jackson (1961), the lynx usually pursues a prey species for a distance of only two or three long jumps. The lynx moves its long-legged body with astonishing quickness and is able to leap 12 to 15 ft (3.6 to 4.6 m). It walks and trots in typical catlike fashion, and runs with a clumsy gallop at speeds no more than 15 mi (24 km) per hour. The short, fast, and leaping rush enables the lynx to catch prey which might otherwise outrun the cat if given an adequate head start. Large "snowshoe-like" feet give the lynx a decided advantage over many other forest creatures in winter, especially if there is a light crust on the snow. The lynx climbs well in trees or on rocky prominences. The animal will take to water, using a powerful swimming stroke which keeps its head high out of the water (Seton, 1953).

The lynx possesses the curiosity attributable to housecats. Its attraction to bright objects or dangling bits of colored material has often led to its downfall, since knowledgeable trappers use this ploy to lure the lynx to snares. The lynx utters various feline purrs, hisses, growls, and mews (Seton, 1953). The latter call is often associated with the female parent attending her kittens. Shrieks and caterwauling are described by persons familiar with the northern forests. Most likely, such yowls or cries are to be expected during the mating season.

The lynx dens in various shelters. An animal having an established home range uses perhaps one major den and occupies others less often in strategic sites in its normal cruising area. Although the lynx may take shelter under rock ledges or in caves, their Michigan lairs are more likely found in hollow trees, under stumps, or in thick brush. Shredded, trampled bark and leafy material serve as bedding. In Alaska, Berrie (1973) found females and kittens using as dens the debris from a blown-down black spruce and the tangle of spruce roots where these structures were washed out on a stream bank. Berrie also described food caches of lynx located within 50 ft (15 m) of a den site. The food (parts of snowshoe hares) was covered with grasses. Directly across the trail from each of the caches were fecal deposits (scats) of the lynx.

ASSOCIATES. As stated earlier, the lynx is not a social creature, except for the female with her litter. Very likely, however, other meat-eating animals of the northern woods devour scraps of food either abandoned or cached by the lynx. Seton (1953) also noted that great horned owls may hover near a hunting cat to pounce on creatures either alarmed or flushed by its activity.

To the ecologist, a most intriguing association (as well as one about which very little is known) exists between the lynx and the bobcat when their ranges interdigitate. Although the lynx may negotiate deep snow more successfully with its huge, broad feet, in other ways the two cats are much alike in size and life requirements. These similar attributes suggest that the lynx and the bobcat may compete for both food and space—a potentially disastrous situation for one of the animals. In presettlement days, the two species were probably separated ecologically in their common living area across a broad sector of what is now the border states and provinces between the United States and Canada. The lynx used boreal coniferous forest while the bobcat was more at home in adjacent deciduous forest. With logging operations and the development of vast areas of deciduous second growth in place of conifers in this border area, the overlap zone between the lynx and bobcat appears to have shifted northward considerably (de Vos, 1964). However, with the return of fairly sizeable numbers of lynx in the early 1960s in such northern states as North Dakota (Adams, 1963), Minnesota (Gunderson, 1978), and Michigan (Harger, 1965), interaction must be greater between these two, especially since the bobcat may have taken over lynx habitat during the latter's absence or great reduction. Even as far north as the Sibley Peninsula on the Ontario shore of Lake Superior, de Vos (1950) commented that the bobcat apparently had replaced the lynx. Certainly an intriguing study in Michigan's Upper Peninsula would be an examination of the ecological and behavioral interaction between these two interesting cats.

REPRODUCTIVE ACTIVITIES. Courtship and mating take place in late winter (mostly in February and March), with litters born about 62 days later, from the last of March to as late as early June (Jackson, 1961; Saunders, 1963b; Seton, 1953). A litter may number two or three (up to six). At birth, the young have closed eyes and folded ear pinnae; are well-furred with overall grayish and brown splotches and stripes on the back, flanks, and limbs; utter low squeals; and weigh about 7 oz (200 g). At nine to 12 days of age, the eyes open. At about 60 days of age, a soft juvenile fur replaces the woolly natal coat and the kittens join their female parent on food-getting forays. At five to six months of age, the permanent canines erupt (Saunders, 1964). At about nine months of age, the adult pelage appears. The female parent is most attentive to the needs of the kittens, retaining them under her care until at least their first winter and possibly until the next mating season.

The physical condition of the breeding female is, of course, dependent on food availability. Recent work in Alberta (Brand and Keith, 1979) indicates that in times of shortage of the major food supply, the snowshoe hare, reproductive effort can be seriously impaired. This may be manifest by (a) decreased rates of pregnancy among the breeding females and (b) decreased litter sizes. Also a problem in lean times is the frequency of post-partum mortality among kittens, resulting from greater susceptibility to disease and predation, including cannibalism, and starvation. This lower productivity allows a predator lynx population to decline along with the prey species.

In captivity, the lynx is known to live for 24 years (Jones, 1979). In the wild, the highest mortality comes when kittens are dispersing in search of home sites. Once established, however, a male or a female may very well survive for several years.

FOOD HABITS. This meat-eating cat relies most heavily on the snowshoe hare for its sustenance; so much so that its population fluctuations are in accord with those of this lagomorph (Brand *et al.*, 1976; Nellis and Keith, 1968; Pulliainen, 1981). There is also a close correlation between numbers of lynxes and an alternate prey species, the ruffed grouse. In most major studies of the food habits of the lynx (Nellis and Keith 1968; Nellis *et al.*, 1972; Saunders, 1963a; etc.), the foods consist primarily of snowshoe hares (69 to 76% of the bulk). Carrion is also important as winter food. Other animals eaten include red squirrel, meadow vole, beaver, red fox, striped skunk, porcupine, other small mammals, and birds. The stomach contents of a lynx taken in Yukon Territory contained two masked shrews, six meadow voles, one long-tailed vole (not found in Michigan), and one savannah sparrow (Youngman, 1975).

When the lynx returned as a viable member of the Michigan meat-eating animal community in the 1950s and early 1960s, there was some concern because of its potential predation on the white-tailed deer. As Seton (1953) ably described, the lynx can prey on white-tailed deer, especially on the summer fawn crop. Although it can down an adult animal if given a chance, most ungulate meat eaten by the lynx is in the form of carrion. The effect of lynx on its more usual Michigan prey, the snowshoe hare and ruffed grouse, has not been appraised.

ENEMIES. The trapper has, of course, been a major force in the reduction of lynx populations. This, coupled with heavy clear-cutting of northern forests, has made profound changes in lynx distribution. In nature, the lynx, especially kittens and dispersing juveniles, might fall victim to gray wolves, coyotes, large owls, or eagles. As an adult, the lynx has little to fear from other creatures in the northern woods.

PARASITES AND DISEASES. Little mention is made in the literature about parasites infesting the lynx; Seton (1953) lists one flea, while Jackson (1961) recorded five different species. The latter author also documents two tapeworms as internal parasites.

MOLT AND COLOR ABERRATIONS. Although there is some difference of opinion, apparently there is a spring molt followed by the growth of a short, brownish summer coat. In autumn, the guard hairs are regrown and are longer to help produce the heavy, silky, winter pelage. The lynx coat can have either a fawn, yellow-orange, or blue tone coloring (Jones, 1923; Schwartz, 1938).

ECONOMIC IMPORTANCE. The long, soft fur of the lynx has long been important in the fur trade. However, in the early days of fur collecting in the Upper Great Lakes Region, pelts of the lynx were much less in demand than those of the popular beaver. British exporting traders at Michilimackinac shipped only 54 "cat" pelts in 1757 as compared with 50,938 beaver pelts (Peterson, 1979). Lynx furs handled by the American Fur Company's Detroit Department (including southern Michigan) between 1835 and 1839 totaled only 100; its Northern Department (including northern Michigan) handled 245 pelts (Johnson, 1969). Although these and subsequent records show no exorbitant annual catches, the population apparently could not be sustained in Michigan. At present, the lynx is under the protective custody of the Michigan Department of Natural Resources. Nevertheless, with long-haired furs in fashion these days and selected lynx pelts bringing as much as $505 at fur auctions in Ontario in 1978, there is some concern that Michigan lynx pelts are occasionally being transported across the northern border for sale. Not only has the pelt been a prize for the trapper and hunter, northern woodlands residents know the lynx flesh to be highly palatable.

COUNTY RECORDS FROM MICHIGAN. Records of lynx gleaned mostly from historic documents show that the species lived in the recent past (nineteenth century) in all parts of the Lower Peninsula and in the Upper Peninsula. Map 63 also shows those Upper Peninsula counties where the lynx has recently become reestablished through natural spread.

LITERATURE CITED

Adams, A. W.
1963 The lynx explosion. North Dakota Outdoors, 26(5): 20–24.

Babcock, R. E., J. Sikorski, L. Holcomb, and T. Norton
1976 A comparison of vertebrate populations in southern Michigan at the time of settlement and the present. Michigan Acad. Sci., Arts and Ltrs., Abstr. of paper presented at 80th Ann. Meeting (mimeo), 5 pp.

Banfield, A. W. F.
1974 The mammals of Canada. Univ. Toronto Press, Toronto. xxiv+438 pp.

Berrie, P. M.
1973 Ecology and status of the lynx in interior Alaska. The World's Cats, 1:4–41.

Brand, C. J., and L. B. Keith
1979 Lynx demography during a snowshoe hare decline in Alberta. Jour. Wildlife Mgt., 43(4):827–849.

Brand, C. J., L. B. Keith, and C. A. Fischer
1976 Lynx responses to changing snowshoe hare densities in central Alberta. Jour. Wildlife Mgt., 40(3):416–428.

Burt, W. H.
1946 The mammals of Michigan. Univ. Michigan Press, Ann Arbor. xv+288 pp.

Cleland, C. E.
1968 Analysis of the fauna of the Indian Point Site on Isle Royale in Lake Superior. Michigan Arch., 14(3–4): 143–146.

de Vos, A.
1950 Timber wolf movements on Sibley Penninsula, Ontario. Jour. Mammalogy, 31(2):169–175.

1964 Range changes of mammals in the Great Lakes Region. American Midl. Nat., 71(1):210–231.

Dice, L. R.
1925 A survey of the mammals of Charlevoix County, Michigan, and vicinity. Univ. Michigan, Mus. Zool., Occas. Papers No. 159, 33 pp.

Dice, L. R., and H. B. Sherman
1922 Notes on the mammals of Gogebic and Ontonagon counties, Michigan, 1920. Univ. Michigan, Mus. Zool., Occas. Papers No. 109, 46 pp.

Doll, A. D., D. S. Balser, and R. F. Wendt
1957 Recent records of Canada lynx in Wisconsin. Jour. Mammalogy, 38(3):414.

Elton, C., and M. Nicholson
1942 The ten-year cycle in numbers of lynx in Canada. Jour. Animal Ecol., 11(2):215–244.

Erickson, A. W.
1955 A recent record of lynx in Michigan. Jour. Mammalogy, 36(1):132–133.

Gunderson, H. L.
1978 A mid-continent irruption of Canada lynx, 1962–63. Prairie Nat., 27:9–24.

Harger, E. M.
1965 The status of the Canada lynx in Michigan. Jack-Pine Warbler, 43(4):150–153.

Jackson, H. H. T.
1961 The mammals of Wisconsin. Univ. Wisconsin Press, Madison. xii+504 pp.

Jones, M. L.
1979 Longevity of mammals in captivity. Internat. Zoo News, 26(3):16–26.

Jones, S. V. H.
1923 Color variations in wild animals. Jour. Mammalogy, 4(3):172–177.

Johnson, D. R.
1969 Returns of the American Fur Company, 1835–1839. Jour. Mammalogy, 50(4):836–839.

Johnson, I. A.
1919 The Michigan fur trade. Michigan Hist. Comm., Lansing. Univ. Ser., vol. 5, vii+201 pp.

Loehler, G. M., M. G. Hornocker, and H. S. Hash
1979 Lynx movements and habitat use in Montana. Canadian Field-Nat., 93(4):441–442.

Mech, L. D.
1977 Record movement of a Canadian lynx. Jour. Mammalogy, 58(4):676–677.
1980 Age, sex, reproduction, and spatial organization of lynxes colonizing northeastern Minnesota. Jour. Mammalogy, 61(2):261–267.

Mumford, R. E.
1969 Distribution of the mammals of Indiana. Indiana Acad. Sci., Monog. No. 1, vii+114 pp.

Nellis, C. H., and L. B. Keith
1968 Hunting activities and success of lynxes in Alberta. Jour. Wildlife Mgt., 32(4):718–722.

Nellis, C. H., S. P. Wetmore, and L. B. Keith
1972 Lynx-prey interactions in central Alberta. Jour. Wildlife Mgt., 36(2):320–329.

Petersen, E. T.
1979 Hunters' heritage. A history of hunting in Michigan. Michigan United Conserv. Clubs, Lansing. 55 pp.

Peterson, R. O.
1977 Wolf ecology and prey relationships on Isle Royale. National Park Serv., Monog. Ser. No. 11, xx+210 pp.

Pruitt, W. O., Jr.
1951 Mammals of the Chase S. Osborn Preserve, Sugar Island, Michigan. Jour. Mammalogy, 32(4):470–472.

Pulliainen, F.
1981 Winter diet of *Felis lynx* L. in SE Finland as compared with the nutrition of other northern lynxes. Zeitschrift für Säugetierkunde, 46(4):249–259.

Saunders, J. K., Jr.
1963a Food habits of the lynx in Newfoundland. Jour. Wildlife Mgt., 27(3):384–390.
1963b Movements and activities of the lynx in Newfoundland. Jour. Wildlife Mgt., 27(3):390–400.
1964 Physical characteristics of the Newfoundland lynx. Jour. Mammalogy, 45(1):36–47.

Schwartz, E.
1938 Blue or dilute mutation in Alaskan lynx. Jour. Mammalogy, 19(3):376.

Seton, E. T.
1953 Lives of game animals. Vol. I, Pt. I. Cats, wolves, and foxes. Charles T. Branford Company, Boston. x+337 pp.

van Zyll de Jong, C. G.
1971 The status and management of the Canada lynx in Canada. Proc. Sympos. Native Cats of North America. Their status and management. U.S. Sport Fisheries and Wildlife, Region 3, v+139 pp.

Wood, N. A.
1914 An annotated check-list of Michigan mammals. Univ. Michigan, Mus. Zool., Occas. Papers No. 4, 13 pp.
1917 Notes on the mammals of Alger County, Michigan. Univ. Michigan, Mus. Zool., Occas. Papers No. 36, 8 pp.
1922 The mammals of Washtenaw County, Michigan. Univ. Michigan, Mus. Zool., Occas. Papers No. 123, 23 pp.

Wood, N. A., and L. R. Dice
1924 Records of the distribution of Michigan mammals. Michigan Acad. Sci., Arts and Ltrs., Papers, 3:425–469.

Youngman, P. M.
1975 Mammals of the Yukon Territory. Nat. Mus. Canada, Publ. Zool. No. 10, 192 pp.

BOBCAT
Felis rufus Schreber

NAMES. The generic name *Felis,* proposed by Linnaeus in 1758, is the Latin word for cat. The specific name *rufus,* proposed by Schreber in 1777, is also a Latin word meaning reddish, appropriately descriptive of the general coloration of the body. In much of the older literature, the bobcat is identified by the scientific name *Lynx rufus.* The name *Lynx* is now generally regarded as merely a subgeneric category of the genus *Felis.* In Michigan, this cat is almost universally known as a bobcat. The names wildcat, bay lynx, and lynx cat have also been used. The bobcat in the Upper Peninsula belongs to the subspecies *Felis rufus superiorensis* (Peterson and Downing), with closest affinities with populations in Wisconsin to the west and perhaps with Ontario to the north. In the Lower Peninsula, the bobcat is classified as the subspecies *Felis rufus rufus* Schreber, with closest relationship to bobcats in adjacent states to the south and east. The presence of these two distinct taxa indicates that the Straits of Mackinac constitutes an effective barrier to crossbreeding between the populations on either side of the Michigan-Huron waterway connector (Table 12).

RECOGNITION. Body size medium large for a felid (Fig. 101), more than twice as large as a large male house cat; length of head and body averaging 29½ in (745 mm) for the larger male and 26¾ in (655 mm) for the smaller female; tail characteristically bobbed short; length of tail vertebrae averaging 6⅛ in (155 mm) for the larger male and 5¾ in (145 mm) for the smaller female; head somewhat small compared to body size; muzzle broad, rounded; nose pad large; eyes large, pupils elliptical in bright light, almost round in dim light;

Figure 101. The bobcat (*Felis rufus*). Photograph by Robert Harrington and courtesy of the Michigan Department of Natural Resources.

ears erect, prominent, generally rounded but having pointed ear tufts conspicuous but short, less than 1 in (25 m) high; vibrissae (whiskers) long, coarse; neck short, muscular; legs long (perhaps comparatively shorter than those of the lynx); tail short, rounded, well-haired; feet large but lacking the furry breadth of the lynx; forefeet with five digits (pollex high on wrist); hindfeet with four digits; claws curved, strong, sharp, retractible; stance digitigrade (walks on tips of bare toe pads); mammary glands consist of two pairs, one abdominal and one inguinal; adult males are approximately 10% larger in external dimensions than females and as much as 34% heavier (Erickson, 1955).

The pelage of the bobcat is soft, fine, and moderately long, shorter and slightly more reddish in summer and longer and slightly more grayish in winter. The upper parts are characterized by irregular dark spotting and splotches, especially along the midline from head to base of tail, against an overall background color of grayish to brownish. The underparts also have blackish spots but against a whitish background. The blackish nose pad, white vibrissae, striped forehead, prominent streaked ruff on each cheek, extending down each side to below the lower jaw, a white spot centered on the blackish dorsal surface of each ear, and short, blackish ear tufts are features of the bobcat's head (Plate 6). The tail has a tip which is blackish above and white below, with subterminal black bars. The legs have irregular spotting aganst a whitish background; on the inside of the front legs, the black markings are in the form of conspicuous bars.

The long, black ear tufts and the large, broad feet distinguish the lynx from the slightly smaller bobcat. The bobcat is also distinguished by its slightly longer tail with tip whitish below and black above; more heavily spotted coat; and blackish dorsal surfaces of the ears, each bearing a large white spot instead of being grayish with black margins.

MEASUREMENTS AND WEIGHTS. Adult specimens of the larger male and the smaller female bobcat measure, respectively, as follows: length from tip of nose to end of tail vertebrae, 33½ to 48¼ in (850 to 1,250 mm) and 29 to 37½ in (740 to 950 mm); length of tail vertebrae, 5⅛ to 7½ in (130 to 170 mm) and 4⅞ to 6½ in (125 to 165 mm); length of hind foot, 6½ to 8¾ in (165 to 220 mm) and 5⅞ to 7⅛ in (150 to 180 mm); height of ear from notch, 2⅝ to 3⅜ in (65 to 85 mm) and 2⅜ to 3 in (60 to 75 mm); weight 14 to 40 lbs (6.4 to 18.3 kg) and 9 to 36 lbs (4.1 to 15.3 kg). A native male taken in Iron County on February 6, 1954, weighed 32½ lbs (14.6 kg), as reported by Erickson (1955).

DISTINCTIVE CRANIAL AND DENTAL CHARACTERISTICS. The skull of the bobcat, like that of the lynx, is strongly constructed (Fig. 102). It is broad with typical catlike short rostrum; has widely-flaring zygomatic arches; possesses a rounded, low braincase with little evidence of a sagittal crest; and is further characterized by a rather narrow interorbital space and narrowly posterior presphenoid. Compared to the skull of the lynx, the bobcat skull is slightly smaller, more inflated in the facial region, and, most conspicuously, the anterior condyloid foramen is confluent (in the same cavity) with the anterior foramen lacerum posterius at the inner posterior end of each auditory bulla (see Fig. 100). These two small openings near the rear end of the underside of the cranium are noticeably separated in the skull of the lynx. The dentition of the bobcat is perhaps slightly less heavy than that of the lynx. The upper carnassial tooth (fourth upper premolar) is shorter, less than ⅝ in (16 mm) long in the bobcat and longer in the lynx. The dental formula for the bobcat is:

$$\text{I (incisors)}\frac{3\text{-}3}{3\text{-}3}, \text{C (canines)}\frac{1\text{-}1}{1\text{-}1}, \text{P (premolars)}\frac{2\text{-}2}{2\text{-}2}, \text{M (molars)}\frac{1\text{-}1}{1\text{-}1} = 28$$

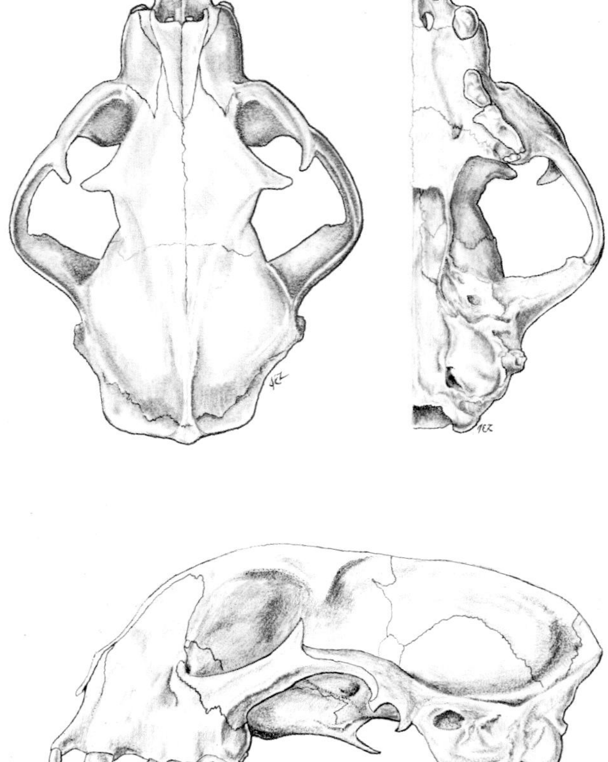

Figure 102. Cranial and dental characteristics of the bobcat (*Felis rufus*).

DISTRIBUTION. The bobcat is a distinctive, short-tailed North American cat distributed as far north as the southern parts of the border provinces of Canada (Nagorsen and Peterson, 1977) southward throughout the United States into Mexico to the southern end of the Mexican Plateau to just below the 18th parallel but avoiding the more tropical coastal sectors (Young, 1958). Despite the inroads of civilization, intense land-use practices, and habitat destruction, the ubiquitous, adaptable bobcat has managed to hold its own, at least in a spotty fashion, in suitable areas in most states.

In Michigan, the bobcat occurred statewide in the early days of settlement (Burt, 1946). There are historic references to the species in most counties in southern Michigan; for example, in Berrien County (in an archaeological site, Cleland, 1966); in Allegan, Huron, Montcalm, Oakland, and Wayne counties (Wood and Dice, 1924); in Ingham, Monroe, and Washtenaw counties (Wood, 1914); in Lenawee, Hillsdale, Sanilac, and Tuscola counties (Burroughs, 1959). However, the bobcat became rare and then disappeared in much of the southern half of the Lower Peninsula during the early years of the twentieth century, at least prior to 1930 (Dearborn, 1932). Nevertheless, late records are known from such southern counties as Kent (1981), Livingston (1952 or 1953) and Shiawassee (1963), according to Marvin Cooley (*in litt.*, December 29, 1980) and T. M. Cooley *et al.* (1982). Today, its survival seems secure in the northern half of the Lower Peninsula and throughout the Upper Peninsula (Map 64). In fact, the second-growth hardwood forests of northern Michigan (in many areas once covered with coniferous trees) appear to favor the bobcat more than the lynx (de Vos, 1964). Bobcats are also known from Bois Blanc and Drummond islands (William

Map 64. Geographic distribution of the bobcat (*Felis rufus*). In pioneer times, this species occurred throughout the entire state.

Barry, *in litt.*, December 20, 1973; Manville, 1950). Moran's (1964) record from Garden Island in Lake Michigan illustrates the bobcat's ability to cross a sizable water barrier, perhaps on winter ice.

HABITAT PREFERENCES. The bobcat has a wide range of habitat preferences, being found in hardwood forests, mountains, and deserts. Its basic need in any situation is some woody cover. In Michigan, the bobcat's primary environment is the temperate hardwoods which originally covered much of southern Michigan. After deciduous second-growth encroached into northern Michigan areas which formerly contained mostly evergreens, the bobcat presumably found considerably more acceptable habitat in the more northern coniferous forests (de Vos, 1964). Today, in northern Michigan, the bobcat may be expected in deciduous forest or mixed hardwoods and conifers. There is some evidence that bobcats tend to avoid leafless deciduous woodlands in winter because of often heavy snow accumulations, increased wind, and low night temperatures in this rather exposed environment, as compared with the well-canopied, evergreen areas (McCord, 1974b). Erickson (1955) found bobcats in Michigan uplands primarily in warmer seasons and in lowland forests in winter. The most used stands were of white and black spruce, white cedar, balsam, alder, willow and poplar followed by swamp situations covered with thick growths of alder-willow or of white cedar. Because the bobcat is often prone to move along wooded waterways, the species may be expected to follow stream systems in more southern parts of the state.

DENSITY AND MOVEMENTS. Unlike the lynx, the bobcat does not seem to undergo pronounced population fluctuations in most of the southern part of its extensive range. There is, however, evidence of cyclic population ups and downs, at least in northern parts of the range (Jackson, 1961). Rollings (1945) suggested a close correlation between the number of bobcats and the snowshoe hare population fluctuations in Minnesota. More quantitative evidence on this matter is needed.

Despite studies by winter tracking and by monitoring radio-collared individuals, population estimates of bobcats have not been well assessed in the Upper Great Lakes Region. Jackson (1961) suggested a density of one animal per 5 sq mi (13 sq km) in suitable range in Wisconsin. Higher densities have been noted in more southern habitats: In a study area in Arizona, Jones and Smith (1979) estimated a density of one bobcat per 1.4 to 1.5 sq mi (3.6 to 4.1 sq km); in Alabama, Miller and Speake (1978) found one animal per 0.3 to 0.5 sq mi (0.9 to 1.3 sq km); in Georgia, Provost *et al.* (1973) calculated one animal per 0.3 to 0.5 sq mi (0.9 to 1.3 sq km).

Most estimates of home range size do not take into account that females use much smaller areas than males (Bailey, 1974). In summarizing studies made in southeastern states, Miller and Speake (1978) reported that summer home ranges of radio-collared males averaged 1.9 sq mi (4.94 sq km); radio-collared females, 0.37 sq mi (0.97 sq km). In Utah, Bailey (1974) found males to move around in areas as much as 2.2 times larger than those used by females; males had an average home range of 16.0 sq mi (42.1 sq km) and females, 7.3 sq mi (19.3 sq km). In Maine, Marston (1942) calculated winter home ranges of bobcats (without distinguishing between sexes) as about 40 sq mi (104 sq km). It was Jackson's (1961) conclusion, perhaps following Seton (1953), that Wisconsin bobcats seldom have home ranges larger than 3.1 sq mi (8.9 sq km). These latter figures might be considered low, since other workers studying bobcats in northern states (Rollings, 1945) suggested home ranges of at least 10 to 15 sq mi (26 to 38 sq km). In Michigan, Erickson (1955) concluded that home ranges might be 10 to 20 sq mi (26 to 52 sq km). As suggested by Bailey (1974), the differences found in widely spaced parts of the country no doubt reflect the local availability of prey species. Home range size of northern bobcats may also be larger in summer and smaller in winter.

Obviously, movements of adult bobcats established in home ranges may be much less extensive than those of transient animals seeking living space. Reports of movements of as much as 50 mi (80 km) are suspected as being those of transient animals (Jackson, 1961). Movements of bobcats from two to 10 mi (3.2 to 16 km) in a single night have been reported in studies from such states as Massachusetts, Minnesota, and Montana (Pollack, 1951b; Robinson and Grand, 1958; Rollings, 1945). In snow tracking Michigan bobcats, Erickson (1955) found the animals made nightly movements of as little as 0.5 mi to as much as 7.25 mi (0.8 to 11.7 km), an average of 3.5 mi (5.6 km).

BEHAVIOR. The bobcat is a rather solitary, retiring animal. Although rarely seen, the creature leaves abundant tracks—in summer, on sand bars, dusty roads and game trails; in winter, in snow, especially across or along roads which traverse white cedar or alder-willow swamps. The sexes associate only briefly during the courtship and mating period. During other seasons, the males maintain their lone existences in home ranges from which other males are excluded by a system of prior rights (Bailey, 1974). Females, often with kittens, also maintain home ranges from which other females are excluded, but these areas are no more than one-quarter to one-half the size of those of males. In fact, two or more female home ranges may be included within the larger range of a single male. The two sexes cruise the same areas but manage to avoid each other and live compatibly.

To discourage transients and communicate with the opposite sex at breeding time, both males and females use scent marking at key localities within their areas. These may be places marked with urine and/or with feces as well as with conspicuous scrapes made by scratching the ground with the hind feet. It is likely that claw marks on trees also play a role in home range maintenance, as the cats scratch noticeable gashes in tree bark (Erickson, 1955; Rollings, 1945).

The bobcat is nocturnal in most activities, often on the move from just after sunset to just prior to sunrise. The senses of vision and hearing, both used in hunting activities, are keenly developed; the olfactory sense is less so. The bobcat can leap considerable distances from a crouch, walks and trots easily, and uses a springing gait for quick rushes. Like other cats, it has a rather awkward bounding gallop. An effective hunter, the bobcat may follow a game trail until prey is sighted or heard, then silently sneak to within pouncing distance (Erickson, 1955; Marston, 1942). This technique is used to bring down white-tailed deer, hares, rabbits, or even mice. Waiting in ambush along trails is a method of "still" hunting (Rollings, 1945). The bobcat is, of course, a good climber, although its downfall has often resulted from its readiness to "tree" when pursued by trail hounds. According to experienced dog handlers, a treed bobcat will jump down and make a run at the approach of a human, unless it is near daylight, at which time the animal is more likely to stay in the tree. The bobcat seems to have no fear of water and easily crosses swamps and streams (Seton, 1953).

The bobcat has the curiosity typical of a playful house cat. It is attracted by colorful objects, bits of paper, and orange peels—items which may be used as trail set lures. A bobcat's calls are also similar to those of the house cat (Young, 1958): spits, growls, puffs, whines, mews, and hisses. Males are apt to scream or caterwaul, especially during the courtship period. Distress calls made by mice, rabbits, turkeys, etc., attract the bobcat and the clever hunter learns how to imitate these calls to decoy the bobcat into shooting distance.

Bobcats use two types of shelters. One is a temporary hiding area for one-day stays when an animal is cruising around its home range. These resting beds may be in brush piles, under rock ledges, inside hollow logs, above ground in tree holes, and below exposed roots (Erickson, 1955). The other type of shelter is usually a substantial den used as a nursery by the female parent over a period of several weeks. Some of the more protective shelters used for temporary occupancy may also qualify for nursery dens. In Minnesota, Rollings (1945) reported a family of four young bobcats denning in a standing, hollow white pine with the opening about 5 ft (1.5 m) aboveground.

ASSOCIATES. The solitary bobcat has little contact, save breeding relationships, with others of its species. Nor does the animal have any real associations with other carnovores. Erickson's (1955) suspicion that Michigan bobcats might be restricted by actions of either the red fox or coyote is probably unwarranted. He did, however, determine that bobcats and coyotes may feed on the same white-tailed deer carrion, even trailing one another to the carcasses. Nevertheless, there was no evidence that the two animals feed on a carcass simultaneously.

The poorly understood relationship between the bobcat and the lynx is also discussed under the account of the latter species. By the turn of the twentieth century, the range of the lynx in Michigan and in other parts of the Upper Great Lakes Region had steadily receded northward due to heavy trapping and basic habitat changes resulting from the logging of the coniferous forests (de Vos, 1964). Although the bobcat traditionally occurred throughout Michigan (Wood, 1914), much of the northern part, where evergreens abounded, was initially lynx country. In the early years of the twentieth century, the bobcat is presumed to have replaced the lynx and become well-established in the second-growth deciduous stands dominating much of the logged-over coniferous areas in northern Michigan. Nevertheless, residual populations of the lynx appeared to persist, probably reinforced by new arrivals spreading southward from north of the Canadian Sault, especially during peak population years near the beginning of each decade. Aided by legal protection from trapping, by mid-century the lynx began to make a comeback (Harger, 1965). How these two cats, similar in size and many habitat requirements, thrive in close association is an intriguing question being asked by wildlife ecologists. Certainly both species share an important role in Michigan's outdoor community.

REPRODUCTIVE ACTIVITIES. Bobcat courtship activities are initiated in January with breeding taking place primarily in February and March (Crowe, 1975b; Fritts and Sealander, 1978a; Gashwiler *et al.*, 1961; McCord, 1974a). In Michigan, Erickson (1955) estimated that most successful matings occur in March, with yearling bobcats becoming responsive somewhat later, in April and May. Reproductive activities also occur in later months; Sheldon (1959) recorded one such case in August in Massachusetts. Late breeders are presumed to be females that failed to become pregnant during their initial estrus cycles. In areas south of Michigan, two litters per year may be expected. In Michigan, most litters are born in May, following a gestation period of about 62 days. Birth occurs in rather secure dens which have all the requirements of a nursery area. However, if threatened, the attentive female parent is quick to move her kittens to other quarters.

A litter may contain from one to four offspring. Erickson (1955) found that an average of 2.5 embroyos develop in Michigan females despite an average of 5.5 eggs ovulated. At birth, the young have closed eyes, are fully covered with spotted, wooly fur, have claws, measure about 10 in (254 mm) long, and weigh about 12 oz (340 g). At nine to ten days of age, the young open their eyes, are highly vocal, and can move around. At about 60 days of age, the young are weaned and very active. The female parent maintains close supervision of her kittens during late summer, autumn, and early winter. She sends them out on their own by the following January, when she prepares again for the breeding season. Some young may also enter the breeding population this first year.

According to Cooley *et al.* (1981) and Hoppe (1980), the young-of-the-year animals make up between 40 to 65% of the Michigan bobcat population between October and March. However, it is likely that the most critical time in their survival comes when they depart from parental care to find an area unoccupied by a bobcat of the same sex in which they can become established. Once established, however, both male and female bobcats have an excellent chance of surviving, perhaps as long as 12 to 14 years (Crowe, 1972, 1975a; Young, 1958). In captivity, the bobcat lives as long as 32 years and four months (Jones, 1979). The age of a bobcat can be estimated by growth characteristics of the canine teeth (Johnson *et al.*, 1981).

FOOD HABITS. More is probably known about the food habits of the bobcat than about any other aspects of its ecology. Studies made in several parts of its extensive range have concerned the identification and quantification of the contents of digestive tracts and fecal deposits, as well as evidence of bobcat kills or carrion use found along snow trails in winter. The diversity of foods used by this cat, and its ability to switch prey species when one declines and another increases, is impressive (Beasom and Moore, 1977; Mahan, 1980; Miller and Speake, 1978). According to Golley *et al.* (1965), the bobcat consumes sufficient food to provide an average of 138 kcal/kg/day of energy. This is expended in the form of feces (9%), urine (8%), weight gain (6%), and metabolism (77%).

It is readily apparent from the wealth of published material about bobcats that this meat eater is highly successful in capturing and eating small rodents and lagomorphs. The eastern cottontail is known to be the chief food in such southern states as Arkansas (Fritts and Sealander, 1978b), Missouri (Korschgen, 1957), Texas (Beasom and Moore, 1977), and Virginia (Progulske, 1955). The snowshoe hare is the major food in such northern states as Maine (Westfall, 1956), Minnesota (Petraborg and Gunvalson, 1962; Rollings, 1945), and Michigan (Dearborn, 1932; Erickson, 1955). The eastern cottontail and the snowshoe hare can be, in combination, major food sources where both species are present, as in New England (Pollack,

1951b), Vermont (Hamilton and Hunter, 1939; Siegler, 1971) and Michigan (Stebler, 1940).

Bobcats eat small mammals at any season of the year. Studies show their diets include such Michigan mammals as fox squirrel, gray squirrel, red squirrel, southern flying squirrel, woodchuck, eastern chipmunk, least chipmunk, beaver, mice of the genus *Peromyscus,* muskrat, meadow vole, bog lemming, woodland vole, southern red-backed vole, meadow jumping mouse, and woodland jumping mouse. Other mammals eaten include opossum, short-tailed shrew, eastern mole, star-nosed mole, raccoon, striped skunk, long-tailed weasel, gray fox, red fox, and house cat. The bobcat also successfully dispatches and eats porcupine (Erickson, 1955; Manville, 1958). Assorted birds including ruffed grouse, reptiles, insects, and even snails are eaten by bobcats but little, if any, vegetable material is consumed.

Another major food of the bobcat is the flesh of the white-tailed deer. Young (1958), in his major work on this cat, pointed out that the bobcat is sufficiently strong to bring down an adult deer. However, it is suspected that most of the bobcat's deer victims are fawns and first-year animals. Such kills have been witnessed by numerous observers (Failing, 1953; Gunvalson, 1962; Marston, 1942). For example, Don Fox (pers. comm.) watched a bobcat jump from a tree and onto the back of a young white-tailed deer on October 22 in Roscommon County. Erickson (1955) recorded three such kills in northern Michigan. Nevertheless, it was Erickson's opinion that the Michigan bobcat is not an important predator on healthy adult white-tail deer; rather, it normally takes weakened individuals but most often eats the species as carrion. Consequently, neither the bobcat nor the lynx are deemed major threats to the Michigan deer herd. The array of available foods, including domestic stock and poultry, effectively relieve both cats of the arduous task of stalking and dispatching white-tailed deer. As carrion, this deer is a major attraction but the bobcat or the lynx should not be automatically blamed for making the initial kill.

ENEMIES. Although the adult bobcat has few adversaries except for the human, young animals can fall victim to red foxes, coyotes, and great horned owls. There are records that a bobcat will occasionally be treed by coyotes (Young, 1958). The mountain lion is also known to dispatch an adult bobcat in much the same way the bobcat kills and eats a house cat.

PARASITES AND DISEASES. Young (1958) listed 20 different kinds of internal parasitic worms (helminths) that infest the bobcat. In Michigan, Manville (1958) found the lumen of the stomach and intestine of a bobcat taken in Montmorency County filled with a tightly-compacted mass of the round (nematode) worm, *Physaloptera praeputialis.* Internal parasites of Michigan bobcats taken in the winter of 1980–81 included tapeworms of the genera *Taenia* and *Spirometra* and roundworms (ascarids) of the genera *Toxocara* and *Toxascaris. Coccidia* and *Capillaria* were also represented (Cooley *et al.,* 1981, 1982). Fleas found on the bobcat include *Cediophyslla simplex, Pulex irritans, Juxta pulix procinus,* and *Hoplopryllus glocialis lynx* (Young, 1958; Pollack, 1951b). Mange is a problem for bobcats (Penner and Parke, 1954). Bobcats are also known to contract tularemia and rabies; the latter is reported in Michigan (Kuta, 1979).

MOLT AND COLOR ABERRATIONS. The bobcats' shorter, redder summer coat appears in spring and is replaced in autumn by a heavier, grayer winter coat. Various color phases have been noted. Young (1958) described the occasional birth of a white (albinistic) or a black (melanistic) individual. A blackish specimen from Florida with a mahogany tint resided for a time in the Philadelphia Zoological Gardens (Ulmer, 1941). Schantz (1939) reported a white-footed bobcat.

ECONOMIC IMPORTANCE. In the early days of settlement, the bobcat was merely looked on as a lynx relative with less valuable fur. To some pioneer families, a dead bobcat was more important as a wholesome dinner and as one less predator on chickens and pigs. In 1822, for example, a bobcat pelt brought only 37½¢, as compared with $1.37 for the more luxurious lynx pelt (Anon., 1962). Pelts purchased between 1835 and 1838 by the American Fur Company (Johnson, 1969) included 3,717 bobcat pelts by the firm's Detroit Department (including southern Michigan) and 154 by the Northern Department (including northern Michigan).

During the years after settlement, the bobcat made few friends in Michigan because of its tendency to raid farmyards for domestic stock and its alleged depredations on populations of white-tailed deer, eastern cottontails, snowshoe hares, and ruffed grouse. The bobcat was easy to trap or trail with hounds and also suffered serious losses as a result of poisoning campaigns in the early years of the twentieth century. Furthermore, a bounty was established on Michigan bobcats in 1935 and, with a few exceptions, remained in effect until 1965. According to Arnold (1971), as many as 1,247 bobcats were bountied for payments of $5,675 in a single year. Beginning in 1965, the entire Lower Peninsula was closed to hunting and trapping of the bobcat while the animal was given no legal protection in the Upper Peninsula until 1976, when it was declared a game species. At that time, a hunting and trapping season was designated in the Upper Peninsula (DNR – Zone 1) from October 25 to March 31 and in six counties (Alpena, Cheboygan, Emmet, Montmorency, Otsego, and Presque Isle) and on Bois Blanc Island in the northern part of the Lower Peninsula (DNR – Zone 2) from January 1 through February 29. In the rest of the state, the season was closed.

Legalizing the bobcat as a game species with a definite harvesting season, but no bag limit, was indeed timely, because at about the same time the fur market price for a prime bobcat pelt jumped from a top of $12 in 1970–1971, to a top of $150 in 1975–1976. In 1978–1979, a quality pelt earned the hunter or trapper as much as $220. In February of 1980, some prime bobcat hides brought $350 at a sale in Newberry but in 1981 the price dropped to around $100. Even so, the harvest remained moderate; the reported catch in 1976–1977 was 341; in 1977–1978, 330; in 1978–1979, 386; in 1979–1980, 597 (474 bobcats from the Upper Peninsula and 123 from open counties in the northern part of the Lower Peninsula). Marketing bobcat hides internationally, as in Canada, has been traditional for many Upper Peninsula hunters and trappers. Beginning in 1977, the Endangered Species Scientific Authority (ESSA) was authorized to regulate international trade in American furs and quotas will probably be set on such business dealings.

Bobcat hunting with trail hounds has become a lively winter sport in snow-covered northern Michigan (Rickey, 1975). Almost half (158) of the 1978–1979 bobcat harvest was taken by this means. Other dog handlers merely enjoy the sport of treeing a snarling bobcat and getting a few photographs (Lowe, 1974). Such an outing is memorable and it assures a close look at one of Michigan's most inconspicuous larger animals.

COUNTY RECORDS FROM MICHIGAN. Records of bobcat specimens in collections of museums, from the literature, from hunters' and trappers' reports, and provided by personnel of the Michigan Department of Natural Resources show that the species once occurred in all counties of the state. At the present time, however, the bobcat is confined to the Upper Peninsula and the northern half of the Lower Peninsula. Bobcats may be expected to range as far southward in the Lower Peninsula as in the line of counties extending east to west from Bay, Midland, Isabella, Mecosta, Newaygo, to Oceana (see Map 64). Larger islands, such as Drummond and Bois Blanc, are also inhabited by the bobcat.

LITERATURE CITED

Anon.
1962 Fur and hide values, 1822. Michigan Conserv., 31(6): 36–37.

Arnold, D. A.
1971 The Michigan bounty system. Michigan Dept. Nat. Res., Wildlife Div., Info. Cir. No. 162, 3 pp.

Bailey, T. N.
1974 Social organization in a bobcat population. Jour. Wildlife Mgt., 38(3):435–446.

Beasom, S. L., and R. A. Moore
1977 Bobcat food habit response to a change in prey abundance. Southwestern Nat., 21(4):451–457.

Burroughs, R. D.
1958 Perry's deer hunting in Michigan, 1838–1855. Michigan Hist., 42(1):35–58.

Burt, W. H.
1946 The mammals of Michigan. Univ. Michigan Press, Ann Arbor. xv+288 pp.

Cleland, C. E.
1966 The prehistoric animal ecology and ethnozoology of the Upper Great Lakes Region. Univ. Michigan, Mus. Anthro., Anthro Papers No. 29, x+294 pp.

Cooley, T. M., S. M. Schmitt, and P. D. Friedrich
1981 Bobcat survey—1980–1981. Michigan Dept. Nat. Res., Wildlife Div., Rept. No. 2894 (mimeo.), 7 pp.
1982 Bobcat survey—1981–82. Michigan Dept. Nat. Res., Wildlife Div., Rept. No. 2116 (mimeo.), 11 pp.

Crowe, D. M.
1972 The presence of annuli in bobcat tooth cementum layers. Jour. Wildlife Mgt., 36(4):1330–1332.

1975a Aspects of ageing, growth, and reproduction of bobcats from Wyoming. Jour. Mammalogy, 56(1):177–198.

1975b A model for exploited bobcat populations in Wyoming. Jour. Wildlife Mgt., 39(2):408–415.

Dearborn, N.
1932 Foods of some predatory fur-bearing animals in Michigan. Univ. Michigan, Sch. Forestry and Conserv., Bull. No. 1, 52 pp.

de Vos, A.
1964 Range changes of mammals in the Great Lakes Region. American Midl. Nat., 71(1):210–231.

Erickson, A. W.
1955 An ecological study of the bobcat in Michigan. Michigan State Univ., unpubl. M.S. thesis, xv+133 pp.

Failing, O.
1953 The bobcat—hunter and hunted. Michigan Conserv., 22(1):7–9.

Fritts, S. H., and J. A. Sealander
1978a Reproductive biology and population characteristics of bobcats (*Lynx rufus*) in Arkansas. Jour. Mammalogy, 59(2):347–353.
1978b Diets of bobcats in Arkansas with special reference to age and sex differences. Jour. Wildlife Mgt., 42(3): 533–539.

Gashwiler, J. S., W. L. Robinette, and O. W. Morris
1961 Breeding habits of bobcats in Utah. Jour. Mammalogy, 42(1):76–84.

Golley, F. B., G. A. Petrides, W. L. Rauber, and J. H. Jenkins
1965 Food intake and assimilation by bobcats under laboratory conditions. Jour. Wildlife Mgt., 29(3):442–447.

Gunvalson, V. E.
1962 Observations on bobcat mortality and bobcat predation on deer. Jour. Mammalogy, 43(3):430–431.

Hamilton, W. J., Jr., and R. P. Hunter
1939 Fall and winter food habits of Vermont bobcats. Jour. Wildlife Mgt., 3(2):99–103.

Harger, E. M.
1965 The status of the Canada lynx in Michigan. Jack-Pine Warbler, 43(4):150–153.

Hoppe, R. T.
1980 Population dynamics of the Michigan bobcat (*Lynx rufus*) with reference to age structure and reproduction. Michigan State Univ., unpubl. M.S. thesis, v+53 pp.

Jackson, H. H. T.
1961 The mammals of Wisconsin. Univ. Wisconsin Press, Madison. xii+504 pp.

Johnson, D. R.
1969 Returns of the American Fur Company, 1835–1839. Jour. Mammalogy, 50(4):836–839.

Johnson, N. F., B. A. Brown, and J. C. Bosomworth
1981 Age and sex characteristics of bobcat canines and their use in population assessment. Wildlife Soc. Bull., 9(3):203–206.

Jones, J. H., and N. S. Smith
1979 Bobcat density and prey selection in central Arizona. Jour. Wildlife Mgt., 43(3):666–672.

Jones, M. L.
1979 Longevity of mammals in captivity. Internat. Zoo News, 23(3):16–26.

Korschgen, L. J.
1957 Food habits of coyotes, foxes, house cats and bobcats in Missouri. Missouri Conserv. Comm., Fish and Game Div., P-R Ser. No. 15, 64 pp.

Kurta, A.
1979 Bat rabies in Michigan. Michigan Acad., 12(2):221–230.

Lowe, K.
1974 Chased, treed and tagged. Michigan Nat. Res., 43(1): 14–17.

Mahan, C. J.
1980 Winter food habits of Nebraska bobcats (*Felis rufus*). Prairie Nat., 12(2):59–63.

Manville, R. H.
1950 The mammals of Drummond Island, Michigan. Jour. Mammalogy, 31(3);358–359.
1958 Odd items in bobcat stomachs. Jour. Mammalogy, 39(3):439.

Marston, M. A.
1942 Winter relations of bobcats to white-tailed deer in Maine. Jour. Wildlife Mgt., 6(4):328–337.

McCord, C. M.
1974a Courtship behavior in free-ranging bobcats. The World's Cats, 2:76–87.
1974b Selection of winter habitat by bobcats (*Lynx rufus*) on the Quabbin Reservation, Massachusetts. Jour. Mammalogy, 55(2):428–437.

Miller, S. D., and D. W. Speake
1978 Status of the bobcat: An endangered species? Proc. Rare and Endangered Wildlife Sympos., Georgia Dept. Nat. Res., Game and Fish Div., Tech. Bull. WL 4, pp. 145–151.

Moran, R. J.
1964 Bobcat found on Lake Michigan Island. Jour. Mammalogy, 45(4):645.

Nagorsen, D., and R. L. Peterson
1977 Two recent bobcat (*Lynx rufus*) specimens from southern Ontario. Canadian Field-Nat., 91(1):98–100.

Penner, L. R., and W. N. Parke
1954 Notoedric mange in the bobcat, *Lynx rufus*. Jour. Mammalogy, 35(3):458.

Petraborg, W. H., and V. E. Gunvalson
1962 Observations on bobcat mortality and bobcat predation on deer. Jour. Mammalogy, 43(3):430–431..

Pollack, E. M.
1951a Food habits of the bobcat in the New England States. Jour. Wildlife Mgt., 15(2):209–213.
1951b Observations on New England bobcats. Jour. Mammalogy, 32(3):356–358.

Progulske, D. R.
1955 Game animals utilized as food by the bobcat in the southern Appalachians. Jour. Wildlife Mgt., 19(2): 249–253.

Provost, E. E., C. A. Nelson, and A. D. Marshall
1973 Population dynamics and behavior in the bobcat. The World's Cats, 1:42–67.

Rickey, D.
1975 Bobcats, coyotes are big game animals! Michigan Out-of-Doors, 29(1):6.

Robinson, W. B., and E. F. Grand
1958 Comparative movements of bobcats and coyotes as disclosed by tagging. Jour. Wildlife Mgt., 22(1):117–122.

Rollings, C. T.
1945 Habits, foods and parasites of the bobcat in Minnesota.. Jour. Wildlife Mgt., 9(2):131–145.

Schantz, V. S.
1939 A white-footed bobcat. Jour. Mammalogy, 20(1):106.

Seton, E. T.
1953 Lives of game animals. Vol. I, Pt. I. Cats, wolves, and foxes. Charles T. Branford Company, Boston, Massachusetts. x+337 pp.

Sheldon, W. G.
1959 A late breeding record of a bobcat in Massachusetts. Jour. Mammalogy, 40(1):148.

Siegler, H. R.
1971 The status of wildcats in New Hampshire. Proc. Sympos. on Native Cats of North America, Their status and management. U.S. Bureau of Sport Fisheries and Wildlife, Region 3, pp. 46–52.

Stebler, A. M.
1940 Cats have their cycles. Michigan Conserv., 9(6):11.

Ulmer, F. A., Jr.
1941 Melanism in the Felidae, with special reference to the genus *Lynx*. Jour. Mammalogy, 22(3):285–288.

Westfall, C. Z.
1956 Foods eaten by bobcats in Maine. Jour. Wildlife Mgt., 20(2):199–200.

Wood, N. A.
1914 An annotated check-list of Michigan mammals. Univ. Michigan, Mus. Zool., Occas. Papers No. 4, 13 pp.

Wood, N. A., and L. R. Dice
1924 Records of the distribution of Michigan mammals. Michigan Acad. Sci., Arts and Ltrs., Papers, 3:425–469.

Young, S. P.
1958 The bobcat of North America. The Stackpole Co., Harrisburg, Pennsylvania. 193 pp.

Even-Toed Hoofed Ungulates

ORDER ARTIODACTYLA

The "even-toed" or "cloven-hoofed" ungulates carry the weight of their bodies on just two toes, the third and fourth, of both fore and hind feet. This condition is termed paraxonic and is commonly shared by this diverse group of animals whose toe tips are sheathed in stout hoofs (the unguligrade stance). Externally, however, four complete and functional digits may be displayed in the pigs, hippopotami, and mouse deer (chevrotains); two complete digits with incomplete remnants of lateral ones in Cervidae (the deer family) and some Bovidae (the cow family); and two complete digits with no lateral remnants in camels, giraffes, and some bovids including the pronghorn.

One characteristic of the families of the camels, cervids, giraffes, and bovids is the cannon bone, a single supportive structure formed by the fusion of the third and fourth metapodials (metacarpals and metatarsals). Unique also is the "double-pulley" arrangement of the astragalus (a bone typical of the tarsal area of the mammalian ankle). The design of this bone allows for a tight articulation distally (toward the foot) with the navicular and cuboid (also anklebones and sometimes fused) and proximally (toward the upper leg) with the tibia. This fit restricts lateral movement and aids in the development of the remarkable running (cursorial) gait of many of these animals. This specialization has been further developed by the elongation of the metapodials (to help lengthen the stride) and the reduction or loss of the clavicle in the shoulder girdle.

The skull of the artiodactyl is strongly built, always with an elongated rostrum (preorbital area). The dentition, either low-crowned (brachyodont) or high-crowned (hypsodont), is characterized by a reduction in the number of upper incisors. The premolars are less complicated in pattern than the molars. The latter reflect the specialization of many even-toed ungulates as herbivores; their teeth having irregular crown patterns, useful in chewing coarse vegetable materials. In some omnivorous groups (pigs), the occlusal surfaces of the molar teeth have rounded cusps (bunodont pattern) while those of the grazers and browsers have detailed patterns exposing crescent-shaped ridges of hard enamel interspersed with valleys of softer dentine (selenodont condition).

To assist in the assimilation of nutrients from vegetable material, many even-toed ungulates (camels, chevrotains, cervids, giraffes, bovids) are ruminants; that is, they have a complicated stomach partitioned into three or four chambers. This stomach contains microorganisms which break down complex plant materials (cellulose) into simple carbohydrates so the ruminant's system can assimilate these products. After vegetable matter is initially chewed and swallowed, the larger food particles are regurgitated and re-chewed. This "cud chewing" aids in further breaking down the fairly rigid cell-walls of the plant materials so that inner cellular proteins can be completely utilized. Even so, these herbivores are obliged to eat great amounts of bulk to survive, while the carnivores need much less bulk since nutrients are more concentrated in flesh than in vegetation.

Horns or antlers, either structured entirely of bone or having a bony core and a keratinous sheath and usually outgrowths of the bones of the forehead (frontals), are present in higher artiodactyls, including the cervids and bovids. Representatives of this order, especially domestic swine, sheep, goats, and cattle, are of major economic importance. All wild species are highly edible, providing necessary "bush" protein for many rural peoples, and sport as well as table delicacies for the public in general. Most of these species are so valued that governments have attempted to regulate their annual harvests, preserved them in national parks, and developed lucrative tourist industries as a result of their presence.

KEY TO THE SPECIES OF ARTIODACTYLA IN MICHIGAN
Using Characteristics of Adult Animals in the Flesh

1a. Bony outgrowths (antlers) from frontal bones in males (sometimes in females) develop through a soft velvet stage to harden by means of ossification and are shed annually; hair not woolly; tail short and lacking terminal brush . . . family Cervidae—deer and allies . . . . 2a.

1b. Bony outgrowths (horns) covered with a horny (keratinous) sheath, usually in both sexes, and persistent through the life of the individual; hair woolly; tail long with terminal bush . . . family Bovidae—BISON *(Bison bison)*, page 620.

2a. Nose not covered with hair all around nostril; bare nose pad borders inside lower edge of nostril; antlers, if present, not palmate; hind leg with distinctive metatarsal gland . . . . 3a.

2b. Nose well-haired all way around nostril; antlers, if present, palmate in shape, hind leg with metatarsal gland absent . . . . 4a.

3a. Size larger, averaging about 500 lbs (227 kg); nose, head, and neck dark brown, in contrast to pale brown back; rump patch large and conspicuously yellowish in color; tail short and colored same as rump . . . WAPITI or ELK *(Cervus elaphus)*, page 566.

3b. Size smaller, averaging less than 200 lbs (91 kg); upper nose (above nose pad) dusky brown with contrasting whitish band above, conspicuously different from darker face, head, and neck; head and neck same general color as back; no large yellow rump patch; tail brownish above and conspicuously white below . . . WHITE-TAILED DEER *(Odocoileus virginianus)*, page 577.

4a. Head, neck, and back overall dark brownish black in color; nose large and arched; snout overhangs mouth; dewlap (bell) hanging from underside of throat; antlers massive with no palmate brow plate (tine) extending down over nose and between eyes; hoofs sharply pointed; bottom length of each hoof more than twice the width . . . MOOSE *(Alces alces)*, page 599.

4b. Head and neck creamy white in contrast to darker back; conspicuous, small, white rump patch; nose not excessively large; snout does not overhang mouth; no dewlap present; antlers less massive with palmate brow plate (tine) extending down over nose and between eyes; hoofs broad and rounded; bottom length of each hoof less than twice the width . . . CARIBOU *(Rangifer tarandus)*, page 612.

KEY TO THE SPECIES OF ARTIODACTYLA IN MICHIGAN
Using Characteristics of the Skulls of Adult Animals

1a. Bony outgrowths (antlers) from frontal bones in males (sometimes in females) develop through a soft velvet stage to complete ossification and are shed annually; lachrymal bone (containing tear duct) separated from adjacent nasal bone on each side of upper rostrum by a large preorbital vacuity (opening in front of eye), at least ⅝ in (15 mm) wide and exposing inner bones . . . family Cervidae—deer and allies . . . . 2a.

1b. Outgrowths (horns) from frontal bones, in both sexes, consist of a bony core covered with a horny (keratinous) sheath and never shed; lachrymal bone in contact with nasal bone on each side of upper rostrum with no large preorbital vacuity present . . . family Bovidae—bison, cattle, sheep, goats . . . . 5a.

2a. Skull larger, more than 20 in (510 mm) long; anterior opening of nasal cavity noticeably long; nasal bones short, length less than half the distance from their anterior tips to the anterior ends of the premaxillaries (bones in front of upper jaw) . . . MOOSE *(Alces alces)*, page 599.

2b. Skull smaller, less than 20 in (510 mm) long; anterior opening of nasal cavity relatively short; nasals long, length much more than half the distance from their anterior tips to the anterior ends of the premaxillary bones . . . . 3a.

3a. Posterior nasal cavity (posterior nares) when viewed on underside of skull divided by substantial, longitudinal, median, bony partition (vomer) for the full length of the cavity . . . 4a.

3b. Posterior nasal cavity (posterior nares) when viewed on underside of skull not divided by a median bony partition . . . WAPITI or ELK *(Cervus elaphus)*, page 566.

4a. Skull larger, more than 14 in (355 mm) long; antlers in male large and oval or flattened in cross-section (palmate), with brow plate (tine) extending forward between eyes and above rostrum; small antlers present in female; width of swollen capsule housing middle ear (auditory bulla) about half the length of the bony tube leading to the external ear opening (meatus); maxillary (upper) canine normally present . . . CARIBOU *(Rangifer tarandus)*, page 612.

4b. Skull smaller, less than 14 in (355 mm) long; antlers in male round in cross-section with no brow plate (tine); width of auditory bulla equal to length of the bony tube leading to the external ear opening (meatus); maxillary canine normally absent . . . WHITE-TAILED DEER *(Odocoileus virginianus)*, page 577.

5a. Skull medium, similar in size to that of white-tailed deer; length of upper jaw tooth-row (premolars and molars) less than 4⅜ in (110 mm) . . . . 6a.

5b. Skull large and massive, length of upper jaw tooth-row (premolars and molars) more than 4⅜ in (110 mm) . . . . 7a.

6a. Premaxillary bones not wedged beween nasals and maxillary bones; lachrymal bone on each side of upper rostrum distinctly concave forming deep pit in front of orbit . . . DOMESTIC SHEEP *(Ovis aries)*.

6b. Premaxillary bones wedged between nasals and maxillary bones; lachrymal bone on each side of upper orbit only slightly concave forming no deep pit in front of orbit . . . DOMESTIC GOAT *(Capra hircus)*.

7a. Skull broad and massive; premaxillary bones do not extend to nasals; each frontal bone projects laterally, just back of orbit, beyond the zygomatic arch (cheek bone) so when viewed from directly above, arch is obscured; horns always present; hind part of cranium (occipital region) projects back at least 1⅛ in (30 mm) beyond posterior edges of horn cores . . . BISON *(Bison bison)* page 620.

7b. Skull relatively narrow and less massive; premaxillary bones extend to nasals; each frontal bone does not project laterally, just back of orbit; horns present or absent; if horns present, occipital region does not project back more than ¾ in (20 mm) beyond posterior edges of horn cores . . . . . . DOMESTIC COW *(Bos taurus)*.

FAMILY CERVIDAE

Members of the deer family number about 16 genera and 37 species. The family is well represented in North and South America, Eurasia, and the islands of Indonesia. Cervids are not native to sub-Sahara Africa or to Australia but some species have been introduced by human design in Australia, New Zealand, the Hawaiian Islands, and other insular areas. The deer family adapts to diverse climates and topography, on arctic prairies and temperate grasslands, in mountains, swamps, and tropical jungles. The species range in size from the small, dog-sized pudu of western South America and musk deer of southern Asia to the massive, horse-like moose.

Like most other Artiodactyla, the Cervidae have developed a fast cursorial gait using an unguligrade stance, with toe tips sheathed in hoofs. For their herbivorous diets, the cervids have brachyodont (low-crowned) jaw teeth with occlusive (chewing) surfaces possessing a selenodon (crescent-shaped) pattern plus a four-chambered, ruminating stomach. There is usually no gall bladder. The most characteristic feature of this family is the presence of antlers in all but two genera. These structures are supported on permanent pedicels of the frontal bones and occur mostly on skulls of males. The growing antler is covered with a "velvet" skin, which is soft to the touch and replete with blood vessels and nerves. This growth, primarily under the control of testicular and pituitary hormones, continues until the structures are full size. Ossification sets in, the antler hardens, blood and nerve supplies are cut off at the pedicel, and the "velvet" is rubbed off about the time courtship begins. Ultimately, decalcification takes place in the pedicel causing the shedding of the antlers. For several months afterwards, the individuals may be antlerless. The rate of growth of these structures is remarkably rapid, causing the biologist to wonder just how the cervids are able to obtain sufficient nutrients to sustain this burden. One also may speculate as to the advantage of such a structure, especially since the horn is a permanent fixture in bovids, and the animals are spared this annual need to produce energy for regrowth.

In Michigan, there are two native cervids, the widespread white-tailed deer and the moose of the northern forest (Figure 103). Two other cervids, the wapiti or elk and the caribou, formerly occurred naturally in the state. Currently, the wapiti thrives after being reestablished.

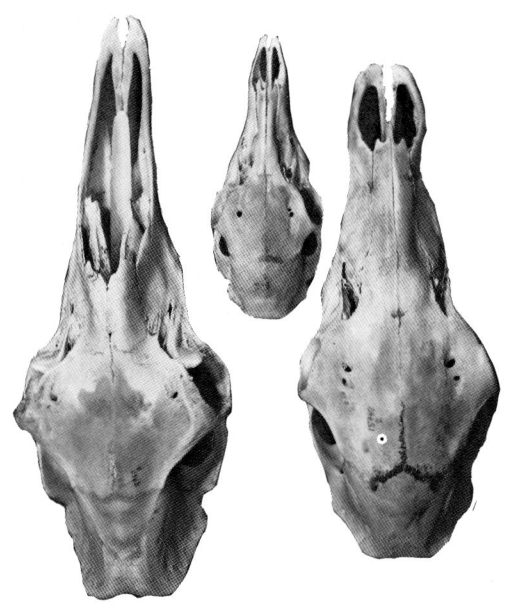

Figure 103. Dorsal aspects of the skulls of three Michigan cervids. From left to right: Moose, white-tailed deer, wapiti.

WAPITI or ELK
Cervus elaphus Linnaeus

NAMES. The generic name *Cervus*, proposed by Linnaeus in 1758, is derived from the Latin word for deer or stag. The specific name *elaphus*, also proposed by Linnaeus in 1758, is from the Greek word *elaphos* also meaning stag. In much of the earlier literature, this large cervid was given the scientific name *Cervus canadensis*. However, it has been demonstrated that North American animals bearing this name and the red deer of Eurasia classified as belonging to *Cervus elaphus* are sufficiently related to be considered conspecific; that is, belonging to the same species. The latter name, the older of the two taxa in terms of usage, applies to both populations (Caughley, 1971; Heptner *et al.*, 1961). The preferred common name wapiti is derived from the Cree word "wapitik" for white deer; also from a Shawnee word meaning white rump. The term elk is of European origin and was first used by early settlers. In Michigan, the name elk is commonly used for this species. Males are often referred to as bulls or stags, females as cows or hinds, and the young as calves. The ancestors of the reestablished wapiti were derived from the Yellowstone National Park area in northwestern Wyoming, from where the subspecies *Cervus elaphus nelsoni* Bailey was described. This name should be applied to the present Michigan wapiti herd. The now-extirpated native wapiti belonged to the subspecies *Cervus elaphus canadensis* Erxleben.

RECOGNITION. Body size large for a cervid, standing about 60 in (1,500 mm) at the shoulder, as much as three times as heavy as a Michigan white-tailed deer; length of head and body averaging 96 in (2,400 mm) for the larger male and 90 in (2,250 mm) for the smaller female; tail averaging for both sexes 5⅞ in (150 mm); head long with muzzle bare; ears large, prominent; preorbital (lachrymal) glands conspicuous; neck long with mane; body thick, sturdy; tail short, well-haired, somewhat wedge-shaped; limbs long, slender; metatarsal glands prominent on the outside of the hocks, oval-shaped, surrounded by stiff yellowish hairs; hoofs bluntly pointed in front, broadly rounded behind; mammary glands consist of two pairs; adult males at least 10% larger in bodily dimensions and weigh almost twice as much as adult females.

Antlers of adult male majestic, widely-branching, arising from large burrs high on the head; each antler consisting of a single, heavy, curved cylindrical beam, as long as 60 in (1,500 mm) sweeping up and back from the head and over the shoulder with several tines; the first tine (brow) extending over face; the second (bez) growing laterally over ear; the third (trez) projecting about one-third of the length of the antler; the fourth (royal) being the longest and stoutest; the final fork producing the fifth tine (sur-royal); the last point being the tip of the main beam; antlers shed no later than April; regrown in late spring and summer; sometimes irregular in form due to the animal's advanced age or glandular disorders (Seton, 1953).

The pelage of the wapiti consists of long, coarse, guard hairs and wooly underfur in winter and shorter guard hairs and no underfur in summer. In general, the coat is darker brown in winter, and tawny in summer. Both males and females have similar coat colors in summer with the females and yearlings somewhat darker in winter. The head, neck, belly, and legs are darker brown in contrast to the paler brown or yellowish gray on the back and sides. Conspicuous is the large buff-colored patch on the rump. The wapiti should not be confused with any other Michigan mammal, wild or domestic, because of its distinctive darkened head, neck and limbs, pronounced mane, obvious buff-colored rump patch, massive size, and in the adult male, the unique and classic antlers (Fig. 104).

MEASUREMENTS AND WEIGHTS. Adult specimens of the male and female measure, respectively, as follows: length from tip of nose to end of tail vertebrae, 92 to 108 in (2,300 to 2,700 mm) and 84 to 96 in (2,100 to 2,400 mm); length of tail vertebrae, 5½ to 6¼ in (140 to 160 mm) and 4¾ to 5¾ in (120 to 145 mm); length of hind foot, 22½ to 25 in (600 to 660 mm) and 22½ to 24½ in (600 to 650 mm); height of ear from notch (same in both sexes), 5¼ to 8 in (130 to 200 mm); height at shoulder (withers), 57 to 69 in (1,400 to 1,700 mm) and 37 to 45 in (1,220 to 1,425 mm); weight, 580 to 1,000 lbs (265 to 454 kg) and 500 to 650 lbs (230 to 300 kg).

DISTINCTIVE CRANIAL AND DENTAL CHARACTERS. The skull of the wapiti is heavy and elongate (Figs. 103, 105). The long upper jaws are typically without incisors but do have large and rounded upper canines (so-called elk's teeth); the preorbital (lachrymal) vacuities form large and conspicuous cavities in front of each orbit; the premaxillary bones make contact with the nasal bones; the cranium is arched; on the ventral side, the vomer is low posteriorly and does not divide the cavity of the posterior nares into two chambers. In the lower jaws, the canines are incisiform, resembling the sharp-edged incisors. The premolars and molars are heavy with crescent-shaped cusp patterns (solenodont) on their chewing surfaces. The skulls of the larger adult male and of the smaller adult female average, respectively, 19¾ in (507 mm) and 16½ in (417 mm) in greatest length and 8¼ in (211 mm) and 7 in (177 mm) in greatest width.

The wapiti skull is shorter than that of the moose and also has rounded instead of palmate antlers, and upper canines. The wapiti skull is longer than

Figure 104. The bull wapiti and harem on an autumn day in Michigan's Pigeon River Country. The grassy opening is bordered by woodlands.
Photograph by Robert Harrington of the Michigan Department of Natural Resources.

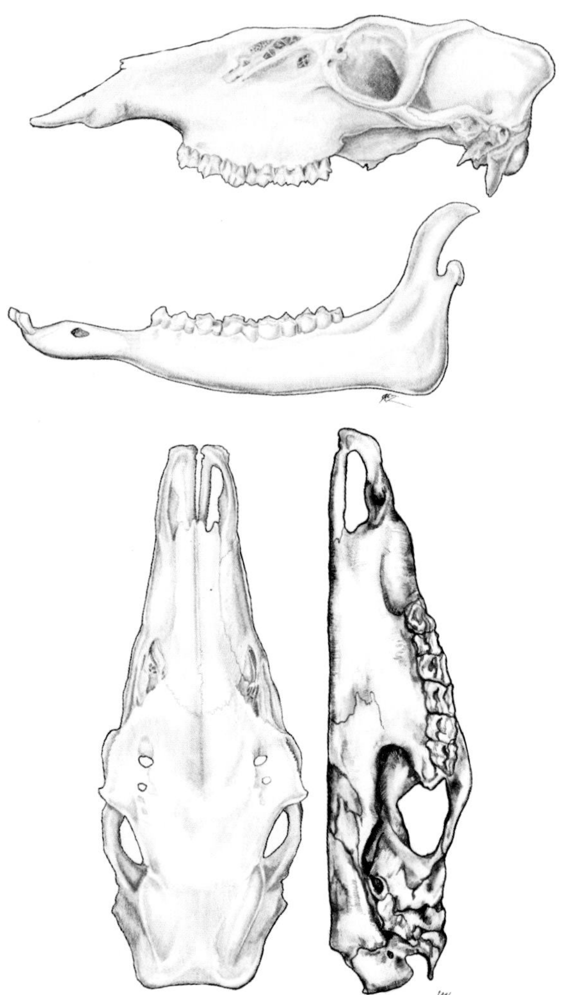

Figure 105. Cranial and dental characteristics of the wapiti, (*Cervus elaphus*).

that of the white-tailed deer and has upper canines, but no complete separation of the posterior nares by the vomer. The wapiti skull differs from that of the extirpated caribou in that the former has rounded instead of palmate antlers, antlers only in males instead of in adults of both sexes, large rounded upper canines instead of small ones (narrow slivers) rarely piercing the gum; and no complete separation of the posterior nares by the vomer. The dental formula for the wapiti is as follows:

I (incisors) $\frac{0\text{-}0}{3\text{-}3}$, C (canines) $\frac{1\text{-}1}{1\text{-}1}$, P (premolars) $\frac{3\text{-}3}{3\text{-}3}$, M (molars) $\frac{3\text{-}3}{3\text{-}3}$ = 34

DISTRIBUTION. The wapiti and its conspecific relative, the Old World red deer, once occurred widely in the Northern Hemisphere, in both Eurasia and North America. Its populations in parts of Europe and eastern North America have been extirpated. In recent years, however, the species has been reestablished in some places in eastern United States and introduced in such exotic places as New Zealand.

In eastern North America, the wapiti occurred in prehistoric times as far east as southern Quebec in the north and southward along the western side of the Allegheny Mountains, probably ranging in some places all the way to the Atlantic coast (Waters and Mack, 1962). Excessive hunting and habitat change as a result of settlement were major factors causing the extirpation of the wapiti—in New York by about 1847, in Pennsylvania by about 1867, in Ohio by about 1838, and in Indiana by about 1830 (Murie, 1951).

In the Lower Peninsula of Michigan, the wapiti

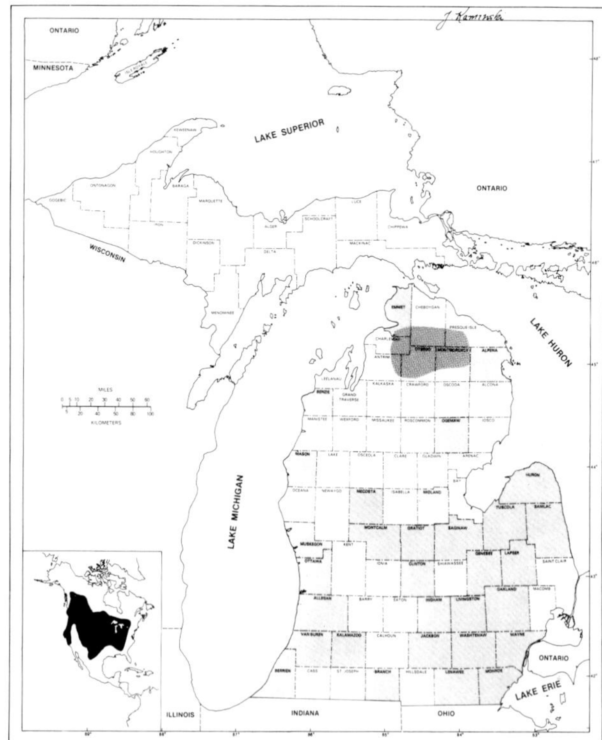

Map 65. Geographic distribution of the wapiti or elk (*Cervus elaphus*). Countries with shading are where actual specimens from the historic and prehistoric record are reported. Present distribution of the reestablished population is shown in dark shading.

is known positively from historic records as having occurred in most if not all counties in the early days of settlement (see Map 65). The Indians' use of the wapiti is shown by numerous bones of the animal found in archaeological sites of the Woodland period in such counties as Allegan, Berrien, Charlevoix, Lapeer, Muskegon, Ottawa, and Saginaw (Cleland, 1966; Lovis, 1973; Martin, 1976, 1979). In addition, antlers have been found in bogs and other post-glacial deposits in such Lower Peninsula counties as Berrien, Branch, Clinton, Crawford, Genesee, Ingham, Jackson, Kalamazoo, Leelanau, Livingston, Mecosta, Midland, Monroe, Montcalm, Oakland, Ogemaw, Van Buren, and Washtenaw (Wilson, 1967). One antler was dredged up by workers who were deepening the ship channel in Saginaw Bay off Bay County. The association of wapiti bones with those of the extinct mastodon (*Mammut americanus*) in a bog deposit in Gratiot County has also been reported (Brown and Cleland, 1969). Such mementos are often preserved in museums, by private collectors and reported in the literature (Wood, 1914). It is assumed that heavily-wooded Upper Peninsula was unattractive to wapiti, although Blois (1839) does place this animal in the western part of the Upper Peninsula early in the nineteenth century. Bones from an archaeological site in Delta County were possibly but not surely those of wapiti, according to Martin (1980). However, there is positive evidence of the occurrence of this cervid in the Wisconsin counties of Iron and Villas which border the western edge of Upper Peninsula Michigan (Schorger, 1954).

In the early period of settlement, wapiti was an important source of human food. Cadillac referred to the species as being in the Detroit area about 1700 (Johnson, 1919). Skeletal remains excavated at Fort Michilimackinac point to use of the animal in that area in the early and middle parts of the eighteenth century (Butsch, 1970). Its presence in that century in southwestern Michigan (Berrien County) is also documented (Baker, 1899). Human settlement in the nineteenth century caused a steady decline of this game species. There are wapiti records for Oakland County as late as 1830 (Hoyt, 1889), and apparently one of the last refuges for the animal was the "thumb" area, especially Huron, Sanilac, and Tuscola counties (Miles, 1861; Murie, 1951), where it survived until at least 1871 (Wood and Dice, 1924). A Cleveland, Ohio, hunter was attracted to this area, and his accounts (Perry, 1899) of hunting wapiti along the watersheds of the Cass and Black rivers in Sanilac and Tuscola counties in 1852 are among the most complete from those early times. Twenty years later, however, in the 1870s, the native wapiti in Michigan was doomed to extinction. Most authors cite 1877 as the year of its final demise (Burt, 1946; Moran, 1973; Murie, 1951).

Following the turn of the century, sentiment apparently built up among citizens to restore this majestic ungulate to the forest glades of Michigan. Besides private citizens' efforts to propagate wapiti obtained from outside the state, there were, according to Moran (1973), three official attempts by the state to reestablish a viable wapiti herd. In March 1915, 23 wapiti imported from Yellowstone National Park in Wyoming were retained in a large enclosure (164 acres) near Houghton Lake. Six others were also brought there from a private herd held at Ionia. Of these, 20 (9 in March 1918 and 11 in February 1919) were released into the wild at the Rayburn Ranch in the Turtle Lake area of Alpena County. These animals apparently disappeared sometime after 1920. In 1918, 7 wapiti (including at least 2 bulls and 3 cows) from this same confined stock were released in Munda Township of Cheboygan County. This small planting thrived and gave rise to the present Michigan herd (Ruhl, 1940; Stephenson, 1942). In 1932, 18 wapiti from various private sources in the state were released but failed to survive in the Houghton Lake State Forest.

The 1918 Cheboygan County release increased and ultimately spread, according to Moran (1973), into parts of Cheboygan, Montmorency, Otsego, and Presque Isle counties by 1958, covering an area of 400 sq mi (1,036 sq km). By the 1960s, a few wapiti also ranged westward into parts of Antrim, Charlevoix, and Emmet counties and southeast into the Au Sable watershed (see Map 65). Hopefully, the wapiti is back in Michigan to stay; problems relating to its livelihood and the effects of controlled hunts held in the late autumn of 1964 and 1965 are discussed under Economic Importance.

HABITAT PREFERENCES. Michigan wapiti thrive in a mosaic pattern of woodlands and openlands much like their relatives in Montana and other western states (Boyce and Hayden-Wing,

1979; Murie, 1951). The wapiti's response to this mixture of habitats including clear-cuts is ". . . a complex function of forage and cover requirements modified by the behavioral patterns of animals in the local environment (Lyon and Jensen, 1980)."

The wapiti range in the northern part of the Lower Peninsula has been characterized (Moran, 1973; Spiegel *et al.*, 1963) as including (1) sandy outwash plains (open savannah); (2) outwash plain-morainic ecotone (edge providing herbaceous forage in openings and tree-shrub browsing in adjacent woody cover); (3) steep morainic slopes (slopes with open pine-hardwood mixtures, when cut, providing new sprouts as forage); (4) morainic uplands (well-drained grassy openings interspersed with forest stands with understory); (5) riverbanks and bottomlands (sources of water and edible grasses and shrubs); and (6) coniferous swamps (white cedar-spruce-balsam stands used for foraging and shelter). Area use has been determined by field observations of wapiti activities and by relating the presence of fecal droppings (pellets) to habitat use. As a consequence, Moran (1973), Knight (1975), and other Michigan observers have concluded that wapiti generally frequent swamp conifers, cuttings, aspen-hardwoods, and upland conifer-hardwoods in winter. Stabilized openings and aspen-hardwood cover are major summer environments. It was Knight's contention that openings are of maximum importance, with small openings most attractive in winter, large ones in spring and autumn, and openings of all sizes in summer. Heavy winter snows, with depths exceeding 18 in (46 mm), appear to restrict wapiti movements. According to Moran (1973), agricultural crops grown in Michigan's wapiti range are sometimes invaded. Such areas, however, may also constitute barriers to dispersal of the animals.

DENSITY AND MOVEMENTS. The principal Michigan wapiti range in 1979 included about 600 sq mi (1,554 sq km) as delineated by a line drawn from the city of Indian River on the northwest, east to Onaway on the northeast, south to Atlanta on the southeast, west to Gaylord on the southwest, and north to Indian River. Through the years, wildlife biologists for the Michigan Department of Natural Resources have carefully recorded the wapiti herd's increase and range expansion. After the initial release of seven animals in 1918, the total population was estimated to be about 200 animals in 1925, 300–400 in 1939, 900–1,000 in 1958, 1,200–1,500 in 1961, and 2,000–3,000, in 1963–1964. Although the total known kill for the hunting seasons in 1964 and 1965 was only 477 animals, the population apparently declined to 1,000–1,250 in the 1966–1967 period. Estimates in the early 1970s were low—200 in 1976 and 255 in 1977—but increasing again to between 400 and 500 in 1979 and 1980. Michigan biologists felt that habitat depletion and some disruption of social structure played roles in this postseason decline, but perhaps to a lesser extent than did illicit poaching and possibly disease (Eveland *et al.*, 1979). Moran (1973) found population densities in the Michigan herd to vary from a low of 4.4 wapiti per sq mi (1.7 per sq km) in 1965–1966 to a high of 11.5 wapiti per sq mi (4.4 per sq km) in 1963–1964. These density figures compare favorably with those obtained for wapiti herds in mountainous parts of western states and provinces.

While movements of wapiti in mountainous regions from summer range to winter range may require extensive travel (Murie, 1951), in Michigan preferred habitats for each season are close together. Michigan workers (Moran, 1973) found collar-marked individuals to rarely move seasonally more than 5 mi (8.3 km) with a maximum of 12 mi (19 km). Mature bulls during the rut may travel more widely. There is evidence, however, that wapiti groups confine themselves to small home ranges correlated with available forage. These areas do not seem to be defended but neither are they used by other such groups (Franklin *et al.*, 1975).

BEHAVIOR. Wapiti are highly social animals and often occur in large herds. Some of the most notable records of this cervid's gregarious nature are for winter congregations in Wyoming's Jackson Hole area (Anderson, 1958). In summer, Michigan wapiti are generally in small bands, with cows (hinds) and their calves forming nursery groups and bulls (stags) in separate associations. Often mature bulls are solitary at this time, but in autumn will gather a selection of cows and their calves in their harems (Fig. 104). These bulls, by their sheer size and aggressive activity, defend their harems from inroads by other males attracted to the females during the autumn rutting period.

After the mating period, the harem system

breaks down and large herds, composed of both sexes and all age groups, with a dominant cow as leader (matriarchy), are formed. These herds are not inclined to "yard" as do white-tailed deer. In spring, the females segregate to give birth to calves while once again the males form bachelor associations. As a result of these seasonal relationships, wapiti herds are not necessarily constant in either size, sex, or age-classes. No special territory is defended in these behavioral interactions (Franklin *et al.,* 1975), except during the autumn rutting season (Struhsaker, 1967).

The herding activity, plus the dominant roles played by harem bulls in the breeding season and often by mature cows at other times of the year, demonstrate both individual and group behavioral characteristics (Altmann, 1952, 1956; Franklin and Lieb, 1979). The animals generally feed in early morning and late evening. They tend to bed down and be inactive during the day and in the middle of the night, when part of their time is spent in chewing their cuds. At all times, the herd as well as its individual members are alert for real or supposed danger, at which time entire groups will be cued to stampede away, although rarely for a long distance. Observers are quick to note the interplay between members of the herd. It is possible that the conspicuous rump patch functions as a signal to promote social interaction (Mc Cullough, 1969). Wapiti feed head down and at a slow walking pace. When trotting, the head is held high with muzzle up and the gait is more stiff than graceful. Wapiti gallop at speeds of more than 30 mi (48 km) per hour and are excellent swimmers, rarely hesitating to cross streams or lakes.

Fieldworkers regard the wapiti as the noisiest of all North American cervids. The newborn calf can utter a loud bleat or squeal, prolonged when danger seems near. Females bark, grunt, and squeal when associating with their calves or when alarmed. These utterances are also used in group interactions. Under stress, wapiti make noisy, tooth-grinding noises. However, the most characteristic call of the wapiti is the bugle (Murie, 1951). The mature bull bugles during the autumn rut, although cows also bugle in late spring during calving time. The bugling bull extends its head and swollen neck forward with antlers laid back to produce this challenging call, making all creatures in the adjacent countryside well aware of his majestic presence, and no doubt subduing and confining his harem while warning away unattached males. The bugle, as described by such observers as Boyd (1978), varies but usually begins as a low-pitched bellow or roar, scaling upward to a high, lucid, trimodal bugle held until the animal needs to take another breath, and subsiding as a grunt or series of coughing sounds. This call, usually occurring between early September and mid-November, may carry as much as a mile (1.6 km) in clear autumn air.

ASSOCIATES. In Michigan, the wapiti has a seemingly close association with the white-tailed deer. Moran's classic study (1973) shows that wapiti and white-tailed deer (1) occur in similar environment; (2) are closely tied to habitats with openings, shrub cover, and young forest growth (early successional stages) which may be maintained by logging, cutting of pulp wood, and management fires; and (3) are highly adaptable browsers and grazers. In contrast to the white-tailed deer, however, the wapiti is (1) more of a wilderness animal especially lacking the white-tailed deer's ability to thrive close to human habitation; (2) perhaps only one-half as productive of offspring; and (3) more mobile especially in winter, whereas white-tailed deer are inclined to confine themselves in heavy snowfall to winter yards in extensive coniferous swamps. The evidence in the literature (Murie, 1951) indicates that the reestablished wapiti and native white-tailed deer may be in competition for food or habitat in such eastern areas as Michigan or Pennsylvania (Hunter, *et al.,* 1979; Moran, 1973). Nevertheless, it is highly suspected that competition for winter food (especially browse) is indeed a factor in Michigan. Studies, using telemetry and other remote-sensing devices, designed to determine interactions between these species for food resources should certainly be attractive to wildlife ecologists.

REPRODUCTIVE ACTIVITIES. In summer, after calves are able to move around well, cows congregate with their offspring in nursery groups. Younger bulls may mix with cows at this season, but older males maintain separate existences either as single individuals or small bachelor groupings. Antlers of these males are in velvet and rapidly completing growth. By mid-August, Michigan bulls initiate the autumn courtship with bugling. At this time, antlers have matured and hardened with males (including spiked yearlings) spending time stripping off velvet coverings by sparring and rubbing their antlers on shrubs and trees, especially conifers.

By early September, the mature bulls by means of their size, aggressive nature, and vigor attract cow-calf groups and drive away the younger males and other mature bulls. Although yearling (15-month old) males are sexually mature, they rarely are allowed to participate in mating at that age (Conaway, 1952). Moran (1973) found rutting activity in Michigan wapiti to occur between September 15 and October 10. His observations showed that the breeding peak occurred by September 20 when "... bugling reached a crescendo and harems were formed up and guarded by dominant bulls." Threshing shrubs and trees, wallowing in urine-tainted mud, following the cows with curled lips and extended heads, shoving male adversaries with antlers, stabbing at them with their forefeet, disregarding food, and bugling noisily are a few of the courtship behavioral patterns described by Murie (1951), Struhsaker (1967), and other observers. In his study of 52 harems, Moran found the cows numbered from 1 to 21 and calves from 1 to 8. The average harem included 10 individuals, 6 cows and 4 calves.

In October, the furor of the rut passes, allowing for the physically exhausted bulls to feed and regain their strength for the coming winter season. The harem breakup is a time for the bulls to wander off singly or once more tolerate other males. The pregnant females, their calves, and young bulls form winter herds with an older cow as lead animal. The bulls' antlers, which began as growth buds from the frontal bones in April, were completely grown, hardened, and rubbed smooth in August and presumably involved in a secondary sexual way in the rutting season, are shed in winter.

With the approach of calving time in May and June, after a gestation period of 249 to 262 days, the pregnant cows separate from the winter herds. Embryonic development of the single calf (rarely two) has been studied to the point that even fetuses can be accurately aged (Morrison, *et al.*, 1959). At birth, the offspring has matted pelage, creamy spots in longitudinal rows on the back and sides in contrast to the overall fawn coloring, stilt-like legs, soft hoofs, and weights about 35 lbs (15.7 kg). At four days of age, the calf has some lower incisors erupting, possesses hardened hoofs, walks erectly, although somewhat wobbly, and drops to "freeze" when alarmed. At seven days of age, the calf has conspicuous lower incisors, shows erupting upper canines and some evidence of cheek teeth, is apt to run instead of freeze when startled, and stands and walks steadily (Johnson, 1951). At about 16 days of age, the calf is sufficiently active for the parent and offspring to join a summer herd. The calf now may nibble tender grass, becoming more dependent on forage by 30 days of age. At more than 60 days of age (usually in September), weaning may be completed, although individual young may be nursed well into winter. In September, the calf's spotted coat is shed and replaced by a juvenile coat of uniform brown.

Wapiti of both sexes may become sexually active as yearlings (approximately 16 months old). In Michigan, for example, Moran (1973) found that about 23% of the 30 yearling females which he examined were pregnant, while 74% of 35 second-year animals also were pregnant. Maximum productivity is expected of cows three or more years of age (Murie, 1951). For mature Michigan cows, Moran found the pregnancy rate was 77%. Although yearling "spike" bulls are capable of breeding and establishing harems (Boyd, 1978), few bulls under four or five years of age, as mentioned previously, have much opportunity to mate.

The wapiti's low annual production, as compared with that of the white-tailed deer, is somewhat offset by the protective maternal care (termed parental investment, Zeveloff and Boyce, 1980) afforded the calves during their first year. Individuals have been successfully aged, at least to the eight-year class, by determination of the time of eruption, development, and attrition of deciduous and permanent canines, premolars and molars (Greer and Yeager, 1967; Quinby and Gaab, 1957). In the wild, Murie (1951) recorded few individuals more than 20 years of age. Of the 476 animals taken in the course of Michigan's two controlled hunts conducted in December of 1964 and 1965 there were only 17 wapiti in the oldest age-class of nine to ten years old. These, with the eight-year old animals, were regarded by Moran (1973) as old; those three to seven years old were prime; those of less than three years were classed as young. Wapiti can live for as long as 25 years in captivity.

FOOD HABITS. Wapiti exhibit a high degree of adaptability in diet, browsing on a variety of woody plants and grazing on grasses, sedges and forbs. The animals spend much time feeding, and consume large amounts of foodstuffs. In the complex stomach of this ruminant, the newly-masticated plant material (full of complex carbohydrates which require bacterial action to break down) is given an opportunity to digest in the rumen. Subsequently, usually when the wapiti have bedded down at noontime or at night, these foods are regurgitated and re-masticated—the chewing of the cud. Then the materials are again swallowed and pass through other stomach chambers to enter the small intestine and move toward elimination; nutrients are extracted by intestinal walls enroute. The wastes from a meal of browse are usually deposited in the form of rounded, cylindrical shaped pellets, resembling those of white-tailed deer but larger, 1⅜ in (35 mm) long. The wastes from a meal of green foods are deposited in a pie-like condition, not unlike the droppings of cattle.

According to Moran (1973), wapiti have defined feeding habits correlated with seasonal changes. In spring, the animals tend to concentrate in openings as soon as green herbaceous vegetation begins to appear. Grasses, sedges, and forbs continue to be the major attraction for wapiti in summer. At frost time in autumn, the animals direct their attention chiefly to woody growth. This diet continues through the winter and early spring although animals will graze where grasses are snow-free on windswept slopes, on streambanks, under trees, and even on green winter field crops on nearby farms.

White cedar, wintergreen, eastern hemlock, staghorn, sumac, and jack pine, in that order, appeared most frequently in 107 rumens of wapiti killed during the December hunting seasons of 1964 and 1965 (Moran, 1973). Other trees eaten included red maple, juneberry, basswood, cherries, striped maple, witch-hazel, staghorn sumac, and aspens. The wapiti diet, especially during heavy snow cover in winter, is partly dependent on edible bark stripped from young trees usually no more than 4 in (100 mm) in diameter at chest height. Little specific information on wapiti food preferences among grasses, sedges and forbs has been obtained in Michigan. Bluegrass and brome are definitely relished. Wapiti also compete with mushroom pickers for these tasty morsels.

It seems important to consider the different needs of the two sexes in both habitat and nutritional studies of any of the cervids because males and females have decidedly different requirements. The cows need foods to maintain their own bodies and to nourish their offspring both *in utero* and as nursing calves. On the other hand, males require unusually large amounts of minerals and other substances to enable them to produce their massive antlers in five months or less, certainly a major metabolic effort. Future studies of the behavioral ecology of wapiti, as well as other cervids, should take into greater account the diverse requirements of the sexes.

ENEMIES. The larger North American carnivores, mountain lion, gray wolf, and bears, have the ability to subdue mature wapiti (Barmore and Stradley, 1971; Murie, 1951; Seton, 1953). With the exception of black bear, Michigan wapiti have no association with these meat-eaters today. Unattended calves and adults weakened from disease or old age could fall victim to coyotes or bobcats. In Otsego County near Vanderbilt, Moran and Ozoga (1965) saw a coyote chasing a wapiti calf. As carrion, however, this cervid serves as a food source for several four-footed carnivores as well as for carrion-eating birds. Moran (1973) and Jenkins (1977) emphasized that the major cause of mortality in the Michigan wapiti herd is illegal hunting. In 1975, for example, 40 animals were known to have been shot.

PARASITES AND DISEASES. The wapiti is infested with both external and internal parasites including ticks (*Dermacenter albipictus, Dermacenter andersoni, Otobius megnini*), a mite (*Psoroptes equi* var. *cervinae*) causing scabies, lice (*Bovicola americanum, Damalinia concavifrons*), tapeworms (*Thysanosoma actinoides, Cysticercus tenuicollis, Cysticercus tarandi, Echinococcus granulosus, Moniezia* sp.), a flatworm (*Fascioloides magna*), nematodes (*Ostertagia* sp., *Marshallagia marshalli, Trichostrongylus* sp., *Cooperia, Nematodirus, Capillatia, Trichuris, Oesophagostomum, Dictyocaulus viviparus, Protostrongylus macrotis, Elaeophora schneideri, Setaria yehi, Parelaphostrongylus tenuis, Wehrdikmansia cervipedis*), protozoans (*Eimeria zurnii, Eimeria wapiti, Sarcocystis* sp., *Trypanosoma cervi, Entamoeba* sp.), and a nasal botfly (*Cephenemyia* sp). Most parasites are known from wapiti in the Rocky Mountain region (Murie, 1951; Worley, 1979). In the Michigan wapiti herd, infestation of the meningeal worm (*Parelaphostrongylus tenuis*) is of special interest. This worm (having a larval stage in a snail or slug) is ingested by the primary cervid host and, after a rather complex migratory route through such areas as the dorsal horn of the spinal cord, matures in the subdural space (Sikarskie, 1978). Its usual host is the white-tailed deer, which seemingly tolerates the parasite without severe results (Anderson, 1972). However, in other cervids (wapiti, moose, caribou, and mule deer) the worm can be highly pathogenic. Moran (1973) attributes some of the poor success of the Michigan wapiti herd to infestations of this worm (see also Tompkins *et al.*, 1977). As more recent arrivals (geologically speaking), the wapiti and the moose appear to lack the resistance of the longer-established white-tailed deer to this parasite. It is also likely that the white-tailed deer, spreading northward after the logging and clearing operation of the nineteenth and early twentieth centuries, came progressively more in contact with wapiti and moose populations in the Upper Great Lakes area and less so with caribou. As suggested by James Sikarskie (pers. comm.), this parasitic roundworm could be helping to define the modern distribution and even the natural selection of cervids at least in northeastern North America. The meningeal worm remains a problem for managers of wapiti herds reestablished in eastern United States, particularly in areas where infested white-tailed deer are already abundant.

Michigan wapiti are not known to have brucellosis (Youatt and Fay, 1961) although it is a problem in animals in the Rocky Mountain area (Thorne *et al.*, 1979).

MOLT AND COLOR ABERRATIONS. The wapiti molts its darker winter coat and grows the paler summer pelage in May and June. In September, the winter coat again appears. Animals with white or spotted white coats are occasionally reported (Murie, 1951).

ECONOMIC IMPORTANCE. The wapiti, although probably never highly abundant in Michigan, offered a sizable and certainly welcome food supply for the early peoples in what is now Michigan. Teeth, bones, sinew, hides, antlers, fat, and other parts were used by both the Indians and early European pioneers. Originally, only animals needed for tribal or family use were taken (Petersen, 1979). However, once the hide trade expanded and Michigan wapiti skins developed commercial value, the animals were more intensely hunted down to their ultimate demise in the 1870s. Between 1835 and 1839, for example, the Detroit Department of the American Fur Company received 231 wapiti skins, many of which may have come from Michigan (Johnson, 1969).

By the turn of the century, interested Michigan citizens realized that the extirpated wapiti might be successfully reestablished. Accordingly, the present population came about after several belated attempts, by both private parties and the state, were made to encourage released animals to thrive. Once established in the four-county area generally known today as the "Pigeon River Country," the herd slowly increased and spread, with distribution ultimately recorded from a six-county area in the extreme northern part of the Lower Peninsula (see Map 65). When the herd became large enough to be easily observable, tourists were attracted, especially by bugling bulls at the time of the autumn rut. At the same time, however, wapiti ventured into neighboring agricultural areas, alienating farmers by their depredations on crops. Overbrowsing of valued trees annoyed foresters and lumbering interests (Moran, 1973). Finally, deer hunters complained that wapiti were competing with their favorite game animal for food and space. Poachers took their toll (Jenkins, 1977). Careful monitoring by state wildlife biologists showed that the population of more than 2,000 individuals in 1963 might very well be cropped by controlled hunts (Blouch and Moran, 1965). The "hunter's choice" wapiti seasons held in December of 1964 and 1965 resulted in a total harvest of 228 cows (48%), 167 bulls (35%) and 81 calves (17%). Following this rather sharp decline in herd size, there was considerable concern for the welfare of Michigan wapiti in the mid-1970s. The herd failed to increase to the population level attained prior to the controlled hunts. Illegal hunting, habitat deterioration, disruption of herd social structure, and perhaps the lethal effects of the meningeal worm were presumed to be responsible. To reverse the downward trend in wapiti numbers, biologists of the Michigan Department of Natural Resources acted to: (1) discourage human disturbances; (2) reduce poaching; and (3) reverse habitat deterioration in the prime range. Happily, the herd slowly began to increase in the late 1970s.

Nevertheless, poaching has remained a difficult problem to control (Jenkins, 1977)—11 animals were reported killed illegally in January and February of 1980. The wapiti's upper canine teeth are valued as watch charms and their antlers "in velvet" are exported to the Orient for medicinal preparations, but the major reason for poaching is presumably for meat. Many wapiti may have been illegally shot merely for the "sport" of it. Another sector of the Michigan citizenry compained that too much money (almost $250,000 annually, Jenkins, 1977) was spent in managing a herd of introduced animals which might never be abundant enough to provide sustained hunting. However, conservationists and others quickly took the side of the wapiti (as well as that of the Kirtland's warbler) and the pristine beauty of the environment when oil and gas exploration was proposed for the Pigeon River Country in 1975. This exploitation of subsurface resources was in the initial stages in 1981. Recent studies, however, show that although seismic activity in pre-drilling explorations can temporarily disturb wapiti behavioral patterns within a distance of 1,000 meters, oil well drilling and pumping seem to have less of a negative influence on the animals (Knight, 1980). In any case, the wapiti seems sufficiently reestablished to remain a Michigan resident for some time to come.

COUNTY RECORDS FROM MICHIGAN. The reestablished wapiti in the extreme northern part of Michigan's Lower Peninsula generally occurs in the following counties: Antrim, Charlevoix, Cheboygan, Montmorency, Otsego, and Presque Isle (see Map 65).

LITERATURE CITED

Altmann, M.

1952 Social behavior of elk *(Cervus canadensis nelsoni)* in Jackson Hole area of Wyoming. Behavior, 4:116–143.

1956 Patterns of herd behavior in free-ranging elk of Wyoming, *Cervus canadensis nelsoni.* Zoologica, Vol. 41, Pt. 2, No. 8, pp. 68–71.

Anderson, C. C.

1958 The elk of Jackson Hole. A review of Jackson Hole elk studies. Wyoming Game and Fish Comm., Bull. No. 10, 184 pp.

Anderson, R. C.

1972 The ecological relationships of meningeal worm and native cervids in North America. Jour. Wildlife Diseases, 8:304–310.

Baker, G. A.

1899 The St. Joseph-Kankakee portage. Its location and use by Marquette, La Salle and the French voyageurs. Northern Indiana Hist. Soc., Publ. No. 1, 48 pp.

Barmore, W. J., and D. Stradley

1971 Predation by black bear on mature elk. Jour. Mammalogy, 52(1):199–202.

Blois, J. T.

1839 Gazetteer of the state of Michigan. Sydney L. Rood & Co., Detroit, Robinson, Pratt & Co., New York.

Blouch, R. I., and R. J. Moran

1965 1964 elk hunt. Michigan Dept. Conserv., Wildlife Dev., Res. and Dev. Rept. No. 30, 24 pp.

Boyce, M. S. and L. D. Hayden-Wing, eds.

1979 North American elk: Ecology, behavior and management. Univ. Wyoming, Laramie, v+294 pp.

Boyd, R. J.

1978 American elk. Pp. 10–29 *in* J. L. Schmidt and D. L. Gilbert, eds., Big game of North America. Ecology and management. Stackpole Books, Harrisburg, Pennsylvania. xv+494 pp.

Brown, J., and C. Cleland

1969 The late glacial and early postglacial faunal resources in midwestern biomes newly opened to human adaptation. Pp. 114–122 *in* R. E. Bergstrom, ed. The Quarternary of Illinois. Univ. Illinois, College of Agric., Sp. Publ. 14, 179 pp.

Butsch, E. A.

1970 The ethnozoology of Fort Michilimackinac. Michigan State Univ., unpubl. M.A. thesis, v+69 pp.

Burt, W. H.

1946 The mammals of Michigan. Univ. Michigan Press, Ann Arbor. xv+288 pp.

Caughley, G.

1971 An investigation of hybridization between free-ranging wapiti and red deer in New Zealand. New Zealand Jour. Sci., 14:993–1008.

Cleland, C. E.

1966 The prehistoric animal ecology and ethnozoology of the Upper Great Lakes Region. Univ. Michigan, Mus. Anthro., Anthro. Papers, No. 29, x+294 pp.

Conaway, C.

1952 The age at sexual maturity in male elk *(Cervus canadensis).* Jour. Wildlife Mgt., 16(3):313–315.

Eveland, J. F., J. L. George, N. B. Hunter, D. M. Forney, and R. L. Harrison

1979 A preliminary evaluation of the ecology of the elk in Pennsylvania. Pp. 145–151 *in* M. S. Boyce and L. D. Hayden-Wing, eds. North American elk: Ecology, behavior and management. Univ. Wyoming, Laramie, v+294 pp.

Franklin, W. L., and J. W. Lieb

1979 The social organization of a sedentary population of North American elk: A model for understanding other populations. Pp. 185–198 *in* M. S. Boyce and L. D. Hayden-Wing, eds. North American elk: Ecology, behavior and management. Univ. Wyoming, Laramie, v+294 pp.

Franklin, W. L., A. S. Mossman, and M. Dole
1975 Social organization and home range of Roosevelt elk. Jour. Mammalogy, 56(1):102–118.

Greer, K. R., and H. W. Yeager
1967 Sex and age indications from upper canine teeth of elk (wapiti). Jour. Wildlife Mgt., 31(3)408–417.

Heptner, V. G., A. A. Nasimovich, and A. G. Bannikov
1961 [Mammals of the Soviet Union, Vol. 1, Artiodactyls and Perissodactyls], in Russian. State Publ., Moscow, 776 pp.

Hoyt, J. M.
1889 History of the town of Commerce. Michigan Pioneer Coll., 14:421–430.

Hunter, N. B., J. L. George, and D. A. Devlin
1979 Herbivore-woody plant relationships on a Pennsylvania clearcut. Pp. 105–111 *in* M. S. Boyce and L. D. Hayden-Wing, eds. North American elk: Ecology, behavior and management. Univ. Wyoming, Laramie, v+294 pp.

Jenkins, D. H.
1977 Managing an elk herd in the face of changing land use and recreational values. Trans. 42nd North American Wildlife and Nat. Res. Conf., pp. 459–464.

Johnson, D. E.
1951 Biology of the elk calf, *Cervus canadensis nelsoni.* Jour. Wildlife Mgt., 15(4):396–410.

Johnson, D. R.
1969 Returns of the American Fur Company, 1835–1839. Jour. Mammalogy, 50(4):836–839.

Johnson, I. A.
1919 The Michigan fur trade. Michigan Historical Comm., Lansing. xii+201 pp.

Knight, J. E.
1975 Use of forest openings by elk in northern Michigan. Michigan State Univ., unpubl. M.S. thesis, v+38 pp.
1980 Effect of hydrocarbon development on elk movements and distribution in northern Michigan. Univ. Michigan, unpubl. Ph.D. dissertation, viii+79 pp.

Lovis, W. A., Jr.
1973 Late Woodland cultural dynamics in the northern Lower Peninsula of Michigan. Michigan State Univ., unpubl. Ph.D. dissertation, xi+346 pp.

Lyon, L. J., and C. E. Jensen
1980 Management implications of elk and deer use of clear-cuts in Montana. Jour. Wildlife Mgt., 44(2):352–362.

Martin, T. J.
1976 A faunal analysis of five Woodland Period archaeological sites in southwestern Michigan. Western Michigan Univ., unpubl. M.A. thesis, viii+200 pp.
1979 Unmodified faunal remains from 20 LP 98. Appendix A, pp. 65–66 *in* W. A. Lovis, ed. The archaeology and physical anthropology of 20 LP 98: A Woodland burial locale in Lapeer County, Michigan. Michigan Arch., 25(1–2), 69 pp.

McCullough, D. R.
1969 The tule elk: Its history, behavior, and ecology. Univ. California Publ. Zool., 88, 209 pp.

Miles, M.
1861 A catalogue of the mammals, birds, reptiles and mollusks of Michigan. Geological Survey of Michigan. First Biennial Rept., pp. 219–241.

Moran, R. J.
1973 The Rocky Mountain elk in Michigan. Michigan Dept. Nat. Res., Wildlife Dev., Res. and Dev. Rept. No. 267, x+93 pp.

Moran, R. J., and J. J. Ozoga
1965 Elk calf pursued by coyotes in Michigan. Jour. Mammalogy, 46(3):498.

Morrison, J. A., C. E. Trainer, and P. L. Wright
1959 Breeding season in elk as determined from known-age embryos. Jour. Wildlife Mgt., 23(1):27–34.

Murie, O. J.
1951 The elk of North America. Stackpole Company, Harrisburg, Pennsylvania. 376 pp.

Perry, O. H.
1899 Hunting expeditions of Oliver Hazard Perry of Cleveland verbatim from his diaries. Privately printed, Cleveland, Ohio viii+246 pp.

Petersen, E. T.
1979 Hunters' heritage. A history of hunting in Michigan. Michigan United Conserv. Clubs, Lansing, 55 pp.

Quinby, D. C., and J. E. Gaab
1957 Mandibular dentition as an age indicator in Rocky Mountain elk. Jour. Wildlife Mgt., 21(4):435–451.

Ruhl, H. D.
1940 Game introductions in Michigan. Trans. 5th North American Wildlife Conf., pp. 424–427.

Schorger, A. W.
1954 The elk in early Wisconsin. Trans. Wisconsin Acad. Sci., Arts and Ltrs., 43:5–23.

Seton, E. T.
1953 Lives of game animals. Vol. III, Pt. I. Hoofed animals. Charles T. Branford Company, Boston. xix+412 pp.

Sikarskie, J. G.
1978 Wildlife lungworm (metastrongyloid) parasites with special consideration to cervids. Michigan State Univ., 8 pp. (mimeo.).

Spiegel, L. E., C. H. Huntly, and G. R. Gerber
1963 A study of the effects of elk browsing on woody plant succession in northern Michigan. Jack-Pine Warbler, 41(2):68–72.

Stephenson, J. H.
1942 Michigan elk. Michigan Conserv., 21(11):8–9.

Struhsaker, T. T.
1967 Behavior of elk *(Cervis canadensis)* during the rut. Zeit. für Tierpsychol., 24(1):30–114.

Thompkins, J. L., J. N. Stuht, and J. E. Knight
1977 Meningeal worm in deer and wapiti of Canada Creek Ranch. Michigan Dept. Nat. Res., Wildlife Div., Rept. no. 2775 (mimeo.), 3 pp.

Thorne, E. T., J. K. Morton, and W. C. Ray
1979 Brucellosis, its effect and impact on elk in western Wyoming. Pp. 212–220 *in* M. S. Boyce and L. D. Hayden-Wing, eds. North American elk: Ecology,

behavior and management. Univ. Wyoming, Laramie, v+294 pp.

Waters, J. H., and C. W. Mack
1962 Note on the former range of *Cervus canadensis* in New England. Jour. Mammalogy, 43(2):266–267.

Wilson, R. L.
1967 The Pleistocene vertebrates of Michigan. Michigan Acad. Sci., Arts and Ltrs., Papers, 52:197–235, 5 figs.

Wood, N. A.
1914 An annotated check-list of Michigan mammals. Univ. Michigan, Mus. Zool., Occas. Papers No. 4, 13 pp.

Wood, N. A. and L. R. Dice
1923 Records of the distribution of Michigan mammals. Michigan Acad. Sci., Arts and Ltrs., 3:424–469.

Worley, D. E.
1979 Parasites and parasitic diseases of elk in the northern Rocky Mountain region. Pp. 206–211 *in* M. S. Boyce and L. D. Hayden-Wing, eds. North American elk: Ecology, behavior and management. Univ. Wyoming, Laramie, v+294 pp.

Youat, W. G., and L. D. Fay
1961 Survey of brucellosis in Michigan wildlife. Jour. American Vet. Med. Assoc., 139(6):677.

Zeveloff, S. I., and M. S. Boyce
1980 Parental investment and mating systems in mammals. Evolution, 34(5):973–982.

WHITE-TAILED DEER
Odocoileus virginianus (Zimmermann)

NAMES. The generic name *Odocoileus,* proposed by Rafinesque in 1832, is derived from two Greek words, *odon* meaning tooth and *koilos* meaning hollow. The specific name, *virginianus,* proposed by Zimmermann in 1777 and again in 1780, means "of Virginia," which the author designated as the locality of this deer in the original description. Although the generic name *Odocoileus* has long been in common usage, some authorities basing their actions on the Law of Priority have applied the older name *Dama,* currently used for the Old World fallow deer. Action by the International Commission on Zoological Nomenclature in Opinion 581 (Bull. Zool. Nomen., 17:267–275, September 6, 1960), using its plenary powers, made an exception to the Rules and validated the use of *Dama* for the fallow deer and *Odocoileus* for the white-tailed deer. This was done partially because of the furor among mammalogists and wildlife biologists over proposed changes in generic names in long usage for "popular" game animals. The common name deer is from an Anglo-Saxon word *deor* or *dior* which originally referred to any wild animal. The underside of the plumed tail provides the reason for the white-tailed part of the common name. In Michigan, this species may be called deer, white-tailed deer, or just whitetail. Males are called bucks; females, does; and young, fawns.

White-tailed deer living in Michigan belong to the subspecies *Odocoileus virginianus borealis* Miller. Recent morphological and protein studies show that there are definable differences between deer from the two peninsulas (Manlove, 1979; Rees, 1969). These differences perhaps reflect the role of the Straits of Mackinac as a partial water barrier isolating these two populations.

RECOGNITION. Body size small when compared with wapiti and moose (Fig. 106); head long; muzzle bare; ears large, prominent; preorbital (lachrymal) gland not conspicuous; antlers on adult males (Fig. 107) moderately-spreading with main beam less than 30 in (750 mm) long, extending slightly backward, outward and then directly forward, curving inward on an almost horizontal plane, bearing several usually unbranched sharp tines (including an inner upright snag near base) rising generally vertically, lacking a brow tine; females antlerless (exceptions noted further in text); neck moderately long, slender (swollen in males during rut); body slender, graceful in appearance; tail with flag-like white fringe; limbs long, slender; metatarsal gland, between ankle and hoof on the outer side of the hind legs, con-

Figure 106. A Michigan buck white-tailed deer (*Odocoileus virginianus*). Photograph by Robert Harrington and courtesy of the Michigan Department of Natural Resources.

spicuously marked by a whitish, elongated patch of hairs longer than surrounding ones and about one in (24 mm) long; tarsal gland on inside of midpart of heel joint, marked by a tuft of coarse hairs; interdigital glands present between hoofs on each foot; dew-claws (actually the remnants of the ancestral second and fifth toes) small and high on the sides of each foot; hoofs delicate, narrow, pointed; mammary glands consist of two pairs located in the inguinal area; adult males slightly larger than adult females.

Both sexes of the white-tailed deer are colored alike. The guard hair of the summer coat is short, sparse, wiry; underfur is lacking. This seasonal pelage is reddish tan on the back and whitish on the belly, throat, eye-ring, inside the ears, and on the underside of the conspicuous tail. The guard hair of the winter coat is long, thick, and somewhat crinkled; underfur is present but sparse. This pelage is grizzled gray on the upperparts. During both seasons, the head is conspicuously marked with a blackish nose, ears fringed with tawny, and a black spot on each side of the white chin.

The characteristic antlers in the male and conspicuous white-plumed tail distinguish the white-tailed deer from all other hoofed mammals, wild or domestic, in Michigan.

MEASUREMENTS AND WEIGHTS. Adult males and females measure, respectively: length from tip of nose to end of tail vertebrae, 75 to 85 in (1,875 to 2,125 mm) and 63 to 79 in (1,575 to 1,975 mm); length of tail vertebrae, 11 to 14 in (275 to 350 mm) and 10 to 13 in (250 to 325 mm); length of hind foot, 20 to 21 in (500 to 525 mm) and 19 to 20½ in (475 to 512 mm); height of ear from notch (same in both sexes), 5½ to 9 in (138 to 225 mm); height at shoulder (withers), 31 to 44 in (775 to 1,100 mm) and 27 to 33 in (675 to 825 mm); weight (depending often on range quality), 150 to 310 lbs (68.6 to 140.9 kg) and 90 to 210 lbs (40.9 to 95.5 kg). Although Michigan bucks are rarely weighed prior to dressing (losing about one-fifth of their weight from blood and visceral removal), animals weighing more than 300 lbs (136 kg) are occasionally reported. A buck taken in 1919 near Trout Creek [=River] in Ontonagon County weighed 354 lbs (161 kg) after being field-dressed. According to Jenkins and Bartlett (1959), this animal may well have weighed as much as 425 lbs (192 kg). Even so, the heaviest Michigan bucks generally occur today

Figure 107. Antler form in the white-tailed deer (*Odocoileus virginianus*).

in the so-called "farm belt" range in the southern part of the Lower Peninsula, although older bucks in remote parts of the Upper Peninsula attain record size as well.

DISTINCTIVE CRANIAL AND DENTAL CHARACTERISTICS. The skull of the white-tailed deer is heavy, elongate, and tapered from a broad cranium forward to a narrow rostrum (Figs. 103 and 108). The extended upper jaws are devoid of incisors and canines; the preorbital (lachrymal) vacuities in front of each orbit are large and conspicuous; each orbit is fully ringed by heavy bone; the width of each auditory bullae is equal to the length of the bony tube leading from it to the external auditory meatus (ear opening); the cranium is prominently elevated; and as seen on the ventral surface, the vomer divides the hind part of the posterior nares into two separate chambers. Each half of the elongated and slender lower jaw contains a full complement of teeth including three incisors and one canine, the latter almost identical to the adjacent incisors in shape (incisiform). The skulls of the larger adult male and smaller adult female average, respectively, 12⅞ in (330 mm) and and 11⅞ in (301 mm) in greatest length and 5¼ in (132 mm) and 4¼ in (108 mm) in greatest width across zygomatic arches (cheek bones).

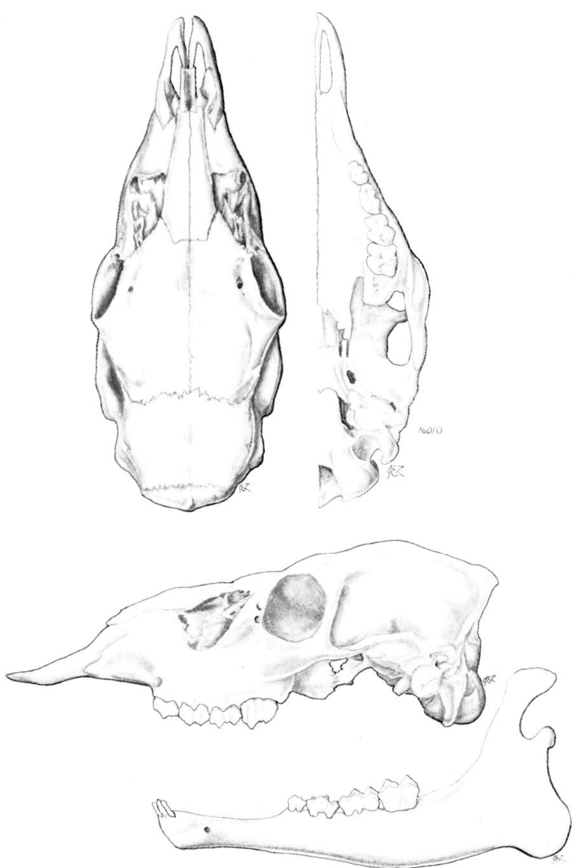

Figure 108. Cranial and dental characteristics of the white-tailed deer (*Odocoileus virginianus*).

The characteristic antlers of the male and the smaller size skulls of both sexes of the white-tailed deer are distinctive when compared to the unique antlers and larger skulls of adult wapiti, moose, and caribou. Skull size of antlerless individuals of the latter three species may be less useful as a means of identification than other characteristics (see Key to the Species, p. 564, also for distinguishing the white-tail deer skull from the skulls of domestic sheep and goat).

Occasionally, white-tailed deer will exhibit cranial or dental anomalies. There are records in Michigan animals of small, peg-like maxillary (upper jaw) canines, extra molar teeth in the lower jaw, mature males failing to grow antlers, females with antlers (often in permanent velvet), and abnormally short lower jaws (Ryel, 1963); Verme and Ozoga, 1966). The usual dental formula for the white-tailed deer is:

$$\text{I (incisors) } \frac{0\text{-}0}{3\text{-}3},\ \text{C (canines) } \frac{0\text{-}0}{1\text{-}1},\ \text{P (premolars) } \frac{3\text{-}3}{3\text{-}3},\ \text{M (molars) } \frac{3\text{-}3}{3\text{-}3} = 32$$

DISTRIBUTION. The white-tailed deer enjoys an unusually large geographic range in the Americas, from the southern edge of the arctic prairie in central Canada, southward into boreal, temperate, and tropical environments as far south as the north bank of the Amazon River in Brazil, northern Bolivia and southern Peru. Despite the fact that this deer provides a necessary protein base for many rural-dwelling people and is a sporting animal for others, it has survived in most parts of this vast intercontinental area, covering about 70° of latitude. In fact, given even minimum protection from overhunting, the species can thrive in very close association with human habitation, as wildlife managers attest.

The white-tailed deer is now found throughout the state (see Map 66), but its numbers in northern

Map 66. Geographic distribution of the white-tailed deer (*Odocoileus virginianus*).

lower Michigan and in the Upper Peninsula are subject to greater fluctuations than those further south. Michigan is geographically almost at the northern edge of the distribution for the species, and the northern boreal habitats can often be only marginally inhabitable to this more southern adapted species.

Even so, the white-tailed deer has been a long-time resident of Michigan. Its bones have been found in association with extinct animals which lived shortly after the glacial melt-back (Holman, 1976), in such counties as Berrien, Genesee, Leelanau, Washtenaw, and Wayne (Wilson, 1967). White-tailed deer remains from the Woodland Culture period have been found occasionally in archaeological sites in the Upper Peninsula counties of Baraga, Delta, and Iron, and abundantly in the southern half of the Lower Peninsula, in Allegan, Berrien, Lapeer, Mackinac, Muskegon, Ottawa, and Saginaw counties (Cleland, 1966; Higgins, 1980; Martin, 1976, 1979, 1980; Rhead and Martin, 1975). After the beginning of the historic period, travelers and settlers in the seventeenth century documented evidence of white-tailed deer at Fort Michilimackinac (Butsch, 1970), in the Detroit area at the time of Cadillac (Johnson, 1919), and in Berrien County (Baker, 1899). These and other data show that although this deer occurred throughout much of the state prior to intensive settlement, its preferred range was apparently in the southern half of the Lower Peninsula.

Prior to 1850, most habitats of the Lower Peninsula north of the Bay City-Muskegon line as well as the Upper Peninsula included mature stands of conifers and hardwoods, which supported few deer. However, as soon as major logging operations got underway in the northern woods (in Michigan and adjacent Ontario), the result was a mosaic of clearings, shrub growth, and young trees which offered diverse and abundant near-ground-level food sources, dramatically increasing the habitat carrying capacity for the species. The white-tailed deer responded, extending its range northward, ultimately almost to Hudson Bay (de Vos, 1964), and increasing greatly in numbers in northern Michigan in the post-Civil War period (Bartlett, 1950).

This northern deer herd, on both sides of the Straits of Mackinac, was thought to have reached its nineteenth century peak between 1875 and 1886, when more than 100,000 deer were killed and shipped out by market hunters (Lorenson, 1973). Although the northern population was depleted drastically through overkill by the 1890s, the habitat, manipulated first by loggers and later by settlers, had become of sufficient quality to support our modern deer herd in all northern counties. The white-tailed deer in the more naturally favorable habitats of the southern part of the Lower Peninsula also suffered drastic setbacks through settlement, especially between 1850 and 1915 (Bartlett, 1938), but all Lower Peninsula counties today provide living space for the white-tailed deer (Map 66). In addition, populations (mostly introductions) on islands in the Great Lakes are thriving.

HABITAT PREFERENCES. The Michigan white-tailed deer lives in two major habitat types: (1) the boreal conifer-hardwood forest environment characteristic of the Upper Peninsula and much of the northern half of the Lower Peninsula, and (2) the temperate hardwood forest environment chiefly in the southern half of the Lower Peninsula. In both of these habitat types, this cervid thrives best in the early stages of forest successions—a mosaic of cultivated and fallow fields, shrub-fringed forest edges, and second-growth, sapling-sized timber. Such cover and food diversity has generally been the result of the human intrusion, where torch, axe, plow, and ultimately the effective enforcement of laws controlling deer harvest have singly or in combination helped produce Michigan's sizable white-tailed deer population (Jenkins and Bartlett, 1959).

In prehistoric times Michigan presumably was covered to a great extent by a canopy of mature tree growth which shaded the leaf-littered forest floor to such an extent that browsing deer could rarely reach the undergrowth. Natural fires, tree-falls, occasional glades, openings at edges of swamps and along lake fronts, and riparian shrub and grass cover along streams did, however, break the ecologic monotony in many places. Indians and later the first European settlers found the white-tailed deer plentiful in the temperate hardwood forests of the southern half of the Lower Peninsula. Here the forest canopy was more open, the soils were more productive, the winter temperatures were relatively milder, and the snowfall was generally lighter—all in marked contrast to the inhospitable conditions of the northern parts of the state (McNeil, 1962), where early settlers found few white-tailed deer (Bartlett, 1938).

As mentioned, intensive logging after 1850 opened these northern forests, diversified the habitats by successional set-back, possibly discouraged large predators, reduced the then resident moose herd, and opened the country to the adaptable white-tailed deer. Although the better habitat for the white-tailed deer is still in the southern half of the Lower Peninsula, where the so-called "corn-fed" bucks thrive in the mixed farm country, the northern forests also remain favorable; the climate in these latitudes continues to be restrictive, however, even with abundant winter food (Potvin *et al.,* 1981). In addition, forest maturation in many areas is gradually reducing the habitat quality. On the bright side, the recent acceleration of an industrially-compatible, short-rotation forest management program (for obtaining wood chips for pulp, energy, chemicals, and pressed-wood products), is bound to insure better white-tailed deer habitat (D. H. Jenkins, pers. comm.).

DENSITY AND MOVEMENTS. The Indian hunter of 1,000 years ago would not only be amazed today at the basic environmental changes wrought by modern human encroachment in Michigan, but also at the abundance of white-tailed deer, especially in the northern sector of the state. Today's populations of white-tailed deer, however, have not been achieved without near-tragic experiences in the past. For example, even though the deer increased markedly in the northern forests during the post-Civil War logging days, they were cut back to scarce status by the 1890s, partly due to overzealous market hunters. The southern Michigan deer herd was also quickly exploited by farmers and other settlers struggling to become established in the raw frontier, and the southern herd virtually disappeared in the late nineteenth century. Covert (1881) and Wood (1922) noted the last record of the species in Washtenaw County in 1879. Citizen willingness to support the enactment and enforcement of laws designed to regulate the harvest of the white-tailed deer allowed for this game animal's remarkable comeback by the 1940s (Jenkins and Bartlett, 1959; McNeil, 1962).

The deer's ability to increase rapidly when its numbers are well below the natural carrying capacity of the habitat is perhaps best illustrated by the population buildup in the fenced Edwin S. George Reserve of The University of Michigan in Livingston County. In 1928, two males and four females obtained from Grand Island in Lake Superior were placed in this 1,146 acres (464 ha) enclosure. A deer drive made five years later (1933) showed the population had increased from the original 6 to 160 animals (Hickie, 1937; McCullough, 1979; O'Roke and Hamerstrom, 1948). Similar efforts have led to the replenishment of the entire Michigan deer herd. In many places, populations have increased beyond the actual carrying capacity of the land, causing local famines. Generally, these have occurred in the boreal forest habitat as a result of a combination of depressed food supplies and severe winters.

To determine harvesting schedules and appraise annual trends, Michigan Department of Natural Resources wildlife biologists estimate white-tailed deer populations (Hawn and Ryel, 1969) by

(1) pellet (fecal deposits) counts in spring in randomly selected areas (Eberhardt and Van Etten, 1956), (2) dead deer surveys also in spring (Whitlock and Eberhardt, 1956), (3) mail surveys of licensed hunters to determine those successful in taking deer (Eberhardt and Murray, 1960), and (4) reconstructing herd characteristics from sex and age data obtained by inspecting a sampling of the annual deer kill in northern Michigan (Hayne and Eberhardt, 1956). Relative population figures and trends aid in monitoring the harvest, by as many as 700,000 licensed hunters, of upwards to 150,000 legally taken white-tailed deer (about 170,000 in 1978) out of a herd of nearly 1,000,000 animals.

Populations of the secretive white-tailed deer, which spend considerable time skulking in wooded terrain, have never been counted with absolute certainty over a large area. Although a herd in a sizable fenced area (Woolf and Harder, 1979) can exist with supplemental foods at a population density in access of one animal per 5 acres (2.1 ha), these numbers are unrealistically high for extensive natural ranges. The Ninth Biennial Report of the Michigan Department of Conservation (1937–1938) suggested an overall population (probably for the autumn season) in the Upper Peninsula as averaging one animal per 35.5 acres (15.4 ha). In selected habitats in Wisconsin, densities appeared comparable; a winter herd using swamp area was estimated at one animal per 40.5 acres (16.4 ha) by Larson *et al.* (1978). In the Huron Mountains in Marquette County, Westover (1971) estimated a winter carrying capacity of one white-tailed deer per 46 acres (18 ha). In the best deer habitat in the southern half of the Lower Peninsula, autumn populations may be as high as one animal per 6.4 to 8.0 acres (2.6 to 3.2 ha), according to Westell (1955). A deer herd count of this magnitude was made in Kalamazoo County's Fort Custer. However, this population was reported to be in excess of the habitat's carrying capacity because a browse line in hardwood stands had become evident and preferred plant species had been severely eaten or eliminated.

Seasonal home ranges used by white-tailed deer vary considerably as a result of such factors as habitat quality, weather conditions, and sex, although early social experience and learning also appear important in habitat selection (Nelson, 1979). These living spaces in Michigan may be no larger than 40 acres (16 ha) in excellent habitat (perhaps in parts of southern Michigan) and as extensive as 300 acres (122 ha) in marginal habitat (perhaps in parts of northern Michigan). Deer become so attached to their living places that they attempt to return if displaced (Hawkins and Montgomery, 1969). To escape danger, these established animals usually run in circular fashion rather than leave the familiar trails of the home territory and flee cross-country. Heavy snow cover in the north may be a major factor causing white-tailed deer to use smaller home ranges in winter than in summer. In southern Michigan where deep snows are the exception rather than the rule, deer may use similar-sized areas throughout the year. Studies of the movements of white-tailed deer equipped with radio transmitter collars are reported for Wisconsin and Minnesota in habitats similar to those in Michigan (Kohn and Mooty, 1971; Larson *et al.,* 1978; Rongstad and Tester, 1969). These studies showed that individual deer may use winter areas ranging from 190 acres (77 ha) to as large as 1,180 acres (486 ha); in summer, from 167 acres (68 ha) to 472 acres (191 ha).

Seasonal movements between winter and summer ranges have also been determined by monitoring the activities of tagged animals (Carlsen and Farmes, 1957; Hoskinson and Mech, 1976; Larson *et al.,* 1978; Nelson, 1979; Rongstad and Tester, 1969; Verme, 1973). In southern Michigan, summer and winter ranges are more apt to overlap (Masek, 1979); whereas, in the Upper Peninsula, Verme (1973) found that deer moved an average of 8.6 mi (13.7 km) from their winter yarding areas to the places where they were shot by hunters in the following November, with seasonal shifts up to 32 mi (53 km) reported. Although movements may depend in part on topography and seasonal availability of foods (Hamerstrom and Blake, 1939), the white-tailed deer, in contrast to other cervids, is generally prone to remain in rather confined areas. Nevertheless, when movements do occur, they generally take place from late March to mid-April and in early December.

BEHAVIOR. The white-tailed deer is usually more solitary than the social wapiti, although this behavior is influenced by seasonal activities. Even so, there is some indication that the conspicuous white-plumed tail erected in times of alarm is a means of maintaining social cohesiveness (Hirth and McCullough, 1977). Moreover, there is a growing body of evidence to show that deer do not necessarily belong to large and homogeneous assemblages but exist as smaller subgroups with little interchange (Dapson *et al.*, 1979; Townsend and Bailey, 1981). It is also recognized that males and females may not compete directly for food and living space in the warm seasons because of their differing needs (antler growth in males and reproduction in females). These matters are discussed by McCullough (1979) in his classic study of white-tailed deer on the Edwin S. George Reserve in Livingston County.

The basic grouping of the white-tailed deer includes the adult doe, her yearling female offspring, and her fawns (Hirth, 1977a; LaGory *et al.*, 1981). Often this matriarchal combination includes females from a fourth generation (Hawkins and Klimstra, 1970). In late summer, these doe-led family groups become conspicuous. At the same time, yearling and older males tend to be solitary or form groups of up to four or more individuals, which, except for the dominant (alpha) male, may not disband even during the rutting season. Autumn and winter are the only seasons when bucks and does are actually in close association. The bucks may utilize "dominance areas" as part of their autumn courtship behavior; however, it is uncertain whether this action can in any way be considered a space-defending action (Smith, 1976). Agggressive behavior by one buck toward another during the rutting period has been described (Cowan and Geist, 1961; Thomas *et al.*, 1965, etc.). Bucks with locked antlers are occasionally found, with the unfortunate individuals dead or dying from exhaustion. There are mounts of bucks with locked antlers from Delta County (collection of Wildlife Div., DNR) and from Hillsdale County (collection of Jens Touborg of Tecumseh).

Once the mating season has passed, the groups of females and fawns remain together through the winter. The first-year male fawns leave the care of their female parent in winter or as late as May or June (McCullough, 1979) to look for home areas and ultimately join small groups of other bucks. In winter in southern Michigan, deer often gather in large numbers at sites of food concentrations, such as fields of standing corn (Masek, 1979). Jenkins (1963) reported groups may contain up to 150 individuals. These winter concentrations demonstrate no real herd organization, except possibly for related female groupings. Again, mature does appear to play the leadership roles, with males apparently having little influence on group action. Such bands may begin to frequent feeding areas in late November and early December with dispersal taking place as late as March. In late spring, each doe bears her young in isolation and nurtures the fawns for a month or two before rejoining her matriarchial family group (Ozoga *et al.*, 1982).

In much of the northern half of the Lower Peninsula and in the Upper Peninsula, the white-tailed deer annually faces severe winter conditions—an unusual challenge for a species which enjoys such widespread distribution in temperate and tropical climates. Factors which certainly influence deer survival in northern boreal Michigan include: (1) sizable snowfall, deep drifts, and snow crusts which will not bear their weight, (2) low ambient temperature, (3) poor quality winter browse, (4) uncertain summer range quality which influences the physical condition of deer entering the winter period, and (5) susceptibility to predation by feral or free-ranging dogs and possibly coyotes. In response to these factors, northern Michigan whitetails seek winter shelter in dense coniferous (often mixed with deciduous) cover in low-elevation swamps. Such an area, called a "yard," is frequented winter after winter by the same individuals or their offspring (Boer, 1978; Krefting and Phillips, 1970; Ozoga and Gysel, 1972; Westover, 1971; Verme, 1965b; Verme and Ozoga, 1971). Yards, many of which are monitored by wildlife biologists, occur as far south as the Deadstream Swamp in Missaukee County and provide the deer herd with wintering space often much less than one-tenth the size of its summer range (Bartlett, 1949). White-tailed deer huddling in dense winter cover may concentrate at the rate of two to three deer per acre (4.9 to 7.4 per ha) in white cedar yards and perhaps one per acre (2.5 per ha) in hardwoods (Davenport *et al.*, 1953). Once yarded, deer form pathways in the deep snow and browse on buds and twigs. Movements elsewhere are made difficult by both low temperatures and snowdrifts; hence, it is possible for the animals to be yarded from the time snowfalls begin in late November to mid-December, until March (or even April) when spring thaws begin.

Because the food supply in confined yards tends to be depleted, the deer's body fat stored in autumn burns up and malnutrition during the winter cold becomes fatal. According to wildlife biologist Louis J. Verme, if a white-tailed deer remains well-nourished, the cold does not necessarily impair the animal. However, there is also evidence that well-fed white-tailed deer may die merely from exposure to severe winter weather (Potvin *et al.*, 1981). Thus, the major limiting factor in the northermost parts of the deer's range is chiefly climate. To conserve energy, yarded deer may generally become inactive, walk slowly, eat less in late winter than in early winter, and exhibit low ecological metabolism (Moen, 1976, 1978; Silver *et al.*, 1969, 1971; Ozoga and Verme, 1970). Apparently the deer's energy is better conserved by remaining more or less inactive and fasting in the shelter of a yard than foraging in exposed areas (Kearney and Gilbert, 1976). Unlike winter concentrations of whitetails in southern Michigan, bucks, does, and fawns in northern groupings have a developed dominance hierarchy (Kabat *et al.*, 1953). Yarded animals have, as Ozoga (1972) described, seven aggressive actions: ear drop, hard look, sidle, rush, snort, strike, and flail. Adult bucks at least two and one-half years old generally dominate all other deer; mature does dominate fawns.

The white-tailed deer is active during the day and night. Some of the best pioneer photographic portraits of Michigan deer were made at night (Gregory, 1930). Nevertheless, this cervid is most apt to be seen in early morning or late afternoon. During the day and in the middle of night, the animals seek bedding areas where they are inactive, chew their cuds, and nurse their young. Deer have good vision and hearing but may depend mostly on a keen sense of smell to warn them of potential danger. Field observers are well aware that if they keep downwind and move cautiously deer can be approached.

The white-tailed deer is the epitome of gracefulness, walking with a springy step, or breaking into a trot or canter. When alarmed, the deer makes long, low strides or bounds of as much as 30 ft (9 m) with white-plumed tail held conspicuously high. Top speed may be as much as 40 mi (64 km) per hour. The white-tailed deer negotiates brushlands, hillsides, tall-grass swamps, and waterways without difficulty. Deer are more apt to swim rivers in summer than in winter (Fuller and Robinson, 1982). The animals prefer to go through fences but will jump them if necessary.

White-tailed deer are quiet animals—almost mute according to many observers. Nevertheless, fawns utter a gentle bleat in response to the female parent's attention and murmuring sounds especially at nursing times. When in danger or restrained, a deer bawls loudly. The buck makes a short *ba* sound while trailing a doe during the rut. But the call most familiar to field persons is a low "blow" or "snort," often given just before or during the animal's first leap after being startled. This sound is also often used by the doe as a warning.

The white-tailed deer's ability to move silently through leaf-littered forests amazes field biologists and hunters. For such a big animal, the deer has a remarkable ability for skulking and remaining inconspicuous. To illustrate, an Upper Peninsula fenced enclosure of 640 acres (260 ha) with an average of 26 deer, including six adult bucks, was hunted with as typical as possible field conditions. According to Van Etten *et al.* (1965), hunters saw only one of these 26 white-tailed deer for each 1.33 hours in the field, while it took an average of 18 hours to see one buck. Obviously, there are more deer in the woods than meet the eye.

ASSOCIATES. Michigan white-tailed deer have perhaps closest relationships with allied cervids—wapiti in the "Pigeon River Country" of the northern part of the Lower Peninsula and an occasional moose in the Upper Peninsula. Association of wapiti and deer, according to Moran (1973), is close because they occupy similar habitats; namely, recently cutover forest, swamp conifers, aspen-hardwoods, and upland conifer-hardwoods. In winter, however, deer tend to restrict their activities to yards while wapiti are much more mobile. According to observations made in Montana (Lyon and Jensen, 1980), wapiti appear to have higher security requirements than deer, preferring smaller, clear-cut openings while deer seem less selective, perhaps because whitetails are better able to elude predators even in large clearings.

Studies, made mostly in Canada, show that moose and white-tailed deer have close association. In Ontario, Kearney and Gilbert (1976) found major habitat overlap in alder growths in spring and summer and pointed out that moose distribution was primarily influenced by food sources and white-tailed deer distribution by shelter factors. Telfer (1972) found a high degree of association between moose and deer from mid-March to early April. The two species were also noted in association in winter spruce habitat in Montana (Singer, 1979) and in newly-burned forest areas in late autumn and early spring in Minnesota (Irwin, 1975). Moose, of course, are able to negotiate deep snow better than the smaller white-tailed deer (Kelsall, 1969). One major concern in the moose-deer association is the lethal effects of the meningeal worm (*Parelaphostrongylus tenuis*) on the moose but not on the deer (Telfer, 1967).

In Michigan, perhaps the most studied association has been between white-tailed deer and snowshoe hare (Bookhout, 1965a, 1965b). These two species appear to feed in common on white cedar and other plants. This overlapping often reduces habitat quality and develops competition for food (Krefting, 1975; Telfer, 1972; 1974). Competition for similar foods could affect over-wintering survival, especially of deer in yarding situations.

Studies made in deer yarding areas in the Upper Peninsula show that the principal small mammalian inhabitants include the masked shrew, the short-tailed shrew, the deer mouse, and the southern red-backed vole (Ozoga and Verme, 1968). Observations have also been made on the inquisitive nature of interactions between deer and sandhill cranes when they forage in close proximity in open areas (Wozencraft, 1979a). Michigan naturalists also report that birds such as the red-breasted nuthatch and cowbird have perched on the backs of white-tailed deer (McNeill, 1967; Tate, 1969).

REPRODUCTIVE ACTIVITIES. Reproductive activities begin in autumn when daylight hours are lessening. At this time, bucks have hardened, velvet-stripped antlers and those 28 months or older develop sexual readiness at least by mid-September in Michigan (Jenkins and Bartlett, 1959). Does (including the spring-born fawns no more than six months of age) on the best range in southern Michigan, generally become receptive to mating in early and middle November. Younger females may breed as much as four weeks later than older animals (Cheatum and Morton, 1946). Does become most attractive to the persistent, belligerent, swollen-necked bucks for at least three days prior to receptiveness and perhaps two days following (Pruitt, 1954; Severinghaus, 1955). During this time, bucks may continually follow does, disregarding nourishment or potential danger. At the time of maximum receptiveness, the doe also becomes highly active (Ozoga and Verme, 1975). Should breeding be unsuccessful in the 24 to 36 hours during which the doe is in estrus, she will become receptive again about 28 days later. Does may conceive while still nursing young (Scanlon *et al.*, 1976), or at least while there is still some milk in the udder. Conceptions as late as March have resulted in spotted fawns in Michigan in November and December (Gordinier, 1948; Verme, 1961). The chances of animals of this age surviving in the Michigan winter are problematical.

Once the rut is virtually completed by late December, females generally separate from the bucks except for feeding assemblages. Pregnant does follow their usual activities until late May or early June when, about 200 days after conception, they separate from their female associates and the fawns are born (Haugen and Davenport, 1950; Verme, 1965a). First-year does (if well nourished) may produce a single fawn; healthy older individuals generally give birth to twins, occasionally triplets. Quadruplet births are rare (Trodd, 1962), but such pregnancies have been reported both in northern and southern Michigan (Dale Fay and Louis J. Verme, pers. comm.). The production of healthy fawns is closely related to the physical condition of the doe throughout the winter months. In the Upper Peninsula, winter food scarcity for does can be a major factor in fetus mortality (Verme, 1965a, 1969, 1977, 1979). Fecundity in well-fed Michigan deer, according to Verme (1965a), amounts to about 1.7 fawns per doe. A Michigan doe's reproductive life can be long; Verme (1962), for example, reported one Mackinac County semi-tame individual producing 20 fawns in ten years.

Fetal development and fawn growth have been well documented (Armstrong, 1950; Severinghaus and Cheatum, 1956; Townsend and Bailey, 1975). At birth, the fawn is marked on the back and flanks with white spots against a reddish background; has fine, silky hair; good vision and upright ear pinnae; stands on long and spindly legs within an hour or so after birth; receives a long, continuous licking by the female parent (Fig. 109); is able to bleat; seems to have little odor; and weighs as little as 90 oz (2,572 g) for females and as much as 120 oz (2,429 g) for males. Within 12 hours, the fawn

Figure 109. The doe and fawn of the white-tailed deer (*Odocoileus virginianus*).

will nurse once or twice daily in a standing position and remain cached in a huddled position between nursings. At three to four days of age, the fawn is able to travel considerable distances. At 21 days of age, the fawn nibbles on grass tips and follows the female parent to join the rest of her family group, with play an obvious part of the behavioral pattern (Miller-Schwarge *et al.*, 1982). Nursing may continue as long as 120 days, although long before, the fawn will begin to eat solid vegetable food. Between 90 and 120 days of age, usually by October, the fawns molt their spotted coats and grow their grayish winter pelage. At about 180 days, small antler buttons can be detected beneath the skin on each side of the male's forehead, in preparation for second summer growth.

Fawn mortality, resulting chiefly from dogs, human-related interference, natural predation, parasites, accidents, and weather, has often been considered a major factor limiting the growth of white-tailed deer populations (Cook *et al.*, 1971; O'Pezio, 1978). Considerable study has been made concerning Michigan fawn diets (Ullrey *et al.*, 1973; Verme and Ozoga, 1980a, 1980b). The Michigan Department of Natural Resources makes a special annual effort to monitor embryo counts and the age and physical condition of does found dead between March 15 and June 1 from various causes, chiefly highway accidents. Summaries of these findings (*e.g.*, Friedrich, 1979) are most useful in determining annual trends in deer herd population characteristics, and, in part, determine recommendations for the annual harvest.

The white-tailed deer has been known to survive for more than 18 years in captivity (Jones, 1979). Does have reportedly thrived in the Upper Peninsula for almost 15 years (Ozoga, 1969). Whitetails have been successfully aged by establishing when various deciduous teeth appear and/or are replaced and when permanent teeth erupt, mature, and become worn (Jenkins and Bartlett, 1959); by counting the annuli in the cementum of the first incisor (Gilbert, 1966; Langenau, 1972; Rice, 1980); and by the condition of the insoluble lens proteins (Ludwig and Dapson, 1977).

The antlers, the crowning glory of the buck white-tailed deer, are considered secondary sexual characteristics and their annual growth and loss are correlated with courtship and breeding. Growing antlers makes great physiological demands on the individual; the cycle is controlled by a hypothalamic-pituitary-gonad endocrine system. Alterations of this growth system, as well as loss of sexual prowess due to old age, may suppress antler development or cause antlers to form in irregular fashion. Castrated bucks may continuously hold antlers in velvet. At least two or three does with antlers in velvet are reported annually in Michigan (Ryel, 1963), as are does with polished antlers, and antlered does which are pregnant (Haugen and Mustard, 1960). An antlered doe, mistaken for a buck, was shot during the hunting season in Eaton County by Clyde Gordinier in November 1977.

The cervid antler begins as a skin-covered button on the forehead (on each frontal bone) of the male fawn during its first winter. Beginning in late April or early May in Michigan, the antlers grow on basal pedicels with a bony dermal core and an epidermal "velvet" skin. The velvety outer covering is soft, contains nerve tissue, and is well supplied with blood vessels to transport the necessary ingredients for bone growth (Fig. 110). When antler growth ends, the velvet disintegrates and is rubbed off (Hirth, 1977b). Michigan bucks may engage in this rubbing routine from as early as late August to as late as early November. Usually the animal pushes its antlers against small trees or even

Figure 110. Antler growth in the male white-tailed deer (*Odocoileus virginianus*).

sprouts of trees, causing bark removal and scarring. Conifers and aspen are often favored for rubbing (de Vos, 1967). During the breeding season, bucks thrash saplings and brush and expose patches of ground by scraping with their hoofs. Urine is also deposited as a marking device at this time (Hirth, 1977a). These rituals apparently serve as a means of communication between bucks and does during courtship activities (Kile and Marchinton, 1977).

Antlers may be shed as early as late December (sometimes even earlier) and as late as early April (Zagata and Moen, 1974). Older bucks may finish shedding earlier than younger ones. During its second summer, the young buck may develop "spike" antlers perhaps as long as 6 in (150 mm). However, with a highly nutritious food supply, this young animal may produce "forked" antlers or even antlers with several points, especially in southern Michigan. The size, symmetry, and number of normal points on each antler are, consequently, dependent on physical well being and proper hormonal balance rather than on age alone (Jenkins and Bartlett, 1959).

FOOD HABITS. Since the Michigan white-tailed deer is very close to the northern-most border of its extensive range, it is expected that food supply would be a factor limiting populations in this marginal area. As wildlife biologists of the Upper Great Lakes Region know (Jenkins and Barlett, 1959), winter forage conditions often make or break the bumper crops of deer, especially in the northern boreal forest habitat. Mild winters appear to help save the northern deer herd from massive die-offs in restrictive yards (Davenport *et al.,* 1953). However, biologists also look for answers to the winter mortality rate in terms of the abundance and nutritive value of summer and autumn foods (Mautz, 1978): If the deer herd can enter the snowy winter carrying heavy fat deposits (using kidney fat as an indicator, Monson *et al.,* 1974) and healthy bodies, it appears likely, although not fully documented, that survival rate can be high. In spring, the physical condition of Michigan deer is monitored by checking fat in femur marrow and in mandibular cavity tissue of animals killed accidentally, primarily by automobiles (Purol *et al.,* 1977; Youatt *et al.,* 1975).

White-tailed deer browse chiefly in the colder months when plant growth is arrested or snow covers the ground, eating buds and twigs of a variety of deciduous shrubs and saplings and needles and leaves of some evergreens. In the warmer months, fruits, grasses, forbs, and the green foliage of shrubs and saplings add variety to the deer's diet. Michigan deer need about 4 lbs (1.8 kg) of nutritious browse and other foodstuffs daily in order to thrive—or at least not lose more than 25% of their body weight during the 100-day winter period (Duvendeck, 1962; Ullrey *et al.,* 1972). There is also some evidence that bucks and does, because of their differing nutritional needs, use slightly separated feeding niches (McCullough, 1979). In early winter, deer may eat every four to six hours. In late winter, at least in snowy northern Michigan, feeding appears concentrated during daylight hours (Ozoga and Verme, 1970). In more southern ranges, deer are generally crepuscular feeders; that is, they feed in early morning and again in late afternoon and early evening.

In the boreal forest habitat of the northern half of the Lower Peninsula and the Upper Peninsula, the choice of food for the white-tailed deer is less diverse and leans more to evergreens (especially in winter), preferrably recent cutovers and sprouting stands of early-stage second growth (Westover, 1971). Wildlife biologists throughout the Upper Great Lakes Region view white cedar (*Thuja occidentalis*) as a major winter food source. Also high on the preferred browse list in this northern area are trembling and big-tooth aspens, ground hemlock, sugar, striped and red maples, willows, yellow birch, white pine, jack pine, basswood, white ash, red-osier dogwood, viburnums, and sumacs (Davenport, 1937; Howard, 1937; King, 1975; Manville, 1947; Stormer and Bauer, 1980; Westell, 1954). In the growing season of summer, deer eat mushrooms, strawberries, brambles, asters and other forbs, grasses, lilies, bush honeysuckle, and a variety of other greenery, with acorns a favorite when available (Duvendeck, 1962). There is also evidence that they have a craving at this time for sodium and are attracted to salt licks and mineral springs (Fraser and Hristienko, 1981). Officials of the Michigan Department of Natural Resources point out that a deer in northern Michigan presumably shifts from a specialist to a generalist in feeding strategies, turning to less nutritious foods including spruce, beech, red pine, balsam, tag alder, and leatherleaf when their usual, more digestible, forage is in short supply. This last-resort diet may fill digestive tracts but offers little sustenance to prevent malnutrition and starvation.

In southern Michigan, on the other hand, the mosaic mix of woodlots, cultivated and fallow fields, forested growth along streams, and succulent greens at marsh edges provide a nutritious variety of summer and winter foods (Masek, 1979; McNeil, 1962). Acorns are important foods, with groves of mature oaks (55 to 80 years old) producing, according to Gysel and Stearns (1968), from 200 to 500 lbs of acorns per acre (225 to 561 kg per ha). Corn and other grains, soybeans, celery and other vegetables, fruit trees, and other agricultural crops also attract deer in southern Michigan but their use of them competes adversely with farming interests. Damage to crops is a direct result of an increasing number of deer in farm and orchard country—an economic matter which requires much more attention. The white-tailed deer also eats such items as lady-bird beetles (Shaw, 1963) and even song birds caught in mist nets set by birdbanders (Allan, 1978). Concretion deposits (bezoars and calculi) have been found in digestive tracts of white-tailed deer (Burt, 1942; Mosby and Cushwa, 1969).

Ecologists and nutritionists with the Michigan Department of Natural Resources and Michigan State University have pioneered in determining dietary requirements, including mineral needs, of the white-tailed deer (Ullrey *et al.*, 1967, 1970, 1971, etc.). In 1960, a rumen fistula, much like those used previously for domesticated bovids, was successfully placed in a tame doe maintained at the Rose Lake Wildlife Research Center (Short, 1962). This device enabled investigators to monitor digestive processes by periodic removal and analysis of rumen material (Short, 1963). The analysis of browse intake and subsequent fecal deposition also provided data on the digestability of browse of white cedar, aspen, jack pine and artificial supplements (Ullrey *et al.*, 1964, 1967, 1975). These intensive investigations point the way to a much better understanding of nutritional strategies of the Michigan deer herd.

ENEMIES. In presettlement Michigan, the white-tailed deer was presumed to be abundant, especially in the southern part of the state, despite the presence of two major predators, the gray wolf and the mountain lion (Young and Goldman, 1944 and 1946). These two large meat-eaters have been removed from the scene, but whitetails must still contend with carnivores such as the coyote, feral dogs, black bear, bobcat, and lynx (Banfield, 1974; Bennett *et al.*, 1943; Marston, 1942; Seton, 1953; Taylor, 1956). The fact that any of these predators can down white-tailed deer (especially fawns) has always been a major concern of the citizenry, especially the license-buying sporting sector. In consequence, Michigan game managers have tried to obtain scientifically indisputable evidence that elaborate predator control campaigns and decades of expensive bounty payments (Arnold, 1971) have been ineffectual in increasing Michigan deer populations. The predators are here to stay and are themselves assets to the outdoor community. In fact, an ever-growing body of citizens would consider the Michigan outdoors even more intriguing should the gray wolf make a comeback, at least in parts of the Upper Peninsula. Hearing a "scream" or sighting an occasional track of a mountain lion would also add a dimension to the Michigan environment. For such an experience today, one must travel to a distant place such as Idaho.

Michigan's present carnivores might best be able to bring down adult white-tailed deer in late winter or early spring when the animals are winter-weakened, especially in northern yarding situations. However, attacks on the summer fawn crop are apt to be much more successful (McCullough,

1979). Even so, most records of deer remains found in the digestive tracts or scats (fecal droppings) of these carnivores are traceable to winter-killed deer carcasses. Spring surveys of dead deer, which have accumulated in the north woods following various winter fatalities, indicate there may be sufficient food so that predators do not need to expend the amount of energy required to stalk live deer. The chances of a coyote, for example, killing a healthy adult deer are poor at best, even though there may be some "pursuit invitation" signal given by the alarmed prey (Hirth and McCullough, 1977). With the coming of spring, rodents, hares, insects, and others become more easily available to meat-eaters, probably relieving some of the danger to the deer herd. Deer (fawns and adults) carcasses also attract red and gray foxes, black bears, ravens, and even gray jays and black-capped chickadees (Field, 1970; Murie, 1936; Ozoga *et al.*, 1982; Ozoga and Harger, 1966; Petraborg and Gunvalson, 1962; Rollings, 1945; Schofield, 1960).

There is considerable evidence indicating that the most serious, four-footed, Michigan white-tailed deer predator is the domestic dog. Although deer-killing dogs are usually classed as stray or feral, numerous well-kept house and yard pets are guilty of such actions. Further, dog owners are more apt to blame coyotes or other wild animals for predation on deer (as well as on sheep, calves or poultry). Predation by dogs, acting alone or in groups, has been studied in many states (Bower, 1953; Causey and Cude, 1980; Gavitt *et al.*, 1974; Kreeger, 1977). In Michigan, reports of dog-killed deer are most prevalent in late winter and early spring when the victims are often less healthy. Officials of the Michigan Department of Natural Resources have reported in the press, for example, 19 deer killed by dogs in the Gladwin area in April 1978, and 14 in the Cadillac area in March 1979. Newborn fawns are also frequent victims of domestic dogs. McNeil (1962) summarized reports from conservation offices for 1962: In the southern half of the Lower Peninsula, dogs accounted for the deaths of at least 322 newborn fawns. The annual kill of white-tailed deer by dogs is estimated by the Michigan Department of Natural Resources at between 2,000 and 5,000 animals.

Although the Michigan deer herd is subject to winter starvation, predation, poaching, and hunter-inflicted wounds, another serious concern to wildlife managers is the high number of deer-car accidents, especially in early autumn in the middle counties of the Lower Peninsula (Allen and McCullough, 1976; Haugen, 1944; Kasul, 1976; McNeil, 1962; Reed and Woodland, 1981; Reilly and Green; 1974; Sicuranza, 1979). Statistics on these collisions are impressive. In Kent County alone in 1979, there were 673 car-deer accidents, almost two per day. Statewide, in 1976, there were 14,349 deer-car accidents; in 1977, 16,315; in 1978, 17,155; in 1979, 16,148; and in 1980, 18,932. A study made in 1966 and 1967 by Allen and McCullough (1976), showed that most of the 2,566 accidents which they investigated occurred at dawn, dusk, or after dark with peaks at sunrise and two hours after dark. Accidents were higher on weekends than during midweek. The fewest collisions occurred in May; the most in November. More females were killed than males, although bucks were more frequently victims during the autumn rutting season. Accidents occurred most commonly when vehicles traveled at speeds of 50 to 59 mi per hour (80 to 95 km/h). In 92% of the deer-car encounters, the deer was killed; in less than 4%, there were human injuries, most resulting from secondary collisions.

PARASITES AND DISEASES. White-tailed deer have internal and external parasites and diseases which often are also characteristic of other closely-related cervids (Davidson *et al.*, 1981). In addition, domestic livestock parasites and diseases can be transmitted to these wild animals (Taylor, 1956). Internal parasites found in deer in Michigan and adjacent parts of the Great Lakes Region include a liver fluke (*Fascioloides magna*), lungworms (*Leptostrongylus alpenae, Protostrongylus coburni,* and *Dictocaulus viviparus*), a stomach worm (*Ostertagia* sp.), a tapeworm larva (*Cysticercus* sp.), a footworm (*Onchocerca cervipedis*), a protozoan (*Trypanosoma* sp.), a whipworm (*Trichurus* sp.), and the important meningeal worm (*Parelaphostrongylus tenuis*). The latter has little pathologic effect on the white-tailed deer but has serious impact on wapiti, moose and caribou (O'Roke and Cheatum, 1950; Severinghaus and Cheatum, 1956; Stuht, 1975; Tompkins *et al.*, 1977; Whitlock, 1939; Youatt and Harte, 1974). Studies of many of these parasites, as well as of cervid ailments, have been conducted by the Wildlife Pathology Laboratory of the Michigan Department of Natural Resources by such workers as Lawrence D. Fay, P. D. Friedrich, John N. Stuht, S. C. Whitlock, and W. G. Youatt.

Other parasities of white-tailed deer include troublesome infestations by the larvae of the nose botfly (*Cephenemyia phobifer*), a tick (*Dermacentor albipictus*), and a biting louse (*Trichodectes tibialis*), Youatt and Harte (1974), Whitlock (1939). Diseases infecting Michigan deer include a serious outbreak of EHD (epizootic hemorrhagic disease) in 1955, deer fibroma, leptospirosis, mycotic hepatitus, encephalitis, and cirrhosis of the liver (Fay *et al.*, 1956; Fay and Youatt, 1959; Stuht, pers. comm.; Stuht and Friedrich, 1978; Wilhelm and Trainer, 1966). Deer also are known to have serious visual problems (Howard *et al.*, 1976).

MOLT AND COLOR ABERRATIONS. The short, thin pelage of summer appears in May and early June in the adult white-tailed deer, and is molted in late August and September to be replaced by the longer, heavier winter coat. This winter coat also replaces, in first-year animals, the spotted coats worn since spring birth. White deer are reported taken almost every hunting season. Sometimes these individuals are true albinos with pink hoofs, nose and eye pigments; others are only partly white with mottled coloration (Ryel, 1963). In the early part of the century, a herd of white deer thrived on Grand Island in Lake Superior just off Munising in Alger County (Shiras, 1918). Although Ryel (1963) recorded no dark (melanistic) deer in Michigan, there have been reports of black individuals in neighboring Wisconsin (Wozencraft, 1979b).

ECONOMIC IMPORTANCE. Of the many natural attributes of Michigan's environment, the white-tailed deer herd is probably among the most appreciated. Through the efforts of Michigan Department of Natural Resources personnel, this animal has added immeasurably to the recreational dimensions of the outdoor public—whether tourists interested in viewing nature (Evans, 1979; Hanson, 1977; Langenau, 1979; Queal, 1968), or dedicated hunters (Mershon, 1923), interested in either or both the quality of the hunt or the quality of the hunting experience (Langenau *et al.*, 1981). The white-tailed deer has also been the center of controversy, with scientific management programs frequently criticized by a well-meaning but often uninformed public. One major disagreement has been the recommendations in recent years that hunting both antlered and antlerless deer be legalized. Bucks only were made legal game during the early days when the deer herd was small and males were deemed more expendable than fawn-producing does. It was assumed that enough bucks would survive to service the protected does during the autumn rut. This was then regarded as the best deer herd-management practice. In reality, however, the law was accepted by the hunters more as a safety measure because, if a hunter had to look for "horns," humans would not be shot.

In the following decades, the sporting public became conditioned to the belief that bucks, especially large ones, were the only valuable trophies. However, once the deer herd increased by means of conservation programs and public adherence to the concept of regulated harvest, the deer population, especially in the north country, reached near carrying capacity proportions. As a result, favored food plants were over-utilized. This pointed to the need for a greater annual kill since the other option was an increasingly larger winter die-off because of cold weather and/or starvation (50,000 perished in 1951; Westell, 1955) or widespread malnutrition with resulting prenatal and postnatal offspring loss. The shooting of antlerless deer (does and fawns) was then legalized to better regulate the size of the herd, especially in problem areas. The hunting public reacted adversely to this change (Moncrief, 1970; Watson, 1972), and the controversy was heated during the 1950s, 1960s, and early 1970s, even entering the political area. Ultimately the scientific arguments of wildlife biologists began to prevail and by the end of the 1970s most of the citizenry began to accept the need for "any deer" hunting regulations. Nevertheless, anti-doe hunting sentiment continues to surface as a result of poor hunting seasons (even locally) and severe winter weather.

Although white-tailed deer in the farming country of the southern half of the Lower Peninsula increased from an estimated 40,000 animals in 1963 to as many as 250,000 in 1977, and in the northern half of the Lower Peninsula in the same years from 380,000 to 500,000, the population in Michigan's Upper Peninsula declined from about 380,000 in 1963 to about 250,000 in 1977. The gradual reduction in habitat quality, as the forest growth advances toward maturity, and severe winters are the causes, rather than overkill of does and fawns, as is widely suspected.

Retarding this successional growth and thus making these northern areas more compatible for a high, sustained deer herd is being accomplished by various habitat manipulations—selective cutting of mature timber, logging of small trees as pulp wood, block cutovers in selected areas of state land, aerial spraying of approved phytocides (Krefting and Hansen, 1969), and controlled burning. Hopefully, a compatible balance will be reached between those citizens who wish to protect the forests so they will reach maturity and those interested in seeing some of this growth arrested so that a brush and sapling stage can prevail, enabling the wildlife which thrives best in the early forest succession stages to abound in at least part of the vast boreal habitat.

The pre-Columbian people probably found white-tailed deer most common in what is now southern Michigan and hunted these animals by such methods as stalking and driving (Weselkov, 1978). The European introduction of firearms facilitated killing the animals and development of the fur trade gave deer commercial value. In the summer of 1767, for example, British traders at Fort Michilimackinac shipped 1,747 deer hides to England (Petersen, 1979). In 1822, at the Astor Trading Post on Mackinac Island shaved deer hides brought 36–37¢; red deer hides, 25–26¢; and gray deer hides, 18–19¢ (Anon., 1962). Between 1835 and 1839, the American Fur Company (Johnson, 1969) processed 51,430 deer skins through their Detroit Department (including southern Michigan) and 22,449 from their Northern Department (including northern Michigan). By the mid-nineteenth century, white-tailed deer were becoming scarce in southern Michigan, although sportsmen like Oliver Hazard Perry of Cleveland, Ohio, found excellent hunting between 1838 and 1855 in Hillsdale, Lenawee, Sanilac, St. Clair, and Tuscola counties (Burroughs, 1958; Perry, 1899). In the post-Civil War period, the professional hunter found ready buyers for meat and hides in logging camps, meat markets, and tanneries. In 1880, for example, Dick Parrish of Roscommon County killed 81 deer in the autumn season alone and, in the same year, well over a million pounds of venison were shipped to city markets on the Mackinaw division of the Jackson, Lansing, and Saginaw Railroad (Petersen, 1953). Even after the turn of the century, the Chicago market was still obtaining part of its venison from Michigan (Oldys, 1910).

In 1859, the first legislative enactment relating to the protection of the white-tailed deer limited the open hunting season to the last five months of each year (Bennett *et al.,* 1966). By 1873, legal hunting was restricted to October, November, and December. In 1895, the season was reduced to November 1 to November 15 with a bag limit of five. In the same year, the deer hunting license (costing 50¢) was first required, and a total of 14,477 licenses were sold. In 1915, the bag limit was reduced to one deer. In 1937, the first bow and arrow season, from November 1 to November 15, was authorized in Iosco and Newaygo counties (Haugen, 1946). Although these constitute only a few of the legal regulations devised to provide sustained, successful hunting of Michigan deer, strict adherence to the laws was slow to be generally accepted by the citizenry. Commercial hunting for out-of-state markets was not really discouraged until 1900 with passage of the Lacey Act, making it a Federal offense to transport illegally-taken wildlife across state lines. The conservation movement now rapidly gained momentum under the leadership, in Michigan, of Walter Barrows, William B. Mershon, and George Shiras, III.

Conservation of renewable wildlife resources became more acceptable through environmental education programs in the schools, greater public awareness of the necessity to preserve breeding stock, employment of competent conservation officers, and demonstration that wildlife populations can be increased by habitat management at such wildlife research centers as Rose Lake in Clinton and Shiawassee Counties, Houghton Lake in Roscommon and Missaukee counties, and Cusino in Alger and Schoolcraft counties.

With passage of the 1938 Pittman-Robertson Act, designed to provide federal aid to wildlife, Michigan began establishing game refuges and public hunting areas (Burroughs, 1946). These, together with properties of the U.S. Forest Service, became bases for the development of long-term programs to restore wildlife and maintain its habitats. Even so, poaching and vandalism continue to be problems (Kesel, 1974). Despite efforts to discourage market hunting, an organized group of poachers was apprehended as recently as January 1979 (Rau, 1979); a total of 54 arrests were made after state and federal conservation officials used undercover agents to gather the evidence needed.

Today, as many as 700,000 licensed gun and

bow hunters annually bag at least 150,000 legally taken white-tailed deer (170,000 in 1978) out of a herd numbering more than 1,000,000 animals. Both in-state and out-of-state hunters support conservation programs with their license-fees; contribute to local economies through purchases of food, fuel, lodging, arms, ammunition, field clothing, boots, and camping gear; and provide business for meat processors, taxidermists and tanners. High school students often are granted time off from classes to hunt deer. Michigan industries also cooperate in allowing their deer-hunting employees time off. In 1980, it was reported that one automobile manufacturer specified in a union contract that the first day of the Michigan deer season be included as one of the employees' paid holidays, as unscheduled absenteeism often increases 300% at this time.

Despite the positive aspects of having a viable deer herd in Michigan, this species can be troublesome and expensive for agriculturalists (Evans, 1979; McNeel and Kennedy, 1959; McNeil, 1962; Queal, 1968). Potato, other vegetable, fruit, grain, and even tree farms (seedling white pines and ornamental yews among others) attract white-tailed deer (Krefting and Arend, 1960; McKee, 1963). While permits may be obtained to shoot deer doing major damage to crops, the problem remains a serious one, especially in the southern half of the Lower Peninsula where the white-tailed deer has increased spectacularly in the past 20 years.

The exertion and excitement of the annual hunting season are surely stressful to the hearts and circulatory systems of many out-of-shape hunters. Nevertheless, fatalities among deer hunters may not be significantly different from a sampling of the non-hunting public for a similar two week period. The gun-related accidents which occur each autumn hunting season are tragic. In the 1975 season, for example, there were 49 such accidents; nine were fatal, although Dr. David H. Jenkins, then Chief of the Wildlife Division of the Michigan Department of Natural Resources, noted in the press that only one out of every 29,000 hunters are involved in these accidents. Further, the accident rate per 100,000 deer hunters has been steadily decreasing for more than 30 years, with notable decreases especially well-documented in 1977, 1978, and 1979. Programs in gun safety and hunting methods have helped considerably in this respect.

COUNTY RECORDS FROM MICHIGAN. Records of specimens of white-tailed deer in collections of museums, from literature, and reported by personnel of the Michigan Department of Natural Resources show that this species occurs in all counties in Michigan (see Map 66). Deer also live on Beaver, Big Summer, Garden, Little Summer, North Fox, North Manitou, Poverty, South Fox, and St. Martins islands in Lake Michigan; on Grand Island in Lake Superior; and on Bois Blanc and Drummond islands in Lake Huron (Cleland, 1966; Hatt *et al.*, 1948; Long, 1978; Manville, 1950; Ozoga and Phillips, 1964; Phillips *et al.*, 1965; Scharf and Jorae, 1980).

LITERATURE CITED

Allan, T. A.
1978 Further evidence of white-tailed deer eating birds in mist nets. Bird Banding, 49(2):184.

Allen, R. E., and D. R. McCullough
1976 Deer-car accidents in southern Michigan. Jour. Wildlife Mgt., 40(2):317–325.

Anon.
1962 Fur and hide values. Michigan Conserv., 31(6):36–37.

Armstrong, R. A.
1950 Fetal development of the northern white-tailed deer. American Midl. Nat., 43(3):650–666.

Arnold, D. A.
1971 The Michigan bounty system. Michigan Dept. Nat. Res., Game Div., Info. Cir., No. 162, 3 pp.

Baker, G. A.
1899 The St. Joseph-Kankakee Portage. Its location and use by Marquette, LaSalle and the French voyageurs. Northern Indiana Hist. Soc., Publ. No. 1, 48 pp.

Banfield, A. W. F.
1974 The mammals of Canada. Univ. Toronto Press, Toronto. xxiv+438 pp.

Bartlett, I. H.
1938 Whitetails. Presenting Michigan's deer problem. Michigan Dept. Conserv., Game Div., 64 pp.
1949 Deer and snow. Michigan Conserv., 18(1):3–5, 23–25.
1950 Michigan deer. Michigan Dept. Conserv., Game Div., 50 pp.

Bennett, C. L., Jr., L. A. Ryel, and L. J. Hawn
1966 A history of Michigan deer hunting. Michigan Dept. Conserv., Res. and Dev. Rept. No. 85, 66 pp.

Bennet, L. J., P. F. English, and R. L. Watts
1943 The food habits of the black bear in Pennsylvania. Jour. Mammalogy, 24(1):25–31.

Boer, A.
1978 Management of deer wintering areas in New Brunswick. Wildlife Soc. Bull., 6(4):200–205.

Bookhout, T. A.
1965a Feeding coactions between snowshoe hares and white-

tailed deer in northern Michigan. Trans. Thirtieth North American Wildlife and Nat. Res. Conf., pp. 321–335.

1965b The snowshoe hare in upper Michigan, its biology and feeding coaction with white-tailed deer. Michigan Dept. Conserv., Res. and Dev. Rept. No. 38, x+191 pp.

Bowers, G. E.
1969 Venison. Upper Peninsula style. Michigan State Univ., Coop. Ext. Serv., Nat. Res. Ser., Ext. Bull. E-657, 20 pp.

Bowers, R. R.
1953 The free-running dog menace. Virginia Wildlife, 14(10):5–7.

Burroughs, R. D.
1946 Game refuges and public hunting grounds in Michigan. Jour. Wildlife Mgt., 10(4):285–296.
1958 Perry's deer hunting in Michigan, 1838–1855. Michigan Hist., 42(1):35–58.

Burt, W. H.
1942 A calculus from the stomach of a deer. Jour. Mammalogy, 23(3):335–336.

Butsch, E. A.
1970 The ethnozoology of Fort Michilimackinac. Michigan State Univ., unpubl. M.S. thesis, v+69 pp.

Carlsen, J. C., and R. E. Farmes
1957 Movements of white-tailed deer tagged in Minnesota. Jour. Wildlife Mgt., 21(4):397–401.

Causey, M. K., and C. A. Cude
1980 Feral dog and white-tailed deer interactions in Alabama. Jour. Wildlife Mgt., 44(2):481–484.

Cheatum, E. L., and G. H. Morton
1946 Breeding season of white-tailed deer in New York. Jour. Wildlife Mgt., 10(3):249–263.

Cleland, C. E.
1966 The prehistoric animal ecology and ethnozoology of the Upper Great Lakes Region. Univ. Michigan, Mus. Anthro., Anthro. Papers, No. 29, x+294 pp.

Cook, R. S., D. O. Trainer and W. C. Glazener
1971 Mortality of young white-tailed deer fawns in South Texas. Jour. Wildlife Mgt., 35(1):47–56.

Covert, A. B.
1881 Natural history. Chapter V, pp. 173–194 *in* History of Washtenaw County, Michigan. Chas. C. Chapman & Co., Chicago. 1,452 pp.

Cowan, I. McT., and V. Geist
1961 Aggressive behavior in deer of the genus *Odocoileus*. Jour. Mammalogy, 42(4):522–526.

Dapson, R. W., P. R. Ramsey, M. H. Smith, and D. F. Urbston
1979 Demographic differences in contiguous populations of white-tailed deer. Jour. Wildlife Mgt., 43(4):889–898.

Davenport, L. A.
1937 Find deer have marked food preference. Michigan Conserv., 7(4):4–6.

Davenport, L. A., D. F. Switzenberg, R. C. Van Etten, and W. D. Burnett
1953 A study of deeryard carrying capacity by controlled browsing. Trans. Eighteenth North American Wildlife Conf., pp. 581–596.

Davidson, W. R., F. A. Hayes, V. F. Nettles, and F. E. Kellogg, eds.
1981 Diseases and parasites of white-tailed deer. Tall Timbers Research Sta., Misc. Publ. No. 7, 450 pp.

de Vos, A.
1964 Range changes of mammals in the Great Lakes Region. American Midl. Nat., 71(1):210–231.
1967 Rubbing of conifers by white-tailed deer in successive years. Jour. Mammalogy, 48(1):146–147.

Duvendeck, J. P.
1962 The value of acorns in the diet of Michigan deer. Jour. Wildlife Mgt., 26(4):371–379.

Eberhardt, L., and R. M. Murray
1960 Estimating the kill of game animals by licensed hunters. Proc. Social Statistics Sect., American Statistics Assoc., pp. 182–188.

Eberhardt, L., and R. C. Van Etten
1956 Evaluation of the pellet group count as a deer census method. Jour. Wildlife Mgt., 20(1):70–74.

Evans, R. A.
1979 Changes in landowner attitudes toward deer and deer hunters in southern Michigan, 1960–1978. University of Michigan, unpubl. M.S. thesis, 40 pp.

Fay, L. D., A. P. Boyce, and W. G. Youatt
1956 An epizootic in deer in Michigan. Trans. Twenty-First North American Wildlife Conf., pp. 173–184.

Fay, L. D., and W. G. Youatt
1959 Results of tests for brucellosis and leptospirosis in deer 1957–1958. Michigan Dept. Conserv., Game Div., Rept. No. 2241 (mimeo.), 3 pp.

Field, R. J.
1970 Scavengers feeding on a Michigan deer carcass. Jack-Pine Warbler, 49(2):73.

Fraser, D., and H. Hristienko
1981 Activity of moose and white-tailed deer at mineral springs. Canadian Jour. Zool., 59(10):1991–2000.

Friedrich, P. D.
1979 Doe productivity and physical condition: 1979 spring survey results. Michigan Dept. Nat. Res., Wildlife Div., Rept. No. 2843 (mimeo.), 2 pp.

Gavitt, J. D., R. L. Downing, and B. S. McGinnes
1974 Effect of dogs on deer production in Virginia. Proc. Twenty-Eighth Ann. Conf. Southeastern Assoc. Game and Fish Comm., pp. 532–539.

Gilbert, F. F.
1966 Aging white-tailed deer by annuli in the cementum of the first incisor. Jour. Wildlife Mgt., 30(1):200–202.

Gordinier, E. J.
1948 September birth of a Michigan white-tailed deer. Jour. Mammalogy, 29(2):184–185.

Gregory, T.
1930 Deer at night in the north woods. Charles C. Thomas, Springfield, Illinois. xiv+211 pp.

Gysel, L. W., and F. Stearns
1968 Deer browse production of oak stands in central lower Michigan. North Central Forest Exp. Sta., Res. Note NC-48, 4 pp.

Hamerstrom, F. N., Jr., and J. Blake
1939 Winter movements and winter foods of white-tailed deer in central Wisconsin. Jour. Mammalogy, 20(2): 206–215.

Hansen, C. S.
1977 Social costs of Michigan's deer habitat improvement program. Univ. Michigan, unpubl. Ph.D. dissertation, 195 pp.

Hatt, R. T., J. Van Tyne, L. C. Stuart, C. H. Pope, and A. B. Grobman
1948 Island life: A study of the land vertebrates of the islands of Lake Michigan. Cranbrook Inst. Sci., Bull. No. 27, xi+179 pp.

Haugen, A. O.
1944 Highway mortality of wildlife in southern Michigan. Jour. Mammalogy, 25(2):177–184.
1946 Deer hunting—Indian style. Michigan Conserv., 15(5):8–11.

Haugen, A. O., and L. A. Davenport
1950 Breeding records of white-tailed deer in the Upper Peninsula of Michigan. Jour. Wildlife Mgt., 14(3): 290–295.

Haugen, A. O., and E. W. Mustard, Jr.
1960 Velvet-antlered pregnant white-tailed doe. Jour. Mammalogy, 41(4):521–523.

Hawkins, R. E., and W. D. Klimstra
1970 A preliminary study of the social organizations of white-tailed deer. Jour. Wildlife Mgt., 34(2):407–419.

Hawkins, R. E., and G. G. Montgomery
1969 Movement of translocated deer as determined by telemetry. Jour. Wildlife Mgt., 35:216–220.

Hawn, L. J., and L. A. Ryel
1969 Michigan deer harvest estimates: Sample surveys versus a complete count. Jour. Wildlife Mgt., 33(4): 871–880.

Hayne, D. W., and L. Eberhardt
1956 Estimating deer kill from counts of deer on automobiles. Unpubl. paper presented at Eighteenth Midwest Wildlife Conf. (mimeo.), 5 pp.

Hickey, P.
1937 Four deer produce 160 in six seasons. Michigan Conserv., 7(1):6–7, 11.

Higgins, M. J.
1980 An analysis of the faunal remains from the Schwerdt Site, a late prehistoric encampment in Allegan County, Michigan. Western Michigan Univ., unpubl. M.A. thesis, v+77 pp.

Hirth, D. H.
1977a Social behavior of white-tailed deer in relation to habitat. Wildlife Monogr., No. 53, 55 pp.
1977b Observations of loss of antler velvet in white-tailed deer. Southwestern Nat., 22(2):269–280.

Hirth, D. H., and D. R. McCullough
1977 Evolution of alarm signals in ungulates with special reference to white-tailed deer. American Nat., 111: 31–42.

Holman, J. A.
1976 Michigan's fossil vertebrates. Michigan State Univ., Publ. Mus., Educ. Bull. No. 2, 54 pp.

Hoskinson, R. L., and L. D. Mech
1976 White-tailed deer migration and its role in wolf predation. Jour. Wildlife Mgt., 40(3):429–441.

Howard, D. R., J. D. Krehbiel, L. D. Fay, J. N. Stuht, and D. L. Whiteneck
1976 Visual defects in white-tailed deer from Michigan: Six case reports. Jour. Wildlife Disease, 12:143–147.

Howard, W. J.
1937 Notes on winter foods of Michigan deer. Jour. Mammalogy, 18(1):77–80.

Irwin, L. L.
1975 Deer-moose relationships on a burn in northeastern Minnesota. Jour. Wildlife Mgt., 39(4):653–662.

Jenkins, D. H.
1963 Deer in 1963. Michigan Dept. Conserv., Game Div., Info. Circ. 133, 21 pp.

Jenkins, D. H., and I. H. Bartlett
1959 Michigan whitetails. Michigan Dept. Conserv., Game Div., 80 pp.

Johnson, D. R.
1969 Returns of the American Fur Company, 1835–1839. Jour. Mammalogy, 50(4):836–839.

Johnson, I. A.
1919 The Michigan fur trade. Michigan Historical Comm., Lansing. xii+201 pp.

Jones, M. L.
1979 Longevity of mammals in captivity. Internatl. Zoo. News, 26(3):16–26.

Kabat, C., N. E. Collias, and R. C. Guettinger
1953 Some winter habits of white-tailed deer and the development of census methods in the Flag Yard of northern Wisconsin. Wisconsin Conserv. Dept., Game Mgt. Div., Tech. Wildlife Bull. No. 7, 32 pp.

Kasul, R. L.
1976 Mortality and movement of mammals and birds on a Michigan interstate highway. Michigan State Univ., unpubl. M.S. thesis, x+39 pp.

Kelsall, J. P.
1969 Structural adaptations of moose and deer for snow. Jour. Mammalogy, 50(2):302–310.

Kearney, S. R., and F. F. Gilbert
1976 Habitat use by white-tailed deer and moose on sympatric range. Jour. Wildlife Mgt., 40(4):645–657.

Kesel, J. A.
1974 Some of the characteristics and attitudes of Michigan deer hunting violators. Michigan State Univ., unpubl. M.S. thesis, vi+42 pp.

Kile, T. L., and R. L. Marchinton
1977 White-tailed deer rubs and scrapes: Spatial, temporal and physical characteristics and social role. American Midl. Nat., 97(2):257–266.

King, D. R.
1975 Estimation of yew and cedar availability and utilization. Jour. Wildlife Mgt., 39(1):101–107.

Kohn, B. E., and J. J. Mooty
1971 Summer habitat of white-tailed deer in north-central Minnesota. Jour. Wildlife Mgt., 35(3):474–487.

Kreeger, T. J.
1977 Impact of dog predation on Minnesota white-tailed deer. Jour. Minnesota Acad. Sci., 43(4):8–13.

Krefting, L. W.
1975 The effect of white-tailed deer and snowshoe hare browsing on trees and shrubs in northern Minnesota. Univ. Minnesota, Agric. Exp. Sta., Forestry Ser., Tech. Bull. 302-1975, 43 pp.

Krefting, L. W., and J. L. Arend
1960 Effect of deer browsing on a young jack pine plantation in northern lower Michigan. Lakes States Forest Exp. Sta., Tech. Notes No. 586, 2 pp.

Krefting, L. W., and H. L. Hansen
1969 Increasing browse for deer by aerial applications of 2,4-D. Jour. Wildlife Mgt., 33(4):784–790.

Krefting, L. W., and R. L. Phillips
1970 Improving deer habitat in upper Michigan by cutting mixed-conifer swamps. Jour. Forestry, 68(11):701–704.

LaGory, K. E., M. K. LaGory, and D. H. Taylor
1981 Evening activities and nearest-neighbor distances in free-ranging white-tailed deer. Jour. Mammalogy, 62(4):752–757.

Langenau, E. E., Jr.
1972 Applications and limitations of aging white-tailed deer by annuli in the cementum of the first lower incisor. Michigan Dept. Nat. Res., Res. and Div. Rept. No. 272 (mimeo.), 13 pp.
1979 Nonconsumptive uses of the Michigan deer herd. Jour. Wildlife Mgt., 43(3):620–625.

Langanau, E. E., Jr., R. J. Moran, J. R. Terry, and D. C. Cue
1981 Relationship between deer kill and ratings of the hunt. Jour. Wildlife Mgt., 45(4):959–963.

Larson, T. J., O. J. Rongstad, and F. W. Terbilcox
1978 Movement and habitat use of white-tailed deer in southcentral Wisconsin. Jour. Wildlife Mgt., 42(1): 113–117.

Long, C. A.
1978 Mammals of the islands of Green Bay, Lake Michigan. Jack-Pine Warbler, 56(2):59–82.

Lorenson, J.
1973 Deer in the north. Michigan Nat. Res., 42(6):22–25.

Ludwig, J. R., and W. Dapson
1977 Use of insoluble lens proteins to estimate age in white-tailed deer. Jour. Wildlife Mgt., 41(2):327–329.

Lyon, L. J., and C. E. Jensen
1980 Management implications of elk and deer use of clear-cuts in Montana. Jour. Wildlife Mgt., 44(2):352–362.

Manville, R. H.
1947 The Huron Mountain deer herd. Jour. Wildlife Mgt., 11(3):253–266.
1948 The vertebrate fauna of the Huron Mountains, Michigan. American Midl. Nat., 39(3):615–640.
1950 The mammals of Drummond Island, Michigan. Jour. Mammalogy, 31(3):358–359.

Manlove, M. N.
1979 Genetic similarity among contiguous and isolated populations of white-tailed deer in Michigan. Michigan State Univ., unpubl. M.S. thesis, 49 pp.

Marston, M. A.
1942 Winter relations of bobcats to white-tailed deer in Maine. Jour. Wildlife Mgt., 6(4):328–337.

Martin, T. J.
1976 A fauna analysis of five Woodland period archaeological sites in southwestern Michigan. Western Michigan Univ., unpubl. M.A. thesis, viii+200 pp.
1979 Unmodified faunal remains from 20 LP 98. Appendix A, pp. 65–66 *in* W. A. Lovis, ed. The archaeology and physical anthropology of 20 LP 98: A Woodland burial locale in Lapeer County, Michigan. Michigan Arch., 25(1–2), 69 pp.
1980 Animal remains from the Winter Site, a Middle Woodland occupation in Delta County, Michigan. Wisconsin Arch., 61(1):91–99.

Masek, J. W.
1979 Deer winter concentration areas in southern Michigan. Michigan State Univ., unpubl. M.S. thesis, vi+48 pp.

Mautz, W. W.
1978 Sledding on a bushy hillside: The fat cycle in deer. Wildlife Soc. Bull., 6(2):88–90.

McCullough, D. R.
1979 The George Reserve deer herd. Population ecology of a K-selected species. Univ. Michigan Press, Ann Arbor. xii+271 pp.

McKee, R.
1963 Deer damage, down on the farm. Michigan Conserv., 32(6):2–7.

McNeel, W., Jr., and J. Kennedy
1959 Prevention of browsing by deer in a pine plantation. Jour. Wildlife Mgt., 23(4):450–451.

McNeil, R. J.
1962 Population dynamics and economic impact of deer in southern Michigan. Michigan Dept. Conserv., Game Div., Rept. No. 2395, x+143 pp.
1967 Cowbirds and white-tailed deer. Jack-Pine Warbler, 45(1):37.

Mershon, W. B.
1923 Recollections of my fifty years hunting and fishing. Stratford Co., Boston. iv+259 pp.

Moen, A. N.
1976 Energy conservation in white-tailed deer in the winter. Ecology, 57(1):192–198.
1978 Seasonal changes in heart rates, activity, metabolism, and forage intake of white-tailed deer. Jour. Wildlife Mgt., 42(4):715–738.

Moncrief, L. W.
1970 An analysis of hunter attitudes toward the state of Michigan's antlerless deer hunting policy. Michigan Dept. Nat. Res., Res. and Dev. Rept. No. 209 (mimeo.), 7 pp.

Monson, R. A., W. B. Stone, B. L. Weber, and F. J. Spadaro
1974 Comparison of Riney and total kidney fat techniques for evaluating the physical condition of white-tailed deer. New York Fish and Game Jour., 21(1):67–72.

Moran, R. J.
1973 The Rocky Mountain elk in Michigan. Michigan Dept. Nat. Res., Wildlife Div., x+93 pp.

Mosby, H. S., and C. T. Cushwa
1969 Deer "madstones" or bezoars. Jour. Wildlife Mgt., 33(2):434–437.

Muller-Schwarze, D., B. Stagge, and C. Muller-Schwarze
1982 Play behavior: Persistence, decrease, and energetic compensation during food shortage in deer fawns. Science, 215(4528):85–87.

Murie, A.
1936 Following fox trails. Univ. Michigan, Mus. Zool., Misc. Publ. No. 32, 45 pp.

Nelson, M. E.
1979 Home range location of white-tailed deer. United States Dept. Agric., Forest Serv., Res. Pap. NC-173, 10 pp.

Oldys, H.
1911 The game market of to-day. Yearbook of Agri. for 1910, Y. B. Sept. 533, pp. 243–254.

O'Pezio, J. P.
1978 Mortality among white-tailed deer fawns on the Seneca Army Depot. New York Fish and Game Jour., 23(1):1–15.

O'Roke, E. C., and E. L. Cheatum
1950 Experimental transmission of the deer lungworm *Leptostrongylus alpenae*. Cornell Vet., 40(3):315–323.

O'Roke, E. C., and F. N. Hamerstrom, Jr.
1948 Productivity and yield of the George Reserve deer herd. Jour. Wildlife Mgt., 12(1):78–86.

Ozoga, J. J.
1969 Some longevity records for female white-tailed deer in northern Michigan. Jour. Wildlife Mgt., 33(4):1027–1028.
1972 Aggressive behavior of white-tailed deer at winter cuttings. Jour. Wildlife Mgt., 36(3):861–868.

Ozoga, J. J., C. S. Bienz, and L. J. Verme
1982 Red fox feeding habits in relation to fawn mortality. Jour. Wildlife Mgt., 46(1):242–243.

Ozoga, J. J., and L. W. Gysel
1972 Response of white-tailed deer in winter weather. Jour. Wildlife Mgt., 36(3):892–896.

Ozoga, J. J., and E. M. Harger
1966 Winter activities and feeding habits of northern Michigan coyotes. Jour. Wildlife Mgt., 30(4):809–818.

Ozoga, J. J., and C. J. Phillips
1964 Mammals of Beaver Island, Michigan. Michigan State Univ., Publ. Mus., Biol. Ser., 2(6):305–348.

Ozoga, J. J., and L. J. Verme
1968 Small mammals of conifer swamp deeryards in northern Michigan. Michigan Acad. Sci., Arts and Ltrs., 53:37–49.
1970 Winter feeding patterns of penned white-tailed deer. Jour. Wildlife Mgt., 34(2):431–439.
1975 Activity patterns of white-tailed deer during estrus. Jour. Wildlife Mgt., 39(4):679–683.

Ozoga, J. J., L. J. Verme, and C. S. Bienz
1982 Parturition behavior and territoriality in white-tailed deer: Impact on neonatal mortality. Jour. Wildlife Mgt., 46(1):1–11.

Perry, O. H.
1899 Hunting expeditions of Oliver Hazard Perry of Cleveland verbatim from his diaries. Privately printed in Cleveland, Ohio. xiii+246 pp.

Petersen, E. T.
1953 The Michigan Sportsmen's Association: A pioneer in game conservation. Michigan Hist., 37(4):355–372.
1979 Hunters' heritage. A history of hunting in Michigan. Michigan United Conserv. Clubs, Lansing, 55 pp.

Petraborg, W. H., and V. E. Gunvalson
1962 Observations on bobcat mortality and bobcat predation on deer. Jour. Mammalogy, 43:430–431.

Phillips, C. J., J. T. Ozoga and L. C. Drew
1965 The land vertebrates of Garden Island, Michigan. Jack-Pine Warbler, 43(1):20–25.

Potvin, F., J. Huot, and F. Duchesneau
1981 Deer mortality in the Pohénégamook wintering area, Quebec. Canadian Field-Nat., 95(1):80–84.

Pruitt, W. O., Jr.
1954 Rutting behavior of the white-tailed deer (*Odocoileus virginianus*). Jour. Mammalogy, 35(1):129–130.

Purol, D. A., J. N. Stuht, and G. E. Burgoyne, Jr.
1977 Mandibular cavity tissue fat as an indicator of spring physical condition for female white-tailed deer in Michigan. Michigan Dept. Nat. Res., Wildlife Div., Rept. No. 2792 (mimeo.), 7 pp.

Queal, L.
1968 Attitudes of landowners toward deer in southern Michigan. Michigan Acad. Sci., Arts and Ltrs., 43:51–69.

Rau, R.
1979 The Michigan massacre. Federal and state authorities make a big bust on a million-dollar poaching ring. Sports Illus., 50(9):49–50.

Reed, D. F., and T. N. Woodard
1981 Effectiveness of highway lighting in reducing deer-vehicle accidents. Jour. Wildlife Mgt., 45(3):721–726.

Rees, J. W.
1969 Morphological variation in the cranium and mandible of the white-tailed deer (*Odocoileus virginianus*): A comparative study of geographical and four biological distances. Jour. Morphol., 128:95–112.

Reilly, R. E., and H. E. Green
1974 Deer mortality on a Michigan highway. Jour. Wildlife Mgt., 38(1):16–19.

Rhead, D., and T. J. Martin
1978 Environment and subsistence at Sand point. Unpubl. rept. presented at Ann. Meet. Central States Anthro. Soc., South Bend, 8 pp.

Rice, L. A.
1980 Influences of irregular dental cementum layers on aging deer incisors. Jour. Wildlife Mgt., 44(1):266–268.

Rollings, C. T.
1945 Habits, foods and parasites of the bobcat in Minnesota. Jour. Wildlife Mgt., 9(2):131–145.

Rongstad, O. J., and J. R. Tester
1969 Movements and habitat use of white-tailed deer in Minnesota. Jour. Wildlife Mgt., 33(2):366–378.

Ryell, L. A.
1963 The occurrence of certain anomalies in Michigan white-tailed deer. Jour. Mammalogy, 44(1):79–98.

Scanlon, P. F., W. F. Murphy, Jr., and D. F. Urbston
1976 Initiation of pregnancy in lactating white-tailed deer. Jour. Wildlife Mgt., 40(2):373–374.

Scharf, W. C. and M. J. Jorae
1980 Birds and land vertebrates of North Manitou Island. Jack-Pine Warbler, 58(1):4–15.

Schofield, R. D.
1960 Determining hunting season waste of deer by following fox trails. Jour. Wildlife Mgt., 24(3):342–344.

Seton, E. T.
1953 Lives of game animals. Vol. III, Pt. I. Hoofed animals. Charles T. Branford Company, Boston. xix+412 pp.

Severinghaus, C. W.
1955 Some observations on the breeding behavior of deer. New York Fish and Game Jour., 2(2):239–241.

Severinghaus, C. W., and E. L. Cheatum
1956 Life and times of the white-tailed deer. Pp. 57–186 *in* W. P. Taylor, ed. The deer of North America. Stackpole Company, Harrisburg, Pennsylvania. xix+668 pp.

Shaw, H.
1963 Insectivorous white-tailed deer. Jour. Mammalogy, 44(2):284.

Shiras, G., III
1918 The albino deer herd of Grand Island. Forest and Stream, 88(3):158–159.

Short, H. L.
1962 The use of a rumen fistula in a white-tailed deer. Jour. Wildlife Mgt., 26(3):341–342.
1963 Rumen fermentations and energy relationships in white-tailed deer. Jour. Wildlife Mgt., 27(2):184–195.

Sicuranza, L. P.
1979 An ecological study of motor vehicle-deer accidents in southern Michigan. Michigan State Univ., unpubl. M.S. thesis, viii+65 pp.

Silver, H., N. F. Colovos, J. B. Holter, and H. H. Hayes
1969 Fasting metabolism of white-tailed deer. Jour. Wildlife Mgt., 33(3):490–498.

Silver, H., J. B. Holter, N. F. Colovos, and H. H. Hayes
1971 Effect of fall temperature on heat production in fasting white-tailed deer. Jour. Wildlife Mgt., 35(1):37–46.

Singer, F. J.
1979 Habitat partitioning and wildfire relationships of cervids in Glacier National Park, Montana. Jour. Wildlife Mgt., 43(2):437–444.

Smith, C. A.
1976 Deer sociobiology—Some second thoughts. Wildlife Soc. Bull., 4(4):181–182.

Stormer, F. A., and W. A. Bauer
1980 Summer forage use by tame deer in northern Michigan. Jour. Wildlife Mgt., 44(1):98–106.

Stuht, J. N.
1975 Morphology of trypanosomes from white-tailed deer and wapiti in Michigan. Jour. Wildlife Diseases, 11:256–262.

Stuht, J. N., and P. D. Friedrich
1978 Distribution and prevalence of skin tumors in Michigan deer. Michigan Dept. Nat. Res., Wildlife Div., Rept. No. 2805 (mimeo.), 3 pp.

Tate, J., Jr.
1969 Red-breasted nuthatch forages on deer's back. Jack-Pine Warbler, 47(1):32.

Taylor, W. P., ed.
1956 The deer of North America. Stackpole Company, Harrisburg, Pennsylvania. xix+668 pp.

Telfer, E. S.
1967 Comparison of moose and deer winter range in Nova Scotia. Jour. Wildlife Mgt., 31(3):418–425.
1972 Association measures applied to moose and white-tailed deer distributions. Jour. Mammalogy, 53(1):196–197.
1974 Vertical distribution of cervid and snowshoe hare browsing. Jour. Wildlife Mgt., 38(4):944–946.

Thomas, J. W., R. M. Robinson, and R. G. Marburger
1965 Social behavior of a white-tailed deer herd containing hypogonadal males. Jour. Mammalogy, 46(2):314–327.

Tompkins, J. L., J. N. Stuht, and J. E. Knight
1977 Meningeal worm in deer and wapiti of Canada Creek Ranch. Michigan Dept. Nat. Res., Wildlife Div., Rept. No. 2775 (mimeo.), 3 pp.

Townsend, T. W., and E. D. Bailey
1975 Parturitional, early maternal, and neonatal behavior in penned white-tailed deer. Jour. Mammalogy, 26(2):347–362.
1981 Effects of age, sex and weight on social rank in penned white-tailed deer. American Midl. Nat., 106(1):92–101.

Trodd, L. L.
1962 Quadruplet fetuses in a white-tailed deer from Espanola, Ontario. Jour. Mammalogy, 43(3):414.

Ullrey, D. E., W. G. Youatt, H. E. Johnson, A. B. Cowan, R. L. Covert, and W. T. Magee
1972 Digestibility and estimated metabolizability of aspen browse for white-tailed deer. Jour. Wildlife Mgt., 36(3):885–891.

Ullrey, D. E., W. G. Youatt, H. E. Johnson, L. D. Fay, and B. E. Brent
1967 Digestibility of cedar and jack pine browse for the white-tailed deer. Jour. Wildlife Mgt., 31(3):448–454.

Ullrey, D. E., W. G. Youatt, H. E. Johnson, L. D. Fay, R. L. Covert, and W. T. Magee
1975 Consumption of artificial browse supplements by penned white-tailed deer. Jour. Wildlife Mgt., 39(4):699–704.

Ullrey, D. E., W. G. Youatt, H. E. Johnson, L. D. Fay, D. B. Purser, B. L. Schoepke, and W. T. Magee
1971 Limitations in winter aspen browse for the white-tailed deer. Jour. Wildlife Mgt., 35(4):732–743.

Ullrey, D. E., W. G. Youatt, H. E. Johnson, L. D. Fay, B. L. Schoepke, and W. T. Magee
1970 Digestible and metabolizable energy requirements for winter maintenance of Michigan white-tailed deer. Jour. Wildlife Mgt., 34(4):863–869.

Ullrey, D. E., W. G. Youatt, H. E. Johnson, L. D. Fay, B. L. Schoepke, W. T. Magee, and K. K. Keahey
1973 Calcium requirements of weaned white-tailed fawns. Jour. Wildlife Mgt., 37(2):187–194.

Ullrey, D. E., W. G. Youatt, H. E. Johnson, P. K. Ku, and L. D. Fay
1964 Digestibility of cedar and aspen browse for the white-tailed deer. Jour. Wildlife Mgt., 28(4):791–797.

Van Etten, R. C., D. F. Switzenburg, and L. Eberhardt
1965 Controlled deer hunting in a square-mile enclosure. Jour. Wildlife Mgt., 29(1):59–73.

Verme, L. J.
1961 Late breeding in northern Michigan deer. Jour. Mammalogy, 42(3):426–427.
1962 Fecundity in a Michigan white-tailed deer. Jour. Mammalogy, 43(1):112–113.
1965a Reproduction studies on penned white-tailed deer. Jour. Wildlife Mgt., 29(1):74–84.
1965b Swamp conifer deeryards in northern Michigan: Their ecology and management. Jour. Forestry, 63: 523–529.
1969 Reproductive patterns of white-tailed deer related to nutritional plane. Jour. Wildlife Mgt., 33(4):881–887.
1973 Movements of white-tailed deer in upper Michigan. Jour. Wildlife Mgt., 37(4):545–552.
1977 Assessment of natal mortality in upper Michigan deer. Jour. Wildlife Mgt. 41(4):700–708.
1979 Influence of nutrition on fetal organ development in deer. Jour. Wildlife Mgt., 43(3):791–796.

Verme, L. J., and J. J. Ozoga
1966 A white-tailed deer jaw with fourth molars. Jack-Pine Warbler, 44(3):148.
1971 Influence of winter weather on white-tailed deer in upper Michigan. Michigan Dept. Nat. Res., Res. and Dev. Rept. No. 237 (mimeo.), 14 pp.
1980a Influence of protein-energy intake on deer fawns in autumn. Jour. Wildlife Mgt., 44(2):305–314.
1980b Effects of diet on growth and lipogenesis in deer fawns. Jour. Wildlife Mgt., 44(2):315–324.

Waselkov, G. A.
1978 Evolution of deer hunting in an eastern woodlands. Midcontinental Jour. Arch., 3(1):15–34.

Watson, M. H.
1972 Characteristics of hunters opposed to Michigan deer management. Michigan Dept. Nat. Res., Res. and Dev. Rept. No. 277 (mimeo.), 12 pp.

Westell, C. E., Jr.
1954 Available browse following aspen logging in lower Michigan. Jour. Wildlife Mgt., 18(2):266–271.
1955 Ecological relationships between deer and forests in lower Michigan. Proc. Soc. American Foresters Meeting, 1955:130–132.

Westover, A. J.
1971 The use of a hemlock-hardwood winter yard in white-tailed deer in northern Michigan. Huron Mt. Foundation, Occ. Papers No. 1, 59 pp.

Whitlock, S. C.
1939 The prevalence of disease and parasites in whitetail deer. Trans. Fourth North American Wildlife Conf., pp. 244–249.

Whitlock, S. C., and L. Eberhardt
1956 Large-scale dead deer surveys: Methods, results and management implications. Trans. Twenty-First North American Wildlife Conf., pp. 555–556.

Wilhelm, A. R., and D. O. Trainer
1966 A serological study of epizootic hemorrhagic disease in deer. Jour. Wildlife Mgt., 30(4):777–780.

Wilson, R. L.
1967 The Pleistocene vertebrates of Michigan. Michigan Acad. Sci., Arts and Ltrs., Papers, 52:197–234.

Wood, N. A.
1922 The mammals of Washtenaw County, Michigan. Univ. Michigan, Mus. Zool., Occ. Papers No. 123, 23 pp.

Woolf, A., and J. D. Harder
1979 Population dynamics of a captive white-tailed deer herd with emphasis on reproduction and mortality. Wildlife Monogr., No. 67, 53 pp.

Wozencraft, W. C.
1979a A note on the behavior of white-tailed deer toward sandhill cranes. Jour. Mammalogy, 60(2):434–435.
1979b Melanistic deer in southern Wisconsin. Jour. Mammalogy, 60(2):437.

Youatt, W. G., L. D. Fay, and H. D. Harte
1975 1975 spring deer survey doe productivity and femur fat analysis April 1–June 1. Michigan Dept. Nat. Res., Wildlife Div., Rept. No. 2743 (mimeo.), 2 pp.

Youatt, W. G., and H. D. Harte
1974 Liver fluke and nose botfly larvae in Michigan white-tailed deer; spring doe deer productivity surveys, 1973–1974. Michigan Dept. Nat. Res., Wildlife Div., Rept. No. 2729 (mimeo.), 9 pp.

Young, S. P., and E. A. Goldman
1944 The wolves of North America. American Wildlife Inst., Washington, D.C., xx + 636 pp.
1946 The puma, mysterious American cat. American Wildlife Inst., Washington, D.C., xvi + 358 pp.

Zagata, M. D., and A. N. Moen
1974 Antler shedding by white-tailed deer in the Midwest. Jour. Mammalogy, 55(3):656–659.

MOOSE
Alces alces (Linnaeus)

NAMES. The generic name *Alces,* proposed by Gray in 1821, is derived from both the Greek and Latin and means elk. The specific name *alces,* proposed by Linnaeus in 1758, has the same derivation. In English-speaking areas of Eurasia the common name for this impressive cervid is elk. However, early European settlers in the New World called the wapiti (or red deer of Eurasia) by the name elk, which led to some confusion. Consequently, the present North American population of the circumboreal species *Alces alces* is commonly called moose, apparently derived from an Algonquian word, *moos,* meaning "eater of twigs" or "he strips off bark." Moose formerly and presently living in Michigan belong to the subspecies *Alces alces americana* (Clinton).

RECOGNITION. Body size large, ungainly with horse-like dimensions (Fig. 111), about 71 in (1,815 mm) at the shoulders (withers) for the male and 69 in (1,780 mm) for the female; head long, massive, rather narrow, gross-looking with a pendulous snout; nose well-haired except for small lower area, flexible anterior end overhanging lower lips; ears large, conspicuous; eyes rather small, recessed; preorbital (lachrymal) glands present; males have massive shovel-shaped (palmate) antlers arising in typical cervid fashion on rugose burrs on frontal bones behind orbits (Fig. 112), with each beam extending laterally for perhaps 12 in (310 mm) where one and perhaps two brow tines branch anteriorly, the beam then flattening, sweeping laterally, and posteriorly as a broad, dished structure with several long, bordering tines; females antlerless except in exceptional circumstances; strip of skin called a dewlap or bell, 6 to 12 in (155 to 310 mm) long, hangs from the throat of both sexes; neck short; slight mane on neck and shoulders; tail short; body heavy, thick, high at the humped shoulders, lower at tapering rump; legs long, slender in comparison to bulky body; metatarsal glands absent; tarsal glands small; dewclaws low on feet; hoofs long, pointed, front ones larger than those on hind feet.

Figure 111. The moose (*Alces alces*).
Sketch by Bonnie Marris.

Figure 112. Details of the massive antlers of the male moose (*Alces alces*).

The two sexes of the moose are similar in color; males may have more brownish foreheads, females more grayish. First-year animals (calves) are reddish brown overall. Adults have coarse, brittle, black-tipped guard hairs about 3 in (77 mm) long over most of the body and as much as 6 in (153 mm) long on the midline of the neck and shoulders. In winter there is a bluish undercoat of fine, slightly curly hairs. Overall coloring of adults in new spring pelage varies from reddish brown to grayish brown to almost black. The face may be grayish; the head, neck, grizzled shoulders, and back are often slightly paler than the flanks; the lower belly and undersides of the legs are whitish. Worn pelage becomes slightly paler; most fading occurs in late summer.

The moose is the largest of the cervids. At close range, the animal appears homely and poorly-proportioned with its drooping muzzle, high and humped shoulders, small rump, and stubby tail. At a distance, however, the moose is graceful and the antlered male has a majestic appearance. The antlerless head of the female moose has a certain grossness (Fig. 113), in sharp contrast to the delicate features of the doe white-tailed deer. Males are called bulls; females, cows; and the young, calves.

Figure 113. The moose (*Alces alces*). Note the prominent, drooping snout emphasized by this portrait of a cow moose by Bonnie Marris.

MEASUREMENTS AND WEIGHTS. Adult specimens of the larger male and the smaller female measure, respectively, as follows: length from tip of nose to end of tail vertebrae, 90 to 104 in (2,300 to 2,800 mm) and 80 to 101 in (2,050 to 2,600 mm); length of tail vertebrae (almost the same in both sexes), 3 to 4¾ in (80 to 120 mm); length of hind foot, 30 to 30¼ in (760 to 835 mm) and 28½ to 31½ in (730 to 810 mm); height of ear from notch (almost the same in both sexes), 10 to 10⅜ in (257 to 265 mm); height at shoulders (withers), 69⅜ to 74⅞ in (1,780 to 1,920 mm) and 67½ to 71⅜ in (1,730 to 1,830 mm); weight, 850 to 1,200 lbs (387 to 545 kg) and 725 to 850 lbs (330 to 396 kg). Field-taken weights of moose are scarce, especially those taken prior to the removal of viscera. Peterson (1974) suggests the viscera may represent about 24% of the total weight of heavy individuals; for less heavy animals the viscera may represent as much as 28% of the total weight.

DISTINCTIVE CRANIAL AND DENTAL CHARACTERISTICS. The skull of the moose is large and heavy (Figs. 103 and 114); the extended

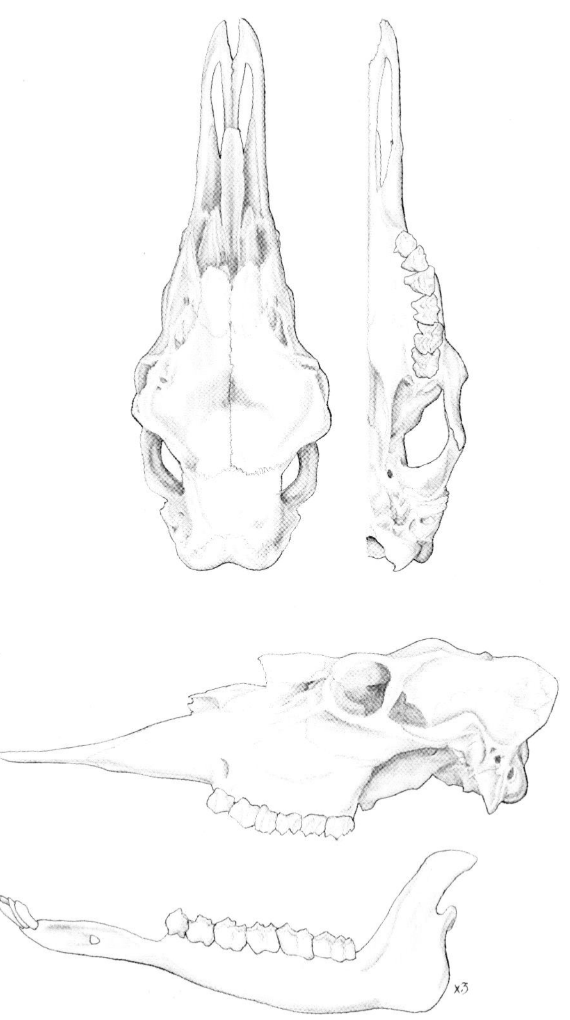

Figure 114. Cranial and dental characteristics of the moose (*Alces alces*).

rostrum consists of elongated and narrow premaxillary bones and noticeably short nasal bones, length of latter less than half the distance from anterior tips to tips of the premaxillaries and almost equal to the distance from the back edge of the nasals to the posterior end of the occiput; vomer in ventral view not dividing aperture of posterior nares; preorbital (lachrymal) vacuities widely open; maxillary canines usually absent; lower canines incisiform in typical cervid fashion; premolars and molars broad and low-crowned. The skulls of the adult male and the adult female average, respectively, 22½ in (577 mm) and 21¾ in (557 mm) in greatest length and 9 in (230 mm) and 8⅝ in (220 mm) in greatest width across the zygomatic arches.

The characteristic palmate antlers of the male, the large size of the skull (in adults of both sexes, more than 19½ in or 500 mm in greatest length), and the elongated rostrum with short nasals are cranial characteristics useful in distinguishing the moose from the wapiti, the white-tailed deer, and the caribou. (Also consult the artificial key to the species of cervids and bovids.) The dental formula for the moose is:

$$\text{I (incisors)}\,\frac{0\text{-}0}{3\text{-}3},\ \text{C (canines)}\,\frac{0\text{-}0}{1\text{-}1},\ \text{P (premolars)}\,\frac{3\text{-}3}{3\text{-}3},\ \text{M (molars)}\,\frac{3\text{-}3}{3\text{-}3} = 32$$

DISTRIBUTION. The moose, like the wapiti, has a circumboreal distribution, occurring (or formerly occurring) widely in northern parts of the Eurasian and North American land masses (Peterson, 1955). Despite its size and edible flesh, the moose has survived in historic times in many parts of its northern range. It has also been successfully introduced in such places as New Zealand (Tustin, 1974).

In North America, the moose occurs in a broad band across Alaska and Canada, primarily south of tree line. The moose ranges southward into boreal parts of northern United States, skirting the Great Plains and into the Rocky Mountain sector as far south as Utah and Colorado. In northeastern United States, this cervid was formerly found as far south as Pennsylvania; presently it lives in northern New England. In the Upper Great Lakes Region, the moose lived in northern Wisconsin, much of Minnesota, and most of Michigan except for the southern and southwestern parts of the Lower Peninsula (Burt, 1946; de Vos, 1964; Peterson, 1955; Schorger, 1942; Wood, 1914).

The moose was present in Michigan in early postglacial times (Holman, 1975; Kelsall and Telfer, 1974; Peterson, 1955), and was an important source of food, clothing, blankets, lacings, bone and antler tools, and other products for the prehistoric peoples (Petersen, 1979). Writers describe various moose hunting methods, including the use of snares (Hickie, 1937). Bones and other evidence of the moose's presence have been found in archaeological sites in Charlevoix County in the Lower Peninsula (Lovis, 1973) and Delta and Mackinac counties in the Upper Peninsula (Janzen, 1968; Martin, 1979; 1980). Moose remains have also been recovered from sites on Summer Island in Lake Michigan (Brose, 1970), Isle Royale in Lake Superior (Cleland, 1968), and Bois Blanc Island in Lake Huron (Cleland, 1966). Radisson reported moose in Michigan at least as early as 1652, when the early fur trade began (Hickie, undated; Krefting, 1974a).

Moose first encountered the onslaught of human settlement in the Lower Peninsula where the primary range was in the north and east (see Map 67), possibly reaching almost to the Ohio line in

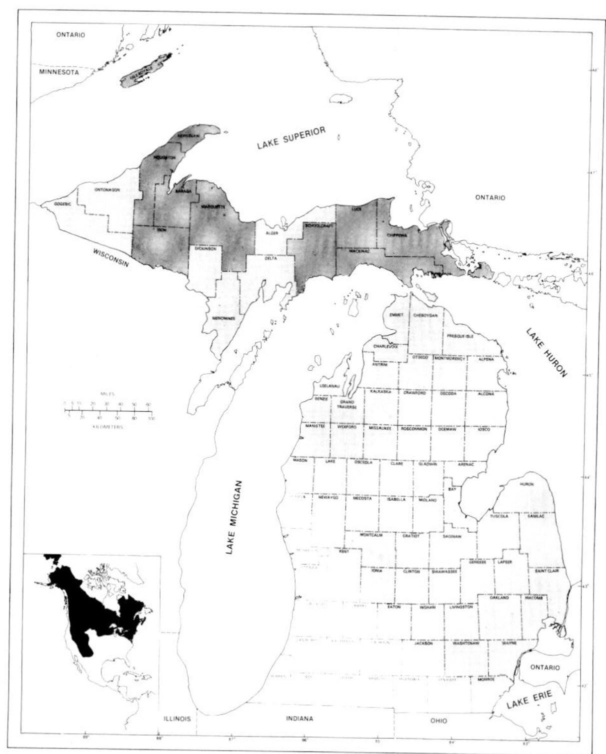

Map 67. Geographic distribution of the moose (*Alces alces*). Moose formerly occurred in areas marked with pale shading and has been found since 1969 in areas marked with dark shading.

extreme southeastern Michigan. Cadillac reported moose in the Detroit area in the early eighteenth century (Johnson, 1919). Moose may have thrived there well into the early nineteenth century; Hoyt (1889) reported moose in Oakland County between 1825 and 1830. Apparently, populations were sustained at this time in the "Thumb" area, especially in Huron, Sanilac and perhaps Tuscola counties. Moose tracks were recorded from along the headwaters of the Black River in Sanilac County on October 4, 1850, by Perry (1899). The carcass of a moose killed in the winter of 1864 in this same county was exhibited in Port Huron; moose were reported at least as late as 1870 in Huron County (Wood and Dice, 1924). In northern parts of the Lower Peninsula, the moose apparently lingered well into the 1880s. An antlered bull moose was reported shot in Missaukee County's Union Township in August, 1881 (Hickie, undated). John Roger's sighting of a moose at Black Lake in Presque Isle County in October, 1883, may be the last recorded evidence for Lower Peninsula moose (Wood and Dice, 1924). Ironically, in 1889, at about the time moose were being extirpated in southern Michigan, the Michigan Legislature passed a law granting the animals full protection.

Moose fared better in the Upper Peninsula, perhaps because higher quality boreal habitat was more extensive and much of the area was neither logged nor settled until after the Civil War. Nevertheless, the moose population slowly declined. Stock in adjacent Wisconsin was also on the wane prior to the turn of the century (Jackson, 1961; Schorger, 1942). In 1870, moose were virtually unknown along the shore of Lake Superior (de Vos, 1964). Wood (1917) reported animals in Alger County in 1878; Dice and Sherman (1922) found evidence of moose at Gogebic Lake in 1885. In the early years of the twentieth century, moose were periodically reported in the eastern part of the Peninsula, primarily in Chippewa, Luce and Schoolcraft counties (Wood and Dice, 1924). The First Biennial Report of the Department of Conservation for 1921–1922 estimated that the population of moose in Chippewa and Luce Counties was well over 1,000 individuals. Apparently, moose have persisted in small numbers thoughout the twentieth century (Hickie, 1937; Krefting, 1974a; Manville, 1948; Wood, 1914). The Department of Natural Resources (previously the Department of Conservation) summarized sightings of moose and/or their identifiable signs (tracks, fecal deposits, etc.) in Biennial Reports. In the Biennial Report for 1931–1932, evidence of moose was recorded in five counties (Alger, Chippewa, Keweenaw, Mackinac, and Schoolcraft); for 1965–1966, in eight counties (Alger, Chippewa, Delta, Dickinson, Luce, Marquette, Mackinac, and Schoolcraft); for 1967–1968, in seven counties; for 1968–1969, in eleven counties (including most of those listed for 1965–1966 plus Baraga, Gogebic, Houghton, Iron, and Ontonagon); for 1971–1972, in eight counties. In 1979–1980, these reports continued to indicate a moderate moose population, again primarily in the eastern counties of the Upper Peninsula. The occasional viewing of calves, as in Mackinac County in September, 1980, indicates that the resident moose population is productive.

Various authors have suggested that these eastern moose populations continue to gain recruits from Ontario when these cervids cross the water gap into Chippewa County. Pruitt (1951) reported that moose were observed swimming from the Canadian mainland to Sugar Island. As long as a viable moose population occurs on the Ontario side of this interlake waterway (Banfield, 1974; Peterson, 1955; Rheaume, 1980), this natural dispersion probably will continue.

To increase the Upper Peninsula moose population, the Michigan Department of Conservation live-trapped moose on Isle Royale for three years. According to the Ninth Biennial Report of the Department, 71 moose (33 males and 38 females) were removed from Isle Royale as follows: in 1934–1935, 11 individuals (1 male and 1 female to the Detroit Zoo, 2 males and 7 females to the Cusino Wildlife Research Station in Alger County); in 1935–1936, 35 individuals (11 males and 13 females to Cusino, 3 males and 3 females to the Escanaba River Tract in Delta County, 3 males and 2 females to Keweenaw Point in the Fort Wilkins area in Keweenaw County); and in 1936–1937, 25 individuals (1 male to Cusino; 7 males and 4 females to Escanaba, 5 males and 8 females to Keweenaw). Five moose (2 males and 3 females) were kept in corrals at Cusino. These live-trapping efforts and the penned animals at Cusino received considerable study (East, 1936; Kellum, 1941; Ruhl, 1940; Verme, 1970). Although the overall effects of these relocations were not totally monitored, they may have added breeding stock to the herd, especially in western areas of the Upper Peninsula (Hickie, undated).

By far the most celebrated Michigan moose herd is found in Isle Royale National Park. Isle Royale, the largest island in Lake Superior, is 45 mi (72 km) long and 8 mi (13 km) wide. Its cluster of about 200 islets and rocky outcrops are only 13 mi (21 km) from the mainland of Ontario and 18 mi (29 km) from Minnesota, even though politically it belongs to the more remote state of Michigan, being about 45 mi (72 km) northwest of Keweenaw Point. Since early times, moose apparently moved over winter ice or possibly swam from the mainland (presumably Ontario) to Isle Royale (Hickie, 1936, 1937; Murie, 1934). Between 1870 and 1900, intensive logging, followed by fires, produced a forage eruption which allowed the rather meager moose population to increase. Beginning in 1929, population ups-and-downs in relation to forest succession and predation by wolf packs were carefully monitored by various observers, especially those from Purdue University and Michigan Technological University in recent years (Aldous and Krefting, 1946; Hickie, 1937; Krefting, 1951, 1974a; Mech, 1966; Murie, 1934; Peterson, 1977; Shelton, 1966; Snyder and Janke, 1976).

HABITAT PREFERENCES. The moose is distributed in North America in close association with northern boreal forests. Park-like stands of coniferous trees with little deciduous undergrowth may be used occasionally as winter retreats but they lack the total environmental quality necessary for year-around moose survival. The moose benefits most from early successional stages of evergreen forests. Most field observers believe that these early seral stages of shrubby growth and saplings stands have historically been best produced and maintained by unrestricted and natural fires and by present-day timber cutting (Eastman, 1974; Irwin, 1975; Krefting, 1974b; McNicol and Gilbert, 1980; Peek *et al.,* 1976; Peterson, 1955). These rather unstable habitats may also be improved by the spruce budworm which, despite its destruction of commercial timber, defoliates and opens closed overhead canopies of balsam fir and white spruce. Ideally, mixed young forest growth has a mosaic pattern providing abundant edge between woody cover and herbaceous plant openings (Peek *et al.,* 1976). Swamps, lakeshores adjacent to subclimax woody vegetation, and forest openings attract moose in summer.

The diverse pattern of well-watered evergreen-deciduous second growth in much of the Upper Peninsula and northern parts of the Lower Peninsula appear of sufficient quality to provide for sustained populations of moose.

DENSITY AND MOVEMENTS. Studies of the moose's seasonal habitat use have been made in the Upper Great Lakes Region, notably in Minnesota and Ontario and on Isle Royale. Investigators have used such methods as aerial counts of either tagged or untagged animals visible against snow backgrounds in winter or in openings in summer (de Vos and Armstrong, undated; Goddard, 1970), pellet (fecal dropping) counts (Krefting, 1974a), and telemetry (van Ballenberghe and Peek, 1971).

Peterson (1955), in his classic work, determined that the moose occurs at densities of approximately one animal per sq mi (0.4 per sq km) in average boreal habitat and at almost three per sq mi (1.2 per sq km) in most favorable habitat. In northeastern Minnesota, Peek *et al.* (1976) found moose using recently logged areas in early winter at densities of 0.74 animals per sq mi (0.3 per sq km). Changes in habitat quality resulting from logging, fires, fire protection, sodium limitations, and browsing (especially on aquatic plants by moose) have caused fluctuations in the carrying capacity of the habitat and affected the density of the moose population on Isle Royale (Belovsky, 1981; Belovsky and Jordan, 1981; Hickie, 1937; Krefting, 1974a; Mech, 1966; Peterson, 1977; Ruhl, 1940; Shelton, 1966; Snyder and Janke, 1976). Density estimates show that this insular population has fluctuated considerably: 1 per sq mi (0.4 per sq km) in 1915, 15 per sq mi (6 per sq km) in 1929, 4 per sq mi (1.6 per sq km) in 1960, 3.5 per sq mi (1.4 per sq km) in 1980.

Despite their size, moose seem to be fairly sedentary; major movements are from winter range to summer range (Goddard, 1970; LeResche, 1974). Most field studies in Minnesota and Ontario (de Vos, 1956; Goddard, 1970; Phillips *et al.,* 1973; Van Ballenberghe and Peek, 1971) show that home ranges of moose vary, depending on habitat, from as small as 0.5 sq mi (0.2 sq km) to as large as 6.9 sq mi (17.9 sq km). One study in northwestern Minnesota showed that cows had slightly larger home ranges than bulls, with both having summer areas as much as five times larger than winter areas (Phillips *et al.,* 1973; Seton, 1953). In boreal forest in northwestern Ontario, radio-collared moose used ranges of 1 to 5 sq mi (2 to 14 sq km) in winter and 2 to 35 sq mi (6 to 90 sq km) in summer (Addison *et al.,* 1980). Bulls, of course, move about in expanded territories during the autumn rut (Murie, 1934).

Movements between winter and summer ranges are necessary, according to LeResche (1974), to provide the animals with optimal physical, biotic, and social environments. In the Upper Great Lakes Region, in contrast to mountainous areas in other parts of its range, the moose may move only minimal distances, especially in confined areas like Isle Royale. The extent and depth of snow cover may trigger these movement responses (Coady, 1974).

BEHAVIOR. The moose is perhaps even more solitary in its habits than the white-tailed deer (Franzmann, 1981). The chief association is that of the cow with her calf; this lasts until the cow is in late pregnancy at which time she usually drives her yearling offspring away. In general, bulls and yearlings may spend much of the year alone. Occasionally, males associate in loose bands in winter and in small groups in summer, but they separate near the end of summer when the older, dominating bulls become belligerent as the rutting period approaches (Altmann, 1959; Dodds, 1958; Hauge and Keith, 1981). Moose are likely to congregate at natural salt licks, concentrations of favored aquatic plants or other places where nutrients may be abundant, and in winter yarding situations, but often ignore each other. Murie (1934) observed moose at a salt lick on Isle Royale and found that older animals tended to dominate younger ones when competing for the minerals (de Vos, 1958). The two sexes associate closely only at the rutting period. According to Lent (1974), there is no real evidence that single bulls form harems of cows.

Moose may be active at any time of day or night, especially in winter, but feeding takes place primarily in the twilight hours of morning and evening. In summer, the animals tend to bed down at midday. Like other cervids, moose are less active in windy and stormy weather (de Vos, 1958).

Although moose are considered quiet creatures, the prolonged, loud wails or moans of the cow during the rutting season are well-known (Kellum, 1941) and may be heard as far away as a mile (1.6 km) on a quiet September day (Lent, 1974). Hunters often use birch bark horns to imitate this call and attract bulls to within gun range. Alarm calls may vary from a dog-like bark and squeal to a mooing sound (Murie, 1934). Males grunt in response to cows' rutting wails. Calves and yearlings utter high plaintive whines to communicate with the female parent; cows may attract the younger animals with low, short, grunts. Moose are apparently more vocal during warm months (including the time of the rut) than during cold months (de Vos, 1958).

Because of its size, the moose does not move as quietly as the graceful white-tailed deer, although the moose's long limbs are advantageous when it moves through brush or wades in lakes and swamps. The moose negotiates uncrusted snow as deep as 30 in (765 mm) without difficulty (Peterson, 1955). However, sharp ice crystals in deep, crusted snow can damage the skin covering on the legs (Kelsall, 1969). The animal's normal gait is a slow walk; it increases its speed with a stiff-legged trot and can gallop at 35 mi (56 km) per hr or more. Surprisingly, the fully-antlered bull moose can move rapidly in thick brush without becoming entangled. The moose rarely jumps, preferring to walk around obstacles. Some ecologists consider the moose semiaquatic, because of its ability as a wader, swimmer, and even diver (Peterson, 1955).

The moose apparently has a keen sense of smell, sharp ears, and dull eyes (Seton, 1955). An observer can approach a moose if the quarry does not smell or hear him. Although cows may stand and defend their calves, and rutting bulls may be belligerent, moose usually avoid humans.

ASSOCIATES. A study of the history of northern Michigan cervid populations indicates that both association and replacement occurred. The Lake Superior area initially may have been the home of the caribou (Hickie, undated). The caribou then gave way to the moose, perhaps no later than the mid-nineteenth century, with the white-tailed deer encroaching and becoming dominant over the moose in the post-Civil War era. In Michigan today, the major moose association with other cervids is with the white-tailed deer, although in many areas in western states the moose shares the environment with the wapiti (McMillan, 1953; Singer, 1979).

In Michigan, moose and white-tailed deer

occupy adjacent areas chiefly in the eastern part of the Upper Peninsula. Based on relationships observed in Minnesota and Ontario (Irwin, 1975; Kearney and Gilbert, 1976), the two cervids probably use similar environments primarily in the warm months, are equally attracted to burned areas, and have a major dietary overlap for browse in autumn. However, moose eat larger-diameter twigs (Peek *et al.*, 1971), Moose also prefer balsam fir and reject white cedar, while deer thrive on white cedar and take balsam fir only as a last resort. Generally, where the two occur together in winter, the white-tailed deer seem more attracted to coniferous habitats; moose to deciduous areas. Kearney and Gilbert (1976) found that distributions of these two species appeared more influenced by habitat characteristics than by habitat types: moose distribution was more related to food availability; white-tailed deer distribution, to shelter availability. In winter, the moose, unlike the white-tailed deer, negotiates deep snow easily and remains highly mobile (Prescott, 1974).

The most negative association between these two cervids is probably infestation by the meningeal worm (*Parelaphostrongylus tenuis*). White-tailed deer tolerate this worm but it causes serious "moose sickness" in moose and "caribou sickness" in caribou (Anderson, 1972; Saunders, 1973). Such infestations have seriously depressed moose populations (Telfer, 1967). Although unreported in Michigan moose to date, this parasite is a potential hazard because it exists in the Michigan white-tail deer herd.

Moose and snowshoe hare often eat the same herbaceous plants in summer and browse species in winter (Dodds, 1960; Telfer, 1974; Wolfe, 1974). Most of this shared food is foliage and twigs near ground level but in winter heavy snow frequently weighs down taller browse plants, placing them at snowshoe hare eating levels. In moose, beaver and porcupine associations, moose may receive the greater benefit because the rodents' gnawing sets back forest and waterside plant successions (Wolfe, 1974). Beaver dams expand surface water in which aquatic plants favored by moose thrive (Hosley, 1949).

REPRODUCTIVE ACTIVITIES. Murie (1934) recorded moose courtship activities as occurring in September on Isle Royale. Early in September, bulls begin removing the withered velvet from their hardened antlers by thrashing them in brush or rubbing them against trees. Males also drive away weaker associating bulls at this time (Dodds, 1958). They grunt, stalk around openings and along shorelines, challenge other bulls (Fig. 115), paw out wallows in which to roll in mud moistened with rain and urine, persistently follow a single cow until she becomes sexually receptive, and finally abandon her to seek out another female (Altmann, 1959; Geist, 1963; Lent, 1974; Murie, 1932; Seton, 1953). The long, drawn-out moan or bleating bawl of the female indicates she is approaching mating readiness. The cows are usually successfully mated

Figure 115. Rutting bull moose joust and in other ways show their belligerent nature during the autumn mating season.

in late September and early October, sometimes as late as December (Coady, 1974). To gain strength for the rut, bulls gain as much as 28% of their body weight prior to courtship; by the end of the rut they have lost as much as 10% of this weight (Franzmann *et al.*, 1978; Verme, 1970). Generally, mature animals, both bulls and cows, breed. A variable number of yearling females (about 16 months old) also mate; males of this age are potentially reproductive but may be excluded by mature bulls (Peterson, 1974, Schwartz *et al.*, 1982).

Pregnancy lasts from 240 to 246 days. As birth approaches, the cow drives away associated yearlings. Calves are born in late May or early June (Murie, 1934). There may be one or two offspring; triplets are rare, although one set was reported from Isle Royale (Mole and Mole, 1976). At birth, the calf has an unspotted, reddish-brown coat, legs are long in comparison to a short body, it stands but has poor balance, has developed vision and hearing, and weighs between 25 to 35 lbs (11.3 to 15.9 kg). Geist (1974) suggested that birth weight is closely related to the quality of the nutrients available to the cow in late pregnancy. The newly-born calf is licked and then suckled, initially with the cow lying down (Stringham, 1974). Within five days after birth, the calf moves around well. Within a month, calves of cows held at the Cusino Wildlife Station in Alger County had tripled their weight (Verme, 1970). By September, calves are usually weaned and have developed their winter coats.

Although moose cows are attentive mothers and keep their calves in close association throughout the first year, the offspring have difficulty surviving. Durward Allen (pers. comm.) suggested that three out of four calves on Isle Royale die in their first year, especially from January through March when deep snow hampers the young animals and makes them more vulnerable to predators (Berg, 1976; Peek *et al.,* 1976; Pimlott, 1959; Simkin, 1965).

Moose are aged by the degree of tooth wear, schedule of tooth eruptions and deciduous tooth replacement, and the number of cementum layers on either incisors or molars (Gasaway *et al.,* 1978; Peterson, 1955; Sergeant and Pimlott, 1959; Wolfe, 1969). North American moose live as long as 22 years (Peterson, 1974).

The rapid development (in a few weeks) of the moose's massive antlers requires abundant, high-quality food. The young male has flat velvety knobs during its first autumn; spike or forked structures appear the second year. In the prime of adulthood, bulls' antlers begin to develop in April and have matured and been cleaned of the dead velvet covering by early September. Presumably, tremendous energy is required to produce these amazing growths, causing them to be questionable assets when related bovids require no similar nutrient supply to maintain permanent horns. The loss of the antlers in early winter (as late as February in the case of younger animals) no doubt relieves overstrained neck muscles but leaves the bulls looking rather unattractive until the new antler growth begins in April. Antler growth can be irregular depending on nutrients and on endocrine changes.

FOOD HABITS. Just as white-tailed deer may have to make adjustments to survive at the northern edge of its natural distribution in the Upper Great Lakes Region, moose also have environmental problems in Michigan because the northern part of the state is just at the southern boundary of its boreal range. The moose is primarily a browser, although it has been called a generalist in food preferences (Belovsky, 1978). In the Upper Great Lakes Region, this large cervid usually prefers second growth lowlands, eating a considerable amount of aquatic plant life in summer and tree saplings and brush in winter (Peek, 1974; Peterson, 1955). According to Kellum (1941), who observed moose confined at the Cusino Wildlife Research Station in Alger County, these animals eat 50 to 60 lbs (22.7 to 27.2 kg) per day in summer and 40 to 50 lbs (18.1 to 22.7 kg) per day in winter. Modern-day workers have also stressed the moose's need for minerals such as copper and sodium (Flynn *et al.,* 1977; Franzmann *et al.,* 1975); sodium is considered a factor in controlling moose populations on Isle Royale (Belovsky, 1978; Jordan *et al.,* 1973). Elongated fecal pellets are passed after meals of woody browse; pie-shaped droppings are a result of meals of aquatic vegetation, other succulents, and leaves.

Most knowledge of moose food preferences in the Michigan area comes from studies conducted in Minnesota and Ontario (de Vos, 1958; McNicol and Gilbert, 1980; Peek *et al.,* 1976; Peterson, 1953, 1955) plus important investigations made on Isle Royale (Aldous and Krefting, 1946; Hickie, 1937; Hosley, 1949; Krefting, 1951; Murie, 1934; Snyder and Janke, 1976). In late spring and summer, moose eat newly-formed leaves of aspen, paper birch, sugar maple, mountain ash, willows, red osier dogwood, beaked hazelnut, and other shrubs and saplings in lesser amounts. At this time of the year, moose are especially attracted to salt licks and mineral springs, motivated chiefly by a need for sodium (Fraser and Hristienko, 1981). In June, moose begin to forage in water, eating sedges, horsetails, rushes, pondweeds and other aquatics or riparian succulents, and forbs may become less important in the diet. In winter, moose are almost entirely woody vegetation browsers (Crete and Audy, 1974). Krefting's (1951) field studies on Isle Royale indicate that three-fourths of the moose's winter diet came from six browse plants: aspen, paper birch, balsam fir, mountain ash, willows and red osier dogwood. Of lesser importance were ground hemlock, juneberry, cherry, sugar maple, and beaked hazelnut. In the careful, long-range monitoring of moose populations on Isle Royale, the impact of moose browsing on favored insular foods has demonstrated some of the deleterious effects on plant diversity and abundance when moose populations exceed the carrying capacity (Snyder and Janke, 1976).

ENEMIES. In its primeval habitat in northern Michigan, the moose was subjected to predation by mountain lion and gray wolf, with black bear, lynx and perhaps wolverine menacing moose calves (Peterson, 1955). Numerous animals, including red fox and fisher, are attracted to winter kills or scraps from remains of individuals killed by gray wolves (de Vos, 1952; Johnson, 1970). Black bear can be a serious threat to moose calves in summer (Franzmann *et al.,* 1980). However, with the exception of the human, the slowly declining gray wolf remains the major moose predator in the Upper Great Lakes Region today (Frenzel, 1974).

This wolf-moose relationship has, of course, been a feature of sustained studies at Isle Royale National Park by scientists from Purdue University (beginning in 1958 under the leadership of Dr. Durward Allen) and more recently by those from Michigan Technological University (Jordan *et al.,* 1971; Mech, 1966; Peterson, 1977; Shelton, 1966; Wolfe, 1977). Moose presumably supplanted caribou on Isle Royale as the major herbivore and fluctuated widely in numbers during the early decades of the twentieth century (Krefting, 1951), sustaining only minor predator exploitation. The gray wolf made its appearance on the island about 1948, and the interaction of these two opposing mammals in a confining, insular situation became a fascinating subject for study. Wolfe (1977) diagnosed the cause of death of 439 moose on Isle Royale between 1950 and 1969; at least 45% were killed by gray wolves, with calves and yearlings comprising 19.3% and 3.5% respectively. Moose in the older categories (age classes 12–17 years) accounted for 29.3% of the kills. Gray wolves also showed a significant preference for female moose. Calves and yearlings are more vulnerable to attack when snow depth reaches more than 30 in (762 mm). In summer, moose often escape by taking to water; gray wolves then turn to beaver and other smaller, island mammals. The balance on Isle Royale between the food requirements for this top carnivore and the productivity of this top herbivore shows that the two can survive compatibly.

PARASITES AND DISEASES. The moose is infested with a variety of diseases and parasites (Anderson and Lankester, 1974; Peterson, 1955). Diseases include several arboviruses, skin tumors, bovine virus diarrhea, leptospirosis, brucellosis, listeriosis, anthrax, necrobacillosis, actinomycosis and others. Internal parasites include two flukes, five tapeworms, and 11 roundworms. Serious parasites in Isle Royale moose populations are the hydatid tapeworm, *Echinococcus granulosus,* and the lungworm, *Dictyocaulus viviparus* (Mech, 1966). The former infests both the liver and the lungs. There are no records, according to Dr. John Stuht of the Michigan Department of Natural Resources, for the meningeal worm, *Parelaphostrongylus tenuis,* in moose from Michigan. This deadly infestation does occur, however, in moose in Ontario and Minnesota (Anderson, 1972; Saunders, 1973). Its presence in Minnesota moose appears to predispose young individuals to wolf predation (Peek *et al.,* 1976).

Ectoparasites reported from moose include a tick, *Dermacentor albipictus* (from Isle Royale, Wilson and Johnson, 1971), the moose fly, *Lyperosiops alcis,* and two nasal botflies, *Cephenemyia jellisoni* and *C. phobifera.*

MOLT AND COLOR ABERRATIONS. In midspring (May) the moose molts its worn, heavy winter coat for a rich brown to almost black summer coat of guard hairs. In autumn, fine, twisted underfur appears along with more guard hairs (de Vos, 1956). An occasional white (albinistic) animal has been reported.

ECONOMIC IMPORTANCE. The long tenure of the moose in most of the Upper Peninsula and in all of the northern and eastern parts of the Lower Peninsula (see Map 67) ended abruptly as a result of human settlement in the nineteenth century. Moose have experienced similar difficulties in the southern part of its geographic range in the coniferous-deciduous ecotone along the Canadian-American border in eastern North America (Karns *et al.,* 1974). Human encroachment in this large border region has not only reduced moose habitat but has resulted in overhunting (at least in the early days before conservation programs could be enacted). Controlled hunts, however, have occurred in recent years in New Brunswick, Nova Scotia, Maine, and Minnesota without loss of basic breeding stock. No such hunts can be included in management plans for the sparse moose populations in Upper Peninsula

Michigan. Nevertheless, the future looks optimistic for the continuing moose population in the Upper Peninsula, probably being augmented by stock moving southward periodically from Ontario or from restocking programs. There is also a growing interest among Michigan citizens for the release of animals in suitable areas in the northern part of the Lower Peninsula. Ultimately, moose should be sufficiently abundant to attract tourists, or perhaps even hunters (Lowe, 1981). Whether such stocks will be adversely affected by the meningeal worm (see Parasites and Diseases) remains to be seen (Margaret Herman and John Stuht, pers. comm.).

Interesting experimental work is being done, mostly in the U.S.S.R., to domesticate this large cervid. Knorre (1974) reported that young individuals have been successfully tamed, and used to draw sleighs and wagons, as mounts and pack animals, and have even been milked. This work is continuing in the hopes that this large browsing animal might become a major provider of meat and dairy products for human populations in forested, boreal environments where grazing domesticated animals, such as cattle, may not thrive.

COUNTY RECORDS FROM MICHIGAN. Records of specimens of moose in collections of museums, from the literature, and reported by personnel of the Michigan Department of Natural Resources and the National Park Service show that this species occurred in historic times throughout the Upper Peninsula and in most counties of the northern and eastern parts of the Lower Peninsula (Map 67). Although specimens have rarely been saved, there are records of moose from the Lower Peninsula for at least eleven counties: Charlevoix, Cheboygan, Genessee, Huron, Kent, Leelanau, Midland, Missaukee, Oakland, Presque Isle, Saginaw, Sanilac, Washtenaw, and Wayne (Burt, 1946; Creaser, 1934; Hatt, 1924; Hickie, undated; Hoyt, 1889; Johnson, 1919; Lovis, 1973; Perry, 1899; Wilson, 1967; Wood and Dice, 1924). Today, moose are apt to be found in any of the counties in the Upper Peninsula, although the most stable residential population is in the eastern counties and, of course, on Isle Royale. Moose are also known from Sugar Island (Pruitt, 1961) and formerly from Drummond Island (Manville, 1950). In 1918, moose were unsuccessfully planted on Grand Island (Ruhl, 1940).

LITERATURE CITED

Addison, R. B., J. C. Williamson, B. P. Saunders, and D. Fraser
1980 Radio-tracking of moose in the boreal forest of northwestern Ontario. Canadian Field-Nat., 94(3):269–276.

Aldous, S. E., and L. W. Krefting
1946 The present state of moose on Isle Royale. Trans. Eleventh North Amer. Wildlife Conf., pp. 296–308.

Altmann, M.
1959 Group dynamics of Wyoming moose during the rutting season. Jour. Mammalogy, 40(3):420–424.

Anderson, R. C.
1972 The ecological relationships of meningeal worm and native cervids in North America. Jour. Wildlife Dis., 8:304–309.

Anderson, R. C., and M. W. Lankester
1974 Infectious and parasitic diseases and arthropod pests of moose in North America. Le Naturaliste Canadien, 101(1–2):23–50.

Banfield, A. W. F.
1974 The mammals of Canada. Univ. Toronto Press, Toronto. xxiv+438 pp.

Belovsky, G. E.
1978 Diet optimization in a generalist herbivore: The moose. Pop. Biol., 14:105–134.
1981 A possible population response of moose to sodium availability. Jour. Mammalogy, 62(3):631–633.

Belovsky, G. E., and P. A. Jordan
1981 Sodium dynamics and adaptations of a moose population. Jour. Mammalogy, 62(3):613–621.

Berg, W. E.
1976 Mortality of moose in northwestern Minnesota. Minnesota Wildlife Res. Quarterly, 36(1):1–10.

Brose, D. S.
1970 The archaeology of Summer Island: Changing settlement systems in northern Lake Michigan. Univ. Michigan, Mus. Anthro., Anthro. Papers No. 41, vii+238 pp.

Burt, W. H.
1946 The mammals of Michigan. Univ. Michigan Press, Ann Arbor. xv+288 pp.

Cleland, C. E.
1966 The prehistoric animal ecology and ethnozoology of the Upper Great Lakes Region. Univ. Michigan, Mus. Anthro., Anthro. Papers, No. 29, x+294 pp.
1968 Analysis of the fauna of the Indian Point Site on Isle Royale in Lake Superior. Michigan Arch., 14(3–4): 143–146.

Coady, J. W.
1974 Influence of snow on behavior of moose. Le Naturaliste Canadien, 101(1–2):417–436.

Creaser, C. W.
1934 The moose (*Alces americanus*) and the water shrew (*Sorex palustris hydrobodistes*), rare mammals of the southern peninsula of Michigan. Michigan Acad. Sci., Arts and Ltrs., Papers, 20:597–598.

Crete, M., and E. Audy
1974 Individual and seasonal variation in the diameter of browsed twigs by moose. North Amer. Moose Conf., 10:145–159.

de Vos, A.
1952 Ecology and management of fisher and marten in Ontario. Ontario Dept. Lands and Forests. Tech. Bull., 90 pp.
1956 Summer studies of moose in Ontario. Trans. Twenty-First North Amer. Wildlife Conf., pp. 510–525.
1958 Summer observations on moose behavior in Ontario. Jour. Mammalogy, 39(1):128–145.
1964 Range changes of mammals in the Great Lakes Region. American Midl. Nat., 71(1):210–231.

de Vos, A., and G. C. Armstrong
n.d. Aerial censusing of moose at Black Bay Peninsula, Ontario. Ontario Dept. Lands and Forests, Tech. Bull., Fish and Wildlife Ser. No. 3, 12 pp.

Dice, L. R. and H. B. Sherman
1922 Notes on the mammals of Gogebic and Ontonagon counties, Michigan, 1920. Univ. Michigan, Mus. Zool., Occas. Papers No. 109, 46 pp.

Dodds, D. G.
1958 Observations of pre-rutting behavior in Newfoundland moose. Jour. Mammalogy, 39(3):412–418.
1960 Food competition and range relationships of moose and snowshoe hare in Newfoundland. Jour. Wildlife Mgt., 24(1):52–60.

East, B.
1936 A moose herd is moved. Michigan transfers thirty-eight animals from Isle Royale to the Upper Peninsula mainland as a conservation measure. Bull. New York Zool. Soc., 39(4):141–149.

Eastman, D. S.
1974 Habitat use by moose of burns, cutovers and forests in northcentral British Columbia. North Amer. Moose Conf., 10:238–256.

Flynn, A., A. W. Franzmann, P. D. Arneson, and J. L. Oldemeyer
1977 Indications of copper deficiency in a subpopulation of Alaskan moose. Jour. Nutrition, 107(7):1182–1189.

Franzmann, A. W.
1981 Alces alces. American Soc. Mammalogists, Mammalian Species, No. 154, 7 pp.

Franzmann, A. W., R. E. LeResche, R. A. Rausch, and J. L. Oldemeyer
1978 Alaskan moose measurements and weights and measurement-weight relationships. Canadian Jour. Zool., 56(2):198–206.

Franzmann, A. W., C. C. Swartz, and R. O. Peterson
1980 Moose calf mortality in summer on the Kenai Peninsula, Alaska. Jour. Wildlife Mgt., 44(3):764–768.

Fraser, D., and H. Hristienko
1981 Activity of moose and white-tailed deer at mineral springs. Canadian Jour. Zool., 59(10):1991–2000.

Frenzel, L. D.
1974 Occurrence of moose in food of wolves as revealed by scat analyses: A review of North American studies. Le Naturaliste Canadien, 101(3–4):467–479.

Gasaway, W. C., D. B. Harkness, and R. A. Rausch
1978 Accuracy of moose age determinations from incisor cementum layers. Jour. Wildlife Mgt., 42(3):558–563.

Geist, V.
1963 On the behavior of the North American moose (*Alces alces andersoni* Peterson, 1950) in British Columbia. Behaviour, 20:377–416.
1974 On the evolution of reproductive potential in moose. Le Naturaliste Canadien, 101(3–4):527–537.

Goddard, J.
1970 Movements of moose in a heavily hunted area of Ontario. Jour. Wildlife Mgt., 34(2):439–445.

Hatt, R. T.
1924 Land vertebrate communities of western Leelanau County, Michigan. Michigan Acad. Sci., Arts and Ltrs, Papers, 3:369–402.

Haughe, T. M., and L. B. Keith
1981 Dynamics of moose populations in northeastern Alaska. Jour. Wildlife Mgt., 45(3):573–597.

Hickie, P. F.
1936 Isle Royale moose studies. Proc. North American Wildlife Conf., pp. 396–398.
1937 A preliminary report on the past and present status of the moose, *Alces americana* (Clinton), in Michigan. Michigan Acad. Sci., Arts and Ltrs., Papers, 22:627–639.
n.d. Michigan moose. Michigan Conserv. Comm., Game Div., 57 pp.

Holman, J. A.
1975 Michigan's fossil vertebrates. Michigan State Univ., Publ. Mus., Educ. Bull. No. 2, 54 pp.

Hosley, N. W.
1949 The moose and its ecology. United States Dept. Interior, Fish and Wildlife Serv., Wildlife Leaflet 312, 51 pp.

Hoyt, J. M.
1889 History of the town of Commerce. Michigan Pioneer Coll., 14:421–430.

Irwin, L. L.
1975 Deer-moose relationships on a burn in northeastern Minnesota. Jour. Wildlife Mgt., 39(4):653–662.

Jackson, H. H. T.
1961 Mammals of Wisconsin. Univ. Wisconsin Press, Madison. xii+504 pp.

Janzen, D. E.
1968 Excavations and survey at Burnt Bluff in 1965. Pp. 61–94 *in* J. E. Fitting. The prehistory of the Burnt Bluff area. Univ. Michigan, Mus. Anthro., Anthro. Papers No. 34, vi+140 pp.

Johnson, I. A.
1919 The Michigan fur trade. Michigan Hist. Comm., Lansing. xii+201 pp.

Johnson, W. L.
1970 Food habits of of the red fox in Isle Royale National Park, Lake Superior. American Midl. Nat., 84(2):568–572.

Jordan, P. A., D. B. Botkin, A. S. Dominski, H. S. Iowendorf, and G. E. Belovsky
1973 Sodium as a critical nutrient for moose of the Isle

Royale. Proc. North Amer. Moose Conf. Workshop, 9:1–28.

Jordan, P. A. D. B. Botkin, and M. L. Wolfe
1971 Biomass dynamics in a moose population. Ecology, 52:147–152.

Karns, P. D., H. Haswell, F. F. Gilbert, and A. E. Petton
1974 Moose management in the coniferous-deciduous ecotone of North America. Le Naturaliste Canadien, 101(3–4):643–656.

Kearney, S. R., and F. F. Gilbert
1976 Habitat use by white-tailed deer and moose on sympatric range. Jour. Wildlife Mgt., 40(4):645–657.

Kellum, F.
1941 Cusino's captive moose. Michigan Conserv., 19(7):4–5.

Kelsall, J. P.
1969 Structural adaptations of moose and deer for snow. Jour. Mammalogy, 50(2):302–310.

Kelsall, J. P., and E. S. Telfer
1974 Biogeography of moose with particular reference to western North America. Le Naturaliste Canadien, 101(1–2):117–130.

Knorre, E. P.
1974 Changes in the behavior of moose with age and during the process of domestication. Le Naturaliste Canadien, 101(1–2):371–377.

Krefting, L. W.
1951 What is the future of the Isle Royale moose herd? Trans. Sixteenth North American Wildlife Conf., pp. 461–472.
1974a Moose distribution and habitat selection in North Central North America. Le Naturaliste Canadien, 101(1–2):81–100.
1974b The ecology of the Isle Royale with special reference to the habitat. Univ. Minnesota, Agric. Exp. Sta., Tech. Bull. 297–1974, Forestry Ser. 15, 75 pp.

Lent, P. C.
1974 A review of rutting behavior in moose. Le Naturaliste Canadien 101(1–2):307–323.

LeResche, R. E.
1974 Moose migrations in North America. Le Naturaliste Canadien, 101(1–2):383–415.

Lovis, W. A., Jr.
1973 Late Woodland cultural dynamics in the northern Lower Peninsula of Michigan. Michigan State Univ. unpubl. Ph.D. dissertation, xi+346 pp.

Lowe, K. S.
1981 Michigan moose hunt? Michigan Out-of-Doors, 35(10):88.

Manville, R. H.
1948 The vertebrate fauna of the Huron Mountains, Michigan. American Midl. Nat., 39(3):615–640.
1950 The mammals of Drummond Island, Michigan. Jour. Mammalogy, 31(3):358–359.

Martin, T. J.
1979 Faunal remains from the Gros Cap Site (20MK6/7), Mackinac County, Michigan. App. E, pp. 133–148 *in* S. R. Martin. Final report: Phase II. Archaeological site examination of the Gros Cap Cemetery area, Mackinac County, Michigan. Michigan Tech. Univ., Cult. Res. Mgt. Rept. No. 6, ix+166 pp.
1980 Animal remains from the Winter Site, a Middle Woodland occupation in Delta County, Michigan. Wisconsin Arch., 61(1):91–99.

McMillan, J. F.
1953 Measures of association between moose and elk on feeding grounds. Jour. Wildlife Mgt., 17(2):162–166.

McNicol, J. G., and F. F. Gilbert
1980 Late winter use of upland cutovers by moose. Jour. Wildlife Mgt., 44(2):363–371.

Mech, L. D.
1966 The wolves of Isle Royale. National Park Serv., Fauna Ser. 7, xiv+210 pp.

Moll, D., and B. K. Moll
1976 Moose triplets on Isle Royale. Trans. Illinois State Acad. Sci., 69(2):151–152.

Murie, A.
1934 The moose of Isle Royale. Univ. Michigan, Mus. Zool., Misc. Publ., No. 25, 44 pp.

Peek, J. M.
1974 A review of moose food habits studies in North America. Le Naturaliste Canadien, 101(1–2):195–215.

Peek, J. M., L. W. Krefting, and J. C. Tappeiner, III
1971 Variation in twig diameter-weight relationships in northern Minnesota. Jour. Wildlife Mgt., 35(3):501–507.

Peek, J. M., D. L. Urich, and R. J. Mackie
1976 Moose habitat selection and relationships to forest management in northeastern Minnesota. Wildlife Monog., No. 48, 65 pp.

Perry, O. H.
1899 Hunting expeditions of Oliver Hazard Perry of Cleveland verbatim from his diaries. Privately printed, Cleveland, Ohio. viii+246 pp.

Petersen, E. T.
1979 Hunters' heritage. A history of hunting in Michigan. Michigan United Conser. Clubs, Lansing. 55 pp.

Peterson, R. L.
1953 Studies of the food habits and the habitat of moose in Ontario. Royal Ontario Mus., Contr. Zool. and Paleo., No. 36, 49 pp.
1955 North American moose. Univ. Toronto Press, Toronto. xi+280 pp.
1974 A review of the general life history of moose. Le Naturaliste Canadien, 101(1–2):9–21.

Peterson, R. O.
1977 Wolf ecology and prey relationships on Isle Royale. National Park Serv., Sci. Monog. Ser., No. 11, xx+210.

Phillips, R. L., W. E. Berg, and D. B. Siniff
1973 Moose movement patterns and range use in northwestern Minnesota. Jour. Wildlife Mgt., 37(3):266–278.

Pimlott, D. H.
1959 Reproduction and productivity of Newfoundland moose. Jour. Wildlife Mgt., 23(4):381–401.

Prescott, W. H.
1974 Interrelationships of moose and deer of the genus Odocoileus. Le Naturaliste Canadien, 101(3–4):493–504.

Pruitt, W. O., Jr.
1951 Mammals of the Chase S. Osborn Preserve, Sugar Island, Michigan. Jour. Mammalogy, 32(4):470–472.

Rheaume, P.
1980 Moose sighting in Chippewa County. Jack-Pine Warbler, 58(4):145.

Ruhl, H. D.
1940 Game introductions in Michigan. Trans. Fifth North American Wildlife Conf., pp. 424–427.

Saunders, B. P.
1973 Meningeal worm in white-tailed deer in northwestern Ontario and moose population densities. Jour. Wildlife Mgt., 37(3):327–330.

Schorger, A. W.
1942 Extinct and endangered mammals and birds of the Upper Great Lakes Region. Trans. Wisconsin Acad. Sci., Arts and Ltrs., 34:23–44.

Schwartz, C. C., W. L. Regelin, and A. W. Franzmann
1982 Male moose successfully breed as yearlings. Jour. Mammalogy, 63(2):334–335.

Sergeant, D. E., and D. H. Pimlott
1959 Age determination in moose from sectioned incisor teeth. Jour. Wildlife Mgt., 23(3):315–321.

Seton, E. T.
1953 Lives of game animals. Vol. III, Pt. I. Hoofed animals. Charles T. Branford Co., Boston. xix+412 pp.

Shelton, P. C.
1966 Ecological studies of beavers, wolves, and moose in Isle Royale National Park, Michigan. Purdue University, unpubl. Ph.D. dissertation, 308 pp.

Simkin, D. W.
1965 Reproduction and productivity of moose in northwestern Ontario. Jour. Wildlife Mgt., 29(4):740–750.

Singer, F. J.
1979 Habitat partitioning and wildfire relationships of cervids in Glacier National Park, Montana. Jour. Wildlife Mgt., 43(2):437–444.

Snyder, J. D., and R. A. Janke
1976 Impact of moose browsing on boreal-type forests of Isle Royale National Park. American Midl. Nat., 95(1): 79–92.

Stringham, S. F.
1974 Mother-infant relations in moose. Le Naturaliste Canadien, 101(1–2):325–369.

Telfer, E. S.
1967 Comparison of moose and deer winter range in Nova Scotia. Jour. Wildlife Mgt., 31(3):418–425.
1974 Vertical distribution of cervid and snowshoe hare browsing. Jour. Wildlife Mgt., 38(4):944–946.

Tustin, K. G.
1974 Status of moose in New Zealand. Jour. Mammalogy, 55(1):199–200.

Van Ballenberghe, V., and J. M. Peek
1971 Radiotelemetry studies of moose in northeastern Minnesota. Jour. Wildlife Mgt., 35(1):63–71.

Verme, L. J.
1970 Some characteristics of captive Michigan moose. Jour. Mammalogy, 51(2):403–405.

Wilson, N., and W. J. Johnson
1971 Ectoparasites of Isle Royale, Michigan. Michigan Ento., 4(4):109–115.

Wilson, R. L.
1967 The Pleistocene vertebrates of Michigan. Michigan Acad. Sci., Arts and Ltrs., Papers, 52:197–234.

Wolfe, M. L.
1969 Age determination in moose from cemental layers of molar teeth. Jour. Wildlife Mgt., 33(2):428–431.
1974 An overview of moose coactions with other animals. Le Naturaliste Canadien, 101(3–4):437–456.
1977 Mortality patterns in the Isle Royale moose population. American Midl. Nat., 99(2):267–279.

Wood, N. A.
1914 An annotated check-list of Michigan mammals. Univ. Michigan, Mus. Zool., Occas. Papers No. 4, 13 pp.
1917 Notes on the mammals of Alger County, Michigan. Univ. Michigan, Mus. Zool., Occas. Papers No. 36, 8 pp.

Wood, N. A., and L. R. Dice
1923 Records of the distribution of Michigan mammals. Michigan Acad. Sci., Arts and Ltrs., 3:425–469.

CARIBOU
Rangifer tarandus (Linnaeus)

NAMES. The generic name, *Rangifer,* proposed by Hamilton-Smith in 1827, is apparently derived from the Old French word *rangifère* meaning reindeer. The specific name, *tarandus,* proposed by Linnaeus in 1758, is from the Greek word *tarandos* also meaning reindeer. The common name for this large cervid in arctic areas and in the Old World is reindeer (Banfield, 1961), originally derived from the word *reino* which is the Lapp name for a young animal. In Michigan and adjacent areas, the common name is caribou, which is derived from the Micmac Indians' word for this mammal's snow-pawing ability (Wright, 1929). Caribou living in Michigan at the time of the first European settlement are classified as having belonged to the subspecies *Rangifer tarandus caribou* (Gmelin). In earlier postglacial times, it is evident that a subspecies belonging to the tundra form existed in Michigan (Banfield, 1961; Cleland, 1965; Hibbard, 1952).

RECOGNITION. Body size medium to large, with height at shoulders (withers) averaging 48 in (1,235 mm) for the largest male and 43 in (1,115 mm) for the smaller female; head large, rather bovine in appearance, with large squared snout not projecting beyond lower jaw (as in the moose); muzzle covered with short hair; nostrils valvular; ears small, broad, well-furred; eyes medium in size; preorbital (lachrymal) glands conspicuous; neck heavy, with underside maned; body stout, tail short and heavily furred; legs slender; tarsal and interdigital glands present; metatarsal glands absent; side toes (dewclaws) placed low on each foot and well developed; hoofs broader than long; in summer, enlarged, fleshy foot pads rest on the ground; in winter, horny rims enlarged so foot pads (well covered with thick hair tufts) are reduced and, as an adaptation for cold, do not rest on the ground. Both sexes grow slender palmate antlers. Female (doe), antlers may be merely spikes or have simple dichotomous branches with a short brow tine and a longer beam. Male (buck or bull) antlers are especially variable and rarely symmetrical (Fig. 116). From the burr on the frontal bone, the beam of the mature buck extends

Figure 116. Palmate antlers of the male caribou (*Rangifer tarandus*).

dorsally and slightly laterally, and then curves upward like a bow so that the tips usually point forward. At the base of this beam, a heavy brow tine points forward and slightly downward over the forehead and eyes where it expands as a compressed palm with small, terminally blunted points. Usually, one brow tine is greatly expanded as described, while its counterpart on the other antler may be only a simple prong. The second tine (bez) extends off the main beam a short distance above the brow tine and sweeps forward in an arc, flattened, palmate, and terminated by blunt tipped protuberances. One bez of the pair may be much more elaborately developed. Distal from the brow and bez tines, the heavy and somewhat flattened antler beam curves in a stylish arc, produces a short back tine, and terminates in a series of digitate or palmate tines. Contrary to popular belief, the flattened, shovel-like brow tine is not used to shovel or scrape snow.

The two sexes are similarily colored and show only slight seasonal differences in coat shade, being somewhat darker in fresh summer pelage and then fading to grayish in winter. The protective body covering consists of long, coarse, hollow, and air-filled guard hairs, and fine, wooly oily underfur. This combination provides a highly insulative pelage. The overall dorsal color ranges from dark cinnamon brown to brownish gray, with darker coloring on the face, chest, and dorsal side of tail. The neck and ventral mane are whitish to pale yellowish. The rump, underside of tail, and belly are white. The white on the feet above the hoofs stands out in contrast to the brown legs.

The caribou is larger than the white-tailed deer but smaller than the heavy moose. The caribou is distinguished from the other cervids in the Upper Great Lakes Region by its stylish, curved, and palmate antlers, the presence of antlers on does, and broad, rounded hoofs.

MEASUREMENTS AND WEIGHTS. Adult specimens of the larger male and the smaller female caribou measure, respectively, a follows: length from tip of nose to end of tail vertebrae, 72 to 88 in (1,840 to 2,260 mm) and 65 to 79 in (1,660 to 2,030 mm); length of tail vertebrae, 4⅜ to 5¾ in (110 to 150 mm) and 3⅞ to 5½ in (100 to 140 mm); length of hind foot, 21⅝ to 26⅛ in (550 to 670 mm) and 19½ to 23⅝ in (500 to 600 mm); height at shoulders (withers), 45 to 51 in (1,170 to 1,300 mm) and 41 to 46 in (1,060 to 1,170 mm); weight, 355 to 400 lbs (161 to 182 kg) and 195 to 260 lbs (89 to 118 kg). Large bucks may weigh as much as 600 lbs (273 kg).

DISTINCTIVE CRANIAL AND DENTAL CHARACTERISTICS. The skull of the caribou is large (Fig. 117); cranium is moderately expanded; rostrum is elongated; premaxillary bones do not extend posteriorly to reach nasals; antler pedicels

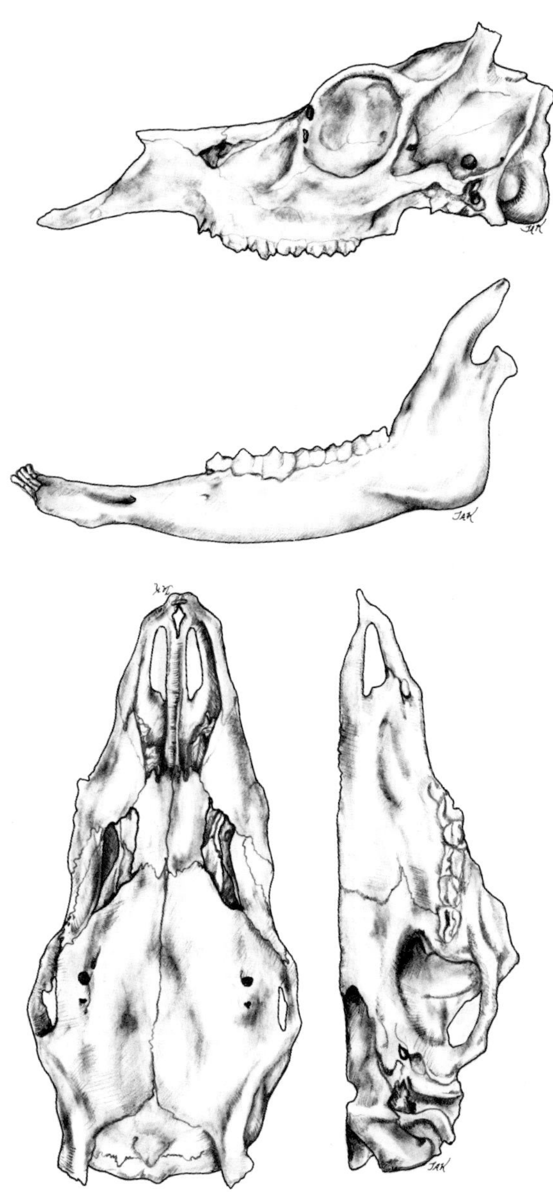

Figure 117. Cranial and dental characteristics of the caribou (*Rangifer tarnadus*).

are posteriorly placed well behind orbital cavities; nasals are characteristically expanded laterally at posterior ends; preorbital (lachrymal) vacuities are moderately large; width of each auditory bulla is only about half the length of the bony tube leading from it to the ear opening (external auditory meatus); and the opening of the posterior nares (viewed on underside of skull) is completely divided by the vertical vomerine plate. The dentition is characterized by small lower incisors; lower premolars and molars are only about half the diameter of upper ones; upper canines are usually present but are somewhat vestigeal and do not pierce the gum. The skulls of the larger adult male and the smaller adult female caribou, respectively, average 16½ in (425 mm) and 15⅛ in (388 mm) in greatest length and 7 in (180 mm) and 6¼ in (160 mm) in greatest width across the zygomatic arches.

The unsymmetrical, palmate antlers found on both sexes, the posterior position of the antler pedicels, the short, proximally-broad nasal bones, the partitioning of the cavity of the posterior nares, and the usual presence of vestigial upper canines are useful, in combination, to distinguish the skull of the caribou from those of other cervids in the Upper Great Lakes Region. The artificial key to the genera and species of Michigan Artiodactyla assists in identifying cranial material. The dental formula for the caribou is:

$$\text{I (incisors) } \frac{0\text{-}0}{3\text{-}3}, \text{ C (canines) } \frac{1\text{-}1}{1\text{-}1} \text{ or } \frac{0\text{-}0}{1\text{-}1}, \text{ P (premolars) } \frac{1\text{-}1}{3\text{-}3},$$
$$\text{M (molars) } \frac{3\text{-}3}{3\text{-}3} = 32 \text{ or } 34$$

DISTRIBUTION. The caribou and the conspecific reindeer have Eurasian-American circumpolar distribution (Banfield, 1961). At the time of the earliest settlement in North America, the natural range of this cervid extended from the arctic southward to beyond the Canadian-American border in most sectors except perhaps the Great Plains (Cringan, 1957). The southern populations, however, declined steadily in the eighteenth and early nineteenth centuries (Bergerud, 1974), with residual populations lingering longest in Maine, Minnesota, and the Rocky Mountain area (Bergerud, 1978; Nelson, 1947; Palmer, 1938). In Wisconsin and Michigan, the caribou was apparently never a common species (Schorger, 1942). Even so, remains (mostly antler parts) of caribou preserved in Michigan deposits show that the cervid inhabited the Upper Peninsula and northern and eastern parts of the Lower Peninsula in both prehistoric and historic periods (see Map 68).

There are indications that, in the earliest of postglacial times, the "tundra" form (Banfield, 1961) of the caribou-reindeer may have briefly inhabited Michigan in some of the bleak, barren environment adjacent to the melting edge of glacial ice. Tundra plants were growing in Lapeer County approximately 13,770 years B.P. (Farrand and Eschman, 1974). Fossils identified as probably belonging to this northern form of *Rangifer tarandus* include a foot bone (phalanx) recovered from a Paleo-Indian occupation dated at 11,200 years B.P. at the Holcomb Site in Macomb County (Cleland, 1965), an antler fragment found in substrate near Fowlerville in Livingston County (Hibbard, 1952), an antler fragment from a peat bog near Minden City in Sanilac County (Burt, 1942), and two antler fragments from a bog in Brandon Township (Sec. 9, T5N, R9E) in Oakland County (MSU 635VP).

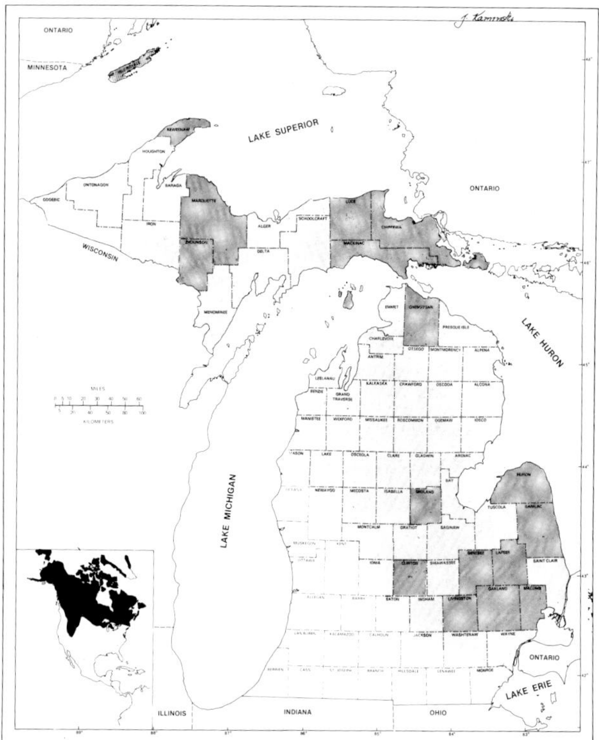

Map 68. Geographic distribution of the caribou (*Rangifer tarandus*). Supposed former distribution is marked with pale shading, with counties suppling actual records marked with dark shading.

The larger "forest" reindeer or woodland caribou apparently made its appearance in what is now Michigan concurrent with the growth of boreal coniferous forest environments. Prehistoric humans of Woodland Culture used these animals for food and other products, as shown by the discovery of caribou remains in association with their occupation areas, including the Juntunen Site in Mackinac County (Cleland, 1966) and the Indian Point Site on Isle Royale (Cleland, 1968). Woodland caribou material with American elm fragments from a bog near Flint in Genesee County was radio-carbon dated at 5,870 ± 200 years ago (Crane, 1956). Other antler fragments were found in Genesee County and also in Cheboygan County (Wilson, 1967). A massive antler fragment, also from the woodland form, was excavated from a bog in Ovid Township in Clinton County in 1958 (MSU 8713); another was obtained in Lapeer County (Mikula, 1964). The early people of Michigan were still hunting caribou long after the first European contacts as shown by an excerpt dated March 2, 1849, from the journal of John H. Pitezel. On that date, while visiting some Indians on Carp Lake in Chippewa County, he wrote, "One of the viands was a great treat—it was *Cariboo* meat, tender and sweet as any venison" (Schorger, 1940).

There are no data to indicate that caribou frequented Michigan's Lower Peninsula much after the middle of the nineteenth century (Miles, 1861). The only dated record is that of Strang (1855), who noted in his report on Beaver Island of December 7, 1853, that "Caribou or reindeer range as far south as here, but visit the islands only on the ice, and very rarely." In the Upper Peninsula, records indicate that caribou appeared as late as 1906 in Dickinson County (Schorger, 1942), and in Luce County (Wood and Dice, 1923) and as late as 1910, perhaps 1912, in Chippewa County (Wood, 1914; Wood and Dice, 1923). Last records for caribou in Michigan are for Isle Royale in 1926 (Mech, 1966).

The scant evidence available shows that caribou may have populated Michigan only during late autumn, winter, and early spring. This conclusion is based on (1) the finding of antlers (carried by both maturing sexes between October and February, when they are dropped, Banfield, 1974); (2) the comments of Strang (1955) that the caribou ranged as far south as Beaver Island on lake ice, indicating that its presence was perhaps only a wintertime phenomenon; (3) the records of early observers both in Wisconsin and northern Michigan listing no actual warm-weather dates for the presence of caribou later than April 18 for Wisconsin (1842 at Fond du Lac, Schorger, 1942) and March 2 for Upper Peninsula Michigan (1849 in Chippewa County, Schorger, 1940); (4) the close correlation of the disappearance of animals from the mainland of the Upper Peninsula in the first decade of the twentieth century with the disappearance of caribou in the Sault Ste. Marie Provincial Forest District of adjacent Ontario (de Vos and Peterson, 1951); and (5) the absence of winter records of caribou on Isle Royale after 1926 because of the decline of the caribou herd in the Port Arthur and Geraldton Forest Districts, adjacent on the south shore of Lake Superior (de Vos, 1964). These data plus the uncertainty expressed by such writers as Blois (1939), Mech (1966), and Schorger (1942) strongly indicate that the now-extirpated Michigan caribou herd might not have been residential but perhaps was visiting from more northern Canadian areas during cold weather. According to the First Biennial Report of the Michigan Department of Conservation for 1921–1922, 225 to 300 caribou on Isle Royale about 1920 may have been a migrating group which returned north.

Supposedly, the forest caribou, although less gregarious and less addicted to long migrational routes than the tundra caribou, was gradually discouraged in this southern sector. Hickie (undated) contended that, in the nineteenth century, there was indeed a succession of browsing animals in the Lake Superior region; the decline of the caribou left the area to the moose; the moose ultimately decreased and the white-tailed deer became the top herbivore (de Vos, 1964; de Vos and Peterson, 1951). Habitat changes and overkill resulting from human encroachment appear to be the principal causes for this succession (Cringan, 1967). The white-tailed deer may have survived successfully because of its greater adaptability in the presence of human settlement.

HABITAT PREFERENCES. Judging from what is known about the so-called "forest" caribou in more northern areas, the caribou which once lived or visited in northern Michigan presumably was closely associated with mature coniferous forests, intermixed with deciduous stands (especially birch) of uneven age, adjacent to swamps and bogs and containing scattered openings and abundant edge. Excessive fires and lumbering for sawlogs or pulpwood were the major factors reducing habitat quality of northern forests for caribou. In short, the caribou was unsuited to the extensive vegetational changes to which the moose and the white-tailed deer were more able to respond.

DENSITY AND MOVEMENTS. Unlike the "tundra" form of the caribou-reindeer, noted for both its gregariousness and migrational patterns, the "woodland" form tends to congregate in small groups and may not make major migrations (Fuller and Keith, 1981). Nevertheless, caribou in northern Michigan apparently did move about considerably. This is evidenced by caribou visits to Isle Royale from the Ontario mainland, and the report by Strang (1855) that caribou reached the Beaver Islands by traveling over winter ice. It is likely that, if caribou did indeed move south into Michigan from Ontario in sizable herds, they were vulnerable to overexploitation: 2,400 animals were reported slaughtered one winter on Manitoulin Island in Ontario (Cringan, 1957).

BEHAVIOR. The Upper Great Lakes Region "forest" form of the caribou frequented mature boreal woodlands, gathered in small groups, apparently required considerable territory for its activities, and probably moved north to south on a seasonal basis (Seton, 1953). During the autumn rut, mature males gather small harems of females. At other times of the year, bands may consist of does and their young or entirely of males. There are some indications, based on more thorough knowledge of the habits of the "tundra" form of the caribou-reindeer, that males assume leadership when bands move from summer to winter areas while pregnant does and their first-year offspring make the initial move to fawning sites in summer range (Kelsall, 1968).

The caribou is primarily diurnal, feeding in early morning and late afternoon, and resting at midday (Harper, 1955). Caribou walk slowly while feeding and may trot or gallop when in a hurry. Pruitt (1960) timed a walking doe and fawn at 5 mi (8 km) per hour; at full speed, animals can travel at more than 40 mi (65 km) per hour. When caribou move at high speed, a clicking noise is produced by tendons slipping over sesamoid bones in the feet. Caribou are excellent swimmers, buoyed somewhat by their long, thick coats.

This large cervid often makes grunting sounds when in a group. When alarmed, the animals may snort. Rutting bucks pant and bellow. The sense of smell appears most developed; caribou apparently can locate food beneath snow by olfaction (Bergerud, 1978). An erect tail stub appears to be a means of communication.

ASSOCIATES. The caribou apparently had a compatible environmental relationship with moose in the Upper Great Lakes Region. The former ordinarily prefers mature forests; the latter, second growth. The same can probably be said for environmental interactions between caribou and white-tailed deer for choices of foods and habitats. However, infestations by the meningeal worm (*Parelaphostrongylus tenuis*) may have been a factor in segregating the caribou and the moose from the white-tailed deer. Studies in Ontario (Anderson, 1972), have shown that this roundworm can be fatal to the caribou and the moose but does not affect the white-tailed deer.

REPRODUCTIVE ACTIVITIES. Each buck caribou of the forest group gathers as many as fifteen does into a harem. By contrast, the tundra caribou do not develop this close association in their large herds. Through October and November, the buck fends off other males by jousting and sparring, eats very little, and ultimately services his does as they become sexually receptive (Seton, 1953). The calves are born to more or less solitary does about 215 to 240 days later, from late May to mid-June.

Normally, a single calf (sometimes twins, McEwan, 1971) is born. The calf is able to stand shortly after birth, has crinkly pelage somewhat reddish brown and unspotted, has black eye-rings and muzzle, receives a good cleaning by licks from the doe, weighs as much as 11 lbs (5 kg), nurses within a few hours, and after several hours can follow along after the doe. Nursing may continue for more than 30 days (sometimes extending into the autumn) although the calf begins to eat solid food by 15 days of age. Growth during the first summer is rapid, although such factors as an overabundance of mosquitoes can seriously affect normal development (Krebs and Cowan, 1962). Females become sexually active in their second autumn (at 16 months of age); males may be excluded by older males from mating until their third year. The age of caribou can be determined by tooth wear, the growth and replacement of various jaw teeth, and the number of annulations in dental cementum (Miller, 1974a, 1974b). Individuals have been known to live for as long as 13 years in the wild, and about 20 years in captivity (Jones, 1979).

Antlers appear on both sexes in their second summer. In bucks, velvet knobs on the frontal bones first appear in March, with rapid growth taking place in midsummer. Full growth may be reached in August, although the beams remain somewhat flexible until ossification is complete in mid-September. By October, the antlers are cleaned of the dried velvet and look polished. The does develop their antlers somewhat later in the season, with growth beginning in June and maturity achieved in October. Bucks shed their antlers by February, while does may retain theirs until April or May. Does occasionally do not grow antlers.

FOOD HABITS. Forest caribou depend to a large extent on ground-growing and tree-growing lichens (Cringan, 1957; Holleman *et al.*, 1979). However, lichens are generally a major food supply for caribou only in late autumn, winter, and early spring. Beginning in late spring, caribou turn to green vegetation, including grasses, mushrooms, sedges, forbs, deciduous shrubs, ferns, and some aquatic plants (Banfield, 1974; Bergerud, 1972; Cringan, 1957). When deciduous forage matures and becomes more fibrous, caribou begin to rely on lichens, frost-killed sedges, and twigs of willow, red osier dogwood, mountain ash, and highbush cranberry. Caribou's broad hoofs are useful for pawing away snow from ground forage.

ENEMIES. Apart from the human, the gray wolf and the lynx appear to be the major predators of both the tundra and the forest forms of the caribou (Bergerud, 1974, 1978). Bergerud (1971) found lynx were apparently the major cause of mortality on calves in Newfoundland. Once caribou attain the age of six months, they are apt to have a much lower mortality rate.

PARASITES AND DISEASES. As stated previously, the meningeal worm (*Parelaphostrongylus tenuis*) can cause a fatal infestation in the caribou (Anderson, 1972). Caribou seem remarkably free of ectoparasites; a louse (*Linognathus tarandi*) is reported from Old World reindeer. Mosquitoes and black flies can be serious pests to caribou, causing major weight loss. Harper (1955) described the serious consequences of the larvae of the warble fly (*Oedemagena tarandi*) and the nostril fly (*Cephenemyia nasalis*).

MOLT AND COLOR ABERRATIONS. In midsummer, the caribou sheds the heavy, worn winter pelage in a patchwork sequence. It is replaced by a short, dark, sleek coat of guard hairs. In late September, a thick undercoat develops along with longer guard hairs. Occasionally, white (albinistic) individuals are reported (Curatolo, 1979).

ECONOMIC IMPORTANCE. The tundra form of the reindeer-caribou has received considerable study in the arctic areas of North America (Bergerud, 1978; Kelsall, 1968, etc.). This migratory animal provides food, clothing, bedding, and utensils for people of the arctic prairies. It is also a lure for southern hunters. The forest caribou has the same kind of game appeal. In early Michigan, the caribou presumably was hunted by the prehistoric people, perhaps when vulnerable during winter movements southward. Judging from early accounts by Europeans, the Michigan caribou may not have been sufficiently common to provide early settlers with a sustaining food supply.

To supplement local meat supplies for arctic people, domesticated reindeer were successfully imported from Siberia to Alaska beginning in 1891 and to Canada beginning in 1907 (Anon., 1940; Brady, 1968; Palmer, 1929). Michigan tried this scheme in 1922, when 60 reindeer (10 bucks and 50 does) were imported from Norway at a total estimated cost of $125,000 (Ruhl, 1940). The animals were first penned at the Mason Game Farm in Ingham County on March 27, 1922. After 12 births, the herd was shipped to Grayling, but did not thrive. One animal lived in the Detroit Belle Isle Zoo until 1927. Caribou brought to Grand Island (in Lake Superior off Munising in Alger County) about 1918 also failed to survive.

COUNTY RECORDS FROM MICHIGAN. Records of specimens of caribou in collections of museums, from the literature, and provided by personnel of the Michigan Department of Natural Resources and the National Park Service show that this now-extirpated species occurred in the postglacial period and into historic times in northern and eastern Michigan (See Map 68). There are records from the following Upper Peninsula Counties: Chippewa, Dickinson, Keweenaw, Luce, Mackinac, and Marquette (Burt, 1946; Cleland, 1966; Manville, 1948; Schorger, 1940; Wood, 1914, 1917; Wood and Dice, 1923); in the Lower Peninsula: Cheboygan, Clinton, Genesee, Huron, Lapeer, Livingston, Macomb, Midland, Oakland, and Sanilac (Burt, 1942; Cleland, 1965; Crane, 1956; Hibbard, 1952; Mikula, 1964; Wilson, 1967). Records of caribou have also been obtained on Beaver and High islands in Lake Michigan (Hatt *et al.*, 1948; Strang, 1855), on Isle Royale in Lake Superior (Cleland, 1968; Mech, 1966), and on Drummond Island in Lake Huron (Manville, 1950). Of the above records, all refer to the "woodland" form of the caribou (*Rangifer tarandus caribou*) except for those from Livingston, Macomb, Oakland, and Sanilac counties. Records from these four counties are for the "tundra" form which existed in earliest postglacial times in what is now Michigan.

LITERATURE CITED

Anderson, R. C.
1972 The ecological relationships of meningeal worm and native cervids in North America. Jour. Wildlife Dis., 8:304–309.

Anon
1940 Canada's reindeer. Canada Dept. Mines and Res., Lands, Parks and Forests Branch, 14 pp.

Banfield, A. W. F.
1961 A revision of the reindeer and caribou, genus *Rangifer*. Nat. Mus. Canada, Bull. No. 117, Biol. Ser., No. 66, vi+137 pp.
1974 The mammals of Canada. Univ. Toronto Press, Toronto. xxiv+438 pp.

Bergerud, A. T.
1971 The population dynamics of Newfoundland caribou. Wildlife Monog., No. 25, 55 pp.
1972 Food habits of Newfoundland caribou. Jour. Wildlife Mgt., 36(3):912–923.
1974 Decline of caribou in North America following settlement. Jour. Wildlife Mgt., 38(4):757–770.
1978 Caribou. Pp. 83–101 *in* J. L. Schmidt, and D. L. Gilbert, eds. Big game of North America. Ecology and management. Stackpole Books, Harrisburg, Pennsylvania. xv+494 pp.

Blois, J. T.
1839 Gazetteer of the State of Michigan. Sydney L. Rood & Co., Detroit, and Robinson, Pratt & Co., New York. xi+418 pp.

Brady, J.
1968 The reindeer industry of Alaska. Alaska Rev. Bus. and Econ. Cond., 5(3):1–20.

Burt, W. H.
1942 A caribou antler from the Lower Peninsula of Michigan. Jour. Mammalogy, 23(2):214.
1946 The mammals of Michigan. Univ. Michigan Press, Ann Arbor. xv+288 pp.

Cleland, C. E.
1965 Barren ground caribou (*Rangifer arcticus*) from an early man site in southeastern Michigan. American Antiquity, 30(3):350–351.
1966 The prehistoric animal ecology and ethnozoology of the Upper Great Lakes Region. Univ. Michigan, Mus. Anthro., Anthro. Papers, No. 29, x+294 pp.
1968 Analysis of the fauna of the Indian Point Site on Isle Royale in Lake Superior. Michigan Arch., 14(3–4): 143–146.

Crane, H. R.
1956 University of Michigan radio-carbon dates I. Science, 124:664–672.

Cringan, A. T.
1957 History, food habits and range requirements of the woodland caribou of continental North America. Trans. Twenty-Second North Amer. Wildlife Conf., pp. 485–500.

Curatolo, J. A.
1979 A sighting of an albino caribou in Alaska and review of North American records. Arctic, 32(4):374–375.

de Vos, A.
1964 Range changes of mammals in the Great Lakes Region. American Midl. Nat., 71(1):210–231.

de Vos, A., and R. L. Peterson
1951 A review of the status of woodland caribou (*Rangifer caribou*) in Ontario. Jour. Mammalogy, 32(3):329–337.

Farrand, W. R., and D. F. Eschman
1974 Glaciation of the Southern Peninsula of Michigan: A review. Michigan Acad., 7(1):31–56.

Fuller, T. K., and L. B. Keith
1981 Woodland caribou population dynamics in northeastern Alberta. Jour. Wildlife Mgt., 45(1):197–213.

Harper, F.
1955 The barren ground caribou of Keewatin. Univ. Kansas, Mus. Nat. Hist., Misc. Publ. No. 6, 164 pp.

Hatt, R. T., J. Van Tyne, L. C. Stuart, C. H. Pope, and A. B. Grobman
1948 Island life: A study of the land vertebrates of the islands of eastern Lake Michigan. Cranbrook Inst. Sci., Bull. No. 27, xi+179 pp.

Hibbard, C. W.
1952 Remains of the barren ground caribou in Pleistocene deposits of Michigan. Michigan Acad. Sci., Arts and Ltrs., Papers, 37:235–237.

Hickie, P. F.
n.d. Michigan moose. Michigan Dept. Conserv., Game Div., 57 pp.

Holleman, D. F., J. R. Luick, and R. G. White
1979 Lichen intake estimates for reindeer and caribou during winter. Jour. Wildlife Mgt., 43(1):192–201.

Jones, M. L.
1979 Longevity of mammals in captivity. Internat. Zoo News, 26(3):16–26.

Kelsall, J. P.
1968 The migratory barren-ground caribou of Canada. Canadian Wildlife Serv., Monog. No. 3, 340 pp.

Krebs, C. J., and I. McT. Cowan
1962 Growth studies of reindeer fawns. Canadian Jour. Zool., 40:963–969.

Manville, R. H.
1948 The vertebrate fauna of the Huron Mountains, Michigan. American Midl. Nat., 39(3):615–640.
1950 The mammals of Drummond Island, Michigan. Jour. Mammalogy, 31(3):358–359.

McEwan, E. H.
1971 Twinning in caribou. Jour. Mammalogy, 52(2):479.

Mech, L. D.
1966 The wolves of Isle Royale. National Park Serv., Fauna Ser. 7, xiv+210 pp.

Mikula, E. J.
1964 Evidence of Pleistocene habitation by woodland caribou in southern Michigan. Jour. Mammalogy, 45(3): 494–495.

Miles, M.
1861 A catalogue of the mammals, birds, and reptiles. Geological Survey of Michigan, First Biennial Rept., pp. 219–241.

Miller, F. L.
1974a Age determination of caribou by annulations in dental cementum. Jour. Wildlife Mgt., 38(1):47–53.
1974b Biology of the Kaminuriak population of barren-ground caribou. Part 2: Dentition as an indicator of age and sex; composition and socialization of the population. Canadian Wildlife Serv., Rept. Ser. No. 31, 88 pp.

Nelson, U. C.
1947 Woodland caribou in Minnesota. Jour. Wildlife Mgt., 11(3):283–284.

Palmer, L. J.
1929 Improved reindeer handling. United States Dept. Agric., Circ. No. 82, 17 pp.

Palmer, R. S.
1938 Late records of caribou in Maine. Jour. Mammalogy, 19(1):37–43.

Pruitt, W. O., Jr.
1960 Locomotor speeds of some large northern mammals. Jour. Mammalogy, 41(1):112.

Ruhl, H. D.
1940 Game introductions in Michigan. Trans. Fifth North American Wildlife Conf., pp. 424–427.

Schorger. A. W.
1940 A record of the caribou in Michigan. Jour. Mammalogy, 21(2):222.
1942 Extinct and endangered mammals and birds of the Upper Great Lakes Region. Trans. Wisconsin Acad. Sci., Arts and Ltrs., 34:23–44.

Seton, E. T.
1953 Lives of game animals. Vol. III, Pt. I. Hoofed Animals. Charles T. Branford Co., Boston. xix+412 pp.

Strang, J. J.
1855 Some remarks on the natural history of Beaver Island, Michigan. Smithsonian Inst., 9th Ann. Rept. for 1854, pp. 282–288.

Wilson, R. L.
1967 The Pleistocene vertebrates of Michigan. Michigan Acad. Sci., Arts and Ltrs., 52:197–234.

Wood, N. A.
1914 Results of the Shiras expeditions to Whitefish Point, Michigan—Mammals. Michigan Acad. Sci., 16th Ann. Rept., pp. 92–98.
1917 Notes on the mammals of Alger County, Michigan. Univ. Michigan, Mus. Zool., Occas. Papers No. 36, 8 pp.

Wood, N. A., and L. R. Dice
1923 Records of the distribution of Michigan mammals. Michigan Acad. Sci., Arts and Ltrs., 3:425–469.

Wright, A. H.
1929 The word "caribou." Jour. Mammalogy, 19(4):353–356.

FAMILY BOVIDAE

Permanent horns, usually present in both sexes, are a major feature of the approximately 110 species of even-toed ungulates belonging to 45 genera of Bovidae. This diverse family of hoofed animals is native to all major land masses except South America and Australia. Domestic cattle, sheep, and goats, have accompanied human occupation of all lands. Although most bovids are adapted to vast open lands of the several continents, many are efficient browsers and live in forest environments. Subsahara Africa has the largest population of these animals, ranging in size from the massive eland to the terrier-sized duiker. The horny sheath covering the bony horn core forms a variety of distinctive bovid projections, from the pointed weapon-like structures of the gemsbok to the dainty and inconspicuous horns of the klipspringer. Bovids lack incisors and canines (the latter sometimes just reduced) in their upper jaws; have generally high-crowned (hypsodont) premolars and molars in their upper jaws; and have cheek teeth adapted for chewing coarse fibrous grasses, forbs, and woody growth. Bovids share most of these features, as well as a complex (ruminating) stomach, with their close relatives, the cervids. The bison, extirpated in Michigan within historic times, is the only native representative of the bovid family. Fossilized remains of bones indicate that, in earlier postglacial times, another bovid, the extinct woodland muskox (*Symbos cavifrons*), also occurred in the state.

BISON
Bison bison (Linnaeus)

NAMES. Both the generic name, *Bison,* proposed by Hamilton-Smith in 1827, and the specific name, *bison,* proposed by Linnaeus in 1758, are derived from the Latin and mean wild ox or buffalo. In Michigan, this large bovid is called either the American buffalo or, more correctly, bison. Bison formerly living in Michigan belonged to the subspecies *Bison bison bison* (Linnaeus).

RECOGNITION. Body size massive, the largest Michigan mammal (Fig. 118), height at withers in larger male, 69 in (1,769 mm) and in smaller female, 60 in (1,538 mm); head broad, flattened, triangular in shape, carried low; nose bare; ears small, inconspicuous; horns extending laterally, upward, those of smaller female less heavy; neck heavy, thickly rising to the high shoulder hump, produced by elongated dorsal processes on the cervical vertebrae; behind shoulder hump, body slim, tapering to rounded rump; tail long, tufted; legs fairly short, hoofs rounded, cow-like; adult males may be 80% heavier and 50% longer than females.

Bison have long, wooly pelage consisting of a matted, thick undercoat covered over by long, coarse guard hairs. The coat is especially shaggy on the head, neck, shoulders and upper forelegs. The body covering posterior to the shoulders is much shorter and not wooly. The anterior parts are darker brown than the posterior body. A pronounced beard may be almost black. The bison differs markedly in external dimensions and form from other Michigan hoofed mammals, including domesticated ones.

MEASUREMENTS AND WEIGHTS. Adult specimens of the male and the female measure, respectively, as follows: length from tip of nose to end of tail vertebrae, 119 to 148 in (3,025 to 3,760 mm) and 77 to 89 in (1,955 to 2,260 mm); length of tail vertebrae, 21 to 32 in (535 to 815 mm) and 18 to 21 in (460 to 535 mm); length of hind foot, 23 to 26 in (585 to 660 mm) and 18 to 22 in (460 to 555 mm); height at shoulder (withers), 65 to 71 in (1,650 to 1,805 mm) and 55 to 63 in (1,400 to 1,600 mm); weight, 1,600 to 2,000 lbs (727 to 909 kg) and 900 to 1,100 lbs (409 to 500 kg).

DISTINCTIVE CRANIAL AND DENTAL CHARACTERISTICS. The skull of the bison is heavy, triangular in shape, and flattened (Fig. 119). It lacks preorbital (lachrymal) vacuities; the

Figure 118. The bison (*Bison bison*).
Sketches by Bonnie Marris.

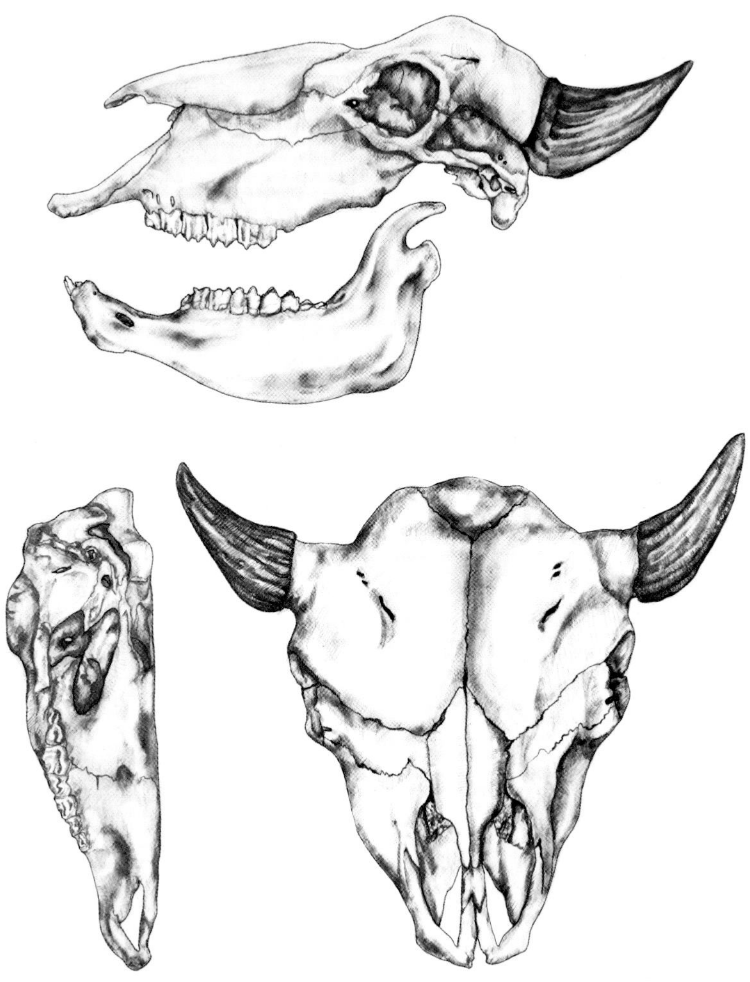

Figure 119. Cranial and dental characteristics of the bison (*Bison bison*).

lachrymal connects to the nasal; and the premaxillary does not extend upwards to contact the nasal. Incisors and canines are absent in the upper jaw; in the lower jaw the canines are incisiform (similar to incisors). The premolars and molars in both jaws are heavy and long, with the crescent-shaped cusp patterns designed for chewing coarse vegetation. The permanent, laterally extending bony horn cores (often the dark horny sheaths are lost in weathered specimens) distinguish bison skulls from those of Michigan cervids. Although the postcranial features of the bison, except for the elongated cervical vertebrae, may be similar to those of domestic cattle (Olsen, 1960), the massive and flattened skull with lateral horn cores, the heavier, longer upper molar tooth row, and the fact that the premaxillary does not contact the nasal are distinctive of the bison. The dental formula for the bison is:

$$\text{I (incisors)} \frac{0\text{-}0}{3\text{-}3}, \text{C (canines)} \frac{0\text{-}0}{1\text{-}1}, \text{P (premolars)} \frac{3\text{-}3}{3\text{-}3}, \text{M (molars)} \frac{3\text{-}3}{3\text{-}3} = 32$$

DISTRIBUTION. The bison has traditionally been known as a mammal of the western plains country, from central Alaska and the western interior of Canada south to northern Mexico. It also occurred in presettlement days in open forest habitat in eastern United States (see Map 69). The gradual extirpation of this massive creature from east to west began with its elimination on the Atlantic side of the Alleghenies by 1730. As settlement of the West continued, it disappeared east of the Mississippi River by about 1810, and then reduced in the remainder of its range to token

Map 69. Former geographic distribution of bison (*Bison bison*). Those counties from where actual records are known are marked with cross-hatched lines.

numbers by 1885. Conservation forces which emerged near the turn of the century carefully documented the bison's demise as a dismal example of wildlife misuse (Allen, 1876; Grinnell, 1892; Hornaday, 1889; McHugh, 1972; Rose, 1951). The bison is included in this report on Michigan mammals because of its occurrence in presettlement times in parts of the southern two tiers of counties in the Lower Peninsula (Map 69), in open country thought to be part of the eastward extending Prairie Province (Transeau, 1935). The earliest records are for the late seventeenth century (for what are now Berrien and Cass counties) within the portage area used by early French travelers moving between the Kankakee River to the south in Indiana and the St. Joseph River to the north in Michigan. Bison lived along the eastern bank of the St. Joseph River south of the present city of Niles in Berrien County and into what is now Indiana in an open area known as *parc aux vaches* or "park of the cows" (Baker, 1899; Engels, 1933). In 1702, Antoine de la Mothe Cadillac, founder of Detroit, wrote, "But 15 leagues from Detroit, at the entrance to Lake Erie, inclining to the south-southwest, are boundless prairies which stretch away for about 100 leagues. It is there that these mightly oxen [= bison], which are covered with wool, find food in abundance" (Burton, 1904). These animals disappeared from the Michigan scene no later than the end of the eighteenth century. Excavated skeletal remains of bison have been found near Norvell in Jackson County (Wood, 1922a), near Scotts in Kalamazoo County, and on Birchwood Beach in Berrien County (Wilson, 1967; Wood, 1922b). A flat bag made of spun bison hair found at an archaeological site in Oceana County (Quinby, 1966) may have been transported there from the south or west. It is also likely that bison could have ranged into the western part of the Upper Peninsula (Martin and Rhead, 1980).

HABITAT PREFERENCES. The paucity of records for bison from southern Michigan allows only educated guesses as to its habitat preferences in this area. Natural and oak openings (Butler, 1978; Veatch, 1927) extending eastward and northward as part of the Prairie Province (Transeau, 1935) from the Illinoian Biotic Province (Dice, 1943) may have attracted bison, perhaps only seasonally, to the Michigan area. Open marshes and park-like stands of deciduous forests also provided grazing areas.

BEHAVIOR. The bison is best known as a denizen of the western plains where immense herds once moved seasonally from one foraging area to another. Southern Michigan may have attracted smaller groups, perhaps family bands containing no more than 15 to 20 individuals. Bulls generally dominate such groups during the mating period. At other seasons cows with their calves and yearling offspring maintain these associations, while bulls become largely solitary. Various group interactions include play, especially by younger individuals, wallowing in dust or mud, milling to discourage biting insects, and the aggressive parcelling of foraging sites (Fuller, 1961; McHugh, 1958; Meagher, 1978).

The bison's senses of sight and smell are highly developed. Its gait is a slow, cow-like, plodding walk. When in a hurry, the bison can trot, or attain speeds of more than 30 mi (48 km) per hour in a lumbering, swaying gallop. They are good swimmers and rarely have difficulty crossing streams and small lakes. They usually forage during the morning and afternoon with midday reserved for rest and cud chewing. Bison may be active on moonlight nights, although normally they remain quiet through the darkness.

Bison are noisy in their group actions (Gunderson and Mahan, 1980). Bulls bellow, especially during the courtship period; cows are more apt to snort and cough. Various visual signals have been identified, such as head-on aggressive moves to assert dominance, broadside displays, threatening nods and head-away submissive postures.

REPRODUCTIVE ACTIVITIES. Bulls normally associate little with cows except during the rut which may begin in late July or early August and continue until at least late September. Fully mature bulls, generally between seven and eight years old, are the breeders, having the strength and aggressiveness to discourage younger (and older) males. Although bulls do not attempt to develop harems, they are attentive to receptive cows, completing their association with one before going to another (Meagher, 1978).

After a gestation period of 270 to 300 days, a single calf (rarely twins) is born between late April and early June. On the day of birth, the calf stands almost immediately on its rather weak legs, nurses within two hours after birth, appears short-necked and spider-legged, has matted pelage somewhat red-brown in color, and in four or more hours is able to follow the cow. At seven days of age, the calf begins to eat grassy foods but may not be totally weaned for several months. Females occasionally breed as yearlings, although many do not mature sexually until their third or fourth year. Males can also mature as yearlings but probably have little opportunity to mate until they are several years older (Meagher, 1978). Bison survive as long as 26 years in captivity (Jones, 1979). Age of individual animals can be estimated by horn growth and by eruption, replacement, and wear of the teeth (Fuller, 1959; Seton, 1953).

FOOD HABITS. This large bovid is chiefly a grazer, eating various grasses, forbs, sedges, and other ground forage. Prairie plants growing on openings in southern Michigan in presettlement days were undoubtedly most available to bison in summer. Although bison can paw away snow cover to find food, Michigan animals might well have moved south to Indiana and Illinois to reach winter foraging areas where snow cover was lighter.

ENEMIES. The gray wolf, mountain lion, and perhaps the black bear were the only top carnivores capable of molesting bison in southern Michigan. In addition to predation, bison were lost through drownings, prairie fires, and freezing. Major mortality in the herd was probably among first year animals.

PARASITES AND DISEASES. In modern times, bison have become afflicted with many of the diseases (*e.g.*, brucellosis) and parasites of domestic cattle. Ticks of the genus *Dermacentor* become pests in the matted pelage. Botfly larvae also attack bison. Internally, the animals may be infested with flukes, tapeworms, and roundworms, including lungworms (*Dictyocaulus hadweni* and *Dictyocaulus viviporus*), stomach worms (*Haemonchus contortus*), and intestinal worms (*Oesophagostomum radiatum*). Lungworms can be fatal in bison (Jackson, 1961).

MOLT AND COLOR ABERRATIONS. Bison undergo two molts each year, one in spring and another in autumn. Occasionally, albinistic (white) and melanistic (dark) individuals are reported (Banfield, 1974).

ECONOMIC IMPORTANCE. Much of the bison range in presettlement times included land with the best soil, now used for growing corn and wheat and grazing domestic livestock. It was a simple matter of replacing a major native resource with human introduced ones. Although early people may have had little opportunity to hunt bison in southern Michigan (Baker, 1899; Petersen, 1979), Michigan merchants were active in the early trade in bison hides, which probably reached them from more western areas via Great Lakes boat traffic. In 1767, British traders at Michilimackinac shipped only 84 bison pelts to foreign customers (Petersen, 1979). In 1822, the John Jacob Astor's fur trading post on Mackinac Island was paying $3.00 for buffalo robes (Anon., 1962). At the height of the western bison slaughter, the Michigan Carbon Works of Detroit, as eloquently described by Barnett (1975), expanded its operation shortly after the firm's founding in 1873 to accommodate the processing of huge shipments of bison bones and hoofs from western states. Bone-char, neat's-foot oil, fertilizer, and other commercial by-products were manufactured. By 1885, this flourishing business was processing about 30 tons of bones and hoofs daily. The supply of bison remains was finally exhausted by 1896. Nevertheless, as Barnett (1975) pointed out, ". . . the contribution of the buffalo to late 19th century Detroit was preserved in the records of the times, and it proved a very significant part of the city's life before the coming of the automobile."

By 1885, there was no space left for the bison except in a few preserves and parks and in the subarctic. A wild source of protein which had nourished native peoples, gray wolves, bears, and mountain lions for uncountable millennia had been depleted. For humans, this also meant the loss of other bison products: skins for clothing, shelter, shields, thong straps and ropes, bedding, and burial garments; horns and bones for tools, ornaments, and weapons; sinews for bowstrings; tallow for waterproofing, leather dressing, medicines, and cooking; and even dung for campfire fuel. Today, token numbers of this once important natural resource remain in scattered western and northern refuges, and zoological gardens. Some ranchers maintain a few animals, occasionally selling surplus ones for slaughter. Several bison meat processors operate in Michigan today (Hammer, 1980).

COUNTY RECORDS FROM MICHIGAN. Records of specimens of bison in collections of museums and from the literature show that this long extirpated species occurred in the following counties: Berrien, Cass, Jackson, Kalamazoo, Monroe, and Wayne (Baker, 1899; Burton, 1904; Wood, 1914, 1922a, 1922b).

LITERATURE CITED

Allen, J. A.
1876 The American bisons, living and extinct. Geol. Serv. Kentucky, Memoir, Vol. I, Pt. II, ix+246 pp.

Anon.
1962 Fur and hide values, 1822. Michigan Conserv., 31(6): 36–37.

Baker, G. A.
1899 The St. Joseph-Kankakee Portage. Northern Indiana Hist. Soc., Publ. No. 1, 48 pp.

Banfield, A. W. F.
1974 The mammals of Canada. Univ. Toronto Press, Toronto. xxiv+438 pp.

Barnett, L. R.
1975 Bones on the Rouge. Michigan Nat. Res., 44(4):10–13.

Burton, C. M., ed.
1904 Cadillac papers. Description of Detroit; advantages found there. Michigan Pioneer and Hist. Coll., 33: 133–151.

Butler, A. F.
1978 Prairies lost, pairies found. Michigan Nat. Res., 47(3): 32–39.

Dice, L. R.
1943 The Biotic Provinces of North America. Univ. Michigan Press, Ann Arbor. viii+78 pp.

Engels, W. L.
1933 Notes on the mammals of St. Joseph County, Indiana. American Midl. Nat., 14(1):1–16.

Fuller, W. A.
1959 The horns and teeth as indicators of age in bison. Jour. Wildlife Mgt., 23(3):342–344.
1961 The biology and management of the bison of West Buffalo National Park. Canadian Wildlife Serv., Mgt. Bull., Ser. 1, No. 16, 52 pp.

Grinnell, G. B.
1892 The last of the buffalo. Schribner's Mag., 12(3):267–286.

Gunderson, H. L., and B. R. Mahan
1980 Analysis of sonagrams of American bison (*Bison bison*). Jour. Mammalogy, 61(2):279–281.

Hammer, M. B.
1980 Michigan's buffalo doctor. Michigan Out-of-Doors, 34(2):32–33, 52.

Hornaday, W. T.
1889 The extermination of the American bison, with a sketch of its discovery and life history. United States Natl. Mus., Rept. 1887, Pt. 2, pp. 367–548.

Jackson, H. H. T.
1961 Mammals of Wisconsin. Univ. Wisconsin Press, Madison. xii+504 pp.

Jones, M. L.
1979 Longevity of mammals in captivity. Internat. Zoo News, 26(3):16–26.

Martin, T. J., and D. K. Rhead
1980 Environment and substance at Sand Point. Michigan Arch., 26(3–4):17–24.

McHugh, T.
1958 Social behavior of the American buffalo (*Bison bison bison*). Zoologica, 43:1–40.
1972 The time of the buffalo. Alfred A. Knopf, New York. xxiv+339+xi pp.

Meagher, M. M.
1978 Bison. Pp. 123–133 *in* J. L. Schmidt and D. L. Gilbert, eds. Big game of North America. Ecology and man-

agement. Stackpole Books, Harrisburg, PA. xv+494 pp.

Olsen, S. J.
1960 Post-cranial skeletal characters of *Bison* and *Bos*. Harvard Univ., Peabody Mus. Arch. and Ethnogr., 35(4), vii+15 pp.

Petersen, E. T.
1979 Hunters' heritage. A history of hunting in Michigan. Michigan United Conserv. Clubs, Lansing. 55 pp.

Quimby, G. I.
1966 The Dumaw Creek Site: A seventeenth century prehistoric Indian village and cemetery in Oceana County, Michigan. Fieldiana: Anthropology, 56(1):1–91.

Roe, F. G.
1951 The North American buffalo. A critical study of the species in the wild state. Univ. Toronto Press, Toronto. xi+991 pp.

Seton, E. T.
1953 Lives of game animals. Vol. III, Pt. II. Hoofed animals. Charles T. Branford Co., Boston, pp. 413–780.

Transeau, E. N.
1935 The prairie peninsula. Ecology, 16(3):423–437.

Veatch, J. O.
1927 The dry prairies of Michigan. Michigan Acad. Sci., Arts and Ltrs., Papers, 8:269–278.

Wood, N. A.
1914 An annotated check-list of Michigan mammals. Univ. Michigan, Mus. Zool., Occas. Papers No. 4, 13 pp.
1922a The mammals of Washtenaw County, Michigan. Univ. Michigan, Mus. Zool., Occas. Papers No. 123, 23 pp.
1922b Notes on the mammals of Berrien County, Michigan. Univ. Michigan, Mus. Zool., Occas. Papers No. 124, 4 pp.

APPENDIX

FUTURE STUDIES OF MICHIGAN MAMMALS

The student of mammalogy constantly looks forward to addressing the array of unanswered questions which are revealed in the course of investigations. Certainly there will never be a time when the last word concerning Michigan's mammals will be written. Naturally, the deeper one probes into the habits of these creatures, the more questions arise. This then is one major objective of this compilation—to stimulate greater inquiry about mammals, their attributes, life styles, and environmental relationships. Following are a few suggested directions for future investigations:

GEOGRAPHIC DISTRIBUTION. Basic distributional patterns of Michigan mammals were mostly well understood by the time of William H. Burt's landmark report (1946). Consequently, today there remains only the need to fill a few minor gaps. Exact ranges are still incomplete, for example, for such species as pygmy, water, and least shrews, some species of bats, and woodland vole in the Lower Peninsula. There is also need to watch for spatial changes in mammalian distributions through time, including west to east movement in the Upper Peninsula of such species as Virginia opossum, eastern cottontail, fox squirrel, prairie subspecies of deer mouse, and least weasel. Also requiring monitoring are distributions in the Lower Peninsula of the prairie vole in southwestern counties and of Virginia opossum and least weasel, for example, as they perhaps expand their ranges into the northern part of the Lower Peninsula. It is anticipated that mammalogists will ultimately obtain evidence of the presence of Franklin's ground squirrel and small-footed bat in the state. The likelihood of the entry of such mammals, previously unreported from either or both peninsulas, has been certainly strengthened by the recent discovery that the smoky shrew has spread from mainland Ontario to Michigan's Sugar Island (Master, 1982).

SPECIATION. We understand only partly the effects of water barriers of the Great Lakes system on speciation of weak-swimming terrestrial mammals. Michigan area mammalian subspecies were diagnosed at a time when only gross morphologic and color distinctions were considered. This evidence of speciation needs re-evaluation supplementing appraisals of the above features with those based on karyological and biochemical characteristics (Ayala, 1976; Smith and Joule, 1981; White, 1977) to assess these physical isolating factors influencing genetic attributes. Such studies might incorporate examinations of terrestrial species having populations living, for example, on (1) both sides of the Straits of Mackinac or (2) on Beaver Island and the adjacent mainland of either the Upper or Lower Peninsulas. More ambitious studies might investigate genetic features of samples of populations of mammals living entirely around the borders of the Great Lakes system not only in Michigan but in Indiana, Illinois, Wisconsin, Minnesota, Ontario, Quebec, Ohio, Pennsylvania, and New York as well. Excellent subjects for such examinations could include masked shrew, short-tailed shrew, woodchuck, gray squirrel, eastern chipmunk, deer mouse, southern bog lemming, meadow vole, meadow jumping mouse, red fox, long-tailed weasel, striped skunk, and even white-tailed deer. Michigan with its temperate and boreal habitats sharply dissected by water barriers is a major focal point in the Great Lakes Region for examining aspects of mammalian speciation by applying traditional and modern methodology. A truly important added dimension is our growing awareness of the geological newness of this environment. We not only have estimates of the time which has elapsed since the meltback of the last glaciation but also when modern environments appeared. Establishing actual timetables for evolutionary changes in area mammals could very well be significant results of these inquiries.

ONTOGENETIC, BEHAVIORAL, AND ECOLOGIC STUDIES. Michigan mammalogists have published many findings from their detailed studies of life history, behavior, physiology, and environmental requirements of both individual species and of communities of mammals. Notable have been the contributions, to mention a few, of Allen (1938, 1943), Blair (1940, 1948), Burt (1940, 1943), Erickson *et al.* (1964), Fitch (1979), Getz (1960), Heidt (1970), Hill (1972), Linduska (1950), Manville (1949), McCullough (1979), Moran (1973), Muul (1968), Nicholson (1941), Stuewer (1943), and Verme and Ozoga (1980). These works and many others have certainly set examples for future work. Of basic importance is the need for further detailed appraisals of food habits of mammals. This often neglected area of study requires meticulous investigation in the sorting out and identifying of insect parts, finely masticated plant tissues, and other food fragments from stomach contents and feces. Such data assist us in understanding nutrient requirements of species as well as how food supplies may be apportioned among several associated species of mammals using the same environments. Without these intimate details, studies of alleged competition between different mammals can surely be stymied. Outdoor laboratories in Michigan provide numerous opportunities for investigating resource uses and of other environmental interactions between such mammals as masked and pygmy shrews, eastern and least chipmunks, meadow and prairie voles, meadow and woodland jumping mice, red and gray foxes, ermine and long-tailed and least weasels, and bobcat and lynx. Modern ecologic, behavioral, and physiological research designs augmented by the use of remote-sensing, data-gathering, electronic devices make such studies most challenging. An added feature of the Michigan area is the presence of species-improverished mammalian populations on islands in the adjacent Great Lakes. Comparison of life styles of these mammals with those on fully-populated adjacent mainland areas can provide insights about species inter-relationships essential to the understanding of mammalian community interactions. Hopefully this brief and surely not all inclusive discussion will serve to illustrate the array of meaningful and exciting investigations yet to be made.

ECONOMICS. There is every evidence that human-dominated environments of Michigan will become progressively less hospitable to many other resident mammals. Present refuges augmented by others in the future will perserve strategically-located natural environments for posterity. However, these areas might very well become isolated by suburbia encroachment, at least in southern Michigan, ecologically separating small populations of both game and none-game species. For some of the species, however, we have yet to determine accurately minimum space requirements essential for sustainable populations. For example, just how large must a southern Michigan woodlot be in order to provide adequately for the survival of fox squirrels, or even white-footed mice? To be sure, small, secretive shrews, rodents, and others co-exist in habitats maintained for the highly visible and important Michigan deer herd. However, it can not be truly said that those environmental factors sustaining the white-tailed deer will also do the same for all other mammals. Nevertheless, the diverse environmental requirements necessary for deer to thrive will perhaps go a long way in preserving living space for Michigan's small mammals. Finally, there must be much more information acquired about successional preferences of Michigan mammals—in old field, brush lands, and several stages of forest growth. Wisely, personnel of the Michigan Department of Natural Resources are joining with those from educational institutions and federal and private agencies to monitor the population status of both small and large species, having concern for their welfare and economic importance even though many Michigan mammals are inconspicuous and do not qualify as game or tourist attractions.

LITERATURE CITED

Allen, D. L.
1938 Ecological studies on the vertebrate fauna of a 500-acre farm in Kalamazoo County, Michigan. Ecol. Monog., 8(3):347–436.
1943 Michigan fox squirrel management. Michigan Dept. Conserv., Game Div., publ. 100, 404 pp.

Ayala, F. J., ed.
1976 Molecular evolution. Sinaver Associates, Sunderland, Mass. 277 pp.

Blair, W. F.
1940 Notes on the home ranges and populations of the short-tailed shrew. Ecology, 20(3):284–288.
1948 Population density, life span, and mortality rates of small mammals in the blue-grass meadow and blue-grass field associations of southern Michigan. American Midl. Nat., 40(2):395–419.

Burt, W. H.
1940 Territorial behavior and populations of some small mammals in southern Michigan. Univ. Michigan, Mus. Zool., Misc. Publ. No. 45, 58 pp.
1943 Territoriality and home range concepts as applied to mammals. Jour. Mammalogy, 24(3):346–352.
1946 The mammals of Michigan. Univ. Michigan Press, Ann Arbor. xv+288 pp.

Erickson, A. W., J. Nellor, and G. A. Petrides
1964 The black bear in Michigan. Michigan State Univ., Agri. Exp. Sta., Res. Bull. 4, 102 pp.

Fitch, J. H.
1979 Patterns of habitat selection and occurrence in the deermouse *Peromyscus maniculatus gracilis*. Michigan State Univ., Publ. Mus., Biol. Ser., 5(6):443–484.

Getz, L. L.
1960 A population study of the vole, *Microtus pennsylvanicus*. American Midl. Nat., 64:392–405.

Heidt, G. A.
1970 The least weasel *Mustela nivalis* Linnaeus. Development biology in comparison with other North American *Mustela*. Michigan State Univ., Publ. Mus., 4(7): 227–282.

Hill, R. W.
1972 The amount of maternal care in *Peromyscus leucopus* and its thermal significance for the young. Jour. Mammalogy, 53(4):774–790.

Linduska, J. P.
1950 Ecology and land-use relationships of small mammals on a Michigan farm. Michigan Dept. Conserv., Game Div., ix+144 pp.

Manville, R. H.
1949 A study of small mammal populations in northern Michigan. Univ. Michigan, Mus. Zool., Misc. Publ. No. 73, 83 pp.

Master, L. L.
1972 Smoky shrew. A new mammal for Michigan. Jack-Pine Warbler, 60(1):28–29.

McCullough, D. R.
1979 The George Reserve deer herd. Population ecology of a k-selected species. Univ. Michigan Press, Ann Arbor. xii+271 pp.

Moran, R. J.
1973 The Rocky Mountain elk in Michigan. Michigan Dept. Nat. Res., Wildlife Div., R. and D. Rept. No. 267, x+93 pp.

Muul, I.
1968 Behavior and physiological influences on the distribution of the flying squirel, *Glaucomys volans*. Univ. Michigan, Mus. Zool., Misc. Publ. No. 134, 66 pp.

Nicholson, A. J.
1941 The homes and social habits of the woodmouse *Peromyscus leucopus novaboracensis* in southern Michigan. American Midl. Nat., 25(1):196–223.

Ryan, J. M.
1982 Distribution and habits of the pygmy shrew, *Sorex (Microsorex) hoyi* in Michigan. Jack-Pine Warbler, 60(2):85–86.

Smith, M. H., and J. Joule, eds.
1981 Mammalian population genetics. Univ. Georgia Press, Athens. 380 pp.

Stuewer, F. W.
1943 Raccoons: Their habits and management in Michigan. Ecol. Monog., 13(2):203–258.

Verme, L. J., and J. J. Ozoga
1980 Influence of protein energy intake on deer fawns in autumn. Jour. Wildlife Mgt., 44(2):305–314.

White, M. J. D.
1977 Animal cytology and evolution. Cambridge Univ. Press, Cambridge, 961 pp.

GLOSSARY

Major features of the mammalian skull and lower jaw, as mentioned in text, are shown in Fig. 120.

Abdominal—Of or pertaining to that part of the body (except the back) between the thorax and the pelvis.

Abdominal pouch—Fur-lined pouch on belly of female Virginia opossum, for carrying newborn young.

Abomasum—The last of the four chambers of the ruminant "stomach," the true stomach of a ruminant.

Aestivation (estivation)—A form of torpor, usually a response to high temperature or scarcity of water.

Albino—Congenital absence of all pigment in hair, skin, eyes, and nails. Thus, the animal's hair is white and the skin and eyes are pink.

Alisphenoid—Pertaining to or designating the right or left bone in front of the squamosal bone and behind the orbitosphenoid bone.

Allopatry—The situation in which two or more populations of species occupy mutually exclusive, but often adjacent, geographic ranges.

Altricial—Young that are relatively poorly developed at birth, usually have scant hair, eyes closed, and little locomotor ability.

Alveolus (plural, alveoli)—A small cavity or pit, as a socket for a tooth.

Amino acids—The primary source of material for growth in animals.

Amnion—The innermost membrane, filled with watery fluid, that encloses the developing embryo.

Anestrus—The quiescent period of the female reproductive cycle, followed by estrus.

Angular process—In lower jaw, posteroventral process below the articular process.

Anterior—Before or toward the front, in place.

Anterior naris (plural, nares)—Opening, out of the skull, of the nasal cavities.

Anterior palatine foramen (plural, foramina)—The incisive foramen, opening through the bony roof of the mouth at the juncture of the premaxillary and maxillary bones.

Antibrachial—Membrane leading from the shoulder to the wrist, on the leading edge of a bat's wing.

Antler—Bony growth (erroneously called horn) on head of deer, wapiti, moose, and caribou shed annually.

Antorbital (process) projection—Forward extended arm of the supraorbital process of the frontal bone in rabbits.

Apocrine gland—A type of gland in which only the apical part of the cell from which the secretion is released breaks down in the process of secretion.

Arid—Describes a habitat with a scanty supply of water (*cf.* mesic).

Auditory bulla (plural, bullae)—A hollow, bony prominence of rounded form (usually formed of the tympanic bone) partly enclosing structures of the middle and inner ear.

Auditory meatus—As used herein, the external auditory meatus which is the opening leading from the external ear to the eardrum (tympanic membrane).

Axilla—The armpit.

Baculum (plural, bacula)—The os penis or bone in the penis (male copulatory organ) of several, but not all, kinds of mammals.

Basioccipital—Pertaining to or designating the

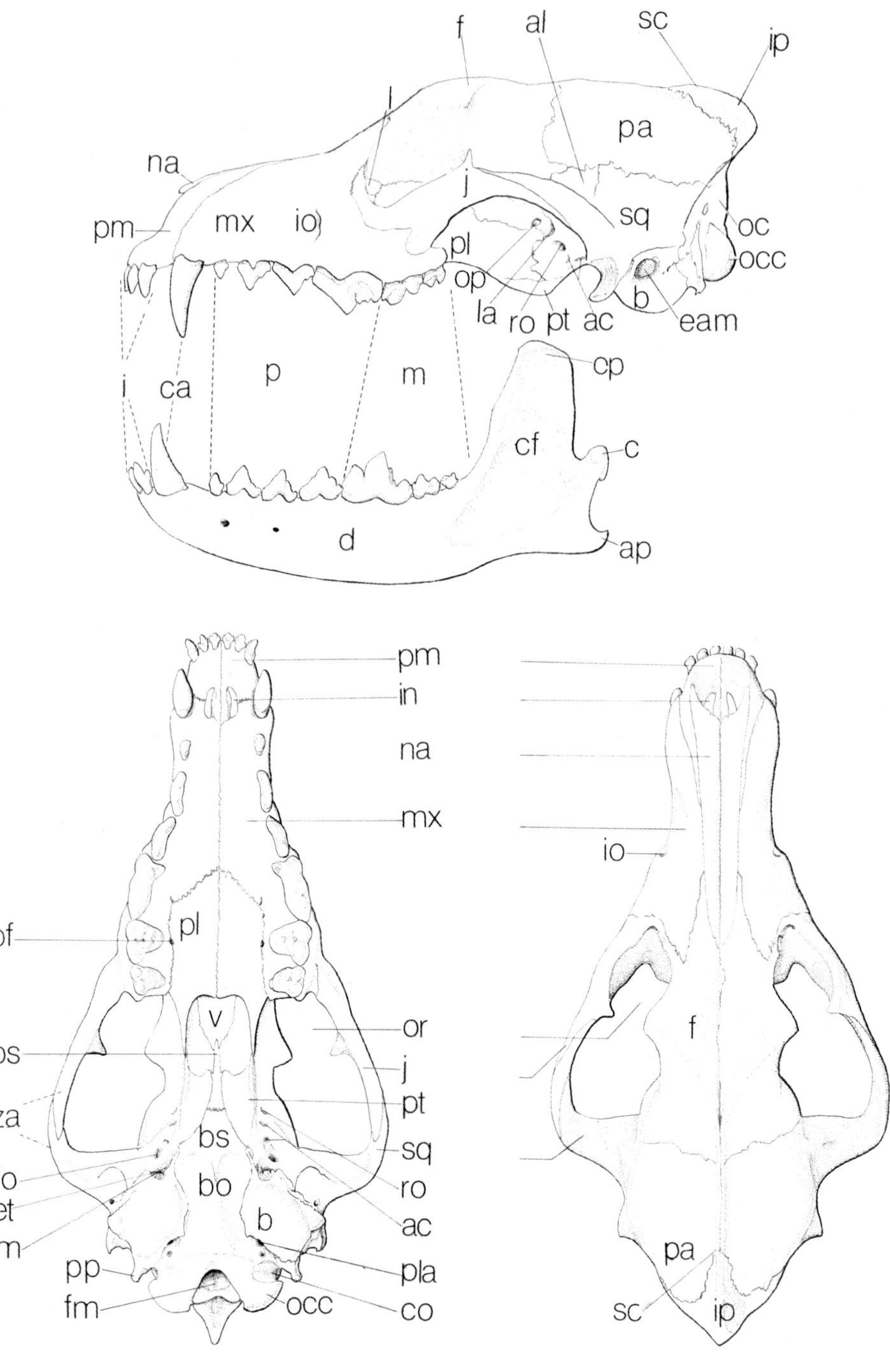

Figure 120. Skull of the coyote (*Canis latrans*), showing bones, major openings (foramina), other features, and teeth. Abbreviations shown are for the following: ac, alisphenoid canal; al, alisphenoid; ap, angular process; b, auditory bulla; bo, basioccipital; bs, basisphenoid; c, condyle; ca, canines; cf, coronoid fossa; co, anterior condyloid foramen; cp, coronoid process; d, dentary; eam, external auditory meatus; et, eustachian tube; fm, foramen magnum; fo, foramen ovale; f, frontal; i, incisors; in, incisive foramen; io, infraorbital foramen; ip, interparietal; j, jugal; l, lacrimal; la, anterior lacerate foramen; lm, medial lacerate foramen; m, molars; mx, maxillary; na, nasal; oc, occipital; occ, occipital condyle; op, optic foramen; or, orbit; p, premolars; pa, parietal; pf, posterior palatine foramen; pl, palatine; pla, post laterate foramen; pm, premaxillary; pp, paroccipital process; ps, presphenoid; pt, pterygoid; ro, foramen rotundum; sc, sagittal crest; sq, squamosal; v, vomer; za, zygomatic arch.

bone (unpaired) immediately in front of the foramen magnum.

Basisphenoid—Pertaining to or designating the bone (unpaired) immediately in front of the basioccipital bone.

Biomass—Living organic material in a habitat.

Biotic succession—The gradual replacement of one community by another in one area in less than geologic time (meaning the changes are not caused by evolution), such changes being largely brought about by the activities of the organisms themselves and excluding changes brought about by invading organisms expanding their geographic ranges.

Brachyodont—Having teeth with low, shallow crowns and well-developed roots, opposed to hypsodont.

Brain case—The part of the skull enclosing the brain.

Browser—Herbivorous animal specializing in eating twigs, leaves, bark and buds of woody plants.

Bunodent—Pertaining to molar teeth which have crowns covered with rounded cusps or tubercules for crushing and grinding food.

Caecum—A blind pouch or saclike extension of the digestive tract.

Calcar—In bats, a process connected with the calcaneum, helping to support the uropatagium (membrane) between the leg and tail.

Canine—Of, pertaining to, or designating the tooth next to the incisors in mammals.

Carnassial—Pertaining to or designating certain teeth in mammals of the order Carnivora, usually adapted for cutting. In these mammals, the last premolars in the upper jaw and first true molars in the lower jaw are the carnassial teeth.

Carnivorous—Eating flesh; preying or feeding on animals, opposed to herbivorous.

Carrying capacity—The maximum number of individuals of a single species which can be maintained in a specified area by the resources in that area for an appreciable (but unspecifiable) period of time. Can also mean the capability of the area to support that number.

Castorum—Waxy secretion of the anal glands of the beaver. Once used extensively in the fixing of perfumes.

Caudal—Of, pertaining to, or like a tail, situated in, near, or directed toward the tail or hind end of the animal.

Cheek teeth—Teeth behind the canines.

Claw—A pointed, curved, horny structure at the end of a toe.

Climax—The final self-perpetuating stage in a sequence of biotic succession. This stage is not replaced by another unless a "disturbance" of some sort occurs.

Commensal—Literally, "common table." An animal, like the house mouse, that depends on another animal, such as man, for food and shelter.

Competition—A relationship between members of the same or different species such that use of resources by one party to the relationship reduces the amount available to the other and vice versa.

Condyle—An articular prominence of a bone.

Conifer—A cone-bearing tree, such as pine, spruce, tamarack or fir.

Conspecific—Mammals or subspecies of mammals which belong to the same species.

Coprophagy—The eating of dung.

Coronoid process—The upward projecting process of the posterior part of the lower jaw.

Cranium—The skull without the lower jaws. In a more restricted sense, the part of the skull which encloses the brain, and, in a more general sense, the skull with lower jaw.

Crepuscular—Of, pertaining to, or like twilight; active in the twilight, as certain bats.

Cursorial—Adapted or modified for running, as are white-tailed deer.

Cusp—One of the bumps or points on the grinding surface of any tooth.

Deciduous—Falling off, shedding, or falling out at maturity, at certain seasons, or at certain ages, of the antlers of deer, hair of mammals, or first (milk) set of teeth of most kinds of mammals.

Deciduous dentition—The first (milk) set of teeth, subsequently replaced by the permanent dentition.

Delayed implantation—A condition in some mammals, such as carnivores, where a fertilized egg will develop to the blastula stage, then lie quiescent, perhaps for months, before implantation and further development.

Dental formula (plural, formulae)—A brief method for expressing the number and kind of teeth of mammals. The abbreviations I (incisor), C (canine), P (premolar), and M (molar) indicate the kinds of teeth in the permanent dentition, and the number in each jaw is written like a fraction, the figure in front of the diagonal line showing the number in the upper jaw, and the figure after, the number in the lower jaw. The dental formula of an adult coyote is I 3/3, C 1/1, P 4/4, M 2/3 × 2 = 42.

Dentine—A calcareous material harder and denser than bone which composes the principal mass of a tooth.

Dentition—The teeth of an animal considered collectively.

Diastema—Any natural toothless interval along a row of teeth.

Digestion—The process of preparing food for absorption and assimilation.

Diphyodont—Having two sets of teeth, the first being the milk teeth and the second the permanent teeth.

Diurnal—Active by day, opposed to nocturnal

Domestic mammal—A mammal that, through direct selection by man, has certain inherent morphological, physiological, or behavioral characteristics by which it differs from its ancestral stock.

Dorsal—Pertaining to or situated on the back of a mammal, opposed to ventral.

Echo-location—A guidance system in which the ear picks up sound waves made by the animal itself which are reflected back after hitting objects. In the case of bats such calls are ultrasonic.

Ecology—The study of the relationships between living things and their environments.

Ecosystem—Abbreviated term for "ecological system," in which the living and the nonliving interact, especially in regard to energy flow and cycling.

Emigration—The act of leaving an area.

Enamel—Of teeth, the hardest substance of the mammalian body, which forms a thin layer that caps or partly covers the teeth.

Enamel loops—Loops formed by enamel ridges bordered by dentine on the grinding surface of tooth.

Epiphysis (plural, epiphyses)—An accessory center of ossification at the ends of the long bones of mammals. When the ossifications of the shaft (diaphysis) and epiphysis meet, further lengthwise growth of the shaft ceases.

Estrus—A period in the reproductive cycle of the female when the animal is most physiologically and psychologically receptive to mating.

Eutherian mammals—Placental mammals, the infraclass Eutheria.

Exoccipital—Pertaining to or designating the right or left bone on the side of the foramen magnum at the back of the cranium.

Fat—A compound of carbon and hydrogen with less oxygen than carbohydrates, main chemicals of storage of energy in animals.

Feces—Intestinal excrement or dung.

Feral—A domestic animal that has reverted to the wild.

Fetus—An unborn, but well-developed young animal.

Foramen (plural, foramina)—A small orifice, opening, or perforation.

Foramen magnum—The large opening at the base of the skull through which the spinal cord passes.

Fossorial—Fitted for digging.

Frontal—Pertaining to or designating the bone (paired) immediately in front of the parietal bone and behind the nasal.

Genitalia—The external sexual organs.

Gestation period—Time between fertilization of egg and birth.

Gland—An organ of secretion or excretion.

Glenoid fossa—The fossa (depression) which accommodates the articular condyle of the lower jaw.

Guard hairs—The stiffer, longer hairs which grow up through the limber, shorter hairs (fur) of a mammals's pelage.

Habitat—A particular area in which one actually finds a given species.

Heat—Period of receptivity in the female when mating may take place.

Herbivorous—Feeding chiefly on grasses or other plant material.

Heterodont—Having the teeth differentiated into incisors, canines, premolars, and molars.

Hibernation—Of an animal, torpidity especially in winter. The body temperature approximates that of the surroundings, the rate of respiration and the heartbeat ordinarily are much slower than in an active mammal.

Home range—The entire area in which an individual moves around.

Hoof—The horny covering which protects the end of the digit and supports the weight of the animal.

Horn—The horny sheath covering an outgrowth of bone from the skull, circular or oval in cross section and pointed at the extremity.

Hyoid—Designating or pertaining to a bone or several connected bones situated at the base of the tongue.

Hypsodont—Having teeth with high or deep crowns and short roots, as in the molar teeth or meadow mice, opposed to brachydont.

Immigration—The act of moving into a new area.

Implantation—The fertilized egg attaching itself to the wall of the uterus.

Incisor—Pertaining to or designating the tooth in front of the canine tooth; those in the upper jaw invariably are in the premaxillary bone.

Infraorbital foramen—An opening through the zygomatic plate from the orbit to the side of the rostrum; for passage of nerves, blood vessels, or muscles.

Inguinal—Pertaining to or in the region of the groin.

Insectivorous—Eating insects; preying or feeding on insects.

Instinctive behavior—Activity traits that are inherited and not learned.

Interfemoral membrane—Thin skin in bats stretching from ankle to ankle including the tail.

Interorbital constriction—The least distance across the top of the skull between the orbits (eye sockets).

Interparietal—Pertaining to or designating the bone (rarely paired) immediately in front of the supraoccipital bone and between the two parietal or temporal bones.

Interspecific—Pertaining to phenomena occurring between members of different species.

Intraspecific—Pertaining to phenomena occurring between members of the same species.

Jugal—Pertaining to or designating the bone in the zygomatic arch which lies between the maxillary and squamosal bones.

Keel—In bats, refers to the free flap of skin attached posteriorly along the calcar.

Lachrymal bone—The bone pierced by the lachrymal duct (tear duct) between the frontal and maxillary bones at the anterior end of the orbit.

Lateral—Of or pertaining to the side.

Learned behavior—Activity traits that are learned and not inherited.

Litter—The two or more young brought forth at one birth by a female mammal.

Longitudinal—Extending along or pertaining to the antero-posterior axis (usually longest) of the body.

Mamma (singular, plural, mammae)—The glandular organ for secreting milk, characteristic of all mammals.

Mammary gland—The milk-producing organ of mammals, inactive in males.

Mandible—In mammals, the dentary bone comprising the right or left half of the lower jaw.

Mane—Pertaining to the long, coarse, black hair on the top of the tail of the gray fox or the long hair on the top of the neck of the weasel and wapiti.

Mange—Skin disease characterized by loss of hair and extensive skin eruptions.

Mast—Tree fruits, such as acorns and beechnuts.

Mastoid—Designating or pertaining to the mastoid bone (paired) or its process. This bone is bounded by the squamosal, exoccipital, and tympanic bones.

Maxilla (plural, maxillae)—The bone (paired) which bears the molar and premolar teeth.

Menopause—The cessation of periodic menstruation and ovulation; the end of reproductive ability in females.

Mental foramen (plural, foramina)—An opening through the extreme mid-anterior (chin) part of each ramus of the lower jaw.

Mesic—Describes an environment with an abundant supply of moisture (*cf.* arid).

Metabolism—The sum of the activities, mainly chemical, of construction and destruction that occurs within living organisms.

Metatarsal gland—Gland on the hind leg of a hoofed animal between the hoof and the hock.

Migration—A periodic movement away from and back to a given area.

Milk Teeth—The baby or deciduous teeth, those that are shed and replaced by the permanent set.

Molar—One of the posterior teeth (three on each side in the upper and lower jaws in coyote, making 12 in all) not preceded by deciduous teeth.

Molariform—Of or pertaining to a tooth which is formed like a molar, as the molars and premolars when the latter are like molars in form.

Molt—In a mammal, the act or process of shedding or casting off the hair.

Mutation—An hereditary change, passed from parent to offspring.

Nail—The horny plate on the upper surface of the end of figures and toes of the human and some other animals; differs from claw and hoof only in shape and form.

Nape—The back part of the neck.

Nares—The openings of the nose.

Nasal—Of or pertaining to the nose, as a nasal bone (paired) on the dorsal surface of the skull at its anterior end.

Niche—Pertaining to the functional role of a species in its environment and the ways it interacts with the biotic and abiotic elements.

Nocturnal—Active by night, opposed to diurnal.

Nose pad—The bare, thickened skin surrounding the nostrils.

Occipital—Pertaining to the posterior part of the skull.

Occlusal—Of or pertaining to the grinding or biting (occluding) surface of a tooth.

Omnivorous—Eating plant and animal matter.

Orbit—The eye socket of the skull.

Oviparous—Reproducing by laying eggs that hatch outside of the body of the mother.

Ovulation—Shedding of a ripe egg (or eggs) from ovary to fallopian tubes.

Palate—Designates the bony roof of the mouth comprised of the two palatine bones, two maxillary bones, and two premaxillary bones.

Parietal—Pertaining to or designating the parietal bone (paired) roofing the brain case. The bone is behind the frontal bone and in front of the occipital bones.

Parietal ridge—A bony ridge arising just back of the orbit and passing back over the brain case to converge posteriorly with its mate from the other side.

Pectoral—Of, pertaining to, or situated or occurring in or on the chest.

Pelage—The covering or coat of a mammal, as of hair, fur, or wool.

Phalanx (plural, phalanges)—The bone in a finger distal to the metacarpus or the bone in a toe distal to the metatarsus.

Photoperiod—The duration of daylight in relation to the duration of darkness over a 24-hour period.

Phylogenetic—Pertaining to the development of an evolutionary lineage.

Pinna (plural, pinnae)—That portion of the ear that sticks out beyond the head and acts as a sound collector.

Placenta—Organ of the developing embryo that provides for its nourishment and the elimination of waste products by interchange through the membrances of the uterus of the mother.

Plantigrade—Walking on the sole with the heel touching the ground, as in humans and bears.

Pollex (plural, pollices)—The first (preaxial) digit of the forelimb, the thumb.

Polyestrus—Having more than one heat or reproductively-receptive period in females per year.

Population—a group of individuals that are interfertile and which regularly contribute germ cells to the formation of fertile offspring.

Posterior—At or toward the hind end of the body.

Precocial—Young that are relatively well developed at birth, usually have much hair, eyes open, and good locomotor ability.

Predator—An animal that kills and consumes animal prey.

Prehensile—Refering to a tail adapted for seizing or grasping limbs or other objects.

Premaxilla (plural, premaxillae)—The bone (paired) in the mammalian skull bearing the incisor teeth of the upper jaw, situated in front of the maxilla.

Premolar—Designating or pertaining to one of the teeth (a maximum of four on each side of the upper and lower jaws of placental mammals, 16 in all) in front of the true molars. When canine teeth are present, premolars are behind the canines. Premolars are preceded by deciduous teeth, and in the upper jaw are confined to the maxilliary bone.

Producers—Organisms that make their own food by transforming "inorganic" raw materials into energy-storing "organic" substances.

Protein—A compound of carbon, hydrogen, oxygen, nitrogen, and usually small amounts of sulfur and other elements.

Rabies—Usually fatal viral disease affecting the nervous system.

Retractile—Capable of being drawn back or in, as with the claws of cats.

Rostrum—The anteriorly projecting part of a mammalian skull in front of the orbits.

Rudimentary—Not fully developed, or represented only by a vestige.

Rumen—An esophageal pouch of certain artiodactyls; the first of the four large digestive chambers of the ruminant "stomach."

Ruminant—A herbivorous mammal that regurgitates and chews a cud and has special modifications of the digestive tract for this process.

Rut—Breeding season, usually in reference to male hoofed mammals.

Sagittal crest—The raised ridge of bone at the juncture of the two parietal bones resulting from the coalescence of the temporal ridges; in old individuals of many species of mammals the crest extends from the middle of the lambdoidal crest anteriorly onto the frontal bones, where it divides into two temporal ridges each of which ends antero-laterally on the posterior edge of the postorbital process of the frontal bone.

Scapula (plural, scapulae)—The shoulder blade, a bone of the pectoral arch.

Scat—A fecal dropping.

Scrotum—A skin-covered sac, actually an extension of the abdominal cavity, that is suspended below the anus and which contains the testes.

Secondary sexual character—External difference, other than the reproductive organs, between the sexes.

Selenodont—Pertaining to molar teeth which have crowns with cusps in the form of crescents for sectioning and grinding vegetation.

Sinus (plural, sinuses)—A cavity in the substance of a bone of the skull which communicates with the nasal cavity and contains air, as the frontal sinus.

Sonar system—A means of locating objects by producing sound waves and then receiving the reflected vibrations or sound waves.

Squamosal—Designating or pertaining to the bone (paired) ventral to the parietal; the posterior root of the zygomatic arch is a process of the squamosal bone.

Supraorbital process—Bony projection of the skull above the eye.

Suture—The line of union in an immovable articulation, like one of the articulations between two bones of the skull.

Sympatry—The occurrence of two or more species in the same area.

Symphysis (plural, symphyses)—Immovable or more or less movable articulation of certain bones in the median plane of the body, as the pubic symphysis.

Synonym—In the zoological sense, a systematic name, as of a subspecies, species, or genus, regarded as incorrectly applied or as incorrect in the form of spelling. It may have been the accepted name which was later rejected in favor of another name because of evidence of priority of the other name or evidence establishing a more natural, assumedly-genetic, classification.

Tactile—Of, pertaining to, or relating to the sense of touch, as a tactile organ.

Tarsal—Of or pertaining to the tarsus.

Tarsus (plural, tarsi)—The ankle; the part of a vertebrate between the metatarsus and the leg.

Terrestrial—Inhabiting the land, rather than water, trees, or air.

Territory—An area, generally surrounding the home, that is defended against other individuals of the same species.

Therian mammals—Mammals of the subclass Theria, including the living marsupials and placental mammals.

Tine—Any of the branches which come off the main beam of an antler.

Torpidity—The condition of being devoid of the power of movement; sluggish in function or action.

Total length—Distance from the tip of the nose to the end of the tail vertebrae when the mammal is laid out straight but not stretched beyond its normal length in this position.

Total length of skull—Greatest length from front to back.

Tragus—Small, fleshy structure, leaf-like and highly developed in bats, projecting upward inside front of ear.

Truncate—Having the end square or even, as if cut off.

Tympanic bulla (plural, bullae)—The hollow, rounded prominence of bone enclosing the three auditory ossicles and related structures of the middle and inner ear. Often referred to as auditory bulla and sometimes as audital bulla.

Type locality—The place where a type specimen was obtained.

Type specimen—The specimen or individual on which the scientific name and accompanying diagnosis of a species or subspecies is based.

Underfur—Thick, soft fur lying beneath the longer and coarser hair of a mammal.

Uroptagium—The interfemoral membrane of a bat, the fold of skin which stretches from the hind legs to the tail.

Vibrissa (plural, vibrissae)—One of the stiff tactile

hairs on the wrist or face of a mammal, as the so-called whiskers of a cat.

Vertebra (plural, vertebrae)—One of the segments of the spinal column or backbone; the first vertebra is immediately behind and in contact with the skull and the last is the segment in the tip of the tail.

Vestigial—Pertaining to a structure or organ that is degenerate or imperfectly developed in the adult.

Viviparous—Reproducing by retaining the young within the mother's uterus. Nourishment for the young is supplied through the placenta.

Width of skull—Greatest width, usually across the zygomatic arches.

Zygoma (plural, zygomata)—The whole zygomatic arch.

Zygomatic arch—The arch of bone which extends along the side of the skull beneath the orbit; formed in most mammals by the union of the jugal bone with the maxillary bone in front and the zygomatic arm of the squamosal bone behind.

Zygomatic plate—The plate of bone comprised of the zygomatic arm of the maxilla.

Index